VENTRAL BRAINSTEM MECHANISMS AND CONTROL OF RESPIRATION AND BLOOD PRESSURE

LUNG BIOLOGY IN HEALTH AND DISEASE

Executive Editor

1. Immunologic and Infectious Reactions in the Lung, *edited by Charles H. Kirkpatrick and Herbert Y. Reynolds*
2. The Biochemical Basis of Pulmonary Function, *edited by Ronald G. Crystal*
3. Bioengineering Aspects of the Lung, *edited by John B. West*
4. Metabolic Functions of the Lung, *edited by Y. S. Bakhle and John R. Vane*
5. Respiratory Defense Mechanisms (in two parts), *edited by Joseph D. Brain, Donald F. Proctor, and Lynne M. Reid*
6. Development of the Lung, *edited by W. Alan Hodson*
7. Lung Water and Solute Exchange, *edited by Norman C. Staub*
8. Extrapulmonary Manifestations of Respiratory Disease, *edited by Eugene Debs Robin*
9. Chronic Obstructive Pulmonary Disease, *edited by Thomas L. Petty*
10. Pathogenesis and Therapy of Lung Cancer, *edited by Curtis C. Harris*
11. Genetic Determinants of Pulmonary Disease, *edited by Stephen D. Litwin*
12. The Lung in the Transition Between Health and Disease, *edited by Peter T. Macklem and Solbert Permutt*
13. Evolution of Respiratory Processes: A Comparative Approach, *edited by Stephen C. Wood and Claude Lenfant*
14. Pulmonary Vascular Diseases, *edited by Kenneth M. Moser*
15. Physiology and Pharmacology of the Airways, *edited by Jay A. Nadel*
16. Diagnostic Techniques in Pulmonary Disease (in two parts), *edited by Marvin A. Sackner*
17. Regulation of Breathing (in two parts), *edited by Thomas F. Hornbein*
18. Occupational Lung Diseases: Research Approaches and Methods, *edited by Hans Weill and Margaret Turner-Warwick*
19. Immunopharmacology of the Lung, *edited by Harold H. Newball*
20. Sarcoidosis and Other Granulomatous Diseases of the Lung, *edited by Barry L. Fanburg*
21. Sleep and Breathing, *edited by Nicholas A. Saunders and Colin E. Sullivan*
22. *Pneumocystis carinii* Pneumonia: Pathogenesis, Diagnosis, and Treatment, *edited by Lowell S. Young*

23. Pulmonary Nuclear Medicine: Techniques in Diagnosis of Lung Disease, *edited by Harold L. Atkins*
24. Acute Respiratory Failure, *edited by Warren M. Zapol and Konrad J. Falke*
25. Gas Mixing and Distribution in the Lung, *edited by Ludwig A. Engel and Manuel Paiva*
26. High-Frequency Ventilation in Intensive Care and During Surgery, *edited by Graziano Carlon and William S. Howland*
27. Pulmonary Development: Transition from Intrauterine to Extrauterine Life, *edited by George H. Nelson*
28. Chronic Obstructive Pulmonary Disease: Second Edition, Revised and Expanded, *edited by Thomas L. Petty*
29. The Thorax (in two parts), *edited by Charis Roussos and Peter T. Macklem*
30. The Pleura in Health and Disease, *edited by Jacques Chrétien, Jean Bignon, and Albert Hirsch*
31. Drug Therapy for Asthma: Research and Clinical Practice, *edited by John W. Jenne and Shirley Murphy*
32. Pulmonary Endothelium in Health and Disease, *edited by Una S. Ryan*
33. The Airways: Neural Control in Health and Disease, *edited by Michael A. Kaliner and Peter J. Barnes*
34. Pathophysiology and Treatment of Inhalation Injuries, *edited by Jacob Loke*
35. Respiratory Function of the Upper Airway, *edited by Oommen P. Mathew and Giuseppe Sant'Ambrogio*
36. Chronic Obstructive Pulmonary Disease: A Behavioral Perspective, *edited by A. John McSweeny and Igor Grant*
37. Biology of Lung Cancer: Diagnosis and Treatment, *edited by Steven T. Rosen, James L. Mulshine, Frank Cuttitta, and Paul G. Abrams*
38. Pulmonary Vascular Physiology and Pathophysiology, *edited by E. Kenneth Weir and John T. Reeves*
39. Comparative Pulmonary Physiology: Current Concepts, *edited by Stephen C. Wood*
40. Respiratory Physiology: An Analytical Approach, *edited by H. K. Chang and Manuel Paiva*
41. Lung Cell Biology, *edited by Donald Massaro*
42. Heart–Lung Interactions in Health and Disease, *edited by Steven M. Scharf and Sharon S. Cassidy*
43. Clinical Epidemiology of Chronic Obstructive Pulmonary Disease, *edited by Michael J. Hensley and Nicholas A. Saunders*
44. Surgical Pathology of Lung Neoplasms, *edited by Alberto M. Marchevsky*
45. The Lung in Rheumatic Diseases, *edited by Grant W. Cannon and Guy A. Zimmerman*
46. Diagnostic Imaging of the Lung, *edited by Charles E. Putman*
47. Models of Lung Disease: Microscopy and Structural Methods, *edited by Joan Gil*

48. Electron Microscopy of the Lung, *edited by Dean E. Schraufnagel*
49. Asthma: Its Pathology and Treatment, *edited by Michael A. Kaliner, Peter J. Barnes, and Carl G. A. Persson*
50. Acute Respiratory Failure: Second Edition, *edited by Warren M. Zapol and Francois Lemaire*
51. Lung Disease in the Tropics, *edited by Om P. Sharma*
52. Exercise: Pulmonary Physiology and Pathophysiology, *edited by Brian J. Whipp and Karlman Wasserman*
53. Developmental Neurobiology of Breathing, *edited by Gabriel G. Haddad and Jay P. Farber*
54. Mediators of Pulmonary Inflammation, *edited by Michael A. Bray and Wayne H. Anderson*
55. The Airway Epithelium, *edited by Stephen G. Farmer and Douglas Hay*
56. Physiological Adaptations in Vertebrates: Respiration, Circulation, and Metabolism, *edited by Stephen C. Wood, Roy E. Weber, Alan R. Hargens, and Ronald W. Millard*
57. The Bronchial Circulation, *edited by John Butler*
58. Lung Cancer Differentiation: Implications for Diagnosis and Treatment, *edited by Samuel D. Bernal and Paul J. Hesketh*
59. Pulmonary Complications of Systemic Disease, *edited by John F. Murray*
60. Lung Vascular Injury: Molecular and Cellular Response, *edited by Arnold Johnson and Thomas J. Ferro*
61. Cytokines of the Lung, *edited by Jason Kelley*
62. The Mast Cell in Health and Disease, *edited by Michael A. Kaliner and Dean D. Metcalfe*
63. Pulmonary Disease in the Elderly Patient, *edited by Donald A. Mahler*
64. Cystic Fibrosis, *edited by Pamela B. Davis*
65. Signal Transduction in Lung Cells, *edited by Jerome S. Brody, David M. Center, and Vsevolod A. Tkachuk*
66. Tuberculosis: A Comprehensive International Approach, *edited by Lee B. Reichman and Earl S. Hershfield*
67. Pharmacology of the Respiratory Tract: Experimental and Clinical Research, *edited by K. Fan Chung and Peter J. Barnes*
68. Prevention of Respiratory Diseases, *edited by Albert Hirsch, Marcel Goldberg, Jean-Pierre Martin, and Roland Masse*
69. *Pneumocystis carinii* Pneumonia: Second Edition, Revised and Expanded, *edited by Peter D. Walzer*
70. Fluid and Solute Transport in the Airspaces of the Lungs, *edited by Richard M. Effros and H. K. Chang*
71. Sleep and Breathing: Second Edition, Revised and Expanded, *edited by Nicholas A. Saunders and Colin E. Sullivan*
72. Airway Secretion: Physiological Bases for the Control of Mucous Hypersecretion, *edited by Tamotsu Takishima and Sanae Shimura*
73. Sarcoidosis and Other Granulomatous Disorders, *edited by D. Geraint James*
74. Epidemiology of Lung Cancer, *edited by Jonathan M. Samet*

75. Pulmonary Embolism, *edited by Mario Morpurgo*
76. Sports and Exercise Medicine, *edited by Stephen C. Wood and Robert C. Roach*
77. Endotoxin and the Lungs, *edited by Kenneth L. Brigham*
78. The Mesothelial Cell and Mesothelioma, *edited by Marie-Claude Jaurand and Jean Bignon*
79. Regulation of Breathing: Second Edition, Revised and Expanded, *edited by Jerome A. Dempsey and Allan I. Pack*
80. Pulmonary Fibrosis, *edited by Sem Hin Phan and Roger S. Thrall*
81. Long-Term Oxygen Therapy: Scientific Basis and Clinical Application, *edited by Walter J. O'Donohue, Jr.*
82. Ventral Brainstem Mechanisms and Control of Respiration and Blood Pressure, *edited by C. Ovid Trouth, Richard M. Millis, Heidrun F. Kiwull-Schöne, and Marianne E. Schläfke*
83. A History of Breathing Physiology, *edited by Donald F. Proctor*
84. Surfactant Therapy for Lung Disease, *edited by Bengt Robertson and H. William Taeusch*

ADDITIONAL VOLUMES IN PREPARATION

The Thorax: Second Edition, Revised and Expanded (in three parts), *edited by Charis Roussos*

Severe Asthma: Pathogenesis and Clinical Management, *edited by Stanley Szefler and Donald Y. M. Leung*

Respiratory Sensation, *edited by Lewis Adams and Abraham Guz*

Mycobacterium avium–Complex Infection: Progress in Research and Treatment, *edited by Joyce A. Korvick and Constance A. Benson*

Alpha 1–Antitrypsin Deficiency: Biology · Pathogenesis · Clinical Manifestations · Therapy, *edited by Ronald G. Crystal*

Adhesion Molecules and the Lung, *edited by Peter A. Ward and Joseph C. Fantone*

Acute Respiratory Failure in Chronic Obstructive Pulmonary Disease, *edited by Jean-Phillippe Derenne, W. A. Whitelaw, and Thomas Similowski*

Pulmonary Rehabilitation, *edited by Alfred P. Fishman*

VENTRAL BRAINSTEM MECHANISMS AND CONTROL OF RESPIRATION AND BLOOD PRESSURE

Edited by

C. Ovid Trouth
Richard M. Millis

College of Medicine
Howard University
Washington, D. C.

Heidrun F. Kiwull-Schöne
Marianne E. Schläfke

Ruhr-University Bochum
Bochum, Germany

Marcel Dekker, Inc. **New York • Basel • Hong Kong**

Library of Congress Cataloging-in-Publication Data

Ventral brainstem mechanisms and control of respiration and blood pressure / edited by C. Ovid Trouth ... [et al.]
p. cm. — (Lung biology in health and disease ; 82)
Includes bibliographical references and index.
ISBN 0-8247-9514-8 (hard cover : alk. paper)
1. Brain stem—Physiology—Congresses. 2. Vasomotor system—Congresses. 3. Respiration—Regulation—Congresses. 4. Blood pressure—Regular—Congresses. I. Trouth, C. Ovid. II. Series: Lung biology in health and disease ; v. 82.
[DNLM: 1. Medulla Oblongata—physiology—congresses. 2. Respiration—physiology—congresses. 3. Blood Pressure—physiology—congresses. 4. Vasomotor System—physiology—congresses. W1 LU62 v.82 1995 / WL 310 V467 1995]
QP377.V46 1995
612.8'28—dc20
DNLM/DLC
for Library of Congress 95-10040
CIP

The publisher offers discounts on this book when ordered in bulk quantities. For more information, write to Special Sales/Professional Marketing at the address below.

This book is printed on acid-free paper.

MARCEL DEKKER, INC.
270 Madison Avenue, New York, New York 10016

Current printing (last digit):
10 9 8 7 6 5 4 3 2 1

Printed in the United States of America

INTRODUCTION

Few topics in physiology have attracted as many great scientists over the centuries as the neural control of respiration. As one ponders why this is so, the views of one contemporary researcher and clinician, Dr. Fred Plum, come to mind: "Breathing, the only autonomic function regulated entirely by skeletal muscle, serves two great homeostatic systems, metabolism and behavior." Dr. Plum further noted that there are "separate neurological structures mediating behavioral and metabolic influences on the respiratory act at almost every level of the brain, from the cerebral hemispheres to the medulla and even the upper cervical spinal cord" (1).

Galen (ca. 129–200), through his pioneering studies of the human body, was the first to elucidate the effect of transection of the spinal cord: all respiration ceases, but the heart continues to beat! It is of further interest that his studies led him to characterize a particular personality type as "sanguine" or "melancholic," expressions that today portray quite different temperaments!

Most likely, the first precise location of the respiratory center may be attributed to Julien Jean Cesar LeGallois, who reported in *Experiences sur le principe de la vie* (1912): "If we open the cranium of a young rabbit, and extract the brain in successive portions, beginning by the anterior part, cutting it into slices, we may then take away the whole cerebrum, then all the cerebellum, and part of the medulla oblongata, without stopping respiration. But, it instantly ceases when we cut the origin of the nerves of the eighth pair." And so, over the centuries, a wealth of discoveries has been made. Sometimes controversies developed, but, undoubtedly, they too contributed to our understanding of the brainstem mechanisms controlling respiratory and circulatory processes.

This volume brings together the experience and the vast knowledge of a great number of experts from the United States and Europe. It is the end product of a symposium held at Howard University in Washington, D.C., and organized by Drs. Trouth, Millis, Kiwull-Schöne, and Schläfke, who are also editors of this monograph. There is no question that this volume constitutes a landmark in the complicated field it addresses. For this reason, it occupies a unique place in the Lung Biology in Health and Disease series. As executive editor of this series, I am grateful to the editors and the authors for the opportunity to include this very important contribution.

Claude Lenfant, M.D.
Bethesda, Maryland

Reference

Plum, F. Neurological integration of behavioral and metabolic control of breathing. In: Breathing. Porter, R., ed. Hering Breuer Century Symposium. London: J & A Churchill, 1970.

FOREWORD

Much has been learned over the past century about how the body operates and how different organ systems are integrated. Although there is more to be learned, certain principles, such as homeostatic control, are well accepted. We have a good appreciation of adaptive processes and how the same functions, such as the body's CO_2 intake, can be subserved by different structures in various species.

Also, over a much shorter time span, we have become quite knowledgeable about cellular mechanisms—about the basic principles of the genetic code and the complexity of intracellular processes—which at this time seem to require amazingly intricate interlocking processes for cells to grow and develop in an active individual. Although, again, there is much that remains unknown.

The relative wealth of information that exists at the cellular level and at the level of the body's behavior contrasts with the scarcity of data concerning the translation of cellular mechanisms to the responses of intact humans or to even less sophisticated multicellular species.

We know very little of the connection between complex cellular mechanisms and the amazing variety of whole-body responses that allow the organism to adapt to equally complex environmental changes. We do not know very much about the relationship of cellular processes to the development of the hormonal and neural connections that play such an important role in the integration of responses of different organ systems.

This volume deals with one small area of the brain, the ventral medulla, which seems to be involved in a multiplicity of physiological responses, including airway responses, secretory behavior, the defense reaction, and especially respiratory and circulatory behavior. This area may prove to be a region that can provide the clues leading to an understanding of the linkages between cellular and whole-body responses.

As a result of the pioneering efforts of Loeschcke and Feldberg and their coworkers (many of whom have made contributions to this volume), it has been established that the ventral medulla plays a significant role in both vasomotor and circulatory responses and possibly in the integrated behavior of these two systems. That this integration takes place commonly is evident. The integrated responses of the whole body to stresses, such as exercise and hypoxia, demonstrate that there is important coordination of the two systems and that to a degree one system can compensate for deficiencies in the other caused by disease affecting the main cellular metabolism. Study of the ventral medulla may provide an opportunity to understand the signals at the cellular level that evoke these responses. Tissue Po_2, or pH, or some complex weighting of the two, in addition to other factors, may be important.

The presentations in this volume record the progress that has been made and indicate the enormous gaps of information that remain. Fortunately, tools necessary to obtain fundamental information are now available. These tools include the reduced preparations that can now be used to study respiratory behavior. Novel genetic techniques now exist that allow the removal or insertion of specific genetic information, leaving the rest of the animal intact so that whole-body responses can be examined. With new imaging techniques, such as magnetic resonance imaging (MRI), positron emission tomography (PET), single-photon emission computed tomography (SPECT), and optical imaging, cellular behavior, or at least the behavior of small groups of cells, can be examined in the intact animal.

Ultimately, we would like to describe the transformation of information from cell to system quantitatively. The advances in computer technology as well as those that have been made in the mathematical analysis of alinear and fuzzy systems make this quantitative description a foreseeable possibility.

The chapters in this book should provide us a better understanding of the possibilities and know-how to frame our questions more sharply so that we can develop further testable hypotheses.

Neil S. Cherniack, M.D.
Case Western Reserve University
Cleveland, Ohio

PREFACE

The idea of an arrangement of brainstem neurons in chemosensitive regions exquisitely sensitive to pH and P_{CO_2} changes in their microenvironment was not new to biomedical researchers and clinicians working in the 1990s. But in the 1960s little was known about the role of the ventral brainstem in regulating ventilatory and vasomotor reflexes. This volume is dedicated to all the biomedical researchers of the past 30 years who have advanced our understanding of the brainstem mechanisms underlying ventilatory and vasomotor control. We have singled out Professors Koepchen and Loeschcke for their outstanding contributions, both being originally influenced by Hans Winterstein's fundamental "reaction theory" to explain the chemical control of breathing by the unique principle of H^+ homeostasis. Their leadership and foresight brought about pioneering experiments in both Germany and the United States that led directly to the first localization of what might be called the chemosensitive components of brainstem cardiorespiratory control mechanisms. Their original hypotheses have provided a structural framework for experiments addressing the significance and diversity of ventral brainstem control phenomena.

The location of vital centers controlling respiration and blood pressure have long been recognized within the reticular formation of the brainstem.

The control functions of these centers have been modeled as pattern generators, and they may have rhythmic characteristics common to those described for locomotor regulation in other parts of the central nervous system. In addition, the reflex activities of these ventilatory and vasomotor control centers are dependent upon numerous peripheral inputs from cranial nerve afferents and higher centers. Significant inputs to these cardiorespiratory control centers have also been traced to discrete regions of the ventral medullary surface, where the central respiratory chemoreceptors are presumed to reside. This anatomical arrangement presents several significant research problems. There appear to be interspecies differences with respect to the localization of the reticular centers and their connections to the ventral medullary surface. It also appears that the control functions of these centers might be observable at several different sites. This could represent redundancy in control of vital functions. However, it also suggests that the observed activity might not be that of the control center per se. In analogy to an electric circuit arrangement, a control switch might lie some distance from either a single electrical appliance or a group of appliances. Although electrical activity might be controlled individually at the appliances, another switch might, at some significant distance, be the critical control trigger for all of the appliances. At the ventral brainstem, the reticular rhythm generators and control centers that have been described all seem to have significant connectivity to the chemosensitive zones of the ventral medullary surface. Studies in the laboratories of Koepchen and Loeschcke were among the earliest to localize what might be found to represent the cardiorespiratory control trigger of the ventral brainstem.

The advances documented in this volume have brought to mind the enormous technical difficulties overcome during the 1960s by Professors Koepchen and Loeschcke in attempting to identify the signposts for chemosensitive control activity at the brainstem level. Their initial hypotheses concerning central respiratory chemosensitivity were insightful and greatly advanced our ability to explain the participation of the brainstem in cardiorespiratory reflex control mechanisms. Techniques had to be developed for maintaining physiological conditions suitable to these studies. This was accomplished by developing electrodes for monitoring the acid–base balance of body fluids and artificial cerebrospinal fluid solutions at and beneath the ventral brainstem surface. The chemosensitive regions had to be demarcated and the morphology of the chemosensitive substrates determined. Recordings of neuronal activity were made from regions where the neuronal density was sparse, requiring tedious experiments to search for a few chemosensitive units. Peripheral inputs from redundant systems, such as carotid and aortic chemoreceptors, had to be separated from their central brainstem components.

This book is based on a symposium held at Howard University to summarize the current knowledge base on ventral brainstem mechanisms. The symposium comprised an international group of researchers whose experience in studying ventral brainstem control mechanisms spanned the past 30 years. Advances in the understanding of neuroregulatory chemicals, cellular transductive, and rhythm generator mechanisms provided a strong infrastructure for workshop discussions on the significance of various research and clinical models for ventral brainstem participation in ventilatory and vasomotor control processes. Each workshop offered an opportunity for scientific enrichment and debate, thereby focusing the writing of the summary chapters.

We are hopeful that this volume will provide an insightful overview for a state-of-the-art scientific summary of our current knowledge of ventral brainstem mechanisms and control of breathing and blood pressure.

The tasks of producing this work have been facilitated by contributions from our colleagues. We wish to give special acknowledgment to the supportive efforts of Dr. Charles H. Epps, Jr., Dean of the Howard University College of Medicine, Dr. LaVal N. Cothran, Chairman of Howard's Department of Physiology and Biophysics, and Pfizer, Inc. This book could not have been produced without the devoted editorial assistance of Drs. Leroy Penix, Nanduri Prabhakar, and David Bernard and of Ms. Deepa Soni, Mr. C.S. Vasudevan, and Ms. Rosalyn Larkin. We have been encouraged in this endeavor by the superb performance of Mr. Paul Dolgert and Ms. Connie Casella and the expert production staff at Marcel Dekker, Inc. We are most grateful for the scientific foresight of Dr. Claude Lenfant, Executive Editor of this Lung Biology in Health and Disease series, and the clarity of each scientist-author's contribution.

C. Ovid Trouth
Richard M. Millis
Heidrun F. Kiwull-Schöne
Marianne E. Schläfke

CONTRIBUTORS

Jonas I. Addae, M.B.B.S., Ph.D. Chair, Department of Physiology, University of the West Indies, St. Augustine, Trinidad and Tobago, West Indies

Andrew M. Allen Howard Florey Institute of Experimental Physiology and Medicine, University of Melbourne, Parkville, Victoria, Australia

Phillip W. Archer, Ph.D. Department of Physiology and Biophysics, College of Medicine, Howard University, Washington, D.C.

David B. Averill, Ph.D. Assistant Professor, Hypertension Center and Departments of Physiology and Pharmacology, Bowman Gray School of Medicine of Wake Forest University, Winston-Salem, North Carolina

Barbara Badio, Ph.D. I.R.T.A. Fellow, Laboratory of Bioorganic Chemistry, National Institutes of Health, Bethesda, Maryland

David G. Bernard, Ph.D.* Postdoctoral Fellow, Department of Physiology and Biophysics, College of Medicine, Howard University, Washington, D.C.

**Present affiliation*: Dartmouth Medical School, Lebanon, New Hampshire.

Gerald E. Bisgard Department of Comparative Biosciences, University of Wisconsin, Madison, Wisconsin

John M. Bissonnette, M.D. Professor of Obstetrics and Gynecology and of Cell Biology and Anatomy, Department of Obstetrics and Gynecology, Oregon Health Sciences University, Portland, Oregon

Donald C. Bolser, Ph.D. Associate Principal Scientist, Department of Allergy, Schering-Plough Research Institute, Kenilworth, New Jersey

P. S. Bolton Division of Human Physiology, Department of Medicine, University of Newcastle, Callaghan, New South Wales, Australia

Ann C. Bonham, Ph.D. Assistant Professor, Division of Cardiovascular Medicine, Departments of Internal Medicine and Pharmacology, University of California at Davis, Davis, California

E. Boudinot Institut Alfred Fessard, Biologie Fonctionnelle du Neurone, CNRS, Gif-sur-Yvette, France

W. Richard Campbell, Ph.D. Assistant Professor, Department of Biology, Morris Brown College, Atlanta, Georgia

Ruy R. Campos, Jr., Ph.D. Assistant Professor, Department of Physiology, Escola Paulista de Medicina, São Paulo, Brazil

Richard W. Chapman Schering-Plough Research Institute, Kenilworth, New Jersey

Clivel George Charlton, Ph.D. Associate Professor of Pharmacology, Division of Basic Pharmaceutical Sciences, College of Pharmacy and Pharmaceutical Sciences, Florida A & M University, Tallahassee, Florida

Neil S. Cherniack, M.D. Dean and Professor of Medicine, School of Medicine, Case Western Reserve University, Cleveland, Ohio

Deepak K. Chugh, M.B.B.S., M.D. Postdoctoral Fellow, Department of Physiology, University of Pennsylvania School of Medicine, Philadelphia, Pennsylvania

John Ciriello, Ph.D. Professor, Department of Physiology, Health Sciences Centre, University of Western Ontario, London, Ontario, Canada

E. Lee Coates, Ph.D. Postdoctoral Fellow, Department of Physiology, Dartmouth Medical School, Lebanon, New Hampshire

Morton I. Cohen, Ph.D. Professor of Physiology and Neuroscience, Department of Physiology and Biophysics, Albert Einstein College of Medicine, Bronx, New York

Sharon K. Coles, Ph.D. Postdoctoral Research Fellow, Division of Pulmonary and Critical Care Medicine, Departments of Medicine, Neuroscience, and Physiology and Biophysics, School of Medicine, Case Western Reserve University, Cleveland, Ohio

Laura A. Collins Research Assistant, Division of Endocrinology/Hypertension, Department of Medicine, School of Medicine, Case Western Reserve University, Cleveland, Ohio

S. L. Cravo, Ph.D. Associate Professor, Department of Physiology, Escola Paulista de Medicina, São Paulo, Brazil

Jay B. Dean, Ph.D. Assistant Professor, Department of Physiology and Biophysics, College of Science and Mathematics, Wright State University School of Medicine, Dayton, Ohio

Frances C. DeGennaro Schering-Plough Research Institute, Kenilworth, New Jersey

Ozra Dehkordi, Ph.D. Assistant Professor, Division of Neurosurgery, Department of Surgery, Howard University Hospital, Washington, D.C.

Michael S. Dekin, Ph.D. Assistant Professor, Division of Pulmonary and Critical Care Medicine, Department of Medicine, University of Medicine and Dentistry of New Jersey–Robert Wood Johnson Medical School, New Brunswick, New Jersey

M. Denavit-Saubié, Ph.D. Director, Institut Alfred Fessard, Biologie Fonctionnelle du Neurone, CNRS, Gif-sur-Yvette, France

Gary C. Dennis, M.D. Associate Professor and Chief of Neurosurgery, Division of Neurosurgery, Department of Surgery, Howard University Hospital, Washington, D.C.

Thomas E, Dick, Ph.D. Associate Professor, Division of Pulmonary and Critical Care Medicine, Department of Medicine, School of Medicine, Case Western Reserve University, Cleveland, Ohio

Robert M. Douglas Predoctoral Fellow, Department of Physiology and Biophysics, College of Medicine, Howard University, Washington, D.C.

Ismail A. Dreshaj, M.D., Ph.D. Department of Pediatrics, School of Medicine, Case Western Reserve University, Cleveland, Ohio

Paul Ernsberger, Ph.D. Assistant Professor, Division of Endocrinology/Hypertension, Departments of Medicine, Pharmacology, and Neuroscience, School of Medicine, Case Western Reserve University, Cleveland, Ohio

Levent Ertugrul, Ph.D. Assistant Professor, Department of Physiology, Faculty of Medicine, University of Istanbul, Istanbul, Turkey

Jay P. Farber, Ph.D. Professor, Department of Physiology, University of Oklahoma Health Sciences Center, Oklahoma City, Oklahoma

Jack L. Feldman, Ph.D. Professor of Neurosciences, Department of Physiological Science, University of California at Los Angeles, Los Angeles, California

J. J. Filiano, M.D. Department of Neurology, Children's Hospital, and Harvard Medical School, Boston, Massachusetts

Eric A. Floyd Research Assistant, Department of Physiology, Meharry Medical College, Nashville, Tennessee

Gregory D. Funk, Ph.D. Research Associate, Department of Physiological Science, University of California at Los Angeles, Los Angeles, California

Owen Gardner Research Assistant, Department of Biological Sciences, Oakwood College, Huntsville, Alabama

Philip J. Gatti, Ph.D. Associate Professor, Department of Pharmacology, College of Medicine, Howard University, Washington, D.C.

Sergio B. Giancola, B.Sc., M.Sc. Department of Physiology, Health Sciences Centre, University of Western Ontario, London, Ontario, Canada

Ashton Gibbons, Ph.D. Professor, Department of Biological Sciences, Oakwood College, Huntsville, Alabama

T. Goto Department of Orthopaedic Surgery, Omiya Red Cross Hospital, Yono City, Saitama, Japan

David Gozal, M.D. Associate Professor of Pediatrics and Physiology, Section of Pediatric Pulmonology, Children's Hospital of Tulane, Tulane University School of Medicine, New Orleans, Louisiana

Antonio R. Granata, M.D., Ph.D. Associate Professor, Department of Pharmacology, University of Tennessee, Memphis, Memphis, Tennessee

Marilyn E. Graves Research Assistant, Division of Endocrinology/Hypertension, Department of Medicine, School of Medicine, Case Western Reserve University, Cleveland, Ohio

Patricia A. Grimes, Ph.D. Associate Professor, Department of Ophthalmology, and Scheie Eye Institute, University of Pennsylvania School of Medicine, Philadelphia, Pennsylvania

P. G. Guertzenstein, M.D., Ph.D.[†] Department of Physiology, Escola Paulista de Medicina, São Paulo, Brazil

Patrice G. Guyenet, Ph.D. Professor, Department of Pharmacology, University of Virginia Health Sciences Center, Charlottesville, Virginia

Brennie E. Hackley, Jr., Ph.D. Adjunct Professor and Chief Scientist, Department of the Army, United States Army Medical Research Institute of Chemical Defense, Aberdeen Proving Ground, Maryland

Gabriel G. Haddad, M.D. Professor of Pediatrics and Cellular and Molecular Physiology, and Director, Section of Respiratory Medicine, Department of Pediatrics, Yale University School of Medicine, New Haven, Connecticut

Ronald M. Harper, Ph.D. Professor of Anatomy and Cell Biology, Brain Research Institute, University of California at Los Angeles School of Medicine, Los Angeles, California

Musa A. Haxhiu, M.D., Ph.D. Professor of Medicine (Pulmonary) and Neuroscience, School of Medicine, Case Western Reserve University, Cleveland, Ohio

Michael S. Hedrick, Ph.D. Postdoctoral Fellow, Department of Comparative Biosciences, University of Wisconsin, Madison, Wisconsin

John A. Hey Schering-Plough Research Institute, Kenilworth, New Jersey

Timothy W. Higenbottam, M.D., F.R.C.P. Consultant Physician and Director, Department of Respiratory Physiology, Papworth Hospital, Cambridge, England

---

[†]Deceased

Yoshiyuki Honda, M.D. Professor of Physiology Emeritus, Departments of Physiology and Chest Medicine, Chiba University School of Medicine, Chiba, Japan

Donald B. Jennings, M.D., C.M., Ph.D. Professor, Department of Physiology, Queen's University, Kingston, Ontario, Canada

Jozef S. Jodkowski, M.D., Ph.D. Visiting Professor, Department of Physiology, Institute of Physiological Sciences, Medical Academy of Warsaw, Warsaw, Poland

Elmond J. Johnson Research Assistant, Department of Biological Sciences, Oakwood College, Huntsville, Alabama

Stephen M. Johnson, M.D., Ph.D. Research Associate, Department of Physiological Science, University of California at Los Angeles, Los Angeles, California

Shereé M. Johnson Research Assistant, Department of Physiology and Biophysics, College of Medicine, Howard University, Washington, D.C.

Yuan-Hsu Kang, Ph.D. Senior Investigator, Pathobiology Division, Naval Medical Research Institute, Bethesda, Maryland

R. Kanjhan Department of Physiology, University of Auckland School of Medicine, Auckland, New Zealand

Masao Katayama, M.D. Visiting Postdoctoral Fellow, Department of Physiology, University of Pennsylvania School of Medicine, Philadelphia, Pennsylvania

F. Kato, Ph.D. Institut Alfred Fessard, Biologie Fonctionnelle du Neurone, CNRS, Gif-sur-Yvette, France

I. A. Kerman Department of Physiology and Biophysics, New York University, New York, New York

Hiroshi Kimura, M.D. Assistant Professor, Department of Chest Medicine, Chiba University School of Medicine, Chiba, Japan

Hannah C. Kinney, M.D. Associate Professor of Neuropathology, Department of Pathology, Harvard Medical School, and Children's Hospital, Boston, Massachusetts

Heidrun F. Kiwull-Schöne, M.D., Ph.D. Professor, Department of Physiology, Ruhr-University Bochum, Bochum, Germany

B. Kruszewska Department of Physiology, University of Auckland School of Medicine, Auckland, New Zealand

Ganesh K. Kumar, Ph.D. Assistant Professor, Department of Biochemistry, School of Medicine, Case Western Reserve University, Cleveland, Ohio

Sukhamay Lahiri, D.Phil. Professor, Department of Physiology, University of Pennsylvania School of Medicine, Philadelphia, Pennsylvania

Aihua Li Department of Physiology, Dartmouth Medical School, Lebanon, New Hampshire

Janusz Lipski, M.D., Ph.D. Professor, Department of Physiology, University of Auckland School of Medicine, Auckland, New Zealand

Arthur D. Loewy, Ph.D. Professor, Department of Anatomy and Neurobiology, Washington University School of Medicine, St. Louis, Missouri

O. U. Lopes, M.D., Ph.D. Full Professor, Department of Physiology, Escola Paulista de Medicina, São Paulo, Brazil

Kevin R. Lynch Department of Biochemistry, University of Virginia Health Sciences Center, Charlottesville, Virginia

Nadine Lyseight Research Assistant, Department of Biological Sciences, Oakwood College, Huntsville, Alabama

Janea Mack Research Assistant, Department of Pharmacology, Meharry Medical College, Nashville, Tennessee

Andrew A. Martin, M.D. Assistant Professor, Division of Pulmonary and Critical Care Medicine, Department of Medicine, University of Medicine and Dentistry of New Jersey–Robert Wood Johnson Medical School, New Brunswick, New Jersey

Richard J. Martin, M.D. Professor, Department of Pediatrics, School of Medicine, Case Western Reserve University, and Rainbow Babies and Childrens Hospital, Cleveland, Ohio

James C. McKenzie, Ph.D. Associate Professor, Department of Anatomy and Howard University Cancer Center, College of Medicine, Howard University, Washington, D.C.

Alecia McLean Research Assistant, Department of Biological Sciences, Oakwood College, Huntsville, Alabama

G. Norris Melville, M.D., Ph.D. Campus Dean and Professor of Physiology, Faculty of Medical Sciences, University of the West Indies, St. Augustine, Trinidad and Tobago, West Indies

Alan D. Miller, Ph.D. Associate Professor, Laboratory of Neurophysiology, Rockefeller University, New York, New York

Martha J. Miller, M.D., Ph.D. Departments of Pediatrics and Medicine, School of Medicine, Case Western Reserve University, and Rainbow Babies and Childrens Hospital, Cleveland, Ohio

Richard M. Millis, Ph.D. Assistant Professor, Department of Physiology and Biophysics, and Instructor, Department of Neurology, College of Medicine, Howard University, Washington, D.C.

Jayoti Mitra, Ph.D. School of Medicine, Case Western Reserve University, Cleveland, Ohio

Anil Mokashi, M.Sc. Research Specialist, Department of Physiology, University of Pennsylvania School of Medicine, Philadelphia, Pennsylvania

Eric L. Moore, Ph.D. Physiologist, Department of Neurotoxicology, United States Army Reserve Institute of Chemical Defense, Aberdeen Proving Ground, Maryland

Tsuguo Morikawa, M.D. Research Fellow, Department of Physiology, Chiba University School of Medicine, Chiba, Japan

M. P. Morin-Surun, Ph.D. Institut Alfred Fessard, Biologie Fonctionnelle du Neurone, CNRS, Gif-sur-Yvette, France

Nicole E. Mourillou Research Assistant, Department of Biology, Xavier University, New Orleans, Louisiana

Eugene E. Nattie, M.D. Professor, Department of Physiology, Dartmouth Medical School, Lebanon, New Hampshire

Judith A. Neubauer, Ph.D. Associate Professor, Division of Pulmonary and Critical Care Medicine, Department of Medicine, University of Medicine and Dentistry of New Jersey–Robert Wood Johnson Medical School, New Brunswick, New Jersey

Patricia J. Ohtake, B.Sc., M.Sc., Ph.D. Department of Physiology, Queen's University, Kingston, Ontario, Canada

Babatunde A. Otulana, M.D. Assistant Professor, Department of Pulmonary and Critical Care Medicine, Howard University Hospital, Washington, D.C.

Yi Pan, M.D., Ph.D.* Department of Physiology and Biophysics, College of Medicine, Howard University, Washington, D.C.

A. Panigraphy Department of Neurology, Children's Hospital, and Harvard Medical School, Boston, Massachusetts

Anthony D. Paul, Ph.D. Chairman, Department of Biological Sciences, Oakwood College, Huntsville, Alabama

O. S. Possas, M.S. Department of Physiology, Escola Paulista de Medicina, São Paulo, Brazil

David E. Potter, Ph.D. Chairman, Department of Pharmacology and Toxicology, Morehouse School of Medicine, Atlanta, Georgia

Nanduri R. Prabhakar, Ph.D., D.Sc. Associate Professor, Department of Physiology and Biophysics, School of Medicine, Case Western Reserve University, Cleveland, Ohio

Ruth Quartey, M.S. Neuroscience Center, Department of Zoology, University of Maryland, College Park, Maryland

L. A. Rava Department of Neurology, Children's Hospital, and Harvard Medical School, Boston, Massachusetts

David M. Rector Interdepartmental Neuroscience Program, Brain Research Institute, University of California at Los Angeles School of Medicine, Los Angeles, California

Radha B. Reddy Department of Physiology and Biophysics, Wright State University School of Medicine, Dayton, Ohio

Stefanie Roder, M.Sc. Department of Physiology, Health Sciences Centre, University of Western Ontario, London, Ontario, Canada

---

**Present affiliation*: St. Louis University Hospital, St. Louis, Missouri.

W. F. Rong Department of Physiology, University of Auckland School of Medicine, Auckland, New Zealand

Diane L. Rosin Department of Pharmacology, University of Virginia Health Sciences Center, Charlottesville, Virginia

Hubert K. Rucker, Ph.D. Associate Professor and Chairman, Department of Physiology, Meharry Medical College, Nashville, Tennessee

Melinda L. Ryan Department of Comparative Biosciences, University of Wisconsin, Madison, Wisconsin

Christoph Schäfer Department of Applied Physiology, Ruhr-University Bochum, Bochum, Germany

Dietmar Schäfer, M.D. Department of Applied Physiology, Ruhr-University Bochum, Bochum, Germany

Thorsten Schäfer, M.D., Ph.D. Department of Applied Physiology, Ruhr-University Bochum, Bochum, Germany

Marianne E. Schläfke, M.D. Head, Department of Applied Physiology, and Professor of Physiology, Ruhr-University Bochum, Bochum, Germany

John W. Severinghaus, M.D. Professor Emeritus, Department of Anesthesia, University of California Medical Center, San Francisco, California

Lucretia M. Sexcius Department of Physiology and Biophysics, College of Medicine, Howard University, Washington, D.C.

Jeffrey C. Smith, Ph.D. Senior Investigator, Laboratory of Neural Control, National Institute of Neurological Disorders and Stroke, National Institutes of Health, Bethesda, Maryland, and University of California at Los Angeles, Los Angeles, California

Winston N. Smith, M.D. Research Assistant, Department of Physiology and Biophysics, College of Medicine, Howard University, Washington, D.C.

Irene C. Solomon, Ph.D. Research Associate, Division of Cardiovascular Medicine, University of California at Davis, Davis, California

Richard A. Stone, M.D. Professor, Department of Ophthalmology, and Scheie Eye Institute, University of Pennsylvania School of Medicine, Philadelphia, Pennsylvania

Ruth L. Stornetta Department of Pharmacology, University of Virginia Health Sciences Center, Charlottesville, Virginia

Ye-Ming Sun, M.D., Ph.D. Division of Pulmonary and Critical Care Medicine, Department of Medicine, University of Medicine and Dentistry of New Jersey–Robert Wood Johnson Medical School, New Brunswick, New Jersey

Koichiro Tatsumi, M.D. Instructor, Department of Chest Medicine, Chiba University School of Medicine, Chiba, Japan

W. Eric Thomas, Ph.D. Professor of Biology and Vice President for Academic Affairs, Department of Natural Sciences and Environmental Health, Mississippi Valley State University, Itta Bena, Mississippi

Pauline Ting, M.D. Associate Professor, Department of Pediatrics, Howard University Hospital, Washington, D.C.

Jason Tomlin Research Assistant, Department of Biological Sciences, Oakwood College, Huntsville, Alabama

A. Jayam Trouth, M.D. Associate Professor, Department of Neurology, College of Medicine, Howard University, Washington, D.C.

C. Ovid Trouth, M.D., Ph.D. Professor of Physiology and Biophysics and Associate Professor of Neurology, College of Medicine, and Graduate Associate Professor, Graduate School of Arts and Sciences, Howard University, Washington, D.C.

Takuya Tsuchihashi, M.D. Second Department of Internal Medicine, Faculty of Medicine, Kyushu University, Fukuoka City, Japan

Beverly Turner Research Assistant, Department of Biological Sciences, Oakwood College, Huntsville, Alabama

Iroka J. Udeinya, Ph.D. Associate Professor, Departments of Anesthesiology and Pharmacology, College of Medicine, Howard University, Washington, D.C.

Paul G. Wagner, Ph.D. Parker B. Francis Postdoctoral Fellow, Division of Pulmonary and Critical Care Medicine, Department of Medicine, University of Medicine and Dentistry of New Jersey–Robert Wood Johnson Medical School, New Brunswick, New Jersey

Stephanie W. Waldrop Department of Pharmacology and Toxicology, Morehouse School of Medicine, Atlanta, Georgia

Julia K. L. Walker, B.Sc., B.Ed., M.Sc. Department of Physiology, Queen's University, Kingston, Ontario, Canada

Lakeischa Webb Research Assistant, Department of Biological Sciences, Oakwood College, Huntsville, Alabama

William L. West, Ph.D. Professor Emeritus, Department of Pharmacology, College of Medicine, Howard University, Washington, D.C.

W. F. White, Ph.D. Pfizer Central Research, Pfizer, Inc., Groton, Connecticut

Christopher G. Wilson Division of Cardiovascular Medicine, Departments of Internal Medicine and Pharmacology, University of California at Davis, Davis, California

Don H. Wood, M.D.[†] Professor, Department of Neurology, Howard University Hospital, Washington, D.C.

Samuel R. Wray, Ph.D., D.Sc. Professor, Neuroscience Adolescent Development and Drug Research Programme, Faculty of Medical Sciences, University of the West Indies, Mona, Jamaica, West Indies

Jerry M. Wright, Ph.D. Senior Staff Fellow, National Institute on Alcohol Abuse and Alcoholism, National Institutes of Health, Bethesda, Maryland

Melville Q. Wyche Jr., M.D. Department of Anesthesiology, College of Medicine, Howard University, Washington, D.C.

B. J. Yates, Ph.D. Assistant Professor, Department of Otolaryngology, Eye and Ear Institute, Pittsburgh, Pennsylvania

---

[†]Deceased.

CONTENTS

AUTONOMIC REGULATION

Part Five DEVELOPMENTAL ASPECTS OF VENTRAL BRAINSTEM PROCESSES

EXPERIMENTAL MODELS FOR CLINICAL PERSPECTIVES

1

The "Reaction Theory" of Hans Winterstein (1879–1963) in the Light of Today's Research on the Ventrolateral Medulla

HEIDRUN F. KIWULL-SCHÖNE

Ruhr-University Bochum
Bochum, Germany

I. Introduction

The nineteenth century was characterized by dynamic development of the sciences, including physiology. In a broad sense, Winterstein's scientific background may be attributed to one of the famous German schools of physiology in Europe initiated by Johannes Müller (1801–1858), who held the chair in physiology in Berlin from 1833 onward. Müller was a predecessor of scientists with such well-known names as du Bois–Reymond, Haeckel, Helmholtz, Henle, Schwann, and Virchow in the first generation and Gad, Heidenhain, Hermann, Heymans, Pflüger, and Verworn in the second generation. There is a direct line of scientific descent from Johannes Müller, Emil du Bois–Reymond, Ludimar Hermann, and Oskar Langendorff to Hans Winterstein (Table 1).

Winterstein's active scientific research period lasted exceptionally long, from 1900 to 1963, and he gained a growing international reputation. The significance of Winterstein's life and work in the history of physiology was appreciated by respiratory physiologists all over the world, whether in a modern textbook (Fig. 1) on acid base balance by Honda (1) in Japan or in the contributions to

Table 1 Scientific Predecessors and Young Collaborators of Hans Winterstein

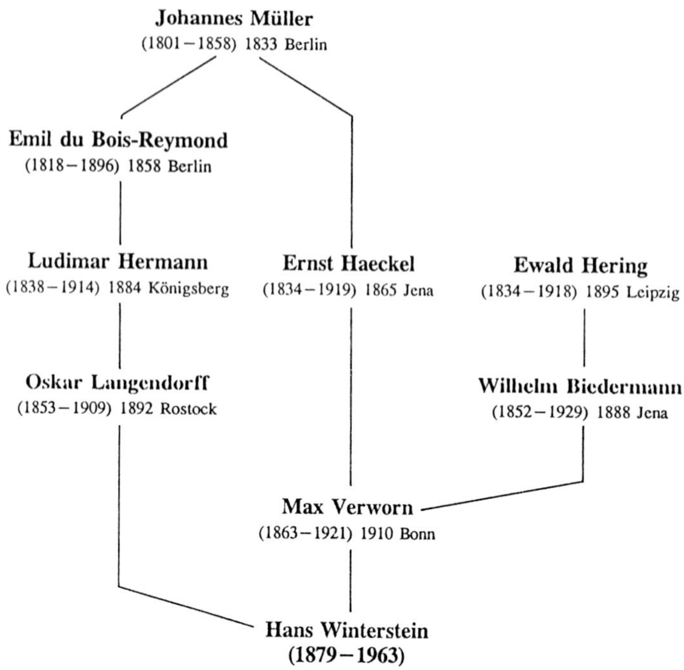

Else Hirschberg, Rostock
Hans H. Weber, Rostock
Harry Rosenbaum, Breslau
Joachim von Ledebur, Breslau
Meliha Inay, Istanbul
Meliha Terzioglu, Istanbul
Nuran Gökhan, Istanbul
Eberhard Frömter, München
Norbert Ott, München
Wolfgang Wiemer, München

the *Handbook of Physiology* made by Kellogg (2) and Perkins (3) in the United States.

The close scientific agreement between Hans Winterstein and Hans H. Loeschcke and their congenial relationship guaranteed that the younger collaborators in Loeschcke's group would be brought up with Winterstein's basic ideas. One of them, C. O. Trouth, now revives this tradition by fascinating contemporary scientists with the reaction theory on the occasion of the present symposium entitled "Ventral Brainstem Mechanisms and Control of Respiration and Blood Pressure."

The following biographical details of Winterstein's life in a restless period of European history mainly refer to Winterstein's autobiography (4), the obituary

酸塩基平衡と呼吸　165

以上の呼吸の化学調節および換気の定量的評価に関し，著者はいくつかの解説[87-92]を述べた．詳細はこれらにゆずる．

IV Hans Winterstein—Reaction theory とその生涯

この章のはじめに述べたように，呼吸の化学刺激の本態は水素イオンであるとする Reaction theory は Winterstein により唱えられ，もっとも強く主張された．これは呼吸調節に関するいわゆる「Winterstein の説」(sog. Wintersteinsche Theorie der Atmungs-regulation) とも呼ばれている．Reaction theory は呼吸生理学の研究の進歩とともに，種々の批判，改変を経て，今日なおその学問的価値を失っていない．また Winterstein その人の生涯[103]も，この説と同様に波乱に富んだ苦難の一生であった．

1) 生い立ち

Winterstein は 1879年 7月 31日プラハにおいて，枢密顧問官の父と貴族出身の母という家庭に生をうけた．生家は旧市内の Tein 教会の近くで，市内を一望に見おろすことのできる眺望のよいところであったという．

医学に志した彼は 1903年 (24歳) まで，Prag, Jena, Göttingen の各大学に学び，就中 Göttingen 大学におけるその頃もっとも高名な生理学者 Verworn に深く傾倒したという．彼の豊富な哲学的思考と魅力ある人格は Winterstein の生理学者としての一生にもっとも大きな影響をおよぼすこととなった．しかしながら，Winterstein は単なる書斎の人ではなく，学生時代 (1900年) すでにその最初の科学論文を発表している．

「Über die Wirkung der Kohlensäure auf das Zentralnerven system. *Arch. Anat. Physiol*. 177-192」

この炭酸ガスの作用の観察が，後年 Reaction theory の提唱につながっていることはこの章の冒頭に述べた．すなわち炭酸ガスの中枢にたいする刺激作用と麻痺作用の 2 元性は，酸としての刺激と CO_2 自体の麻酔作用に由来するものではないかという考えかたに発展していくことになった．

2) Rostock 時代

若くして俊才を謳われた Winterstein は，Göttingen 大学卒業後 3 年の1906年に Rostock 大学において学位を得た．Rostock はバルト海に面した小ハンザ都市で，その大学も小さな規模のものであったが，Winterstein は 25～47歳まで (1905～1927年) の長期間，学問的にもっとも充実した時期を送ることになった．

最初の Reaction theory の構想は，1910年 Wien の国際生理学会の席上，「Die Wirkung

Figure 1 First page of the article “Hans Winterstein—reaction theory and his life” in a Japanese textbook on basic and clinical medicine. (From Ref. 1 by courtesy of Professor Dr. Y. Honda, Shinko-Koeki Medical Book Publisher, Tokyo.)

by Weber and Loeschcke (5), and the historical description of the Physiological Institute in Rostock by Burmeister (6).

II. Curriculum Vitae

On July 31, 1879, during the Austro-Hungarian monarchy of the Hapsburgs, Hans Winterstein was born in Prague, where he spent his childhood and school years. In 1897 he started his medical studies in Prague, just missing Ewald Hering, who since 1870 had been successor to Purkinje in Prague but in 1895 moved to Leipzig to the vacant chair of the late Carl Ludwig.

As a young student, Winterstein was strongly impressed with Max Verworn's seemingly conclusive and comprehensive textbook *Allgemeine Physiologie* (*General Physiology*) (Jena, 1895). He therefore continued his studies in Jena (Germany) and later followed Verworn to Göttingen. Under the guidance of Verworn in Jena, Winterstein was introduced to *Kohlensäure* (carbonic acid)

Q. B. F. F. F. Q. S.!
SUMMIS AUSPICIIS
AUGUSTISSIMI IMPERATORIS AC REGIS
FRANCISCI JOSEPHI I.
NOS
OSCARUS LENZ,
PHILOSOPHIAE DOCTOR, GEOGRAPHIAE C. R. PROFESSOR PUBLICUS ORDINARIUS, UNIVERSITATIS CAROLINAE FERDINANDEAE GERMANICAE PRAGENSIS H. T.
RECTOR MAGNIFICUS,
JOANNES GAD,
MEDICINAE DOCTOR, PHYSIOLOGIAE C. R. PROFESSOR PUBLICUS ORDINARIUS,
ORDINIS MEDICORUM H. T. DECANUS,
ET
Carolus Hugo Huppert,
Med.UDr; a consiliis aulicae regiae aulae nec non chemiae mediae c.r. professor publ. ord.,
PROMOTOR RITE CONSTITUTUS,
IN VIRUM CLARISSIMUM
JOANNEM WINTERSTEIN,
Pragensem
POSTQUAM EXAMINIBUS LEGITIMIS LAUDABILEM CUM DOCTRINAM TUM FACULTATEM ARTIS MEDICAE PROBAVIT
DOCTORIS UNIVERSAE MEDICINAE
NOMEN ET HONORES POTESTATEMQUE ARTEM TAM MEDICAM CHIRURGICAMQUE QUAM OPHTHALMICAM ATQUE OBSTETRICIAM EXERCENDI CONTULIMUS IN EIUSQUE REI FIDEM HASCE LITTERAS UNIVERSITATIS SIGILLO SANCIENDAS CURAVIMUS.
DATUM PRAGAE, DIE XXV. Martii ANNI MDCCCCIII.
Dr. phil. Oscarus Lenz
Geographiae P.P.O.
h.t. Rector

Figure 2 Hans Winterstein's medical doctor's certificate from the Carl-Ferdinand's University in Prague, 1903. (Reproduction by courtesy of PD Dr. N. Ott.)

research, leading to his first scientific publication in 1900 entitled "Über die Wirkung der Kohlensäure auf das Centralnervensystem" ("On the Effect of Carbonic Acid on the Central Nervous System") (7).

In 1903 Winterstein received his doctoral degree in the medical sciences. The impressive certificate was written in Latin on handmade paper headed by the name of the emperor of Austria-Hungary, Franz Joseph I, followed by those of the Rector Magnificus of Carl-Ferdinand's University in Prague and the Dean of the Medical Faculty (Fig. 2). Incidentally, this dean was the well-known physiologist Johannes Gad, who a few years earlier had cauterized discrete areas of the floor of the fourth ventricle by tiny glass beads to show their role in respiration (8).

After a short time as Verworn's research assistant in Göttingen, Winterstein transferred to Langendorff in Rostock, who was well known for his isolated heart preparations. In Winterstein's own words, "It was a rather fortunate providence that led me from the somehow superficial but brilliant world, rich in ideas, of Verworn to the unpretentious but thorough research problems of Langendorff." In 1906 Winterstein finished his "Habilitationsschrift" on tissue respiration entitled "Der Mechanismus der Gewebsatmung" ("Mechanism of Tissue Respiration").

Figure 3 Neo-Renaissance building raised in 1878 for the medical faculty of the University of Rostock. Since then, as throughout Winterstein's time, the building has housed the Institute of Physiology. (Picture taken a few years before German reunification.)

In 1911, 2 years after Langendorff's death, Winterstein accepted the Professorial Chair of Physiology in Rostock. The Physiological Institute was situated on the first floor of a beautiful neo-Renaissance building, constructed in 1878, in the Gertrudenstrasse, which still stands today (Fig. 3). Winterstein, working in close proximity to the anatomist Elze and the pharmacologist Trendelenburg, formed a most fruitful association with them. Before long, Winterstein became a leading expert on the control of breathing in mammals.

Winterstein presented his "reaction theory" for the first time as a consistent concept at the Eighth International Congress of Physiological Sciences in Vienna in 1910. A comprehensive publication supporting the importance of the hydrogen ion concentration of the blood in respiratory control appeared in 1911 in Pflügers Archive (9); it was entitled "Die Regulierung der Atmung durch das Blut" ("Regulation of Respiration by the Blood") (Fig. 4). Winterstein developed a novel technique to measure the blood pH directly by means of a platinum-hydrogen electrode connected to an ionometer, which can still be seen in the historical collection of the Rostock Institute (Fig. 5). Besides, because of his

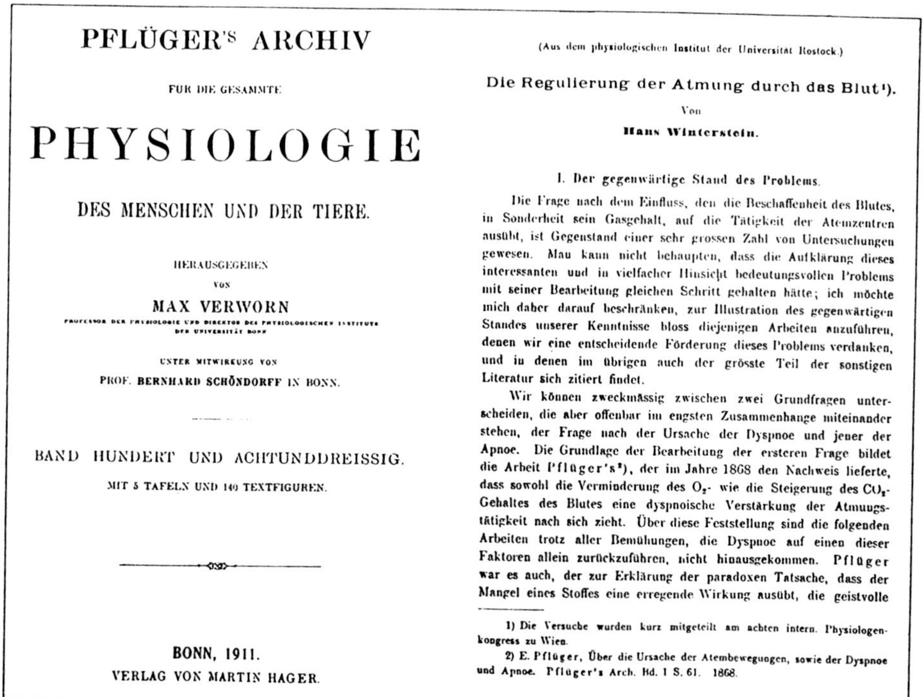

PFLÜGER'S ARCHIV

FÜR DIE GESAMMTE

PHYSIOLOGIE

DES MENSCHEN UND DER TIERE.

HERAUSGEGEBEN

VON

MAX VERWORN

PROFESSOR DER PHYSIOLOGIE UND DIREKTOR DES PHYSIOLOGISCHEN INSTITUTS DER UNIVERSITÄT BONN

UNTER MITWIRKUNG VON

PROF. BERNHARD SCHÖNDORFF IN BONN.

BAND HUNDERT UND ACHTUNDDREISSIG.

MIT 5 TAFELN UND 140 TEXTFIGUREN.

BONN, 1911.

VERLAG VON MARTIN HAGER.

(Aus dem physiologischen Institut der Universität Rostock.)

Die Regulierung der Atmung durch das Blut[1]).

Von

Hans Winterstein.

I. Der gegenwärtige Stand des Problems.

Die Frage nach dem Einfluss, den die Beschaffenheit des Blutes, in Sonderheit sein Gasgehalt, auf die Tätigkeit der Atemzentren ausübt, ist Gegenstand einer sehr grossen Zahl von Untersuchungen gewesen. Man kann nicht behaupten, dass die Aufklärung dieses interessanten und in vielfacher Hinsicht bedeutungsvollen Problems mit seiner Bearbeitung gleichen Schritt gehalten hätte; ich möchte mich daher darauf beschränken, zur Illustration des gegenwärtigen Standes unserer Kenntnisse bloss diejenigen Arbeiten anzuführen, denen wir eine entscheidende Förderung dieses Problems verdanken, und in denen im übrigen auch der grösste Teil der sonstigen Literatur sich zitiert findet.

Wir können zweckmässig zwischen zwei Grundfragen unterscheiden, die aber offenbar im engsten Zusammenhange miteinander stehen, der Frage nach der Ursache der Dyspnoe und jener der Apnoe. Die Grundlage der Bearbeitung der ersteren Frage bildet die Arbeit Pflüger's[2]), der im Jahre 1868 den Nachweis lieferte, dass sowohl die Verminderung des O_2- wie die Steigerung des CO_2-Gehaltes des Blutes eine dyspnoische Verstärkung der Atmungstätigkeit nach sich zieht. Über diese Feststellung sind die folgenden Arbeiten trotz aller Bemühungen, die Dyspnoe auf einen dieser Faktoren allein zurückzuführen, nicht hinausgekommen. Pflüger war es auch, der zur Erklärung der paradoxen Tatsache, dass der Mangel eines Stoffes eine erregende Wirkung ausübt, die geistvolle

1) Die Versuche wurden kurz mitgeteilt am achten intern. Physiologenkongress zu Wien.

2) E. Pflüger, Über die Ursache der Atembewegungen, sowie der Dyspnoe und Apnoe. Pflüger's Arch. Bd. 1 S. 61. 1868.

Figure 4 First version of the reaction theory. Original publication 1911 in Pflüger's Archive of "All Human and Animal Physiology," edited by Max Verworn, a former teacher of Winterstein's. (Courtesy of Springer-Verlag, Berlin-Heidelberg-New York.)

Figure 5 Platinum-hydrogen electrode and ionometer developed by Winterstein for blood pH measurement. Historical collection of the Physiological Institute in Rostock. (Photo taken by W. Kimenkowski with kind permission of Professor Dr. C. Pfeiffer.)

enormous productivity, Winterstein, on commission from the Naples Zoological Research Station, undertook to edit the eight volumes of the *Handbuch der Vergleichenden Physiologie* (*Handbook of Comparative Physiology*) (Jena: G. Fischer, 1910–1925).

His time in Rostock ended in 1927, when he was invited to take the prestigious chair in Breslau that was previously held by Heidenhain and Purkinje. Winterstein left behind his portrait, dedicated "To my place of work in the years 1905–1927 in grateful memory" (Fig. 6), now adjoining the portraits of Stannius and Langendorff in Rostock. In his memoirs, Winterstein remembers with some emotion the torchlight procession with which the students in Rostock honored him upon his departure.

In the fateful year 1933, Winterstein was exposed to growing intrigues by the German government because of his Jewish descent. He was therefore glad to be invited to the University of Istanbul, Turkey, first as a visiting professor and later as full professor and chairman. This was due to the ingenious political decisions of Kemal Pascha (Atatürk), calculated to bring western standards of science to Turkey by offering refuge to some of the outstanding Jewish scientists being expelled from Germany.

Figure 6 Portrait dedicated by Hans Winterstein to the Physiological Institute in Rostock on the occasion of his call to Breslau 1927. (Photo kindly made available by Professor Dr. C. Pfeiffer.)

The next 20 years were characterized by the intensive development of programs to establish research and teaching facilities in physiology at the University of Istanbul. Interestingly, among the young talents once again attracted to Winterstein, there were a considerable number of Turkish women scientists (Meliha Inay, Meliha Terzioglu, Nuran Gökhan; see Table 1). Winterstein did not retire from his professorship in Istanbul until he was 74 years old, and for a long time he hesitated to return to the Federal Republic of Germany.

Nevertheless, in 1956, Winterstein decided to move to Munich, partly because Professor Wagner, then director of the Physiological Institute in Munich, offered him adequate research facilities. Once more, Winterstein started fruitful scientific work and gathered a group of young doctoral candidates around him. They all built successful careers and are still active leaders in physiology or clinical medicine (e.g., E. Frömter, N. Ott, W. Wiemer).

In 1963, in his eighty-fifth year, Winterstein died—only a few weeks after having finished his scientific work at the Institute of Physiology in Munich.

III. Winterstein's Scientific Life Work

A. Philosophical Background

Hans Winterstein may be regarded as one of the last representatives of the "classic period" of German and European physiology, scholars characterized by their encyclopedic knowledge and interest in medicine, biology, and sciences. He integrated the phenomena of life into an enclosed and unique world view of sciences, being nurtured by the philosophical position of the positivism and empiricism represented by Ernst Mach (1838–1916) and the Wiener Kreis (Vienna Circle) in the second half of the nineteenth century. Thus, he attempted the explanation of phenomena by strict economy of thinking. Winterstein's general intellectual world is best delineated in his comprehensive review (10) on the chemical control of breathing, which emerged late during his stay in Istanbul and was published 1955. One of the most characteristic features of his way of thinking—in terms of cognition theory—is the critical reflection on definitions of such terms as *excitability, respiration, regulation,* or *explanation.*

Winterstein distinctly disapproved of the use of the term *excitability* in the chemical control of respiration since statements about any changed excitability of the centers would not explain mechanisms of chemical drives.

In defining the term *respiration*, Winterstein strictly differentiates the processes of cell metabolism—referring to Lazare Spallanzani's *Mémoires sur la respiration* (translated into French from his manuscript edited by Jean Senebier, Genève: Paschoud, 1803)—from those processes concerning gas exchange in lung and tissues.

Most important is Winterstein's definition of *regulation* in respiratory physiology: "Pulmonary ventilation is not regulated but regulates, at most being controlled by the regulated variable." Winterstein has clearly illustrated a general problem arising with the investigation of biological control systems. The more sensitive the regulator, the smaller the necessary changes to start it. Thus, it depends on the sensitivity of the method to show the changes controlling the regulator at all. "Observing a cyclist, he drives in a straight line not showing any deviations, which have to be there to elicit the regulatory movements necessary to maintain the balance." A biological reaction to unmeasurable changes of a variable has been designated as a *paradoxon* by several investigators. According to Winterstein, this has to be understood in much the same way as the described "cyclist-paradoxon."

Winterstein's definition of *explanation* (11) gives a typical example of his basic philosophical attitude:

> It appears to be generally accepted that it is the task of exploration of nature to explain the observed phenomena. But there appears to be no clarity on what is the nature of this "explaining." This clarity will certainly not be gained by taking the in itself unclear and controversial term "causality." To my mind, the aim of all research has been defined in unsurpassable clearness by Mach [Ernst Mach 1838–1916, physicist, psychologist, philosopher in Prague and Vienna], who sees this in representing facts mentally in the most simple and economical way. If this mental representation is not possible, likewise nothing is "explained."

Winterstein's reaction theory, based on this way of thinking, must be understood as his main contribution to respiratory physiology. In his very first scientific study of 1900, Winterstein, still as a student, found that carbonic acid had both excitatory and paralyzing effects on warm- and cold-blooded animals (7). By logical deduction, he postulated that these controversial observations cannot be attributed to the direct influence of one and the same compound. The postulate can only be accomplished by considering the ambiguity of CO_2 as carbonic acid on the one hand and as narcotic gas on the other. Thus, the respiratory system presents itself as an ideal model for studying the highly sensitive excitatory effect of carbonic acid.

B. First and Second Version of the Reaction Theory, 1911 and 1921

As early as 1911, while holding the chair in physiology at the University of Rostock, Winterstein proposed his first reaction theory, whereby the well-known responses of pulmonary ventilation to low oxygen and high CO_2 were reduced to one and the same mechanism, namely the acidification of blood by either lactic or carbonic acid.

The experiments leading to the first version of the reaction theory were published 1911 in Pflügers Archive (9): newborn rabbits were continuously perfused from the aorta with O_2-saturated Ringer's solution, resulting in complete apnea but maintained reflexes. Rhythmic respiration could be elicited by either adding CO_2 or hydrochloric acid to the perfusion fluid.

Having previously observed the specific paralyzing property of CO_2 (7), Winterstein concluded that the excitatory effect might be due to its nature as an acid. Since hypoxia also may lead to acid production, a unique explanation was given, namely that "neither the lack of oxygen nor the CO_2 tension as such, but solely the hydrogen-ion concentration in the blood performs the chemical control of breathing. . . ." Thereby, the acid reaction of the blood was thought to determine the excitability of the respiratory centers themselves.

This idea was strongly corroborated by the investigations of Hasselbalch (12). By changing the diet of normal humans, the hydrogen ion concentration in urine could be varied over a wide range. Nevertheless, there was no considerable decrease in blood pH upon changing the diets from vegetable to meat, since the CO_2 tension in the alveoli (and hence in the blood) was strongly reduced by increased pulmonary ventilation. Winterstein's theory was supported by this finding, since ventilation could not have been driven by (blood) CO_2. Winterstein (13) designed a series of experiments especially adapted to his question in rabbits, using slow intravenous infusion of different strong mineral and weak organic acids. It could be shown that the resultant considerable rise in ventilation was accompanied by a rise in blood hydrogen ion concentration (cH) but a fall in alveolar CO_2 tension.

In the following period, a growing number of objections and arguments arose against the reaction theory, provided by the seemingly paradoxical excitation of ventilation by sodium bicarbonate ($NaHCO_3$), the ventilatory depression by ammonium chloride (NH_4Cl), and the hypoxic dyspnea despite concomitant blood alkalinity. The main objections were based on the observations that the same change in blood pH caused greater drives to pulmonary ventilation when induced by CO_2 inhalation than if induced by fixed acid infusion; Gray's "multiple factor theory" later summarized these objections. According to Gray (14), the reaction theory does not stand the quantitative test; rather, carbonic acid has to be postulated as being an autonomous factor in respiratory control.

Winterstein's reply was impassioned:

> It is not the task of cH to keep ventilation constant, but the task of ventilation to keep cH constant. With each breath the restrained CO_2 release causes a relatively important augmentation of cH . . . leading to a correspondingly greater excitation of breathing. The gradually occurring slight metabolic drift of reaction correspondingly induces a hardly perceptible enhancement of breathing.

Another convincing argument in favor of a specific effect of carbonic acid came from repeated observations of different investigators that bicarbonate application excited pulmonary ventilation, although the cH of the blood did not undergo augmentation but diminution.

This conflict was solved in a very elegant way, as Winterstein became aware of Jacobs's now classic experiments with starfish eggs (15): Unfertilized starfish eggs were lightly stained in neutral red and introduced into different solutions of the same pH. $NaHCO_3$ saturated with CO_2 caused a bright pink by intracellular acidification, and a mixture of NH_4Cl and some NH_4OH caused a yellow staining by intracellular alkalinization. With clear-cut logical arguments deduced from these experiments, Winterstein explained even paradoxical reactions to $NaHCO_3$ or NH_4Cl by appropriate $[H^+]$ changes expected within nerve cells of the respiratory center, separated by differentially permeable membranes or barriers. This is an early example of how experiments at the cellular level may help in understanding problems of physiological systems at higher levels of complexity. Winterstein commented on these results as follows (10): "It is self-evident that the cells of the respiratory centres will behave in much the same way as . . . the starfish eggs of Jacobs. . . . The seemingly contradictory effect of bicarbonate solutions to the reaction theory is thus simply explained." And, in summary:

> The basic idea of the reaction theory that the excitatory effect of carbonic acid is not specific, but is based on its nature as an acid, has been proved to hold true in its entirety. We have, however, to modify the theory saying that we add to the cH of the blood the cH of the centres. Furthermore, we learnt to understand what the frequently observed seemingly specific effect of carbonic acid is based on. It is based on the "specific" way in which it brings to bear its "unspecific" acid nature, in that it is able to go through all tissues and cell membranes easily in the form of carbon dioxide and thus to appear at sites and—changing back into HCO_3^- and H^+—there to take effect, where other acid ions are not able to get to.

Typically, Winterstein developed these arguments by logical deduction, not performing experiments or using skilful methods if not necessary. Gesell and Hertzman (16) later proved that the acid-base status of the blood did indeed differ from that in the cerebrospinal fluid (CSF) by continuously and concomitantly measuring pH in arterial and venous blood as well as in the CSF of the fourth ventricle.

The second version of the reaction theory was published in 1921 in Pflügers Archive (17). Whereas the seemingly specific role of CO_2 for the respiratory system could have been easily contributed to differing permeation properties in its ionized and unionized form, Winterstein was persistently faced with the problem of incorporating the well-known ventilatory responses to hypoxia into the reaction theory. Thereby, it has to be kept in mind that the peripheral chemoreceptors had not yet been discovered. Winterstein repeatedly investigated acid production during oxygen lack to prove that blood cH was the common factor in respiratory control under these conditions also. But his direct measurements of cH in the blood during hypoxia revealed the weakness of his theory: the cH of the blood was not enhanced; quite the contrary—it was lowered. Winterstein explained the hypoxic dyspnea by an accumulation of *Erstickungsstoffen* (asphyxiating substances) in the respiratory centers themselves without any change in the blood condition. Therefore, in his second version of the reaction theory, Winterstein sharply distinguished two kinds of hyperpnea: hematogenic hyperpnea caused by a rise in blood cH, and centrogenic hyperpnea resulting from increased cH in the centers themselves. In summary, the new version of the reaction theory then read: "The chemical control of pulmonary ventilation is mediated by the reaction prevailing in the centres." This statement was later expanded to "The reaction in the centres depends on local metabolism, the number of H^+ ions in the surrounding extracellular fluid and the permeation capability of compounds determining the reaction" (18). Thus, before the discovery of the peripheral chemoreceptors, the second version of the reaction theory included an explanation for the origin of the hypoxic dyspnea as being "centrogenic."

C. Third and Fourth Versions of the Reaction Theory, 1949 and 1955

The discovery of the arterial chemoreceptors by J. Fr. Heymans and C. Heymans (19) left no doubt that the hyperpnea caused by lack of oxygen was by no means central but was rather entirely reflexogenic in origin, so that, concerning the effect of hypoxia, apparently any basis was taken from the reaction theory. Being faced with this important discovery, Winterstein tried to extend the reaction theory to the peripheral chemoreflex control of breathing.

Based on the observation that changes in the CO_2 tension and the cH as well as in the O_2 tension in the surrounding medium determined the excitation level of the glomera, the same complex of questions as to their adequate stimulus arose as earlier for central chemosensitive respiratory control. The crucial experiment to prove the acid theory and explain the excitatory mechanism of the arterial chemoreceptors appeared to Winterstein to have been that by von Euler, Liljestrand, and Zotterman (20). These authors investigated whether the carotid body chemoreceptors were differentially sensitive either to oxygen lack or to carbonic acid accumulation. They detected that when there was stepwise diminution in the number of functioning chemoreceptor fibers down to some few, the discharge frequency was reduced for lack of oxygen in much the same way as for rise in CO_2, and—what is even more significant—that the effect of a small amount of intravenous alkali transiently but completely abolished the chemoreceptor impulses elicited by both anoxia and hypercapnia.

In Winterstein's opinion, "These experiments are perfect evidence at least for that there are the same chemoreceptor fibres and the same kind of excitation mechanism that take action in the reflexogenic excitation of breathing due to both lack of oxygen and carbonic acid excess, and that this mechanism consists in a control by cH" (10). Thus, to Winterstein, "The reaction theory appeared to be resuscitated in a surprising manner, in its third version (21), in which the expression 'centrogenic' is just replaced by 'glomerogenic.' "

However, the third version of the reaction theory also very soon needed revision and completion. New observations on metabolic acidosis in animals with either intact or denervated arterial chemoreceptors (22) led to the more inclusive concept of the reaction theory in its fourth version: "With the discovery of the glomerogenic control of breathing the important problem arose which are the relative contributions in the interaction of reflexogenic and directly centrally elicited impulses. . ." (10). In chemoreceptor-deprived animals, acid injection invariably led to a fall in CO_2 tension caused by enhancement of pulmonary ventilation. On the other hand, dogs fed with ammonium chloride for several days developed a shift of blood cH toward acidity but a shift of cerebrospinal fluid cH toward alkalinity. This, after all, could explain that the elicited augmentation of pulmonary ventilation was small compared to the strong rise in blood cH: "Obviously the respiratory response is the algebraic sum of the two opposing factors, the driving influence of the rise in blood cH and the attenuating influence of the fall in CSF cH. . . ." (23). Winterstein and Gökhan (23) demonstrated furthermore that H^+ ions were unable to penetrate the blood-brain barrier, and they concluded that the cH of the blood takes effect directly only via the chemoreceptors, since after their elimination blood acidosis no longer had any respiratory exciting effect. Since inhalation of CO_2 still elicited hyperpnea even after elimination of all chemoreceptors, they finally concluded: "The chemical control of pulmonary ventilation by carbonic acid or cH, respectively, occurs both centrally and glomerogenically, the portion of each being still controversial. . . ."

The fact that CO_2 inhalation took effect directly on the respiratory centers seemed to prove a specifically respiratory excitatory effect of carbonic acid. The solution of this seeming contradiction to the reaction theory had just resulted, however, from investigations of the effects of NH_4Cl and $NaHCO_3$, showing that the transmission of effects from blood into the centers occurred through the undissociated molecules NH_3 and CO_2, since ions cannot pass the blood-brain barrier.

Considering these facts, the reaction theory in its fourth version (10) reads as follows:

The chemical control of breathing occurs

a) haematogenic through the blood cH, being effective in a two-fold manner: first, in that it determines the cH of the respiratory centres by undissociated CO_2 passing the blood-brain barrier, and second, reflexogenic via the chemoreceptors, which in turn happens in a two-fold manner,
   1) in that it directly (extraglomerally) influences the cH of the glomera
   2) in that it—as we will see later—through the Bohr-effect determines the O_2 tension of the blood and thus indirectly the excitation status of the chemoreceptors.

b) glomerogenic, in that the tonus of the chemoreceptors depends on the intraglomeral cH, which—as just mentioned—is determined by the blood cH, but predominantly by the O_2 pressure in the blood. Finally, add

c) centrogenic (in a narrow sense) the control of impulses, which is primarily (and autochthonously) generated by changes in the excitatory state occurring in the centres themselves, being enhanced or lowered by manifold influences. The most important among them will be haunting us later on (as temperature, sleep, anaesthesia, pregnancy).

Up to now, no phenomenon is known that would contradict the reaction theory in this version.

D. Interaction with Loeschcke's Work

The rather complex fourth version of the reaction theory must have placed Winterstein in something of a dilemma, the "economy of thinking" principle apparently being abandoned. A complex biological feedback system with multiple interactions and connections could obviously no longer be represented mentally in that most simply way required originally. Nevertheless, also in this final version of the reaction theory, a unique H^+ sensitivity both at peripheral chemoreceptor and central level was adhered to as the basic mechanism of phenomena in the control of breathing.

Upon his return to Munich, Winterstein pursued at least two main lines of interest expressed by the fourth version of the reaction theory. First, together with his newly recruited graduate students, he carried out studies on the quantitative contributions and interactions of different autonomic nervous feedback systems involved in the chemical control of breathing. Second, he placed much emphasis on the pharmacological exploration of central chemosensitive mechanisms beyond the blood-brain barrier via the cerebrospinal fluid (CSF) and summarized the results in a further comprehensive review (24).

Because Winterstein had meanwhile become aware of Loeschcke's fourth ventricle perfusion experiments (25), scientific agreement and congeniality laid the foundation for their ensuing close relationship. These experiments lent rather suggestive support to H^+ and not CO_2 as being the adequate stimulus for central respiratory chemosensitivity. The historical concept of a CO_2-sensitive center was defeated in some respect when it was shown that low pH/low P_{CO_2} solutions did stimulate ventilation but that iso-pH/high P_{CO_2} solutions did not. Thereby the fluid had to pass the lateral apertures and reached the ventral surface of the medulla oblongata. From these and further perfusion experiments (26), two main conclusions were drawn, namely that (a) the central respiratory chemosensitivity uniquely responded to changes in the H^+ ion activity in its environment and (b) it consisted in a chemoreceptorlike structure located at the ventral surface of the medulla oblongata, which was not identical with the integrative neurons of the respiratory centers themselves.

Thus, Winterstein (27) set another example for continuity and conciseness of his thinking by adopting "the discovery of new sensory areas for the chemical control of breathing" for his reaction theory:

> The significance of these ideas in terms of "economy of thinking" I have already pointed to on several occasions: The exceptional position which was attributed to the respiratory centre as the only centre provided with chemical excitability is falling away. It becomes a nerve centre like all the others, having the task to transfer and to forward the excitation being conducted to it by manifold nervous pathways.

Winterstein died in Munich on August 18, 1963, at the age of 85, having continued his active scientific work in physiology up to only a few weeks before. A few months later, a symposium entitled "Cerebrospinal Fluid and the Regulation of Ventilation" was held at the Downstate Medical Center of the State University of New York, organized by Chandler McC. Brooks, Frederick F. Kao, and Brian B. Lloyd. Winterstein contributed a great deal to the topic of this symposium and stimulated ensuing research on central chemosensitive control of breathing.

IV. Posthumous Appreciation: The Symposium on Cerebrospinal Fluid, New York, 1964

A. Scientific Background

Winterstein's posthumous contribution to the symposium was twofold, along the main lines of his last interests: first, in referring to experiments just completed on the interaction of intracranial and extracranial respiratory mechanisms, by Wiemer, Winterstein, Kiwull, and Ott (28), and, second, in having stimulated and encouraged most contributors to adopt his lifelong effort and prove the validity of the reaction theory, at least for the central chemosensitive subsystem beyond the blood-brain barier. Thus, Perkins, in his historical introduction to the symposium, expressed appreciation for Winterstein's life work and devoted a great part of his text to the description of the formation of the reaction theory (29).

To validate Winterstein's reaction theory beyond the blood-brain barrier, the possible localization of central chemoreceptors as well as the role of CSF acid-base balance in the chemical control of breathing were examined. Presentations addressing these main questions were given by Frencl, Leusen, Loeschcke, Mitchell, and Severinghaus by means of classic in vivo whole-system approaches.

Fencl as well as Loeschcke attempted to localize central H^+-sensitive areas suitable to fulfil the requirements of a unifying reaction theory. Mitchell postulated a more complex although unifying concept of interactive peripheral and central chemoreception, and Leusen as well as Severinghaus gave new, although controversial, insights into mechanisms of nonventilatory CSF acid-base regulation providing the tentative adequate stimulus for central chemosensors.

Leusen postulated changes in cerebral blood flow and in brain metabolism as additional mechanisms for $[HCO_3^-]$ adjustment (30), whereas Severinghaus postulated an active transport, based on the evidence that the blood-brain barrier was capable of generating significant electrochemical gradients for bicarbonate (31,32).

This elicited discussion on signal transmission from blood to brain tissue. Loeschcke hesitated to accept an active uphill transport of bicarbonate between CSF and blood, since HCO_3^- was observed to move rather rapidly in the same direction in both compartments under conditions of acute respiratory and metabolic acidosis. Generally, strong arguments in favor of the reaction theory arose, but also a fair amount of its weakness became evident at this symposium.

B. Central Versus Peripheral Chemoreception

Mitchell attempted an interactive multiple factor type of unifying theory for the chemical control of breathing (33), referring mainly to the second version of Winterstein's reaction theory (17). There was general agreement that the pH of the CSF, although lacking buffers, was better regulated than the blood pH. The question arose, whether variations in CSF pH and ventilation were uniquely correlated, as proposed by Winterstein, or whether there were additional stabilizing influences. The respiratory response to increased mock CSF P_{CO_2} at constant end-tidal P_{CO_2} was found to be only about 40% of the respiratory response to inhaled CO_2; thus 60% should be mediated by peripheral chemoreceptors (26). Furthermore, a lowered blood pH even at constant CSF pH was accompanied by a lowered P_{CO_2}. Therefore, the source of this ventilatory drive in metabolic acidosis should arise from outside the central nervous system (CNS), possibly at peripheral chemoreceptors.

To test this hypothesis, awake dogs were subjected to a few days of ammonium chloride acidosis, whereby—after arterial chemodenervation—metabolic acidosis indeed failed to stimulate ventilation. These observations led Mitchell (33) to postulate an extended interactive but unifying concept of the chemical regulation of respiration: Acute acidosis was thought to stimulate carotid and aortic H^+-receptors, to increase ventilation, and to decrease arterial and CSF P_{CO_2}. CSF alkalinity would first decrease the medullary H^+ receptor activity; but after active transport of HCO_3^- across the blood-brain barrier, the medullary H^+ receptor would be reset to normal activity at a lower than normal CO_2 tension. The adjustment of CSF $[HCO_3^-]$ would thus contribute to respiratory adaptation under any condition of respiratory and metabolic acid-base disturbances, whereby a major contribution of carotid and aortic bodies to the chemical control of breathing was assumed.

Since then and up to now, there is still controversy about the relative contribution of peripheral chemoreceptors to respiratory control in metabolic acidosis. Winterstein's own experiments in rabbits using reversible cold blockade of the carotid sinus nerves (34) showed that the share of the carotid chemoreceptors was unimportant if not entirely lacking in the ventilatory response to acute metabolic (and respiratory) acidosis.

Loeschcke's contribution to the symposium (35) most unequivocally supported Winterstein's reaction theory by demarcating chemosensitive areas at the lateral apertures of the fourth ventricle. According to Winterstein, the influence of CO_2 on ventilation was explained by permeation of CO_2 through the blood-brain barrier, possibly determining $[H^+]$ in the environment of "some superficial structures in the neighborhood of the lateral aperture of the fourth ventricle."

To explain those influences of blood $[H^+]$ on ventilation that were unexplained by the peripheral chemoreflexes, transient responses of tidal volume and of choroid plexus surface pH were studied in response to intravenous injection of hydrochloric acid (cp. 36). Since instead of hysteresis a seemingly unique function was observed, Loeschcke postulated an important central mechanism sensitive to CSF $[H^+]$ accounting for the "metabolic" respiratory drive of blood $[H^+]$, which remained after sectioning the chemoreceptor nerves. The ventilatory drive elicited by a central $[H^+]$ receptor in the region of the lateral aperture would serve homeostasis of the freshly produced CSF during metabolic and respiratory acid-base changes.

C. Central Chemoreception of Cerebrospinal Fluid Composition

Fencl likewise attempted a unifying theory of central respiratory control by using ventriculo-cisternal perfusion combined with inhaled CO_2 (37, cp. 38). The main question was, whether there existed central chemoreceptors and whether these were superficially located at the ventrolateral medulla. The presented results appeared to support the view originally expressed by Winterstein's third version of his reaction theory (21) "that central chemical control of breathing involves the acidity of fluid in immediate contact with respiratory neurons, which may be located at some distance from cerebrospinal fluid in the large cavities."

Respiratory responses to inhaled CO_2 were measured in unanesthetized goats during ventriculo-cisternal perfusion with normal, low, and high $[HCO_3^-]$ in mock CSF. Lowering $[HCO_3^-]$ and pH in CSF increased alveolar ventilation and therefore decreased arterial and CSF P_{CO_2}. Related to the same decrease in CSF pH, this was only about 15% of the response to inhaled CO_2. It was essential to find a unique function between pulmonary ventilation and a tissue fluid pH value, which could be estimated along a gradient between CSF and blood composition for all the conditions of CSF $[HCO_3^-]$ and inhaled CO_2 studied. Since large ventilatory responses occurred in response to inhaled CO_2 even when $[H^+]$ in cranial CSF was maintained constant, the necessity for special H^+ receptors at the medullary surface was questioned, but generally it was concluded that the locus of excitability should be closer to cerebral capillaries than to the fluids of the large cavities. An interesting discussion arose between Loeschcke and Fencl. In the experiments presented, goats with high bicarbonate at the same CSF pH were breathing more. This was just the opposite to what was found in perfusion experiments, where proportionate $[HCO_3^-]$ and P_{CO_2} increases in the CSF depressed ventilation (25,26,39,40). The essential reply by Fencl that "our P_{CO_2} was set by the goat" was not quite recognized. Obviously, at the same CSF pH and P_{CO_2}, ventilatory responses to inhaled CO_2 are completely different from those to hypercapnic superfusion.

Leusen (30) referred to several publications of Winterstein with special emphasis on the third version of the reaction theory (21) and later studies of ammonium chloride acidosis (23). The first systematic perfusions of the cerebral ventricles with bicarbonate and nonbicarbonate buffers were performed by Leusen (39,40) to prove Winterstein's "liquorogenic theory," postulating a determinant role for the fluids surrounding the respiratory centers. Up to that time, the greater augmentation of ventilation by CO_2 inhalation than by fixed acid infusion for the same change in blood pH was generally accepted and ascribed to easier penetration of CO_2 through the blood-brain barrier. Unexpectedly, with an increased $[HCO_3^-]$ in CSF at constant $[H^+]$ due to proportionate increase of P_{CO_2}, ventilation transiently rose but stabilized at a lower level, whereas with lower $[HCO_3^-]$ and proportionate decrease of P_{CO_2}, respiratory activity transiently fell but stabilized at a higher level. Although these findings did not support the liquorogenic version of the reaction theory, Leusen transferred the role of $[H^+]$ by postulating some $[H^+]$ gradient across the receptor cell membrane due to different penetration abilities of HCO_3^- and CO_2.

However, a more striking evidence against the reaction theory was the progressive and regular increase in ventilation in response to perfusing the cerebrospinal space with mock CSF at normal pH, in which the normal bicarbonate was replaced by Tris buffer (41). Changing this perfusion from a more alkaline to a more acid solution produced practically no further increase in ventilation. These observations were comparable with results obtained in earlier experiments, when solutions were used containing either borate buffer or no buffer at all (39,40).

The impression was thus confirmed that the bicarbonate concentration at the cerebral tissue level played a crucial role in fixing the basic activity of the respiratory centers. The rather provocative observations of Leusen were not sufficiently considered at the symposium, but there is now growing evidence of a special role for bicarbonate.

V. On the Way to Today's Research on the Ventrolateral Medulla

A. Search for the Central Chemosensor by Loeschcke's Group

Localization of Neuronal Activities and Cholinergic Mechanisms

The problems that arose at the Brooks Symposium have been followed up in a diverging manner. One main road was trod by Loeschcke and collaborators in Bochum, their experimental work on central chemosensitivity being reviewed and summarized by Schläfke (42), Loeschcke (43), and Kiwull-Schöne (44).

Localization experiments—using focal cooling, natural stimuli, electrical stimuli, and single-unit recording in vivo—were combined with histological studies. Three bilateral areas at the ventral medullary surface involved in central chemosensitivity were circumscribed more precisely, at the "caudal," "intermediate," and "rostral" areas (42,43). Medial of the hypoglossal root and corresponding to the caudal area, a rather superficial peak distribution of neurons (between 150 and 200 μm below the surface) was found, from which it was possible to drive ventilation by electrical microstimulation before histological mapping (45,46). These neurons confirmed later by electron microscopy (47, 48) were characterized by a random firing pattern, reminiscent of peripheral chemoreceptors, and were sensitive to both inhaled CO_2 and pH/[HCO_3^-] changes of the superfused CSF (49,50).

A functional localization of chemosensitive substrates was achieved by a modified version of the early perfusion experiments (25), carried out by Berndt et al. (51–54), now directly superfusing the ventral medullary surface in combination with CO_2 inhalation. The central chemosensors would thus be accessible to CO_2 of the blood and to H^+ of the CSF. Regarding the diffusion laws for CO_2 and HCO_3^- in brain tissue, the extracellular pH value was calculated at different depths below the ventral medullary surface. The highest correlation between pulmonary ventilation and extracellular pH was found in a depth of 250 to 300 μm. This apparently supported a "unique function (reaction) theory" and the idea of rather superficially located chemosensitive structures, distant from the respiratory centers as such.

Neurophysiological studies on ventral medullary surface neurons have been carried out in all three areas in vivo, as reviewed by Schläfke (42): among a total number of 185 units, 45% responded to changes in surface pH. Detailed studies of quantitative relationships between neuronal discharge rates and lowered surface pH in the "intermediate area" revealed comparable results (55): Among 49 neurons, 31 significantly increased their impulse rate to any kind of acidification, irrespective of whether it was induced by CO_2 inhalation, intravenous acid injection, or local superfusion with low HCO_3^- buffer solutions. There were, however, two strikingly different sensitivity ranges, whereby about one-fifth of the cells showed a sensitivity as low as reported for brain slices (about +35% per 0.1 units decrease in pH), whereas the majority showed an approximately tenfold higher sensitivity (+300%/−0.1 pH), reaching the order of magnitude of pulmonary ventilation in response to medullary surface pH-changes induced by CO_2-inhalation (56–58). The neurons were likewise activated in different ways, but coordination between sensitivity and mode of acidification is still unknown.

After the pioneer work of Fukuda and Honda (59), in vitro brainstem slice recordings were continued under different respiratory and metabolic acid-base conditions. More than 50% of the cells investigated in superficial slices of the ventral medulla were found to be H^+-sensitive, but only 13% of those in the dorsal medulla as in whole brain. The H^+-sensitive neurons in the ventral medullary surface slices showed an average rise of impulse activity by not more than about 25% in response to a decrease in pH of 0.1 units (60), whereby CO_2 and H^+ responses were not differentiated.

In order to furnish proof of a cholinergic mechanism underlying central H^+ chemosensitivity, pharmacological experiments with cholinergic agonists and antagonists were conducted on intact animals in vivo and on brainstem slices in vitro.

In vivo, it could be shown that the spatial distribution of the respiratory responses to locally applied acetylcholine (ACh) or nicotine coincided rather accurately with that of the respiratory responses to locally varied pH or electrical microstimulation (42,45,61). These responses could be diminished or abolished by prior application of cholinergic blocking agents, whereas inhibition of ACh-esterase (AChE) enhanced resting ventilation without a change in CO_2-sensitivity (62). Recently, these findings were extended by Trouth et al. (50). The H^+ and CO_2 responses of non–respiratory modulated neurons in the "caudal area" were depressed by topically applied atropine and enhanced by ACh, suggesting cholinergic involvement at muscarinic receptor sites. Naloxone potentiating the excitatory responses to low pH, ACh, and physostigmine pointed to cholinergic and opioid interaction.

In vitro, it could be demonstrated that H^+-sensitive excitation of neurons depended on intact cholinergic synaptic transmission (63). Thus, three possible roles of H^+ were discussed, either increasing the sensitivity of cholinoceptive neurons to ACh, or augmenting the release of ACh from presynaptic terminals due to greater ionization of calcium, or inhibiting AChE activity.

Since then, an enormous number of studies have been carried out dealing with putative transmitters involved in central and peripheral chemoreception by others. Some recent attempts by Prabhakar should be mentioned (64), since, when he was a young graduate student, his views were decisively shaped by Loeschcke.

Signal Transmission from Blood to Brain ECF

In addition to the theoretical calculations of brainstem extracellular fluid (ECF) pH by Berndt et al. (51), direct measurements were attempted by Ahmad and Loeschcke (65–67). Based on the morphological evidence of free access of substances from the subarachnoid to the intercellular spaces (47,48), ECF pH and P_{CO_2} could be measured with sufficient accuracy immediately at the ventral surface of the medulla by balanced noninvasive macroelectrodes. Following on-and-off step changes of inspired CO_2, the time courses of pulmonary ventilation and medullary (but not cortex) surface pH were closely correlated with each other without hysteresis, pointing to acid-base changes in the ECF environment of the central chemosensor as adequate stimulus (65).

Since the ECF does not contain considerable amounts of protein, an increment in P_{CO_2} was expected to reduce pH (according to the Henderson-Hasselbalch relationship) with almost no change in $[HCO_3^-]$. However, marked increases in ECF $(HCO_3^-]$ were calculated from simultaneous measurements of pH and P_{CO_2}. Adjacent glial cells containing carbonic anhydrase (68) were thus assumed to serve as cellular buffer for the ECF, in much the same way as red cells in contact with blood plasma (Hamburger shift). To support the idea of cellular buffering, a (third) chloride electrode was applied, whereby in the ECF of the ventral medulla a one-to-one monoexponential increase in [HCO_3^-] and decrease in [Cl^-] could be demonstrated in response to step rises in end-tidal P_{CO_2} (66).

By analogy, a nonelectrogenic carrier-mediated HCO_3^-/Cl^- exchange across the blood-brain barrier was postulated for metabolic acid-base changes in the blood. Indeed, correspondingly fast inverse changes of both anions in response to a sudden bicarbonate load at constant P_{CO_2} could be demonstrated. It was suggested that rapid signal transmission from blood to brain should provide rapid adjustment of pulmonary ventilation by central chemosensitivity during both respiratory and metabolic acid-base disturbances (67).

The topic of the last doctoral thesis initiated by Loeschcke was to compare signal transmission from blood to brainstem ECF with ventilatory responses during acute respiratory and metabolic acidosis. In his thesis, Shams studied the ventilatory reactions to inhaled CO_2 and to intravenously infused sulfuric acid. Unexpectedly, pulmonary ventilation during respiratory acidosis reacted about 15 times more sensitively to the same decrease in ECF pH than during metabolic acidosis (58). This coincided at least qualitatively with our own findings (56,57), when metabolic acidosis was not caused by infusion of strong mineral acids but instead by endogenously accumulated lactic acid. In spite of a considerable reduction in ECF pH, no ventilatory drive was thereby elicited at all, although the CO_2 sensitivity was not disturbed.

From their early perfusion experiments, Loeschcke et al. (25) had not recognized the rather low H^+-sensitivity of ventilation to a decrease in CSF [HCO_3^-] at constant P_{CO_2}. Tidal volume (V_T) increased only by 12% per 0.1 unit fall in CSF pH upon low [HCO_3^-] perfusion, being quite in the order of magnitude observed later on for ventilatory responses to ECF pH changes induced by intravenous acid infusions (58,69,70). Furthermore, it had been concluded from the superfusion studies that only H^+ ions, and not CO_2, were able to drive lung ventilation, since tidal volume was depressed by a rise in superfused CSF P_{CO_2} at constant [HCO_3^-]. However, a CO_2 drive in addition to that of ECF H^+ could not generally be ruled out, since inhaling CO_2 to keep CSF (or ECF) pH constant at enhanced [HCO_3^-] rather increased than decreased pulmonary ventilation markedly (37,38,58).

Thus, Loeschcke (43), trying to explain these discrepancies, suggested that H^+ ions, unlike CO_2, might not under all circumstances have free access to the receptor, e.g., possibly the synaptic cleft of cholinergic neurons.

B. Persisting Objections Against the Reaction Theory

Hypoxia and Endogenous Lactacidosis

Before the discovery of the peripheral chemoreceptors, the difficulty of reconciling hypoxic dyspnea and blood alkalinity was solved by assuming "asphyxiating substances" in the centers according to the second version of the reaction theory (17). After the discovery of the peripheral chemoreceptors, Winterstein (22) claimed that the hypoxic ventilatory depression and CO_2 rise observed in chemoreceptor-deprived animals was an "experimentum crucis" against the traditional CO_2 theory, for once not considering the role of hypoxia-induced acid metabolites for central chemosensitivity any more.

To explore the role of hypoxia-induced lactic acid, we recorded the ventral medullary surface pH and pulmonary ventilation in chemodenervated cats after repeated periods of hypoxia (56,57). The distinct acidification of ECF (by about 0.15 pH units), which persisted during posthypoxic recovery, unexpectedly did not lead to any ventilatory response at all (Fig. 7). On the other hand, a strong and reproducible rise in ventilation (by about +360%/−0.1 pH units) could be observed when ECF pH was reduced by CO_2 inhalation. Under the same experimental condition, we found extraordinarily high concentrations of lactic acid in the arterial blood of chemodenervated rabbits but likewise no ventilatory responses (71).

To explain the discrepancy between the lack of ventilatory reaction to endogenous brainstem lactacidosis and the high sensitivity of pulmonary ventilation against CO_2-induced changes in ECF pH, the unique significance of brainstem ECF [H^+] for central chemoreception had to be questioned. Kiwull-Schöne and Kiwull (56) speculated about taking the intracellular pH as reference value, so that, assuming chemosensory structures were affected by the intra/extracellular H^+ gradient, the simultaneous decrease of pH in both compartments would remain ineffective as respiratory drive.

More recently, this concept was substantiated by similar observations of Xu et al. (72) in awake goats, showing "anomalous" hypoxic acidification of medullary ventral surface, no ventilatory reaction to CO-hypoxia, but maintained responses of ECF pH and ventilation to inhaled CO_2. These observations were discussed to be compatible with the hypothesis that the chemosensor cell might be stimulated by a fall of the transmembrane H^+ gradient, whereby internally generated (hypoxic) acidosis would increase the gradient and therefore depress rather than stimulate the chemoreceptors.

For hypercapnia, there is indeed experimental evidence of a decreased transmembrane H^+ gradient due to more effective buffering in the intracellular than in the extracellular compartment (73,74). For anaerobic intracellular lactic acid production, however, direct evidence of an increased transmembrane H^+ gradient is still lacking.

To explain the small or even absent ventilatory response to the significant fall in brainstem ECF pH following fixed acid infusion (57,58,69–71), a necessary implication of this rather attractive hypothesis would be that extracellular metabolic acidosis should lead to a less pronounced diminution of the gradient than respiratory acidosis. The same argument is valid for the rather strong metabolic brainstem acidification (by 0.250 to 0.730 pH units) reached by super-

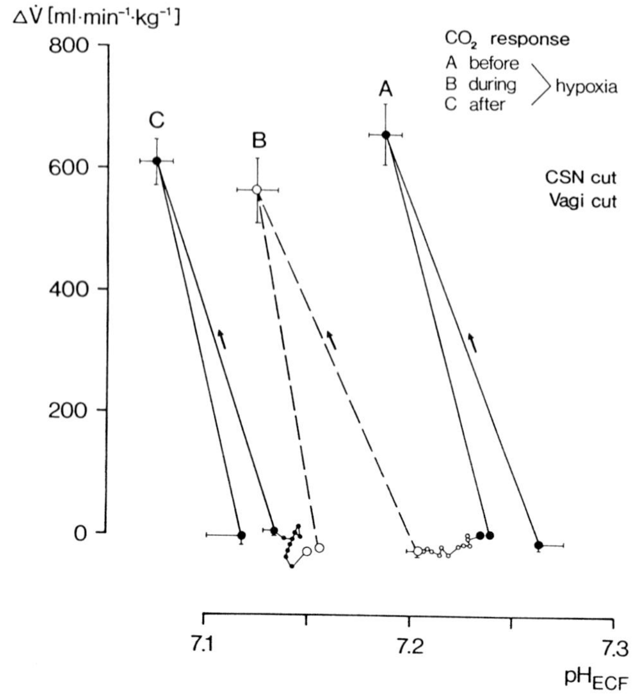

Figure 7 Lack of ventilatory reaction to hypoxia-induced metabolic brainstem acidosis despite maintained CO_2 sensitivity. Changes in pulmonary ventilation ($\Delta\dot{V} \pm$ SEM) from baseline level (304 ± 15 mL/min^{-1}/kg^{-1}) as a function of the extracellular fluid (ECF) pH at the ventral medullary surface in 11 carotid chemodenervated cats during hyperoxia (●—●) and hypoxia (○—○) before and after step rises in Pa_{CO_2} of about 1.3 kPa (A,B,C). (Slightly modified from Ref. 57, Fig. 1, courtesy of Plenum Press, New York.)

fusion with isocapnic low-[HCO_3^-] mock CSF, leading to a rise in tidal volume of only 10% to 15%/−0.1 pH units in cats and rats (43,75).

Thus, the unifying transmembrane H^+-gradient hypothesis would only explain the rather low central chemosensitivity to metabolic acid-base changes in brainstem ECF, if these were less effectively buffered inside the cell than hypercapnia.

Breathing Pattern Differences

Another argument against a unique central chemoreception for respiratory and metabolic acidosis concerns the different type of breathing pattern. The now classic spirometric recordings during perfusion of the fourth ventricle with low [HCO_3^-] CSF by Loeschcke et al. (25) demonstrate a nearly exclusive rise in tidal volume.

Similarly, the ventilatory response to acute and chronic metabolic acidosis in chemodenervated and vagotomized rabbits exclusively consisted in a tidal volume rise, respiratory rate being even reduced (71,76). On the other hand, the central chemosensitive breathing pattern response to inhaled CO_2 in these rabbits included both tidal volume and respiratory frequency in a rather definite proportion.

According to Hey et al. (77), the type of breathing pattern response can be quantified by the linear relationship between ventilation ($\dot{V}$) and tidal volume (V_T), $\dot{V} = m\ (V_T\text{-}k)$. The $\dot{V}/V_T$ relationships we found for acute respiratory acidosis and chronic metabolic acidosis are compared in Figure 8: The CO_2-responses are characterized by rather high values of $m \pm SD = 42.0 \pm 2.3\ min^{-1}$

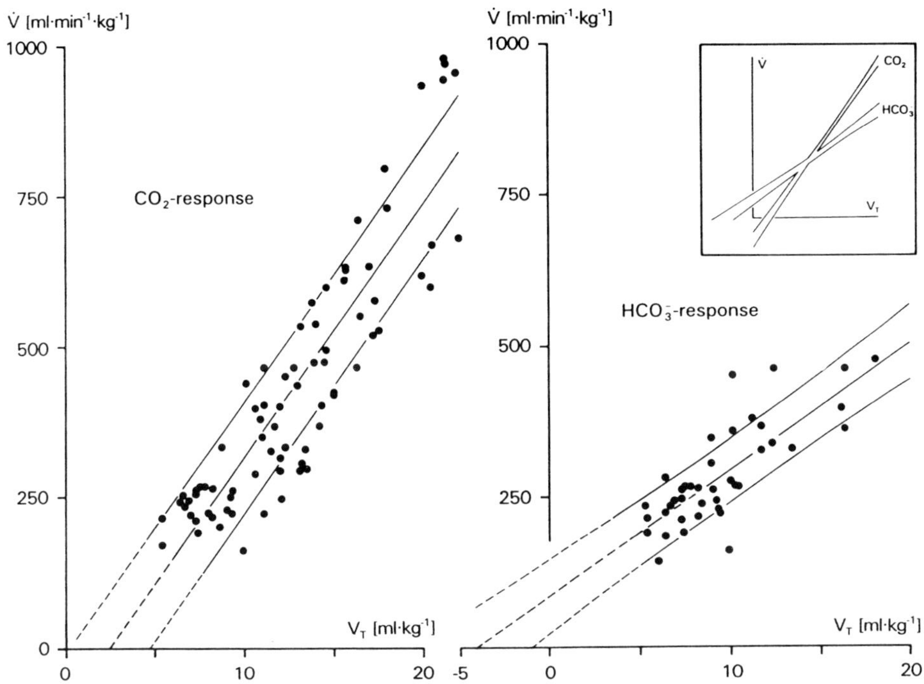

Figure 8 Quantitative comparison of breathing pattern during respiratory and metabolic acidosis. $\dot{V}/V_T$ relationship in anesthetized chemodenervated and vagotomized rabbits during CO_2 inhalation ($n = 80$, left-hand diagram) and during chronic metabolic acidosis in the range between 13 and 28 m*M* [HCO_3^-] ($n = 41$, right-hand diagram). Linear regression analysis $\pm$SD (or $\pm$SEM in the small insert) showed slopes (m) and intercepts (k) with the V_T axis to be significantly different in the two groups ($P < 0.0002$, unpaired t-test). (Data from Ref. 76.)

and positive values of $k \pm SE = 2.5 \pm 0.5$ mL/kg, pointing to an increasing involvement of respiratory rate, whereas the responses to lowered arterial [HCO_3^-] (in the range between 28 and 13 m*M*) consist in a pure tidal volume response and even decreasing frequency, thus showing smaller values of m (20.8 ± 2.6 min^{-1}) and negative values of k (-4.2 ± 1.1 mL/kg). Interestingly, the same principle can be demonstrated when re-evaluating the experiments of Pappenheimer et al. (38) in awake goats, using ventricular-cisternal perfusion combined with CO_2 inhalation. Breathing pattern responses to CO_2 inhalation are likewise characterized by high values of m (~ 41 min^{-1}) and positive values of k (~ 6 mL/kg), whereas those to changed CSF [HCO_3^-] at constant P_{CO_2} yield smaller values of m (~ 12 min^{-1}) and negative values of k (~ -18 mL/kg).

This consistently different quality in the breathing pattern response to either respiratory or metabolic acid-base changes suggests different underlying central chemosensory control mechanisms.

VI. Today's Research on the Ventrolateral Medulla

A. NMR Studies of Whole-Brain Intracellular pH

One promising type of modern approach to explore the still undefined nature of central chemosensitivity is noninvasive nuclear magnetic resonance (NMR) spectroscopy. Under the topic of chemical respiratory control, whole-brain intracellular pH (pH_i) was estimated by NMR spectroscopy in anaesthetized animals (74,78) and conscious humans (73,79). Based on the NMR pH measurements in brain of conscious humans, Lassen (80) raised the crucial question of whether the central chemoreceptors were sensitive to intracellular pH (pH_i) rather than to extracellular pH (pH_e). Conclusions were drawn from comparing effects of hypercapnia and oral acetazolamide: since during hypercapnia the brain pH_i was diminished and pulmonary ventilation concomitantly enhanced but following acetazolamide application neither variable was changed, pH_i but not pH_e was assumed to be the central chemoreceptor stimulus. This rather "economic" supposition may be taken as some resuscitated intracellular reaction theory.

On the other hand, the simultaneous determination of intracellular and extracellular brain pH at different levels of arterial P_{CO_2} by Portman et al. (74) lent support to the transmembrane H^+ gradient hypothesis (72) at least for hypercapnia, since better buffering of CO_2 inside the cell should markedly decrease in the pH gradient between the extracellular and the intracellular space and may influence central respiratory control. As pointed out above, the experimental proof of the H^+ gradient hypothesis for endogenous and exogenous metabolic acidosis is still lacking.

Thus, in spite of this very elegant noninvasive technique to determine intracellular brain pH, the adequate stimulus for central chemosensitive control of breathing is not yet finally determined, but unifying theories are supported.

B. Reduced Central Chemosensitive Systems

Another main line of modern approaches to the problem of central chemosensitivity is the development of reduced chemosensitive subsystems in vitro to eliminate complex whole-system interactions.

The isolated "brainstem–spinal cord preparation", including cranial and phrenic nerve roots to record respiratory motor output, is usually taken from neonatal rats and superfused with oxygen-enriched mock CSF at reduced temperature. Besides rhythmogenesis, this preparation serves to explore central chemosensitivity and differential effects of respiratory and nonrespiratory acidosis.

Harada et al. (81) found maintained respiratory responses directly related to extracellular $[H^+]$, when $[HCO_3^-]$ in the superfusate was changed at constant P_{CO_2}. The average pH sensitivity of phrenic minute activity was 15%/0.1 pH units and both phrenic amplitude and frequency were involved. Unexpected was the distinct transient rise in phrenic discharge amplitude alone, elicited by superfusion with high CO_2 at constant pH, pointing to some specific CO_2 sensing mechanism. Furthermore, there was no response to extracellular acidification by HCl in HEPES buffer free of CO_2/HCO_3^-, comparable to in vivo findings by Leusen (30,39,40). Central chemoreceptors eliciting strikingly different breathing pattern responses appeared to be independently sensitive to either H^+/HCO_3^- or CO_2.

Contrarily, a pure frequency response to hypercapnic superfusion was found by Okada et al. (82) in a similar preparation. However, depth profiles of pH and P_{O_2} have shown that the superfused brainstem-spinal cord preparations are more or less anoxic and acidic except for a small layer below the surface (83), possibly causing this functional variability.

Thus, it is reasonable to perfuse the vertebral arteries additionally, as done in preparations of adult guinea pigs by Morin-Surun et al. (84). This better supplied preparation did react to high CO_2 at constant pH in the superfusate by increasing again the burst amplitude of hypoglossal nerve discharge. However, if the same solution was delivered via the vertebral arteries, respiratory frequency was accelerated. Nonrespiratory acidosis at constant P_{CO_2} affected the frequency throughout, disregarding the mode of application. In agreement with Harada et al. (81), this preparation also revealed results in favor of different central chemosensitive subsystems, some sensitivity to CO_2 at ventral surface affecting burst amplitude, but no specific H^+-sensitivity.

The next step was to reduce the system to progressively thin brainstem slices. An extensive study of respiratory and nonrespiratory acid-base effects on the neuronal activity in this slices of the ventral medullary surface layer was performed by Fukuda (85). Only a small number of the neurons studied (12%) responded as expected from the reaction theory, namely to the both kinds of pH reduction either by high CO_2 at normal $[HCO_3^-]$ or by low $[HCO_3^-]$ at normal P_{CO_2}. The majority (72%) was activated by high CO_2 even at normal pH.

A neuronal subpopulation of about 30% did respond to low pH only when induced by high CO_2, another of about 40% only when low pH was induced by low [HCO_3^-]. Thus, this slice approach, in much the same way as the isolated brainstem preparation, did not support the unifying reaction theory, but CO_2 and HCO_3^- appeared to act rather differently on chemosensitive elements.

Another study by Jarolimek et al. (86) on ventral medullary slices in the rat revealed that 43% of the investigated neurons were activated by low bicarbonate acidification, but in all cases only transiently. Increased P_{CO_2} at constant pH, however, generally depressed the activity, except for a few chemosensitive neurons, in which the response to the same extracellular pH reduction was potentiated by hypercapnia. This observation, pointing at least to some prominent role of CO_2 over fixed acids, led these authors to suggest a role for intracellular pH in the central chemosensitive process.

The next promising approach to explore the nature of central chemosensitivity is based on organotypic tissue cultures of small but defined brainstem regions, adequately supplied by oxygen and substrates, and disconnected from extended neuronal networks. Results available up to now are, however, contradictory (87,88).

Bingmann et al. (87) performed intracellular recordings in neurons of cultured (horizontal) medullary slice explants under different respiratory and nonrespiratory acid-base conditions in the superfusate. He found 62 neurons with periodic (respirationlike) activity not confined to any dorsal or ventral region. This periodic activity consisted in depolarizing (D-) waves of 100 to 300 ms duration, generating short trains of action potentials. In part of the neurons, occurrence of these periodic events was tested for decreased [HCO_3^-] at constant P_{CO_2}, for changes in P_{CO_2} at constant [HCO_3^-] and for HCl administration. Periodic activity changed with all kinds of acid-base deviation from the control pH (7.4) with an average sensitivity of about 7%/0.1 pH units. This appeared to be consistent with the hypothesis of a unique dependence on H^+ activity of medullary neurons involved in central respiratory chemosensitivity.

Neubauer et al. (88) came to the opposite conclusion. They likewise used explant tissue cultures of ventral and dorsal medulla. All ventral brainstem neurons showed steady pattern firing. About half of the dorsal neurons showed either "steady" or "bursting" firing pattern. Part of both kinds of neurons were investigated for chemosensitive behavior. About one-third of both ventral and dorsal neurons increased firing frequency in response to hypercapnic acidosis (in one representative response +10%/0.1 pH unit). In some neurons tested for lowered [HCO_3^-], no reaction was found. Thus, chemosensitive neurons appeared to be present in explant tissue cultures of both the ventral and dorsal medulla, but these cells responded to changes in pH only in association with changes in CO_2 tension.

The role of carbonic anhydrase has also been questioned in this respect. Martin and Neubauer (89) studied the effect of acetazolamide on CO_2 sensitivity of cultured medullary neurons using the perforated-patch technique for whole-cell recording. Respiratory acidosis produced an overshooting depolarization and spike discharge, more rapidly adapting after the addition of acetazolamide. It appeared that in these neurons, carbonic anhydrase was required to sustain the membrane responses to CO_2-induced acidosis. Unfortunately, nothing is known about concomitant intracellular pH changes in these neurons.

E. The Peripheral Chemoreceptors Model for H^+ Sensitivity

There are recent publications dealing with intra/extracellular pH relationships in cultured carotid body type I cells under acid-base conditions comparable to those studied in the central chemosensitive system. The idea to take the peripheral chemoreceptors as a model for H^+-chemosensitivity arose as early as the third and fourth version of the reaction theory (10,21). The acid theory to explain the peripheral chemoreception was followed up by Torrance (90), initially postulating the extracellular pH as being the unique and adequate stimulus during hypoxia, hypercapnia, and acidosis. Thereby, the question arose analogous to that raised earlier for central chemosensitivity, namely, whether arterial CO_2 was sensed independently of pH (91) or not (92).

More recent experiments, comparing membrane permeating and nonpermeating carbonic anhydrase inhibitors, led to the replacement of the extracellular theory of peripheral chemoreception by an intracellular one (93), suggesting that intracellular carbonic anhydrase and a rise in $[H^+]$ were involved at least in the quick response to CO_2.

Indeed, isolated carotid body type I cells, stained with a pH-sensitive fluorescent dye, exhibited intracellular acidification upon changing extracellular nonbicarbonate buffer to isohydric CO_2/HCO_3^- solution. The rate of intracellular acidification, probably due to CO_2/H^+ conversion, was considerably slowed by the membrane permeant carbonic anhydrase inhibitor acetazolamide (94). Likewise, switching from bicarbonate-free to prolonged CO_2/HCO_3^- buffer perfusion of the isolated carotid body in vitro caused biphasic responses of sinus nerve discharge. Thereby, intracellular carbonic anhydrase inhibition, affecting only the peak-response but not the persisting response, substantiated that the CO_2 response of the peripheral chemoreceptors most probably is mediated by a rise of intracellular $[H^+]$, independent of extracellular $[H^+]$ (95).

The possible role of intracellular $[H^+]$ for chemoreception in metabolic acidosis has been investigated in isolated glomus cells by varying either external $[HCO_3^-]$ or P_{CO_2} independently. Wilding et al. (96), using a pH sensitive fluorrescent dye, made the fundamental observations that the steady state relationship between intra- and extracellular pH was the same under different respiratory and nonrespiratory acid-base conditions as well as in the presence and absence of extracellular $CO_2/[HCO_3^-]$. The exceptional sensitivity of the intracellular to the extracellular pH suggested pH_i of the glomus cell to be a link in the chemoreceptor's response to external acidity.

With the same technique, Buckler et al. (97) investigated dynamic responses of intracellular pH responses to extracellular hypercapnic acidosis, isocapnic acidosis, and isohydric hypercapnia. Hypercapnic acidosis led to a rapid and isocapnic acidosis to a slower reduction in pH_i, reaching the same steady-state level. Isohydric hypercapnia, however, although leading to rapid transient reductions in pH_i, did not significantly affect steady-state pH_i over a wide range of P_{CO_2}. By comparing these dynamics of intracellular pH to those of carotid sinus nerve discharge recorded elsewhere in situ under comparable acid-base conditions (92), the authors indirectly concluded that peripheral chemoreceptors should uniquely respond to intracellular pH.

However, there are no direct experiments to support intracellular $[H^+]$ as being the adequate stimulus for peripheral chemoreceptor activation. The only report on the resting potential and intracellular pH in glomus cells measured simultaneously by microelectrodes is that of He et al. (98): independent of pH_e, the membrane potential was hyperpolarized by low pH_i, whereas depolarization was correlated with a lower pH_i–pH_e difference, much in agreement with the membrane H^+ gradient hypothesis for central chemosensitivity proposed by Xu et al. (72).

Thus, it would be useful to extend the techniques developed for pH_i measurement in isolated glomus cells to the reduced preparations of central chemosensitive structures, to verify or to discard unifying intracellular H^+ or transmembrane H^+ gradient theories.

VII. Open Questions and Future Perspectives

The history of research on chemical control of breathing is characterized by partial acceptance of the unique role of $[H^+]$ sensitivity but also by repeated objections against the rather reductive reaction theory. By means of whole-system approaches, several research groups have substantiated unifying theories, earlier concerning central nervous extracellular fluid $[H^+]$ (25,26), including within tissue $[H^+]$ gradients (38,51–54). Since, however, techniques for direct measurement of brainstem ECF pH became available, its unique role in ventilatory responses to inhaled CO_2 and to intravenous acid infusion has been challenged (57,58,68,70,71).

With the possibility to determine whole-brain pH_i by NMR techniques, the unique role of extracellular $[H^+]$ as adequate stimulus for central chemosensors has been replaced by intracellular $[H^+]$ (80). For instance, the behavior of pH_i in glomus cells could serve as a model for central chemosensitive neurons and explain the repeatedly observed respiratory responses to proportionate rise in CSF P_{CO_2} and $[HCO_3^-]$ in vivo (30,39,40) and in vitro (81,85,88) by CO_2 more easily penetrating cell membranes and acidifying the cytosol.

If, for the central chemosensitive process, CO_2 had to be converted to intracellular H^+, then for quick CO_2 reactions of the respiratory system carbonic anhydrase would be required. Carbonic anhydrase appeared indeed to act at different sites in the region of medullary chemoreceptors, on either side of the blood-brain barrier (99). More recently, focal application of small amounts of acetazolamide (AZ) to different brainstem sites was performed by Coates et al. (100). The induced slope reduction and slowing of the respiratory response to CO_2 step rises speak in favor of H^+ as being the adequate stimulus for detection of transient changes. Since respiratory responses to AZ were not uniquely correlated with medullary surface pH, this altogether led to the assumption that central chemoreceptors responded to changes in intracellular rather than extracellular fluid pH.

However, this assumption of an attenuated intracellular acidification cannot explain why considerable respiratory effects can be elicited by microinjections of AZ at "widespread sites" of the medulla (100,101) and by systemic carbonic anhydrase inhibition in servoventilated animals (57,102). Likewise, another important observation is not compatible with the intracellular H^+ theory, namely the lack of ventilatory reaction to posthypoxic brainstem lactacidosis in spite of a maintained CO_2 response (56,57). This led to transmembrane H^+ gradients being considered earlier (56) and substantiated more recently (72,103).

Transmembrane H^+ gradients have also been proposed by Severinghaus (103) to explain the respiratory drive by carbonic anhydrase inhibition: The seeming paradox that systemic administration of AZ does acidify brain ECF at constant brain tissue P_{CO_2} with no fall in ICF pH (78) could be due to a rise in brain tissue $[HCO_3^-]$, opposed by metabolically generated carbonic acid, which, better buffered inside the cell and diffusing out, would lead to a greater decrease in extracellular than intracellular pH. This, in agreement with the transmembrane H^+ gradient hypothesis, could account for respiratory stimulation by AZ due to extracellular carbonic acidosis in much the same way as for the lack of stimulation due to intracellular hypoxic acidosis (103).

Interestingly, there is one report on simultaneous recording of intracellular pH and membrane potential pointing directly to a possible role for the transmembrane H^+ gradient in glomus cell depolarization (98).

There are, however, several findings on central chemosensitive systems in vivo and in vitro that are not compatible with either intracellular or transmembrane $[H^+]$ theories. The intracellular $[H^+]$ or transmembrane H^+ gradient in peripheral chemoreceptor type I cells, when taken as a model for central H^+ sensitivity, could not generally explain the differing respiratory responses observed. The intra/extracellular pH relationship in the glomus cell model remained the same under all respiratory and metabolic acid-base conditions and was independent of whether bicarbonate or nonbicarbonate buffer was used. Thus, neither the differing effects of metabolic and respiratory acidosis in isolated brainstem-spinal cord preparations, slices, and tissue cultures (81,85,88) would be explained, nor the lacking effects of acid nonbicarbonate buffers on respiratory responses (81), unless central chemosensitive neurons underwent transmembrane H^+ distributions different from those in glomus cells. Furthermore, it is also inexplicable in the light of this theory why respiration in vivo does not react to acid nonbicarbonate buffers at the ventral medullary surface (30,39,40) and why, in spite of predicted intracellular CO_2/[H^+] conversion, local high CO_2/saline injections in central chemosensitive areas remain ineffective (101).

There is persisting doubt about unifying hypotheses throughout, based on strikingly different sensitivities to respiratory and metabolic blood or CSF acidosis observed in vivo, either during ventriculo-cisternal perfusion and medullary superfusion (26,38,51–54) or during intravenous acid infusion (57,58, 69,70,71). Likewise, respiratory pattern analysis of CO_2 responses and acid responses in vivo (76) led to the postulate of different central chemosensitive subsystems for CO_2 and [H^+] or [HCO_3^-]. Furthermore, the observation that AZ did not affect the sensitivity of the steady state CO_2 response (57,100,102) but did affect the apnoeic threshold and breathing pattern (102) does not rule out a possible role of CO_2 independent of [H^+].

Experimental evidence also from reduced central chemosensitive systems pointed to the possibility of at least two different subsystems of central chemosensitivity. This has been proposed for medullary slices by Fukuda (85) and for explant tissue cultures by Neubauer et al. (88) as well as for isolated brainstem–spinal cord preparations by Harada et al. (81) and Morin-Surun et al. (84), the latter approaches also showing characteristic differences in breathing pattern. Intracellular pH measurement in addition to neuronal activity in reduced brainstem preparations would be necessary to verify or falsify the significance of intracellular H^+ or the transmembrane H^+ gradient as unique stimulus for central CO_2 and acid sensitivity and, if falsified, to substantiate the involvement of at least two different subsystems, tentatively a more sensitive blood-related CO_2 system and a less sensitive but rather effective CSF-related acid-base system.

In spite of the considerable development of experimental techniques up to now, the exact nature of central chemosensitivity is still undefined. This may be due to inherent dangers in highly sophisticated techniques that sometimes the problem gets lost. Winterstein, although remaining up to date with new techniques throughout his scientific life, never took a method as an end in itself but always kept close to the primary question. Originating from Winterstein's theory there is still the problem whether a unique stimulus of central chemosensitivity or CO_2 different from H^+ elicit the drives to pulmonary ventilation during respiratory and metabolic acidosis.

Acknowledgments

Thanks are expressed to my friends and colleagues who have supported this work by providing material: Professor Dr. Y. Honda, Chiba; Professor Dr. C. Pfeiffer, Rostock; PD Dr. N. Ott, Meschede; Professor Dr. W. Wiemer, Essen; and W. Kimenkowski, Bochum.

It is gratefully acknowledged that Frau Dr. G. Loeschcke left Winterstein's collection of scientific reprints to the Institute of Medical History, Bochum (Prof. Dr. I. Müller), supporting research on the modern history of physiology. W. Kimenkowski, in preparing his thesis on Winterstein's life and work, has contributed fruitful discussion to the present article.

It was important for me that Professor C. Pfeiffer, who holds the physiology chair in Rostock at the present time, preserved the memory of Hans Winterstein along with the great academic tradition of the institute.

Last but not least, I have to thank my husband, Professor Dr. P. Kiwull, for his expert help with preparing the manuscript. He started his work in physiology close to the present topic together with Winterstein's last collaborators, Wiemer and Ott. The skillful assistance of S. Adler and C. Bräuer with text and figures is likewise gratefully acknowledged.

References

1. Honda Y. Acid-Base Balance: Review in Basic and Clinical Medicine. 2nd ed. Tokyo: Shinko-Koeki Medical Book Publisher, 1984.
2. Kellogg RH. Central chemical regulation of respiration. In: Handbook of Physiology: Vol. 1, Sec. 3. Respiration. Washington DC: American Physiological Society, 1964:507–534.
3. Perkins JF Jr. Historical development of respiratory physiology. In: Fenn WO, Rahn H, eds. Handbook of Physiology: Vol. 1, Sec. 3. Respiration. Washington DC: American Physiological Society, 1964:1–62.
4. Winterstein H. Skizzen aus meinem Leben. In: Ärzte unserer Zeit in Selbstdarstellungen. Hippokrates 1962; 33:79–83.
5. Weber HH, Loeschcke HH. In Memoriam Hans Winterstein. Ergebn Physiol 1964; 55:1–27.
6. Burmeister J. Geschichte der Physiologie an der Universität Rostock. Wiss Z Universität Rostock 1989; N-Reihe 38/8:35–41.
7. Winterstein H. Über die Wirkung der Kohlensäure auf das Centralnervensystem. Arch Anat Physiol 1990 (suppl):177–192.
8. Gad J. Über das Atemzentrum in der Medulla oblongata. Arch Anat Physiol 1893; 175–184.
9. Winterstein H. Die Regulierung der Atmung durch das Blut. Pflügers Arch 1911; 138:167–184.
10. Winterstein H. Die chemische Steuerung der Atmung. Ergebn Physiol 1955; 48: 328–528.
11. Winterstein H. Kausalität und Vitalismus vom Standpunkt der Denkökonomie: 2. Aufl. Berlin: Springer, 1928.
12. Hasselbalch KA. Neutralitätsregulation und Reizbarkeit des Atemzentrums in ihren Wirkungen auf die Kohlensäurespannung des Blutes. Biochem Z 1912; 46:403.
13. Winterstein H. Neue Untersuchungen über die physikalisch-chemische Regulierung der Atmung. Biochem Z 1915; 70:45–73.
14. Gray JS. Pulmonary Ventilation and Its Physiological Regulation. Springfield, IL: Charles C. Thomas, 1950.

15. Jacobs M. The influence of ammonium salts on cell reaction. J Gen Physiol 1922; 2:181–187.
16. Gesell R, Hertzman AB. The regulation of respiration: IV. Tissue acidity, blood acidity and pulmonary ventilation: A study of the effects of semipermeability of membranes and the buffering action of tissues with the continuous method of recording changes in acidity. Am J Physiol 1926; 78:610–629.
17. Winterstein H. Die Reaktionstheorie der Atmungsregulation. Pflügers Arch 1921; 187:293.
18. Winterstein H. Atmungsregulation und Reaktionsregulation. Naturwissenschaften 1923; 625:644.
19. Heymans JF, Heymans C. Sur les modifications directes et sur la régulation réflexe de l'activité du centre respiratoire de la tête isolée du chien. Arch Int Pharmacodyn 1927; 33:273.
20. von Euler US, Liljestrand G, Zotterman Y. The excitation mechanism of the chemoreceptors of the carotid body. Skand Arch Physiol 1939/40; 83:132.
21. Winterstein H. The "reaction theory" of respiratory regulation. Experientia (Basel) 1949; 5:221.
22. Winterstein H. Die Atmung ohne Chemoreceptoren. Arch Int Pharmacodyn 1950; 83:80.
23. Winterstein H. Gökhan N. Ammoniumchlorid-Acidose und Reaktionstheorie der Atmungsregulation. Arch Int Pharmacodyn 1953; 93:212.
24. Winterstein H. The actions of substances introduced into the cerebrospinal fluid and the problem of intracranial chemoreceptors. Pharmacol Rev 1961; 13:71–107.
25. Loeschcke HH, Koepchen HP, Gertz KH. Über den Einfluss von Wasserstoffionenkonzentration und CO_2-Druck im Liquor cerebrospinalis auf die Atmung. Pflügers Arch 1958; 266:569–585.
26. Mitchell RA, Loeschcke HH, Massion WH, Severinghaus JW. Respiratory responses mediated through superficial chemo-sensitive areas on the medulla. J Appl Physiol 1963; 18:523–533.
27. Winterstein H. Die Entdeckung neuer Sinnesflächen für die chemische Steuerung der Atmung. Naturwissenschaften 1960; 47:99–133.
28. Wiemer W, Winterstein H, Kiwull P, Ott N. Interaction of intracranial and extracranial respiratory mechanisms. In: Brooks Ch McC, Kao FF, Lloyd BB, eds. Cerebrospinal Fluid and the Regulation of Ventilation. Oxford: Blackwell Scientific Publications, 1965:303–330.
29. Perkins JF Jr. The relation of the cerebrospinal fluid to respiration. A brief historical introduction. In: Brooks Ch McC, Kao FF, Lloyd BB, eds. Cerebrospinal Fluid and the Regulation of Ventilation. Oxford: Blackwell Scientific Publications, 1965:7–42.
30. Leusen I. Aspects of the acid-base balance between blood and cerebrospinal fluid. In: Brooks Ch McC, Kao FF, Lloyd BB, eds. Cerebrospinal Fluid and the Regulation of Ventilation. Oxford: Blackwell Scientific Publications 1965:55–105.
31. Severinghaus JW, Mitchell RA, Richardson BW, Singer MM. Respiratory control at high altitude suggesting active transport regulation of CSF pH. J Appl Physiol 1963; 18:1155–1166.

32. Severinghaus HW. Electrochemical gradients for hydrogen and bicarbonate ions across the blood-CSF barrier in response to acid base balance changes. In: Brooks Ch McC, Kao FF, Lloyd BB, eds. Cerebrospinal Fluid and the Regulation of Ventilation. Oxford: Blackwell Scientific Publications, 1965:247–267.
33. Mitchell RA. The regulation of respiration in metabolic acidosis and alkalosis. In: Brooks Ch McC, Kao FF, Lloyd BB, eds. Cerebrospinal Fluid and the Regulation of Ventilation. Oxford: Blackwell Scientific Publications, 1965:109–140.
34. Wiemer W, Ott N, Winterstein H. Reflektorische und zentrale Anteile der Hyperpnoe bei HCl-Acidose. Z Biol 1964; 114:299–308.
35. Loeschcke HH. A concept of the role of intracranial chemosensitivity in respiratory control. In: Brooks Ch McC, Kao FF, Lloyd BB, eds. Cerebrospinal Fluid and the Regulation of Ventilation. Oxford: Blackwell Scientific Publications, 1965: 183–210.
36. Loeschcke HH, Sugioka K. pH of cerebrospinal fluid in the cisterna magna and on the surface of the choroid plexus of the 4th ventricle and its effect on ventilation in experimental disturbances of acid-base-balance. Pflügers Arch 1969; 312:161–188.
37. Fencl V, Heisey SR, Held D, Pappenheimer JR. Role of cerebrospinal fluid in the respiratory response to CO_2 as studied in unanaesthetized goats. In: Brooks Ch McC, Kao FF, Lloyd BB, eds. Cerebrospinal Fluid and the Regulation of Ventilation. Oxford: Blackwell Scientific Publications, 1965:141–149.
38. Pappenheimer JR, Fencl V, Heisey SR, Held D. Role of cerebral fluids in control of respiration as studied in unaesthetized goats. Am J Physiol 1965; 208: 436–450.
39. Leusen IR. Chemosensitivity of the respiratory center: Influence of CO_2 in the cerebral ventricles on respiration. Am J Physiol 1954; 176:39–44.
40. Leusen IR. Chemosensitivity of the respiratory center: Influence of changes in the H^+ and total buffer concentrations in the cerebral ventricles on respiration. Am J Physiol 1954; 176:45–51.
41. Van Vaerenbergh P, van der Mijnsbrugge K, Leusen I. The influence on respiration of the substitution of bicarbonate by tris buffer in the cerebral ventricles. Arch Int Pharmacodyn Ther 1962; 138:334–337.
42. Schläfke ME. Central chemosensitivity: A respiratory drive. Rev Physiol Biochem Pharmacol 1981; 90:171–244.
43. Loeschcke HH. Central chemosensitivity and the reaction theory. J Physiol (Lond) 1982; 332:1–24.
44. Kiwull-Schöne H. Chemical signals and the control of breathing. In: Lüttgau HCh, Necker R, eds. Biological Signal Processing. VCH Weinheim 1989:112–121.
45. Trouth CO, Loeschcke HH, Berndt J. Histological structures in the chemosensitive regions on the ventral surface of the cat's medulla oblongata. Pflügers Arch 1973; 339:171–183.
46. Trouth CO, Loeschcke HH, Berndt J. A superficial substrate on the ventral surface of the medulla oblongata influencing respiration. Pflügers Arch 1973; 339:135–152.
47. Dermietzel R. Central chemosensitivity, morphological studies. In: Loeschcke HH, ed. Acid-Base Homeostasis of the Brain Extracellular Fluid and the Respiratory Control System. Stuttgart: Thieme Verlag 1976:52–65.

48. Trouth CO, Odek-Ogunde M, Holloway JA. Morphological observations on superficial medullary CO_2-chemosensitive areas. Brain Res 1982; 246:35–45.
49. Trouth CO, Patrickson JW, Holloway JA, Wright LE. Neurophysiological studies on superficial medullary chemosensitive area for respiration. Brain Res 1982; 246: 47–56.
50. Trouth CO, Millis RM, Bernard DG, et al. Cholinergic-opioid interactions at brainstem respiratory chemosensitive areas in cats. Neuro Toxicology 1993; 14:459–468.
51. Berndt J, Berger W, Mückenhoff K. Untersuchungen zum zentralen chemosensiblen Mechanismus der Atmung: I. Die Bestimmung des extracellulären pH im Gehirngewebe. Pflügers Arch 1972; 332:127–145.
52. Berndt J, Berger W, Berger K. Schmidt M. Untersuchungen zum zentralen chemosensiblen Mechanismus der Atmung: II. Die Steuerung der Atmung durch das extracelluläre pH im Gewebe der Medulla oblongata. Pflügers Arch 1972; 332: 146–170.
53. Berndt J, Berger W, Berger K, Schmidt M. Untersuchungen zum zentralen chemosensiblen Mechanismus der Atmung: III. Die Wirkung starker Änderungen des Liquor-pH (pH 5.4–7.7) bei Katzen vor und nach Vagotomie. Pflügers Arch 1972; 332:171–182.
54. Berndt J, Berger W, Berger K, Schmidt M. Untersuchungen zum zentralen chemosensiblen Mechanismus der Atmung: IV. Beobachtungen an anaesthesierten und dezerebrierten Katzen. Pflügers Arch 1972; 332:184–197.
55. See WR. Untersuchungen zur Konvergenz von atemwirksamen Afferenzen an chemosensiblen Neuronen des mittleren Feldes (Area S) der ventralen Oberfläche der Medulla oblongata der Katze. Habilitationsschrift, Abteilung für Naturwissenschaftliche Medizin der Ruhr-Universität Bochum, 1984.
56. Kiwull-Schöne H, Kiwull P. Hypoxic modulation of central chemosensitivity. In: Schläfke ME, Koepchen HP, See WR, eds. Central Neurone Environment and the Control Systems of Breathing and Circulation. Berlin-Heidelberg-New York: Springer-Verlag, 1983:88–95.
57. Kiwull-Schöne H, Kiwull P. Hypoxia and the "reaction theory" of central respiratory chemosensitivity. Adv Exp Med Biol 1992; 316:347–357.
58. Shams H. Differential effects of CO_2 and H^+ as central stimuli of respiration in the cat. J Appl Physiol 1985; 58:357–364.
59. Fukuda Y, Honda Y. pH sensitive cells at ventrolateral surface of the rat medulla oblongata. Nature 1975; 256:317–318.
60. Fukuda Y, Loeschcke HH. Effect of H^+ on spontaneous neuronal activity in the surface layer of the rat medulla oblongata. Pflügers Arch 1977; 371:125–134.
61. Dev NB, Loeschcke HH. Topography of the respiratory and circulatory responses to acetylcholine and nicotine on the ventral surface of the medulla oblongata. Pflügers Arch 1979; 379:19–27.
62. Dev NB, Loeschcke HH. A cholinergic mechanism involved in the respiratory chemosensitivity of the medulla oblongata of the cat. Pflügers Arch 1979; 379: 29–36.
63. Fukuda Y, Loeschcke HH. A cholinergic mechanism involved in the neuronal excitation by H^+ in the respiratory chemosensitive structures on the ventral medulla oblongata of rats in vitro. Pflügers Arch 1979; 379:125–135.

64. Mitra J, Prabhakar NR, Overholt JL, Cherniack NS. Respiratory effects of N-methyl-D-aspartate on the ventrolateral medullary surface. J Appl Physiol 1989; 67:1814–1819.
65. Ahmad HR, Loeschcke HH. Transient and steady state response of pulmonary ventilation to the medullary extracellular pH. Pflügers Arch 1982; 395:285–292.
66. Ahmad HR, Loeschcke HH. Fast bicarbonate-chloride exchange between brain cells and brain extracellular fluid in respiratory acidosis. Pflügers Arch 1982; 395: 293–299.
67. Ahmad HR, Loeschcke HH. Fast bicarbonate-chloride exchange between plasma and brain extracellular fluid at maintained P_{CO_2}. Pflügers Arch 1982; 395:300–305.
68. Maren Th H. Carbonic anhydrase: Chemistry, physiology, and inhibition. Physiol Rev 1967; 47:595–781.
69. Eldridge FL, Kiley JP, Millhorn DE. Respiratory responses to medullary hydrogen ion changes in cats: Different effects of respiratory and metabolic acidoses. J Physiol (Lond) 1985; 358:285–297.
70. Teppema LJ, Barts PWJA, Folgering HTh, Evers JAM. Effects of respiratory and (isocapnic) metabolic arterial acid-base disturbances on medullary extracellular fluid pH and ventilation in cats. Respir Physiol 1983; 53:379–395.
71. Kiwull-Schöne H, Kiwull P. Lack of ventilatory reaction to endogenous and exogenous metabolic brain-stem acidosis. In: Honda Y, Miyamoto Y, Konno K, Widdicombe JG, eds. Control of Breathing and its Modeling Perspective. New York: Plenum Press 1992:143–146.
72. Xu FD, Spellman MJ Jr, Sato M, et al. Anomalous hypoxic acidification of medullary ventral surface. J Appl Physiol 1991; 71:2211–2217.
73. Jensen KE, Thomsen C, Henriksen O. In vivo measurement of intracellular pH in human brain during different tensions of carbon dioxide in arterial blood: A 31P-NMR study. Acta Physiol Scand 1988; 134:295–298.
74. Portman MA, Lassen NA, Cooper Th G, et al. Intra- and extracellular pH of the brain in vivo studied by 31P-NMR during hyper- and hypocapnia. J Appl Physiol 1991; 71:2168–2172.
75. Tojima H, Kuriyama T, Fukuda Y. Differential respiratory effects of HCO_3^- and CO_2 applied on ventral medullary surface of rats. J Appl Physiol 1991; 70:2217–2225.
76. Kiwull-Schöne H, Kiwull P. Central chemical control of breathing pattern during metabolic and respiratory acid-base changes. Proc XXXII Int Congr Physiol Sci, Glasgow, Aug. 1–6, 1993.
77. Hey EN, Lloyd BB, Cunningham DJC, et al. Effects of various respiratory stimuli on the depth and frequency of breathing in man. Respir Physiol 1966; 1:193–205.
78. Bickler PhE, Litt L, Banville DL, Severinghaus JW. Effects of acetazolamide on cerebral acid-base balance. J Appl Physiol 1988; 65:422–427.
79. Vorstrup S, Jensen KE, Thomsen C, et al. Neuronal pH regulation: Constant normal intracellular pH is maintained in brain during low extracellular pH induced by acetazolamide- 31P NMR study. J Cerebr Blood Flow Metab 1989; 9:417–421.
80. Lassen NA. Is central chemoreceptor sensitive to intracellular rather than extracellular pH? Clin Physiol 1990; 10:311–319.
81. Harada Y, Kuno M, Wang Y Zh. Differential effects of carbon dioxide and pH on central chemoreceptors in the rat in vitro. J Physiol (Lond) 1985; 368:679–693.

82. Okada Y, Mückenhoff K, Scheid P. Hypercapnia and medullary neurons in the isolated brain stem–spinal cord of the rat. Respir Physiol 1993; 93:327–336.
83. Okada Y, Mückenhoff K, Holtermann G, et al. Depth profiles of pH and P_{O_2} in the isolated brain stem-spinal cord of the neonatal rat. Respir Physiol 1993; 93: 315–326.
84. Morin-Surun MP, Schäfer T, Boudinot E, et al. Perfused isolated adult brainstem: A novel approach to study central chemosensitivity. Ventral Brainstem Mechanisms and Control of Respiration and Blood Pressure, Washington, Nov. 5–6, 1993, this volume.
85. Fukuda Y. Difference between actions of P_{CO_2} and low $[HCO_3^-]$ on neurons in the rat medullary chemosensitive areas in vitro. Pflügers Arch 1983; 398:324–330.
86. Jarolimek W, Misgeld U, Lux HD. Neurons sensitive to pH in slices of the rat ventral medulla oblongata. Pflügers Arch 1990; 416:247–253.
87. Bingmann D, Baker RE, Ballantyne D. Rhythm generation in brainstem cultures grown in a serum-free medium. Neurosci Lett 1991; 132:167–170.
88. Neubauer JA, Consalves SF, Chou W, et al. Chemosensitivity of medullary neurons in explant tissue cultures. Neuroscience 1991; 45:701–708.
89. Martin A, Neubauer JA. Effect of acetazolamide on CO_2 chemosensitivity of cultured medullary neurons. ISAC, Dublin, Aug. 9–13, 1994.
90. Torrance RW. A comparison of central and peripheral chemoreceptors. In: Loeschcke HH, ed. Acid-Base Homeostasis of the Brain Extracellular Fluid and the Respiratory Control System. Stuttgart: Georg Thieme, 1976:95–103.
91. Pokorski M, Lahiri S. Aortic and carotid chemoreceptor responses to metabolic acidosis in the cat. Am J Physiol 1983; 244:R652–R658.
92. Gray BA. Response of the perfused carotid body to changes in pH and P_{CO_2}. Respir Physiol 1968; 4:229–245.
93. Hanson MA, Nye PCG, Torrance RW. The exodus of an extracellular bicarbonate theory of chemoreception and the genesis of an intracellular one. In: Belmonte C, Pallot DJ, Acker H, Fidone S, eds. Arterial Chemoreceptors. Leicester: University Press 1981:403–416.
94. Buckler KJ, Vaughan-Jones RD, Peers C, Nye PCG. Intracellular pH and its regulation in isolated type I carotid body cells of the neonatal rat. J Physiol (Lond) 1991; 436:107–129.
95. Iturriaga R, Lahiri S, Mokashi A. Carbonic anhydrase and chemoreception in the cat carotid body. Am J Physiol 1991; 261:C565–C573.
96. Wilding TJ, Cheng B, Roos A. pH regulation in adult rat carotid body glomus cells—Importance of extracellular pH, sodium, and potassium. J Gen Physiol 1992; 100:593–608.
97. Buckler KJ, Vaughan-Jones RD, Peers C, et al. Effects of extracellular pH, P_{CO_2} and HCO_3^- on intracellular pH in isolated type-I cells of the neonatal rat carotid body. J Physiol (Lond) 1991; 444:703–721.
98. He Sh-F, Wei JY, Eyzaguirre C. Influence of intracellular pH on the membrane potential of cultured carotid body glomus cells. Brain Res 1993; 601:353–357.
99. Hanson MA, Nye PCG, Torrance RW. The location of carbonic anhydrase in relation to the blood-brain barrier at the medullary chemoreceptors of the cat. J Physiol (Lond) 1981; 320:113–125.

100. Coates EL, Li A, Nattie EE. Acetazolamide on the ventral medulla of the cat increases phrenic output and delays the ventilatory response to CO_2. J Physiol (Lond) 1991; 441:433–451.
101. Coates EL, Li A, Nattie EE. Widespread sites of brainstem ventilatory chemoreceptors. J Appl Physiol 1993; 75:5–14.
102. Kiwull-Schöne H, Kiwull P. Effects of acetazolamide on the central respiratory CO_2 response and apneic threshold. Pflügers Arch 1992; 420(suppl):R129.
103. Severinghaus JW. Invited editorial on widespread sites of brain stem ventilatory chemoreceptors. J Appl Physiol 1993; 75:3–4.

2

Control and Rhythm

MARIANNE E. SCHLÄFKE

Ruhr-University Bochum
Bochum, Germany

This symposium is dedicated to Hans H. Loeschcke (1912–1986) and Hans-Peter Koepchen (1924–1993), who initiated the work on ventral brainstem mechanisms and control functions through their original joint discovery leading up to their varying views on control and rhythm, the result of their lives' work.

Both Loeschcke and Koepchen, disciples of the famous Hermann Rein school (1), were devoted to "system physiology" (24). Their main fields of interest originated in Göttingen: Loeschcke focused on acid-base homeostasis of the brain and respiratory control, whereas Koepchen's focus was on rhythm and cardiorespiratory physiological control.

I. Hans H. Loeschcke

A. The Years Before 1958

Loeschcke was the son of the famous lung pathologist who chaired the Department of Pathology at the University of Greifswald. His delightful mother was the founder and director of a private grammar school in Mannheim, where Hans began his schooling. Later, in 1938, as a 26-year-old physician, Loeschcke completed his residency training in internal medicine at the Hermann Rein's

Institute. In addition, while there he produced four original publications on the microanatomy of the lymphatic and vascular system, on the Vater-Pacini corpuscles, and on infectious diseases.

During the "golden age" at the University of Göttingen—when Aschoff, Gauer, and Loeschcke enjoyed great academic freedom in the Rein laboratory—a most exciting and fruitful period in their departmental lives evolved in which a flood of now classic papers on respiratory and circulatory regulation were published. A movie from 1938 still exists in which Loeschcke is seated in a self-made contrifuge that is being accelerated to threefold gravity by his younger colleague Gauer. Many experiments were performed on men and on pregnant women, including the scientists' wives, and Loeschcke's medical qualifications obviously proved to be very useful. Many years later, when we used to sit together in his laboratory, nightly observing the endless paper from polygraph recordings of cat data, Loeschcke used to recount jokes about their self-experiments of those days. He would describe that there were seven spontaneously breathing doctors who were the subjects in the paper about the influence of muscular exercise on respiratory augmentation by partial neuromuscular blockade (with tubocurarine). This heroic investigation was devoted to Hans Winterstein on the occasion of his eightieth birthday.

Loeschcke soon developed a particular interest in the effect of hypoxia and high-altitude adaptation. In the summer of 1939, he participated in a scientific expedition to the Jungfraujoch in Switzerland to investigate high-altitude adaptation. There he worked together with Becker-Freyseng, Luft, and Opitz and a series of fundamental studies came out in the early 1940s.

The war interrupted Loeschcke's work and he received prompt orders to act as an junior (house) physician. In the spring of 1942, he fortunately got a new command and was reassigned to the institute. In May 1942, he completed his second thesis in Göttingen, with the study entitled "Adaptation of the Cardiovascular System and Oxygen Partial Pressure Profile in the Brain During Acute Hypoxia." This was also the year in which he married Dr. Gertrud Freitag. In 1948 he was promoted to a senior physician. This was also the year in which he published a paper with his wife called "On the Exchange of Lactic Acid Between Arterial Blood and Brain Tissue and Its Variation by Hypoxia" (2). In 1949, Loeschcke became an associate professor in Göttingen.

During this period, Loeschcke's primary interest was in methods research, e.g., on continuous blood gas measurements, heart-lung pumps, blood cell counting, regulation and metabolism during hypoxia, regulation of blood formation, respiratory regulation and acid-base balance during pregnancy, and the effect of steroid hormones on ventilation. From autumn of 1949 to the autumn of 1950, Loeschcke was a fellow at the Rockefeller Foundation in Philadelphia. Under the direction of C. F. Schmidt, he worked along with Lambertsen on the effects of increased oxygen pressure on respiration and cerebral blood flow. Back in Göttingen, he continued his research on steroid hormones and on gas diffusion as well as on drugs and ventilation in both humans and dogs.

In the early 1950s, Loeschcke received several requests to accept a chair: in 1951 from Leipzig, in 1954 from Istanbul, and also in 1954 from Rostock. Both Istanbul and Rostock were places of Winterstein's activity; nonetheless, Loeschcke refused these positions. Instead, he accepted an invitation from Alexander von Muralt as a visiting professor for a year at the Institute in Bern, Switzerland. There, he continued the work on the influence of cerebrospinal fluid on ventilation that he had started in Göttingen and which was the result of his interest not only in Winterstein's ideas but also in Leusen's approach and methods specificities. This research became more and more his main field upon his return to Göttingen. Aside from his wife, Gertrud Loeschcke who was his first doctoral fellow, Hans-Peter Koepchen was his best-known doctoral fellow.

B. Some Milestones After 1958

In 1963 Loeschcke was appointed to the extraordinary chair of clinical physiology at Göttingen, and in 1964 he accepted the position as chair in physiology at Ruhr University, where he worked very successfully with a large group of international fellows and numerous visiting scientists from all over the world until his illness in the year of his retirement in 1981. In early 1970 Loeschcke initiated the Sonderforschungsbereich 114, called Biological Signal Perception and Processing, an institution of the DFG (German Research Society); he was generously supported in this for 15 years. The list of his associates is long and includes names like Ahmad; Beck from Torrance's lab; Davis; De Lattre from Bargeton's lab; Dev; Folgering; Fukuda from Honda's lab; Goto from Hukuhara's lab; Gürtzenstein; Lipski, and Majcherczyk from Trzebski's lab; Mueller from Grodins's lab; Trouth and others and many of the foreign classic representatives spent some time in Bochum or were frequent guests in the department, like Bargeton, Bouveroi, Cunningham, von Euler, Fencl, Grodins, Guenard, Haxhiu, Honda, Hukuhara Jr., Kimura, Kreutzer, Kumazawa, Lahiri, Leusen, Lloyd, Paintal, Ponten, Purves, Remmers, Severinghaus, Torrance, Trzebski, Ward, Weyne, Wright and others. In 1974 Hans H. Loeschcke received an honorary doctorate from the University Rene Descartes, Paris.

Loeschcke enjoyed a most pleasant time during his sabbatical in 1973, which Cloyd Ovid Trouth had organized for him. Trouth had finished his Ph.D. with a comprehensive paper on respiratory and cardiovascular function as well as the neurohistology of the ventral brainstem under the guidance of Loeschcke in the various labs in Bochum; now both could work in Trouth's laboratory in Kingston, Jamaica. Loeschcke was therefore relieved, for a while, from the manifold faculty duties, and he could enjoy the beauties of the lovely home country of his former doctoral student. As far as I remember, the work with

Trouth in Jamaica was on the cerebellum, and its role in the cardiorespiratory system has only been documented as an abstract by the German Physiological Society. And I am convinced that there is still a mine of valuables unpublished and classified as "top secret" in Trouth's safe.

C. Some Evaluations

The main scientific work of Loeschcke and his cybernetic concept are documented in the last symposium to which he could contribute personally in spite of his illness. The symposium was entitled "Central Neurone Environment and the Control Systems of Breathing and Circulation" and was held in October 1981 (Fig. 1). In November of the same year he should have delivered the Annual Review Lecture of the Physiological Society in Oxford. Unfortunately, he could not write the manuscript but he allowed us to use the last official drawing of his concept on the central chemoreceptor mechanism, which he had developed for his review lecture published in the Journal of Physiology (3). Figure 2 is a modified version of this; Koepchen, whom we implored to give a speech in honor of Loeschcke at this symposium, formulated the preface to Loeschcke's last symposium in 1982 as follows:

Figure 1 Photo of Hans H. Loeschcke and H. P. Koepchen from the Symposium on Central Neurone Environment and the Control System of Breathing and Circulation, held in Bochum, 1981.

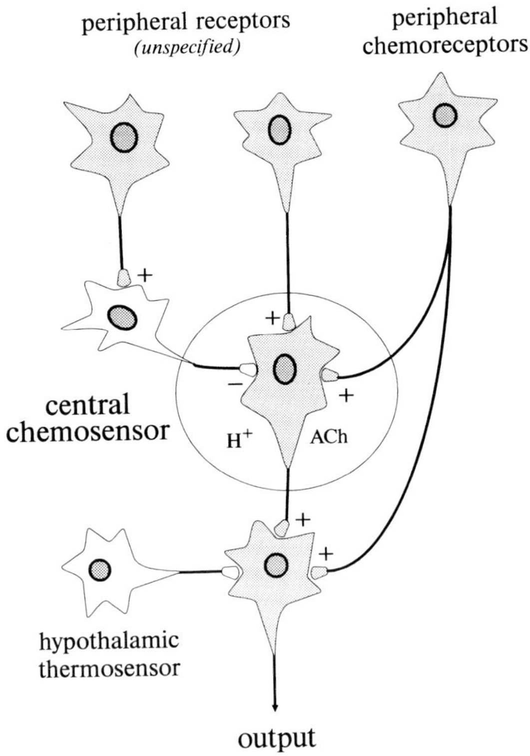

Figure 2 Loeschcke's last diagram of central chemosensitivity "with its principal interconnections. Inhibitory and excitatory afferents both unspecific (sympathetic and fibers from the integument and from muscle) and specific fibers from the peripheral chemoreceptors are supposed to feed in synaptically to a central chemosensitive neurone. These synapses are believed to be cholinergic and may be affected by applied acetylcholine. Hydrogen ions affect cholinergic synaptic transmission. Some fibers from the peripheral chemoreceptors bypass the region of central chemosensitivity as do the fibers from the hypothalamic thermosensor. There is a final convergence on an output neuron which drives the respiratory motoneurons" (3).

Loeschcke's discovery of ventral medullary substrates forming an essential drive for the ventilatory as well as for the circulatory control systems, and the elaboration of this concept during the last 25 years have profoundly influenced the concepts in this field. In an age of proceeding specialization on partial mechanisms, his approach has always emphasized the integrative aspects of the control systems, especially the role of the respiratory control system for the ionic homeostasis of the brain and the close interrelation between neuronal respiratory and cardiovascular control. This general intention implies the combination of physicochemical and neurophysiological approaches as well as the study of peripheral function of respiration and circulation, topics which are usually handled separately according to the different methods (4).

The zenith of Hans Loeschcke's scientific life is reflected in the symposium entitled "Acid-Base Homeostasis of the Brain Extracellular Fluid and the Respiratory Control System" (5) organized by him in March 1975. On the basis of the results of the three main projects—neuroanatomy, neurophysiology and function, and mathematical modeling—and with his main coworkers (Ahmad, Berger, Berger, Bernolt, Dermietzel, Dev, Fukuda, Kiwull, Kiwull-Schöne, Majcherczyk, Marek, Middendort, Mückenhoff, Mueller, Natsui, Plaas-Link, Pokorski, Prabhakar, Schläfke, See, Shams, Shimada, Tok, Trouth, Woidke, and Loeschcke), he had created the modern concept of the respiratory control system subserving the precision of brain extracellular fluid homeostasis. In accordance with Pappenheimer's theory (6), Loeschcke considered a unique pH-sensitive mechanism as the input for the respiratory drive system. In contrast to Pappenheimer and as the result of 17 years of research, he proposed a special receptor site (Fig. 3). The condition of free access from the intercellular compartment to the CSF space was guaranteed for the ventral medullary sur-

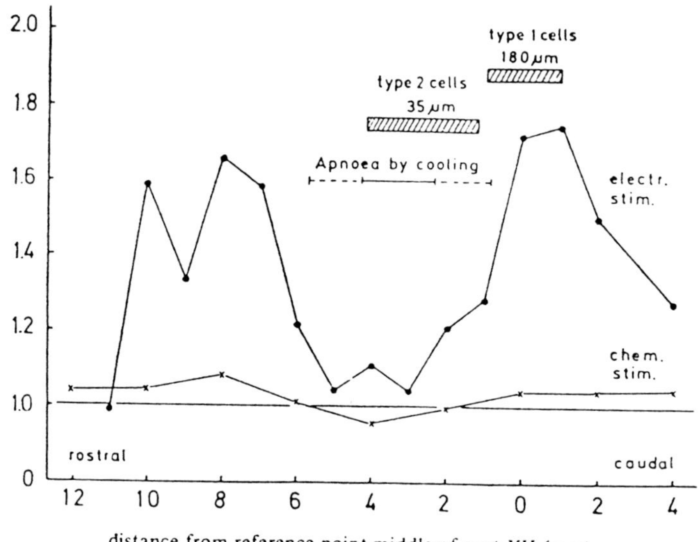

Figure 3 The "areas": Loeschcke emphasized that the chemosensitive function is located within the rostral (M for Mitchell) and the caudal (L for Loeschcke) area, and that the intermediate area (S for Schläfke) is not chemosensitive. In his drawing he combined the results obtained by electrical stimulation (26), local cold block (27), the histological findings of Trouth et al. (28), type 1 cells, those of Petrovisky (29), and of Schläfke et al. (30) regarding type 2 cells. As the local recording techniques gave evidence for a dense populatiion of pH-responsive neurons below the intermediate area as well as (from the central chemosensitive mechanism can be abolished completely) Schläfke had joked about the typical female fate of her area just being less important (31).

face layer, as Dermietzel (the same symposium) could demonstrate by electron microscopy. Loeschcke described the process of acid-base homeostasis in respiratory as well as in nonrespiratory acidosis using the noninvasive pH electrode technique to measure the pH on the ventral medullary surface. The time course of ventilation and of brain extracellular pH followed step changes of inspired CO_2. The tidal volume, especially after denervation of the peripheral chemoreceptors, followed the pH closely.

Tidal volume and pH measured on the cortical surface, however, were unrelated. Winterstein's and Loeschcke's theory claims that there is a receptor in the brain tissue which responds to H^+ ions. The transmission from blood to the extracellular fluid is not the same for the three variables of the Henderson-Hasselbalch equation. The choroid plexus and the endothelium of the brain capillaries form the boundaries and have to be passed in parallel. It is to be expected that the arteriovenous CO_2 difference of the plexus blood is low on account of the high blood flow. The CO_2 of the secreted plexus fluid comes into equilibrium with the CO_2 of the brain surface. This partially explains the higher acidity of cerebrospinal fluid (CSF) in comparison to blood plasma. Free exchange between CSF and cerebral extracellular fluid (ECF) reduces the differences in composition between the two fluids. The local extracellular pH is only partly under the influence of the CSF because the tissue is perfused by the blood. Loeschcke assumed that the local P_{CO_2} was determined by local metabolism, CO_2 binding capacity of the blood, blood flow, and distance to the capillaries. The average tissue P_{CO_2} in the brain under steady-state conditions has been proposed as the algebraic means of venous and arterial P_{CO_2} of cerebral blood plus 1 torr.

Bicarbonate in response to acid-base changes in CSF varies by only 40% as much as in blood plasma. The HCO_3^- exchange between blood and brain is a fast process occurring simultaneously with an exchange of Cl^- in the opposite direction. The exchange possibly uses a protein carrier through an anion exchange channel of the endothelium of the brain capillaries. When CO_2 is inhaled, the increase of bicarbonate concentration in ECF is greater than in blood plasma, even if the plasma bicarbonate is kept constant. Ahmad and Loeschcke used electrochemical methods to demonstrate that HCO_3^- in this case is exchanged for Cl^- in a one-to-one fashion between ECF and cells. Extracellular pH depends on anion exchange between plasma and extracellular fluid (passing the endothelium) on the one hand and on anion exchange between brain tissue cells and extracellular fluid on the other. In the ECF, all five compartments—namely red cells, plasma, endothelium, extracellular fluid, and glial cells—must be considered for the pH kinetics. This is true for metabolic as well as for respiratory acidosis. Cunningham, a beloved friend of the Loeschckes, in his enthusiasm for Loeschcke's excellent work, named this exchange mechanism the Bochumer shift in analogy to the Hamburger shift in blood.

A mathematical model made use of the bicarbonate relationship between plasma and CSF (Fig. 4) (7). Hereby it was shown that the ventilatory response

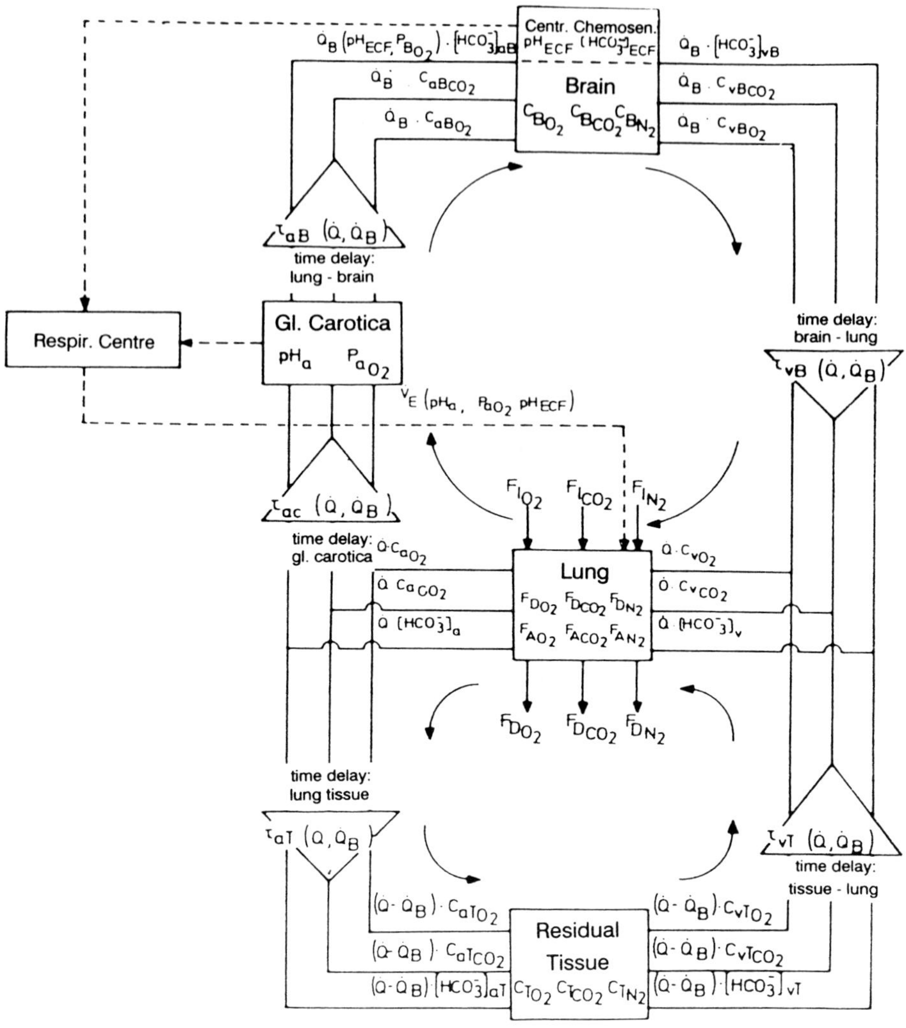

Figure 4 Mathematical simulation of the respiratory system, the result of a cooperation between Loeschcke and Ewald, a former fellow of N. Wiener (32).

to respiratory acidosis is much higher than that to metabolic acidosis. The increase in ventilation, however, is more effective in regulating blood pH in metabolic acidosis than in respiratory acidosis. In the light of these experimental and mathematical approaches, Loeschcke stated that the extracellular pH in the brain is the main chemical signal determining ventilation. This pH depends on the tissue P_{CO_2} and the bicarbonate concentration in the tissue. The distribution of bicarbonate between plasma and CSF under given conditions can be experimentally determined. There are processes preventing a full equilibrium, and any assessment of the extracellular pH should take into account the exchange process between the five compartments.

Differences in the ventilatory responses when comparing mild and high degrees of metabolic acidosis with the responses to CO_2 inhalation have been reported by Shams (8). CO_2 responses during mild metabolic acidosis have a much higher slope than during severe metabolic acidosis. During the process of this investigation, which was still surveyed by Loeschcke, he assumed that the sensing mechanism resides in a compartment that is initially accessible to H^+-ions, but under the continued application of acid further entrance of H^+-ions is prevented, whereas the lipid-soluble acid CO_2 penetrates as before. A synaptic gap would have been a good model in Loeschcke's eyes. Shams (8) and Dempsey and Forster (9) doubted the validity of these assumptions and so also the classic one-receptor theory. Their conclusions are substantiated by findings of Teppema et al. (10), indicating that changes in pH caused by hypercapnia give rise to a greater ventilatory response than the same changes in pH caused by isocapnic changes in arterial pH. Kiwull-Schöne and Kiwull (11) observed in peripherally chemodenervated cats that the response of ventilation to CO_2 was maintained at any level of medullary pH during hypoxia. However, hypoxia-induced decreases in pH had no effect on respiration. Shams directed interests toward the significance of the intracellular pH for the one-receptor theory, since the pH difference between the extracellular and the intracellular compartment might be dependent upon the CO_2 level. So the classic one-receptor theory will have to be challenged by new intracellular pH recording techniques.

When Loeschcke left the department in 1982, his coworkers hoped that he would return in order to pursue some more collaborative experimental approaches in accordance with his progressive ideas. Unfortunately his illness caused him too much trouble.

Those who were present at his last candlelight ceremony will never forget the beauty of Lake Constance reflecting his wonderful white house on the other side of the village.

Working in the lab with Hans Loeschcke was a most exciting way of learning. Nobody knows Loeschcke who has not spent a few days in his lab with him and, the day after, discussed the results of the experiment very critically with him. His broad scientific background, his technical skills, his physicochemical and mathematical expertise combined with his surprising practical skills, and his experience as an internist helped to make him a fantastic teacher who motivated us and trained us to be adventurers in the sciences. He was the best "master" we could ever have.

II. Hans-Peter Koepchen

Koepchen was the first son of four children of a trustworthy and popular country doctor. For young Koepchen, his father was the ideal and role model. His thoroughness and sensitivity in his dealings with people was evident in his private surroundings as well as in his institute in the international scientific society. Young Koepchen's most attractive trait was his unlimited capacity to grasp the crucial point in a scientific argument and to lead the discussion in a way that impressed those who were associated with him. A congress on cardiorespiratory research would be painfully dull without the comments from Koepchen.

One personal story should be mentioned because it truly characterizes Koepchen's personality. When he was a young student in Göttingen in 1945, he met a Mrs. Muller emerging from the cafeteria of the university—her daughter was studying music there—the only place at that terrible time where one could get a cheap meal. She had lost her son in the war during the previous year. Koepchen decided to replace him and stood by this decision and its consequences for the rest of his life. This explains why two families signed the death notice of the unmarried Koepchen—they were the Koepchen and Braig families. Carmen Braig, born Muller, was his congenial "sister" and her husband his congenial "brother-in-law." Many scientists know Mrs. Muller, who cared for Koepchen's guests and accompanied him at symposiums. And some may know the Braigs, who were professional musicians in the famous Berlin Orchestra and who played the violin at symposium in Berlin when it was organized by Hans-Peter Koepchen, who himself was a violinist. They accompanied the freedom-loving Koepchen in a most wonderfully understanding and warm way up to the end of his unusual and shining life.

Koepchen joined the Department of Physiology in Göttingen as a student during the period when Loeschcke was investigating the effect of steroid hormones on ventilation. His thesis, completed in 1952, was entitled "Further Investigations on the Effect of Steroid Hormones on Lung Ventilation." His first publication with Loeschcke, "On the Effect of Adrenocorticotropic Hormone on Lung Ventilation of Man" appeared in 1954 (12). During that period Koepchen also worked with Bartels, Kramer, Thurau, Kanzow, and von Baumgarten on methods of blood gas measurement, control of heart rate, muscular exercise, interrelationship between respiration blood pressure, blood pressure waves, and neurophysiological mechanisms of vagal afferent and their influence on respiratory neurons. They developed techniques in extra- and intracellular recording and stereotactic techniques in those years.

In 1959, Koepchen completed his second thesis, entitled "On Rhythmic Variability of Circulatory Measures and their Relationship to the Respiratory Rhythm." During this time until 1965, when he received his first chair, he wrote 20 original papers, one monograph on blood pressure rhythms, and 33 congress and symposium contributions on such topics as cardiovascular and respiratory regulation and interactions, cardiovascular and respiratory rhythm interactions and coordination, vasomotor control, the motor system, and on neurophysiological methods. In 1965 he became the head of Germany's first Department of Neurovegetative Physiology in Munich, and in 1969 he accepted the chair in physiology at the Free University of Berlin, which he held until his retirement in 1992. He was chairman of the committee for the Max Planck Institute for System Physiology in Dortmund from 1975 to 1985. From 1986 to 1988 he was guest professor of the University of Bari. He became president of the International Society for Neurovegetative Research from 1984 to 1989, belonged to the commission of the Autonomic Nervous System of the International Union of Physiological Sciences (IUPS) and was its honorary chairman until 1990. In 1991 he became chairman of the IUPS Commitee for Integrative Physiology. In 1985 he started a joint research program with a famous and sophisticated hospital for sports medicine, supported by the Deutsche Sporthilfe in Lüdenscheid, entitled "The Neurovegetative State in Humans." After his retirement, he held a guest professorship at the Max Planck Institute for System Physiology in Dortmund. In spite of his illness, which began about 2 years before his retirement, Koepchen was extremely active until the last months of his life. He completed various monographs and papers and succeeded in creating a network connection between institutions in Jülich, Karlsruhe, Dortmund, Berlin, and Lüdenscheid, with the aim of promoting a true interdisciplinary research on neurovegetative regulations, theoretical analysis, and application on humans under various physiological and pathophysiological conditions. Koepchen's already successful program was interrupted by his much too early death. In addition to about 200 publications and 8 monographs, Koepchen leaves the best German textbook on respiration that has ever been written. Unfortunately, his contribution to his international textbook on human physiology, edited together with Greger, Mommaerts, and Windhorst, remains unfinished.

A. Some Evaluations

One of the most significant symposiums on central control of breathing, with the presentations of many original data and stimulating concepts, was one organized by Curt von Euler and Hugo Lagercrantz at the Wenner-Gren Center in Stockholm in 1978. After many expressions of thanks to the organizer and at the very end of this great symposium, Koepchen asked for another minute, went to the blackboard, and developed a scheme reviewing the complete symposium. Everyone could find his own contribution incorporated in Koepchen's general review of the control system, and everyone found his work confirmed but was gently made to feel insecure as well. His cautiously dosed, keen-witted humor made his soberly presented view a breathtaking experience. His students loved this about him—his typical way of teaching.

Koepchen's ideas on rhythm and neurovegetative control have now been accepted as the international standard knowledge and form the basis of the worldwide approaches in research on control and rhythm. Symposiums organized by him attest to his integrated view on neurovegetative and motor control systems. His special concern was to bring together the specialized research groups on respiration and circulation as well as clinicians and mathematicians in order to follow-up on the aim of a general biological principle which was already expressed by Goethe. Koepchen translated him in one of his last papers: "It would be highly desirable, in order to accelerate progress in the whole of physiology, not to lose sight of the interaction between all parts of a living organism. Only by recognizing that in an organic body all parts together act on the single part and, in turn, each single part influences all others, can we hope to fill in step by step the gaps of physiology." The symposiums entitled "Central Rhythmic and Regulation" (13), Central Interactions Between Respiratory and Cardiovascular Control Systems" (14), Neurovegetative Control Systems" (15), Cardiorespiratory and Motor Coordination" (16), and "Rhythms in Physiological Systems" (17) are valuable documentations of his message.

Koepchen's theoretical approaches on integrative neurovegetative and motor control made use of his experimental studies on spontaneous coordination or rhythmic activities in anaesthetized dogs and conscious men, microelectrode recordings, and local cooling experiments in the brainstem of anaesthetized dogs and cats. Analysis of his findings and logical deductions result in the statement that central control systems must be partly identical and that usual criteria of labeling central neurons according to their coordination to specific control systems are necessary but not sufficient (Fig. 5) (18). Instead the picture of a multifunctional common network emerges forming the substrate for the observed common central rhythmicity (Fig. 6) (19). In this network specificity is a quantitative property of neurones. Figure 7 presents Koepchen's "simple model" illustrating that slight inhomogeneity leads to graded input-output specificity in the network which changes with changing conditions. The degree of neuronal specificity increased in the direction of the outputs (20).

Together with his friend Haken, Koepchen attempted to follow up his ideas on physiological rhythmicity and synergetics. The symposium in common on rhythms in physiological systems (21) Koepchen introduced his chapters as follows:

> There is a broad spectrum of physiological rhythms in many different systems extending over a large range of frequencies. The medium ultradian frequency range is characterized by relatively great variability of frequency, intense mutual interaction and interference with the homeostatic feedback regulatory circuits and behavioral influences. In spite of the resulting multiform phenomenology, certain basic rules of "relative" sliding coordination between the subunits of one system, and between the systems themselves can be identified. They are revealed through certain kinds of mutual entrainment, the occurrence of multiple integer frequency relations, instability of phases and amplitudes during transitory states, rhythm-related excitability cycles, and through variable periods of synchronization and desynchronization. This principle is illustrated by respiratory, vascular, cardiovascular, and motor systems and their interaction found in anaesthetized animals and in man. The fact that the same general kind of rhythm coordination is found, irrespective of the particular system and frequency range, justifies the hope of a successful analysis and understanding of rhythmicity as a basic principle of physio-

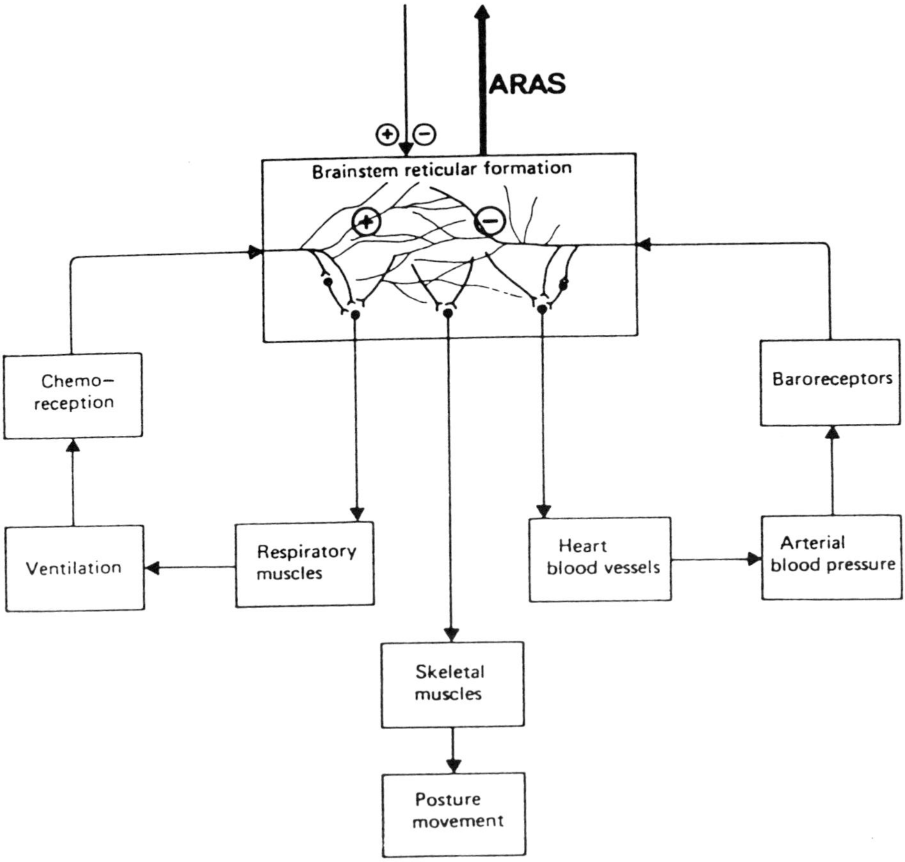

Figure 5 Scheme of the functional interdependencies between the homeostatic feedback control systems of respiration and circulation and the general activating brainstem system influencing vigilance, skeletomotor tone, ventilation, and blood pressure (18).

> logical self-organization. In view of their general validity, irrespective of the particular physical or living system under investigation, the concepts and analytical tools of synergetics will play a key role in the future of this whole subject.

Figure 8 is taken from this paper, showing the scale of frequencies and their range of variation of relevant human physiological rhythms in comparison with rhythms of the outer world. The great variability and overlapping of frequencies in the cardiorespiratory and motor systems in evident (21).

In his paper on physiological rhythmicity and music in medicine (22), Koepchen uses this "scale of human life" (Fig. 8).

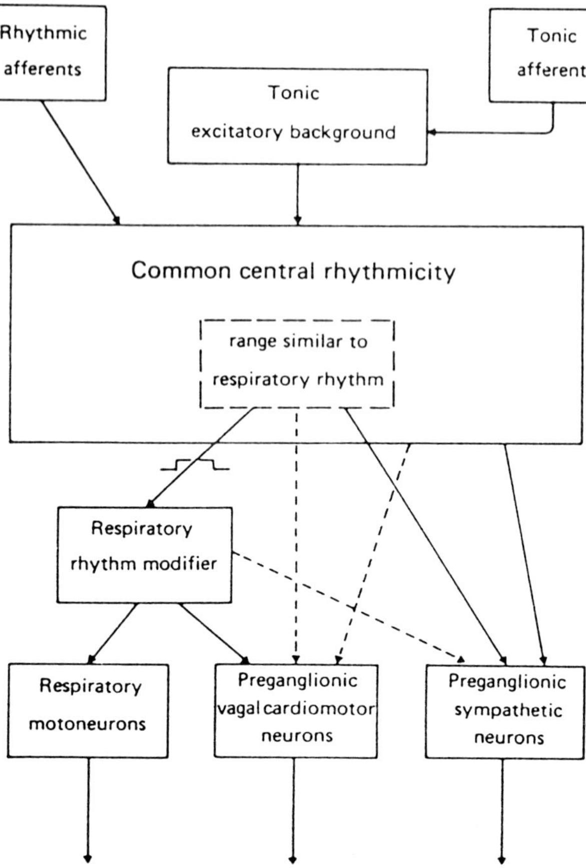

Figure 6 Schematic illustration of the supposed functional organization of common cardiorespiratory rhythmicity. Dotted lines indicate weaker actions. "The respiratory rhythm generator, i.e., that part of the network which directly projects on respiratory motoneurons, does not represent the origin of respiration-related rhythmicity in cardiovascular outputs. In contrast, both rhythms are induced by a superimposed, more differentiated prerhythm, the time course of which is reflected more directly in the cardiovascular than in the respiratory innervation. This prerhythm is transformed by a special rhythm transformer near the respiratory output into the characteristic respiratory pattern seen in the phrenic and respiratory motoneurons. The transformer makes the oscillation extremely nonlinear—it sharpens it—thus producing the physiological kind of respiratory innervation pattern. These components of the network of which the excitatory and inhibitory synaptic interactions lead to the known distinct phases of clear-cut inspiratory activity, and silence belong to the final rhythm transformer and not to the primary rhythm generator. The rhythm transformer has a higher threshold, i.e., needs a higher level of excitation, than the autonomic output to produce the rising phase of activity" (33).

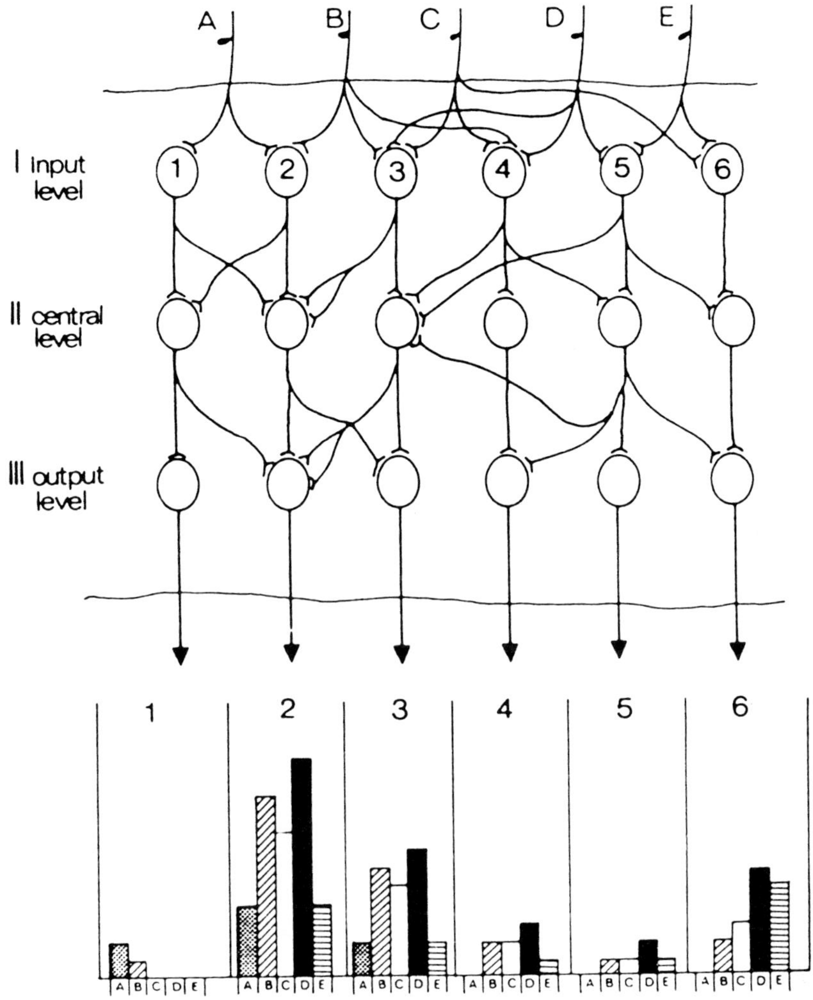

Figure 7 Model of a simple network with five different inputs (A–E) and six different outputs (1–6), illustrating that slight inhomogeneity in intracentral diverging and converging connections leads to a markedly uneven distribution of excitation between the outputs. The model contains excitatory connections only and is based on the assumption that all inputs deliver the same amount of excitation and that pure additive interaction occurs at the synapses. The columns below the network indicate the relative amount of activity reaching the respective output from the different sources (20).

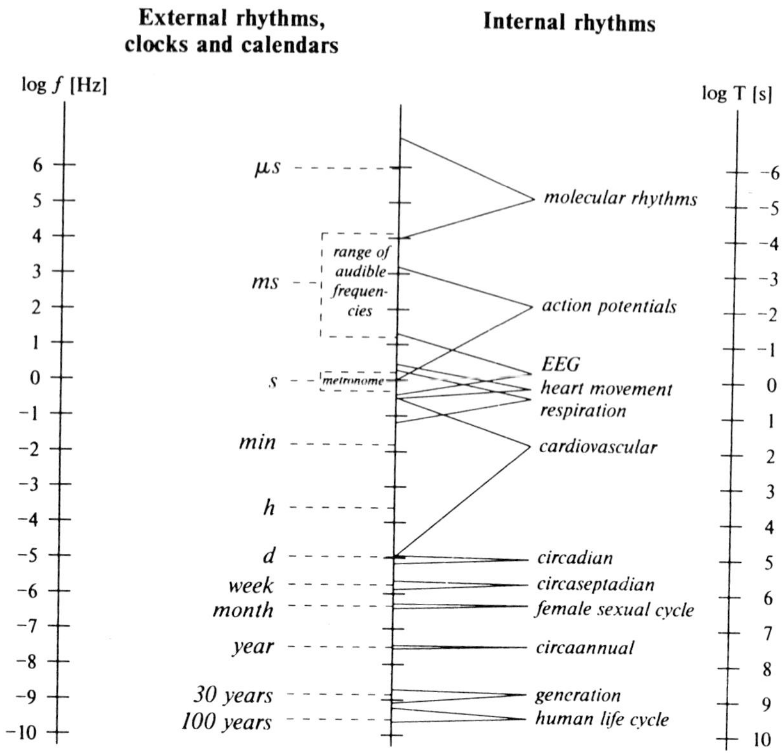

Figure 8 Scale of frequencies and their range of variation of relevant human physiological rhythms in comparison with rhythms of the outer world. Note the great variability and overlapping of frequencies in the cardiorespiratory and motor systems (21).

The rhythms, which are of particular importance to music are found in a medium range. At one end we find the EEG rhythm overlapping with that of the action potentials; at the other end, the cardiovascular parameters. The middle of this medium range between 10 Hz and 10^{-2} Hz is the proper domain of musical rhythms in the usual sense of the word. It is here that we find the inner rhythms of the heartbeat with all its variations, those of the respiratory system, and the frequencies of most of our normal movements.

In the context of neurovegetative rhythmicity for the medium frequency range it is remarkable that a similar coordination was demonstrated in the relationship between the respiratory, cardiovascular, and motor systems and in the relationship between the individual components of these systems. This also signifies the close functional connection between our inner world of neurovegetative rhythmicity and our motor activities which are used for interaction with the outer world and to express our feelings.

Koepchen develops general relationships between physiological function and music and asks the following questions:

> What is the relationship between general biological rhythmicity and music medicine? What kind of special phenomena can be found and evaluated in the realm of cardiorespiratory and motor rhythms? What is the relationship to homeostasis, stress, and relaxation? What are the possible applications to the concrete problems of music medicine?

Again, he uses a scheme (Fig. 9) of bidirectional relations between the components of brain function and those of music, illustrating the role of neurovegetative rhythmicity in the frame of these relations.

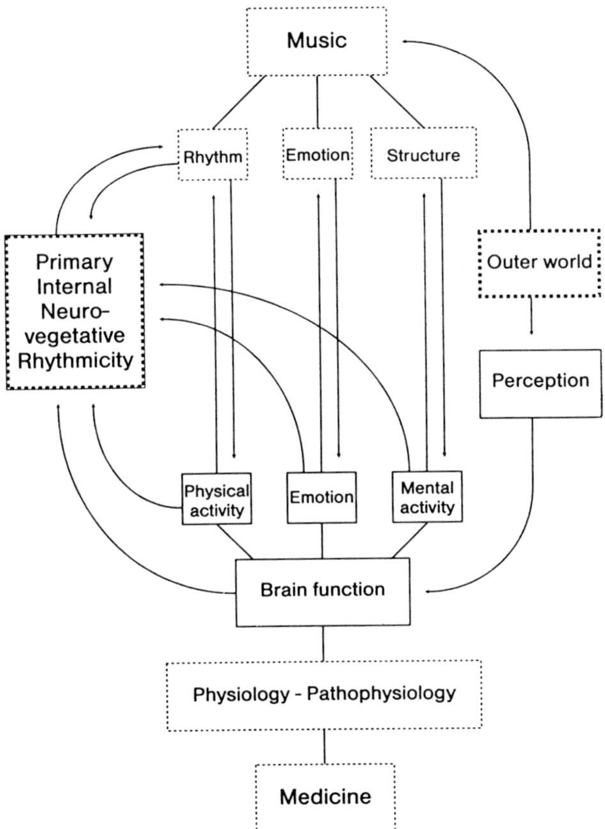

Figure 9 Scheme of bidirectional relations between the components of brain function and those of music, illustrating the role of neurovegetative rhythmicity in the frame of these relations. Music therapy is based on these interrelations (22).

Music at an archaic level comes to man from surrounding nature. We can detect music in birdsong, the rustling of trees in the wind, the roar of the surging seas, or even within a higher spiritual sense, in the harmony of the universe as Kepler did. All of this becomes music only through the meditation of the human mind. Music in the narrower sense is a creation of man. The physiological roots of this creation can be found in his body as well as in his mind. In a certain sense man has projected the essential components of his brain and bodily functions into his creation called music.

Like music, brain function is an entirety but we can differentiate a few main components in both which correspond to each other. In our brain we find the organization for the use of the motor system, the function of abstraction and thinking and for the function of the involuntary neurovegetative control systems and closely linked emotions. The elements of music can be subdivided into rhythm, emotion, and structure which is expressed in the architecture of melody or of a polyphonic fugue, for example. The subdivisions correspond to the components of brain function: abstract mental activity and neurovegetative rhythms induced by the brain correspond to musical rhythm, and finally man's emotional experience corresponds to the emotional content of music.

On the background of these activities at that period, Loeschcke and Koepchen worked intensively together for a while, which is documented, among others, in four papers that were published in 1958 (Figs. 10 to 13). The prominent role of the base of the brainstem as the site of action for respiratory and cardiovascular effects of H^+ ions and other chemical stimuli has been discovered (Figs. 14 to 16). These papers introduced a turn in the research on central control of homeostatic mechanisms, and it is rather strange that they are rarely

Pflügers Archiv, Bd. 266, S. 569—585 (1958)

Aus dem Physiologischen Institut der Universität Göttingen

Über den Einfluß von Wasserstoffionenkonzentration und CO_2-Druck im Liquor cerebrospinalis auf die Atmung*

Von

H. H. Loeschcke, H. P. Koepchen und K. H. Gertz

Mit 5 Textabbildungen

(Eingegangen am 16. Dezember 1957)

Figures 10–13 Title pages of original papers by Loeschcke and Koepchen from 1958 on the localization of central chemosensitivity as a H^+-sensitive mechanism separated from the respiratory "centers."

Pflügers Archiv, Bd. 266, S. 586—610 (1958)

Aus dem Physiologischen Institut der Universität Göttingen

Über das Verhalten der Atmung und des arteriellen Druck bei Einbringen von Veratridin, Lobelin und Cyanid in den Liquor cerebrospinalis*

Von

H. H. Loeschcke und H. P. Koepchen

Mit 11 Textabbildungen

(Eingegangen am 16. Dezember 1957)

Figure 11

Pflügers Archiv, Bd. 266, S. 611—627 (1958)

Aus dem Physiologischen Institut der Universität Göttingen

Beeinflussung von Atmung und Vasomotorik durch Einbringen von Novocain in die Liquorräume*

Von

H. H. Loeschcke und H. P. Koepchen

Mit 14 Textabbildungen

(Eingegangen am 16. Dezember 1957)

Figure 12

Pflügers Archiv, Bd. 266, S. 628—641 (1958)

Aus dem Physiologischen Institut der Universität Göttingen

Versuch zur Lokalisation des Angriffsortes der Atmungs- und Kreislaufwirkung von Novocain im Liquor cerebrospinalis *

Von

H. H. LOESCHCKE und H. P. KOEPCHEN

Mit 7 Textabbildungen

(Eingegangen am 16. Dezember 1957)

Figure 13

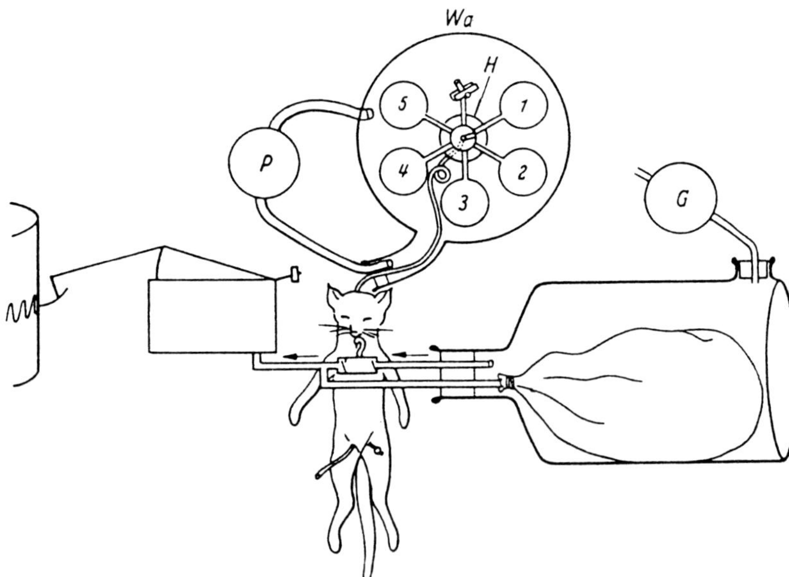

Figures 14–15 Experimental setup for the perfusion of the ventricular system of the cat.

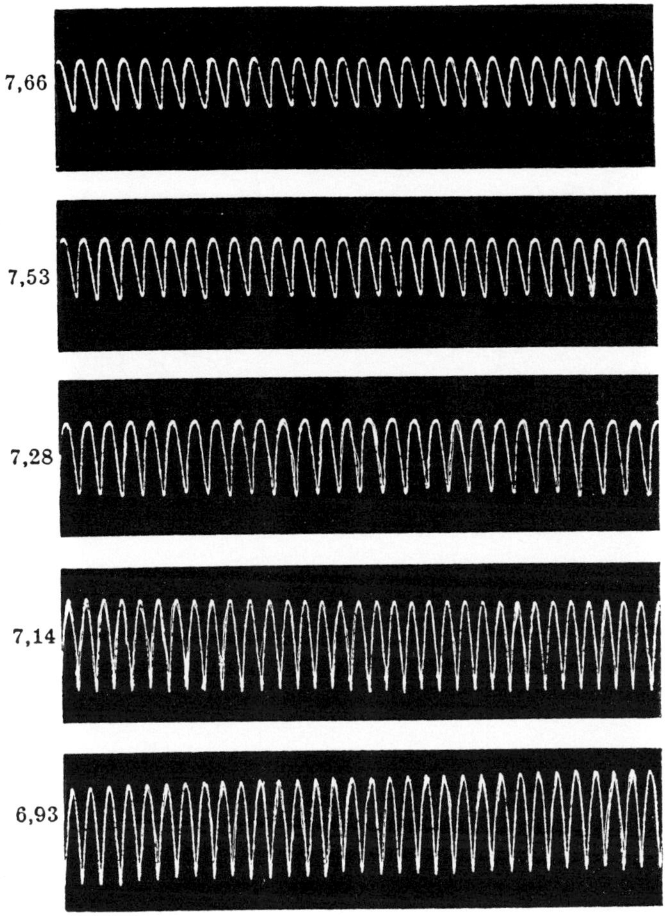

Figure 15

quoted today. In 1961, an invitation from Mitchell and Severinghaus brought Loeschcke to join the cardiovascular institute, where the San Francisco group had been searching for chemical senses within the fourth ventricle at that time. Loeschcke took his findings from Göttingen and became a coauthor in the famous paper on the ventral medullary surface H^+-ion sensitive area in 1963 (23).

Two outstanding scientists, the one responsible for our present concept on the acid-base homeostasis by the central chemosensitive mechanism (3) and the other "creator" of the term *cardiorespiratory system*, who opened our eyes to the integrative approach to physiological rhythms and synergetics (17).

An almost unearthly discipline regarding self-control and a highly developed sense of justice characterizes both Loeschcke and Koepchen. They were

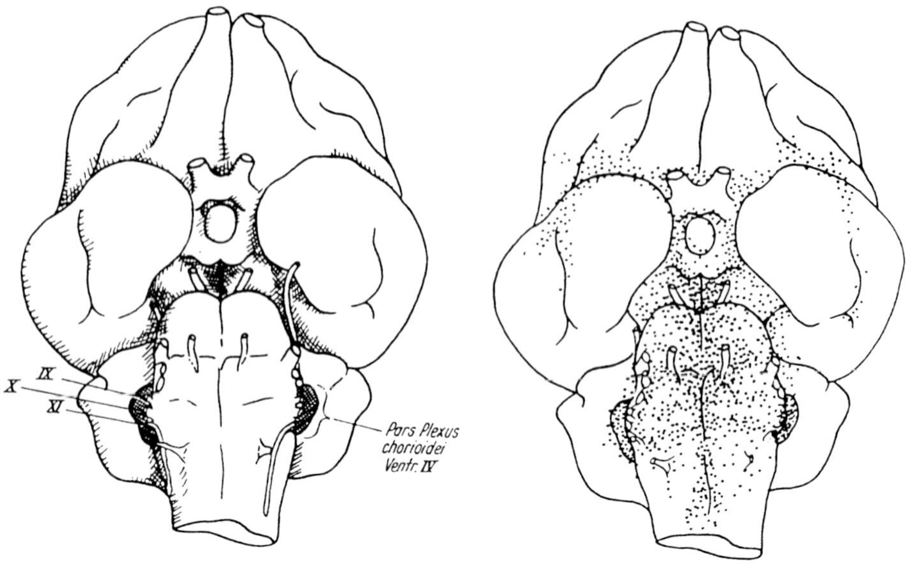

Figure 16 Distribution of eosin solution infused into the fourth ventricle. The dye reaches the ventral surface through the recessus lateralis (25).

indebted to the virtue of honesty and devoted to freedom. Both were authorities without being authoritarian. Loeschcke and Koepchen grew up in the tradition of physiology between Albrecht von Haller (1708–1777) and Hermann Rein (1898–1953). The primary general message of both scientists is expressed in two traditional lectures: The Göttinger Universitätsreden 1979 and the invited main lecture of the German Physiological Society in 1984.

> We come to the view for the today and tomorrow. Besides the detailed analysis we see the cleaning up of complex processes in complicated systems as one of our tasks, thereby it will be necessary to develop theoretical methods in order to understand the tertiary structures. Here one sees new approaches, e.g., in the common consideration of different, but linked feed back control systems by Albers or in the description of hypercycles by Eigen (24).
>
> A further very important postulate came out for the future organization of research: A real picture of the functional connections between central, vegetative, and somatic control systems cannot be obtained, when the systems continue to be worked on in an isolated way, and this isolation is continuously strengthened by further specialization. The mere survey over the nervous control of only one functional system is difficult for the single scientist already, a research prefect on the

overlapping systems can only be handled successfully by intensive cooperation of different laboratories. This, however, at present is incompatible with the organization of research as well as with the main fields of today's granted research (19).

Sir Andrew Huxley wrote on August 3, 1993: "Professor Koepchen was an important figure in the world of physiology, and his passing will be sorely felt by all those who knew him." Those who knew him best headlined his death notice with his confirmation verse: I will bless you and you may be a blessing (Genesis 12,2). May this honest and convinced obituary be integrated in what happens to us when listening to the melody "Air," D-major, BWV 1068/2 by Johann Sebastian Bach in memory of Hans H. Loeschcke and Hans-Peter Koepchen.

References

1. Kiwull-Schöne H. The "reaction theory" of Hans Winterstein (1879–1963) in the light of today's research on the ventrolateral medulla. This volume.
2. Loeschcke GC, Loeschcke HH. On the exchange of lactic acid between arterial blood and brain tissue and its variation by hypoxia. Pflügers Arch 1958; 248:521–538.
3. Loeschcke HH. Central chemosensitivity and the reaction theory. J Physiol 1982; 332:1–24.
4. Schläfke ME, Koepchen HP, See WR, eds. Central Neuron Environment and the Control System of Breathing and Circulation. Berlin, Heidelberg, New York: Springer-Verlag, 1983.
5. Loeschcke HH. Acid Base Homeostasis of the Brain Extracellular Fluid and the Respiratory Control System. Stuttgart: Georg Thieme, 1976.
6. Pappenheimer JR, Fencl V, Heisey R, Held D. Role of cerebral fluid in the control of respiration as studied in unanesthetized goats. Am J Physiol 1965; 208:436–450.
7. Middendorf T, Loeschcke, HH. Cooperation of the peripheral and central chemosensitive mechanisms in the control of the extracellular pH in brain in non-respiratory acidosis. Pflügers Arch 1978; 375:257–260.
8. Shams H. Differential effects of CO_2 and H^+ as central stimuli of respiration in the cat. J Appl Physiol 1985; 58:357–364.
9. Dempsey JA, Forster HV. Mediation of ventilatory adaptations. Physiol Rev 1982; 62:262–346.
10. Teppema LJ, Barts PWJA, Folgering HT, Evers JAM. Effects of respiratory and (isocapnic) metabolic arterial acid-base disturbances on medullary extracellular fluid pH and ventilation in cats. Respir Physiol 1983; 53:379–395.
11. Kiwull-Schöne H, Kiwull P. Hypoxia and the "reaction theory" of central respiratory chemosensitivity. Adv Exp Med Biol 1992; 316:347–357.
12. Koepchen HP, Loeschcke GC, Loeschcke HH. On the effect of adrenocorticotropic hormone on lung ventilation in man. Pflügers Arch 1954; 258:475–480.
13. Umbach W, Koepchen HP, eds. Central Rhythm and Regulation. Stuttgart: Hippokrates, 1974.
14. Koepchen HP, ed. Central Integration Between Respiratory and Cardiovascular Control Systems. Berlin, Heidelberg, New York: Springer-Verlag, 1980.

15. Koepchen HP, McC Brooks C, Koizumi K. Neurovegetative Control Systems. Amsterdam, New York, Oxford: Elsevier, 1988.
16. Koepchen HP, Huopaniemi T, eds. Cardiorespiratory and Motor Coordination. Berlin, Heidelberg, New York: Springer-Verlag, 1991.
17. Haken H, Koepchen HP, eds. Rhythms in Physiological Sciences. Berlin, Heidelberg, New York: Springer-Verlag, 1991.
18. Koepchen HP, Abel HH, Klüssendorf D. Integrative neurovegetative and motor control: Phenomena and theory. Functional Neurology 1987; 2:389–406.
19. Koepchen HP. Zentrale Kardiorespiratorische Integration: Ein Beitrag zum Zentrenproblem. In: Bromm B, Lübbers DW, eds. Hauptvorträge der Jahrestagung der deutschen Physiolgischen Gesselschaft. Stuttgart: Gustav Fischer, 1984:147–166.
20. Koepchen HP. Respiratory and cardiovascular "centres": Functional entirety or separate structures? In: Schläfke ME, Koepchen HP, See WR, eds. Central Neuron Environment and the Control System of Breathing and Circulation. Berlin, Heidelberg, New York: Springer-Verlag, 1983:221–237.
21. Koepchen HP. Physiology of rhythms and control systems: An integrative approach. In: Haken H, Koepchen HP, eds. Rhythms in Physiological Systems. Berlin, Heidelberg, New York: Springer-Verlag, 1991:3–20.
22. Koepchen HP, Droh R, Spintge R, et al. Physiological rhythmicity and music in medicine. In: Spingte R, Droh R, eds. Music Medicine. Saint Louis, MO: MMB Music, 1992:39–70.
23. Mitchell RA, Loeschcke HH, Massion WH, et al. Respiratory responses mediated through superficial chemosensitive areas on the medulla. J Appl Physiol 1963; 18: 522–533.
24. Loeschcke HH. Zum Selbstverständnis der Physiologie zwischen Albrecht von Haller und Hermann Rein. Göttinger Universitätsreden 1979; 64:13–23.
25. Loeschcke HH, Koepchen HP. Versuch zur Lokalisation des Angriffsortes der Atmungs- und Kreislaufwirkung von Novocain im Liquor cerebrospinalis. Pflügers Arch 1958; 266:628–641.
26. Loeschcke HH, De Lattre J, Schläfke ME, Trouth CO. Effects on respiration and circulation of electrically stimulating the ventral surface of the medulla oblongata. Resp Physiol 1970; 10:184–197.
27. Schläfke ME, Loeschcke HH. Lokalisation eines an der Regulatiion von Atmung und Kreislauf beteiligten Gebietes an der ventralen Oberfläche der Medulla oblongata durch Kälteblockade. Pflügers Arch 1967; 297:201–220.
28. Trouth CO, Loeschcke HH, Berndt J. Histological structures in the chemosensitive regions on the ventral surface of the cat's medulla oblongata. Pflügers Arch 1973; 339:171–183.
29. Petrovicky R. Über die Glia marginalis und oberflächliche Nervenzellen im Himstamm der Katze. Z Anat Antw 1968; 127:221–231.
30. Schläfke ME, Kille JF, Folgering H, et al. Breathing without central chemosensitivity. In: Umbach W, Koepchen HP, eds. Central Rhythm and Regulation. Stuttgart: Hippokrates, 1974:97–104.
31. Schläfke ME, Pokorski M, See WR, et al. Chemosensitive neurons on the ventral medullary surface. Bull Physio-Pathol Respir 1975; 11:277–284.

32. Middendorf T, Loeschcke HH. Mathematische Simulation des Respirations Systems. J Math Biol 1976; 3:149–177.
33. Koepchen HP, Abel HH, Klüssendorf D, Lazar H. Respiratory and cardiovascular rhythmicity. In: Von Euler C, Lagercrantz H, eds. Neurobiology of the Control of Breathing. New York: Raven Press, 1986:303–312.

3

Functional Assessment of the Ventral Medullary Surface by Optical Imaging

DAVID GOZAL

Children's Hospital of Tulane
Tulane University School of Medicine
New Orleans, Louisiana

DAVID M. RECTOR
and RONALD M. HARPER

Brain Research Institute
University of California Los Angeles School of Medicine
Los Angeles, California

I. Introduction

Determining the role of ventral medullary regions in mediating cardiorespiratory control has been a difficult challenge for neuroscience. Afferent and efferent projections in the ventral medulla are just now being defined, but the functional organization of neurons within these regions remains largely unexplored (1). In this chapter, we will outline, using recently developed procedures, response patterns over relatively large areas of the ventral medulla that have the potential to describe local neural interactions.

Anatomical descriptions and functional knowledge of the ventral medullary surface (VMS) can be contrasted with other neural systems serving specific functions, such as the visual system. It is now relatively easier to understand neural mechanisms underlying vision, since the major projection pathways have been well described and a substantial body of data exists on single neuron discharge along major portions of the visual system. Moreover, a wide variety of pathological entities exists, allowing for precise localization of dysfunctional sites within the visual system. Extensive insights into visual neural processing have required high-resolution anatomical and neurophysiological strategies, which have provided a rich description of neuronal interactions, columnar organization of specific visual functions, and reorganization of neural regions under developmental circumstances. Comparable descriptions for neural processes underlying blood pressure or respiratory control are lacking, and we must strive to provide the basic descriptions of the neural interactions that mediate such control.

Several techniques have contributed to the astonishing body of information available on neural mechanisms underlying vision. A major contribution has come from single neuron studies, which provided evidence of interactions between neuronal fields and columnar structural organization for receptive neurons (2). Additional information developed from functional anatomical studies using autoradiographic, immunohistochemical, or other tools that delineate responsive neural elements. Finally, the emergence of positron emission tomography (PET) and functional magnetic resonance imaging (FMRI) suggests that these visualization procedures may be useful in defining interactive elements within neural systems (3,4).

Although single neuron recordings have been extraordinarily useful in delineating functional brain pathways, those procedures are extremely arduous. At most, only a few neurons can be recorded simultaneously. The limitation of sampling ensures difficulty in examining interactions between neurons and guarantees that only a very small proportion of neurons from the site of interest will ever be represented. Thus, studies will likely examine neurons from limited areas rather than from widespread regions over an area. Recording from single neurons over long time periods in exposed neural areas is difficult, because tissue and vascular movements disrupt microelectrode tip positioning from cell bodies. Wide-area sampling techniques, using voltage- or calcium-sensitive dyes, are available and have joined the arsenal for examining spread of neural excitation, particularly in in vitro preparations. Dyes, however, are usually toxic, by production of reactive oxygen singlets with subsequent cellular damage, and have a finite lifetime within an experiment (5). Thus, for examination of cardiorespiratory control mechanisms that involve successive challenges or require long-term recordings, use of such dyes is limited.

Functional anatomical procedures also possess certain limitations, especially for determination of dynamic patterns. Autoradiographic procedures obviously preclude differentiation of conditions within a single animal except where bilateral representation can be incorporated into the protocol; certainly, differentiation of serial conditions is precluded. Such anatomical procedures often impose major time resolution consequences on function as well. In contrast, PET has the advantage of surveying large areas of the brain simultaneously and allows for assessment of topographical distribution of specific neurotransmitters during activation paradigms. Several minutes of continuous sampling are required, however, restricting temporal resolution. Moreover, PET provides only limited spatial resolution and involves ionizing radiation. Functional magnetic resonance imaging (FMRI) offers great promise in both temporal and spatial resolution; however, that procedure is at an early stage of development, and the neural mechanisms underlying FMRI remain largely unknown (6–8).

II. Assessment of Neural Activity by Scattered Light

An association between transmitted or reflected light changes with cellular discharge has long been known in neurophysiology (9). Indeed, at the turn of the last century, increased cell size and transparency were reported during activation (10). Since these early observations, Cohen has demonstrated a close temporal relationship between transparency of a giant squid axon and electrical events of the neural membrane (11). The mechanisms leading to altered transparency closely follow increased cell body size associated with the formation of action potentials. Some of the increased transparency characteristics with cell swelling most likely derive from light scattering within the cell body; other characteristics may depend on surface membrane properties. The latter properties might act in an analogous fashion to increasing the size of an inflatable colored balloon, which alters transparency of the balloon walls (Fig. 1). If an external illuminating source is applied over neural tissue, any change in light transmission through cell bodies or their processes will affect light reflectance

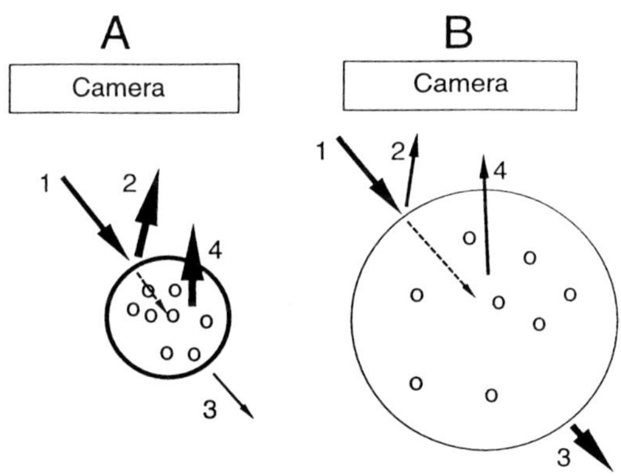

Figure 1 Schematic diagram of potential mechanisms operating in light reflectance and scattering associated with neural activation. Quiescent cell activity (A) is associated with diminished cell volumes. Incident light (1) will reflect more from a less transparent cell body (2), resulting in decreased light transmission (3); scatter will also be enhanced in the smaller volume (4). Cell discharge (B) is accompanied by cell swelling, increased transparency, and light transmission (3), resulting in decreased reflectance (2) and scatter (4).

and scatter (12,13). These light reflectance changes can be recorded by photosensitive elements. Moreover, if spatial integrity in recording elements is preserved, a topographical representation of neural activity can be demonstrated. These principles have been successfully used by subsequent investigators to visualize neural activity. Thus, MacVicar and colleagues have described patterns of neural excitation in the hippocampal slice following electrical stimulation (14), and spread of stimulated or epileptic activity in the whole brain in both animals and humans (15–17). Similarly, Grinwald and his colleagues at the Rockefeller University and Weitzman Institute described visual cortical dominance columns and other subtle interactive properties of cortical neurones within the visual system of the in vivo cat and monkey, using optical signals (18–20).

The choice of light wavelength to illuminate the neural tissue beneath the optical probe will determine which particular cellular aspects can be imaged. The number of potential physical changes associated with cellular activity is large, ranging from blood flow alterations to protein conformation changes; several of these physical changes are associated with particular light bandwidths.

It is widely accepted that regional cerebral blood flow is tightly coupled to neural tissue function and metabolism (21). Although the mechanisms responsible for coupling are not well understood, a vast array of vasoactive substances (e.g., $CO_2/H+$, $K+$) has been implicated (22). Moreover, neuronal mediation of regional blood flow increases has also been proposed in light of the close anatomical relationship between the microvasculature and surrounding nerve fibers (23). Local capillary flow changes can occur rapidly with neural stimulation, with peak flows being measured within 2 to 3 s (24). Upon activation, onset of oxygen delivery antecedes the blood volume increase (19), leading some investigators to relate the changes found in reflected light, using slow scanning cameras, to localized vascular phenomena indicative of regional activation of metabolic processes inherent to neuronal activation. Light absorption by hemoglobin plays some role in overall light capture from the neural surface, and this relationship should not be overlooked. However, the coefficients of light absorption by oxyhemoglobin and deoxyhemoglobin vary depending on the light wavelength employed. Peak absorptions occur between 420 and 430 nm, with abrupt decreases in light absorption with increasing wavelengths. At 660 to 700 nm, the wavelengths used in our studies, a decrease in light absorption by a factor of ~ 500 would be expected (25), thus achieving improved signal-to-noise ratios. Furthermore, as described earlier, neural activity elicits light transmission changes in a blood-free preparation, and such changes are blocked by ionic flow inhibitors, suggesting that cell membrane characteristics induce the principal reflectance changes following cell discharge (14). Indeed, cell swelling closely tracks action potentials during neuronal activation and evokes parallel dynamic changes in light scattering properties (26–28).

The imaging procedure typically employs a very low light level camera, of the type used in astronomy, usually using slow-scan techniques to measure the minuscule changes in light reflected or scattered by illumination of neural tissue at specific bandwidths. These slow-scan astronomical cameras have limitations of temporal resolution (1 to 3 s), although recent advancements have increased sampling rates to every 200 ms (29). The cost of the camera is relatively high, and the field of view is limited to surface structures, such as the neocortex. In addition, since the cameras are lens-coupled, significant light losses occur. To minimize light loss, we developed a coherent image conduit that is directly coupled to the sensing element. Because structures that mediate long-term effects on respiration affect both surface and subsurface sites, we needed an optical procedure that could examine activity on both superficial and deep areas. Understanding VMS neural mechanisms requires the ability to examine both rapid and long-lasting phenomena for prolonged recording periods over a large area of neural tissue. The imaging device allows acquisition at variable sampling rates (16 ms to 3 s), and thus can examine both fast- and slow changing interactions between neural areas (30–32). The procedure consists of attaching a coherent optic fiber bundle containing approximately 10,000 fibers directly to the sensing element of a miniaturized charge-coupled device (CCD) camera (Fig. 2). Light at specific bandwidths is conveyed to the neural tissue, either

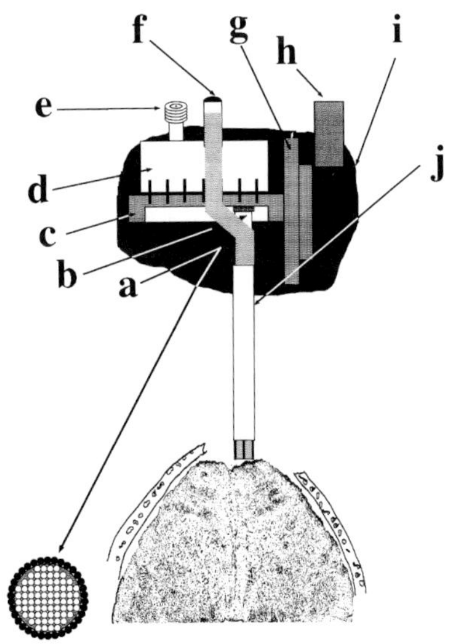

Figure 2 Schematic diagram of the camera device for acquisition of subtle light changes associated with cellular discharge: (a) illuminating fibers circumferentially arranged around the coherent optical probe (b); (c) charge-coupled device; (d) mounting plate for video connector (e); (f) illuminating light-emitting diode (LED); (g) camera circuitry; (h) pin connector for camera circuitry synchronizing pulses; (i) embedding protective material; (j) opaque tube surrounding optic probe to contain stray light.

through the cladding surrounding the optic fibers or through additional fibers attached to the perimeter of the coherent bundle. The optic probe can be placed anywhere in the brain; since the sensing element is very small, the device can be fixed to the head of a freely moving animal with a sealed skull. The capability of sealing the skull, of course, greatly minimizes brain movements from cardiovascular or other potential noise sources; we can synchronize image grabbing to the cardiac pulse or other physiological signals as well so as to minimize movement artifacts (33). A substantial contribution to the size reduction comes from recently developed camera-on-a-chip from the University of Edinburgh, a development that greatly reduced the ancillary circuitry needed for image acquisi-

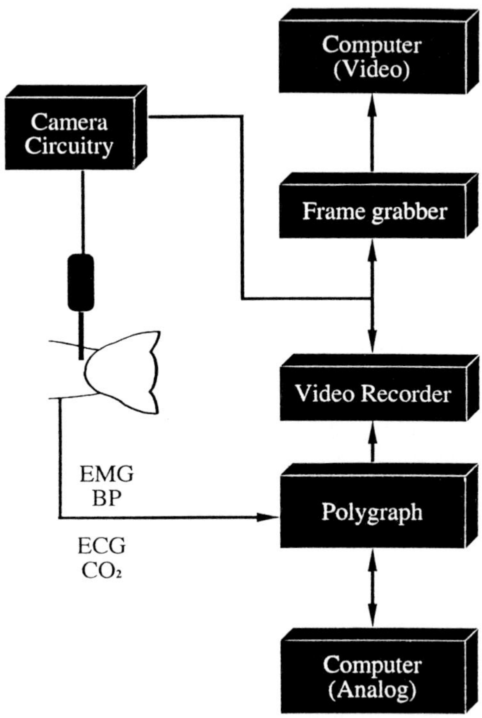

Figure 3 Schematic diagram of recording apparatus. The coherent optic probe is placed on the VMS of the supine animal. The light images are captured by a frame grabber; timing of capture is controlled by the camera circuitry. The captured images are stored on a video computer mass storage device and simultaneously recorded on a videotape recorder. Electrophysiological signals (EMG = electromyograph of the diaphragm; BP = arterial pressure; ECG = electrocardiogram; CO_2 = end-tidal carbon dioxide tension) are amplified by the polygraph and stored on a separate computer mass storage device; they are also multiplexed with the video signals on the video recorder.

tion. Images can be collected, subtracted from a reference to show changes, and displayed to show patterns of activity over time. Typically, images are collected by a frame grabber attached to a computer and are simultaneously recorded on videotape. Electrophysiological signals—such as electrocardiogram (ECG), electroencephalogram (EEG), blood pressure, respiratory patterning, and electromyographic activity (EMG)—are also recorded and stored on a second computer, or, more recently, multiplexed on a second video channel and stored with the images (Fig. 3).

A sample of the type of images that can be collected and processed is shown in Fig. 4. This image represents a digital subtraction of the averages of 50 images collected from the caudal portions of the ventral medullary surface prior to and following administration of 5% CO_2 to an anesthetized, spontaneously breathing cat. Changes in mean reflectance occurring between conditions are not homogeneous across different subregions of the probe, illustrating the different topographical responsivity to a stimulus within a neural structure. The tracings shown in Fig. 5 represent the individual values of scattered light acquired from three different subregions of the optical probe plotted over time. A substantial increase in activity (decreased reflectance) is found immediately upon application of a mild poikylocapnic hypoxic stimulus (12% O_2) in all three

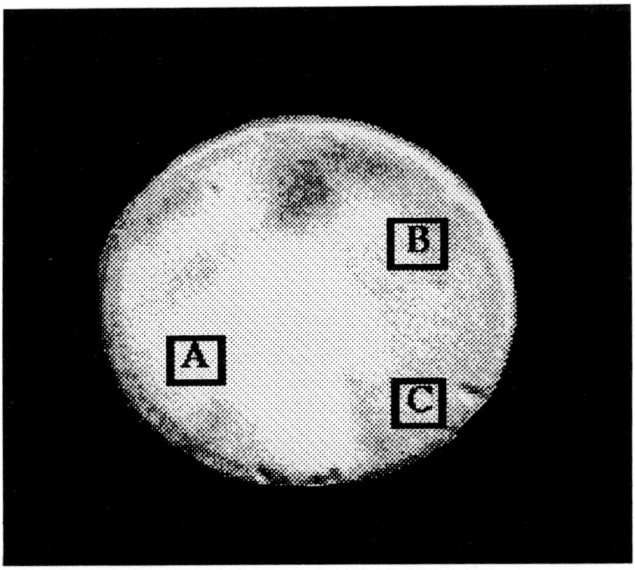

Figure 4 Gray-scale-encoded image from the caudal VMS, with three separate areas, A, B, and C, depicted. The image was derived from an average of 50 images during hypercapnic challenge (5% CO_2), and subtracted from a similar average image obtained while the animal was breathing room air.

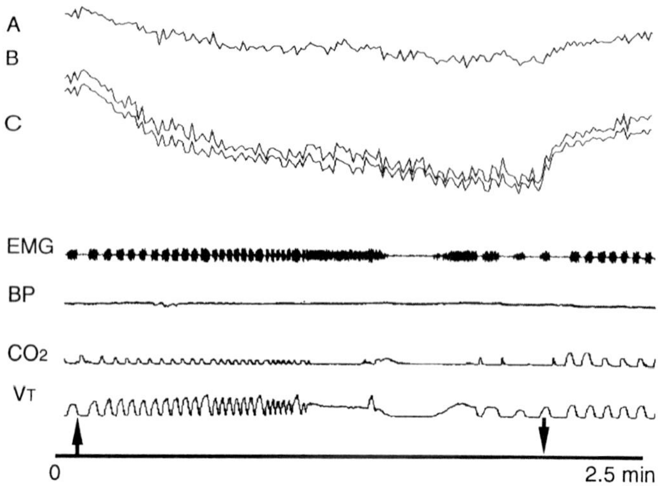

Figure 5 Time course of overall light reflectance from three separate areas of an image derived from the caudal VMS following 12% hypoxia (arrows). Note increase in activity in all areas (decreased reflectance) occurring immediately upon onset of administration of hypoxic gas, while a longer latency is required for the increase in ventilation and subsequent alveolar hypocapnia. Furthermore, two areas (B and C) decline in close accord, while the third area (A) declines only slightly. The percent decrease in light scatter during peak neural activation measured 6.3%. EMG = diaphragm electromyogram; BP = arterial blood pressure; CO_2 = end-tidal carbon dioxide tension; V_T = tidal volume.

subregions, followed by an expected increase in ventilation and subsequent alveolar hypocapnia. However, the magnitude and time course of changes in neural activity are very similar in two subregions and different in the third subregion, further illustrating regional differences in neural response of the VMS to ventilatory challenges. Thus, activity from a large number of neurons can be sampled simultaneously and interactions between subregions of the neural site examined can be studied.

III. Ventral Medullary Surface Respiratory-Induced Optical Activity

The initial studies examined neural activity in the VMS during ventilatory challenges. Eleven cats underwent general anesthesia with pentobarbital sodium. The ventrolateral medulla was exposed by a ventral surgical approach; the basioccipital bone and the underlying dura from the level of the styliform process rostrally to the level of the jugular foramen caudally, and from the midline to tympanic bulla laterally, were removed. Arterial pressure, end-tidal CO_2, costal diaphragmatic EMG, and ECG were continuously monitored in the spontaneously breathing, eucapnic animals. Hypoxic challenges consisted of up to 2 min exposure to 6%, 9%, and 12% O_2 balanced in N_2. Forty raw frames were synchronized to cardiac R waves and inspiratory offsets in each condition and averaged; pixels were pseudocolored. In all animals, 12% O_2 (mild hypoxia) induced an increase in activity in the caudal VMS (immediately rostral to the hypoglossal rootlet), while decreased activity was observed in the rostral position. Increased activity appeared with deeper levels of hypoxic stimuli in the rostral area (defined as immediately caudal to the facial nerve exit). Thus, differential patterns of response to mild hypoxic challenge were found within the VMS in the adult animal. Evidence for similar changes in light scattering were not identified in cortical surfaces. The VMS regional differences disappeared with accentuated levels of hypoxia. Dose-dependent responses were observed in intrinsic optical activity in both caudal and rostral VMS regions when exposed to 6%, 9%, and 12% O_2 (34).

Hypercapnic challenges were delivered to the preparation with the probe in similar VMS positions, as described for the hypoxic studies above. After gathering baseline images, sets of ventilatory hypercapnic stimulation with CO_2 were delivered at concentrations of 3%, 5%, and 10% balanced in O_2 with a duration of 2 min and intervals of 30 min. Hypercapnic stimulation induced regional discrete patterns of increases and declines of neuronal activity on the ventral medulla. Neuronal activity in caudal regions tended to increase. On the other hand, although rostral regions occasionally exhibited clusters of increased and decreased patterns, neuronal activity in rostral regions tended to decrease. The changes in neuronal activity induced by CO_2 stimulation were dose-dependent. With mild CO_2 stimulation, decreased activity occurred in rostral regions. With an increase of CO_2 concentration, the extent of change became greater; that is, activity diminished even further. Neurons near the cortical surface, however, did not display consistent reflectance changes during stimulation with 5% CO_2. These results suggest that patterns elicited in the ventral medulla represent responses from respiratory-related neurons rather than general effects caused by CO_2 effects on brain tissue. This study indicates that optical imaging procedures reveal topographical patterns of increased and decreased activity on the ventral medullary surface during ventilatory CO_2 stimulation (35).

We used rapid-image-acquisition optical procedures to examine VMS reflectance characteristics in order to establish whether fast-occurring neural phenomena develop during ventilatory challenges. Scattered 660 nm light was digitized continuously at 16-ms intervals from regions of the rostral VMS during a baseline period, 5% CO_2 in O_2 hypercapnic challenge, and 30-min recovery. During room air breathing, we consistently found frequency peaks, calculated by fast Fourier procedures, within the range of 8 to 13 Hz, which were not present in neocortical surfaces. During hypercapnia, both the power and frequency of these peaks increased, returning to baseline values within several minutes of cessation of the stimulus (36). During mild hypoxia, similar increases in spectral power occurred. However, during moderate to severe hypoxia, multiple peaks emerged over a wide frequency band (36). These results suggest the presence of discrete neural populations within the VMS which are sensitive to a variety of ventilatory challenges.

A. Blood Pressure Elevation Elicits Optical Changes in the Ventral Medullary Surface

We followed the time course and extent of neural activity following transient blood pressure elevation over the VMS using optical procedures. The VMS was exposed in 10 adult cats, as previously described. Light at 700 nm, reflected through an image probe, was digitized at 2- to 3-s intervals during baseline and following administration of 10, 20, and 40 mg/kg intravenous phenylephrine. Phenylephrine elicited a marked dose-dependent blood pressure elevation, profound respiratory slowing, diaphragmatic EMG amplitude diminution, and a substantial and widespread decline in neural activity in the adult cats (37).

B. Optical Assessment of VMS During Development

In contrast to adults, kittens aged 10 to 30 days did not display the caudal-rostral differentiation of neural activity observed during mild hypoxia in the VMS (38). However, accelerated initial response slopes in 20- and 30-day kittens relative to 10-day and adult animals emerged in overall scattered light measurements (39). We speculate that the age-dependent slower VMS neural responses coincide with maturational contributions of peripheral chemoreceptor function. Similar studies performed in developing acutely instrumented kittens using graded hyperoxic hypercapnic mixtures did not disclose significant topographical differences in the optical patterns when compared to similar studies in mature animals. However, age-related differences in temporal characteristics of optical responses developed, similar to those found during hypoxic challenges (39).

IV. Summary and Conclusions

A significant issue underlying functional mechanisms within the VMS is the nature of interactions between neural elements as they relate to cardiorespiratory control. Optical imaging procedures allow examination of relatively widespread regions of the VMS. Gross time resolution images reveal patterns of activity that indicate discrete regional activation of structures to ventilatory challenge, while fast temporal resolution studies disclose the presence of a range of relatively high-frequency components in the VMS. These higher-frequency components are remarkably similar to components found with single neuron discharge in the VMS; for example, shifts in cell discharge frequencies within the 8- to 15-Hz range accompany a switch from breathing room air to hypercapnia (40–42), a frequency shift that we also observed in optical patterns. The caudal-rostral differentiation within the VMS found with 12% O_2 suggests a regionalization of functional response to hypoxia and may indicate that caudal regions play a more prominent role in the neural relays responsible for mediating ventilatory response to such stimuli. In contrast, hypercapnia elicited highly variable interindividual patterns of response, which may indicate the heterogeneity of the topographical distribution of cells mediating the response to changes in chemical drive. The widespread diminution of cell activity following pharmacological pressor stimulation is consistent with single neuron discharge studies and may mediate the overall decline in respiratory activity that characteristically accompanies transient elevation in blood pressure. Changing patterns of optical activity during postnatal development concur with previous studies of ventilatory function and cellular discharge, suggesting that different neural recruitment strategies are employed by VMS neural elements during maturation.

The optical imaging data suggest that the neural mechanisms underlying cardiorespiratory and pressor control within the VMS may be revealed with far fewer arduous studies than were required for comparable levels of understanding in the mammalian visual system.

Acknowledgments

Research from the authors' laboratory was supported by HL-22418 and HD-22506. David Gozal is supported by a Parker B. Francis Family Foundation Fellowship for Pulmonary Research. David M. Rector is the recipient of a predoctoral fellowship, NIDR DE 07212. The University of Glasgow Technology Transfer Program kindly provided advanced models of Charge Coupled Device integrated circuits.

References

1. Millhorn DE, Eldridge F. Role of ventrolateral medulla in regulation of respiratory and cardiovascular system. J Appl Physiol 1986; 61:1249–1263.
2. Hubel DH, Wiesel TN. Receptive fields, binocular interactions and functional architecture in the cat's visual cortex. J Physiol (Lond) 1962; 160:106–154.
3. Fox PT, Mintun MA, Raichle ME, et al. Mapping human visual cortex with positron emission tomography. Nature 1986; 323:806–809.
4. Schneider W, Noll DC, Cohen JD. Functional topographic mapping of the cortical ribbon in human vision with conventional MRI scanners. Nature 1993; 365:150–153.
5. Pooler JP. Photodynamic alteration of sodium current in lobster axon. J Gen Physiol 1972; 60:367–387.

6. Belliveau JW, Kennedy DN, McKinstry RC, et al. Functional mapping of human visual cortex by magnetic resonance imaging. Science 1991; 254:716–718.
7. Blamire AM, Ogawa S, Ugurbil K, et al. Dynamic mapping of the human visual cortex by high-speed magnetic resonance imaging. Proc Natl Acad Sci USA 1992; 89:11069–11073.
8. McCarthy G, Blamire AM, Rothman DL, et al. Echo-planar magnetic resonance imaging studies of frontal cortex activation during word generation in humans. Proc Natl Acad Sci USA 1993; 90:4952–4956.
9. Hodge CF. A microscopical study of changes due to functional activity in nerve cells. J Morphol 1892; 7:95–168.
10. Mann G. Histological changes induced in sympathetic, motor, and sensory nerve cells by functional activity. J Anat Physiol 1894; 29:100–108.
11. Cohen LB. Changes in neuron structure during action potential propagation and synaptic transmission. Physiol Rev 1973; 53:373–418.
12. Cohen LB, Keynes RD, Hille B. Light scattering and birefringence changes during nerve activity. Nature 1968; 218:438–441.
13. Cohen LB, Keynes RD, Landowne D. Changes in axon light scattering that accompany the action potential: Current dependent components. J Physiol (Lond) 1972; 224:727–752.
14. MacVicar BA, Hochman D. Imaging of synaptically evoked intrinsic optical signals in hippocampal slices. J Neurosci 1991; 11:1458–1469.
15. Federico P, Borg SG, MacVicar BA. Imaging of seizure spread across cortex of the perfused guinea-pig brain. Soc Neurosci Abstr 1992; 18:909.
16. Hochman DW, Ojemann GA, Haglund MM. Imaging of stimulation and functionally evoked intrinsic optical changes in monkey and human cortex. Soc Neurosci Abstr 1992; 18:583.
17. Haglund MM, Ojemann GA, Hochman DW. Optical imaging of epileptiform and functional activity in human cerebral cortex. Nature 1992; 358:668–671.
18. Grinvald A, Lieke EE, Frostig RD, et al. Functional architecture of cortex revealed by optical imaging of intrinsic signals. Nature 1986; 324:361–364.
19. Frostig RD, Lieke EE, Ts'o DY, Grinvald A. Cortical functional architecture and local coupling between neuronal activity and the microcirculation revealed by in vivo high resolution optical imaging of intrinsic signals. Proc Natl Acad Sci USA 1990; 87:6082–6086.
20. Bonhoeffer T, Grinvald A. Iso-orientation domains in cat visual cortex are arranged in pinwheel-like patterns. Nature 1991; 353:429–431.
21. Sokoloff L. Relationship among local functional activity, energy metabolism, and blood flow in the central nervous system. Fed Proc 1981; 40:2311–2316.
22. Kuschinsky W, Wahl M, Bosse O, Thurau K. Perivascular potassium and pH as determinants of local pial arterial diameter in cats: A microapplication study. Circ Res 1972; 31:240–247.
23. Lou HC, Edvinsson L, Mackenzie ET. The concept of coupling bloodflow to brain function: Revision required? Ann Neurol 1987; 22:289–297.
24. Lindauer U, Villringer A, Dirnagl U. Characterization of CBF response to somatosensory stimulation: Model and influence of anesthetics. Am J Physiol 1993; 264:H1223–H1228.
25. Stryer L. In: Biochemistry. New York: Freeman, 1981:52.

26. Tasaki I, Iwasa K. Further studies of rapid mechanical changes in squid giant axon associated with action potential production. Jpn J Physiol 1982; 32:505–518.
27. Tasaki I, Kusano K, Byrne PM. Rapid mechanical and thermal changes in the garfish olfactory nerve associated with propagated impulse. Biophys J 1989; 55:1033–1040.
28. Tasaki I, Byrne PM. Rapid structural changes in nerve fibers evoked by electric current pulses. Biochem Biophys Res Comm 1992; 188:559–564.
29. Grinvald A, Frostig RD, Lieke E, Hildesheim R. Optical imaging of neural activity. Physiol Rev 1988; 68:1285–1365.
30. Rector DM, Harper RM. Imaging of hippocampal neural activity in freely behaving animals. Behav Brain Res 1991; 42:143–149.
31. Rector DM, Poe GR, Harper RM. Assessment of dorsal hippocampal optical properties concurrent with electrophysiological measures at high temporal and spatial resolution. Soc Neurosci Abstr 1992; 18:918.
32. Rector DM, Burk P, Harper RM. A data acquisition system for long term monitoring of physiological and video signals. Electroencephalogr Clin Neurophysiol 1993; 87:380–384.
33. Rector DM, Poe GR, Harper RM. Imaging of hipoccampal and neocortical neural activity following intravenous cocaine administration in freely behaving cats. Neuroscience 1993; 54:633–641.
34. Gozal D, Dong XW, Rector DM, Harper RM. Optical imaging of the ventral medullary surface of cats: Hypoxia-induced differences in neural activation. J Appl Physiol 1993; 74:1658–1665.
35. Dong XW, Gozal D, Rector DM, Harper RM. Ventilatory CO_2-induced optical activity changes of the cat ventral medullary surface. Am J Physiol 1993; 265:R494–R503.
36. Gozal DM, Rector DM, Harper RK, et al. Assessment of ventral medullary surface activity during graded hypoxia in the cat with high temporal resolution optical imaging. Soc Neurosci Abst 1993; 19:1490.
37. Harper RM, Gozal D, Aljadeff G, et al. Pressor-induced responses of the cat ventral medullary surface. Am J Physiol. In press.
38. Gozal D, Dong X-W, Rector DM, et al. Optical imaging of the ventral medullary surface of developing kittens during ventilatory challenges. J Dev Physiol. In press.
39. Rector DM, Gozal D, Dong X-W, et al. Maturational differences in 660 nm light scattering from ventral medullary surface of kittens during graded hypercapnia and hypoxia. Soc Neurosci Abst 1993; 19:1490.
40. Cakar L, Terzioglu M. Localization of CO_2-sensitive unit in the rostral medullary chemosensitivity area of the cat. In: Schläfke ME, Koepchen HP, See WR, eds. Central Neurone Environment and the Control Systems of Breathing and Circulation. Berlin: Springer-Verlag, 1983:52–60.
41. Pokorski M. Neurophysiological studies on central chemosensor in medullary ventrolateral areas. Am J Physiol 1986; 230:1288–1295.
42. Arita H, Yamabayashi H. Locations of central chemoreceptors. In: Takashima T, Cherniack NS, eds. Advances in the Biosciences: Control of Breathing and Dyspnea. New York: Pergamon Press, 1991:35–41.

4

Identification of Brainstem H^+ Ion–Chemosensitive Regions by c-fos Immunocytochemistry

ROBERT M. DOUGLAS, C. OVID TROUTH, YI PAN,* JAMES C. McKENZIE and WINSTON N. SMITH

College of Medicine
Howard University
Washington, D.C.

NICOLE E. MOURILLON

Xavier University
New Orleans, Louisiana

G. NORRIS MELVILLE

University of the West Indies
St. Augustine, Trinidad and Tobago, West Indies

I. Introduction

Electrical or chemical stimulation of discrete regions on the ventrolateral medullary surface (VMS) of the medulla oblongata can modulate the output of the central regulators of vasomotor and respiratory activity. Three such chemosensory regions have been described in the cat (1–5) and two in the rat (6–9). These regions are hypothesized to be the loci of central cardiorespiratory chemosensitive neurons (Fig. 1). The topical application of procaine to (10,11) and destruction of (12,13) these areas in cats abolishes carbon dioxide (CO_2) sensitivity. The abolition of CO_2 sensitivity has also been noted following topical application of the excitotoxic glutamate analog kainic acid (KA) to the ventrolateral medulla (VLM) of rats (14,15).

Neurons located within a depth of 200 μm of the VMS have been described by Schläfke et al. (16) and Kille and Schläfke (17) beneath the intermediate chemosensitive area (area S) and by Trouth et al. (5) beneath the caudal chemosensitive area (area L) in cats. These neurons have also been documented by electron microscopic examination of the medulla oblongata (18–20). Recent

**Present affiliation*: St. Louis University Hospital, St. Louis, Missouri.

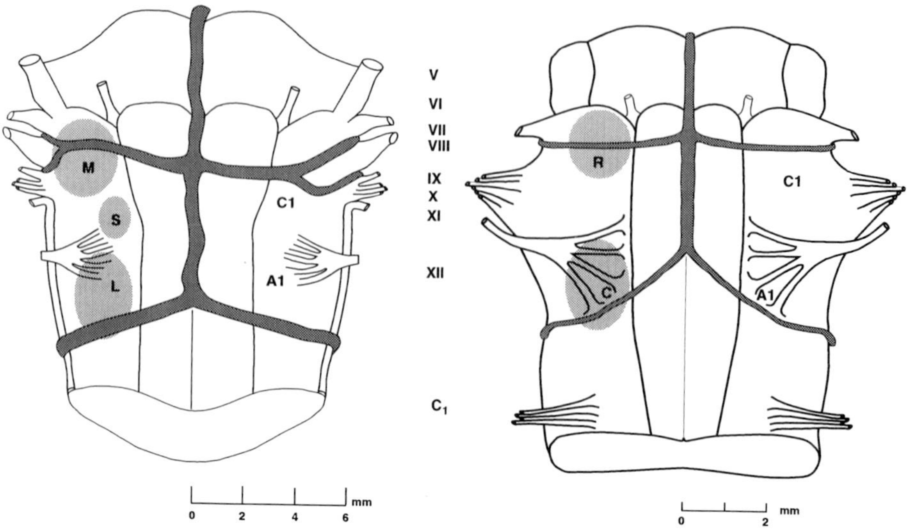

Figure 1 Ventral surface of the medulla oblongata of the cat (left) and the rat (right). In the cat, the three respiratory chemosensitive areas described by Mitchell et al. (3), M, and Schläfke et al. (1), S and L, are indicated on the left half of the diagram and represent the rostral, intermediate, and caudal chemosensitive areas, respectively. In the rat, the two chemosensitive zones are indicated on the left side of the diagram by the symbols R and C, representing the rostral and caudal chemosensitive areas, respectively, as described by Mitra et al. (J Appl Physiol 1989; 67:1814). In both right hemimedullas, the symbols C_1 and A_1 represent the vasopressor and vasodepressor regions of the medulla oblongata, respectively. V–XII = cranial nerves.

immunocytochemical studies with antibodies to tyrosine hydroxylase (TH) (21, 22), choline acetyltransferase (ChAT) (23–25), and c-fos (15,26–29) have also demonstrated superficially located neurons in the chemosensitive zones of the medulla oblongata. However, the anatomical substrate subserving CO_2 or hydrogen ion (H^+) chemosensitivity has not been clearly elucidated.

It is believed that c-fos and other immediate-early genes play a role in signal transduction, functioning as third messengers that couple acute events at the cell surface to long-term changes in gene expression (30). Therefore, despite an incomplete understanding of the mechanism of action of c-fos, it has been successfully used as a metabolic marker in tracing neuronal pathways (31–35). Sato et al. (29) recently described a discrete distribution of c-fos–positive cells in the VLM that indicated response to inhaled CO_2. They identified a population of c-fos–positive cells within 150 μm of the VMS, with a significant proportion of those within 50 μm of the surface. Erickson and Millhorn (26) demonstrated that stimulation of the carotid sinus nerve and hypoxia resulted in the distribution of c-fos along the VMS, within the VLM, in the nucleus of the solitary tract (NTS) bilaterally, and within the area postrema (AP) and raphe pallidus.

The advent of c-fos immunocytochemistry has therefore provided a tool that may circumvent the dilemma of isolating those central chemosensitive elements responsive to respiratory stimuli such as CO_2 and H^+. The present study utilizing the immunocytochemical identification of the c-fos protein was undertaken to localize central chemosensitive neurons within the VLM modulating respiratory and cardiovascular activity.

II. Expression of c-fos Protein in the Brainstem of the Rat in Response to the Topical Application of Kainic Acid (1%)

In spontaneously breathing animals ($n = 5$), pledgets containing KA (1%, 47 m*M*) were topically and unilaterally applied to the VMS in the caudal chemosensitive zone (area L) for approximately 2 min. The pledgets were then removed and the VMS thoroughly rinsed with mock cerebrospinal fluid (mCSF) at pH 7.4. KA induced transient increases in tidal volume (V_T) and mean arterial blood pressure (MABP). These increases were followed by a return of the MABP to baseline levels; however, V_T was depressed to apnea within 30 min. On induction of apnea, the animals were placed on a mechanical ventilator for approximately 2 h to ensure adequate translation of c-fos mRNA to c-fos protein. The brains of the animals were then perfused via the intracardiac route with 4% paraformaldehyde in 0.1 *M* phosphate buffer, pH 7.3; postfixed overnight; cryoprotected in glycerol buffer over a 48-h period; and 40-μm cryostat sections were prepared. The tissue was then processed for c-fos protein immunoreactivity utilizing the avidin-biotin peroxidase complex technique of Uemura et al. (36). Evaluation of the tissue revealed a limited expression of the c-fos protein in neuronal elements along the VMS in area L but a robust response within the AP—with very little or scant staining in the remainder of the neuraxis (Fig. 2).

The c-fos protein expression may have been scant in the medulla because KA appears to be region-specific and sometimes unpredictable in its ability to induce c-fos protein production (37–39). It has been noted that in KA-induced seizures, c-fos mRNA is maximally and protractedly induced, yet the c-fos protein demonstrates a low level of induction—also over a protracted time period. This suggests that the c-fos protein is rapidly degraded or that there is failure in the translative process (40).

The demonstration of c-fos protein in the AP raises many intriguing possibilities. The AP is known to be involved in chemosensory transduction, e.g., the chemoreceptor trigger zone (41,42), and in visceromotor function due to its reciprocal connectivity to the NTS (43–45), and it is one of those unique central nervous system (CNS) structures, a circumventricular organ, that exists

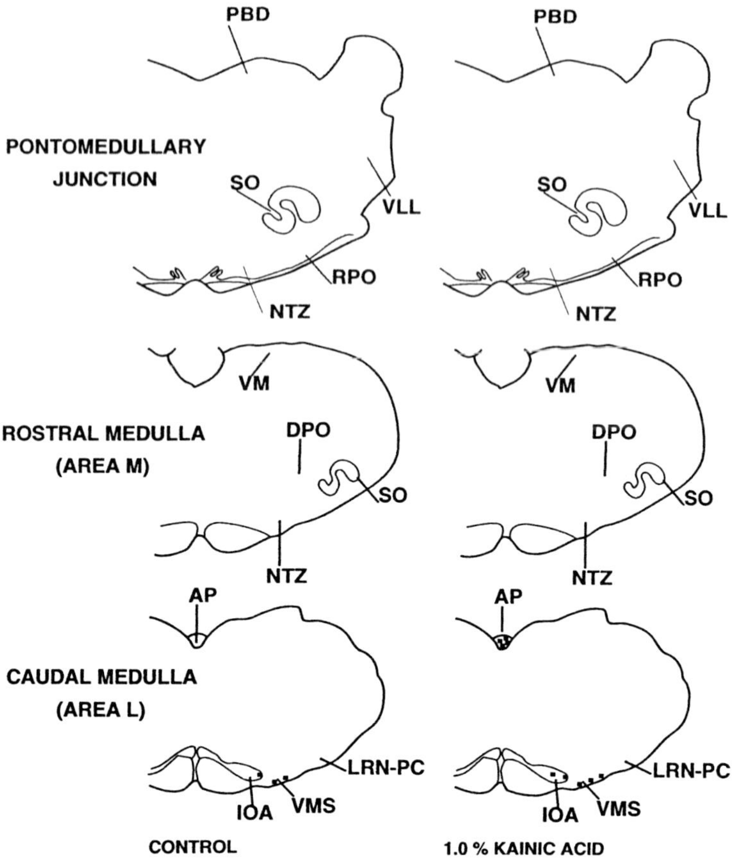

Figure 2 Diagrammatic representation of c-fos immunoreactivity in the brainstem of the rat in response to the unilateral topical application of 1% KA to the cVMS of the medulla oblongata. Each dot represents a single neuron. Reaction product of c-fos can be visualized along the VMS, occasionally in the ventrolateral IO, and dramatically within the AP. VMS = ventrolateral medullary surface; IO = inferior olivary nucleus; AP = area postrema; DPO = dorsal periolivary region; VM = medial vestibular nucleus; VLL = nucleus of the lateral lemniscus.

outside of the blood-brain barrier (46). As a result, c-fos expression within the AP could represent a blood pressure response, a ventilatory response, or a metabolic response caused by the compromised condition of the animal. The suggested participation of the AP in the central control of respiration (47) and in the respiratory component of emesis (48) and the stability of the MABP during KA-induced apnea argue for the AP being a component of the respiratory chemoreflex apparatus operating within the mammalian brainstem.

III. Expression of c-fos Protein in the Brainstem of the Rat in Response to the Topical Application of Acidic Mock Cerebrospinal Fluid

In another series of studies, we applied pledgets containing mCSF at pH 7.4 (control) and pH 7.2 (acidic) to the caudal VMS (cVMS) of spontaneously breathing, chloralose-urethane anesthetized rats ($n = 5$) for 2 h in order to allow suf-

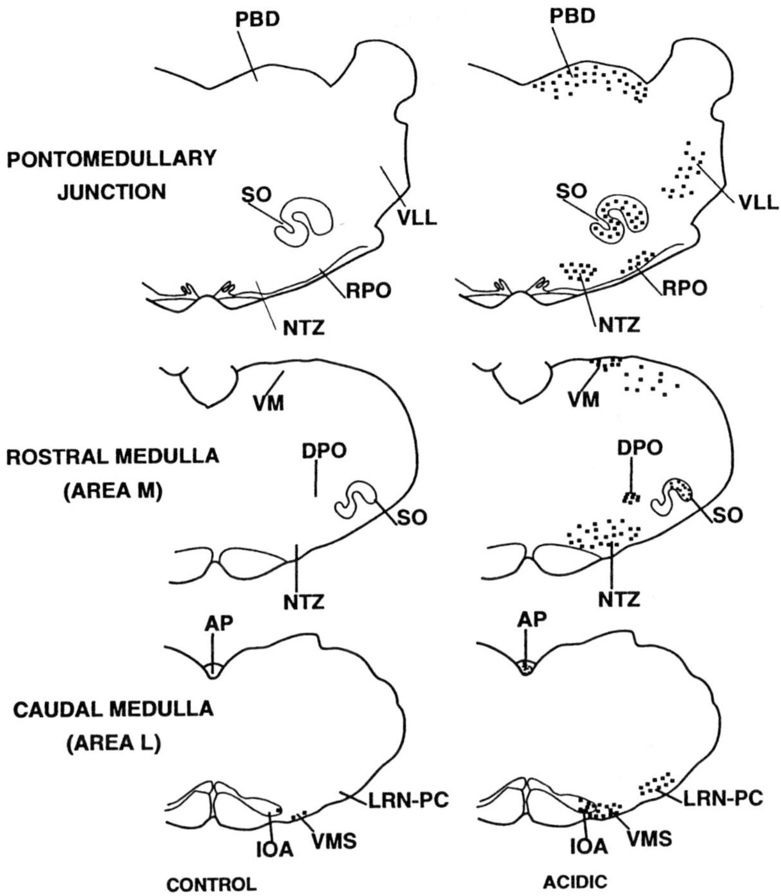

Figure 3 Diagrammatic representation of c-fos immunoreactivity in the brainstem of the rat in response to the topical application of acidic mCSF (pH 7.2) to the cVMS of the medulla oblongata. Control = pH 7.4. Each dot represents a single neuron. Reaction product of c-fos was conspicuous in the IO and LRN-PC at the caudal VMS, in the NTZ and RPO, at the rostral VMS and the caudal pons. IO = inferior olivary nucleus; LRN-PC = parvocellular region of the lateral reticular nucleus; NTZ = nucleus of the trapezoid body; RPO = rostral periolivary group; SO = superior olivary nucleus.

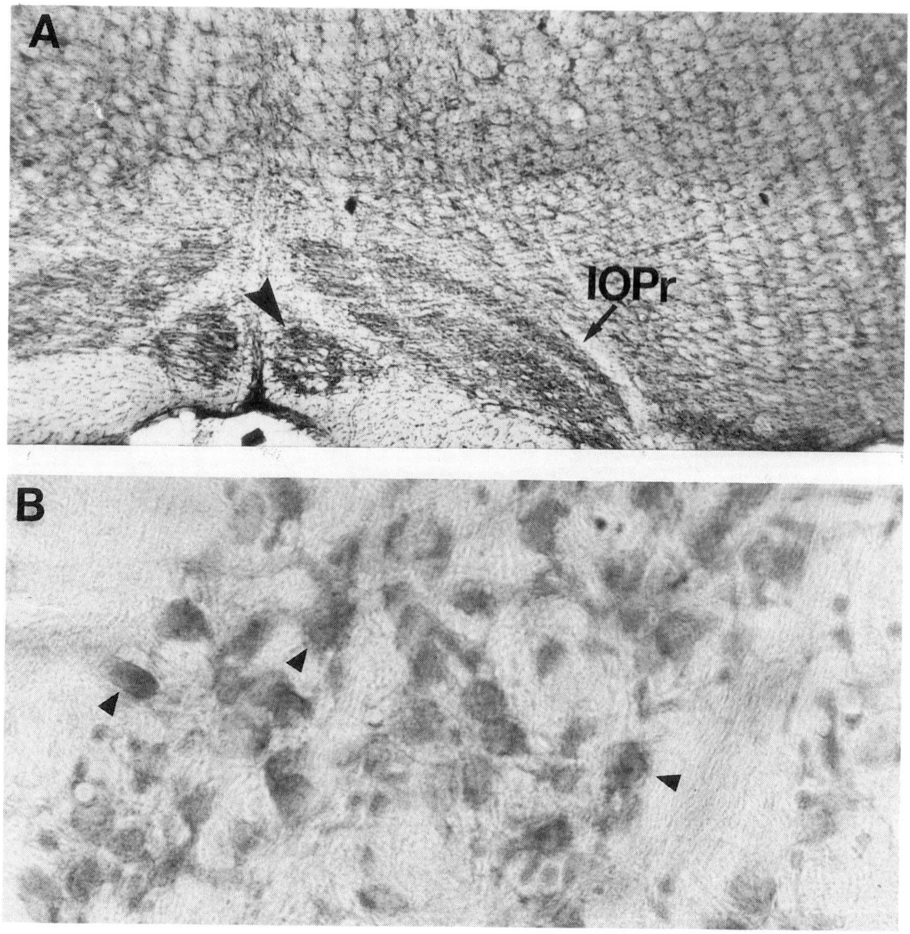

Figure 4 Distribution of c-fos protein reaction product at the VMS of the rat medulla oblongata in response to acidic mCSF (pH 7.2). (A) c-fos immunoreactivity is visualized in the IOPr and along the VMS ($\times 10$). Ventromedial to the pyramids, c-fos reaction product in response to acidic mCSF is indicated by the large arrowhead. This region has been variously described in the rat as (a) inferior olive, medial nucleus (57) and (b) part of nucleus raphe pallidus (58). This location is reminiscent of the arcuate nucleus described in humans and cats (59), which has been implicated in sudden infant death syndrome (SIDS) by Filiano et al. (59) and "Ondine's curse" (primary alveolar hypoventilation) by Folgering et al. (60). The relationship between nucleus raphe pallidus, arcuate nucleus, and VMS cells is rather close, wherein the topographic proximity as well as developmental, chemical, and functional similarities argue for a common function (59). These three groups are believed to originate ontogenetically from neuronal precursors that migrate ventromedially from the rhombic lip and differentiate into medullary arcuate nucleus, inferior olivary complex, griseum pontis, and serotoninergic cells of nucleus raphe pallidus and PGCL (61). Cresyl violet counterstain. (B) Here, c-fos–positive cells

ficient time for the translation of the c-fos protein from the induced c-fos mRNA. The animals were then rapidly perfused and processed for c-fos immunoreactivity. Adjusted to pH 7.4, mCSF produced a light stain at the VMS in the caudal chemosensitive zone and also in what appear to be a few neurons at the most ventral part of the inferior olivary nucleus (IO). There was scant or no c-fos immunoreactivity throughout the rest of the neuraxis (Fig. 3). It can be postulated that temperature- or pressure-sensitive units within the VMS could have been responsible for the induction of low levels of c-fos at the VMS.

Topical application of mCSF at pH 7.2 (acidic) to the cVMS (area L) for a 2-h period induced the production of c-fos protein in many brainstem structures. At the level of the cVLM—i.e., at the exit of the hypoglossal rootlets—c-fos reaction product was noted in the IO bilaterally, in the lateral reticular nucleus (LRN) unilaterally, and along the cVMS (Fig. 4). Immunoreactivity of c-fos was also detected in the parvocellular region of the LRN (LRN-PC) bilaterally, within the AP, and with slight responses in the NTS and the hypoglossal nucleus (nXII) (Fig. 5). The most dramatic responses at this level were noted in the inferior olive, the LRN-PC, and along the VMS—i.e., within nuclei that are most superficial. These results correspond well with the reports of Sato et al. (29), who described cells along the VMS that were c-fos–responsive to 1 h of hypercapnia and that resided ventral to the LRN and the nucleus paragigantocellularis lateralis (PGCL), and with the work of Erickson and Millhorn (26), in which carotid sinus nerve stimulation and hypoxia induced a similar pattern of c-fos immunoreactivity, particularly in a group of cells ventral to the IO and lateral to the parapyramidal cell groups.

At the level of the rostral medulla, in the vicinity of the facial nucleus and nerve (CN VII), c-fos immunoreactivity was noted in the nucleus of the trapezoid body (NTZ), the medial and lateral periolvary nuclei (MVPO and LVPO), the superior periolivary nucleus (SPO), the lateral and medial superior olives (LSO and MSO) and the dorsal periolivary region (DPO) (Fig. 6). Immunoreactivity of c-fos was also detected along the rostral VMS, although with a sparser representation as compared with the caudal VMS, and in the PGCL in this region. There was a distinct reaction in the ventral and dorsal cochlear nuclei, but such a reaction also existed in control sections. There was also a response in the cerebellum in the lateral (dentate) and interposed (interpositus) cerebellar nuclei. Notably, changes in ventilation and blood pressure have been elicited in cats upon stimulation of the surface and interior of the cerebellum (49,50).

---

can be distinguished in the nucleus of the trapezoid body (NTZ) in response to the topical application of acidic mCSF to the cVMS (×40). Arrowheads point to c-fos–positive cells. Cresyl violet counterstain. IOPr = inferior olive, principal nucleus; VMS = ventral medullary surface.

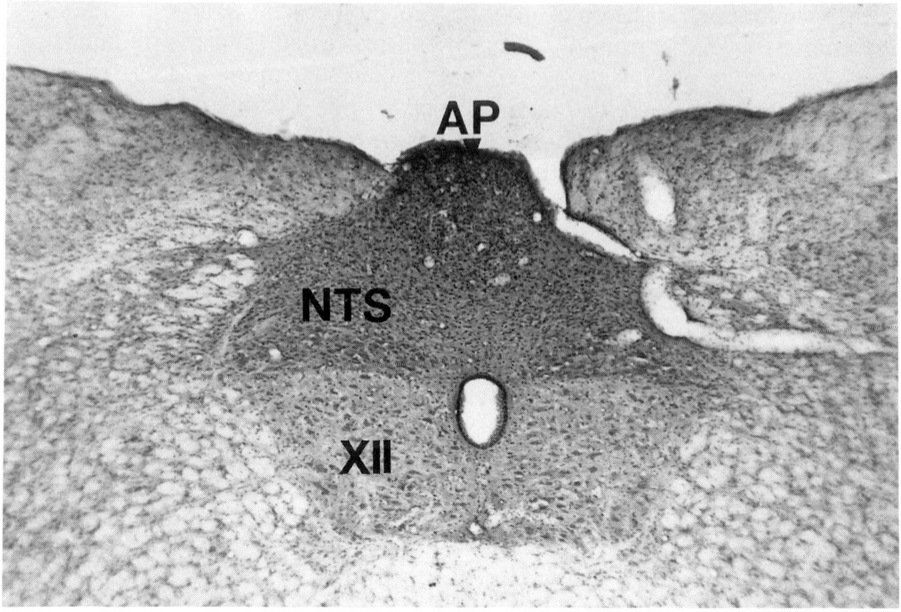

Figure 5 Distribution of c-fos protein reaction product at the caudal VMS of the rat medulla oblongata in response to acidic mCSF (pH 7.2) (×10). Cresyl violet counterstain. Here, c-fos immunoreactivity is visualized in the AP, with slight responses in the NTS. AP = area postrema, NTS = nucleus of the solitary tract.

The dorsal periolivary region, at the level of the rostral medulla, appears to coincide with the location of the expiratory modulated neurons of the Bötzinger complex, which is considered the rostral component of the ventral respiratory group (VRG) (51–53). The existence of c-fos–positive cells ventral to the facial motor nucleus is reminiscent of the retrotrapezoid nucleus (RTN), which has extensive efferent projections to both the dorsal respiratory group (DRG) and VRG (53,54). The rostral ventrolateral medulla (rVLM) also contains the rostral chemosensitive zone (area M), as described by Mitchell et al. (3), which is sensitive to the acid-base changes of the surrounding CSF. The RTN could therefore serve as the anatomical substrate for chemosensitivity residing at area M.

At the pontomedullary border, c-fos immunoreactivity was detectable in the NTZ, the ventral pontine nuclei, more rostrally in the rostral periolivary region (RPO), and within the nuclei of the lateral lemniscus (VLL and DLL). At midbrain regions, light staining was noted at the dorsal aspects of the superior and inferior colliculi; however, a more robust response was noted in the red nucleus. The demonstration of c-fos in cerebellar nuclei and the red nucleus raises the possibility that these neurons are involved in the modulation of the

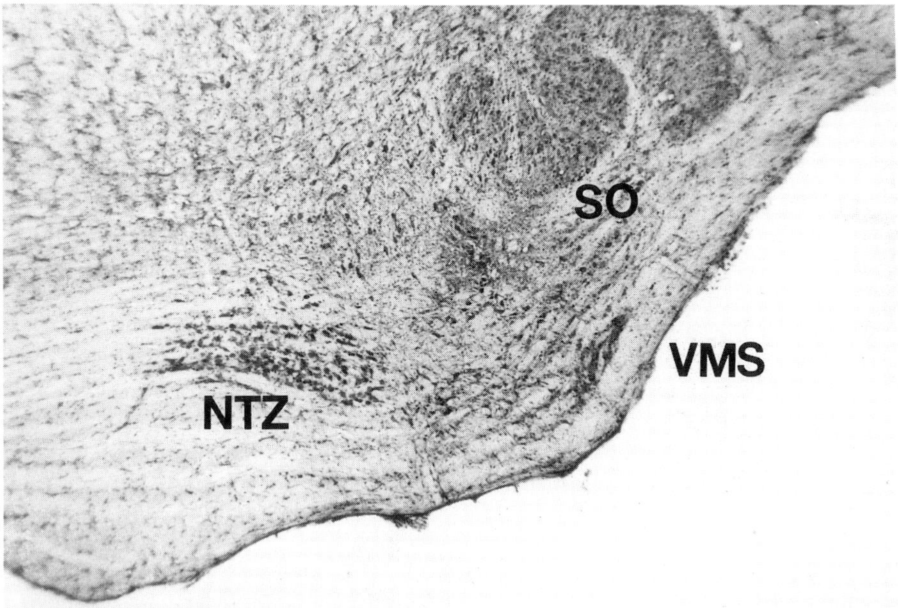

Figure 6 Distribution of c-fos protein reaction product at the rostral VMS of the rat medulla oblongata in response to acidic mCSF (pH 7.2). Here, c-fos immunoreactivity can be distinctly visualized in the NTZ, the SO, and along the VMS (×10). Cresyl violet counterstain. NTZ = nucleus of the trapezoid body; SO = superior olivary nucleus; VMS = ventrolateral medullary surface.

central respiratory efferent pathway. There was also a response in the interpeduncular nucleus, but this also appeared in control sections and was therefore utilized, with the dorsal cochlear nucleus, as an internal positive control.

IV. Summary

Physiological investigations have indicated that neurons located on the VMS subserve central respiratory chemosensitivity and vasomotor control (55). The proto-oncogene c-fos and its concomitant protein are expressed rapidly, transiently, and polysynaptically within neurons in response to synaptic activation, neuronal bursting, noxious stimulation, and voltage-gated calcium entry into the cell (56). The present study was conducted to identify cells within the rat medulla oblongata that express the c-fos protein in response to altered physiological pH on the VMS. In spontaneously breathing, chloralose-urethane anesthetized rats, filter-paper pledgets soaked in mCSF at a pH of 7.2 were applied to the caudal chemosensitive area (area L) for approximately 2 h. Topical application of acidic mCSF to area L induced a consistent and identifiable distribution of c-fos protein immunoreactivity. At the level of the caudal medulla, c-fos protein reaction product was observed within the inferior olivary nucleus, lateral reticular nucleus, nucleus of the solitary tract, area postrema, and bilaterally along the VMS. At the level of the rostral medulla, c-fos was noted in the nucleus of the trapezoid body, the PGCL, the medial and lateral periolivary nuclei, the superior periolivary nucleus, the medial and lateral superior olives, the dorsal periolivary region, and the cochlear nucleus. At the level of the pons, c-fos was noted in the NTZ, ventral tegmental pontine nuclei, the rostral periolivary region, and within the nucleus of the lateral lemniscus. Cells responsive to c-fos–positive cells were also detected within the red nucleus and cerebellar nuclei. These observations are suggestive of the possible involvement of the inferior olivary nucleus, the superior and lateral olivary nuclei, the nucleus of the trapezoid body, and cells along the VMS operating at the caudal (area L) and rostral (area M) regions of the ventrolateral medulla in the higher-order processing within central respiratory chemoreceptor reflex pathways.

This work was supported by NIH/NIGMS Grant #S069M08016.

References

1. Schläfke ME, Loeschcke HH. Lokalisation eines an der Regulation von Atmung und Kreislauf beteiligten Gebietes an der ventra Oberflache der Medulla oblongata durch Kalteblockade. Pflügers Arch 1967; 297:201–220.
2. Loeschcke HH, DeLattre J, Schläfke ME, Trouth CO. Effects on respiration and circulation of electrically stimulating the ventral surface of the medulla oblongata. Resp Physiol 1970; 10:184–197.
3. Mitchell RA, Loeschcke HH, Massion WH, Severinghaus JW. Respiratory responses mediated through superficial chemosensitive areas on the medulla. J Appl Physiol 1963; 18:523–533.
4. Mitchell RA, Loeschcke HH, Severinghaus JW, et al. Regions of respiratory chemosensitivity on the surface of the medulla. Ann NY Acad Sci 1963; 109:661–681.
5. Trouth CO, Loeschcke HH, Berndt J. A superficial substrate on the ventral surface of the medulla oblongata influencing respiration. Pflügers Arch 1973; 339:135–152.
6. Pan Y, Bernard DG, Holloway JA, Trouth CO. Histological studies on brainstem respiratory chemosensitive areas in rats. Neurosci Abstr 1989; 15:99.
7. Mitra J, Prabhakar NR, Overholt JL, Cherniack NS. Respiratory and vasomotor responses to focal cooling of the ventral medullary surface (VMS) of the rat. Resp Physiol 1988; 74:35–48.
8. Laha PK, Nayar U, Chhina GS, Singh B. Carbon dioxide sensitivity of the central chemosensitive mechanisms: An exploration by direct stimulation in rats. Pflügers Arch 1977; 367:241–247.
9. Malcolm JL, Sarelius IH, Sinclair JD. The respiratory role of the ventral surface of the medulla studied in the anesthetized rat. J Physiol 1980; 307:503–515.
10. Berndt J, Berger W, Trouth CO. Respiratory and circulatory effects of 100 mEq/l potassium or 2% procaine in the cerebrospinal fluid of cats. Pflügers Arch 1970; 321:346–363.

11. Schwanghart F, Schroeter R, Kleussendorf D, Koepchen HP. The influence of novocaine block of superficial brain stem structures on respiratory and reticular neurons. In: Umbach W, Koepchen HP, eds. Central Rhythmic and Regulation. Stuttgart: Hippokrates, 1974:104–110.
12. Schläfke ME, Kille JF, Loeschcke HH. Elimination of central chemosensitivity by coagulation of a bilateral area on the ventral surface in awake cats. Pflügers Arch 1979; 378:231-241.
13. Schläfke ME, See WR, Herker-See A, Loeschcke HH. Respiratory response to hypoxia and hypercapnia after elimination of central chemosensitivity. Pflügers Arch 1979; 381:241–248.
14. McAllen RM, Neil JJ, Loewy AD. Effects of kainic acid applied to the ventral surface of the medulla oblongata on vasomotor tone, the baroreceptor reflex and hypothalamic autonomic responses. Brain Res 1982; 238:65–76.
15. Douglas RM, Bernard DG, Pan Y, et al. Cardiorespiratory effects of kainic acid on brainstem chemosensitive zones. Neurosci Abstr 1992; 18:828.
16. Schläfke ME, Kille J, Folgering H, et al. Breathing without chemosensitivity. In: Umbach W, Koepchen HP, eds. Central Rhythmic and Regulation. Stuttgart: Hippokrates, 1974:97–104.
17. Kille J, Schläfke ME. Histological investigations in the chemosensitive areas on the ventral surface of the medulla oblongata. Pflügers Arch ges Physiol 1974; 347(suppl):R76.
18. Dermietzel R, Willenberg I, Leibstein A. Histochemical studies on the central chemosensitive fields of the medulla oblongata. In: Schläfke ME, Koepchen HP, See WR, eds. Central Neurone Environment and the Control Systems of Breathing and Circulation. Berlin: Springer-Verlag, 1983:251–256.
19. Trouth CO, Odek-Ogunde M, Holloway JA. Morphological observations on superficial medullary CO_2-chemosensitive areas. Brain Res 1982; 246:35–45.
20. Bada FJ. Identification of a naloxone-sensitive site in the brainstem influencing respiratory CO_2 chemosensitivity. Ph.D. dissertation, Howard University, Washington, DC, 1990.
21. Foster GA, Schulzberg M, Goldstein M, Hökfelt T. Ontogeny of phenylethanolamine N-methyltransferase and tyrosine hydroxylase-like immunoreactivity in presumptive adrenaline neurones of the foetal rat central nervous system. J Comp Neurol 1985; 236:348–381.
22. Pilowsky PM, Jiang C, Lipski J. An intracellular study of respiratory neurons in the rostral ventrolateral medulla of the rat and their relationship to catecholamine-containing neurons. J Comp Neurol 1990; 301:604–617.
23. Satoh K, Armstrong DM, Fibiger HC. A comparison of the distribution of central cholinergic neurons as demonstrated by acetylcholinesterase pharmacohistochemistry and choline acetyltransferase immunohistochemistry. Brain Res Bull 1983; 11:693–720.
24. Ruggiero DA, Giuliano R, Anwar M, et al. Anatomical substrates of cholinergic-autonomic regulation in the rat. J Comp Neurol 1990; 292:1–53.
25. Jones BE, Beaudet A. Distribution of acetylcholine and catecholamine neurons in the cat brainstem: A choline acetyltransferase and tyrosine hydroxylase immunohistochemical study. J Comp Neurol 1992; 261:15–32.
26. Erickson JT, Millhorn DE. Fos-like protein is induced in neurons of the medulla oblongata after stimulation of the carotid sinus nerve in awake and anesthetized rats. Brain Res 1991; 567:11–24.

27. Badoer E, McKinley MJ, Oldfield BJ, McAllen RM. Distribution of hypothalamic, medullary and lamina terminalis neurons expressing fos after hemorrhage in conscious rats. Brain Res 1992; 582:323–328.
28. Li Y-W, Dampney RAL. Expression of c-fos protein in the medulla oblongata of conscious rabbits in response to baroreceptor activation. Neurosci Lett 1992; 144: 70–74.
29. Sato M, Severinghaus JW, Basbaum AI. Medullary CO_2 chemoreceptor neuron identification by c-fos immunocytochemistry. J Appl Physiol 1992; 73:96–100.
30. Morgan JI, Curran T. Stimulus-transcription coupling in neurons: role of cellular immediate-early genes. Trends Neurosci 1989; 12:459–462.
31. Hunt SP, Pini A, Evan G. Induction of c-fos-like protein in spinal cord neurons following sensory stimulation. Nature 1987; 328:632–634.
32. Morgan JI, Cohen DR, Hempstead JL, Curran T. Mapping patterns of c-fos expression in the central nervous system after seizure. Science 1987; 237:192–197.
33. Sagar SM, Sharp FR, Curran T. Expression of c-fos protein in brain: Metabolic mapping at the cellular level. Science 1987; 240:1328–1331.
34. Dragunow M, Faull RLM. The use of c-fos as a metabolic marker in neuronal pathway tracing. J Neurosci Methods 1989; 29:261–265.
35. Bullitt E. Expression of c-fos-like protein as a marker for neuronal activity following noxious stimulation in the rat. J Comp Neurol 1990; 296:517–530.
36. Uemura Y, Kowall NW, Moskowitz MA. Focal ischemia in rats causes time-dependent expression of c-fos protein immunoreactivity in widespread regions of ipsilateral cortex. Brain Res 1991; 552:99–105.
37. Szekely AM, Barbaccia ML, Costa E. Activation of specific glutamate receptor subtypes increases c-fos proto-oncogene expression in primary cultures of neonatal rat cerebellar granule cells. Neuropharmacology 1987; 26:1779–1782.
38. Le Gal La Salle G. Long-lasting and sequential increase of c-fos oncoprotein expression in kainic acid-induced status epilepticus. Neurosci Lett 1988; 88:127–130.
39. Popovici Th, Barbin G, Ben Ari Y. Kainic acid-induced seizures increase c-fos-like protein in the hippocampus. Eur J Pharmacol 1988; 150:405–406.
40. Morgan JI, Curran T. Stimulus-transcription coupling in the nervous system: Involvement of the inducible proto-oncogenes fos and jun. Annu Rev Neurosci 1991; 14:421–451.
41. Borison HL, Brizzee KR. Morphology of emetic chemoreceptor trigger zone in cat medulla oblongata. Proc Soc Exp Biol Med 1951; 77:38–42.
42. Borison HL. Area postrema: Chemoreceptor trigger zone for vomiting—Is that all? Life Sci 1974; 14:1807–1817.
43. Palkovits M, Záborsky L. Neuroanatomy of central cardiovascular control. Nucleus tractus solitarii: Afferent and efferent neuronal connections in relation to the baroreceptor reflex arc. Prog Brain Res 1977; 47:9–34.
44. Van der Kooy D, Koda LY. Organization of the projections of a circumventricular organ: The area postrema in the rat. J Comp Neurol 1983; 219:328–338.
45. Shapiro RE, Miselis RR. The central neural connections of the area postrema of the rat. J Comp Neurol 1985; 234:344–364.
46. Leslie RA. Comparative aspects of the area postrema: Fine-structural considerations help to determine its function. J Cell Mol Neurobiol 1986; 6:95–120.
47. Masland WS, Yamamoto WS. Abolition of ventilatory response to inhaled CO_2 by neurological lesions. Am J Physiol 1962; 203:789–795.

48. Fukuda H, Koga T. The Bötzinger complex as the pattern generator for retching and vomiting in the dog. Neurosci Res 1991; 12:471–485.
49. Loeschcke HH, Trouth CO. Effects on ventilation and circulation of topical electrical stimulation of the cerebellum. Pflügers Arch 1974; 347:112.
50. Patrickson J, Trouth CO, Holloway JA, Brooks R. Effects of electrical stimulation of the cerebellum on ventilation and circulation in the cat. Neurosci Abstr 1976; 1:116.
51. Fedorko L, Merrill EG. Axonal projections from the rostral expiratory neurones of the Bötzinger complex to medulla and spinal cord in the cat. J Physiol 1984; 350:487–496.
52. Speck DF. Bötzinger complex region role in phrenic-to-phrenic inhibitory reflex of cat. J Appl Physiol 1989; 67:1364–1370.
53. Smith JC, Morrison DE, Ellenberger HH, et al. Brainstem projections to the major respiratory neuron populations in the medulla of the cat. J Comp Neurol 1989; 281:69–96.
54. Connelly CA, Ellenberger HH, Feldman JL. Respiratory activity in retrotrapezoid nucleus in cat. J Physiol 1990; 258:L33–L44.
55. Loeschcke HH. Review lecture: Central chemosensitivity and the reaction theory. J Physiol 1982; 332:1–24.
56. Sheng M, Greenberg ME. The regulation and function of c-fos and other immediate early genes in the nervous system. Neuron 1990; 4:477–485.
57. Paxinos G, Watson C. In: The Rat Brain in Stereotaxic Coordinates, 2nd ed. Australia: Academic Press, 1986.
58. Palkovits M, Brownstein MJ. In: Maps and Guide to Microdissection of the Rat Brain. New York: Elsevier, 1988.
59. Filiano JJ, Choi JC, Kinney HC. Candidate cell populations for respiratory chemosensitive fields in the human infant medulla. J Comp Neurol 1990; 293:448–465.
60. Folgering H, Kuyper F, Kille JF. Primary alveolar hypoventilation (Ondine's curse syndrome) in an infant without external arcuate nucleus. Case Report Bull Eur Physiopathol Resp 1979; 15:659–665.
61. Essick CR. The corpus ponto-bulbare: A hitherto undescribed nuclear mass in the human hindbrain. Am J Anat 1907; 7:119–135.

5

Central and Peripheral Sites of Action of Antitussive Drugs in the Cat

DONALD C. BOLSER, FRANCES C. DeGENNARO, RICHARD W. CHAPMAN, and JOHN A. HEY

Schering-Plough Research Institute
Kenilworth, New Jersey

I. Introduction

There are a variety of symptoms of pulmonary disease: chest tightness or pain, cough, dyspnea, hemoptysis, hypopnea, sneeze, tachypnea, and wheeze. Of these symptoms, cough is the most common (1). While cough normally has an important defensive function in maintaining airway integrity, it is often clinically important to inhibit this reflex. Cough can be inhibited by a variety of pharmacological agents (2). These agents exert their antitussive effects primarily by inhibition of the responsiveness of one or more neural elements of the reflex pathway for cough (2). For example, the most commonly used antitussive drugs, codeine and dextromethorphan, inhibit cough by a central site of action (3). These drugs presumably inhibit one or more components of the brainstem pattern generator for cough (4,5). Conversely, the peripheral μ-opioid agonist BW 443C inhibits cough at a peripheral site, presumably by reducing the responsiveness of rapidly adapting pulmonary receptors to cough-provoking stimuli (6).

We have recently shown that the GABA-B receptor agonists baclofen and 3-aminopropylphosphinic acid (3-APPi) have antitussive activity in several animal models of cough (7). However, whether these drugs inhibit the reflex pathway for cough at central or peripheral sites is unknown. We investigated the site of action of baclofen and 3-APPi and several standard antitussive drugs in a model of cough in the anesthetized cat. The antitussive potency of these drugs was compared after intravenous (IV) and intraarterial (IA) injection of each compound. Intraarterial injections were made into the vertebral artery, which supplies the brainstem (3). The brainstem is the site of action of centrally acting antitussive drugs (3,8). The ratio of the effective doses to inhibit cough by 50% (ED_{50}) IV and IA was used to differentiate central and peripheral sites of action of these drugs (3).

II. Methods

Cats (2.5 to 5.0 kg) were anesthetized with pentobarbital sodium (35 $mgkg^{-1}$, i.p.) and given supplemental anesthetic as necessary (5 mgkg, i.v.). Atropine sulfate (1.0 mgkg, i.v.) was administered to block reflex tracheal secretions. The trachea was cannulated to allow access to the intrathoracic airway. The femoral vein and femoral artery were cannulated for administration of drugs IV and for measurement of arterial blood pressure. In some animals, a catheter was inserted into the left axillary artery and advanced until the catheter tip was at the branch of the left vertebral artery (3). The omocervical, pericardiophrenic, and costocervical branches were ligated. Proper placement of this catheter was confirmed postmortem.

Electromyograms (EMG) from the diaphragm and rectus abdominis muscles were recorded with bipolar silver wire electrodes. The EMG were amplified, filtered (0.5 to 10 kHz), and integrated with a resistance-capacitance circuit (100 ms time constant). The integrated EMG were displayed on a chart recorder.

Cough was defined as a burst of EMG activity in the diaphragm immediately following a burst of EMG activity in the rectus abdominis muscle (7). Coughing was produced by probing the intrathoracic trachea continuously with a thin flexible polyethylene cannula for approximately 10 s per trial. Control values were generated by averaging the number of coughs during five consecutive stimulus (probing) trials obtained after vehicle administration. One minute elapsed between probing trials. Stimulus trials were applied at 1-min intervals after each dose of drug for a total of five stimulus trials between doses. The cough response after each dose of drug was determined by averaging the number of coughs observed during these five stimulus trials. Five min elapsed between each dose of drug.

The antitussive activity of morphine, codeine, carbetapentane, dextromethorphan, BW 443C, baclofen, and 3-APPi was evaluated from cumulative dose-response relationships after IV administration and, in separate groups of animals, IA administration of each drug. An effective dose ratio (EDR) was generated for each drug. The EDR was defined as the ED_{50} for IV activity divided by the ED_{50} for IA activity of the drug. This ratio will be 20 or greater for a centrally acting antitussive drug and less than 20 for a peripheral antitussive drug (3).

A. Compounds

Compounds used in this study included atropine sulfate, morphine sulfate, carbetapentane (Sigma Chemical Co., St. Louis, MO), codeine sulfate (Mallinckrodt, St. Louis, MO), BW 443C (kind gift of Wellcome Research Laboratories, Beckenham, Kent), dextromethorphan hydrobromide, baclofen (Research Biochemicals Inc., Natick, MA), and 3-APPi (prepared at Schering-Plough Research Institute, Kenilworth, NJ). All drugs were dissolved in physiological saline. Doses were calculated as their free base.

B. Statistics

Effective doses for 50% inhibition (ED_{50}) of cough frequency were obtained by regression analysis of dose-response relationships. Data were expressed as means ± SE.

III. Results

In animals that received drugs IV, mechanical stimuli applied to the trachea elicited 7.1 ± 0.5 coughs per stimulus trial during the control period. Likewise, in animals that received drugs IA, mechanical stimuli applied to the trachea elicited 9.6 ± 0.6 coughs per stimulus trial during the control period. Figure 1 shows cumulative dose-response relationships for antitussive activity after IV and IA administration of two reference standards—morphine (Fig. 1A) and carbetapentane (Fig. 1B)—as well as the GABA-B agonists baclofen (Fig. 1C), and 3-APPi (Fig. 1D). Each dose-response relationship was generated from a separate group of animals. Table 1 shows the EDR for all drugs used in this study.

Morphine was considerably more potent when given by the IA route ($ED_{50} = 0.002$ mgkg^{-1}) than by the IV route ($ED_{50} = 0.04$ mgkg^{-1}) and had an EDR of 20 (Fig. 1, Table 1). Likewise, dextromethorphan was more potent when given IA ($ED_{50} = 0.013$ mgkg^{-1}) than IV ($ED_{50} = 0.23$ mgkg^{-1}) and had an EDR of 18 (Table 1). In contrast, carbetapentane was only slightly more potent when given by the IA route ($ED_{50} = 0.16$ mgkg^{-1}) than when given by the IV route ($ED_{50} = 0.4$ mgkg^{-1}) and had an EDR of 3 (Fig. 1, Table 1). Codeine was much more potent when given by the IA route ($ED_{50} = 0.013$ mgkg^{-1}) than by the IV route ($ED_{50} = 0.34$ mgkg^{-1}). The EDR for codeine was 26 (Table 1). The peripherally acting opioid BW 443C was slightly more potent when given by the IA route ($ED_{50} = 0.06$ mgkg^{-1}) than by the IV route ($ED_{50} = 0.17$ mgkg^{-1})

A

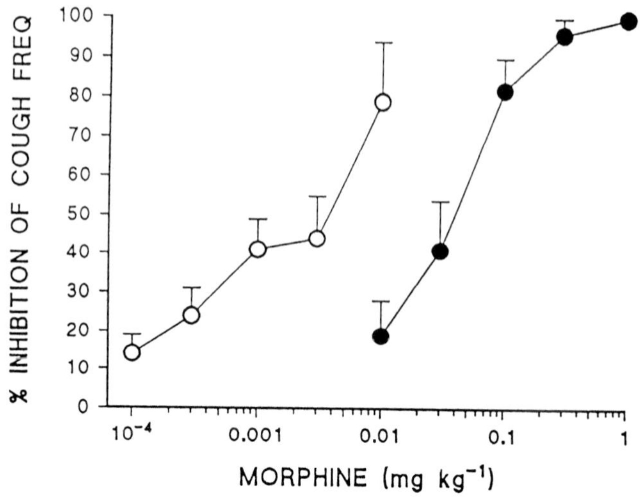

B

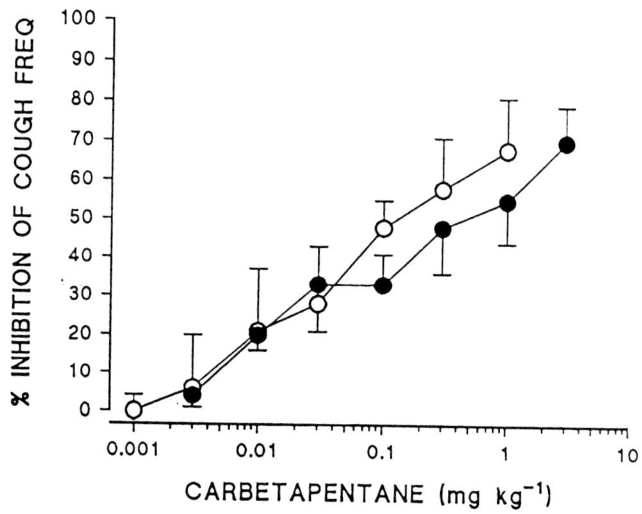

Figure 1 Antitussive effects of IV and IA administration of morphine, carbetapentane, baclofen, and 3-APPi in the cat. Cumulative dose-response relationships for IV (●) and IA (○) administration of morphine (A), carbetapentane (B), baclofen (C), and 3-APPi (D) are shown.

C

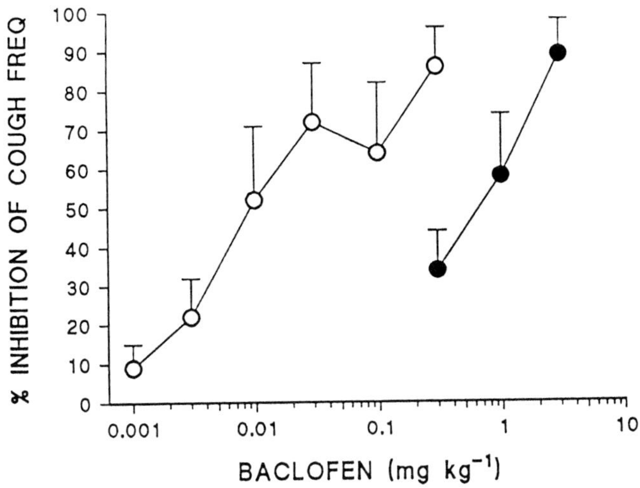

D

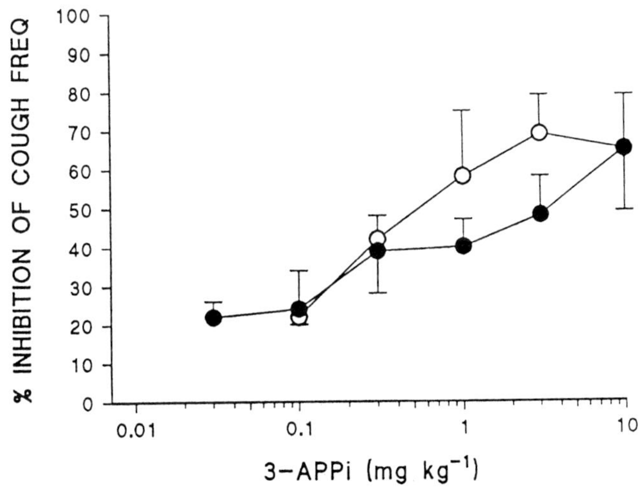

and had an EDR of 3 (Table 1). Baclofen was much more potent when given by the IA route (ED_{50} = 0.016 $mgkg^{-1}$) than by the IV route (ED_{50} = 0.63 $mgkg^{-1}$) and had an EDR of 39 (Fig. 1, Table 1). Conversely, 3-APPi was only slightly more potent by the IA (ED_{50} = 0.87 $mgkg^{-1}$) route than by the IV route (ED_{50} = 2.3 $mgkg^{-1}$) and had an EDR of 3.0 (Fig. 1, Table 1).

Table 1 Effective Dose Ratio (EDR) for Inhibition of Cough by Standard Antitussives and GABA-B Agonists in Cats

Compounds	Inhibition of cough (ED_{50}, $mgkg^{-1}$) IV	IA	EDR (ED_{50} IV/ED_{50} IA)
Morphine	0.04	0.002	20
Dextromethorphan	0.23	0.013	18
Carbetapentane	0.40	0.160	3
Codeine	0.34	0.013	26
BW 443C	0.17	0.060	3
Baclofen	0.63	0.016	39
3-APPi	2.30	0.870	3

IV. Discussion

The use of effective dose ratios generated from IV and IA (vertebral artery) dose-response relationships has been demonstrated to differentiate between centrally and peripherally active antitussive drugs (3). The EDR for centrally active antitussive drugs is generally high (greater than 20), because very small doses of these drugs are necessary to inhibit cough when injected into the vertebral artery. This artery supplies the brainstem, which is the primary site where the central neural elements generating cough are located (3). In contrast, much larger doses of peripheral antitussive drugs are necessary to inhibit cough when given in the vertebral artery because these drugs must recirculate in the venous system to reach their site of action. EDR of peripheral antitussive drugs are much less than 20 (3). In these studies, the EDR for baclofen was 39 and that of the standard centrally acting antitussive drugs codeine, morphine, and dextromethorphan ranged from 18 to 26. This observation indicates that baclofen had a central site of action to inhibit the cough in the cat. The EDR for 3-APPi was only 3 and was similar to that of the peripheral μ-opioid agonist BW443C. These findings indicate that the antitussive activity of 3-APPi in the cat occurs through a peripheral site of action.

The results of this study indicate that the antitussive activity of baclofen has a central component and that of 3-APPi is due to a peripheral site of action, with both drugs acting at GABA-B receptors. These conclusions confirm and extend a previous suggestion that baclofen inhibits cough at a central site and 3-APPi inhibits cough at a peripheral site (9). However, this suggestion was based primarily on the idea that 3-APPi penetrates the central nervous system (CNS) to a limited extent (10). The data presented provide direct evidence to support the conclusion that baclofen can inhibit cough by a central site of action and 3-APPi inhibits cough by a peripheral site of action. The mechanisms by which baclofen elicits its central antitussive effect is probably due to inhibition of one or more components of the central reflex pathway for cough. Conversely, 3-APPi probably elicits its peripheral antitussive effect by inhibiting vagal sensory afferent responsiveness to mechanical stimulation of the trachea in the cat. Similar mechanisms of action have been suggested for the peripheral antitussive drugs BW443C (11) and benzonatate (8).

The fact that morphine, codeine, and dextromethorphan had EDR that were high (18 or greater) is consistent with the classic view that these drugs inhibit cough by a central site of action (3,5). On the other hand, the low EDR for carbetapentane is consistent with a peripheral site of action for this drug. Carbetapentane is classically thought of as a centrally acting antitussive drug (12,13). Our data suggest that this view of the site of action of carbetapentane should be reevaluated.

The slope of the dose-response relationships and efficacy of 3-APPi appeared to be lower than the relationships for baclofen. There are at least two hypotheses that could account for this observation. First, peripheral GABA-B receptors may have only a limited inhibitory influence on the cough reflex in this species. In the guinea pig, the efficacies of baclofen and 3-APPi are the same (7). It is possible that this difference in efficacy of 3-APPi is not related as much to the different species used as it is to different sensory receptor types inhibited by 3-APPi in each model. 3-APPi may be more effective in reducing the responsiveness of pulmonary C fibers than rapidly adapting receptors. Although it is possible that the pharmacokinetics of 3-APPi may limit access of this drug to its site of action, we favor an explanation including either species differences or differential sensory receptor sensitivities to 3-APPi.

References

1. Karlsson J-A, Sant'Ambrogio G, Widdicombe JG. Afferent neural pathways in cough and reflex bronchoconstriction. J Appl Physiol 1989; 65:1007–1023.
2. Braman SS, Corrao WM. Cough: Differential diagnosis and treatment. Clin Chest Med 1987; 8:177–188.
3. Chou DT, Wang SC. Studies on the localization of central cough mechanism: Site of action of antitussive drugs. J Pharmacol Exp Ther 1975; 194:499–505.
4. Kase Y. Antitussive agents and their sites of action. Trends Pharmacol Sci 1980; 1:237–239.
5. Braga PC. Centrally acting opioid drugs. In: Braga PC, Allegra L, eds. Cough. New York: Raven Press, 1989:109–143.
6. Adcock JJ. Peripheral opioid receptors and the cough reflex. Respir Med 1991; 85 (suppl A):43–46.
7. Bolser DC, Aziz SM, DeGennaro FC, et al. Antitussive effects of GABA-B agonists in the cat and guinea pig. Br J Pharmacol 1993; 110:491–495.
8. Korpas J, Tomori Z. Cough and other Respiratory Reflexes. New York: Karger, 1979.

9. Chapman RW, Hey JA, Rizzo CA, Bolser DC. GABA-B receptors in the lung. Trends Pharmacol Sci 1993; 14:26–29.
10. Hills JM, Howson W. The GABA-B profile of a series of phosphinic acids-agonist and antagonist activity in a range of peripheral tissue. In: Erdö SL, ed. GABA Outside the CNS. New York: Springer-Verlag, 1992:249–260.
11. Adcock JJ, Schneider C, Smith TW. Effects of codeine, morphine and a novel pentapeptide BW443C, on cough, nociception and ventilation in the unanesthetized guinea pig. Br J Pharmacol 1988; 93:93–100.
12. Guffanti EE. Centrally acting nonopioid drugs. In: Braga PC, Allegra L, eds. Cough. New York: Raven Press, 1989:147–194.
13. Jaffe JH, Martin WR. Opioid analgesics and antagonists. In: Goodman Gillman A, Goodman LS, Rall TW, Mured F, eds. The Pharmacological Basis of Therapeutics. New York: Macmillan, 1985:491–531.

6

The Role of the Ventral Medulla in Hypoxic Respiratory Depression and Sympathetic Excitation

NEIL S. CHERNIACK, PAUL ERNSBERGER, JAYOTI MITRA, and MUSA A. HAXHIU

School of Medicine
Case Western Reserve University
Cleveland, Ohio

I. Introduction

In the absence of afferent inputs from peripheral chemoreceptors, systemic hypoxia triggers central mechanisms that depress breathing and produce sympathetic excitation (1–5). Several explanations have been proposed for the depressive actions of central nervous system (CNS) hypoxia on respiration. Hypoxic reduction of respiratory activity has, for example, been attributed to increased cerebral blood flow or changes in energy supply that interfere with the function of respiratory neurons (6–8).

Since hypoxia has a global action affecting different areas of the CNS (9, 10) and causing the release of either inhibitory or excitatory neurotransmitters, there may be specific sites in the brain where neurotransmitter release produces both sympathetic excitation and respiratory depression even before energy stores are exhausted. Our recent data suggest, as described below, that one of these sites could be the ventrolateral portion of the medulla oblongata (5,11, 12), a region that plays a pivotal role in the regulation of breathing and arterial blood pressure (13–18). There the excitation and depression may occur by a single mechanism at specific neuronal groups, which projects by multisynaptic pathways to vasomotor and respiratory neurons within the ventrolateral medulla; or it is possible that respiratory depression and sympathetic activity occur by independent mechanisms.

The depression of breathing and the sympathetic excitation response may serve a common purpose. It may act on an emergency basis—like the diving reflex—to prevent further loss of available oxygen store while preserving the O_2 delivered to vital organs when O_2 is absent in the environment.

In this contribution we discuss some previous work on the effects of hypoxia of the CNS on breathing activity and sympathetic tone as well as our own recent studies on cardiorespiratory responses to hypoxia confined to caudal brainstem structures.

II. Hypoxic Depression of Respiration

Hypoxic depression of respiration occurs invariably in peripherally chemodenervated animals, but it can also be observed in animals and humans with intact peripheral chemoreceptors.

In an early study of hypoxic depression, we showed a depressive effect of systemic hypoxia on phrenic nerve activity in anesthetized, paralyzed, and artificially ventilated dogs (with chemoreceptors intact). These animals were exposed in separate trials to asphyxia, progressive hypercapnia under hyperoxic conditions, and hypoxia while hypercapnia was prevented by TRIS buffer infusion (19). The paralyzed dogs were mechanically ventilated with a constant volume so the effect of pulmonary stretch receptors and muscle proprioreceptors were minimal. After an initial increase in fictive breathing, a clear depressive effect of hypoxia on phrenic activity could be observed during both isocapnic and hypercapnic conditions, while hypercapnia alone did not. The level of hypoxia at which phrenic nerve activity decreased from its peak was higher during hypercapnia than during isocapnic hypoxia. Tris buffer infusion, while it maintained isocapnia, regularly produced an increase in arterial pH despite the occurrence of hypoxia. In light of more recent experiments by Neubauer et al. (20) showing that prevention of acidosis largely prevented respiratory depression of cats exposed to a progressive increase in carbon monoxide level, it is possible that the Tris buffer alkalosis contributed to the greater resistance of the isocapnic animals to hypoxic depression.

Neubauer and coworkers have described different phases of hypoxic respiratory depression (20). An initial phase of depression is due to hypoxic augmentation of cerebral blood flow, which reduces P_{CO_2} (and $[H^+]$) at central chemoreceptors. In a subsequent phase, the release of inhibitory neurotransmitters such as GABA, enkephalins, and adenosine (17,21–23) probably further diminish respiration, since the administration of the appropriate antagonist at least partially reverses the respiratory depression, even though the results reported in different studies have not been entirely consistent. When positive effects have been seen, it is unclear whether these agents act on bulbar respiratory neurons or on neurons upstream that project to them. During this stage of hypoxic depression, excitatory responses to increments in CO_2 are maintained. Finally, the last stage of hypoxic depression of neuronal metabolism may be sufficiently severe that neuronal death is imminent.

It has been proposed that hypoxic depression of breathing, by decreasing neuronal metabolism, may serve a useful function by increasing the potential for survival during hypoxic stress. It could be that a beneficial object is also attained by the central excitatory effect of hypoxia on vasomotor activity. Differential changes in sympathetic discharge, by promoting perfusion of vital organs (analogous to the diving reflex), could have survival value.

III. Regional Differences in the Brain in Sensitivity to Hypoxic Depression

Respiratory neurons seem to be mainly concentrated in the medulla in two well-described aggregations of nerve cells in its ventral and dorsal portions (24–26). These groups are, in turn, reciprocally connected to a dorsally located pontine respiratory group in the parabrachial regions (27–29). In addition, evidence exists that, in several species, neurons sensitive to CO_2 are located superfical to the ventral respiratory group, particularly in rostral medullary regions. These respiratory neurons seem to form a population that is distinct from the vasomotor neurons (30–32). Hypoxic depression may not affect all respiratory outputs uniformly. Data obtained in the goat and the piglet show differential actions of hypoxia on upper airway dilating versus respiratory pumping muscles (16,33,34). Recently, we determined the contribution of brainstem hypoxia to the response of upper airway and chest wall muscles during early life. We perfused the brainstem through a vertebral artery intermittently with blood from an extracorporeal circuit in nine newborn piglets, ages 1 to 5 days. Brainstem perfusions were performed with hypoxemic blood (arterial P_{O_2} 32 $\pm$ 6 to 38 $\pm$ 8 Torr) with different levels of brainstem P_{CO_2} (28 $\pm$ 2, 37 $\pm$ 4, and 56 $\pm$ 5 torr) while systemic normocapnic hyperoxia was maintained (arterial P_{CO_2} 36 $\pm$ 3 to 40 $\pm$ 6 torr, arterial P_{O_2} 345 $\pm$ 73 to 392 $\pm$ 37 torr). Electromyograms (EMG) of alae nasi (AN), external intercostal (EI) muscles, and diaphragm (DIA) were recorded. Normocapnic hypoxia of the brainstem induced a sustained increase in AN EMG ($P<0.01$, analysis of variance) and depression of EI and DIA EMG without a transient increase. These contrasting responses were also observed during hypocapnic and hypercapnic hypoxia of the brainstem and were not affected by inputs from the peripheral chemoreceptors or rostral cerebral structures that were not exposed to hypoxia. Thus, brainstem hypoxia elicits a selective rather than a generalized respiratory muscle depression (16). It may be that similar differences occur in regional sympathetic activity.

Studies by Korner stress the importance of higher brain centers in modulating the interplay of baroreceptor, lung stretch receptor, and peripheral chemoreceptor output in setting vasomotor and respiratory response during hypoxia (35). To a certain extent the adverse actions of hypoxia on nerve tissue seem to depend on differences in metabolic rates among brain tissues and vary with maturation (36,37).

Nonetheless, experiments in animals indicate that hypoxic depression of ventilation does not require brain tissue above the pons or lower mesencephalon. Brain transection experiments suggest that hypoxic depression arises from stimulation of the caudal mesencephalon or dorsolateral pons (38,39). These studies raise the possibility that hypoxia, rather than depressing, might excite certain brain areas that then have an inhibitory effect on breathing. It is possible that hypoxia, by stimulating these same areas, excites vasomotor- and inhibits respiratory-related neurons. Severe hypoxia is known to cause the release in brain slices of excitatory and/or inhibitory amino acids, which could form the basis of a mechanism for the excitatory and inhibitory effects of hypoxia (40–42).

The effects of central hypoxia on nerve activity require more time to develop than the effects of hypoxic stimulation through peripheral chemoreceptors, suggesting the involvement of relatively slow metabolic processes (8,43). Certain enzymes involved in the metabolism of neurotransmitters—like choline acetyl transferase, tyrosine, and tryptophan hydroxylase—are very sensitive to the effects of hypoxia (44). Much of the information on the cellular and membrane effects of hypoxia comes from studies of hippocampal slices (8).

Acidosis occurs during hypoxia because of augmented glycolysis. Recent studies of respiratory activity suggest that the depressive effects of hypoxia are largely abrogated if lactic acid production is reduced and acidosis prevented by systemic dichloracetate (DCA) (20). It has been suggested that acidosis arising from hypoxia may interfere with the breakdown of neurotransmitters such as GABA, since the enzymes involved may be particularly sensitive to pH changes. GABA, for example, has inhibitory effects on ventilation when GABA acts on the structures near the ventral surface of the medulla (45).

IV. Effects of Hypoxia on the Vasomotor System

Brain ischemia, studied more extensively than hypoxia alone, produces sympathetic excitation, which can be eliminated by lesions in the rostral ventrolateral medulla (RVLM). However, ischemic hypoxia is complicated by simultaneously occurring loss of nutrient delivery and by hypercapnia (20,40,46,47). Hypercapnia may not contribute significantly to the ischemic reflex, since elimination of the excitatory effects of CO_2 by cooling of the intermediate area of Schläfke on the VMS, which overlaps a portion of the RVLM, does not prevent the pressor actions of cerebral ischemia (47).

The frequency content of sympathetic nerve activity is altered during cerebral ischemia. First there is accentuation of the normal 2-, 6-, and 10-Hz rhythmic fluctuations in the frequency content of sympathetic nerve, which arises, in large part, from the hindbrain, but later these cyclic changes become desynchronized. This desynchronization has been attributed to the central effects of hypoxia (48).

There are relatively fewer studies of the effect of CNS hypoxia alone on the cardiovascular system. Several investigators have observed an increase in sympathetic activity in heart rate and cardiac contractility with hypoxia in chemodenervated animals (4,49,50), although this is sometimes preceded by a transient decrease in activity. Hypoxia in peripherally chemodenervated cats acutely transected at C_1 increases the activity of preganglionic sympathetic nerve fibers (4). This suggest that hypoxia can excite preganglionic sympathetic neurons when input from the brain is eliminated. What remains unclear is whether the excitatory effects elicited by hypoxia are due to a direct action on the preganglionic neurons or are secondary to input from extracarotid chemoreceptors. In addition to the carotid and aortic bodies, oxygen-sensitive tissue is present in the thorax and abdomen, which is supplied by sympathetic afferents (1,51–55).

Pulmonary neuroepithelial bodies, composed of innervated clusters of amine- and peptide-containing cells, are widely distributed throughout the airway mucosa of human and animal lungs; they resemble chemoreceptors (such as carotid body, or taste buds) and may function as hypoxia-sensitive airway sensors (51). Recently, Youngston et al. cultured neuroepithelial bodies isolated from rabbit fetal lungs and identified voltage-activated potassium, calcium, and sodium currents using the whole-cell patch-clamp technique. Upon exposure to hypoxia, there was a reversible reduction (25% to 30%) in the outward potassium current, with no change in inward currents. In addition, it was demonstrated the expression of an oxygen-binding protein (b-cytochrome, NADPH oxidase) on the plasma membrane of these cells. The identification of an oxygen-sensing mechanism (namely the presence of an O_2-sensitive potassium channel coupled to an O_2 sensor protein) in the cells of pulmonary neuroepithelial bodies indicates that they are transducers of the hypoxia stimulus and hence may function as airway chemoreceptors in the regulation of respiration (55). It is possible that a spinal reflex involving these afferents contributes to hypoxic preganglionic neuron excitation.

Investigations by Arita and coworkers and Mitra et al. support a supraspinal pressor site sensitive to hypoxia (3,12,49). They found pressor responses to small injections of solutions equilibrated with N_2 injected into the vertebral artery in cats. Respiratory depression occurred simultaneously with blood pressure increases. Decerebration had no effect on the responses, suggesting that the vasopressor and respiratory inhibiting effects of N_2 were due to actions confined to hindbrain neurons.

V. Effects of Hypoxia on the Respiratory Modulation of Sympathetic Activity

A. The Effect of Hypoxic Loading of the Caudal Brainstem on Respiratory and Sympathetic Outflow

Sympathetic activity often has a pronounced respiratory modulation effect, particularly in nerves supplying the vasculature and the heart (2,56). Phasic increases occur in many sympathetic nerves synchronous with inspiration; these grow greater when respiration is stimulated (56). Haselton and Guyenet have reported respiratory modulation of the baroreceptor-sensitive neurons in the RVLM of vagotomized rats (57). Both hypoxia and hypercapnia heighten sympathetic activity, in part indirectly by increasing respiratory activity (due to spillover of central respiratory drive potentials to vasomotor neurons) but also in part independently of respiratory effects (58). Studies of the central effects of hypoxia in cats show that it is the inspiratory-related sympathetic activity that declines with respiratory depression (8). However, tonic activity of the sympathetic CNS generally grows greater in cats with central hypoxia, suggesting a more direct effect on vasomotor activity (59).

The effects of systemic hypoxia are complex and responses may differ from the site of action within the CNS. Because of that, we have studied the effect of hypoxia confined to the caudal brainstem using different techniques to localize hypoxic loading.

B. Cardiorespiratory Changes Induced by Vertebral Artery Injectiion of Sodium Cyanide

To assess the effects of hypoxia confined to the structures of the medulla on respiratory activity and sympathetic outflow, we utilized sodium cyanide (NaCN) loading (3,11,12).

Sodium cyanide–induced hypoxia was produced by administration of NaCN unilaterally into a vertebral artery at the spinomedullary junction or by topical application of NaCN. Unlike inhalation of hypoxic gas mixtures, local applications of sodium cyanide have the advantage of avoiding the systemic effects of hypoxia and limiting its effect to specific areas of the brain. The responses of phrenic and sympathetic activity in 14 anesthetized, paralyzed, and mechanically ventilated cats to different doses were compared.

Three dose levels of NaCN were used for the intravertebral injections (1, 10, and 20 μg) in animals ventilated with air. Similar doses, given systemically, had no effect on phrenic activity, sympathetic discharge, or blood pressure. Intravertebral injection of NaCN reduced integrated phrenic nerve amplitude, increased tonic sympathetic activity, and raised blood pressure, as shown in Figure 1. Phrenic amplitude decreased with increasing NaCN dose. Occasionally,

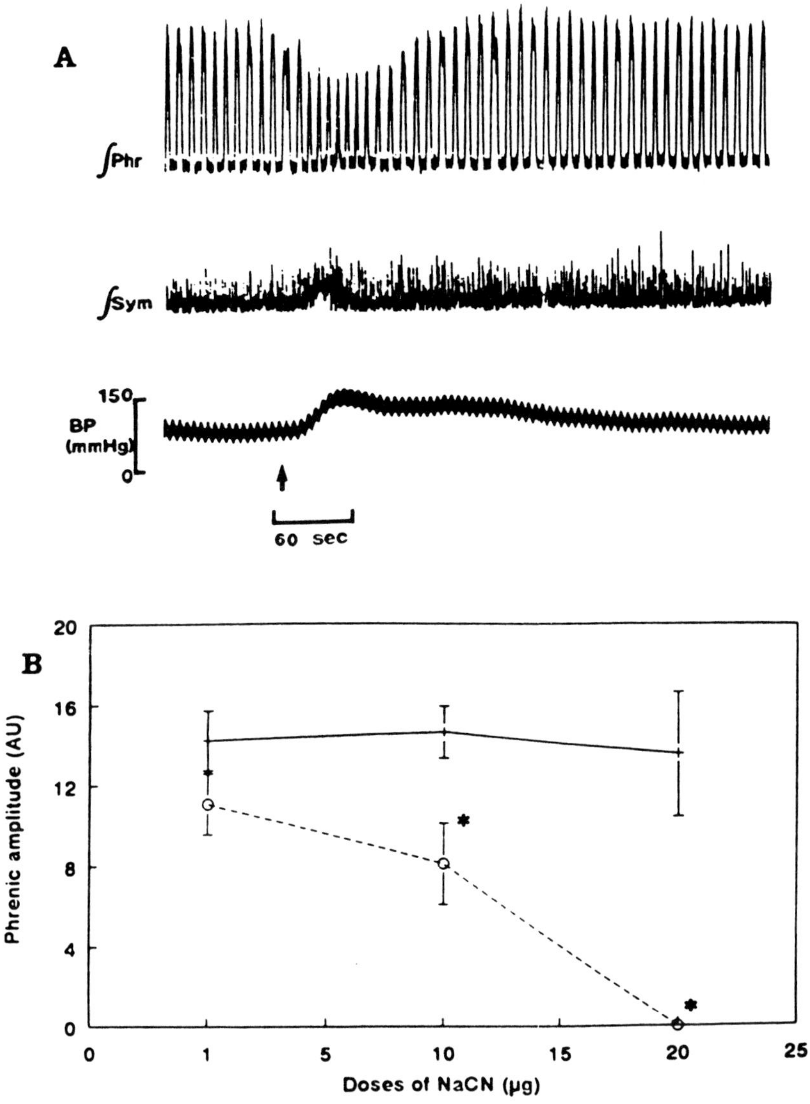

Figure 1 Panel A: Effect of intravertebral injection of NaCN (10 μg) in a single animal on integrated phrenic (Phr) and cervical sympathetic (Sym) activity. ↑: injection of NaCN; BP: arterial blood pressure. (Modified from Ref. 12.) Panels B, C, and D: Mean changes in phrenic nerve amplitude, sympathetic tone, and blood pressure. *: $P<0.05$.

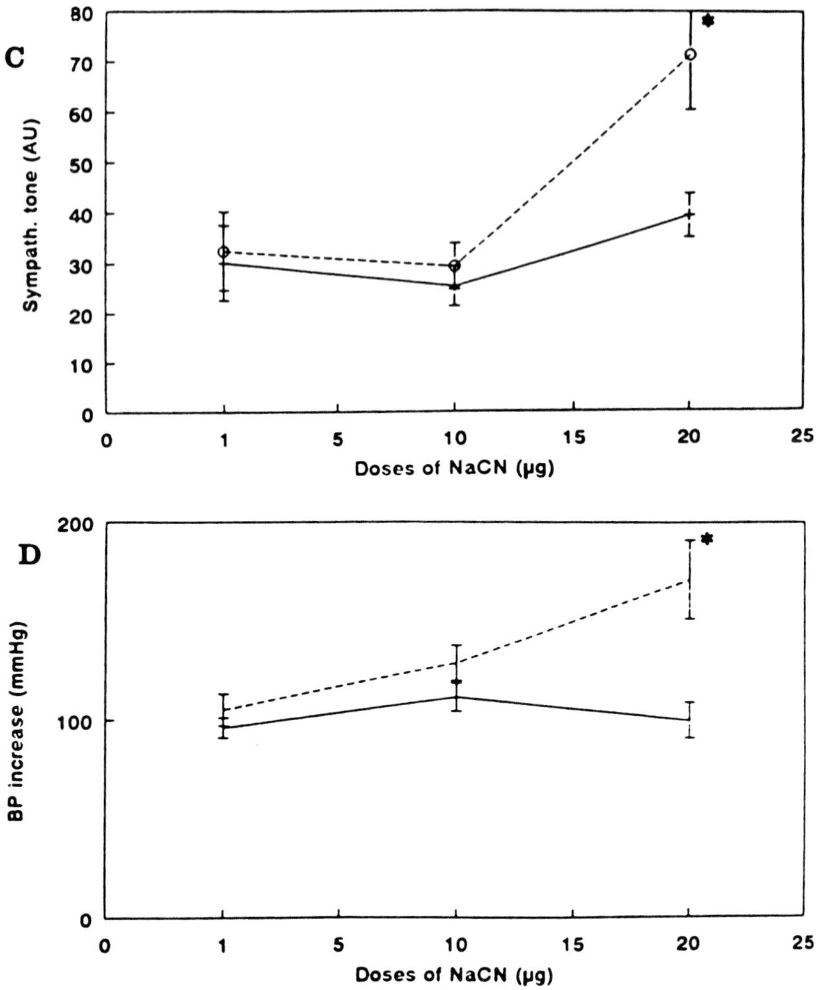

respiratory frequency increased with depression of phrenic amplitude, but this was not a consistent finding. All animals became apneic after receiving the higher dose (20 μg) of NaCN. The average times for the onset of phrenic depression and apnea were 4.0 ± 1.0, and 8.0 ± 2 s ($P < 0.05$), respectively, following injection (for all doses). The duration of apnea varied considerably (20.0 ± 10.0 s). The depression of phrenic nerve activity produced by NaCN was less in animals breathing 7% CO_2 in O_2 ($P < 0.05$).

It is known that phasic respiratory modulation of sympathetic nerve activity increases and decreases in parallel with changes in phrenic nerve activity (56,60). However, during hypoxic loading, we found that as phrenic nerve discharges decreased, the tonic discharge in sympathetic nerve increased, growing progressively with increasing dosages of NaCN. Although the increase in sympathetic activity was not statistically significant between the 1- and 10-μg dose levels, it was significant when the 10- and 20-μg dose levels were compared ($P < 0.05$). These changes were associated with an increase in arterial blood pressure and were the same with 7% CO_2 breathing.

From these experiments, it cannot be determined which neuronal population is responsible for respiratory depression, sympathetic excitation, and increased sympathetic nerve discharge.

C. Effect of NaCN Applied to VLM Surface and Microinjected into the Superficial Layer of the Ventral Medulla on Respiratory Activity

To ascertain the role of the ventrolateral medulla (VLM) in hypoxic depression of breathing, we examined the respiratory response to local application of cyanide to superficial structures of the ventrolateral medulla (Fig. 2).

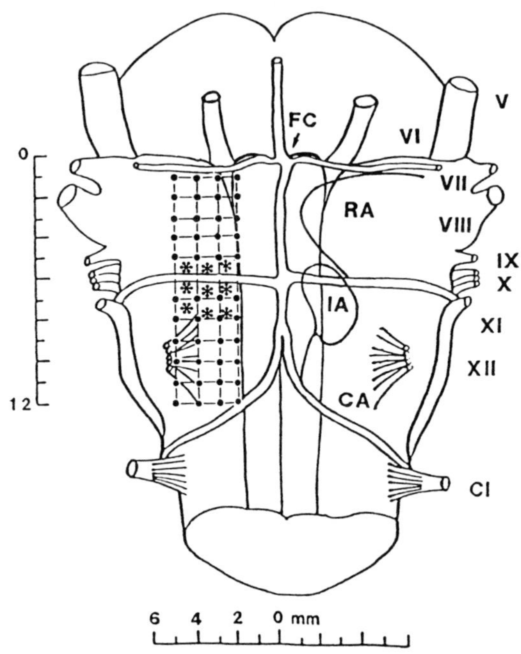

Figure 2 Diagrammatic sketch of cat's ventral medulla shows grid mark. Each dot in the grid represents an injection site. Typical response was observed when NaCN was applied 5 to 8.0 mm caudal to FC and 3.0 to 5.0 mm lateral to the midline.

In these experiments, cats were sedated with ketamine (10 mg/kg IV) and anesthetized with α-chloralose (60 mg/kg IV). To avoid superposition of secondary effects on the immediate respiratory changes induced by interventions, animals were mechanically ventilated with O_2.

To determine whether changes of inspiratory discharge are related to overall inhibition of respiratory output or are caused by inhibition through tonically active expiratory neurons, we recorded EMG activity of inspiratory and expiratory chest wall pumping muscles (11). Topical application of cyanide (10 to 100 μg) on the intermediate area of the ventral medullary surface (VMS) as described by Schläfke et al. (29) produced a consistent respiratory depression. In each animal, the electrical activity of the phrenic nerve and of all respiratory muscles studied began to decrease within 20 s from the application of cyanide. The maximal respiratory depressive nerve response occurred ~1 min after application of cyanide. The decrease in activity of the expiratory muscles exceeded that of the parasternal intercostal, which was in turn more pronounced than that of the diaphragm, but differences did not reach statistical significance ($P > 0.05$). However, doses of cyanide that had no effect on diaphragm activity and arterial blood pressure could cause a significant inhibition of transversus abdominis muscle activity. After the VMS was washed with warm buffered saline, the activity of abdominal expiratory muscle returned to its control level of activity later than that of the diaphragm and the parasternal intercostal muscle by ~60 to 90 s. An example of respiratory responses to topical application of cyanide in one cat (panel A) and results of individual animals (panel B) are presented in Figure 3. Topical application of cyanide to the VMS was associated with a decrease of phrenic nerve activity from 24 ± 3 to 10 ± 3 angstrom units (AU) ($P < 0.05$; $n = 4$) and diaphragm activity from 22 ± 1 to 7 ± 3 AU ($P < 0.005$; $n = 7$); inspiratory intercostal activity decreased from 22 ± 2 to 4 ± 1 AU ($P < 0.05$; $n = 6$) and expiratory activity of the transversus abdominis from 22 ± 7 to 1 ± 1 AU ($P < 0.05$; $n = 5$).

Respiratory depression was accompanied by a concomitant increase in arterial blood pressure. The peak pressure response occurred earlier than the maximal change in respiratory muscle activity and subsequently returned to the preapplication level. On average, mean arterial blood pressure increased by 12% ± 3% ($P < 0.05$).

Results of this study showed for the first time that cyanide acting locally at the VMS inhibits both diaphragm and intercostal muscle activity. Effects on intercostal electrical activity were greater than the effects on diaphragm activity. Expiratory intercostal and abdominal muscle activities were very sensitive to cyanide applied to the intermediate area of the VMS. Doses of cyanide that induced only small alterations in diaphragm activity caused appreciable decreases in expiratory activity. Our findings indicate that cyanide acting on the VMS influences respiratory-related activity, but they do not permit us to determine whether these changes are due entirely to an action of cyanide on

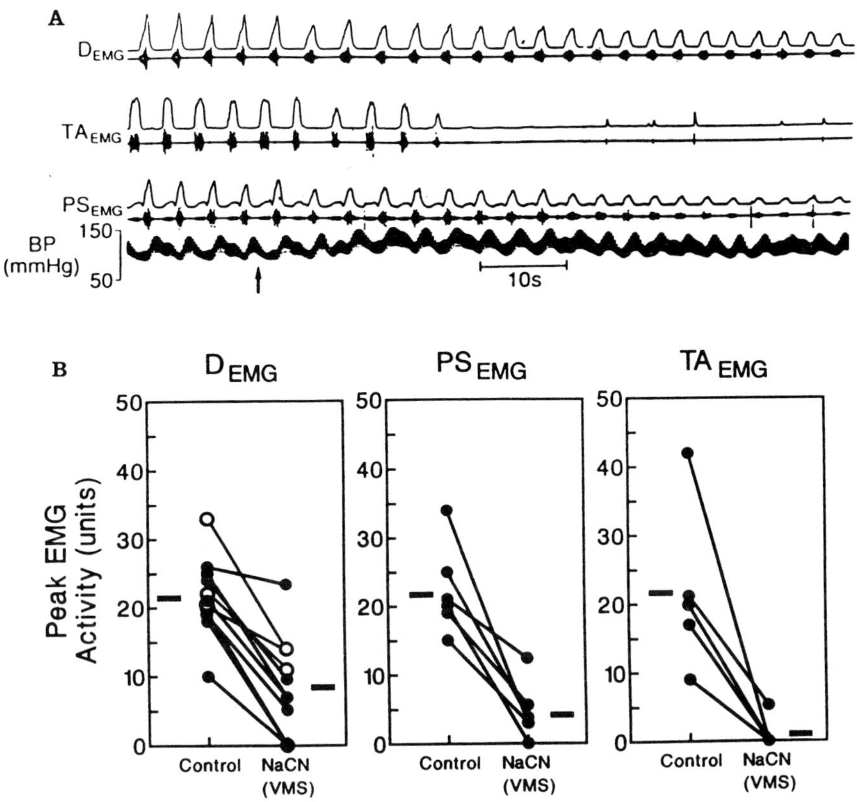

Figure 3 An example (panel A), and individual data (panel B) of responses of electrical activity of phrenic nerve (○) and electromyographic activity of diaphragm [(—),A], parasternal intercostal (B), and transversus abdominis muscle (C) to local application of 100 μg of NaCn on intermediate area of ventral surface of medulla (panel B). Horizontal bars, mean values. (Modified from Ref. 12.)

chemoreceptors or whether a high density of capillaries, a high local blood flow, and the structure of the VMS facilitate the access of cyanide to deeper portions of the medulla.

To further study the role of these structures in hypoxia-induced respiratory depression and sympathetic excitation and to localize the site of action, NaCN was microinjected into the ventrolateral medulla in sites which were 1 to 12 mm caudal to the foramen cecum and 2 to 5 mm lateral to the midline (3). Injections were made unilaterally at a depth of 1 mm below the ventral surface. The central chemoreceptors have been reported to be located in the region explored. On the basis of focal cooling and the response to chemicals applied

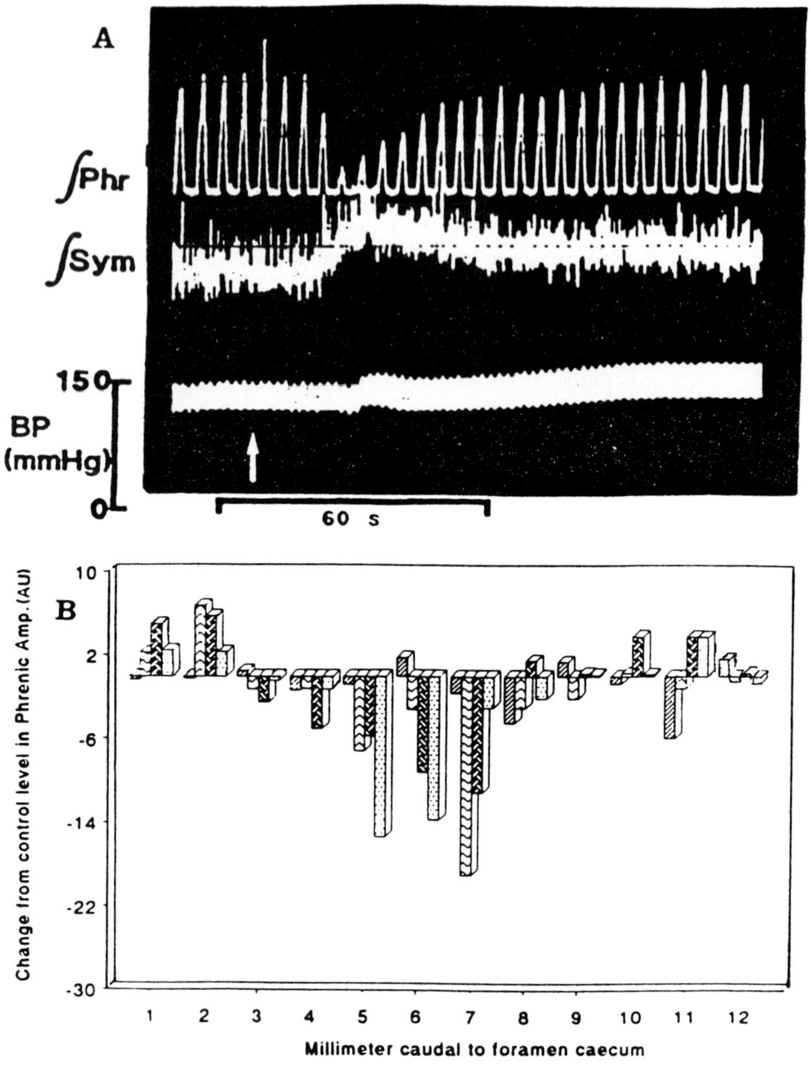

Figure 4 Panel A: Typical response to NaCN injection (100 ng/100 nl) into the ventrolateral medulla of integrated phrenic (Phr) and sympathetic (Sym) nerves, and arterial blood pressure (BP). Panels B, C, and D: Mean changes in phrenic amplitude, sympathetic tone, and blood pressure. ↑: time of injection. (Modified from Ref. 3.)

topically (using the foramen cecum as a reference), this area has been subdivided into three segjments, i.e., rostral (R), intermediate (I), and caudal (C). In our experiment, the R segment roughly corresponds to 1.0 to 4.0 mm caudal; the I segment, 5.0 to 8.0 mm; and the C segment 9.0 to 12.0 mm caudal to FC, as shown in Figure 2.

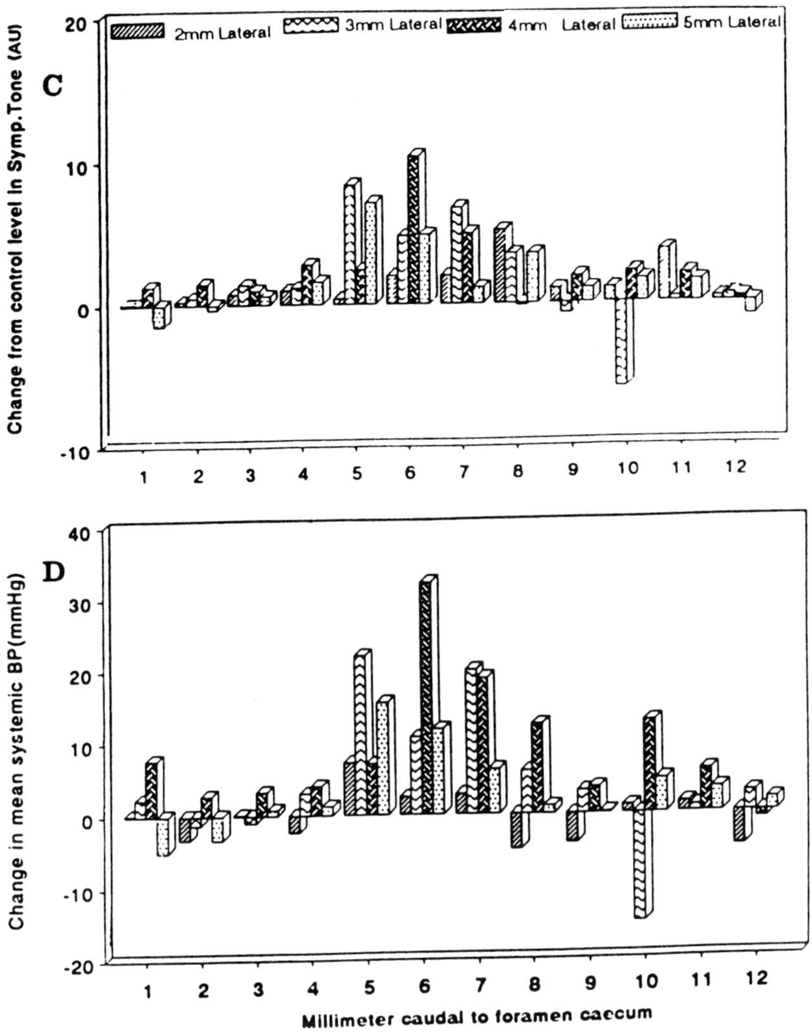

Data were obtained from 269 injections at 48 sites near the ventral medullary surface in 17 adult cats. Statistical analysis showed that there were significant differences along the rostrocaudal axis in phrenic activity, sympathetic discharge, and blood pressure following injection. The responses at rostral and caudal sites were in general less than response in the intermediate area. The null hypothesis of no overall rostrocaudal difference with respect to all three measured parameters was strongly rejected at $P < 0.0001$, using Wilk's lambda statistic. However, the same analysis revealed only a marginal difference ($P \sim$

0.07) in vasomotor and respiratory responses to injections along the medial-to-lateral axis.

The predominant respiratory response to sodium cyanide microinjections into the VMS was depression of breathing. Usually the onset of depression was abrupt, occurring within 10 ± 5 s of the injection and recovered gradually over 45 ± 15 s. Figure 4, panel A, shows the effect of NaCN injection in one animal. Depression of breathing was obtained consistently from an area 5.0 to 7.0 mm caudal to the foramen cecum and 3.0 to 5.0 mm lateral to the midline ($P<0.05$). This region corresponds to the intermediate area. Respiratory responses to the injection of NaCN more rostral and more caudal also caused depression, but to a lesser degree. The distribution of cyanide following microinjection was determined by using ^{14}CN. As shown in Figure 5, cyanide was confined within 1 mm of the injection site after 7 min. Less than 10% of the injected cyanide was covalently bound to proteins.

As in phrenic nerve activity, there were no changes in sympathetic tone when NaCN was injected in either the most rostral or most caudal areas. Significant increases in sympathetic tone were obtained only between 5.0 to 8.0 mm caudal to the foramen cecum and 2.0 to 5.0 mm lateral to the midline ($P<0.05$). In general, the onset latency and duration of response was similar to that of phrenic nerve activity. Figure 4 summarizes the mean phrenic (B) and the mean sympathetic (C) responses from the regions explored in these studies. Note that sympathetic activity increases most prominently in the intermediate area. Comparison of panels B and C show that there is considerable overlap in the regions involved in sympathetic excitation and respiratory depression. As in respiratory depression and sympathetic excitation, significant BP elevation was obtained between 5.0 to 8.0 mm caudal and 2.0 to 5.0 mm lateral (area I) with cyanide injection (D).

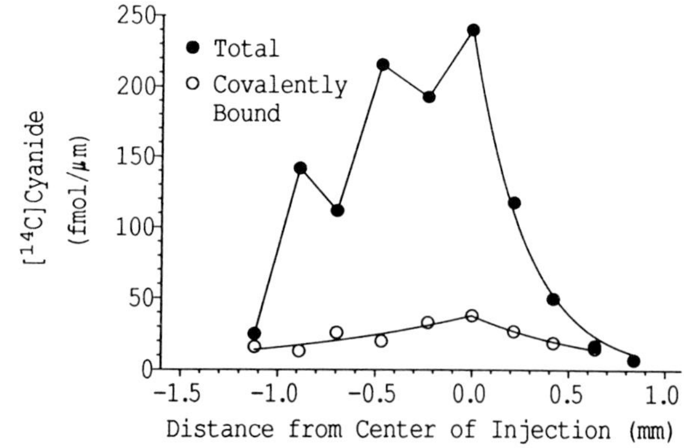

Figure 5 Distribution of [^{14}C]cyanide following microinjection into the rostral ventrolateral medulla.

The major findings of these studies are that a significant depression of respiration and increases in sympathetic tone and blood pressure could be obtained by the topical application or injection of NaCN into the intermediate area of the ventral medulla. These responses are similar to those elicited by central hypoxia produced by intravertebral injections of hypoxic saline or NaCN (12) or topical application of NaCN on intermediate areas (11). Hence the origin of central hypoxic responses may be the ventrolateral medulla, particularly the intermediate region.

The excitotoxin kainic acid, when applied superficially on the rostral but not in the intermediate ventrolateral medulla, depressed phrenic activity irreversibly following an initial increase in BP (61). In contrast, our greatest phrenic depression and rise in BP with NaCN applications were obtained from the intermediate area, and both effects were reversible, indicating two different sites and mechanisms.

Although more direct actions of cyanide on nervous tissue have been proposed, it is generally believed that NaCN acts by blocking tissue oxidation, thereby causing histotoxic hypoxia (62). Aitken and Braitman (1989) studied the effects of NaCN on isolated hippocampal slice preparation and found that cyanide in lower doses, like hypoxia (63), reversibly blocks synaptic transmission (43).

The mechanisms through which hypoxia produces respiratory depression and sympathetic excitation are still speculative. Recently it has been shown that cyanide, like hypoxia, increases the release of several neurotransmitters, i.e., GABA (64), glutamate (65), and catecholamines (66). Whether the respiratory and vasomotor changes in response to NaCN injection are also mediated through neurotransmitter release is not known.

It has been shown in the past that respiratory and vasomotor control can be dissociated functionally under certain experimental conditions (67). Since maximum sympathetic response was obtained from the intermediate area and sympathoexcitatory neurons that project to the spinal cord can be found underlying the intermediate area (68), it is possible that these neurons are affected directly by hypoxia to excite sympathetic response.

Microinjection of NaCN into the RVLM in the rat raises systemic blood pressure (69). Morphological examination and study of cerebral ischemia have led some investigators to suggest that neurons in the RVLM may be sensitive to hypoxia (57). Our results support this hypothesis. However, a contribution from surrounding regions cannot be ruled out. Isolated regions scattered outside the intermediate area when injected with NaCN elicited only small, nonsignificant increases in sympathetic tone and blood pressure following NaCN injection. Both in the rostral and caudal medulla we also found small regions (1.0 to 2.0 mm and 10.10 to 12.0 mm caudal to FC, 3.0 to 5.0 mm lateral) where NaCN caused a slow augmentation of phrenic nerve activity with little or no change in systemic blood pressure. Also in the caudal medulla (10.0 mm caudal, 3.0 mm lateral) there was a region where NaCN lowered both sympathetic tone and blood pressure. This site may be part of the vasodepressor area, since in this general region others have found a blood pressure–lowering response to electrical or chemical stimulation (70).

D. Response of Single Neurons in the Medulla to Systemic and Brainstem Hypoxia

To better characterize the responses of ventral-medullary neurons with different functional characteristics, we examined the behavior of respiratory modulated and nonmodulated neurons (including bulbo-sympathetic neurons) during hypoxia caused either by ventilation with 10% O_2 in N_2 or intravertebral infusion of deoxygenated saline. Experiments were performed in 10 anesthetized, vagotomized, sinus-denervated, paralyzed, and mechanically ventilated cats.

We found that the effects of systemic hypoxia on different types of neurons in the ventral medulla were varied and ranged from excitation in some cases to inhibition in others.

In our experiment, all inspiratory modulated neurons were inhibited by hypoxia but excited by hypercapnia. These neurons also increased their discharge frequency in response to proprio- or nociceptive stimuli. The findings of our study confirm the results of other investigators (71,72).

Expiratory modulated neurons behaved nonuniformly during hypoxia. One group became silent, as observed by others (73), while the other became tonic as phrenic nerve activity disappeared. To our knowledge, the tonic discharge we observed in some expiratory modulated neurons during hypoxia has not been described previously. It is possible that these tonic expiratory neurons were motoneurons supplying the muscles of the oropharyngeal region. Muscles of the oropharynx can discharge tonically during expiration (74).

There were no clear differences in the response to hypoxia of the neurons that were activated by peripheral sensory inputs versus those that were not. Some neurons were inhibited by hypoxia while others were excited. The excitation may have been caused by disinhibition rather than by direct excitation.

Neurons that decreased their activity or became silent during the apnea produced by hypoxia could still be excited as before, indicating that the polysynaptic pathways were functional. The amplitude of the evoked activity during the early part of respiratory arrest (10 s) was similar to that during the control period. However, in the later part of respiratory arrest (90 s), the evoked amplitude decreased to less than 50% of control values. Since these sensory neurons could still be activated during the early phase of central apnea, it seems likely that the decrease in discharge rate was caused by an active inhibition. Direct depression of neuronal excitability may occur in the later phase.

Similar observations on the neuronal excitability of phasic inspiratory modulated neurons during hypoxia have been reported by Richter et al. (75). They found that during hypoxia induced central apnea, "respiratory neurons were still excitable by stimulus-evoked orthodromic and antidromic action potentials." This suggests that central apnea initially may be due to withdrawal of excitatory drive input and not to neuronal failure (75). This idea is further reinforced by observations that during central apnea, electrical stimulation of the sinus nerve or inhalation of high CO_2 initiates rhythmic phrenic nerve activity (76).

Twenty-nine percent of antidromically activated RVL neurons (bulbosympathetic) increased their discharge—an example is shown in Figure 6—while 71% decreased or became silent during apnea. However, intermediolateral stimulation could still excite the latter group during the early phase of apnea. Sympathetic nerve tonic activity during the later part of apnea could not be correlated with the discharge rate of the identified bulbosympathetic neurons. Moreover, sympathetic tone remained high even when the activity of these neurons decreased. This supports the idea that the rise in sympathetic tone during hypoxia does not depend exclusively on the medullary neurons and indicates that there are multiple sites for the hypoxic excitation of sympathetic activity, including spinal neurons. The spinal cord is known to be more resistant to hypoxia than the medulla (77). We have reported that sympathetic tone can be augmented in spinal animals by intrathecal administration of sodium cyanide (11). Moreover, we observed that central hypoxia produced by infusion of hypoxic saline into the vertebral artery and topical application of cyanide on the ventral surface of the medulla augments sympathetic activity.

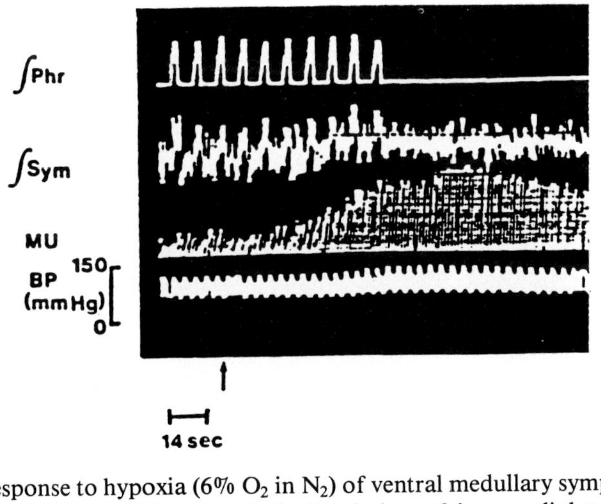

Figure 6 Response to hypoxia (6% O_2 in N_2) of ventral medullary sympathetic neuron activity of which increased by electrical stimulation of intermediolateral cell colum. ↑: Beginning of hypoxia; Phr: integrated nerve activity; Sym: integrated sympathetic activity; Mu: raster display of medullary sympathetic unit discharge (per second); BP: arterial blood pressure.

This indicates that neurons in the medulla or supramedullary structure(s) are capable of contributing to hypoxic sympathoexcitation, and these tonic influences may affect sympathetic nerves differently (18).

Neurons that did not show respiratory modulation and were sensitive (78) or insensitive to hypercapnia (5% CO_2 in O_2) and sympathetic stimuli responded to hypoxia. However, responses were variable; some could be excited and some inhibited by hypoxic challenge (Fig. 7).

Our study shows that both respiratory bulbospinal sympathetic and nonrespiratory neurons within the ventrolateral medulla that are depressed by hypoxia maintain their excitability during the early phase of respiratory depression to an adequate stimulus, as previously demonstrated for inspiratory neurons. Our results also suggest that the sympathetic activation that occurs with hypoxia probably arises from multiple sites, including the ventral medulla and spinal cord. These findings could be partially explained by an increase in H^+ concentration, which is in agreement with earlier studies (79).

VI. In Vitro Studies on the Electrophysiological Effects of O_2 Deprivation in Ventrolateral Medullary Neurons

Expression of c-fos has been used to identify cells activated by specific stimuli such as CO_2. Data obtained from experiments in which the effect of CO_2 on c-fos expression have been assessed demonstrate that CO_2-responsive neurons are present in the ventrolateral medulla, including the parapyramidal region (27–29), and that the axonal targets of these cells include brainstem and spinal cord neurons. Conceivably these neurons are among the sites inhibited by hypoxia, leading to depression of respiratory activity. Recently we assessed the effects of O_2 reduction on the electrophysiological characteristics of parapyramidal

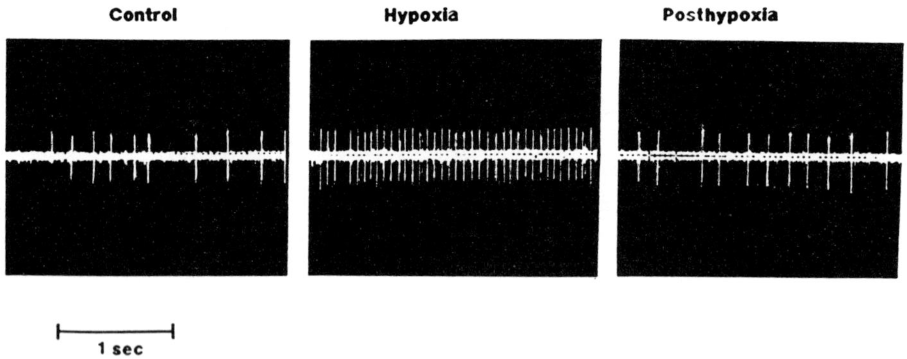

Figure 7 Response to hypoxia of ventral medullary neuron, activity of which did not change by hypercapnia (5% CO_2 in O_2) or electrical stimulation of intermediolateral cell column.

neurons in neonatal (7 to 20 days) and adult rats. Intracellular recordings were obtained in vitro from parapyramidal cells of the ventral half of medullary slices in control conditions (95% O_2 and 5% CO_2) and after exposure to anoxia (95% N_2 + 5% CO_2). Anoxia abolished all spontaneous activity of the cells, which were found to fire spontaneously at a rate of about 3 spikes/s. However, an electrical stimulus could generate action potentials, whose shape or size did not appear to differ significantly from the control. Anoxia decreased the somatic firing threshold. A transient decrease in the input resistance was also observed in anoxia. Complete recovery was obtained by switching to 95% O_2 and 5% CO_2. Subsequent anoxic insults did not produce a significant change in the resting potential. Neonatal cells survived in anoxia for >45 min, whereas survival was up to 30 min in adults. These results indicated that O_2 deprivation produces depolarization and abolishes spontaneous activity of ventromedullary cells. However, their excitability was not altered (37).

VII. Neurotransmitters Involved in Hypoxia-Induced Respiratory Depression and Sympathetic Excitation

An increasingly accepted idea is that moderately severe hypoxia affects the activity of neurotransmitters in the CNS, tipping the balance between inhibitory and excitatory influences on respiratory and sympathetic brainstem neurons (21,64–66). These effects may be local and/or global. Hypoxia may act on local areas of the ventrolateral medulla to inhibit respiration but excite sympathetic activity. Sympathetic excitation might then occur as a result of a disinhibition (a release of inhibitory effects) on vasomotor neurons. Alternatively, an overabundance of excitatory neuromodulations might also explain the CNS effects of hypoxia. Indeed, even with respect to breathing, a number of excitatory effects of central hypoxia, such as tachypnea, have been described during CNS hypoxia, particularly in unanesthetized animals.

Hypoxia may release inhibitory neurotransmitters, like GABA, opiates, dopamine, adenosine, and endothelin. GABA is the predominant inhibitory transmitter in the brain, and the depression of respiratory neuronal activity during hypoxia is a potentially adaptive response mediated principally by the effects of of acidosis and increased GABA levels (3,12,21,66). Activation of GABAergic receptors has been specifically implicated in hypoxic depression in neonates (66).

Opiate receptors may participate in hypoxic depression in newborn neonates (80–82). Systemically administered, the opiate antagonist naloxone stimulates respiration in hypoxic neonates and can prevent apnea. Endorphin and enkephalin levels are sharply elevated in hypoxic neonates. The role of endogenous opiates in the cardiovascular responses to hypoxia remains largely unexplored, even though opiates have been implicated in vasomotor control, particularly in the RVLM and caudal VLM (CVLM) (17). However, the role of endogenous opiates in the response of adult organisms to hypoxia can be questioned. For example, naloxone only slightly attenuates respiratory depression in adult cats exposed to CO (22). Inconsistent results with naloxone in adult CNS hypoxia may be caused by the existence of multiple opiate receptor subtypes. The opiate receptor subtypes that mediate effects on respiration have not been identified with certainty, although μ-receptor subtypes appear to predominate.

Dopaminergic mechanisms also may play a role in hypoxic depression of respiration. It is known that dopamine is expressed by glomus cells in the carotid body and in several regions of brainstem. When it is released, it may cause respiration depression.

Adenosine has been implicated by a number of investigators in hypoxic depression of respiration, particularly in neonates (21,23,83,84). The adenosine antagonist theophylline is a respiratory stimulant used clinically to reverse hypoxic depression of respiration, while adenosine agonists depress respiration in normoxia. Brain levels of adenosine rise up to sevenfold during severe hypoxia (23). Brain adenosine interacts extensively with GABA, since nearly all putative adenosine neurons also synthesize GABA. Adenosine may be important as a neuromodulator regulating the activity of neurotransmitters during hypoxia.

Since theophylline (adenosine antagonist) reverses hypoxic depression and stimulates breathing (83), we investigated its effect on central hypoxia produced by vertebral artery injections of NaCN or saline equilibrated with N_2. Experiments were done in chloralose-anesthetized, paralyzed, vagotomized, and sinus-denervated cats ventilated with room air. Recordings were made from the N_2 phrenic and preganglionic cervical sympathetic nerves, along with blood pressure. The vertebral artery on one side was cannulated for injections of NaCN (10 and 20 μg) and hypoxic saline (97% N_2 in 3% CO_2). Systemic administration of theophylline (5 mg/kg) either reduced the depression of phrenic amplitude (22 $\pm$ 5 AU %) or the duration of apnea (14 $\pm$ 7 s) caused either by NaCN or hypoxic saline but had little or no effect on the changes produced in sympathetic tone and blood pressure. We conclude that the central respiratory depression produced by NaCN or hypoxic saline may be in part due to the release of adenosine. However, the increase in BP and sympathetic tone seems not be due to inhibition of interneurons by adenosine.

Endothelin-1 (ET_1) is a member of a family of peptides found in endothelial and epithelial cells. This 21–amino acid peptide also is present in the brain, in a nonvascular distribution associated with specific nuclei (85). Recently, we have identified endothelin receptors in isolated cell membranes from the ventral medulla as the nonselective (ET) subtype which binds each of the endothelin isopeptides with high affinity (Fig. 8). In addition, we determined the effects of chronic hypobaric hypoxia (0.5 atm for 21 days) on medullary ET receptors and examined the possible role of ET in medullary cardiorespiratory control. For visualizing ET receptors, 15-μm sections of rat medulla were preincubated 0.5 h, incubated 12 min with 50 pM [^{125}I]ET_1, washed 1 h, and either exposed to film or cross-linked in paraformaldehyde vapor and processed for

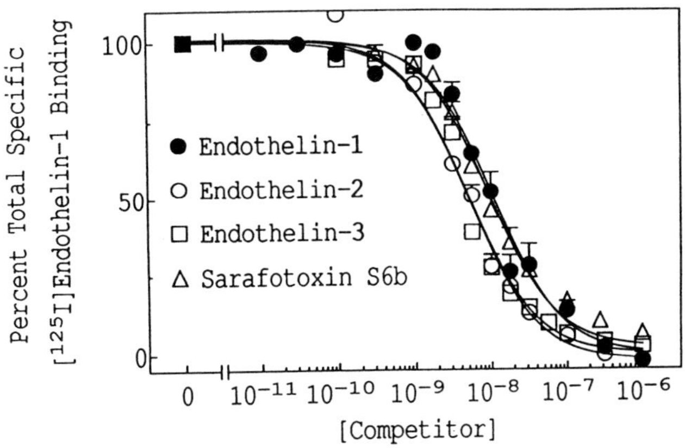

Figure 8 Dose-dependent inhibition of specific [^{125}I] endothelial binding in ventrolateral medulla.

emulsion autoradiography. Nonspecific binding was defined with 1 μM ET_1. Autoradiogram showed that ET receptors were enriched in VLM regions implicated in cardiorespiratory control: ventral surface, parapyramidal, retrotrapezoid, and rostroventrolateralis area. Autoradiographic grains were associated with neurons as well as glia and major vessels. In biochemical studies, intact

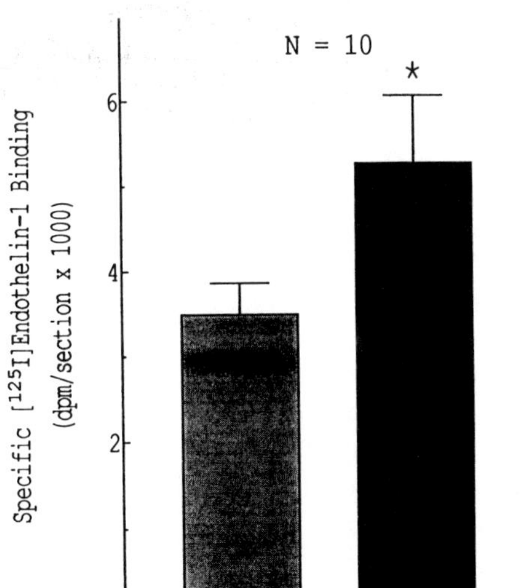

Figure 9 Effects of hypoxia on endothelin binding sites in rat medulla.

medullary sections from rats exposed to chronic hypoxia showed a higher density of ET binding sites than matched controls (Fig. 9; 150 ± 18% of control, $P < 0.05$). Regulation of ET receptors by hypoxia suggests a role in respiratory control. In vivo studies in anesthetized rats ($n = 15$) showed that bilateral microinjections into superficial layers of ventral medulla with high ET receptor density, in doses of (1 to 10 pmol), produced expiratory apnea associated with an initial increase followed by a decrease in arterial blood pressure. Changes occurred within 30 s and lasted longer than 30 min. These findings are in agreement with recent studies (86). We conclude that ET receptors in the VLM (1) are upregulated by chronic hypoxia; (2) when activated, reduce respiratory activity; and (3) might be involved in hypoxic depression of respiration.

VIII. Conclusion

In conclusion, our studies suggest that structures within the ventrolateral medulla are involved in regulation of respiratory and sympathetic activity during hypoxia. Hypoxia acting on ventrolateral medullary neurons induces qualitatively different site-specific responses: generally it causes respiratory depression and sympathetic excitation. These responses may be of a protective nature and their value could be of particular importance to the in utero life of the fetus. During prolonged hypoxia, increase in activity of respiratory pumping muscles may lead to depletion of O_2 stores without increasing O_2 supply; decrease in their activity would conserve energy supplies necessary for maintenance of ionic gradients. In addition, increase in sympathetic activity may provide critical blood flow and redistribution of cardiac output during hypoxia.

Acknowledgments

The authors wish to thank Cecily Lewis for her secretarial assistance in typing this manuscript. Support for this research was provided by the National Heart, Lung and Blood Institute, Grant No. HL 25830.

References

1. Bean JW. Intestine in reflex chemical control of breathing. Am J Physiol 1952; 171: 522–527.
2. Berman SM, Gebber GL. Sympathetic nerve rhythm of brain stem origin. Am J Physiol 1980; 239:R42–R47.
3. Mitra J, Dev NB, Trivedi R, et al. Intramedullary sodium cyanide injection on respiratory and vasomotor responses. Respir Physiol 1993; 93:71–82.
4. Rohlicek CV, Polosa C. Hypoxic responses of sympathetic preganglionic neurons in aortic-denervated cats. Am J Physiol 1983; 244:H681–H686.

5. Wasicko MJ, Melton JE, Neubaur JA, et al. Cervical sympathetic and phrenic nerve responses to progressive brain hypoxia. J Appl Physiol 1990; 68:53–58.
6. Santiago TV, Guerra E, Neubauer JA, Edelman NH. Correlation between ventilation and brain blood flow during sleep. J Clin Invest 1984; 73:497–506.
7. Santiago TV, Neubauer JA, Edelman NH. Correlation between ventilation and brain blood flow during hypoxic sleep. J Appl Physiol 1986; 60:295–298.
8. Whittingham TS, Lipton P. Cerebral synaptic transmission during anoxia is protected by creatine. J Neurochem 1981; 37:1618–1621.
9. Davenport HW, Brewer G, Chambers AH, Coldschmidt S. The respiratory responses to anoxemia of unanesthetized dogs with chronically denervated aortic and carotid chemoreceptors and their causes. Am J Physiol 1947; 148:406–417.
10. Hayes K, Weaver CL. Tonic sympathetic excitation and vasomotor control from pontine reticular neurons. Am J Physiol 1992; 263:H1567–H1575.
11. Haxhiu MA, Erokwu B, van Lunteren E, et al. Central and spinal effects of sodium cyanide on respiratory activity. J Appl Physiol 1993; 74:574–579.
12. Mitra J, Dev NB, Romaniuk JR, et al. Cardiorespiratory changes induced by vertebral artery injection of sodium cyanide in cats. Respir Physiol 1992; 87:49–61.
13. Dev NB, Loeschcke HH. Topography of the respiratory and circulatory responses to acetylcholine and nicotine on the ventral surface of the medulla. Pflügers Arch 1979; 379:19–27.
14. Giuliano R, Ruggiero DA, Morrison S, et al. Cholinergic regulation of arterial pressure by the C1 areas of the rostral ventrolateral medulla. J Neurosci 1989; 9:923–942.
15. Haxhiu MA, Mitra J, van Lunteren EN, et al. Hypoglossal and phrenic responses to cholinergic agents applied to ventral medullary surface. Am J Physiol 1988; 247:R939–R944.
16. Cattarossi L, Haxhiu-Poskurica B, Haxhiu MA, Carlo WA. Response of upper airway and chest wall muscles to selective brain stem hypoxia in the newborn. J appl Physiol 1993; 74:2443–2449.
17. Schläfke ME. Central chemosensitivity: A respiratory drive. Rev Physiol Biochem Pharmacol 1981; 90:171–244.
18. Yardley CP, Stein RD, Weaver LC. Tonic influences from the rostral medulla affect sympathetic nerves differentially. Am J Physiol 1989; 256:R323–R331.
19. Cherniack NS, Edelman NH, Lahiri S. Hypoxia and hypercapnia as respiratory stimulants and depressants. Respir Physiol 1970/71; 11:113–126.
20. Neubauer JA, Simone A, Edelman NH. Role of brain lactic acidosis in hypoxic depression of respiration. J Appl Physiol 1988; 65:1324–1331.
21. Hedner Th, Hedner J, Wessberg P, Jonason J. Regulation of breathing in the rat: Indications for a role of central adenosine mechanisms. Neurosci Lett 1982; 33: 147–151.
22. Neubauer JA, Posner MA, Santiago TV, Edelman NH. Naloxone reduces ventilatory depression of brain hypoxia. J Appl Physiol 1987; 63:699–706.
23. Winn HR, Rubio R, Berne RM. Brain adenosine concentrations during hypoxia in rat. Am J Physiol 1981; 241:H235–H242.
24. Feldman JL. Neurophysiology of breathing in mammals. In: Montcastle VB, Bloom FE, Geiger SR, eds. Handbook of Physiology, Section: The Nervous System Regu-

latory System of the Brain, Vol. IV. Bethesda, MD: American Physiological Society, 1986:463–524.
25. Loeschcke HH. Central chemosensitivity and the reaction theory. J Physiol (Lond) 1982; 332:1–24.
26. Smith JC, Morrison DE, Ellenberger HE, et al. Brainstem projections to the major respiratory neuron populations in the medulla of the cat. J Comp Neurol 1989; 281:69–96.
27. Haxhiu MA, Erokwu B, Martin RJ. Medullary chemosensory neurons can be identified by expression of c-fos protein during early postnatal life. Pediatr Res 1992; 31:1842.
28. Saphir R, Rapoport E. Cardiovascular responses of the cat to mesenteric intra-arterial administration of nicotine, cyanide and venous blood. Circ Res 1969; 25: 713–724.
29. Sapru HN, Punnen S, Willette RN. Role of enkephalins in ventrolateral medullary control of blood pressure. In: Buckley JD, Ferrario CM, eds. Brain Peptides and Catecholamines in Cardiovascular Regulation. New York: Raven Press, 1987: 153–167.
30. McAllen RM. Location of neurons with cardiovascular and respiratory function, at the ventral surface of the cat's medulla. Neuroscience 1986; 18:43–49.
31. McAllen RM. Action and specificity of ventral medullary vasopressor neurons in the cat. Neuroscience 1986; 18:51–59.
32. McAllen RM, Dampney RA. Vasomotor neurons in the rostral ventrolateral medulla are organized topographically with respect to type of vascular bed but not doy region. Neurosci Lett 1990; 110:91–96.
33. Hutt DA, Parisi RA, Santiago TV, Edelman NH. Brain hypoxia preferentially stimulates genioglossal EMG responses to CO_2. J Appl Physiol 1989; 66:51–56.
34. Martin RJ, van Lunteren E, Haxhiu MA, Carlo WA. Upper airway muscle and diaphragm responses to hypoxia in the piglet. J Appl Physiol 1990; 68:672–677.
35. Korner PI. Integrative neural cardiovascular control. Physiol Rev 1971; 51:312–357.
36. Haddad GG, Mellis RB. Hypoxia and respiratory control in early life. Am Rev Respir Physiol 1988; 46:629–643.
37. Patel M, Durand D, Haxhiu MA. Anoxic response of medullary parapyramidal neurons. Soc Neurosci 1993; 19:989.
38. Martin-Body RL. Brain transections demonstrate the central origin of hypoxic ventilatory depression in carotid body-denervated rats. J Physiol 1988; 407:41–52.
39. Martin-Body RL, Johston BM. Central origin of the hypoxic depression of breathing in the newborn. Respir Physiol 1988; 71:25–32.
40. Hansen AJ. Effect of anoxia on ion distribution in the brain. Physiol Rev 1985; 65:101–148.
41. Meldrum B. Possible therapeutic applications of antagonists of excitatory amino acid neurotransmitters. Clin Sci 1985; 68:113–122.
42. Rothman SM, Olney JW. Glutamate and the pathophysiology of hypoxic-ischemic brain damage. Ann Neurol 1986; 19:105–111.
43. Aitken PG, Braitman DJ. The effects of cyanide on neuronal and synaptic function in hippocampal slices. Neurotoxicology 1989; 10:239–248.

44. Siesjö BK, Johannsson H, Ljunggren B, Norberg K. Brain dysfunction in cerebral hypoxia and ischemia. In: Plum F, ed. Brain Dysfunction in Metabolic Disorders. New York: Raven Press, 1974:75–112.
45. Yamada KA, Hamosh P, Gillis RA. Respiratory depression produced by activation of GABA receptors in hindbrain in cat. J Appl Physiol 1981; 51:1278–1286.
46. McGillicuddy JE, Kindt GW, Raisis JE, Miller CA. The relation of cerebral ischemia, hypoxia and hypercarbia to the Cushing response. J Neurosurg 1978; 48:730–740.
47. Prabhakar NR, Mitra J, van de Graaf W, et al. Effect of focal cooling of central chemosensitive areas on cerebral ischemic response. Am J Physiol 1986; 251:R295–R302.
48. Kocsis B, Fedina L, Pasztor E. Two-phase change of sympathetic rhythms in brain ischemia, Cushing reaction, and asphyxia. Am J Physiol 1989; 256:R120–R132.
49. Arita H, Ischikawa K, Ochiishi MJ. Pressor response and depression evoked by intravertebral arterial injections of hypoxic saline. J Auton Nerv Syst 1989; 28: 27–33.
50. Downing ES, Mitchell JH, Wallace AG. Cardiovascular responses to ischemia, hypoxia, and hypercapnia of the central nervous system. Am J Physiol 1963; 204: 881–887.
51. Lauweryns JM, Cokelaere M, Deleersnyder M, Liebens M. Intrapulmonary neuroepithelial bodies in newborn rabbits. Cell Tiss Res 1977; 182:425–440.
52. Lew WYW, Longhurst JL. Substance P, 5-hydroxytryptamine, and bradykinin stimulate abdominal visceral afferents. Am J Physiol 1986; 250:R465–R473.
53. Long S, Duffin J. The neuronal determinants of respiratory rhythm. Prog Neurobiol 1986; 27:101–182.
54. Sato J, Severinghaus JM, Basbaum AI. Medullary CO_2 chemoreceptor neuron identification by c-fos immunochemistry.
55. Youngston Chr, Nurse C, Yeger H, Cutz E. Oxygen sensing in airway chemoreceptors. Nature 1993; 365:153–155.
56. van Lunteren E, Mitra J, Prabhakar NR, et al. Ventral medullary surface inputs to cervical sympathetic respiratory oscillations. Am J Physiol 1987; 252:R1032–R1038.
57. Haselton JR, Guyenet PG. Electrophysiological characterization of putative C′ adrenergic neurons in the rat. Neuroscience 1986; 30:199–214.
58. Kubin LA, Trzebski A, Lipski J. Split medulla preparation in the cat: Arterial chemoreceptor reflex and respiratory modulation of the renal sympathetic nerve activity. J Auton Nerv Syst 1985; 12:211–215.
59. Miller MJ, Tenney SM. Hypoxia-induced tachypnea in carotid-deafferented cats. Respir Physiol 1975; 23:31–39.
60. Numao Y, Koshiya N, Gilbey MP, Spyer KM. Central respiratory drive-related activity in sympathetic nerves of the rat: The regional differences. Neurosci Lett 1987; 81:279–284.
61. Nattie EE, Mills W, Ou LC, St. John WM. Kainic acid on the rostral ventrolateral medulla inhibits output and CO_2 sensitivity. J Appl Physiol 1988; 65:1523–1524.
62. Isom GE, Way JL. Effect of oxygen on cyanide intoxication: Cytochrome oxidase in vitro. Toxicol Appl Pharmacol 1984; 74:57–62.

63. Czéh G, Somjen GG. Hypoxic failure of synaptic transmission in the isolated spinal cord, and the effects of divalent cations. Brain Res 1990; 527:224–233.
64. Tursky T, Sajter V. The influence of potassium cyanide poisoning on τ-aminobutyric acid level in rat brain. J Neurochem 1962; 9:519–523.
65. Patel M, Ardelt BK, Yim GKW, Isom GE. Cyanide induces Ca^{+2} dependent-independent release of glutamate from mouse brain slices. Neurosci Lett 1991; 131: 42–44.
66. Kanthasamy AG, Maduh EU, Peoples RW, et al. Calcium mediation of cyanide-induced catecholamine release: Implications for neurotoxicity. Toxicol Appl Pharmacol 1991; 110:275–282.
67. Guyenet PG, Brown DL. Unit activity in nucleus paragigantiocellularis lateralis during cerebral ischemia in the rat. Brain Res 1986; 364:301–314.
68. Amendt K, Czachurski J, Dembowsky K, Seller H. Neurons within the "chemosensitive area" on the ventral surface of the brainstem which project to the intermediolateral column. Pflügers Arch 1978; 375:289–292.
69. Sun M-K, Jeske IT, Reis DJ. Cyanide excites medullary sympathoexcitatory neurons in rats. Am J Physiol 1992; 262:R182–R189.
70. Granata AQ, Numao Y, Kumada M, Reiks DJ. A1 noradrenergic neurons tonically inhibit sympathoexcitatory neurons of C1 area in rat brainstem. Brain Res 1986; 377:127–146.
71. Koizumi K, Ushiyama J, McBrooks C. Muscle afferents and activity of respiratory neurons. Am J Physiol 1961; 200:679–684.
72. St John WM, Bianchi AL. Responses of bulbospinal and laryngeal respiratory neurons to hypercapnia and hypoxia. J Appl Physiol 1985; 59:1201–1207.
73. St John WM, Wang SC. Response of medullary respiratory neurons to hypercapnia and isocapnic hypoxia. J Appl Physiol 1977; 43:812–821.
74. Grelot L, Bianchi AL, Iscoe S, Remmers JE. Expiratory neurons of the rostral medulla: Anatomical and functional correlates. Neurosci Lett 1988; 89:140–145.
75. Richter DW, Bischoff A, Anders K, et al. Response of the medullary respiratory network of the cat to hypoxia. J Physiol (Lond) 1991; 443:231–356.
76. Parisi RA, Melton JE, Wasicko MJ, et al. Phrenic responsiveness to carotid sinus nerve stimulation during progressive brain hypoxia. FASEB J 1988; 2:A1507.
77. Van Harreveld A, Biersteker PA. Acute asphyxiation of the spinal cord and of other sections of the nervous system. Am J Physiol 1964; 206:8–14.
78. Arita H, Kogo N, Ichikawa K. Locations of medullary neurons with non-phasic discharges excited by stimulations of central and/or peripheral chemoreceptors and activation of nociceptors in cat. Brain Res 1988; 442:1–10.
79. Seller H, König S, Czachurski J. Chemosensitivity of sympathoexcitatory neurons in the rostroventrolateral medulla of the cat. Pflügers Arch 1990; 416:735–741.
80. Chernick V, Madansky DL, Lawson EE. Naloxone decreases the duration of primary apnea with neonatal asphyxia. Pediatr Res 1980; 14:357–359.
81. Chernick V, Warshaw JB, Kiley JP. Developmental neurobiology of respiratory control. Am Rev Respir Dis 1989; 139:1295–1301.
82. Langercrantz H. Neuromodulators and respiratory control during development. Trends Neurosci 1987; 10:368–372.
83. Darnall AR, Aminophylline reduces hypoxic ventilatory depression: Possible role of adenosine. Pediatr Res 1985; 19:706–710.

84. Easton PA, Anthonisen NR. Ventilatory responses to sustained hypoxia after pretreatment with aminophylline. J Appl Physiol 1988; 74:1445–1450.
85. Giad A, Gibson SJ, Herrero MT, et al. Topographical localization of endothelin mRNA and peptide immunoreactivity in neurons of the human brain. Histochemistry 1991; 95:303–314.
86. Kuwai T-I, Cao W-H, Unekawa M, et al. Endothelin-sensitive areas in the ventral surface of the rat medulla. J Auton Nerv Syst 1991; 37:149–158.

7

Central Chemoreceptor Location and the Ventrolateral Medulla

EUGENE E. NATTIE, AIHUA LI, and E. LEE COATES

Dartmouth Medical School
Lebanon, New Hampshire

I. Introduction

Historically, central chemoreceptors have been located at or within a few hundred micrometers of the ventral surface of the medulla (1–6). Mitchell et al. (3,4) stimulated ventilation by bathing the subarachnoid region of the ventrolateral medulla (VLM) directly with acidic solutions (Fig. 1). The direct application of acidic or basic artificial cerebrospinal fluid to specific locations on the VLM surface using small cotton pledgets or a superperfusion apparatus delineated a rostral chemosensitive area, referred to as Mitchell's area (M), and a caudal chemosensitive area, referred to as Loeschcke's area (L) (4,5) (Fig. 2). Tonic nonrespiratory modulated medullary units located just beneath the VLM surface at these chemosensitive areas have been shown in vivo (7–10) and in vitro (11–15) to increase firing with increased CO_2. However, it is difficult to prove that these units are actually chemoreceptive and responsible for changes in breathing. An intermediate area (IMA), lying between the M and L areas and referred to as Schläfke's area (S), was also described (Fig. 2). The IMA was not thought to be chemosensitive, but cooling of this area produced depression of ventilation and of chemosensitivity (6,16–18). This paper will

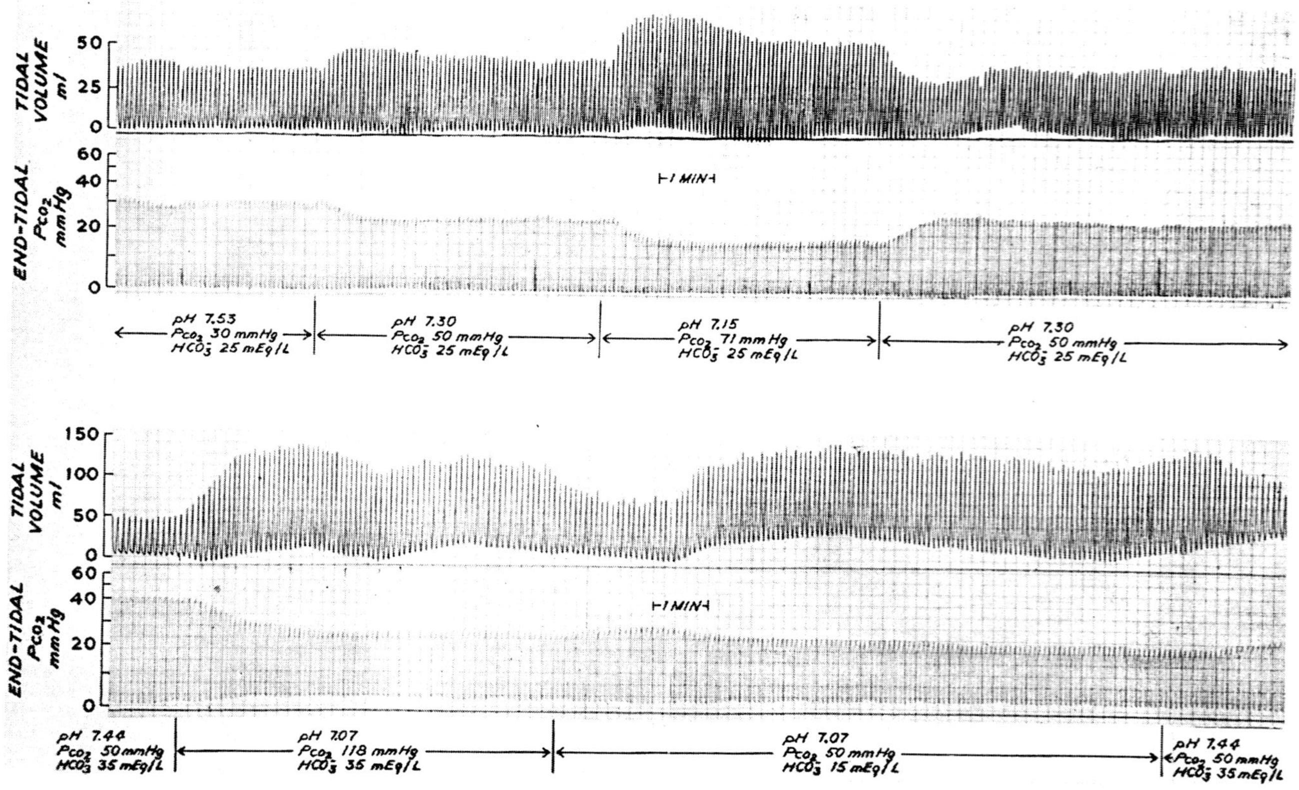

Figure 1 Effects on ventilation and end-tidal P_{CO_2} of changing the hydrogen ion concentration and P_{CO_2} of artificial cerebrospinal fluid introduced into the subarachnoid space of two anesthetized cats with vagal and carotid sinus nerve section. (From Ref. 3.)

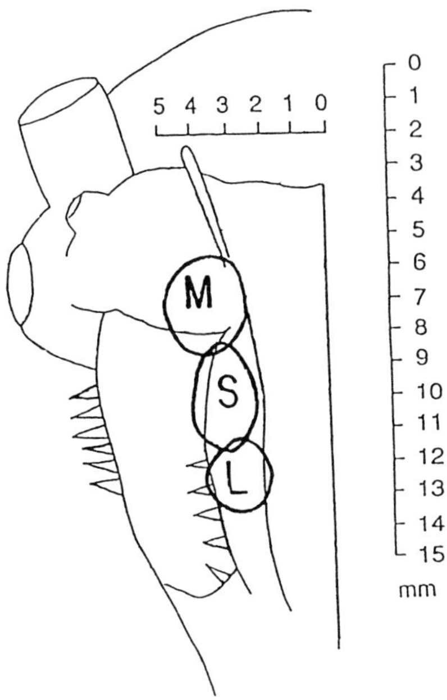

Figure 2 The topographical locations of the rostral (Mitchell's or M) and caudal (Loeschcke's or L) chemosensitive areas and the intermediate (Schläfke's or S) area in the cat. The vertical rule is in millimeters caudal to the foramen cecum; the other in millimeters lateral to the midline.

expand on these traditional concepts on central chemoreceptor location and on the role of the IMA.

II. Central Chemoreceptors

This set of locations for central chemoreceptors, at or within a few hundred micrometers of the VLM surface at rostral and caudal sites, has received widespread but not universal acceptance. In rats, for example, some have found ventilatory responses to VLM acid application (19) and others have not (20). One problem is that substances applied to the VLM surface can actually be transported more deeply within the medulla by numerous blood vessels that penetrate from the ventral surface (7). For example, both large (21) and small radiolabeled molecules (22–24) applied to the VLM surface have been shown to penetrate within minutes to a depth of up to 2 mm. These observations suggest that the precise location of central chemoreceptors cannot be accurately deduced from surface application studies.

Others have argued for the presence of central chemoreception at sites well distant from the surface of the VLM. In vivo studies have shown that a high proportion of respiratory-modulated medullary single units in the pons and and dorsal medulla (25,26) increase their firing rate with systemic hypercapnia. Injections of CO_2-laden saline into the vertebral artery (27–29) have shown the presence of CO_2-responsive tonic units located near the rostral aspect of the ventral respiratory group (VRG) and ventral to the nucleus tractus solitarius (dorsal respiratory group, DRG) as well as near the VLM surface. In these experiments however, the increase in unit discharge may not reflect chemosensitivity per se, it may merely reflect a respiratory system response—i.e., the unit is part of the respiratory control system responding to the chemoreceptor stimulation. In vitro studies have also shown cells responsive to acidic stimulation at more dorsal locations near the nucleus tractus solitarius (13,14,30,31) as well as in parts of the slice not usually associated with respiratory control (14). These in vitro studies indicate that there are many neurons within the brainstem that are excited by CO_2/H^+. Whether any of these responses represent ventilatory chemoreception is unknown, since no measure of ventilatory output could be included.

Non-electrophysiological approaches have also suggested that many parts of the brainstem participate in the response to hypercapnia. The immunohistochemical localization of the protein product of the c-fos oncogene describes neurons activated by a particular stimulus and has been applied to the study of chemoreception. Rats exposed to 13% to 15% CO_2 show fos-like immunoreactivity within 150 μm of the VLM surface at the rostral and caudal chemosensitive areas and at the IMA as well as more dorsally near the NTS (32). Similar studies have shown fos-like immunoreactivity at the locus ceruleus (33) and in the region ventral and ventromedial to the facial nucleus (34). This approach does not distinguish between receptor neurons and any other neurons activated in the response to hypercapnia.

We have searched for a method to produce a focal region of acidosis in the brainstem that might result in a respiratory system response. For example, we used a microdialysis system to dialyze focal brainstem regions with hypercapnic solutions. This approach did stimulate respiratory output but was limited by nonspecific effects, which were related to the size of the dialysis probe. In a study designed initially for a different reason, we noted that acetazolamide (AZ) applied to the surface of the VLM in anesthetized cats under isocapnic conditions increased phrenic amplitude and decreased medullary tissue pH (35). Surprisingly, we also found that microinjections (10 to 100 nL) of AZ unilaterally at putative VLM chemoreceptor sites also brought about large increases in phrenic activity (35). This observation suggested that we might use AZ injections as a probe to search for central chemoreceptor locations.

To determine the size of the region of tissue acidosis produced by AZ microinjections, we made 1-nL AZ (5×10^{-6} *M*) injections in the medulla of anesthetized rats and measured pH at varying distances from the AZ injection center (36). The measured pH change following AZ injection was normalized to that produced in each animal by a 20 mm Hg increase in end-tidal P_{CO_2}. At the center of the AZ injection, the average tissue pH change was equivalent to that produced by a 36 mm Hg increase in end-tidal P_{CO_2}. The medullary pH change produced by AZ rapidly decreased with radial distance away from the injection center, such that at distances greater than 350 μm, tissue pH was unaffected. We then used this injection volume with AZ concentrations of 5×10^{-6} to 10^{-5} *M* as a probe to search for the locations of central chemoreceptor sites, knowing that the region of tissue acidosis would have a radius of 350 μm or less.

We made such 1-nL AZ injections to produce focal brainstem regions of decreased pH in anesthetized cats and rats and measured the response using the phrenic nerve as an index of the whole respiratory control system response to the focal pH change. Figure 3 shows a typical response. The nondiffusible fluorescent microbeads included with each injection mark the injection site, in this case in the region just dorsal to the caudal chemosensitive area. The prior measured tissue pH changes indicate that the entire region affected by the combination of the subsequent AZ diffusion from the injection site and the AZ effects on tissue pH lies within 350 μm of the injection center. While many neurons could be affected, the region of decreased tissue pH is small enough to allow the use of this approach to probe the brainstem for the location of central chemoreceptor sites.

Figure 3 also shows the time course of the increase in phrenic nerve activity, measured with constant end-tidal P_{CO_2}, which resulted from this injection. The response to increased systemic P_{CO_2} is shown and indicates the maximum response to stimulation of all central chemoreceptors in this cat. Note that at 60 min following the AZ injection, which stimulated only the chemoreceptors within the region of decreased pH produced by the injection, the phrenic nerve activity has increased to a level that is a large fraction of the maximum observed with systemic CO_2. Overall in these experiments, 26 of 57 (46%) injections in 30 cats produced a significant increase in phrenic activity, which returned to baseline within 120 min. For these 26 positive injections, on average, the normalized phrenic activity increased from 35% of maximum to 56% of maximum following the AZ injection. The sites of these positive injections are shown on Fig. 4, and they can be grouped into three regions: first, sites within 800 μm of the VLM surface at the rostral and caudal chemosensitive areas and at the IMA; second, sites near the nucleus tractus solitarius; and third, sites more rostrally located in the vicinity of the locus ceruleus. Similar findings were obtained in rats.

The use of 1-nL AZ injections to produce focal regions of decreased pH results in the stimulation of a sufficient population of chemoreceptors to produce a measurable control system output—i.e., a phrenic activity. This fortunate outcome allows us to know that the AZ injection did stimulate a ventila-

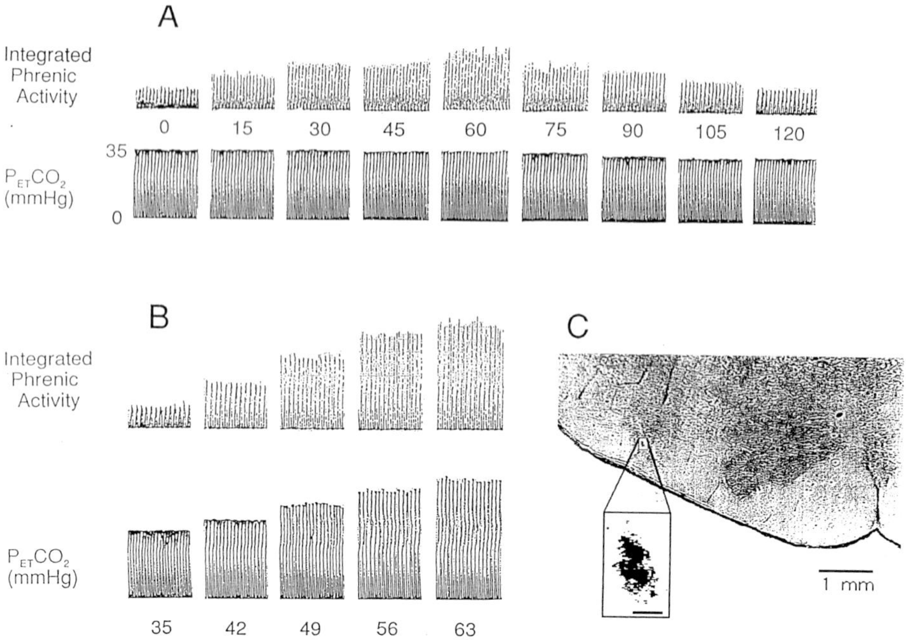

Figure 3 (A) Changes in phrenic activity at times (shown in numbers below tracings) following a 1 nL injection of acetazolamide into the region shown on panel C. Note that end-tidal P_{CO_2} is kept constant. (B) The response of the phrenic to increased end-tidal P_{CO_2} (numbers below bottom tracing reflect mean end-tidal P_{CO_2}). (C) Cross section of medulla showing site of fluorescent beads marking this injection lateral to the inferior olivary nucleus. Inset shows a higher-power image of the fluorescent beads. Bar is 100 μm. Note that the response to this single AZ injection is maximum at 60 min and that this maximum is a significant fraction of that observed with systemic hypercapnia. (From Ref. 37.)

tory chemoreceptor site; it did not merely excite a group of medullary neurons of unknown function. The results indicate that central chemoreception is located at widespread sites within the brainstem. In fact, there may well be other sites. For example, hypercapnia excites neurons in hypothalamic slices (37), and the approach of the AZ microinjection probe can be used to ask if local acidosis within the hypothalamus results in ventilatory stimulation. Such work is in progress.

One may ask if the AZ could possibly have an effect independent from that of the focal tissue acidosis. The presence of many ineffective injection sites and the absence of any response to injections of an inactive AZ analogue are helpful controls for the specificity of the AZ injection. Inhibition of brain carbonic anhydrase using acetazolamide administered systemically (38) and via

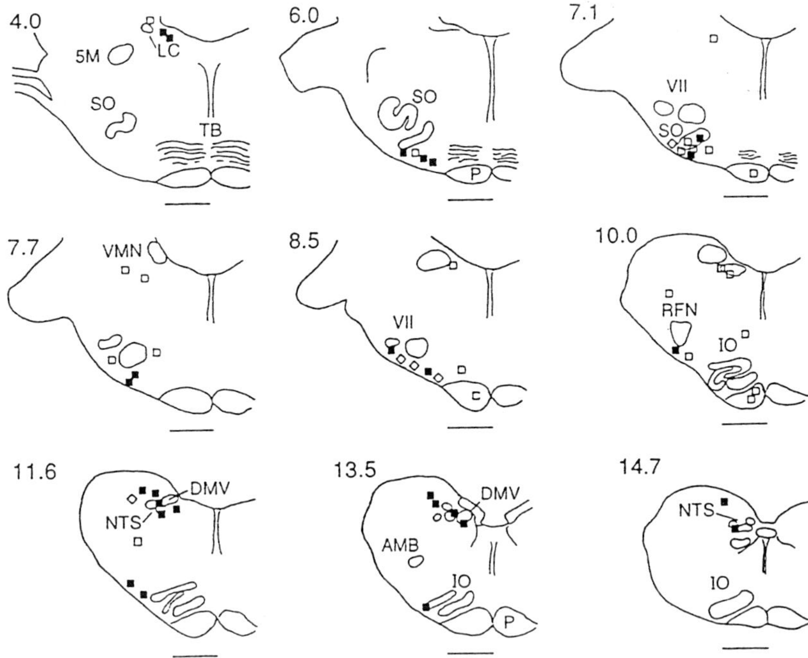

Figure 4 A series of medullary cross sections of the cat showing microinjection sites determined by localization of the fluorescent beads injected with the AZ. Numbers at the top left of each section refer to millimeters caudal to the interaural plane. The solid squares represent AZ injection sites associated with an increased phrenic amplitude; open squares sites at which the AZ injection had no effect; and open diamonds sites of the injection of an AZ analogue which had no effect. Note the distribution of positive AZ injection sites along the VLM surface, in the vicinity of the NTS, and in the vicinity of the LC. Calibration bars are 2 mm. Abbreviations: AMB, nucleus ambiguus; DMV, dorsal motor nucleus of the vagus; LC, locus ceruleus; 5M, motor trigeminal nucleus; NTS, nucleus tractus solitarii; P, pyramidal tract; RFN, retrofacial nucleus; TB, trapezoid body; VMN, medial vestibular nucleus. (From Ref. 36.)

local application on, or microinjection into, the VLM (35,36) increases ventilatory output and decreases medullary tissue pH. In response to increased CO_2, the phrenic output at any level of CO_2 is increased; i.e., the response is shifted upward. This reflects the additive effects of the greater baseline levels of ventilatory output produced by AZ injection and the stimulation of the hypercapnia (35,38). The slope of the steady-state phrenic nerve response to hypercapnic stimulation is unaffected by AZ treatment, although responses to step increases in P_{CO_2} are delayed in reaching the steady state (35,39,40). These data suggest that the CO_2/H^+ transduction process is unaffected by AZ, although the time course is slowed.

The exact function of carbonic anhydrase within the central nervous system is unknown. Carbonic anhydrase is located in cells within the VLM chemosensitive areas (41) and in neurons cultured from the medulla (42), many of which can be excited by hypercapnia (43). Recent in vitro evidence points to an important role for the enzyme in pH mediated events. Accidental dialysis of intracellular carbonic anhydrase via a whole cell patch pipette removes the pH sensitivity of some neurons and addition of the enzyme to the pipette restores it (44). Chesler and Kaila (45) have proposed that carbonic anhydrase is necessary for activity-dependent changes in extra- and intracellular pH in the nervous system. For example, in crayfish, the use of carbonic anhydrase inhibitors affects the GABA-induced increase in extracellular pH and decrease in intracellular pH. These activity-dependent changes in neuronal intra- and extracellular pH presumably modulate the effectiveness of neurotransmitters and perhaps neuromodulators.

Thus, while carbonic anhydrase appears to be required for some single neurons to be excited by hypercapnia in vitro and for activity-dependent pH changes to take place at the synaptic level, in contrast, the steady-state ventilatory response to hypercapnic stimulation of central chemoreceptors measured in vivo does not appear to require the presence of this enzyme. We conclude that the effect of AZ injections on phrenic activity is most likely due to the focal changes in tissue pH.

The location of central chemoreceptors at many sites together with the unexpectedly large response to any single injection when expressed as percent of the maximum response observed with stimulation of all central chemoreceptor sites leads to the following hypothesis. Central chemoreceptors exist at many locations within the brainstem to sense local tissue or cell pH changes. Acidosis at any of these locations can result in a ventilatory response, which would tend to correct or minimize this focal acidosis. Central chemoreceptors, then, would exist as a guardian of medullary pH balance, using the ventilatory system to regulate local pH quickly. We predict that studies in progress will show other brainstem locations with central chemoreceptor function.

III. The Intermediate Area

The IMA or Schläfke's area (3–6), located 3 to 4 mm lateral to the midline and rostral to the rootlets of the twelfth cranial nerve (Fig. 2), is remarkable in that surface cooling using temperatures that block neuronal function but not that of passing axons results in apnea (16,17). The use of a cooling probe at the IMA and, just rostrally, at the level of the caudal aspect of the rostral chemosensitive area also results in apnea and in decreased CO_2 sensitivity (18). Initially the IMA was proposed to be an integrative site not directly involved in chemoreception (46). The use of 1-nL injections of AZ to probe for central chemoreceptive locations (36) showed that chemoreception is present at regions all along the VLM, including the IMA. However, this does not detract from the significance of the effects of IMA cooling on ventilatory output; it suggests that neurons within this medullary region may have many respiratory control functions. The probable anatomical substrates of the IMA include the retrotrapezoid nucleus, the retrofacial nucleus, the subretrofacial region, the nucleus paragigantocellularis lateralis, and perhaps aspects of the rostral VRG, including the Bötzinger and pre-Bötzinger regions.

Many recent studies have examined the possible role of these structures in the effects observed during IMA cooling. Within the retrofacial nucleus, lesions in the cat (47,48) and procaine injections in the rabbit (49) decrease phrenic nerve activity, sometimes to apnea, and CO_2 sensitivity. At sites approximately within the rostral VRG, 1.5 mm deep to the VLM surface, bilateral injections of glutamate receptor antagonists produce apnea in anesthetized cats (50,51) and rats (52,53). Small injections of excitatory amino acids at low concentrations (e.g., 1 to 100 mM for glutamate) within the dorsomedial and ventrolateral VRG also produced phrenic inhibition (54). The nucleus paragigantocellularis lateralis (PGCL) appears to be of particular importance. This structure in cat (55) and rat (56) lies essentially just dorsal to the center of the IMA. Afferent inputs converge there from a variety of sites (57,58), suggesting involvement of PGCL neurons in pain, analgesia, cardiovascular control, respiratory control, exteroceptive sensation, and arousal. A major source of efferents from the PGCL is the locus ceruleus, which itself has widespread inputs to the higher nervous system and is thought to be primarily involved in arousal and vigilance (59).

Studies in the in vitro neonatal rat brainstem preparation in a region that probably lies just deep to the IMA have provided evidence for the site of respiratory rhythmogenesis, at least for this rhythm in this preparation. Neurons located within a small region of the VLM just medial to the caudal aspect of the facial nucleus and close to the surface of the VLM (60) and a region just caudal to the level of the retrofacial nucleus (61) have been proposed for this function. This is a difficult preparation in that it has a clearly demarcated hypoxic central core of tissue, with the viable outside ring of medullary tissue being of the order of 800 μm in thickness (62). So the importance of these findings for respiratory control in an adult with a completely viable brainstem is uncertain. Yet it is striking that both of these proposed regions for respiratory rhythmogenesis lie approximately within those anatomical regions which are substrates for the IMA in adult mammals, and that cooling or lesioning at sites within these regions in adult mammals have dramatic effects on ventilatory control.

Recent work in my laboratory has focused on one of these neuronal groups lying dorsal to the IMA, the retrotrapezoid nucleus (RTN) (Fig. 5). Initial attempts in my laboratory to ask if neuronal cell bodies, not axons of passage, lying near to the VLM surface were involved in chemoreception involved the use of kainic acid, an excitatory amino acid analogue and neurotoxin. Applica-

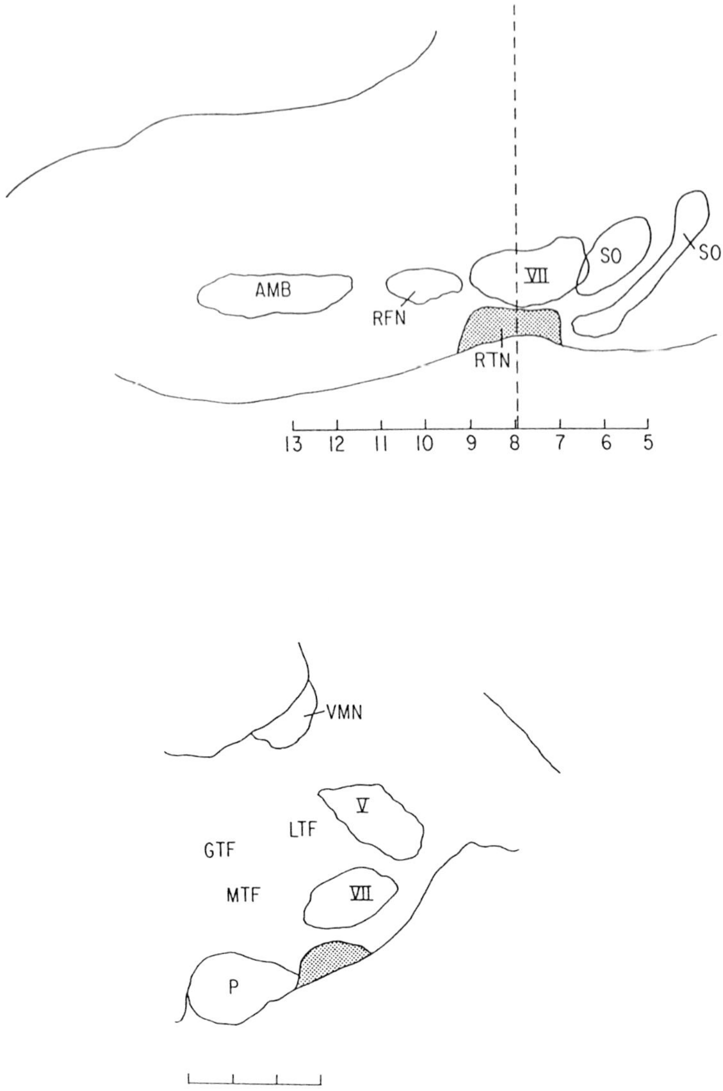

Figure 5 The location of the retrotrapezoid nucleus (RTN). The top panel is a parasaggital section 3.7 mm from the midline; the bottom panel a cross section at the dotted line in the top panel. The stippled area represents the RTN. The rules are in millimeters. Abbreviations not in Fig. 4 are: VII, facial nucleus; SO, superior olivary nucleus; LTF, lateral tegmental field; GTF, gigantocellular tegmental field; MTF, magnocellular tegmental field. (Modified from Fig. 1 in Ref. 69.)

tion of kainic acid to the VLM surface at the level of the rostral aspects of the IMA and caudal aspect of area M (63) and, subsequently, its microinjection within 800 μm of the VLM surface specifically at one anatomical location (64) results in decreased phrenic activity, often to apnea, and an absent response to hypercapnia (Fig. 6). These effects last for many hours. The specific site of greatest effect was just ventral and ventromedial to the facial nucleus, a location that appears to be beneath the topographical border of area M and the rostral aspect of the IMA. Electrolytic as well as chemical lesions produced in decerebrate as well as anesthetized cats have similar effects (65) as they do in the anesthetized rat (82).

Following injection of retrograde tracers into the DRG and VRG, Smith et al. (66) described labeling in this same region, indicating the presence of monosynaptic anatomical connections between DRG and VRG and this location. They named this site lying ventral and ventromedial to the facial nucleus the retrotrapezoid nucleus. Studies have now shown the presence of respiratory-modulated single units in the retrotrapezoid nucleus of cat (67) and rat (68), some of which increase their firing rate with hypercapnia (8). The fact that rather small amounts of destruction produced unilaterally at this location can

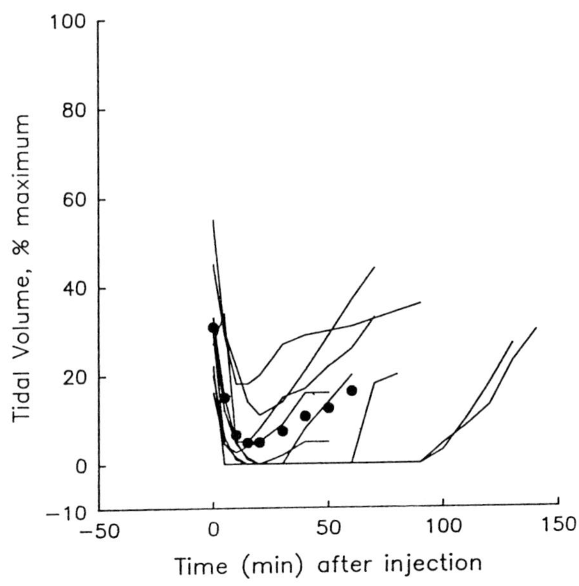

Figure 6 The individual (thin lines) and average (solid symbols) responses of integrated phrenic amplitude ("tidal volume") to single, unilateral 10-nL injections of kainic acid (4.7 m*M*) into the RTN of the anesthetized cat. (Data from Ref. 64.)

result in long-lasting apnea and absent chemosensitivity suggests an important role for these respiratory neurons. The medial aspects of this region lying ventral to the facial nucleus can overlap with portions of the PGCL and the parapyramidal region of the midline raphe. Here we refer to the region of interest as the RTN, realizing that the precise identification of neurons in this area is a task for future experiments.

We have examined the effects of glutamate receptor antagonists injected into the RTN of the anesthetized cat (69) and rat (Nattie E.E., and Li A.; unpublished observations). Unilateral 10-nL injection of kynurenate (a nonspecific glutamate receptor antagonist), CNQX (a non-N-methyl-D-aspartate [NMDA] receptor antagonist), or AP-5 (an NMDA receptor antagonist) into the RTN results in a decrease in the amplitude of the integrated phrenic signal (Fig. 7) and in the slope of the CO_2 response. These effects have a dose dependence, with an effective dose for kynurenate of 100 m*M*; for CNQX and AP-5, of 10 m*M*. We concluded that endogenous glutamate input to RTN neurons is present under eucapnic conditions and that this input is necessary for expression of the normal CO_2 response. The source of this input is unknown. We do not attribute the decrease in CO_2 sensitivity to direct inhibition of the chemoreceptor sites in this region. Instead, we feel that it is due to inhibition of neurons that provide a tonic excitation of the respiratory system and whose function is necessary for the expression of a normal CO_2 response.

Direct injection (10 nL) into the RTN of glutamate itself (70) at the appropriate dose increases the amplitude of the integrated phrenic signal. A 100-m*M* dose produces this excitatory response, a 10-m*M* dose has no effect (Fig. 8), and a 1*M*-dose depresses phrenic amplitude, probably via depolarization block. In response to the 100-m*M* dose, we found, surprisingly and unexpectedly, that the duration of the phrenic stimulation was related to the duration of the glutamate injection (Fig. 8). In anesthetized cats (70) and rats (83) unilateral 10-nL RTN glutamate injection at 100 m*M* stimulates phrenic amplitude transiently if injected over 30 ms to 3 s. When the same volume and concentration is injected over 60 s, the phrenic amplitude is increased for 40 to 50 min. We compare this prolonged phrenic response to the long-term facilitation observed after repeated trains of carotid sinus nerve stimulation (71), suggesting that the 60 s of continuous glutamate injection mimics the release of glutamate in the RTN during the repeated stimulation of the nerve. The mechanism is uncertain. RTN single units do fire longer with 60-s versus 30-ms injections (70), with one unit observed having a sustained increase in firing rate over 40 min. We have suggested and are evaluating the possible role of metabotropic glutamate receptors in this long-term response.

We have observed a similar long-term stimulation of phrenic amplitude following unilateral RTN injection of thyrotropin releasing hormone (TRH) (0.5 m*M*; 10 nL) but this effect does not require long injection durations (72). This preliminary observation suggests an important role for TRH as a neuromodulator in the RTN region.

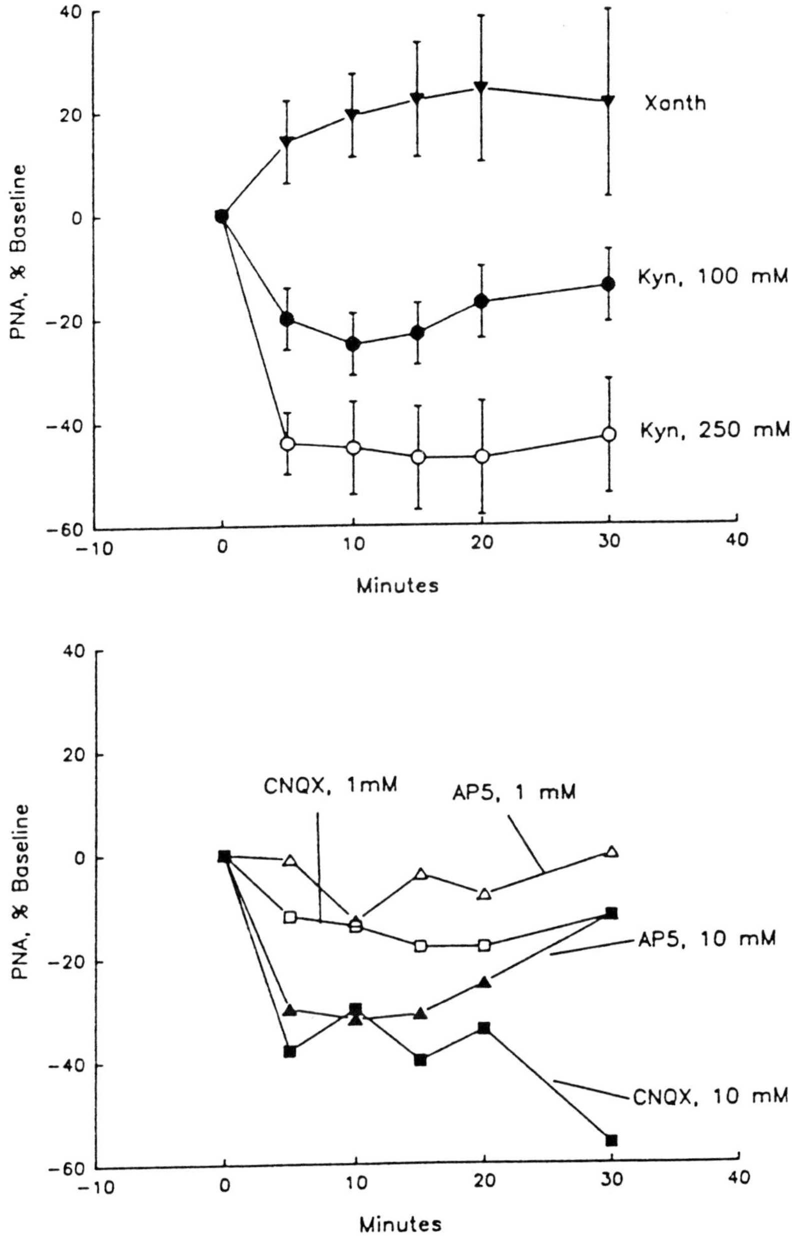

Figure 7 Phrenic amplitude (mean ± SEM; $N = 5$) of anesthetized cats following unilateral RTN injection (10 nL) of xanthurenic acid (an analogue of kynurenate without glutamate receptor antagonist properties) and kynurenate (top panel) and the non-NMDA antagonist CNQX or the NMDA receptor antagonist AP-5 (bottom panel). (From Ref. 69.)

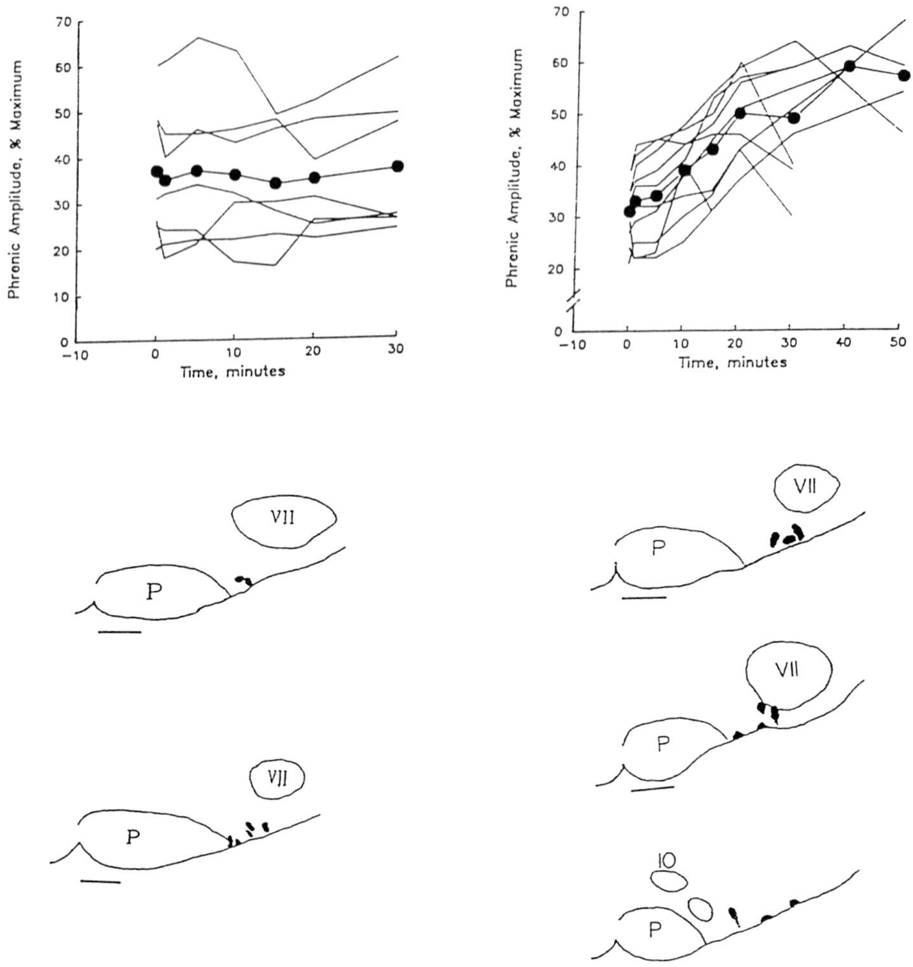

Figure 8 The individual (thin lines) and average (solid symbols) responses of phrenic amplitude in the anesthetized cat to the injection (10 nL) into the RTN of glutamate, 10 m*M* shown at left, 100 m*M* shown at right. At the bottom are panels depicting the location of the injections as marked by the use of fluorescent microbeads. The two sections at left represent the rostral and caudal aspects of the facial nucleus (VII); the sections at right, the rostral and caudal aspects of the facial nucleus and just caudal to the facial nucleus at the rostral level of the inferior olivary nucleus. (From Ref. 70.)

The studies cited above, which used surface cooling and lesions to examine the role of the IMA and the more specific anatomical sites beneath the IMA, were performed in anesthetized or decerebrate animals. Studies have also been conducted in conscious animals. Coagulation of the IMA bilaterally in cats (73, 74) results in a chronic loss of chemosensitivity in these animals as well as resting hypoventilation. These findings have been related to "Ondine's curse" and to the sudden infant death syndrome (75). Forster and colleagues (76) have begun studies with cooling of the IMA in conscious goats. They have found severe depression of ventilation at rest, in response to hypercapnia, and in exercise during IMA cooling (Forster HV; personal communication). In conscious dogs, periods of cluster breathing can be seen after unilateral lesioning of the retrotrapezoid region by kainic acid injections (77). In my laboratory, we have found it difficult to keep rats alive after lesioning the retrotrapezoid nucleus bilaterally, an experience shared by Dormer et al. (77) and Forster (personal communication) in their experiments. These studies suggest an important role for this region of the VLM in the control of breathing, which is demonstrable in the more physiological conditions of a chronic, conscious animal preparation.

Neurons accessible from the IMA are very important in the control of breathing but are also involved in many other physiological functions. The focus of this chapter is on chemoreception and the control of breathing, but it is important to realize that neurons important in other physiological control systems lie close to and are, at some sites, probably intermingled with those whose primary purpose is in the control of breathing. For example, the subretrofacial pressor region (78), also called the C1 region because of the presence of many catecholamine-containing neurons (79), includes a large portion of the PGCL. This pressor region projects to the intermediolateral column of the spinal cord and is thought to be important in the origin and maintenance of cardiovascular sympathetic tone. Conditional pacemaker neurons, proposed as a source of cardiovascular sympathetic tone, have been reported here (80), and the link in which a portion of cardiovascular sympathetic tone is phasically related to phrenic nerve output apparently occurs here (69,81). The neurons of this rostral pressor region lie just caudal and dorsal to the retrotrapezoid nucleus, within the region ventral to the retrofacial nucleus. These neurons are accessible to surface manipulations at the IMA, and such studies often had blood pressure as well as respiratory effects, especially with manipulation of the caudal IMA. Hypercapnia increases cardiovascular sympathetic tone and blood pressure (69,81) as well as ventilation. The role of central chemoreceptors in these changes is unknown.

IV. Conclusions

1. Central chemoreceptors, as located by the use of tiny injections of acetazolamide to produce a small region of focal acidosis, are present in at least three brainstem sites; at or just deep to the VLM surface at the traditional rostral and caudal chemosensitive areas and the IMA, in the vicinity of the nucleus tractus solitarius, and in the region of the locus ceruleus.
2. Stimulation at one chemoreceptor site results in an increase in phrenic amplitude, which is 35% to 56% of that observed with maximum stim-

ulation of all central chemoreceptor sites. This suggests that central chemoreceptors serve to regulate regional brainstem pH.

3. Central chemoreceptors are present within the region just deep to the IMA. However, the IMA, at least the RTN portion of the IMA, appears to have many important functions in the control of breathing.
4. Unilateral RTN lesions result in depression of respiratory output and abolish central chemosensitivity. This suggests that neurons in the RTN are required to maintain normal eupneic ventilation and to allow expression of the CO_2 response.
5. Glutamate receptors of both the NMDA and non-NMDA subtypes are involved in normal RTN function presumably sensing glutamate released by afferents or interneurons.
6. Stimulation of the RTN by glutamate over long time periods (60 s) or TRH (30 ms) can result in prolonged stimulation of phrenic amplitude, raising the possibility that the RTN plays a role in long-term facilitation.

Acknowledgment

The work of the authors is supported by NIH HL 28066.

References

1. Leusen IR. Chemosensitivity of the respiratory center: Influence of CO_2 in the cerebral ventricles on respiration. Am J Physiol 1954; 176:39–44.
2. Loeschcke HH, Koepchen HP, Gertz KH. Über den Einfluss von Wasserstoffionenkonzentration und CO_2 -Druck im Liquor cerebrospinalis auf die Atmung. Pflügers Arch 1958; 266:565–585.
3. Mitchell RA, Loeschcke HH, Massion WH, Severinghaus JW. Respiratory responses mediated through superficial chemosensitive areas on the medulla. J Appl Physiol 1963; 18:523–533.
4. Mitchell RA, Loeschcke HH, Severinghaus JL, et al. Regions of respiratory chemosensitivity on the surface of the medulla. Ann NY Acad Sci 1963; 109:661–681.
5. Loeschcke HH. Central chemosensitivity and the reaction theory. J Physiol (Lond) 1982; 332:1–24.
6. Schlaefke ME, See WR, Loeschcke HH. Ventilatory response to alterations of H^+ ion concentration in small areas of the ventral medullary surface. Respir Physiol 1970; 10:198–212.
7. Lipscomb WT, Boyarski LL. Neurophysiological investigations of medullary chemosensitive areas of respiration. Respir Physiol 1972; 16:362–376.
8. Nattie EE, Fung ML, Li A, St. John WM. Responses of respiratory modulated and tonic units in the retrotrapezoid nucleus to CO_2. Respir Physiol 1993; 94:35–50.
9. Pokorski M. Neurophysiological studies on central chemosensor in medullary areas. Am J Physiol 1976; 230:1288–1295.
10. Euler C von, Soderberg U. Medullary chemosensitive receptors. J Physiol (Lond) 1952; 118:545–554.

11. Fukuda Y, Honda Y. pH-sensitive cells at ventro-lateral surface of rat medulla oblongata. Nature 1975; 256:317–318.
12. Fukuda Y, Honda Y. pH sensitivity of cells located at the ventrolateral surface of the cat medulla oblongata in vitro. Pflügers Arch 1976; 364:243–247.
13. Fukuda Y, Loeschcke HH. Effect of H^+ on spontaneous neuronal activity in the surface layer of the rat medulla oblongata in vitro. Pflügers Arch 1977; 371:125–134.
14. Miles R. Does low pH stimulate central chemoreceptors located near the ventral medullary surface? Brain Res 1983; 271:349–353.
15. Jarolimek W, Misgeld U, Lux HD. Neurons sensitive to pH in slices of the rat ventral medulla oblongata. Pflügers Arch 1990; 416:247–253.
16. Cherniack NS, von Euler C, Homma I, Kao FF. Graded changes in central chemoreceptor input by local temperature changes on the ventral surface of the medulla. J Physiol (Lond) 1979; 287:191–211.
17. Millhorn DE, Eldridge FL, Waldrop TG. Effects of medullary area $I_{(S)}$ cooling on respiratory response to chemoreceptor inputs. Respir Physiol 1982; 49:23–39.
18. Budzinska K, von Euler C, Kao FF, et al. Effects of graded focal cold block in rostral areas of the medulla. Acta Physiol Scand 1985; 124:329–340.
19. Hori T, Roth GI, Yamamoto WS. Respiratory sensitivity of rat brainstem surface to chemical stimuli. J Appl Physiol 1970; 28:721–724.
20. Malcolm JL, Sarelius IH, Sinclair JD. The respiratory role of the ventral surface of the medulla studied in the anesthetized rat. J Physiol (Lond) 1980; 307:503–515.
21. Borison HL, Borison R, McCarthy LE. Brain stem penetration by horseradish peroxidase from the cerebrospinal fluid spaces in the cat. Exp Neurol 1980; 69: 271–289.
22. Keeler JR, Shults CW, Chase TN, Helke CJ. The ventral surface of the medulla in the rat: Pharmacological and autoradiographical localization of GABA-induced cardiovascular effects. Brain Res 1984; 297:217–224.
23. Nattie EE, Mills JW, Ou LC, St. John WM. Kainic acid on the rostral ventrolateral medulla inhibits phrenic output and CO_2 sensitivity. J Appl Physiol 1988; 65:1525–1534.
24. Yamada KA, McAllen RM, Loewy AD. GABA antagonists applied to the ventral surface of the medulla oblongata block the baroreceptor reflex. Brain Res 1984; 297:175–180.
25. Cohen MI. Discharge patterns of brainstem respiratory neurons in relation to carbon dioxide tension. J Neurophysiol 1968; 31:142–165.
26. St. John WM, Wang SC. Response of medullary respiratory neurons to hypercapnia and isocapnic hypoxia. J Appl Physiol 1977; 43:812–821.
27. Arita H, Kogo N, Ichikawa K. Rapid and transient excitation of respiration mediated by central chemoreceptor. J Appl Physiol 1988; 64:1369–1375.
28. Arita H, Kogo N, Ichikawa K. Locations of medullary neurons with non-phasic discharges excited by stimulation of central and/or peripheral chemoreceptors and by activation of nociceptors in cat. Brain Res 1988; 442:1–10.
29. Kogo N, Arita H. In vivo study on medullary H^+-sensitive neurons. J Appl Physiol 1990; 69:1408–1412.
30. Dean JB, Lawing WL, Millhorn DE. CO_2 decreases membrane conductance and depolarizes neurons in the nucleus tractus solitarii. Expl Brain Res 1989; 76:656–661.

31. Dean JB, Bayliss DA, Erickson JT, et al. Depolarization and stimulation of neurons in nucleus tractus solitarii by carbon dioxide does not require chemical synaptic input. Neuroscience 1990; 36:207–216.
32. Sato M, Severinghaus JW, Basbaum A. Medullary CO_2 chemoreceptor neuron identification by c-fos immunocytochemistry. J Appl Physiol 1992; 73:96–100.
33. Haxhiu MA, Erokwu B, Prabhakar NR, et al. Locus coeruleus neurons express c-fos immunoreactivity upon stimulation of central chemosensory system. Soc Neurosci Abstr 1992; 18:828.
34. Teppema LJ, Berkenbosch A, Veening JG, Olievier CN. Hypercapnia induces c-FOS expression in neurons of retrotrapezoid nucleus in cat. IUPS Abstr 1993; #282.21.
35. Coates EL, Li A, Nattie EE. Acetazolamide on the ventral medulla of the cat increases phrenic output and delays the ventilatory response to CO_2. J Physiol (Lond) 1991; 441:433–451.
36. Coates EL, Li A, Nattie EE. Widespread sites of brainstem ventilatory chemoreceptors. J Appl Physiol 1993; 75:5–14.
37. Dillon GH, Waldrop TG. Intracellular and extracellular recordings of posterior hypothalamic neurons in vitro: Response to hypoxia and hypercapnia. Soc Neurosci Abstr 1991; 17:200.
38. Teppema LJ, Rochette F, Demedts M. Effects of acetazolamide on medullary extracellular pH and PCO_2 and on ventilation in peripherally chemodenervated cats. Pflügers Arch 1990; 415:519–525.
39. Hanson MA, Nye PCG, Torrance RW. The location of carbonic anhydrase in relation to the blood-brain barrier at the medullary chemoreceptors of the cat. J Physiol (Lond) 1981; 320:113–125.
40. Toyjimia H, Kuriyama T, Fukuda Y. Delayed ventilatory response to CO_2 after carbonic anhydrase inhibition with acetazolamide administration in the anesthetized rat. Jpn J Physiol 1988; 38:55–65.
41. Ridderstråle Y, Hanson M. Histochemical study of the distribution of carbonic anhydrase in the cat brain. Acta Physiol Scand 1985; 124:557–564.
42. Neubauer J. Carbonic anhydrase and sensory function in the central nervous system. In: Dodgson SJ, ed. The Carbonic Anhydrases: Cellular Physiology and Molecular Genetics. New York: Plenum Press, 1991.
43. Neubauer JA, Gonsalves SF, Chou W, et al. Chemosensitivity of medullary neurons in explant tissue cultures. Neuroscience 1991; 45:701–708.
44. Dean JB. Brainstem whole-cell responses to CO_2 after intracellular dialysis with carbonic anhydrase II. FASEB J 1993; 17:A403.
45. Chesler M, Kaila K. Modulation of pH by neuronal activity. Trends Neurosci 1992; 15:396–402.
46. Schläfke ME. Central chemosensitivity: A respiratory drive. Rev Physiol Biochem Pharm 1981; 90:171–172.
47. St. John WM, Hwang Q, Nattie EE, Zhou D. Functions of the retrofacial nucleus in chemosensitivity and ventilatory neurogenesis. Respir Physiol 1989; 76:159–172.
48. Nattie EE, Blanchford C, Li A. Retrofacial lesions: Effects on CO_2-sensitive phrenic and sympathetic activity. J Appl Physiol 1992; 73:1317–1325.
49. Zhang F, Wu Z, Li Y. Effect of blocking medial area of nucleus retrofacialis on respiratory rhythm. Respir Physiol 1991; 85:73–81.

50. Abrahams TP, Hornby PJ, Walton DP, et al. An excitatory amino acid(s) in the ventrolateral medulla is (are) required for breathing to occur in the anesthetized cat. J Exp Pharmacol Ther 1991; 259:1388–1395.
51. McManigle JE, Panico WH, Pineo S, et al. Evidence for a role for tonic release of an excitatory amino acid at the caudal-subretrofacial area in the control of ventilation. Soc Neurosci Abstr 1991; 17:201.
52. Jung R, Bruce EN, Katona PG. Cardiorespiratory responses to glutaminergic antagonists in the caudal ventrolateral medulla of rats. Brain Res 1991; 564:286–295.
53. Dillon GH, Welsh DE, Waldrop TG. Modulation of respiratory reflexes by an excitatory amino acid mechanism in the ventrolateral medulla. Respir Physiol 1991; 85:55–72.
54. McCrimmon DR, Feldman JL, Speck DF. Respiratory motoneuron activity is altered by injections of picomoles of glutamate into cat brainstem. J Neurosci 1986; 6:2384–2392.
55. Taber E. The cytoarchitecture of the brain stem of the cat: I. Brain stem nuclei. J Comp Neurol 1961; 116:27–70.
56. Andrezik JA, Chan-Palay V. The nucleus paragigantocellularis (PGCL): Definition and afferents. Anat Rec 1977; 187:524–525.
57. Van Bockstaele EJ, Pieribone VA, Aston-Jones G. Diverse afferents converge on the nucleus paragigantocellularis in the rat ventrolateral medulla: Retrograde and anterograde tracing studies. J Comp Neurol 1989; 290:561–584.
58. Van Bockstaele EJ, Akaoka H, Aston-Jones G. Brainstem afferents to the rostral (juxtafacial) nucleus paragigantocellularis: Integration of exteroceptive and interoceptive sensory inputs in the ventral tegmentum. Brain Res 1993; 603:1–18.
59. Aston-Jones G, Ennis VA, Pieribone VA, et al. The brain nucleus locus coeruleus: Restricted afferent control of a broad efferent network. Science 1986; 234:734–737.
60. Onimaru H, Arata A, Homma I. Localization of respiratory rhythm-generating neurons in the medulla of brainstem-spinal cord preparations from newborn rats. Neurosci Lett 1987; 78:151–155.
61. Smith JC, Ellenberger HH, Ballanyi K, et al. Pre-Bötzinger complex: A brainstem region that may generate respiratory rhythm in mammals. Science 1991; 254:726–729.
62. Brockhaus J, Ballanyi K, Smith JC, Richter DW. Microenvironment of respiratory neurons in the in vitro brainstem-spinal cord of neonatal rats. J Physiol (Lond) 1993; 462:421–445.
63. Nattie EE, Mills JW, Ou LC, St. John WM. Kainic acid on the rostral ventrolateral medulla inhibits phrenic output and CO_2 sensitivity. J Appl Physiol 1988; 65:1525–1534.
64. Nattie EE, Li A. Fluorescence location of RVLM kainate microinjections that alter the control of breathing. J Appl Physiol 1990; 68:1157–1166.
65. Nattie EE, Li A, St. John WM. Lesions in retrotrapezoid nucleus decrease ventilatory output in anesthetized or decerebrate cats. J Appl Physiol 1991; 71:1364–1375.
66. Smith JC, Morrison DE, Ellenberger HH, et al. Brainstem projections to the major respiratory neuron populations in the medulla of the cat. J Comp Neurol 1989; 281: 69–96.
67. Connelly CA, Ellenberger HH, Feldman JL. Respiratory activity in retrotrapezoid nucleus in cat. Am J Physiol (Lung Cell Molec Physiol) 1990; 258:L33–L44.

68. Pearce RA, Stornetta RL, Guyenet PG. Retrotrapezoid nucleus in the rat. Neurosci Lett 1989; 101:138–142.
69. Nattie EE, Gdovin M, Li A. Retrotrapezoid nucleus glutamate receptors: Control of CO_2 sensitive phrenic and sympathetic output. J Appl Physiol 1993; 74:2958–2968.
70. Nattie EE, Li A. Retrotrapezoid nucleus glutamate injections: Long-term stimulation of phrenic activity. J Appl Physiol 1994; 76:760–772.
71. Millhorn DE, Eldridge FL, Waldrop TG. Prolonged stimulation of respiration by a new central mechanism. Respir Physiol 1980; 41:87–103.
72. Cream CL, Beeman H, Li A, Nattie EE. Ventilatory response to microinjection of TRH in the rostral ventrolateral medulla in the rat. FASEB J 1993; 7:A459.
73. Schläfke ME. Elimination of central chemosensitivity by coagulation of a bilateral area on the ventral medullary surface in awake cats. Pflügers Arch 1979; 379:231–241.
74. Burghardt F, Schläfke ME. Loss of central chemosensitivity: an animal model to overcome respiratory insufficiency. J Auton Nerv Syst 1986; suppl:105–109.
75. Schläfke ME. Central chemosensitivity. Adv Biosci 1991; 79:27–33.
76. Pan LG, Forster HV, Erickson BK, Lowry T. Effect of transient cooling of the ventral surface of the medulla (VLM) in awake goats. FASEB J 1991; 5:A665.
77. Dormer KJ, Bedford TG. Cardiovascular control by the rostral ventrolateral medulla in the conscious dog. In: Ciriello J, Caverson MM, Polosa E, eds. The Central Neural Organization of Cardiovascular Control, Progress Brain Research 81. Amsterdam: Elsevier, 1981:265–277.
78. Dampney R. The subretrofacial nucleus: Its pivotal role in cardiovascular regulation. News Physiol Sci 1990; 5:63–66.
79. Reis DJ, Ross C, Granata AR, Ruggiero DA. Role of C1 area of rostroventrolateral medulla in cardiovascular control. In: Buckley JP, Ferrario CM, eds. Brain Peptides and Catecholamines in Cardiovascular Regulation. New York: Raven Press, 1987:1–14.
80. Sun M-K, Young BS, Hackett JT, Guyenet PG. Reticulospinal pacemaker neurons of the rat rostral ventrolateral medulla with putative sympathoexcitatory function: An intracellular study "in vitro." Brain Res 1988; 442:229–239.
81. Guyenet P, Darnall RA, Riley TA. Rostral ventrolateral medulla and sympathorespiratory integration in the rat. Am J Physiol (Regul Integrat Comp Physiol 28) 1990; 259:R1063–R1074.
82. Nattie EE, Li A. Retrotrapezoid nucleus lesions decrease phrenic activity and CO_2 sensitivity in rats. Respir Physiol 1994; 97:63–77.
83. Nattie EE, Li A. Rat retrotrapezoid nucleus iono- and metabotropic glutamate receptors and the control of breathing. J Appl Physiol 1995; 78 (in press).

8

Hypoxic Ventilatory Depression and Ventral Medullary Acidosis

JOHN W. SEVERINGHAUS

University of California Medical Center
San Francisco, California

This paper concerns the possible relationship of the acid secretion found to be localized primarily over the CO_2-chemosensitive regions of the ventral medullary surface to the phenomenon of hypoxic ventilatory depression (HVD), or "roll-off."

HVD is a phenomenon in which, between the fifth and twenty-fifth min of acute steady isocapnic hypoxia in a normal human at sea level, about half of the hypoxic ventilatory stimulus disappears (1–5). HVD does not reduce the slope of the CO_2 response curve (2,4,6), and brainstem hypoxia has been shown not to depress central response to a constant stimulus from peripheral chemoreceptors (7). Despite extensive studies of HVD, the mechanism is not clear. Some evidence has supported adenosine (8) or GABA (5) as responsible. It can be prevented by either systemic or ventral medullary topical use of dichloroacetate, which blocks the degradation of pyruvate dehydrogenase and thus increases pyruvate entry into the citric acid cycle in mitochondria (9). HVD is not blocked by naloxone, ruling out endorphins as the major cause (10). HVD persists 5 to 25 min after hypoxia, shorter recovery time being seen with hyperoxia (1).

In studies at altitude ($P_b \simeq 488$ torr), we observed HVD of subjects breathing ambient air, with S_aO_2 equal to 80% to 90% (11). After 25 min at $S_pO_2 \simeq$ 75%, when ventilation was apparently depressed by HVD, subjects had a normal or above-normal HVR (computed as the differential slope) when S_pO_2 was rapidly lowered from 75% to 65%. A second study (12) confirmed that HVD is more significant when assessed as a fall of ventilation at constant S_aO_2 and P_{CO_2} than when measured as a change of HVR slope, tested with and without preoxygenation.

HVD was computed in two ways: (1) The reduction in HVR slope was tested beginning with ambient saturation. $P_{ET}CO_2$ was increased for 6 to 8 min to a predefined control P_{CO_2} (yielding a standard minute ventilation during hyperoxia of 140 $mL \cdot kg^{-1} \cdot min^{-1}$) while keeping S_pO_2 constant at its ambient value; then S_aO_2 was suddenly reduced to about 70% to 75%, and the slope $\Delta\dot{V}_I/\Delta S_aO_2$ was computed. (2) The difference in ventilation was determined between that found chronically while breathing ambient air and that found after 25 min of oxygenation, when S_aO_2 was abruptly lowered to the ambient level, and P_{CO_2} was set at the level observed while breathing ambient air.

Hypoxic depression failed to significantly reduce HVR slope when measured without preoxygenation on any altitude day. However, if all altitude data were combined, HVR slope averaged 18% lower without preoxygenation ($P<0.05$). The difference of ventilation with and without preoxygenation at the same S_aO_2 and P_{CO_2} diminished as S_aO_2 improved with acclimatization. An example of the mean preoxygenated HVR and ambient HVR data of day 2 is shown in Figure 1. The vertical dashed lines indicate the decrement of ventilation attributable to HVD. This supports the proposal of Van Beek et al. (7) that HVD is an alteration of central CO_2 chemosensitivity set point and not a decrease in either peripheral or central chemosensitivity.

The ventral medullary surface becomes acidic during hypoxia (13). The magnitude of this acid is greater over the ventral medulla than over pons, spinal cord, or cortex in goats with implanted electrodes (13). Xu et al. mapped the acidity of the ventral medulla in cats, showing that its magnitude was closely related to the known areas of chemosensitivity (14). The degree of acid shift over the rostral area (15) averaged -0.12 ± 0.09 pH units at a mean S_aO_2 of 48% ± 10%, whereas the surface of adjacent brainstem areas, spinal cord, and cortex pH fell only 0.02 ± 0.01. The slope $\Delta pH/\Delta S_aO_2$, was three times greater on the rostral area than on cortex.

The mechanism of acid generation appears to be lactic acid, resulting from local cellular hypoxia (9). Sato et al. (16) have been able to stain cells in this ventral surface chemoreceptor by use of a c-fos immunocytochemical method after stimulating the cells with 1 h of 13% to 15% CO_2 inhalation in rats. Increased numbers of c-fos stained cells were found in the areas identified as topically chemosensitive by Loeschcke, Mitchell, and their coworkers long

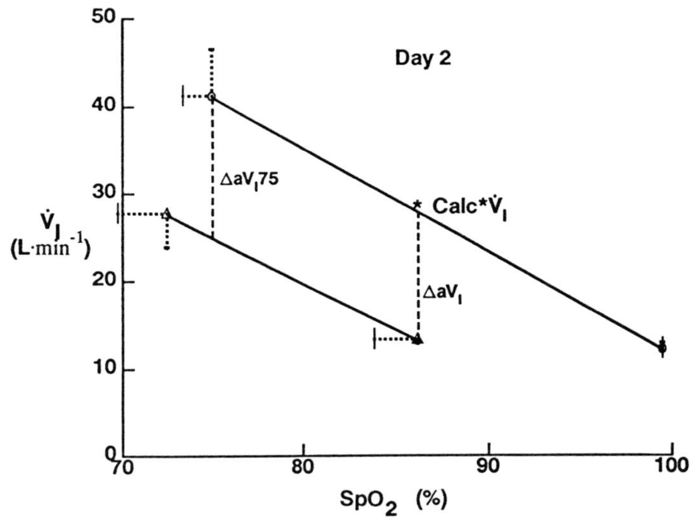

Figure 1 The mean ventilatory response to isocapnic step hypoxia with preoxygenation (circles) and without (triangles) on day 2 at altitude. Calc*$\dot{V}_I$ is the ventilation expected with step reduction to ambient S_pO_2. The vertical dashed lines indicate the estimated decrements of ventilation produced by HVD at ambient S_pO_2 and at 75% S_pO_2. All data obtained at the same $P_{ET}CO_2$ (defined after 20 min of hyperoxia as setting $\dot{V}_I$ to 140 mL·kg^{-1}·min^{-1}).

ago, together with increased staining of the NTS (nucleus tractus solitarius), but surprisingly, no general labeling in other medullary areas thought to be involved in respiratory integration.

Using the altitude data described above, we estimated the shift of P_{CO_2} setpoint due to ambient hypoxia compared with that found after preoxygenation to be 9.2 ± 2.1 torr on the second day at altitude when mean S_aO_2 was 86.2%, 6.1 ± 2.4 torr on day 6 with S_aO_2 = 89.4% ± 0.9%, and only 4.5 ± 1.3 torr on day 12 with S_aO_2 = 91.0% ± 0.6%. This shift of set point P_{CO_2} was calculated at ambient S_pO_2 as the ratio of the ambient-induced ventilatory decrease divided by the slope of the CO_2 response, HCVR.

The topographic identity of these CO_2-stimulated cells, this hypoxic acid generation, and the classic chemosensitive region suggests a connection between hypoxic ventilatory depression, especially when measured as an upward shift of the CO_2 set point, and intracellular acidosis of chemoreceptor cells. Xu et al (14) speculated as follows:

> The data indicating easily detectable acid production from a very few scattered cells suggest either a very high local metabolic rate of the specific chemosensor

> cells or an unusual sensitivity to hypoxia of other non-neuronal cells in the region. This may indicate that the CO_2 and acid detection process involves significant [metabolic] work.
>
> These observations fault the hypothesis that the normal stimulation of medullary respiratory chemoreceptors is ECF acidification. Hypoxia, which fails to stimulate ventilation centrally . . . causes prompt acidification of the ECF specifically over the MVS regions in which topically applied acid stimulates ventilation.
>
> We speculate that these chemoreceptor neurons respond to changes in their transmembrane H^+ gradient, akin to glass pH electrodes. In such a model an electrogenic inward flux of H^+ depolarizes (stimulates) the chemoreceptor, whereas outward H^+ flux hyperpolarizes and quiets (stabilizes). If the source of this local hypoxic acid is the chemoreceptor cells, the intracellular fluid pH presumably falls first and most, and thus depresses neuronal activity. Both increased P_{CO_2} and reduced ECF HCO_3^- acidify the ECF more than intracellular fluid and stimulate ventilation.

If these chemoreceptor cells have an unusually high metabolic rate, it may be possible to detect local heating by infrared surface scanning in response to hypercapnia or hypoxia.

In the context of the upward shift of CO_2 set point produced by hypoxic ventilatory depression, we may then ask how acid production can produce an upward shift which resembles an increase in ECF HCO_3^- concentration. The hypothesis suggested by Xu et al. is consistent with a steady state of outward flux of H^+ (or the equivalent movement of strong cations and anions), tending to stabilize these cells until the P_{CO_2} rises sufficiently to normalize the transcellular gradients.

References

1. Easton PA, Slykerman LJ, Anthonisen NR. Recovery of the ventilatory response to hypoxia in normal adults. J Appl Physiol 1988; 64:521–528.
2. Georgopoulos D, Walker S, Anthonisen NR. Effect of sustained hypoxia on ventilatory response to CO_2 in normal adults. J Appl Physiol 1990; 68:891–896.
3. Kiwull-Schöne H, Kiwull P. Hypoxic modulation of central chemosensitivity. In: Schläfke ME, Köpchen HP, See WR, eds. Central Neurone Environment. Berlin: Springer-Verlag, 1983:88–95.
4. Melton JE, Neubauer JA, Edelman NH. CO_2 sensitivity of cat phrenic neurogram during hypoxic respiratory depression. J Appl Physiol 1988; 65:736–743.
5. Melton JE, Neubauer JA, Edelman NH. GABA antagonism reverses hypoxic respiratory depression in the cat. J Appl Physiol 1990; 69:1296–1301.
6. Maruyama R, Yoshida A, Fukuda Y. Differential sensitivity to hypoxic inhibition of respiratory processes in the anesthetized rat. Jpn J Physiol 1980; 39:857–871.
7. Van Beek JHGM, Berkenbosch A, DeGoede J, Olievier CN. Effects of brain stem hypoxaemia on the regulation of breathing. Respir Physiol 1984; 57:171–188.
8. Georgopoulos D, Holtby SG, Berezanski D. Anthonisen NR. Aminophylline effects on ventilatory response to hypoxia and hyperoxia in normal adults. J Appl Physiol 1989; 67:1150–1156.

9. Neubauer JA, Simone A, Edelman NH. Role of brain lactic acid in hypoxic depression of respiration. J Appl Physiol 1988; 65:1324–1331.
10. Kagawa S, Stafford MJ, Waggener TB, Severinghaus JW. No effect of naloxone on hypoxia-induced ventilatory depression in adults. J Appl Physiol 1982; 52:1030–1034.
11. Sato M, Severinghaus JW, Powell FL, et al. Augmented hypoxic ventilatory response in men at altitude. J Appl Physiol 1992; 73:101–107.
12. Sato M, Severinghaus JW, Bickler P. Time course of augmentation and depression of hypoxic ventilatory responses at altitude. J Appl Physiol 1994;76:313–316.
13. Xu F, Spellman MJ Jr, Sato M, et al. Anomalous hypoxic acidification of medullary ventral surface. J Appl Physiol 1991; 71:2211–2217.
14. Xu F, Sato M, Spellman MJ Jr, et al. Topography of cat medullary ventral surface hypoxic acidification. J Appl Physiol 1992; 73:2631–2637.
15. Mitchell RA, Loeschcke HH, Massion WH, Severinghaus JW. Respiratory responses mediated through superficial chemosensitive areas on the medulla. J Appl Physiol 1963; 18:661–668.
16. Sato M, Severinghaus JW, Basbaum AI. Medullary CO_2 chemoreceptor neuron identification by c-fos immunocytochemistry. J Appl Physiol 1992; 73:96–100.

9

The Ventral Medullary Network and Airway Control

MUSA A. HAXHIU and
NEIL S. CHERNIACK

School of Medicine
Case Western Reserve University
Cleveland, Ohio

ARTHUR D. LOEWY

Washington University School of Medicine
St. Louis, Missouri

I. Introduction

An adequate gas exchange in the lungs and sufficient supply of O_2 to tissues depends on the patency of the airways, as they serve as a conduit for the tidal movement of gases entering and leaving the lung.

The airways, extending from nares to alveoli, have a complex geometry. Their size and resistance to airflow may be influenced by numerous factors, including changes in neural drive to the upper airway dilating muscles, airway vasculature, and airway smooth muscle.

The chemosensitive elements located in the vicinity of the ventral medullary surface play a critical role in the regulation of breathing and blood pressure (1–7). Neurons localized in this chemosensitive region are also involved in the control of airway size and in coordination of airway, ventilatory, and cardiovascular defense responses (8–13). This article summarizes the findings of our studies on the importance of the structures near the ventrolateral medullary surface in the regulation of airway geometry and discusses the neuroanatomical and functional findings of neuronal circuitry involved in the control of parasympathetic outflow to the airways.

II. The Role of Ventrolateral Medulla in Regulation of Nasal Patency

The volume of blood in the mucosal vascular bed largely determines nasal caliber. This blood volume fluctuates synchronously with breathing and is predominantly regulated by the sympathetic nervous system (14–16). The nasal vasculature also has nerve supplies, cholinergic and noncholinergic; but the significance of these innervations in the regulation of nasal resistance is not clear (17,18).

In anesthetized cats, we examined whether structures located near the surface of ventrolateral medulla are involved in the control of nasal patency and in coordination of changes in nasal caliber with changes in respiratory activity. The influences of skeletal muscle activation on the size of nasal passages were eliminated by paralyzing the animals. Changes in the pressure across the nose at a constant flow were used to indicate modulations in nasal caliber (for details, see Refs. 19 and 20).

In the first series of seven experiments in cats, we examined the importance of the ventrolateral medullary surface in the response of nasal resistance to elevated CO_2. When the animal was ventilated with 100% O_2 at end-tidal P_{CO_2}, close to the level found during spontaneous breathing (end-tidal P_{CO_2} = 35 to 40 torr), phasic oscillations in nasal resistance could be observed. Hyperventilation with 100% O_2 to apnea increased nasal resistance and abolished respiratory-related changes in nasal pressure (Fig. 1, panel 1). Ventilating the lungs with 6% CO_2 and 94% O_2 resulted in the reappearance of phrenic nerve activity associated with a concomitant fall in nasal pressure. The nasal pressure, on averge, decreased by 31% (range, 20% to 61%). Prior administration of lidocaine to the intermediate area of the ventral surface of medulla blocked the CO_2-induced nasal pressure changes, indicating that the nasal vasculature response to hypercapnia is mediated through the ventrolateral medullary structures (19).

In the second series of 15 experiments, we assessed the importance of the ventrolateral medulla in the regulation of nasal patency by topical application of *N*-methyl-D-aspartic acid (NMDA), a synthetic excitatory amino acid that acts on soma-dendritic membranes (20). In all animals studied, application of NMDA to the intermediate area of the ventral surface, while the cat was ventilated with 100% O_2 at rate sufficient to produce neural apnea, induced the appearance of phrenic nerve activity and a decrease in nasal pressure. Lower concentrations of NMDA often caused a decrease in nasal pressure without affecting phrenic nerve activity (Fig. 1, panel 2). Individual data for the effects of NMDA applied to intermediate area of the ventrolateral medullary surface (VMS) are presented in Figure 1, panel 3.

The response to NMDA could be diminished or abolished by the application to the ventral medullary surface of the NMDA antagonist 2-amino-5-

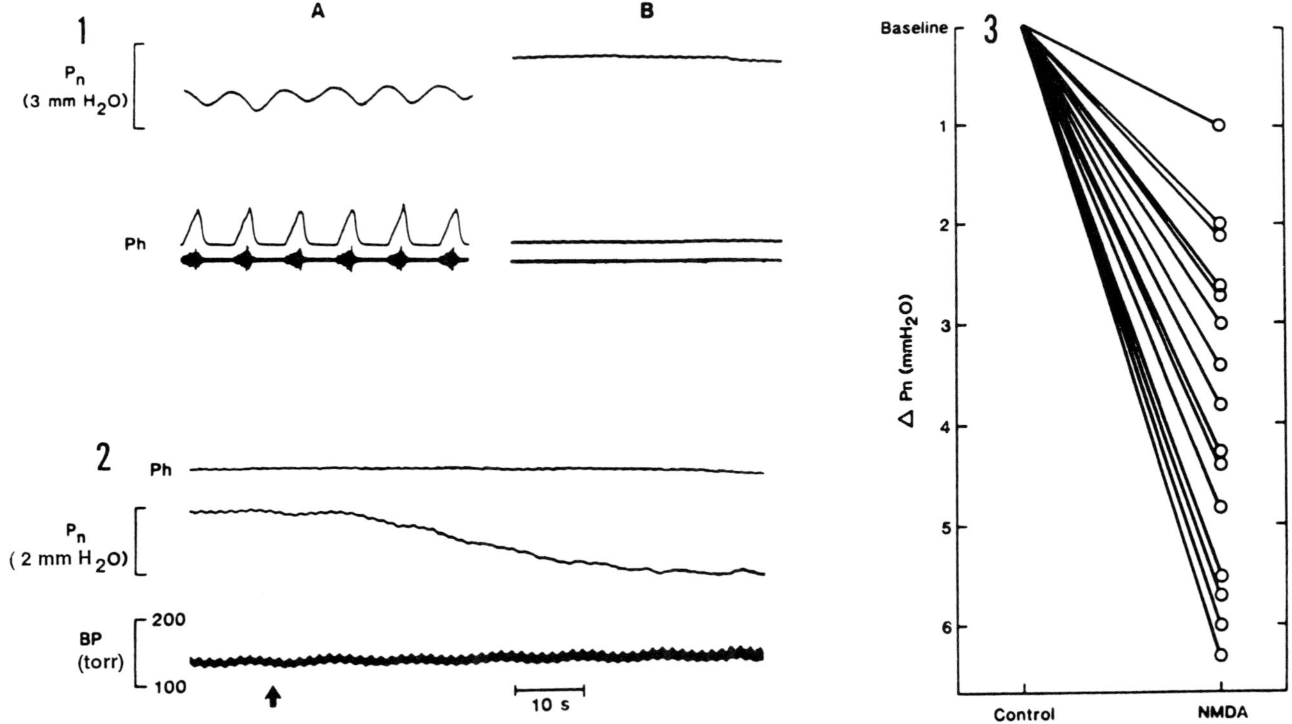

Figure 1 Panel 1: Tracings from a representative cat, paralyzed and artificially ventilated with 100% O_2 at tidal pCO_2 of 35 torr (A) and 22 torr (B), showing nasal pressure (Pn), mass, and moving average of phrenic nerve activity (Ph). Panel 2: Example of the effect of NMDA applied to left intermediate area of ventral medullary surface (VMS) on nasal pressure (Pn), phrenic nerve activity (Ph), and arterial blood pressure (BP) in a paralyzed cat artificially ventilated with 100% O_2 at apneic end-tidal pCO_2. Panel 3: Nasal pressure response (ΔPn) to local application of NMDA in each studied animal. (Modified from Ref. 20.)

phosphonovalerate (2-APV) or the local anesthetic lidocaine. Carotid sinus denervation and posthypothalamic decerebration did not alter the nasal and phrenic nerve responses to NMDA; however, cervical sympathetic denervation decreased but did not abolish these responses, both in intact and in bilaterally adrenalectomized animals. These data suggest that pathways other than the sympathetic participate in the changes in nasal resistance induced by activation of neurons located within the intermediate area.

In a third series of seven experiments, we examined the role of cholinergic agents in the regulation of nasal patency by topical application of nicotine on the intermediate area of the ventral medullary surface. In six of these animals, studies were performed before and after application of lidocaine to the field. Application of nicotine resulted in a 62% fall in nasal resistance (range = 42% to 59%). After lidocaine administration, the same dose of nicotine caused only a 4% decrease (range 0% to 10%) in nasal pressure. However, application of hexamethonium increased nasal pressure by 42% (19). The relevance of these findings will be discussed in relation to changes induced by interventions to the VMS on the other segments of the airways.

III. A Role for the Ventral Surface of the Medulla (VMS) in Regulation of Pharyngeal Patency

In the obstructive sleep apnea syndrome, the pharyngeal passage seems to be the predominant site of airway closure (21–24). During inspiration, this compliant structure is subjected to negative intra-airway pressure. Alterations in its caliber may result from differences in the activity of upper airway dilating and chest wall pumping muscles.

Motoneurons of the hypoglossal nerve that innervate the genioglossus muscle exhibit phasic discharge coincident with breathing (25,26). While the phrenic nerve activity of anesthetized or awake unsedated cats typically rises steadily during inspiration and then declines abruptly, the activity pattern of the hypoglossal nerve is characterized by a rapid rise, peaking in early inspiration, followed by a plateau or downward slope during the remainder of inspiration.

The early peaking pattern of hypoglossal nerve activity, which closely resembles the pattern of inspiratory airflow, may serve to dilate or stiffen the upper airways before they are subjected to the greatest transmural forces produced by inspiratory contraction of the diaphragm. Changes in this pattern or any other change that produces an imbalance in the activities of upper airway dilating and chest wall pumping muscles may induce the occurrence of obstructive apneas during sleep.

The pattern of electrical activity of the hypoglossal nerve or the genioglossus muscle that it innervates is shaped by vagally mediated inhibition, which preferentially depresses the activity of hypoglossal motoneurons in midinspiration. Thus, after vagotomy or when lung inflation is prevented, the pattern of hypoglossal motoneuron discharge changes and reaches maximum near the end of inspiration. In addition to pulmonary stretch receptors, other afferent inputs (i.e., from the central chemosensory system) may also modulate within-breath discharge patterns of upper airway dilating muscles (like the genioglossus) and of cranial nerves (like the hypoglossus).

In cats, we investigated the effects of changes in central respiratory drive produced by activation of central cholinergic and GABAergic neurons located in the vicinity of the ventrolateral medulla (26–29) on the phasic pattern and the activity of the hypoglossal and phrenic nerves after vagotomy, eliminating lung mechanoreceptor influences. Experiments were performed in anesthetized, paralyzed, vagotomized, and artificially ventilated animals. The effects of cholinergic agents (acetylcholine, carbachol, methacholine, physostigmine) were

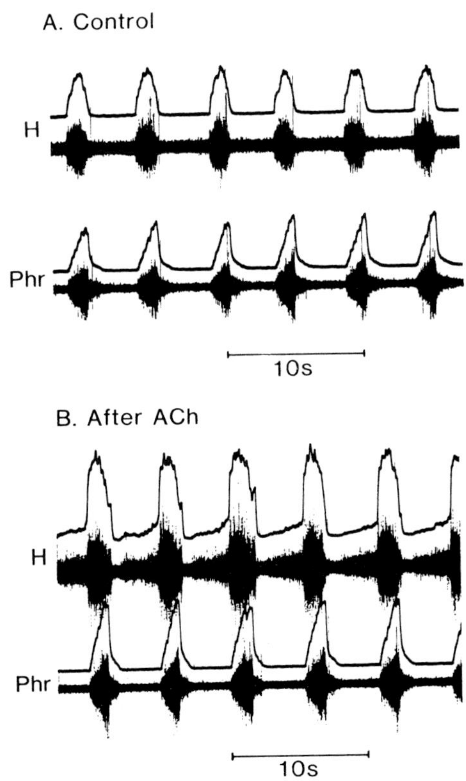

Figure 2 The effect of acetylcholine (ACh) applied to intermediate area of the VMS on hypoglossal (H) and phrenic nerve (Phr) activity in an anesthetized, paralyzed, and vagotomized cat artificially ventilated with 7% CO_2 in O_2. (A) Control period; (B) after application of ACh. (From Ref. 28.)

examined in 19 animals. All the cholinergic agents applied to the intermediate area increased tonic and phasic hypoglossal activity significantly. Atropine applied topically to the same medullary areas blocked the respiratory effects of locally administered acetylcholine. In addition, after the administration of cholinergic agents, hypoglossal nerve discharge began earlier and peaked sooner than did phrenic nerve activity. The maximal amplitude of hypoglossal activity was attained while phrenic nerve discharge was still augmenting, shown in Figure 2. The effects of the drugs were relatively greater on hypoglossal nerve activity than on phrenic discharge. Similar effects were obtained by administration of nicotine.

On the other hand, cooling of the intermediate areas of the VMS and topical application of $GABA_A$ receptor agonists (GABA, muscimol) or benzodiaze-

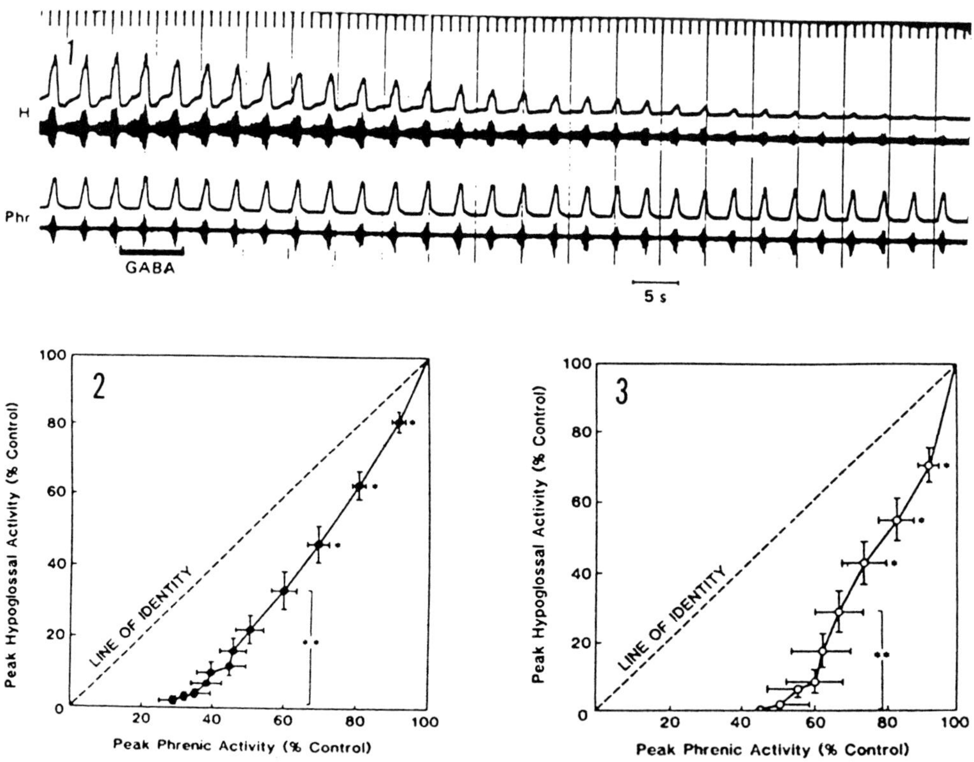

Figure 3 Panel 1: An example of the effect of GABA on Phr and H activities. Panels 2 and 3: relationships between mean changes of phrenic and hypoglossal nerve activities induced by local cooling (20°C) (panel 2) and by topical application of $GABA_A$ receptor agonists (GABA and muscimol) on intermediate area of VMS (panel 3). Data are expressed as percent of control ($\bar{X} \pm$ SEM; *: $P < 0.05$; **: $P < 0.001$). (Modified from Ref. 29.)

pines depressed respiratory activity. The depressive effects were more pronounced on hypoglossal than on phrenic nerve discharge (Fig. 3). Furthermore, the reduction in hypoglossal nerve activity was associated with a change in the pattern of hypoglossal nerve dischrage, which often had its onset later than that of the phrenic nerve and ended near the end of phrenic inspiratory activity.

In addition, the effect of neurons near the ventral medullary surface on respiratory reflex responses to pulmonary C-fiber stimulation was studied. In eight experiments, activation of ventral medullary neurons close to the surface and just lateral to the pyramidal tract by topical application of NMDA significantly increased phrenic activity; NMDA application decreased the duration of apnea induced by administration of phenyldiguinaide or capsaicin into the right atrium (14.6 ± 2.65: 4.1 ± 1.75; $P<0.05$). Likewise, cooling the intermediate area to 20°C markedly prolonged the capsaicin- or phenyldiguinaide-induced hypoglossal nerve apnea compared to the phrenic nerve activity. Cooling the caudal areas had less effect, while cooling the rostral area had the least effect. Cooling of the intermediate area to 15°C had no effect on the increase in laryngeal constrictor activity of recurrent laryngeal nerve produced by capsaicin, which occurred during phrenic apnea (30).

IV. The Ventral Medullary Network and Control of Bronchial Caliber

Airway smooth muscle tone can be modulated centrally by the parasympathetic outflow generated within the medulla oblongata. In part these changes in airway caliber seem to be related to changes in respiratory activity (31). Since structures near the VMS can produce profound effects on respiration, it seems reasonable to believe that they also might be capable of modifying airway smooth muscle tone. In the first series of experiments, we examined the effects of agents applied to VMS on tracheal tone: NMDA and nicotine, which stimulate respiratory activity, and gamma aminobutyric acid (GABA), which depresses respiratory activity. In chloralose-anesthetized, paralyzed, artificially ventilated cats, airway smooth muscle tone was assessed by measuring pressure changes in a rostral bypassed segment of the trachea, while phrenic nerve activity was monitored simultaneously (10).

Phasic oscillations in tracheal segment pressure were recorded at an end-tidal CO_2 pressure close to the level found during spontaneous breathing (PET_{CO2} = 35 to 40 torr). Usually tracheal pressure began to increase near the end of inspiration, reached a peak in early expiration, and started to decrease when postinspiratory activity of the phrenic nerve disappeared (Fig. 4, panel 1). The magnitude of the oscillations varied from cat to cat and was inversely related to phrenic nerve frequency. Hypercapnia increased peak phrenic activity and mean tracheal pressure. Hypocapnia induced by hyperventilation of the cat with O_2 decreased phrenic nerve activity and tracheal tone and abolished the

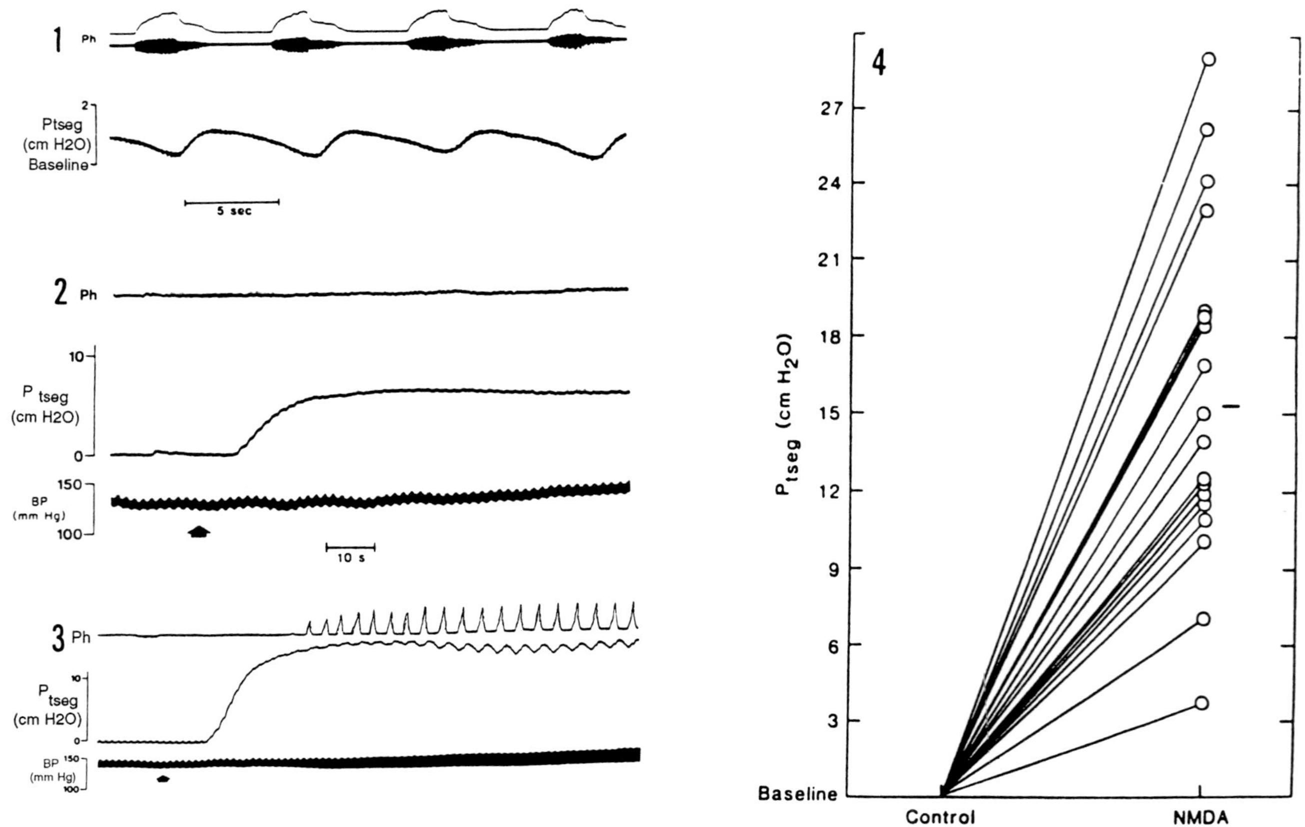

Figure 4 Panel 1: Tracheal pressure (Ptseg) and phrenic nerve activity (Phr) in paralyzed, artificially ventilated cat at end-tidal CO_2 above apneic point. Panels 2 and 3: Responses to local application of lower (panel 2) and higher (panel 3) concentration of NMDA in a cat ventilated with 100% O_2 to neural apnea. Panel 4: individual data of all studied cats. ↑: NMDA application; BP: arterial blood pressure. (Modified from Ref. 33.)

phasic oscillations in the pressure of the tracheal segment; cooling the VMS caused similar changes. Vagotomy reduced both mean tracheal tone and abolished the oscillations in tracheal pressure.

Application of pledgets containing NMDA produced tracheal constriction without the appearance of phasic phrenic nerve activity (Fig. 4, panel 2) (33). Higher concentrations raised tracheal tone and reactivated phrenic nerve activity (Fig. 4, panel 3). Application to the entire VMS either 2-APV, (10^{-6} mol) or the local anesthetic lidocaine (2%) 60 s before the application of pledgets containing NMDA prevented the increase in tracheal tone and the elevation of phrenic activity. Intravenous administration of a cholinergic antagonist, atropine methyl nitrate (0.5 mg/kg), blocked tracheal responses to NMDA but did not affect the increase in phasic phrenic nerve activity. These findings suggest that stimulation of neurons near the surface of the VMS within the intermediate area increases the parasympathetic outflow to the airways.

The responses to nicotine applied topically on the intermediate area in doses of 2×10^{-8} to 2×10^{-7} mol were determined in 11 artificially ventilated cats. Nicotine, like NMDA, produced an increase in the tracheal segment pressure and a reappearance of phrenic nerve activity (Fig. 5). It should be noted that the increase in tracheal smooth muscle tone seemed to precede the increase in phrenic nerve activity.

The tracheal constrictor response to medullary application of nicotine was abolished by local administration of 2% lidocaine to the VMS even when the carotid sinus nerve was intact. Furthermore, in one cat, focal cooling of the intermediate area to 20°C completely abolished the effect of nicotine applied to the caudal area, while rewarming to 37°C restored the increased tracheal tone. In addition, the prior application to the VMS of the nicotinic antagonist hexamethonium bromide, performed in two cats, eliminated the tracheal response to topically administered nicotine.

The effect of GABA on tracheal pressure was also studied in nine animals. When GABA was applied on the VMS of anesthetized cats ventilated with 7% CO_2 in O_2, tracheal segment pressure fell (Fig. 5, panel 1) so that, on the average, it was only 4 ± 1 cm H_2O greater than its baseline level during O_2 ventilation (Fig. 5, panel 2). Phrenic nerve activity decreased at the same time. The decrease in tracheal pressure and phrenic nerve activity were both statistically significant ($P < 0.01$). These effects of GABA on tracheal pressure were blocked by the prior application of bicuculline methiodide in four cats studied. Similarly, administration of benzodiazepines and cooling of the intermediate area of the VMS diminished the response to elevated CO_2 (8,32).

V. Influences of the VMS on Airway Responses to Stimulation of Peripheral Chemoreceptors

The intermediate area of the VMS may integrate central chemoreceptor information with that from peripheral chemoreceptors (34).

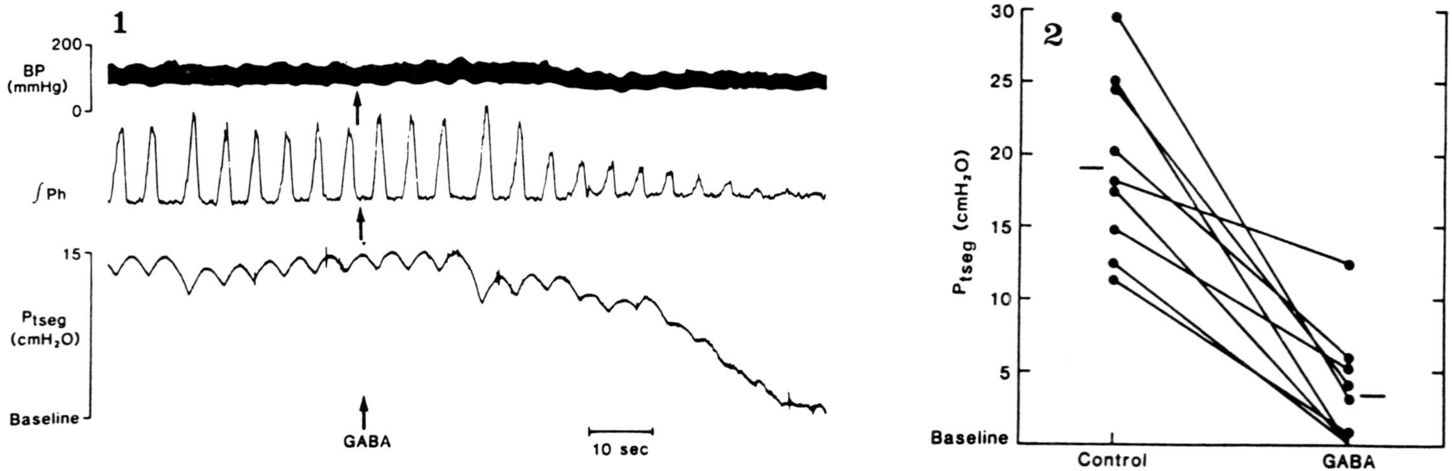

Figure 5 An example (panel 1) and single data (panel 2) of the effects of GABA topically applied to the intermediate area of the VMS on tracheal pressure (Ptseg), phrenic nerve activity (Phr), and arterial blood pressure (BP) in paralyzed and artificially ventilated cats with 7% CO_2 in O_2 (panel 2). ↑: GABA application. (Modified from Ref. 10.)

In anesthetized animals, carotid stimulatiion results in airway constriction which is abolished by vagotomy (35,36). In series of experiments, we determined the influence of the intermediate area of the VMS on respiratory and airway responses to carotid body stimulation by hypoxia (11). Anesthetized, paralyzed, and artificially ventilated cats were examined to minimize the reflex effects of changes in pulmonary mechanoreceptor activity and of PCO_2 on tracheal tone. Respiratory activity was assessed from the moving averaged phrenic electroneurogram, and tracheal constriction (as measured with a saline-filled balloon placed in the rostral cervical trachea) was used as an index of airway responses.

Anesthetized, paralyzed cats were hyperventilated with 100% oxygen to produce phrenic neural apnea. Switching the inspired gas from 100% O_2 to 12% O_2 increased tracheal pressure in 11 of 12 cats but caused phrenic activity to reappear in only 6 of the animals. Ventilation with 6% O_2 increased tracheal tone significantly prior to the appearance of phrenic activity. Neither the tracheal pressure nor the phasic phrenic electroneurogram responded to hypoxia after the carotid sinus nerves were cut. When the intermediate area of the VMS was cooled to 20°C prior to ventilation with the hypoxic gases, both tracheal and phrenic responses were significantly diminished. Both returned to the same levels after rewarming. Cooling of the intermediate area also significantly diminished the increase in tracheal pressure and phrenic nerve activity caused by 6% O_2. Cooling of the intermediate area blunted tracheal and phrenic responses to carotid body stimulation by sodium cyanide (NaCN).

These results support the idea that the intermediate area modulates the parasympathetic reflex responses of the trachea (8,9). The pathways through which the intermediate area modulates the respiratory response to carotid body stimulation is not entirely clear. Schläfke et al. (1979) found that coagulation of the intermediate area reduced the ventilatory response to hypoxia by 50% in spontaneously breathing animals (5). Millhorn et al. (1982) also studied the effect of carotid sinus nerve (CSN) stimulation at various intermediate area temperatures. When baseline phrenic activity was held constant by varying the CO_2 at different intermediate area temperatures, the effect of CSN stimulation on minute phrenic activity was reduced by 50% when the intermediate area was cooled to 20°C. They concluded that the intermediate area is part of a common pathway from both central and peripheral chemoreceptors and that it is involved in the integration of both inputs (34).

From the results of these studies, we cannot state conclusively whether cooling of the intermediate area removed an excitatory effect of the central chemoreceptors on phrenic nerve discharge even when the CO_2 level was below the apneic threshold. There is a possibility that tonically active neurons in the intermediate area project to both the respiratory controller and to the airway-related preganglionic vagal neurons and, in turn, that respiratory related neurons modulate parasympathetic outflow to the airways.

VI. Influence of the VMS in Tracheal Gland Secretion

Many stimuli that affect smooth muscle tone also influence airway submucosal gland function. For example, changes in airway tone induced by hypoxia or reflexly initiated by stimulation of sensory nerve endings are associated with an increase in fluid secretion from tracheal submucosal glands (37–40). Because structures near the VMS can produce profound effects on cholinergic outflow to the airway smooth muscle, it seems reasonable that they also might be capable of modifying tracheal secretions. Therefore, in another study, we evaluated the possibility that structures near the VMS may influence secretion of fluid from tracheal submucosal glands (41,42).

We examined the changes in the number of hillocks of secretion appearing from submucosal glands in an exposed field of tracheal epithelium (1.2 cm^2) coated with tantalum dust before and after interventions on the VMS. Experiments were performed in α-chloralose-anesthetized dogs paralyzed and ventilated with 40% O_2. Stimulation of nicotinergic receptors by application of a pledget containing nicotine in 11 dogs caused a significant elevation in tracheal gland secretion in the subsequent 60 s, compared with a control period in which buffered saline was applied (Fig. 6). Prior application of lidocaine or hexamethonium bromide to the VMS blocked the effect of topically applied nicotine. The central effects of nicotine were diminished by atropine methylnitrate given intravenously. In addition, lidocaine application to the VMS or focal cooling of intermediate areas to between 20 and 15°C significantly decreased the rate of secretion reflexly produced by capsaicin-induced stimulation of pulmonary C-fiber receptors or by mechanical stimulation of the carina and larynx. These findings suggest that the ventral medulla contains cells near its surface that influence tracheal fluid secretion and modulate reflex responses of airway submucosal glands, probably by altering the level of general excitation within central respiratory integrating circuits.

VII. Transneural Labeling Studies of Ventral Medullary Circuitry Involved in Regulation of Parasympathetic Outflow to the Airways

The pathways connecting ventral medullary neurons with the airway-related vagal preganglionic cell bodies have not been described. This is because most contemporary neuroanatomical methods, due to their inherent properties, cannot provide information regarding the location or chemical nature of the central neurons that innervate specific functional classes of autonomic preganglionic neurons. Recently, we have circumvented this problem by using the viral transneuronal labeling method that utilizes a weakened strain of pseudorabies virus (PRV), a neurotropic herpes virus, to produce controlled infections that

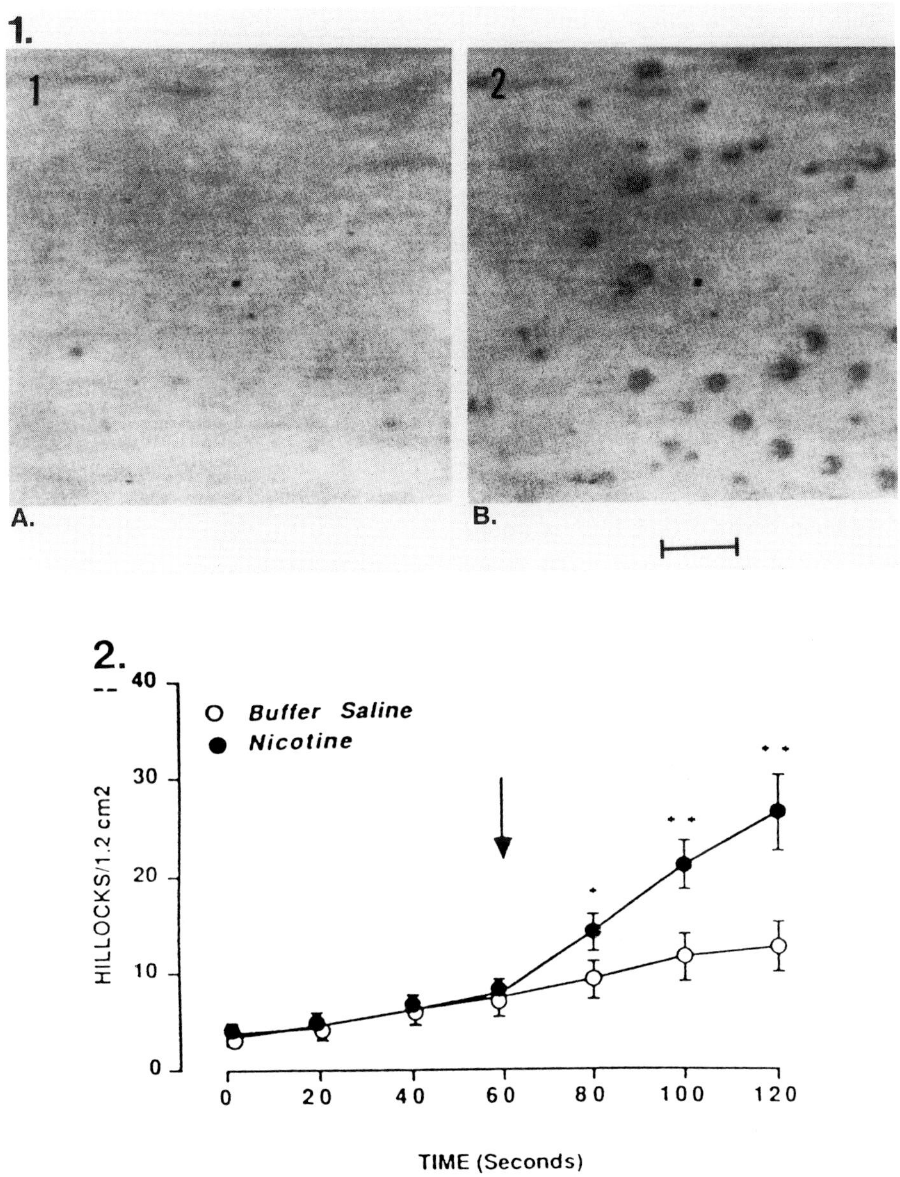

Figure 6 Panel 1: Video images of the effects of topical application of nicotine to intermediate area of VMS on tracheal submucosal gland secretion. (A) tracheal epithelium at the end of 60 s control period; (B) tracheal epithelium 60 s after topical application of nicotine. Horizontal bar, 1 mm. Panel 2: Average results ($\bar{X} \pm$ SEM) of the effects of nicotine on submucosal gland secretion. (Modified from Ref. 42.)

spread in a hierarchical manner within functionally related sets of neurons (43). Since this virus travels preferentially in a retrograde direction (i.e., from axon terminal to the cell body), it is possible to use it as a cell body marker of hierarchical chains of central nervous system (CNS) neurons that regulate autonomic motor systems. The utility of this method rests on experimental evidence indicating that the Bartha strain of PRV has the properties of a specific retrograde transneuronal marker (44–47).

The viral transneuronal labeling method works in the following manner. An injection of PRV (or other types of herpesvirus) is made in a particular end organ; then it is taken up and transported in a retrograde fashion to the cell bodies of the first-order autonomic preganglionic neurons that innervate this tissue. An active infection occurs within these neurons. Next, the infection spreads to the second-order neurons that synapse on these infected cells. In principle, if enough time is allowed for this infectious process to develop more fully, it is possible to label (i.e., infect) third- (or even higher) level neurons that form a specific CNS motor network. However, in practice, only a few reports have verified that this technique can be used to label neurons beyond those of the second-order level, because as the infection proceeds, there is a high probability that a nonspecific central viremia will develop. The infected neurons are easily visualized with standard immunohistochemical techniques, and this technique has been used to localize the CNS cell groups that innervate a variety of different autonomic and somatic motor outflows.

Since most functional investigations of the airway conducting system have used the trachea as a model system, we decided to study the CNS innervation of this organ. Furthermore, the rationale for selecting this particular airway segment stems from the fact that all the organs that form the tracheobronchial tree originate from a common embryonic structure, the respiratory diverticulum, and thus are likely to retain certain basic similar patterns in both their peripheral and central innervation.

Experiments were performed in C_8-level spinal rats in order to eliminate the possibility that PRV injected into the tracheal wall entered the CNS via the sympathetic nervous system. The results were compared with the findings of experiments in which cholera toxin β-subunit (CTb) was used as the retrograde marker to determine the location of the parasympathetic preganglionic neurons that innervate the trachea (for details, see Ref. 43).

The results showed that PRV-labeled first-order parasympathetic preganglionic neurons were observed in the same medullary sites as found in the CTb experiments—namely, in a compact part of the nucleus ambiguus, the area ventral to it, and the rostralmost part of medial nucleus tractus solitarius. In the ventral medulla, an extensive rostral caudal network of PRV-labeled neurons extended from the pontomedullary border to the decussation of the pyramidal tract. Within this network, five superficial individual cytoarchitectonic groupings were recognized: the lateral medullary surface, the lateral parapyramidal nucleus, the parapyramidal nucleus, the raphe pallidus, and the retrotrapezoid nucleus. In addition, two deeper-lying ventral medullary nuclei were labeled as well: the gigantocellular and lateral paragigantocellular reticular nuclei.

Many of these neurons were located literally on or next to the VMS (Fig. 7). This network of PRV-labeled neurons were continuous with a more medial

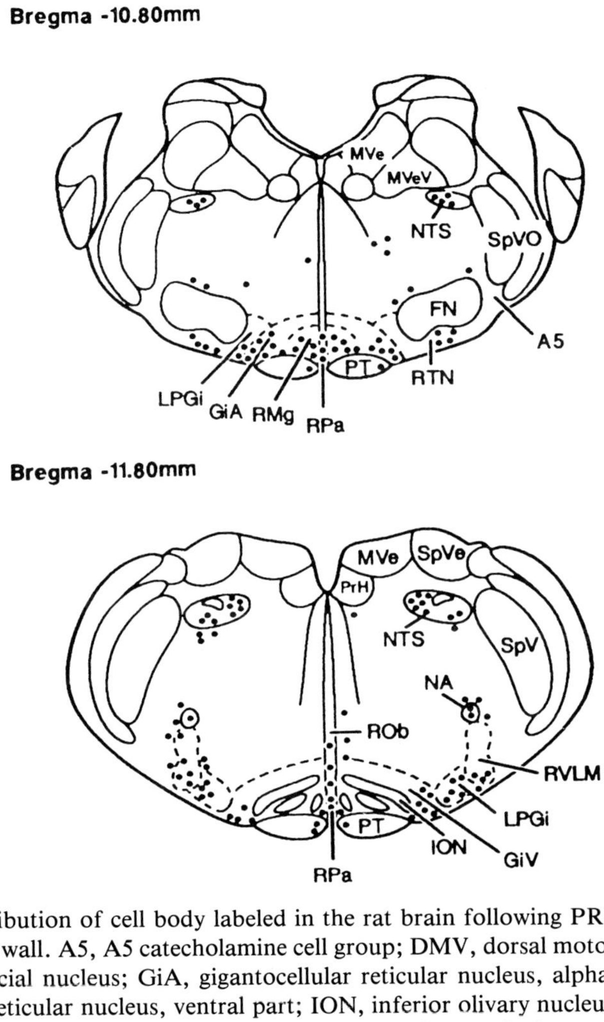

Figure 7 Distribution of cell body labeled in the rat brain following PRV injections into the tracheal wall. A5, A5 catecholamine cell group; DMV, dorsal motor nucleus of vagus F; FN, facial nucleus; GiA, gigantocellular reticular nucleus, alpha part; GiV, gigantocellular reticular nucleus, ventral part; ION, inferior olivary nucleus; LPG, lateral paragigantocellular reticular nucleus; MdD, medullary reticular nucleus; MVe, medial vestibular nucleus; MVeV, medial vestibular nucleus, ventral part; NA, nucleus ambiguus; NTS, nucleus tractus solitarius; PT, pyramidal tract; ROb, raphe obscurus nucleus; RPa, raphe pallidus nucleus; RTN, retrotrapezoid nucleus; RVLM, rostroventrolateral reticular nucleus; SpV, spinal trigeminal nucleus; SpVO, spinal trigeminal nucleus, oral part; XII, hypoglossal nucleus. (Modified from Ref. 44.)

Bregma -13.24mm

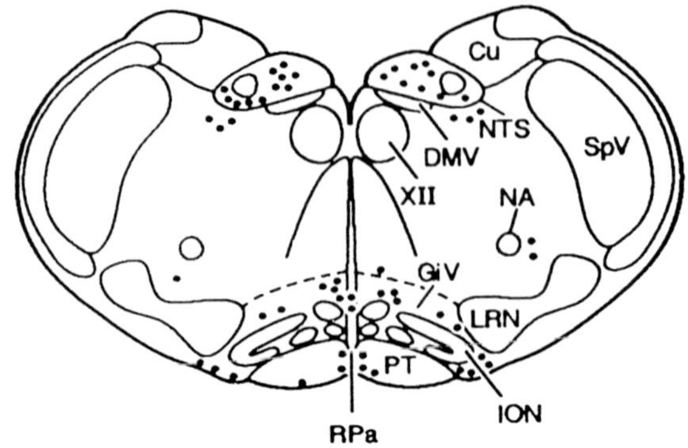

Bregma -14.30mm

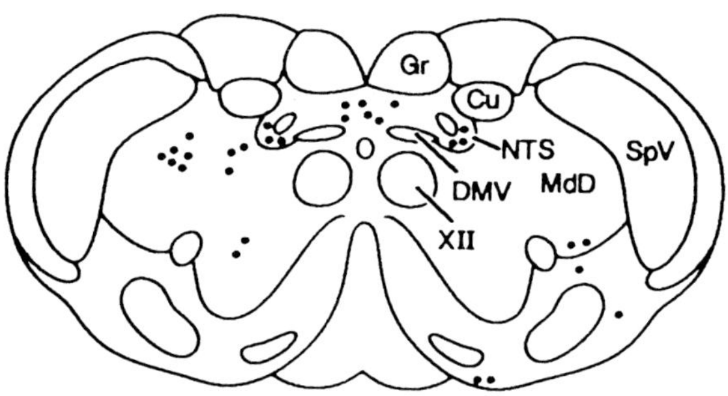

Figure 7 Continued

group of neurons that are concentrated just lateral to the pyramidal tract in the region of the ventral medulla where the hypoglossal nerve exits the brain; the second group of PRV-labeled neurons has also sometimes been termed the lateral parapyramidal nucleus, nucleus interfascicularis hypoglossi, or external arcuate nucleus. The PRV-labeled neurons extend to the midline as a third group of the superficial neurons that outlines the pyramidal tract and has been referred to as the parapyramidal nucleus (equivalent to the arcuate nucleus). The fourth group was the nucleus raphe pallidus. A fifth group of neurons was identified in the area ventral to the facial nucleus, a region that has been termed the retrotrapezoid nucleus. This area contained relatively few PRV-labeled neurons.

The two groups of deeper-lying PRV, labeled ventral medullary neurons, were located in the lateral paragigantocellular and gigantocellular reticular nuclei. Of particular interest was the fact that some of these neurons possessed dendrites that projected to the medullary surface (Fig. 8).

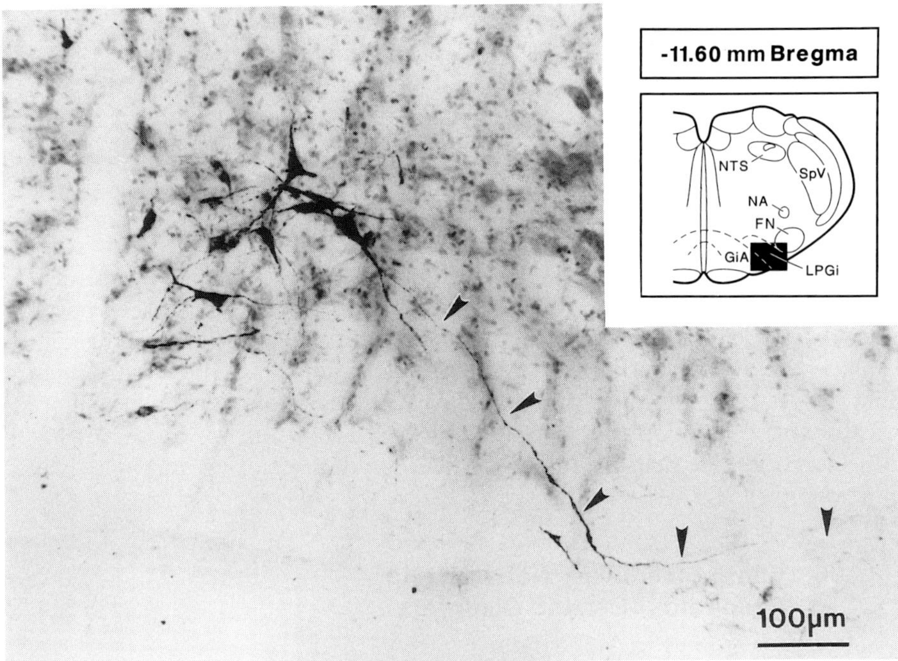

Figure 8 Transneurally labeled neurons within lateral paragigantocellular neurons with dendrites reaching the VMS. (From Ref. 43.)

One of the most striking findings of these studies was the observation that the neurons projecting to the tracheal parasympathetic preganglionic neurons arise predominantly from the same regions of the ventral medulla known to be involved in cardiovascular or respiratory regulatory functions. While these studies do not establish whether the same individual neuron located in any one of these sites is actually involved in the control of three types of functions (i.e., cardiovascular, respiratory, and airway regulation), the coincidence of the neuronal labeling in these sites is impressive and highly suggestive that each of these regions may be involved in multifunctional regulatory processes. In addition, it is possible some of these individual neurons regulate all three motor systems and serve as a central synergistic regulator of vital visceral functions.

The caudal raphe complex—which consists mainly of the raphe pallidus, raphe obscurus, and nearby parapyramidal nucleus—contains neurons involved in cardiovascular, respiratory, and airway control mechanisms. All three regions provide direct projections to the cervical ventral horn and intermediolateral cell column (48,49) as well as to the tracheal parasympathetic preganglionic neurons (43). Considerable evidence has been collected indicating that the raphe obscurus nucleus is involved in the regulation of cardiovascular and respiratory functions (50,51); evidence linking this cell group to the control of airway functions is not yet available. In addition, there is evidence that the raphe pallidus nucleus is also involved in cardiovascular and respiratory regulatory functions (21). To date, no study has examined the role these neurons may play in airway functions.

Since the ventral medullary surface has been implicated as part of a central chemosensory system and since chemical activation of this region alters nasal and pharyngeal patency (19,20,25–29), airway tone and secretions (7, 8), respiration (52), blood pressure, and sympathetic activity (2), of particular significance is our observation that some neurons lying in the paragigantocellular reticular formation send dendrites to the superficial medullary surface (Fig. 8). At this time, it is impossible to provide a detailed functional scenario for this cell group, but like the other cell groups in this area, it is likely to be composed of neurons or sets of neurons that integrate multiple visceral functions.

VIII. Ventral Medullary Network and Respiratory-Related Disorders

The findings of these studies show that neurons located in the vicinity of the ventrolateral surface of the medulla oblongata are involved in the regulation of airway size and secretions as well as respiratory activity of the thoracic and upper airway muscles.

Stimulation of VMS structures induces reciprocal changes in upper airway diameter (the nose, the pharynx, the larynx), and the caliber of tracheobronchial conduit. With an increase in respiratory drive, a concomitant, simultaneous increase in upper airway patency and a decreased resistance to airflow leads to a reduction in the work of breathing. The elevation of tracheobronchial smooth muscle tone with increased respiration may allow dead space to be modulated to maximize gas exchange. Phasic changes in airway tone and nasal pressures, reaching a peak in the early phase of expiration—like the postinspiratory activity of inspiratory pumping muscles—may oppose rapid emptying of the lung, and that assists in the regulation of functional residual capacity, requiring little metabolic energy expenditure.

The ventral medullary network probably functions as a single unit in reflex responses, such as exercise or the response to exposure to irritant gases when parallel in nasal, laryngopharyngeal, tracheobronchial, respiratory, and circulatory changes occur sufficiently to supply tissues with O_2 or to protect the lung and organism from harmful effects of toxic agents.

This region has also been implicated as a vital center in a few neuropathological studies dealing with fatal human syndromes (53). For example, one condition, referred to as central alveolar hypoventilation or "Ondine's curse" syndrome, is a rare form of sleep apnea in which there is a loss of automatic control of ventilation. In one case report, neurons were absent in the external arcuate nucleus (i.e., the region of the ventral surface of the medulla near the exit zone of the hypoglossal rootlets). The CNS lesion(s) producing this syndrome probably affects additional neural systems as well, because even when these patients are treated by phrenic pacing, airway collapse is a potential complication, suggesting that some type of central disorder may also be involved which results in a decrease in excitatory drive to the motor outflow controlling the upper airway dilator muscles—namely, the genioglossus muscle (54). Related neuropathological observations have been made in autopsy material from two children who died from sudden infant death syndrome (55). In this report, a decreased number of neurons was found in the arcuate nucleus (i.e., the superficial neurons lying on the outer portion of the pyramidal tract), a region that is probably functionally similar to the more lateral superficial medullary neurons that has been called the external arcuate nucleus and is probably equivalent to the nucleus interfascicularis hypoglossal or lateral parapyramidal nucleus. These neuropathological findings are of considerable interest in light of the data presented here, which indicate that this region is a major source of inputs to upper airways dilating muscles as well as tracheobronchial smooth muscles. Because pharyngeal dilating muscles are more dependent on the activity of these neurons than the chest wall muscles, it could be expected that lesions in this area will increase the likelihood of obstructive apneas. However, it is important to point out that this superficial medullary region may not be the only CNS region involved in these often fatal clinical entities. Evidence exists in the clinical literature that lesions of other medullary regions can also produce failure of automatic control of ventilation (56,57). In addition, higher CNS regions may be involved as well. In any event, it is clear that increased attention must be directed to the analysis of this region of the brain in neurological disorders affecting respiration and airway patency during sleep. In infants suffering from respiratory viral infections, propagation of virion from the airways into the respiratory-related neurons and neurons controlling autonomic function may explain, in some cases, the association of airway viral infections and sudden infant death syndrome.

Further support of the idea that this superficial region of the ventral medulla oblongata is of fundamental importance to airway function comes from the observation that this region contains a particularly high concentration of high-density 5-HT_{1A} binding sites (58). While attention has been directed mainly to localize the central site of action where 5-HT_{1A} agonists act to lower blood pressure and heart rate (59), the selective 5-HT_{1A} agonist called 8-OH-DPAT [(8-hydroxy-2-(di-n-propylamino)-tetralin] is a drug that also acts to suppress the cough reflex (60). This drug probably operates directly at ventral medullary surface rather than at the level of the nucleus ambiguus, because the latter region lacks this type of 5-HT-receptor (61) and microinjections of this drug into the region of the dorsal vagal nucleus (which surely also involved the NTS as well) did not affect the cough reflex (61). The findings presented and other studies strongly suggest that the VLM plays an important role in regulation of airway tone (62–68).

References

1. Cherniack NS, von Euler C, Homma I, Kao FF. Graded changes in central chemoreceptor input by local temperature changes on the ventral surface of the medulla. J Physiol (Lond) 1979; 287:191–211.
2. Baradziej S, Trzebski A. Specific areas of the ventral medulla controlling sympathetic and respiratory activities and their functional synchronization in the rat. Prog Brain Res 1989; 81:193–204.
3. Mitchell RA, Loeschcke HH, Massion WH, Severinghaus JW. Respiratory responses mediated through superficial chemosensitive area of the medulla. J Appl Physiol 1963; 18:523–533.
4. Loeschcke HH. Central chemosensitivity and the reaction theory. J Physiol 1982; 332:1–24.
5. Schläfke M, See WR, Herker-See A, Loeschcke HH. Respiratory responses to hypoxia and hypercapnia after elimination of central chemosensitivity. Pflügers Arch 1979; 381:241–278.
6. Schläfke ME. Central chemosensitivity: A respiratory drive. Rev Physiol Biochem Pharmacol 1981; 90:171–244.
7. Haxhiu MA, Mitra J, van Lunteren E, et al. Influence of central chemoreceptor afferent inputs on respiratory muscle activity. Am J Physiol 1985; 249:R266–R273.
8. Deal EC Jr, Haxhiu MA, Norcia MP, et al. Influence of structures near the ventral surface of the medulla in reflex airway responses to CO_2. J Appl Physiol 1986; 60:1091–1097.
9. Haxhiu MA, Deal EC, Norcia MP, et al. Influence of ventral surface of medulla on reflex tracheal constriction. J Appl Physiol 1986; 61:791–796.
10. Haxhiu MA, Deal EC, Norcia MP, et al. Medullary effects of nicotine and GABA on tracheal smooth muscle tone. Respir Physiol 1986; 64:351–363.
11. Deal EC Jr, Haxhiu MA, Norcia MP, et al. Cooling the intermediate area of the ventral medullary surface affects tracheal responses to hypoxia. Respir Physiol 1987; 69:335–345.
12. Hilton SM, Marshall RJ, Timms RJ. Ventral medullary relay neurons in the pathway from the defense areas of the cat and their effect on blood pressure. J Physiol (Lond) 1983; 345:149–166.
13. Hilton SM, Redfern WS. A search for brain stem cell groups integrating the defence reaction in the rat. J Physiol 1986; 378:213–228.
14. Anggard A. Capillary and shunt blood flow in the nasal mucosa of the cat. Acta Otolaryngeol 1978; 78:418–422.
15. Eccles R, Wilson H. The autonomic innervation of the nasal blood vessels of the cat. J Physiol (Lond) 1974; 238:549–560.
16. Eccles R. Sympathetic control of nasal erectile tissue. Eur J Respir 1983; 64:150–154.

17. Uddman R, Sundler F. Innervation of the upper airways. In: Widdicombe J, ed. Clinics in Chest Medicine: The Upper Airways. Vol. 7. Philadelphia: Saunders, 1986:201–210.
18. Widdicombe JG. The physiology of the nose. In: Widdicombe J, ed. Clinics in Chest Medicine. The Upper Airways, Vol. 7. Philadelphia: Saunders, 1986:157–170.
19. Strohl KP, Norcia MP, Wolin AD, et al. Nasal and tracheal response to chemical and somatic afferent stimulation in anesthetized cats. J Appl Physiol 1988; 65: 870–877.
20. Haxhiu MA, Strohl KP, Norcia MP, et al. A role for the ventral surface of the medulla in regulation of nasal resistance. Am J Physiol 1987; 253 (Reg. Integr. Comp. Physiol. 22):R494–R500.
21. Remmers J, deGroot EWJ, Sauerland EK, Anch AM. Pathogenesis of upper airway occlusion during sleep. J Appl Physiol 1978; 44:931–938.
22. Cherniack NS. Sleep apnea and its causes. J Clin Invest 1984; 73:1501–1506.
23. Cherniack NS. Respiratory dysrhythmias during sleep. N Engl J Med 1981; 305: 325–330.
24. Strohl KP, Cherniack NS, Gothe B. Physiological basis for therapy of sleep apnea. Am Rev Respir Dis 1986; 134:791–802.
25. Haxhiu MA, van Lunteren E, Mitra J, Cherniack NS. Responses to chemical stimulation of upper airway muscles and diaphragm in awake cats. J Appl Physiol 1984; 56:397–403.
26. Haxhiu MA, Mitra J, van Lunteren E, et al. Hypoglossal and phrenic responses to cholinergic agents applied to ventral medullary surface. Am J Physiol 1984; 247:R939–R944.
27. Haxhiu MA, van Lunteren E, Van de Graaff WB, et al. Action of nicotine on the respiratory activity of the diaphragm and genioglossus muscles and the nerves that innervate them. Respir Physiol 1984; 57:153–169.
28. Haxhiu MA, Cherniack NS, Mitra J, et al. Nonvagal modulation of hypoglossal neural activity. Respiration 1992; 59:65–71.
29. Haxhiu MA, Mitra J, van Lunteren E, et al. Responses of hypoglossal and phrenic nerves to decreased respiratory drive in cats. Respiration 1986; 50:130–138.
30. Mitra J, Prabhakar NR, Haxhiu MA, Cherniack NS. The effects of hypercapnia and cooling the ventral medullary surface on capsaicin induced respiratory reflexes. Respir Physiol 1985; 60:377–382.
31. Mitchell RA, Herbert DA, Baker DG. Inspiratory rhythm in airway smooth muscle tone. J Appl Physiol 1985; 58:911–920.
32. Haxhiu MA, van Lunteren E, Cherniack NS, Deal EC. Benzodiazepines acting on ventral surface of medulla cause airway dilation. Am J Physiol 1989; 257:R810–R815.
33. Haxhiu MA, Deal EC, Norcia MP, et al. Effect of N-methyl-D-aspartate applied to the ventral surface of the medulla on the trachea. J Appl Physiol 1987; 63:1268–1274.
34. Millhorn OE, Eldridge FL, Waldrop TG. Effects of medullary area $I_{(S)}$ cooling on respiratory response to chemoreceptor inputs. Respir Physiol 1982; 49:23–29.
35. Nadel JA, Widdicombe JG. Effect of changes in blood gas tensions and carotid sinus pressure on tracheal volume and total lung resistance to airflow. J Physiol 1962; 163:13–33.

36. Nadel JA, Barnes PJ. Autonomic regulation of the airways. Annu Rev Med 1984; 35:451–467.
37. Davis B, Chinn R, Gold J, et al. Hypoxemia reflexly increases secretion from tracheal submucosal glands in dogs. J Appl Physiol 1982; 3:1416–1419.
38. Davis B, Roberts AM, Coleridge HM, Coleridge JCC. Reflex tracheal gland secretion evoked by stimulation of bronchial C-fibers in dogs. J Appl Physiol 1982; 53:985–991.
39. Coleridge HM, Coleridge JCG. Reflexes evoked from tracheobronchial tree and lungs. In: Handbook of Physiology. Sec. 3: The Respiratory System. Vol. 11. Control of Breathing, Part 1. Baltimore, MD: American Physiological Society, 1986: 395–429.
40. Yu J, Schultz HD, Goodman J, et al. Pulmonary rapidly adapting receptors reflexly increase airway secretion in dogs. J Appl Physiol 1989; 67:782–687.
41. Haxhiu MA, van Lunteren E, Cherniack NS. Central effects of tachykinin peptide on tracheal secretion. Respir Physiol 1991; 86:405–414.
42. Haxhiu MA, van Lunteren E, Cherniack NS. Influence of ventrolateral surface of medulla on tracheal gland secretion. J Appl Physiol 1991; 71:1663–1668.
43. Haxhiu MA, Jansen ASP, Cherniack NS, Loewy AD. CNS innervation of airway-related parasympathetic preganglionic neurons: A transneuronal labeling study using pseudorabies virus. Brain Res 1993; 618:115–134.
44. Strack AM, Sawyer WB, Platt KB, Loewy AD. CNS cell groups regulating the sympathetic outflow to the adrenal gland as revealed by transneuronal cell body labeling with pseudorabies virus. Brain Res 1989; 491:274–296.
45. Strack AM, Sawyer WB, Hughes JH, et al. A general pattern of CNS innervation of the sympathetic outflow demonstrated by transneuronal pseudorabies viral infections. Brain Res 1989; 491:156–162.
46. Strack AM, Loewy AD. Pseudorabies virus: A highly specific transneuronal cell body marker in the sympathetic nervous system. J Neurosci 1990; 10:2139–2147.
47. Rotto-Percelay DM, Wheeler JG, Osorio FA, et al. Transneuronal labeling of spinal interneurons and sympathetic preganglionic neurons after pseudorabies virus injections in the rat medial gastrocnemius muscle. Brain Res 1992; 574:291–306.
48. Loewy AD. Raphe pallidus and raphe obscurus projections to the intermediolateral cell column in the rat. Brain Res 1981; 222:129–133.
49. Loewy AD, McKellar S. Serotonergic projections from the ventral medulla to the intermediolateral cell column in the rat. Brain Res 1981; 211:146–152.
50. Holtman JR, Anastasi NC, Norman WP, Dretchen KL. Effect of electrical and chemical stimulation of the raphe obscurus on phrenic nerve activity in the cat. Brain Research 1986; 362:214–220.
51. Holtman JR, Marion LJ, Speck DF. Origin of serotonin-containing projections to the ventral respiratory group in the rat. Neuroscience 1990; 37:541–552.
52. Connelly CA, Ellenberger HH, Feldman JL. Respiratory activity in retrotrapezoid nucleus in the cat. Am J Physiol 1990; 258:L33–L44.
53. Folgering H, Kuyper F, Kille JF. Primary alveolar hypoventilation (Ondine's curse syndrome) in an infant without external arcuate nucleus: Case report. Bull Eur Physiopathol Resp 1979; 15:659–665.
54. Olson TJ, Woodson GE, Heldt GP. Upper airway function in Ondine's curse. Arch Otolaryngol Head Neck Lung 1992; 118:310–312.

55. Filiano JJ, Kinney HC. Arcuate nucleus hypoplasia in the sudden infant death syndrome. J Neuropathol Exp Neurol 1992; 51:394–404.
56. Kinney HC, Filiano JJ, Herper RM. The neuropathology of the sudden infant syndrome: A review. J Neuropathol Exp Neurol 1992; 51:115–126.
57. Giangaspero F, Schiavina M, Sturani C, et al. Failure of automatic control of ventilation (Ondine's curse) associated with viral encephalitis of the brainstem: A clinicopathologic study of one case. Clin Neuropathol 1988; 7:234–237.
58. Thor KB, Blitz-Siebert A, Helke CJ. Discrete localization of high-density 5-HT_{1A} binding sites in the midline raphe and parapyramidal region of the ventral medulla oblongata of the rat. Neurosci Lett 1990; 108:249–254.
59. Gillis RA, Hill KJ, Kirby JS, et al. Effect of activation of central nervous system serotonin 1A receptors on cardiorespiratory function. J Pharmacol Exp Ther 1985; 248:99–108.
60. Thor KB, Blitz-Siebert A, Helke CJ. Autoradiographic localization of 5HT, binding sites in autonomic areas of the rat dorsomedial medulla oblongata. Synapse 1992; 10:217–227.
61. Kamei J, Mori J, Igarashi H, Kasuya Y. Effects of 8-hydroxy-2-(di-n-propylamino) tetralin, a selective agonist of 5-HT_{1A} receptors, on the cough reflex in rats. Eur J Pharmacol 1991; 203:253–258.
62. McCallister LW, Connelly JC, Kaufman MP. Stimulation of H fields of Forel decrease total lung resistance in dogs. J Appl Physiol 1988; 65:2156–2163.
63. Padrid PhA, Haselton JR, Kaufman MP. Role of caudal ventrolateral medulla in reflex and central control of airway caliber. Physiology 1991; 71:2274–2289.
64. Haselton JR, Solomon IC, Motekaitis AM, Kaufman MP. Bronchomotor vagal preganglionic cell bodies in the dog: An anatomic and functional study. J Appl Physiol 1992; 73:1122–1129.
65. Connelly JC, McCallister LW, Kaufman MD. Stimulation of the caudal ventrolateral medulla decreases total lung resistance in dogs. J Appl Physiol 1987; 63:912–917.
66. Dreshaj IA, Jakupi M, Karahoda N, et al. Influence of ventral surface of medulla on tracheal and cardiovascular responses to stimulation of pulmonary c-fiber receptors. Acta Biol Med Exp 1990; 15:017–025.
67. Haxhiu MA, Deal EC, Trivedi RD, et al. Tracheal and phrenic neural responses to neurotensin applied to the ventral medulla. Am J Physiol 1988; 255:R780–R786.
68. Stein JR, Widdicombe JG. Nervously-mediated changes in tracheal volume on medullary stimulation of dogs. Resp Physiol 1976; 9:348–355.

10

The Role of the Ventral Brainstem in Vestibulosympathetic and Vestibulorespiratory Reflexes

B. J. YATES

Eye and Ear Institute
Pittsburgh, Pennsylvania

P. S. BOLTON

University of Newcastle
Callaghan, New South Wales, Australia

T. GOTO

Omiya Red Cross Hospital
Yono City, Saitama, Japan

I. A. KERMAN

New York University
New York, New York

ALAN D. MILLER

Rockefeller University
New York, New York

I. Introduction

Vestibular receptors located in the inner ear detect linear and angular acceleration imposed on the head and thus provide signals to the central nervous system that indicate head position and the direction and velocity of head movements (1,2). By combining vestibular inputs with signals from neck receptors, the central nervous system can distinguish head and whole-body movements and thus produce appropriate compensatory reflexes (3,4); examples are vestibuloocular reflexes that assist in maintaining fixed gaze during head rotations and vestibulospinal reflexes that compensate for postural perturbations (1,2).

Disturbances in homeostasis can also occur during movement and changes in posture, and there is evidence to suggest that the vestibular system participates in correcting these perturbations. For example, vestibulosympathetic reflexes are presumably involved in compensating for orthostatic hypotension (5). Vestibular effects on respiratory outflow have also been reported (6–9). Vestibular actions on expiratory and inspiratory motoneurons could assist in adjustments of breathing during changes in posture that place adverse mechanical restraints on respiratory muscles. In addition, such vestibulorespiratory reflexes could aid in cardiac return during movements that might lead to orthostatic hypotension (10,11) or be involved in the production of movements and the maintenance of postures that require the participation of respiratory muscles (12–14).

This review will first describe evidence indicating that the vestibular system influences both sympathetic and respiratory outflow; it will then discuss the role of the ventral brainstem in producing these effects.

II. Vestibular Influences on Sympathetic and Respiratory Outflow

Electrical stimulation of the vestibular nerve in the cat produces evoked potentials bilaterally in the phrenic, abdominal, intercostal, and sympathetic nerves (5,9). Examples are shown in Figure 1. Such potentials can typically be elicited by short stimulus trains (two or three shocks) at intensities less than half those that produce current spread to the closest nontarget nerve, the facial nerve. Vestibulorespiratory responses in the phrenic nerve usually have a latency of 7 to 11 ms; the latency for the lumbar abdominal nerves is 10 to 18 ms. Furthermore, the amplitudes of the responses are a sizable fraction of the amplitude of ongoing spontaneous respiratory discharges (9). Thus, the vestibular system appears to make a significant contribution to the excitability of respiratory motoneurons via a fairly direct neural pathway. In contrast, vestibulosympathetic reflexes recorded from the splanchnic nerve have much longer

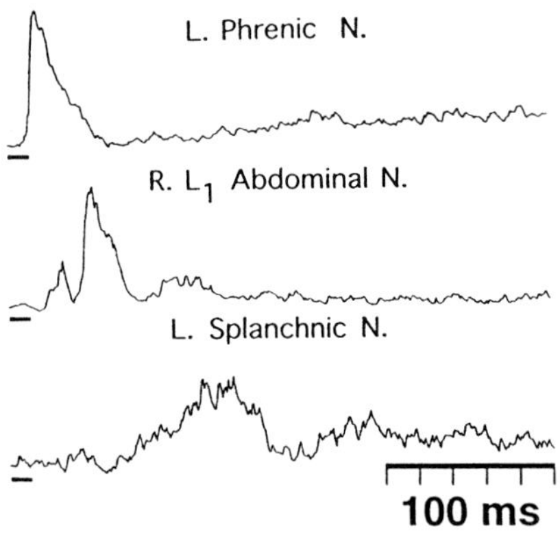

Figure 1 Evoked potentials recorded simultaneously from the left phrenic, right abdominal, and left splanchnic nerves following stimulation of the left vestibular nerve in a decerebrate and paralyzed cat. Fifty-four sweeps were averaged. The potentials were elicited by a train of five shocks (3 ms intershock interval; time of delivery indicated by bar). The amplification is the same for each waveform.

latencies, typically of the order of 60 to 100 ms (5,9). This is not to say, however, that the central processing time for vestibulosympathetic interactions is long, since the recording site was several centimeters distal to the spinal cord in these studies and the conduction time along the unmyelinated and finely myelinated fibers of the splanchnic nerve is lengthy (15).

Effects on respiratory and sympathetic outflow produced by stimulation of the vestibular nerve can be abolished by making electrolytic or chemical lesions affecting the medial vestibular nucleus or the adjacent parts of the inferior nucleus (9,16–18). Thus, it is clear that the autonomic activity produced by vestibular nerve stimulation is the result of activation of vestibular afferents and is *not* due to stimulus spread to nontarget nerves. Furthermore, it is likely that the medial and/or inferior vestibular nuclei are involved in producing the responses.

Evidence that the vestibular system influences sympathetic outflow also comes from studies employing natural stimulation of the vestibular system. Caloric stimulation of the labyrinth in rabbits has been reported to elicit effects on blood pressure that are abolished by lesions placed in the vestibular nuclei (16,17). Furthermore, Doba and Reis (19) showed that transection of the vestibular nerves in paralyzed, chloralose-anesthetized cats significantly impairs compensation for orthostatic hypotension produced by 30° or 60° head-up tilt, suggesting that the vestibular system is involved in correcting posturally related changes in blood pressure. This finding is supported by our recent experiments in decerebrate cats. We delivered sinusoidal head rotations in multiple vertical planes (in animals with upper cervical dorsal root rhizotomies to remove inputs from neck receptors) and found that maximal sympathetic outflow recorded from the splanchnic nerve was elicited by nose-up pitch, the direction most likely to produce orthostatic hypotension (20). Since the position of the trunk was not altered, it seems unlikely that the sympathetic reflexes elicited by head rotation were due to nonvestibular inputs from such sources as baroreceptors or abdominal receptors.

III. The Role of the Ventral Brainstem in Producing Vestibulo-Sympathetic Reflexes

Many of the bulbospinal neurons in the cat relaying descending signals to sympathetic preganglionic neurons are found in two regions located partly or entirely in the ventral brainstem: the subretrofacial rostral ventrolateral medulla (21–23) and the caudal medullary raphe nuclei (24–26). Over two-thirds of the neurons in these areas with slowly conducting ($\leqslant$ 10 m/s for the subretrofacial rostral ventrolateral medulla and $\leqslant$ 5 m/s for the raphe nuclei) projections to the thoracic spinal cord responded to electrical stimulation of the vestibular nerve (27,28). The responses of one such neuron are illustrated in the left panel of Figure 2; Figure 3A shows the location in the ventral brainstem of slowly

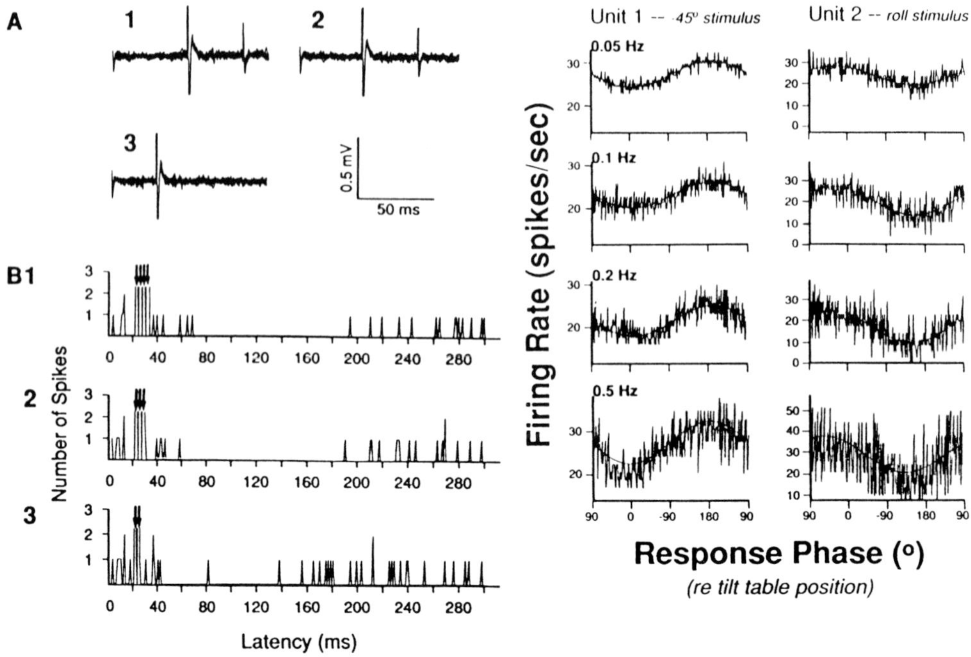

Figure 2 Left panel: Responses of a subretrofacial neuron with a slowly conducting projection to the thoracic spinal cord to stimulation of the vestibular nerve. (A) Collision test showing antidromic driving from the upper thoracic spinal cord. The conduction velocity was 3 m/s. (B) Poststimulus time histograms showing the responses of the same cell to stimulation of the vestibular nerve. Arrows denote the times when stimuli were delivered, and responses to stimulus trains of three different lengths are shown. (From Ref. 27.) Right panel: Histograms showing responses of raphe neurons to sinusoidal natural vestibular stimulation in vertical planes. The traces are averages of binned spike data indicating the response to several frequencies of tilt; the stimulus frequency tested is indicated above each pair of waveforms. Superimposed on each trace is the sinusoid fit to the fundamental of the response. Unit 1 is a raphespinal neuron that responded best to tilt near the plane of the left anterior and right posterior semicircular canals (−45° in our orientation scheme). The stimulus amplitude was 7.5° in all runs. The response phase remained near table position at all frequencies and response gain only increased by a factor of 1.6 as frequency was increased from 0.05 to 0.5 Hz. This behavior is similar to that of otolith afferents. Unit 2 was not tested for a projection to the spinal cord. Like unit 1, it had response phases near table position and almost equivalent response gains across stimulus frequencies and was also classified as mainly having otolith input. (From Ref. 30.)

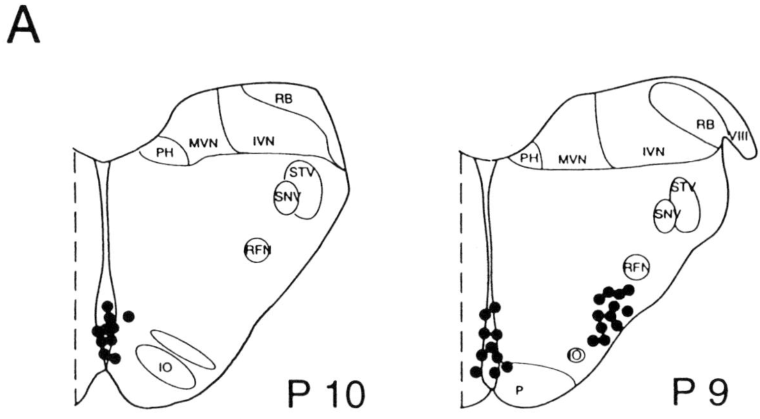

Figure 3 (A) Locations of cells with slowly conducting projections to the thoracic spinal cord that responded to stimulation of the vestibular nerve. The numbers below each drawing indicate the distance (in millimeters) posterior to stereotaxic zero. The laterally located subretrofacial neurons had conduction velocities $\leqslant$ 10 m/s, whereas the medially placed raphe neurons had conduction velocities between 1 and 4 m/s. (Data from Refs. 27 and 28.) (B) Neurons that responded to natural vestibular stimulation. The medially located raphe neurons had projections to the spinal cord, whereas the laterally located subretrofacial neurons were not tested for a spinal projection. Filled circles indicate cells that were classified as receiving inputs predominantly from otolith receptors, and open circles denote cells with other inputs (from semicircular canals or convergent inputs from otoliths and canals). (Data from Refs. 29 and 30.) *Abbreviations*: EC, external cuneate nucleus; IO, inferior olivary nucleus; IVN, inferior vestibular nucleus; LVN, lateral vestibular nucleus; MVN, medial vestibular nucleus; P, pyramid; PH, prepositus hypoglossi; RB, restiform body; RFN, retrofacial nucleus; SA, stria acustica; SNV, spinal trigeminal (Vth) nucleus; STV, spinal trigeminal tract; VII, facial nucleus, VIII, vestibulocochlear nerve.

conducting bulbospinal neurons with vestibular input. Most slowly conducting subretrofacial-spinal neurons (21) and many slowly conducting raphespinal neurons (25,26) make direct connections with sympathetic preganglionic neurons; the presence of vestibular inputs to a large fraction of these cells suggests that the vestibular system is likely to have an important role in cardiovascular control.

Subretrofacial and raphespinal neurons were also tested for responses to natural vestibular stimulation. Sinusoidal rotations in multiple vertical planes as well as horizontal rotations were employed in these experiments; the animals used were baroreceptor-denervated, vagotomized, and had a cervical spinal

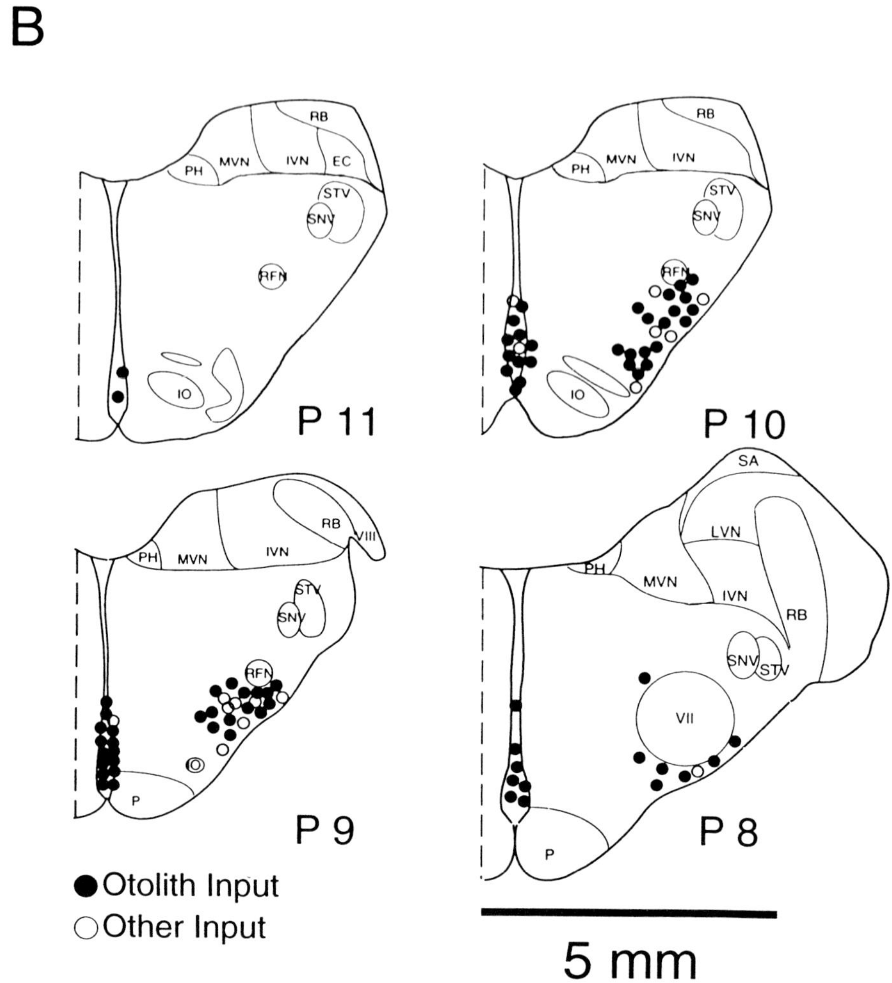

Figure 3 Continued

transection to prevent inputs from tilt-sensitive receptors outside of the labyrinth from influencing the studied neurons. The responses of two raphe neurons to natural vestibular stimulation are shown in the right panel of Figure 2; Figure 3B shows the locations in the ventral brainstem of subretrofacial and raphespinal neurons that responded to natural vestibular stimulation. The majority of neurons had response properties typical of cells with inputs from otolith organs; only a few had responses similar to those of afferents from semicircular canals (29,30). Figure 4A indicates the types of vertical vestibular in-

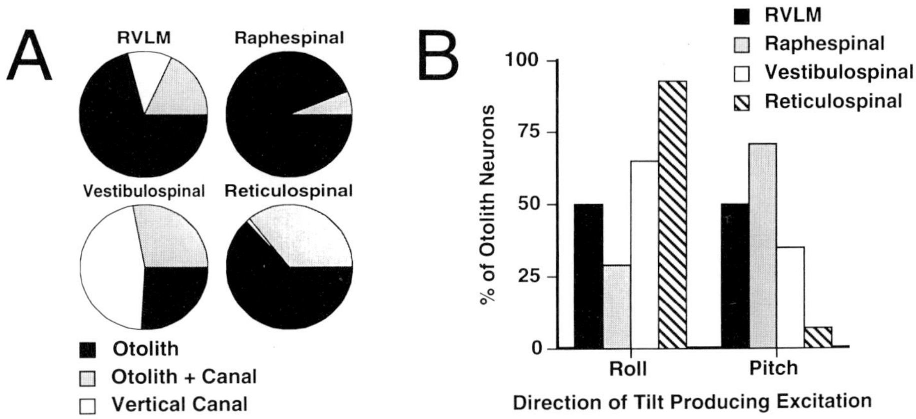

Figure 4 (A) Comparison of the predominant type of vertical vestibular inputs to neurons in the subretrofacial rostral ventrolateral medulla (RVLM), raphespinal neurons, vestibulospinal neurons, and medial pontomedullary reticulospinal neurons. (RVLM data are from Ref. 29, raphespinal data are from Ref. 30, reticulospinal data are from Ref. 31, and vestibulospinal responses represent the pooled data of Refs. 32 and 33.) (B) Direction of tilt producing maximal excitation of subretrofacial RVLM, raphespinal, medial reticulospinal, and vestibulospinal neurons that were classified as receiving predominant otolith input. *Pitch* refers to nose-up or nose-down rotations, whereas *roll* indicates ear-down tilt.

puts (i.e., from the anterior and posterior semicircular canals and otolith organs) to raphespinal neurons and cells in the subretrofacial rostral ventrolateral medulla and, for comparison, shows the types of inputs to vestibulospinal and medial reticulospinal neurons which presumably are involved in motor control. The otolith-dominant pattern of vertical vestibular inputs to the cardioregulatory areas is shared by medial reticulospinal neurons (31); in contrast, many vestibulospinal neurons receive inputs from the vertical semicircular canals (32,33). Furthermore, raphe and subretrofacial neurons (29, 30) differ from both vestibulospinal and reticulospinal neurons (34–36) in lacking a significant input from the horizontal semicircular canals. The otolith-dominant pattern of vestibular inputs to cardioregulatory areas in the ventral brainstem is in keeping with the idea that these regions are involved in compensation for posturally related changes in blood pressure. Because otolith receptors respond to linear accelerations, including gravity, they signal static head position; in contrast, semicircular canals are primarily activated during angular head movements and provide information concerning the direction and velocity of those movements (1,2). Compensation for postural changes requires specific information about static body position, and it would be expected that such compensation calls for selective inputs from otolith organs.

Figure 4B shows the direction of vertical tilt producing maximal excitation of subretrofacial, raphespinal, vestibulospinal, and medial reticulospinal neurons classified as receiving predominant otolith input. A large fraction of neurons in cardioregulatory areas in the ventral brainstem responded best to nose-up or nose-down tilt (pitch). In contrast, vestibulospinal and reticulospinal neurons that mainly had otolith inputs responded best to roll (ear-down tilt). The preference of raphespinal and subretrofacial neurons for activation by pitch rotations is consistent with their presumed involvement in compensation for orthostatic hypotension, which can result from nose-up body rotations in the cat (which has a long longitudinal axis).

The differences in the responses of subretrofacial and raphespinal neurons from those of both reticulospinal and vestibulospinal neurons to natural vestibular stimulation suggest that cardioregulatory areas in the ventral brainstem receive their vestibular input through specific circuits and not through diffuse connections with the vestibular nuclei or reticular formation. This is further evidence to suggest that vestibular inputs to subretrofacial and raphe neurons serve a particular physiological function and are *not* present just to modulate the background firing rate of cells in the ventral brainstem.

IV. The Roles of the Ventral Brainstem in Producing Vestibulorespiratory Reflexes

The neural underpinnings of vestibular influences on respiration have not been as well investigated as the circuitry producing vestibulosympathetic reflexes. However, recently we tested neurons with inspiratory-related activity that were located in the ventrolateral reticular formation near nucleus ambiguus for their responses to electrical stimulation of the vestibular nerve. The locations of the neurons studied are shown in Figure 5. Only 2/21 cells responded at all to stimulation of the vestibular nerve on either side, and the effects on these two neurons were extremely weak, despite the fact that large evoked potentials were simultaneously recorded from the phrenic nerve. Thus, the ventral respiratory group does not appear to play a substantial role in transmitting vestibular signals to inspiratory motoneurons. Furthermore, the dorsal respiratory group (located in the ventrolateral part of nucleus solitarius) also fails to receive appreciable vestibular input (37), suggesting that neurons besides those in the two major groups that are involved in imparting rhythmic activity on inspiratory motoneurons (the dorsal and ventral respiratory groups) must be responsible for vestibular effects on phrenic outflow. Although the descending pathways responsible for vestibulorespiratory reflexes are yet to be determined, they may include vestibulospinal and/or reticulospinal tracts, which are known to make synapses on motoneurons in the vicinity of respiratory motoneurons (1,2).

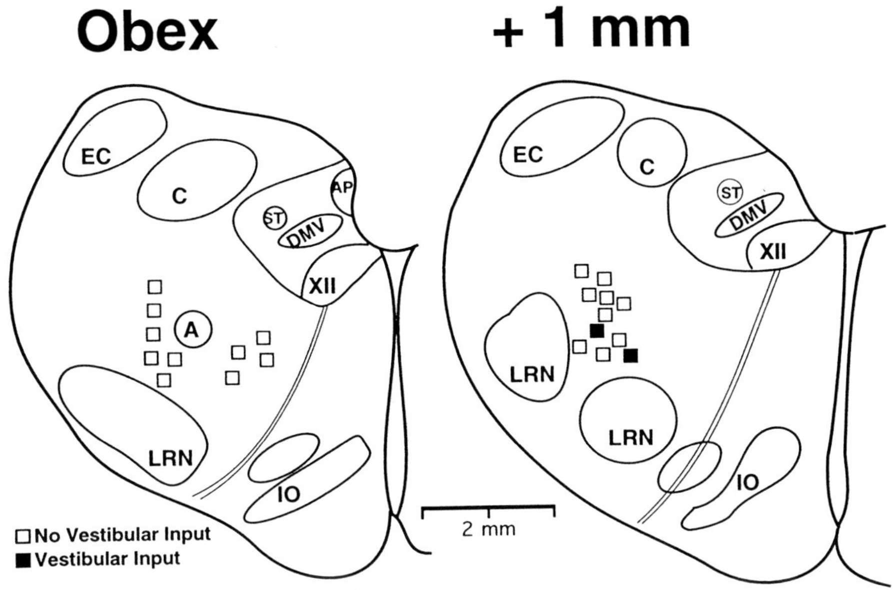

Figure 5 Locations of neurons with inspiratory-related activity that were tested for responses to vestibular nerve stimulation. *Abbreviations*: A, nucleus ambiguus; AP, area postrema; C, cuneate nucleus; DMV, dorsal motor nucleus of the vagus; EC, external cuneate nucleus; IO, inferior olivary nucleus; LRN, lateral reticular nucleus; ST, solitary tract; XII, hypoglossal nucleus.

V. Summary

The vestibular system influences both sympathetic and respiratory outflow. Two regions of the cat ventral brainstem, the subretrofacial rostral ventrolateral medulla and ventral portions of the caudal medullary raphe nuclei, appear to be involved with relaying vestibular signals to sympathetic preganglionic neurons. These areas receive their vestibular inputs mainly from otolith organs, and many ventral brainstem neurons respond better to nose-up or nose-down rotations (pitch) than to ear-down rotations (roll). Thus, the responses of neurons in ventral medullary cardioregulatory areas to natural vestibular stimulation are consistent with their participation in compensation for orthostatic hypotension. In contrast, ventrally located neurons near nucleus ambiguus with inspiratory-related activity receive little vestibular input, suggesting that other neuronal groups must be involved in generating vestibulorespiratory reflexes.

Acknowledgments

We thank Drs. Robert Schor and Victor Wilson for their helpful comments on the manuscript. This work was supported by National Institutes of Health grants DC00693 (to B.J.Y.) and NS20585 (to A.D.M.), and a career scientist award from the Irma T. Hirschl Trust (to B.J.Y.).

References

1. Goldberg JM, Fernández C. The vestibular system. In: Brookhart JM, Mountcastle VB, eds. Handbook of Physiology. Sec. 1: The Nervous System. Vol. III: Sensory Processes, Part 2. Washington, DC: American Physiological Society, 1984:977–1022.
2. Wilson VJ, Melvill Jones G. Mammalian Vestibular Physiology. New York: Plenum Press, 1979.
3. Boyle R, Pompeiano O. Convergence and interaction of neck and macular vestibular inputs on vestibulospinal neurons. J Neurophysiol 1981; 45:852–868.
4. Ezure K, Wilson VJ. Interaction of tonic neck and vestibular reflexes in the forelimb of the decerebrate cat. Exp Brain Res 1984; 54:289–292.
5. Yates BJ. Vestibular influences on the sympathetic nervous system. Brain Res Rev 1992; 17:51–59.
6. Megirian D. Vestibular control of laryngeal and phrenic motoneurons of cat. Arch Ital Biol 1968; 106:333–342.
7. Cobbold AF, Megirian D, Sherrey JH. Vestibular evoked activity in autonomic motor outflows. Arch Ital Biol 1968; 106:113–123.
8. Tang PC, Gernandt BE. Autonomic responses to vestibular stimulation. Exp Neurol 1969; 24:558–578.
9. Yates BJ, Jakus J, Miller AD. Vestibular effects on respiratory outflow in the decerebrate cat. Brain Res 1993; 629:209–217.
10. Lloyd TC. Effect of inspiration on inferior vena caval blood flow in dogs. J Appl Physiol 1983; 55:1701–1708.
11. Moreno AH, Burchell AR, Burke JH. Respiratory regulation of splanchnic and systemic venous return. Am J Physiol 1967; 213:455–465.
12. Davis PR, Troup JDG. Pressure in the trunk cavities when pulling, pushing and lifting. Ergonomics 1964; 7:465–474.
13. De Troyer A. Mechanical role of the abdominal muscles in relation to posture. Resp Physiol 1983; 53:341–353.
14. Grillner S, Nilsson J, Thorstensson A. Intra-abdominal pressure changes during natural movements in man. Acta Physiol Scand 1978; 103:275–283.
15. Cabot JB. Sympathetic preganglionic neurons: cytoarchitecture, ultrastructure and biophysical properties. In: Loewy AD, Spyer KM, eds. Central Regulation of Autonomic Functions. New York: Oxford University Press, 1990:44–67.
16. Spiegel EA, Démétriades TD. Der Einfluss des Vestibular-apparates auf das Gefässsystem. Pflügers Arch ges Physiol 1922; 196:185–188.
17. Spiegel EA. Effect of labyrinthine reflexes on the vegetative nervous system. Arch Otolaryngol 1946; 44:61–72.

18. Uchino Y, Kudo N, Tsuda K, et al. Vestibular inhibition of sympathetic nerve activities. Brain Res 1970; 22:195–206.
19. Doba N, Reis DJ. Role of the cerebellum and vestibular apparatus in regulation of orthostatic reflexes in the cat. Circ Res 1974; 34:9–18.
20. Yates BJ, Miller AD. Properties of sympathetic reflexes elicited by natural vestibular stimulation: implications for cardiovascular control. J Neurophysiol 1994; 71: 2087–2092.
21. Barman SM, Gebber GL. Axonal projection patterns of ventrolateral medullospinal sympathoexcitatory neurons. J Neurophysiol 1985; 53:1551–1566.
22. Dampney RAL. The subretrofacial nucleus: Its pivital role in cardiovascular regulation. News Physiol Sci 1990; 5:63–67.
23. Dampney RAL, Goodchild AK, McAllen RM. Vasomotor control by subretrofacial neurons in the rostral ventrolateral medulla. Can J Physiol Pharmacol 1987; 65: 1572–1579.
24. Morrison SF, Gebber GL. Classification of raphe neurons with cardiac-related activity. Am J Physiol 1982; 243:R49–R59.
25. Morrison SF, Gebber GL. Raphe neurons with sympathetic-related activity: Baroreceptor responses and spinal connections. Am J Physiol 1984; 246:R338–R348.
26. Morrison SF, Gebber GL. Axonal branching patterns and funicular trajectories of raphespinal sympathoinhibitory neurons. J Neurophysiol 1985; 53:759–772.
27. Yates BJ, Yamagata Y, Bolton PS. The ventrolateral medulla of the cat mediates vestibulosympathetic reflexes. Brain Res 1991; 552:265–272.
28. Yates BJ, Goto T, Bolton PS. Responses of neurons in the caudal medullary raphe nuclei of the cat to stimulation of the vestibular nerve. Exp Brain Res 1992; 89: 323–332.
29. Yates BJ, Goto T, Bolton PS. Responses of neurons in the rostral ventrolateral medulla of the cat to natural vestibular stimulation. Brain Res 1993; 601:255–264.
30. Yates BJ, Goto T, Kerman I, et al. Responses of caudal medullary raphe neurons to natural vestibular stimulation. J Neurophysiol 1993; 70:938–946.
31. Bolton PS, Goto T, Schor RH, et al. Response of pontomedullary reticulospinal neurons to vestibular stimuli in vertical planes. Role in vertical vestibulospinal reflexes of the decerebrate cat. J Neurophysiol 1992; 67:639–647.
32. Kasper J, Schor RH, Wilson VJ. Response of vestibular neurons to head rotations in vertical planes. I. Response to vestibular stimulation. J Neurophysiol 1988; 60: 1753–1764.
33. Wilson VJ, Yamagata Y, Yates BJ, et al. Response of vestibular neurons to head rotations in vertical planes. III. Response of vestibulocollic neurons to vestibular and neck stimulation. J Neurophysiol 1990; 64:1695–1703.
34. Peterson BW, Fukushima K, Hirai N, et al. Responses of vestibulospinal and reticulospinal neurons to sinusoidal vestibular stimulation. J Neurophysiol 1980; 43: 1236–1250.
35. Shimazu H, Precht W. Tonic and kinetic responses of cat's vestibular neurons to horizontal angular acceleration. J Neurophysiol 1965; 28:991–1013.
36. Schor RH, Cotter LA, Yates BJ. Responses of cat pontomedullary reticular neurons to horizontal rotation. Soc Neurosci Abstr 1993; 19:1492.
37. Yates BJ, Grélot L, Kerman IA, et al. Organization of vestibular inputs to nucleus tractus solitarius and adjacent structures in cat brain stem. Am J Physiol 1994; 267:R974–R983.

11

The Role of the Caudal Ventrolateral Medulla in the Reflex Bronchodilation Arising from the Hindlimb

IRENE C. SOLOMON

University of California at Davis
Davis, California

I. Introduction

The caudal ventrolateral medulla plays an important role in mediating the reflex airway dilation evoked by stimulation of thin fiber afferents arising from the hindlimb of anesthetized dogs. This paper focuses on the central neural pathways and neurotransmitters that are likely to play a role in causing the reflex airway dilation arising from the hindlimb. Particular attention has been paid to two areas. The first is the dorsal horn, the site of the first synapse in the reflex arc causing these autonomic effects. The second is the caudal ventrolateral medulla, whose participation in this reflex arc also appears to be vital.

II. Reflex Responses Arising from Hindlimb Afferents

Static contraction of the hindlimb muscles of anesthetized cats and dogs has been shown to increase arterial pressure, heart rate, and ventilation (1–3). In cats, McCloskey and Mitchell (2) demonstrated that these contraction-induced cardiovascular and ventilatory responses resulted from the activation of group

III and IV muscle afferents but not from the activation of group I and II afferents. Similarly, Tibes (4) found that the cardiovascular and ventilatory responses to dynamic contraction of the hindlimb muscles of anesthetized dogs were due to stimulation of group III and IV afferents.

Contraction of the hindlimb muscles in both anesthetized cats and dogs has also been shown to dilate the airways. Longhurst (5) demonstrated that static contraction of the hindlimb muscles of anesthetized cats reflexly relaxes tracheal smooth muscle. Similarly, Kaufman and Rybicki (6) showed that both static and repetitive twitch contractions of the gracilis muscles of anesthetized dogs reflexly decreased tracheal smooth muscle tone. Kaufman et al. (7) extended their previous findings to include total lung resistance as a functional index of airway caliber by demonstrating that both types of contractions also reflexly decreased total lung resistance. The reflex decreases in total lung resistance evoked by contraction were unaffected by propranolol but were blocked by atropine, suggesting that the contraction induced airway dilation resulted from the withdrawal of a tonic cholinergic outflow to airway smooth muscle.

The afferent arm of this reflex arc is believed to comprise group III and IV muscle afferents but not group I and II afferents. Stimulation of group I and II afferents with succinylcholine has been shown to have no effect on tracheal smooth muscle tone, whereas stimulation of group III and IV afferents with capsaicin has been shown to reflexly decrease tracheal smooth muscle tone (8,9). Rybicki and Kaufman (10) demonstrated that electrical stimulation of the gracilis muscle afferents at current intensities that activate group I and II afferents elicited no change in total lung resistance, whereas stimulation of these muscle afferents at current intensities that activate group III and IV afferents decreased total lung resistance. This reflex decrease in total lung resistance evoked by electrical stimulation of muscle afferents was due to a withdrawal of tonic cholinergic input to airway smooth muscle.

III. The Dorsal Horn

Thin fiber hindlimb muscle afferents enter the spinal cord via the dorsal roots and synapse in the superficial laminae of the dorsal horn (11,12). The neurotransmitters released onto second-order dorsal horn neurons from contraction-activated group III and IV hindlimb afferents is unknown; however, potential candidates include glutamate and substance P.

Primary afferents synapsing in the dorsal horn are known to contain glutamate (13,14), and neurons in the superficial laminae of the dorsal horn have been shown to be depolarized by glutamate (15). The role of glutamate in the spinal transmission of the cardiovascular and ventilatory responses evoked by stimulation of hindlimb afferents, however, has received little attention. Recently, Hill et al. (16) demonstrated that the reflex cardiovascular responses to static contraction of the hindlimb of chloralose-anesthetized cats were unaffected by intrathecal injection of *N* methyl-D-aspartate (NMDA) receptor antagonists but were attenuated by intrathecal injection of a non-NMDA receptor antagonist.

Considerable evidence has been accumulated suggesting that substance P plays a role in the spinal transmission of the cardiovascular and ventilatory responses evoked by stimulation of hindlimb afferents. In anesthetized cats, both electrical stimulation of group III and IV afferents in the sciatic nerve and static contraction of hindlimb muscles have been shown to evoke release of substance P in the dorsal horn (17–19). Further, intrathecal injections of either a peptide antagonist to substance P (20,21) or an antibody to this peptide (22) attenuated the reflex pressor and ventilatory responses to static contraction of the hindlimb muscles in anesthetized cats. Similarly, microinjection of a substance P antagonist into the gray matter of the spinal cord has also been shown to attenuate the reflex pressor response to static contraction (23).

Hill et al. (16) have extended their findings regarding the role of substance P and glutamate receptors in the spinal transmission of the reflex cardiovascular responses to static contraction of the hindlimb muscles in anesthetized cats. These investigators reported that the reflex pressor response to static contraction was nearly abolished by intrathecal injection of a substance P antagonist and a non-NMDA receptor antagonist. They speculated that substance P and glutamate are the major neurotransmitters in the first synapse of the exercise pressor reflex arc. The role of the dorsal horn in the reflex arc causing airway dilation when hindlimb afferents are stimulated remains to be determined.

IV. The Ventrolateral Medulla

The role played by neurons in the ventrolateral medulla in the regulation of cardiovascular function has been examined extensively. For example, both electrical and chemical stimulation of the caudal ventrolateral medulla (CVLM) have been shown to evoke depressor responses in anesthetized animals (24,25), although, in a few instances, chemical stimulation of this area has been shown to evoke a pressor response (26,27).

The ventrolateral medulla has also been shown to play a role in the control of airway tone. For example, Haxhiu et al. (28) demonstrated that topical application of nicotine to the intermediate area in cats evoked tracheal constriction, whereas application of GABA evoked tracheal dilation. Haselton et al. (29) showed that microinjection of DL-homocysteic acid into the lateral aspect of the rostral ventrolateral medulla increased total lung resistance and tracheal smooth muscle tension in chloralose-anesthetized dogs. In addition, the increase in total lung resistance evoked by DL-homocysteic acid was not affected by β-adrenergic blockade but was abolished by muscarinic blockade.

The cells of origin of the vagal efferent fibers whose activation causes constriction of airway smooth muscle appear to be located in the nucleus ambiguus.

In cats, McAllen and Spyer (30) demonstrated that cells in the nucleus ambiguus could be antidromically invaded from the pulmonary branches of the vagus nerves. Additionally, cell body labeling in the nucleus ambiguus following application of horseradish peroxidase to the cut central ends of the pulmonary branches of the vagus has been reported in cats (31) and dogs (32,33). Similarly, Wallach et al. (34) found cell body labeling in this nucleus when horseradish peroxidase was applied to the cut central end of the superior laryngeal nerve in the dog.

The role of the CVLM in the control of airway tone has also been examined. Connelly et al. (35) demonstrated that both electrical and chemical stimulation of the CVLM significantly decreased total lung resistance (TLR) in chloralose-anesthetized dogs. The bronchodilation evoked by stimulation of the CVLM was unaffected by β-adrenergic blockade but was abolished by muscarinic blockade. These investigators reported that the CVLM contains a pool of neurons whose excitation evokes a bronchodilation by withdrawing cholinergic input to the airways.

V. Hindlimb Afferents and the Ventrolateral Medulla

Electrophysiological evidence links input from hindlimb afferents with the ventrolateral medulla. Iwamoto and Kaufman (36) demonstrated that activation of group III and IV afferents by intra-arterial injection of capsaicin and by static muscular contraction of the triceps surae muscles in cats stimulated cells in the CVLM. Similarly, Bauer et al. (37) showed that cells in the ventrolateral medulla responded to muscular contraction of the hindlimb of anesthetized cats and that many of these cells projected to the intermediolateral cell column. Moreover, these cells exhibited a discharge related to either sympathetic nerve activity or the cardiac cycle. These cells described by Bauer et al. (37) appeared to be located more rostrally than those described by Iwamoto and Kaufman (36) and may play a role in the control of sympathetic outflow.

Although the activity of neurons in the ventrolateral medulla increased in response to activation of hindlimb afferents, it is not clear from the above studies whether these neurons play a role in mediating the reflex-induced cardiovascular and ventilatory responses evoked by stimulation of hindlimb afferents. There is, however, evidence suggesting that the CVLM plays an important role in the mediation of reflex-induced cardiovascular and ventilatory responses arising from the hindlimb. For example, electrolytic destruction of the CVLM has been shown to attenuate the reflex pressor responses to either electrical stimulation of afferent fibers in the sciatic nerve (38) or to static muscular contraction (39). Stornetta et al. (40), however, suggested that the attenuation of these reflex pressor responses resulted from destruction of fibers of passage in the CVLM.

An intact CVLM appears to be necessary for the full expression of the bronchodilation evoked reflexly from hindlimb afferents. In anesthetized cats, topical application of lidocaine to the ventrolateral medullary surface has been shown to abolish the reflex tracheal dilation evoked by electrical stimulation of the sciatic nerve (41). Similarly, chemical inactivation of cell bodies in the CVLM has been shown to attenuate the reflex decrease in TLR evoked by stimulation of hindlimb afferents. Using chloralose-anesthetized dogs, Padrid et al. (42) demonstrated that microinjection of either ibotenic acid or cobalt chloride into the CVLM attenuated the bronchodilation evoked by static muscular contraction and by electrical stimulation of C fibers in the sciatic nerve. Ibotenic acid destroyed cell bodies but not axons (43), whereas cobalt chloride prevented neurotransmitter release by blocking presynaptic calcium channels (44). These findings suggested that cells in the CVLM served as interneurons in the reflex arc that dilates the airways in response to afferent stimulation arising from the hindlimb.

Although the findings of Padrid et al. (42) implicated cell bodies located in the CVLM in the reflex control of airway caliber, the pharmacological mechanism by which hindlimb afferents evoked these reflex bronchomotor effects was not investigated. We therefore examined the role played by excitatory amino acids in the CVLM in the reflex airway dilation arising from the hindlimb of anesthetized dogs. In preliminary experiments, using the broad-spectrum glutamate antagonist kynurenic acid, we found that blockade of glutamatergic receptors located in the CVLM reversibly attenuated the reflex bronchodilation evoked by electrical stimulation of C fibers in the sciatic nerve (45). We also found that these injections of kynurenic acid potentiated the pressor response to sciatic nerve stimulation. From this work, however, it is not clear whether NMDA or non-NMDA receptors were involved. To distinguish which of the two excitatory amino acid receptor subtypes in the CVLM were responsible for evoking this reflex airway dilation, we examined the role played by NMDA and non-NMDA receptors in the CVLM in the reflex airway dilation arising from the hindlimb.

Decreases in TLR were evoked by electrical stimulation of the sciatic nerve at current intensities shown to recruit C fibers and by static contraction of the gastrocnemius muscles of both hindlimbs in chloralose-anesthetized dogs. Once we established control levels for the reflex airway dilation evoked by these maneuvers, we bilaterally microinjected 50 nL of a NMDA or non-NMDA receptor antagonist into the CVLM. We then repeated stimulation at 15- to 30-min intervals for 2 h and recorded the bronchomotor and cardiovascular effects. We used ($\pm$)-3-(2-Carboxypiperazin-4-yl)-propyl-1-phosphonic acid (CPP; 25 m*M*) and ($\pm$)-2-Amino-5-phosphono-valeric acid (AP5; 50 m*M*) to block NMDA receptors and 6-Cyano-7-nitroquinoxaline-2,3-dione (CNQX; 39 μM) to block non-NMDA receptors (46).

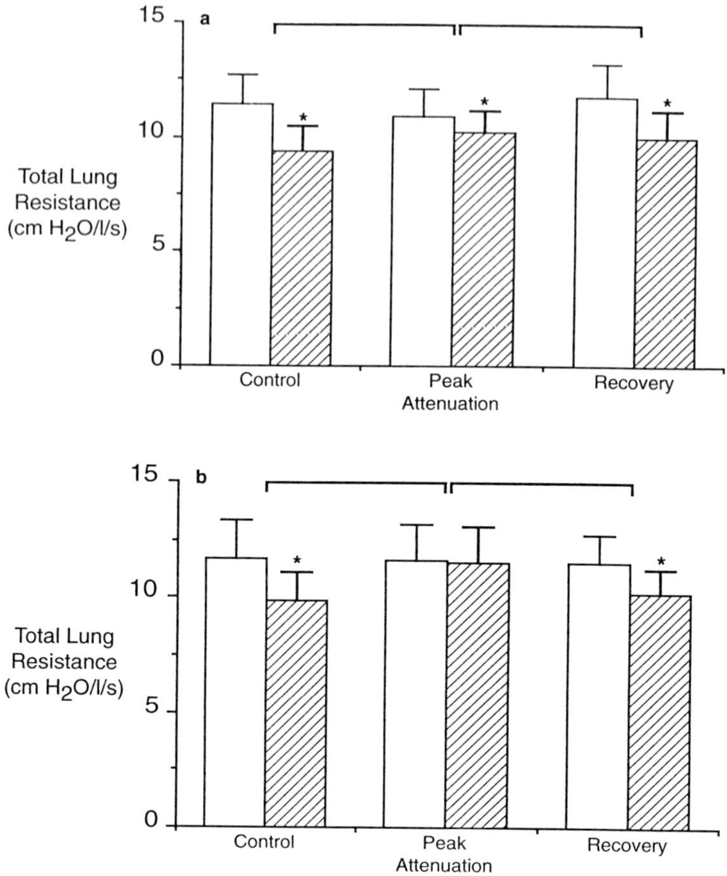

Figure 1 Effects of bilateral microinjection into the CVLM of either CPP (25 m*M*; 50 nL) or AP5 (50 m*M*, 50 nL) on the reflex airway dilation evoked by (a) electrical stimulation of the sciatic nerve ($n = 9$) and by (b) static contraction of both gastrocnemius muscles ($n = 5$). Blockade of NMDA receptors in the CVLM reversibly attenuated the reflex decrease in total lung resistance evoked by both maneuvers. Data are reported as means ± SE. Asterisks represent statistically significant differences ($P < 0.05$) between baseline (open bar) and its corresponding stimulation effect (hatched bar). Horizontal brackets connect control, peak attenuation, and recovery values that are significantly different ($P < 0.05$) from each other.

We found that bilateral microinjection into the CVLM of either CPP (25 m*M*; 50 nL) or AP5 (50 m*M*; 50 nL) reversibly attenuated the decrease in total lung resistance that was evoked by electrical stimulation of C fibers in the sciatic nerve (Fig. 1a). These injections also augmented the pressor response to stimulation. Similarly, bilateral microinjection of AP5 (50 m*M*; 50 nL) into the

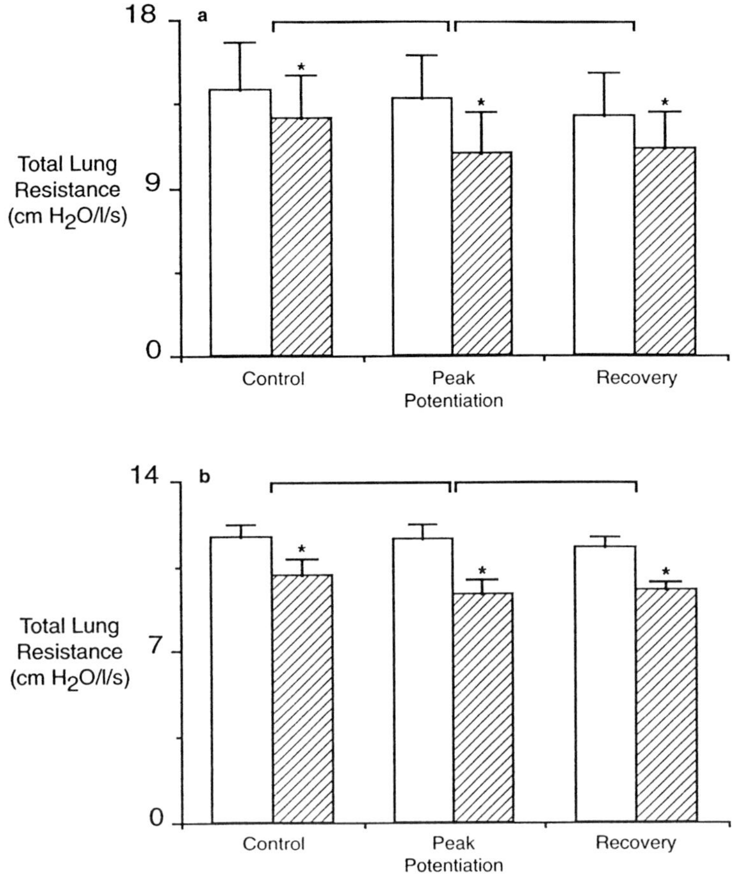

Figure 2 Effects of bilateral microinjection into the CVLM of CNQX (39 μM, 50 nL) on the reflex airway dilation evoked by (a) electrical stimulation of the sciatic nerve ($n = 6$) and by (b) static contraction of both gastrocnemius muscles ($n = 5$). Blockade of non-NMDA receptors in the CVLM reversibly augmented the reflex decrease in total lung resistance evoked by both maneuvers. Data are reported as means ± SE. Asterisks represent statistically significant differences ($P < 0.05$) between baseline (open bar) and its corresponding stimulation effect (hatched bar). Horizontal brackets connect control, peak potentiation, and recovery values that are significantly different ($P < 0.05$) from each other.

CVLM reversibly attenuated or abolished the decrease in total lung resistance elicited by static contraction of both gastrocnemius muscles (Fig. 1b). In contrast, bilateral microinjection into the CVLM of CNQX (39 μM; 50 nL) reversibly augmented the reflex decrease in total lung resistance that was evoked by either sciatic nerve stimulation (Fig. 2a) or by static muscular contraction (Fig. 2b).

Bilateral microinjections of xanthurenic acid (100 mM; 50 nL) into the CVLM had no effect on the decrease in total lung resistance that was evoked by sciatic nerve stimulation, and bilateral microinjection of either AP5 or CNQX into the dorsal medulla did not simulate the effects reported above. These findings suggest that NMDA receptors, but not non-NMDA receptors, located in the CVLM play the predominant role in mediating the reflex-induced bronchodilation arising from the hindlimb.

VI. Summary

Sensory input arising from hindlimb afferents has been shown to have marked reflex effects on the cardiovascular and respiratory systems. Both static contraction and electrical stimulation of hindlimb afferents can evoke increases in blood pressure, heart rate, and ventilation, as well as dilate the airways. The afferents responsible for inducing these reflex cardiovascular and respiratory effects appear to be group III and IV afferents. The CVLM has been shown to participate in the reflex airway dilation evoked by stimulation of these thin fiber afferents arising from the hindlimb. The pharmacological mechanism in the CVLM by which hindlimb afferents evoke this reflex airway dilation may in part be due to stimulation of excitatory amino acid receptors, specifically NMDA receptors.

References

1. Coote JH, Hilton SM, Perez-Gonzalez JF. The reflex nature of the pressor response to muscular exercise. J Physiol 1971; 215:789–804.
2. McCloskey DI, Mitchell JH. Reflex cardiovascular and respiratory responses originating in exercising muscle. J Physiol 1972; 224:173–186.
3. Fisher ML, Nutter DO. Cardiovascular and reflex adjustments to static muscular contraction in the canine hindlimb. Am J Physiol 1974; 226:648–655.
4. Tibes U. Reflex inputs to the cardiovascular and respiratory centers from dynamically working canine muscles. Circ Res 1977; 41:332–341.
5. Longhurst JC. Static contraction of hindlimb muscles in cats reflexly relaxes tracheal smooth muscle. J Appl Physiol 1984; 57:380–387.
6. Kaufman MP, Rybicki KJ. Muscular contraction reflexly relaxes tracheal smooth muscle in dogs. Respir Physiol 1984; 56:61–72.
7. Kaufman MP, Rybicki KJ, Mitchell JH. Hindlimb muscular contraction reflexly decreases total pulmonary resistance in dogs. J Appl Physiol 1985; 59:1521–1526.
8. Coleridge JCG, Coleridge HM, Roberts AM, et al. Tracheal contraction and relaxation initiated by lung and somatic afferents in dogs. J Appl Physiol 1982; 52:984–990.
9. Kaufman MP, Ordway GA, Longhurst JC, Mitchell JH. Reflex relaxation of tracheal smooth muscle by thin-fiber muscle afferents in dogs. Am J Physiol 1982; 243:R383–R388.

10. Rybicki KJ, Kaufman MP. Stimulation of group III and IV muscle afferents reflexly decreases total pulmonary resistance in dogs. Respir Physiol 1985; 59:185–195.
11. Pomeranz B, Wall PD, Weber WV. Cord cells responding to fine myelinated afferents from viscera, muscle, and skin. J Physiol 1968; 199:511–532.
12. Hoheisel U, Mense S. Response behaviour of cat dorsal horn neurons receiving input from skeletal muscle and other deep somatic tissues. J Physiol 1990; 426: 265–280.
13. Graham LT, Shank RP, Werman R, Aprison MH. Distribution of some synaptic transmitter suspects in cat spinal cord. J Neurochem 1967; 14:465–172.
14. Johnson JL, Aprison MH. The distribution of glutamic acid, a transmitter candidate, and other amino acids in the dorsal sensory neuron of the cat. Brain Res 1970; 24:285–292.
15. Schneider SP, Perl ER. Comparison of primary afferent and glutamate excitation of neurons in the mammalian spinal dorsal horn. J Neurosci 1988; 8:2062–2073.
16. Hill JM, Pickar JG, Kaufman MP. Blockade of non-NMDA receptors attenuates the reflex pressor response to static contraction. Am J Physiol (Heart Circ Physiol) 1994; 266:H1769–H1776.
17. Go VLW, Yaksh TL. Release of substance P from the cat spinal cord. J Physiol 1987; 391:141–167.
18. Duggan AW, Hope PJ, Lang CW, Williams CA. Sustained isometric contraction of skeletal muscle results in release of immunoreactive neurokinins in the spinal cord of the anesthetized cat. Neurosci Lett 1991; 122:191–194.
19. Wilson LB, Fuchs IE, Matsukawa K, et al. Substance P release in the spinal cord during the exercise pressor reflex in anesthetized cats. J Physiol 1993; 460:79–90.
20. Kaufman MP, Kozlowski GP, Rybicki KL. Attenuation of the reflex pressor response to muscular contraction by a substance P antagonist. Brain Res 1985; 333: 182–184.
21. Hill JM, Pickar JG, Kaufman MP. Attenuation of reflex pressor and ventilatory responses to static contraction by an NK-1 receptor antagonist. J Appl Physiol 1992; 73:1389–1395.
22. Kaufman MP, Rybicki KJ, Kozlowski GP, Iwamoto GA. Immunoneutralization of substance P attenuates the reflex pressor response to muscular contraction. Brain Res 1986; 377:199–203.
23. Wilson LB, Wall PT, Matsukawa K, Mitchell JH. The effect of spinal microinjections of an antagonist to substance P or somatostatin on the exercise pressor reflex. Circ Res 1992; 70:213–222.
24. Blessing WW, Reis DJ. Inhibitory cardiovascular function of neurons in the caudal ventrolateral medulla of the rabbit: Relationship to the area containing A1 noradrenergic cells. Brain Res 1982; 253:161–171.
25. Willette RN, Barcas PP, Krieger AJ, Sapru HN. Vasopressor and depressor areas in the rat medulla: Identification by microinjection of L-glutamate. Neuropharmacology 1983; 22:1071–1079.
26. Gordon FJ, McCann LA. Pressor responses evoked by microinjections of L-glutamate into the caudal ventrolateral medulla of the rat. Brain Res 1988; 457:251–258.
27. Iwamoto GA, Brtva RD, Waldrop TG. Cardiorespiratory responses to chemical stimulation of the caudalmost ventrolateral medulla in the cat. Neurosci Lett 1991; 129:86–90.

28. Haxhiu MA, Deal EC, Norcia MP, et al. Influence of ventrolateral surface of medulla on reflex tracheal constriction. J Appl Physiol 1986; 61:791–796.
29. Haselton JR, Padrid PA, Kaufman MP. Activation of neurons in the rostral ventrolateral medulla increases bronchomotor tone in dogs. J Appl Physiol 1991; 71:210–216.
30. McAllen RM, Spyer KM. Two types of vagal preganglionic motoneurons projecting to the heart and lungs. J Physiol (Lond) 1978; 282:353–364.
31. Kalia M, Mesulam MM. Brainstem projections of sensory and motor components of the vagus complex of the cat: II. Laryngeal, tracheobronchial, pulmonary, cardiac and gastrointestinal branches. J Comp Neurol 1980; 193:467–508.
32. Bennett JA, Kidd C, Latif AB, McWilliam PN. A horseradish peroxidase study of vagal motoneurons with axons in cardiac and pulmonary branches of the cat and dog. Q J Exp Physiol 1981; 66:145–154.
33. Haselton JR, Solomon IC, Motekaitis AM, Kaufman MP. Bronchomotor vagal preganglionic cell bodies in the dog: An anatomic and functional study. J Appl Physiol 1992; 73:1122–1129.
34. Wallach JH, Rybicki KJ, Kaufman MP. Anatomical localization of the cells of origin of efferent fibers in the superior laryngeal and recurrent laryngeal nerves of dogs. Brain Res 1983; 261:307–311.
35. Connelly JC, McCallister LW, Kaufman MP. Stimulation of the caudal ventrolateral medulla decreases total lung resistance in dogs. J Appl Physiol 1987; 63: 912–917.
36. Iwamoto GA, Kaufman MP. Caudal ventrolateral medullary cells responsive to static muscular contraction. J Appl Physiol 1987; 62:149–157.
37. Bauer RM, Waldrop TG, Iwamoto GA, Holzwarth MA. Properties of ventrolateral medullary neurons that respond to muscular contraction. Brain Res Bull 1992; 28: 167–178.
38. Ciriello J, Calaresu FR. Lateral reticular nucleus: A site of somatic and cardiovascular integration in the cat. Am J Physiol 1977; 233:R100–R109.
39. Iwamoto GA, Kaufman MP, Botterman BR, Mitchell JH. Effects of lateral reticular nucleus lesions on the exercise pressor reflex in cats. Circ Res 1982; 51:400–403.
40. Stornetta RL, Morrison SR, Ruggiero DA, Reis DJ. Neurons of rostral ventrolateral medulla mediate somatic pressor reflex. Am J Physiol 1989; 256:R448–R462.
41. Strohl KP, Norcia MP, Wolin DA, et al. Nasal and tracheal responses to chemical and somatic afferent stimulation in anesthetized cats. J Appl Physiol 1988; 65: 870–877.
42. Padrid PA, Haselton JR, Kaufman MP. Role of caudal ventrolateral medulla in reflex and central control of airway caliber. J Appl Physiol 1991; 71:2274–2282.
43. Schwarcz R, Hökfelt T, Fuxe K, et al. Ibotenic acid-induced neuronal degeneration: A morphological and neurochemical study. Exp Brain Res 1979; 37:199–216.
44. Kretz R. Local cobalt injection: a method to discriminate presynaptic axonal from post synaptic activity. J Neurosci Meth 1984; 11:129–135.
45. Solomon IC, Motekaitis AM, Kaufman MP. Caudal ventrolateral medullary neurons mediate reflex and central bronchomotor responses by a glutamatergic mechanism. FASEB J 1993; 7:A400.
46. Honoré T, Davies SN, Drejer J, et al. Quinoxalinediones: Potent competitive non-NMDA glutamate receptor antagonists. Science 1988; 241:701–703.

12

Central Effects and Interactions of the Renin–Angiotensin System and Vasopressin on Ventilation and Pa_{CO_2}

DONALD B. JENNINGS, JULIA K. L. WALKER, and PATRICIA J. OHTAKE

Queen's University
Kingston, Ontario, Canada

This review presents an hypothesis related to the roles of angiotensin II (ANG II) and vasopressin (AVP) in respiratory control. In the conscious dog, alterations in NaCl dietary intake affect ventilation and gas exchange in relation to chronic changes in strong ions in body fluids, osmolality, and hormones (62). During acute lowering of osmolality, increased ventilation and lowered Pa_{CO_2} were correlated with activation of the renin-angiotensin system and increases in plasma renin activity (PRA) (69). Subsequent experiments indicated that exogenous increases in plasma ANG II levels in conscious dogs stimulate respiration, presumably via central circumventricular organ(s) (CVO) (56). Furthermore, we found that (1) blocking ANG II receptors with saralasin, which does not cross the blood-brain barrier but accesses CVO, prevented ventilatory stimulation during acute hypotension (72) and (2) AVP inhibition of the renin-angiotensin system in normal dogs modulates ventilatory stimulation by ANG II (74). In both of the latter experiments, saralasin block of ANG II receptors prevented ventilatory stimulation under circumstances in which the *systemic* renin-angiotensin system was not activated. We speculate that both the renal and brain renin-angiotensin systems are involved in respiratory control and that this may be important for respiratory control during adaptations and disease.

I. Introduction

Medullary respiratory centers are sites of integration for numerous modulating factors affecting the homeostatic regulation of H^+ by ventilatory mechanisms. Circumventricular organs (CVO), including those of the hypothalamus, are among the important sensory groups with reflex neuronal connections to medullary respiratory centers. An extensive literature has been published, within the last decade or so, on the integrative role of CVO in the homeostatic regulation of body water and its ionic constituents and in circulatory control. These sensory receptors are stimulated or inhibited by multiple hormonal systems and by changes in osmolality. This chapter develops the hypothesis that CVO are influenced by changes in osmolality, angiotensin II (ANG II), and vasopressin (AVP) to coordinate brainstem respiratory control of Pa_{CO_2}, and hence H^+ homeostasis, with other physicochemical properties of the *milieu interieur.*

In order to put our hypotheses for hormonal regulation of ventilation and Pa_{CO_2} into perspective, the current state of knowledge on CVO and their relations to hormonal systems and osmolality are reviewed. To assess the effects of hormones on respiratory control, hormonal effects on blood pressure and arterial baroreflex modulation of respiratory control are also taken into account.

II. Homeostasis of Body Fluids and the Circumventricular Organs

Several recent review articles and monographs describe CVO and their roles in the homeostatic regulation of the composition of body fluids (1–12). CVO are specialized organs in the midline of the central nervous system at the blood-brain interface. They have extensive fenestrated capillary networks with a loose glial cell apposition to capillaries. Based on these structural characteristics, CVO lack a blood-brain barrier and have a high permeability to plasma solutes; they have been called the "windows of the brain" (13). CVO receive and transduce hormonal messages from the blood for transmission to other centers protected by the blood-brain barrier. They are also accessed by active mediators in the brain interstitial fluid and cerebrospinal fluid (CSF), and by neurotransmitters of efferent nerves from other brain centers (8).

CVO with a potential role in respiratory control include the subfornical organ (SFO) and the organum vasculosum of the lamina terminalis (OVLT) of the hypothalamus and the area postrema (AP) of the medulla (Fig. 1). The SFO and OVLT are in the walls of the third ventricle and the AP protrudes into the fourth ventricle. CVO are separated from the CSF by specialized ependymal cells that modulate the exchange of chemical messengers between the CSF and the neurons of CVO, either by acting as a variable barrier for diffusion or by transporting neuromodulators (reviewed in Refs. 14–17). Another CVO, the neurohypophysis of the pituitary gland, releases AVP into the blood

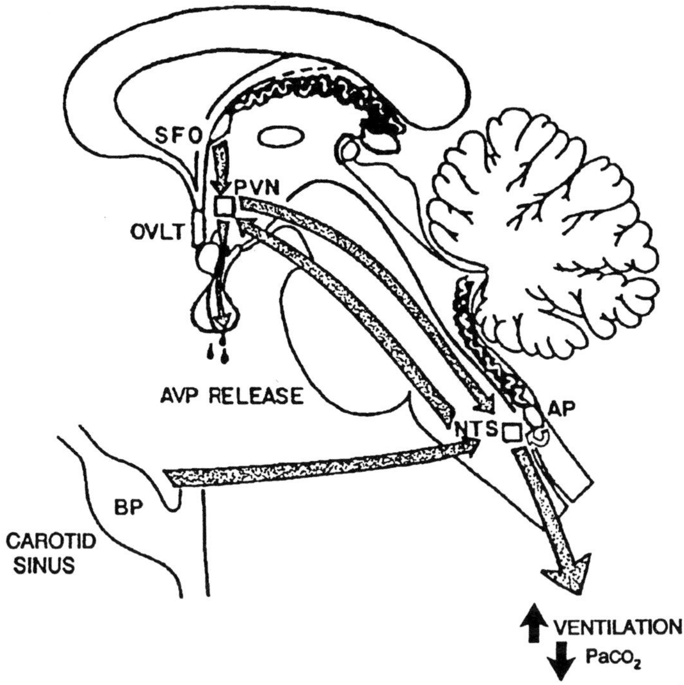

Figure 1 Schematic of the brain indicating the general locations of circumventricular organs (CVO): the subfornical organ (SFO), organum vasculosum of the lamina terminalis (OVLT) and area postrema (AP). Hypothalamic CVO have reciprocal connections with each other and stimulate vasopressin (AVP) release from the neurohypophysis. CVO also have reciprocal connections with the paraventricular nucleus (PVN) of the hypothalamus, which, in turn, has reciprocal connections with the nucleus of the tractus solitarius (NTS) and respiratory control. AVP acts at the AP, which has input to the NTS and reciprocal connections with the PVN. Carotid sinus baroreceptor afferents synapse at the NTS before the baroreceptor respiratory reflex relays to the posterior hypothalamus, where it inhibits hypothalamic facilitatory neurons that synapse on NTS respiratory neurons. (Modified from Ref. 72.) See text for discussion. (With permission of the American Physiological Society.)

and, as will be subsequently described, AVP has important implications for respiratory control.

Hypothalamic CVO have extensive connections with other hypothalamic centers such as the paraventricular nucleus (PVN); the PVN, in turn, has reciprocal connections with medullary centers (Fig. 1). Thus, CVO have indirect connections with respiratory centers of the pons (lateral parabrachial nucleus) and the medulla (nucleus of the tractus solitarius, or NTS, and the ventrolateral medulla, or VLM (reviewed in Refs. 4, 8, and 12). Hypothalamic CVO sense water concentration (osmolality), ion concentrations, and hormones involved in the regulation of water and electrolyte balance—ANG II, AVP, and atrial/brain natriuretic peptides (ANP, BNP). Therefore, CVO are perfectly situated to coordinate the acidity of body and brain fluids relative to their ion concentrations and osmolality via modulation of ventilation and $PaCO_2$.

III. Interactions Between CVO and Osmotic and Hormonal Neuromodulators

CVO are stimulated by increases in osmolality and ANG II in both the systemic circulation and the brain. For example, neuronal activity in the SFO is stimulated by both increased ANG II and hypernatremia (18). CVO regulate a number of responses, including AVP release from the neurohypophysis, drinking behavior, salt appetite, and arterial blood pressure and heart rate (reviewed in Refs. 9, 10, and 17). Because hypothalamic CVO are potentially activated by the same stimuli, our understanding of the contributions of individual CVO to these physiological regulations is still evolving. Some believe that the SFO is the primary sensor responsible for ANG II–induced AVP release and drinking behavior, whereas these same responses are evoked at the OVLT due to changes in osmolality (11). In keeping with this view is the finding in dogs that lesions of the OVLT result in impaired regulation of osmolality; lesions of the SFO do not impair regulation of osmolality but abrogate drinking responses to ANG II (reviewed in Ref. 11). ANP, which has its highest degree of binding in the brain at CVO (19), opposes ANG II at the SFO by counteracting ANG II's stimulatory effect on drinking behavior, salt appetite, blood pressure, and AVP release (20).

Electrical and ANG II stimulation of the SFO affects circulatory control via PVN connections to medullary centers (3,4); most importantly for our hypotheses, electrical stimulation of the SFO also evokes a respiratory response (21). The AP, another important CVO for the control of blood pressure (2,4), has been stated to be unimportant for respiratory control (reviewed in Ref. 1).

IV. ANG II Within the Brain and CVO

Neurons of CVO bind many peptides, including ANG II, AVP, and ANP (reviewed in Ref. 10); precise locations of individual receptors have recently begun to be identified with specific receptor antibodies. Peptides, such as ANG II, do not normally cross the blood-brain barrier from either the blood or CSF side when administered within physiological levels (9,22–24). Neuronal pericarya of the SFO and OVLT both bind ANG II at ANG II AT_1 receptors or a subtype thereof (10,25–32). In contrast, the AP binds ANG II only weakly (25,33) near its borders (31) and not in association with cell bodies (28). Despite this,

AP neurons respond to circulating ANG II (34). The combined observations suggest that ANG II may be acting presynaptically at the AP to affect its activity.

ANG II is a neurotransmitter of some afferents and efferents between the SFO and the PVN (35), and neurons of the PVN have extensive receptor binding for ANG II (27,36). The extensive distribution of ANG II within nerve fibers and on neuronal membranes of centers involved in respiratory and cardiovascular control—including the nucleus parabrachialis, the NTS, the Bötzinger complex, the Kölliker-Fuse area, the nucleus ambiguus, and the nucleus retroambiguus—led to the suggestion that ANG II has a role in respiratory control (36). Indeed, iontophoretic application of ANG II excites some NTS neurons with respiratory-related activity as well as some related interneurons (37). Furthermore, afferents in the carotid sinus (38) and vagus (39) nerves, which are potentially involved in respiratory reflexes, influence the number of ANG II receptor binding sites on NTS neurons as well as other medullary centers. For example, denervation of the carotid sinus nerves leads to decreased ANG II binding at the NTS (38).

In contrast to ANG II, there appears to be little binding of AVP in the SFO, OVLT, and PVN, but a very dense receptor binding of AVP occurs at the AP (40,41). These observations suggest that neurons of hypothalamic CVO have ANG II as a major hormonal mediator, while neurons of the AP of the medulla have AVP as a major hormonal mediator.

V. The Brain Renin-Angiotensin System

ANG II can be generated in the interstitial fluids of the brain (reviewed in Refs. 10 and 12). Many years ago, renin was identified in brain tissue (42); it is present primarily in the synaptosomes of nerve terminals (43), although some may be located in glia (10). The presence of mRNA for both renin and angiotensinogen (AOGEN) in the brain has been established (10,12). AOGEN is most likely produced in nonneuronal tissue, including glia of the SFO (29), but about 95% of AOGEN is in the extracellular fluid of the brain, where it constitutes about 3% of the CSF protein (12). There is a negative feedback between the peripheral and central renin-angiotensin systems. The brain renin-angiotensin system is stimulated by increased blood levels of ANG II, whereas the renal renin-angiotensin system is inhibited by increased brain ANG II (reviewed in Refs. 10 and 17).

In the SFO, angiotensin converting enzyme (ACE) has been identified on the external membrane surfaces of ependymal cells and glia and within organelles of neuronal processes (43,44). Thus, CVO appear to have the appropriate structural organization, following release of renin from nerve endings (10), to produce ANG II in the interstitial fluid around neuronal cells and synaptic terminals. Dehydration is a physiological state in which brain ANG II may play a role; under this condition, brain levels of ANG II, as measured in the CSF of the third ventricle of conscious dogs, increase (45).

VI. Inhibition of Renal and Brain Renin-Angiotensin Systems by AVP

Although ANG II acts centrally to release AVP into the systemic circulation, AVP, in turn, inhibits the renin-angiotensin system. Systemically administered AVP acts at arterial baroreceptor afferent endings to reset the arterial baroreceptor reflex and inhibit the renal release of renin; in the dog carotid sinus nerve denervation blocks this reflex (46). Systemic AVP also acts centrally, via the NTS of the medulla of rabbits, to inhibit sympathetic renal renin release; the most likely central CVO mediating the effect of AVP in the rabbit was the AP (47). Brain receptors for AVP are AVP V_1 receptors, or a subtype thereof and similar to AVP receptors which elicit constriction of vascular smooth muscle (reviewed in Refs. 40 and 41).

There is evidence that systemic AVP inhibits the brain renin-angiotensin system. Brattleboro rats, which lack systemic AVP due to genetic abnormalities of the hypothalamus, have increased activation of the brain renin-angiotensin system (48). Administration of systemic AVP in Brattleboro rats inhibits the brain renin-angiotensin system and lowers the level of ACE in specific hypothalamic centers (48).

AVP from neurons of the central nervous system is released into brain extracellular fluids and CSF during dehydration and hemorrhage in the dog (49). Therefore, AVP produced in the brain may act within the central nervous system to affect responses involving angiotensin. For example, an AVP blocker administered into the cisternal CSF reduced the pressor response to systemically administered ANG II by a vasopressinergic pathway inside the blood-brain barrier; the AP was unlikely to be involved, since the vasopressin blocker was not effective when administered via the vertebral artery or intravenously (50).

VII. Respiratory Effects of ANG II and AVP and Interactions with Arterial Baroreceptors

The confounding effects of ANG II and AVP in increasing arterial blood pressure and evoking baroreceptor respiratory reflexes can obscure independent effects on respiratory control, so that assessments are inconclusive (51). Increases in arterial pressure inhibit, and decreases in arterial pressure augment, respiration via baroreceptor reflexes (reviewed in Ref. 52). Anesthetized dogs require relatively large increases in carotid sinus pressure to change ventilation (Fig. 2; Ref. 53). Other cardiac and arterial baroreceptors do not interact in the anesthetized dog with the ventilatory response to changes in isolated carotid

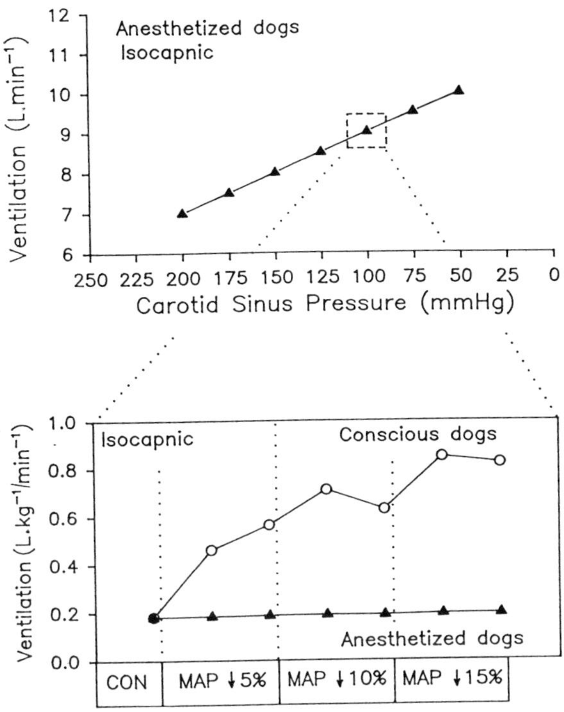

Figure 2 The upper panel represents changes in ventilation, at constant PCO_2, in anesthetized dogs as pressure is regulated bilaterally in isolated carotid sinuses. (Modified from Ref. 53). The lower panel demonstrates changes in ventilation, corrected to isocapnia, in conscious dogs as MAP is lowered by 5%, 10%, and 15% using a vasodilator (open circles). (From Ref. 52.) For comparison, changes in ventilation for the same changes in MAP in anesthetized dogs (solid triangles) are extracted from the upper panel. See text for discussion. (With permission of the American Physiological Society.)

sinus pressure (54). In comparison with anesthetized dogs, conscious dogs have extremely sensitive arterial baroreceptor respiratory reflexes, especially when the effects of changes in PCO_2 are taken into account (Fig. 2; Ref. 52). Dillon et al. (55) demonstrated that reflex respiratory responses to carotid sinus nerve baroreceptor activity involve a GABAergic inhibitory projection from the NTS to the posterior hypothalamus. The posterior hypothalamic center for the baroreceptor respiratory reflex augments ventilatory drive to the NTS (55). Baroreceptor reflexes affect only the threshold and not the slope of the ventilatory response to CO_2 in conscious and anesthetized dogs (52,53).

When mean arterial pressure (MAP) is increased by 15 to 20 mm Hg by intravenous infusion of AVP, ventilation is depressed (Fig. 3; Ref. 56). However,

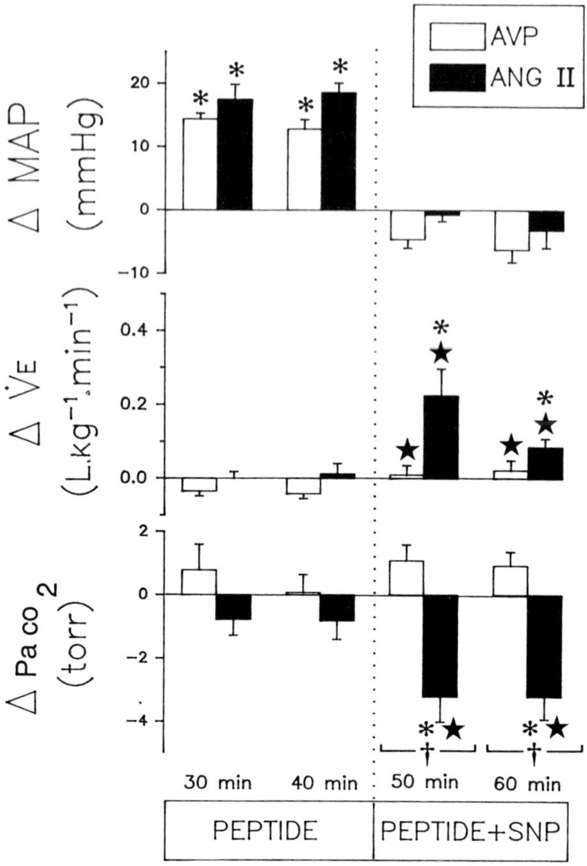

Figure 3 Changes in mean arterial pressure (MAP), expired minute ventilation ($\dot{V}_E$), and Pa_{CO_2} from six conscious dogs following infusions of angiotensin II (ANG II; solid bars) and vasopressin (AVP; open bars) at doses to increase MAP 15 to 20 mm Hg over 20 min, at 30 and 40 min of the protocol. Subsequently, MAP was titrated back to control levels with sodium nitroprusside. *, significantly different from the control period of the same protocol. +, significant difference between protocols. ★, significant difference from period of raised MAP, same protocol. (Modified from Ref. 56.) See text for discussion. (With permission of Elsevier Science Publishers.)

Pa_{CO_2} does not change during AVP-induced respiratory depression because CO_2 production also decreases (56). Baroreceptor-induced respiratory depression during AVP infusion is reversed by returning MAP to control levels (56). It is of interest that AVP has another effect on respiratory control, independent from arterial pressure, in that it depresses the slope of the ventilatory response to increased P_{CO_2} (56).

In contrast to AVP, a comparable increase in MAP resulting from intravenous infusion of ANG II does not affect resting ventilation or $PaCO_2$, and when MAP is reduced to control levels by simultaneous infusion of a vasodilator, ventilation increases and $PaCO_2$ falls (Fig. 3; Ref. 56). Intraventricular injection of ANG II into the brain of conscious rabbits also stimulates respiratory evaporative heat loss at a neutral ambient temperature (57). In anesthetized dogs, pharmacological doses of ANG II given intravenously resulted in increased respiration when the effects of arterial pressure were taken into account (58,59). The effects of ANG II on respiration in dogs are centrally mediated because they are not prevented by sinus nerve denervation; moreover, intracarotid injection of ANG II does not affect carotid chemoreceptor afferent activity (59). ANG II does not affect carotid baroreceptor afferent activity (60), which could influence respiratory control. Based on our earlier analysis of the central effects of systemic ANG II, increased ANG II likely acts at one or more CVO, probably the SFO (21), to stimulate ventilation.

VIII. Regulation of Ventilation and $PaCO_2$ During Changes in Osmolality

$PaCO_2$ and arterial hydrogen ion concentration ($[H^+]a$) in conscious dogs are positively correlated with plasma osmolality under steady-state conditions during low and high dietary intakes of NaCl (61,62). This relation between osmolality and respiratory regulation of $PaCO_2$ indicates that CVO osmoreceptors may be involved in respiratory regulation; the lower the osmolality, the greater the stimulation to breathe. During heat stress in conscious animals, infusion of hypoosmotic solutions into the blood vessels of the brain, or into the CSF, enhances panting; infusion of hyperosmotic solutions inhibits panting (63–65). Ventilation decreases and $PaCO_2$ increases with increased plasma osmolarity in heat stressed men (66). Infusion of ANP in conscious dogs on chronic low- and high-NaCl diets disrupts the relation between respiratory control of $PaCO_2$ and osmolality (67), presumably via ANP receptors at CVO (68). If respiratory regulation associated with changes in osmolality has the same mechanism proposed for the regulation of drinking and AVP release (11), our earlier analysis indicates the most likely CVO to sense osmotic change and affect respiratory control is the OVLT.

IX. Stimulation of Ventilation During Endogenous Increases in ANG II

Despite the positive relation between chronic adaptations of osmolality and $PaCO_2$ (62), during acute reductions of osmolality in conscious dogs increases in ventilation and lowering of $PaCO_2$ were associated primarily with increases in plasma renin activity (PRA) (Fig. 4; Ref. 69). For the experiments of acute

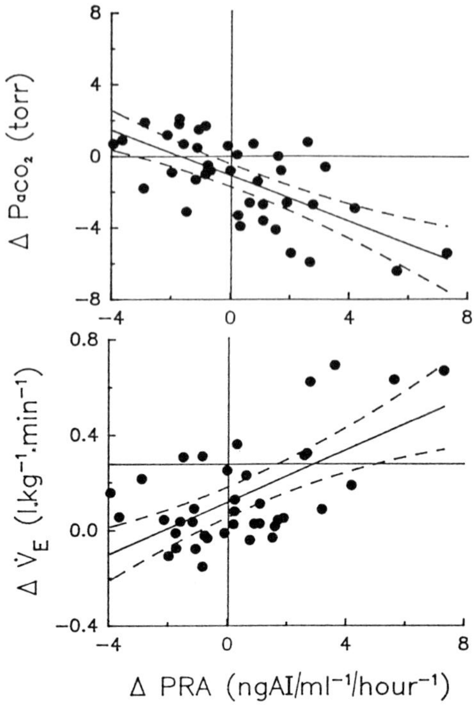

Figure 4 Changes in $PaCO_2$ and $\dot{V}E$ in seven conscious dogs following the acute lowering of plasma osmolality (~ 10 mOsm) by a gastric water load. In one of the protocols AVP was administered prior to the water load, resulting in inhibition of the renal renin-angiotensin system and a decrease in plasma renin activity (PRA) despite decreased osmolality. Thus, in the figure, PRA both increases and decreases from control levels. (Modified from Ref. 69.) See text for discussion. (With permission of Elsevier Science Publishers.)

lowering of osmolality, the prediction of ventilatory outcome was improved by including osmolality with PRA in equations of stepwise multiple regression; however, change in osmolality by itself did not predict ventilatory changes (69). PRA is a very good predictor of plasma ANG II levels (70,71). Because PRA/ANG II predicted change in ventilation during acute lowering of osmolality, we carried out the studies of intravenous infusions of ANG II in conscious dogs (plasma levels about twice normal) and stimulated ventilation and lowered $PaCO_2$ when baroreceptor reflexes were eliminated by returning MAP to control levels with sodium nitroprusside (Fig. 3; Ref. 56). ANG II infusion in conscious dogs affects only the threshold and not the slope of the ventilatory response to CO_2 (56). These combined studies indicated that systemic ANG II can contribute to the control of $PaCO_2$ in awake dogs.

We subsequently investigated the importance of endogenous ANG II in respiratory control in awake dogs during acute hypotension, a state in which ventilation increases in association with activation of the renin-angiotensin system (52). When the ANG II receptor blocker saralasin was infused intravenously in conscious dogs during relatively small decreases in MAP (~10, 15, and 20 mm Hg), ventilatory stimulation was reduced and $PaCO_2$ did not decrease, as it had in the unblocked studies (Fig. 5; Ref. 72). As previously reviewed, there is no evidence that ANG II acts at peripheral chemoreceptors or baroreceptors. Since intravenous saralasin does not cross the blood-brain barrier (24), the baroreceptor disinhibition (excitation) of ventilation is unblocked dogs must have occurred via ANG II-mediated stimulation at CVO. Based on the previous analyses of the potential central site(s) of action of ANG II, we speculate that endogenously released ANG II acted at the SFO to evoke the increased ventilation and hypocapnia during acute hypotension (72). Whatever the precise mechanism, ANG II, under physiological conditions, clearly acts as a mediator of ventilatory adaptations.

We also tested the hypothesis that AVP, by inhibiting the renin-angiotensin system, modulates the control of respiration by ANG II. The injection of the AVP V_1 receptor antagonist 1,β-mercapto-β,β-cyclopentamethylene propionic acid 2(O)-methyl-tyrosine arginine vasopressin (PMP; Ref. 73) into normal conscious dogs increased ventilation and lowered $PaCO_2$ (Fig. 6; Ref. 74). This AVP V_1 receptor blockade–induced hyperventilation was prevented or reversed by simultaneous administration of the ANG II receptor antagonist saralasin (Fig. 6). This finding is remarkable because it indicates that the normal basal level of AVP is an important modulator of respiration via ANG II. In contrast, under normal conditions in our conscious dogs, ANG II receptor blockade with saralasin did not affect ventilation (72), indicating that normal plasma levels of ANG II are not modulating respiration.

Surprisingly, following AVP V_1 receptor blockade, despite the ANG II-mediated increase in ventilation, there was no increase in PRA compared to control studies in which respiration does not change (Fig. 6). Because activation of the renal renin-angiotensin system could not account for respiratory stimulation, we surmise that the brain renin-angiotensin system must have been activated by AVP V_1 receptor blockade. AVP did not contribute to vasomotor tone in the dogs because MAP did not decrease but, in fact, increased (Fig. 6). The finding that saralasin abrogated the increase in MAP following AVP V_1 receptor block, is additional evidence that a brain ANG II was stimulating the circulation, as well as respiration, via CVO. Since the systemically administered ANG II receptor antagonist could not cross the blood-brain barrier, the site of action for brain ANG II must have been a CVO; both the blood and brain sides of the CVO are blocked by a systemic ANG II receptor blocker (24).

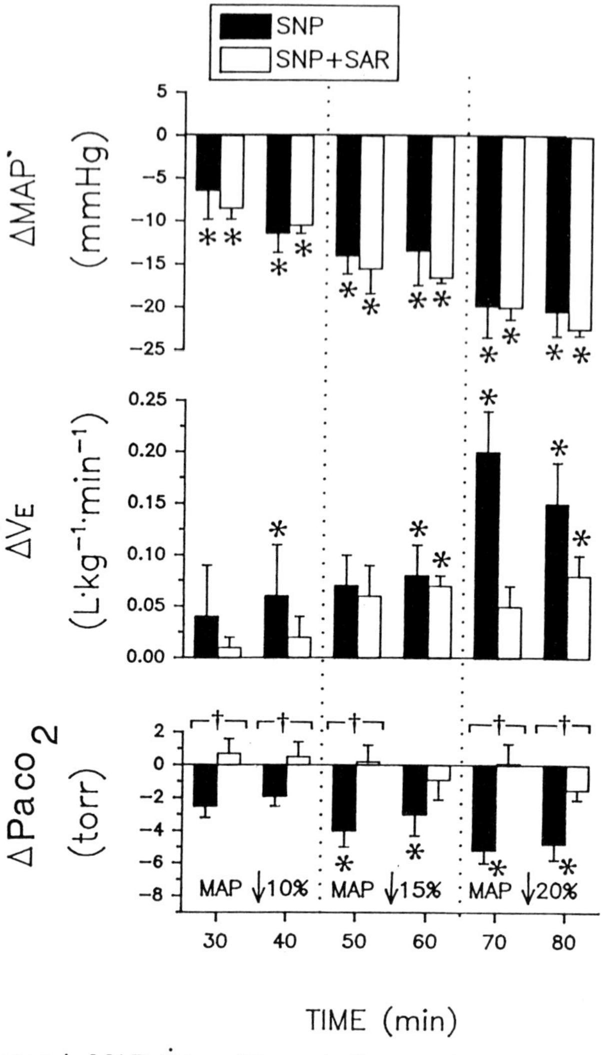

Figure 5 Changes in MAP, $\dot{V}_E$, and Pa_{CO_2} in five conscious dogs in each of two protocols in which acute hypotension was induced. In one protocol, they were unblocked during hypotension (solid bars) and PRA was progressively increased. (From Ref. 72.) In the second protocol, they were infused with the ANG II receptor blocker seralasin during the hypotension (open bars). MAP was lowered by 10%, 15%, and 20% in each protocol by titrated infusions of sodium notroprusside. (Modified from Ref. 72.) See text for discussion. (With permission of the American Physiological Society.)

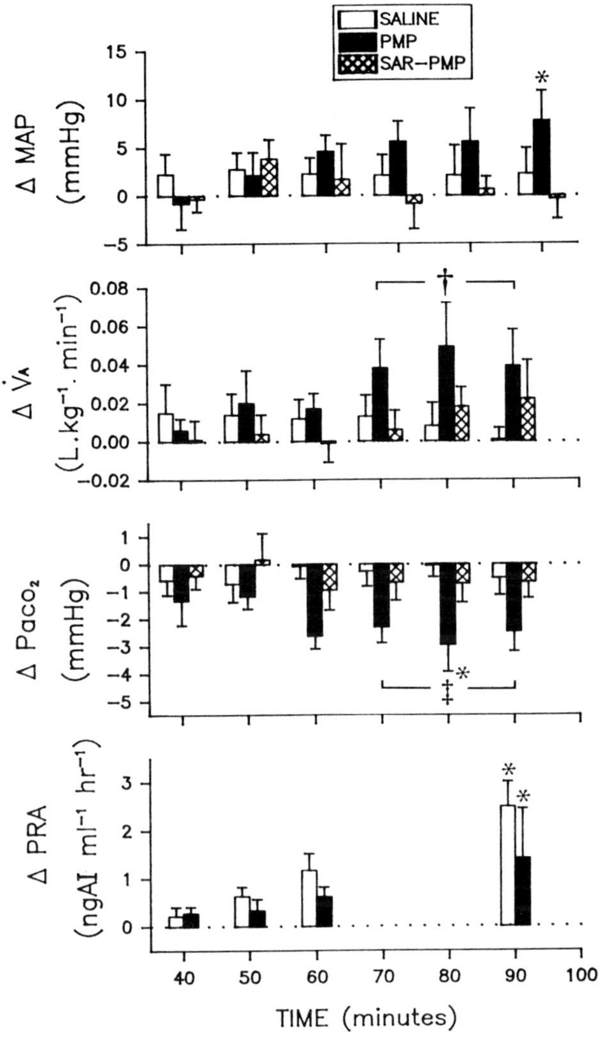

Figure 6 Changes in MAP, alveolar ventilation ($\dot{V}A$), $PaCO_2$, and plasma renin activity (PRA) in six conscious dogs following injection of an AVP V_1 receptor blocker at 30 min to disinhibit the renin-angiotensin system. In one protocol, the dogs had only the AVP V_1 receptor block (solid bars). In the second protocol, there was a combined blockade; a continuous infusion of the ANG II receptor blocker saralasin was initiated 20 min prior to the AVP receptor blockade (cross-hatched bars). Control studies are represented by the open bars. (Modified from Ref. 74.) See text for discussion. (With permission of the American Physiological Society.)

X. Potential Importance of Up- and Downregulation of ANG II Receptors for Respiratory Control

Saavedra (10) provides an extensive review of factors affecting upregulation and downregulation of hypothalamic ANG II receptors. For example, at the SFO, glucocorticoids and α-adrenergic agents upregulate and estrogens downregulate ANG II receptors. Most importantly for this review, sodium (Na^+) depletion and increased chloride (Cl^-) downregulate ANG II receptors at the SFO, and vice versa; in contrast, ANG II receptors at the PVN are unaffected by alterations in the concentrations of Na^+ and Cl^- (75–78). The potential variations in ANG II receptor numbers during high- and low-salt diets could be important for the relative significance of systemic and brain ANG II in respiratory control in these different states.

During a low-NaCl diet, the peripheral renin-angiotensin system in dogs can be activated about ninefold (79), but ventilation is depressed relative to that during a high-NaCl diet, in association with an increase in the strong ion difference ([SID])[1] and a relative metabolic alkalosis (62,81). The question is: Does an increased ANG II during a low-NaCl diet oppose the depressant effects of the metabolic alkalosis on respiratory control (81)? The literature indicates that the brain renin-angiotensin system is depressed in dogs on a low-NaCl diet (82) but activated in rats (83). As indicated in the preceding paragraph, ANG II receptors could be up- or downregulated by changes in ion concentrations during NaCl depletion, although the opposing effects of changes in Na^+ and Cl^- leave the outcome unclear.

Table 1 provides preliminary data on ventilation and Pa_{CO_2} in one of our conscious dogs prior to and following ANG II receptor blockade with saralasin during low- (<5 mEq·d^{-1}) and high- (~ 120 mEq·d^{-1}) NaCl special diets, as described in Ref. 62. Data from the same dog during a normal chow diet (~ 50–60 mEq NaCl·d^{-1}) are also presented in Table 1. PRA was severalfold higher in the dog during the low-salt diet compared with the high-salt diet and the normal-chow diet. Although we expected ventilation to be depressed by ANG II receptor blockade during the low-NaCl diet, this did not occur. ANG II receptor blockade also did not affect ventilation and Pa_{CO_2} in this dog during the normal chow diet, a response typical of normal dogs (72). However, on the high-NaCl diet, ANG II receptor blockade depressed ventilation compared to the normal- and low-NaCl diets (Table 1). Although these findings are based on one dog, they support the idea that up- and downregulation of ANG II receptors is an important point of control for respiratory adaptive outcomes.

[1]From physicochemistry, [SID] is the difference between the concentrations of the strong cations and strong anions; a decrease in [SID] acts to make the solution acidic, and vice versa (80). For practical purposes, in plasma and CSF the [SID] = ($[Na^+] + [K^+]$) − ($[Cl^-]$ + [lactate$^-$]).

Table 1 Measurements Prior to and Following Angiotensin II Receptor Blockade with Saralasin in a Dog on Low-, Normal-, and High-NaCl Diets

Diet	Infusion	PRA ($ngAI \cdot ml^{-1} \cdot h^{-1}$)	$\dot{V}A$ ($L \cdot kg^{-1} \cdot min^{-1}$)	Pa_{CO_2} (torr)	Osmolality ($mOsm \cdot kg\ H_2O^{-1}$)	MAP (mm Hg)
Low-NaCl diet, $< 5\ mEq \cdot d^{-1}$	Control	8.3	0.123	36.0	294	128
	Saralasin	15.0	0.124	35.1	294	132
	Δ	6.7	0.001	−0.9	0	4
Normal-NaCl diet, $\sim 60\ mEq \cdot d^{-1}$	Control	2.0	0.165	41.0	304	108
	Saralasin	1.2	0.168	42.3	304	111
	Δ	−0.8	0.003	1.3	0	3
High-NaCl diet, $\sim 120\ mEq \cdot d^{-1}$	Control	3.2	0.160	35.5	298	119
	Saralasin	8.3	0.094	37.7	294	121
	Δ	5.1	−0.066	2.2	−4	2

From the data in Table 1, ANG II receptors at CVO may have been downregulated during the low-NaCl diet and upregulated during the high-NaCl diet; paradoxically, low circulating levels of ANG II appeared to be tonically stimulating respiration during the high-salt diet, whereas very high levels of ANG II during the low-salt diet exerted no tonic effect on ventilation. This is consistent with the definitive evidence obtained in rats on the effects of NaCl diets on ANG II receptor regulation within the SFO (75,76).

XI. Summary of Our Current Hypothesis for Respiratory Control by Hormones

As outlined in Figure 7, we speculate that increases in systemic or brain ANG II stimulate respiration, lowering P_{CO_2} in body fluids. From our evidence, we believe that AVP normally inhibits the renin-angiotensin system, presumably in the brain, thereby modulating ventilation. We also speculate that the SFO mediates these ventilatory responses to ANG II and that the AP mediates the

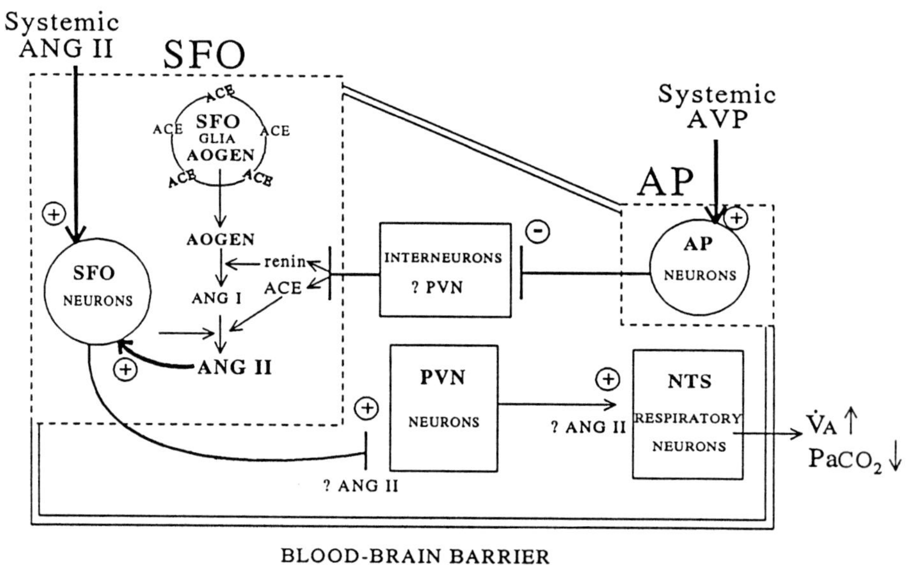

Figure 7 A speculative schematic, based on our analysis of the literature, of the effects of ANG II and AVP at CVO and potential neural interconnections evoking respiratory responses. Known elements of the organization of the brain renin-angiotensin system are included. AOGEN, angiotensinogen; ACE, angiotensin converting enzyme. The lack of a blood-brain barrier at CVO is depicted by the dotted lines. See text for discussion.

effects of AVP on the brain renin-angiotensin system. Future experiments are required to verify these hypotheses. We find that endogenous ANG II plays a role in respiratory stimulation during modest decreases in arterial pressure (72) and probably during acute lowering of osmolality (69). AVP also depresses the ventilatory responses to increased PCO_2, whereas ANG II has no interaction with hypercapnia on respiratory control (72).

XII. How Is Hormonal and Osmolality Control of Respiration Coordinated with Chemical Control During Respiratory Adaptations?

Living processes depend on the optimal function of proteins. The charge state, and thus function, of some proteins is influenced by small changes in the pH of body fluids (84). These proteins are said to be "pH dependent." More accurately, these proteins are "charge state dependent," where charge state is determined by the equilibrium between the pK of the protein and the pH of body fluids (84).

It has been proposed that stimulation of respiration is controlled by chemoreceptor proteins which are "charge state dependent" (84,85). Thus, a stimulus to breathe may be a factor which alters either the pH of body fluids or the pK of a chemoreceptor protein (86). Physicochemical factors in body fluids such as temperature (84), osmolality and strong ions (61) affect the pK of proteins and will therefore change the optimal pH for body fluids.

Variability of NaCl in the diet results in variability of [Na^+] and [Cl^-], [SID], and osmolality (61), factors which affect protein pK. A high NaCl intake results in a decreased [SID] which acts to make body fluids relatively acidic (80), and vice versa (61). We hypothesized, based on experimental evidence, that the [SID] is the stimulus to central chemoreceptors which controls $PaCO_2$, and that PCO_2 acts independently as a stimulus to ventilation (62,81,86). How hormonal influences are coordinated with chemical control for optimal regulation of PCO_2 and H^+ has yet to be determined. However, since ANG II and AVP are hormones which affect strong ions and osmolality of body fluids (thereby affecting the pK of proteins), it seems appropriate that these same hormones can simultaneously influence pH through direct effects on respiratory control, so that charge state of proteins remains constant.

The renin-angiotensin system is stimulated under many conditions when ventilatory drive increases, e.g., during exercise (87,88), acute hypercapnia (72) and adaptation to high altitude (89). Although ANG II can potentially stimulate ventilation (56,72), it does not contribute to ventilatory drive during acute hypercapnia (72) nor during a low salt diet (Table 1). As well, it appears that up- or down-regulation of receptors is critical in the ANG II contribution to respiratory drive (Table 1). Abnormalities of the renin-angiotensin system and vasopressin during cardiorenal diseases, and especially during respiratory failure and chronic obstructive pulmonary disease (90–92), raise important questions. First, how do alterations in osmolality and ions in body fluids during these diseases alter the optimal pH for regulatory purposes? Second, when do the hormonal changes play a significant role in respiratory control during disease states?

Acknowledgments

The editorial comments of Drs. A. Ferguson and S. Iscoe were helpful to the authors in writing this manuscript. Julia Walker was supported by an Ontario Graduate Student Award and a graduate student award of the Ontario Thoracic Society. Patricia Ohtake was supported by a graduate student award of the Canadian Heart and Stroke Foundation. The research was funded by grants from the Medical Research Council of Canada and the Ontario Thoracic Society.

References

1. Borison HL. Area postrema: Chemoreceptor circumventricular organ of the medulla oblongata. Prog Neurobiol 1989; 32:351–390.
2. Ferguson AV. The area postrema: A cardiovascular control centre at the blood-brain interface? Can J Physiol Pharmacol 1991; 69:1026–1034.
3. Ferguson AV. Neurophysiological analysis of mechanisms for subfornical organ and area postrema involvement in autonomic control. In: Ermisch A, Landgraf R, Ruhle H-J, eds. Progress in Brain Research. Vol. 91. New York: Elsevier Science Publishers, 1992:413–421.
4. Ferguson AV, et al. Circumventricular organs and cardiovascular homeostasis. In: Kunos G, Ciriello J, eds. Central Neural Mechanisms in Cardiovascular Regulation. Vol. 2. New York: Birkhauser, 1993:80–101.
5. Gross PM. Circumventricular Organs and Body Fluids. Vol. I. Boca Raton, FL: CRC Press, 1987.
6. Gross PM. Circumventricular Organs and Body Fluids. Vol. II. Boca Raton, FL: CRC Press, 1987.
7. Gross PM. Circumventricular Organs and Body Fluids. Vol. III. Boca Raton, FL: CRC Press, 1987.
8. Johnson AK, Gross PM. Sensory circumventricular organs and brain homeostatic pathways. FASEB J 1993; 7:678–686.
9. Simpson JB. The circumventricular organs and the central actions of angiotensin. Neuroendocrinology 1981; 32:248–256.
10. Saavedra JM. Brain and pituitary angiotensin. Endocr Rev 1992; 13:329–380.
11. Thrasher TN. Role of forebrain circumventricular organs in body fluid balance. Acta Physiol Scand 1989; 136(suppl 583):140–150.
12. Wright JW, Harding JW. Regulatory role of brain angiotensins in the control of physiological and behavioral responses. Brain Res Rev 1992; 17:227–262.
13. Gross PM, Weindl A. Peering through the windows of the brain. J Cereb Blood Flow Metab 1987; 7:663–672.

14. Dellmann HD. Subfornical organ—Fine structure. In: Gross PM, ed. Circumventricular Organs and Body Fluids. Vol. I. Boca Raton, FL: CRC Press, 1987:3–25.
15. McKinley MJ, et al. The Organum vasculosum of the lamina terminalis (OVLT) and the antroventral third ventricular (AV3V) region. In: Gross PM, ed. Circumventricular Organs and Body Fluids. Vol. 1. Boca Raton, FL: CRC Press, 1987: 111–130.
16. Miselis RR, et al. The area postrema. In: Gross PM, ed. Circumventricular Organs and Body Fluids. Vol. II. Boca Raton, FL: CRC Press, 1987:185–207.
17. Phillips MI. Functions of angiotensin in the central nervous system. Annu Rev Physiol 1987; 49:413–435.
18. Gutman MB, et al. Effects of plasma angiotensin II and hypernatremia on subfornical organ neurons. Am J Physiol 1988; 254:R746–R754.
19. Gibson TR, et al. Autoradiographic localization and characterization of atrial natriuretic peptide binding sites in the rat central nervous system and adrenal gland. J Neurosci 1986; 6:2004–2011.
20. Samson WK. Atrial natriuretic factor and the central nervous system. Endocrinol Metab Clin North Am 1987; 16:145–161.
21. Ferguson AV, et al. Effects of subfornical organ stimulation on respiration in the anesthetized rat. Can J Physiol Pharmacol 1989; 67:1097–1101.
22. Harding JW, et al. Inability of [^{125}I]Sar^1,Ile^8-angiotensin II to move between the blood and cerebrospinal fluid compartments. J Neurochem 1988; 50:554–557.
23. Volicer L, Loew CG. Penetration of angiotensin II into the brain. Neuropharmacology 1971; 10:631-636.
24. Van Houten M, et al. [Sar^1,Ala^8]angiotensin II in cerebrospinal fluid blocks the binding of blood-borne [^{125}I]angiotensin II to the circumventricular organs. Neuroscience 1983; 10:1421–1426.
25. Aldred GP, et al. Distribution of angiotensin II receptor subtypes in the rabbit brain. Regul Pept 1993; 44:119–130.
26. Van Houten M, et al. Radioautographic localization of specific binding sites for blood-borne angiotensin II in the rat brain. Brain Res 1980; 186:480–485.
27. Israel A, Plunkett LM. Quantitative autoradiographic characterization of receptors for angiotensin II and other neuropeptides in individual brain nuclei and peripheral tissues from single rats. Cell Mol Neurobiol 1985; 5:211–222.
28. Phillips MI, et al. Immunohistochemical mapping of angiotensin AT_1 receptors in the brain. Regul Pept 1993; 44:95–107.
29. Lippold A, et al. Cellular localization of angiotensin type 1 receptor and angiotensinogen mRNAs in the subfornical organ of the rat brain. Neurosci Lett 1993; 150:153–158.
30. Pfister J, et al. Immunohistochemical demonstration of angiotensin II receptors in rat brain by use of an anti-idiotypic antibody. Regul Pept 1993; 44:109–117.
31. Lind RW, et al. Organization of angiotensin II immunoreactive cells and fibers in the rat central nervous system. Neuroendocrinology 1985; 40:2–24.
32. Speth RC, et al. Angiotensin II receptor localization in the canine CNS. Brain Res 1985; 326:137–143.
33. Mendelsohn FAO, et al. Localization of angiotensin II receptor binding in rabbit brain by in vitro autoradiography. J Comp Neurol 1988; 270:372–384.
34. Papas S, et al. Electrophysiological evidence that systemic angiotensin influences rat area postrema neurons. Am J Physiol 1990; 258:R70–R76.

35. Bains JS, et al. Angiotensin II actions in paraventricular nucleus: Functional evidence for neurotransmitter role in efferents originating in the subfornical organ. Brain Res 1992; 599:223–229.
36. Aguirre JA, et al. Immunocytochemical study of angiotensin-II fibres and cell bodies in the brainstem respiratory areas of the cat. Brain Res 1989; 489:311–317.
37. Sessle BJ, Henry JL. Angiotensin II excites neurones in cat solitary tract nuclei which are involved in respiration and related activities. Brain Res 1987; 407:163–167.
38. Saavedra JM, Alexander N. Angiotensin II receptors in adrenal gland, pituitary gland and brain of sino-aortic denervated rats. J Hypertens 1986; 4(suppl 5):S154–S157.
39. Diz DI, Ferrario CM. Bidirectional transport of angiotensin II binding sites in the vagus nerve. Hypertension 1988; 11(suppl I):I-139–I-143.
40. Phillips PA, et al. Vasopressin receptors in rat brain and kidney: Studies using a radio-iodinated V_1 receptor antagonist. J Hypertens 1988; 6(suppl 4):S550–S553.
41. Tribollet E, et al. Localization and pharmacological characterization of high affinity binding sites for vasopressin and oxytocin in the rat brain by light microscopic autoradiography. Brain Res 1988; 442:105–118.
42. Ganten D, et al. Renin in dog brain. Am J Physiol 1971; 221:1733–1737.
43. Paul M, et al. Localization of renin (EC 3.4.23) and converting enzyme (EC 3.4.15.1) in nerve endings of rat brain. Brain Res 1985; 334:315–324.
44. Pickel VM, et al. Dual peroxidase and colloidal gold-labelling study of angiotensin converting enzyme and angiotensin-like immunoreactivity in the rat subfornical organ. J Neurosci 1986; 6:2457–2469.
45. Simon-Oppermann C, et al. Independent osmoregulatory control of central and systemic angiotensin II concentrations in dogs. Am J Physiol 1986; 250:R918–R925.
46. Gregory LC, et al. Effect of baroreceptor denervation on the inhibition of renin release by vasopressin. Endocrinology 1988; 123:319–327.
47. Suzuki S, et al. Central nervous system mechanisms involved in inhibition of renal sympathetic nerve activity induced by arginine vasopressin. Circ Res 1989; 65:1390–1399.
48. Saavedra JM, Chevillard C. Vasopressin-reversible increase in angiotensin-converting enzyme in specific hypothalamic nuclei of Brattleboro rats. Brain Res 1982; 246:157–160.
49. Szczepanska-Sadowska E, et al. Vasopressin in blood and third ventricle CSF during dehydration, thirst, and hemorrhage. Am J Physiol 1983; 245:R549–R555.
50. Michelini LC, et al. Arginine vasopressin modulates the central action of angiotensin II in the dog. Hypertension 1983; 5(suppl V):V-94–V-100.
51. Dempsey JA, et al. Hormones and neurochemicals in the regulation of breathing. In: Fishman AP, Cherniack NS, Widdicombe JG, Geiger SR, eds. Handbook of Physiology. Sec. 3. The Respiratory System. Vol. II. Control of Breathing. Part 1. Bethesda, MD: American Physiological Society, 1986:181–221.
52. Ohtake PJ, Jennings DB. Ventilation is stimulated by small reductions in arterial pressure in the awake dog. J Appl Physiol 1992; 73:1549–1557.
53. Brunner MJ, et al. Interaction of carotid chemoreceptor and baroreceptor reflexes in anesthetized dogs. Am J Physiol 1988; 254:R1–R10.

54. Brunner MJ, et al. Carotid sinus baroreceptor reflex control of respiration. Circ Res 1982; 51:624–636.
55. Dillon GH, et al. Hypothalamic GABAergic modulation of respiratory responses to baroreceptor stimulation. Respir Physiol 1991; 85:289–304.
56. Ohtake PJ, Jennings DB. Angiotensin II stimulates respiration in awake dogs and antagonizes baroreceptor inhibition. Respir Physiol 1993; 91:335–351.
57. Lin MT. Effects of angiotensin II on metabolic, respiratory and vasomotor activities as well as body temperatures in the rabbit. J Neural Transm 1980; 49:197–204.
58. Alexander IE, Lumbers ER. The effects of angiotensin on respiratory patterns of anaesthetized dogs. Respir Physiol 1981; 46:261–270.
59. Potter EK, McCloskey DI. Respiratory stimulation by angiotensin II. Respir Physiol 1979; 36:367–373.
60. Schmid PG, et al. Different aspects of vasopressin and angiotensin II on baroreflexes. Fed Proc 1985; 44:2388–2392.
61. Anderson JW, Jennings DB. H^+ homeostasis, osmolality, and body temperature during controlled NaCl and H_2O intake. Am J Physiol 1988; 255:R97–R104.
62. Anderson JW, Jennings DB. Osmolality, NaCl dietary intake, and regulation of ventilation by CO_2. Am J Physiol 1988; 255:R105–R112.
63. Baker MA, Doris PA. Control of evaporative heat loss during changes in plasma osmolality in the cat. J Physiol (Lond) 1982; 328:535–545.
64. Baker MA, Dawson DD. Inhibition of thermal panting by intracarotid infusion of hypertonic saline in dogs. Am J Physiol 1985; 249:R787–R791.
65. Turlejska E, Baker MA. Elevated CSF osmolality inhibits thermoregulatory heat loss responses. Am J Physiol 1986; 251:R749–R754.
66. Senay LC Jr. Increased blood osmolarity and its effect on respiration of dehydrating; men. Pflügers Arch 1969; 309:167–175.
67. Jennings DB, et al. Atrial natriuretic factor (ANF) affects the relation between plasma osmolality and the ventilatory response to CO_2 (abstr). Fed Proc 1987; 45: 658.
68. Samson WK. Central nervous system actions of atrial natriuetic factor. Brain Res Bull 1988; 20:831–837.
69. Anderson JW, et al. Acute changes in osmolality and renin and respiratory control of arterial P_{CO_2} and $[H^+]$. Respir Physiol 1990; 80:1–16.
70. Ben LK, et al. Role of the renin-angiotensin system in the control of vasopressin and ACTH secretion during the development of renal hypertension in dogs. Hypertension 1984; 6:35–41.
71. Morganti A, et al. Comparative measurements of immunoreactive renin, plasma renin activity and angiotensin II in human plasma. Clin Exp Theory Pract 1987; A9:1367–1381.
72. Ohtake PJ, et al. Renin-angiotensin system stimulates respiration during acute hypotension but not during hypercapnia. J Appl Physiol 1993; 74:1220–1228.
73. Manning M, Sawyer WH. Synthesis and receptor specificities of vasopressin antagonists. J Cardiovasc Pharmacol 1986; 8(suppl 7):S29–S35.
74. Walker JKL, Jennings DB. Angiotensin mediates stimulation of respiration after vasopressin V_1 receptor blockade. J Appl Physiol 1994; 76:2517–2526.
75. Ray PE, et al. Down-regulation of angiotensin II receptors in subfornical organ of young male rats by chronic dietary sodium depletion. Brain Res 1990; 510:303–308.

76. Ray PE, et al. Different effects of sodium or chloride depletion on angiotensin II receptors in rats. Am J Physiol 1990; 258:R1008–R1015.
77. Feldstein JB, et al. Sodium increases angiotensin II receptors in neuronal cultures from brains of normotensive and hypertensive rats. Brain Res 1986; 370:265–272.
78. Thomas WG, Sernia C. Regulation of rat brain angiotensin II (AII) receptors by intravenous AII and low dietary Na^+. Brain Res 1985; 345:54–61.
79. Jennings DB, et al. Infusion of atrial natriuretic peptide in dogs on low and high NaCl diets: Plasma levels and disappearance times. In: Brenner BM, Laragh JH, eds. American Society of Hypertension Symposium Series. Vol. 3. New York: Raven Press, 1989:227–235.
80. Stewart PA. How to understand acid-base: A quantitative acid-base primer for biology and medicine. New York: Elsevier, 1981.
81. Jennings DB. Breathing for protein function and $[H^+]$ homeostasis. Respir Physiol 1993; 93:1–12.
82. Brosnihan KB, et al. Effects of chronic sodium depletion on canine brain renin and cathepsin D activities. Hypertension 1982; 4:604–608.
83. Genain CP, et al. Distribution of renin activity and angiotensinogen in rat brain: Effects of dietary sodium chloride intake on brain renin. J Clin Invest 1985; 76: 1939–1945.
84. Reeves RB, Rahn H. Patterns of vertebrate acid-base regulation. In: Wood SC, Lenfant C, eds. Lung Biology in Health and Disease. Vol. 13. Evolution of Respiratory Processes. A Comparative Approach. New York: Marcel Dekker, 1979: 225–252.
85. Nattie EE. The alphastat hypothesis in respiratory control and acid-base balance. J Appl Physiol 1990; 69:1201–1207.
86. Jennings DB. The physicochemistry of H^+ and respiratory regulation: roles of Pa_{CO_2}, strong ions and their hormonal regulators. Can J Physiol Pharmacol 1994. In press.
87. Convertino VA, et al. Plasma volume, renin, and vasopressin responses to graded exercise after training. J Appl Physiol 1983; 54:508–514.
88. Freund BJ, et al. Hormonal and vascular fluid responses to maximal exercise in trained and untrained males. J Appl Physiol 1987; 63:669–675.
89. Milledge JS. Renin-aldosterone system. In: West JB, Lahiri S, eds. High Altitude and Man. Bethesda, MD: American Physiological Society, 1984:47–57.
90. Farber MO, et al. Studies on plasma vasopressin and the renin-angiotensin-aldosterone system in chronic obstructive lung disease. J Lab Clin Med 1977; 90:373–380.
91. Farber MO, et al. Hormonal abnormalities affecting sodium and water balance in acute respiratory failure due to chronic obstructive lung disease. Chest 1984; 85: 49–54.
92. Farber MO, et al. Abnormalities of sodium and H_2O handling in chronic obstructive lung disease. Arch Intern Med 1982; 142:1326–1330.

13

Central and Peripheral Respiratory Control Mechanisms After Lung Denervation in Humans

BABATUNDE A. OTULANA

Howard University Hospital
Washington, D.C.

TIMOTHY W. HIGENBOTTAM

Papworth Hospital
Cambridge, England

I. Introduction

More is known about the mechanics of respiration than about the processes controlling the regulation of breathing. The functions of the diaphragm and other accessory muscles of respiration and the efferent innervation pathways to these muscles are well elucidated. However, the neural regulation of breathing and the interplay between the chemical and neural factors in stimulated ventilation are areas of pulmonary and neural physiology that continue to generate controversies. For instance, it is often asked: Are there important roles for intrapulmonary receptors in the control of ventilation during exercise or hypercapnia? Are there receptors—chemosensitive or mechanosensitive—in the pulmonary circulation that are involved in the hyperpnea of exercise?

Centrally, the location of neurons that generate the rhythm of breathing in mammals is unknown. Recent location of rhythmogenic neurons in the ventral medulla (1) may pave the way for understanding this process. Much still requires to be understood about the feedback mechanisms involving intrapulmonary receptors, pulmonary afferents, and such central neurons that regulate the pattern of breathing in humans.

It is generally accepted that the carotid bodies are the main peripheral chemoreceptors responsible for hypoxic ventilatory chemosensitivity (2,3), while the central chemoreceptors located in the ventrolateral surface of the medulla are believed to mediate augmented ventilation in response to hypercapnic or hydrogen ion load (4). However, although no distinct pulmonary chemoreceptors have yet been described in mammals, some investigators have postulated that hyperpnea of exercise might be due partly to an increase in the carbon dioxide load to the lungs (5,6). And although it is well accepted that afferent information from lower airways are of little relevance in the regulation of resting ventilation (7), it has been suggested that certain feedback mechanisms between the lower airways and the central respiratory control areas may be important during stimulated ventilation (8). Studies in dogs in which the pulmonary and systemic circulations have been separated (9,10) have demonstrated that increased pulmonary arterial CO_2 tension can produce hyperpnea, thus appearing to support this theory. Furthermore, the involvement of vagus as the afferent pathway for this mechanism was demonstrated in these studies when vagotomy was shown to abolish the respiratory responses.

There have been understandable difficulties in studying the role of pulmonary afferents in control of breathing in humans. Previous experimental attempts in humans included vagal blockade by pharmacological means (11) or airway anesthesia by means of inhaled anesthetic agents (12). Technically, vagal blockade is invasive and induced airway anesthesia may not produce complete blockade. With these drawbacks, one such study (8) in which the vagus and glossopharyngeal nerves were blocked bilaterally in a conscious subject suggested that vagal afferents are required for the rise in respiratory rate and total ventilation to a hypercapnic stimulus.

Heart-lung transplantation provides a human model, which may not be hindered by any of those factors that confound other experimental approaches such as airway anesthesia or vagal blockade. Combined heart-lung transplantation (HLT) was first carried out in humans in 1981 by Reitz and others (13). The surgical operation involves placing the donor heart and lung en bloc into the recipient's thorax in the normal anatomical position (13). The recipient's right atrium and ascending aorta are sutured to the donor heart, while the trachea is anastomosed just above the carina. As a result, heart-lung transplantation results in all neural connections to the lungs being severed. Although animal studies have shown vagal reinnervation of reimplanted lungs, physiological (14) and histochemical (15) studies in human recipients have consistently established that afferent reinnervation of the lungs does not occur after HLT. The heart also loses its innervation except for the cuff of right atrium retained from the recipient's heart, which is sutured to the donor heart (13). Therefore the recipient's sinoatrial node does not control the heart rate. This absence of cardiac innervation has also been confirmed histochemically in heart transplants (16) and is responsible for the high resting heart rate and delayed cardiac response to exercise in heart transplant patients (17). Reinnervation also does not occur in the transplanted heart (16). Studies of static pressure-volume relationship of transplanted lungs have shown normal elastic properties soon after the operation (18); by 1 year after operation, the lungs have usually returned to their pretransplant volume (19).

Given the alteration in the neural supply to the heart and lungs described above, HLT patients provide a convenient model for delineating the effect of cardiopulmonary denervation on respiratory drive. A number of stimuli can be used to assess the responsiveness of the respiratory drive. Among them are hypoxia, exercise, and hypercapnia. One important index of response to these challenges is the minute ventilation ($\dot{V}_E$)—the volume of air leaving the lungs per minute. Described in this section are data from studies in which the ventilatory responses of HLT patients to hypercapnic stimulus and to submaximal exercise challenge were assessed.

II. Hypercapnic Ventilatory Response

The intensity of the afferent stimulus to respiration is one of the factors that controls the minute ventilation. An important stimulus for driving the ventilation is alveolar CO_2 tension, which, in turn, is correlated with the arterial CO_2 tension. Ventilatory response to CO_2 is influenced by the tension of oxygen in the inspired air. In order to eliminate the hypoxic chemoreceptor drive to ventilation, therefore, it is important to carry out CO_2 challenge with patients rebreathing a hyperoxic gas mixture.

We recently studied the hypercapnic ventilatory response in 16 patients, using a modified Read method (20), before and after heart-lung transplantation. At the time of the first testing (prior to the transplant operation), 10 of the patients had cystic fibrosis, and 1 patient each had Eisenmenger syndrome, ventricular septal defect, bronchiectasis, pneumoconiosis, emphysema, and histiocytosis-X. Seventeen age- and sex-matched nonsmoking normal subjects were studied by the same method, and they served as controls. Demographics and spirometric data on the patients and the controls are shown in Table 1. At the time of post-HLT hypercapnic challenge, all patients were free from acute or chronic lung rejection and from pulmonary infection as confirmed by pulmonary function testing (21) and on transbronchial biopsy (22). The control subjects also had normal pulmonary function data at the time of the study. Post-HLT patients were receiving cyclosporine and azathioprine for immunosuppression (23).

For the testing, a volume of 100% oxygen approximately equal to the subject's vital capacity was placed in a 10-L rebreathing bag. Subjects breathed through a rubber mouthpiece, which was connected to a two-way valve via a Fleisch pneumotachograph to measure flow rates. The flow rates were converted to volumes by a micromanometer and a respiratory integrator and re-

Table 1 Characteristics of the Patients and Control Subjects Whose Ventilatory Response to CO_2 Was Studied

	Pre-HLT	Post-HLT	Controls
n	16	16	17
Male	12	12	12
Female	4	4	5
Age (years)	29.6 ± 10	31.1 ± 11	30.1 ± 8
Height (m)	1.69 ± 0.11	1.69 ± 0.11	1.75 ± 0.09
Weight (kg)	51.1 ± 13.3	54.6 ± 13.7	70.0 ± 11.9
FEV_1/VC (%)	61.0 ± 27	85.0 ± 16	81.0 ± 10

Numbers are mean ± SD.

corded on a chart recorder. Expired gas was analyzed breath by breath for end-tidal oxygen and carbon dioxide with a mass spectrometer. Integration of inspiratory flow signals yielded inspiratory time (TI), tidal volume (VT), and expiratory time (TE).

The mean responses of the patients to hypercapnic challenge are shown in Table 2. Ventilatory responses to hypercapnia were similar in post-HLT and normal controls but reduced in pre-HLT patients. Correcting for the tension of CO_2 in the expired breath, both the respiratory rate response (Δf) and the tidal volume response (ΔVT) became normal in the patients after the HLT operation, although chronic cardiopulmonary disease had blunted the responses before transplantation in these same individuals. The VT/VI ratio, a measure of inspiratory flow, also became normalized after HLT. We inferred from these data that chronic denervation of the lungs does not substantially alter ventilatory drive from hypercapnia.

III. Response to Exercise

The metabolic demands of exercise require a coordinated increase in ventilation, pulmonary gas diffusion, cardiac output, and blood flow to the periph-

Table 2 Ventilatory Response Parameters in Patients and Control Subjects

	Pre-HLT	Post-HLT	Controls
Age (years)	29.6 ± 10.6	31.1 ± 11.2	30.1 ± 8.2
FEV_1/ VC (%)	61.0 ± 27.0	85.0 ± 16.0	81.0 ± 10.0
$\Delta f/\Delta CO_2$ (breaths/min/kPa)	1.11 ± 4.85	2.79 ± 3.1	2.8 ± 3.73
$\Delta V_T/\Delta CO_2$ (L/kPa)	−0.72 ± 3.37	0.33 ± 0.31	0.40 ± 0.38
$\Delta \dot{V}_E/\Delta CO_2$ (L/min/kPa)	4.52 ± 6.38	8.43 ± 5.3	11.36 ± 7.3[a]
$\Delta V_T/T_I/\Delta CO_2$ (L/s/kPa)	0.17 ± 0.25	0.35 ± 0.2[a]	0.42 ± 0.25[a]

[a] $P < 0.05$ compared to pre-HLT.
Δf = respiratory rate response; ΔCO_2 = change in CO_2 tension in expired breath; ΔVT = tidal volume response; ΔVE = minute volume response; ΔVT/TI = change in ratio of tidal volume: inspiratory time.

eral muscles (24). Apart from central neural control, complete innervation of the heart and lungs was initially thought to be necessary for an adequate cardiopulmonary response to exercise (25,26). Studies in experimental animals with deafferented hearts (27) and in human heart transplant recipients (17) have demonstrated a reduced cardiac output response to muscular exercise. In HLT patients, however, there has been less consensus on the ventilatory response to exercise (28,29). While there are reports of increased exercise ventilatory responses (28), a normal level of ventilation compatible with the exercise intensity has also been described (29) in recipients of HLT. Furthermore, it is uncertain whether exercise response in these patients is limited by the cardiac or the pulmonary denervation.

In order to clarify the influence of pulmonary innervation on exercise response, we administered a steady-state exercise test consisting of cycling for 6 min at 50 W workload and 60 rpm to 14 HLT recipients (8 males, 6 females), mean age 35.6 ($\pm$9 SD) years, and 14 normal untrained adults (8 males, 6 females), mean age 27.9 ($\pm$7.6) years.

The HLT patients were over 1 year posttransplant and had had stable lung function for at least 1 month prior to exercise study. That they were free of significant lung rejection or opportunistic infection was confirmed on histology of transbronchial biopsy carried out prior to the exercise study (30).

In the HLT patients, mixed venous blood samples for oxygen and carbon dioxide tensions were obtained through a pulmonary artery catheter put in place prior to exercise, and cardiac output by thermodilution was measured. Arterial blood gases and systemic arterial pressures were also measured. Ventilation-perfusion distribution was estimated in the HLT patients by calculating the physiological dead space (V_D/V_T) using the modified Bohr equation and the physiological shunt ($\dot{Q}_S/\dot{Q}_T$) by the Riley method, using the alveolar gas equation (31).

In the control subjects, cardiac output and heart rate were measured noninvasively with an impedance cardiograph, and oxygen saturation was recorded at baseline and each minute of the exercise using a pulse oximeter. In all subjects, continuous expired gas analysis was carried out using an on-line mass spectrometer.

The study showed that HLT patients had normal cardiac output at rest but that the heart rate was moderately elevated in comparison to the mean value in normal subjects ($P = 0.003$) (Table 3). Resting blood gas measurements were also normal in the HLT patients (Table 4).

On exercise, the mean increase in minute ventilation ($\dot{V}_E$) was higher in the HLT patients than in controls: 231.6 ($\pm$101.8)% and 153.6 ($\pm$44)% respectively ($P = 0.03$). This difference in ventilatory response was also reflected in greater reduction in ventilatory equivalent for carbon dioxide ($\dot{V}_E/V_{CO_2}$) in the normal subjects compared to the HLT patients. The increase in $\dot{V}_E$ during exercise in HLT patients was achieved from an increase of 124% in tidal volume

Table 3 Resting Ventilatory and Hemodynamic Measurements in HLT and Normal Subjects Studied with Exercise

	$\dot{V}_E$ (L/min)	f (per min)	HR (beats/min)	$\dot{Q}_T$ (L/min)
HLT patients ($n = 14$)				
Mean	8.1	14.6	85	6.0
SD	1.3	3.5	12	1.2
Normal controls ($n = 14$)				
Mean	9.8	15.8	71	7.1
SD	1.2	2.6	13	2.0

Abbreviations as defined in the text.

and 58% in respiratory rate (f). The normals by comparison showed a 109% increase in tidal volume and a 33% increase in respiratory rate.

The cardiac output increased with exercise in the HLT group from a mean 5.95 ($\pm$1.2) to 10.86 ($\pm$1.7) L/min ($P = 0.003$). The increase in the normal subjects was from 7.05 ($\pm$2.0) to 14.0 ($\pm$2.8) L/min ($P < 0.001$). Expressing these increases as cardiac index to take the lower body weight of the HLT patients into consideration, the rise in cardiac index was significantly higher in the normal subjects ($P = 0.002$). In the HLT patients, increase in cardiac output was due mainly to an augmentation of the stroke volume during exercise (mean increase 47.0 ± 29%), with less pronounced increase in heart rate (mean increase 25.1 ± 10%). By contrast, the normal subjects had comparable fractional increases in both heart rate (47.9 ± 22.5%) and stroke volume (64.3 ± 80.6%).

Exercise produced significant changes in mixed venous O_2 saturation, CO_2 tension, and hydrogen ion concentration [H^+] in the HLT patients, but the corresponding arterial measurements were unaltered (Table 5). The normal subjects maintained an unchanged SaO_2 from rest (mean 98.8 ± 0.3%) to exercise (mean 98.1 ± 0.2%). The physiological dead space (V_D/V_T) was normal

Table 4 Resting Blood Gas Data in HLT Patients Studied with Exercise ($n = 14$)

	SaO_2 (%)	$PaCO_2$ (kPa)	Arterial pH	$S\bar{v}O_2$ (%)	$P\bar{v}CO_2$ (kPa)	Venous pH
Mean	97.1	4.9	7.39	73	5.5	7.38
SD	1.6	0.6	0.04	7.0	0.6	0.03

SaO_2 = Arterial O_2 saturation; $PaCO_2$ = arterial CO_2 tension; $S\bar{v}O_2$ = mixed venous O_2 saturation; $P\bar{v}CO_2$ = mixed venous CO_2 tension.

Table 5 Mean Changes in Blood Gases and Hydrogen Ion in HLT Patients on Exercise

Arterial			
	SaO_2 (%)	$PaCO_2$ (kPa)	$[H^+]$ (mmol/L)
Mean change	1.2 (1.1)	0.26 (0.4)	1.0 (2)
P value	0.5	0.3	0.6
Mixed venous			
	$S\bar{v}O_2$ (%)	$P\bar{v}CO_2$ (kPa)	$[H^+]$ (mmol/L)
Mean change	29.7 (9)	0.79 (0.5)	5.5 (3)
P value	<0.001	0.03	<0.01

Abbreviations as for Table 4.

at rest in the HLT patients (mean 0.25 ± 0.14) and declined appropriately with exercise (mean 0.16 ± 0.09). Similarly, physiological shunt ($\dot{Q}S/\dot{Q}T$) was normal at rest (mean 5.09 ± 3.7%) and decreased to 1.26 ± 1.1% on exercise.

IV. Pulmonary Afferents and Control of Ventilation

Data from the two studies described above confirm the generally held view that afferent innervation to the lungs is not crucial for ventilation in humans at rest (7,32). Feedback loops from the airways or the lung parenchyma that were once thought to be important in stimulating breathing (8) were not shown to be of major influence on the total ventilation in the CO_2-rebreathing study described above. However, similar CO_2-rebreathing studies in HLT patients reported in the literature have produced diverse results. Sanders et al. (33), while finding comparable VT and VT/VI responses between normal and HLT patients, demonstrated lower overall ventilatory response to hypercapnia in HLT patients. In their study, HLT patients and normal subjects had similar ventilatory responses to hypoxia. In the study by Duncan et al. (34), the overall ventilatory response in HLT patients (who had normal spirometry) and control subjects was similar. A trend toward slower, deeper respiration was shown in the male HLT patients compared to controls but not in the female subjects. Overall, the authors did not show any difference in pattern of breathing between the HLT patients and normal controls.

The hypercapnic challenge study described in this section represents a unique situation in which the same patients were challenged before and after heart-lung transplantation. The abnormal ventilatory response seen in the patients pretransplantation allows an understanding of how abberations of pulmonary function and chronic cardiorespiratory insufficiency could affect chemosensitivity to CO_2 stimulus. After transplantation, however, and in the absence of pulmonary function impairment—as may result from acute or chronic lung rejection and lung infection (21)—the ventilatory response to hypercapnia returns to normal. It is possible that blunted ventilatory response or altered breathing pattern in HLT patients reported in other studies was due to abnormal pulmonary mechanics resulting from residual rejection or infection in the experimental subjects.

The response of the HLT patients to exercise as reported here also shed some light on the role of pulmonary afferents in the control of breathing. Despite the chronic denervation of the lungs, no significant difference was seen in the resting ventilation and hemodynamics between the HLT and the control subjects. These data, again, support the belief that intrapulmonary receptors and pulmonary afferents are of no consequence in the regulation of the pattern of breathing at rest (7,32).

With exercise, some important differences in ventilatory response between the HLT and normal subjects emerged. The increase in minute ventilation ($\dot{V}_E$) was higher in the HLT group than in the normal subjects. Similar findings have been reported in other studies of exercise in HLT patients (28,35). Additionally, Sciurba et al. reported a disparity in the increase in tidal volume and respiratory rate at maximal exercise in HLT patients (36). The authors attributed the lack of appropriate increase in respiratory rate at maximal exercise to the loss of intrapulmonary receptors. This finding has not been duplicated in other studies.

The slight elevation of the resting heart rate in the HLT group revealed the loss of the cardiodeceleratory role of the vagal reflexes (17), but this had no apparent influence on myocardial contractility, as resting cardiac output was normal. Cardiac response to exercise was reduced in the HLT patients as compared to the normal subjects. This failure to achieve adequate stroke volume on exercise is also seen in heart transplant patients, suggesting that the cause is related to the effect of loss of cardiac innervation on myocardial contractility (17).

The physiological shunt and dead space were normal both at rest and on exercise in the HLT patients, showing that ventilation-perfusion imbalance was not a factor in the hyperventilation.

How then can the excessive hyperventilation on exercise in the HLT patients be explained? Before implicating a defect of ventilatory control in HLT patients per se, a number of simpler possibilities ought to be considered. Mild anemia, a common finding in transplant patients due to chronic immunosuppression, is known to be associated with hyperventilation on exercise (37,38). Anemia limits exercise by reducing the capacity for oxygen transport, which is needed to meet increased metabolic demands (38). Chronic deconditioning as a cause of exercise hyperventilation is another possibility in transplant patients (39), who—in most cases—have had long sedentary lives prior to their operation.

Although anemia, deconditioning, and other similar factors may contribute to the increased ventilation in HLT patients, the exercise response data in this section and in other reports stimulate interesting speculations that appear to support some long-standing hypotheses. Such theories include those that favor the controlling influence of mixed venous CO_2 on hyperventilation (40,41). The original concept of humoral regulation of ventilation by a chemoreceptor system in the venous circulation was postulated by Zuntz and Geppert in 1886 (42). Animal experiments (40) and human studies (43) have supported the existence of such receptors. For instance, double-bypass studies in puppies (44) demonstrated the presence of chemosensitive receptors in the right atrium and the venae cavae. Evidence of a role for these receptors in the regulation of breathing is, however, lacking. In HLT patients, cardiac innervation exists only in the right atrium of the recipient, resulting in complete denervation of the remaining parts of the heart. If indeed chemoreceptors in the right atrium or the venae cavae can sense changes in mixed venous P_{CO_2} or $[H^+]$, can this explain the greater than normal exercise hyperpnea observed in HLT patients? Carbon dioxide loading of the venous system has been shown to stimulate hyperventilation in a time-dependent fashion (45,46) with no significant differences in $\dot{V}_E/\Delta Pa_{CO_2}$ for inhaled and intravenous carbon dioxide. There is no consensus as to where the receptors sensitive to the change in P_{CO_2} are located. In CO_2 infusion studies in animals, very high doses of CO_2 infused into the pulmonary artery produced marginal increase in ventilation (9).

The increase in P_{CO_2} and $[H^+]$ along with substantial decrease in mixed venous oxygen saturation on exercise in HLT patients seemed to reflect the inadequate cardiac output response to exercise. Compared to reported figures in normal subjects (35,47), these mixed venous blood changes were more profound than expected, for the level of exercise. The absence of significant changes in the arterial blood oxygen saturation or CO_2 tension suggests a cardiac rather than respiratory cause for the mixed venous blood changes. More studies are needed to define a threshold for sensing these changes and to assess their contribution, if any, to control of ventilation in mammals.

V. Conclusion

Heart-lung transplant patients provide a model, unconfounded by technical difficulties of experimental vagal blockade, for studying the contribution of pulmonary afferents to the control of breathing in humans. Available data confirm that pulmonary afferents and feedback loops involving airway receptors contribute very little to total ventilation and pattern of breathing at rest. Similarly, response to hypercapnic overload does not depend on intact pulmonary innervation. Central coordination of such a response is unaffected by lung denervation. Increased ventilatory response to exercise in HLT patients—seen in the study reported here and similar studies—may be multifactorial in origin but raises a question of a possible role for venous receptors in ventilatory control.

References

1. Smith JC, Elleberger HH, Ballanyi K, et al. Pre-Botzinger complex: A brainstem region that may generate respiratory rhythm in mammals. Science 1991; 254:726–729.
2. Lugliani R, Whipp BJ, Seard C, et al. Effect of bilateral carotid-body resection on ventilatory control at rest and during exercise in man. N Engl J Med 1971; 285:1105–1111.
3. Honda YS, Watanabe I, Hashizume Y, et al. Hypoxic chemosensitivity in asthmatic patients two decades after carotid body resection. J Appl Physiol 1979; 46: 632–638.
4. Mitchell RA, Berger AJ. Neural regulation of respiration. Am Rev Respir Dis 1975; 111:206–224.
5. Wasserman K, Whipp BJ, Castagna J. Cardiodynamic hyperpnea: Hyperpnea secondary to cardiac output increase. J Appl Physiol 1974; 36:457–464.
6. Wasserman K, Whipp BJ, et al. CO_2 flow to the lungs and ventilatory control. In: Dempsey JA, Reed CE, eds. Muscular Exercise and the Lung. Madison, WI: University of Wisconsin Press, 1977:103–132.
7. Guz A, Noble MIM, Trenchard D, et al. Studies on the vagus nerves in man: Their role in respiratory and circulatory control. Clin Sci 1964; 27:293–304.
8. Guz A, Noble MIM, Widdicombe JG, et al. The effect of bilateral block of vagus and glossopharyngeal nerves on the ventilatory response to CO_2 of conscious man. Respir Physiol 1966; 1:206–210.
9. Sheldon MI, Green JF. Evidence for pulmonary chemosensitivity: Effects on ventilation. J Appl Physiol 1982; 1192–1197.
10. Green JF, Sheldon MI. Mechanisms of hyperpnea induced by changes in pulmonary blood flow. 1984; 56:1418–1422.
11. Berglund E, Kurhoff AK, Loefstroem B, Oequist L. A study of the effect of unilateral vagus nerve block in a dyspnoeic patient. Scand J Respir Dis 1971; 52:34–38.
12. Winning AJ, Hamilton RD, Shea SA, et al. The effect of airway anaesthesia on the control of breathing and the sensation of breathlessness in man. Clin Sci 1985; 68: 215–225.
13. Reitz BA, Wallwork JL, Hunt SA, et al. Heart-lung transplantation: Successful therapy for patients with pulmonary vascular disease. N Engl J Med 1982; 306: 557–564.
14. Higenbottam T, Jackson M, Woolman P, et al. The cough response to ultrasonically nebulised distilled water in heart-lung transplantation patients. Am Rev Respir Dis 1989; 140:58–61.
15. Springall DR, Polak JM, Howard L, et al. Persistence of intrinsic neurones and possible phenotypic changes after extrinsic denervation of human respiratory tract by heart-lung transplantation. Am Rev Respir Dis 1990; 141:1538–1546.
16. Rowan RA, Billingham ME. Myocardial innervation in long-term heart transplant survivors: A quantitative ultrastructural survey. J Heart Transplant 1988; 7:448–452.

17. Savin WM, Haskell WL, Schroeder JS, Stinson EB. Cardiorespiratory responses of cardiac transplant patients to graded, symptom-limited exercise. Circulation 1980; 62:55–60.
18. Glanville AR, Theordore J, Harvey J, Robin E. Elastic behaviour of the transplanted lung: Exponential analysis of static pressure-volume relationship. Am Rev Respir Dis 1988; 137:308–312.
19. Otulana BA, Mist BA, Wallwork J, Higenbottam TW. The effect of recipient lung size on lung physiology after heart-lung transplantation. Transplantation 1989; 48: 625–629.
20. Read DJ. A clinical method for assessing the ventilatory response to carbon dioxide. Australas Ann Med 1967; 16:20–32.
21. Otulana BA, Higenbottam T, Scott J, et al. Lung function associated with histologically diagnosed acute lung rejection and pulmonary infection in heart-lung transplant patients. Am Rev Respir Dis 1990; 142:329–332.
22. Higenbottam T, Stewart S, Penketh A, Wallwork J. The diagnosis of lung rejection and opportunistic infection by transbronchial lung biopsy. Transplant Proc 1987; 19:3777–3778.
23. Higenbottam T, Otulana BA, Wallwork J. Transplantation of the lung. Eur Respir J 1990; 3:594–605.
24. Wasserman K, Hansen JE, Sue DY, Whipp B. Principles of Exercise Testing and Interpretation. Philadelphia: Lea & Febiger, 1987:255.
25. Dempsey JA, Vidruk EH, Masterbrook SM. Pulmonary control systems in exercise. Fed Proc 1980; 39:1498–1505.
26. Stone HL, Liang IYS. Cardiovascular response and control during exercise. Am Rev Respir Dis 1984; (suppl):S13–S16.
27. Donald DE, Milburn SE, Shepherd JT. Effect of cardiac denervation on the maximal capacity for exercise in the racing greyhound. J Appl Physiol 1964; 19:849–952.
28. Banner N, Guz A, Heaton R, et al. Ventilatory and circulatory responses at the onset of exercise in man following heart or heart-lung transplantation. J Physiol (Lond) 1988; 399:437–449.
29. Sciurba FC, Owens GR, Sanders MH, et al. Evidence of an altered pattern of breathing during exercise in recipients of heart-lung transplants. N Engl J Med 1988; 319: 1186–1192.
30. Higenbottam TW, Stewart S, Penketh ARL, Wallwork J. Transbronchial lung biopsy for the diagnosis of rejection in heart-lung transplant patients. Transplantation 1988; 46:532–540.
31. Riley RL, Cournand A. "Ideal" alveolar air and the analysis of ventilation-perfusion relationship in the lungs. J Appl Physiol 1949; 1:825–847.
32. Otis AB, Fenn WO, Rahn H. Mechanisms of breathing in man. J Appl Physiol 1950; 2:592–607.
33. Sanders MH, Owens GR, Sciurba FC, et al. Ventilation and breathing pattern during progressive hypercapnia and hypoxia after human heart-lung transplantation. Am Rev Respir Dis 1989; 140:38–44.
34. Duncan SR, Kagawa FT, Starnes VA, et al. Hypercarbic ventilatory responses of human heart-lung transplant recipients. Am Rev Respir Dis 1991; 144:126–130.
35. Karlson H, Lindborg B, Linnarsson D. Time courses of pulmonary gas exchange and heart rate changes in supine exercise. Acta Physiol Scand 1975; 95:329–340.

36. Sciurba FC, Owens GR, Sanders MH, et al. Evidence of an altered pattern of breathing during exercise in recipients of heart-lung transplants. N Engl J Med 1988; 319: 1186–1192.
37. Pappenheimer JR. Standardization of definitions and symbols in respiratory physiology. Fed Proc 1950; 9:602–605.
38. Woodson RD, Wills RE, Lenfant C. Effect of acute and established anemia on O_2 transport at rest, submaximal and maximal work. J Appl Physiol 1978; 44:36–43.
39. Otulana BA, Higenbottam TW, Wallwork J. Causes of exercise limitation after heart-lung transplantation. J Heart Transplant 1992; 11:S244–S251.
40. Armstrong BW, Hurt HH, Blide RW, Workman JM. The humoral regulation of breathing. Science 1961; 133:1897–1906.
41. Duke NH, Green JH, Heffron PF, Stubbens VWJ. Pulmonary chemoreceptors. Q J Exp Physiol 1963; 48:164–175.
42. Zuntz W, Geppert J. Ueber die Natur der Normalen Atemreize und den Orthrer Wirkung. Pflügers Arch 1886; 38:337–338.
43. Riley RL, Ross RS, Armstrong B. Pulmonary arterial PCO_2 and ventilation in man (abstr). Fed Proc 1961; 20:431.
44. Kollmeyer KR, Kleinman LI. A respiratory venous chemoreceptor in the young puppy. J Appl Physiol 1975; 38:819–826.
45. Fordyce WE, Grodins FS. Ventilatory responses to intravenous and airway CO_2 administration in anaesthetised dogs. J Appl Physiol 1980; 48:337–346.
46. Linton RAF, Miller R, Cameron IR. Ventilatory response to CO_2 inhalation and intravenous infusion of hypercapnic blood. Respir Physiol 1976; 26:383–394.
47. Cruz JC, Rahn H, Farhi LE. Mixed venous PO_2, PCO_2, pH and cardiac output during exercise in trained subjects. J Appl Physiol 1969; 27:431–434.

14

Modulation of Pedunculopontine Tegmental Nucleus and Alterations in Glutamate-Evoked Discharge Rate of Nucleus Ambiguus Cholinoceptive Neurons in Rats

HUBERT K. RUCKER and ERIC A. FLOYD

Meharry Medical College
Nashville, Tennessee

W. ERIC THOMAS

Mississippi Valley State University
Itta Bena, Mississippi

ERIC L. MOORE

United States Army Reserve Institute of Chemical Defense
Aberdeen Proving Ground, Maryland

I. Introduction

The pedunculopontine tegmental nucleus (PPTn) is part of a heterogenous mesopontine complex that provides diffuse ascending projections to diencephalic and forebrain structures as well as descending projections to the brainstem and possibly spinal cord (1). Rye et al. (2) described the PPTn in the rat, based on cytoarchitecture, as including all of the darkly stained magnocellular neurons in the field extending from the substantia nigra and retrorubral field rostral to parabrachial nucleus caudally. The complex is bordered medially by the ascending limb of the superior cerebellar peduncle and laterally by the lateral lemniscus and associated nuclei. The dorsal and ventral extent of the nuclear complex is delimited by the cuneiform nucleus and pontine tegmental field, respectively. At the caudal extreme, the PPTn merges with the laterodorsal tegmental nucleus (LDTn). While admixed with smaller noncholinergic cells, most, if not all, of the large multipolar cells in this area are immunopositive

The experiments described in this report were conducted according to the "Guide for Care and Use of Laboratory Animals," National Research Council, NIH Publication No. 86-23, revised 1985. The opinions or assertions contained herein are the private views of the authors and are not to be construed as official or as reflecting the views of the Army or the Department of Defense.

for choline acetyltransferase (ChAT). ChAT-positive cells in PPTn and LDTn (PPTn/LDTn) have been designated as the Ch-5 and Ch-6 cholinergic cell groups by Mesulam and co-workers (3)

While attempts to formulate a unitary model for central cholinergic function have been replete with inconsistencies, the hypothesis has been advanced that PPTn/LDTn may be considered a functional unit with an important role in behavioral state regulation (4–7). Ascending cholinergic projections from the PPTn/LDTn may control, in part, shifts in thalamocortical processing associated with different arousal states (reviewed in 8). This conclusion is supported by the finding that PPTn/LDTn (presumably cholinergic) neurons with antidromically identified thalamic projections exhibit "precursor signs" of increased discharge rates during transitions from EEG-synchronized to EEG-desynchronized, or rapid eye movement (REM), sleep state in the chronically implanted cat model (9). Likewise, presumptive experimental evidence supports a role for descending cholinergic projections from PPTn/LDTn to the medulla oblongata in the behavioral manifestations of REM sleep (10). Cholinergic agonists applied to selective sites within the medial pontine reticular formation (MPRF) invoke muscle atonia, EEG desynchronization, pontogeniculo-occipital (PGO) waves, rapid eye movements, and hippocampal theta rhythms. The PPTn/LDTn is considered to be the endogenous source of cholinergic afferents to the MPRF (reviewed in 10). More recently, Lydic and co-workers have reported a series of findings which also argue for the involvement of PPTn/LDTn in state-dependent respiratory depression (6,11–16). While those studies seem to have provided a link between PPTn-related acetylcholine (ACh) release within the MPRF and respiratory depression, it was acknowledged that it is also possible that PPTn/LDTn neurons might provide parallel cholinergic input to pontomedullary respiratory neurons as well as to the MPRF (6).

The evidence is also good that cholinergic mechanisms are involved in cardiovascular and respiratory regulation in the lower brain stem. Earlier studies localized the circulatory and/or respiratory effect of cholinergic substances to chemosensitive zones on the ventral surface of the medulla oblongata (17–19). Subsequent iontophoretic studies have demonstrated that the majority of cardioinhibitory (20,21) and respiratory-related neurons (22) studied in the ventrolateral medulla (nucleus ambiguus area) are also cholinoceptive. Additionally, a number of other neuroactive substances are known to be present in the ventrolateral medulla (see 23,24 for review). However, very little is known about their role and synaptic interactions. It is likely that most of these neuroactive substances, including ACh, are modulatory to central chemosensitivity and to the central integration in the medulla of cardiovascular and respiratory reflex afferents.

On the other hand, the excitatory amino acid glutamate may play a more key role in the integration of cardiovascular and respiratory reflexes (25). Glutamate is thought to mediate fast synaptic transmission at most excitatory synapses within the central nervous system. Glutamate appears to be released by central terminals of cardiopulmonary afferent fibers in the nucleus of the solitary tract (NTS) (26,27). Furthermore, central glutamate metabolism and glutamate release from glial elements have been proposed as additional mechanisms of the central respiratory response to hypoxia (25). This latter point is controversial as other investigators have not observed an increase in brain glutamate during hypoxia after peripheral chemodenervation (28). However, several studies using cats, dogs, and rats are consistent in demonstrating that excitatory amino acid receptor antagonists alter baro- and chemoreceptor-induced respiratory and cardiovascular responses (29–31). Moreover, immunohistochemical and microinjection studies suggest that glutamate activity is likely to be widespread in medullary cardiorespiratory integration areas (32), including a recently demonstrated NTS to nucleus ambiguus (AMB) pathway (23).

Therefore, in the current chapter, we propose to present evidence for PPTn/LDTn (cholinergic) modulation of glutamate elicited postsynaptic responses in AMB. Our experimental model combined glutamate microdialysis within PPTn with cholinergic and glutamate microiontophoresis in the AMB. The relationship of these findings to the hypothesis that descending mesopontine cholinergic fibers contribute to state-dependent respiratory depression (6) will be discussed.

II. The Rat Pedunculopontine Tegmental Nucleus

A. Functional Considerations

The rat mesopontine tegmentum contains large cholinergic and smaller noncholinergic cells (3). As discussed, cholinergic afferents from PPTn have been implicated in a number of physiological functions, including the production of components of REM sleep, regulation of thalamic activity, as well as stereotypical motor activity (10). Rye and co-workers have attempted to ascribe this latter function to noncholinergic neurons of the "midbrain extrapyramidal area" (1,2); however, this view has recently been challenged (32–34). In in vivo extracellular recording studies, although the cholinergic identity of the neuron was questionable, correlations between mesopontine neuronal firing and locomotor behavior (35,36) and sleep–wake states (4,36) have in fact been reported.

Recording from the nucleus parabrachialis lateralis of unanesthetized cats, Saito and co-workers (36) were able to identify three populations of cells based on their activity patterns during the sleep–wake cycle. In a population of 84 single units, 31% exhibited a marked reduction or cessation of firing during REM or paradoxical sleep. These were termed PS-off cells. A majority (55%) of the units isolated increased their firing during REM sleep and were termed PS-on cells. Six of the PS-on cells were phase-locked with PGO waves and fired in bursts of 3–5 action potentials 5–25 ms before the PGO wave was recorded in the thalamus. PS-on cells, therefore, might participate in the generation of thalamocortical PGO waves. The question then arises as to whether the firing patterns of mesopontine neurons are reflections of synaptic input or arise from intrinsic membrane properties. In other words, are PPTn/LDTn neurons the "pattern generator" for thalamocortical oscillations? This question has been addressed, in part, by in vitro studies.

Based on the electrophysiological properties of PPTn/LDTn neurons in brain slice preparations from guinea pigs, three types of neurons also have been described (37). Type I neurons make up approximately 10% of the populations and are characterized by low-threshold spikes (LTS). These LTS are manifested as rebound bursts after hyperpolarizing current pulses. A voltage-dependent calcium current underlies the LTS. The conductance was inactivated at the resting membrane potential and de-inactivated at hyperpolarized membrane potentials. It is unlikely that type I neurons are cholinergic due to their small size and the apparent lack of projections outside the tegmentum. Type II neurons fired spontaneously and exhibited a prominent late after-hyperpolarization (AHP) in contrast to the rebound excitation of type I neurons. Several conductances contributed to the firing behavior of type II neurons. Action potentials were preceded by two-component (fast and sustained) tetrodotoxin (TTX)-sensitive depolarizing prepotentials. Transient outward A-currents were apparently responsible for the delay in returning to the baseline potential following an action potential. In addition, high-threshold cadmium-sensitive (calcium) spikes were uncovered after TTX blockade of fast action potentials. Morphologically, type II and type III neurons were described as large fusiform multidendritic putatively cholinergic cells. Type III cells were also spontaneously active and exhibited A-currents as well as LTS. It was noted that these two currents may provide a sensitive mechanism for controlling rebound excitability in light of their opposing actions and susceptibility to neurochemical modulation. It was also acknowledged that type II and type III neurons might represent two functional states of the same neuron. Similar results were reported from rat brain slices by Kang and Kitai (38).

In summary, electrophysiological data are consistent with state-dependent phasic and tonic discharge patterns of mesopontine neurons. While correlations between state-dependent in vivo discharge behavior and the in vitro findings relative to the intrinsic membrane conductances are not straightforward, it is apparent that these neurons exhibit properties compatible with regulation of behavioral states, including the sleep–wake cycle.

B. Transmitter-Specific Afferents

Ultimately, the discharge patterns of mesopontine neurons must reflect the interplay of intrinsic membrane conductances and afferent inputs. However, the neurochemicals employed by afferent pathways to PPTn/LDTn have not been thoroughly delineated. Reportedly, the rat PPTn/LDTn receive multiple inputs from limbic structures, basal forebrain, oculomotor nuclei, ascending sensory systems, substantia nigra and the brainstem reticular formation (2,5, 38). Afferents from the limbic system include massive inputs from the lateral-hypothalamus-zona incerta area and the midbrain central gray, with lesser inputs from the medial prefrontal cortex. The afferents from the lateral hypothalamic area, a region associated with regulating autonomic functions, appear to contain α-melanocyte–stimulating hormone (5). Little is known about projections from the midbrain central gray, although they originate in areas found to be rich with neuropeptides, including somatostatin, substance P, vasoactive intestinal peptide, and opioid peptides (39). It has therefore been proposed that neuropeptides are involved in the long-term modulation of state control (40). On the other hand, the synaptic input from the prefrontal cortex is thought to be via excitatory amino acid non-NMDA receptors (41). Reportedly, glutamate is also contained within an efferent subpopulation of PPTn/LDTn neurons (42). Considering the interconnections between ipsi- and contralateral mesopontine neurons (5), it is reasonable to assume that cholinergic mesopontine neurons might also receive an excitatory synaptic input from intrinsic mesopontine glutamatergic neurons. Moreover, studies utilizing excitatory neurotoxins suggest that this region contains a high density of NMDA receptors (43). Inhibitory GABAergic inputs appear to arise, in part, from the substantia innominata of the basal forebrain (44) as well as the substantia nigra pars reticulata (38,45). Other neurotransmitters inhibitory to PPTn/LDTn neurons include serotonin from the dorsal and median raphe nuclei (5,46) and noradrenaline from the locus ceruleus (5,47). Moreover, carbachol has been demonstrated to reliably suppress the firing of at least one type of mesopontine neuron, presumably via a muscarinic autoreceptor (4,48).

Hence, PPTn/LDTn receive inputs from areas known to be involved in behavioral and autonomic regulation. Additionally, the postsynaptic effects of the afferent neurochemicals are likely to be consistent with known state-dependent pharmacology and firing behavior of PPTn/LDTn neurons (36,40). For example, the increased tonic firing of some putative cholinergic PPTn/LDTn neurons during arousal and REM sleep appears to be enabled by a decrease in monoaminergic (noradrenaline and serotonin) afferent tone. The state-dependent excitatory drive to tonic PPTn/LDTn neurons likely comes from peptides and glutamate. Phasic or burst firing PPTn/LDTn neuronal behavior might reflect rebound excitation in type I and possibly type II neurons subsequent to GABAergic-mediated hyperpolarization.

C. Efferent Projections of PPTn/LDTn

Widely distributed diffuse cholinergic and noncholinergic projections ascend and descend from the mesopontine tegmentum (1–3,8,34,39). While this chapter focuses on descending effects on PPTn/LDTn modulation, ascending projections in the rat are reportedly distributed to the thalamus, basal forebrain, thalamus, limbic cortex, and perhaps the basal ganglia (reviewed in 2). Descending projections from PPTn/LDTn have been described in the cat (49,50) and rat (1,51–57). Descending projections were found mainly distributed to pontomedullary reticular nuclei. In the rat, Woolf and Butcher (57) utilized retrograde tracer with ChAT histochemistry and described mesopontine projections to reticular spinal nuclei, which included the raphe magnus nucleus, lateral reticular nucleus, medullary reticular nucleus, and the oral and caudal pontine reticular nuclei. These latter projections, along with specific efferents to the gigantocellular reticular nucleus, were confirmed by *Phaseolus vulgaris*–leukoagglutinin anterograde tracer studies (51). These projections appear to be involved in REM sleep induction and the muscle atonia of REM sleep (58). Additionally, a ChAT immunoreactive pathway has been described from PPTn to the rostral ventrolateral medulla (RVLM), which has implications for cardiovascular regulation. Rye and co-workers (1) also described pathways from the mesopontine tegmentum, which innervated autonomic control areas in the medulla. The NTS and caudal VLM (CVLM), including AMB and adjacent cardiorespiratory control neurons, appear to be innervated by the ventrolateral branch of Probst's tract. It was believed that most of these fibers originated in the predominantely noncholinergic parabrachial nucleus. However, it would appear that the CVLM is innervated by some cholinergic PPTn fibers (see Table 2, case 1 in Ref. 1).

As mentioned, it is likely that some of the noncholinergic cells that contribute to the efferent output of PPTn/LDTn are glutamatergic (42). Furthermore, PPTn/LDTn neurons exhibit immunoreactivity for a number of peptides including substance P, corticotropin releasing factor, bombesin, and atriopeptin, at least in the rat (59,60,61). The terminal field neuronal effects of PPTn/LDTn stimulation are not well understood. Stimulation of PPTn and LDTn, under the appropriate conditions, elicited a short latency excitation of thalamic neurons followed by a long latency and prolonged depolarization superimposed by a barrage of action potentials (see Ref. 9). While the early depolarization is thought to be mediated by nicotinic receptors (NAchR), the slow depolarization may be via muscarinic receptor (MAChR) mechanisms and/or peptide modulation.

III. Nucleus Ambiguus

A. Functional Considerations

In the rat, the AMB and associated ventroambigual (periambigual) region appear as aggregations of small to large, mainly multipolar neurons located ventral to the parvocellular reticular nucleus and dorsal to the lateral reticular nucleus (62). The cell column extends from the rostral "retrofacial," or Botzinger, complex to the caudal "retroambigual" nucleus. Presumably, this cell column represents a developmental ventral descent of the visceral efferent column of the medulla. In most mammals, these motor and premotor neurons are integral to cardiorespiratory control including the control of laryngeal and bronchial musculature.

Both electrophysiological (63) and morphological (64,65) studies confirm that cardioinhibitory fibers originate from the dorsal motor nucleus of the vagus (DMV) and AMB—mainly from the AMB. Standish and co-workers (64) have proposed functional differences between them, in that DMV neurons have primarily unmyelinated axons that mediate slow vagal effects, whereas AMB neurons are myelinated and mediate fast effects. This latter study also demonstrated the ventricular innervation of the vagus. The AMB neurons are also in close proximity to the caudal ventrolateral group of cardiovascular neurons (CVLM), which are premotor to sympathetic preganglionic neurons (66).

The ventral respiratory group (VRG) is also associated with the AMB cell column. The caudal AMB-retroambigual complex exhibits primarily expiratory activity and most of these neurons project to the thoracic spinal cord, where they innervate internal intercostal motoneurons (66,67). Others appear to be bulbar output neurons innervating laryngeal, bronchial, and/or esophageal musculature.

B. Neurochemical Modulation of Ambiguus Neurons

Precise information regarding the localization, postsynaptic effect, and physiological impact of neurochemicals on autonomic reflexes within the ventrolateral medulla is not available (see 68,69). Nevertheless, it appears likely that reflex excitatory synaptic interactions in AMB are mediated by excitatory amino acids, and these synaptic effects are in turn modulated by classical neurotransmitters and neuropeptides.

The most direct evidence comes from exogenous iontophoretic application or microinjection of putative neurotransmitter agonists or antagonists to observe their cellular and/or autonomic effects. Excitatory amino acids have been routinely used to locate cardioinhibitory neurons in the AMB by their ability to induce bradycardia (20,21). In addition, recent evidence suggests that NMDA receptors might participate in solitario-ambigual synaptic transmission (70). Somatostatin may be colocalized with glutamate within this pathway (71), and somatostatin enhances glutamate-evoked excitation of AMB motoneurons. It is also likely that the AMB region receives glutamatergic efferents from the mesopontine tegmentum (42). Reportedly cardioinhibitory AMB neurons are also excited by dopamine, enkephalins, substance P, and acetylcholine (21,72), whereas they appear to be inhibited by glycine (73) and possibly serotonin and noradrenaline.

Although more variable, pharmacological studies of the VRG mirror the responses of cardioinhibitory AMB neurons (63). Bradley and Lucy (22) reported that the iontophoretic administration of glutamate resulted in excitation of inspiratory, expiratory, phase-spanning, and nonrespiratory units in the region of the AMB. Glutamate often converted phasic activity to tonic activity. In the same study, acetylcholine administration caused excitation in 80% of the neurons studied; no inhibition was reported. However, two different time courses of the response to ACh were noted: a short latency onset and offset, and a slow onset with persistent excitation after the iontophoretic current was turned off. This latter finding is particularly significant as it seems to suggest that ACh might modulate or enable response plasticity in cardiorespiratory reflex circuitry. However, the underlying mechanisms are not straightforward. In the forebrain, the M1 muscarinic receptor is primarily associated with response plasticity (74,75). In the case of AMB, the actions of cholinergic antagonists seem to suggest the presence of both muscarinic and nicotinic receptors. Zhang et al. (76) and Wang et al. (77) have provided further evidence for an AMB cholinergic nicotinic synapse, mediating information transfer in brainstem vagal motoneurons. Moreover, M2 seems to be the predominant muscarinic type in the rat brainstem, presumably subserving an autoreceptor role (78). The source of cholinergic afferents in these studies is left unclear.

Therefore, we designed a series of experiments to test the hypothesis that activity within the PPTn is capable of modulating glutamate-evoked neuronal activity in AMB, the underlying assumption being that glutamate microiono-phoresis mimics the effect of visceral reflex interneurons on AMB neurons. The specific aim of these experiments was (a) to test the effect of pharmacological modulation of PPTn on AMB neurons, (b) to assess the effect of muscarinic antagonist (i.e. atropine), and (c) to look for persistent effects of PPTn stimulation.

Studies were performed on urethane-anesthetized (1.5 g/Kg, IP) male Sprague-Dawley rats. Neuronal activity was recorded via 3 *M* NaCl-filled center barrels of 7 barreled glass multipipettes. For microiontophoresis, the multipipettes also contained acetylcholine chloride (0.5 *M*) atropine sulfate (0.025 *M*), sodium glutamate (0.2 *M*), and artificial cerebrospinal fluid (aCSF). Single-unit AMB activity was discriminated and collected via a F. Haer counter that reset each 5 s. Data were converted to a rate per second display in software and analyzed as peak firing rates. The remaining barrels were filled with the retrograde traces, rhodamine, and/or fluorescein latex microspheres. The rhodamine-labeled latex microspheres were deposited at the end of the recording session; fluorescein was used as a nonspecific marker and was placed in the medial reticular formation. The AMB was approached using stereotaxic coordinates, and recording sites were selected based on the ability of high-current (~100 nA) delivery of glutamate to effect changes in heart rate. Microdialysis of glutamate (10 mM), GABA (10 mM), or aCSF was through a Carnegie Medicin CMA/12 microdialysis probe (1-mm membrane tip) directed at the caudal PPTn. Perfusion rate was 2 μL/min with in vitro transfer factor ranging from 36% to 54%.

Results are summarized for 26 neurons that responded with excitation to direct application of ACh with an atropine-sensitive change in firing rate. During glutamate microdialysis into PPTn, 42% (n = 11) of the AMB neurons exhibited an increase in net peak evoked response compared to the glutamate-evoked response during aCSF microdialysis into PPTn. 54% (n = 14) of the

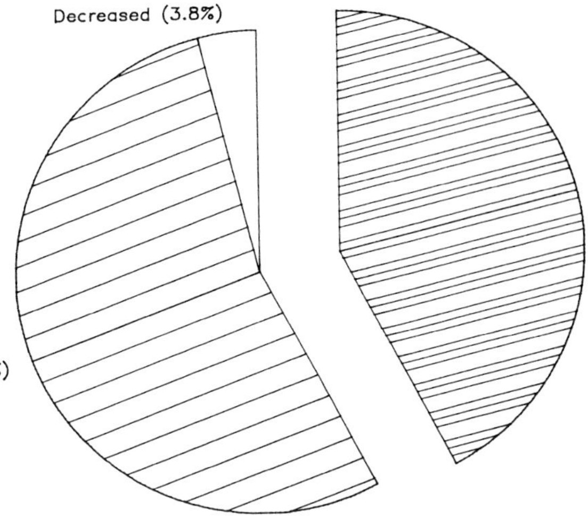

Figure 1 Differential effect of glutamate-evoked activity in nucleus ambiguus neurons (n = 26) during glutamate microdialysis into PPTn as compared to microdialysis of aCSF. Exploded slice represents 42.3% of the neurons (n = 11), which exhibited a net increase in peak glutamate-evoked activity.

neurons demonstrated no change in net evoked response, and only 4% (n = 1) showed a decrease in net evoked response (Fig. 1). A typical experiment is depicted in Figures 2A and 2B. In Figure 2, panel A, the increase in glutamate-evoked discharge in AMB during glutamate microdialysis is illustrated, along with the progressive attenuation by simultaneous application of atropine. Atropine attenuated the response in each of the 11 neurons exhibiting PPTn facilitation. Atropine was equally effective in blocking the excitation produced by ACh in all 26 neurons tested (data not shown). Panel B of Figure 2 is a continuation of panel A (lapsed time between graphs is 30 min). It demonstrates the recovery after atropine antagonism and the progressive increase in background and evoked discharge with continuous glutamate microdialysis into PPTn. It should be noted that both remained elevated for up to 30 min after glutamate microdialysis was terminated in three neurons. The data are summarized for the 11 facilitated neurons in Figure 3. During PPTn aCSF microdialysis, the glutamate-evoked response tended to be more robust as compared to ACh. GABA microdialysis into PPTn was ineffective in altering glutamate-evoked activity in AMB.

IV. Discussion

The literature reviewed and data presented in this chapter strongly suggest that descending projections from the PPTn/LDTn complex might be involved in

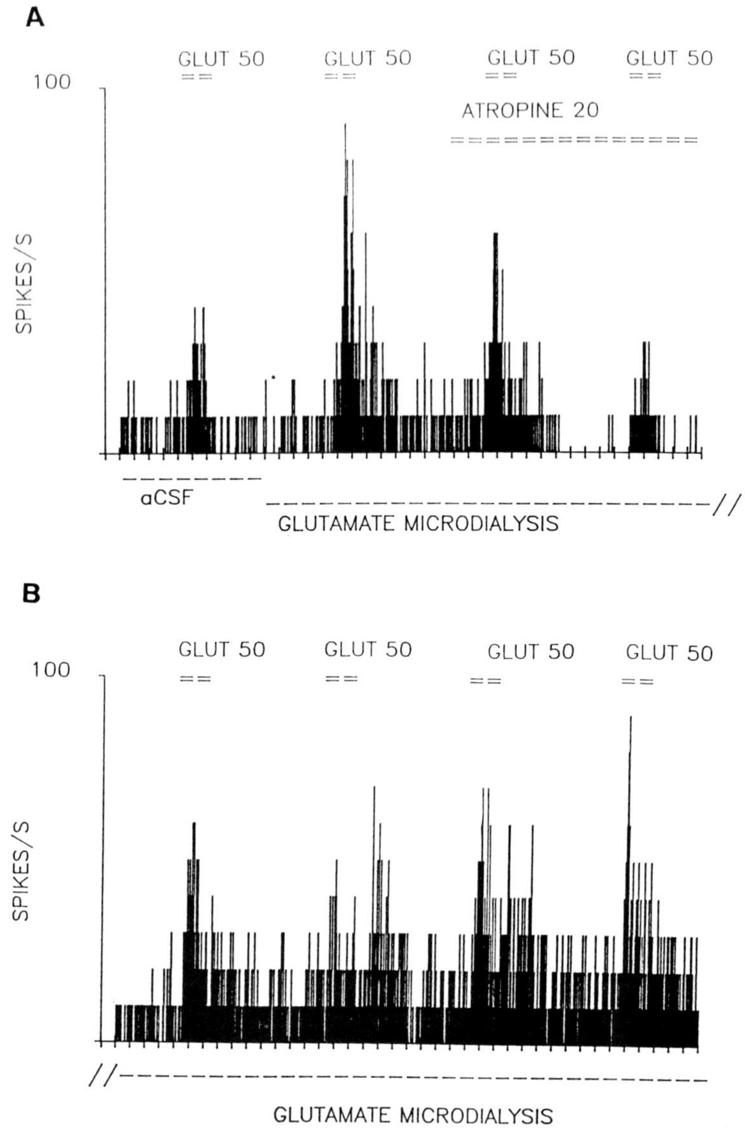

Figure 2 Discharge behavior of a typical facilitated unit during experimental protocol. (A) From left to right, discharge behavior of a single unit to two 50-s test pulses of 50-nA glutamate in the presence of aCSF, 50-nA glutamate after 200 s of PPTn microdialysis of 10-m*M* glutamate, and the progressive attenuation of glutamate-evoked activity by simultaneous application of 20-nA atropine sulfate. (B) Continuation of panel A after 30 min lapsed time. Figure illustrates progressive buildup of glutamate-evoked discharge and background firing with continuous PPTn glutamate microdialysis. Time between tic marks on x-axis equal 50 s, bin equal 5 s.

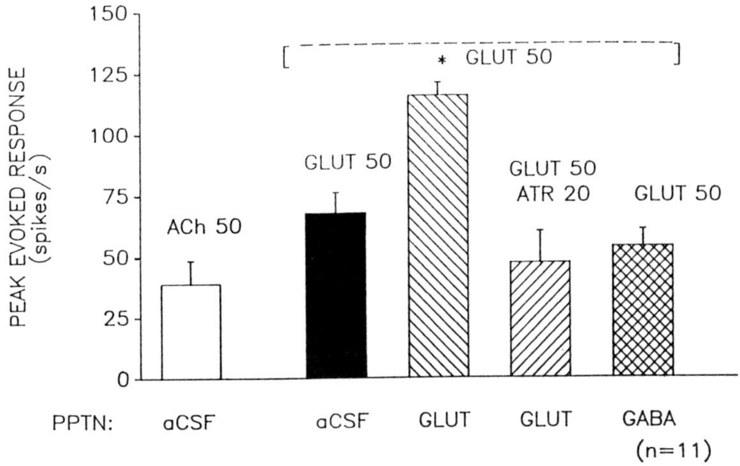

Figure 3 Summary of average net peak evoked activity for facilitated neurons (n = 11). Bottom legend provides substance microdialyazed into PPTn while top legend indicates substance(s) applied by microiontophoresis into the AMB. Average peak evoked activity was as follows: 50-nA ACh in presence of aCSF, 38.8 ± 9.7 spike/s (mean ± SEM); 50-nA glutamate with aCSF, 67.8 ± 8.3 spike/s; 50-nA glutamate in presence of 10-m*M* glutamate microdialysis, 115.6 ± 5.3 spike/s; 50-nA glutamate with 20-nA atropine in presence of glutamate microdialysis, 47.2 ± 13.0 spike/s; and 50-nA glutamate in presence of 10-m*M* GABA microdialysis, 53.8 ± 7.0. Treatments were randomized. $^*P < 0.05$ (ANOVA with Bonferoni-corrected post hoc comparisons).

state-dependent regulation of visceral sensory-motor integration in the CVLM. Presumptive cholinergic and noncholinergic cells in this complex exhibit electrophysiological characteristics consistent with a role in behavioral state modulation of neuronal activity up and down the neuraxis. State control seems to reflect an interplay between cholinergic activity and the other neurochemicals, specifically noradrenaline and serotonin, both within the PPTn/LDTn and at target sites. Neurons within and around the AMB, premotor, and motor efferents to visceral effectors are integral to cardiorespiratory reflex integration and are overwhelmingly cholinoceptive. Cholinergic mechanisms have been implicated in state-dependent modulation of neuronal activity throughout the central nervous system. Finally, our studies reveal that the discharge rate of a significant number of AMB neurons can be facilitated by putative pharmacological (glutamate microdialysis) upregulation of PPTn firing.

However, the current study is not without limitations. Because the retrograde tracing studies failed to provide a conclusive link between mesopontine cholinergic neurons and AMB (Rucker et al., unpublished), the origin of cholinergic afferents to AMB remain unclear. It remains to be demonstrated directly that PPTn glutamate microdialysis increases ACh release in the cortex. In addition, detailed pharmacological studies are needed to clarify the receptor type(s) involved in the observed responses.

Nevertheless, if we assume that descending cholinergic mechanisms were, at least in part, responsible, then several points can be addressed. During the transition from waking to slow-wave sleep (SWS) to REM sleep, subpopulations of PPTn neurons exhibit tonic firing during waking, become silent during SWS, and display bursting activity during REM sleep (58). Therefore our glutamate microdialysis should have induced PPTn firing behavior consistent with waking and/or REM sleep, and GABA microdialysis should have mimicked PPTn behavior during SWS. We saw no effects with PPTn GABA microdialysis. However, the increase in AMB neuron firing during PPTn glutamate microdialysis is consistent with the increased cardioinhibitory tone during waking and REM sleep. Furthermore, the decreased respiratory tone of REM sleep is no doubt linked to the atonia-related hyperpolarization of respiratory motor neurons in the lower brainstem, which has been linked to cholinergic mechanisms (11–16). This phenomenon may involve cholinergic activation of inhibitory interneurons, and is therefore also consistent with our current findings. Further speculation at establishing correlations between PPTn behavior and cardiorespiratory regulation in the lower brainstem is clearly not warranted at the present time; however, further study in this area is indicated.

Acknowledgments

This work was supported in part by the following awards: ONR N00014-91-J-1372, NIH GM08037, and NSF HRD 9106096.

References

1. Rye DB, Lee HJ, Saper CB, Wainer BH. Medullary and spinal efferents of the pedunculopontine tegmental nucleus and adjacent mesopontine tegmentum in the rat. J Comp Neurol 1988; 269:315–341.
2. Rye DB, Saper CB, Lee HJ, Wainer BH. Pedunculopontine tegmental nucleus of the rat: Cytoarchitecture, cytochemistry, and some extrapyramidal connections of the mesopontine tegmentum. J Comp Neurol 1987; 259:483–528.
3. Mesulam M-M, Mufson EJ, Wainer BH, Levy AI. Central cholinergic pathways in the rat: An overview based on an alternative nomenclature (Ch1-Ch6). Neuroscience 1983; 10:1185–1201.
4. Sakai K, Mansari ME, Jouvet M. Inhibition by carbacol microinjections of presumptive cholinergic PGO-on neurons in freely moving cats. Brain Res 1990; 527:213–223.
5. Semba K, Fibiger HC. Afferent connections of the laterodorsal and the pedunculopontine tegmental nuclei in the rat: A retro- and antero-grade transport and immunohistochemical study. J Comp Neurol 1992; 323:387–410.

6. Lydic R, Baghdoyan HA. Pedunculopontine stimulation alters respiration and increases ACh release in the pontine reticular formation. Am J Physiol 1993; 264: R544–R554.
7. Semba K, Reiner PB, Fibiger HC. Single cholinergic mesopontine tegmental neurons project to both the pontine reticular formation and the thalamus in the rat. Neurosci 1990; 38:643–654.
8. McCormick DA. Cholinergic and noradrenergic modulation of thalamocortical processing. TINS 1989; 12:215–221.
9. Steriade M, Buzsaki G. Parallel activation of thalamic and cortical neurons by brainstem and basal forebrain cholinergic system. In: Steriade M, Biesold D, eds. Brain Cholinergic Systems. New York: Oxford University Press, 1990:3–62.
10. Green RW, McCarley RW. Cholinergic neurotransmission in the brainstem: Implications for behavioural state control. In: Steriade M, Biesold D, eds. Brain Cholinergic Systems. New York: Oxford University Press, 1990:224–235.
11. Gilbert KA, Lydic R. Parabrachial neuron discharge is altered during the carbachol induced REM sleep-like state. Neurosci Lett 1990; 120:241–244.
12. Lee LH, Veino J, Lydic R. Fluorescent dye tracing reveals reciprocal connections between respiratory nuclei and the medial pontine reticular formation (abstr). Neurosci Abstr 1989; 18:124.
13. Lydic R, Baghdoyan HA. Cholinoceptive pontine reticular mechanisms cause state-dependent changes in respiration. Neurosci Lett 1989; 102:211–216.
14. Lydic R, Baghdoyan HA, Lorinc Z. Microdialysis of cat pons reveals enhanced acetylcholine release during state-dependent respiratory depression. Am J Physiol 1991; 261:R766–R770.
15. Lydic R, Baghdoyan HA, Wertz R, White DP. Cholinergic reticular mechanisms influence the state-dependent ventilatory response to hypercapnia. Am J Physiol 1991; 261:R738–R746.
16. Lydic R, Baghdoyan HA, Swillich CW. Cholinoceptive reticular mechanisms cause state-dependent hypotonia in posterior cricoarytenoid muscles of the larynx. FASEB J 1989; 3:1625–1631.
17. Guertzenstein PG. Blood pressure effects obtained by drugs applied to the ventral surface of the brainstem. J Physiol (Lond) 1973; 229:395–408.
18. Dev NB, Loeschcke HH. Topography of the respiratory and circulatory responses to acetylcholine and nicotine on the ventral surface of the medulla oblongata. Pflügers Arch 1979; 379:19–27.
19. Dev NB, Loeschcke HH. A cholinergic mechanism involved in the respiratory chemosensitivity of the medulla oblongata in the cat. Pflügers Arch 1979; 379:29–36.
20. Gilbey MP, Jordan D, Richter DW, Spyer KM. Synaptic mechanisms involved in the inspiratory modulation of vagal cardio-inhibitory neurons in the cat. J Physiol (Lond) 1984; 356:65–78.
21. Agarwal SK, Calaresu FR. Enkephalins, substance P and acetylcholine microinjected into the nucleus ambiguus elicit vagal bradycardia in rats. Brain Res 1991; 563:203–208.
22. Bradley PB, Lucy AP. Cholinoceptive properties of respiratory neurones in the rat medulla. Neuropharmacology 1983; 22:853–858.
23. Eldridge FL, Millhorn DE. Central regulation of respiration by endogenous neurotransmitters and neuromodulators. Ann Rev Physiol 1981; 43:121–135.

24. Gardiner SM, Bennett T. Brain neuropeptides: Actions on central cardiovascular control mechanisms. Brain Res Rev 1989; 14:79–116.
25. Kazemi H, Hoop B. Glutamic and -aminobutyric acid neurotransmitters in central control of breathing. J Appl Physiol 1991; 70:1–7.
26. Reis DJ, Granata AR, Perrone MH, Talman WT. Evidence that glutamic acid is the neurotransmitter of baroreceptor afferents terminating in the nucleus tractus solitarius. J Auton Nerv Syst 1981; 3:321–334.
27. Meeley MP, Underwood MD, Talman WT, Reis DJ. Content and in vitro release of endogenous amino acids in the area of the nucleus of the solitary tract of the rat. J Neurochem 1989; 53:1807–1817.
28. Kazemi H, Chiang CH, Hoop B. Role of medullary glutamate in the hypoxic ventilatory response. In: Lahiri S, ed. Comroe Memorial Symposium—Chemoreceptors and Reflexes in Breathing. New York: Oxford University Press, 1989:233–242.
29. Ang RC, Hoop B, Kazemi H. Role of glutamate as the central neurotransmitter in the hypoxic ventilatory response. J Appl Physiol 1992; 72:1480–1487.
30. Vardhan A, Kachroo A, Sapru HN. Excitatory amino acid receptors in commissural nucleus of the NTS mediate carotid chemoreceptor responses. Am J Physiol 1993; 264:R41–R50.
31. Koshiya N, Huangfu D, Guyenet PG. Ventrolateral medulla and sympathetic chemoreflex in the rat. Brain Res 1993; 609:174–184.
32. Hall WC, Fitzpatrick D, Klatt LL, Raczkowski D. Cholinergic innervation of the superior colliculus in the cat. J Comp Neurol 1989; 287:295–514.
33. Kang Y, Kitai ST. Electrophysiological properties of pedunculopontine neurons and their postsynaptic responses following stimulation of substantia nigra reticulata. Brain Res 1990; 535:79–95.
34. Spann BM, Grofova I. Cholinergic and non-cholinergic neurons in the rat pedunculopontine tegmental nucleus. Anat Embryo 1992; 186:215–227.
35. Garcia-Rill E, Skinner RD. Modulation of rhythmic function in the posterior midbrain. Neuroscience 1988; 27:639–54.
36. Saito H, Sakai K, Jouvet M. Discharge patterns of the nucleus parabrachialis lateralis neurons of the cat during sleep and waking. Brain Res 1977; 134:59–72.
37. Leonard CS, Llinas RR. Electrophysiology of mammalian pedunculopontine and laterodorsal tegmental neurons in vitro: Implications for the control of REM sleep. In: Steriade M, Biesold D, eds. Brain Cholinergic Systems. New York: Oxford University Press, 1990:205–223.
38. Kang Y, Kitai ST. Electrophysiological properties of pedunculopontine neurons and their postsynaptic responses following stimulation of substantia nigra reticulata. Brain Res 1990; 535:79–95.
39. Satoh K, Fibiger HC. Cholinergic neurons of the laterodorsal tegmental nucleus: Efferent and afferent connections. J Comp Neurol 1986; 253:277–302.
40. Jones BE. Paradoxical sleep and its chemical/structural substrates in the brain. Neuroscience 1991; 40:637–656.
41. Highfield D, Grant SJ. Electrophysiological evidence for an excitatory amino acid pathway from medial prefrontal cortex to lateral dorsal tegmental nucleus and rostral locus coeruleus (abstr). Soc Neurosci Abstr 1989; 15:644.
42. Clements JR, Grant S. Glutamate-like immunoreactivity in neurons of the laterodorsal tegmental and pedunculopontine nuclei in the rat. Neurosci Lett 1990; 120: 70–73.

43. Inglis WL, Dunbar JS, Winn P. Barbiturate anaesthesia reduces the neurotoxic effects of quinolinate but not ibotenate in the rat pedunculopontine tegemental nucleus. Neurosci Lett 1993; 156:78–82.
44. Semba K, Reiner PB, McGeer EG, Fibiger HC. Brainstem projecting neurons in the rat basal forebrain: Neurochemical, topographical, and physiological distinctions from cortically projecting cholinergic neurons. Brain Res Bull 1989; 22:501–509.
45. Childs JA, Gale K. Neurochemical evidence for a nigrotegmental GABAergic projection. Brain Res 1983; 258:109–114.
46. Luebke JL, Greene RW, Semba K, Kamondi A, McCarley RW, Reiner PB. Serotonin hyperpolarizes cholinergic low-threshold burst neurons in the rat laterodorsal tegmental nucleus in vitro. Proc Natl Acad Sci 1992; 89:743–747.
47. Williams JA, Reiner PB. Noradrenaline hyperpolarizes identified rat mesopontine cholinergic neurons in vitro. J Neurosci 1993; 13:3878–3883.
48. El Mansari M, Sakai K, Jouvet M. Responses of presumed cholinergic mesopontine tegmental neurons to carbachol microinjections in freely moving cats. Brain Res 1990; 83:115–123.
49. Shiromani PJ, Lai YY, Siegel JM. Descending projections from the dorsolateral pontine tegmentum to the paramedian reticular nucleus of the caudal medulla in the cat. Brain Res 1990; 517:224–228.
50. Quattrochi JJ, Mamelak AN, Madison RD, Macklis JD, Hobson JA. Mapping neuronal inputs to REM sleep induction sites with carbachol-fluorescent microspheres. Science 1989; 245:984–986.
51. Grofova I, Keane S. Descending brainstem projections of the pedunculopontine tegmental nucleus in the rat. Anat Embryol 1991; 184:275–290.
52. Jones BE. Immunohistochemical study of choline acetyltransferase-immunoreactive processes and cells innervating the pontomedullary reticular formation in the rat. J Comp Neurol 1990; 295:485–514.
53. Semba K. Aminergic and cholinergic afferents to REM sleep induction regions of the pontine reticular formation in the rat. J Comp Neurol 1993; 330:543–556.
54. Semba K, Reiner PB, Fibiger HC. Single cholinergic mesopontine tegmental neurons project to both the pontine reticular formation and the thalamus in the rat. Neuroscience 1990; 38:643–654.
55. Skinner RD, Kinjo N, Ishikawa Y, Biedermann JA, Garcia-Rill E. Locomotor projections from the pedunculopontine nucleus to the medioventral medulla. Neuroreport 1990; 1:207–210.
56. Yasui Y, Cechetto DF, Saper CB. Evidence for a cholinergic projection from the pedunculopontine tegmental nucleus to the rostral ventrolateral medulla in the rat. Brain Res 1990; 517:19–24.
57. Woolf NJ, Butcher LL. Cholinergic systems in the rat brain: IV. Descending projections of pontomesencephalic tegmentum. Brain Res Bull 1989; 23:519–540.
58. Garcia-Rill E. The pedunculopontine nucleus. Prog Brain Res 1991; 36:363–389.
59. Crawley J, Olschowka J, Diz D, Jacobowitz D. Behavioral investigation of the coexistence of substance P, corticotropin releasing factor, and acetylcholinesterase in lateral dorsal tegmental neurons projecting to the medial frontal cortex of the rat. Peptides 1985; 6:891–901.
60. Standaert D, Saper C, Rye D, Wainer B. Colocalization of atriopeptin-like immunoreactivity with choline acetyltransferase and substance P-like immunoreactivity

in the pedunculopontine and laterodorsal tegmental nuclei in the rat. Brain Res 1986; 382:163–168.

61. Sutin E, Jacobowitz D. Immunocytochemical localization of peptides and other neurochemicals in the rat laterodorsal tegmental nucleus and adjacent area. J Comp Neurol 1988; 270:243–270.
62. Bystrzycka EK, Nail BS. Brain stem nuclei associated with respiratory, cardiovascular and other autonomic functions. In: Paxinos G, ed. The Rat Nervous System: Volume 2, Hindbrain and Spinal Cord. Orlando, Florida: Academic Press, 1985:95–110.
63. McAllen RM, Spyer KM. The location of cardiac vagal preganglionic motorneurones in the medulla of the cat. J Physiol (Lond) 1976; 258:187–204.
64. Standish A, Enquist LW, Schwaber JS. Innervation of the heart and its central medullary origin defined by viral tracing. Science 1994; 263:232–235.
65. Izzo PN, Deuchars J, Spyer DM. Localization of cardiac vagal preganglionic motorneurons in the rat: Immunocytochemical evidence of synaptic inputs containing 5-hydroxytryptamine. J Comp Neurol 1993; 327:572–583.
66. Richter DW, Spyer KM. Cardiorespiratory control. In: Loewy AD, Spyer KM, eds. Central Regulation of Autonomic Function. New York: Oxford University Press, 1990:189–207.
67. Richerson GB, Getting PA. Medullary respiratory neurons in the guinea pig: Localization and firing patterns. Brain Res 1992; 591:79–87.
68. Mueller RA, Lundberg DBA, Breese GR, et al. The neuropharmacology of respiratory control. Pharmacol Rev 1982; 34:255–285.
69. Gardiner SM, Bennett T. Brain neuropeptides: Actions on central cardiovascular control mechanism. Brain Res 1989; 14:79–116.
70. Wang YT, Bieger D, Neuman RS. Activation of NMDA receptors is necessary for fast information transfer at brainstem vagal neurons. Brain Res 1991; 567:260–266.
71. Wang YT, Zhang M, Neuman RS, Bieger D. Somatostatin regulates amino acid receptor mediated fast excitatory postsynaptic potential components in vagal motoneurons. Neuroscience 1993; 53:7–9.
72. Chitravanshi VC, Calaresu FR. Dopamine microinjected into the nucleus ambiguus elicits vagal bradycardia in spinal rats. Brain Res 1992; 583:308–311.
73. Chitravanshi VC, Agarwal SK, Calaresu FR. Microinjection of glycine into the nucleus ambiguus elicits tachycardia in spinal rats. Brain Res 1991; 566:290–294.
74. Pirch J, Rigdon G, Rucker H, Turco K. Basal forebrain modulation of cortical cell activity during conditioning. In: Napier TC, Kalivas PW, Hanin I, eds. The Basal Forebrain: Anatomy to Function. New York: Plenum, 1991:219–231.
75. Pirch JH, Turco K, Rucker HK. A role for acetylcholine in conditioning-related responses of rat frontal cortex neurons: Microiontophoretic evidence. Brain Res 1992; 586:19–26.
76. Zhang M, Wang YT, Vyas DM, et al. Nicotinic cholinoceptor-mediated excitatory postsynaptic potentials in rat nucleus ambiguus. Exp Brain Res 1993; 96:83–88.
77. Wang YT, Neuman RS, Bieger D. Nicotinic cholinoceptor-mediated excitation in ambigual motorneurons of the rat. Neuroscience 1991; 40:759–767.
78. Miyoshi R, Kito S, Shimoyama M. Quantitative autoradiographic localization of the M1 and M2 subtypes of muscarinic acetylcholine receptors in the monkey brain. Jpn J Pharmacol 1989; 51:247–255.

15

Integrative Aspects of Ventral Brainstem Function

The Intermediate Area and the Ventral Medulla

EUGENE E. NATTIE

Dartmouth Medical School
Lebanon, New Hampshire

I. Introduction

Part I of this book includes a number of presentations that describe the physiological roles of neurons located within the ventral aspect of the medulla oblongata. Many of these neurons have access from the surface of the ventrolateral medulla (VLM) and a surprising number of physiological systems appear to require input from these neurons for maintenance or modulation of their normal function. For the most part, experimental studies in each system (e.g., respiration, blood pressure regulation) have been performed in isolation without regard for other physiological systems. The outline below, which includes many of the papers of Part I, shows in brief the diversity of function represented within small regions of the ventral medulla and points out evidence indicating the importance of the ventral medullary neurons in these functions. The final (and speculative) section of the chapter asks why so many vital functions involve neurons with easy access to the VLM surface and what the functional significance of these neurons is.

II. The Intermediate Area

Cooling of the ventral medullary surface using temperatures that block function of neurons but not of passing axons can result in apnea and in decreased blood pressure (1–3). The most sensitive surface location for the respiratory effects, the intermediate area (IMA), lies adjacent to a similar area associated with potent blood pressure effects (see Ref. 4, this volume, for further discussion). Additional studies using surface application of neuroactive substances further described potent effects of a variety of these agents on breathing and blood pressure (see Refs. 4–6).

The specific neuroanatomical locus within the medulla of these effects due to perturbations of the VLM surface has been controversial. The blood supply to the brainstem arises at the ventral surface from penetrating arteries, which, together with their accompanying perivascular spaces, can allow rather deep penetration within the medulla of surface perturbations. This has been shown by evaluation of the rate and depth of penetration of small radiolabeled and larger molecules applied to the VLM surface (see Ref. 4, this volume, for discussion).

III. Possible Neuronal Substrates of the IMA

Some possible anatomical structures affected by these IMA surface experiments include the retrotrapezoid nucleus, the retrofacial nucleus, the subretrofacial region (including the C1 group), the *nucleus paragigantocellularis lateralis* (*PGCL*), midline and parapyramidal raphe neurons, the lateral reticular nucleus, and perhaps aspects of the rostral portion of the ventral respiratory group including the Bötzinger and pre-Bötzinger regions (see Ref. 4, this volume, for discussion). Here I will give a brief overview of some of these specific structures as they relate to this chapter.

The *retrotrapezoid nucleus* was identified by injection of retrograde anatomical tracers into major groups of respiratory neurons, which are located more dorsally. Labeling of a novel site lying ventral and ventromedial to the facial nucleus (within a few hundred micrometers of the VLM surface) led to this new anatomical definition (7). The retrotrapezoid nucleus was then found to contain neurons with respiratory-modulated as well as tonic firing patterns, both types increasing their firing rate with hypercapnia (see Ref. 4). Destruction of the retrotrapezoid nucleus region unilaterally can result in depression of respiratory output (including apnea) and absent chemosensitivity, suggesting an important role for these respiratory neurons (see Ref. 4). Such lesions have little effect on blood pressure.

Bilateral (8) and unilateral (9) electrolytic and chemical lesions within the *retrofacial nucleus* of the cat can also result in decreased respiratory output and chemosensitivity. The retrofacial and subretrofacial regions have neurons with respiratory-modulating firing patterns, some of which have processes that extend to the surface of the VLM (10,11). The retrofacial region has known respiratory involvement, while the subretrofacial region is important in blood pressure regulation (see below). At quite discrete sites located more caudally and deeper, bilateral injections of glutamate receptor antagonists can also produce apnea in anesthetized animals (12,13).

The PGCL (14,15) lies between the VLM surface and the nucleus ambiguus. It extends rostrally from the lateral reticular nucleus to the region medial to the facial nucleus lying just lateral to the inferior olive. It is essentially just beneath the center of the IMA. Retrograde tracing experiments have shown that afferent inputs converge there from a variety of sites, including the spinal cord, caudal lateral medulla, contralateral PGCL, nucleus of the solitary tract, A1 area, lateral parabrachial nucleus, Kölliker-Fuse nucleus, periaqueductal gray, and supraoculomotor nucleus (14). Studies of retrograde transport from injections limited to the juxtafacial portion of PGCL (15) show a different but also diverse pattern of afferent sources, including the dorsal column nuclei, inferior colliculus, paralemniscus, paramedian reticular formation, paraolivary reticular formation, retrofacial PGCL, caudal medullary reticular formation, and dorsal periaqueductal gray. The PGCL seems to be a source of convergence of many afferent inputs involved in pain, analgesia, cardiovascular control, respiratory control, exteroceptive sensation, and arousal. A major source of efferents from the PGCL is the locus ceruleus, which itself has widespread inputs to the higher nervous system and is thought to be primarily involved in arousal and vigilance (16).

Parapyramidal neurons of the midline raphe lie at sites accessible to surface applications at the IMA (17). While the raphe has been studied largely in regard to cardiovascular control (see Ref. 18, for example), there is evidence that this region is involved in respiratory control as well (19,20). Parapyramidal cells have also been implicated in the control of breathing by localized surface cooling experiments (21).

Thus, there are a number of defined anatomical structures, accessible to substances applied to the surface of the VLM, which are likely sites of physiological effects on respiratory and blood pressure regulation. The nomenclature used for these structures has evolved more from an anatomical than a functional rationale. From the studies cited above (and in Ref. 4), it appears that a rostral-to-caudal column of neurons exists in the VLM with involvement in the control of respiration and blood pressure (and in other functions discussed below). Cardiorespiratory neurons in the column appear to be distributed among a number of these (and other) anatomically designated structures. Much more detailed examination of the lateral tegmental field has led to a similar conclusion: that respiratory control neurons of the ventral respiratory group are distributed among a number of discrete anatomical nuclei (22,23).

IV. Central Chemoreceptors; CO_2

The traditional view of central chemoreception has focused on the VLM as the chemoreceptor locus, based largely on studies that used alterations in surface pH as the localizing stimulus. Some workers have criticized this interpretation and much of the evidence for and against this idea is reviewed in Ref. 4.

In my laboratory, we have used microinjections of 1 nL acetazolamide to produce focal regions of reversible tissue acidosis (see Refs. 4 and 24 for details). The results indicate that central chemoreception is located at widespread sites within the brainstem, including (1) the nucleus tractus solitarius, (2) the vicinity of the locus ceruleus, and (3) within 800 μm of the VLM surface at sites probably accessible from the VLM surface, as originally proposed. Ongoing work suggests the involvement of other sites as well (25).

Of interest in the role of the VLM in chemoreception are the measurements of tissue pH profiles during hypercapnia induced by injection of hypercapnic saline into the vertebral artery (26). Surprisingly, at some sites the pH decreased significantly and at others it did not change. The sites with large pH changes, proposed as the location of central chemoreceptors, were located in the vicinity of the nucleus tractus solitarius, within the nucleus ambiguus, and, again, near the VLM surface as originally proposed.

In this volume, data was presented from brainstem staining for the c-fos gene product following application of acidic pledgets to the caudal VLM surface (Ref. 27, this volume). The stained regions were surprising in their number and distribution throughout the caudal to rostral axis of the brainstem. While it is difficult to evaluate the specificity of such c-fos experiments, the widespread nature of neuronal stimulation from such a localized surface application emphasizes the number of connections of neurons accessible from the VLM surface. An alternative approach, optical imaging following systemic changes in CO_2, O_2, and blood pressure, indicated a rather widespread involvement of VLM neurons with altered function (Ref. 28, this volume). This approach has value in discerning patterns of involvement of neurons within a finite distance of the VLM surface and suggests discrete patterns of neuron involvement that are specific for the different stimuli.

In respect to the VLM, central chemoreception exists within a few hundred micrometers of the VLM surface at sites from the level of the hypoglossal nerve rootlets rostrally to the level of the trapezoid body. Stimulation by CO_2 or H^+ at these sites can stimulate respiratory output and large numbers of medullary neurons. A small number of observations from my lab suggests that some central chemoreceptor sites may be specific for blood pressure regulation, others for respiration (24).

V. Central Chemoreception; O_2

While the primary transduction of a hypoxic stimulus that increases respiration occurs at the carotid body, there are some interesting effects of hypoxia which are localized to the VLM. In the ventilatory acclimatization to chronic hypoxia, the role of central chemoreceptors and the VLM has been controversial. One hypothesis described the prolonged increase in ventilation seen with acclimatization to acidic changes in cerebral fluids (29). It seems clear that large-cavity cerebrospinal fluid pH is not acidic under these conditions (30). But VLM tissue pH in some regions does become acidic under hypoxic conditions (30). For example, in anesthetized cats, 5 min of isocapnic hypoxia such as to decrease arterial oxygen saturation to 48% results in regional VLM acidosis. The pH was decreased at rostral and caudal VLM regions but not at the IMA (31). This is reminiscent of the regional VLM pH changes observed following vertebral artery injection of hypercapnic saline (see above). It is possible that such changes participate in the increase in respiration observed in acclimatization. Respiration in adult animals is not stimulated by hypoxia (and the presumed VLM regional acidosis) following denervation of the peripheral chemoreceptors. But it seems reasonable to hypothesize that central chemoreceptor stimulation by local hypoxia-induced acidosis may potentiate the hypoxic ventilatory response in animals with intact peripheral chemoreceptors. The suggestion that the response to acute peripheral chemoreceptor stimulation involves an input from neurons in the VLM region (see Ref. 32, this volume, for discussion) adds support to this hypothesis.

There is evidence suggesting a possible role for VLM neurons in sensing changes in metabolism, perhaps via tissue hypoxia. First, the stimulation of respiration produced by infusion of 2,4-dinitrophenol can be blocked by cooling or destruction of the VLM at the IMA (33). Second, immunohistochemistry has localized the enzyme involved in nitric oxide metabolism, NADPH diaphorase, in a 2-mm long column of cells lying lateral to the inferior olive and medial to the C1 cell group in the rostral VLM (RVLM) (34). Third, specifically identified RVLM neurons involved in blood pressure regulation have been described that can be stimulated by local injections of sodium cyanide and by the production of hypoxia (arterial PO_2 = 27 mm Hg) even following removal of peripheral chemoreceptor tissue (35). Nearby respiratory neurons were not excited by these stimuli. These neurons excited by hypoxia appear to be the source of the central ischemic pressor response.

VI. Airways

Neurons in the caudal VLM stimulated by activation of group III and IV hindlimb afferents are involved in the decrease in airway resistance observed in such conditions (see Ref. 36, this volume). Neurons in more rostral sites can be stimulated by muscle contraction and produce a pressor response (Ref. 36, this volume). The VLM neurotransmitters involved may include glutamate, as microinjections of glutamate receptor antagonists attenuate these bronchodilator and pressor responses (Ref. 36, this volume).

Other airway responses can be affected by VLM neurons. Application to the VLM surface at the IMA of agents that block or stimulate the function of accessible neurons can alter (1) CO_2-sensitive respiratory phasic changes in nasal patency, (2) hypoglossal nerve activity affecting upper airway patency, (3) the pharyngeal response to pulmonary irritant receptor stimulation, (4) the tone of tracheal smooth muscle, and (5) the secretion of tracheal glands (32).

VII. Breathing

The studies discussed above, which showed profound effects on respiratory output and chemosensitivity with surface cooling at the IMA and lesions of more specific anatomical sites beneath the IMA, were performed in anesthetized or decerebrate animals. Of great interest are the results of similar studies in conscious animals. Conscious cats following bilateral coagulation of the IMA have decreased respiration and CO_2 retention and decreased responsivity to CO_2 (37,38). Cooling of the IMA in conscious goats also depresses ventilation at rest, in response to hypercapnia, and in exercise (Ref. 39 and H. V. Forster, personal communication). In conscious dogs, after unilateral lesioning of the retrotrapezoid region by kainic acid injections, periods of cluster breathing have been observed (40). In humans with neurological disease of the brainstem resulting in unilateral lesions of the RVLM, ventilation and CO_2 sensitivity are decreased and apneas during sleep are more common (41). And in victims of sudden infant death syndrome, some show abnormally few neurons in a region of the rostral ventrolateral medulla (42). Changes in airway caliber related to VLM neuronal function have not been evaluated in conscious animals.

Neurons located within a small region of the VLM just medial to the caudal aspect of the facial nucleus, close to the surface (43), and within a region just caudal to the level of the retrofacial nucleus—the pre-Bötzinger region (44,45)—have been proposed to be the site of respiratory rhythm generation, at least in the neonatal rat brainstem preparation. This is a difficult preparation in that it has a hypoxic central core with a viable outside ring of tissue approximately 800 μm in thickness (46). Yet the similarity of the anatomical location of these proposed regions for respiratory rhythmogenesis and those in adult mammals where cooling or lesioning have dramatic effects on respiratory output is striking.

VIII. Blood Pressure

Surface manipulations at and near to the IMA often had blood pressure as well as respiratory effects. Subsequent studies identified a subretrofacial pressor region (47), also called the C1 region (48), which includes a large portion of the PGCL. Neurons of this pressor region project to the intermediolateral column of the spinal cord, may contain catecholamines, and are thought important in the origin and maintenance of sympathetic tone to the cardiovascular system. Conditional pacemaker neurons, thought to be cardiovascular in function, have been reported here (49), and the link between respiratory and cardiovascular sympathetic output, whereby a large portion of cardiovascular sympathetic tone is phasically related to phrenic output, occurs here (50). The neurons in this rostral pressor region lie just caudal and dorsal to the retrotrapezoid nucleus, within the region ventral to the retrofacial nucleus. Within the RVLM, there appears to be a functional organization of the neurons into discrete groups that have specific vascular bed connections (51,52). This rostral pressor region can also be divided anatomically into three subregions: the midline raphe, PGCL, and rostroventrolateral reticular nucleus (53). This RVLM pressor region also has anatomical connections to the periaqueductal gray (18), the lateral parabrachial nucleus, the locus ceruleus, and the lateral hypothalamic area (54) and the amygdaloid complex (55). It receives widespread afferents (see Ref. 56, for example) and it can modulate cardiovascular reflexes at the nucleus tractus solitarius (57). Bilateral lesions of the C1 region can result in decreased blood pressure (to spinal levels) and abolish pressor sympathetic outflow (48,58).

Neurons responsive to serotonin receptor agonists (producing a hypotensive effect) lie in the midline raphe and at the parapyramidal region of the RVLM (59). And a more caudal region of the VLM appears to be involved in depressor responses (60,61). A single study suggests involvement of neurons in the midline raphe in the control of pulmonary blood flow. Injections of a glutamate agonist and toxin, ibotenic acid, into the raphe of the rat results in increased pulmonary artery pressure, hypoxemia, acidosis, hypothermia, and death without any evidence of pulmonary edema (62).

IX. Nociception

Neurons in the rostral ventromedial medulla are part of a network involved with control of pain transmission. They are located in the nucleus raphe magnus, the nucleus PGCL, and the adjacent reticular network, and they have been characterized by their projections to spinal cord and their responses to painful stimuli and opiate analgesic agents. Noxious stimuli to the extremities result in withdrawal of the paw and either increased (on cells) or decreased (off cells) firing rates of these neurons (63). The off cells are excited by opiate analgesic agents, suggesting that they have an inhibitory effect on pain transmission (63). Of particular interest is the fact that increased CO_2 can increase the threshold for painful stimuli, possibly via an endogenous opiate-related mechanism (64). This effect could well be occurring at this region of the VLM, where nociceptive and respiratory neurons are in close proximity. Of similar interest is the observation that application of opiates onto the VLM surface at the rostral aspect of the IMA can result in respiratory depression (65). And the hypotensive and respiratory depressant responses to cocaine application on the VLM surface at more caudal locations can be prevented in part by prior treatment at these sites with an anticholinesterase (see Ref. 66, this volume for discussion).

A specific role of PGCL neurons, which are easily accessible from the VLM surface, in the transmission of peripheral pain to the locus ceruleus has been demonstrated. Injections of lidocaine into the ventromedial PGCL attenuate the response of single neurons within the locus ceruleus to foot shock in the anesthetized rat (67).

X. Vestibular System

The stimulation of vestibular afferents has effects on autonomic and respiratory output (see Ref. 68, this volume). Neurons in the VLM rostral pressor region and the more caudal depressor region with known connections to the sympathetic preganglionic neurons respond with an increased firing rate to vestibular stimulation. There is a specificity to the VLM connections in that their input comes largely from otolith organs and they have a greater response to pitch than to roll.

XI. Why Are So Many Functions Attributed to Neurons So Close to Each Other and to the VLM Surface? What Is the Functional Significance of These Rostral VLM Neurons?

These are speculative questions without proven answers at present. In respect to the proximity to the VLM surface, one thought is that the projections to the surface of many neurons lying within the VLM allows them to sense or sample the cerebrospinal fluid (CSF) coming into contact with the surface as it flows by in its course through the ventricular system. There are substances in the CSF that might reflect activity at other more central sites within the neuroaxis or in blood via entry through regions having a less stringent blood-brain barrier (see Ref. 69, this volume, for example). These could be sensed at the VLM surface. This seems to me to be a rather unlikely possibility, given the incredibly diverse neuronal signaling possibilities in both the anatomical and synaptic realms. Also, the turnover for CSF is rather slow, giving this idea of sensing important stimuli via CSF a rather phlegmatic potential in respect to speed of response. Alternatively, if substances in CSF reflect some time average of central nervous system events or state, then this type of signaling system could be of value, allowing medullary autonomic neurons to monitor higher cerebral events in a slow, stately manner.

A second possibility to explain the location of these neurons near the VLM surface relates to the blood supply to the brainstem, which enters from the ventral surface via penetrating arteries. Any investigator who has examined the ventral medullary surface under the operating microscope can vouch for the impressive number of blood vessels both on the surface and entering into the parenchyma. These could provide a means for nearby neurons within the VLM to sample what is happening in the blood. This possible explanation would have to include an evaluation of the significance of the blood-brain barrier, which is intact in this region.

Third, the location of all of these important physiological effects to neurons lying close together within the VLM may just be coincidence. The entire brainstem is very compact, and other brainstem regions—e.g., the nucleus tractus solitarius (NTS) (see Ref. 70, for example) and the lateral tegmental field (22, 23)—are known to have many functions expressed within a small anatomical region.

I argue here that compelling evidence supports the hypothesis that neurons within the VLM have a particularly important role in the control of respiration and blood pressure. In respiratory regulation, specific lesions of small VLM regions and cooling of larger surface areas (cited above) have dramatic effects on respiratory output and reflexes even in chronic conscious preparations. Similar or larger lesions in other medullary (71) and pontine (72) regions known to be involved in respiratory regulation do not have such effects. In blood pressure regulation, again, specific VLM lesions dramatically reduce cardiovascular sympathetic outflow, even in chronic conscious preparations (see Ref. 58). The most effective lesion location for the blood pressure effects is not the same as that for the respiratory effects, although they are in proximity. While lesion studies do have problems and recent reviews (70) have downplayed any special significance of the RVLM in blood pressure regulation, I find the hypothesis for a specific and significant role for these RVLM cardiorespiratory neurons to be of interest and worthy of continued investigation.

What are the physiological roles of these RVLM neurons that are important in respiratory and blood pressure regulation? Early work on the IMA emphasized the word *integration*—that the IMA is an integrative area (3,37). There is ample evidence that these neurons receive afferents from many central and peripheral sources (cited above). In fact, single neurons located near these RVLM sites have been shown to receive multiple afferent inputs, which converge on them (73). However, most cardiovascular and respiratory peripheral afferents have their primary synapse in the more dorsally located NTS. The functional relationship between the RVLM and the NTS is unclear. Recent evidence (cited above) has shown that RVLM neurons can modulate specific reflexes at the NTS. Thus, RVLM neurons may integrate afferent information, and they can modify reflexes at other medullary sites. This region of the VLM also contains CO_2/H^+ chemoreceptor function, as shown by studies from my own lab using 1-nL injections of acetazolamide to produce focal acidosis (as well as other studies reviewed in Ref. 4), and it contains hypoxia-sensitive neurons (cited above).

I propose that, in addition to receiving afferent information, modifying reflexes at other sites, and containing chemosensitive elements, the cardiovascular and respiratory neurons of the rostral VLM provide a tonic input or drive to the neuronal networks involved in these respective control systems, an input that sustains these networks in an active state. This is not a new suggestion (74). And recent work in conscious goats with cooling of the rostral VLM surface shows decreased respiratory output in eucapnic, hypercapnic, hypoxic, and exercise conditions but no disruption of rhythm generation (Forster H. V., personal communication), observations supporting this hypothesis. Recent work from my lab has shown significant long-term stimulation of phrenic activity following single injections of glutamate (100 m*M*; 10 nL) into the RTN of the cat (75), again supporting this hypothesis. An analogy for this hypothesis in a better understood neural network is the driver or P cell for the pyloric rhythm of the crustacean stomatogastric ganglion (76). Specific destruction of these cells reduces the rhythmic output of this network into tonic activity. In mammals, there is evidence that afferent input is required for the respiratory control system to maintain a rhythmic output and that there is a tonic respiratory drive independent of the afferent determined rhythmic output (77,78). Is it possible that the source of this tonic drive is these rostral VLM neurons?

Acknowledgment

The author is supported by NIH RO1 HL28066.

References

1. Cherniack NS, von Euler C, Homma I, Kao FF. Graded changes in central chemoreceptor input by local temperature changes on the ventral surface of the medulla. J Physiol (Lond) 1979; 287:191–211.
2. Millhorn DE, Eldridge FL, Waldrop TG. Effects of medullary area $I_{(s)}$ cooling on respiratory response to chemoreceptor inputs. Respir Physiol 1982; 49:23–39.
3. Schläfke ME, See WR, Loeschcke HH. Ventilatory response to alterations of H^+ ion concentration in small areas of the ventral medullary surface. Respir Physiol 1970; 10:198–212.
4. Nattie EE, Li A, Coates EL. Central chemoreceptor location and the ventrolateral medulla (this volume).
5. Millhorn DE, Eldridge FL. Role of ventrolateral medulla in regulation of respiratory and cardiovascular systems. J Appl Physiol 1986; 61:1249–1263.
6. Bruce EN, Cherniack NS. Central chemoreceptors (brief review). J Appl Physiol 1987; 62:389–402.
7. Smith JC, Morrison DE, Ellenberger HH, et al. Brainstem projections to the major respiratory neuron populations in the medulla of the cat. J Comp Neurol 1989;281: 69–96.

8. St. John WM, Hwang Q, Nattie EE, Zhou D. Functions of the retrofacial neurons in chemosensitivity and ventilatory neurogenesis. Respir Physiol 1989; 76:159–172.
9. Nattie EE, Blanchford C, Li A. Retrofacial lesions: Effects on CO_2 sensitive phrenic and sympathetic activity. J Appl Physiol 1992; 73:1317–1325.
10. Bianchi AL, Grélot L, Iscoe S, Remmers JE. Electrophysiological properties of rostral medullary respiratory neurones in the cat: An intracellular study. J Physiol (Lond) 1988; 407:293–310.
11. Pilowsky PM, Jiang C, Lipski J. An intracellular study of respiratory neurons in the rostral ventrolateral medulla of the rat and their relationship to catecholamine-containing neurons. J Comp Neurol 1990; 301:604–617.
12. Abrahams TP, Hornby PJ, Walton DP, et al. An excitatory amino acid(s) in the ventrolateral medulla is (are) required for breathing to occur in the anesthetized cat. J Exp Pharm Ther 1991; 259:1388–1395.
13. Jung R, Bruce EN, Katona PG. Cardiorespiratory responses to glutaminergic antagonists in the caudal ventrolateral medulla of rats. Brain Res 1991; 564:286–295.
14. Van Bockstaele EJ, Pieribone VA, Aston-Jones G. Diverse afferents converge on the nucleus paragigantocellularis in the rat ventrolateral medulla: Retrograde and anterograde tracing studies. J Comp Neurol 1989; 290:561–584.
15. Van Bockstaele EJ, Akaoka H, Aston-Jones G. Brainstem afferents to the rostral (juxtafacial) nucleus paragigantocellularis: Integration of exteroceptive and interoceptive sensory inputs in the ventral tegmentum. Brain Res 1993; 603:1–18.
16. Aston-Jones G, Ennis VA, Pieribone VA, et al. The brain nucleus locus coeruleus: Restricted afferent control of a broad efferent network. Science 1986; 234:734–737.
17. Helke CJ, Thor KB, Sasek CA. Chemical neuroanatomy of the parapyramidal region of the ventral medulla in the rat. In: Ciriello J, Caverson MM, Polosa E, eds. The Central Organization of Cardiovascular Control. New York: Elsevier, 19898: 17–28.
18. Wang WH, Lovick TA. The inhibitory effect of the ventrolateral periaqueductal grey matter on neurones in the rostral ventrolateral medulla involves a relay in the medullary raphe nuclei. Exp Brain Res 1993; 94:295–300.
19. Holtman JR Jr, Anastasi NC, Norman WP, Dretchen KL. Effect of electrical stimulation of the raphe obscurus on phrenic nerve activity in the cat. Brain Res 1986; 362:214–220.
20. Holtman JR Jr, Marion LJ, Speck DF. Origin of serotonin-containing projections to the ventral respiratory group in the rat. Neuroscience 1990; 37:541–552.
21. Fukuda Y, Tojima H, Tanaka K, Chiba T. Respiratory suppression by focal cooling of ventral medullary surface in anesthetized rats: Functional and neuroanatomical correlate. Neurosci Lett 1993; 153:177–180.
22. Ellenberger HH, Feldman JL, Shan W-Z. Subnuclear organization of the lateral tegmental field. II: Catecholamine neurons and ventral respiratory group. J Comp Neurol 1990; 294:212–222.
23. Ellenberger HH, Feldman JL. Subnuclear organization of the lateral tegmental field. I: Nucleus ambiguus and ventral respiratory group. J Comp Neurol 1990; 294: 202–211.
24. Coates EL, Li A, Nattie EE. Widespread sites of brainstem ventilatory chemoreceptors. J Appl Physiol 1993; 75:5–14.
25. Bernard DG, Li A, Nattie EE. Ventilatory chemoreceptors in the midline brainstem raphe of rats. Soc Neurosci Abst 1994; 20:959.

26. Arita H, Ichikawa K, Kuwana S, Kogo N. Possible locations of pH-dependent central chemoreceptors: Intramedullary regions with acidic shift of extracellular fluid pH during hypercapnia. Brain Res 1989; 485:285–293.
27. Douglas RM, Trouth CO, Melville GN, et al. Localization of c-FOS protein in rat brainstem in response to the topical application of acidic mock cerebrospinal fluid (this volume).
28. Gozal D, Rector DM, Harper RM. Functional assessment of the ventral medullary surface by optical imaging (this volume).
29. Fencl V, Gabel RA, Wolfe D. Composition of cerebral fluids in goats adapted to high altitude. J Appl Physiol 1979; 47:508–513.
30. Musch TI, Dempsey JA, Smith CA, et al. Metabolic acids and $[H^+]$ regulation in brain tissue during acclimatization to chronic hypoxia. J Appl Physiol 1983; 55: 1486–1495.
31. Xu F, Sato M, Spellman MJ Jr, et al. Topography of cat medullary ventral surface hypoxic acidification. J Appl Physiol 1992; 73:2631–2637.
32. Haxhiu MA, Loewy AD, Cherniack NS. Ventral medullary network and airway control (this volume).
33. Prabhakar NR, Mitra J, Adams EM, Cherniack NS. Involvement of ventral medullary surface in respiratory responses induced by 2,4-dinitrophenol. J Appl Physiol 1989; 66:598–605.
34. Iadecola C, Faris PL, Hartman BK, Xu Y. Localization of NADPH diaphorase in neurons of the rostral ventral medulla: Possible role of nitric oxide in central autonomic regulation and oxygen chemoreception. Brain Res 1993; 603:173–179.
35. Sun MK, Reis DJ. Hypoxia selectively excites vasomotor neurons of rostral ventrolateral medulla in rats. Am J Physiol (Regul Integrat Comp Physiol 35) 1994; 266:R245–R256.
36. Solomon IC. Role of the caudal ventrolateral medulla in the reflex bronchodilation arising from the hindlimb (this volume).
37. Schläfke ME. Elimination of central chemosensitivity by coagulation of a bilateral area on the ventral medullary surface in awake cats. Pflügers Arch 1979; 379:231–241.
38. Burghardt F, Schläfke ME. Loss of central chemosensitivity: An animal model to overcome respiratory insufficiency. J Auton Nerv Syst 1986; (suppl):105–109.
39. Pan LG, Forster HV, Erickson BK, Lowry T. Effect of transient cooling of the ventral surface of the medulla (VLM) in awake goats. FASEB J 1991; 5:A665.
40. Dormer KJ, Bedford TG. Cardiovascular control by the rostral ventrolateral medulla in the conscious dog. In: Ciriello J, Caverson MM, Polosa E, eds. The Central Neural Organization of Cardiovascular Control, Progress Brain Research 81. Amsterdam: Elsevier, 1981:265–277.
41. Guz A. Brain-stem lesions in man with loss, or reduction of, ventilatory responses to CO_2: Implications of breathing at rest and on exercise. Intern Union Physiol Sci Abst 1993; 32:25.
42. Filiano JJ, Choi JC, Kinney HC. Arcuate nucleus hypoplasia in the sudden infant death syndrome. J Neuropathol Exp Neurol 1992; 51:394–403.
43. Onimaru H, Arata A, Homma I. Localization of respiratory rhythm-generating neurons in the medulla of brainstem-spinal cord preparations from newborn rats. Neurosci Lett 1987; 78:151–155.

44. Smith JC, Ellenberger HH, Ballanyi K, et al. Pre-Bötzinger complex: A brainstem region that may generate respiratory rhythm in mammals. Science 1991; 254: 726–729.
45. Smith JC, Ballanyi K, Richter DW. Whole-cell patch-clamp recording from respiratory neurons in mammalian brainstem in vitro. Neurosci Lett 1992; 134:153–156.
46. Brockhaus J, Ballanyi K, Smith JC, Richter DW. Microenvironment of respiratory neurons in the in vitro brainstem-spinal cord of neonatal rats. J Physiol (Lond) 1993; 462:421–445.
47. Dampney R. The subretrofacial nucleus: Its pivotal role in cardiovascular regulation. News Physiol Sci 1990; 5:63–66.
48. Reis DJ, Ross C, Granata AR, Ruggiero DA. Role of C1 area of rostroventrolateral medulla in cardiovascular control. In: Buckley JP, Ferrario CM, eds. Brain Peptides and Catecholamines in Cardiovascular Regulation. New York: Raven Press, 1987:1–14.
49. Sun M-K, Young BS, Hackett JT, Guyenet PG. Reticulospinal pacemaker neurons of the rat rostral ventrolateral medulla with putative sympathoexcitatory function: an intracellular study "in vitro." Brain Res 1988; 442:229–239.
50. Guyenet P, Darnall RA, Riley TA. Rostral ventrolateral medulla and sympatho-respiratory integration in the rat. Am J Physiol (Regul Integrat Comp Physiol 28) 1990; 259:R1063–R1074.
51. Dean C, Seagard JL, Hopp FA, Kampine JP. Differential control of sympathetic activity to kidney and skeletal muscle by ventral medullary neurons. J Auton Nerv Syst 1992; 37:1–10.
52. McAllen RM, Dampney RAL. Vasomotor neurons in the rostral ventrolateral medulla are organized topographically with respect to type of vascular bed but not body region. Neurosci Lett 1990; 110:91–96.
53. Zagon A. Innervation of serotonergic medullary raphe neurons from cells of the rostral ventrolateral medulla. Neuroscience 1993; 55:849–867.
54. Agarwal SK, Calaresu FR. Supramedullary inputs to cardiovascular neurons of rostral ventrolateral medulla in rats. Am J Physiol (Regul Integrat Comp Physiol) 1993; 265:R111–R116.
55. Roder S, Ciriello J. Innervation of the amygdaloid complex by catecholaminergic cell groups of the ventrolateral medulla. J Comp Neurol 1993; 332:105–122.
56. Ermirio R, Ruggeri P, Molinari C, Weaver LC. Somatic and visceral inputs to neurons of the rostral ventrolateral medulla. Am J Physiol (Regul Integrat Comp Physiol) 1993; 265:R35–R40.
57. Roder S, Ciriello J. Convergence of ventrolateral medullary and aortic baroreceptor inputs in nucleus of the solitary tract. Cand J Physiol Pharm 1993; 71:365–373.
58. Cochrane KL, Nathan MA. Cardiovascular effects of lesions of the rostral ventrolateral medulla and the nucleus reticularis parvocellularis in rats. J Auton Nerv Syst 1993; 43:69–81.
59. Helke CJ, McDonald CH, Phillips ET. Hypotensive effects of 5-HTIA receptor activation: Ventral medullary sites and mechanisms of action in the rat. J Auton Nerv Syst 1993; 42:177–188.
60. Cravo SL, Morrison SF. The caudal ventrolateral medulla is a source of tonic sympathoinhibition. Brain Res 1993; 621:133–136.

61. Willette RN, Barcas PP, Krieger AJ, Sapru HN. Vasopressor and depressor areas in the rat medulla: Identification by microinjection of L-glutamate. Neuropharmacology 1983; 22:1071–1079.
62. Carruth MK, Fowler AA, Fairman RP, et al. Respiratory failure without pulmonary edema following injection of a glutamate agonist into the ventral medulla of the rat. Brain Res Bull 1992; 28:365–378.
63. Fields HL, Barbaro NM, Heinricher MM. Brain stem neuronal circuitry underlying the antinociceptive action of opiates. In: Fields HL, Besson J-M, eds. Progress Brain Research 77. Amsterdam: Elsevier, 1988:245–257.
64. Gamble GD, Milne RJ. Hypercapnia depresses nociception: Endogenous opioids implicated. Brain Res 1990; 514:198–204.
65. Taveira Da Silva AM, Dias Souza J, Quest JA, et al. Central nervous system site of action for the respiratory depressant effect of diacetylmorphine (heroin) in the cat. J Clin Invest 1983; 72:1209–1217.
66. Dehkordi O, Dennis GC, Millis RM, et al. Effects of physostigmine on cocaine-induced cardiorespiratory responses at the caudal ventrolateral medullary surface (this volume).
67. Chiang C, Aston-Jones G. Response of locus coeruleus neurons to footshock stimulation is mediated by neurons in the rostral ventral medulla. Neuroscience 1993; 53:705–715.
68. Yates BJ, Bolton PS, Goto T, et al. The role of the ventral brainstem in vestibulosympathetic and respiratory reflexes (this volume).
69. Jennings DB, Walker JKL, Ohtake PJ. Central effects and interactions of the renin-angiotensin system and vasopressin on ventilation and $PaCO_2$ (this volume).
70. Spyer KM. Central nervous mechanisms contributing to cardiovascular control. J Physiol (Lond) 1994; 474:1–20.
71. Speck DF, Beck ER. Respiratory rhythmicity after extensive lesions of the dorsal and ventral respiratory groups in the decerebrate cat. Brain Res 1989; 482:387–392.
72. St John WM, Glasser RL, King RA. Rhythmic respiration in awake vagotomized cats with chronic pneumotaxic center lesions. Respir Physiol 1972; 15:233–244.
73. Blair RW. Convergence of sympathetic, vagal, and other sensory inputs onto neurons in feline ventrolateral medulla. Am J Physiol (Heart Circ Physiol) 1991; 260: H1918–H1928.
74. Koepchen HP. Respiratory and cardiovascular "centres": Functional entirety or separate structures? In: Schlaefke ME, Koepchen HP, See WR, eds. Central Neurone Environment and the Control System of Breathing and Circulation. Berlin: Springer-Verlag, 1983:221–237.
75. Nattie EE, Li A. Retrotrapezoid nucleus glutamate injections: Long-term stimulation of phrenic activity. J Appl Physiol 1994; 76:760–772.
76. Nagy F, Bal T, Cardi P. Dynamic re-wiring of CPG circuits in a simple nervous system. In: Eeckman FH, ed. Analysis and Modeling of Neural Systems. Boston: Kluwer, 1992:339–351.
77. Horner RL, Kozar LF, Phillipson EA. Tonic respiratory drive in the absence of rhythm generation in the conscious dog. J Appl Physiol 1994; 76:671–680.
78. Sears TA, Berger AJ, Phillipson EA. Reciprocal tonic activation of inspiratory and expiratory motoneurons by chemical drive. Nature 1982; 299:728–730.

16

Effect of CMA and MPA, Synthetic Progesterones, on Control of Respiration in Humans

YOSHIYUKI HONDA, TSUGUO MORIKAWA, HIROSHI KIMURA, and KOICHIRO TATSUMI

Chiba University School of Medicine
Chiba, Japan

I. Introduction

Loeschcke and his associates were the first to have experimentally demonstrated augmented ventilation by progesterone administration in male and postmenopausal female subjects (1,2). They further tried to correlate this finding with central respiratory chemosensitivity by perfusing the ventrolateral surface of the medulla with progesterone-containing solution (3). The Pa_{CO_2} ventilation (tidal volume) response curve was shown to shift to the left and the slope to increase. The possibility of the involvement of peripheral chemosensitivity was also examined, but the cross-circulation technique used by Mei et al. produced negative results (4). However, Hannhart et al. (5,6) demonstrated in pregnant cats that the carotid sinus nerve (CSN) discharge response to progressive hypoxia was enhanced, suggesting a directly stimulating influence of progesterone plus estrogen on the carotid body. However, this does not necessarily mean the presence of augmented peripheral chemoreceptor activity under natural conditions, because ventilation substantially increases and arterial blood should be hyperoxemic rather than hypoxemic in pregnancy (7).

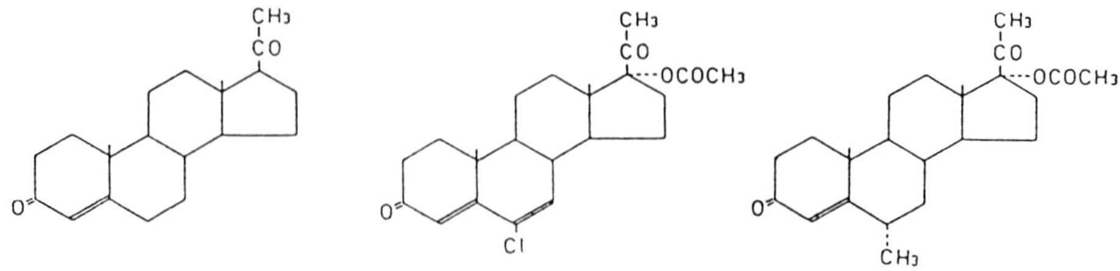

Figure 1 The chemical structures of progesterone and the synthetic progesterones used in the present study.

II. Synthetic Progesterone

In our study, we used mainly a synthetic progesterone, chlormadinone acetate (CMA), which has 10 times the progestational activity of medroxyprogestrone acetate (MPA) (8). MPA is an agent that has often been used in North America for respiratory studies and treatment. For comparison with CMA, MPA was also tested in our study. Since MPA is 15 times more active than natural progesterone (9), CMA is considered to be as much as 150 times stronger than natural progesterone. The presence of chlorine in the compound induces potentiation of luteinizing activity. Figure 1 compares the chemical structures of progesterone, CMA, and MPA.

III. Effect of Synthetic Progesterones on Respiratory Activities

A. Outline of Study Methods

First, 16 normal male subjects (25 ± 2.7 years of age) were administered CMA, MPA, and placebo in a randomized double-blind crossover maneuver (10). CMA and MPA were administered orally at a dose of 25 mg twice a day for 7 days. The placebo tablet, identical in appearance to the test drugs, contained lactose and was also given twice a day for 7 days. To make sure that the drug effect had sufficiently worn off, an interval of more than 2 weeks was allowed between each drug administration run. Just before and again on the last day of drug administration, the subjects were examined between 9 and 10 A.M. for the following: body weight, oral temperature, heart rate, venous blood sampling to measure various hormones and neuropeptides, arterial blood drawn to measure blood gas and acid-base balance, pulmonary function, ventilatory pattern, metabolic rate measured by $\dot{V}O_2$ and $\dot{V}CO_2$, and finally hypercapnic and hypoxic ventilatory responses (HCVR and HVR) by Read's rebreathing method and the progressive isocapnic hypoxia test respectively. To evaluate the magnitude of the ventilatory drive, both the amount of minute ventilation and height of occlusion pressure were measured. For the latter item, we used $P_{0.2}$ which was the occluded inspiratory airway pressure obtained at 0.2 s after the onset of inspiration; this was found to be more reproducible than the conventionally used $P_{0.1}$ (11). Furthermore, a number of patients with prostatic hypertrophy (12), chronic obstructive pulmonary disease (COPD) (13–15), and

Table 1 Body Weight, Body Temperature, Heart Rate, Pulmonary Function, Arterial Blood Gases, Alveolar Ventilation and Metabolic Rate with Placebo, CMA, and MPA Treatments ($n = 16$)[a]

	Placebo	CMA	MPA
Body weight, kg	67.7 ± 1.4	67.4 ± 1.5	67.4 ± 1.4
Body temp, °C	36.5 ± 0.1	36.8 ± 0.1[b]	36.7 ± 0.1
Heart rate, beats/min	70.1 ± 1.9	67.9 ± 1.4	68.3 ± 1.5
Pulmonary function			
%VC	104.7 ± 3.2	103.6 ± 2.8	101.6 ± 3.5
$FEV_{1.0\%}$	86.1 ± 1.6	86.3 ± 1.4	85.9 ± 1.3
Arterial blood			
pH	7.39 ± 0.01	7.40 ± 0.06	7.40 ± 0.01
$PaCO_2$, mm Hg	39.4 ± 0.7	35.4 ± 0.7[c]	36.5 ± 0.8[c]
PaO_2, mm Hg	91.9 ± 1.4	95.4 ± 1.2	9.4.4 ± 1.5
$[HCO_3^-]$, m*M*/L	23.1 ± 0.3	21.0 ± 0.3[c]	21.8 ± 0.4[c]
Ventilation			
$\dot{V}_A$, L/min	4.36 ± 0.20	5.40 ± 0.30[c]	5.07 ± 0.25[c]
V_T, liters	0.59 ± 0.03	0.63 ± 0.04	0.63 ± 0.04
f, breaths/min	15.8 ± 1.4	17.1 ± 1.7	17.3 ± 1.6
V_T/T_I, L/s	0.41 ± 0.02	0.43 ± 0.02	0.44 ± 0.02
$P_{0.2}$, cm H_2O	3.7 ± 0.2	3.8 ± 0.3	4.3 ± 0.3
V_D/V_T	0.32 ± 0.02	0.28 ± 0.03	0.30 ± 0.0]2
Metabolic rate			
$\dot{V}CO_2$, mL/min	194 ± 7	218 ± 10	213 ± 10
$\dot{V}O_2$, mL/min	247 ± 11	273 ± 14	274 ± 12

[a]Values are mean ± SE. CMA, chlormadinone acetate; MPA, medoxyprogesterone acetate; %VC, vital capacity of the predicted value in %; $FEV_{1.0\%}$, forced expiratory volume in 1 s in percent for forced vital capacity: $\dot{V}_A$, minute alveolar ventilation; V_T, tidal volume; f, breathing frequency; T_I, inspiratory duration; $P_{0.2}$, mouth occlusion pressure at 0.2 s after onset of inspiratory effort; V_D/V_T, ratio of dead space to tidal volume.

[b]Significant difference between placebo and MPA or CMA data, $P<0.05$.

[c]Significant difference between placebo and MPA or CMA data, $P<0.01$.

From reference 9: reproduced by permission.

sleep apnea syndrome (SAS) (16) were treated with CMA; a beneficial effect was seen in a substantial number of cases. The underlying pathophysiological mechanisms for this improvement were also pursued.

B. Ventilatory Pattern and Ventilatory Responses

Table 1 represents the physical condition, pulmonary function data, blood gas, and metabolic rate at rest after drug administration. Compared with the placebo run, the acid-base status of the arterial blood clearly indicates the presence of respiratory alkalosis by CMA and MPA ingestion, and this was derived from increases in both V_T and f, resulting in significant augmentation in $\dot{V}_A$. On the other hand, no definite changes were detected in body weight, heart rate, pulmonary function, or metabolic rate.

$P_{ET}CO_2$ ventilation response curves shifted to the left and their slopes tended to increase with synthetic progesterones (Fig. 2). When the subjects were loaded with inspiratory flow resistance in an amount of 16 cm H_2O/L/s at a flow rate of 1.0 L/s, the drug effect was better demonstrated. Furthermore, the response curves evaluated in terms of occlusion pressure were found to better represent the augmented response by CMA and MPA (Fig. 3).

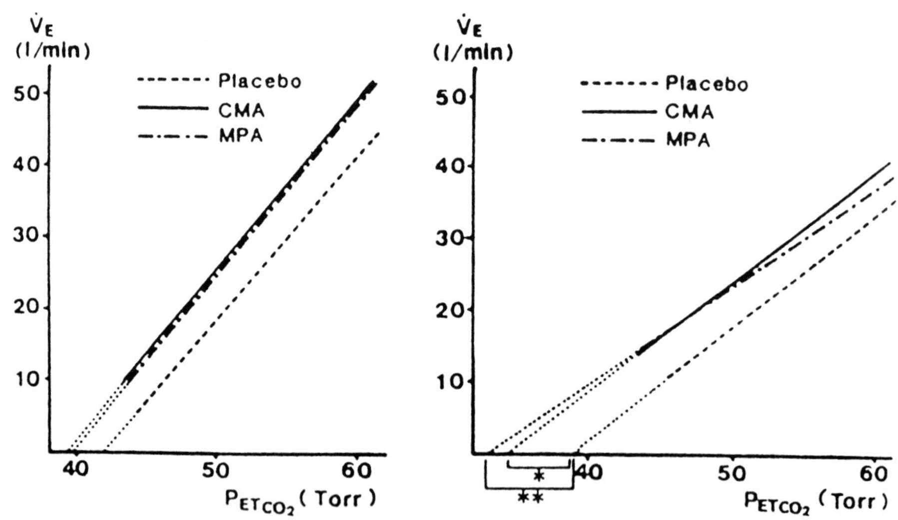

Figure 2 Mean ventilatory response ($\dot{V}_E$) to CO_2 in unloaded (left) and loaded (right) conditions. *$0.05 < P < 0.01$, **$P < 0.05$: differences in horizontal intercept of $\dot{V}_E$ end-tidal P_{CO_2} ($P_{ET}CO_2$) response line between placebo and chlormadinone acetate (CMA) or medroxyprogesterone acetate (MPA) data are of borderline significances at a 10% level and significant at a 5% level, respectively. (From Ref. 9.)

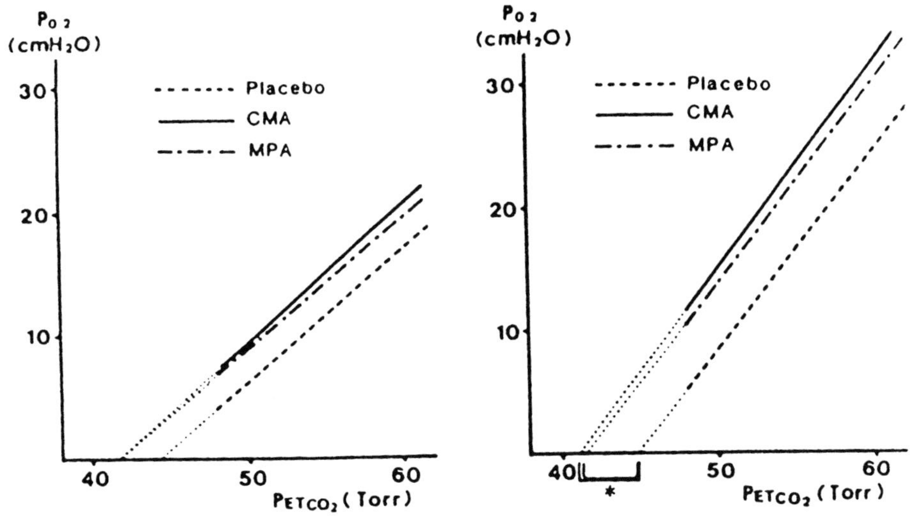

Figure 3 Mean occlusion pressure response to CO_2 in unloaded (left) and loaded (right) conditions. $^*0.05<P<0.1$: difference in horizontal intercept of occlusion pressure at 0.2 s after inspiratory effort end-tidal P_{CO_2} ($P_{0.2}$–$P_{ET}CO_2$) response line between placebo and chlormadinone acetate (CMA) or medroxyprogesterone acetate (MPA) data are of borderline significance at a 10% level. (From Ref. 9.)

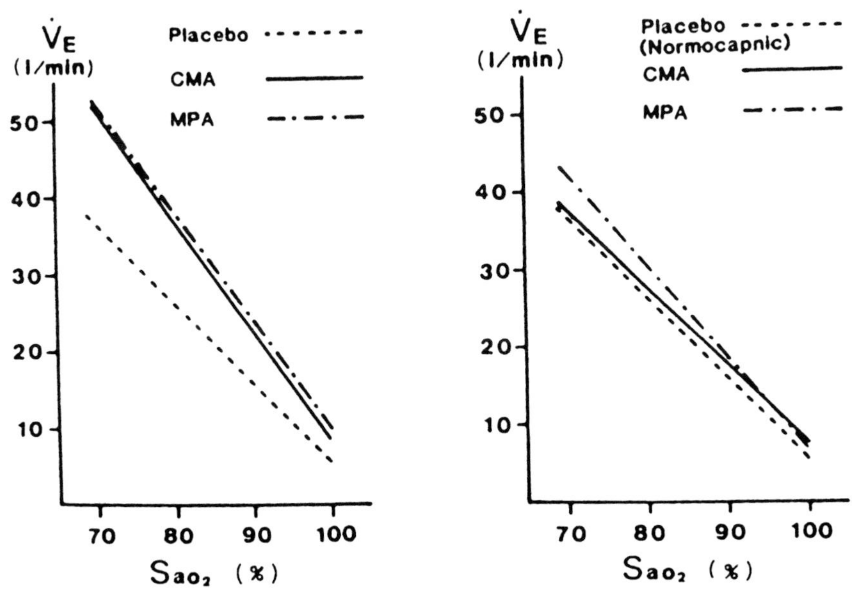

Figure 4 Mean ventilatory response ($\dot{V}_E$) to hypoxia in normocapnic (left) and hypocapnic (right) conditions. Dashed line representing placebo data is the normocapnic response curve because Pa_{CO_2} was not decreased in this condition. CMA, chlormadinone acetate; MPA, medroxyprogesterone acetate; Sa_{O_2}, arterial O_2 saturation. (From Ref. 9.)

SaO_2 ventilation and occlusion pressure response curves exhibited a shift to the right and increased slope (Figs. 4 and 5). This tendency was pronounced when tested under normocapnic conditions and was distinct when evaluated by the occlusion pressure.

C. Load Compensation Response

Figure 6 illustrates the $P_{ET}CO_2$ ventilation response curves before and after CMA in 10 patients with prostatic hypertrophy without cardiorespiratory complication. Augmented ventilatory drive response to flow-resistive loading (i.e., increased occlusion pressure response) is the most pronounced feature demonstrated in this figure. The degree of increased occlusion pressure in response to flow-resistive loading in 12 COPD patients was plotted on the ordinate axis and termed the load compensation ratio (LCR) (Fig. 7). The magnitude of LCR is known to be correlated with higher central nervous system (CNS) activity

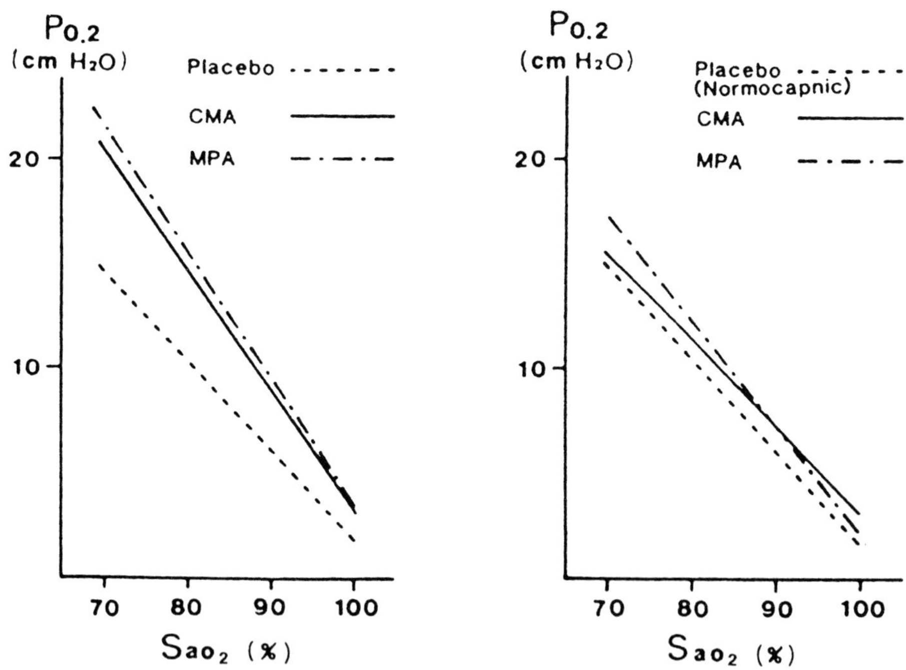

Figure 5 Mean occlusion pressure response ($P_{0.2}$) to hypoxia in normocapnic (left) and hypocapnic (right) conditions. Placebo responses indicated by dashed lines in both figures are normocapnic, as explained in Figure 4. (From Ref. 9.)

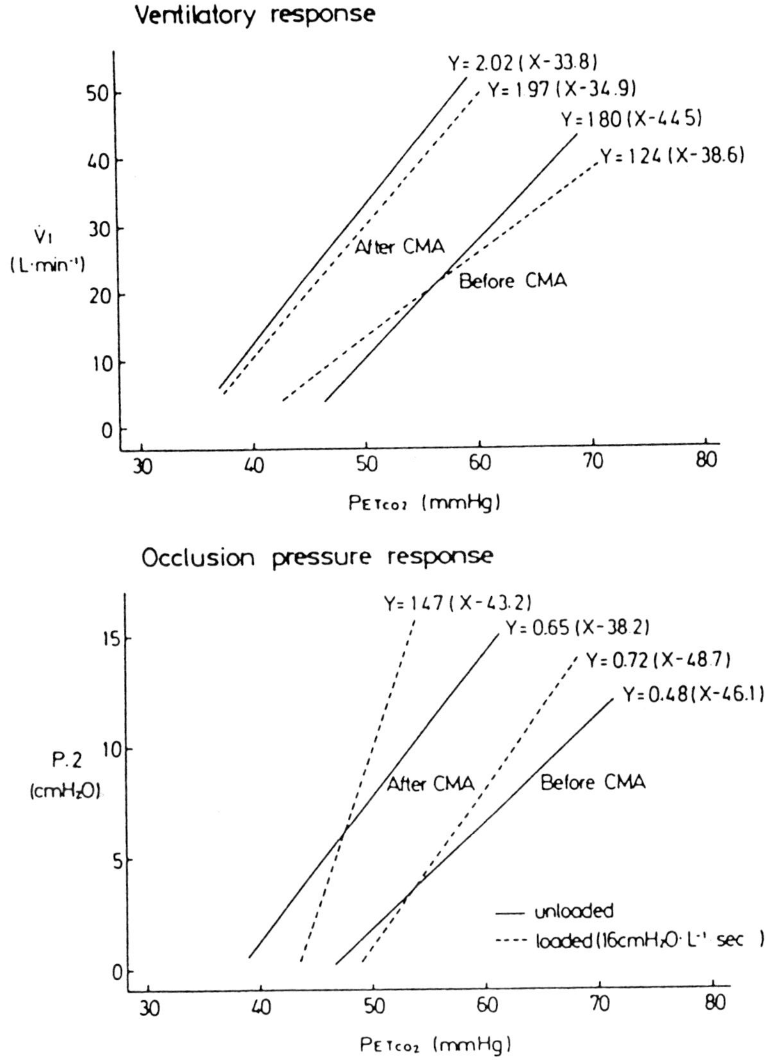

Figure 6 Mean ventilatory and occlusion pressure response lines to CO_2. Solid line, response line without loading; dashed line, response line with loading; $\dot{V}_I$, inspired minute ventilation; CMA, chlormadinone acetate; $P_{ET}CO_2$, end-tidal P_{CO_2}; $P_{0.2}$, occlusion pressure at 0.2 s. (From Ref. 11.)

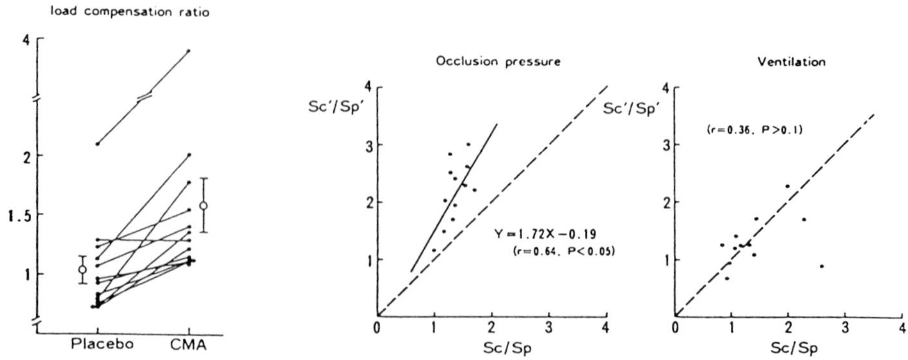

Figure 7 Left panel: Changes of inspiratory flow-resistive load compensation with placebo and CMA therapy. The change of load compensation is expressed as the ratio of loaded to unloaded $\Delta P_{0.2}/\Delta P_{CO_2}$. Closed circles and solid lines indicate individual data; open circle with vertical bar represents mean ± SE. Middle and right panels: Relationship between chemical and neuromechanical drives with placebo against CMA conditions. Subscripts p and c indicate placebo and CMA, respectively; prime indicates the condition with flow-resistive loading; Sp and Sp′ are the slopes of ventilatory and occlusion pressure responses in unloaded and loaded condition, respectively, with placebo treatment; Sc and Sc′ are the same as above with CMA therapy. The middle panel shows the relationship of occlusion pressure response to CO_2, and the right panel that of ventilatory response to CO_2. (From Ref. 12.)

and is effectively and disproportionately augmented by CMA, as shown in the middle section of this figure. Since occlusion pressure independently assesses the ventilatory drive regardless of alterations in mechanical properties of the airways, we consider LCR to be a good indicator for evaluating the ventilatory effect of CMA. In 9 SAS patients, parallel increases in both hypercapnic ventilation and occlusion pressure responses were observed in 7 out of 9 subjects (Fig. 8), and this was found to be a useful indication of improvement in clinical symptoms (16).

IV. Hormones and Neuropeptides After Administration of Synthetic Progesterone

Table 2 shows the level of certain hormones and neuropeptides after placebo, CMA, and MPA administration. Both TSH and ACTH showed a tendency to be increased by synthetic progesterones, and particularly for augmentation of TSH was shown to be a significant level by MPA. This is in contrast to the levels of T_3 and T_4 as shown in Figure 9. Figure 10 illustrates a possible mechanism

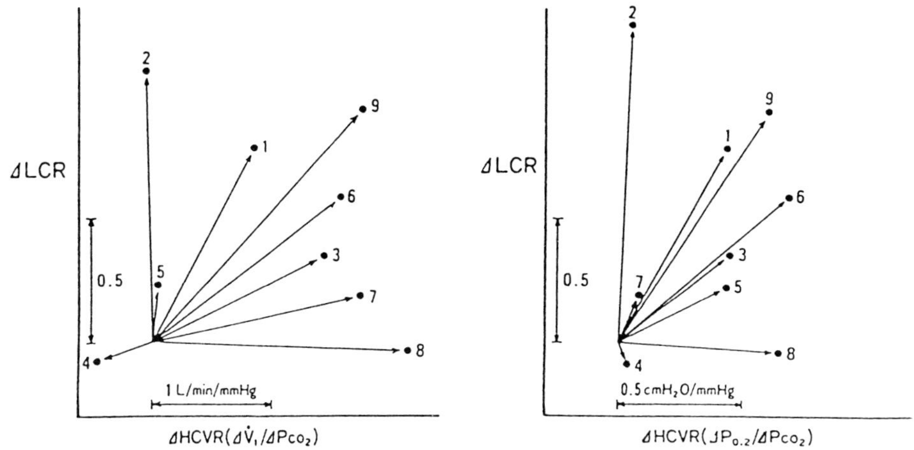

Figure 8 Relationship between the amount of change in load compensation ratio (ΔLCR) and that in hypercapnic ventilatory response (ΔHCVR) demonstrated by a vector. ΔHCVR in the left and right panels shows the change in hypercapnic ventilatory and occlusion pressure responses, respectively. Numbers are patient numbers. (From Ref. 15.)

Table 2 Effect of Placebo, CMA and MPA Administration on Certain Hormones and Neuropeptides

		Placebo ($n = 13$)	CMA ($n = 12$)	MPA ($n = 13$)
TSH	(μU/ml)	1.98 ± 0.21	2.45 ± 0.36	2.72 ± 0.32*
T_3	(ng/ml)	156.9 ± 8.4	156.2 ± 4.5	164.2 ± 8.0
T_4	(μg/ml)	9.31 ± 0.43	9.18 ± 0.32	10.12 ± 0.44
ACTH	(pg/ml)	4.70 ± 3.24	8.65 ± 4.71	6.75 ± 3.53
Cortisol	(ng/ml)	160.4 ± 19.8	141.7 ± 19.0	174.7 ± 19.9
Adrenaline	(ng/ml)	0.040 ± 0.013	0.032 ± 0.007	0.035 ± 0.008
Noradrenaline	(ng/ml)	0.25 ± 0.03	0.25 ± 0.02	0.26 ± 0.03
Dopamine	(ng/ml)	<0.2	<0.2	<0.2
Serotonin	(ng/ml)	76.9 ± 9.1	75.5 ± 8.7	81.5 ± 9.8
β-endorphin	(pg/ml)	17.3 ± 3.1	17.1 ± 2.4	20.3 ± 2.5

Values are mean ± SE.
*$P < 0.05$: difference between placebo and MPA data is significant at 5% level.

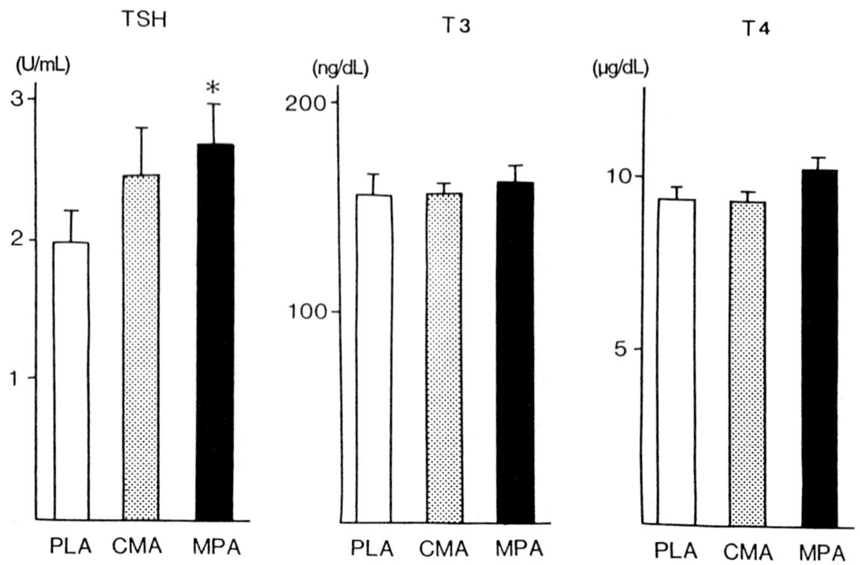

Figure 9 TSH, T_3, and T_4 levels after placebo, CMA, and MPA administration, respectively, for 7 days. Vertical bars indicate the magnitude of SE. * indicates $P<0.05$.

to explain the above data. Since T_3 and T_4 did not show consistent change, feedback control in TSH secretion at the pituitary level, which is considered to be the major control mechanism, may not be playing an effective role in this situation. Therefore, the elevated TSH observed in this study may have been of hypothalamic origin—i.e., thyrotropin releasing hormone (TRH). Since TRH is now known to induce a potent analeptic and hyperpneic effect in many animal species (17–31), the enhanced ventilatory chemosensitivity and load compensation described above may, at least in part, be explained by this hormone.

It must be mentioned that Bayliss and Milhorn (32) localized the anatomic substrate in the ventromedial nucleus of the hypothalamus responsible for gene expression of the progesterone receptors that induced augmented ventilatory and reproductive activities in the cat. Since TRH is known to exhibit interaction with various neuropeptides such as norepinephrine, GABA, and serotonin (23), the coexistence of increased progesterone and TRH secretion may possibly be associated in the enhancement of their biological activities. It is also known that the hyperpneic effect of progesterone is abolished by decerebration but not by decortication (32) and that TRH-induced hyperventilation is effectively attenuated after interfollicular decerebration (27). When these data are taken together, it seems very likely that the hypothalamus is responsible for the hyperpnea induced by progesterone.

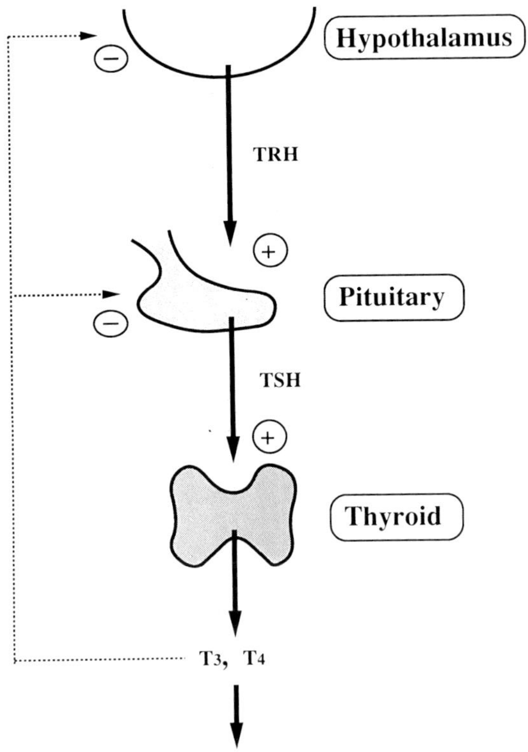

Figure 10 Schematic diagram explaining the elevated TSH level observed after administration of synthetic progesterone. Since no increase in T_3 and T_4 levels was detected, feedback control for TSH secretion by thyroid hormones at the pituitary level may not be effectively operative. Therefore, TSH elevation may have been induced by augmented thyrotropin releasing hormone (TRH).

V. Conclusion

Potent synthetic progesterones, chlormadinone acetate (CMA) and medroxyprogesterone acetate (MPA), induced augmented hypercapnic and hypoxic ventilatory as well as occlusion pressure responses. Furthermore, an enhanced compensation response to flow-resistive loading was demonstrated.

A significant increase in TSH was found, indicating augmented TRH release. This was assumed to have originated from an associated activation of the area of the ventromedial nucleus in the hypothalamus, where a genomic effect on progesterone receptors was shown to develop.

Since TRH is known to exert analeptic and hyperpneic influence, the observed CMA and MPA responses may, at least in part, be explained by an augmentation of this hormone.

References

1. Heerhaber I, Loeschcke HH, Westphal U. Eine Wirkung des Progesterons auf die Atmung. Plügers Arch 1948; 250:42–55.
2. Döring GK, Loeschke HH, Ochwadt B. Weitere Untersuchungen über die Wirkung der Sexualhormone auf die Atmung. Pflügers Arch 1950; 252:216–230.
3. Tok G, Loeschcke HH. Untersuchung über zentrale Wirkung von Progesteron auf die Atmung und Vasomotrik bei Katzen. Atemw-Lungenkrkh Jahrgang 1981; 7:148–153.
4. Mei SS, Gort D, Kao FF. The investigation of respiratory effect of progesterone in cross-circulated dogs (abstr). Fed Proc 1977; 36:489.
5. Hannhart B, Pickett CK, Weil JV, Moore LG. Influence of pregnancy on ventilatory and carotid body neural output responsiveness to hypoxia in cats. J Appl Physiol 1989; 67:797–803.
6. Hannhart B, Pickett CK, Moore LG. Effects of estrogen and progesterone on carotid body neural output responsiveness to hypoxia. J Appl Physiol 1990; 68:1909–1916.
7. Dempsey JA, Olson EB Jr, Skatrud JB. Hormones and neurochemicals in the regulation of breathing. In: Cherniack NS, Widdicombe JG, eds. Handbook of Physiology. Sec 3: The Respiratory System. Vol II: Control of Breathing, Part 1. Bethesda, MD: American Physiological Society, 1986:181–221.
8. Brennan DM, Kraay KJ. Chlormadinone acetate, a new highly active gestation-supporting agent. Acta Endocrinol 1963; 44:367–379.
9. Tausk M. Pharmacology of orally active progestational compounds: Animal studies. Tausk M. ed. Pharmacology of the Endocrine System and Related Drugs. Sec 48, Vol 2, Chap 28. Oxford: Pergamon Press, 1972:35–216.
10. Morikawa T, Tanaka Y, Maruyama R, et al. Comparison of two synthetic progesterones on ventilation in normal males: CMA vs MPA. J Appl Physiol 1987; 63: 1610–1615.
11. Yoshida A, Hayashi F, Sasaki K, et al. Analysis of pressure profile in the occluded airway obtained at the beginning of inspiration in steady state hypercapnia. Am Rev Respir Dis 1981; 124:252–256.
12. Kimura H, Yahashi F, Yoshida A, et al. Augmentation of CO_2 drives by chlormadinone acetate, a synthetic progesterone. J Appl Physiol 1984; 56:1627–1632.
13. Tatsumi K, Kimura H, Kunitomo F, et al. Effect of chlormadinone acetate on ventilatory control in patients with chronic obstructive pulmonary disease. Am Rev Respir Dis 1986; 133:552–557.
14. Tatsumi K, Kimura H, Kunitomo F, et al. Effect of chlormadinone acetate on sleep arterial oxygen desaturation in patients with chronic obstructive pulmonary disease. Chest 1987; 91:688–692.
15. Kimura H, Tatsumi K, Kuriyama T, et al. Effect of chlormadinone acetate, a synthetic progesterone, on restoring impaired load compensation in chronic obstructive pulmonary disease. Tohoku J Exp Med 1986; 149:119–132.

16. Kimura H, Tatsumi K, Kunitomo F, et al. Progesterone therapy for sleep apnea syndrome evaluated by occlusion pressure responses to exogenous loading. Am Rev Respir Dis 1989; 139:1198–1206.
17. McCann MJ, Hermann GE, Roger RC. Thyrotropin-releasing hormone: Effects on identified neurons of the dorsal vagal complex. J Auton Nerv Syst 1989; 26:107–112.
18. Mueller RA, Towle AC, Breese GR. The role of vagal afferents and carbon dioxide in the respiratory response to thyrotropin-releasing hormone. Regulat Peptide 1985; 10:157–166.
19. Homma I, Oouchi M, Ichikawa S. Facilitation of inspiration by intracerebroventricular injection of thyrotropin-releasing hormone in rabbits. Neurosci Lett 1984; 44:265–269.
20. Sharp T. Tullochi F, Bennett GW, et al. Analeptic effects of centrally injected TRH and analogues of TRH in the pentobarbitone-anaesthetized rat. Neuropharmacology 1984; 23:339–348.
21. Mueller RA, Towle AC, Breese GR. Supersensitivity to respiratory stimulatory effect of TRH in 5,7-dihydroxytryptamine-treated rats. Brain Res 1984; 298:370–373.
22. Hedner J, Hedner T, Wessberg P, et al. Effects of TRH and TRH analogues on the central integration of breathing in the rat. Acta Physiol Scand 1983; 117:427–437.
23. Mueller RA, Lundsberg DBA, Breese GR, et al. The neuropharmacology of respiratory control. Pharmacol Rev 1982; 34:255–285.
24. Hender T, Hender J, Jonason J, Lundsberg D. Respiratory effects of TRH in preterm rabbits. Pediatr Res 1982; 16:543–548.
25. Fraber JF, Connors AF, Gisolfi CV, et al. Effects on breathing of putative neurotransmitters in the rostral hypothalamus of the rat. Brain Res Bull 1981; 6:13–17.
26. Hender J, Hender T, Jonason J, Lundsberg D. Central respiratory stimulant effect by thyrotropin releasing hormone in the rabbit. Neurosci Lett 1981; 25:317–320.
27. Yamamoto Y, Lagercrantz H, Von Euler C. Effects of substance P and TRH on ventilation and pattern of breathing in newborn rabbits. Acta Physiol Scand 1981; 113:541–543.
28. Koivusalo F, Paakkari J, Leppäluoto, Kappanen H. The effect of centrally administered TRH on blood pressure, heart rate and ventilation in rat. Acta Physiol Scand 1979; 106:83–86.
29. Carino MA, Smith JR, Weick BG, Horita A. Effects of thyrotopin releasing hormone (TRH) microinjected into various brain areas of conscious and pentobarbital-pretreated rabbits. Life Sci 1976; 19:1687–1692.
30. Metcalf G, Myers RD. A comparison between the hypothermia induced by intraventricular injections of thyrotropin releasing hormone, noradrenaline or calcium ions in unaesthetized cats. Br J Pharmacol 1976; 58:489–495.
31. Myers RD, Metcalf G, Rice JC. Identification by microinjection of TRH-sensitive sites in the cat's brain stem that mediate respiratory, temperature and other autonomic changes. Brain Res 1977; 126:105–115.
32. Bayliss DA, Milhorn DE. Central neural mechanisms of progesterone action: Application to the respiratory system. J Appl Physiol 1982; 73:393–404.

17

α_2-Adrenergic Receptors Rather Than Imidazoline Binding Sites Mediate the Sympatholytic Effect of Clonidine in the Rostral Ventrolateral Medulla

A New Look at the Evidence

PATRICE G. GUYENET, KEVIN R. LYNCH, DIANE L. ROSIN, and RUTH L. STORNETTA

University of Virginia Health Sciences Center
Charlottesville, Virginia

ANDREW M. ALLEN

Howard Florey Institute of Experimental Physiology and Medicine
University of Melbourne
Parkville, Victoria, Australia

I. Introduction

Under anesthesia, the generation of sympathetic vasomotor tone depends primarily on the output of the rostral ventrolateral medulla (RVL) in the form of a largely monosynaptic excitatory drive from bulbospinal vasomotor presympathetic neurons (VPN) to spinal preganglionic neurons (SPGN) (1). These VPNs fall into two main categories, adrenergic (C1) cells and nonadrenergic, probably glutamatergic cells (1–8). Collectively, these neurons perform a dual function (1). The first is sympathetic tone generation that appears to rely in large part on the intrinsic pacemaker properties of some of these cells [suggested so far only in the case of the nonadrenergic components of the pathway (9–11) but certainly not excluded in the case of the C1 cells]. The second major role of these neurons is synaptic integration of a wide variety of vasomotor reflexes and central command inputs (1).

Under anesthesia, RVL is thought to be a major though surely not unique central nervous system (CNS) site where clonidine and related centrally acting hypotensive agents produce their sympatholytic effect (12). This notion originates from the observation that intraparenchymal injection of clonidine into

RVL produces considerable hypotension, an effect that is receptor-mediated, since it is antagonized by local injection of agents with alpha-2 adrenergic receptor (α_2-AR) antagonist activity (12,13).

The purpose of the present review is twofold. The first is to propose a refined version of the α_2-adrenergic receptor theory of the hypotensive action of clonidine in RVL (α_2-AR theory) and to document the evidence in its support recently obtained in our laboratories. This theory (Fig. 1) states that (a) C1 adrenergic cells, like CNS noradrenergic neurons, have somatic α_2-adrenergic inhibitory receptors ("autoreceptors"); (b) clonidine and its congeners act as agonists of these receptors, thereby triggering neuronal inhibition by means presumably identical to those described in the locus ceruleus and neurons elsewhere (14,15); and (c) since these cells most likely subserve a sympathoexcitatory role (1,5,16), their inhibition produces disfacilitation of sympathetic preganglionic neurons, hence reduction in sympathetic tone and hypotension.

The second purpose of this paper is to analyze the evidence that supports the competing point of view (i.e., that the hypotensive action of clonidine in RVL has to do with "imidazoline receptors" instead of bona fide α_2-ARs) (17–22). The concept of imidazoline receptors originates from the identification in brain and elsewhere of several types of nonadrenergic binding sites with high affinity for a class of α_2-AR agonists and antagonists characterized by an imidazole or imidazoline structure. The most extreme view has been adopted by Bousquet and collaborators (17), who believe that the central hypotensive effect of clonidine bears no relation at all to the α_2-AR agonist property of this drug and is entirely due to interaction with these hypothetical receptors. These authors postulate that clonidine mimics the action of an endogenous agonist of these receptors dubbed "endazoline" (18). These views have gained wider acceptance recently (21–23), seemingly without being subjected to much critical evaluation. The second half of the paper is therefore of an editorial nature and consists of a brief critical analysis of the imidazoline theory of the sympatholytic action of clonidine (henceforth abbreviated imidazoline theory).

II. Classification of α_2-ARs

There is no excellent evidence that α_2-ARs are encoded by three separate genes (these will be called α_{2A}, α_{2B}, and α_{2C}, after Refs. 24 and 25). For a given species, each of the three (paralogous) gene products displays a slightly different pharmacological profile (25). In addition, there are differences between the pharmacological profile of orthologous gene products (26). The difference in affinity exhibited by orthologous human and rodent A subtypes of α_2-AR for yohimbine (and its diastereoisomer rauwolscine) has led some (27) to suggest that this gene encodes a "D" subtype in rat and mouse. We favor a classification scheme based on comparison of protein sequences; by this criterion, the so-

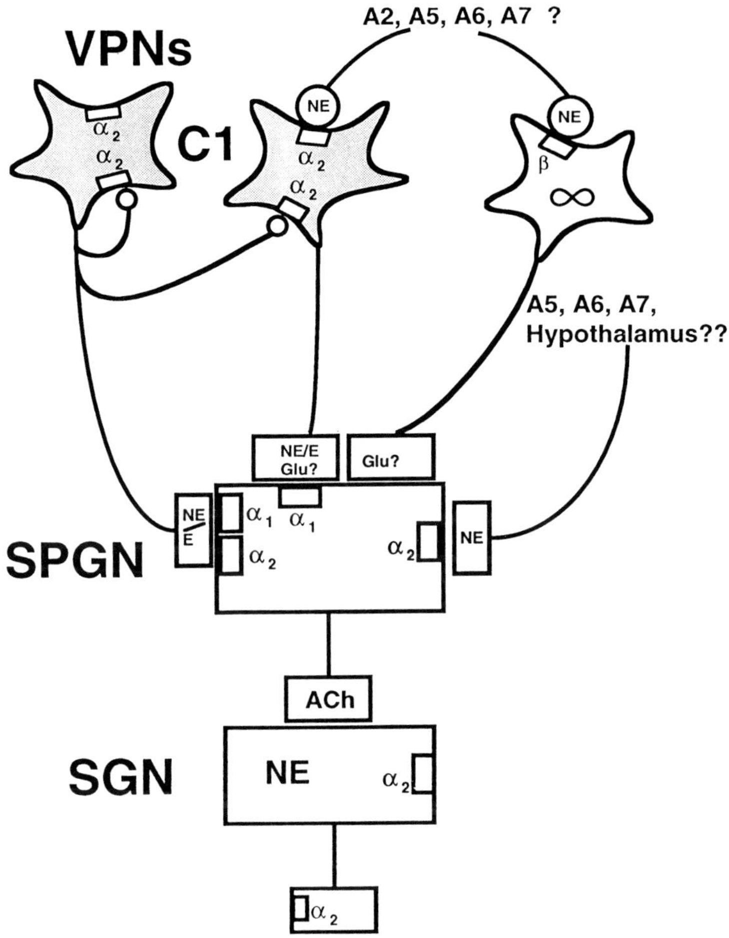

Figure 1 Presumed location of α_2-ARs principally responsible for the sympatholytic effects of clonidine under anesthesia. The extreme sensitivity of the sympathetic outflow to clonidine is attributed to the agonist activity of this drug at α_2-ARs located at the various sites illustrated on the figure. Activation of these receptors at each site appears to produce cumulative levels of inhibition of the sympathetic vasomotor tone. Some of the effect of clonidine is due to the direct inhibition of RVL VPNs (presympathetic cells), probably the adrenergic ones (C1 cells, stippled). The effects of catecholamines in RVL are complex since the nonadrenergic VPNs (pacemaker cells, identified by the infinity sign) are excited via β-adrenergic receptors, an effect which would tend to counteract the α_2-AR mediated inhibition of the adrenergic cells. Inhibitory α_2-ARs are also located on sympathetic preganglionic neurons (SPGN) and sympathetic postganglionic neurons (SGN). The nature of the afferent input to α_2-ARs is unknown. Some of the possible sources of catecholaminergic innervation of these receptors are represented. A2-A7 are clusters of CNS noradrenergic cells.

called D subtype in rodents is clearly orthologous to the A subtype in humans. Minor species differences in binding profile are common with G protein–coupled receptors and should not be used for classifying receptor subtypes.

III. Anatomical and Electrophysiological Evidence for the Existence of Alpha-2 Adrenergic Autoreceptors on C1 Cells

A. Prior Anatomical Evidence

Conventional ligand binding studies in thin tissue slices or in tissue punches have suggested that RVL contains an intermediate concentration of α_2-ARs (28, 29). The presence of mRNA coding for the α_{2A}-AR gene has also been detected in RVL by hybridization histochemistry (30).

B. Development of Subtype-Specific Antibodies to α_{2A}-Adrenergic Receptors (Summary of Data from Ref. 31)

Two of us (KRL and DLR) have developed a polyclonal antiserum directed against a 47–amino acid fragment of the third intracellular loop of the A subtype of the α_2-AR (clone RG 20α; rat orthologous gene product to human α_2C10) (24,27). A 141-bp fragment of the α_{2A}-AR gene was subcloned into the *Bam* HI-*Hind* III site of pGEX.KG expression vector, which positioned the fragment downstream from the gene encoding a 26-kDa helminthic glutathione-S-transferase (GST). Fusion proteins expressed in *Escherichia coli* HB101 were purified over reduced glutathione agarose beads and injected into rabbits. Antisera were affinity-purified. The specificity of the affinity-purified antibody was characterized in the following way: (a) the antibody showed immunoreactivity on Western blots with membrane extracts from COS cells transfected with the α_{2A} but not the α_{2B} or α_{2C} subtype of AR; (b) immunoreactivity in these blots was completely eliminated by preincubation with fusion protein but was intact after preincubation with GST; (c) using the avidin-biotin HRP reaction, formaldehyde-fixed COS cells transfected with DNA encoding α_{2A}-AR stained strongly with the antiserum while control cells transfected with either the α_{2B}- or α_{2C} DNAs did not stain at all; (d) the staining in fixed, transfected COS cells was eliminated by preincubation with fusion protein but not GST.

C. Distribution of α_{2A}-AR–Like Immunoreactivity (α_{2A}-AR-LIR) Using Subtype-Specific Antibody (Summary from Refs. 31 and 32)

Specific α_{2A}-AR-LIR in brain (eliminated by preincubation of antibody with fusion protein but not with GST) appeared as submicron punctate spherical structures. Punctate immunoreactivity was prevalent especially in the autonomic areas of the brainstem and spinal cord (NTS, ventrolateral medulla, lateral tegmental field, raphe nuclei, dorsolateral pons, midbrain central gray and locus coeruleus) [for details, see figures in 31,32]. The distribution of these receptors within the rostral medulla is shown in Figure 2.

Double labeling for tyrosine hydroxylase (TH) and α_{2A}-AR-LIR revealed the presence of α_{2A}-AR-LIR in all noradrenergic (A6, A5, A1, A2, A7 groups) and adrenergic cells (C1 and C3) of the lower brainstem (example in Fig. 3). Triple labeling experiments (including fluorescent microbeads used as a retrograde marker from cord, and immunostaining for TH- and α_{2A}-AR-LIR) revealed that virtually every C1 adrenergic cell with projection to the thoracic cord exhibited punctate α_{2A}-AR-LIR within its cell body. The probable intracellular location of the α_{2A}-AR-LIR within C1 adrenergic cells was identified by confocal microscopy in fluorescence-stained material (Texas red avidin biotin labeling of α_{2A}-LIR) double labeled for TH (fluorescein-tagged secondary antibody detection system). Since α_{2A}-LIR could be located within TH-IR cells, it can be inferred that the protein is manufactured by the catecholaminergic neurons themselves.

Using double and triple fluorescent labeling, α_{2A}-AR-LIR was detected also within most serotonergic cells of the medullary raphe, including all those with spinal projections. In contrast, very few serotonergic cells of the dorsal raphe contained α_{2A}-LIR.

In summary, (a) punctate α_{2A}-AR LIR was found within all adrenergic and noradrenergic cells of the lower brainstem; (b) for reasons unknown (pos-

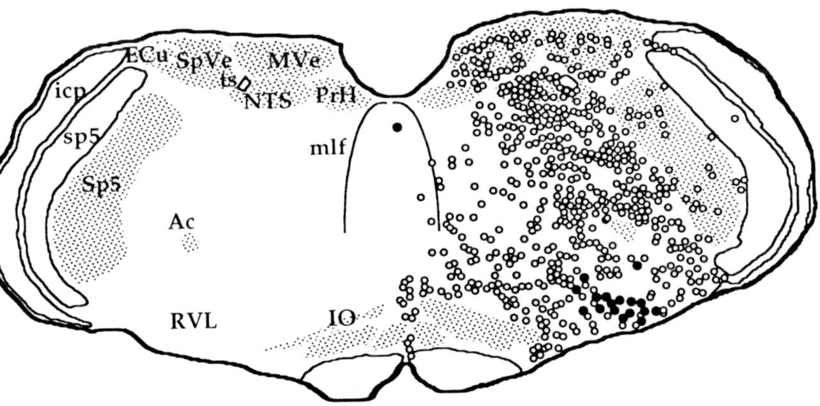

Figure 2 Location of α_{2A}-AR in the rostral medulla of the rat. Computer-assisted plot of the distribution of α_{2A}-AR-LIR in a coronal section at the level of the rostral medulla of the rat. Each open circle represents one cell with clusters of α_{2A}-LIR. Closed circles indicate C1 adrenergic cells (cells double-labeled with TH-LIR and α_{2A}-LIR). Abbreviations: Ac, nucleus ambiguus, compact division; IO, inferior olive; RVL rostral ventrolateral medulla.

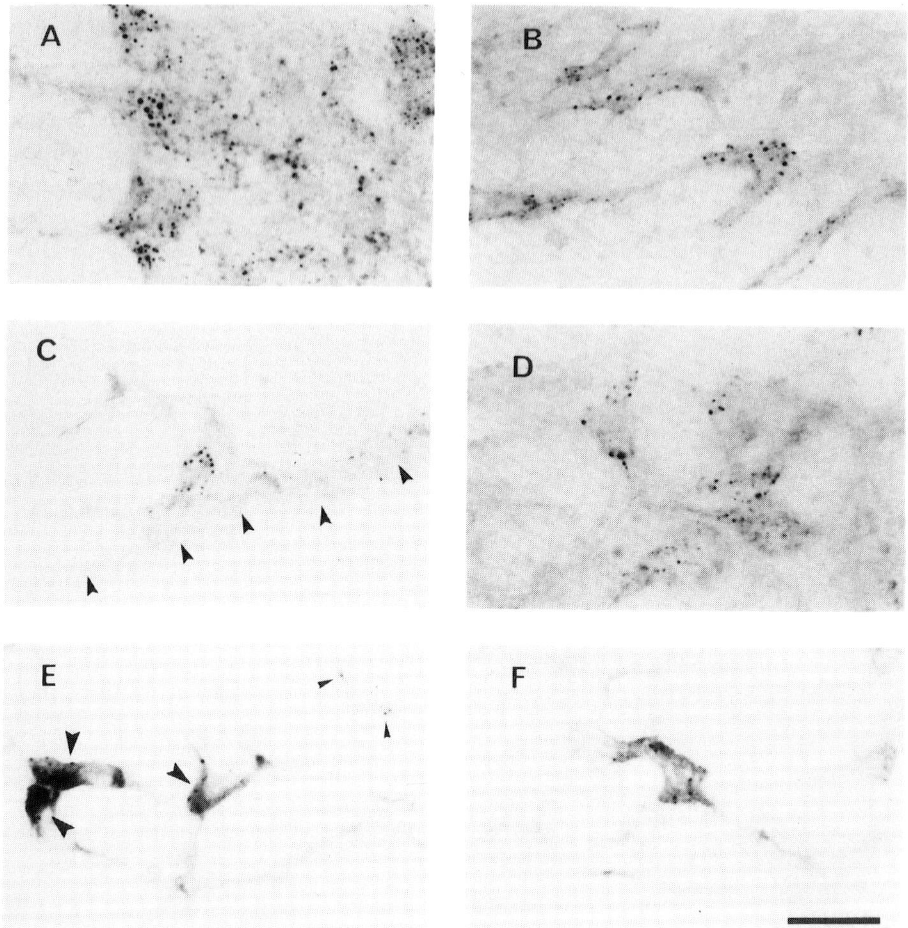

Figure 3 α_{2A}-AR-LIR in the lower medulla oblongata. Photomicrographs depicting punctate appearance of immunoreactive material detected with subtype-specific polyclonal antibody raised against a 47–amino acid stretch of the i3 loop of the rat α_{2A}-adrenergic receptor. Areas represented are raphe pallidus (panel A), parapyramidal area (ventromedial rostral medulla, panel B), subependymal location on the ventral surface of the medulla (panel C), and RVL (panel D, see location in Fig. 2). Panels E and F represent two examples of tyrosine-hydroxylase-immunoreactive cells in RVL (C1 adrenergic cells with presumed vasomotor presympathetic function) containing α_{2A}-LIR (dark dots). The one shown in F is wrapped around a capillary, a frequently reported feature of this type of cells.

sibly a question of sensitivity), our antiserum does not seem to detect the presence of the α_{2A}-AR in cell membranes and therefore does not permit localization of all receptors; (c) the presence of the receptor within all noradrenergic cells of the locus ceruleus suggests that the adrenergic autoreceptors responsible for inhibition of these cells may belong to the α_{2A} subtype (clone RG20α); and (d) the presence of these receptors within all other noradrenergic and the adrenergic cells suggests that all these cells may, like the locus ceruleus, also have somatodendritic autoreceptors of the α_{2A}-AR variety (14,15).

Finally, the presence of α_{2A}-LIR in serotoninergic cells of the medulla supports prior physiological work by others suggesting that a portion of the hypotensive effect of clonidine and α-methyl DOPA could be mediated by inhibition of the medullary serotonergic system (33,34).

D. Electrophysiological Evidence for α_2-AR-Mediated Inhibition of Bulbospinal C1 Adrenergic Cells by Clonidine (from Ref. 35)

RVL of urethane-anesthetized rats was found to contain two classes of tonically active bulbospinal VPNs. These cells share the characteristics of being barosensitive and pulse-synchronous. One type has a slow rate of discharge (mean of 6 ± spikes/s, $N = 29$), a slow axonal conduction velocity (2.0 ± 0.3 m/s), and is powerfully inhibited by low doses of IV clonidine (83 ± 11% inhibition with 5 to 10 μg/kg). The other has a much higher discharge rate (20 ± 1 spikes/s, $N=47$), a higher axonal conduction velocity (4.1 ± 0.2 m/s), and is insensitive to IV clonidine up to 20 μg/kg (maximally effective hypotensive dose in rat).

Prior work from this laboratory indicates that the clonidine-insensitive cells are probably not adrenergic and have intrinsic pacemaker discharges (7, 10), while the clonidine-sensitive neurons may be adrenergic (2).

Every RVL VPN that was inhibited by clonidine IV exhibited the following properties: (a) they were also inhibited by iontophoretic application of α-methyl norepinephrine [α-methyl NE (69% ± 3%)], (b) they were inhibited to the same extent by iontophoretically applied clonidine (51% ± 7%); (c) in each case where this was tested, the inhibitory effects of iontophoretically applied clonidine and α-methyl NE were both antagonized by idazoxan (an α_2-AR antagonist of the imidazoline class) and piperoxane (an α_2-AR blocker devoid of imidazoline structure) (15); (d) the effect of α-methyl NE was unaffected by the α_1-AR antagonist WB 4101; (e) the inhibition of RVL VPNs by IV clonidine could be reversed (50% to 100%) by local iontophoretic application of the antagonist idazoxan suggesting that most of the effect of IV clonidine on these cells is due to a local action within RVL; and (f) the inhibition produced by iontophoretic application of α-methyl NE could be attenuated by systemic administration of the non-imidazoline α_2-AR antagonist yohimbine.

In short, the iontophoretic results obtained with the presumed C1 cells are virtually indistinguishable from those obtained previously in the locus ceruleus using the same in vivo techniques (36,37). In the latter case, the postsynaptic location of the α_2-AR has been verified subsequently (15).

E. Conclusions

Both the neuroanatomical and the neurophysiological approaches described above have limitations. The rabbit antibody, while subtype-specific, (a) does not reveal the location of all receptors, most particularly it does not appear to stain visibly the receptors associated with the plasma membrane, which obviously must be the most significant functionally, and (b) by virtue of being subtype-specific does not recognize α_{2B}- or α_{2C}-AR that may also be present in C1 cells. As for the iontophoretic approach, it cannot ascertain that the responses investigated are due to postsynaptic receptors, although this is most generally the case. Finally while it is probable that the clonidine-sensitive VPNs of RVL are C1 adrenergic cells (2), this has not been established definitively by intracellular staining in vivo. More detailed neurophysiology is therefore needed to confirm that C1 cells have somatodendritic α_2-AR. However, despite those limitations, the results are fully compatible overall with the scheme represented in Figure 1, especially with the notion that clonidine inhibits RVL VPNs via an action on bona fide α_2-ARs and not imidazoline-selective receptors.

IV. Mechanism of Hypotensive Action of Clonidine: Imidazoline Binding Sites or Bona Fide Alpha-2 Adrenergic Receptors?

The imidazoline theory relies on the following categories of evidence: (a) the brain, including RVL, contains nonadrenergic binding sites with high affinity for clonidine and related imidazolines (nonadrenergic imidazoline binding sites, or NAIBS); (b) various partially purified brain extracts compete with clonidine for these binding sites; (c) a protein with high affinity for clonidine has been isolated and its distribution mapped by histochemistry; (d) intraparenchymal microinjections of catecholamines into RVL reportedly do not produce hypotension, while clonidine does; (e) the hypotension produced by microinjection of a variety of drugs into RVL correlates better with the apparent affinity of these agents for imidazoline binding sites than for α_2-ARs; (f) the α_2-AR antagonist yohimbine is sometimes found to be a weak antagonist of the hypotensive effect of clonidine, while α_2-AR antagonists with affinity for NAIBS are much better antagonists; and (g) the sympatholytic agent rilmenidine purportedly exerts its effect exclusively via imidazoline sites.

Each of these observations will be analyzed briefly from three perspectives, namely (a) whether they provide relevant or pharmacologically adequate evidence to support an imidazoline theory, (b) whether some of the odd pharmacological findings used to support the imidazoline theory could be reinterpreted in a manner consistent with the α_2-AR theory in the light of other or more recent work, and (c) by contrasting the overall consistency of the α_2-AR theory of action of clonidine with the gaps and inconsistencies of the imidazoline hypothesis.

A. The Nonadrenergic Imidazoline Binding Sites (NAIBS)

Multiple nonadrenergic binding sites with high affinity for various classes of α_2-AR agonists and antagonists characterized by an imidazoline structure have now been delineated by radioligand methods (22,38,39). These are sometimes referred to as imidazoline-preferring "receptors" (IPR), but are more precisely termed NAIBS (nonadrenergic imidazoline binding sites), since these binding sites have yet to be proven receptors. The clonidine-preferring site (I-1 site) (a) has a high affinity for clonidine and para-aminoclonidine, (b) is competed for by the α_2-AR agonists moxonidine and cirazoline, various imidazoles (imidazole acetic acid, histamine, cimetidine), and the antagonist idazoxan but not by the α_2-AR agonist (and hypotensive agent) guanabenz (19); and (c) appears to be the binding site present in small quantities in the brain of rat and cow and possibly in high quantities in human RVL. Idazoxan, an α_2-AR antagonist and imidazoline derivative, binds to α_2-ARs and to a class of NAIBS (I-2 sites) that appear unrelated to the I-1, clonidine-prefering sites. The idazoxan-preferring sites have (a) very low (K_is $> 10\ \mu M$) affinity for catecholamines (by definition); (b) virtually no affinity for nonimidazoline α_2-adrenergic antagonists such as yohimbine and rauwolscine ($K_i > 10\ \mu M$); (c) virtually no affinity for at least one powerful imidazoline-type α_2-AR antagonist, 2-methoxy-idazoxan [RX 821002 (40)]; (d) high affinity for cirazoline ($K_i = 1$ to $50\ nM$) and, in contrast to the clonidine sites; (e) relatively low affinity for either clonidine, para-aminoclonidine (Ki 200 nM to $>10\ \mu M$), or imidazole derivatives (histamine, cimetidine, imidazole acetic acid); and (f) high affinity for guanabenz. I-2 sites are present in astrocytes rather than neurons (41) and in the bovine adrenal medulla (42,43).

A few guiding principles are helpful in evaluating the potential significance of this cornucopia of binding data. The first is that the existence of a binding site, however high its affinity for a ligand, does not define a receptor. In theory, very high affinity binding sites may be found by pure serendipity on any protein or other macromolecule. Second, a large variety of experimental observations can provide legitimate evidence that a given biological response is due to interaction with a given binding site (44). For example, (a) obviously at least one biological response triggered by drugs that interact with the binding site should be found; (b) agonists and antagonists should be identifiable; (c) there should be a high degree of correlation between the absolute and relative potency of a series of antagonists, as determined by Schild-plot analysis, and the K_is of the same ligands for the binding site; (d) the presence of an endogenous ligand would lend support but is not required for receptor definition; (e) if responses in simple effector systems (e.g., isolated cells or membranes) are examined, the K_{act} of agonists should be either similar to the K_d of the drug for its binding site (if action is proportional to receptor occupancy) or lower (if spare receptors are present), but almost never larger. Finally, the definitive characterization of the agonist or antagonist binding properties of a receptor usually await the cloning of the gene and the availability via gene expression systems of amounts of the receptor sufficiently large for detailed study (24).

Obviously an immense literature that cannot even begin to be described here indicates that the above-mentioned criteria for receptor identification have all been met fully in the case of α_2-ARs (24,45). In contrast, as will be shown later, not a single one of the basic criteria for identification of a functional receptor has been fulfilled to date in the case of the NAIBS. In addition, a technical consideration of primary importance in evaluating the significance of the clonidine preferring sites of the medulla oblongata is that most binding experiments purporting to demonstrate the existence of NAIBS using clonidine or para-aminoclonidine as radioligands are conducted under what appears to be very unfavorable conditions for this type of work. First, the commonly used ligands clonidine or para-aminoclonidine are both agonists of α_2-ARs. The kinetics of interaction of agonists with G protein–interactive receptors can be complex, especially in the absence of a stable GTP analogue to uncouple receptors from the G proteins, a condition rarely used in this field. Second, in a typical experiment on bovine brain medulla membranes, the ratio between the specific binding of clonidine to NAIBS and its nonspecific binding appears diminutive (28). For example, "specific" binding to both α_2-ARs and NAIBS (displaceable by 10μM phentolamine) all together represents only 55% to 65% of the total radioactivity recovered on the filters. Typically 75% of this "specific" binding is displaced by catecholamines and thus due to α_2-AR binding, the remainder being attributed to NAIBS (28). Thus simple arithmetic indicates that "specific" binding of the radioligand to NAIBS typically represents a number of dpm equivalent only to between 20% and 30% of the nonspecific binding. These unfavorable conditions (signal/noise ratio of 1.3) may explain the disconcerting discrepancies in the published estimates of the affinity of clonidine or para-aminoclonidine for the clonidine-preferring site of the ventrolateral medulla [from less than 1 nM in the bovine brain (28) to 283 nM in human RVL (46)]. In contrast, the K_ds of clonidine or para-aminoclonidine for α_2-ARs, which can be determined under reasonably favorable conditions because of the higher affinity and capacity of the binding, only vary within narrow limits (1 to 20 nM). The relatively low affinity of clonidine for the NAIBS of RVL in humans (46) argues strongly against the involvement of NAIBS in the hypotensive effect of clinically effective doses of this agent. On the other hand, this low affinity may be congruent with the fact that concentrations of clonidine in excess of 1 μM are required to produce the few cellular or biochemical effects of clonidine so far attributed to interaction of the drug with other sites than α_2-AR (10,41,47–49). This includes a non-α_2-AR–mediated inhibition of nonadrenergic RVL VPNs in vitro (10), an effect that could conceivably account for some of the hypotensive effect produced by injections of high doses of clonidine into RVL.

In summary, (a) at present there is no convincing cell biological evidence that NAIBS belong to neuronal receptors involved in synaptic transmission or, for that matter, that they belong to important regulatory molecules, (b) at present, there is no convincing evidence that relevant concentrations of imidazoline-type α_2-AR agonists or antagonists (1 to 50 nM) exert any cellular effect attributable to interaction with binding sites other than α_2-ARs.

B. Endogenous Substances with Affinity for Imidazoline Binding Sites and Clonidine Binding Protein

The isolation of a clonidine-binding protein that is distinct from α_2-ARs is consistent with, and extends previous evidence for, the existence of nonadrenergic binding sites for imidazoline-type drugs in the CNS (50). Identification of this protein could be an important step in favor of the imidazoline theory if (a) it could be determined that this substance is a receptor and (b) if the binding of clonidine could be shown to change its activity. Presently available information does not permit an opinion on whether the protein is a structural protein or an enzyme and whether the binding of clonidine affects its properties in any way. Therefore, at this point, the existence of a clonidine-binding protein can neither support nor invalidate the imidazoline theory.

Low-molecular-weight substances present in partially purified brain extracts compete with clonidine for imidazoline binding sites (clonidine-displacing substances, CDS) in brain including RVL (51). The group at Cornell University has demonstrated that CDS has affinity for antibodies raised against clonidine and concludes that CDS must share some structural similarity with clonidine. This substance also produces hypotension upon microinjection into RVL (21) and stimulates adrenal chromaffin cells to secrete catecholamines (52). Anticlonidine antibodies have also been obtained by Bousquet et al. (53) and immunoreactivity was detected in human brain. However the substance partially purified by Bousquet's group ("endazoline") produces a pressor effect when injected in the lower brainstem (54). These inconsistencies could be due to the low degree of purification of CDS and endazoline and thus the data is too preliminary at this stage to support convincingly any one theory of action of clonidine.

C. Do Catecholamines Produce Hypotension upon Microinjection into RVL?

The lack of hypotension following microinjection of catecholamines into RVL is often contrasted with the powerful effect of clonidine injections into the same area (e.g., Ref. 55). This negative result is presented as a major piece of evidence that clonidine cannot produce hypotension by activating adrenergic receptors (55). However, these claims should not obscure the fact that such negative findings are not uniformly obtained. We will argue here, using two specific studies as examples (13,56), that the apparent lack of effect of catecholamines could very well be due to the use of insufficiently high concentrations of these substances by the authors reporting lack of effects. Indeed Granata et al. (56) initially reported that a substantial dose-dependent hypotensive effect was produced by injections of α-methyl NE into RVL. Specifically, this study describes dose-dependent hypotension following injection of α-methyl NE into RVL with an ED_{50} of around 2 to 3 nmol per side and a maximum drop of 55 mm Hg in arterial pressure at saturating doses (>10 nmol). Moreover this effect, just like that of clonidine, was antagonized by the α_2-AR antagonist phentolamine. According to these results, the α_2-AR–mediated hypotension produced by microinjection of α-methyl NE into RVL is virtually identical to that produced by clonidine in the rat when the catecholamine is administered in saturating doses. In addition, endogenous release of catecholamines via intraparenchymal injections of the indirect sympathomimetic amine tyramine in RVL was shown also by the same group to produce powerful (-60 mm Hg) hypotensive effects that depended, as expected, on the integrity of the local noradrenergic innervation (56). The same group of investigators later reversed their original interpretation, which was supportive of an α_2-AR adrenergic mechanism of action of clonidine in RVL (56). This retraction concluded a second study wherein injection into RVL of a fixed dose of 1 nmol of α-methyl NE was found to produce a smaller effect than that of 1 nmol of clonidine (-12 mm Hg, compared to -40 mm Hg for clonidine). The hypotension produced by α-methyl NE (and two other catecholamines, NE and adrenaline) was disregarded despite its statistical significance (19). However the effect produced by a 1 nmol dose of the catecholamine was essentially the same as that found in the first study (56), indicating clearly that 1 nmol did not represent a saturating dose. Therefore, the conclusion that catecholamines are ineffective in RVL while clonidine is effective appears to have rested in this case on an inappropriate comparison between a maximally effective dose of clonidine (1 nmol) and a minimally effective dose of the catecholamine α-methyl NE (also 1 nmol), both of which could very well have acted on RVL α_2-AR receptors. In support of this mundane interpretation, it is very well established that the K_{act} of clonidine for hyperpolarization of neurons in vitro is much lower than that of catecholamines. For example, in locus ceruleus slices, the threshold concentration for NE-induced inhibition is 10 μM without uptake inhibitor and 300 nM with desmethylimipramine (15), indicating first of all that reuptake lowers considerably the concentration of NE at the level of the receptors. In the same preparation, the threshold dose of clonidine is 3 nM (i.e., 100-fold lower than that of NE in the presence of uptake inhibitor and 3000-fold lower than NE alone). The effects of both drugs on LC neurons are proven to be due to stimulation of somatodendritic α_2-ARs and they are blocked by yohimbine and piperoxane, two drugs devoid of affinity for imidazoline binding sites (15). The approximately 50-fold difference in K_{act} between NE and clonidine is borne out by the 40-fold difference in the dissociation constants of NE and clonidine for α_{2A}-AR [633 n*M* for NE, 16 n*M* for clonidine (25)]. Moreover, not only is clonidine more potent, but its effect lasts a great deal longer than that of catecholamines because it is neither degraded as quickly nor taken up by nerve terminals. Thus, the use of identical doses of clonidine and catecholamines in microinjection studies represents a basically faulty paradigm if the object is to rule out the α_2-AR theory of action of clonidine in RVL.

In addition, catecholamines also activate β-adrenergic receptors in RVL, which results in an increased firing of nonadrenergic VPNs in vitro and appears to elevate AP in vivo (10,56). Therefore these opposite actions of catecholamines could reduce the hypotensive effect produced by microinjection of these agents into RVL, while the presence of β receptors would not affect the hypotensive effects of clonidine. Explanations of that sort could well account for the peculiarly small effect of NE injection into RVL of the cat (55).

In short, the sometimes reported lack of hypotensive effect of catecholamines (NE and α-methyl NE) microinjection into RVL, especially in rat, is likely due to the use of insufficiently high doses of these labile compounds in comparison with those of the much more powerful drug clonidine. An additional contributing factor is that catecholamines may activate other receptors within RVL (e.g., β receptors) that exert opposing effects. Thus, these variable and occasionally negative results do not provide solid evidence against the theory that α_2-AR agonists produce sympathoinhibition in RVL.

D. Are α_2-AR Antagonists Without Imidazoline Structure Really Unable to Antagonize the Effect of Clonidine in RVL?

The argument rests on two kinds of evidence: (a) the purported inability of yohimbine to antagonize the hypotensive effect of clonidine and (b) the inability of SKF 86466 (an α_2-AR antagonist devoid of affinity for NAIBS) to reverse the hypotensive effect of clonidine injections into RVL.

The Yohimbine Case

To evaluate these results, some background remarks on the potency of yohimbine relative to other commonly used α_2-AR antagonists is necessary. First, yohimbine by the i.v. route always appears to be a relatively weak α_2-AR antagonist in rats. For example, in adrenaline turnover studies, the dose required to fully antagonize the effect of 0.1 mg/kg i.p. of clonidine is between 1 and 10 mg/kg, a dose of 1 mg/kg producing only partial antagonism (57). In neurophysiological experiments on the locus ceruleus in vivo, yohimbine also must be administered IV in the 1- to 5-mg/kg range to antagonize the effect of a very small dose (20 μg/kg IV) of clonidine (58). The reason for this relatively low potency by the IV route may simply be poor penetration through the blood-brain barrier since, in vitro, the antagonist potency of this nonimidazoline α_2-AR antagonist is reasonably high in the same locus ceruleus system. With the very accurate Schild plot approach, the K_is of yohimbine, idazoxan, phentolamine, and piperoxan in antagonizing clonidine hyperpolarization in locus ceruleus were 14 n*M*, 9 n*M*, 20 n*M*, and 49 n*M* (15). These values are in fairly close agreement with binding data that generally reveal about a threefold difference in K_i between idazoxan and yohimbine for the rodent α_{2A}-AR (25). Yohimbine has been shown recently to have lower affinity for the A subtype of α_2-adrenergic receptor (RG20) than for the other subtypes in rats (42.6 n*M* for A/RG20 subtype versus 6.9 for B/RNG type and 2.6 for the C/RG10) (25). Thus, in the rat, the affinity of yohimbine for the A subtype is about 3 times lower than that of idazoxan and 50 times lower than the most potent α_2-AR antagonist with an imidazoline structure, atipamezole. The relatively low affinity of yohimbine for the A subtype is a peculiarity of the rat that seems to be due to a critical single amino acid change in the antagonist binding region of the α_{2A}-AR gene product (26). Our own results (see part A of this review) suggest that the A subtype of receptor is present in both locus ceruleus and C1 adrenergic cells of RVL. Therefore, in the rat, a relatively low potency of yohimbine as antagonist of the hypotensive effect of clonidine in RVL is not an unexpected result in the context of the α_2-AR theory depicted in Figure 1. The question is therefore to establish whether or not the relevant cardiovascular literature supports the following expectations: (1) yohimbine does antagonize the hypotensive effect of clonidine in all species investigated and (2) yohimbine is a weaker antagonist in rats, especially via the IV route.

The first point has been answered affirmatively by many authors, beginning with Schmitt and Laubie almost twenty years ago (see references in Ref. 59). These authors reported that, in the cat, the sympatholytic effect of clonidine is quite well antagonized by yohimbine as well as by other drugs, such as piperoxane, that have alpha blocking activity but do not have an imidazoline structure. Piperoxane (an α_2-AR selective antagonist relatively specific of α_2-AR) is able to block also the α_2-AR–mediated inhibitory effect of catecholamines on the locus ceruleus (15) (where α_2-AR mediation is not controversial) and on RVL presympathetic cells (presumed C1 cells) (35). Timmermans et al. have established, again in cats, that the hypotension produced by the central action of clonidine (intravertebral injections) is not merely attenuated but blocked by pretreatment of the animals with as low as 10 μg/kg IV of yohimbine or rauwolscine (59). Both these substances lack any significant affinity for NAIBS.

In rats, doses of yohimbine IV in the 1- to 5-mg/kg range need to be used to reverse the sympatholytic effect of clonidine (Guyenet, unpublished results), consistent with the higher doses required to prevent inhibition of both locus ceruleus or RVL neurons by clonidine in this species (35,58). However, very low doses of yohimbine i.c.v. (1 to 5 μg/kg) prevent the hypotensive effect of clonidine in spontaneously hypertensive rats (60).

In the rabbit, intracisternal yohimbine antagonizes the central hypotensive effect produced by clonidine or clonidine-like drugs (e.g., rilmenidine) very well also (61).

Therefore, positive evidence for antagonism of the central hypotensive effect of clonidine by at least three α_2-AR antagonists devoid of action on NAIBS has been obtained in three different animal species.

The Case of SKF 86466

SKF 86466 is an α_2-AR antagonist with the following characteristics: (a) it is devoid of imidazoline structure, (b) it has a 20-fold α_2/α_1 adrenergic receptor selectivity ratio, and (c) its potency as α_2-AR antagonist is relatively low (Ki ~ 40 n*M*) (62) in comparison with such drugs as idazoxan or even yohimbine. The inability of SKF 86466 to antagonize the hypotensive effect of clonidine microinjected into RVL is also given as evidence that clonidine cannot exert its effect via bona fide α_2-ARs (19). However, the evidence rests on the use of a single intraparenchymal dose of SKF 86466 that is compared to the same dose of idazoxan (an α_2-AR antagonist with imidazoline structure). The K_i of idazoxan for α_2-AR in RVL is at least 10 times higher than that of SKF 86466 (3.6 versus 35 n*M*) (19), therefore 1 nmol of each of these compounds does not represent equieffective blocking doses vis-à-vis α_2-ARs. Hence, the lack of effect of the SKF compound in antagonizing clonidine in these experiments may, once again, be due to the use of an ineffective dose with regard to α_2-AR blockade. Quite possibly, injection of higher doses may have been impractical for solubility reasons or because the drug exerts some unrelated depressant effect when microinjected into the brain, as suggested by the authors' admission that SKF 86466 produces a "rapid fall in blood pressure" (19) when injected by itself into RVL in a dose of 3 nmol. Regardless of the reason, negative evidence of this sort remains inconclusive.

E. Efficacy of Hypotensive Drugs Injected into RVL Correlates Better with Affinity for Imidazoline Binding Sites Than for α-Adrenergic Binding Sites

Ernsberger and colleagues compared the hypotensive effect produced by microinjection into RVL in vivo of a single dose of a variety of drugs and the K_is of the same drugs for either α_2-AR or imidazoline binding sites (13). The authors concluded that the hypotensive effect of the drugs correlates better with their affinity for imidazoline binding sites. The significance of these correlations is limited for at least two reasons. One is that the hypotensive effect of catecholamines is compared with that of non- or very slowly metabolized substances like clonidine, naphazoline, and oxymetazoline. As already discussed above (experiments of Williams et al. in tissue slices) (15), the rapid degradation and uptake of catecholamines could underestimate the K_{act} of catecholamines for α_2-ARs (estimated by measuring a hypotensive effect in this case) by several orders of magnitude. Second, the authors included in their correlation the drug imidazole acetic acid (IAA), which is also a powerful agonist of $GABA_A$ receptors that binds these receptors with 50 n*M* affinity (63). No evidence was provided by Ernsberger et al. (13) that IAA produced hypotension in RVL by mechanisms related either to α_2-ARs or imidazoline binding sites. The possibility exists that many of the other drugs used in the same study might have also had neuronal effects unrelated to α_2-ARs or NAIBS. The IAA case illustrates the poor predictability of this type of in vivo correlation.

In short, it is suggested here that the validity of this correlative approach in vivo is considerably weakened by an overwhelming burden of uncertainties and unverified assumptions regarding the pharmacokinetics and pharmacodynamics of the drugs microinjected into RVL.

The Case of Rilmenidine

Rilmenidine, a centrally acting hypotensive agent used clinically in Europe, has a chemical structure characterized by an oxazoline frame that is distinct from that of other α_2-AR agonists (64). Binding of rilmenidine to cortical membranes reveals that the drug has an affinity for α_2-ARs which is about 17 times lower than that of clonidine (K_d: 87 versus 5 n*M*, after Ref. 65; K_d: 17.3 n*M* versus 1 n*M*, after Ref. 66) and a low affinity (K_d: 250 n*M*) for a still unidentified NAIBS (66). The moderate affinity of rilmenidine for α_2-ARs (roughly one-tenth to one-twentieth clonidine) is in excellent agreement with the following observations: (a) the hypotensive potency of rilmenidine via the IV route in rats is about 20-fold less than that of clonidine via the IV route [EC_{50} for rilmenidine: 100 μg/kg (65), EC_{50} for clonidine IV: 5 μg/kg IV (67)], (b) the same potency ratio is found in rabbits following the IV or IC route [17-fold difference (68)], and (c) the same ~ 20-fold potency ratio is observed after microinjection into RVL of rats (65) or rabbits (54).

In addition, the effects of ICV rilmenidine are antagonized by both yohimbine and idazoxan (54). Figure 2 of this study can be read in Schild-plot fashion and illustrates with great clarity that the antagonist potency of yohimbine is only threefold lower than that of the imidazoline idazoxan. To be precise, the figure indicates that after ICV injection of 5 nmol/kg of yohimbine, the dose of rilmenidine required to produce the same hypotension as in absence of antagonist must be raised 30-fold, while after pretreatment with the same dose of idazoxan (5 nmol/kg), the dose of rilmenidine has to be increased by about 100-fold. This identifies a threefold difference in potency between yohimbine and idazoxan, which agrees quite well with the slightly higher potency of idazoxan for α_2-ARs generally found in binding data and other systems.

In awake rabbits, the hypotensive effect of rilmenidine depends on the integrity of both noradrenergic and serotonergic systems, just like that of clonidine (68). Like clonidine, rilmenidine decreases the catechol oxidation current in RVL at doses slightly lower than in the locus ceruleus (69). The reason for this potency difference is not explained and this phenomenon conflicts with neurophysiological results, since there is no difference between the dose of clonidine necessary to inhibit the discharges of locus ceruleus cells and those of RVL VPNs (5 to 20 μg/kg IV). Since (a) neither the type or receptor nor the transduction mechanism responsible for the effect of clonidine on the catechol oxidation current in locus ceruleus and RVL is known and (b) the relationship between catechol overflow and cellular discharge rate is also unknown, the electrochemical data cannot provide solid evidence against the theory that either clonidine or rilmenidine inhibits RVL cell discharges via bona fide α_2-ARs.

The final piece of evidence that has been interpreted as suggesting that rilmenidine acts via a receptor other than an α_2-AR is that the hypotensive effect of injection of rilmenidine into RVL is reversed by injections of idazoxan and not by injections of SKF 86466, an α_2-AR antagonist devoid of affinity for NAIBS (65) [see section IV.D for a pharmacological profile of this drug]. The behavior of rilmenidine is again identical to that of clonidine in this respect (19). As discussed before, the apparent lack of effect of SKF 86466 is likely due to a combination of factors including the fact that this drug is a weaker antagonist of α_2-ARs than idazoxan (19) and that, at the high dose required to achieve some α_2-AR antagonism, it produces hypotensive effects of its own (Ref. 19, page 414).

In summary, the available pharmacological data in animals suggest that the cardiovascular effects of rilmenidine are virtually indistinguishable from those of clonidine. K_Ds from ligand binding experiments and potency measurements relative to blood pressure effects are completely compatible with the notion that the drug exerts its hypotensive effect via an agonist action at CNS α_2-ARs, for which it has an affinity on the order of one-tenth to one-twentieth that of the prototypical agent clonidine.

F. Are There Really Clonidine-Preferring Imidazoline Binding Sites in RVL? How Many and of What Type?

The original claim by Bousquet's team (70) that the RVL of humans contained only imidazoline binding sites and no α_2-adrenergic receptors was the most puzzling aspect of this field (70,71). In these original studies, the human RVL reportedly contained 65 fmol/mg protein of clonidine-preferring NAIBS with K_d in the 9 nM range. However a more recent report from the same laboratory indicates that human RVL membranes actually contain significant amounts of α_2-AR (22 fmol/mg membrane protein). This density is small but well within the usual 10 to 150 fmol/mg range found for most G protein–coupled receptors in brain membranes. In addition, this type of receptor density is certainly not unexpected for a reticular formation area that is criss-crossed by major fiber tracts and therefore likely to have low neuronal density. This low but reasonable value is to be contrasted with the enormous density of the newly described clonidine-preferring NAIBS of human RVL (1700 fmol/mg protein) (46). The reported K_d of clonidine for this site is so large (280 n*M*, determined by displacement of radiolabeled clonidine by unlabeled ligand) that the very existence of this low-affinity site can legitimately be questioned on technical grounds. Indeed, according to Ref. 44, it may not be theoretically possible to analyze by filtration assay the equilibrium binding characteristics of a tritiated ligand to a receptor if the K_d is in the range of 280 n*M* as described unless the on-rate for the binding reaction is unusually slow (considerably below the normal 10^7 to 10^8 $M^{-1} \times min^{-1}$ found for all G protein–coupled and other membrane receptors). Based on (a) the rather invariant rate of association (10^7 to 10^8 $M^{-1} \times min^{-1}$ at 22 to 37°C) for ligands to their membrane receptors (e.g., Ref. 25) and (b) the authors' K_d value of 300 n*M*, it can be calculated that the $t_{1/2}$ for dissociation of labeled clonidine from its NAIBS in RVL should be from 1 to 10 s (optimistic hypothesis) and is therefore virtually incompatible with a filtration assay. It could be argued that there appear to be other types of NAIBS in the medulla of humans and that these might be the relevant receptors. However these seem to be idazoxan-preferring (I2) sites with very low affinity for clonidine (46) and therefore lacking the properties to account for the sympatholytic effect of the latter drug. In addition, more contradictory evidence comes from the assertion of Bousquet and colleagues that, in rat brainstem, clonidine binds exclusively to α_2-AR (70).

In summary, it can be concluded at this point (a) that α_2-ARs are present in the RVL of all species examined including the human, (b) that there are few, if any, high-affinity clonidine-preferring NAIBS in the human, maybe none in the rat, few in bovine medulla, and that their characterization in this species is still relatively preliminary due to the extremely marginal conditions used to determine their presence in binding experiments.

The collective weight of this evidence once more argues in favor of the α_2-AR theory and against the imidazoline hypothesis.

V. Conclusion

In our opinion, the imidazoline theory of the central sympatholytic effect of clonidine will not be acceptable until (a) some neuronal imidazoline receptor responsive to low nanomolar concentrations of clonidine is characterized fully, (b) the presence of this receptor is detected on relevant RVL vasomotor neurons, and (c) activation of this receptor is found to inhibit these neurons. At the time this is being written, the imidazoline theory still rests on indirect and/or controversial experimental results. This paper has attempted to illustrate with specific examples that many of the apparently discrepant pharmacological results that have been interpreted as supporting the imidazoline theory of the hypotensive effect of clonidine can be easily reconciled with the α_2-adrenergic hypothesis. Admittedly, there still remain a very few apparently unexplainable pharmacological results, mostly those obtained by electrochemical detection of the catechol overflow in brain (72,73). However, the variable recorded in these experiments is only remotely related to the unit discharge of the catecholaminergic cells. Regulation of catecholamine synthesis in cell bodies by α_2-AR agonists may involve receptor subtypes or transduction mechanisms quite different from those involved in cell hyperpolarization and inhibition.

In contrast to the imidazoline hypothesis, the α_2-AR hypothesis meets fully the three criteria listed previously (i.e., responsiveness to low doses of clonidine, presence of receptors on relevant cells in RVL, and inhibition of these cells by α_2-AR agonists).

In summary, the theory that clonidine lowers arterial pressure by virtue of its agonistic interactions with α_2-ARs is backed by an overwhelming wealth of fully congruent neuroanatomical, neurophysiological, cell biological, and pharmacological results. The new anatomical and electrophysiological data provided by our laboratory are completely consistent with the notion that the mechanism of action of low doses of clonidine at the level of the RVL is exactly the same as elsewhere (i.e. activation of inhibitory α_2-ARs). Furthermore, it is proposed that the sympatholytic effect of clonidine injections into RVL may simply be due to the presence of inhibitory autoreceptors on the C1 adrenergic cells, and that the receptor involved in the action of clonidine in RVL is probably the α_{2A} subtype. The physiological role of these α_2-ARs remains to be identified.

References

1. Guyenet PG. Role of the ventral medulla oblongata in blood pressure regulation. In: Loewy AD, Spyer KM, eds. Central Regulation of Autonomic Functions. New York: Oxford University Press, 1990:145–167.
2. Haselton JR, Guyenet PG. Electrophysiological characterization of putative C1 adrenergic neurons in the rat. Neuroscience 1989; 30:199–214.
3. Morrison SF, Callaway J, Milner TA, Reis DJ. Glutamate in the spinal sympathetic intermediolateral nucleus: Localization by light and electron microscopy. Brain Res 1989; 503:5–15.
4. Morrison SF, Ernsberger P, Milner TA, et al. A glutamate mechanism in the intermediolateral nucleus mediates sympathoexcitatory responses to stimulation of the rostral ventrolateral medulla. Prog Brain Res 1989; 81:159–169.
5. Morrison SF, Milner TA, Reis DJ. Reticulospinal vasomotor neurons of the rat rostral ventrolateral medulla: Relationship to sympathetic nerve activity and the C1 adrenergic cell group. J Neurosci 1988; 8:1286–1301.
6. Reis DJ, Ruggiero DA, Morrison SF. The C1 area of the rostral ventrolateral medulla oblongata: A critical brainstem region for control of resting and reflex integration of arterial pressure. Am J Hypertens 1989; 2:363S–374S.

7. Sun M-K, Hackett JT, Guyenet PG. Sympathoexcitatory neurons of rostral ventrolateral medulla exhibit pacemaker properties in the presence of a glutamate-receptor antagonist. Brain Res 1988; 438:23–40.
8. Sun M-K, Young BS, Hackett JT, Guyenet PG. Reticulospinal pacemaker neurons of the rat rostral ventrolateral medulla with putative sympathoexcitatory function: An intracellular study in vitro. Brain Res 1988; 442:229–239.
9. Sun M-K, Stornetta RL, Guyenet PG. Morphology of rostral medullary neurons with intrinsic pacemaker activity in the rat. Brain Res 1991; 556:61–70.
10. Sun MK, Guyenet PG. Excitation of rostral medullary pacemaker neurons with putative sympathoexcitatory function by cyclic AMP and beta-adrenoceptor agonists "in vitro." Brain Res 1990; 511:30–40.
11. Sun MK, Young BS, Hackett JT, Guyenet PG. Reticulospinal pacemaker neurons of the rat rostral ventrolateral medulla with putative sympathoexcitatory function: an intracellular study in vitro. Brain Res 1988; 442:229–239.
12. Punnen S, Urbanski R, Krieger AJ, Sapru HN. Ventrolateral medullary pressor area: Site of hypotensive action of clonidine. Brain Research 1987; 422:336–346.
13. Ernsberger P, Giuliano R, Willette RN, et al. Hypotensive action of clonidine analogues correlates with binding affinity at imidazole and not alpha-2-adrenergic receptors in the rostral ventrolateral medulla. J Hypertens 1988; 6(suppl):S554–S557.
14. Aghajanian GK, Wang YY. Common alpha 2- and opiate effector mechanisms in the locus coeruleus: Intracellular studies in brain slices. Neuropharmacology 1987; 26:793–799.
15. Williams JT, Henderson G, North RA. Characterization of alpha 2-adrenoceptors which increase potassium conductance in rat locus coeruleus neurones. Neuroscience 1985; 14:95–101.
16. Reis DJ, Morrison S, Ruggiero DA. The C1 area of the brainstem in tonic and reflex control of blood pressure: State of the art lecture. Hypertension 1988; 11:I8–13.
17. Bousquet P, Feldman J, Tibirica E, et al. New concepts on the central regulation of blood pressure: Alpha 2-adrenoceptors and "imidazoline receptors" (review). Am J Med 1989; 87:10S–13S.
18. Bousquet P, Feldman J, Tibirica E, et al. From alpha 2-adrenergic to endazoline receptors [Fre]. Arch Mal Coeur 1989; 82(spec no 5):11–14.
19. Ernsberger P, Giuliano R, Willette RN, Reis DJ. Role of imidazole receptors in the vasodepressor response to clonidine analogs in the rostral ventrolateral medulla. J Pharmacol Exp Ther 1990; 253:408–418.
20. Tibirica E, Feldman J, Mermet C, et al. An imidazoline-specific mechanism for the hypotensive effect of clonidine—A study with yohimbine and idazoxan. J Pharmacol Exp Ther 1991; 256:606–613.
21. Reis DJ, Ernsberger PR, Meeley MP. Imidazole receptors and their endogenous ligand in the rostral ventrolateral medulla: relationship to the action of clonidine on arterial pressure. In: Kunos G, Ciriello J, eds. Central Neural Mechanisms in Cardiovascular Regulation. Vol. 1. Boston: Birkhauser, 1991:55–68.
22. Hieble JP, Ruffolo RR Jr. Imidazoline receptors: Historical perspective. Fund Clin Pharmacol 1992; 6(suppl 1):7S–13S.
23. Reis DJ, Regunathan S, Meeley MP. Imidazole receptors and clonidine-displacing substance in relationship to control of blood pressure, neuroprotection, and adrenomedullary secretion. Am J Hypertens 1992; 5:51S–57S.

24. Harrison JK, Pearson WR, Lynch KR. Molecular characterization of α1- and α2-adrenoceptors. Trends Pharmacol Sci 1991; 62:67.
25. Harrison JK, D'Angelo DD, Zeng D, Lynch KR. Pharmacological characterization of rat α_2-adrenergic receptors. Molecular Pharmacology 1991; 40:407–412.
26. Link R, Daunt D, Barsh G, et al. Cloning of two mouse genes encoding alpha-2 adrenergic receptor subtypes and identification of a single aminoacid in the mouse 2-C10 homolog responsible for an interspecies variation in antagonist binding. Mol Pharmacol 1992; 42:16–27.
27. Lanier SM, Downing S, Duzic E, Homcy CJ. Isolation of rat genomic clones encoding subtypes of the alpha 2-adrenergic receptor: Identification of a unique receptor subtype. J Biol Chem 1991; 266:10470–10478.
28. Ernsberger P, Meeley MP, Mann JJ, Reis DJ. Clonidine binds to imidazole binding sites as well as alpha 2-adrenoceptors in the ventrolateral medulla. Eur J Pharmacol 1987; 134:1–13.
29. Unnerstall JR, Kopaitic TA, Kuhar MJ. Distribution of alpha-2 agonist binding sites in the rat and human central nervous system: Analysis of some functional, anatomic correlates of the pharmacological effects of clonidine and related adrenergic agents. Brain Res Rev 1984; 7:69–101.
30. Nicholas AP, Pieribone V, Hokfelt T. Distributions of messenger RNAs for alpha-2 adrenergic receptor subtypes in rat brain—An in situ hybridization study. J Comp Neurol 1993; 328:575–594.
31. Rosin DL, Zeng D, Stornetta RL, et al. Histochemical localization of α_{2A}-adrenergic receptors in catecholaminergic and other brainstem neurons in the rat. Neuroscience 1993; 56:139–155.
32. Guyenet PG, Stornetta RL, Riley T, et al. Alpha$_{2A}$-adrenergic receptors are present in lower brainstem catecholaminergic and serotonergic neurons innervating sympathetic preganglionic cells. Brain Res 1993; 638:285–294.
33. Chalmers JP, Pilowsky PM, MInson JB, et al. Central serotonergic mechanisms in hypertension. Am J Hypertens 1988; 1:79–83.
34. Drolet G, Aslanian V, Minson J, Morris M, et al. Differences in the central hypotensive actions of alpha-methyldopa and clonidine in the spontaneously hypertensive rat: contribution of neurons arising from the B3 and the C1 areas of the rostral ventrolateral medulla. J Cardiovasc Pharmacol 1990; 15:118–123.
35. Allen AM, Guyenet PG. α_2-Adrenoceptor-mediated inhibition of bulbospinal barosensitive cells of rat rostral medulla. Am J Physiol 1993; 265:R1065–1075.
36. Marwaha J, Aghajanian GK. Relative potencies of alpha-1 and alpha-2 antagonists in the locus ceruleus, dorsal raphe and dorsal lateral geniculate nuclei: An electrophysiological study. J Pharmacol Exp Ther 1982; 222:287–293.
37. Cedarbaum JM, Aghajanian GK. Catecholamine receptors of locus coeruleus neurons: Pharmacological characterization. Eur J Pharmacol 1977; 44:375–385.
38. Bousquet P. Too early to subtype imidazoline receptors (letter; comment). Trends Pharmacol Sci 1990; 11:7.
39. Kamisaki Y, Ishikawa T, Takao Y, et al. Binding of [H-3] para-aminoclonidine to two sites, alpha-2-adrenoceptors and imidazoline binding sites—distribution of imidazoline binding sites in rat brain. Brain Res 1990; 514:15–21.
40. Langin D, Paris H, Lafontan M. Binding of [H3] idazoxan and of its methoxy derivative [3H]RX8210 in human fat cells: [3H] idazoxan but not [3H]RX821002 labels non alpha-2 adrenergic binding sites. Mol Pharmacol 1990; 37:876–885.

41. Regunathan S, Feinstein DL, Reis DJ. Expression of non-adrenergic imidazoline sites in rat cerebral cortical astrocytes. J Neurosci Res 1993; 34:681–688.
42. Regunathan S, Meeley MP, Reis DJ. Expression of non-adrenergic imidazoline sites in chromaffin cells and mitochondrial membranes of bovine adrenal medulla. Biochem Pharmacol 1993; 45:1667–1675.
43. Wang H, Regunathan S, Meeley MP, Reis DJ. Isolation and characterization of imidazoline receptor protein from bovine adrenal chromaffin cells. Mol Pharmacol 1992; 42:792–801.
44. Rall TW. General aspects of ligand-binding phenomena. In: Berne RM, Rall TW, Rubio R, eds. Regulatory Function of Adenosine. The Hague: Martinus Nijhoff, 1993:63–76.
45. Ruffolo RR Jr, Nichols AJ, Stadel JM, Hieble JP. Pharmacologic and therapeutic applications of alpha 2-adrenoceptor subtypes. Annu Rev Pharmacol Toxicol 1993; 33:243–279.
46. Greney H, Bricca G, Dontenwill M, et al. [3H]idazoxan and [3H]clonidine binding in the human nucleus reticularis lateralis (NRL). Fund Clin Pharmacol 1992; 6(suppl 1):51s.
47. Regunathan S, Meeley MP, Reis DJ. Effect of clonidine on second messenger systems in rat adrenal gland. Life Sci 1990; 47:2127–2133.
48. Pineda J, Ugedo L, Garcia-Sevilla JA. 2-adrenoceptor-independent effect of clonidine on locus coeruleus neurons in rats. Fund Clin Pharmacol 1992; 6/suppl 1:51s.
49. Regunathan S, Evinger MJ, Meeley MP, Reis DJ. Effects of clonidine and other imidazole-receptor binding agents on second messenger systems and calcium influx in bovine adrenal chromaffin cells. Biochem Pharmacol 1991; 42:2011–2018.
50. Ruggiero DA, Milner TA, Wang H, et al. Imidazoline receptor-associated binding protein in central nervous system. Soc Neurosci Abstr 1992; 18(part 1):591.
51. Ernsberger PR, Meeley MP, Reis DJ. An endogenous clonidine-like substance binds preferentially to imidazole binding sites in the ventrolateral medulla labelled by 3H-para-aminoclonidine. J Hypertens 1986; 4(suppl 5):S109–S111.
52. Regunathan S, Meeley MP, Reis DJ. Clonidine-displacing substance from bovine brain binds to imidazoline receptors and releases catecholamines in adrenal chromaffin cells. Mol Pharmacol 1991; 40:884–888.
53. Dontenwill M, Molines A, Bricca G, et al. Production and characterization of an iminoimidazolidine specific monoclonal antibody using para-aminoclonidine as antigen. Life Sci 1992; 50:1859–1868.
54. Feldman J, Tibirièa E, Bricca G, et al. Evidence for the involvement of imidazoline receptors in the central hypotensive effect of rilmenidine in the rabbit. Br J Pharmacol 1990; 100:600–604.
55. Bousquet P, Feldman J, Schwartz J. Central cardiovascular effects of alpha adrenergic drugs: Differences between catecholamines and imidazolines. J Pharmacol Exp Ther 1984; 230:232–236.
56. Granata AR, Numao Y, Kumada M, Reis DJ. A1 noradrenergic neurons tonically inhibit sympathoexcitatory neurons of C1 area in rat brainstem. Brain Res 1986; 377:127–146.
57. Scatton B, Pelayo F, Ducobovitch ML, et al. Effect of clonidine on utilization and potassium-evoked release of adrenaline in rat brain areas. Brain Res 1979; 176: 197–201.

58. Engberg G, Svensson TH. Amphetamine-induced inhibition of central noradrenergic neurons: A pharmacological analysis. Life Sci 1979; 24:2245–2254.
59. Timmermans PBMWM, Schoop AMC, Kwa HY, Van Zwieten PA. Characterization of α-adrenoceptors participating in the central hypotensive and sedative effects of clonidine, using yohimbine, rauwolscine and corynanthine. Eur J Pharmacol 1981; 70:7–11.
60. Tibirica E, Feldman J, Bousquet P. Differences in the ability of yohimbine to antagonize the hypotensive effect of clonidine in normotensive and spontaneously hypertensive anesthetized rats. J Pharmacol Exp Ther 1988; 244:1062–1066.
61. Feldman J, Tibirica E, Bricca G, et al. Evidence for the involvement of imidazoline receptors in the central hypotensive effect of rilmenidine in the rabbit. Br J Pharmacol 1990; 100:600–604.
62. Hieble JP, DeMarinis RM, Fowler PJ, Matthews WD. Selective alpha-2 adrenoceptor blockade by SK&F 86466: In vitro characterization of receptor selectivity. J Pharmacol Exp Ther 1986; 236:90–96.
63. Enna SJ, Snyder SH. Influence of ions, enzymes and detergents on gamma-aminobutyric acid-receptor binding in synaptic membranes of rat brain. Mol Pharmacol 1977; 13:442–453.
64. Verbeuren TJ, Xuan ATD, Koenig-Berard E, Vitou P. Rilmenidine. Cardiovasc Drugs Rev 1990; 8:56–70.
65. Gomez RE, Ernsberger P, Feinland G, Reis DJ. Rilmenidine lowers arterial pressure via imidazole receptors in brainstem C1 area. Eur J Pharmacol 1991; 195: 181–191.
66. King PR, Gundlach AL, Jarrott B, Louis WJ. Nature of [3H]rilmenidine binding to membranes of rat cerebral cortex. Am J Hypertens 1992; 5:64S–68S.
67. Sun MK, Guyenet PG. Effect of clonidine and gamma-aminobutyric acid on the discharges of medullo-spinal sympathoexcitatory neurons in the rat. Brain Res 1986; 368:1–17.
68. Head GH, Burke S. Importance of central noradrenergic and serotonergic pathways in the cardiovascular actions of rilmenidine and clonidine. J Cardiovasc Pharmacol 1993; 18:819–826.
69. Tibirica E, Feldman J, Mermet C, Monassier L, et al. Selectivity of rilmenidine for the nucleus reticularis lateralis, a ventrolateral medullary structure containing imidazoline-preferring receptors. Eur J Pharmacol 1991; 209:213–221.
70. Bricca G, Dontenwill M, Molines A, et al. The imidazoline preferring receptor: Binding studies in bovine, rat and human brainstem. Eur J Pharmacol 1989; 162: 1–9.
71. Bricca G, Dontenwill M, Molines A, et al. Evidence for the existence of a homogeneous population of imidazoline receptors in the human brainstem. Eur J Pharmacol 1988; 150:401–402.
72. Tibirica E, Feldman J, Mermet C, et al. Selectivity of rilmenidine for the nucleus-reticularis lateralis, a ventrolateral medullary structure containing imidazoline-preferring receptors. Eur J Pharmacol 1991; 209:213–222.
73. Tibirica E, Feldman J, Mermet C, et al. An imidazoline-specific mechanism for the hypotensive effect of clonidine: A study with yohimbine and idazoxan. J Pharmacol Exp Ther 1991; 256:606–613.

18

Ocular Hypotension Mediated via the CNS

Imidazoline Receptors and the RVLM

W. RICHARD CAMPBELL

Morris Brown College
Atlanta, Georgia

STEPHANIE W. WALDROP and DAVID E. POTTER

Morehouse School of Medicine
Atlanta, Georgia

Summary

1. Moxonidine (MOX) injected intracerebroventricularly (ICVT) into the third ventricle of NZW rabbits induced bilateral, ocular hypotension (> 6.0 mm Hg) that persisted for over 2 h; the effect was significantly inhibited by pretreatment with the putative imidazoline (I_1) receptor antagonist efaroxan, injected ICVT into the third ventricle.
2. MOX, injected unilaterally into the rostral ventrolateral medulla (RVLM), induced bilateral ocular hypotension (> 6.0 mm Hg) that persisted for over 2 h; the effect was significantly inhibited by unilateral efaroxan pretreatment into the RVLM.
3. UK-14,304 injected into the third ventricle induced bilateral ocular hypotension (> 6.0 mm Hg) that persisted for over 2 h; the effect was significantly inhibited by pretreatment with the α-2 antagonist rauwolscine.
4. UK-14,304 injected unilaterally into the RVLM induced no change in intraocular pressure (IOP).

5. The RVLM region of the brainstem plays a role in the modulation of ocular hydrodynamics, as demonstrated by MOX-induced ocular hypotension when injected into the RVLM.
6. I_1 receptors, located generally within the central nervous system (CNS) and specifically within the RVLM region of the brainstem, play a role in MOX-induced ocular hypotension, as suggested by the ability of the putative imidazoline I_1 receptor antagonist efaroxan to inhibit MOX-induced ocular hypotension when pretreated into the third ventricle or the RVLM.
7. Alpha-2 receptors in the RVLM region of the brainstem are not the predominant mediators of ocular hypotension induced by UK-14,304.

I. Introduction

Glaucoma is a leading cause of blindness, particularly in African Americans (1), and represents a major public health problem (2). One of the characteristic risk factors most consistently associated with the disease is a chronic increase in intraocular pressure (IOP) (3). An intact sympathetic nervous system seems to be required for some aspects of IOP regulation, and drugs that suppress sympathoadrenal function are effective antiglaucoma agents. Another manipulation, such as superior cervical ganglionectomy in rabbits, eliminates the elevation of IOP at night (4), indicating that intact ocular sympathetic innervation is required for maintanence of the normal circadian rhythm of IOP in this species (5,6). In patients with open-angle glaucoma, blocking of the stellate ganglion with local anesthesia will lower IOP (7,8). A number of other clinical reports have postulated a relationship between the central nervous system (CNS) and the control of IOP (9,10). Past workers sought a center in the brain, the stimulation of which would affect IOP selectively. Working mostly in the cat, they tried to stimulate discrete nuclei that would affect IOP without parallel changes in systemic blood pressure or extraocular muscle tone—alterations with secondary IOP effects (Refs. 11, 12–16; for review, see Ref. 16). Stimulation in some areas, such as the posterior hypothalamus, induced an IOP change simultaneous with a change in blood pressure, while stimulation in other areas, such as the dorsal hypothalamus and the ventral thalamus, induced changes in IOP without parallel blood pressure changes (17).

A number of substances have been introduced into the cerebrospinal fluid by intraventricular perfusion or by single intraventricular injections in order to study their effects on IOP. Effects on IOP have been demonstrated independent of actions on related physiologic parameters such as blood pressure (16). These substances include hypo- and hyperosmotic agents, prostaglandins, arachidonic acid, calcium, clonidine, substance P, thyrotropin-releasing hormone, and vasopressin (10,18–23). Administration of agents into the cerebrospinal fluid, however, delivers these agents to many areas of the brain; therefore multiple sites of action are possible. Like the studies with diencephalic stimulation, the intraventricular perfusion and injection studies have not identified a specific brain locus for the regulation of aqueous humor dynamics, nor have they identified an integrating neural mechanism for IOP regulation (16).

In recent years, therapeutic and research interests in alpha (α)-2 agonists as potential antiglaucoma drugs has increased. Whether given systemically or applied topically to eyes of animals (24–30) or of normal and glaucomatous humans (31–33), α-2 adrenoceptor agonists have been demonstrated to lower IOP. When perfused intracisternally in conscious rabbits (23), clonidine has been demonstrated to lower IOP. More recently, the centrally mediated systemic antihypertensive actions of clonidine, classified chemically as an imidazoline and functionally as an α-2 adrenergic agonist, have been attributed to interactions at imidazoline-preferring receptors (IPR) (34) rather than to actions on α-2 adrenergic receptors (35,36). The centrally mediated antihypertensive actions of other imidazolines—e.g., moxonidine (MOX)—when injected into the rostral ventrolateral medulla (RVLM) area of the brainstem, have also been attributed, in part, to actions on an imidazoline-preferring receptor (37,38). Recently, MOX and another imidazoline-structured α-2 agonist, oxymetazoline (39), have been shown by this laboratory to lower IOP bilaterally when applied topically to the eyes of rabbits (27,29,40,41) or when injected intracerebroventricularly (ICVT) in rabbits (27–29). Additionally, the α-2 agonist UK-14,304 (UK; brimonidine) (42), defined chemically as an imidazolidine, has been shown by this laboratory and others to induce ocular hypotension when applied topically to the eyes of rabbits (25,29) or injected into the lateral ventricle of the rabbit (29). The purposes of this study were to (1) further assess the contribution of CNS mechanisms to the ocular actions of MOX and UK, (2) determine the magnitude of involvement of central imidazoline (I_1) receptors and α-2 adrenoceptors in the ocular hypotensive responses to MOX and UK, and (3) identify a specific brain locus and integrating neural mechanism for IOP regulation. Efaroxan, which has been classified as a relatively selective I_1 imidazoline receptor antagonist (36), and rauwolscine, a relatively selective α-2 adrenoceptor antagonist (43), were selected to help characterize the centrally mediated ocular hypotensive actions of MOX and UK.

II. Materials and Methods

A. Experimental Animals

New Zealand White (NZW) male rabbits (2 to 4 kg), utilized in IOP experiments, were maintained on a reversed 12-h dark/light cycle. All animals were housed individually at normal room temperature and maintained on Purina Rabbit Chow and tap water ad libitum. Animal care and treatment in this investigation were in accordance with the NIH and ARVO guidelines on the use of animals for research and were approved by the Institutional Animal Care and Use Committee of Morehouse School of Medicine.

B. Implantation of Chronic Indwelling Cannulas

Male NZW rabbits (2 to 4 kg) were anesthetized with ketamine/xylazine, 25/3 mg/kg (i.m.), respectively, and placed in a stereotaxic apparatus (Kopf). A small incision was made to the dorsal cranial surface to reveal the skull sutures lambda and bregma, which were used to align the skull and served as stereotaxic reference points. Trephination of the skull was performed to produce entry ports for the stainless steel guide cannula and three anchor screws. After the anchor screws were placed and the guide cannula was stereotaxically implanted cranioplastic dental cement was built up around the guide cannula and the anchor screws to solidly hold the cannula in place. The surface of the hardened dental cement cap was filed to remove any rough edges that could irritate the skin of the animal. The guide cannula was sealed with a removable dummy cannula to prevent any foreign material from entering the guide cannula and subsequently the CNS. The incision was closed using stainless steel wound clips, which are removed after 1 week. The animals were allowed a minimum of 5 days to recuperate prior to experimentation. Coordinates for cannula placement were obtained using the *Atlas of the Rabbit Brain and Spinal Cord* (44). Verification of injection sites occurred via macroscopic and microscopic methods. In each case a dye, trypan blue, was injected into the injection site to aid in visualizing the area.

C. Intraocular Pressure and Pupil Diameter Measurements

IOP was measured in millimeters of mercury (mm Hg) using an applanation pneumatonograph (Digilab Inc., Division of Bio-Rad, Cambridge, MA), which was calibrated manometrically for rabbit eyes. Pupil diameter (PD) was measured horizontally in millimeters (mm), under constant illumination utilizing an Optistick (Allergan, Irvine, CA). Before tonometry, 25 μL of 0.05% tetracaine (Sigma, St. Louis, MO) was applied to the corneas to minimize any discomfort to the animal. Two baseline readings were taken at -0.5 and 0 h, after which the test drug and vehicle were administered as described below. Subsequent IOP and PD measurements were taken at 0.5, 1, 2, 3, and 4 h postdrug. At the end of each day's measurement, stability of the tonometer was confirmed using the verifier supplied by Digilab.

D. Drug Preparation and Administration

The drugs used and their suppliers were moxonidine (4-chloro-*N*-[4,5 dihydro-1H-imidazol-2-yl]-6-methoxy-2-methyl-pyridinamine hydrochloride, Beiersdorf, Hamburg, Germany); UK-14,304 (brimonidine) 5-bromo-*N*-(4,5-dihydro-1H-imidazol-2-yl)-6-quinoxalinamine (Research Biochemicals, Inc.); efaroxan (2-ethyl-2-[imidazolin-2-yl]-2,3-dihydrobenzofuran) hydrochloride; and rauwolscine (17 alpha-hydroxy-20 alpha yohimban-16 beta-carboxylic acid methy ester hydrochloride) (Sigma Chemical Co., St. Louis, MO; Fig. 1). The salt of each

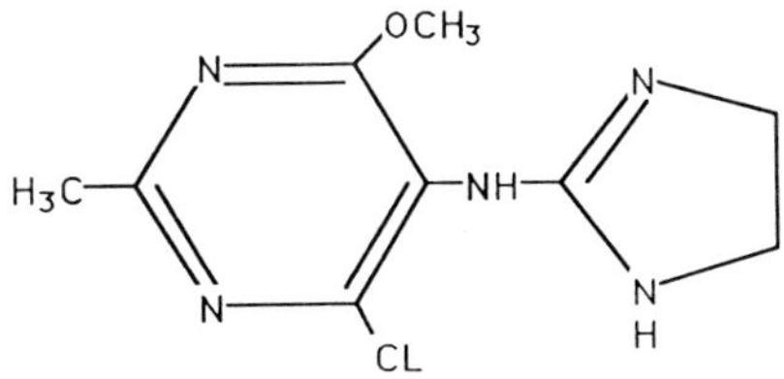

MOXONIDINE
(Imidazoline I-1 / Alpha-2 Agonist)

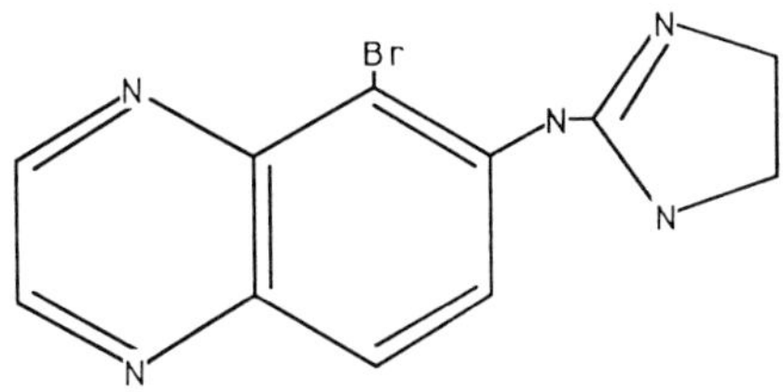

UK-14,304-18
(Alpha-2 Agonist)

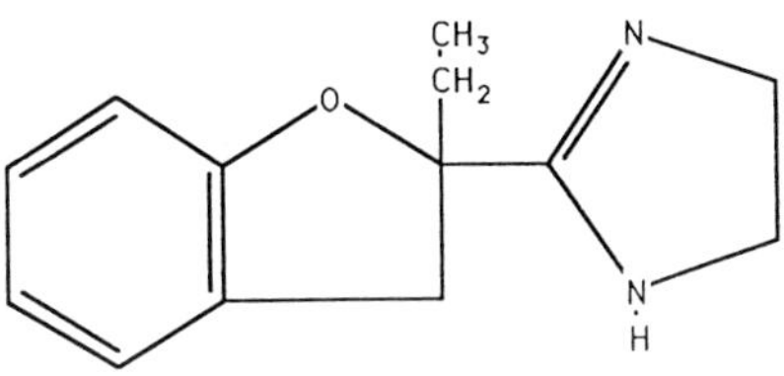

EFAROXAN
(Centrally & Peripherally Acting I-1 & Alpha-2 Antagonist)

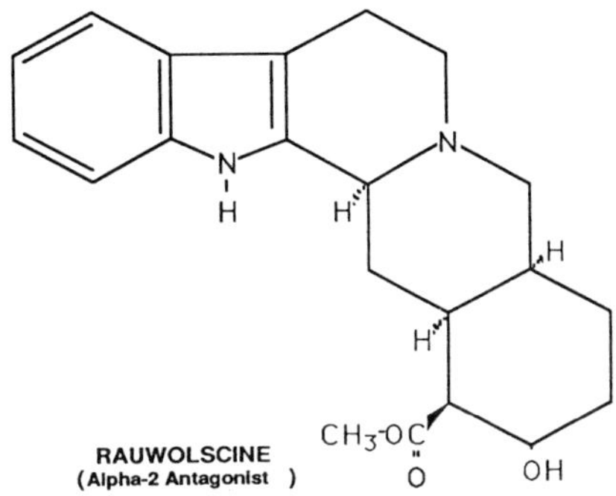

RAUWOLSCINE
(Alpha-2 Antagonist)

Figure 1 Chemical structures of the agonists (top) and antagonists (bottom) used in this study. Each agent has traditionally been described as a potent and selective alpha-2 adrenoceptor agent. Moxonidine and efaroxan are also currently described as imidazoline-preferring receptor agents.

agent was dissolved in distilled water on each day of the experiment; concentrations are expressed as the salt.

Drug and vehicle solutions used in IOP experiments were administered, using Hamilton microliter syringes (1) intracerebroventricularly (ICVT) into the third ventricle in 20-μL volumes and (2) into the RVLM in 1.0-μL volumes, using cannulas (26 g) and injectors (33 g) supplied by Plastic Products, Inc. MOX and UK were administered at time $t = 0$; efaroxan and rauwolscine were administered 30 min prior to the addition of MOX or UK. Since the drugs were salts, saline was used as placebo treatment for control.

E. Presentation and Statistical Analysis of Data

Student's t-test for paired data was used to evaluate drug-induced changes. A probability level of 0.05 was selected as the minimum level of significance; data presented in graphs represent the mean ± standard error (SE).

III. Results

Third Ventricle

Single doses of MOX, UK, efaroxan, and rauwolscine were used in experiments to assess the activity of drugs administered into the CNS, using IOP and PD as the surrogate endpoints of activity. MOX (0.9 μg, 20 μL, $n=4$), injected ICVT into the third ventricle, lowered IOP over a 4-h period as measured by applanation pneumatonometry (Fig. 2). From a starting IOP of 28.8 $\pm$ 0.7 mm Hg, MOX lowered IOP bilaterally 6.1 $\pm$ 0.9 ($P<0.001$) and 6.7 $\pm$ 0.8 ($P<$ 0.001) mm Hg at 30 and 120 min, respectively, as compared to saline controls. IOP returned to control levels after 4 h. MOX (0.9 μg) had no effect on PD at this dose (data not shown) and produced no observable change in rabbit behavior. Efaroxan, an I_1 imidazoline receptor antagonist, injected (3.0 μg, 20 μL, ICVT, $n=4$) 30 min prior to ICVT MOX, inhibited the bilateral IOP reductions at 30 and 120 min by 54% and 99% ($P<0.001$), respectively (Fig. 2). Rauwolscine, an α-2 receptor antagonist, injected (3.0 μg, 20 μL, ICVT, $n=4$) 30 min prior to MOX, inhibited the bilateral reductions at 30 and 120 min by 28% and 56% ($P<0.05$), respectively (Fig. 2). UK-14,304 (0.9 μg, 20 μL, ICVT, $n=4$),

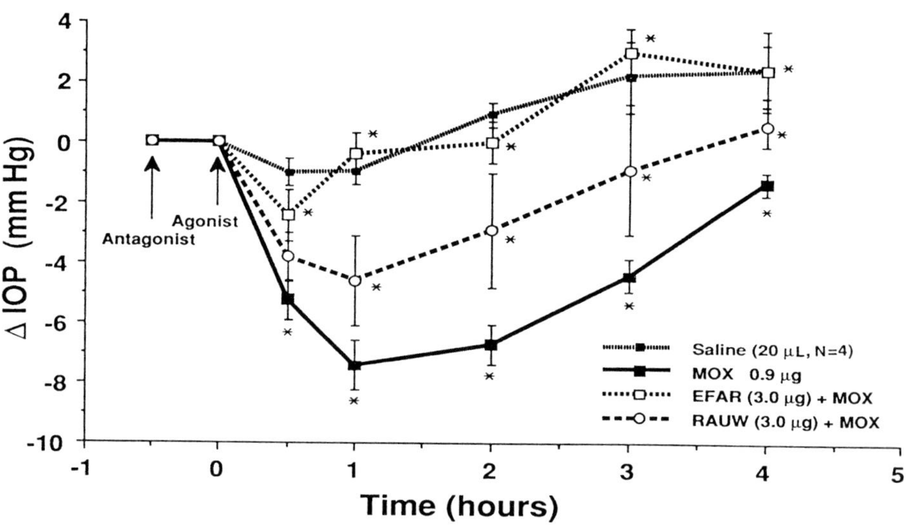

Figure 2 The effect of ICVT injection of MOX into the third ventricle on IOP, with and without efaroxan (ICVT) and rauwolscine (ICVT), as compared to saline controls ($n=4$, for all groups). Starting IOP for treatment groups were MOX alone, 28.8 $\pm$ .7 mm Hg; efaroxan plus MOX, 29.8 $\pm$ .9 mm Hg; rauwolscine plus MOX, 29.3 $\pm$ .5 mm Hg. Individual significance levels were determined by Student's t-test: MOX was compared to saline controls; efaroxan plus MOX data were compared to MOX alone; rauwolscine plus MOX data were compared to MOX alone; * = $P<0.05$.

from a starting pressure of 31.7 ± 0.49 mm Hg, lowered IOP bilaterally, 4.8 ± 0.4 ($P<0.001$) and 6.2 ± 1.7 ($P<0.05$) mm Hg, at 30 and 120 min, respectively, as compared to saline controls (Fig. 3). UK (0.9 μg, ICVT) had no observable effect on PD at this dose (data not shown) and produced no observable change in rabbit behavior. The antagonist rauwolscine (3.0 μg, ICVT, $n=4$), given 30 min prior to UK, inhibited UK-induced ocular hypotension by 55% at 30 min ($P<0.001$) and by 95% at 120 min ($P<0.001$) (Fig. 3). Efaroxan (3.0 μg, 20 μL, ICVT, $n=4$), given ICVT 30 min prior to UK (0.9 μg), had no effect on UK-induced ocular hypotension (Fig. 3). Neither saline (20 μL, ICVT, $n=4$), efaroxan (3.0 μg, 20 μL, ICVT, $n=4$), nor rauwolscine (3.0 μg, 20 μL, ICVT, $n=4$) alone produced any significant change in IOP (Fig. 4).

Rostral Ventrolateral Medulla (RVLM)

Single doses of MOX and UK (0.9 μg/1.0 μL) were unilaterally injected in the RVLM region and possible changes in IOP were evaluated. From a starting pressure of 29.0 ± 0.9 mm Hg ($n=4$), MOX lowered IOP, −3.8 ± .4 ($P<0.05$) mm Hg at 30 min and −6.9 ± 1.3 ($P<0.001$) mm Hg at 120 min. The I-1

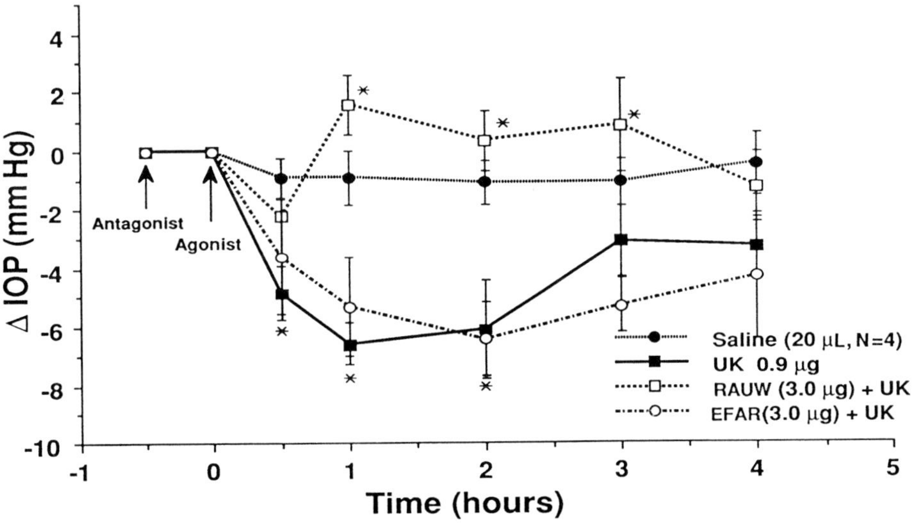

Figure 3 The effect of ICVT injection of UK into the third ventricle on IOP with and without efaroxan (ICVT) and rauwolscine (ICVT) as compared to saline controls ($n=4$, for all groups). Starting IOP for treatment groups were UK alone, 31.7 ± .5 mm Hg; efaroxan plus UK, 31.6 ± .8 mm Hg; rauwolscine plus UK, 32.0 ± .7 mm Hg. Individual significance levels were determined by Student's *t*-test: UK was compared to saline controls; efaroxan (ICVT) plus UK data were compared to UK alone; rauwolscine (ICVT) plus UK data were compared to UK alone; * = $P<0.05$.

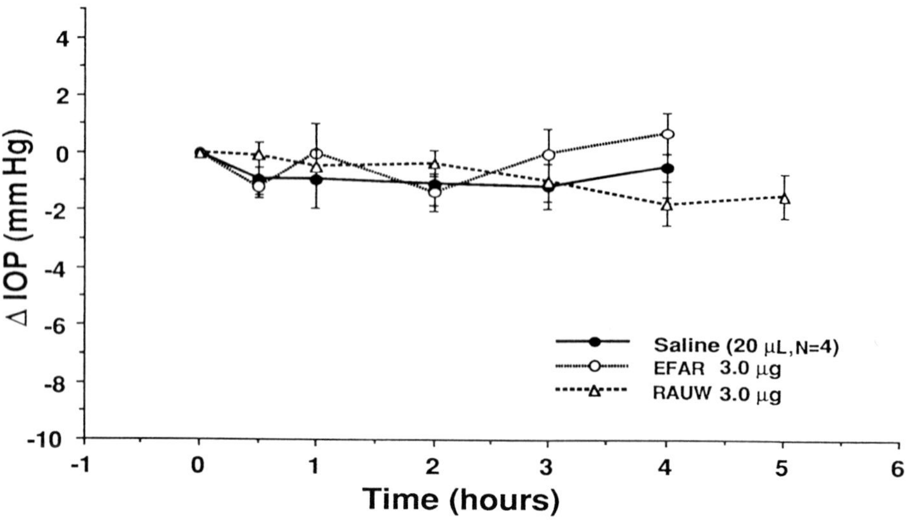

Figure 4 The effects on IOP of injection of saline (20 μL, ICVT), efaroxan (3.0 μg, 20 μL, ICVT) and rauwolscine (3.0 μg, 20 μL, ICVT), into the third ventricle (3V), ($n=4$, for all groups). Starting IOP for treatment groups were saline, 27.5 ± .9 mm Hg; efaroxan, 26.8 ± .9 mm Hg; rauwolscine, 28.7 ± .3 mm Hg.

antagonist efaroxan, injected unilaterally into the RVLM region, inhibited the MOX-induced hypotensive response (100% at 30 min and 60% at 120 min) more than did the α-2 antagonist rauwolscine (1% at 30 min and 31.4% at 120 min) (Fig. 5). There was no observable change in PD as a result of MOX (0.9 μg) injection into the RVLM region. When UK was injected into the RVLM region, IOP was not lowered significantly (Fig. 5). From an initial IOP reading of 33.8 ± 0.4 mm Hg, the change following UK was only −0.9 mm Hg at 30 min and only −1.8 mm Hg at 120 min. Saline (1.0 μg, $n=4$) alone did not produce any significant change in IOP (Fig. 5) when injected unilaterally into the RVLM region.

IV. Discussion

The objectives of this preliminary study were to assess further the role of centrally located α-2 adrenoceptors and I_1 imidazoline-preferring receptors (IPR) in the mediation of changes in intraocular pressure. Drugs were administered into the third cerebral ventricle and into the RVLM region of the brainstem. MOX and UK-14,304, given ICVT into the third ventricle, significantly lowered IOP in conscious NZW male rabbits. MOX, but not UK, significantly lowered IOP when injected unilaterally into the RVLM region of the brainstem.

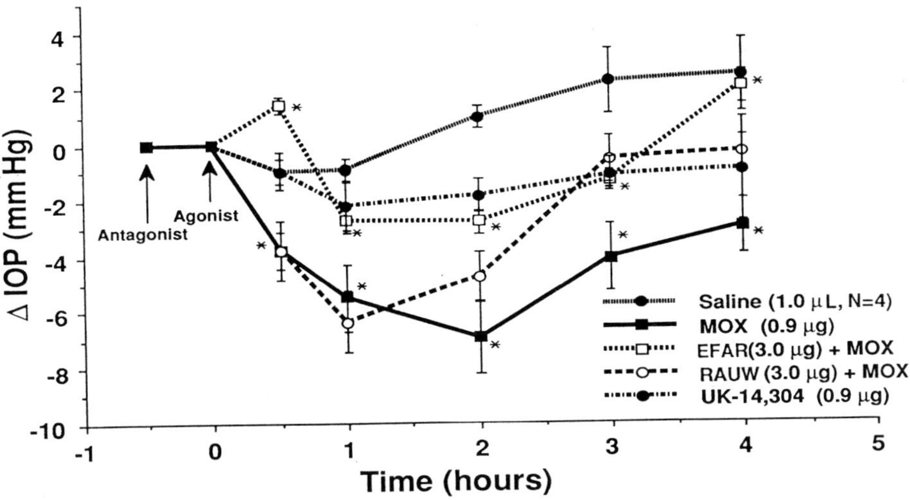

Figure 5 The effect on IOP of unilateral injection of (1) MOX into the RVLM region with and without efaroxan and rauwolscine and (2) UK alone, into the RVLM region, as compared to saline controls ($n = 4$ for all groups). Starting IOP for treatment groups were MOX alone, 29.9 ± .9 mm Hg; efaroxan plus MOX, 27.8 ± 1.3 mm Hg; rauwolscine plus MOX, 29.9 ± .5 mm Hg; UK alone, 33.8 ± .4 mm Hg. Individual significance levels were determined by Student's *t*-test: MOX was compared to saline controls; efaroxan plus MOX data were compared to MOX alone; rauwolscine plus MOX data were compared to MOX alone; * = $P < 0.05$.

MOX is a clinically utilized second-generation antihypertensive agent that was originally classified as an alpha-2 adrenergic agonist (45). The recent findings of several laboratories have questioned the limitations of this classification, however, and have proposed the idea that MOX may activate an IPR (28, 34,37,38,46–48) rather than an α-2 receptor. This IPR has been divided into subtypes, I_1 and I_2 (49) as suggested by binding experiments (37,38,46) and by functional alterations in cardiovascular parameters (48). Based on the recent binding studies of Ernsberger and coworkers, MOX has been shown to be one of the most selective IPR (I_1) agonists available at this time. To determine the actions of MOX on cardiovascular parameters, workers have injected MOX into the brainstem area called the RVLM (48) in different animal species. MOX (2 to 10 nmol), an imidazoline, was able to significantly lower blood pressure in rabbits (48), while α-methylnorepinephrine (0.1–10 μg/kg), a catecholamine, injected into rats was unable to affect blood pressure in this brainstem region, whereas clonidine produced dose-dependent hypotension (36). Therefore, not only MOX but a number of imidazoline-structured compounds, including clonidine and rilmenidine, another second-generation antihypertensive agent, have been shown to preferentially induce systemic hypotension when injected into the RVLM. Moreover, nonimidazoline compounds traditionally classified as α-2 agonists, such as α-methylnorepinephrine (36) and guanabenz (37), are much less effective in reducing systemic blood pressure when administered into the RVLM. UK-14,304 (brimonidine), classified as a relatively selective α-2 adrenergic agonist (42), has been demonstrated to lower IOP bilaterally when applied unilaterally to the eyes of rabbits (25,29) or when injected ICVT into the anterior lateral ventricle of rabbits (29).

In order to ascertain the receptor type(s) and site(s) of action that mediate(s) the ocular hypotension induced by centrally administered MOX and UK, relatively selective antagonists were used to inhibit the effects of these agonists. Efaroxan, described as a relatively selective I_1 imidazoline-preferring antagonist (37), and rauwolscine, described as a relatively selective α-2 adrenoceptor antagonist (43), were used to evaluate their effects on both MOX and UK. Efaroxan was able to significantly inhibit the ocular hypotension induced by MOX, whereas rauwolscine was a much less effective antagonist when injected ICVT into the third ventricle prior to MOX. The decrease in IOP induced by ICVT MOX was inhibited markedly by pretreatment with ICVT efaroxan, whereas rauwolscine was able to inhibit ICVT MOX only to a moderate degree. The ocular hypotension induced by UK (0.9 μg), administered ICVT, was significantly inhibited by ICVT pretreatment with rauwolscine. In contrast, efaroxan was relatively ineffective against UK. The marked inhibition by ICVT efaroxan of the actions of centrally applied MOX on IOP indicates that central I_1 imidazoline receptors function to mediate ocular hypotension. The ability of ICVT rauwolscine to inhibit the IOP effects of UK markedly demonstrates that central α-2 receptors also play a role in ocular hypotension. It should be noted that there was a previous suggestion of a requirement for an intact sympathetic nervous system in order for MOX to induce ocular hypotension when applied topically to the eyes of rabbits (40). However, this former study was not designed to identify multiple site(s) of action for MOX and did not utilize centrally mediated responses to assess the activity of MOX.

To explore a possible central site of action of α-2 and imidazoline I_1 receptors in mediating ocular hypotension, MOX and UK were injected unilaterally, in a smaller volume (1 μL), into the RVLM region of the brainstem. The RVLM is a critical component of an area in the brainstem called the medullary intermediate reticular zone (MIRZ). The MIRZ contains premotor autonomic and respiratory neurons as well as interneurons (50). The RVLM is a critical region of the MIRZ that contains vasomotor (51–53), cardiovagal (54), and respiratory neurons (55) forming a cardiorespiratory network, which controls inputs to spinal, respiratory, and preganglionic motoneurons (56). The MIRZ and its associated RVLM are components of the central autonomic network (CAN) (57). The CAN is a network that receives and integrates viscerosensory, humoral, and exteroceptive information; in turn, it controls and coordinates preganglionic, neurosecretory, and respiratory outputs. The main output of the CAN is mediated through preganglionic sympathetic and parasympathetic neurons (for a review, see Ref. 57). While a number of midbrain regions in the hypothalamus and thalamus have been shown to induce alterations in IOP upon electrical stimulation (17), there is a paucity of reports describing modulation of IOP by specific sites in the brainstem. MOX injected unilaterally into the RVLM region significantly lowered IOP. This MOX-induced ocular hypotension was significantly and preferentially inhibited by unilateral pretreatment with efaroxan into the RVLM. Efaroxan was able to inhibit MOX-induced ocular hypotension significantly (approximately 60%); whereas rauwolscine inhibited the MOX-induced decrease in IOP by a modest amount (approximately 30%). UK injected unilaterally into the RVLM did not induce a change in IOP.

The inhibition by efaroxan, injected into the third ventricle and RVLM, of MOX-induced ocular hypotension implicates the involvement of a centrally mediated I_1 imidazoline receptor mechanism in the ocular action of MOX. The inhibition of UK-induced ocular hypotension by rauwolscine, when injected into the third ventricle, demonstrates that central α-2 adrenoceptors also play a role in ocular hydrodynamics. The inability of UK to induce a decrease in IOP when injected into the RVLM region suggests that this region contains a predominance of IPR rather than α-2 receptors. Additionally however, recent data from our laboratory (27,28) have suggested that ICVT MOX can also induce ocular hypotension via an α-2 adrenergic mechanism. For example, the relatively selective α-2 adrenergic receptor antagonists L-659,066 (58) and L-657,743 (59) were able to significantly inhibit the MOX-induced decrease in IOP when injected into the anterior lateral ventricle, whereas rauwolscine again only partially inhibited the ICVT MOX-induced decrease in IOP. It is not known however, if L-659,066 and L-657,743 also have IPR antagonist activity. This latter possibility is currently being explored.

V. Conclusion

In conclusion, the findings of this study support the hypotheses that (1) activation of central α-2 adrenoceptors and imidazoline I_1 receptors influences aqueous humor dynamics; (2) MOX acts primarily via imidazoline-1 receptors to lower IOP when injected into the third ventricle and RVLM region; (3) UK-14,304 acts primarily via α-2 receptors to lower IOP when injected into the third ventricle but was ineffective in changing IOP when injected into the RVLM region; and (4) imidazoline I_1 receptors are present in the RVLM region, as demonstrated by the ability of MOX but not UK-14,304 to induce ocular hypotension and by antagonism of MOX effects by efaroxan.

Acknowledgment

This research was supported, in part, by EY06338, from the National Eye Institute.

References

1. Sommer A, Tielsch JM, Katz J, et al. Relationship between intraocular pressure and primary open angle glaucoma among white and black Americans. Arch Ophthalmol 1991; 109:1090–1095.
2. Wilson MR. Primary open-angle glaucoma: Magnitude of the problem in the United States. J Glaucoma 1992; 1:64–68.
3. Quigley HA. Open angle glaucoma: Medical progress. N Engl J Med 1993; 328: 1097–1106.
4. Gregory DS, Aviado DG, Sears ML. Cervical ganglionectomy alters the circadian rhythm of intraocular pressure in New Zealand rabbits. Curr Eye Res 1985; 4:1273.
5. Rowland JM, Potter DE, Reiter RJ. Circadian rhythm in an intraocular pressure model. Curr Eye Res 1981; 1:169–173.
6. Braslow RA, Gregory DS. Adrenergic decentralization modifies the circadian rhythm of intraocular pressure. Invest Ophthalmol Vis Sci 1987; 28:1730.
7. Miller SJH. Stellate ganglion block in glaucoma. Br J Ophthalmol 1953; 37:70.
8. Endo Y. Fluctuation of intraocular pressure by blockage of the stellar ganglion. Acta Soc Ophthalmol Jpn 1961; 70:926.
9. Waitzman MB. Hypothalamus and ocular pressure. Surv Ophthalmol 1971; 16:161.
10. Schmerl E, Steinberg B. The role of the ciliary and superior cervical ganglion in ocular tension. Am J Ophthalmol 1949; 32:947.
11. Schmerl E, Steinberg B. Separation of diencephalic centers concerned with pupillary motility and ocular tension. Am J Ophthalmol 1950; 33:1379.
12. Nagai M, Ban T, Kurotsu T. Studies on the changes of intraocular pressure induced by electrical stimulation of the hypothalamus. Med J Osaka U 1951; 2:87.
13. von Sallmann L, Lowenstein O. Responses of intraocular pressure, blood pressure and cutaneous vessels to electric stimulation of the diencephalon. Am J Ophthalmol 1955; 39:11.
14. Gloster J, Greaves DP. Effect of diencephalic stimulation upon intra-ocular pressure. Br J Ophthalmol 1957; 41:513.
15. Gloster J. Responses to the intraocular pressure to diencephalic stimulation. Br J Ophthalmol 1960; 44:649.
16. Stone RA, Kuwayama Y. The nervous system and intraocular pressure. In: Ritch R, Shields MB, and Krupin T, eds. The Glaucomas. St. Louis, Mosby, 1989:257–279.
17. von Sallmann L. The role of the central nervous system in the regulation of the intraocular pressure. Doc Ophthalmol 1959; 13:93.
18. Krupin T, Podos SM, Becker B. Alteration of intraocular pressure after third ventricle injections of osmotic agents. Am J Ophthalmol 1973; 76:948.
19. Krupin T, Oestrich CJ, Podos SM, Becker B. Increased intraocular pressure after third ventricle injections of prostaglandins E1 and arachidonic acid. Am J Ophthalmol 1976; 81:346.
20. Krupin T, Grove JC, Gugenheim SM, Oestrich CJ, Podos SM, Becker B. Increased intraocular pressure and hypothermia following injection of calcium into the rabbit third ventricle. Exp Eye Res 1978; 27:129.
21. Krupin T, Koloms BA, Klutho I, Webb G, Becker B. Increased intraocular pressure and hypothermia following injection of substance P into the rabbit third ventricle. Exp Eye Res 1982; 34:319.
22. Krupin T, Webb G, Barbosa A, Gulli B, Levine J, Becker B. Central effects of thyrotropin-releasing hormone and arginine vasopressin on intraocular pressure in rabbits. Invest Ophthalmol Vis Sci 1984; 25:932.

23. Liu JHK, Neufeld AH. Study of central regulation of intraocular pressure using ventriculocisternal perfusion. Invest Ophthalmol Vis Sci 1985; 26:136–143.
24. Allen RC, Langham ME. The intraocular pressure response of conscious rabbits to clonidine. Invest Ophthalmol 1976; 15:815–823.
25. Burke JA, Potter DE. Ocular effects of a relatively selective $\alpha 2$ agonist (UK 14,304-18) in cats, rabbits and monkeys. Curr Eye Res 1986; 5:665–676.
26. Campbell WR, Potter DE. The central effects of moxonidine on intraocular pressure and its antagonism by L-659,066, in the rabbit. Am Coll Neuropsychopharmacol 1992; :143.
27. Campbell WR, Wilson CJ, Potter DE. Ocular hypotensive effects of moxonidine and oxymetazoline: Involvement of central imidazoline (I1) receptors. Invest Ophthalmol Vis Sci (ARVO suppl) 1993; 34:928.
28. Campbell WR, Potter DE. The central effects of moxonidine on intraocular pressure and its antagonism by L-659,066 and L-657,743 in the rabbit. J Prog Neuro-Psychopharmacol & Biol Psychiatry 1994; Vol 18:1051–1061.
29. Campbell WR, Long S, Strothers S, Potter DE. Centrally-induced ocular hypotension in the rabbit: Involvement of alpha-2 and imidazoline receptor subtypes. Soc Neurosci Abstr 1993; 19:1790.
30. Campbell WR, Potter DE, Potential role of imidazoline (I1) receptors in modulating aqueous humor dynamics. Int Symp Ocular Pharm, Novi, Michigan, #53.
31. Makabe R. Ophthalmologische Untersuchungen und Dichlorophenylaminoimidazolin. Dtsch Med Woschenschr 1966; 91:1686–1696.
32. Hodapp E, Kolker AE, Kass MA, et al. The effect of topical clonidine on intraocular pressure. Arch Ophthalmol 1981; 99:1208–1212.
33. Robin AL, Pollack IP, House B, Eager C. Effects of AL02145 on ocular pressure following argon laser trabeculoplasty. Arch Ophthalmol 1987; 105:646–650.
34. Lehman J, Koenig-Berard E, Vitou P. The imidazoline-preferring receptor. Life Sci 1989; 45:1609–1615.
35. Karppanen H, Paakkari I, Paakkari P. Studies on the nature of cerebral receptors mediating the hypotensive effects of clonidine in rats. In: Yellin TO, ed. Histamine Receptors. New York, S.P. Medical and Scientific, 1977:25.
36. Bousquet P, Feldman J, Schwartz J. Central cardiovascular effects of alpha adrenergic drugs: differences between catecholamines and imidazolines. J Pharmacol Exp Ther 1984; 230:232–236.
37. Ernsberger P, Westbrooks KL, Christen MO, Schafer SG. A second generation of centrally acting antihypertensive agents act on putative I1-imidazoline receptors. J Cardio Pharm 1992; 20(suppl. 4):S1–S10.
38. Ernsberger P, Damon TH, Graff LM, et al. Moxonidine, a centrally acting antihypertensive agent, is a selective ligand for I1-imidazoline sites. J Pharmacol Exp Ther 1993; 264:172–182.
39. Bylund DB, Ray-Prenger C, Murphy TJ. Alpha-2A and alpha-2B adrenergic receptor subtypes: Antagonists binding in tissues and cell lines containing only one subtype. J Pharmacol Exp Ther 1988; 245:600–607.
40. Ogidigben MJ, Potter DE. Moxonidine: Effects on intraocular pressure and pupil diameter and contraction of the nictitating membrane. Invest Ophthalmol Vis Sci (ARVO suppl) 1991; 32:867.
41. Potter DE, Ogidigben MJ, Chu TC. Oxymetazoline: An atypical alpha-2 agonist with ocular hypotensive activity. Invest Ophthalmol Vis Sci (ARVO suppl) 1992; 33:1116.

42. Cambridge D. UK-14,304-18, a potent and selective α2-agonist for the characterization of α-adrenoceptor subtype. Eur J Pharmacol 1981; 72:413–415.
43. Perry BD, U'Prichard DC. [^{3}H]Rauwolscine (α-yohimbine), a specific antagonist radioligand for brain α2-adrenergic receptors. Eur J Pharmacol 1981; 76:461–464.
44. Shek JW, Wen GY, Wisniewski HM. Atlas of the rabbit brain and spinal cord. New York, Karger, 1986.
45. Armah IB, Stenzel W. BE 5895, a new clonidine-type antihypertensive aminopyrimidine derivative. Naunyn-Schmiedeberg's Arch Pharmacol 1981; 316:R42.
46. Ernsberger P, Giuliano R, Willette RN, et al. Hypotensive action of clonidine analogues correlates with binding affinity at imidazoline and not alpha-2-adrenergic receptors in the rostral ventrolateral medulla. J Hypertens 1988; 6(suppl 4):S554–557.
47. Ernsberger P, Biuliano R, Willette RN, Reis DJ. Role of imidazole receptors in the vasodepressor response to clonidine analogs in the rostral ventrolateral medulla. J Pharmacol Exp Ther 1990; 253:408–418.
48. Haxhiu MA, Dreshaj I, Erokwu B, et al. Vasodepression elicited in hypertensive rats by the selective I1-imidazoline agonist moxonidine administered into the rostral ventrolateral medulla. J Cardiovasc Pharmacol 1992; 20(suppl 4):S11–S15.
49. Reis DJ, Regunathan S, Wang H, et al. Imidazoline receptors in the nervous sytem. Fund Clin Pharmacol 1992; 6(suppl 1):23S–29S.
50. Tork I, McRitchie DA, Rikard-Bell GC, Paxinos G. Autonomic regulatory centers in the medulla oblongata. In: Paxinos G, ed. The Human Nervous System, San Diego, CA: Academic Press, 1990.
51. Guyenet PG, Haselton JR, Sun M-K. Sympathoexcitatory neurons of the rostroventrolateral medulla and the origin of the sympathetic vasomotor tone. Prog Brain Res 1989; 81:105–116.
52. Arango V, Ruggiero DA, Callaway JL, et al. Catecholaminergic neurons in the ventrolateral medulla and nucleus of the solitary tract in the human. J Comp Neurol 1988; 273:224–240.
53. McAllen RM, Dampney RAL. Vasomotor neurons in the rostral ventrolateral medulla are organized topographically with respect to type of vascular bed but not body region. Neurosci Lett 1990; 110:91–96.
54. Loewy AD, Spyer KM. Vagal preganglionic neurons. In: Loewy AD, Spyer KM, eds. Central Regulation of Autonomic Functions, New York: Oxford University Press, 1990:68–87.
55. Feldman JL. Neurophysiology of breathing in mammals. In: Mountcastle VB, Bloom FE, Greiger SR, eds. Handbook of Physiology. Sec. 1: The Nervous System. Vol. 4. Bethesda, MD: American Physiological Society, 1986:466–524.
56. Richter DW, Spyer KM. Cardiorespiratory control. In: Loewy AD, Spyer KM, eds. Central Regulation of Autonomic Functions. New York: Oxford University Press, 1990:189–207.
57. Benarroch EE. The central autonomic network: Functional organization, dysfunction and perspective. Mayo Clin Proc 1993; 68:988–1001.
58. Clineschmidt BV, Pettibone DJ, Lotti VJ, et al. A peripherally acting alpha-2 adrenoceptor antagonist: L-650,066. J Pharmacol Exp Ther 1988; 245:32–40.
59. Pettibone DJ, Clineschmidt BV, Lotti VJ, et al. Pharmacological profile of a new potent and specific α2-adrenoceptor antagonist, L-657,743. Naunyn-Schmiedeberg's Arch Pharmacol 1987; 336:169–175.

19

Imidazoline I_1 Receptors in the Ventrolateral Medulla and Their Role in Cardiorespiratory Control

PAUL ERNSBERGER, LAURA A. COLLINS, MARILYN E. GRAVES, ISMAIL A. DRESHAJ, and MUSA A. HAXHIU

School of Medicine
Case Western Reserve University
Cleveland, Ohio

I. The RVLM as the Site of Action for Central Antihypertensive Agents

The probable site of the hypotensive action of clonidine is within the rostral ventrolateral medulla (RVLM). Localization of clonidine's action to the medulla oblongata arises from experiments showing that transection of the brainstem at the pons preserves the response to clonidine and its analogues, whereas transection of the spinal cord above the first thoracic segment abolishes it (1,2). Human patients with cervical spinal cord transection also fail to show a vasodepressor response to clonidine and manifest only the direct pressor action on blood vessels (3). Furthermore, clonidine lowers blood pressure when given into the vertebral arteries at doses that are inactive when given into the carotid artery, even when the basilar artery is clipped to limit the injection to pontomedullary structures (4). Interestingly, clonidine does not lower heart rate when given by this route, suggesting that the bradycardiac and vasopressor actions of clonidine are mediated by distinct neuronal sites.

Within the medulla of the rat, direct microinjection of clonidine or related analogues into the RVLM lowers blood pressure, but microinjections as little as 2 mm distant to the pressor area of the RVLM are without effect (5,6).

However, delivery of clonidine into the caudal ventrolateral medulla (CVLM) has been reported to lower blood pressure as effectively as RVLM microinjection, although with a longer delay to reach maximum effect (7). The site of action for clonidine has been mapped in the cat, a species with a larger medulla that lends itself more readily to localization studies (8). The largest effect of clonidine was observed in an intermediate area 2 mm caudal to the RVLM pressor area and at the same dorsoventral and lateral coordinates (8). Clonidine was assumed to stimulate neurons in this region, which in turn inhibit RVLM pressor neurons. A major French group has also noted that the active site for clonidine in the cat is caudal to the region from which sympathoexcitatory RVLM neurons have been recorded (9). Another group found two active zones for clonidine microinjections within the RVLM region, one located within the pressor zone and another area more caudal and medial (10). The two sites appear to be independent, since lesion of one allows an effect of clonidine at the other site (10). Thus, it appears likely that the action of clonidine is not confined to the region containing the cell bodies of tonically active sympathoexcitatory neurons and may involve other nearby groups of neurons.

Microinjection of clonidine into the nucleus tractus solitarius (NTS) will also lower blood pressure; 100-fold higher concentrations are required relative to the RVLM (11). Lesions of the NTS have no effect on the response to clonidine (12,13), but animals with RVLM lesions show only pressor responses to clonidine (14). Moreover, localized microinjection of the imidazoline α_2-antagonist idazoxan into the RVLM prevents the vasodepressor effect of systemic clonidine (8,15,16) or rilmenidine (2). Thus, administration of an antagonist specifically into the RVLM, leaving intact the NTS and other potential sites of action, completely abolishes the hypotensive action of centrally acting imidazolines.

Clonidine elicits bradycardia by two mechanisms, one mediated by sympathoinhibition and the other mediated by an increase in vagal tone. There is some evidence that the vagal component of the heart rate responses might be mediated within the nucleus ambiguus. Microinjection of either norepinephrine or clonidine into the nucleus ambiguus elicits bradycardia with little effect on blood pressure (17). This bradycardiac response is abolished by vagotomy. In contrast, microinjection of clonidine into the nearby RVLM lowers blood pressure with little effect on heart rate, and this response is unaffected by vagotomy (17). Furthermore, microinjection of idazoxan into the nucleus ambiguus blocks the bradycardic action of clonidine without affecting the blood pressure response, whereas delivery of idazoxan into the RVLM blocks the hypotensive action without affecting bradycardia (16). Most studies show that microinjection of idazoxan in the RVLM blocks the bradycardic effect of imidazoline agents (8,15), but the spread of antagonist into the nucleus ambiguus cannot be ruled out.

II. NTS α_2-Adrenergic Receptors as a Site of Action for Central Antihypertensive Agents

Activation of central α_2-adrenergic receptors elicits falls in blood pressure and heart rate. The action of the nonimidazoline central antihypertensive agents—specifically α-methyldopa, guanfacine, and guanabenz—is almost certainly attributable to an action of α_2-adrenergic receptors. The most likely site of action for α_2-agonist action in the brainstem is the NTS. Microinjection of minute doses (0.02 to 0.3 nmol) of norepinephrine, epinephrine, or α-methylnorepinephrine, a selective α_2-agonist which is the active metabolite of α-methyldopa, into this brain region lowers blood pressure and heart rate (1,18–21). Blood pressure and heart rate are usually decreased by about the same proportion and there is no evidence for a dissociation of blood pressure and heart rate effects. The dose-dependent fall in blood pressure and heart rate elicited by α-methylnorepinephrine microinjected into the NTS is shown in Figure 1 [solid symbols; data from Zandberg and colleagues (20)]. Vagotomy or treatment with atropine actually potentiates the hypotensive action of catecholamines within the NTS, indicating that α_2-receptor activation inhibits sympathetic activity.

The imidazoline α_2-agonist clonidine is much less effective than the catecholamines, having no effect at 1.25 nmol and decreasing blood pressure only 20 mm Hg at a 5-nmol dose, whereas epinephrine lowers blood pressure 24 mm Hg at a 600-fold lower dose (20). Higher doses of clonidine, up to 80 nmol, actually increase blood pressure, as shown in Figure 2. Note the contrast with the potent action of α-methylnorepinephrine shown in Figure 1. Oxymetazoline is completely ineffective in lowering blood pressure when administered into the NTS in doses up to 20 nmol (20). In another study, clonidine was inactive at doses up to 30 nmol, whereas the effect of epinephrine was half-maximal at 3 nmol (22). The ineffectiveness of clonidine and oxymetazoline in lowering blood pressure from within the NTS is probably due to their status as partial agonists. Clonidine and oxymetazoline often act as α_2-adrenergic antagonists, particularly in the presence of high concentrations of endogenous catecholamines (23), such as might be found in the NTS region. BHT 920, a noncatecholamine full α_2-agonist, is about as potent as α-methylnorepinephrine in lowering blood pressure when given into the NTS (21). Interestingly, clonidine is able to decrease heart rate when microinjected into the NTS (Fig. 2), implying that blood pressure and heart rate are independently regulated within the medulla.

A mapping study showed the active zone for α-methylnorepinephrine to be within 0.5 mm of the caudal tip of the area postrema in proximity to the solitary tract (24). This corresponds to the area eliciting the greatest depressor response when stimulated. When delivered within the NTS, α-methylnorepin-

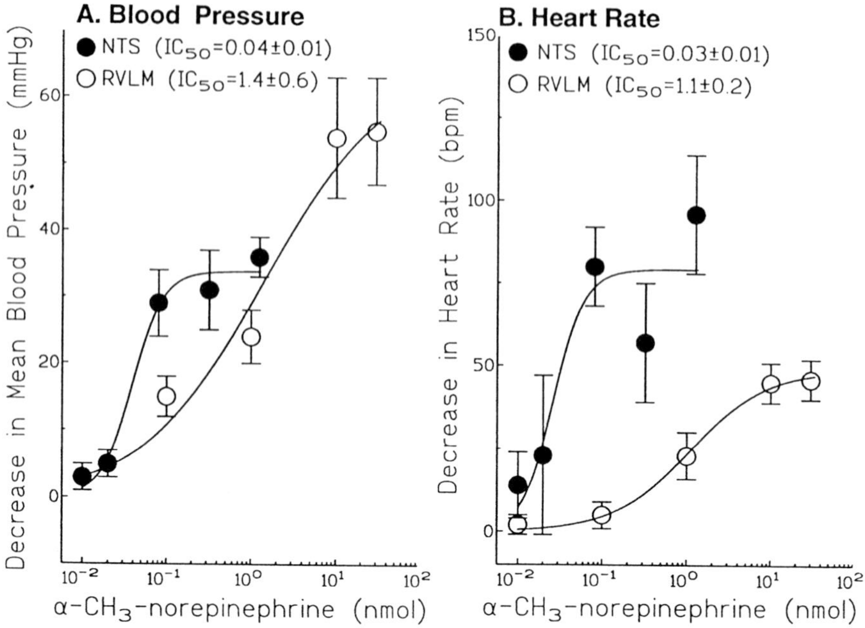

Figure 1 Dose-dependent cardiovascular actions of α-methylnorepinephrine upon microinjection into the RVLM or the NTS. Data for NTS (solid circles) were reanalyzed from Zandberg and colleagues (20) and represent the decrease in mean blood pressure (panel A) or heart rate (panel B) at 5 min after bilateral microinjection in a volume of 500 nL. Data for the RVLM (open circles) were reanalyzed from Granata and colleagues (44) and represent peak decreases in blood pressure (panel A) or heart rate (panel B) after bilateral microinjection in a volume of 100 nL. IC_{50} values were determined by nonlinear curve fitting to a logistic equation by using the program InPlot (152) (Graphpad Software, San Diego, CA). The logistic function accounted for at least 92% of the total variance in each case. α-Methylnorepinephrine is at least 30-fold more potent in the NTS than in the RVLM ($P<0.01$).

ephrine (0.5 nmol) not only decreases blood pressure and heart rate but also potentiates the depressor response to stimulation of the aortic depressor nerve (25). The α_2-adrenergic antagonist yohimbine has the opposite effect, even at very low doses, attenuating the response to aortic nerve stimulation while abolishing the effect of α-methylnorepinephrine (25). Microinjection of yohimbine into the NTS increases blood pressure in doses as low as 10 pmol, whereas the imidazoline α_2-antagonist idazoxan is 25-fold less potent (19). These data demonstrate a potent α_2-adrenergic receptor mechanism for regulating blood pressure within the NTS, which is integrated into the baroreflex and is tonically active, as indicated by the potent pressor response to antagonists.

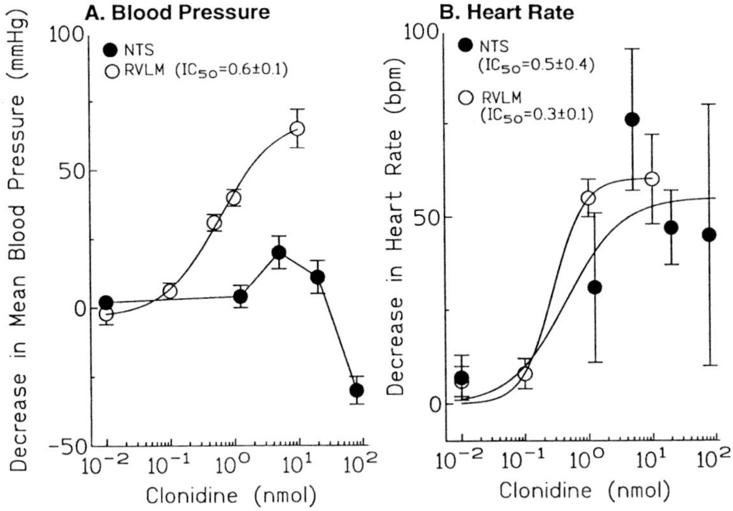

Figure 2 Dose-dependent cardiovascular actions of clonidine upon microinjection into the RVLM or the NTS. All details as in Figure 1. Clonidine had dose-dependent action on blood pressure only within the RVLM. The IC_{50} values for clonidine for bardycardiac actions were identical in the NTS and the RVLM.

III. Central Nervous System α_2-Adrenergic Receptors as Mediators of Central Nervous System Depression

While activation of α_2-adrenergic receptors is an effective means of lowering blood pressure, it also causes sedation and other signs of central nervous system depression, which are the major side effects of centrally acting antihypertensive agents (26). Agents that act within the central nervous system to inhibit sympathetic nervous system activity have long been in use as antihypertensive agents (26). Clonidine and α-methyldopa represent the first generation of centrally acting antihypertensive agents but have fallen out of use owing to a high incidence of side effects such as sedation and dry mouth. Extensive efforts to separate the vasodepressor actions from sedative effects were unsuccessful. Guanabenz and guanfacine were developed as more potent and selective α_2-agonists; indeed, in binding studies, they show 3- to 10-fold higher affinity at α_2-adrenergic receptors than clonidine, yet they are up to 10-fold less potent than clonidine as antihypertensive agents and elicit the same profile of side effects (26,27).

The sedative action of α_2-agonists may be mediated within the locus ceruleus, a brain region critical in the control of alertness and sleep-wake cycles, because direct microinjection of clonidine or other α_2-adrenergic agonists into the locus ceruleus elicits sedation (28,29). Microinjection of pertussis toxin, an agent that functionally uncouples α_2-adrenergic receptors, into the locus ceruleus attenuates the sedative action of clonidine and the altering action of the α_2-antagonist yohimbine (28). The activity of neurons in the locus ceruleus neurons is potently inhibited by clonidine and by nonimidazoline α_2-agonists, whereas rilmenidine, a central antihypertensive agent with reduced sedative action, inhibits activity in the locus ceruleus only at very high doses (30). Clonidine and other imidazolines have no effect on blood pressure when applied directly into the locus ceruleus, showing that this region is involved only in sedative side effects and not the vasodepressor action of these agents (20,31).

The evidence that α_2-adrenergic receptors mediate sedation is strong, since the alkaloid antagonists yohimbine and rauwolscine potently and completely reverse the sedation elicited by clonidine (32,33), whereas higher doses are required to reverse the hypotensive action (9,32). Profound sedation is elicited by α_2-agonists that are not imidazolines, including the antihypertensive agents α-methyldopa, guanabenz, and guanfacine. In fact, the potent α_2-agonist dexmedetomidine, which has little affinity for imidazoline-specific binding sites (Ernsberger, unpublished data), induces profound sedation and near-anesthesia without lowering blood pressure (34,35). The nonimidazoline selective α_2-agonist azepexole is more potent than clonidine in facilitating sedation and anesthesia (35) but is at least 100-fold less potent in inhibiting sympathetic activity (30,36). α_2-Adrenergic receptors have also been implicated in the sedation produced by ethanol (37). Participation of imidazoline-specific mechanisms in ethanol-induced central nervous system depression has been ruled out (37).

IV. Evidence for Nonadrenergic Actions of Imidazolines

Thus far, we have reviewed evidence that imidazoline α_2-agonists such as clonidine act specifically within the RVLM to lower blood pressure, that nonimidazoline α_2-adrenergic agonists act specifically within the NTS to lower blood pressure and heart rate, and that α_2-agonists act within the locus ceruleus and cerebral cortex to elicit central nervous system (CNS) depression. This difference in the site of action of different chemical classes of α_2-agonists implies that an additional mechanism must be involved in the action of imidazoline α_2-agonists within the RVLM. This additional mechanism would by definition be independent of the stimulation of α_2-adrenergic receptors. In this section, we will review historical evidence for nonadrenergic actions of imidazolines and the characterization of I_1-imidazoline binding sites, which provide a potential molecular mechanism for the nonadrenergic actions of imidazolines.

Clonidine is a potent drug with multiple physiological effects (23,38). Early doubts that these diverse actions were entirely due to activation of α_2-adrenergic receptors were raised when it was found that clonidine is a weak partial agonist or even an antagonist in many α_2-receptor systems (23). Furthermore, it was noted that α-antagonists as well as α-agonists lower blood pressure when administered into brainstem ventricles (39–41). For example, phentolamine, an imidazoline that acts as an antagonist at α_1- and α_2-adrenergic receptors, elicits a centrally mediated fall in blood pressure by inhibiting sympathetic outflow (40–43). The effect of clonidine is additive (43) or somewhat less than additive (42) with phentolamine, in contrast to the antagonism observed at α_2-adrenergic receptors. Phentolamine also lowered blood pressure (by 14 mm Hg) upon microinjection into the RVLM of normotensive rats (44).

In the 1970s, Karppanen noted interactions between clonidine and imidazole compounds such as cimetidine and imidazole-4-acetic acid (IAA) and proposed that clonidine may act via brainstem "imidazole receptors" (45–47). The first direct evidence for a nonadrenergic action of imidazolines was reported by Bousquet and colleagues, who used microinjection to administer clonidine analogs directly into their site of action in the rostral ventrolateral medulla (RVLM) (18,48). Cirazoline, an imidazoline that is an α_2-adrenergic antagonist, showed a clonidine-like vasodepressor action (48). In contrast, α-methylnorepinephrine, a potent α_2-agonist that lacks an imidazole ring, had no effect on blood pressure. These findings led to the proposal that the vasodepressor action of clonidine within the RVLM was mediated not by α_2-adrenergic receptors but by a novel receptor specific for imidazolines.

A major line of evidence supporting the assumption that clonidine acted specifically through α_2-adrenergic receptors was that [^{3}H]clonidine specifically labeled α_2-adrenergic receptors in brain and platelet membranes (23). However, [^{3}H]clonidine binding assays in brain had always used cerebral cortex membranes, due to its very high density of α_2-adrenergic receptors (23). When [^{3}H]clonidine binding assays were conducted using membranes from the RVLM, it was found that [^{3}H]clonidine labeled not only α_2-adrenergic receptors but also a novel population of nonadrenergic binding sites that were insensitive to catecholamines and other adrenergic agents with a phenylethylamine structure and were specific for imidazoles and imidazolines (2,5,49–57). I_1-Imidazoline binding sites have been described in the medulla oblongata of cow (5,50,51,58, 59), rat (54,60), and human (58). These sites have also been characterized in the peripheral nervous system, including adrenomedullary chromaffin cells and the carotid body (59,61), and in clonal neuron-like cell lines such as NG108-15 neuroblastoma-glioma (62) and PC12 pheochromocytoma cells (63). Primary cultures of astrocytes lack these imidazoline sites, although they express a moderate density of α_2-adrenergic receptors (64). The lack of I_1-imidazoline sites in glia implies that, within the brain, they are predominantly neuronal.

I_1-Imidazoline binding sites are distinct not only from adrenergic and histaminergic receptors (50,51) but also from imidazoline recognition sites labeled with [^{3}H]idazoxan. Imidazoline sites labeled by [^{3}H]idazoxan show 100-fold lower affinity for clonidine and do not recognize imidazoles such as cimetidine, which have a moderately high affinity for I_1-imidazoline sites (65–68). Unlike the nonadrenergic sites labeled by [^{3}H]clonidine, [^{3}H]idazoxan sites show high affinity for guanidines and thus have been termed "imidazoline-guanadinium receptive sites" (65–67,69). We have proposed the terminology of I_1- and I_2-imidazoline sites for the binding sites labeled by [^{3}H]clonidine and [^{3}H]idazoxan, respectively (52,54,70), and these terms were recently adopted by consensus (56).

The binding of [^{3}H]*p*-aminoclonidine and [^{3}H]clonidine to I_1-imidazoline sites shows properties characteristic of known neurotransmitter receptors, since radioligand binding is rapid, specific, saturable, reversible, and of high affinity (5,50,51,54,58–60). Furthermore, the pharmacological profile of these binding sites is distinct from that of any known receptor. Although many of the compounds that bind to I_1-imidazoline sites also bind to α_1- or α_2-adrenergic receptors, I_1-sites fail to bind many other agents active at these receptors, including phenylethylamine agonists and the nonimidazoline antagonists such as phenoxybenzamine and SK&F86466. Similarly, although I_1-imidazoline sites bind cimetidine and other agents active at histamine H2 receptors, I_1-sites fail to bind histamine agonists and antagonists lacking an imidazole ring, such as dimaprit, ranitidine, tiotidine, lupitidine, methapyraline, and thiazolylethlamine (Ernsberger, unpublished data). Further evidence that I_1-imidazoline sites are distinct from histamine receptors is that there are no histamine-containing neurons or cell bodies in the RVLM (71,72), despite the presence of a high density of I_1-imidazoline sites.

The heterogeneity of α_2-adrenergic receptors has recently been established by radioligand binding and molecular cloning (73–76). Of the four or more proposed subtypes, the best known are α_{2A}- and α_{2B}-adrenergic receptors, which can be distinguished on the basis of the relative potencies of oxymetazoline, which is α_{2A}-selective, and prazosxin, which is α_{2B}-selective (73,76). Prazosin also has high affinity for α_{2C} receptors. The α_{2A}-receptor predominates in most regions of the brain (73,77), including the NTS and the RVLM (78). The α_{2C} subtype is mainly found in cerebral cortex (78). Consistent with the predominance of the α_{2A} subtype, we noted very low affinity of prazosin at α_2-adrenergic receptor binding sites in the bovine RVLM. Because all of the α_2-adrenergic receptor subtypes show high affinity for epinephrine, SK&F 86466, and other nonimidazoline agonists and antagonists, whereas I_1-imidazoline sites show micromolar or lower affinity for these agents, it is clear that I_1-imidazoline sites do not represent another subtype of α_2-adrenergic receptors but are a distinct class of putative membrane receptors.

The endogenous ligand for imidazoline receptors is unknown. The catecholamines can be excluded, since they do not bind to these sites. Histamine is an imidazole neurotransmitter, but the ligand specificity of imidazoline receptors is distinct from that of histamine H_1, H_2, or H_3 receptors. The probable endogenous ligand at imidazoline receptors is a clonidine-displacing substance (CDS) isolated from the brain and originally described as an endogenous α-agonist (79). CDS is a competitive inhibitor at both imidazoline and α_2-adrenergic sites, but exhibits 30-fold higher affinity for imidazoline sites (51,62). CDS may be structurally related to clonidine, since it cross-reacts with anti-clonidine antibodies (80). Several laboratories have major initiatives dedicated to purifying and identifying this substance (56).

Independent replication of many of the key findings supporting the existence of I_1-imidazoline receptors have been reported by laboratories on four continents. For example, a recent article replicates the basic findings characterizing imidazoline receptors in bovine RVLM and extends these results to the human RVLM (58). Other reports characterize binding sites in rat brain (60), human platelets (81) and bovine adrenal medulla (82) that appear indistinguishable from I_1-imidazoline sites of the bovine RVLM.

V. Evidence That an I_1-Imidazoline Receptor in the RVLM Mediates a Vasodepressor Response

The primary findings supporting a role of I_1-imidazoline receptors in the action of imidazoline vasodepressor agents within the RVLM are summarized in Table 1. An essential first criterion, the presence of I_1-imidazoline sites in the RVLM, is well established. The vasodepressor action of cirazoline (48) and phentolamine, two imidazolines which are α_2-antagonists, is difficult to reconcile with the hypothesis that all of the actions of imidazolines are mediated by adrenergic mechanisms. Participation of α_1-adrenergic receptors in the effects of cirazoline and phentolamine is unlikely, since the α_1-agonist methoxamine and the α_1-antagonist prazosin had no effect on blood pressure following microinjection into the cat RVLM (83,84). This implies a nonadrenergic action of imidazolines in the RVLM.

Further evidence against a purely α_2-adrenergic mechanism in the RVLM is the finding that phenylethylamine α-agonists are relatively weak in eliciting vasodepression in the RVLM relative to imidazolines, despite the fact that phenylethylamines, including the catecholamines epinephrine and norepinephrine, are full agonists whereas the imidazolines are only partial agonists and therefore can act as antagonists in the presence of an endogenous agonists. α-Methylnorepinephrine, a potent full agonist at α_2-adrenergic receptors, had no significant effect on blood pressure at doses up to 40 nmol when microinjected into the cat RVLM (48). Follow-up studies showed that epinephrine, norepinephrine, and phenylephrine were also completely ineffective up the same high dose (83). Comparable experiments in the rat have yielded inconsistent results. In one study in spontaneously hypertensive rats, α-methylnorepinephrine was completely ineffective microinjected into the RVLM, despite the robust response to clonidine in the same region (85). α-Methylnorepinephrine did elicit a moderate hypotension when administered into the nearby parapyramidal regions

Table 1 Evidence Supporting a Role for I_1-Imidazoline Receptors in Vasodepressor Actions in the RVLM

1. I_1-Imidazoline binding sites are localized to the RVLM, as shown by membrane binding studies (5,50,51,57,91) and by autoradiographic visualization (91,121) (Fig. 5).
2. Imidazolines that are antagonists at α_2-adrenergic receptors can elicit vasodepressor responses within the RVLM, including cirazoline (48) and phentolamine (40–44).
3. α_2-Adrenergic antagonists fail to increase blood pressure when administered into the RVLM and may actually decrease pressure (5,84,87,100) (Figs. 6 and 9), implying that α_2-adrenergic receptors in the RVLM are not activated under resting conditions. In contrast, the I_1/α_2 antagonist efaroxan raises pressure when administered by itself into the RVLM (87) (Fig. 9), implying that I_1-imidazoline receptors are activated under resting conditions.
4. α_2-Adrenergic full agonists having a phenylethylamine or guanidine structure are either completely inactive (48,83,85,86) or relatively weak (5,44) (Fig. 1) in eliciting vasodespression in the RVLM.
5. Upon microinjection into the NTS, phenylethylamine full α_2-agonists are extremely potent in eliciting vasodepression (1,18–21) (Fig. 1), whereas imidazoline partial α_2-agonists, including clonidine, are very weak (11,20,22) (Fig. 2). Unlike the RVLM, microinjection of α_2-antagonists into the NTS elicit elevations in blood pressure. The pharmacological profile of vasodepressor responses in the NTS closely corresponds to an α_2-adrenergic receptor type and is clearly distinct from the pharmacolocial profile in the RVLM.
6. Second-generation central antihypertensives such as rilmenidine and moxonidine have reduced affinity at α_2-adrenergic receptors and attenuated sedative side effects, but they retain effectiveness in lowering blood pressure (2,53,90,91).
7. Binding affinity at I_1-imidazoline sites predicts clinical antihypertensive efficacy, whereas α_2-adrenergic binding affinity or α_2-efficacy are unrelated (57) (Figs. 3 and 4).
8. Binding affinity at I_1-imidazoline sites predicts vasodepressor activity within the RVLM, whereas α_2-adrenergic binding affinity or α_2-efficacy are unrelated (5,90).
9. Cimetidine, which has no affinity for α_2-adrenergic receptors, elicits vasodepression when microinjected into the RVLM with a potency consistent with its affinity at I_1-imidazoline sites (5).
10. The vasodepressor actions of imidazolines (clonidine, rilmenidine, or moxonidine) given either intravenously or by RVLM microinjection, can be prevented or reversed by RVLM microinjection of antagonists with high affinity for I_1-imidazoline sites (idazoxan, efaroxan), whereas nonimidazoline antagonists (SK&F 86466, yohimbine, rauwolscine) have little effect (2,5,9,87,100) (Figs, 5, 6, and 9).

containing the B3 serotonergic neurons (85). A study in normotensive rats found modest (10 to 18 mm Hg) vasodepressor responses to bilateral microinjection of 1 nmol of epinephrine, norepinephrine, α-methylnorepinephrine, and guanabenz (5). Another study found guanabenz to be inactive when applied to the ventral surface of the brainstem, in contrast to the potent hypotensive effect of clonidine applied in this manner (86). A single study found α-methylnorepinephrine to lower blood pressure and heart rate effectively in the RVLM (44). However, the doses required are 30-fold higher than those needed to elicit blood pressure and heart rate responses in the NTS (20) (Fig. 1). Because multiple injections were tested in each rat in the study reporting large responses to α-methylnorepinephrine in rats (44), and because the micropipette must pass through the NTS to target the RVLM, minor seepage along the micropipette tract (3% of total volume) could account for the effect observed.

There are major differences in pharmacological activity of α-adrenergic agonists between the NTS and the RVLM, with phenylethylamines such as α-methylnorepinephrine much more potent in the NTS and imidazolines such as clonidine being much more potent in the RVLM (Figs. 1 and 2). The contrast between NTS and RVLM is even greater for epinephrine, which is active in the NTS after unilateral injection of 0.02 nmol (20), whereas 50-fold higher doses are required bilaterally within the RVLM to elicit a comparable response (5). Oxymetazoline is completely inactive in the NTS at doses up to 20 nmol (20), whereas in the RVLM oxymetazoline is nearly as potent as clonidine (5). The relative effectiveness of different α-agonists in the NTS is fully consistent with an α_2-adrenergic receptor mechanism, with an order of potency of epinephrine > norepinephrine > α-methylnorepinephrine and with reduced effectiveness of partial agonists such as clonidine and oxymetazoline. α_2-Antagonists also have contrasting effects in the NTS and RVLM. When microinjected into the NTS, yohimbine even in very low doses (<1 nmol) induces prolonged increases in blood pressure (19,21,22,25). In contrast, yohimbine and idazoxan had no effect in low doses and actually lowered pressure when microinjected into the cat RVLM (84), and SK&F 86466 induced a dose-dependent decrease in blood pressure in normotensive rats (5) and in hypertensive rats (87). The α_2-adrenergic hypothesis of central vasodepressor action cannot account for the differences between RVLM and NTS sites of injection. Furthermore, the failure of α_2-adrenergic antagonists to elevate pressure when applied within the RVLM implies that there is not a tonically active depressor system coupled to α_2-receptors in this region, in contrast to a highly active α_2-adrenergic receptor system of the NTS.

One might argue that the catecholamines are relatively weak when microinjected into the RVLM because they are subject to uptake and degradation, whereas the imidazolines are not. However, this cannot account for the high potency of these agents upon microinjection into the NTS, an area with high concentration of noradrenergic terminals containing uptake sites and degradative enzymes for catecholamines. Myelinated axons make up the bulk of the brain tissue within the RVLM and only a low density of neurons is present relative to the NTS (88,89). The primary effect of uptake and degradation appears to be upon duration of action. In the NTS, α-methylnorepinephrine has a longer duration of action relative to norepinephrine (20) and is less susceptible to degradation. Nonetheless, norepinephrine is more potent than α-methylnorepinephrine in lowering blood pressure. In all probability, the high concentrations of catecholamine applied by microinjection (>1 m*M* in most cases) greatly exceed the capacity of local uptake and degradation mechanisms within the small (>1 mm^3) sphere of tissue affected by the injection.

The clinical application of I_1-imidazoline receptor pharmacology has been realized with the introduction of two selective I_1-agonists for the treatment of essential hypertension. Rilmenidine, now on the market in France, and moxonidine, now available in Germany and soon to be introduced throughout Europe, represent a second generation of centrally acting antihypertensive agents that cause greatly diminished side effects of sedation and dry mouth relative to first-generation agents such as clonidine, guanfacine and guanabenz (2,53, 90,91). The reduced side effects from rilmenidine and moxonidine appear to be due to lower α_2-adrenergic affinity. Controlled studies in laboratory animals also support the contention that rilmenidine and moxonidine have reduced sedative side effects. As reviewed in Section III above, the locus ceruleus is a likely site for the sedative action of α_2-agonists. Rilmenidine is 60-fold less potent than clonidine in inhibiting the activity of locus ceruleus neurons, whether given intravenously or by iontophoresis (92). Rilmenidine does not prolong barbiturate-induced sleeping time in rats and can antagonize the sedative action of clonidine (93). Rilmenidine was 80-fold less potent than clonidine in reducing locomotor activity in rats (94). Moxonidine was over 20-fold less potent than clonidine in a locomotor activity test in mice (95).

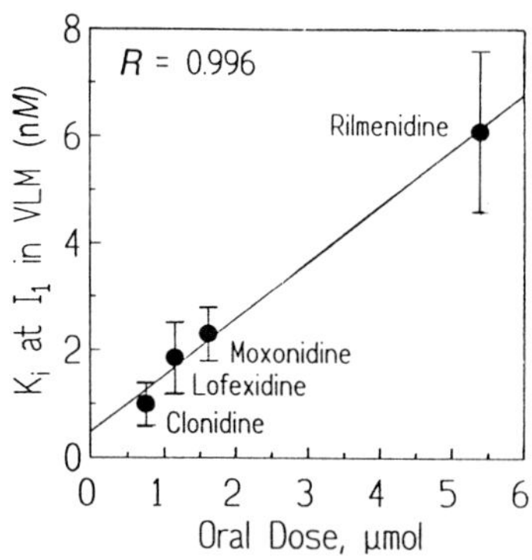

Figure 3 Relationship between affinity at I_1-imidazoline sites in the bovine VLM and therapeutic potency in hypertensive humans. Clinical doses were estimated from long-term clinical trials reporting effective control of human essential hypertension. The affinity of each agent at I_1-imidazoline binding sites in the VLM is highly correlated with clinically effective dose ($P<0.01$). The four compounds are similar in their chemical properties (hydrophobicity, pK_a) and their pharmacokinetics. (From Ref. 57.)

We compared binding affinity at I_1-imidazoline sites to the dose required to treat human essential hypertension for the four imidazoline compounds that have been subjected to controlled clinical trials (57). As shown in Figure 3, the effective dose for control of essential hypertension is tightly correlated with binding affinity at bovine RVLM I_1-imidazoline sites. Rilmenidine has a lower affinity for imidazoline sites in the RVLM than clonidine (2,96,97), which is consistent with its reduced potency clinically. Lofexidine and moxonidine are intermediate with respect to both binding affinity and clinical efficacy. The slope of this correlation is very close to one (1.05 ± 0.07), implying a one-to-one correspondence of binding affinity and clinical potency. This correlation implicates the putative I_1-imidazoline receptor in the therapeutic actions of these agents and also implies that there are few species differences in the characteristics of I_1-imidazoline sites, in agreement with a direct comparison of RVLM I_1-imidazoline sites with those of human platelets (152).

We compared binding affinity at α_2-adrenergic sites to the clinically effective dose for seven compounds. Dosages for six of the compounds were from clinical trials (57), whereas azepexole data were from a dosing study in normotensive humans (98). Unlike I_1-imidazoline affinity, the α_2-affinity of central antihypertensive agents was not correlated with clinical efficacy, as shown in Figure 4. This result is difficult to reconcile with the currently accepted notion that α_2-adrenergic receptors alone are responsible for central antihypertensive actions. It might be argued that differences in blood-brain-barrier penetration or pharmacokinetics might obscure a relationship between α_2 potency and effi-

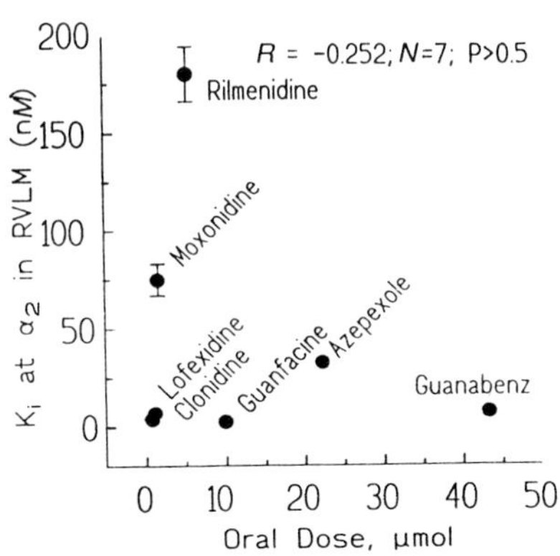

Figure 4 Relationship between affinity at α_2-adrenergic receptor sites in the VLM and clinical potency. Error bars are not visible for four of the compounds because the standard error was smaller than the size of the symbol. Other details as in Figure 3. (Modification of an original figure in Ref. 57.)

cacy. However, all seven of the comparison agents readily gain access to the central nervous system (57). In fact, of the seven compounds, guanfacine shows the best penetration of the blood-brain barrier, the highest accumulation within the central nervous system, and the longest half-life (21 h), yet it is 10-fold less potent than clonidine despite 10-fold higher affinity at α_2-adrenergic receptors (26,27). Furthermore, the α_2-affinity of the centrally acting antihypertensives varied over an almost 100-fold range, in contrast to the 6-fold range of I_1-affinities for this class of agent. This result implies that a high affinity at I_1-imidazoline receptors (1 to 6 n*M*) may be a property that distinguishes potent centrally acting antihypertensive agents.

A correlation between vasodepressor actions and I_1-imidazoline binding affinity has been reported for controlled laboratory studies as well as for clinical antihypertensive potency data. The ability of a series of imidazoline, imidazole, and phenylethylamine agents to lower blood pressure and heart rate when microinjected into the RVLM of normotensive rats was correlated with binding affinity at I_1 imidazoline receptors (5,90). In the original series of six compounds, there were significant correlations of I_1-imidazoline affinity with vasodepressor and bradycardiac responses ($r=0.84$ and 0.89, respectively) (5). Addition of moxonidine to make a total of seven compounds maintained the correlation with vasodepressor response ($r=0.93$, $P=0.0025$) (90). The further addition of rilmenidine to the series decreased the correlation slightly ($r=0.79$, $P=0.02$), owing to rilmenidine's low potency compared to that predicted from its I_1 affinity. The potency of rilmenidine in humans, however, closely corresponds to the value predicted from its I_1-imidazoline affinity (see Fig. 3), suggesting that the low efficacy of rilmenidine in the RVLM of rats is a peculiarity of that species.

In contrast to the strong correlation of cardiovascular actions with I_1-imidazoline affinity, α_2-adrenergic affinity or efficacy is unrelated to functional potency upon administration into the RVLM. The correlation of α_2-affinity and vasodepressor responses was not significant for the original series of 11 α_2-agonists ($r=-0.05$) (5) or with the addition of moxonidine and rilmenidine ($r=-0.17$) (90). In fact, there was a nonsignificant tendency for compounds with the lowest affinity at α_2-adrenergic receptors to have the greatest effect on blood pressure. α_2-Affinity was also unrelated to bradycardiac potency ($r=0.10$) (5). There was also no relationship between intrinsic efficacy at α_2-adrenergic receptors and cardiovascular effects in the RVLM, because partial agonists such as clonidine and oxymetazoline were more effective than full agonists such as epinephrine, norepinephrine, and α-methylnorepinephrine.

The one-to-one correspondence of binding affinity and both clinical and experimental potency is an important finding, because it fulfills a major criterion for the identification of a binding site as a functional receptor. On the basis of these data, the Nomenclature Committee of the International Union of Pharmacology has accepted the terminology *I_1-imidazoline receptor* for inclusion in the 1994 handbook of receptor nomenclature (*Trends in Pharmaco-*

logical Sciences Supplement, 1994). However, these correlational data are not direct tests of the relative participation of I_1-imidazoline and α_2-adrenergic receptors in the vasodepressor response of imidazolines mediated within the RVLM. We have used selective antagonists in an attempt to separate the relative roles of these receptor types. In one study, clonidine (0.5 nmol) was microinjected into the RVLM, followed 4 min later by administration via the same route of vehicle or antagonist (5). Some 6 min after the second microinjection, rats receiving vehicle showed a maintained vasodepressor response of over 30 mm Hg (Fig. 5; open bar). The selective α_2-antagonist SK&F 86466 (1 nmol) did not reverse the effect of clonidine (stippled bar). However, idazoxan (1 nmol), an α_2-antagonist with a high affinity for I_1-imidazoline receptors was able to completely and rapidly reverse the effect of clonidine (solid bar). Because both SK&F 86466 and idazoxan are potent α_2-antagonists, the major difference be-

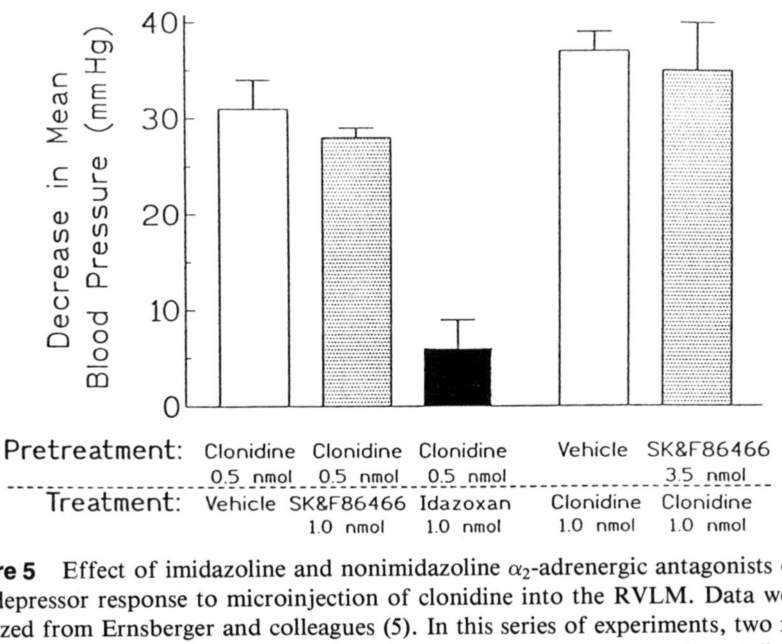

Figure 5 Effect of imidazoline and nonimidazoline α_2-adrenergic antagonists on the vasodepressor response to microinjection of clonidine into the RVLM. Data were reanalyzed from Ernsberger and colleagues (5). In this series of experiments, two microinjections were made. In the first experiment, clonidine was microinjected first (0.5 nmol) followed 4 min later by either normal saline vehicle (open bar), the selective α_2-antagonist SK&F 86466 (1.0 nmol; stippled bar), or the α_2/I_1-antagonist idazoxan (1.0 nmol; solid bar). In the second experiment, a higher dose of SK&F 86466 was tested. Pretreatment by microinjection of vehicle (open bar) or SK&F 86466 (3.5 nmol) was followed by microinjection of clonidine (1.0 nmol). Idazoxan completely reversed the effect of clonidine, whereas SK&F 86466 neither reversed nor prevented the action of clonidine, even when given at a dose 3.5-fold higher than that used for idazoxan.

tween them is the higher affinity of idazoxan for I_1-imidazoline sites. Thus, it appears that I_1-imidazoline receptors mediate most if not all of the action of clonidine within the RVLM. To test for the possibility that 1 nmol of SK&F 86466 was insufficient to achieve full blockade of α_2-adrenergic receptors, the selective α_2-antagonist was administered at a dose of 3.5 nmol. SK&F 86466 rapidly lowered blood pressure by 27 ± 2 mm Hg and microinjection of clonidine (1 nmol) lowered blood pressure an additional 35 ± 5 mm Hg, which did not differ from the effect of clonidine following pretreatment with vehicle (37 ± 2 mm Hg) (5) (Fig. 5). The fact that α_2-antagonists lower blood pressure when microinjected into the RVLM and that the action of clonidine is additive with that of an α_2-antagonist further argues against a primary role of α_2-adrenergic receptors in control of cardiovascular function by the RVLM.

Initial tests of the effects of selective antagonists on the action of imidazolines within the RVLM have used microinjection for the delivery of both agonist and antagonist. However, local microinjection introduces high local concentrations of the drug, whose effects may not be entirely representative of the actions of systemically administered agents. Furthermore, in clinical practice, centrally acting drugs are distributed throughout the organism and not confined to a specific nucleus. Therefore, in some recent studies, we have examined the ability of locally injected antagonists within the RVLM to prevent the effect of systemically administered drug. In addition, we have used second-generation central antihypertensive agents with selectivity for I_1-imidazoline sites relative to α_2-adrenergic receptors. Microinjection of the selective α_2-antagonist SK&F 86466 into the RVLM in either a low dose (1 nmol) or a high dose (10 nmol) did not significantly attenuate the effect of intravenous rilmenidine (0.5 mg/kg). In contrast, the α_2/I_1 antagonist idazoxan completely abolished the effect of intravenous rilmenidine even at a low dose (1 nmol) (2). More recently, we have shown that microinjection of the selective I_1-imidazoline antagonist efuroxan (10 nmol) into the RVLM completely prevented the hypotensive action of intravenous moxonidine (40 μg/kg) and appeared to unmask a direct vasoconstrictor effect on the vasculature (87). These studies implicate I_1-imidazoline receptors within the RVLM in the action of systemically administered imidazolines and confirm the RVLM as the site of action of second-generation central antihypertensives as well as the prototype drug clonidine.

Accumulating evidence supports the role of I_1-imidazoline receptors in vasodepressor actions of imidazolines. Intracisternal administration of a high dose of rauwolscine (300 nmol) completely abolished the action of the α_2-selective agonist S 8350 (30 nmol) but only attenuated the effect of clonidine (3 nmol) by one-third (99). Elegant support for a role of nonadrenergic mechanisms in the action of clonidine was provided by Nosjean and Guyenet (100). Intravenous clonidine (5 μg/kg) produced a significant fall in blood pressure relative to baseline (Fig. 6). Microinjection of rauwolscine (2 nmol) into the RVLM, like other α_2-antagonists, elicited a fall in blood pressure (see above). Subsequent injec-

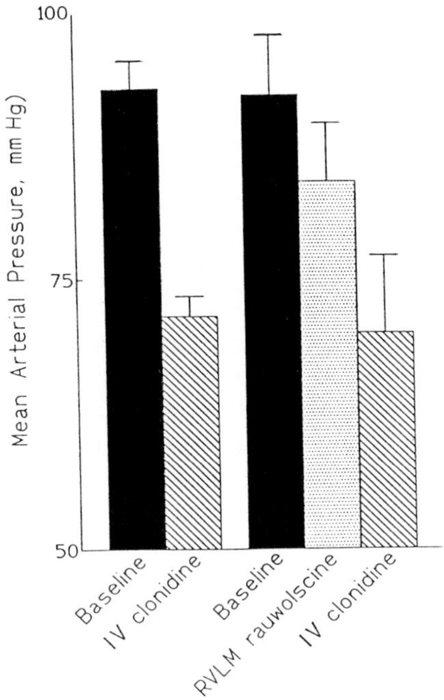

Figure 6 Vasodepressor action of intravenous clonidine (20 μg/kg) with or without blockade of α_2-adrenergic receptors in the RVLM by local microinjection of rauwolscine (16 nmol). (Data are plotted from Ref. 100.) The fall in mean arterial pressure elicited by clonidine was not affected by pretreatment with rauwolscine in the RVLM. Rauwolscine pretreatment did, however, block the effect of the serotonin 5-HT_{1a} agonist 8-OH-DPAT (not shown). A much lower dose of the imidazoline antagonist idazoxan (2 nmol) microinjected into the RVLM completely blocked the effect of intravenous clonidine (not shown).

tion of clonidine produced a depressor response indistinguishable from the control response. Rauwolscine was able to block to the responses to systemic 8-hydroxy-DPAT, a serotonin 5-HT_{1a} agonist, demonstrating that rauwolscine is an effective antagonist for 5-HT_{1a}-serotonergic as well as α_2-adrenergic receptors. In contrast to rauwolscine, idazoxan completely blocked the action of clonidine after microinjection into the RVLM (100) (data not shown).

Much of the support for a role of I_1-imidazoline receptors in the action of clonidine comes from studies in anesthetized rats. However, one study of fourth-ventricle administration of drugs in conscious, freely moving rats showed that the selective α_2-agonist BHT 920 actually increased blood pressure and plasma catecholamine levels, whereas the α_2/I_1 agonist clonidine lowered blood pressure and catecholamines (101). The α_2-antagonist rauwolscine (10 μg) did not abolish the action of clonidine (1.0 μg), ruling out a role of α_2-adrenergic receptors (101). An independent study using intracisternal application in conscious spontaneously hypertensive (SHR) rats determined dose-response curves for a series of imidazoline compounds with similar α_2-affinity but varying affinity for I_1-imidazoline sites. Vasodepressor potency as defined by ED_{50} values was correlated with I_1 but not α_2 affinity (102). These findings imply that the role of I_1-imidazoline receptors in cardiovascular regulation may be more pronounced in conscious animals.

In vivo electrochemical studies support a role for nonadrenergic receptors in the actions of imidazolines. Intravenous clonidine or rilmenidine inhibited catecholamine activity in the RVLM in parallel with a fall in blood pressure (9,30). Catecholaminergic activity in the locus ceruleus was also inhibited, but higher doses of the selective I_1-agonist rilmenidine were required to inhibit the locus coerleus relative to the RVLM (30). Intracisternal administration of the α_2/I_1 antagonist idazoxan (5 nmol) completely prevented the fall of both the blood pressure and catecholamine activity. In contrast, the α_2/5-HT antagonist yohimbine (5 nmol) did not alter either effect of clonidine (9). The superiority of idazoxan was not due to more effective α_2-blockade, because catecholamine activity in the locus ceruleus was reversed by yohimbine but not idazoxan, showing that yohimbine is actually more effective in antagonizing a true α_2-adrenergic effect. These data support the idea that there is a mechanism in the RVLM which is selectively affected by imidazoline agonists and antagonists and shows a distinct specificity relative to the classical α_2-adrenergic receptor system expressed in the locus ceruleus.

The ability of clonidine, a mixed I_1/α_2 agonist, to elicit parellel decreases in blood pressure and RVLM catecholamine activity (9) is not observed for dexmedetomidine, a specific α_2-agonist (103). Unlike clonidine, which transiently elevates blood pressure but then produces prolonged hypotension, dexmedetomidine raises blood pressure persistently when given intravenously because the direct vasoconstrictor action on vascular α_2-adrenergic receptors overwhelms any centrally mediated fall in blood pressure. This is not due to a failure to penetrate the brain, because dexmedetomidine enters the CNS readily and elicits profound sedation at very low doses (35). When administered intracerebroventricularly to anesthetized rats, dexmedetomidine only transiently lowers blood pressure. In contrast, catecholamine activity in the RVLM is profoundly depressed, falling to one-third of baseline within 45 min (103). Mean blood pressure measured simultaneously was decreased less than 10 mm Hg. Thus, in contrast to the correlated and parallel decreases in blood pressure and RVLM catecholamine activity elicited by the imidazolines clonidine and rilmenidine, the selective α_2-agonist dexmedetomidine decreases catecholamine activity to a much greater extent than it decreases blood pressure. The selective effect of α_2-adrenergic agonists on catecholamine activity is probably due to an α_2-autoreceptor mechanism present on adrenergic neurons in the RVLM. These α_2-autoreceptors might not be involved in controlling sympathetic outflow.

VI. New Tests of the I_1-Imidazoline Receptor Hypothesis

The discovery of selective agonists and antagonists binding with high affinity to I_1-imidazoline sites suggests new tools enabling direct tests of the hypothesis that I_1-imidazoline sites are functional receptors. We used these selective agents in autoradiographic studies to separately visualize α_2-adrenergic and I_1-imidazoline binding sites in the medulla and for functional studies in medulla and carotid body in vivo. For autoradiography, slide-mounted brainstem sections were preincubated for 1 h, then incubated 1 h with 0.5 n*M* [^{125}I]*p*-iodoclonidine on adjacent sections; I_1 and α_2 were selectively blocked with 10 μM epinephrine and moxonidine, respectively. Autoradiography of I_1-imidazoline and α_2-receptors in rat medulla and their interaction with moxonidine is shown in Figure 7. Panel A shows total [^{125}I]*p*-iodoclonidine binding in the presence of vehicle alone. Total binding of [^{125}I]*p*-iodoclonidine is diffusely distributed throughout the medulla. Panel B shows an adjacent section incubated with 10 μM epinephrine in order to mask α_2-adrenergic sites. The horizontal arrow in Panel B indicates the NTS. Labeling in the NTS was almost completely eliminated by epinephrine, indicating that [^{125}I]*p*-iodoclonidine labels mainly α_2-adrenergic receptors in this dorsal region. The oblique arrow indicates the RVLM. Labeling in the RVLM was maintained in the presence of epinephrine, suggesting that a non-adrenergic site is mainly present in this region. Panel C shows another nearly adjacent section incubated with 100 n*M* moxonidine to mask I_1-imidazoline sites. The oblique arrow indicates the RVLM, which shows a loss of labeling in the presence of moxonidine, particularly on the left side of the section. The remaining labeling in the RVLM region presumably represents α_2-adrenergic binding sites. Labeling in the dorsal medulla is preserved. Panel D shows an adjacent section incubated with 10 μM BDF-6143 to mask both I_1 and α_2 sites. The remaining labeling is defined as nonspecific.

These autoradiograms imply that I_1-imidazoline sites, defined as [^{125}I]*p*-iodoclonidine binding sites labeled in the presence of epinephrine to block α_2 sites, are absent in the NTS but are widely distributed across reticular areas of the medulla, including the RVLM. A similar pattern was observed in the pons (not shown). The locus ceruleus in the dorsal pons expressed mainly α_2, while I_1 were present in A5 and ventral tegmental areas. We conclude that I_1-imidazoline sites are localized to cardiorespiratory regions of the reticular formation.

Direct microinjection of substances into brain tissue is powerful technique for testing the sites and mechanism of action of CNS drugs. However, in most studies, interpretation of the results is severely limited by a lack of knowledge of the brain region receiving the drug by diffusion. To address this issue, we characterized the pharmacokinetics of [^{3}H]moxonidine microinjected into the RVLM. Microinjection of 4 nmol [^{3}H]moxonidine elicited a typical vasodepressor response at 8 min after injection (not shown). The distribution of [^{3}H]moxonidine in the rostral-caudal plane was determined by scintillation counting of alternate transverse sections through the medulla. As shown in

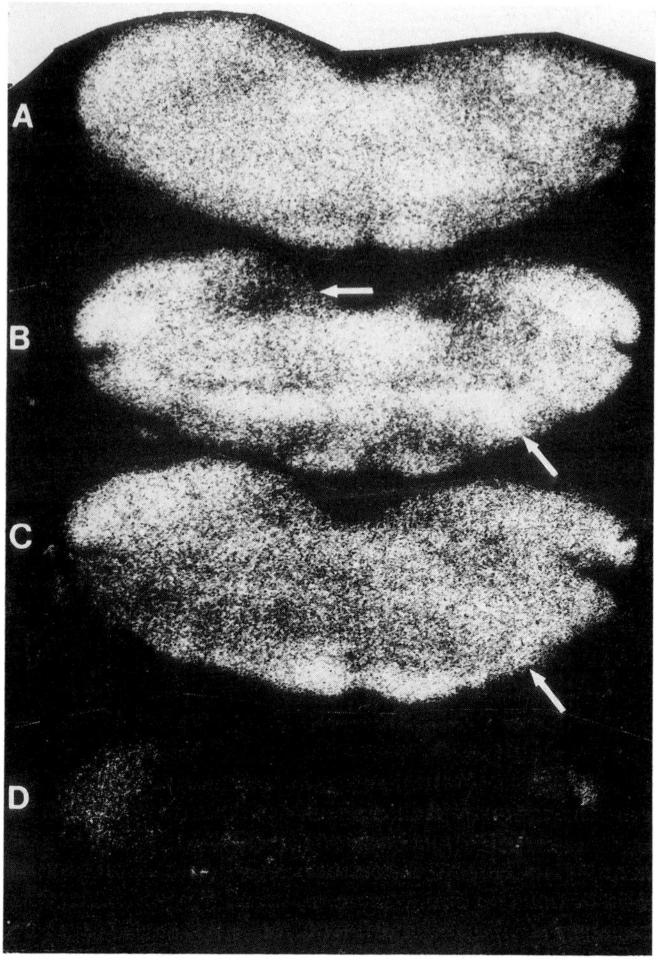

Figure 7 Autoradiography of [^{125}I]*p*-iodoclonidine binding sites in the medulla of an SHR rat. Shown are photographic prints made under identical conditions using the autoradiograms as negatives. Light areas correspond to the highest intensity of labeling. Adjacent 15 μm sections of rat brainstem were incubated with 0.5 n*M* [^{125}I]*p*-iodoclonidine under one of four different conditions. The section in panel A was incubated with antioxidant vehicle alone, and shows the total binding of [^{125}I]*p*-iodoclonidine. Binding sites were distributed throughout the section. The section below in panel B was incubated with 10 *μM epinephrine to mask* α_2-receptors, and thus shows the distribution of I_1 binding. The upper arrow indicates the area of the NTS, where [^{125}I]*p*-iodoclonidine labeling was almost completely eliminated by epinephrine. The labeling of the RVLM, indicated by the second arrow, was not inhibited by epinephrine. The section in panel C was incubated with 100 n*M* moxonidine to selectively mask I_1-imidazoline sites and thus shows the distribution of α_2-adrenergic sites. Labeling in the RVLM, indicated by an arrow, was inhibited by 100 n*M* moxonidine. The section in panel D was incubated with 10 μ*M* BDF-6143 to block both I_1 and α_2 binding, and thus shows the distribution of nonspecific sites. (From Ref. 91.)

Figure 8, at the peak of the vasodepressor response [^{3}H]moxonidine had diffused less than 1 mm from the site of injection. In fact, 90% of the total of 8 nmol [^{3}H]moxonidine from the bilateral microinjections was recovered within 0.6 mm of the center of the injection site. The concentration of [^{3}H]moxonidine declined exponentially as a function of distance, as expected for simple diffusion. Nonlinear curve fitting showed that [^{3}H]moxonidine concentration could be expressed as $37 \pm 2 \cdot e^{-d(2.1 \pm 0.2)}$, where d is distance in millimeters, with a goodness of fit (r^2) of 0.92. Recovery of injected radioactivity was 102%, ruling out any significant leakage to the peripheral circulation.

The volume of the diffusion sphere is approximately 1 μL (1 mm^3). Thus, the average concentration of [^{3}H]moxonidine in the sphere is 4 m*M*, but the concentration on the margins of the sphere is at least an order of magnitude lower. At the threshold dose of 50 pmol moxonidine, the predicted average concentration in the diffusion sphere would be 50 μM and the margin of the sphere would show concentrations in the low micromolar range. These concentrations of moxonidine are still 100 to 1000-fold higher than the concentrations required for receptor occupancy or those achieved in clinical pharmacotherapy. The requirement for these high concentrations is puzzling. However, it is likely that a "critical mass" of RVLM sympathoexcitatory neurons must be bathed by a compound before effects on systemic blood pressure can be detected. In previous pharmacokinetic studies, a diffusion spread of about 1 mm across the extent of the RVLM was found to be required for effects of the metabolic toxin sodium cyanide (104) and the muscarinic antagonist scopolamine (105). The histological basis for this phenomenon might be the sparse distribution of neurons in the RVLM interspersed with heavily myelinated fibers, which may absorb lipophilic drugs. Dendrites of RVLM neurons also cover a large area, and a critical proportion of dendrites might have to be affected before net sympathetic discharge is altered. In microinjection studies, the critical factor may be drug concentration in the leading edge of the diffusion sphere. One prediction from this hypothesis would be that the same quantity of drug administered in a larger volume should be more effective, due to the coverage of a greater area.

Recent in vivo studies of I_1-imidazoline receptor function have used SHR rats which were anesthetized with urethane (1 g/kg IP), paralyzed with gallamine, and ventilated. The RVLM was localized by testing for a blood pressure rise >30 mm Hg from 1 nmol glutamate. Simultaneous bilateral microinjection of the selective I_1-imidazoline agonist moxonidine into the rostral RVLM of the anesthetized, spontaneously hypertensive rat elicits falls in blood pressure at doses as low as 50 pmol (6,106). At a threshold dose of 50 pmol, moxonidine lowered mean blood pressure from 190 ± 11 to 182 ± 15 mm Hg but heart rate only from 392 ± 15 to 383 ± 12 bpm ($n = 6$).

Radioligand binding studies in RVLM membranes identified efaroxan and SK&F86466 as a selective antagonists at I_1-imidazoline and α_2-adrenergic receptors, respectively (K_i at I_1 versus α_2: 0.12 ± 0.14 n*M* versus 6 ± 1 n*M* for efaroxan and 93 ± 150 μM versus 35 ± 8 n*M* for SK&F86466). To test whether

I_1 or α_2 mediate the in vivo action of moxonidine in the RVLM, the I_1/α_2 antagonist efaroxan (4 nmol) or the α_2-blocker SK&F86466 (10 nmol) was microinjected into the RVLM 15 min prior to 4 nmol moxonidine (Fig. 9). Efaroxan elevated blood pressure itself and abolished the action of moxonidine (up to 40 nmol), whereas α_2-blockade with SK&F86466 slightly lowered blood pressure when given alone and insignificantly attenuated moxonidine's effect, ruling out a major role of α_2-adrenergic receptors in the action of moxonidine. The effect of moxonidine on heart rate was completely unaffected by α_2-blockade with SK&F 86466 but was eliminated by efaroxan (Fig. 9). Although efaroxan is slightly more potent at α_2-adrenergic sites than SK&F 86466 (6 ± 1 versus 35 ± 8 nM), SK&F 86466 was given at a higher dose (10 versus 4 nmol). These studies support a predominant role for I_1-imidazoline receptors in the action of moxonidine within the RVLM but do not rule out minor contributions from activation of α_2-adrenergic receptors.

Both I_1-imidazoline agonists and α_2-adrenergic agonists are capable of lowering blood pressure by inhibiting central sympathetic outflow. The primary difference between these agents is the additional effects of α_2-receptor stimulation, which includes nonspecific depression of CNS activity similar to drugs of the sedative or anesthetic classes. Sedatives and anesthetics tend to depress respiration, so an indirect measure of sedative action is respiratory depression. Clonidine has been shown to reduce phrenic nerve activity (107), slow the

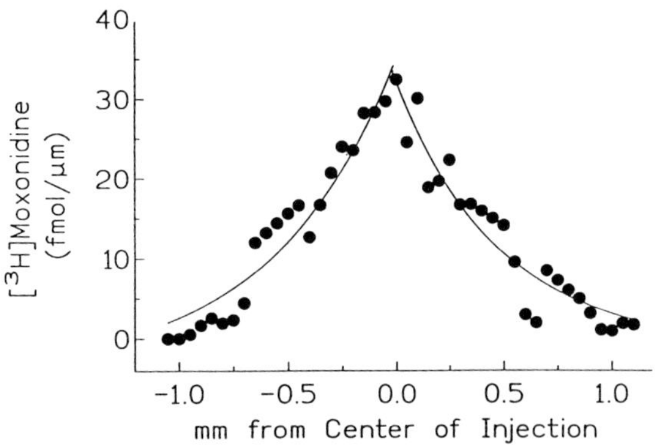

Figure 8 Distribution of [^{3}H]moxonidine in the RVLM region of an SHR rat 8 min after bilateral microinjection of a 4-nmol dose in a volume of 80 nL. After a maximum fall in blood pressure was stabilized, the medulla was rapidly removed and frozen in isopentane chilled on dry ice. Cryostat sections were alternately used for anatomical reference or directly counted for [^{3}H] activity in a scintillation counter. Distance from the site of injection is shown on the *x* axis, where positive numbers are rostral to the injection and negative numbers are caudal. The content of [^{3}H]moxonidine is expressed as fmol per micron of coronally sectioned brain. (From Ref. 87.)

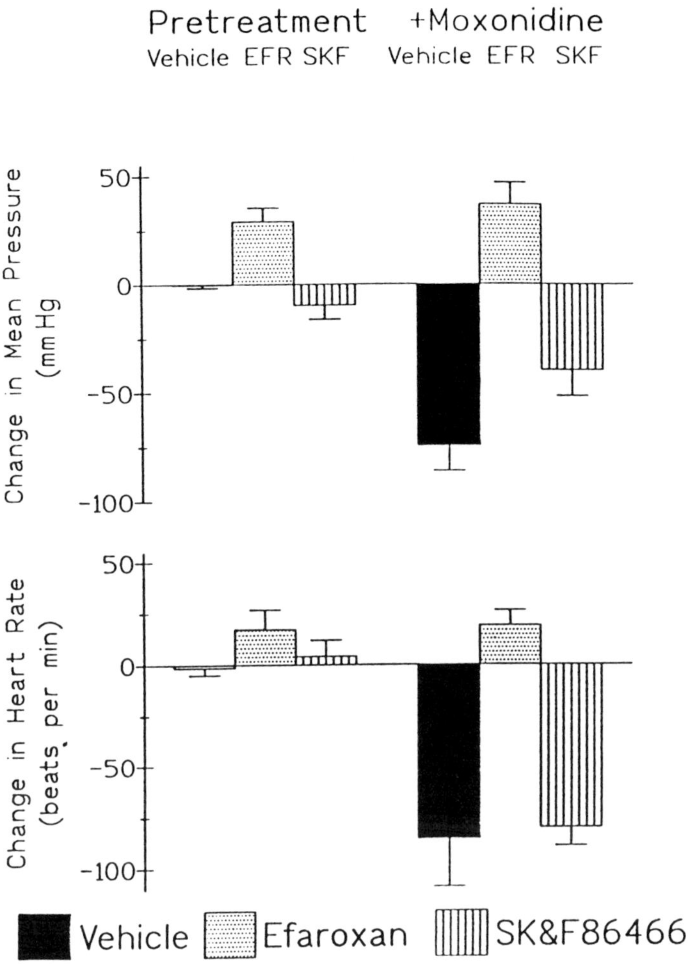

Figure 9 Effect of prior administration into the RVLM of vehicle ($N=9$ rats), 4 nmol efaroxan ($N=6$), or 10 nmol SK&F 86466 ($N=9$) on the blood pressure (top) and heart rate (bottom) responses to simultaneous bilateral microinjection of moxonidine (4 nmol) into the same location 15 min later. Data are expressed as net change from baseline value obtained prior to any treatment and are mean ± standard error. (From Ref. 87.)

respiratory rate (108), increase arterial carbon dioxide levels (109), attenuate respiratory reflex responses to hypoxia (110), decrease chemosensory discharge from carotid body afferents (61), and reduce other indices of respiratory activity (111–117). The pharmacology of the respiratory depressant effect of clonidine has been examined in several of these studies, and α_2-adrenergic rather than I_1-imidazoline receptors appear to mediate respiratory depression, since the depressant effect of clonidine is mimicked by the nonimidazoline α_2-agonists guanfacine, guanabenz, or α-methylnorepinephrine and readily blocked by low doses of yohimbine. We tested the respiratory effect of imidazolines to determine the specificity of the sympathoinhibition elicited by centrally acting imidazolines. We compared the effects of clonidine, which has equal affinity for I_1 and α_2 sites, with the effects of moxonidine, which is 30-fold selective for I_1 over α_2 sites. Moxonidine (20 μmol/kg IV) almost abolished cervical sympathetic activity (3 ± 2 versus 17 ± 3 U); lowered blood pressure $23 \pm 2\%$ and heart rate $19 \pm 3\%$ in anesthetized, paralyzed, peripherally denervated cats; but did not affect the rate of rise in phrenic activity (23 ± 6 vs. 22 ± 6 U/s) or the carbon dioxide threshold for hypoglossal or phrenic activation. Clonidine (10 μmol/kg IV), while having similar effects on sympathetic activity and blood pressure, nearly abolished the hypoglossal response to CO_2 and inhibited phrenic activity. We conclude that α_2-adrenergic but not I_1-imidazoline receptors mediate respiratory depression, particularly on the upper airway (54,108).

In addition to its role in cardiovascular regulation, the RVLM also contains elements of the central chemoreceptors (118,119) and may contribute to pressor responses to hypercapnia and hypoxia. In the RVLM, I_1-imidazoline receptors may participate in respiratory regulation (54). To test for possible involvement of I_1-imidazoline receptors in peripheral chemosensory responses, we tested for the presence of I_1-imidazoline and α_2-adrenergic receptors in the carotid body. During development, cells of the carotid body originate from chromaffin cells migrating from the neural crest (120). Chromaffin cells within the adrenal medulla have recently been found to express a high density of I_1-imidazoline sites (57). We hypothesized that the peripheral chemoreceptors in the carotid body may also express I_1-imidazoline sites. The small mass of the carotid body (less than 1 mg) demanded development of a sensitive binding assay using ^{125}I. $[^{125}I]$*p*-Iodoclonidine binding was only partially inhibited by a high concentration (100 μM) of the endogenous α_2-agonist (−)-epinephrine ($37 \pm 3\%$ inhibition). A similar fraction of total specific $[^{125}I]$*p*-iodoclonidine binding was blocked by the specific α_2-antagonist SK&F 86466 ($44 \pm 3\%$ inhibition at 10 μM). In contrast, clonidine, which binds to both I_1-imidazoline and α_2-adrenergic sites, inhibited the total population of specific sites. These findings imply that only about 40% of $[^{125}I]$*p*-iodoclonidine binding sites in the cat carotid body are α_2-adrenergic receptor sites. Imidazole-4-acetic acid, which binds to I_1 but not α_2 sites (5,50), inhibited $72 \pm 6\%$ of the specific $[^{125}I]$*p*-iodoclonidine binding sites at a concentration of 100 μM, consistent with the presence of specific I_1-imidazole binding. A total of 1.5 fmol of specific binding sites per carotid body were labeled by 0.6 nM $[^{125}I]$*p*-iodoclonidine, of which 0.9 fmol were insensitive to epinephrine and thus nonadrenergic. The remaining 0.6 fmol per carotid body had the characteristics of an α_2-adrenergic receptor.

The role of I_1-imidazoline receptors in control of chemosensory discharge was determined in cats pretreated with 0.5 mg/kg SK&F 86466 IV, which blocks carotid body α_2-receptors (61). During α_2-blockade, clonidine stimulated chemosensory discharge under hypoxia when injected into the carotid body (20 μg/kg) or when given intravenously (100 μg/kg). A two-way analysis of variance by clonidine treatment and oxygen tension showed significant effects of clonidine and of oxygen level as well as a significant interaction between clonidine treatment and oxygen. The latter finding indicates that clonidine's facilitory effect was greater under hypoxia than under normoxic or hyperoxic conditions.

These data indicate that I_1-imidazoline sites as well as α_2-adrenergic receptors are expressed within the carotid bodies of cats. Functional studies of chemosensory discharge suggest that I_1-imidazoline receptors may facilitate the response to hypoxia, whereas, in contrast, α_2-adrenergic receptors inhibit chemosensory function, as previously reported (61). The opposing effect of I_1-imidazoline and α_2-adrenergic receptor activation supports the contention that these are distinct classes of receptor.

VII. Alternatives to the I_1-Imidazoline Receptor Hypothesis

The hypothesis that an I_1-imidazoline receptor participates in central antihypertensive actions is in seeming opposition to a large body of work in neuropharmacology which operated under the assumption that only α_2-adrenergic receptors in the CNS are involved. Thus, inevitably, many research findings must be reexamined in light of this new hypothesis. The arguments favoring the hypothesis that imidazolines act solely on α_2-adrenergic receptors within the RVLM to lower blood pressure are summarized in Table 2. The presence of α_2-adrenergic receptors in the RVLM has been established by membrane binding studies (5, 50,51,57), autoradiography (91,121) (Fig. 7C), in situ hybridization (78), and electrophysiology (Guyenet et al., this volume). However, based on the relative density of receptors, the NTS must be identified as the most likely location for α_2-adrenergic regulation of cardiovascular function, since it expresses considerably higher density of α_2-adrenergic receptors (see Figs. 7A and 7C). Furthermore, microinjection studies support the hypothesis that α_2-adrenergic receptors in the NTS subserve a tonically active vasodepressor function and that the NTS is the most likely CNS site of action for specific α_2-agonists (see Section II, above).

The primary argument in favor of the hypothesis that clonidine and other imidazoline centrally acting antihypertensives reduce blood pressure through an activation of α_2-adrenergic receptors is that the hypotensive effect of clonidine can be blocked by α_2-adrenergic antagonists. However, there are several weaknesses in these experiments that have become apparent as more knowledge has accumulated. First, many of the antagonists have significant affinity for I_1-imidazoline sites. This is obviously true for the imidazoline antagonists such as idazoxan or tolazoline, which have roughly equal affinity for I_1-imidazoline and α_2-adrenergic receptors binding sites (90). Piperoxan, an α_2-antagonist

Table 2 Evidence Supporting a Role for α_2-Adrenergic Receptors in Vasodepressor Actions in the RVLM

1. α_2-Adrenergic receptor binding sites are localized to the RVLM, as shown by membrane binding studies (5,50,51,57), by autoradiographic visualization (91,121) (Fig. 7B), by in situ hybridization of mRNA (78) and electrophysiological studies (Guyenet, this volume).
2. α_2-Adrenergic agonists with little or no affinity for I_1-imidazoline sites, such as α-methylnorepinephrine, elicit vasodepressor responses within the RVLM (44).
3. α_2-Adrenergic/serotonergic antagonists with little or no affinity for I_1-imidazoline sites, such as yohimbine and rauwolscine, are able to block or reverse the effect of clonidine when administered into the circulation of the cerebral ventricles (1,32,127).

that potently blocks the action of clonidine (122–124), shows an affinity (K_i) of 40 n*M* for both I_1- and α_2-adrenergic receptor binding sites (Ernsberger, unpublished data).

Yohimbine and its diastereoisomers rauwolscine and corynanthine have been widely used to characterize α_2-adrenergic receptors. Yohimbine and rauwolscine are equally potent at α_2-adrenergic receptors, while corynanthine is much weaker and is actually selective for α_1-adrenergic receptors (32). Rauwolscine is equally potent at serotonin receptors as at α_2-adrenergic receptors, and yohimbine is only moderately selective for α_2 relative to serotonin receptors (125,126). Yohimbine, rauwolscine, and corynanthine all show about 1 μM affinity for I_1 sites (50). Yohimbine is only 5-fold more potent than corynanthine in reversing the hypotensive effect of clonidine, whereas yohimbine is 100-fold more potent than corynanthine in blocking clonidine's sedative action (32). Yohimbine is also 100-fold more potent than corynanthine at α_2-adrenergic receptor binding sites (32,50). These results imply that the sedative action of clonidine is mediated by α_2-adrenergic receptors, but that a major component of the hypotensive action of clonidine may be mediated by a receptor with equal affinities for yohimbine and corynanthine. Consistent with this conclusion, in another study, yohimbine and corynanthine were found to be equally effective in blocking the antihypertensive action of clonidine in renal hypertensive rats (127). Moreover, yohimbine antagonizes the hypotensive effect of clonidine in a biphasic manner, suggesting that more than one receptor is involved (128). Interestingly, the vasodepressor action of clonidine is not inhibited by several α_2-antagonists that do not bind to I_1 sites, such as ergot alkaloids (123,124) or phenoxybenzamine (42,123). All of these results are compatible with roles for an additional mechanism besides α_2-adrenergic agonist actions in central antihypertensive efficacy.

A second weakness of studies showing reversal of the action of clonidine and other central antihypertensive agents by α_2-antagonists is that the antagonists were administered systemically, into the cerebral circulation, or into the brain ventricles. Systemic administration of α-adrenergic antagonists greatly complicates the interpretation of results. Blockade of systemic α-adrenergic receptors, even if partial, will decrease the importance of the sympathetic nervous system in maintaining blood pressure. Typically, blood pressure may be maintained during α-adrenergic blockade by increased production of pressor hormones. Following blockade of vascular α_1- and/or α_2-adrenergic receptors, centrally mediated sympathoinhibition may lose effectiveness. Thus, systemic administration of the specific α_1-adrenergic antagonist prazosin (1 mg/kg) attenuates the hypotensive response to clonidine nearly as effectively as yohimbine (1 mg/kg) (127,129). This does not mean that clonidine's action is mediated by α_1-adrenergic receptors but rather that the antagonist has interfered with its mechanism of action downstream of the receptor. Probably by a similar mechanism, intravenous administration of the β-adrenergic antagonist propranolol can prevent or reverse the vasodepressor action of clonidine, an effect also observed clinically (130).

Comparable drug interactions at sites distal to the original site of action can occur when drugs are administered throughout the CNS through the brain ventricles or the vertebral arteries. Centrally acting antihypertensives may interact with a complex neural pathway that involves several synapses, so that when antagonists are administered globally into the brainstem, it cannot be assumed that their blocking action is direct. For example (Table 3), the hypotensive effect of clonidine can be effectively antagonized by the serotonin antagonist methysergide or by lesions of serotonin neurons (131–133), by the opiate antagonist naloxone (134–136), by the γ-aminobutyric acid antagonist bicuculline (137–139), and even by ethanol (140). The blockade of clonidine's hypotensive action by each of these antagonists cannot be mediated by a direct interaction with the receptor upon which clonidine acts. A more likely explanation is that clonidine activates a complex pathway within the brainstem, originating in the RVLM, which involves multiple neurotransmitters. Interference with any of these modulatory neurotransmitters may attenuate the action of clonidine. For example, clonidine acts within the B3 serotonergic cell group to decrease serotonin activity (141). Both α_2 and I_1 sites appear to be present in the B3 region in autoradiographic studies, so it remains to be determined which receptor type (or both) mediates interaction with serotonergic systems. Several studies have implicated GABA neurons in the vasodepressor action of clonidine. Pretreatment with the GABA antagonist bicuculline completely prevents the hypotensive effect of clonidine when both drugs are given intracerebroventricularly (142). Furthermore, clonidine enhances GABA release in the medulla and the hypothalamus but not in the cerebellum (143).

VIII. A Local Circuit Model of the Action of Imidazoline Agents in the RVLM

The ability of antagonists at GABA, opiate, and serotonin receptors to block the vasodepressor action of clonidine implies that imidazolines interact with a neuronal circuit using multiple transmitters. A potential minimal model of a local circuit within the RVLM is shown in Figure 10. Rather than an exogenous imidazoline drug, the putative endogenous ligand CDS is included in the model as the stimulus for I_1-imidazoline receptors. The location of the cell bodies for neurons providing CDS input to the RVLM is not known. However, the hypothalamus contains the highest levels of CDS in the brain (144), and there are strong hypothalamic projections to the RVLM region, suggesting a hypothalamic source of CDS in the RVLM. In this model, I_1-imidazoline receptor stimulation acts not by directly inhibiting sympathoexcitatory neurons in the RVLM but by stimulating inhibitory interneurons containing opiates, GABA, or serotonin. The RVLM is densely innervated by enkephalin-containing nerve terminals, some of which may arise from nearby enkephalin cell bodies (89). Serotonergic neurons might also activate local GABA neurons, which have been characterized within the RVLM (145–147). Activation of these interneurons would then inhibit reticulospinal neurons, providing tonic control of sympathetic neurons in intermediolateral (IML) column of the spinal cord.

As shown in Table 3, blockade of opiate, GABA, or serotonin receptors attenuates the action of clonidine. In the proposed model, blockade of these inhibitory neurotransmitters would interrupt the pathway activated by imidazolines. Parapyramidal neurons also provide direct serotonin and substance P to the IML (148). This would provide an additional pathway for serotonergic neurons to participate in the action of imidazolines. As mentioned in Section I, clonidine may act not directly in the RVLM but in a region slightly caudal, which exerts an inhibitory control over the RVLM (8). This might be the location of the postulated inhibitory interneurons stimulated by imidazolines.

The model outlined in Figure 10 represents hypotheses that are readily testable. The existence of local circuits within the RVLM, particularly inhibitory loops involving GABA, have been postulated for a decade (145,147) but seldom considered. As one example, we predict that local blockade of GABA, serotonin, or opiate receptors within the RVLM may interfere with the action of imidazolines and other drugs with central cardiovascular actions.

Given the potential for an interaction between imidazoline vasodepressor agents and a local circuit within the RVLM, the experimental design most relevant to the direct actions of imidazoline antihypertensive agents with their receptors is that wherein antagonists are directly applied within the RVLM and agonists are given either by the same route or systemically. With this experimental design, the specific α_2-adrenergic antagonists SK&F 86466 and rauwolscine have no effect on the hypotensive action of clonidine or rilmenidine (5,100,149). In contrast, administration into the RVLM of idazoxan, an antagonist at I_1-imidazoline as well as α_2-adrenergic receptors, abolishes the action of clonidine or rilmenidine (5,15,100,149). Comparable findings have recently been obtained with moxonidine (87) (Fig. 9). These data show that I_1-imidazoline receptors are primarily responsible for the vasodepressor action of centrally acting antihypertensives with an imidazoline structure.

Table 3 Antagonists That Block the Depressor Action of Clonidine

Antagonist	Receptor	References
Methysergide	Serotonin	131
5,7-DHT	Serotonin	132,133
Bicuculline	GABA	137–139
Ethanol	GABA?	140
Naloxone	Opiate	134–136
Prazosin	α_1-Adrenergic	127,129
Propranolol	β-Adrenergic	130

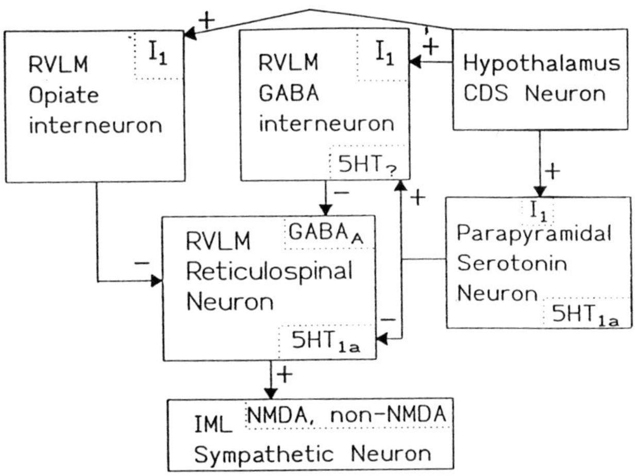

Figure 10 Hypothetical model of a local circuit of interneurons within the RVLM mediating the vasodepressor action of imidazolines. The endogenous ligand for I_1-imidazoline receptors is assumed to be clonidine-displacing substance (CDS). Because levels of CDS are highest in the hypothalamus relative to other brain areas and the hypothalamus projects heavily to the RVLM, we have placed putative "CDS neurons" in the hypothalamus. Stimulation of I_1-imidazoline receptors in the RVLM region, whether by endogenous ligand or exogenous drug, may activate inhibitory interneurons. Inhibitory GABA, opiate, and serotonin inputs to cardiovascular neurons in the RVLM have been identified, and in this model they are stimulated by I_1-imidazoline agonists, which, in turn, leads to inhibition of RVLM reticulospinal neurons. Serotonin neurons might also innervate local GABA interneurons. Reduced activity of RVLM cardiovascular reticulospinal neurons reduces activity of sympathetic neurons in the intermediolateral (IML) column of the spinal cord, thus resulting in sympathoinhibition. Parapyramidal neurons containing serotonin and substance P also project directly to the spinal cord, providing a separate pathway for modulating sympathetic drive.

Electrophysiological studies of the effects of I_1-imidazoline and α_2-adrenergic agonists and antagonists have yielded paradoxical results that do not seem to correlate with cardiovascular actions. One possible explanation for these anomalous results is that the neurons recorded from are not those responsive to imidazolines. As noted above, the active zone for effects of imidazolines on blood pressure within the RVLM appears to be caudal to the location from which sympathoexcitatory reticulospinal neurons have been recorded. Furthermore, as speculated above, imidazolines may act on local circuits of neurons rather than having a direct action on the final output neurons of the RVLM. In the cat RVLM, only about half the neurons are inhibited by clonidine, even at doses that silence sympathetic nerve discharge (150). All of the sympathoexcitatory neurons, even those insensitive to clonidine, could be inhibited by systemic 8-OH-DPAT, a serotonin 5-HT_{1A} agonist. None of the sympathoexcitatory neurons were inhibited by iontophoretic clonidine or norepinephrine, ruling out a direct action of clonidine on RVLM sympathoexcitatory neurons. One study compared the effects of iontophoresed clonidine and norepinephrine on neurons in the region of the RVLM (151). Norepinephrine produced a large, rapid inhibition, whereas clonidine produced a much smaller effect with a latency of 60 s (versus 10 s for norepinephrine). Of the 43 responsive RVLM neurons, 30 were inhibited by both norepinephrine and clonidine, but the other 13 neurons showed opposing patterns, including 7 neurons that were excited by clonidine (151). These results are consistent with clonidine's status as a weak partial agonist at α_2-adrenergic receptors and is also consistent with the hypothesis that excitatory responses to clonidine may be elicited from certain neurons in the RVLM region.

Inhibition of spontaneously active units in the RVLM may also be mediated by α_2-adrenergic receptors, as reported by Guyenet in this volume. However, norepinephrine and α-methylnorepinephrine are equal to clonidine as inhibitors of neuronal firing but are weaker than clonidine in eliciting a cardiovascular response. Similarly, α_2-adrenergic antagonists are powerful stimulators of RVLM units but have no cardiovascular action except a paradoxical vasodepressor effect. Furthermore, activation of α_2-receptors elicits global CNS depression. RVLM neurons are less sensitive to α_2-adrenergic inhibition than neurons in the NTS and locus ceruleus, which conflicts with the overwhelming evidence that the RVLM is the sole of site of action of the imidazoline vasodepressor agents (see Section I, above). Therefore, the I_1-imidazoline receptor hypothesis of the central vasodepressor actions of imidazolines in the RVLM shows the greatest consistency with the overall body of data accumulated to date.

IX. Summary

The RVLM is the site of action for antihypertensives of the imidazoline class, including clonidine and the second-generation agents rilmenidine and moxonidine. Microinjection of minute doses of these agents into the RVLM elicits vasodepression, whereas larger doses are required for effects elsewhere in the brain or in the periphery. Furthermore, the RVLM mediates the depressor effect of systemic imidazolines, since microinjection of antagonists into the RVLM reverses the effect of intravenous agonists. Nonimidazoline antihypertensives appear to act mainly in the NTS region rather than the RVLM. Antagonists against many receptors can attenuate the depressor action of imidazolines, including not only α-antagonists but also GABA and serotonin blockers. This suggests that imidazolines may act on a multisynaptic pathway utilizing several transmitters, rather than acting directly on reticulospinal neurons in the RVLM. This hypothesis is consistent with electrophysiological evidence that reticulospinal neurons are largely unaffected by clonidine (see Guyenet, this volume).

A novel binding site specific for imidazolines and related compounds has been characterized within the RVLM. These I_1-imidazoline sites are distinct from receptors for catecholamines, histamine, and other known neurotransmitters. Accumulating evidence implicates I_1 sites in the vasodepressor actions of imidazolines: (1) Some imidazoline α_2-antagonists lower blood pressure when microinjected into the RVLM. (2) Nonimidazolines—which are potent full agonists at α_2-receptors, such as α-methylnorepinephrine and guanabenz—are 100-fold less efficacious than the imidazolines in the RVLM. (3) Binding affinity at I_1 sites in the RVLM correlates with clinical antihypertensive efficacy ($r=0.996$, $N=4$) and with vasodepressor activity within the RVLM ($r=0.79$, $N=8$), whereas α_2-affinity correlates with neither measure of efficacy ($r=0.24$ and -0.17, respectively). (4) The vasodepressor action of imidazolines (clonidine, rilmenidine, or moxonidine), given either intravenously or by RVLM microinjection, can be prevented or reversed by RVLM microinjection of imidazoline antagonists (idazoxan, efaroxan), whereas nonimidazoline antagonists (SK&F 86466, yohimbine) have little effect.

References

1. Kobinger W. Central α-adrenergic systems as targets for hypotensive drugs. Rev Physiol Biochem Pharmacol 1978; 81:39–100.
2. Gomez RE, Ernsberger P, Feinland G, Reis DJ. Rilmenidine lowers arterial pressure via imidazole receptors in brainstem C1 area. Eur J Pharmacol 1991; 195:181–191.
3. Kooner JS, Birch R, Frankel HL, et al. Hemodynamic and neurohormonal effects of clonidine in patients with preganglionic and postganglionic sympathetic lesions: Evidence for a central sympatholytic action. Circulation 1991; 84:75–83.
4. Rudolph CD, Kaplan SL, Ganong WF. Sites at which clonidine acts to affect blood pressure and the section of renin, growth hormone and ACTH. Neuroendocrinology 1980; 31:121–128.
5. Ernsberger P, Giuliano R, Willette RN, Reis DJ. Role of imidazole receptors in the vasodepressor response to clonidine analogs in the rostral ventrolateral medulla. J Pharmacol Exp Ther 1990; 253:408–418.

6. Haxhiu MA, Dreshaj I, Erokwu B, et al. Vasodepression elicited in hypertensive rats by the selective I_1-imidazoline agonist moxonidine administered into the rostral ventrolateral medulla. J Cardiovasc Pharmacol 1992; 20(suppl) 4:S11–S15.
7. McAuley MA, Macrae IM, Reid JL. The cardiovascular actions of clonidine and neuropeptide-Y in the ventrolateral medulla of the rat. Br J Pharmacol 1989; 97: 1067–1074.
8. Gatti PJ, Hill KJ, Da Silva MT, Norman WP, Gillis RA. Central nervous system site of action for the hypotensive effect of clonidine in the cat. J Pharmacol Exp Ther 1988; 245:373–380.
9. Tibiriça E, Feldman J, Mermet C, et al. An imidazoline-specific mechanism for the hypotensive effect of clonidine: A study with yohimbine and idazoxan. J Pharmacol Exp Ther 1991; 256:606–613.
10. Williams CA, Roberts JR, Freels DB. Changes in blood pressure during isometric contractions to fatigue in the cat after brain stem lesions: Effects of clonidine. Cardiovasc Res 1990; 24:821–833.
11. Bousquet P, Feldman J, Bloch R, Schwartz J. The nucleus reticular lateralis: A region highly sensitive to clonidine. Eur J Pharmacol 1981; 69:389–392.
12. Antonaccio MJ, Halley J. Clonidine hypotension: Lack of effect of bilateral lesions of the nucleus solitary tract in anaesthetized cats. Neuropharmacology 1977; 16:431–433.
13. Sved AF. Clonidine can lower blood pressure by inhibiting vasopressin release. Eur J Pharmacol 1985; 109:111–116.
14. Bousquet P, Feldman J, Velly J, Bloch R. Role of the ventral surface of the brain stem in the hypotensive action of clonidine. Eur J Pharmacol 1975; 34:151–156.
15. Punnen S, Urbanski R, Krieger AJ, Sapru HN. Ventrolateral medullary pressor area: Site of hypotensive action of clonidine. Brain Res 1987; 422:336–346.
16. Gurtu S, Pant KK, Sinha JN, Bhargava KP. Role of α_2-adrenoceptors of nucleus ambiguus & lateral medullary pressor area in hypotensive & bradycardiac effects of clonidine. Indian J Med Res 1986; 83:429–434.
17. Gurtu S, Sharma DK, Sinha JN, Bhargava KP. Evidence for involvement of α_2-adrenoceptors in the nucleus ambiguus in baroreflex-mediated bradycardia. Naunyn-Schmiedeberg's Arch Pharmacol 1983; 323:199–204.
18. Bousquet P, Schwartz J. Alpha-Adrenergic drugs: Pharmacological tools for the study of the central nervous system. Biochem Pharmacol 1983; 32:1459–1465.
19. Sved AF, Tsukamoto K, Schreihofer AM. Stimulation of α_2-adrenergic receptors in nucleus tractus solitarius is required for the baroreceptor reflex. Brain Res 1992; 576:297–303.
20. Zandberg P, De Jong W, De Wied D. Effect of catecholamine-receptor stimulating agents on blood pressure after local application in the nucleus tractus solitarii of the medulla oblongata. Eur J Pharmacol 1979; 55:43–56.
21. Kubo T, Kihara M, Hata H, Misu Y. Cardiovascular effects in rats of α_1 and α_2 adrenergic agents injected into the nucleus tractus solitarii. Naunyn-Schmied Arch Pharmacol 1987; 335:274–277.
22. Kubo T, Misu Y. Pharmacological characterisation of the α-adrenoceptors responsible for a decrease of blood pressure in the nucleus tractus solitarii of the rat. Naunyn-Schmiedeberg's Arch Pharmacol 1981; 317:120–125.
23. Bylund DB, U'Prichard DC. Characterization of α_1- and α_2-adrenergic receptors. Int Rev Neurobiol 1983; 24:343–431.

24. Zandberg P, De Jong W. α-Methylnoradrenaline-induced hypotension in the nucleus tractus solarii of the rat: A localization study. Neuropharmacology 1977; 16:219–222.
25. Kubo T, Goshima Y, Hata H, Misu Y. Evidence that endogenous catecholamines are involved in α_2-adrenoceptor-mediated modulation of the aortic baroreceptor reflex in the nucleus tractus solitarii of the rat. Brain Res 1990; 526:313–317.
26. Weber MA, Graettinger WF, Cheung DG. Centrally acting sympathetic inhibitors. In: Laragh JH, Brenner BM, eds. Hypertension: Pathophysiology, Diagnosis, and Management. New York: Raven Press, 1990:2251–2261.
27. Mosqueda-Garcia R. Guanfacine: A second generation α_2-adrenergic blocker. Am J Med Sci 1990; 299:73–76.
28. De Sarro GB, Bagetta G, Ascioti C, et al. Effects of pertussis toxin on the behavioural and ECoG spectrum changes induced by clonidine and yohimbine after their microinjection into the locus coeruleus. Br J Pharmacol 1980; 96:59–64.
29. De Sarro GB, Ascioti C, Froio F, et al. Evidence that locus coeruleus is the site where clonidine and drugs acting at α_1- and α_2-adrenoceptors affect sleep and arousal mechanisms. Br J Pharmacol 1987; 90:675–685.
30. Tibiriça E, Feldman J, Mermet C, et al. Selectivity of rilmenidine for the nucleus reticularis lateralis, a ventrolateral medullary structure containing imidazoline-preferring receptors. Eur J Pharmacol 1991; 209:213–221.
31. Miyawaki T, Kawamura H, Komatsu K, Yasugi T. Chemical stimulation of the locus coeruleus: Inhibitory effects on hemodynamics and renal sympathetic nerve activity. Brain Res 1991; 568:101–108.
32. Timmermans PBMWM, Schoop AMC, et al. Characterization of α-adrenergic receptors participating in the central hypotensive and sedative effects of clonidine using yohimbine, rauwolscine, and corynanthine. Eur J Pharmacol 1981; 70:7–15.
33. Timmermans PBMWM, Van Zwieten PA. Alpha-adrenoceptors: Classification, localization, mechanisms, and targets for drugs. J Med Chem 1982; 25:1389–1401.
34. Zornow MH. Ventilatory, hemodynamic and sedative effects of the α_2-adrenergic agonist, dexmedetomidine. Neuropharmacology 1991; 30:1065–1071.
35. Hayashi Y, Maze M. Alpha$_2$ adrenoceptor agonists and anaesthesia. Br J Anaesth 1993; 71:108–118.
36. Van Zwieten PA, Timmermans PBMWM. Centrally mediated hypotensive activity of B-HT 933 upon infusion via the cat's vertebral artery. Pharmacology 1980; 21:327–332.
37. Durcan MJ, Lister RG, Linnoila M. Evidence for central alpha-2 adrenoceptors, not imidazoline binding sites, mediating the ethanol-attenuating properties of alpha-2 adrenoceptor antagonists. J Pharmacol Exp Ther 1991; 258:276–582.
38. Aronstam RS, Daly JW, Spande TF, et al. Interaction of gephyrotoxin and indolozidine alkaloids with the nicotinic acetylcholine receptor-ion channel complex of Torpedo electroplax. Neurochem Res 1986; 11:1227–1240.
39. Tsoucaris-Kupfer D, Bogaievsky D, Schmitt H. Further investigations on the interaction between α-adrenoceptor antagonists and the central hypotensive effect of clonidine in rats, rabbits and dogs. Clin Exp Pharmacol Physiol 1980; 7:355–366.
40. Vollmer RR, Buckley JP. Central cardiovascular effects of phentolamine in chloralose-anesthetized cats. Eur J Pharmacol 1977; 43:17–25.
41. Ito A, Schanberg SM. Maintenance of tonic vasomotor activity by alpha and beta adrenergic mechanisms in medullary cardiovascular centers. J Pharmacol Exp Ther 1974; 189:392–404.

42. Bogaievsky D, Bogaievsky Y, Tsoucaris-Kupfer D, Schmitt H. Blockade of the central hypotensive effect of clonidine by alpha-adrenoreceptor antagonists in rats, rabbits and dogs. Clin Exp Pharmacol Physiol 1974; 1:527–534.
43. Finch L, Harvey CA, Hicks PE, Owen DAA. Clonidine-induced hypotension: further evidence for a central interaction with histamine H_2 receptor antagonists in the rat. Neuropharmacology 1978; 17:307–313.
44. Granata AR, Numao Y, Kumada M, Reis DJ. A1 noradrenergic neurons tonically inhibit sympathoexcitatory neurons of the C1 area in rat brainstem. Brain Res 1986; 377:127–146.
45. Karppanen H, Paakkari I, Paakkari P, et al. Possible involvement of central histamine H_2-receptors in the hypotensive effect of clonidine. Nature 1976; 259:587–588.
46. Karppanen H. Interrelationships between clonidine and histaminergic mechanisms. Trends Pharmacol Sci 1981; 2:35–37.
47. Karppanen H, Paakkari I, Paakkari P. Further evidence for central histamine H_2-receptor involvement in the hypotensive effect of clonidine in the rat. Eur J Pharmacol 1977; 42:299–302.
48. Bousquet P, Feldman J, Bloch R, Schwartz J. Central cardiovascular effects of alpha-adrenergic drugs: Differences between catecholamines and imidazolines. J Pharmacol Exp Ther 1984; 230:232–236.
49. Ernsberger P, Meeley MP, Reis DJ. An endogenous clonidine-like substance binds preferentially to imidazole binding sites in the ventrolateral medulla labelled by ^{3}H-para-aminoclonidine. J Hypertens 1986; 4(suppl 5):S109–S111.
50. Ernsberger P, Meeley MP, Mann JJ, Reis DJ. Clonidine binds to imidazole binding sites as well as α_2-adrenoceptors in the ventrolateral medulla. Eur J Pharmacol 1987; 134:1–13.
51. Ernsberger P, Meeley MP, Reis DJ. An endogenous substance with clonidine-like properties: selective binding to imidazole sites in the ventrolateral medulla. Brain Res 1988; 441:309–318.
52. Ernsberger P, Westbrooks KL, Graff LM. Imidazoline binding sites labeled by clonidine analogs (I_1 sites): Putative G-protein-linked receptors. FASEB J 1991; 5:A1066.
53. Ernsberger P, Gomez RE, Cannata MA. Comparison of moxonidine and rilmenidine: Two second-generation centrally acting antihypertensives selective for I_1-imidazoline receptors (abstr). Fundam Clin Pharmacol 1992; 6:49s.
54. Ernsberger P, Haxhiu MA, Damon TH, Wendorff LA. I_1-Imidazoline receptors and brainstem autonomic control: Membrane binding, autoradiographic and functional studies (abstr). FASEB J 1992; 6:A1874.
55. Ernsberger P. Heterogeneity of imidazoline binding sites: Proposed I_1 and I_2 subtypes (abstr). Fundam Clin Pharmacol 1992; 6:55s.
56. Michel MC, Ernsberger P. Keep an eye on the I site: Imidazoline-preferring receptors. Trends Pharmacol Sci 1992; 13:369–370.
57. Ernsberger P, Damon TH, Graff LM, et al. Moxonidine, a centrally acting antihypertensive agent, is a selective ligand for I_1-imidazoline sites. J Pharmacol Exp ther 1993; 264:172–182.
58. Bricca G, Dontenwill M, Molines A, et al. The imidazoline preferring receptor: Binding studies in bovine, rat and human brainstem. Eur J Pharmacol 1989; 162: 1–9.

59. Ernsberger P, Kou YR, Prabhakar NR. Imidazole binding sites in the ventrolateral medulla (VLM) and carotid body are labeled by ^{125}I-p-iodoclonidine (^{125}I-PIC): Possible chemosensory role (abstr). FASEB J 1990; 6:A716.
60. Kamisaki Y, Ishikawa T, Takao Y, et al. Binding of [^{3}H]*p*-aminoclonidine to two sites, α_2-adrenoceptors and imidazoline binding sites: Distribution of imidazoline binding sites in rat brain. Brain Res 1990; 514:15–21.
61. Kou YR, Ernsberger P, Cragg PA, et al. Role of α_2-adrenergic receptors in the carotid body response to isocapnic hypoxia. Respir Physiol 1991; 83:353–364.
62. Ernsberger P, Meeley MP, Reis DJ. Neuroblastoma-glioma hybrid cells contain clonidine-displacing substance. Eur J Pharmacol 1989; 174:135–138.
63. Ernsberger P. I_1-Imidazoline binding sites in PC12 cells are up-regulated by nerve growth factor (NGF) (abstr). Soc Neurosci Abstr 1992; 18:622.
64. Ernsberger P, Iacovitti L, Reis DJ. Astrocytes cultured from specific brain regions differ in their expression of adrenergic binding sites. Brain Res 1990; 517:202–208.
65. Michel MC, Insel PA. Are there multiple imidazoline binding sites? Trends Pharmacol Sci 1989; 10:342–344.
66. Brown CM, MacKinnon AC, McGrath JC, et al. α_2-Adrenoceptor subtypes and imidazoline-like binding sites in the rat brain. Br J Pharmacol 1990; 99:803–809.
67. Coupry I, Atlas D, Podevin R-A, et al. Imidazoline-guanidinium receptive site in renal proximal tubule: Asymetric distribution, regulation by cations and interaction with an endogenous clonidine displacing substance. J Pharmacol Exp Ther 1990; 252:293–299.
68. Wikberg JES, Uhlén S, Chhajlani V. Medetomidine stereoisomers delineate two closely related subtypes of idazoxan (imidazoline) I-receptors in the guinea pig. Eur J Pharmacol 1991; 193:335–340.
69. Bidet M, Poujeol P, Parini A. Effect of imidazolines on Na^+ transport and intracellular pH in renal proximal tubule cells. Biochim Biophys Acta Bio-Membr 1990; 1024:173–178.
70. Ernsberger P, Sims C, Douglas JG. I_1-imidazoline sites in rat renal medulla: Binding properties and up-regulation by subpressor angiotensin infusion (abstr). Hypertension 1991; 18:380.
71. Steinbusch HWM, Mulder AH. Localization and projections of histamine-immunoreactive neurons in the central nervous system of the rat. In: Ganellin CR, Schwartz JR, eds. Frontiers in Histamine Research. Oxford: Pergamon, 1985:119–128.
72. Airaksinen MS, Panula P. The histaminergic system in the guinea pig central nervous system: An immunocytochemical mapping study using an antiserum against histamine. J Comp Neurol 1988; 273:163–186.
73. Bylund DB. Subtypes of α_2-adrenoceptors: Pharmacological and molecular biological evidence converge. Trends Pharmacol Sci 1988; 9:356–361.
74. Lorenz W, Lomasney JW, Collins S, et al. Expression of three α_2-adrenergic receptor subtypes in rat tissues: Implications for α_2 receptor classification. Mol Pharmacol 1990; 38:599–603.
75. Lanier SM, Downing S, Duzic E, Homcy CJ. Isolation of rat genomic clones encoding subtypes of the α_2-adrenergic receptor: Identification of a unique receptor subtype. J Biol Chem 1991; 266:10470–10478.
76. Bylund DB, Ray-Prenger C. Alpha-2A and alpha-2B adrenergic receptor subtypes: Attenuation of cyclic AMP production in cell lines containing only one receptor subtype. J Pharmacol Exp Ther 1989; 251:640–644.

77. Harrison JK, Pearson WR, Lynch KR. Molecular characterization of α_1- and α_2-adrenoceptors. Trends Pharmacol Sci 1991; 12:62–67.
78. Nicholas AP. Pieribone V, Hökfelt T. Distributions of mRNAs for alpha-2 adrenergic receptor subtypes in rat brain: An in situ hybridization study. J Comp Neurol 1993; 328:575–594.
79. Atlas D, Burstein Y. Isolation and partial purification of a clonidine-displacing endogenous brain substance. Eur J Biochem 1984; 144:287–293.
80. Meeley MP, Towle AC, Ernsberger P, et al. Clonidine-specific antisera recognize an endogenous clonidine-displacing substance in brain. Hypertension 1989; 13: 341–351.
81. Piletz JE, Andorn AC, Unnerstall JR, Halaris A. Binding of [^{3}H]-*p*-aminoclonidine to α_2-adrenoceptor states plus a non-adrenergic site on human platelet plasma membranes. Biochem Pharmacol 1991; 42:569–584.
82. Molderings GJ, Moura D, Fink K, et al. Binding of [^{3}H]clonidine to I_1-imidazoline sites in bovine adrenal medullary membranes. Naunyn Schmiedebergs Arch Pharmacol 1993; 348:70–76.
83. Bousquet P, Feldman J, Schwartz J. The central hypotensive actions of α-adrenoceptor agonists: A structure-activity relationship study. Trends Autonom Pharmacol 1985; 3:213–223.
84. Bousquet P, Feldman J. The blood pressure effects of alpha-adrenoceptor antagonists injected in the medullary site of action of clonidine: The nucleus reticularis lateralis. Life Sci 1987; 40:1045–1050.
85. Estes ML, Block CH, Barnes KL. The canine nucleus tractus solitarii: Light microscopic analysis of subnuclear divisions. Brain Res Bull 1989; 23:509–517.
86. Scholtysik G, Laurener H, Eichenberger E, et al. Pharmacological actions of the antihypertensive agent N-amino-2-(2,6-dichlorophenyl)acetamidehydrochloride (BS 100-141). Arzneim-Forsch 1975; 25:1483–1491.
87. Haxhiu MA, Dreshaj I, Schäfer SG, Ernsberger P. Selective antihypertensive action of moxonidine is mediated mainly by I_1-imidazoline receptors in the rostral ventrolateral medulla. J Cardiovasc Pharmacol 1994; 24(suppl. 1):S1–S8.
88. Ernsberger P, Arango V, Reis DJ. A high density of muscarinic receptors in the ventrolateral medulla of the rat is revealed by correction for autoradiographic efficiency. Neurosci Lett 1988; 85:179–186.
89. Milner TA, Pickel VM, Reis DJ. Ultrastructural basis for interactions between central opioids and catecholamines. I. Rostral ventrolateral medulla. J Neurosci 1989; 9:2114–2130.
90. Ernsberger PR, Westbrooks KL, Christen MO, Schäfer SG. A second generation of centrally acting antihypertensive agents act on putative I_1-imidazoline receptors. J Cardiovasc Pharmacol 1992; 20(suppl) 4:S1–S10.
91. Ernsberger P, Haxhiu MA, Graff LM, et al. A novel mechanism of action for hypertension control: Moxonidine as a selective I_1-imidazoline agonist. Cardiovasc Drugs Ther 1994; 8:27–41.
92. Dresse A, Scuvée-Moreau J, Seutin V. Action of rilmenidine on locus ceruleus and dorsal raphe cells in vivo and in vitro. Am J Cardiol 1988; 61:32D–34D.
93. Van Zwieten PA. Pharmacology of the alpha$_2$-adrenoceptor agonist rilmenidine. Am J Cardiol 1988; 61:6D–14D.
94. Koenig-Bérard E, Tierney C, Beau B, et al. Cardiovascular and central nervous system effects of rilmenidine (S 3341) in rats). Am J Cardiol 1988; 61:22D–31D.

95. Armah BI, Hofferber E, Stenzel W. General pharmacology of the novel centrally acting antihypertensive agent moxonidine. Arzneim-Forsch 1988; 38:1426–1434.
96. Bricca G, Dontenwill M, Molines A, et al. Rilmenidine selectivity for imidazoline receptors in human brain. Eur J Pharmacol 1989; 163:373–377.
97. Feldman J, Tibiriça E, Bricca G, et al. Evidence for the involvement of imidazoline receptors in the central hypotensive effect of rilmenidine in the rabbit. Br J Pharmacol 1990; 100:600–604.
98. Reid JL, Rubin PC, Howden CW. Central α_2-adrenoceptors and BP regulation in man: Studies with guanfacine (BA 100-141) and azepexole (B-HT 933). Br J Pharmacol 1983; 15:463S–470S.
99. Deckert V, Lachaud V, Parini A, Elghozi J-L. Contribution of α_2-adrenoceptors to the central cardiovascular effect of clonidine and S 8350 in anaesthetized rats. Clin Exp Pharmacol Phsyiol 1991; 18:401–408.
100. Nosjean A, Guyenet PG. Role of ventrolateral medulla in sympatholytic effect of 8-OHDPAT in rats. Am J Physiol Regul Integr Comp Physiol 1991; 260:R600–R609.
101. King KA, Pang CCY. Differential cardiovascular effects of central clonidine and B-HT 920 in conscious rats. Can J Physiol Pharmacol 1988; 66:1455–1460.
102. Lapp CA, Buccafusco JJ. Comparison of the antihypertensive actions of α-adrenergic agonists after intracisternal injection (abstr). FASEB J 1991; 5:A1411.
103. Hong M, Milne B, Loomis C, Jhamandas K. Stereoselective effects of central α_2-adrenergic agonist medetomidine on in vivo catechol activity in the rat rostral ventrolateral medulla (RVLM). Brain Res 1992; 592:163–169.
104. Mitra J, Dev NB, Trivedi R, et al. Intramedullary sodium cyanide injection on respiratory and vasomotor responses in cats. Respir Physiol 1993; 93:71–82.
105. Giuliano R, Ruggiero DA, Morrison S, et al. Cholinergic regulation of arterial pressure by the C1 area of the rostral ventrolateral medulla. J Neurosci 1989; 9: 923–942.
106. Haxhiu MA, Dreshaj I, Erokwu B, Ernsberger P. Vasodepression elicited in hypertensive rats and normal cats by the selective I_1-imidazoline agonist moxonidine administered into the rostral ventrolateral medulla (RVLM) (abstr). Fundam Clin Pharmacol 1992; 6:62s.
107. McCrimmon DR, Lalley PM. Inhibition of respiratory neural discharges by clonidine and 5-hydroxytryptophan. J Pharmacol Exp Ther 1982; 222:771–777.
108. Udvary E, Schäfer SG, Vegh A, Szekeres L. Haemodynamic effects of a new centrally acting antihypertensive agent moxonidine as compared to clonidine. Eur J Pharmacol 1990; 183:2064.
109. Smith BD, Baudendistel LJ, Gibbons JJ, Schweiss JF. A comparison of two epidural α_2-agonists, guanfacine and clonidine, in regard to duration of antinociception, and ventilatory and hemodynamic effects in goats. Anesth Analg 1992; 74: 712–718.
110. Kou YR, Ernsberger P, Cherniack NS, Prabhakar NR. Role of α_2-adrenergic receptors in the carotid body response to hypoxia (abstr). FASEB J 1990; 4:A716.
111. Bamford OS, Hawkins RL, Blanco CE. Effects of clonidine on breathing movements and electrocortical activity in the fetal lamb. Am J Obstet Gynecol 1990; 163:661–668.
112. Bailey PL, Sperry RJ, Johnson GK, et al. Respiratory effects of clonidine alone and combined with morphine, in humans. Anesthesiology 1991; 74:43–48.

113. Penon C, Ecoffey C, Cohen SE. Ventilatory response to carbon dioxide after epidural clonidine injection. Anesth Analg 1991; 72:761–764.
114. Benhamou D, Veillette Y, Narchi P, Ecoffey C. Ventilatory effects of premedication with clonidine. Anesth Analg 1991; 73:799–803.
115. Ooi R, Pattison J, Feldman SA. The effects of intravenous clonidine on ventilation. Anaesthesia 1991; 46:632–633.
116. Errchidi S, Monteau R, Hilaire G. Noradrenergic modulation of the medullary respiratory rhythm generator in the newborn rat: An in vitro study. J Physiol (Lond) 1991; 443:477–498.
117. Jarvis DA, Duncan SR, Segal IS, Maze M. Ventilatory effects of clonidine alone and in the presence of alfentanil in human volunteers. Anesthesiology 1992; 76: 899–905.
118. Prabhakar NR, Mitra J, Adams EM, Cherniack NS. Involvement of ventral medullary surface in respiratory responses induced by 2,4-dinitrophenol. J Appl Physiol 1989; 66:598–605.
119. Neubauer JA, Gonsalves SF, Chou W, et al. Chemosensitivity of medullary neurons in explant tissue cultures. Neuroscience 1991; 45:701–708.
120. Heath D, Khan Q, Smith P. Histopathology of the carotid bodies in neonates and infants. Histopathology 1990; 17:511–520.
121. Ernsberger P, Collins LA. Autoradiography of I_1-imidazoline and α_2-adrenergic sites in rat brainstem: Reticular localization of I_1-sites (abstr). Soc Neurosci Abstr 1993; 19:1694.
122. Schmitt H, Frenard S. Evidence for an alpha-sympathomimetic component in the effects of catapressan on vasomotor centres: antagonism by piperoxane. Eur J Pharmacol 1971; 14:98–100.
123. Schmitt H, Fenard S. Action of alpha-adrenergic blocking drugs on the sympathetic centres and their interactions with the central sympatho-inhibitory effect of clonidine. Arzneim-Forsch 1973; 23:40–45.
124. Laubie M, Delbarre B, Bogaievsky D, et al. Pharmacological evidence for a central alpha-sympathomimetic mechanism controlling blood pressure and heart rate. Circ Res 1976; 38(suppl II):II-35–II-41.
125. Dwoskin LP, Neal BS, Sparber SB. Evidence for antiserotonergic properties of yohimbine. Pharmacol Biochem Behav 1988; 31:321–326.
126. Schoeffter P, Hoyer D. 5-Hydroxytryptamine 5-HT_{1B} and 5-HT_{1D} receptors mediating inhibition of adenylate cyclase activity: Pharmacological comparison with special reference to the effects of yohimbine, rauwolscine and some β-adrenoceptor antagonists. Naunyn-Schmiedeberg's Arch Pharmacol 1989; 340:285–292.
127. Beckett PJ, Finch L. The $alpha_1$- and $alpha_2$-adrenoceptor involvement in the central cardiovascular action of clonidine in the conscious renal hypertensive cat. Eur J Pharmacol 1982; 82:155–160.
128. Tibiriça E, Feldman J, Bousquet P. Differences in the ability of yohimbine to antagonize the hypotensive effect of clonidine in normotensive and spontaneously hypertensive anesthetized rats. J Pharmacol Exp Ther 1988; 244:1062–1066.
129. Timmermans PBMWM, Lam E, VanZwieten PA. The interaction between prazosin and clonidine at alpha-adrenoceptors in rats and cats. Eur J Pharmacol 1979; 55:57–66.
130. Garvey HL, Woodhouse BL. Reversal of clonidine-induced hypotension by beta-adrenoceptor blocking drugs. Eur J Pharmacol 1980; 65:55–62.

131. Drolet G, Aslanian V, Minson J, et al. Differences in the central hypotensive actions of α-methyldopa and clonidine in the spontaneously hypertensive rat: Contribution of neurons arising from the B3 and the C1 areas of the rostral ventrolateral medulla. J Cardiovasc Pharmacol 1990; 15:118–123.
132. Head GA, Burke S. Importance of central noradrenergic and serotonergic pathways in the cardiovascular actions of rilmenidine and clonidine. J Cardiovasc Pharmacol 1991; 18:819–826.
133. Chalmers J, Pilowsky P. Brainstem and bulbospinal neurotransmitter systems in the control of blood pressure. J Hypertens 1991; 9:675–694.
134. Farsang C, Kunos G. Naloxone reverses the antihypertensive effect of clonidine. Br J Pharmacol 1979; 67:161–164.
135. Van Giersbergen PLM, Tierney SAV, Wiegant VM, De Jong W. Possible involvement of brain opioid peptides in clonidine-induced hypotension in spontaneously hypertensive rats. Hypertension 1989; 13:83–90.
136. Hieble Jp, Ruffolo RR Jr. Therapeutic applications of agents interacting with α-adrenoceptors. Prog Basic Clin Pharmacol 1991; 8:180–220.
137. Czyzewska-Szafran H, Jastrzebski Z, Remiszewska M, Wutkiewicz M. Effect of clonidine on blood pressure and GABAergic mechanisms in spontaneously hypertensive rats. Eur J Pharmacol 1991; 198:115–120.
138. Marmo E, Filippelli W, Marrazzo R, et al. Participation of GABAergic mechanisms in the hypotensive and bradycardic effects of clonidine: Experimental study in conscious normotensive and hypertensive rats. Neuropharmacology 1987; 26: 1525–1528.
139. Marmo E, Rossi F, Berrino L, et al. Participation of GABA-ergic mechanisms in the hypotensive effect of clonidine. Curr Ther Res 1991; 39:269–279.
140. Abdel-Rahman AA. Reversal by ethanol of the hypotensive effect of clonidine in conscious spontaneously hypertensive rats. Hypertension 1989; 14:531–541.
141. Aslanian V, Renaud B. Changes in serotonin metabolism in the rat raphe magnus and cardiovascular modifications following systemic administration of clonidine and other central α_2-agonists: An in vivo voltammetry study. Neuropharmacology 1989; 28:387–395.
142. Gillis RA, DiMicco JA, Williford DJ, et al. Importance of CNS GABAergic mechanisms in the regulation of cardiovascular function. Brain Res Bull 1980; 5(suppl 2):303–315.
143. Pittaluga A, Torelli R, Raiteri M. Clonidine differentially modulates the release of endogenous GABA in various rat brain areas. Pharmacol Res 1991; 24:189–196.
144. Meeley MP, Ernsberger P, Char LK, et al. Regional and subcellular distribution of clonidine-displacing substance in brain (abstr). Soc Neurosci Abstr 1989; 15: 1179.
145. Ruggiero DA, Meeley MP, Muhammed A, Reis DJ. Newly identified GABAergic neurons in regions of the ventrolateral medulla which regulate blood pressure. Brain Res 1985; 339:171–177.
146. Pickel VM, Chan J, Milner TA. Cellular substrates for interactions between neurons containing phenylethanolamine N-methyltransferase and GABA in the nuclei of the solitary tracts. J Comp Neurol 1989; 286:243–259.
147. Willette RN, Barcas PP, Krieger AJ, Sapru HN. Endogenous GABAergic mechanisms in the medulla and the regulation of blood pressure. J Pharmacol Exp Ther 1984; 230:34–39.

148. Sasek CA, Wessendorf MW, Helke CJ. Evidence for co-existence of thyrotropin-releasing hormone, substance P and serotonin in ventral medullary neurons that project to the intermediolateral cell column in the rat. Neuroscience 1990; 35: 105–119.
149. Gomez R, Ernsberger P, Feinland G, Reis DJ. Rilmenidine acts upon imidazole receptors in the rostral ventrolateral medulla to elicit hypotension and bradycardia. Eur J Pharmacol 1990; 183:168–169.
150. Clement ME, McCall RB. Effects of clonidine on sympathoexcitatory neurons in the rostral ventrolateral medulla. Brain Res 1991; 550:353–357.
151. Anderson C, Stone TW. On the mechanism of action of clonidine: effects on single central neurons. Br J Pharmacol 1974; 51:359–365.
152. Piletz JE, Sletten K. Nonadrenergic imidazoline binding sites on human platelets. J Pharmacol Exp Ther 1993; 267:1493–1502.

20

GABA and TRH Modulate the Release of Neurotransmitter from Premotor Respiratory Neurons Via a Common Mechanism

PAUL G. WAGNER, YE-MING SUN, and MICHAEL S. DEKIN

University of Medicine and Dentistry of New Jersey–Robert Wood Johnson Medical School
New Brunswick, New Jersey

The respiratory central pattern generator (CPG) is capable of producing a normal respiratory rhythm but also exhibits variations and adaptations to changes in the environment. The ability of the respiratory CPG to vary its activity resides in the plasticity of its individual elements. While many mechanisms are utilized by the central nervous system to produce such plasticity, one of the most common is to selectively adjust the strength of the monosynaptic connections between CPG neurons. This adjustment in synaptic strength is accomplished via the modulation of neurotransmitter release from CPG neurons. Though never tested directly, presynaptic inhibition and facilitation have been postulated to be responsible for changes observed in respiratory output. Using a spinal cord slice preparation in which monosynaptic connections between premotor respiratory neurons and phrenic motoneurons were isolated, we showed that GABA and thyrotropin-releasing hormone (TRH) reciprocally modulate the release of neurotransmitter via a presynaptic mechanism. In addition, we have identified a K^+ channel (gK_{OR}) in the membranes of cultured premotor neurons that is modified by baclofen and TRH. This channel exhibits outward rectification, is insensitive to blockade by Ba^{++} ions, and is second-messenger-dependent. The properties of gK_{OR} are similar to those of the S channel in invertebrates, which plays a role in presynaptic inhibition and facilitation. We propose that gK_{OR} has a similar role in the respiratory CPG. This channel may also participate in other respiratory phenomena like poststimulus potentiation and acclimatization to high altitude.

I. Introduction

Rhythmic breathing movements in mammals are initiated and maintained by an organized group of neurons called a central pattern generator (CPG), which is located in the lower brainstem (1). Conceptually, the respiratory CPG consists of a rhythm generator coupled in series to a burst pattern generator (2). The rhythm generator is made up of pacemaker neurons located in the pre-Botzinger complex and initiates inspiratory activity (3). The burst generator comprises nonpacemaker inspiratory premotor neurons from the medullary brainstem regions known as the dorsal (nucleus tractus solitarius) and ventral (nucleus ambiguus) respiratory groups, which innervate phrenic motoneurons (4–6).

The respiratory CPG is capable of producing a normal respiratory pattern but requires additional inputs to display the variations observed in day-to-day breathing behavior (7) as well as long-term adaptations (8). The ability of the respiratory CPG to vary its activity resides in the plasticity of its individual elements. While many mechanisms are utilized by the central nervous system to produce such plasticity, one of the most common is to selectively adjust the strength of the monosynaptic connections between CPG neurons. This adjustment in synaptic strength can be accomplished via the modulation of neurotransmitter release from CPG neurons.

The modulation of neurotransmitter release from respiratory neurons has never been directly tested in vivo; however, such modulation has been postulated to mediate several adaptations of the respiratory CPG circuit (9). One example involves the neuropeptide thyrotropin releasing hormone (TRH), which has been shown to enhance the excitability of motoneurons (10). Part of this action has been attributed to presynaptic facilitation (11,12). TRH is known to have potent analeptic effects on the respiratory CPG, and nerve terminals containing TRH have been localized within the phrenic motor nucleus (13,14). Another example of this type of regulation is the postulated presynaptic inhibition of phrenic motoneurons produced by the selective $GABA_B$ agonist (±)baclofen (15,16).

Based on these observations, we tested the hypothesis that GABA and TRH reciprocally modulate the release of neurotransmitter from premotor respiratory neurons. To do this, we used a spinal cord slice preparation from neonatal rats in which monosynaptic connections between premotor respiratory neurons and phrenic motor neurons could be isolated. We then studied the effects of (±)baclofen and TRH on monosynaptic excitatory postsynaptic potentials (EPSP). We found that TRH and baclofen reciprocally altered the amplitude of the EPSP. In parallel studies using a cell culture system in which premotor respiratory neurons were isolated, we observed that baclofen and TRH reciprocally modulated the activity of an outward-rectifying potassium conductance. The properties of this K^+ channel are consistent with its being involved in the modulation of neurotransmitter release. As such, this channel could be an important presynaptic site of action for TRH and GABA.

II. Methods

A. Spinal Cord Slice Preparation

Rats 3 to 5 days old were anesthetized with ether and hypothermia then quickly laminectomized to expose the high cervical spinal cord. A section of the cord containing the phrenic motor nucleus (C3–5) was removed with the dorsal roots attached (2 to 3 mm). Transverse slices (500 μm) were cut with a vibratome and transferred to a recording chamber, where they were perfused with oxygenated Ringer's solution at room temperature. To prevent swelling during the slicing procedure, a modified Ringer's solution in which sucrose replaced NaCl was used (17). This solution contained (in m*M*) 252 sucrose, 5 KCl, 1.3 $MgSO_4$, 2 $CaCl_2$, 2.5 NaH_2PO_4, 26 $NaHCO_3$, 10 glucose (pH 7.4; aerated with 95% O_2–5% CO_2). The sucrose-modified Ringer's solution was also used during a 1-h recovery period after the slices were placed in the chamber.

Following recovery, the bath was switched to a normal Ringer's solution that contained (in m*M*) 125 NaCl, 2.5 KCl, 2 $CaCl_2$, 1 $MgCl_2$, 10 Na-Hepes, 30 glucose (pH 7.4; aerated with 100% O_2). All experiments were done using an axoclamp 1A amplifier (Axon Instr., Burlingame, CA) and data were stored on digital tape (Sony DT-120). Whole cell patch recording pipettes (3 to 5 Mohm) were filled with (in m*M*) 130 K^+ gluconate, 10 EGTA, 10 HEPES, 1 $CaCl_2$, 1 Mg-ATP (pH 7.2; 270–280 mOsm). To isolate the monosynaptic EPSP, a high-Ca^{++}–high Mg^{++} Ringer was used for recording and contained (in m*M*) 125 NaCl, 2.5 KCl, 4 $CaCl_2$, 2 m*M* $MgCl_2$, 10 Na-Hepes, 30 glucose (pH 7.4; aerated with 100% O_2). This modified Ringer's solution also contained 0.5 m*M* $BaCl_2$ to block the postsynaptic actions of GABA and TRH on inward-rectifying K^+ channels (18,19) and 1 m*M* bicuculline to block inhibitory post-synaptic potentials. In addition, all EPSP were measured at a membrane potential of -80 mV to avoid contamination from Ba^{++}-insensitive outward-rectifying K^+ channels (20). EPSP were elicited using bipolar electrodes (100 μm in diameter) to stimulate the ventral (or lateral) funiculus, where axons of premotor respiratory neurons from the ventral and dorsal respiratory groups are located (5). Antidromic spikes were induced using a second bipolar electrode placed on the ventral root. To avoid potential problems with EPSP variability, averaged EPSP were calculated using 25 to 30 individual events (stimulated at 0.5 Hz). Averaging was done using the CODAS data acquisition and processing system (DATAQ Instr., Akron, OH). The amplitudes of the averaged EPSP were expressed as mean ± SD. All drugs were applied directly to the bath.

B. Cell Culture Preparation

All experiments used labeled premotor respiratory bulbospinal neurons from the regions of the dorsal and ventral respiratory group of 3- to 4-day-old Sprague-Dawley rat pups (20). Cell-attached patch recordings were done on cultured neurons perfused with Ringer's solution at a rate of 1 to 2 ml/min. The normal Ringer's solution contained (in m*M*) 130 NaCl, 5.4 KCl, 1 $CaCl_2$, 1 $MgCl_2$, 10 HEPES, and 30 glucose (pH 7.4). To isolate (±)baclofen-activated channels and zero the membrane potential, a modified Ringer's solution was employed that contained (in m*M*) 140 KCl, 5 CsCl, 1 $BaCl_2$, 1 $MgCl_2$, 0.2 $CdCl_2$, 0.2 $NiCl_2$, 0.5 4-AP, 0.0005 tetrodotoxin, 10 HEPES, and 30 glucose (pH 7.4). All experiments were done at room temperature with an Axopatch 2A amplifier and were analyzed with p-Clamp software. Outward currents are shown as downward deflections, and membrane potential values are reported as inside with respect to out. Except where noted, patch pipettes (5 to 7 MΩ) contained the same saline solution as the bath. Current-voltage relationships were plotted using the amplitudes of single-channel currents measured at different voltages. Regression lines were calculated by the method of least squares and used to calculate the single-channel conductance.

TRH and bicuculline were obtained from Research Biochemicals Inc. (Natick, MA). All other drugs and chemicals were obtained from Sigma Chemical Co. (St. Louis, MO).

III. Results

To isolate the presynaptic effects of baclofen and TRH, it was necessary to block their postsynaptic actions. In postsynaptic membranes, baclofen potentiates an intrinsic Ba^{++}-sensitive K^+ current (18,21) while TRH inhibits a similar current (19). In contrast, presynaptic inhibition due to baclofen has been demonstrated to be insensitive to blockade by Ba^{++} (22,23). In our experiments, therefore, we added 500 μM $BaCl_2$ to our Ringer's solution to block the inward-rectifier current in phrenic motoneurons. Figure 1 shows averaged compound monosynaptic EPSP recorded from phrenic motoneurons. The EPSP were evoked by stimulating in the region of the ventral funiculus, where premotor inspiratory axons from the VRG are located. Each trace represents the average of 30 individual events elicited at a rate of 0.5 Hz. The membrane potential of the phrenic motoneuron was current-clamped at −80 mV. EPSP had a reversal potential of 0 mV and could be blocked by CNQX, indicating that they were due to the activation of a non-NMDA glutamate activated conductance (data not shown). The averaged EPSP for control, 1 μM TRH, and 1 μM TRH

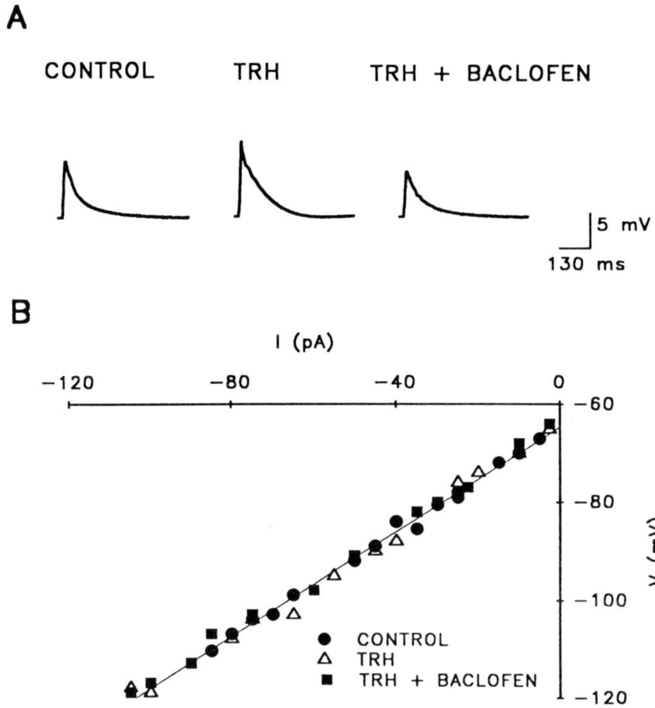

Figure 1 Effects of TRH and baclofen on the amplitude of the mEPSP. (A) Each trace is an averaged EPSP of 30 individual events. TRH caused an increase in the amplitude of the EPSP while the addition of baclofen antagonized the increase caused by TRH. (B) Voltage-current relationships generated before and after the addition of each drug.

plus 100 μM ($\pm$)baclofen are illustrated in Figure 1A. The addition of TRH to the perfusate caused an obvious increase in the amplitude of the averaged EPSP. In the presence of both TRH and ($\pm$)baclofen, the averaged EPSP amplitude decreased below the amplitude of control. Neither TRH nor baclofen had an effect on the time course of the EPSP.

Voltage-current plots for this experiment were determined using short hyperpolarizing pulses from a membrane potential of -60 mV and are illustrated in Figure 1B. The curve for control was linear, indicating that inward rectification was effectively blocked by the addition of Ba^{++} to the Ringer's solution. Also, the slopes of the voltage-current relationships in the presence of TRH alone and TRH plus ($\pm$)baclofen were not altered (Fig. 1B; control = 534 Mohm, TRH = 535 Mohm, and TRH + baclofen = 535 Mohm). Thus, the enhancement of the EPSP by TRH and the reversal of this increase by (±)baclofen were likely due to a presynaptic mechanism of action.

While the above experiments suggest a role of baclofen and TRH in presynaptic inhibition and facilitation, it is not clear whether these drugs had a common site of action or if they acted on separate processes that were mutually antagonistic. To address this question, we used the patch-clamp technique combined with a cell culture preparation that allowed us to isolate and directly visualize premotor respiratory bulbospinal neurons. Figure 2 is an example of one of these neurons from the region of the ventral respiratory group (nucleus ambiguus). The cell was pyramidal in shape and displayed several processes (Fig. 2A). Under epifluorescence illumination, rhodamine-labeled latex microspheres were observed clustered in the cell body (Fig. 2B).

A cell-attached patch recording of spontaneous single-channel activity from a DRG neuron at rest exposed to normal Ringer's solution is shown in the top trace of Figure 2C. The amount of activity varied from cell to cell but typically contained both inward and outward currents. This activity was abolished (Fig. 2C, middle) by perfusing the cells with a modified Ringer's solution (see methods). The addition of baclofen (100 μM) to the bath caused the appearance of large outward currents despite the presence of Ba^{++}, Cs^{+}, and 4-AP (Fig. 2C, bottom). The appearance of these baclofen-activated channels also demonstrated that these currents were second-messenger-gated, since baclofen was added to the bath and had no direct access to the channel that was isolated under the patch pipette.

The outward direction of the current suggested that the channel was permeable to potassium. To identify the type of channel and examine its biophysical properties in more detail, we determined the current-voltage relationship of the channel for different concentrations of $[KCl]_o$ in the patch pipette (Fig. 3). All reductions in $[KCl]_o$ were made by stoichiometric replacement with NaCl in the modified Ringer's solution. In addition, the membrane potential was zeroed using 140 mM K^+ in the bath. Figure 3A shows recordings of the baclofen-activated channel at different membrane potentials with $[K^+]_o$ in the pipette equal to 140 mM. Single-channel currents were outward at membrane potentials positive to 0 mV (the reversal potential) and increased in amplitude as the cell was depolarized. In contrast, the amplitude of the currents were small and did not change in an ohmic fashion at membrane potential levels negative to 0 mV. These data indicate that this baclofen-activated potassium channel (gK_{OR}) exhibited outward rectification. When $[K^+]_o$ in the pipette was reduced to 25 mM, the reversal potential was shifted to more negative membrane potentials (Fig. 3B). The inward currents were smaller than with the higher concentration of K^+ but still exhibited outward rectification. In Figure 3C, currents were recorded with $[K^+]_o$ = 5.4 mM in the pipette. Outward currents were observed at membrane potentials greater than −65 mV, but no inward currents could be resolved at more negative potential levels. The effects of altering $[K^+]_o$ in

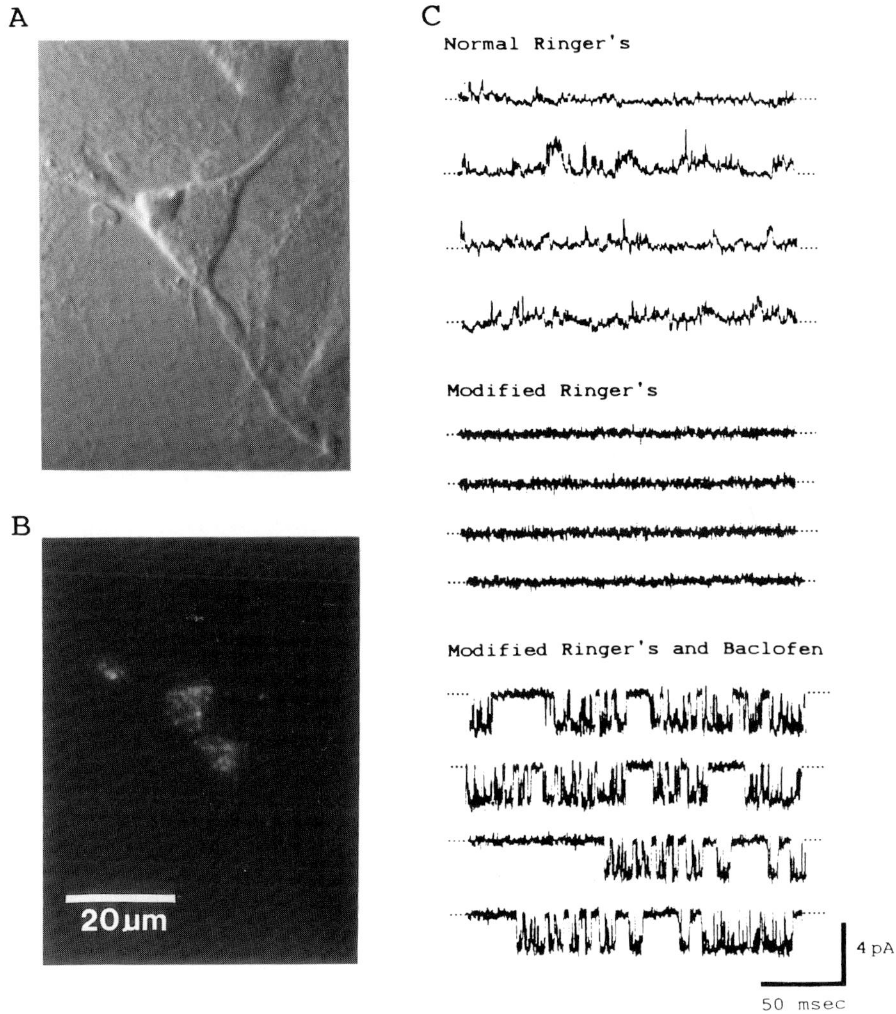

Figure 2 (A) Normaski image of a labeled neuron in culture. (B) Same cell as in A uncer epifluorescence illumination (scale applies to both A and B). (C) Cell-attached recordings under different ionic conditions (see text for details). Dotted lines in each trace represent the closed state of the channel. Downward deflections represent outward currents. (From Ref. 20.)

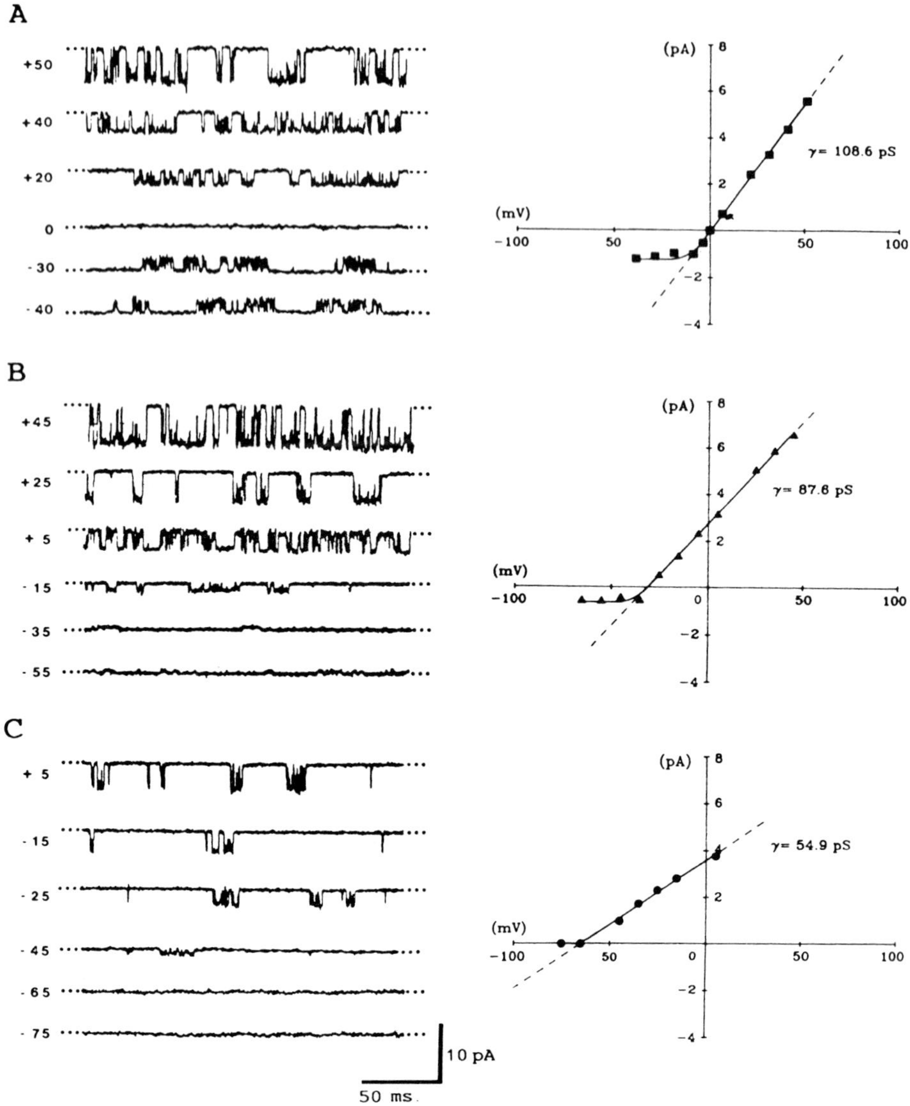

Figure 3 Cell-attached recordings (left) are shown of baclofen activated channels with different concentrations of K^+ in the pipette (A = 140 m*M*, B = 25 m*M*, and C = 5.4 m*M*). I-V plots for these data are shown on the right. Solid lines were drawn by eye. Values for single-channel conductances (γ) were calculated from the slope of the regression lines (----) fit to the linear portion of each curve. (From Ref. 20.)

the pipette on the current-voltage (I-V) relationships are shown to the right of the channel recordings in Figure 3. Outward rectification of the single channel current is very obvious in these plots. With 140 mM K^+ in the pipette, the I-V plot was linear between -10 and $+50$ mV and flat at membrane potentials negative to -10 mV. The I-V curves for $[K^+]_o = 25$ and 5.4 mM had similar shapes but were shifted to the left. The single-channel conductance (γ) values that were calculated from the linear portions of each curve are shown for each graph. The reversal potentials for these plots were 0 ± 1.4 mV for 140 mM ($n = 5$), -31.8 ± 7.1 mM for 25 mM ($n = 4$), and -63.7 ± 4.6 mV for 5.4 mM ($n = 3$). These changes in reversal potential are consistent with a channel that is predominantly selective for potassium (20).

The biophysical properties of gK_{OR} are similar to those of the S channel found in sensory neurons of the marine mollusc *Aplysia* (24). The S channel is inhibited by phosphorylation that is mediated by a cAMP-dependent pathway. We therefore tested the effects of cAMP on gK_{OR} by exposing neurons to the lipophilic cAMP analogue 8-bromo-cAMP (Fig. 4A). This was done by adding 100 μM (±)baclofen alone (Fig. 4A, top) to stimulate gK_{OR} and then exposing the neurons to both 100 μM (±)baclofen and 1 mM 8-bromo-cAMP (bottom).

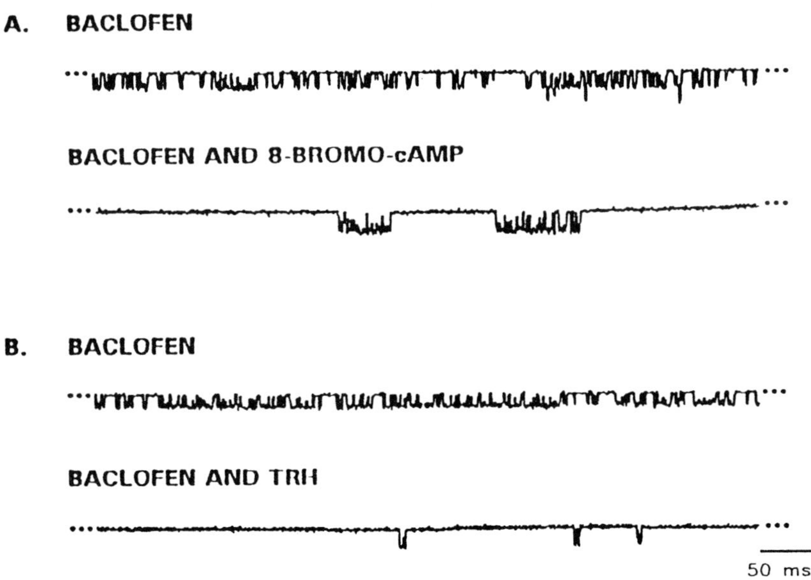

Figure 4 Effects of 8-bromo-cAMP and TRH on the activity of gK_{OR}. (A) The effect of 100 μM (±)baclofen alone (top) and after the addition of 1 mM 8-bromo-cAMP (bottom, Vm = 20 mV, $[K^+]_o = 140$ mM). (B) Effects of 100 μM (±)baclofen alone (top) and after the addition of 1 μM TRH (bottom, Vm = 0 mV, $[K^+]_o = 25$ mM).

The mean open time probability of gK_{OR} was reduced in the presence of 8-bromo-cAMP (55.1% in control versus 13.3% in 8-bromo-cAMP). Since our data from the spinal cord slices suggested that baclofen and TRH were acting antagonistically, we tested the ability of TRH to inhibit the activity of gK_{OR}. An example of an experiment in which gK_{OR} channels activated by 100 μM (±)baclofen were exposed to 1 μM TRH is shown in Figure 4B. The top trace shows gK_{OR} channels activated by exposing the neurons to 100 μM baclofen. The mean open time probability for this example was 36.1%. The bottom trace shows the effect of adding 1 μM TRH to the perfusate in the presence of baclofen. As was the case with 8-bromo-cAMP, TRH caused a dramatic decrease in the activity of gK_{OR} (mean open probability = 0.9%).

Analysis of the effects of 8-bromo-cAMP and TRH on gK_{OR} indicated that the normal channel opening and closing kinetics were not altered during inhibition, but rather that the mean open time probability of the channel was reduced. Table 1 shows typical results from the analysis of one experiment in which the action of 8-bromo-cAMP was tested and another in which TRH was used. The mean open time of gK_{OR} after 8-bromo-cAMP (4.2 ms) was comparable to that of control (3.9 ms). The closing kinetics of the channel were described by two time constants. The fast time constant (τ_{CF}) represented the short closed durations that occurred during bursts of openings and which gave the appearance that the channel was flickering. The slower time constant (τ_{CS}) reflected the gaps between bursts of openings. As with the open times, there were no significant differences between the fast components during control and after 8-bromo-cAMP. However, the slower component showed a dramatic increase following the application of 8-bromo-cAMP. This increase in the interval between bursts of openings (τ_{CS}) resulted in a significant decrease in the mean open time probability (Table 1). The data shown in Table 1 for the analysis of an experiment in which TRH was tested were similar to those of 8-bromo-cAMP (i.e., no differences in open time or τ_{CF} but a dramatic increase in τ_{CS}). We concluded from these data that cAMP and TRH did not have a direct effect on the opening and closing kinetics of gK_{OR} but rather decreased the probability that gK_{OR} would enter into the active state. In addition, these data suggested that the inhibition of gK_{OR} by TRH was mediated by cAMP. This conclusion was further supported by the fact that Rp-cAMP, a lipophilic competitive inhibitor

Table 1 Open and Closed Times

	Control	8-bromo-cAMP	Control	TRH
Mean open time (ms)	3.90	4.20	1.58	1.71
Closed time (msec)				
Fast time constant (τ_{CF})	0.70	0.66	1.11	1.12
Slow time constant (τ_{CS})	4.50	135.0	5.01	25.6

of cAMP as its protein kinase A (PKA) binding site, prevented the inhibition of gK_{OR} caused by TRH (25).

IV. Discussion

In this report, we have demonstrated that baclofen and TRH reciprocally modulated the release of neurotransmitter from premotor respiratory neurons. We have also shown that baclofen and TRH antagonistically regulated the activity of an outward-rectifying K^+ channel (gK_{OR}) in the membranes of cultured premotor respiratory neurons. The potential regulation of neurotransmitter release by GABA and TRH acting through gK_{OR} at the premotor-phrenic motoneuron synapse is reminiscent of the heterosynaptic modulation of neurotransmitter release by FMRFamide and serotonin acting on the S channel in the *Aplysia* gill-siphon withdrawal reflex (26,27). The biophysical properties of gK_{OR} and the S channel are very similar. Both these channels are resistant to Ba^{++} blockade, are second-messenger-dependent, have a single channel conductance of 60 pS ($[K^+] = 10$ mM), and exhibit outward rectification. In *Aplysia* sensory neurons, the outward-rectifying properties of the S channel allow it to shunt membrane currents during depolarization and enhance the repolarization of calcium action potentials. Consequently, less calcium crosses the membrane and neurotransmitter release is reduced. This mechanism could, in part, account for the depression of compound monosynaptic EPSP during exposure to baclofen (Fig. 1). Likewise, inhibition of gK_{OR} by TRH would broaden the calcium action potential, allow more Ca^{++} influx, and increase the amount of neurotransmitter released.

The demonstration of heterosynaptic modulation of neurotransmitter release within the respiratory CPG suggests that mechanisms of synaptic plasticity characterized in invertebrate preparations such as *Aplysia* are important in the control of breathing. Such modulation could participate in respiratory phenomena such as poststimulus potentiation (i.e., respiratory afterdischarge) (9) as well as longer-term adaptations such as acclimatization to high altitude (8). In fact, it has been hypothesized that the cellular mechanisms responsible for short-term potentiation in *Aplysia* sensory neurons (e.g., presynaptic facilitation and cAMP) underlie the potentiation of respiratory activity observed following a facilitatory stimulus (8,9). The demonstration that such cellular mechanisms are present in the respiratory CPG provides us with a firm basis for pursuing such hypotheses at the level of the whole organism.

References

1. Feldman JL. In: Bloom FE, ed. Handbook of Physiology; Section 1: The Nervous System; Vol. IV. Washington, DC: American Physiological Society, 1986:424–463.

2. Feldman JL, Smith JC. Cellular mechanisms underlying modulation of breathing pattern in mammals. Ann NY Acad Sci 1989; 563:114–130.
3. Smith JC, Ellenberger HH, Ballanyi K, et al. Pre-Botzinger complex: A brainstem region that may generate respiratory rhythm in mammals. Science 1991; 254:726–729.
4. Sather K, Hilaire G, Monteau R. Dorsal and ventral respiratory groups of neurons in the medulla of the rat. Brain Res 1987; 419:87–96.
5. Monteau R, Hilaire G. Spinal respiratory motoneurons. Prog Neurobiol 1991; 37: 83–144.
6. Speck DF, Feldman JL. The effects of microstimulation and microlesions in the ventral and dorsal respiratory groups in medulla of cat. J Neurosci 1982; 2:744–757.
7. Dekin MS. Comparative neurobiology of invertebrate motor networks: Implications for the control of breathing in mammals. In: Haddad GG, Faber J, eds. Developmental Neurobiology of the Lung. New York: Marcel Dekker, 1991:95–143.
8. Eldridge FL, Milhorn DE. Oscillation, gating, and memory in the respiratory control system. In: Cherniack NS, Widdecombe JG, eds. Handbook of Physiology—Respiration II: Part I, Control of Breathing. Bethesda, MD: American Physiological Society, 1986:93–114.
9. Wagner PG, Eldridge FL. Development of short-term potentiation of respiration. Resp Physiol 1991; 83:129–140.
10. Clarke RW, Ford TW, Harris SM, Taylor JS. Thyrotropin-releasing hormone, cholecystokinin and endogenous opioids in the modulation of spinal reflexes in the rabbit. Neuropharmacology 1988; 27:1279–1284.
11. Lacey G, Nistri A, Rhys-Maitland ER. Large enhancement of excitatory post-synaptic potentials and currents by TRH in frog spinal motoneurones. Brain Res 1989; 488:80–88.
12. Behbehani MM, Pun RYK, Means ED, Anderson DK. TRH has profound pre-synaptic action on cultured spinal cord neurons. Synapse 1990; 6:169–174.
13. Hokfelt T, Fuxe K, Johansson O, et al. Thyrotropin-releasing hormone containing nerve terminals in certain brainstem nuclei and in the spinal cord. Neurosci Lett 1975; 1:133–139.
14. Johansson O, Hokfelt T, Pernow B, et al. Immunohistochemical support for three putative transmitters in one neuron: Coexistence of 5-hydroxytryptamine, substance P and thyrotropin releasing-like immunoreactivity in medullary neurons projecting to the spinal cord. Neuroscience 1981; 6:1857–1881.
15. Lalley NA, Harrison NL, Teyler TJ. Effects of baclofen and GABA on different types of medullary respiratory neurons. Brain Res 1986; 376:392–395.
16. Lalley PM. Biphasic effects of baclofen on phrenic motoneurons: Possible involvement of 2 types of GABA receptors. JPET 1983; 226:616–624.
17. Aghajanian GK, Rasmussen K. Intracellular studies in facial nucleus illustrating a simple new method for obtaining viable motoneurons in adult rat brain slices. Synapse 1989; 3:331–338.
18. Andrade R, Malenka RC, Nicoll RA. A G-protein couples serotonin and $GABA_B$ receptors to the same channels in the hippocampus. Science 1986; 234:1261–1265.
19. Bayliss DA, Viana F, Berger AJ. Mechanisms underlying excitatory effects of TRH on rat hypoglossal motoneurons in vitro. J Neurophysiol 1992; 68:1733–1745.

20. Wagner PG, Dekin MS. $GABA_b$ receptors are coupled to a barium-insensitive outward rectifying potassium conductance in premotor respiratory neurons. J Neurophysiol 1993; 69:286–289.
21. Innis RB, Nestler EJ, Aghajanian GK. Evidence for G protein mediation of serotonin and $GABA_B$ induced hyperpolarization of rat dorsal raphe neurons. Brain Res 1988; 459:27–36.
22. Allerton CA, Boden DR, Hill RG. Actions of $GABA_B$ agonist baclofen on neurons in deep dorsal horn of the rat spinal cord in vitro. Br J Pharmacol 1989; 96:29–38.
23. Lambert NA, Harrison NL, Teylor YJ. Baclofen-induced disinhibition in area CA1 of rat hippocampus is resistant to extracellular barium. Brain Res 1991; 547: 349–352.
24. Siegelbaum SA, Camardo JS, Kandel ER. Serotonin an cAMP close single K^+ channels in *Aplysia* sensory neurones. Nature (Lond) 1982; 299:413–417.
25. Dekin MS, Sun Y, Wagner PG. Thyrotropin-releasing hormone (TRH) inhibits outward rectifying $GABA_B$ activated K^+ channels via a protein kinase C dependent pathway. Neurosci Abstr 1993; 19:1194.
26. Kandel ER, Schwartz JH. Molecular biology of learning: Modulation of transmitter release. Science 1982; 218:433–443.
27. Hawkins RD, Kandel ER, Siegelbaum SA. Learning to modulate transmitter release: Themes and variations in synaptic plasticity. Annu Rev Neurosci 1993; 16: 625–665.

21

The Role of Glycine in the Pressor Response to NMDA Injected into the Subretrofacial Nucleus (SRFN) of the Cat

BARBARA BADIO

National Institutes of Health
Bethesda, Maryland

PHILIP J. GATTI

College of Medicine
Howard University
Washington, D.C.

This study examined the role of glycine in the pressor response to the glutamate agonist *N*-methyl-D-aspartic acid (NMDA) microinjected into the subretrofacial nucleus (SRFN) of the chloralose-anesthetized cat. NMDA evoked a dose-dependent increase in arterial blood pressure upon injection into the SRFN. This response could be repeated 30 min later. Bilateral injection of the glycine antagonist 3-amino-1-hydroxy-2-pyrrolidone (HA-966) into the SRFN 5 min before the second NMDA injection significantly reduced the pressor response observed following NMDA injection. HA-966 had no effect of its own on resting levels of arterial blood pressure or heart rate. HA-966 did not attenuate the pressor response seen following the injection of bromowillardiine, a non-NMDA receptor agonist, into the SRFN. The results suggest that there is a glycine site associated with NMDA receptors located on neurons in the SRFN and that activation of this site is important for the full expression of the NMDA-induced increase in arterial pressure in the cat.

I. Introduction

The *N*-methyl-D-aspartic acid (NMDA) receptor has been the most extensively studied subtype of the excitatory amino acid (EAA) receptors (1). Many studies have shown that there are a number of functional sites on the NMDA receptor complex for endogenous as well as exogenous ligands. These include (a) a transmitter site that binds glutamate, NMDA, and competitive antagonists such as 3-(2-carboxypiperazin-4-yl)-propyl-1-phosphonic acid (CPP); (b) a cotransmitter site that binds glycine; (c) a polyamine-sensitive site; (d) a site within the channel that binds phencyclidine, MK-801, and related compounds; (e) a voltage dependent Mg^{++} binding site; and (f) and an inhibitory divalent cation site that binds Zn^{++} (2,3).

The glycine site on the NMDA receptor was discovered in 1987, when Johnson and Ascher observed an increase in the frequency of NMDA-induced channel opening in the presence of glycine (4). They also found that this action of glycine was not blocked by the classic glycine antagonist strychnine. These findings were later confirmed by other researchers (5,6). Later studies by Foster and Kemp (7) showed that 3-amino-1-hydroxy-2-pyrrolidone (HA-966) antagonized NMDA receptors through a selective interaction with the glycine site. Therefore, HA-966 was classified as a noncompetitive antagonist of the NMDA receptor but a competitive antagonist of the glycine site on the NMDA receptor.

Since the discovery of a strychnine-insensitive glycine binding site on the NMDA receptor, its role in the function of the receptor has been the focus of many studies. Kleckner and Dingledine (5) suggested that glycine was absolutely necessary for NMDA receptor activation by glutamate. Therefore, glycine was believed to be a cotransmitter with glutamate at the NMDA receptor. However, later studies by Mayer and coworkers (6) proposed that glycine served predominantly as a positive modulator of the NMDA receptor by reducing the desensitization of NMDA receptor activation. Still, the definitive role of glycine in NMDA receptor function is still uncertain.

Mitra and coworkers (8) were the first to demonstrate that application of NMDA to the intermediate area on the ventral medullary surface of the cat evoked an increase in arterial blood pressure. This pressor response is the result of stimulating neurons located in the subretrofacial nucleus located beneath the intermediate area (9). In this study, we demonstrate that the antagonist of the glycine site located on the NMDA receptor, HA-966, reduced NMDA-induced responses in the subretrofacial nucleus (SRFN) of the cat.

II. Methods

Eighteen adult cats of either sex weighing 2 to 4 kg were used. The animals were sedated with an intramuscular injection of ketamine/acepromazine and then anesthetized with a 70 mg/kg solution of reagent-grade alpha-chloralose. This dose of ketamine did not attenuate responses seen with NMDA (10). A femoral artery and vein were cannulated for the measurement of arterial blood pressure and the systemic administration of drugs, respectively. Mean arterial pressure (MAP) was calculated from systolic and diastolic pressures. Body temperature was monitored rectally and maintained between 37 and 38°C using a heat lamp. Heart rate (HR) was determined directly by counting pulse tracings from blood pressure recordings using the fast trace (25 mm/s).

The trachea was cannulated and the animals were artificially ventilated with room air. Recordings were made continuously on a Narco four-channel recorder (desk model, type DMP-4A).

A longitudinal midline incision was then made in the ventral surface of the neck. The trachea and esophagus were retracted to expose the prevertebral muscles. The prevertebral muscles were removed from the basal plate of the skull. The basal plate was then removed to create an opening of approximately 15 by 9 mm over the medullary surface. The arachnoid and dura membranes were cut and reflected to expose the ventral surface of the medulla (VMS) and the region of the intermediate chemosensitive area. Bone wax and gel foam were used to control bleeding. An aspirator was used when needed to prevent the accumulation of blood and cerebrospinal fluid on the VMS.

Animals were then placed in a Kopf cat stereotaxic head holder in the supine position. A micromanipulator was attached to the head holder and a micropipette was fastened to the manipulator. Each division on the pipette represented 1 μL. Each microliter division was further subdivided into 50 nL divisions using formaline tape (type 9008B) attached to the pipette. Drugs were filtered using a syringe filter (0.25 μm, Fisher Scientific Co.). Micropipettes were then filled from the tip, using a vacuum.

The solutions (50 nL) were injected into the SRFN using a nanopump (WPI, Model PV800) with a pressure regulator. The SRFN in the brainstem of the cat is located 1.5 mm caudal to the caudal edge of the trapezoid body or 6.5 mm caudal to the foramen cecum, 4 mm lateral to the basilar artery, and 1 mm below the VSM.

Control values were obtained using the agonist, NMDA, or bromowillardiine (BW). Following the first injection, the animal was allowed to stabilize. Bilateral injections of either saline or the antagonist, HA-966 were made 30 min later. Readings were taken before and after injection of the antagonist to determine the effect of the antagonist on the resting level of arterial pressure if any. A second injection of NMDA or BW was made 5 to 10 min after injection of the antagonist.

Drugs used included *N*-methyl-D-aspartic acid and HA-966 (Research Biochemicals Inc., Natick, MA); bromowillardiine (Tocris Neuramin, Bristol, England); alpha-chloralose (K&K Laboratories, Cleveland, OH); ketamine (Upjohn Company, Kalamazoo, MI).

Statistical tests included the paired and unpaired Student's *t*-test for comparing means and linear regression analysis for the analysis of the dose-response studies. The criterion for statistical significance was $P<0.05$. All values are expressed as the mean ± SEM.

Table 1 Effect of Microinjection of NMDA into the Subretrofacial Nucleus of Cat on Mean Arterial Pressure[a]

NMDA	MAP_B (mm Hg)	MAP_A (mm Hg)	Difference (mm Hg)
0.1 m*M*	138 ± 10	150 ± 15	12 ± 5
1 m*M*	144 ± 13	191 ± 9	47 ± 5*
10 m*M*	122 ± 7	191 ± 7	69 ± 6*

[a]Values represent the mean ± SEM; $N = 4$; significantly different from response before NMDA at $*P < 0.01$ using the paired Student's *t*-test. *Abbreviations*: MAP_B = mean arterial pressure before injection of NMDA; MAP_A = mean arterial pressure after injection of NMDA; Difference = change in arterial pressure.

III. Results

Table 1 shows the effect on mean arterial pressure of unilateral injection of various concentrations of NMDA (0.1, 1.0, and 10 m*M*) into the SRFN. The results show a dose-dependent increase in mean arterial pressure. There was a significant ($P < 0.01$) increase in MAP at concentrations of 1 and 10 m*M* NMDA. The onset of the pressor response caused by injection of NMDA into the SRFN occurred immediately at all concentrations. The time to peak effect of arterial pressure was approximately 1 to 3 min and pressor response lasted for approximately 10 to 15 min. No significant effect on HR was observed. For example, prior to the injection of the 10-m*M* concentration of NMDA, HR was 210 ± 12.2 bpm; following injection, HR was 195 ± 19.4 bpm. Desensitization of the receptors did not occur when arterial pressure was allowed to stabilize. NMDA increased MAP to a similar level when reinjected 30 min after the initial dose (Fig. 1).

Bilateral injection of 10 m*M* HA-966 into the SRFN produced no significant change in mean arterial pressure (MAP_B = 129 ± 7.5 mm Hg; MAP_A = 126 ± 8.4 mm Hg). This suggests that blockade of the glycine sites does not affect resting level of arterial pressure.

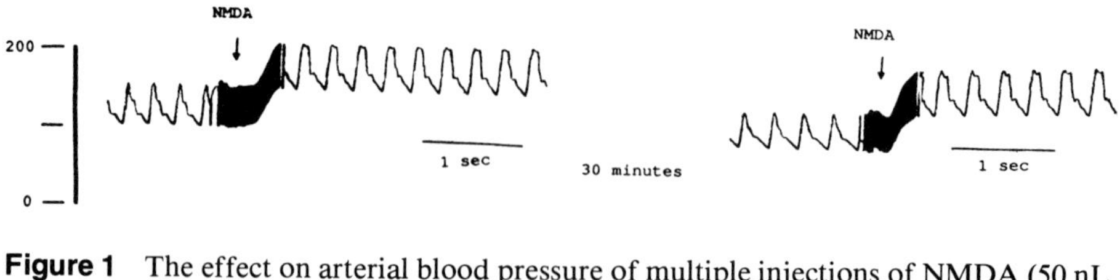

Figure 1 The effect on arterial blood pressure of multiple injections of NMDA (50 nL of a 10-m*M* solution) into the SRFN of the cat.

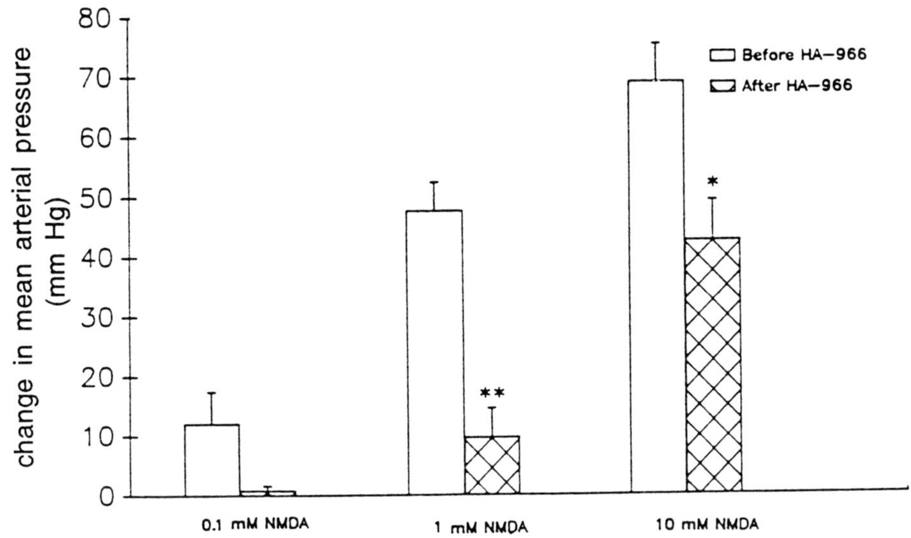

Figure 2 The effect on the pressor response to various doses of NMDA of bilateral injection of HA-966 (50 nL of a 10-m*M* solution) into the SRFN. $N=4$ for each dose. $*=P<0.05$ and $**$ $P<0.01$ using the paried Student's *t*-test.

Figure 2 shows the effect of bilateral injection of HA-966 on NMDA-induced increase in MAP. Using each animal as its own control, a second injection of NMDA was made into the SRFN. As shown in Figure 2, 10 m*M* HA-966 caused a significant reduction in the pressor response due to injection of 1 and 10 m*M* NMDA into the SRFN. With time, the effect of HA-966 gradually wore off resulting in a gradual increase in the response to NMDA.

Table 2 Effect of Bilateral Injection of HA-966 on the Pressor Response to Bromowillardiine Injected into the Subretrofacial Nucleus of the Cat[a]

	MAP before BW (mm Hg)	MAP after BW (mm Hg)	Difference (mm Hg)
In the absence of HA-966	133 ± 13	176 ± 11	43 ± 8
In the presence of HA-966	139 ± 12	186 ± 9	47 ± 8

[a]Bromowillardiine (1 m*M*) was injected unilaterally into the subretrofacial nucleus before and after bilateral injection of 10 m*M* HA-966. $N=3$; Data represent the mean ± SEM.

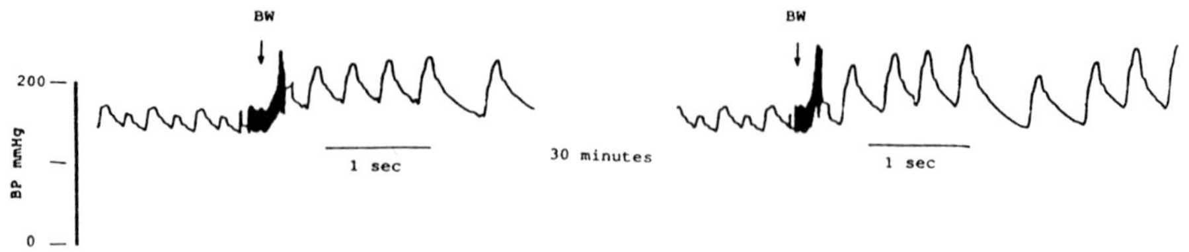

Figure 3 The effect on arterial blood pressure of multiple injections of BW (50 nL of a 1-m*M* solution) into the SRFN of the cat.

To test the selectivity of HA-966 for the NMDA receptor, experiments were done using a nonselective non-NMDA receptor agonist, bromowillardiine (BW), Table 2. Control values were obtained with injection of BW into the subretrofacial nucleus. Like NMDA, BW did not cause desensitization of the receptors when injected twice into the same site 30 min following the initial dose (Fig. 3). BW caused a significant increase in MAP ($P < 0.05$). The BW-induced pressor response was not significantly affected following the bilateral injection of the glycine antagonist HA-966.

IV. Discussion

The purpose of this study was to examine the effect of pharmacological manipulation of the glycine site located on the NMDA receptor. We attempted to determine whether a competitive antagonist of the strychnine-insensitive glycine site, HA-966, would modify the NMDA-induced pressor response in the SRFN of the cat.

The results show that microinjection of different concentrations of NMDA produced dose-dependent increases in MAP with no consistent change in HR. This effect was repeatable after 30 min. Thus, we used each animal as its own control. In the presence of HA-966, the pressor response to NMDA was significantly reduced at the two higher doses. The lowest dose appeared to be attenuated, but this effect was not statistically significant. This is consistent with previous studies suggesting that a glycine site is associated with the NMDA receptor complex (11–15). It also suggests that glycine binding to the NMDA receptor is necessary for the full expression of agonist-induced action. This effect of HA-966 was selective for the NMDA receptor, since the pressor response to BW, a non-NMDA receptor agonist, was not affected by this dose of compounds like HA-966.

HA-966 (10 m*M*) injected bilaterally into the SRFN produced no effect on arterial pressure. Since this dose of HA-966 blocked the pressor effect of exogenous NMDA, this finding might indicate that there is little or no tonic release of glutamate in the SRFN acting on NMDA receptors involved in cardiovascular control. This possibility has been suggested by Abrahams (16). Abrahams suggests that non-NMDA receptors in the SRFN are more important than NMDA receptors in controlling cardiovascular functions at this site. This conclusion was based on studies utilizing antagonists of both NMDA and non-NMDA receptors. It was noted in the study that non-NMDA receptors antagonists 6-Cyano-7-nitroquinoxaline-2,3-dione (CNQX) and gamma-D-Glutamylaminomethylsulphonic acid (GAMS) produced a greater fall in arterial pressure per se when injected into the SRFN than the competitive NMDA receptor antagonist CPP. Thus, glutamate is being tonically released at this site, and it interacts primarily with non-NMDA receptors. It can be postulated that the NMDA receptors in this area might be involved in the reflex control of MAP but not in maintaining resting arterial pressure. This has yet to be demonstrated.

The results obtained in this investigation demonstrates the necessary involvement of glycine in the pressor response to exogenously administered NMDA in the SRFN of the cat. It also shows that the NMDA receptors can be regulated in vivo using compounds that modulate the NMDA receptor by binding to specific binding sites on that receptor. This might provide a new avenue for the control of some of the excitotoxic actions of EAAs that act through the NMDA receptor. Compounds like NMDA might be of potential benefit in neurological diseases such as Alzheimer's disease, Huntington's disease, and AIDS dementia.

Acknowledgment

This work was supported by the American Heart Association/Nation's Capital Affiliate.

References

1. Collingridge GL, Lester RA. Excitatory amino acid receptors in the vertebrate central nervous system. Pharmacol Rev 1989; 40:143–210.
2. Monaghan DT, Bridges RJ, Cotman CW. The excitatory amino acid receptors: Their classes, pharmacology, and distinct properties in the function of the central nervous system. Annu Rev Pharmacol Toxicol 1989; 29:365–402.
3. Williams K, Romano C, Dichter MA, Molinoff PB. Modulation of the NMDA receptor by polyamines. Life Sci 1991; 48:469–498.
4. Johnson JW, Ascher P. Glycine potentiates the NMDA response in cultured mouse brain neurons. Nature 1987; 325:529–531.
5. Kleckner NW, Dingledine R. Requirement for glycine in activation of NMDA-receptors expressed in Xenopus oocytes. Science 1988; 241:835–837.
6. Mayer ML, Westbrook GL, Vyklicky L. Sites of antagonist action on N-methyl-D-aspartic acid receptors studied using fluctuation analysis and a rapid perfusion technique. J Neurophysiol 1988; 60:645–663.
7. Foster AC, Kemp JA. HA-966 antagonizes N-methyl-D aspartate receptors through a selective interaction with the glycine modulatory site. J Neurosci 1988; 9:2191–2196.

8. Mitra J, Prabhakar NR, Overholt JR, Cherniak NS. Respiratory and vasomotor effects of excitatory amino acids on the ventral medullary surface. Brain Res Bull 1987; 18:681–684.
9. Badio B, Gatti PJ. Investigation into the role of glycine in the pressor response to N-methyl-D-aspartic acid (NMDA) injected into the subretrofacial nucleus (SRFN) of the cat. Soc Neurosci 1992; 18:1016.
10. Badio B. An investigation into the role of glutamate receptors in the control of the circulation. PhD dissertation, Howard University, Washington, DC, 1993.
11. Drejer J, Sheardown M, Nielsen EO, Honore T. Glycine reverses the effect of HA-966 on NMDA responses in cultured rat cortical neurons and in chick retina. Neurosci Lett 1989; 98:333–338.
12. Stirling JM, Cross AJ, Green AR. The binding of [3H]-thienyl cyclohexylpiperidine ([3H]-TCP) to the NMDA-phencyclidine receptor complex. Neuropharmacology 1989; 28:1–7.
13. Watson GB, Bolanowski MA, Baganoff MP, et al. Glycine antagonist action of 1-aminocyclobutane-1-carboxylate (ACBC) in Xenopus oocytes injected with rat brain mRNA. Eur J Pharmacol 1989; 167:291–294.
14. Wood PL, Emmett MR, Rao TS, et al. In vivo modulation of the N-methyl-D-aspartate receptor complex by D-serine: Potentiation of ongoing neuron A1 activity as evidenced by increased cerebellar cyclic GMP. J Neurochem 1989; 53:979–981.
15. Procter AW, Stratmann GC, Francis PT, et al. Characterization of the glycine modulatory site of the N-methyl-D-aspartate receptor ionophore complex in human brain. J Neurochem 1991; 56:299–310.
16. Abrahams TP. Evidence for hind brain excitatory amino acid systems controlling cardiovascular and respiratory function. PhD dissertation, Georgetown University, Washington, DC, 1991.

22

Significance of Peptidases in Regulation of Respiration

Role of Neutral Endopeptidase in the Peripheral and Central Chemoreceptors

GANESH K. KUMAR and MUSA A. HAXHIU

School of Medicine
Case Western Reserve University
Cleveland, Ohio

I. Introduction

Adjustments of ventilatory patterns in response to environmental perturbations are largely mediated by neural pathways associated with the peripheral and central chemoreceptors. The peripheral chemoreceptors, primarily carotid bodies, are responsible for the increased ventilation during hypoxia (1–5), whereas neurons near the ventrolateral surface of the medulla oblongata, often referred to as "central chemoreceptors," play crucial roles in mediating ventilatory responses to hypercapnia (6–10). The precise mechanisms underlying the transduction of chemical signals to ventilatory changes are not fully understood, but neuroactive substances are thought to be involved in the transmission of natural stimuli.

In recent years, several neuropeptides have been identified in the carotid body (11–20) and in the neurons of the ventrolateral medullary surface (VMS) (21–24). Several physiological studies have demonstrated that neuropeptides such as substance P (25–28) and enkephalins (25,29–31) influence the ventilatory responses by acting either on carotid bodies or on the VMS neurons. Given their abundance and possible physiological role(s) in chemoreceptor tissues, detailed information on inactivation and/or regulation of neuropeptides will be of great value in further elucidating their roles in chemical control of breathing. In general, the physiological actions of neuropeptides are terminated by hydrolysis involving specific cell-surface peptidases (32–34). In this chapter, a brief review of our recent results in cats on the characterization and physiological significance of endopeptidase(s) involved in the hydrolysis of neuropeptides in the carotid body and in the VMS will be presented.

II. The Carotid Chemoreceptors

The carotid body, along with its numerous interglomic fibers, contains groups of glomus cells (also known as type I cells) that are of neural crest origin and sustentacular cells (referred as type II cells) resembling glia in their morphology (35–37). Type I cells along with the apposing nerve endings are believed to constitute the chemoreceptor complex (25). Immunocytochemical studies have shown that several neuropeptides occur in association with certain structures of the carotid body. For example, as shown in Table 1, substance P-like immunoreactivity was found to occur both in type I cells and in the nerve fibers of the cat carotid body (13,14). Also, the interstitial region between certain type I cells was reported to be positive for substance P-like immunoreactivity (13). By contrast, peptides like enkephalin and atrial natriuretic peptide occur exclusively in type I cells (18,25,35), whereas galanin and CGRP are colocalized to certain nerve fibers (16). Other examples of colocalization of neuropeptides in the carotid body include substance P/neurokinin A (14) and substance P/CGRP (19). These results suggest that the mammalian carotid body contains several neuropeptides occurring in specific cellular compartments and frequently colocalized with other neuropeptides.

Table 1 Neuropeptides of the Mammalian Carotid Body

Neuropeptides	Species	Cellular localization
Enkephalin	Cat, rabbit, human, rat	Type I cells
Substance P	Cat, rabbit, human, rat	Nerve fibers and type I cells
CGRP	Human, guinea pig, rat	Mostly nerve fibers
Galanin	Human, guinea pig, rat	Nerve fibers
ANP	Rabbit	Type I cells
VIP	Cat	Nerve endings apposed to blood vessels and ganglion cells
Neuropeptide Y	Human	ND
Neuropeptide YY	Human	ND

Abbreviations: CGRP, calcitonin gene related peptide; ANP, atrial natriuretic peptide; VIP, vasoactive intestinal polypeptide; ND, not determined.

A. Peptidases of the Carotid Body

We have investigated the role of peptidases in the inactivation of neuropeptides in the carotid body. For this purpose, cell-free extracts of the cat carotid bodies ($n = 24$) were made in 50 m*M* HEPES buffer, pH 7.4. Synthetic peptide substrates containing either 4-methoxy-2-naphthylamine or 4-methyl-coumaryl-7-amide chromogenic group were used for monitoring the activities of peptidases in the chemoreceptor tissue. A comparison of activities suggested that aminopeptidase, carboxypeptidase, trypsin, chymotrypsin, and neutral endopeptidase (NEP)-like activities occur in the carotid body. Notably, NEP-like activity (3.8×10^3 pmol/h/mg protein), occurring at several orders of magnitude higher than other peptidases, appears to be the major endopeptidase in the chemoreceptor tissue. However, other neuropeptide-degrading activities—like proline-specific endopeptidase and angiotensin converting enzyme—were not detected. It is possible that these activities occur at levels well below the detection limit of our assay.

In accordance with the above observations, our in vitro studies (38) demonstrated that hydrolysis of substance P by cell-free extracts of the carotid body occurs via initial cleavage of phe^7-phe^8 peptide bond that was inhibited ($\sim 90\%$) by metal ion chelators and by phosphoramidon, a well known inhibitor of NEP. It has been shown by others that neuropeptides like enkephalin (32), atrial natriuretic peptide (39), and endothelin (40) are rapidly hydrolyzed by NEP. It is possible that the broad in vitro specificity displayed by NEP may be reflected in vivo as well (41). It is, therefore, reasonable to suggest that NEP-like activity is involved in the degradation of substance P and possibly other peptides in the carotid body.

B. Biochemistry of Neutral Endopeptidase-like Activity of the Carotid Body

Neutral endopeptidase (NEP; E.C.3.4.24.11), also known as enkephalinase, has been shown to occur widely in several peripheral tissues (42–46) and in many regions of the central nervous system (47–49). NEP is a cell-surface metalloenzyme and hydrolyzes several small neuropeptides (33). It cleaves the amino terminal side of the peptide bonds involving hydrophobic amino acids with enzyme specificity resembling that of the bacterial protease thermolysin (50).

Unlike that of the kidney and brain tissues (39,47–49) the NEP-like activity of the carotid body occurs in two forms: a common membrane-bound form and a unique soluble active form (51). A third NEP form, which represents less than 10% of the total activity, could be extracted only in the presence of neutral detergents. A soluble NEP in addition to the membrane-bound form has been described in lymphoid tissues (46).

NEP-like activity in the cell-free extracts of the carotid body was analyzed for its inhibition and immunological characteristics. While metal-ion chelators like EDTA and 1,10-phenanthroline fully inhibited NEP-like activity of the carotid body in the concentration range of 10^{-3} *M*, both phosphoramidon and thiorphan inhibited this activity at much lower concentrations (10^{-9} to 10^{-7} *M*). Other protease inhibitors, even at 10^{-6} *M* concentrations, did not show any significant inhibitory effects.

The immunological properties of NEP-like activity of the carotid body was assessed using two antibody preparations: a human monoclonal antibody (hNEP) and a rat polyclonal antibody (rNEP). NEP-like activity of the cat carotid body was quantitatively immunoprecipitated by hNEP, whereas rNEP effected only partial precipitation (~20%). These observations suggest that inhibition characteristics of NEP-like activity of the cat carotid body are quite similar to NEP characterized previously from other tissues (32), but the carotid body NEP appears to be unique with respect to epitope or folded structures.

C. Physiological Role of NEP-like Activity of the Carotid Body

The functional significance of the occurrence of NEP-like activity of the carotid body was assessed in cats in vivo (51). The chemoreceptor responses of the carotid body to hypoxia (12% O_2 + N_2; 36 ± 8 mm Hg) and hypercapnia (7% CO_2 in 93% O_2; 60 ± 3 mm Hg) were monitored before and after administration of phosphoramidon, an inhibitor of NEP, close to the carotid body. In comparison to the prephosphoramidon control, the hypoxic response of the carotid body was markedly potentiated (~59%) following administration of phosphoramidon (for 3 min at a concentration of 46 μM/min). In seven experiments, phosphoramidon consistently potentiated the hypoxic response in a reversible manner (51). By contrast, the hypercapnic response to phosphoramidon was inconsistent and was comparable to that of the prephosphoramidon control value. Taken together, these results suggest that inhibition of NEP-like activity of the carotid body by phosphoramidon might have caused an elevation of certain endogenous peptide substrate levels, possibly an excitatory neuropeptide like substance P, thus influencing the hypoxic response. Furthermore, the absence of a consistent effect of phosphoramidon suggests that peptidergic mechanism may not contribute significantly to the hypercapnic responses in the peripheral chemoreceptor tissue. Alternatively, if peptidergic mechanisms involving certain neuropeptides are crucial for the expression of hypercapnic response in the carotid body, then peptidase(s) other than NEP may be involved in their inactivation.

D. Cellular Localization of NEP-like Activity of the Carotid Body

The precise localization of NEP-like activity to one or another cell structure of the carotid body will significantly enhance our understanding of the functional roles of NEP in the chemoreceptor tissue. Toward this goal, we have initiated immunolabeling studies using thin sections of the cat carotid body.

The carotid bodies were removed from anesthetized animals spontaneously breathing room air. Prior to cryosectioning, the tissues were immersion-fixed overnight in 0.2 *M* glutaraldehyde and 3% paraformaldehyde in 0.1 *M* phosphate buffer, pH 7.2 at 4°C. The sections were immunolabeled by incubating first with hNEP monoclonal antibody preparation, followed by a secondary antibody conjugated with FITC. Appropriate controls lacking one or the other antibody were included to assess specificity of NEP labeling. Fluorescent microscopic analysis revealed a "punctate" type of labeling along the interstitial region of type I and II cells of the carotid body. NEP immunolabeling of either intracellular components or cell membranes was less pronounced.

Toward further understanding NEP distribution at the ultrastructural level, sections of the carotid body immunolabeled with hNEP and gold-conjugated protein A were viewed by electron microscopy. In well-preserved sections, specific NEP immunolabeling at interstitial regions of certain type I cells was consistently observed, confirming the results of the light microscopic studies. In view of the potential limitations of immunolabeling techniques that include antigen accessibility, antibody specificity, and epitope recognition (sequence-specific versus conformational), the possibility remains that not all the NEP cellular sites have been identified. Therefore it is deemed essential to explore alternative autoradiographic methods that utilize radiolabeled inhibitors (47) in order to further elucidate NEP distribution in the carotid body.

The above-described intercellular distribution at least represents one of the NEP locations in the carotid body. Previously, Chen et al. (13) demonstrated that certain interstitial regions in the carotid body are positive for substance P-like immunoreactivity. It is conceivable that the intercellular NEP form of the carotid body by virtue of its close proximity may be involved in the inactivation of substance P of the interstitial origin.

E. Regulation of NEP-like Activity of the Carotid Body

Effect of Hypoxia

Previously, we demonstrated that following 1 h of sustained hypoxia there is an increase in substance P and neurokinin-A levels in the cat carotid body (14). This increase may arise from an enhanced de novo synthesis of tachykinins or by a mechanism involving their decreased degradation.

Given the fact that NEP is the major endopeptidase of the cat carotid body, we have assessed the changes, if any, in NEP activity of the carotid body ($n = 7$) during 1 h of sustained hypoxia (PO_2 41 ± 2 mm Hg) under isocapnic conditions (PCO_2 37 ± 3 mm Hg). Compared to room air controls, hypoxia, on average, caused ~30% decrease in NEP activity of the carotid body. However, NEP activity in the superior cervical ganglion was unaffected by acute hypoxia (52). In the rabbit, it has been shown that during intermittent hypoxia both the substance P and enkephalin content of the carotid body decreased (53).

Taken together, these results suggest that the effects of hypoxia on neuropeptide levels and on the activity of NEP in the carotid body are quite complex and that differential regulatory mechanisms may be evoked, depending on the severity and duration of the hypoxic stress, in a tissue-specific manner.

Effect of Anions

Multivalent anions like PO_4^{3-} are shown to inhibit NEP purified from bovine brain (54). We have tested the possibility that cellular phosphates may serve as modulators of NEP activity in the carotid body. The effects of inorganic phosphates and citrates were examined by assessing the rate of in vitro substance-P degradation by cell-free extracts of the carotid body while maintaining the reaction pH at 7.4. The hydrolysis of substance P was monitored by reverse-phase high-performance liquid chromatography (HPLC), as described in Ref. 38. From the results shown in Table 2, it is evident that inorganic phosphates inhibited the hydrolysis of substance P. A maximum inhibition was observed at 10 m*M* PO_4^{3-} (~45%), which may be attributed to anion binding at certain regulatory sites, possibly to the metal center on NEP. By contrast, in the presence of citrate (1 m*M*), more than 90% of substance P hydrolysis was inhibited. This inhibition may be due to the chelation of zinc at NEP active sites. Although the effects of hypoxia on the inorganic phosphates of the carotid body are not fully understood, it is reasonable to hypothesize that phosphate anions are involved in the downregulation of NEP by acute hypoxia via mechanism(s) yet to be defined.

III. The Ventrolateral Medullary Surface

The ventrolateral medullary surface is considered to be the putative chemosensory region mediating ventilatory responses to CO_2 (6–10). Morphological studies have demonstrated the existence of peptidergic neurons and nerve fibers

Table 2 Effect of Anions on the Hydrolysis of Substance P by Carotid Body Extracts

Sample	Rate of substance P hydrolysis: fmol/min/mg protein	Rate of substance P hydrolysis: Percent of control
1. Cell extract + SP (control)	$54 \pm 4 \times 10^3$	100
2. Experiment #1 + phosphoramidon	$4 \pm 1 \times 10^3$	7
3. Experiment #1 + EDTA	—	0
4. Experiment #1 + 1 m*M* PO_4^{3-}	$49 \pm 3 \times 10^3$	91
5. Experiment #1 + 5 m*M* PO_4^{3-}	$42 \pm 2 \times 10^3$	78
6. Experiment #1 + 10 m*M* PO_4^{3-}	$30 \pm 5 \times 10^3$	56
7. Experiment #1 + 1 m*M* citrate	$5 \pm 1 \times 10^3$	9

in the ventrolateral medulla of the brainstem (23,24). It is likely that these peptidergic neurons, in addition to the pharmacologically demonstrated cholinergic and GABAergic (55,56) neuronal pathways, may contribute to regulation of respiratory-related activity and parasympathetic outflow to the airways (57).

A. Evidence that NEP-like Activity Is Involved in the Degradation of Neuropeptides of the VMS

We have investigated the peptidases of the VMS that might be involved in the inactivation of neuropeptides like substance P. Slices (200 μm in thickness) of the VMS of the cat brainstem, 6 to 9 mm caudal from foramen cecum and 2 mm lateral to midline, were used for cell-free extract preparation. A membrane-enriched fraction of the VMS was prepared by following the procedure of Maeda et al. (58). The protein content of the membrane-enriched fraction of VMS was determined by the Rose Bengal method (59).

NEP-like activity was determined by following the hydrolysis of the synthetic substrate glutaryl-alanyl-alanyl-phenylalanyl-4-methoxy-2-naphthylamide, as outlined in the modified procedure of Sonnenberg et al. (60). The membrane-enriched fraction of the VMS rapidly hydrolyzed the synthetic substrate at a rate of 2.8×10^3 pmol/h/mg protein. More than 90% of the activity was inhibited by phosphoramidon and thiorphan at 10^{-8} *M* concentrations. Moreover, metal ion chelators, in the concentration range of 0.1 to 0.5 m*M*, fully inhibited this activity. These results provide evidence that the measured activity indeed represents NEP but not other endopeptidases.

The ability of a membrane-enriched fraction of the VMS to hydrolyze neuropeptides in vitro was examined using substance P as a model peptide. The degradation products of substance P that are generated following incubation with a membrane-enriched fraction of the VMS were analyzed by reverse-phase HPLC. The peptide profiles were compared with the elution profiles of intact substance P (1-11) and its fragments (1-5; 1-7; 1-9; 5-11, and 8-11). The results showed that substance P is hydrolyzed at a rate of 31.5 pmol/min/mg protein. Significant accumulation of peptide fragments of substance P was not observed, suggesting that multiple peptidases, associated with membrane fraction of the VMS, are involved in the hydrolysis of substance P.

Toward identifying the major peptidases of the VMS, the effect of various protease inhibitors on substance P hydrolysis was determined. The results are summarized in Table 3. Phosphoramidon or thiorphan at 10^{-6} *M* concentration only partially inhibited (~45%) the hydrolysis, whereas a slightly higher protection (~55%) was observed in the presence of either EDTA or 1,10-phenanthroline. However, the hydrolysis was only marginally affected by either chymostatin or leupeptin. Based on the above results, it is concluded that more than one peptidase including NEP-like activity is involved in the hydrolysis of substance P in the VMS. Thus, NEP along with other endopeptidases may

Table 3 Effect of Protease Inhibitors on the Hydrolysis of Substance P by Membrane Fraction of the VMS

Sample	Rate of hydrolysis (pmol/min/mg protein)
VMS membrane + SP	31.5 ± 6
VMS membrane + phosphoramidon + SP	17.4 ± 7
VMS membrane + thiorphan + SP	16.9 ± 4
VMS membrane + EDTA + SP	14.1 ± 3
VMS membrane + 1,10-phenanthroline + SP	13.6 ± 7
VMS membrane + chymostatin + SP	30.5 ± 3
VMS membrane + leupeptin + SP	31.0 ± 1

be involved in the inactivation of substance P–mediated chemosensory responses of the VMS.

B. Physiological Role of NEP-like Activity of the VMS

Previous studies have suggested that structures within 1 mm of the VMS are involved in the regulation of cholinergic outflow to the trachea (55). Topical application of tachykinins, like substance P and neurokinin-A, to the intermediate area of the VMS increased airway smooth muscle tone (57). Application of either phosphoramidon or thiorphan (10^{-5} to 10^{-3} *M*) significantly potentiated the airway and phrenic nerve responses to tachykinins. Furthermore, under similar experimental conditions, the tracheal and phrenic nerve responses to hypoxia and hypercapnia are also augmented (57). These results suggest that NEP may serve as an important modulator of tachykinin induced effects in certain structures located in the vicinity of the VMS and thus affecting respiratory activity and parasympathetic outflow to the airways.

C. Immunolabeling of NEP-like Activity Within Ventrolateral Medullary Structures

The distribution of NEP-like activity in the ventral medullary aspect was examined by fractionation studies involving cell-free extracts and also by immunocytochemistry. More than 95% of NEP-like activity of the VMS is associated with the membrane-enriched fraction, whereas NEP-like activity was not discernible in the soluble fraction.

A polyclonal NEP primary antibody was used to label NEP in the VMS, and the antibody binding sites were subsequently identified using the avidin-biotin peroxidase complex method (61). Light-microscopic analysis of immunolabeled serial sections indicated that NEP-like immunoreactivity is located 50 to 350 μm beneath the surface of the ventrolateral medulla. Colocalization studies with glial fibrillary acidic protein indicated that astroglial cells of this region also harbor NEP.

D. Summary

In summary, our studies on the peptidases of the carotid body and of the ventrolateral aspect of medulla oblongata areas suggest that NEP-like activity is present and participates in the regulation of tachykinin effects in these chemosensory structures. However, a few differences in NEP of these two structures were noted. In the VMS, NEP is exclusively associated with the membrane structures of astroglia and neurons, whereas an additional soluble form in near equal proportion occurs in the carotid body. While NEP is the major peptidase regulating peptide levels in the carotid body, peptidases in addition to NEP also contribute to tachykinin metabolism in the VMS. Finally, inhibition of NEP selectively potentiated the hypoxic response of the carotid body. By contrast, both hypoxia and hypercapnia induced changes in phrenic activity, and parasympathetic outflow to the airways is affected by NEP inhibition in the VMS. It is anticipated that future studies on NEP regulation under a variety of conditions that influence respiratory drive may unravel new roles for the neural endopeptidase in chemoreceptor functions.

Acknowledgments

This work was supported in part by grants from American Lung Association to G.K.K. The authors thank Dr. Cherniack for his support and encouragement and Dr. Prabhakar for valuable discussions. We also thank Dr. Moracic and Ms. Beegen for their excellent assistance in NEP immunolabeling studies.

References

1. Eyzaguirre C, Zapata P. Perspectives in carotid body research. J Appl Physiol 1984; 57:931–957.
2. Fidone SJ, Gonzalez C. Initiation and control of chemoreceptor activity in the carotid body. In: Cherniack NS, Widdicombe JG, eds. Hand Book of Physiology, Section 3, Vol II. Bethesda, MD: American Physiological Society, 1986:247–312.
3. Weil JV. Ventilatory control at high altitude. In: Cherniack NS, Widdicombe JG, eds. Hand Book of Physiology, Section 3, Vol II. Bethesda, MD: American Physiological Society, 1986:703–727.
4. Fitzgerald RS, Lahiri S. Reflex responses to chemoreceptor stimulation. In: Cherniack NS, Widdicombe JG, eds. Hand Book of Physiology, Sec 3, Vol II. Bethesda, MD: American Physiological Society, 1986:313–362.
5. Schramm CM, Grunstein M. Respiratory influence of peripheral chemoreceptor stimulation in maturing rabbits. J Appl Physiol 1987; 63:1671–1680.
6. Cherniack NS, von Euler C, Homma I, Kao FF. Graded changes in central chemoreceptor input by local temperature changes on the ventral surface of the medulla. J Physiol (Lond) 1979; 287:191–211.
7. Loeschcke HH. Central chemosensitivity and the reaction theory. J Physiol (Lond) 1982; 332:1–24.

8. Millhorn DE, Eldridge F. Role of ventrolateral medulla in regulation of respiratory and cardiovascular system. J Appl Physiol 1986; 61:1249–1263.
9. Schlafke ME. Central chemosensitivity: A respiratory drive. Rev Physiol Biochem Pharmacol 1981; 90:171–244.
10. Cherniack NS, Prabhakar NR, Haxhiu MA, Runold M. Excitatory and inhibitory influences on the ventilatory augmentation caused by hypoxia. In: Lahiri S, Cherniack NS, Fitzgerald RS, eds. Response and Adaptation to Hypoxia: Organ to Organelle. New York: Oxford University Press, 1991:107–121.
11. Lundberg JM, Hokfelt T, Fahrenkrug J, et al. Peptides in the cat carotid body (glomus caroticum): VIP-, enkephalin- and substance P-like immunoreactivity. Acta Physiol Scand 1979; 107:279–281.
12. Jacobowitz DM, Helke CJ. Localization of substance P immunoreactive nerves in the carotid body. Brain Res Bull 1980; 5:195–197.
13. Chen IL, Yates RD, Hansen JT. Substance P-like immunoreactivity in rat and cat carotid bodies: Light and electron microscopic studies. Histol Histopathol 1986; 1:203–212.
14. Prabhakar NR, Landis S, Kumar GK, et al. Substance P and neurokinin A in the cat carotid body: Localization, exogenous effects and changes in content in response to arterial PO_2. Brain Res 1989; 481:205–214.
15. Kummer W, Gibbins IL, Heym C. Peptidergic innervation of arterial chemoreceptors. Arch Histol Cytol 1989; 52(suppl):361-364.
16. Ichikawa H, Helke CJ. Distribution, origin and plasticity of galanin-immunoreactivity in the rat carotid body. Neuroscience 1993; 52:757–767.
17. Scraggs M, Smith P, Heath D. Glomic cells and their peptides in the carotid body of the human fetus. Pediatr Pathol 1992; 12:823–834.
18. Wang ZZ, He L, Stensaas LJ, et al. Localization and in vitro actions of atrial natriuretic peptide in the cat carotid body. J Appl Physiol 1991; 70:942–946.
19. Kummer W, Habeck JO. Substance P and calcitonin gene related peptide-like immunoreactivities in the human carotid body studied at light and electron microscopical level. Brain Res 1991; 554:286–292.
20. Oomori Y, Ishikawa K, Satoh Y, et al. Neuropeptide-Y immunoreactive chief cells in the carotid body of young rats. Acta Anat (Basel) 1991; 140:120–123.
21. Cuello AC, Kanazawa I. The distribution of substance P immunoreactive fibers in the rat central nervous system. J Comp Neurol 1978; 178:129–156.
22. Armstrong DM, Pickel VM, Joh TH, et al. Immunocytochemical localization of catecholamine synthesizing enzymes and neuropeptides in area prostrema and medial nucleus tractus solitarius of rat brain. J Comp Neurol 1981; 196:505–517.
23. Helke CJ, Neil JJ, Massari VJ, Loewy AD. Substance P neurons project from the ventral medulla to the intermediolateral cell column and ventral horn in the rat. Brain Res 1982; 243:147–152.
24. Liebstein AG, Dermietzel R, Willenberg IM, Dauschert J. Mapping of different neuropeptides in the lower brain stem of the rat: With special reference to the ventral surface. J Auton Nerv Syst 1985; 14:299–313.
25. Eyzaguirre C, Fidone SJ. Transduction mechanisms in carotid body: Glomus cells, putative neurotransmitters and nerve endings. Am J Physiol 1980; 239:C135–C152.
26. Prabhakar NR, Runold M, Yamamoto Y, et al. Effect of substance P antagonist on the hypoxia-induced and carotid chemoreceptor activity. Acta Physiol Scand 1984; 121:301–303.

27. Yamamoto Y, Lagercrantz H. Substance P: A putative mediator of the hypoxic drive. In: von Euler C, Lagercrantz H, eds. Neurobiology of the Control of Breathing. New York: Raven Press, 1986:97–100.
28. Prabhakar NR, Cao H, Lowe JA III, Snider RM. Selective inhibition of the carotid body sensory response to hypoxia by the substance P receptor antagonist CP-96,345. Proc Natl Acad Sci 1993; 30:10041–10045.
29. Florez J, Mediavilla A. Respiratory and cardiovascular effects of Met-enkephalin applied to the ventral surface of the brain stem. Brain Res 1977; 138:585–596.
30. Pokorski M, Lahiri S. Effects of naloxone on carotid body chemoreception and ventilation in the cat. J Appl Physiol 1981; 51:1533–1538.
31. Monti-Bloch J, Eyzaguirre C. Effects of methionine-enkephalin and substance P on the chemosensory discharge of the cat carotid body. Brain Res 1985; 338:297–307.
32. Matsas R, Kenny AJ, Turner AJ. The metabolism of neuropeptides: The hydrolysis of peptides, including enkephalins, tachykinins and their analogues by endopeptidase-24.11. Biochem J 1984; 223:433–440.
33. Turner AJ. Endopeptidase-24.11 and neuropeptide metabolism. In: Turner AJ, ed. Neuropeptides and Their Peptidases. Chichester, England: Ellis-Horwood, 1987: 183–201.
34. Erdos EG, Skidgel RA. Neutral endopeptidase 24.11 (enkephalinase) and related regulators of peptide hormones. FASEB J 1989; 3:145–151.
35. McDonald DM. Peripheral chemoreceptors: Structure-function relationships in the carotid body. In: Hornbein TF, ed. Lung Biology in Health and Disease: The Regulation of Breathing. New York: Marcel Dekker, 1981; 17:105–319.
36. Verna A. Ultrastructure of the carotid body in the mammals. Int Rev Cytol 1979; 60:271–330.
37. Pallot DJ. The mammalian carotid body. In: Advances in Anatomy Embryology and Cell Biology. Berlin: Springer-Verlag, 1987.
38. Kumar GK, Prabhakar NR, Cherniack NS. In vitro degradation of substance P by the carotid body proteases. In: Eyzaguirre C, Fidone SJ, Fitzgerald RS, et al., eds. Arterial Chemoreception. New York: Springer-Verlag, 1990:137–142.
39. Stephenson SL, Kenny AJ. The hydrolysis of a human atrial natriuretic peptide by pig kidney microvillar membranes is initiated by endopeptidase-24.11. Biochem J 1987; 243:183–187.
40. Vijayaraghavan J, Scili AG, Carretero OA, et al. The hydrolysis of endothelins by neutral endopeptidase-24.11 (enkephalinase). J Biol Chem 1990; 265:14150–14155.
41. Turner AJ, Matsas R, Kenny AJ. Are there neuropeptide-specific peptidases? Biochem Pharmacol 1985; 34:1347–1356.
42. Llorens C, Schwartz JC. Enkephalinase activity in rat peripheral organs. Eur J Pharmacol 1981; 69:113–116.
43. Ronco P, Pollard H, Galceran M, et al. Distribution of enkaphalinase (membrane metalloendopeptidase, E.C.3.4.24.11) in rat organs. Lab Invest 1988; 58:210–217.
44. Kioussi C, Matsas R. Endopeptidase-24.11, a cell-surface peptidase of central nervous system neurons is expressed by Schwann cells in the pig peripheral nervous system. J Neurochem 1991; 57:431–440.
45. Barnes K, Bourne A, Cook PA, et al. Membrane peptidases in the peripheral nervous system of the pig: Their localization by immunohistochemistry at light and electron microscopic levels. Neuroscience 1991; 44:245–261.

46. Bowes MA, Kenny AJ. Endopeptidase-24.11 in pig lymph nodes: Purification and immunocytochemical localization in reticular cells. Biochem J 1986; 236:801–810.
47. Waksman G, Hamel G, Delay-Goyet P, Roques BP. Neuronal localization of the neutral endopeptidase "enkephalinase" in rat brain revealed by lesions and autoradiography. EMBO J 1986; 5:3163–3166.
48. Malfroy B, Swerts JP, Llorens C, Schwartz JC. Regional distribution of a high affinity enkephalin degrading peptidase ("enkephalinase") and effects of lesions suggest localization in the vicinity of opiate receptors in brain. Neurosci Lett 1979; 11:329–334.
49. Matsas R, Turner AJ, Kenny AJ. Immunocytochemical localization of endopeptidase-24.11 (enkephalinase) in the pig nervous system. Neuroscience 1986; 18:991–1012.
50. Pozsgay M, Michaud C, Liebman M, Orlowski M. Substrate and inhibitor studies of thermolysin-like neutral endopeptidase from kidney membrane fractions: Comparison with bacterial thermolysin. Biochemistry 1986; 25:1292–1299.
51. Kumar GK, Runold M, Ghai RD, et al. Occurrence of neutral endopeptidase activity in the cat carotid body and its significance in chemoreception. Brain Res 1990; 517:341–343.
52. Kumar GK, Cherniack NS, Prabhakar NR. Biochemical analysis of neutral endopeptidase activity of the cat carotid body. In: Acker J, Trzebski A, O'Regan RG, eds. Chemoreceptors and Chemoreceptor Reflexes. New York: Plenum Press, 1990:99–102.
53. Hansen G, Jones L, Fidone S. Physiological chemoreceptor stimulation decreases enkephalin and substance P in the cat carotid body. Peptides 1986; 7:767–769.
54. van Amsterdam JGC, van Buuren KJH, de Jong AM, Soudijn W. Inhibitors of calf-brain enkephalinase A and B. Life Sci 1983; 33:109–112.
55. Haxhiu MA, Deal EC, Norcia MP, et al. Medullary effects of nicotine and GABA on tracheal smooth muscle tone. Respir Physiol 1986; 64:351–363.
56. Haxhiu MA, Deal EC, Norcia MP, et al. Effect of N-methyl-D-aspartate applied to the ventral surface of the medulla on the trachea. J Appl Physiol 1987; 63:1268–1274.
57. Haxhiu MA, Deal EC, van Lunteren E, Cherniack NS. Central medullary effects of tachykinin peptides on airway tone. J Auton Nerv Syst 1989; 28:105–116.
58. Maeda T, Balakrishnan K, Mehdi SQ. A simple and rapid method for the preparation of plasma membranes. Biochim Biophys Acta 1983; 731:115–121.
59. Elliot JI, Brewer JM. The inactivation of yeast enolase by 2,3-butanedione. Arch Biochem Biophys 1978; 190:351–357.
60. Sonnenberg JL, Sakane Y, Jeng AY, et al. Identification of protease 3.4.24.11 as the major atrial natriuretic factor degrading enzyme in the rat kidney. Peptides 1988; 9:173–180.
61. Moracic V, Haxhiu MA. Neutral endopeptidase in the ventral medulla, an immunochemistry study. Acta Biol Med Exp 1989; 14:173–176.

23

Inhibitory and Excitatory Effects of Nitric Oxide on Respiratory Responses to Hypoxia

NANDURI R. PRABHAKAR, NEIL S. CHERNIACK, and MUSA A. HAXHIU

School of Medicine
Case Western Reserve University
Cleveland, Ohio

I. Introduction

Nitric oxide (NO) is a gas molecule with free radical properties. Till recently, NO was considered a noxious gas harmful to the body. It is now established that mammalian cells synthesize NO. The enzyme nitric oxide synthase (NOS) catalyzes the formation of NO from L-arginine (1). Two types of NOS have been purified: a constitutive Ca^{2+}-calmodulin-dependent type and an inducible Ca^{2+}-independent form (2). NOS is a heme-containing enzyme and requires NADPH, tetrahydrobiopterin, and molecular oxygen for its activity (2). Both neuronal and nonneuronal cells synthesize NO, as revealed by histobiochemical analysis of NOS (3,4). Once formed, NO diffuses rapidly into neighboring cells and is inactivated to nitrites in the presence of oxygen. Many of the actions of NO are mediated by activating the enzyme soluble guanylate cyclase (1,3,4).

Nitric oxide (NO) has been implicated as a chemical messenger in many physiological systems. For example, NO generated by vascular endothelium seems to be important for maintaining arterial blood pressure (3,4). Nitric oxide (NO) has been implicated in immune function of macrophages and platelet aggregation (3). In the nervous system, NO may function as an intercellular messenger (4,5). For instance, in central neurons, NO mediates the neuronal stimulation by excitatory amino acids and is considered a "retrograde chemical messenger" (5). In the peripheral nervous system, stimulation of autonomic nerves leads to inhibition of certain smooth muscles. Neither cholinergic nor adrenergic blockers affect the inhibition, whereas blockade of NO synthesis prevents the neuronally induced inhibition (4,5). Thus, NO is an inhibitory chemical that mediates noncholinergic and nonadrenergic transmission in the peripheral nervous system.

Although the roles of NO have been extensively investigated in many physiological systems, relatively little is known on its effects on respiratory control. Recently we reported that hypoxia inhibits NOS activity in the carotid bodies (6). The finding that low oxygen (or, more appropriately, reduced oxygen availability) decreases NO synthesis prompted us to examine the role of endogenous NO on the ventilatory response to hypoxia. For this purpose, we examined the distribution of NOS activity in carotid chemoreceptors and in respiratory-related neurons in the brainstem. The physiological relevance of endogenous NO was examined by monitoring chemoreceptor activity and respiratory responses to hypoxia using an inhibitor of NO synthesis.

II. Distribution of NO Synthase in Carotid Body and Brainstem Neurons

It is now established that neuronal NADPH-diaphorase is a nitric oxide synthase (7,8). Therefore, NADPH-diaphorase histochemistry provides a simple method to localize the distribution of NOS in the neuronal tissues (8). Using this method, we examined the distribution of NOS in cat carotid bodies and brainstem. The methodological details are as follows. Animals were anesthetized with pentobarbital sodium (40 mg/kg; IP) and the trachea was intubated. A femoral artery and a vein were cannulated for monitoring of arterial blood pressure and systemic administration of fluids. The carotid bifurcation was exposed and the tissues were covered with cotton gauze soaked in saline. In the experiments involving the brainstem, the animal's head was fixed to a stereotaxic frame and the dorsal surface of the brain was exposed. Anesthetized animals were mechanically ventilated with a positive-pressure respirator. Heparin (1000 U/kg) was administered IV 15 min before the beginning of perfusion. After opening the thorax, the ascending aorta was cannulated and connected to reservoirs containing perfusion medium. An opening was made in the right atrium for drainage of fluids. Heparinized saline was perfused for 15 min until the blood vessels were free of blood. Thereafter the perfusion medium was switched to 4% paraformaldehyde in 0.1 *M* phosphate buffer, pH = 7.3, for 30 min at room temperature. The tissues were removed, rinsed with phosphate buffer, cryostat-protected with 30% sucrose in 0.1 *M* phosphate buffer, and stored at $-70°C$. Cryostat sections 8 μ thick were cut and thaw-mounted on gelatin-coated slides. The sections were allowed to air-dry for 30 min and were then either stained immediately or stored frozen at −20°C for staining at a later time. To demonstrate NADPH-diaphorase reaction, sections were incubated in 0.1 *M* phosphate buffer, pH = 7.4, containing 0.3% Triton X-100, 0.1 mg/ml nitroblue tetrazolium, and 1.0 mg/ml β-NADPH at 37°C for 60 min. Following the reaction, the sections were rinsed in phosphate buffer, pH = 7.4, followed by distilled water, and air-dried. After placing the cover slips, the sections were examined under the microscope. As controls, sections were incubated with the oxidized form of NADP along with nitroblue tetrazolium.

The results of the histochemical studies were as follows. In the cat carotid body, many nerve plexuses innervating the glomus tissue were positive for NADPH-diaphorase (Fig. 1). The positive staining was consistently seen in all the sections examined. On the other hand, NOS activity was not apparent in glomus cells. By contrast, in presence of the oxidized form of NADP (negative control), no staining was observed. Recently, Wang et al. examined the distribution of NOS in rat carotid bodies (9). Using antibodies specific for the constitutive form of the enzyme, these investigators also observed NOS reactivity

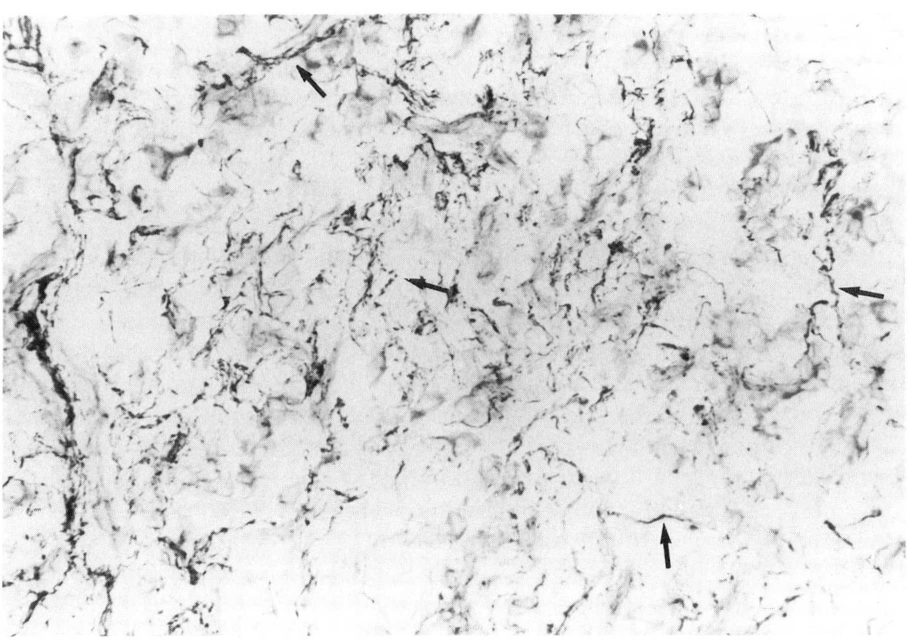

Figure 1 NADPH-diaphorase histochemistry of the cat carotid body. Note that many nerve fibers were positive for NADPH-diaphorase. Some examples of nerve fibers are denoted by arrows.

in nerve fibers innervating the glomus tissue. Furthermore, these authors reported that chronic sectioning of sinus nerve abolished, whereas sympathectomy had no influence on, NOS-positive fibers, suggesting that they are of petrosal ganglion origin. We have not performed similar experiments with chronic ablation of sinus nerve, but in consideration of the study of Wang et al. it is conceivable that in the cat carotid body, NOS-positive fibers are also of sinus nerve origin. Thus, in the carotid body, the primary source of NO production resides in the sinus nerve fibers.

Chemoreceptor afferents are known to project in the n. tractus solitarius (NTS). Distinct NOS reactivity was evident in NTS neurons. No reaction product was observed in subcoerules and in the Kölliker-Fuse nucleus. On the other hand, in the ventrolateral medulla, in the area that corresponds to respiratory-related neurons, NOS reactivity was evident in a small number of neurons and nerve fibers. These results demonstrate that in the cat, NOS activity is present in the carotid bodies, which are sensors for hypoxic stimulus, and the brainstem neurons, which integrate the chemoreceptor afferent input. These observations suggest that neuronal substrates related to ventilatory responses to hypoxia are capable of generating NO.

III. Endogenous NO and Carotid Chemoreceptor Activity

In order to examine the influence of endogenous NO on sensory activity of the carotid body, we performed the following physiological studies. Several analogues of L-arginine have been shown to inhibit NO synthesis and provide valuable tools to examine the role of endogenous NO in various physiological systems (3). We examined the effects of L*w*-Nitro-L-arginine (L-NNA), a potent selective inhibitor of NOS (10), on sensory activity of the carotid body in vitro. The isolated carotid body preparation was chosen because, in intact animals, inhibitors of NO synthesis may affect arterial blood pressure, which may confound the interpretation of the results. The experimental procedures for recording the carotid body activity in vitro were described in detail elsewhere (6, 11). Briefly, they are as follows. The carotid bifurcation together with sinus nerve was dissected from anesthetized cats. After removing the connective tissue, carotid bifurcation was placed in a perspex chamber having two compartments, A and B, separated by a partition having a small opening. The carotid sinus nerve was pulled through the small opening and placed on a platform in chamber B and covered with warm mineral oil. The common carotid artery in chamber A was cannulated and perfused and superfused with Kreb's solution at a rate of 2 to 4 mL/min and at a temperature of 36 ± 1°C. Chamber A was closed with a lid to prevent the exposure of the carotid body to atmospheric air. The composition of the bathing medium was as follows (in m*M*): NaCl = 110, sucrose = 54, KCl = 5, glucose = 5.5, $MgCl_2$ = 0.5, HEPES = 5, $CaCl_2$ = 2.2, and pH = 7.38. The reservoirs containing superfusion medium were bubbled with 21% O_2. Action potentials were recorded from thin filaments dissected from the sinus nerve. Stopping the perfusion (i.e., hypoxia for 1 to 2 min) augmented the sensory activity, suggesting that the action potentials were primarily of chemoreceptor origin. Known amounts of analogues of arginine to be tested were added to the reservoirs containing the bathing medium.

Addition of L-NNA to the perfusate increased the sensory discharge in a dose-dependent manner (Fig. 2). Maximum increase was seen at 100 μM of L-NNA and plateaued with higher doses (i.e., 300 μM). The increase in sensory discharge at any given dose of L-NNA commenced within 2.4 ± 0.8 min (excluding 2.5 min—i.e., the time taken for the chemical to reach the tissue from the reservoir). The activity increased progressively. We examined the selectivity of L-NNA by comparing the relative efficacy of L- and D-arginine (300 to 500 μM) to reverse the effects. Examples from two different experiments are depicted in Figures 3A and B. As can be seen, L-NNA-induced chemosensory excitation was reversed in the presence of L-arginine (300 μM), whereas such a reversal was not seen in the presence of comparable doses of D-arginine (300 μM). Average results showed that L- but not D-arginine significantly reversed L-NNA-induced chemoreceptor excitation ($P<0.01$; $n=7$ carotid bodies).

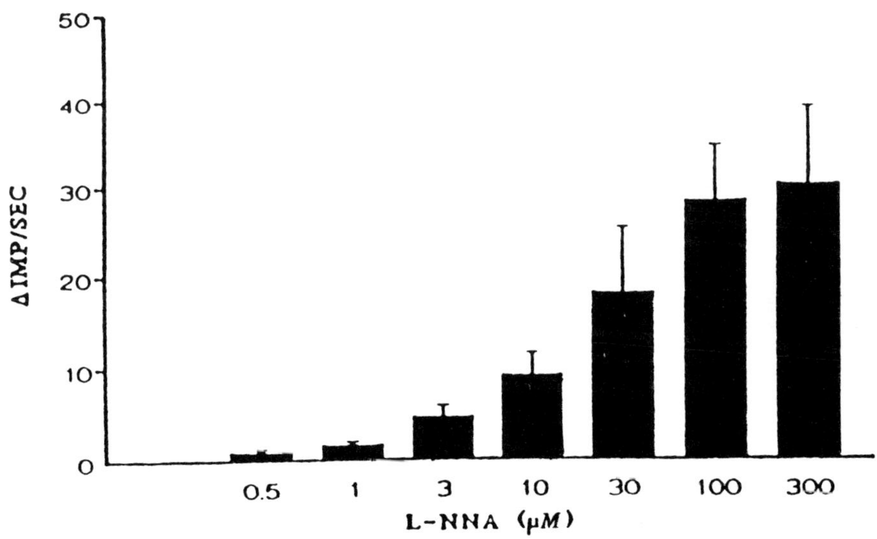

Figure 2 Effect of Lw-Nitro-L-arginine (L-NNA) on baseline activity of the isolated in vitro carotid body (perfusate: P_{O_2} = 140 ± 7 mm Hg; P_{CO_2} = 3 ± 1 mm Hg; pH = 7.4; temp = 36.5°C). Changes in chemoreceptor act are expressed as delta imp/s (i.e., test-control). Data presented are mean ± SEM from 10 carotid bodies. Note the dose-dependent increase in sensory activity with L-NNA, which reached a plateau at 100 μM.

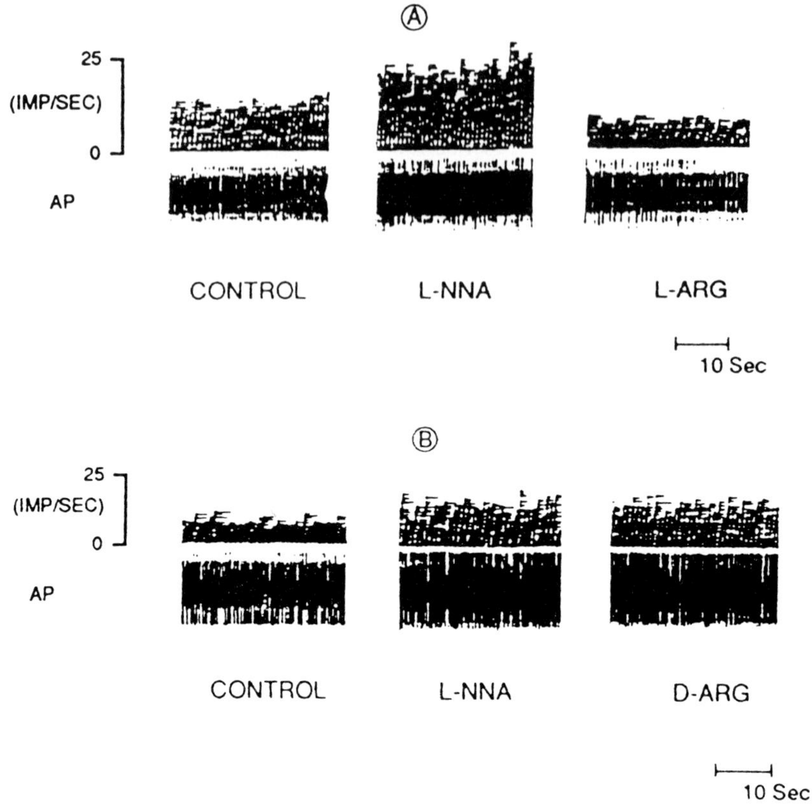

Figure 3 Examples of experiments illustrating the effects of L- and D-arginine on L-NNA-induced chemosensory activity. Chemosensory response to L-NNA (100 *μM*) was recorded for 10 min. Then either L-arginine (300 *μM*) (A) or D-arginine (B) was added to the perfusing medium containing L-NNA. Representative tracings were taken during tenth min of perfusion with L-NNA and L- or D-arginine plus L-NNA-containing medium. Note that L- but not D-arginine reduced the sensory excitation by L-NNA. Abbreviations: (imp/s) = rate meter output; AP = action potentials of the sinus nerve.

The effects of L-NNA on carotid body response to hypoxia were examined in six carotid bodies. Because higher doses (100 *μM*) of L-NNA markedly increased the baseline activity, we chose to examine the effects of lower doses of L-NNA (10 *μM*) on hypoxic response. Prior to L-NNA, lowering the Po_2 of the medium from 142 ± 12 mm Hg to 36 ± 9 mm Hg increased the chemoreceptor activity on average by +22 ± 6 impulses per second. In the presence of 10 *μM* of L-NNA, hypoxia increased the chemoreceptor activity by +36 ± 9 impulses per second (a 64% increase). This potentiation in hypoxic response was associated with a 26% elevation of baseline activity compared to controls. Furthermore, in the presence of L-NNA chemoreceptor, the response to hypoxia commenced on average by 40 ± 3 s earlier compared to controls.

These results demonstrate that (a) L-NNA, an inhibitor of nitric oxide synthesizing enzyme, increases the baseline carotid body activity, (b) the effects of the inhibitor are reversed by L- but not by D-arginine, and (c) lower doses of L-NNA potentiated the carotid body response to hypoxia. The increased chemosensory activity by L-NNA, we believe, is due to selective inhibition of NOS activity. This notion is supported by the finding that L- but not D-arginine reversed the effects of L-NNA. The results with L- and D-arginine also demonstrate the enantiomer selectivity of L-NNA. The observation that lower doses of L-NNA potentiated the hypoxic response and augmentation of baseline activity by NOS inhibitor suggests that endogenous NO is inhibitory to the carotid body activity. The effects of NO in the carotid body resemble its actions in the peripheral nervous system, which are inhibitory, rather than those in central neurons, where it causes excitation (4). Furthermore, the stimulatory actions of L-NNA were seen under normoxia, indicating that NO is continuously produced under basal conditions in the chemoreceptor tissue and tonically inhibits chemosensory activity. Some of the inhibitory actions of NO at the carotid body are coupled to levels of cyclic guanosine monophosphate (cGMP) (6). The site of action of NO cannot be ascertained from the present results. They may include either type I cells, sustentacular cells, vascular endothelial cells, or a combination of all.

IV. Influence of NOS Inhibitor on Ventilatory Response to Hypoxia—Evidence for the Central Excitatory Effects of NO

Thus far, the studies described above indicate that endogenous NO is inhibitory to the carotid body sensory activity. If, endogenous NO is inhibitory to the chemoreceptor activity, then systemic administration of NOS inhibitor should potentiate the hypoxic ventilatory drive in animals with intact peripheral chemoreceptors. To test this possibility, the following experiments were performed on anesthetized, paralyzed cats. The electrical activity of the efferent phrenic nerve was used as an index of central respiratory drive. Briefly, the phrenic nerve on one side was isolated at the C5 level, cut, and desheathed. Electrical activity from the cut central end was recorded by bipolar electrodes using standard electrophysiological techniques. The action potentials were fully rectified and integrated using 100-ms time constant. To avoid the interference of lung stretch receptor input with the respirator, all experiments were performed after bilateral vagotomy (including the aortic nerve). The protocols in this series of experiments were as follows. Control phrenic nerve activity was recorded while the animals were ventilated with 100% O_2. Arterial P_{CO_2} was maintained between 35 and 40 mm Hg by adjusting the rate and stroke volume of the respirator. Thereafter, the inspiratory port of the respirator was switched to reservoir bags containing 12% O_2 balanced N_2. The duration of hypoxic challenge was 10 min. After the control respiratory responses to hypoxia were recorded, L-NNA was administered intravenously; the dose was 30 to 50 mg/kg. Hypoxic challenge was repeated 10 to 40 min after administration of L-NNA.

An example of the phrenic nerve response to hypoxia before and after systemic administration of L-NNA is shown in Figure 4A. Before L-NNA, hypoxia increased phrenic activity and the increased respiratory drive persisted for the entire duration of hypoxic challenge. In this animal, there was an increase in basal respiratory activity during hyperoxia after L-NNA. Ventilating the animal with low Po_2 further increased the phrenic nerve activity. But this increase was followed by a gradual decline in respiratory activity. By the fifth minute of hypoxia, rhythmic phrenic activity disappeared (neural apnea). Average data normalized as percent of controls (hyperoxia = 100%) are summarized in Figure 4B. The results from the five cats studied showed that L-NNA had no effect on the magnitude of respiratory excitation by hypoxia ($P > 0.05$). However, the peak increase in respiratory excitation was reached 1 min earlier compared to pre-L-NNA controls. The increase in phrenic activity was followed by a gradual decrease, and by the fifth minute of hypoxia, average amplitude of phrenic nerve activity was 50% below the controls.

The effects of L-NNA on respiratory response to hypoxia was examined after sectioning carotid sinus nerves bilaterally in four additional cats. An example of an experiment illustrating the effect of hypoxia on phrenic nerve activity before and after L-NNA in a peripherally chemodenervated and vagotomized cat is shown in Figure 5A. Before L-NNA, hypoxia resulted in a progressive decrease in respiratory activity and there was no initial respiratory stimulation. L-NNA decreased baseline respiratory activity during hyperoxia compared to pre-L-NNA controls. Ventilating with a low-Po_2 gas mixture resulted in a complete cessation of rhythmic respiratory activity within 5 min after the onset of hypoxia. In other words, L-NNA enhanced hypoxic ventilatory depression in the peripherally chemodenervated animal. Such a potentiation of hypoxic ventilatory depression by L-NNA was seen in all of the four cats studied. Average results are summarized in Figure 5B. The data are presented as a percent of hyperoxic controls because of substantial reduction in baseline activity after L-NNA (40% reduction). As can be seen, before L-NNA, hypoxia resulted in a progressive decline in respiratory activity. After 10 min of hypoxia, phrenic nerve activity was reduced by 80% of the normoxic controls. By contrast, after L-NNA, rhythmic phrenic activity disappeared by fifth minute of hypoxia. Administration of L-NNA increased arterial blood pressure by 20 to 40 mm Hg in all cats studied. However, during 10 min of hypoxic challenge, arterial blood pressure decreased by 15 to 20 mm Hg. L-NNA had no consistent effects on blood pressure changes during hypoxia. Moreover, there was no

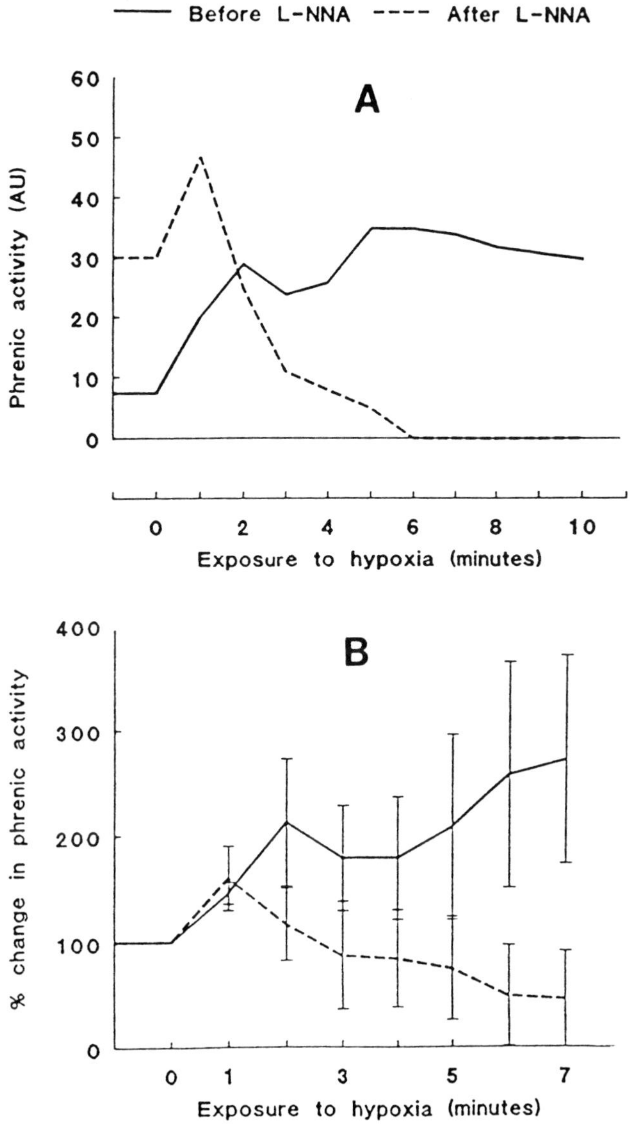

Figure 4 The effects of L-NNA on respiratory response to hypoxia in anaesthetized, paralyzed, and vagotomized cats with intact carotid body chemoreceptors. Time course of the efferent phrenic nerve activity in response to hypoxia from a single experiment is given in top panel A and the average results from five experiments are presented in panel B. The data presented in panel B are mean ± SEM. L-NNA increased basal respiratory activity during hyperoxia and peak respiratory stimulation by hypoxia occurred earlier. Increased respiratory activity was not sustained after L-NNA.

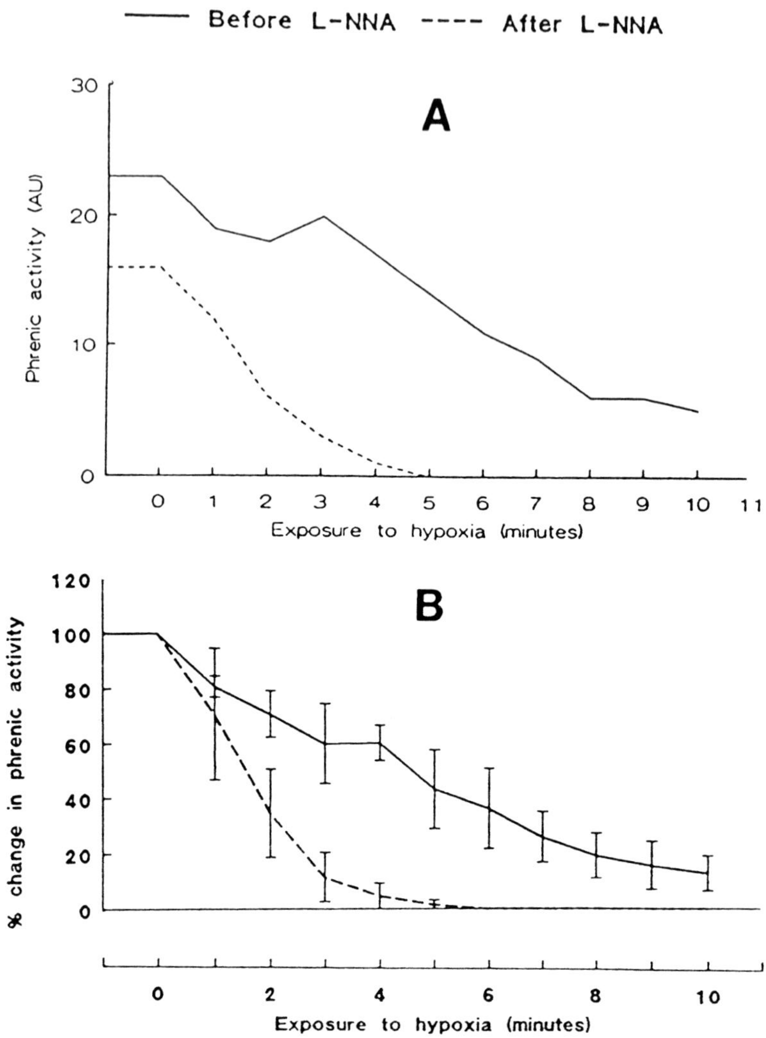

Figure 5 The effects of L-NNA on respiratory response to hypoxia in anaesthetized, paralyzed, vagotomized and denervated cats. Time course of the efferent phrenic nerve activity during hypoxia from a single experiment is given in top panel A, and the average results from four experiments are presented in panel B. The data presented in panel B are mean ± SEM. L-NNA decreased respiratory activity during hyperoxia and potentiated hypoxic ventilatory depression.

causal relation between blood pressure changes and the respiratory responses to hypoxia before and after L-NNA.

The results described above demonstrate that systemic administration of L-NNA not only stimulated baseline respiratory activity, but also preceded the respiratory excitation by 1 min. Taken together, these observations suggest that L-NNA stimulated carotid bodies, resulting in enhanced respiratory stimulation by hypoxia. This assumption is supported by the finding that L-NNA did not stimulate respiration in peripherally denervated animals. Thus, both in vivo and in vitro experiments suggest that endogenous NO is inhibitory to the carotid body activity.

Interestingly after L-NNA, respiratory stimulation by hypoxia was no longer maintained. Rather, it declined progressively with hypoxia. Hypoxic respiratory depression was more pronounced after L-NNA in peripheral chemodenervated animals. It has been reported that NO is excitatory to central neurons (5). It is likely that the enhanced ventilatory depression by hypoxia is due to inhibition of NO synthesis in the central neurons that mediate hypoxic ventilatory drive. Present observation that NOS is present in nTS neurons may support such a notion. Alternatively, blockade of NO synthesis may have promoted central hypoxia by way of causing cerebral vasoconstriction.

V. Summary and Conclusions

The present results provide convincing evidence that endogenous NO can potentially influence hypoxic ventilatory drive. Interestingly, NO seems to have completely opposite effects on peripheral versus central structures that are related to hypoxic ventilatory drive. On one hand, endogenous NO seems to tonically inhibit carotid body sensory discharge; on the other, it appears to exert excitatory influence on brainstem respiratory-related neurons. Thus, the fact that hypoxia inhibits NO synthesis may have potential implications on the expression of ventilatory response to hypoxia. The magnitude of the ventilatory response to low P_{O_2} depends on the fine balance between the peripheral inhibitory versus central excitatory actions of NO. As recently suggested (12), ventilatory and circulatory adaptations to environmental oxygen and CO_2 depend on the balance of excitatory and inhibitory neurochemicals. The present results not only provide experimental support for such a notion but also suggest that such a fine balance can be brought out by a single chemical such as NO by virtue of its diametrically opposing effects on central versus peripheral neurons. Future studies on the mechanism(s) and site(s) of action of NO may unravel hitherto its unrecognized roles in respiratory control.

Acknowledgments

Our thanks go to Mrs. Cheryl D. Gilliam-Walker for her secretarial assistance with the manuscript. This work was supported by grants from National Institutes of Health; Heart, Lung and Blood Institute HL-45780, HL-25830 and a Research Career Development Award HL-02599 (N.R.P.).

References

1. Knowles RG, Palacios M, Palmer RMJ, Moncada S. Formation of nitric oxide from L-arginine in the central nervous system: A transduction mechanism for stimulation of the soluble guanylate cyclase. Proc Natl Acad Sci USA 1989; 86:5159–5162.
2. Knowles RG, Palacios M, Palmer RMJ, Moncada S. Kinetic characteristics of nitric oxide synthase from rat brain. Biochem J 1990; 269:207–210.
3. Moncada S, Palmer RMJ, Higgs EA. Nitric oxide: Physiology, pathophysiology, and pharmacology. Pharmacol Rev 1991; 43:109–141.
4. Snyder SH. Nitric oxide: First in a new class of neurotransmitters. Science 1992; 257:494–496.
5. Garthwaite J, Charles SL, Chess-Williams R. Endothelium-derived relaxing factor release on activation of NMDA receptors suggests a role as intercellular messenger in the brain. Nature (Lond) 1988; 336:385–388.
6. Prabhakar NR, Kumar GK, Chang CH, et al. Nitric oxide in the sensory function of the carotid body. Brain Res 1993; 625:16–22.
7. Dawson TM, Bredt DS, Fotuhi M, et al. Nitric oxide synthase and neuronal NADPH diaphorase are identical in brain and peripheral tissues. Proc Natl Acad Sci USA 1991; 88:7797–7801.
8. Vincent SR, Kimura H. Histochemical mapping of nitric oxide synthase in the rat brain. Neuroscience 1992; 46:755–784.
9. Wang ZZ, Berdt DS, Snyder SH, et al. Nitric oxide synthase in the carotid body Soc Neurosci (Abstr). 1992; 18:1197.
10. Burnett AL, Lowenstein CJ, Bredt DS, et al. Nitric oxide: A physiological mediator of penile erection. Science 1992; 257:401–403.
11. Prabhakar NR, Cao H, Lowe JA III, Snider RM. Selective inhibition of the carotid body sensory response to hypoxia by the substance P receptor antagonist CP-96,345. Proc Natl Acad Sci USA 1993; 90:10041–10045.
12. Cherniack NS, Prabhakar NR, Haxhiu MA, Runold M. Excitatory and inhibitory influences on the ventilatory augmentation caused by hypoxia. In: Lahiri S, Cherniack NS, Fitzgerald RS, eds. Response and Adaptation to Hypoxia. New York: Oxford University Press, 1991:107–121.

24

Nitric Oxide in Carotid Body Chemoreception

DEEPAK K. CHUGH, PATRICIA A. GRIMES, MASAO KATAYAMA, ANIL MOKASHI, RICHARD A. STONE, and SUKHAMAY LAHIRI

University of Pennsylvania School of Medicine
Philadelphia, Pennsylvania

Since nitric oxide (NO) is produced endogenously in the peripheral blood vessels and nerves, its extensive vasculature and innervation make the carotid body a likely target organ for the synthesis and function of NO. We studied the localization of nitric oxide synthase (NOS) and also the effect of NOS inhibitor on the carotid chemosensory responses. To localize NO, we used NOS immunoreactivity; to test its function, we used the NOS-inhibitor L-NAME (25 to 300 μM) in vitro. Also, we tested the reversal of the L-NAME effect by sodium nitroprusside (SNP) (0.5 to 10 μM), which is a source for NO. L-NAME slowly stimulated the sensory discharge, attaining the peak response in a few minutes. The stimulatory effect was totally eliminated by SNP, which presumably relieved the vasoconstriction. Consistent with this interpretation was the finding that NOS-positive nerve fibers in the carotid body were mostly associated with the blood vessels. The highly diffusible NO could have also directly affected glomus cell function. In either case, NO exerts an inhibitory effect on chemosensory activity.

I. Introduction

Nitric oxide (NO) has appeared in biology as if from nowhere. It is present in blood vessels, peripheral nerves to the blood vessels, and certain neurons in the central nervous system. These contain a constitutive type enzyme, nitric oxide synthase (NOS), and its formation depends on demand (1,2). This simple molecule is a free radical gas that clearly triggers a signal transduction cascade. Once generated, it can diffuse away and influence nearby tissues. In vivo, this effect is limited by its reaction with oxyhemoglobin. There is evidence also that superoxide removes NO. In addition, its affinity and reactions with molecular oxygen are well known. These limit the reaction of NO, but the most important target of NO is the soluble guanylate cyclase for biological responses. The stimulatory effect occurs because the reaction of NO with the heme group of soluble guanylate cyclase (3) increases the rate of cyclic guanosine monophosphate (cGMP) synthesis. This pathway of NO effect has been found to be endothelium-dependent. This results in the relaxation of smooth muscle cells (2). Inhibition of soluble guanylate cyclase with methylene blue blocks the relaxation, 8-bromo-cGMP mimics it, and inhibitors of cGMP phosphodiesterase potentiate it. Thus, NO generated in one cell produces a second messenger in a neighboring cell.

In endothelium-independent vascular systems, the response occurs through the neural pathway (1). It is likely that both the depolarization of nerves and an increase in Ca^{2+} entry stimulate NOS in the nerve terminals, which results in stimulation of soluble guanylate cyclase in neighboring smooth muscle cells.

In the brain, particularly in the cerebellum, which contains the highest concentration of cGMP, glutamate stimulates *N*-methyl-D-aspartate receptors and enhances both NOS activity and NO formation while increasing cGMP concentration. NOS inhibitor N^G-monomethyl-L-arginine acetate completely inhibits this reaction and cGMP production (1).

The carotid body has all the attributes needed to produce NO. It is a highly vascular organ that contains a large number of endothelial cells, and it has nerves that supply the glomus cells, blood vessels, and ganglia. Thus the NO signal pathway could operate through neural tissue or through glomus cells and blood vessels. On both counts, the effects on NO when cGMP production is increased are expected to be inhibitory. Accordingly, an inhibitor of NOS would stimulate the chemosensory nerve discharge if NO production were a continual process. On the other hand, a direct application of an NO source would inhibit the discharge rate.

We tested these predictions in this study by using an in vitro cat carotid body and thus confirmed our original predictions. Because of the importance of localization and identification of the cells and tissues, we sought to localize NOS by using immunocytochemistry in the carotid body of the cat.

II. Methods

A. Immunochemistry for NOS

Carotid bodies, with a connecting segment of carotid artery, were removed from anesthetized cats and fixed in vitro by intraarterial perfusion with either 4% paraformaldehyde or periodate-lysine-paraformaldehyde (PLP). The tissues were postfixed by immersion for an additional 2-h period in the same fixatives, cryoprotected with 30% sucrose, and frozen and sectioned in a cryostat at 16 μm thickness. Portions of the glossopharyngeal nerve, including the petrosal ganglion, were fixed by immersion in either solution for 2 h and then were similarly processed and sectioned.

Sections of PLP-fixed tissues were incubated with a rabbit polyclonal antibody against NOS (4), and immunoreactivity was localized with a biotinylated secondary antibody and a streptavidin–Texas red conjugate. Specificity of immunostaining was established by omitting the primary antiserum from the incubation procedure.

B. Chemosensory Recording In Vitro

The carotid bodies (CB) in anesthetized cats were vascularly isolated, and the carotid artery, proximal to the carotid sinus, was cannulated. The CB was flushed with oxygenated and heparinized perfusate and excised with the majority of the carotid sinus nerve. The perfusion was started immediately in a thermostated (36 to 37°C) chamber, similar to the one previously described (5). Modified Tyrode solution containing CO_2-HCO_3^- (pH = 7.38 to 7.40 at 37°C) was used for perfusion and superfusion of the carotid body preparation. The composition of Tyrode solution was as follows (m*M*): 112 NaCl, 21.4 $NaHCO_3$, 4.7 KCl, 2.2 $CaCl_2$, 1.1 $MgCl_2$, 22.0 sodium glutamate, 5.0 *N*-2-hydroxyethyl piperazine-*N*-2-ethanesulfonic acid (HEPES), and 5.0 glucose. HEPES was included to raise the buffering power of the bicarbonate solution. Perfusate Po_2 was maintained constant or changed as desired. Next the preparation was covered with a layer of warm paraffin oil. The carotid sinus nerve was then desheathed, placed on a platinum electrode, and lifted into the oil away from the superfusate. The perfusate level covering the carotid body was fixed by adjusting the height of the effluent suction needle. All of the veins and one or two very small arteries were left open so that the perfusate flow was rapid at a finite perfusion pressure. The perfusate flowed by gravity at a pressure difference of 80 to 100 Torr through a water-jacketed glass tubing at a temperature of 37°C. We also measured perfusion pressure, in four preparations, by connecting the external carotid artery to a calibrated transducer and recording its output continuously. This provided an estimate of change of the perfusate pressure during experimental maneuver.

To study the effects of NO, we used the NOS inhibitor L-nitroarginine-methyl ester (L-NAME), which decreases the production of endogenous NO, and sodium nitroprusside (SNP), which raises the supply of exogenous NO, among other associated products. The following protocols consisting of several types of perturbations were used: (a) perfusion with L-NAME (25 to 200 μM) during normoxia; (b) mixture of L-arginine (50 to 500 μM) and L-NAME (50 and 100 μM) during normoxia; (c) SNP (0.5 to 10 μM) during L-NAME administration; (d) SNP perfusion with and without hypoxia; and (e) interruption of perfusate flow with and without SNP (1 to 2 μM) to test if NO blocked response to severe hypoxia.

Separate perfusates were placed in separate bottles and were equilibrated with a continuous flow of gases (O_2, CO_2, and N_2 mixtures). The time lag for the medium to reach the carotid body was less than 5 s. Perfusates for the same gas composition were treated with the same gas mixture from the same source. Samples of the equilibrated perfusates were collected anaerobically in 5-mL glass syringes, and P_{O_2}, P_{CO_2} and pH were checked by measuring them with a blood gas analyzer at 37°C (PHM 73, Radiometer Inc.).

C. Data Analysis

All results are expressed as a mean ± standard error of the mean (SEM). Statistical analysis was evaluated by a paired *t*-test. *P* values less than 0.05 were considered significant.

III. Results*

A. NOS Localization

NOS-like immunoreactive (NOS-LI) nerve fibers and ganglion cells were present in all carotid bodies examined (Fig. 1). Positively stained nerve fibers, both smooth and varicose, coursed throughout the structure, usually in connective tissue septa and in association with blood vessels. Occasionally NOS-LI nerves appeared to lie close to or among type I glomus cells, but specific innervation of these cells could not be detected. Some parts of the NOS-LI innervation of the carotid body were derived from endogenous NOS-positive ganglion cells. The number of positive ganglion cells seen in the carotid body and its closely adjacent tissues ranged from 10 to 30, depending on the animal examined; not all of the parasympathetic ganglion cells within a given organ were NOS-positive. Small nerve bundles of unknown origin in the proximity of the carotid body often contained several NOS-positive axons. The main trunk of the glossopharyngeal nerve distal to the petrosal ganglion also contained a small population of immunoreactive axons, though examination of a limited number of sections from 2 petrosal ganglia demonstrated no positive ganglion cells.

*Adapted from Ref. 23.

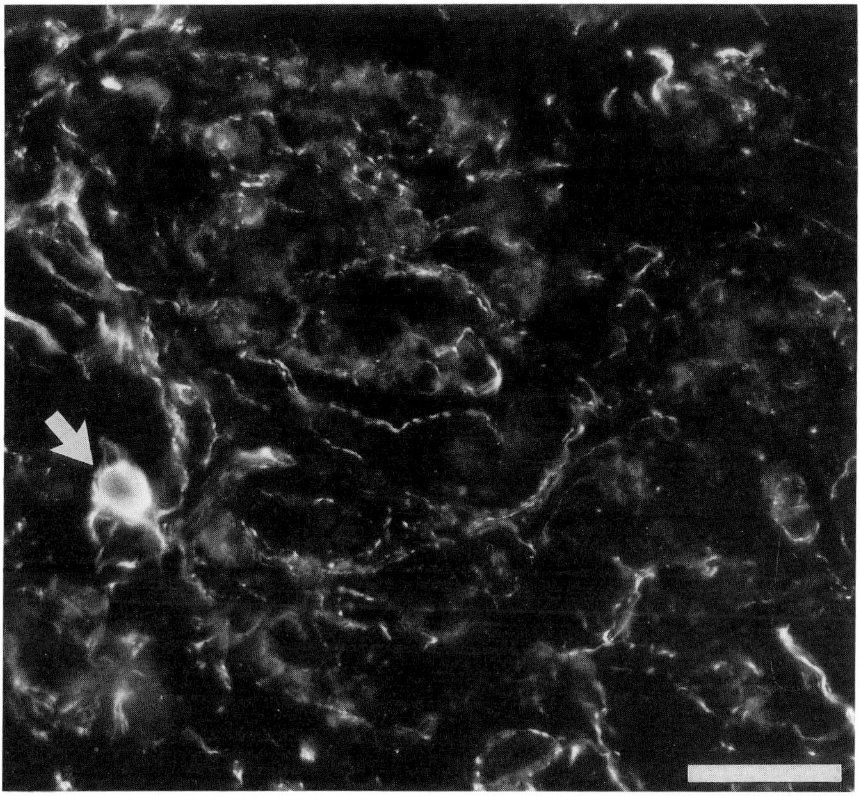

Figure 1 NOS-like immunoreactive nerve fibers, both smooth and varicose, can be seen coursing throughout the organ in association with blood vessels, with connective tissue septa, and at places in close proximity of type I glomus cells. The arrow points to a NOS-positive ganglion cell. (Magnification bar = 50 μm.)

B. Chemosensory Discharge

Figure 2 shows the effects of 100 μM L-NAME during normoxia ($Po_2 = 145$ Torr) on the chemosensory activity and arterial perfusion pressure. Following L-NAME perfusion, after a delay, there was a gradual rise in the chemosensory activity. Three sections of the record over a 17-min period are shown. The perfusion pressure remained fairly stable (95 Torr) throughout the period of increased sensory activity. Figure 2B shows another example of a more rapid response to L-NAME (100 μM) during normoxia ($Po_2 = 122$ Torr). Its withdrawal was followed by a transient decline in the chemosensory activity, but

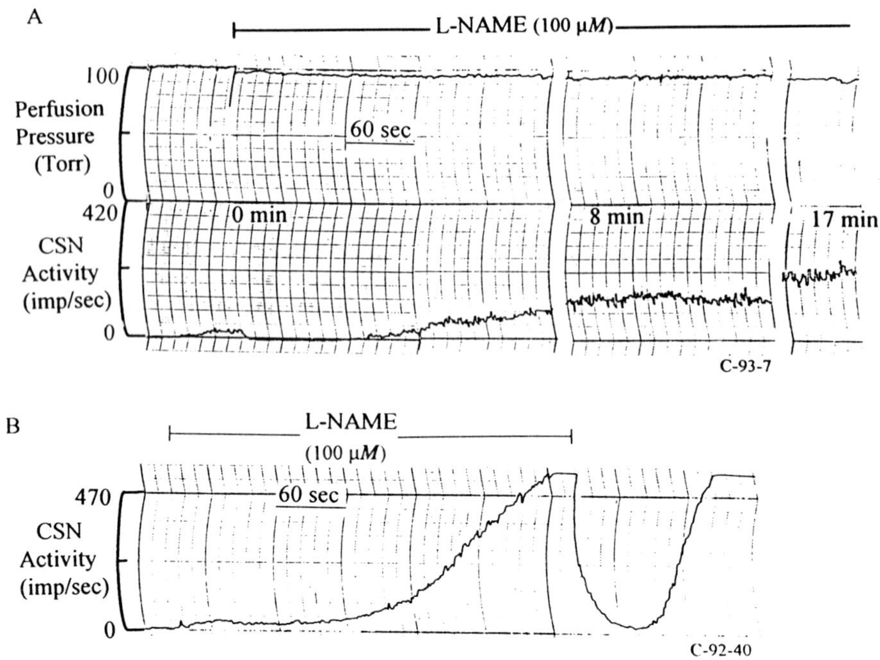

Figure 2 Examples of L-NAME (100 μM)-induced increase in carotid chemosensory activity (during normoxia). The effect developed slowly, more slowly in panel A than in panel B. Withdrawal of L-NAME in panel B was followed by only a transient decrease in the activity. Perfusion pressure remained fairly stable throughout. (Note that the peak activity in panel B reached the set limit of the pen recorder.)

the stimulation returned even after withdrawal of L-NAME. However, after prolonged perfusion, the drug was washed out and activity subsided.

The dose-dependent (25 to 200 μM) effects of L-NAME on the chemosensory (CSN) peak activity is shown in Figure 3. The peak response in the activity increased with the dose but appeared to manifest saturation, although the increase appeared to be maximum at the dose level of 100 to 200 μM.

Figure 4 shows that SNP consistently blocked the stimulatory effect of L-NAME. Panels A and B show that L-NAME (50 μM and 100 μM, respectively) stimulated the activity and that the subsequent administration of SNP promptly and persistently blocked the stimulatory effect during normoxia ($Po_2 = 122$ Torr).

Panel C in Figure 4 shows that 1 μM SNP blocked the oscilaltory activity that developed in the presence of L-NAME (50 μM) and L-arginine (500 μM) during normoxia ($Po_2 = 143$ Torr).

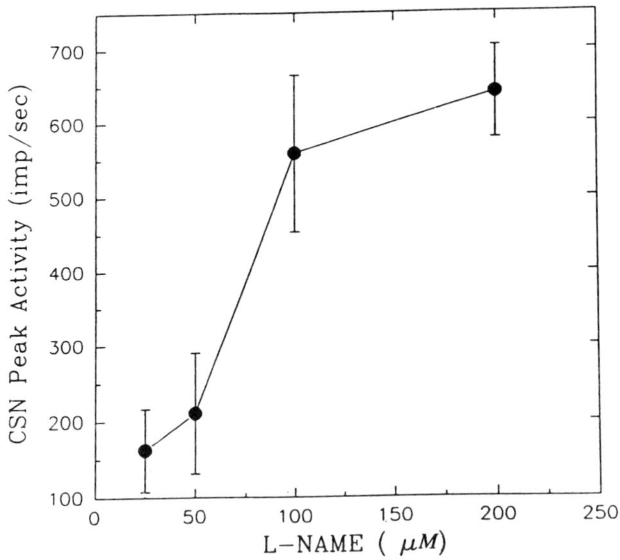

Figure 3 The dose-dependent effect of L-NAME during normoxia. A total of 14 observations were made and plotted as mean ± SEM. The increase in the activity was dose-related.

Blockade of excitation was seen in 11 tests; the effect of L-NAME (100 *μM*) decreased from 467 ± 27 to 61 ± 9 imp/s in response to 1 *μM* SNP.

C. Effect of SNP on the Response to Hypoxia

SNP showed no discernible effect on the baseline normoxic chemosensory activity, but it did attenuate the chemosensory response to hypoxia. An example of graded attenuation is shown in Figure 5: (A) hypoxia alone ($Po_2 = 50$ Torr); (B) 0.5 *μM* SNP plus hypoxia; and (C) 1 *μM* SNP plus hypoxia. The suppressant effect of SNP was graded. The effect of SNP was readily reversed. Large doses of SNP (> 200 *μM*) did not eliminate the response to hypoxia further (not shown).

IV. Discussion

The important findings of this study are as follows: NOS inhibitor L-NAME stimulated carotid chemosensory discharge; SNP strikingly and persistently attenuated the chemosensory excitation by L-NAME; and nerve fibers in the carotid body contain the NOS.

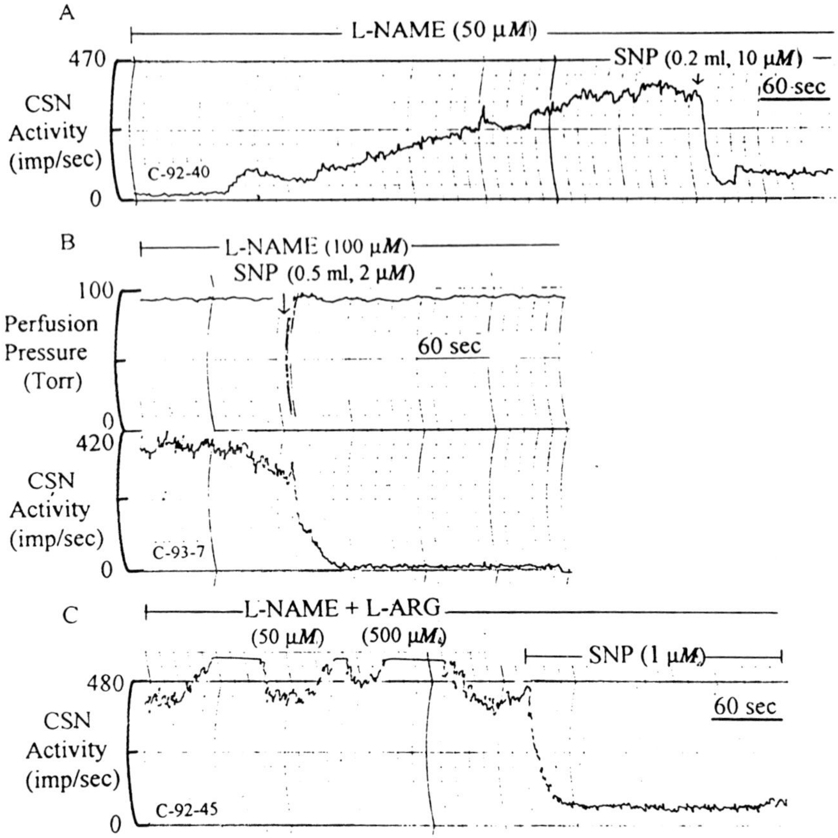

Figure 4 Effects of SNP on the L-NAME-stimulated activity. Panels A, B, and C show the stimulatory effects of L-NAME, which were promptly inhibited by SNP.

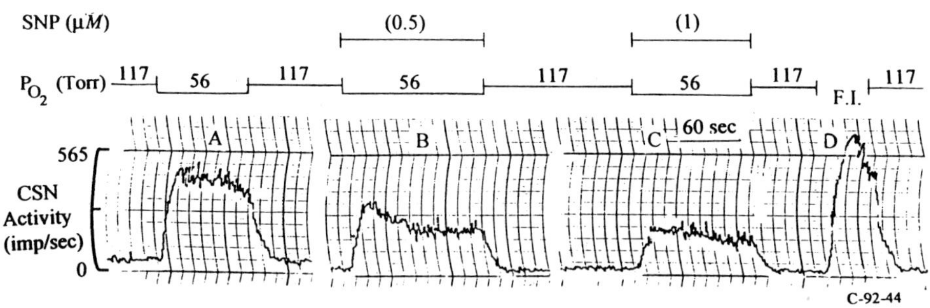

Figure 5 Effects of SNP on the chemosensory responses to hypoxia: (A) hypoxia only; (B) hypoxia and 0.5 μM SNP; (C) hypoxia and 1 μM SNP; and (D) flow interruption (FI) after 1 μM SNP. Larger SNP doses diminished the response to hypoxia but did not block the response to the perfusate flow interruption.

The excitation of chemosensory nerve activity by the NOS inhibitor indicates that a basal release of NO in the carotid body is a part of the regulation of chemosensory nerve activity. NOS was localized mostly in the nerves and in and around the blood vessels. A few resident ganglion cells and their processes also showed a positive reaction. However, not all of the fibers in the carotid sinus nerve stained positive, and the innervation at the glomus cell level could not be traced using light microscopy. Wang et al. (6,7) reported NOS localization in the petrosal ganglion cells and in CB nerves down to glomus cells. Using diaphorase technique, Prabhakar et al. (8) found that it is primarily nerve fibers which stain most positively. Our impression is that the vascular varicose nerves are by far the most strongly NOS-positive. NO seems to be of neural origin, and the slow development of chemosensory excitation that follows NOS inhibition is consistent with the effects of NO presence on blood vessels. The scenario is that the constitutive NO production keeps the vessels dilated. In its absence, constriction occurs, which lowers the CB tissue, P_{O_2}, and sensory excitation.

SNP, an exogenous source for NO, bypassed NOS and accordingly overrode the effect of L-NAME. The attenuating effect of SNP on the chemosensory response to hypoxia presumably means that cGMP production was augmented, thereby inhibiting the discharges (9). The effect need not be hypoxia-specific only. There is evidence that cGMP stimulates Ca^{2+}-ATPase of plasma membrane and reduces $[Ca^{2+}]_i$ (10,11). Since Ca^{2+}-ATPase is generally ubiquitous in the plasma membrane of eukaryotic cells, it is possible that it also operates in the glomus cells. cGMP-dependent protein kinase may also participate, as it does elsewhere (12). Since SNP did not block the effect of extreme hypoxia, it is reasonable to assume that cGMP is not the critical mediator of O_2 chemoreception. The scheme here for the NO effect is that cGMP level in the carotid body is normally high and thus keeps the sensory activity low. Hypoxia decreases cGMP level by inhibiting O_2-dependent NO production and hence releases the sensory mechanism from the inhibition. Hypoxia has been reported to decrease cGMP concentration in the entire CB, which includes smooth muscle, glomus, and other cells (9). However, observations on the cGMP in the carotid body and the sensory discharge are not always consistent. For example, SNP which causes a several fold increase in cGMP concentration during normoxia hardly manifests any effect on the baseline sensory discharge (13).

Pharmacological effects of L-NAME may not be just anti-NOS, because L-arginine did not reverse its effect in this and other studies. Barer et al. (14) reported that the constrictive effect of L-NAME on the pulmonary blood vessels of rats was not reversed by L-arginine and was persistent. More recently Buxton et al. (15) reported that L-NAME exhibits properties of a muscarinic receptor antagonist. This could have contributed to the observed mixed effects of L-NAME. It is important to note, however, that SNP eliminated all of the effects of L-NAME, suggesting that the muscarinic receptor effect is subject to SNP modification.

A. Perspectives

NO production by NOS requires O_2 as a substrate (2). According to Rengasamy and Johns (16), hypoxia ($Po_2 = 28 \pm 5$ Torr) significantly but reversibly decreased V_{max} of NOS in bovine cerebellum homogenate. Thus hypoxia can cause significant reduction in the production of NO and consequently lead to serious physiological dysfunction. At high altitudes, for example, lower levels of NO would lead to pulmonary and systemic hypertension, impotence, distended gut, etc. However, evidence for these NO-related disorders at high altitudes has not been specifically investigated. Guanylate cyclase, which contains a heme group, may also be Po_2-dependent functionally. Because of possible O_2/NO competition, a low Po_2 may augment NO affinity for guanylate cyclase, but NO is expected to have a very high affinity for the heme group anyway (17), and a low Po_2-effect on the NO affinity may not take it farther. However, taken together, NO at low Po_2 is expected to augment guanylate cyclase activity, which would, in turn, increase cGMP level in the CB during hypoxia. This is contrary to a decreased cGMP level in the CB during acute hypoxia (9). Enhanced cGMP production through H_2O_2-related free radicals is another mechanism by which hypoxia could decrease H_2O_2 and hence decrease cGMP concentration (18).

Finally, since NO is inhibitory to the chemosensory response and its release from the peripheral nerve is augmented during neural discharge (1), an explanation for the efferent inhibition of the chemosensory discharge (19) may be found in the pathways involving L-arginine-NOS-NO. Lahiri et al. (20,21) found an augmented efferent inhibition of the chemosensory discharge in chronically hypoxic cats. This was a dopaminergic-related effect. It is possible that NO stimulates dopamine release, as it secretes insulin from pancreatic beta cells (22) and may cause dopaminergic inhibition of the chemosensory response to hypoxia (21). Nitric oxide biology in the carotid body is yet another factor in the complex area of neurotransmission.

Acknowledgment

Supported in part by HL-43413-05. The grant T32-HL-07027-18 provided a research training fellowship to D. K. Chugh.

References

1. Bredt DS, Snyder SH. Nitric oxide, a novel neural messenger. Neuron 1992; 8:3–11.
2. Moncada S, et al. Nitric oxide: Physiology, pathophysiology and pharmacology. Pharmacol Rev 1991; 43:109–142.
3. Knowles RG, et al. Formation of nitric oxide from L-arginine in the central nervous system: A transduction mechanism for stimulation of the soluble guanylate cyclase. Proc Natl Acad Sci USA 1989; 86:5159–5162.

4. Bredt DS, et al. Localization of nitric oxide synthase indicating a neuronal role for nitric oxide. Nature 1990; 347:768–770.
5. Iturriaga R, et al. In-vitro perfused-superfused cat carotid body for physiological and pharmacological studies. J Appl Physiol 1991; 70:1393–1400.
6. Wang ZZ, et al. Nitric oxide synthase in the carotid body (abstr). Soc Neurosci Abstr 1992; 18:1197.
7. Wang ZZ, et al. Physiological role of nitric oxide (NO) in the carotid body (abstr). FASEB J Abstr 1993; 7:A431.
8. Prabhakar NR, et al. Nitric oxide and oxygen chemoreception of the carotid body (abstr). Soc Neurosci Abstr 1992; 18:1197.
9. Wang WJ, et al. Effects of hypoxia on cyclic nucleotide formation in rabbit carotid body in vitro. Neurosci Lett 1989; 105:164–168.
10. Furakawa KI, Nakumara H. Cyclic GMP regulation of the plasma membrane: Ca-Mg-ATPase in vascular smooth muscle. J Biol Chem 1981; 256:3263–3270.
11. Carafoli E. Intracellular calcium homeostasis. Annu Rev Biochem 1987; 56:395–433.
12. Nishizuka Y. Studies and perspectives of protein kinase C. Science 1986; 233:305–312.
13. Haque U, et al. Nitric oxide, a novel chemical messenger in carotid body (abstr). FASEB J Abstr 1993; 7:A 2492.
14. Barer G, et al. Endothelial control of the pulmonary circulation in normal and chronically hypoxic rats. J Physiol 1993; 463:1–16.
15. Buxton ILO, et al. N^G-Nitro L-arginine methyl ester and other alkyl esters of arginine are muscarinic receptor antagonists. Cir Res 1993; 72:387–395.
16. Rengasamy A, Johns RA. Characterization of endothelium-derived relaxing factor/nitric oxide synthase from bovine cerebellum and mechanism of modulation by high and low oxygen tensions. J Pharmacol Exp Ther 1991; 259:310–316.
17. Keilin D. The History of Cell Respiration and Cytochrome. London: Cambridge University Press, 1970:329–332.
18. Wolin MS, et al. Association of pulmonary artery photorelaxation with H_2O_2 metabolism by catalase. Am J Physiol 1991; 261:H 1141–1147.
19. O'Regan RG, Majcherczyk S. Control of peripheral chemoreceptors by efferent nerves. In: Acker H, O'Regan RG, eds. Physiology of the Arterial Chemoreceptors. Amsterdam: Elsevier, 1983:257–298.
20. Lahiri S, et al. Efferent inhibition of carotid body chemoreception in chronically hypoxic cats. Am J Physiol 1983; 245:R 678–683.
21. Lahiri S, et al. Dopaminergic inhibition of carotid body chemoreceptors in chronically hypoxic cats. Am J Physiol 1984; 247:R 24–28.
22. Schmidt HHHW, et al. Insulin secretion from pancreatic beta cells caused by L-arginine-derived nitrogen oxide. Science 1992; 255:721–723.
23. Chugh DK, et al. Nitric oxide-related inhibition of carotid chemosensory nerve activity in the cat. Resp Physiol 1994; 97:147–156.

25

Ventilatory Responses to L-Arginine and Nitric Oxide Control Mechanisms

RICHARD M. MILLIS, IROKA J. UDEINYA, MELVILLE Q. WYCHE JR., and WILLIAM L. WEST

College of Medicine
Howard University
Washington, D.C.

The amino acid L-arginine is the precursor for synthesis of nitric oxide (NO), also identified as the endothelial derived relaxing factor (EDRF), which plays a significant role in the maintenance of resting vasomotor tone and arterial blood pressure in humans. However, in anesthetized rats, administration of L-arginine and the L-arginine antagonist *N*-monomethyl-L-arginine did not affect arterial blood pressure. Therefore, the current studies were designed to determine the effects of L-arginine on ventilation in anesthetized rats, hypothesizing that blood pressure effects might be masked reflexly by NO-induced hyperventilation. In rats anesthetized with chloralose-urethane, arterial blood pressure, heart rate, tidal volume, respiratory frequency, and minute ventilation were continuously monitored for a 50-min interval. In one control group of 10 rats, isovolumetric intravenous doses of physiological saline (vehicle) and D-arginine (controls) were given. In an experimental group of 10 rats, 1.0 mg L-arginine was given and, in a second experimental group, 1.0 mg hemoglobin was administered prior to administration of L-arginine (1.0 mg). D-arginine, vehicle, and L-arginine failed to produce significant changes in heart rate and arterial blood pressure. Increases of minute ventilation were associated with administration of L-arginine: 1.0 mg L-arginine in 0.5 mL saline increased minute ventilation by approximately 15%; hemoglobin effectively inhibited the ventilatory responses to L-arginine. L-arginine increases minute ventilation by a mechanism that might involve NO.

I. Introduction

The L-arginine infusion test has long been established as a means of assessing release of growth hormone and insulin, although release of glucagon and prolactin may also be regulated by circulating L-arginine (1). More recent studies have shown that L-arginine intravenously administered to human subjects also produces a physiologically significant decrement in arterial blood pressure (2–4). It has been presumed that this response results from L-arginine–induced synthesis of an endothelium-derived relaxation factor (EDRF) (5–8), which decreases resting vasomotor tone. The vasoactive principle is thought to be nitric oxide (NO) (9–11), the activity of which is inhibited by oxyhemoglobin (Hb) (9,12–14). D-arginine lacks the capacity for stimulating synthesis of NO (12); therefore the absence of D-arginine–induced vasoreactivity has been used as a reliable control for the vasomotor effects of L-arginine (12).

Changes in arterial blood pressure associated with changes in ventilation are commonly encountered in research on cardiorespiratory control mechanisms (15–17). Changes in arterial blood pressure might be a primary effect or a baroreceptor reflex response secondary to changes in ventilation (16). The close linkage of vasomotor and ventilatory reflexes is further supported by the findings of ventilatory-dependent changes in sympathetic nerve activity in both cats (16) and humans (17). Indeed, vasomotor control centers have been colocalized with ventilatory control centers at the surface of the ventrolateral medulla oblongata, where the central respiratory chemoreceptors are presumed to reside (18,19).

The inability of researchers to produce consistent decrements of arterial blood pressure in laboratory rats following the administration of L-arginine to stimulate NO synthesis has been attributed to the usage of anesthesia (2); however, blood pressure changes resulting from arteriolar relaxation might be masked by ventilation-induced cardiovascular responses. Therefore, we hypothesized that when L-arginine is administered to anesthetized rats, decrements in blood pressure are inhibited by NO-induced hyperventilation.

The present study has shown that the intravenous administration of L-arginine to anesthetized rats increases minute ventilation dose-dependently by a mechanism which might involve the synthesis of NO. A possible role for NO in respiration in the present study is indicated by the fact that Hb, a specific inhibitor of NO activity (9), inhibited the respiratory effects of administered L-arginine. Administration of D-arginine, a compound that does not result in NO synthesis, did not affect respiration.

II. Methods

Rats (Sprague-Dawley, 250 to 300 g) were anesthetized with chloralose-urethane (40 mg/kg glucochloralose, 200 mg/kg urethane). Physiological measurements were made on tracheotomized animals with indwelling femoral arterial and jugular venous catheters. The catheterized rats were sealed in a plethysmographic chamber at constant body temperature (37 to 38°C) maintained by water circulation under the animals. Blood pressure and respiratory parameters (tidal volume, minute ventilation, and frequency) were monitored by a blood pressure transducer (Cobe strain gauge, supplied by Buxco Electronics Inc., Sharon, CT) and by a plethysmographic chamber low-range differential pressure transducer (Validyne, Model MP45-1) and were recorded using a Buxco Physiological Monitoring System. The mean arterial pressure and heart rate were computed electronically from the pressure pulse; tidal volume, respiratory rate, and minute ventilation from the plethysmographic airflow waveforms by a microcomputer software program (Buxco). Following both electronic and mechanical calibrations, a 10-min period was provided for stabilization of plethysmographic chamber temperature and all subsequent measurements were made during a 50-min interval. Statistical significance of differences between groups was evaluated by Duncan's multiple range test for repeated measures and was determined at $P \leqslant 0.05$.

III. Results

In one control group of 10 rats, isovolumetric (0.5 mL) intravenous doses of physiological saline (vehicle) and D-arginine (controls) were given. In an experimental group of 10 rats, a 1.0-mg dose of L-arginine was given; in a second experimental group, 1.0 mg L-arginine was administered immediately after a 1.0-mg dose of hemoglobin. Significant changes in systolic, diastolic, and mean arterial blood pressures were not detected after intravenous administration of two different doses of L-arginine to anesthetized rats. The same isovolumetric doses of physiological saline (vehicle) and D-arginine (control) failed to produce significant changes in either arterial blood pressure or ventilation. Figure 1 shows that L-arginine increased minute ventilation significantly to levels about 13% above controls during the 50-min experimental interval. Hb (1.0 mg) effectively inhibited the ventilatory responses to L-arginine, decreasing minute ventilation significantly—about 15% below baseline control values.

IV. Discussion

The infusion of L-arginine into humans has been associated with release of growth hormone, prolactin, insulin, and glucagon (1) as well as decrements in

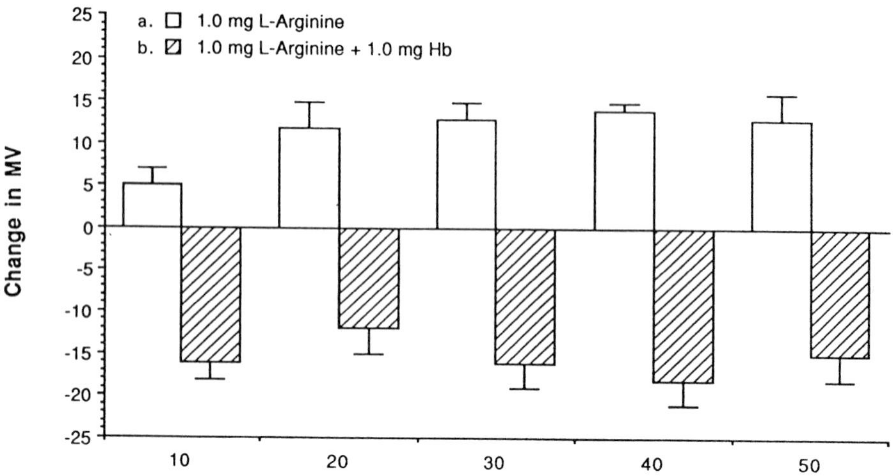

Figure 1 Ventilatory response to L-arginine and hemoglobin. The percent changes from baseline control values in minute ventilation (MV) are shown over a 50-min experimental interval, following intravenous infusion of either (a) 1.0 mg L-arginine alone or (b) 1.0 mg L-arginine immediately after 1.0 mg hemoglobin (Hb). Statistically significant increases above control resulted from L-arginine, and statistically significant decreases below control were produced by L-arginine after hemoglobin.

resting vasomotor tone and arterial blood pressure (2–4). The effects of endothelium-derived NO on peripheral arteriolar tone in humans has been assumed from the decrement in forearm blood flow following infusion of *N*-monomethyl L-argtinine (L-NMMA), a specific inhibitor of NO synthetase, into the brachial arteries of healthy volunteers (20). Other researchers have demonstrated increased mean arterial blood pressure and reversal of acetylcholine-induced systemic hypotension in anesthetized rabbits given L-NMMA (8). The demonstration that intravenous infusion of L-arginine into healthy and hypertensive humans produced a rapid onset of systemic hypotension in both groups (2) was not without controversy, however, since L-arginine has also been thought to release histamine from human skin, and histaminic effects might have accounted for the systemic hypotension associated with L-arginine (21). However, the antihistamine chlorpheniramine was subsequently shown to exert no discernible effect on the L-arginine-induced hypotension (4) in the same subjects originally reported (2). Tachycardia and increased cardiac output have been reported to accompany the decreased peripheral resistance and systemic hypotension associated with L-arginine infusions in humans (3).

Consistent decrements in arterial blood pressure have not been found in anesthetized laboratory animals (2); however, a dose-dependent atropine-sensitive hypotensive effect of very short duration has been produced by L-arginine in unanesthetized rats (22). Oral ingestion of L-NMMA, a selective inhibitor of NO synthesis, in conscious vasopressin-deficient rats produced marked systemic hypertension without bradycardia (23). The fact that the changes in systemic blood pressure were found to be associated with significant constriction in the renal, mesenteric, and hindquarters vasculature (23) suggested that NO-mediated, local control of vasomotor tone in several different vascular beds might contribute to L-arginine–induced systemic hypotension, a response that seems to be independent of mediation by baroreceptor mechanisms.

The present studies demonstrate L-arginine–induced hyperventilation in anesthetized rats in the absence of systemic hypotension. In healthy humans, inhalation of NO has recently been reported to produce pulmonary vasodilation without systemic hypotension (24). Furthermore, NO has been implicated in the control of airway tone by acting as a bronchodilator in guinea pig lungs (25). Additionally, NO aerosolized in the inspired gas of rabbits has been reported to modulate increased airways resistance (26), and treatment of status asthmaticus by inhalation of NO in humans might be possible in the future. Decrements in either the tonus of the airways or of the pulmonary vasculature may stimulate ventilation reflexly. Increased pulmonary blood flow is considered to be a reliable stimulus for hyperventilation in humans (27–30). Hence, in the present studies, the observed hyperventilation following L-arginine administration to anesthetized rats may have resulted from a possible increase in pulmonary blood flow.

Other explanations are also possible, such as L-arginine–mediated stimulation of central nervous system (CNS) cardiorespiratory control sites wherein electrical (15) and chemical (31) stimuli produce increases in both ventilation and arterial blood pressure. Studies using direct application of drugs on the ventral medullary surface suggest a pharmacological basis for colocalization of vasomotor and ventilatory control mechanisms (20,21). More recently, a ventral medullary pathway of catecholamine-containing neurons with synaptic input directly to neurons within the Bötzinger complex, the ventral respiratory group, and respiratory motoneurons has provided a physiological basis for the association between vasomotor control and respiratory pattern generator mechanisms (32). Vasomotor control has been reported to be influenced by NO mechanisms at the ventral medullary surface (33). Although NO synthetase activity has been localized in ventral medullary neurons, double-labeling techniques have failed to identify this enzyme in the adrenergic C1 group neurons thought to be involved in vasomotor control at the ventral medullary surface (34). Hence, putative NO activity at the ventral medullary surface cannot be ignored, and there may be a relationship between NO neuroregulation and the dissociation of ventilatory and vasomotor responses reported herein. The effects of L-arginine administration to anesthetized rats might provide a fortuitous model for studying the linkage between NO and cardiorespiratory control mechanisms at the ventral medullary sites where vasomotor and ventilatory control neurons are intermingled.

Acknowledgment

The authors wish to acknowledge the expert editorial assistance of Dr. DeLee F. Minner in preparation of the manuscript.

References

1. Visek WJ. Arginine needs, physiological state and usual diets. A reevaluation. J Nutr 1986; 116:36–46.
2. Nakaki T, et al. L-arginine-induced hypotension. Lancet 1990; 336:696.
3. Nakaki T, et al. L-Arginine-induced changes in haemodynamics in man. In: Moncadas S, Marletta MA, Hibbs JB Jr, Higgs EA, eds. The Biology of Nitric Oxide: Vol I. Physiological and Clinical Aspects. London: Portland Press, 1992:351–352.
4. Hishikawa K, et al. L-arginine-induced hypotension. Lancet 1991; 337:683–684.
5. Forchgott RF, Zawadski JV. The obligatory role of endothelial cells in the relaxation of arterial smooth muscle by acetylcholine. Nature (Lond) 1980; 288:373–376.
6. Furchgott RF. The role of endothelium in the responses of vascular muscle to drugs. Annu Rev Pharmacol Toxicol 1984; 24:175–197.
7. Griffith TM, et al. The nature of endothelium-derived vascular relaxant factor. Nature (Lond) 1984; 308:645–647.
8. Rees DD, et al. Role of endothelium-derived nitric oxide in the regulation of blood pressure. Proc Natl Acad Sci USA 1989; 86:3375–3378.
9. Palmer RMJ, et al. Nitric oxide release accounts for the biological activity of endothelium-derived relaxing factor. Nature (Lond) 1987; 327:524–526.
10. Ignarro LJ, et al. Endothelium-derived relaxing factor produced and released from artery and vein is nitric oxide. Proc Natl Acad Sci USA 1987; 84:9265–9269.
11. Kelm M, Schrader J. Nitric oxide release from the isolated guinea pig heart. Eur J Pharmacol 1988; 155:317–321.
12. Palmer RMJ, et al. Vascular endothelium cells synthesize nitric oxide from L-arginine. Nature (Lond) 1988; 333:664–666.
13. Auch-Schwelk W, et al. Pressure natriuresis in rats during blockade of the L-arginine/nitric oxide pathway. Hypertension 1992; 19:333–338.
14. Braun M, Schror K. Prostaglandin D_2 relaxes bovine coronary arteries by endothelium-dependent nitric oxide-mediated cGMP formation. Circ Res 1992; 71:1305–1313.
15. Trouth CO, et al. A superficial substrate on the ventral surface of the medulla oblongata influencing respiration. Pflügers Arch 1973; 339:135–152.
16. Boczek-Funcke A, et al. Respiratory-related activity patterns in preganglionic neurones projecting into the cat cervical sympathetic trunk. J Physiol (Lond) 1992; 457:277–296.

17. Seals DR, et al. Respiratory modulation of muscle sympathetic nerve activity in intact and lung denervated humans. Circ Res 1993; 72:440–454.
18. Feldberg W, Guertzenstein PG. Vasodepressor effects obtained by drugs acting on the ventral surface of the brain stem. J Physiol (Lond) 1976; 258:337–355.
19. Millhorn DE, Eldridge FL. Role of ventrolateral medulla in regulation of respiratory and cardiovascular systems. J Appl Physiol 1986; 61:1249–1263.
20. Vallance P, et al. Effects of endothelium-derived nitric oxide on peripheral arteriolar tone in man. Lancet 1989; 2:997–1000.
21. Paton WDM. L-arginine-induced hypotension. Lancet 1990; 336:1016–1017.
22. Cernadas MR, et al. Hypotensive effect of L-arginine in conscious unrestricted rats. Lancet 1990; 336:1017.
23. Gardiner SM, et al. Regional haemodynamic changes during oral ingestion of N^G-monomethyl-L-arginine or N^G-nitro-L-arginine methyl ester in conscious Brattleboro rats. Br J Pharmacol 1990; 101:10–12.
24. Frostell CG, et al. Inhaled nitric oxide selectively reverses human hypoxic pulmonary vasoconstriction without causing systemic vasodilation. Anesthesiology 1993; 78:427–435.
25. Dupuy P, et al. Bronchodilator action of inhaled nitric oxide in guinea pigs. J Clin Invest 1992; 90:421–428.
26. Hogman M, et al. Nitric oxide and airways. Eur Resp J 1993; 6:163–165.
27. Huszczuk A, et al. Ventilatory responses to partial cardiopulmonary bypass at rest and exercise in dogs. J Appl Physiol 1986; 61:575–583.
28. Miyamoto Y, et al. Dynamics of cardiac, respiratory, and metabolic function in men in response to step work load. J Appl Physiol 1982; 52:1198–1208.
29. Wasserman K, et al. Cardiodynamic hyperpnea: Hyperpnea secondary to cardiac output increase. J Appl Physiol 1974; 36:457–464.
30. Weiler-Ravell D, et al. Control of breathing at the start of exercise as influenced by posture. J Appl Physiol 1983; 55:1460–1466.
31. Feldberg W. The ventral surface of the brain stem: A scarcely explored region of pharmacological sensitivity. Neuroscience 1976; 1:427–441.
32. Sun Q-J, et al. Close appositions between tyrosine hydroxylase immunoreactive boutons and respiratory neurons in the rat ventrolateral medulla. J Comp Neurol 1994; 340:1–10.
33. Shaboval LN, et al. Nitric oxide influences ventrolateral medullary mechanisms of vasomotor control in the cat. Neurosci Lett 1991; 132:47–50.
34. Iadecola C, et al. Localization of NADPH diaphorase in neurons of the rostral ventral medulla: Possible role of nitric oxide in central autonomic regulation and oxygen chemoreception. Brain Res 1993; 603:173–178.

26

Retrograde Neuronal Tracer Injected into the Diaphragm Labeled Neurons in the Brainstem: A Possible Direct Brain Connection for the Diaphragm

CLIVEL GEORGE CHARLTON

College of Pharmacy and Pharmaceutical Sciences
Florida A&M University
Tallahassee, Florida

JANEA MACK

Meharry Medical College
Nashville, Tennessee

The diaphragm is under both voluntary and involuntary controls. It has a multifunctional role, serving primarily in respiration but also in coughing, sneezing, defecating, yawning, swallowing, and vocalization. This means that the diaphragm is intricately controlled, which should be reflected in the complexity of its neuronal innervations. To determine whether the neuronal innervations of the diaphragm reflect its multifunctional role, true blue (TB) was injected into the diaphragm of rats with intact nerves or with unilateral transected phrenic or vagal nerves. Horseradish peroxidase (HRP) was also injected into the crushed vagus nerve. Following the injection of TB into the diaphragm, neurons were bilaterally labeled in the phrenic motor nucleus (PMN) as well as in the nucleus ambiguus (NA), in the dorsal motor nucleus of the vagus nerve (DMNV), in the area postrema (AP), and in the caudal regions of the mesencephalic trigeminal nucleus (MTN). Transection of the phrenic nerve did not affect the labeling in the brain, which shows that the tracer was not transneurally transported via the PMN. Tissues prepared from rats with transected vagus nerves showed labeled neurons in the contralateral NA, DMNV, and AP and in the bilateral PMN and MTN to the transected vagus nerve. HRP microinjected into the crushed vagus nerve labeled nerve endings in the diaphragm. Choline acetyl transferase (ChAT)-like immunoreactivity was observed in blood vessels. As controls, the injection of TB into the thoracic or abdominal cavity did not result in labeling the brain neurons. The findings suggest that direct projections from the brain to the diaphragm exist. Axons originating in the NA, DMNV, and AP traverse the vagus nerve to the diaphragm. The NA-DMNV connections may corroborate the finding of blood vessel–associated cholinergic fibers in the diaphragm (1) and may explain the ChAT-like activity seen in blood vessels. It indicates that neurons in the NA may regulate respiratory functions not only by modulating breathing but logically also by modulating vascular perfusion mechanisms. Based on the known functions of the NA, AP, and MTN, the labeled neurons in these areas may be involved in neuronal circuitry that control the reflex involvement of the diaphragm in functions like yawning, coughing, talking, vomiting, and gasping.

I. Introduction

The diaphragm is the principal muscle of respiration. During contraction, its dome descends, the base of the thorax expands, intrathoracic pressure decreases, and abdominal pressure increases. These activities facilitate the expansion of the lung and cause inspiration. Besides respiration, the diaphragm also serves other functions. It contracts strongly during expulsive efforts, probably participating in all efforts requiring the rigidity of the thoracoabdominal system (2,3). Thus the diaphragm serves and maintains functional relationship with the systems that control coughing, sneezing, laughing, vocalization, defecation, yawning, and swallowing. During swallowing, for example, inspirational neurons instantly stop their discharge (4); vomiting is preceded by rhythmic spasmodic respiratory movement; and the expulsion phase of vomiting is brought about by the squeezing action of the diaphragm as well as the abdominal muscles. Most of the diaphragm contracts during both retching and expulsion, but the crural region around the esophagus is virtually inactive (5). This crural relaxation is necessary to augment the relaxed intrinsic smooth muscle sphincter mechanism at the gastroesophageal junction (6). Since the coastal region continues to contract when the crural region relaxes (6), it suggests that these two regions may be under different neural control mechanisms. Furthermore, the fact that the diaphragm participates in these various functions means that there may be still unresolved neuronal control mechanisms for the diaphragm in addition to that directly involving the phrenic motor nucleus (PMN). The normal neural mechanism that influences diaphragmatic functions is generally accepted to involve brainstem respiratory neuronal groups, notably the ventral respiratory group, with which the nucleus ambiguus (NA) is associated. The brain neurons, in turn, influence the PMN, which innervates the diaphragm via the phrenic nerve. In spite of this well-known pathway for the control of breathing, other salient but direct projections from the brain to the diaphragm may exist, especially in light of the findings (7) that the dorsal and ventral respiratory neuronal groups do not have the appropriate firing pattern to activate phrenic motor neurons during vomiting. Neuronal connections from the brain to the diaphragm may make up the circuitry through which the diaphragm is modulated during its participation in expulsion and elimination functions—for example, vomiting, coughing, sneezing, defecation, and vocalization. Such circuitry will not be spontaneously active, neither will the activity of the circuitry be evoked during normal breathing. It means that the connections may elude identification using procedures that will normally identify active respiratory neuronal circuitry. Interestingly, respiratory electrical activity in the diaphragm of cats, dogs, monkeys, and rabbits following bilateral phrenicectomy has been reported (8). Such observation, however, was not supported (9) when it was shown that sectioning of the phrenic nerve roots in rats caused the respiratory electrical activity in the diaphragm to disappear. In his review, Felix (10) had maintained that there is some extraphrenic motor innervation of the diaphragm, but this too was not supported (11–14). More recently, however, documentation of the autonomic innervation of the diaphragm has been reported (1), using neuronal tracing, cholinesterase activity, and fluorescence microscopy for catecholamine. It was suggested that fibers enter the diaphragm from the gut wall. In the present study, neuronal projections to the diaphragm were determined by injecting retrograde neuronal tracers into the diaphragm and looking for labeled cell bodies in the spinal cord and brain.

II. Materials and Methods

Sprague-Dawley male rats were kept in an animal room maintained at a 12-h light/12-h dark cycle; they were given free access to food and water. For the injection of neuronal tracers into the diaphragm, at least seven rats were anesthetized with chloral hydrate (400 mg/kg IP). Each rat was placed on an operating board and a longitudinal incision was made into the abdomen. The intestine was gently placed on the external abdomen, lateral to the incision, and was kept covered and moistened. The liver was gently pushed aside with cotton-tipped applicators to expose the diaphragm, and the diaphragm was injected at about 4 to 5 adjacent sites with about 8 μL of 2% True Blue (TB) (Sigma Chemical Co., St. Louis, MO) using a 10 μL Hamilton syringe. The sites of injection were blotted to remove traces of TB at the surface of the needle tract. The intestines were then repositioned, the wound closed, and the animals allowed to recover in a warmed cage. To determine whether the transport of TB from the diaphragm occurred via the phrenic nerve and/or the vagus nerve, these nerves were transected in two groups of rats prior to the injection of TB into the diaphragm. In another group of rats, the right vagus nerve was exposed and crushed with a plastic-tipped forceps. Horseradish peroxidase (HRP) coupled to cholera toxin B-chain was prepared according to McIlhinney et al. (15) and about 3.0 μL injected into the crushed vagus nerve. As controls, the injection of TB into the thoracic or abdominal cavity was done in two groups of rats to determine whether the labeled brain neurons were due to spillage and uptake of the marker by vagal fibers innervating the esophagus, lungs, and gut. The diaphragms of rats not injected with HRP were used as the HRP control.

For the TB experiments, the rats were reanesthetized with 400 mg/kg chloral hydrate at 4 to 5 days postinjection. They were then perfused transcardially with cold phosphate-buffered saline (PBS), followed by 4% paraformaldehyde (PF) in PBS. The spinal cord and brain were removed, postfixed for about 1 h in cold 4% PF, and placed in 15% sucrose-PBS for 1 day. The tissues were frozen in powdered dry ice and stored at −78°C. Sections 30 μm thick were prepared using a cryostat microtome and mounted on gelatin chrome-alum coated slides. In some experiments, the diaphragm was removed, processed, and examined to verify the injection sites. The sections were viewed with a Zeiss microscope equipped with epifluorescent illumination and with a filter-set G365 for TB visualization. Photomicrographs of the labeled neurons were prepared.

For the HRP experiments, the rats were reanesthetized after 3 days and perfused with 2% glutaraldehyde and 1% PF in PBS, pH 7.4. The diaphragm was removed and sections prepared. The sections were treated for the determination of HRP according to the modification of the methods of Graham and Kornovsky (16), Mesulam (17), and McIlhinney et al. (15). Sections were soaked in 0.1 *M* phosphate buffer, pH 7.4, then preincubated in a solution of 25 μg/mL of tetramethylbenzidine (TMB) and 1 mg/mL of sodium nitroprusside in 0.05 *M* sodium acetate buffer, pH 3.3, for 30 min. This was followed by incubation, three consecutive changes, in a similar solution containing 0.006% hydrogen peroxide for 10 min each. The slices were then stabilized for 5 min in 0.9% sodium nitroprusside made up in 50% ethanol and 0.05 *M* sodium acetate buffer. They were then rinsed in buffer and lightly stained in 0.04% neutral red in 50% ethanol, rinsed, dried, and cover-slipped, using permount, and examined under a bright field microscope for TMB reaction products.

Immunohistochemical localization of ChAT was done in diaphragm slices. The procedure for tissue preparation was similar to that used for the detection of TB, in addition to a modification of the indirect immunohistochemical procedure of Coons (18). The slide-mounted slices were preincubated in 0.3% Triton X-100 in PBS, pH 7.4, for three 5-min periods, then incubated in a similar buffer containing ChAT antiserum (developed in rabbit or normal rabbit serum) as controls for 1 h at room temperature and about 16 h at 4°C. The slices were washed in 0.2% Triton X-100 buffer for three 5-min periods and incubated in a humidity box in the dark for about 30 min in the buffer containing the fluorescent-labeled goat antiserum, developed against rabbit IgG. The slides were then washed for 5 min in buffer and for two consecutive 5-min periods in PBS. The sections were drained and cover-slipped using Fluoromount (Fisher Scientific Co.), and viewed with an epifluorescent microscope.

III. Results

Following the injection of TB into the diaphragm, the examination of cervical spinal cord sections showed that bilateral labeling of neurons occurred in the PMN. Figure 1 shows the labeled neurons in a cross-section tissue slice (A) and a longitudinal tissue slice (B). Figure 1A also highlights the periphery of the spinal ventral gray matter. Neurons were also labeled bilaterally in the nucleus ambiguus (NA), as shown in Figure 1C and D. D shows a low magnification of the NA within the unlabeled brainstem reticular formation. Labeled cells were also identified in the dorsal motor nucleus of vagus (DMNV) and

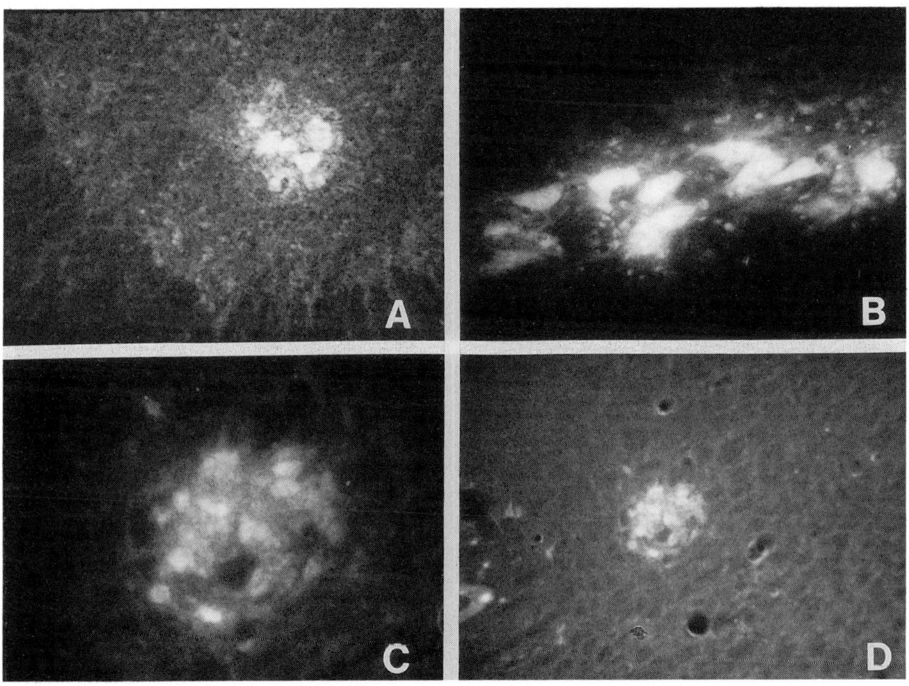

Figure 1 Photomicrographs of neurons labeled in the phrenic motor nucleus (PMN) (A and B) and the nucleus ambiguus (NA) (C and D) following the injection of True Blue (TB) into the diaphragm. (A) Cross section of the PMN; (B) Longitudinal section at a higher magnification. (D) Lower magnification of (C), highlighting the unlabeled reticular formation surrounding the NA.

the area postrema (AP) (Figure 2A and B) as well as in the caudal regions of the mesencephalic trigeminal nucleus (MTN) ventral to the superior cerebral peduncle and adjacent to the locus ceruleus (Fig. 2C and D). Cells were also labeled in the dorsal root ganglia.

Following the injection of TB into the diaphragm in rats with unilaterally transected phrenic nerves, neurons were labeled unilaterally in the PMN, on the side corresponding to the intact phrenic nerve. Bilaterally, labeled neurons were seen in the NA, DMNV, AP, and MTN (Table 1). The results showed that the labeling of brain neurons following the injection of TB into the diaphragm was not by way of the PMN. When the vagus nerve was transected

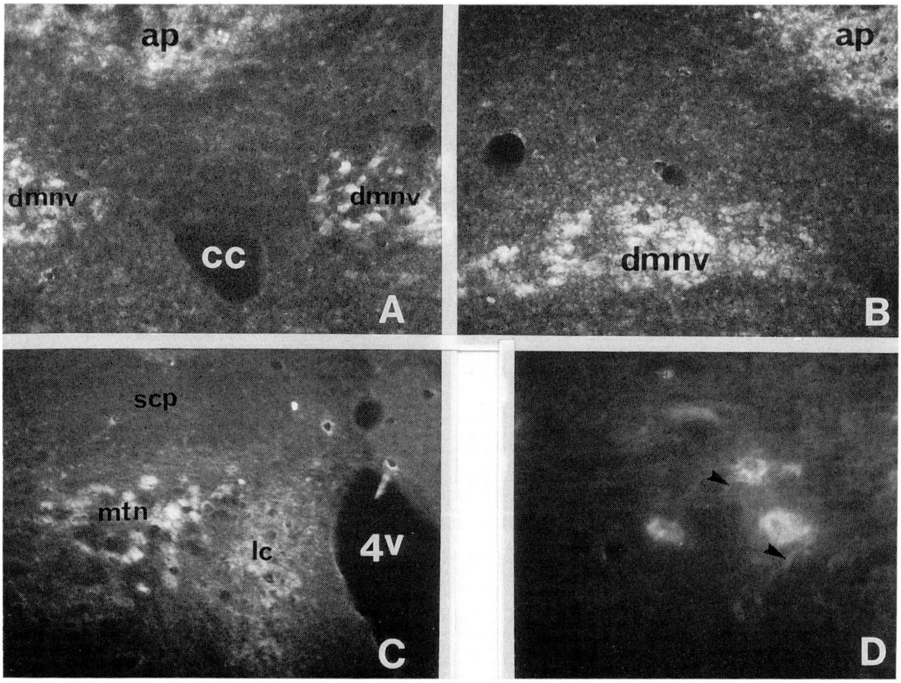

Figure 2 Photomicrographs of neurons labeled in the dorsal motor nucleus of vagus (DMNV) and area postrema (AP) (A and B) and in the mesencephalic trigeminal nucleus (MTN) (C and D) following the injection of TB into the diaphragm. (A) The bilateral DMNV and the ventral AP are highlighted; (B) The entire mediolateral extent of the DMNV is highlighted. Essential landmarks for these neurons are the central canal (cc) seen in (A), and the superior cerebral peduncle (scp), the locus ceruleus (lc) and the fourth ventricle (4v) seen in (C). (D) Higher magnification, identifying the cell nucleus and labeled processes (arrowheads) of neurons labeled in the MTN. The visibility of the lc is due to autofluorescence and not to the TB tracer.

Table 1 Nuclei Labeled Postinjection of Retrograde Neuronal Tracer into Rats with Transected Phrenic Nerves

Nuclei	Ipsilateral	Contralateral
Phrenic motor nucleus	No	Yes
Nucleus ambiguus	Yes	Yes
Dorsal motor nucleus of vagus	Yes	Yes
Mesencephalic trigeminal nucleus	Yes	Yes
Area postrema	Yes	Yes

and rats were injected into the diaphragm with TB, labeled neurons were seen in the bilateral PMN and the MTN but only in the ipsilateral NA, DMNV, and AP; that is, labeling occurred on the same side as the intact vagus nerve (Table 2). This suggests that the labeling of the NA, DMNV, and AP occurs by way of the vagus nerve. The injection of TB into the thoracic cavity or abdominal cavity did not cause TB labeling of these neurons.

To determine the types of terminals that may be associated with the likely vagal projections to the diaphragm, HRP was injected into the crushed vagus nerve, with the objective of labeling nerve endings. Following the injection of HRP into the crushed vagus nerve and processing of slices of the diaphragm for HRP reaction products, it was shown that reaction products occurred in nerve ending–like structures (Fig. 3A, B, and C). Neutral red-stained nerve processes can be seen coursing among the muscle fibers. Some of the fibers bifurcate proximal to their terminals, which contain the HRP reaction products. Large blood vessels (Fig. 3D, b) also seem to contain HRP-reactive products, but not the smaller blood vessels (Fig. 3C, b). These results suggest that the vagus nerve projects to the diaphragm. The determination of ChAT-like immunoreactivity in slices from the diaphragm revealed the occurrence of ChAT-like immunoreactivity in larger blood vessels (Fig. 4A,b) but not in small thin-walled blood vessels (Fig. 4B, b). Nerve fibers containing ChAT-like immunoreactive substances were also detected (Fig. 4D).

Table 2 Nuclei Labeled Postinjection of Neuronal Tracer into Rats with Transected Vagus Nerves

Nuclei	Ipsilateral	Contralateral
Phrenic motor nucleus	Yes	Yes
Nucleus ambiguus	No	Yes
Dorsal motor nucleus of vagus	No	Yes
Mesencephalic trigeminal nucleus	Yes	Yes
Area postrema	No	Yes

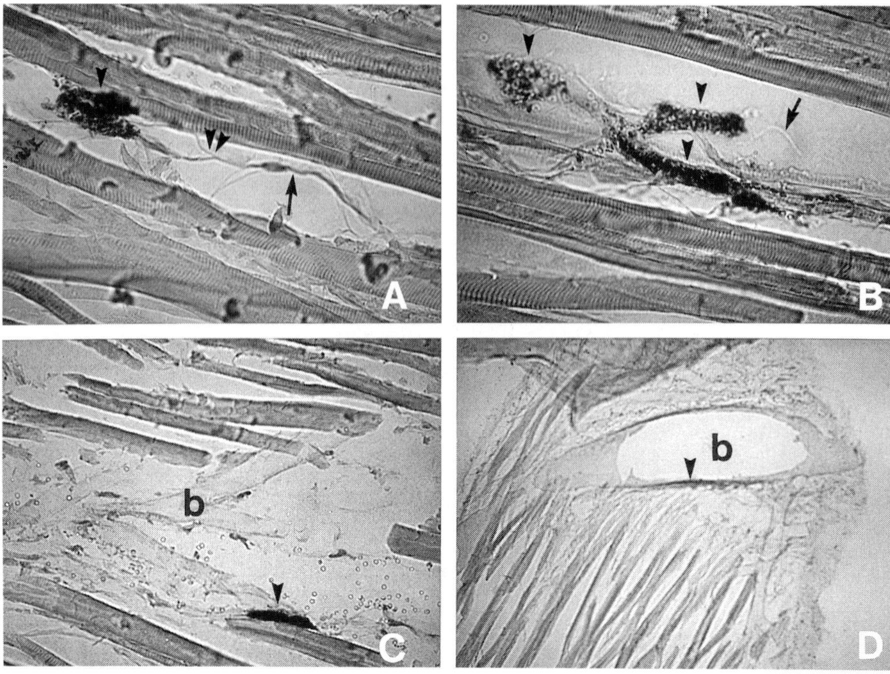

Figure 3 Nerve endings (A, B and C, arrowheads) and a blood vessel (D, arrowhead) containing HRP reaction products following the injection of HRP-conjugate into the crushed vagus nerve and assayed for HRP reaction products in the rat's diaphragm. Small blood vessels (e.g., in panel C did not show HRP reaction products). Nerve fibers (arrows) proximal to the HRP-containing nerve endings and a bifurcated fiber (double arrowheads) can be seen in panels A and B.

IV. Discussion

The labeling of brain neurons following the injection of TB into the diaphragm suggests that direct projections from the brain to the diaphragm exist, probably similar to those observed for the liver, kidney, lung, and intestines. Axons originating in the NA, DMNV, and AP apparently traverse the vagus nerve to the diaphragm. Clearly, the labeling was not due to transsynaptic transport via the phrenic nerve, because transection of the phrenic nerve, which prevented the labeling of neurons in the PMN on the transected side, did not interfere with the labeling of brain neurons. Since the unilateral transection of the vagus nerve prevented the labeling of neurons in the NA, DMNV, and AP ipsilateral to the transected vagus nerve, we may conclude that neurons located in these nuclei send projections to the diaphragm via the vagus nerve. The labeled

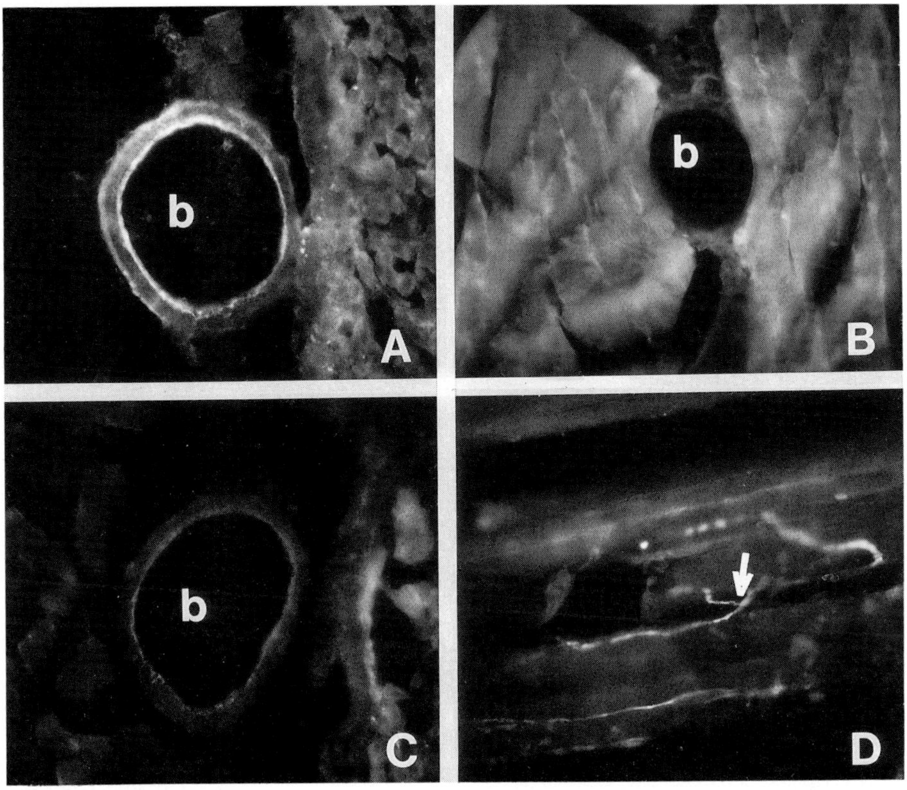

Figure 4 (A) Choline acetyl transferase (ChAT)-like immunoreactivity in the rat's diaphragm. (B) ChAT-like immunoreactivity in a large arterylike blood vessel is highlighted; (C) Significant reduction of ChAT-like immunoreactivity in a control diaphragm slice reacted with normal rabbit serum instead of the immune serum. The photomicrograph in panel B was overexposed to demonstrate the absence of ChAT-like immunoreactivity in vein-like small blood vessels. (D) ChAT-like immunoreactivity in nerve fiber-like processes. These processes resemble and may be identical to similar fibers that were shown to contain HRP-like reaction products; note the bifurcation in panel D (arrow) and in Fig. 3A (double arrowheads).

neurons in the caudal MTN were not affected by either transection, which means that the route of transport or the mechanism for labeling the MTN was not through the phrenic nerve or the vagus nerve.

The NA and DMNV are known to contain cholinergic vagal cell bodies; therefore the neurons labeled in these nuclei are probably cholinergic neurons, which send fibers to the diaphragm. The functions for these brain-to-diaphragm connections are yet to be identified. However, blood vessels associated cholinergic fibers in the diaphragm had been reported (1), and in this study ChAT immunoreactivity was also identified in large artery-like blood vessels. This means that the vagal projections to the diaphragm may be cholinergic and serve to regulate large-vessel blood perfusion, because HRP-like reaction products were confined to large but not to small, thin-walled vein-like blood vessels. By virtue of the brain neuronal localization, these connections seem to be different from the cholinergic sympathetic vasodilator system, originating from neurons in the sympathetic chain. Because these projections may originate in the NA, the ventral respiratory nucleus, which includes the NA, may control respiratory functions, not only by modulating breathing but logically also by regulating vascular perfusion mechanisms.

Vagal nerve endings not associated with blood vessels were also seen. The HRP-labeled nerve endings seen in the diaphragm are distinct from the somatic nerve endings coming from the phrenic nerve, because HRP was specifically injected into the vagus nerve. It is of interest that the HRP-labeled nerve fibers were similar to the nerve fibers that contain ChAT-like immunoreactivity; note, for example, that both labeled fibers bifurcate. The HRP-labeled nerve endings may be presynaptic vagal endings located in the diaphragm. Although HRP-reactive products were also found in the nodose ganglion following the injection of HRP into the crushed vagus nerve, this does not, by itself, support the occurrence of sensory endings in the diaphragm, because fibers originating in the nodose ganglia and traversing the vagus nerve may have picked up the HRP from the injection site. The vagal endings in the diaphragm could also serve a motor function. It is intriguing to propose that these fibers may serve to directly modulate the diaphragm during physiological operations that demand the incorporation of the diaphragm—for example, swallowing, vomiting, talking, and sneezing. Since the NA innervates most laryngeal muscles, the NA may directly coregulate the functions of the larynx and diaphragm. For example, it may cause the abductor laryngeal muscle to close the larynx during coughing and at the same time regulate the diaphragm function to facilitate the coughing reflexes. Similarly, the NA may be involved in regulating the laryngeal muscles during speech as well as the participation of the diaphragm in speech. A dual function may also be expected of the AP, because the chemoreceptive emetic trigger zone is located in the AP. Thus the AP may control the role of the diaphragm in emesis.

The DMNV-to-diaphragm connection could help to control the function of the diaphragm directly. Thus it could facilitate the coordination of the respiratory functions with cardiovascular functions via the vagus nerve. The neurons labeled in the area of the MTN are also located within the vicinity of the so-called pneumotaxic region (19), and the MTN nucleus labeled in this study could have been the specific respiratory-responsive region. The MTN control muscles that play a principal role in gasping, in which the diaphragm is intimately involved. Finally, the labeling of neurons in the brain following the injection of TB into the diaphragm may not be exclusive only to the diaphragm, because preliminary results showed that the injection of TB into other skeletal muscles—for example, the triceps—caused labeling of neurons in the brain. If the tracers are transported retrogradely, and not transsynaptically, direct connections from the brain to skeletal muscles exist, the function of which is more likely to be of a general nature, like the regulation of blood perfusion. Such direct neuronal connections from the brain to skeletal muscles would help to explain the highly discrete and localized manner in which peripheral blood flow can be regulated.

References

1. Azanza Maria J, Garin P. The autonomic innervation of the rat diaphragm. Gen Pharmacol 1986; 17:109–112.
2. Agostoni E, Sant'Ambrogio G, Del Portillo Carrasco H. Electromyography of the diaphragm in man and transdiaphragmatic pressure. J Appl Physiol 1960; 15:1093–1097.
3. Agostoni E. Diaphragm activity and thoracoabdominal mechanisms during positive pressure breathing. J Appl Physiol 1962; 17:215–220.
4. Hukuhara T, Okada H. Effects of deglutition upon the spike discharges of neurones in the respiratory center. Jpn J Physiol 1956; 6:162–166.
5. Monges H, Salducci J, Naudy B. Dissociation between the electrical activity of the diaphragmatic dome and crura muscular fibers during esophageal distention, vomiting and eructation: An electromyographic study in the dog. J Physiol (Paris) 1978; 74:541–554.
6. Altschuler SM, Boyle JT, Nixon TE, et al. Simultaneous reflex inhibition of lower esophageal sphincter and crural diaphragm in cats. Am J Physikol 1985; 249:G586–G591.
7. Miller AD, Tan LK, Lakos SF. Activity of inspiratory neurons in the dorsal and ventral respiratory groups during fictive vomiting (abstr). Soc Neurosci Abstr 1987; 4:1638.
8. Rosenblueth A, Alanis J, Pilar G. The accessory motor innervation of the diaphragm. Arch Int Physiol Biochem 1961; 69:19–25.
9. Sant'Ambrogio G, Frazer DT, Wilson MF, Agostoni E. Motor innervation and pattern of activity of cat diaphragm. J Appl Physiol 1963; 18:43–46.
10. Felix W. Anatomische, experimentelle und klinische Untersuchungen uber den Phrenicus und uber die Zwerchfellinnervation. Dtsch Z Chir 1922; 171:283–397.
11. Kiss F, Ballon HC. Contribution to the nerve supply of the diaphragm. Anat Rec 1929; 41:285–298.
12. Muller Botha GS. The anatomy of the phrenic nerve termination and the motor innervation of the diaphragm. Thorax 1957; 12:50–56.
13. Ogawa T. Studies on the phrenic nerve and the diaphragm of the dog. Amer J Surg 1959; 97:744–748.
14. Wilson AS. Studies on the innervation of the diaphragm in cats and rodents. J Anat (Lond) 1963; 97:482–483.

15. McIlhinney RJA, Bacon SJ, Smith AD. A simple and rapid method for the production of cholera B-toxin coupled to horseradish peroxidase for neuronal tracing. J Neurosci Methods 1988; 22:189–194.
16. Graham RC, Kornovsky MJ. The early stage of absorption of injected horseradish peroxidase in the proximal tubules of mouse kidney: Ultrastructural cytochemistry by a new technique. J Histochem Cytochem 1966; 14:291–302.
17. Mesulam MM. Tetramethylbenzidine for horseradish peroxidase neurohistochemistry: A non-carcinogenic blue reaction product with superior sensitivity for visualizing neural afferents and efferents. J Histochem Cytochem 1978; 26:106–117.
18. Coons AH. General Cytochemical Methods. New York: Academic Press, 1958.
19. Johnson FH, Russel GV. The locus coeruleus as a pneumotaxic center. Anat Rec 1952; 112:348.

27

Cardiovascular and Respiratory Neurons in the Rostral Ventrolateral Medulla of the Rat

JANUSZ LIPSKI, R. KANJHAN, B. KRUSZEWSKA, and W. F. RONG

University of Auckland School of Medicine
Auckland, New Zealand

I. Introduction

There is now considerable evidence that the rostral ventrolateral medulla (RVLM) is a key brainstem site for the integration of many autonomic and somatic functions, including the control of blood pressure and respiration (e.g., Refs. 1–3 and 6). This region is located just caudal to the facial nucleus and extends from the compact formation of the nucleus ambiguus (retrofacial nucleus) to the ventrolateral medullary surface (4). The RVLM is known to contain a heterogenous neuronal population, but the boundaries between various anatomical and functional components have not been clearly defined. Two neuronal groups in this region have attracted particular attention: (a) presympathetic neurons and (b) respiratory neurons of the Bötzinger complex. The RVLM presympathetic neurons project to the intermediolateral cell column in the spinal cord and are believed to be responsible, at least in part, for generating the resting vasomotor tone and for changes in vasomotor output following activation of a variety of cardiovascular reflexes, including the arterial baroreceptor reflex (e.g., Refs. 1, 5, and 6). The Bötzinger complex consists of propriobulbar and bulbospinal respiratory (mainly expiratory) neurons located ventral to the compact formation of the nucleus ambiguus, which are known to play an important role in the network generating respiratory movements (e.g., Refs. 2, 7, and 8). Several previous electrophysiological studies (e.g., Refs. 9–12) reported that these two types of neurons (sympathoexcitatory presympathetic and respiratory) can be found in close proximity, but no detailed maps based on microelectrode recordings have been published. The main aim of the present study was to examine the degree of anatomical and functional overlap between these neurons. In addition, we examined the relationship between these two neuronal groups and the catecholaminergic neurons of the C1 group. These adrenergic neurons have previously been implicated in cardiovascular control (e.g., Refs. 1, 4–6, 11, 13–21).

II. Methods

Experiments were conducted on pentobarbital anesthetized (Nembutal, 80 mg/kg IP, followed by 4 to 6 mg/h IV) adult male Wistar rats ($n = 21$). The animals were paralyzed with pancuronium bromide (Pavulon, 0.4 mg/h), and artificially ventilated with O_2-enriched air. The tidal volume and frequency (60 to 70/min) of ventilation were adjusted to ensure that the end-tidal CO_2 was between 4% and 6% and that the phrenic discharge remained at an approximately constant level throughout the experiment. A bilateral pneumothorax was made and an expiratory load of 2 to 3 cm H_2O was applied. Tracheal pressure (peak less than 10 cm H_2O), arterial blood pressure (mean more than 70 mm Hg measured in the femoral artery), and body temperature (35 to 37°C) were monitored. The dorsal surface of the medulla and spinal cord (C1 to T2 segments) was exposed. Phrenic nerve activity was recorded with standard bipolar electrodes either directly (after amplification), or was full-wave rectified and integrated (time constant, 50 ms). Bipolar stimulating electrodes were placed on one or both aortic depressor nerves, which were dissected several millimeters distal from their junction with the superior laryngeal nerve and cut peripherally (interelectrode distance, 2 mm). The aortic nerves were stimulated at a frequency of 1 Hz with trains of three rectangular pulses 5 ms apart. The duration of the pulses was 0.2 ms and the amplitude less than 1.5 V (less than 10 μA). The stimulus intensity was adjusted to five times the threshold for evoking a depressor response during stimulation of the nerve at 100 Hz for 10 s (the threshold stimulus to elicit the depressor response in individual animals was 0.1 to 0.3 V). Stimulating electrodes were also placed on the mandibular branch of the facial nerve and inside the esophagus, for identification of the facial nucleus and the compact formation of the nucleus ambiguus, respectively (see "Results," below). Glass microelectrodes filled with 5% pontamine sky blue in 0.5 *M* sodium acetate (resistance, 4 to 8 MΩ) were used for extracellular recording from RVLM neurons. Some recording points were marked by iontophoretic injection of the dye (7 μA for 10 min). Two monopolar stimulating electrodes were placed bilaterally in the lateral funiculus at T2 segment for antidromic identification of RVLM neurons projecting to this level (stimulus duration, 0.2 ms; amplitude, up to 1 mA). An additional bipolar stimulating electrode was used for antidromic identification of Bötzinger neurons. It was placed in the region of the nucleus ambiguus (at approximately obex level), either ipsilateral or contralateral to the recording microelectrode.

After recording, the animals were perfused with a fixative (4% formaldehyde and 0.1% glutaraldehyde in 0.1 *M* phosphate buffer). Coronal sections of the medulla 50 μm thick were cut with a Vibratome and counterstained with neutral red. The recording positions marked with pontamine sky blue were drawn on the outline of a coronal section of the medulla taken 200 μm caudal to the caudal border of the facial nucleus, using a normalization procedure that enabled us to compensate for the variation in the distance between the compact formation and both the ventral medullary surface and the midline between individual animals. First, using the histological section containing the deposit of pontamine sky blue, measurements were made of the distance between the center of the compact formation of the nucleus ambiguus and both the ventral surface (*Y* coordinate) and the midline (*X* coordinate). Second, the position of dye deposits was measured and expressed as percentages of these two coordinates. Finally, the recording points were drawn on the outline of a standard section, taking into account their relative distance from the *X* and *Y* axes.

In two animals, the distribution of tyrosine hydroxylase–immunoreactive neurons in the RVLM was examined in 50-μm sections using two different antityrosine hydroxylase antibodies: a monoclonal antibody (Boehringer Mannheim), and a rabbit antibody (Eugene Tech). A standard detection system was used, consisting of biotinylated sheep antimouse (or antirabbit) secondary antibody, ExtrAvidin peroxidase, and diaminobenzidine intensified by nickel (Sigma).

III. Results

Microelectrode tracking was carried out at 75-μm intervals in the mediolateral direction and at 100-μm intervals in the rostrocaudal direction. The explored area extended 800 μm caudally from the caudal border of the facial nucleus, from 1.3 to 2.1 mm lateral to the midline, and ventrally from the compact formation of nucleus ambiguus to near the medullary surface. The location of the facial nucleus was identified by recording antidromic field potentials evoked by stimulation of the ipsilateral mandibular branch of the facial nerve. The caudal boundary of the nucleus could easily be mapped, as described previously (22). The compact formation of the nucleus ambiguus was mapped by recording antidromic field potentials evoked by stimulation of the esophageal wall (23). The motoneurons in this part of the nucleus ambiguus display no spontaneous activity (23). When the microelectrode was advanced just ventral (100 to 300 μm)

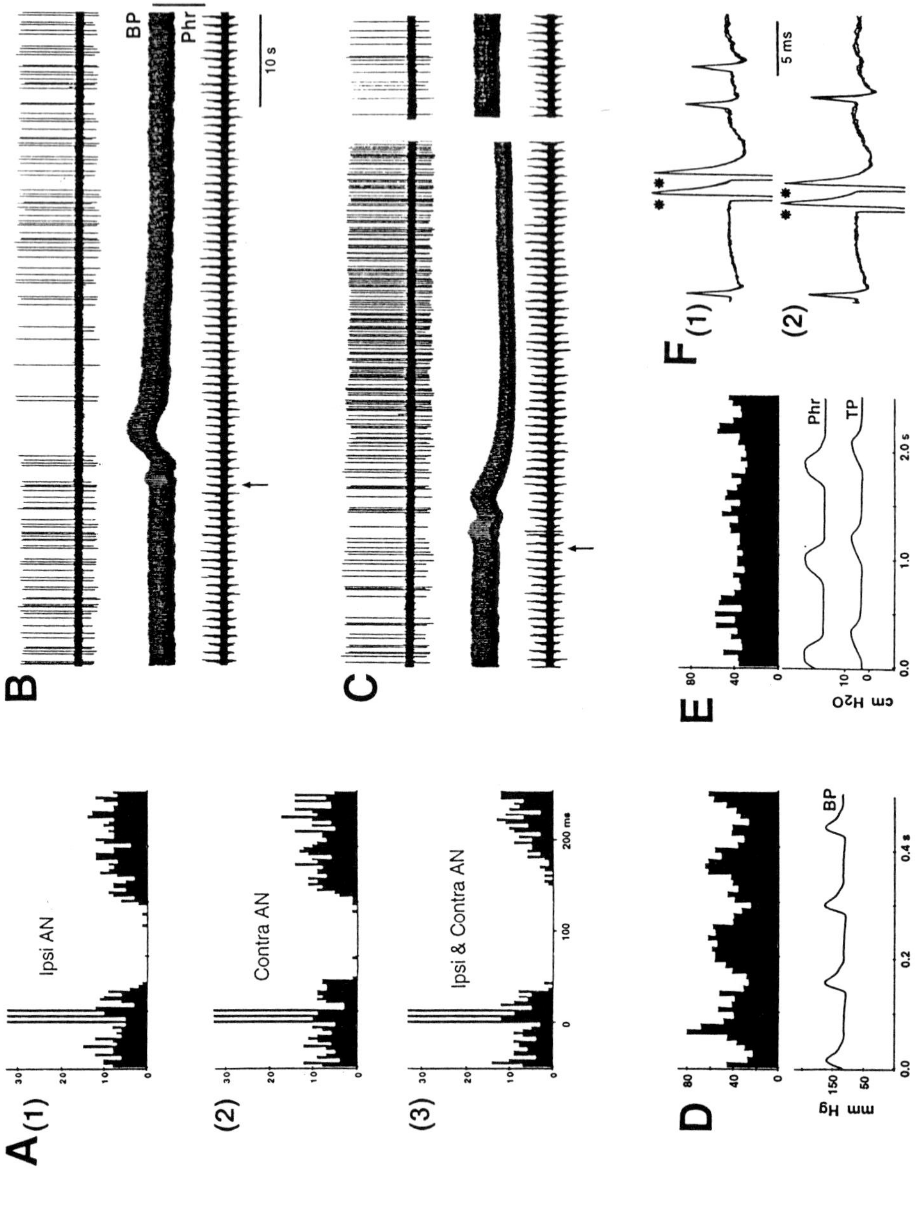
A(1)
Ipsi AN
(2)
Contra AN
(3)
Ipsi & Contra AN
30
20
10
0
0
100
200 ms
B
BP
Phr
10 s
C
D
80
40
0
mm Hg
150
50
BP
0.0
0.2
0.4 s
E
80
40
0
cm H2O
10
0
Phr
TP
0.0
1.0
2.0 s
F(1)
(2)
5 ms

to the compact formation, single units with respiratory modulation, which were thought to be the motoneurons of the semicompact formation of the nucleus ambiguus were recorded (see "Discussion," below). None of these units responded antidromically following stimuli applied through electrodes placed in the spinal cord (T2) or in the medulla. Advancing the microelectrode below the compact and semicompact formations resulted in recording of units which could often be classified into one of two categories: presympathetic neurons or respiratory neurons of the Bötzinger complex.

A. RVLM Presympathetic Neurons

Thirty neurons were classified as presympathetic on the basis of the following criteria: (a) irregular firing pattern, (b) inhibition after stimulation of the aortic depressor nerve(s), (c) response to pharmacologically induced changes of the arterial blood pressure, and (d) axonal projection to the thoracic spinal cord.

The presympathetic neurons showed an irregular pattern of activity, with the frequency ranging from 2 to 46 spikes/s (18 ± 12.3, mean ± SD). Cardiac rhythmicity in the firing was usually not obvious, at least in animals with mean blood pressures less than 100 mm Hg. It was confirmed in all nine tested neurons (all recordings with mean blood pressure >100 mm Hg) with histograms triggered by arterial systole (Fig. 1D). Similarly, respiratory modulation of the firing frequency was not apparent when the activity was monitored by the audio amplifier, or inspected on the chart recorder, at end-tidal CO_2 less than 6%.

Figure 1 Identification of a presympathetic neuron in the rostral ventrolateral medulla. (A) Peristimulus histograms of unit responses to stimulation of the ipsilateral (1), contralateral (2), and (3) both aortic depressor nerves with trains of three pulses. The first impulse in the train was applied at time 0. Each histogram obtained with 200 sweeps. (B) Response of the unit (top trace) to increase in blood pressure (BP; cal. bar, 0 to 100 mm Hg) evoked by IV administration of phenylephrine (arrow). Note no change in activity of the phrenic nerve (Phr). (C) Response to decrease in blood pressure evoked by IV administration of sodium nitroprusside (arrow). Traces as for B. Note a 30-s gap in the record. (D) Histogram triggered by arterial systole (BP) showing unit activity during cardiac cycle, obtained with 500 sweeps. (E) Histogram of unit activity during respiratory cycle triggered with integrated activity of the phrenic nerve (Phr), obtained with 300 sweeps. Note 1:1 relationship between bursts of phrenic nerve and changes in the tracheal pressure (TP) indicating cycles of the ventilator. (F) Antidromic activation following stimulation in the T2 spinal segment (ipsilateral side). Three superimposed sweeps in each record. Top trace (1), paired stimuli (asterisks) delivered 8.5 ms after spontaneous spikes. Bottom trace (2), paired stimuli delivered 8.0 ms after spontaneous spikes. Note in 2 absence of the antidromic response following the first of the two stimuli due to collision and antidromic responses following the second stimulus.

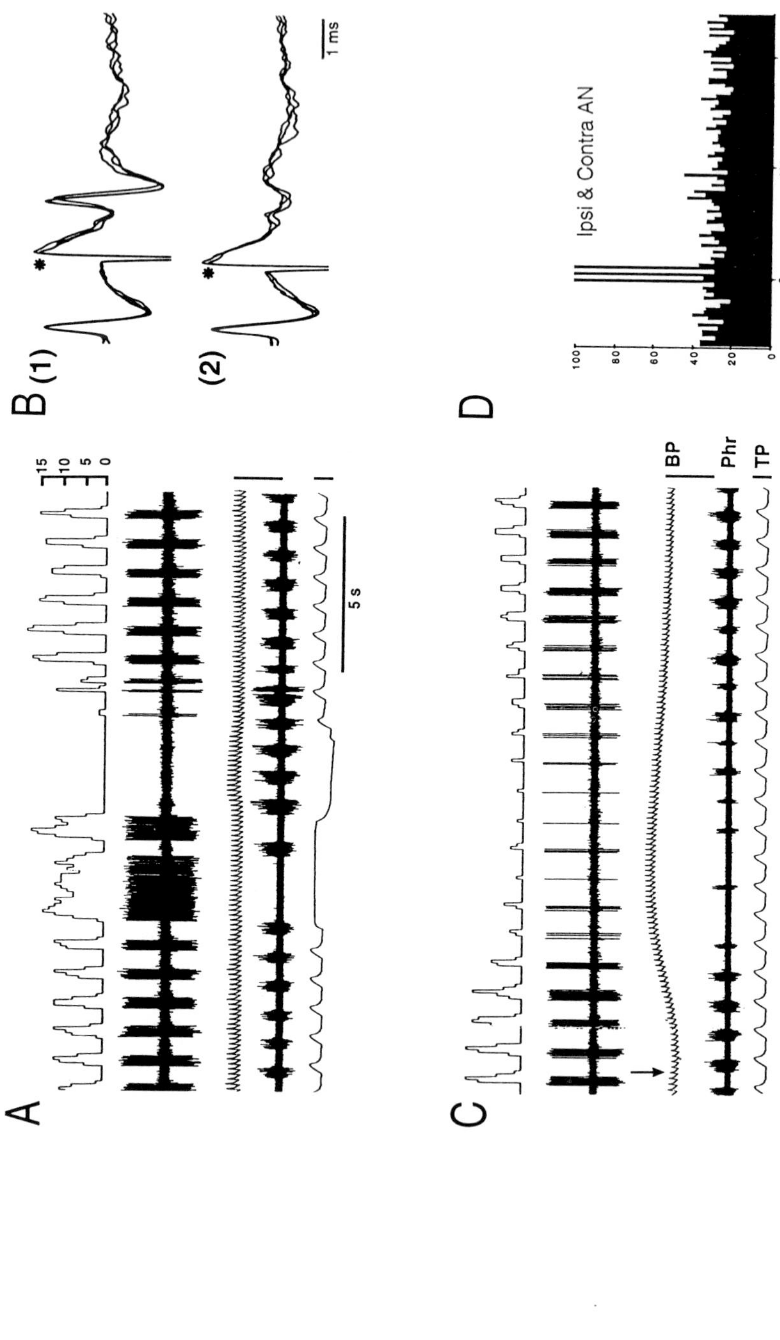
A
15
10
5
0
5 s
B (1)
(2)
1 ms
C
BP
Phr
TP
D
Ipsi & Contra AN
100
80
60
40
20
0
0
100
200 ms

However, a weak respiratory modulation could usually be detected in histograms triggered with bursts of activity of the phrenic nerve (Fig. 1E). Some presympathetic neurons ($n = 13$) were also tested with maintained lung inflations, with pressure of approximately 10 cm H_2O. In contrast to the Bötzinger neurons which were strongly affected by such inflations (see below), none of the tested presympathetic neurons was affected (not illustrated).

Electrical stimulation of the aortic depressor nerve(s) resulted in a transient inhibition of cell firing. This inhibition was observed after stimulation of ipsilateral, contralateral, or (when both nerves were dissected) both aortic nerves and was always confirmed by poststimulus time histograms (Fig. 1A). The latency was 37.9 ± 9.3 ms (mean ± SD) following stimulation of the ipsilateral aortic nerve and 41.9 ± 7.4 ms after stimulation of the contralateral nerve. The duration of inhibition was 59.2 ± 29.8 ms (ipsilateral stimulation) and 70.1 ± 34.4 ms (contralateral stimulation). Nineteen of these cells were tested with intravenous administration of phenylephrine (5 to 10 μg). All were inhibited during the increase in arterial blood pressure at least 40 mm Hg above baseline (Fig. 1B). In addition, seven cells were excited (in all cases the baseline blood pressure was >110 mm Hg) when the blood pressure was lowered with sodium nitroprusside (15 to 25 μg, IV) (Fig. 1C).

Only cells that responded antidromically (threshold range, 15 to 750 μA) following stimulation of the (ipsilateral) spinal cord (Fig. 1F) were included in this study. All antidromic responses were confirmed by the collision test (24). The mean conduction velocity (single point stimulation and assumed linear axonal trajectory) was 5.0 ± 2.3 m/s (range, 1.6 to 10.6 m/s).

B. Respiratory Neurons of the Bötzinger Complex

Twenty-nine units were classified as respiratory neurons of the Bötzinger complex on the basis of two criteria: (a) distinct ("on-off" type) respiratory activity,

Figure 2 Identification of a Bötzinger neuron. (A) Activity in relation to phrenic discharge and lung inflations. Traces from above—frequency of firing (number of spikes per 90 ms bin), unit activity, arterial blood pressure (cal. bar, 0 to 100 mm Hg), activity of the phrenic nerve, and tracheal pressure (cal. bar, 0 to 10 cm H_2O). Note that the unit fires during each ventilator-induced lung inflation. The tonic activity induced by maintained lung inflation ceases during the burst of phrenic activity. (B) Antidromic responses to stimulation in the region of contralateral nucleus ambiguus. Three superimposed sweeps in each record. Top trace (1), antidromic action potentials following stimuli delivered 2.2 ms after spontaneous spikes (stimulus artifacts marked by asterisk). Bottom trace (2), no antidromic response due to collision when stimuli applied 2.0 ms after spontaneous spikes. (C) Response to increase in blood pressure evoked by IV administration of phenylephrine (arrow). Traces as in A. (D) Peristimulus histogram illustrating lack of response to stimulation of both aortic depressor nerves with train of three pulses (1000 sweeps).

with action potentials during expiration and (b) axonal projections to contralateral medulla or caudal parts of the ipsilateral nucleus ambiguus. Only expiratory neurons that could be excited antidromically (latency less than 1.5 ms; mean threshold, 65 μA) following stimuli applied in the region of the nucleus ambiguus (on the ipsi- or contralateral side) were considered to belong to the Bötzinger complex.

These expiratory neurons showed decrementing (postinspiratory) plateaus or (only in a few cases) augmenting patterns of expiratory activity. When the rhythmic activity of the phrenic nerve was inhibited by maintained lung inflation (Hering-Breuer reflex), most Bötzinger neurons ($n=23$) displayed continuous firing which was instantly inhibited during the burst(s) of the phrenic nerve occurring toward the end of the inflation (Fig. 2A). Six Bötzinger neurons were inhibited by lung inflations (not illustrated). None of the tested Bötzinger neurons could be excited antidromically following stimulation of the spinal cord at T2 level.

Thirteen of the 29 Bötzinger neurons were tested with stimulation of the ipsi- and/or contralateral aortic nerve(s), with stimulation parameters similar to those used for identification of presympathetic neurons. No inhibitory response was detected in any of the computed poststimulus histograms (Fig. 2D). Eight units were tested with intravenous administration of phenylephrine. The firing of all these units was inhibited during the rise in blood pressure (Fig. 2C).

C. Distribution of Recording Sites and Relationship to the C1 Group

Recording sites from 20 presympathetic and 18 Bötzinger neurons were marked with extracellular deposits of pontamine sky blue. Figure 3A shows that the majority of recording points were ventral and ventromedial to the compact formation of nucleus ambiguus. However, the two groups were largely separate, with Bötzinger neurons generally more lateral and presympathetic neurons more medial. There was also a tendency for Bötzinger neurons to be more dorsal (closer to the compact formation), while the presympathetic neurons were closer to the ventral medullary surface. There was only a relatively small area of overlap between the recording sites of these two groups of neurons.

Figure 3B shows the location of catecholaminergic cell bodies detected in the RVLM with the monoclonal antityrosine hydroxylase antibody. Tyrosine hydroxylase–immunoreactive neurons extended from the compact formation to near the inferior olive. The lateral part of the area containing these neurons overlapped with the recording sites of presympathetic neurons. However, few catecholaminergic neurons lay within the area of Bötzinger neurons. A similar relationship between the distribution of tyrosine hydroxylase–immunoreactive neurons and the recording positions of presympathetic and Bötzinger neurons was observed when the polyclonal antibody was used (not illustrated).

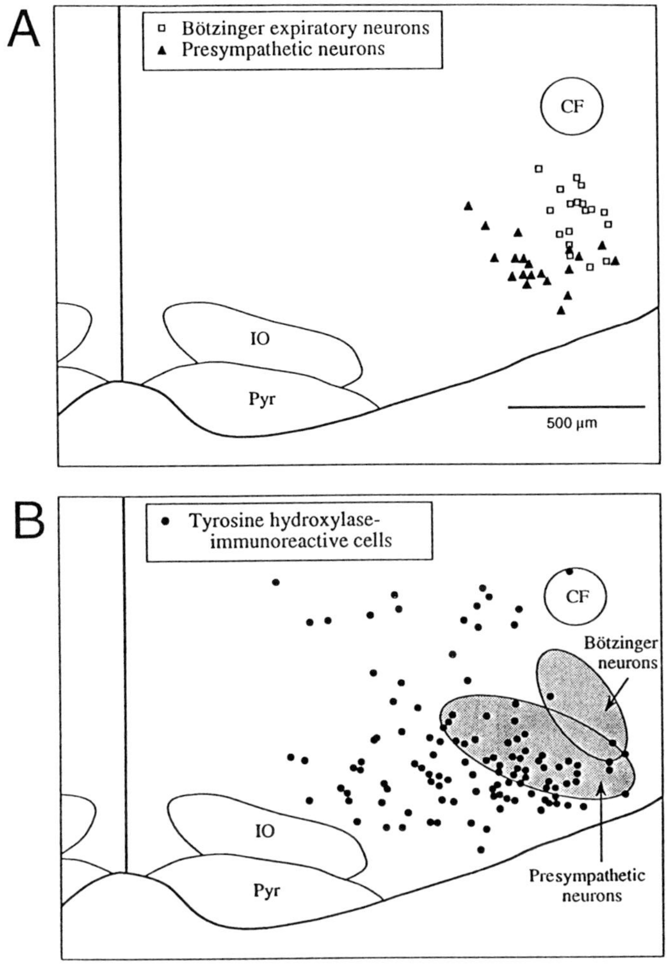

Figure 3 (A) Location of presympathetic neurons (triangles) and Bötzinger expiratory neurons (squares) mapped in relation to the compact formation of nucleus ambiguus, projected onto a transverse section of the medulla 200 μm caudal to the caudal end of the facial nucleus. The rostrocaudal extent for the two groups of neurons (with reference to the caudal border of the facial nucleus) was from 200 μm rostral to 450 μm caudal for presympathetic, and 100 μm rostral to 650 μm caudal for Bötzinger neurons. (B) Location of tyrosine hydroxylase-immunoreactive cells (circles) and projected areas of recording from presympathetic and Bötzinger neurons. The distribution of cells immunoreactive for tyrosine hydroxylase is a composite of four 50-μm coronal sections taken between the caudal border of the facial nucleus and 450 μm caudally. Abbreviations: IO, inferior olive; Pyr, pyramidal tract; CF, compact formation of nucleus ambiguus.

IV. Discussion

The main finding of this study is that presympathetic and Bötzinger expiratory neurons form two functionally distinct neuronal populations, with only a small degree of anatomical overlap. Both groups of neurons have been extensively studied before, but usually in separate studies conducted under different experimental conditions (e.g., differences in species, anaesthesia, monitoring of ventilation). This, to our knowledge, is the first report comparing the two groups of neurons under standardized conditions.

A. Properties of Presympathetic and Bötzinger Neurons

The presympathetic neurons recorded in our study showed many features previously described in the rat by others (e.g., Refs. 10–14, 22, 25–29). These included inhibition following pressor response to phenylephrine, excitation during fall in blood pressure evoked by nitroprusside, cardiac rhythm at appropriate levels of blood pressure, and antidromic activation following stimulation in the rostral thoracic segments. In addition, they showed short-latency (less than 50 ms) inhibition following a burst of stimuli applied to the aortic depressor nerves. The inhibitory responses to aortic nerve stimulation have been reported (28), but it was not stated whether the responses were evoked from the ipsi- or contralateral aortic nerve. Our study demonstrates a convergence of inhibitory inputs from both nerves. Previous experiments with a lesion of the nucleus of the solitary tract on one side and another in the RVLM on the opposite side (15) suggest that the inhibition of presympathetic neurons from the contralateral aortic nerve is mediated by primary afferents crossing the midline. However, this interpretation is not supported by the finding that the aortic nerve in the rat projects predominantly to the ipsilateral nucleus of the solitary tract (30). Therefore the pathway for this crossed inhibition remains to be established. In addition, we demonstrated a convergence of inhibitory input onto presympathetic neurons from the aortic nerve and baroreceptor afferents originating in the carotid sinus and/or the heart. This conclusion is based on the observation that in spite of cutting both aortic nerves, all tested neurons were inhibited during the pressor response evoked by phenylephrine.

Another previously reported feature of RVLM presympathetic neurons, both in the rat and the cat, is their respiratory modulation, with increased firing often occurring in the rat during expiration (10,11,26,31). However, the present study shows that this modulation is highly probablistic in nature during normocapnia. This contrasts with the strong "on-off" respiratory (expiratory) activity characteristic of Bötzinger neurons. The expiratory neurons of the Bötzinger complex, previously studied mainly in the cat, are inhibitory interneurons that through extensive axonal arborizations, monosynaptically inhibit several other subpopulations of respiratory neurons in the medulla and spinal cord (e.g., Refs. 7, 32, 33). There are only a few reports describing these neurons in the rat (34,35), but a recent intracellular study conducted in this species concluded that many features of these neurons in the rat are similar to those previously described in the cat (8). The main problem with identification of Bötzinger neurons is that they show patterns of activity identical to those observed in motoneurons located in the adjacent medullary region (in the semicompact formation of the nucleus ambiguus) which innervate the striated muscles of the upper airways (8,36). However, in contrast to Bötzinger neurons, which have axonal arborizations projecting to both the ipsi- and contralateral sides of the medulla, these motoneurons have no axon collaterals (8,33). This feature allowed us to distinguish between these two types of neurons using antidromic stimulation in the medulla. This distinction was not made in many previous studies (e.g., Refs. 9, 10–12, 26, 34).

Similar to the response observed in presympathetic neurons, the activity of Bötzinger neurons was inhibited when the arterial blood pressure was increased with phenylephrine (but see Refs. 12, 22, and 25, reporting little or no response of unidentified respiratory neurons to increase in blood pressure). However, unlike the presympathetic neurons, Bötzinger neurons did not respond to stimulation of the aortic depressor nerve(s) with a short train of pulses. Therefore the responses to blood pressure increase and aortic nerve stimulation must be due to different central pathways. The baroreceptor-induced inhibition of presympathetic neurons is believed to be mediated through an oligosynaptic pathway involving second-order neurons in the nucleus of the solitary tract, and inhibitory (GABAergic) interneurons in the caudal ventrolateral medulla which directly inhibit presympathetic neurons in the RVLM (16,26,28). The inhibitory response of Bötzinger neurons is probably mediated through a polysynaptic pathway, which requires a greater degree of temporal facilitation. We speculate that the inhibition of Bötzinger neurons is secondary to inhibition of some other elements in the reticular formation which provide excitation to the expiratory (and presumably inspiratory) medullary neurons responsible for the respiratory motor output. If this is the case, the response of the Bötzinger neurons during increase of the arterial blood pressure would result from disfacilitation rather than direct inhibition.

B. Location

Several maps of spinally projecting, barosensitive rat RVLM neurons have been published (14,25,27,29). The distribution of presympathetic neurons recorded in our study was similar. The recording sites were within the lateral part of the nucleus paragigantocellularis lateralis and the ventromedial part of the rostroventrolateral reticular nucleus, as defined by Paxinos and Watson (interaural, −2.8 to −3.3 mm) (37). Maps illustrating the distribution of Bötzinger neurons (in the coronal plane) have previously been published for the cat (38–40) but not the rat. Our map indicates that the perikarya of these respiratory neurons only partially overlap with the location of presympathetic neurons and that the majority of expiratory neurons are located dorsolateral to the main concentration of presympathetic neurons.

The distribution of RVLM tyrosine hydroxylase–immunoreactive cell bodies (C1 adrenergic cells containing phenylethanolamine *N*-methyltransferase) was generally similar to that reported in the rat in previous studies (e.g., Refs. 4, 16, 17, 35, 41–43). It has been reported that some of these catecholaminergic cells project rostrally (e.g., to the dorsolateral pons, locus ceruleus, and hypothalamus) (11,18,19,43) and/or to the thoracic spinal cord, where they make synaptic connections with preganglionic sympathetic neurons (Refs. 20,21; but see Ref. 44). It has been estimated that approximately half of all spinally projecting RVLM neurons are tyrosine hydroxylase–immunoreactive (19). However, there is no direct evidence that barosensitive presympathetic neurons recorded in our and other extracellular studies contain this enzyme (cf. Refs. 42, 45). In some previous studies, an inhibitory response to microelectrophoretic or systemic administration of clonidine was taken as an indication that the responding neurons were catecholaminergic (11,13,46). In addition, it has been shown that clonidine-sensitive neurons had spinally projecting axons with slower conduction velocities than neurons which did not respond to this drug (11,13, 46,47). The response to clonidine was not tested in our study, but the relatively fast conduction velocity of spinal axons (5.0 ± 0.4 m/s) suggests that most of these neurons were noncatecholaminergic. However, the conduction velocity cannot be used as the only criterion, as it has been argued that even RVLM neurons with myelinated, faster-conducting axons (up to 5.3 m/s) can be tyrosine hydroxylase–immunoreactive (14). Our map (Fig. 3B) confirms previous observations that there is a large overlap between the location of the catecholaminergic and presympathetic neurons (14,27) (see also Ref. 42). However, tyrosine hydroxylase–immunoreactive cell bodies are also located in more medial aspects of the RVLM where no presympathetic neurons were found in our study.

Examination of the distribution of recording sites for the Bötzinger group revealed that these respiratory neurons are located mainly lateral to the C1 neurons and that there is only a small overlap in the location of these two neuronal groups (as well as between the Bötzinger neurons and presympathetic neurons; see above). This conclusion is supported by results of a study which used retrograde labeling of RVLM neurons (17), as well as by a previous study conducted in our laboratory demonstrating that none of the RVLM expiratory neurons labeled intracellularly with Lucifer yellow were immunoreactive for tyrosine hydroxylase (35).

V. Summary and Conclusions

This study was undertaken to examine the degree of functional and anatomical overlap between two major neuronal groups in the rostral ventrolateral medulla

(RVLM): presympathetic (sympathoexcitatory) neurons and expiratory neurons of the Bötzinger complex. Recordings from both types of neurons were made in anesthetized rats with extracellular microelectrodes, and recording sites were marked with deposits of a dye. In addition, immunohistochemistry was used to locate tyrosine hydroxylase-immunoreactive cell bodies (C1 adrenergic neurons) in this medullary region.

Presympathetic neurons were inhibited by a burst of stimuli applied to the aortic depressor nerve(s) and by a pressure rise evoked by IV administration of phenylephrine. They displayed cardiac rhythm when the mean arterial blood pressure was above 100 mm Hg but showed little or no response to changes of lung volume. All ($n = 30$) responded antidromically following stimulation in the T2 spinal segment. Bötzinger neurons ($n = 29$) were identified by on-off expiratory activity and antidromic responses to stimulation within the nucleus ambiguus but not to stimulation at the T2 segment. They showed no inhibition following stimulation of the aortic nerve(s) but were inhibited by phenylephrine and strongly affected by changes of lung volume. The recording positions for presympathetic and Bötzinger expiratory neurons were largely separate, with presympathetic neurons mainly located ventromedial to the Bötzinger neurons. There was a large overlap between the location of presympathetic neurons and the lateral part of the C1 adrenergic group. On the other hand, there was only a small overlap between the location of these adrenergic neurons and Bötzinger neurons.

These results indicate that the RVLM cardiovascular (presympathetic) and respiratory (Bötzinger) neurons form two electrophysiologically distinct neuronal populations, with only a small degree of anatomical overlap. Thus it is unlikely that these neurons are elements of a common cardiorespiratory network of the type proposed by Richter and Spyer (48). A large overlap between the location of the C1 adrenergic and the presympathetic neurons is consistent with results of previous studies and suggests that some presympathetic neurons may be catecholaminergic. Further studies combining electrophysiology, anatomy, immunohistochemistry, and pharmacology are needed to examine the relationship among presympathetic, Bötzinger, and other chemically identified neurons in the RVLM.

References

1. Guyenet PG. Role of ventral medulla oblongata in blood pressure regulation. In: Loewy AD, Spyer KM, eds. Central Regulation of Autonomic Functions. New York: Oxford University Press, 1990:145–167.
2. Duffin J, et al. Breathing rhythm generation: Focus on the rostral ventrolateral medulla. NIPS. In press.
3. Lovick TA, Li P. Integrated function of neurons in the rostral ventrolateral medulla. In: Ciriello J, Caverson MM, Polosa C, eds. Progress in Brain Research, vol. 81. New York: Elsevier, 1989:223–232.

4. Ruggiero DA, et al. Central control of the circulation by the rostral ventrolateral reticular nucleus: Anatomical substrates. In: Ciriello J, Caverson MM, Polosa C, eds. Progress in Brain Research, vol 81. New York: Elsevier, 1989:49–79.
5. Reis DJ, et al. The C1 area of the rostral ventrolateral medulla oblongata: A critical brainstem region for control of resting and reflex integration of arterial pressure. Am J Hypertens 1989; 2:363S–374S.
6. Dampney RAL. The subretrofacial vasomotor nucleus: Anatomical, chemical and pharmacological properties, and role in cardiovascular regulation. Prog Neurobiol 1994; 42:197–227.
7. Ezure K. Synaptic connections between medullary respiratory neurons and considerations on the genesis of respiratory rhythm. Prog Neurobiol 1990; 35:429–450.
8. Bryant TH, et al. Expiratory neurons of the Bötzinger complex in the rat: A morphological study following intracellular labelling with biocytin. J Comp Neurol 1993; 335:267–282.
9. Huangfu D, et al. Rostral ventrolateral medullary neurons projecting to locus coeruleus have cardiorespiratory inputs. Brain Res 192; 598:67–75.
10. Haselton JR, Guyenet PG. Central respiratory modulation of medullary sympathoexcitatory neurons in rat. Am J Physiol 1989; 256:R739–R750.
11. Haselton JR, Guyenet PG. Electrophysiological characterisation of putative C1 adrenergic neurons in the rat. Neuroscience 1989; 30:199–214.
12. Sun M-K, Spyer KM. Nociceptive inputs into rostral ventrolateral medulla-spinal vasomotor neurons. J Physiol 1991; 436:685–700.
13. Sun M-K, Guyenet PG. Effect of clonidine and γ-aminobutyric acid on the discharges of medullo-spinal sympathoexcitatory neurons in the rat. Brain Res 1986; 368:1–17.
14. Morrison SF, et al. Reticulospinal vasomotor neurons of the rat rostral ventrolateral medulla: Relationship to sympathetic nerve activity and C1 adrenergic cell group. J Neurosci 1988; 8:1286–1301.
15. Granata AR, et al. Brain stem area with C1 epinephrine neurons mediate baroreflex vasodepressor responses. Am J Physiol 1985; 248:H547–H567.
16. Chalmers J, et al. Amino acid neurotransmitters in the central control of blood pressure and in experimental hypertension. J Hypertens 1992; 10(suppl 7):S27–S37.
17. Ellenberger HH, et al. Subnuclear organisation of the lateral tegmental field of the rat: II. Catecholamine neurons and ventral respiratory group. J Comp Neurol 1990; 294:212–222.
18. Haselton JR, Guyenet PG. Ascending collaterals of medullary barosensitive neurons and C1 cells in rats. Am J Physiol 1990; 258:R1051–R1063.
19. Tucker DC, et al. Organisation of central adrenergic pathways: I. Relationships of ventrolateral medullary projections to the hypothalamus and spinal cord. J Comp Neurol 1987; 259:591–603.
20. Ross CA, et al. Rostral ventrolateral medulla: Selective projections to the thoracic autonomic cell column from the region containing C1 adrenaline neurons. J Comp Neurol 1984; 228:168–185.
21. Milner TA, et al. Phenylethanolamine N-methyltransferase-containing terminals synapse directly on sympathetic preganglionic neurons in the rat. Brain Res 1988; 448:205–222.

22. Brown DL, Guyenet PG. Electrophysiological study of cardiovascular neurons in the rostral ventrolateral medulla in rats. Circ Res 1985; 56:359–369.
23. Kruszewska B, et al. An electrophysiological and morphological study of esophageal motoneurons in the rat. Am J Physiol 1994; 266:R622–R632.
24. Lipski J. Antidromic activation of neurons as an analytic tool in the study of the central nervous system. J Neurosci Meth 1981; 4:1–32.
25. Brown DL, Guyenet PG. Cardiovascular neurons of brain stem with projections to spinal cord. Am J Physiol 1984; 247:R1009–R1016.
26. Koshiya N, et al. Ventrolateral medulla and sympathetic chemoreflex in the rat. Brain Res 1993; 609:174–184.
27. Morrison SF, Reis DJ. Reticulospinal vasomotor neurons in the RVL mediate the somatosympathetic reflex. Am J Physiol 1989; 256:R1084–R1097.
28. Jeske I, et al. Identification of baroreceptor reflex interneurons in the caudal ventrolateral medulla. Am J Physiol 1993; 264:R169–R178.
29. Wang WH, Lovick TA. Inhibitory serotoninergic effects on rostral ventrolateral medullary neurons. Pflügers Arch 1992; 422:93–97.
30. Ciriello J. Brainstem projections of aortic baroreceptor afferent fibres in the rat. Neurosci Lett 1983; 36:37–42.
31. McAllen RM. Central respiratory modulation of subretrofacial bulbospinal neurons in the cat. J Physiol (Lond) 1987; 388:533–545.
32. Lipski J, Merrill EG. Electrophysiological demonstration of the projection from expiratory neurons in the rostral medulla to contralateral dorsal respiratory group. Brain Res 1980; 197:521–524.
33. Jiang C, Lipski J. Extensive monosynaptic inhibition of ventral respiratory group neurons by augmenting neurons in the Bötzinger complex in the cat. Exp Brain Res 1990; 81:639–648.
34. Ezure K, et al. Distribution of medullary respiratory neurons in the rat. Brain Res 1988; 455:262–270.
35. Pilowsky PM, et al. An intracellular study of respiratory neurons in the rostral ventrolateral medulla of the rat and their relationship to catecholamine-containing neurons. J Comp Neurol 1990; 301:604–617.
36. Bieger D, Hopkins DA. Viscerotopic representation of the upper alimentary tract in the medulla oblongata of the rat: The nucleus ambiguus. J Comp Neurol 1987; 262:545–562.
37. Paxinos G, Watson C. The Rat Brain in Stereotaxic Coordinates. 2nd ed. New York: Academic Press, 1986.
38. Merrill EG, et al. Origin of the expiratory inhibition of the nucleus tractus solitarius inspiratory neurones. Brain Res 1983; 263:43–50.
39. Fedorko L, Merrill EG. Axonal projections from the rostral expiratory neurons of the Bötzinger complex to medulla and spinal cord in the cat. J Physiol (Lond) 1984; 350:487–496.
40. Otake K, et al. Morphology of expiratory neurons of the Bötzinger complex: A HRP study in the cat. J Comp Neurol 1987; 258:565–579.
41. Kalia M, et al. Rat medulla oblongata: III. Adrenergic (C1 and C2) neurons, nerve fibres and presumptive terminal processes. J Comp Neurol 1985; 233:333–349.
42. Sun M-K, et al. Rostral ventrolateral medullary neurons with intrinsic pacemaker properties are not catecholaminergic. Brain Res 1988; 451:345–349.

43. Guyenet PG, Young B. Projections of nucleus paragigantocellularis lateralis to locus coeruleus and other structures in the rat. Brain Res 1987; 406:171–184.
44. Llewellyn-Smith IJ, et al. There are few catecholamine- or neuropeptide Y-containing synapses in the intermediolateral cell column of rat thoracic spinal cord. Clin Exp Pharmacol Physiol 1991; 18:111–116.
45. Granata AR, Kitai S. Intracellular analysis in vivo of different barosensitive bulbospinal neurons in the rat rostral ventrolateral medulla. J Neurosci 1992; 12:1–20.
46. Huangfu D, et al. A5 noradrenergic unit activity and sympathetic nerve discharge in rats. Am J Physiol 1991; 261:R393–R402.
47. Allen AM, Guyenet PG. α_2-Adrenoreceptor-mediated inhibition of bulbospinal barosensitive cells of rat rostral medulla. Am J Physiol 1993; 265:R1065–R1075.
48. Richter DW, Spyer KM. Cardiorespiratory control. In: Loewy AD, Spyer KM, eds. Central Regulation of Autonomic Functions. New York: Oxford University Press, 1990:189–201.

28

Effects of Intracellular Dialysis on CO_2/H^+ Chemosensitivity in Brainstem Neurons

JAY B. DEAN and RADHA B. REDDY

Wright State University School of Medicine
Dayton, Ohio

I. Introduction

Since the early 1960s, two areas of the ventrolateral medulla oblongata (VLM) have been accepted as the sites for central chemoreception of CO_2 and/or H^+ after it was reported that localized application of acid or CO_2-saturated solution to VLM stimulated the cardiorespiratory systems (1–3). The theory that the VLM is the only site for central chemoreception is currently being debated as accumulating evidence demonstrates that other areas outside of the VLM are chemosensitive and may function in chemoreception (4–10). One such area outside of the VLM that demonstrates chemosensitive properties and is hypothesized to function in central CO_2/H^+ chemoreception is the nucleus tractus solitarii (4,6,8,9,11–14). We have been studying neurons in the nucleus tractus solitarii and the adjacent dorsal motor nucleus of vagus, referred to collectively as the solitary complex (SC), for several years using single-unit recording (SUR) and intracellular recording (ICR) techniques in brain slices to identify the membrane and synaptic mechanisms underlying CO_2/H^+ chemosensitivity (6,11,12,15).

Recently, we expanded our experimental approach to include the tight-seal whole-cell recording (WCR) technique of Hamill et al. (16), adapted for thick slices (17). The physical relationship between the patch pipette, tightly sealed membrane, and ruptured membrane patch results in a low series resistance (R_S) situation that favors current passing and voltage clamping compared to the higher R_S characteristics of ICR (18). A potential problem with WCR is that these same physical characteristics of the whole-cell configuration that facilitate electrophysiological measurements also permit rapid diffusional exchange of ions and large molecules between the cytoplasm and pipette solution (19,20). The consequences of intracellular dialysis or "washout" are that mechanisms for specific neuronal functions may be altered or impaired due to removal of obligatory endogenous substances (21–24).

The present study was conducted to determine what effects intracellular dialysis with a "standard" pipette solution during WCR has on the responsiveness of neurons in the SC to hypercapnia. Comparing CO_2-chemosensitivity in SC neurons during WCR to the less invasive recording procedures of SUR and ICR, where washout is not a major concern (18), would establish the validity of the WCR technique for studying chemosensory mechanisms in SC neurons. Whole-cell responses to hypercapnia were also studied using a standard pipette solution containing the cytosolic isozyme carbonic anhydrase II (CA II) in the event that washout removed CA II from SC neurons (13,25) and affected CO_2 chemosensitivity. Carbonic anhydrase, which catalyzes the reversible CO_2-HCO_3^- reaction, is thought to be important in determining the rate and/or magnitude of the CO_2 response of the peripheral chemoreceptors (26–28) and central chemoreceptors (13,25,29,30).

Neurons tested for CO_2 chemosensitivity during earlier SUR and ICR (11, 12) studies have been included in this report for comparison with our results from recent ICR and WCR experiments.

II. Methods

Procedures, solutions, and gas mixtures utilized for testing neuronal CO_2/H^+ chemosensitivity have been described in detail (11,12). Briefly, transverse brainstem slices (400 μm thick) were prepared from an adult rat in chilled (4 to 6°C) conditions using either a guillotine-type tissue chopper or vibratome. Slices were incubated for $\geq$1 h in oxygenated, bicarbonate-buffered nutrient medium maintained at 28 to 30°C. Control conditions were 5% CO_2, 95% O_2, medium pH 7.4, and 35 to 37°C. The test for chemosensitivity involved increasing the CO_2 level in the gas mixture saturating the air over the slice by 2% to 10% (balance O_2) for 2 to 5 min causing medium pH to drop by 0.1 to 0.47 units. Carbonic anhydrase II loading experiments in the whole-cell configuration used longer periods of hypercapnia ranging from 2 to 15 min.

Single-unit recordings were made with 3 to 5 MΩ microelectrodes filled with 3 *M* NaCl. Intracellular recordings were made using 70 to 150 MΩ microelectrodes filled with three different solutions: (a) 3 *M* potassium acetate; (b) 0.5 *M* KCl, 0.1 *M* TRIS buffer, 1% biocytin, pH 7.4 to 7.6; and (c) 0.1 *M* LiCl, 3% lucifer yellow, back-filled with 0.1% LiCl. Patch pipettes for WCR had DC resistances of 3 to 6 *M*Ω and were filled with (in millimolar) 130 or 140 potassium gluconate, 10 EGTA, 10 HEPES, 1 $MgCl_2$, 1 $CaCl_2$, 2 ATP, pH 7.25 to 7.3. In some WCR experiments, pipette tips were filled with the above solution containing 0.1 to 0.5% lucifer yellow or biocytin. Carbonic anhydrase II (10 and 20 μM; Sigma C2522, C3640) was added to the pipette solution in approximately one-half of the WCR. Pipette solution containing CA II was prepared freshly for each experiment and kept chilled between pipette fillings.

As in our previous studies (6,11,12), neurons that were depolarized and increased their firing rate during hypercapnia were classified as CO_2-excited if the excitatory response returned toward the prehypercapnic firing rate and membrane potential once CO_2 was lowered. Neurons were classified as CO_2-inhibited if they were hyperpolarized and reduced their firing rate during hypercapnia with recovery toward the prehypercapnic levels after CO_2 was reduced. All other neurons were classified as CO_2-insensitive if they demonstrated no significant change in membrane potential and/or firing rate during hypercapnia. This latter group also included neurons that demonstrated changes in excitability over time that did not correlate with the onset and removal of hypercapnia; e.g., unidirectional increase or decrease in excitability.

III. Results

Single-unit and ICR experiments, when pooled together ($n = 158$ cells), showed that 34% of the neurons tested in the SC were depolarized and/or increased their firing rate during 2 to 5 min of hypercapnia. Chemosensitive neurons were observed during ICR experiments using all three microelectrode solutions. Figure 1 shows that excitatory responses during hypercapnia were similar when recorded extracellularly (panel A) or intracellularly (panels B and C) and typically began within 0.5 to 1 min after increasing the CO_2 concentration in the gas mixture aerating the slice chamber. Firing rate responses gradually increased during hypercapnia or else reached a plateau without any firing rate adaptation to hypercapnia. Excitatory responses typically reversed during the first 0.5 to 2 min after returning CO_2 to control level. Excitatory responses were observed at all levels of CO_2 used. The majority of neurons tested in SC during SUR and ICR were CO_2-insensitive (60%), whereas the smallest population of neurons was CO_2-inhibited (6%).

Whole-cell recordings in SC using the standard HEPES-buffered pipette solution without CA II added ($n = 37$) gave dramatically different results than

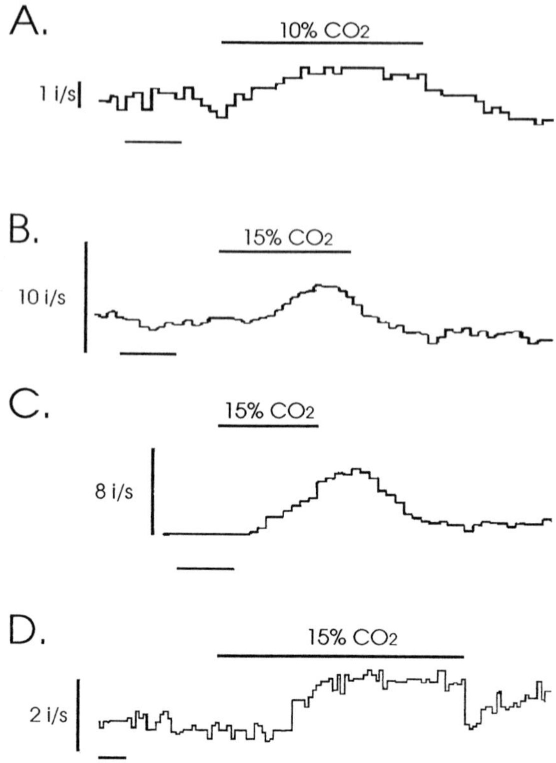

Figure 1 Integrated firing rate responses measured in four CO_2-excited neurons in SC using three different recording techniques. (A) Single-unit recording using 3 *M* NaCl. (B and C) Intracellular recordings in two neurons using 3 *M* potassium acetate. (D) Whole-cell recording using HEPES-buffered pipette solution containing 10 μM CA II. The time calibration in each panel is 1 min. Notice the compressed time scale in panel D relative to panels A through C. Firing rate was integrated using 10-s bins.

SUR and ICR experiments. Ninety-five percent of the neurons tested were identified as CO_2-insensitive. Inhibitory responses to increased CO_2 were still observed during WCR (5%); however, no neuron tested in SC responded to CO_2 with a reversible depolarization and/or increased firing rate. Many neurons studied with patch clamp pipettes showed unidirectional increases or decreases in excitability during the first 30 min of WCR that did not correlate with exposure to hypercapnia.

Whole-cell recordings that dialyzed neurons for ≥10 min with CA II ($n = 36$) prior to hypercapnia evoked reversible depolarization and increased firing rate responses in 20% of the neurons tested. Figure 1, panel D, shows an example of an SC neuron that was dialyzed with 10 μM CA II for 10 min prior to testing CO_2 chemosensitivity. The firing rate began to increase after ~3.75 min of hypercapnia and temporarily recovered within the first 0.5 min after CO_2 removal. During the posthypercapnic period, however, firing rate gradually increased as the neuron depolarized until the recording was lost. All of the CA II–loaded neurons that were depolarized during hypercapnia demonstrated excitatory changes atypical of SUR and ICR. These differences included longer latencies for CO_2 to increase firing rate (e.g., Fig. 1D), transient excitation to CO_2 onset and removal, large membrane depolarizations during hypercapnia or during the posthypercapnia period (e.g., Fig. 1D). At least one of these anomalies was observed in each CO_2-excited neuron dialyzed with CA II. CA II–loaded neurons were inhibited more frequently (33%) during hypercapnia compared to those experiments employing SUR, ICR (6%) or WCR with CA II–free solution (5%). Forty-seven percent of the CA II–loaded neurons were classified as CO_2-insensitive.

IV. Discussion

Approximately one-third of the neurons in SC studies under thick-slice conditions using SUR and ICR techniques were stimulated or depolarized by hypercapnia and the accompanying acidosis. No CO_2-excited neurons were encountered, however, when conventional WCR were made from SC neurons under identical conditions using CA II–free pipette solution. The small population of CO_2-inhibited neurons was still observed during WCR. We concluded that WCR, and its inherent problem of intracellular dialysis, removed intracellular substances critical for mediating CO_2-induced depolarization in chemosensitive neurons. Conventional ICR, or the perforated patch configuration of WCR that enables tight-seal recording while maintaining a metabolically intact neuron (31,32), is obviously better suited for the study of cellular mechanisms underlying CO_2 chemosensitivity in the SC. Preliminary reports of perforated-patch recordings in the VLM indicate that perforated patch recordings are indeed appropriate electrophysiological procedures for studying neuronal chemosensitivity in brainstem neurons (33,34).

Not all neuronal responses to physiological stimuli in SC were disrupted by WCR and washout. Parallel experiments in SC neurons, not reported here, showed that the four categories of hypoxia-induced responses observed during ICR were also observed during WCR (35).

What intracellular substances were removed during WCR and prevented CO_2-excited neurons from responding to hypercapnia? Carbonic anhydrase catalyzes the hydration of CO_2 and dehydration of carbonic acid and has been identified in neurons of the SC as well as the VLM (13,25). The need for CA II to be added to the pipette solution in order to observe CO_2 excitation during WCR suggests that cytosolic CA participates in CO_2-induced depolarization and that it was one of the substances washed out during WCR. Pusch and Neher (20) have shown that large cytoplasmic enzymes will undergo diffusional exchange with the pipette solution with a time constant of several minutes depending on their molecular weight and the R_S of the patch pipette and cell membrane configuration. Carbonic anhydrase II (molecular weight ~30,000) would be washed out from some neurons in SC (13,25) during the initial period of WCR unless cytosolic levels of CA were maintained by the addition of CA II to the pipette solution that was dialyzing the neuron.

The fact that CO_2-excitation was observed during WCR only when CA II was supplied illustrates the important role of CA II in CO_2-excitation and implies that CO_2 acts on certain brainstem neurons in SC by causing an intracellular acidosis (36,37). Cytosolic CA would directly influence the rate of CO_2 hydration and carbonic acid dehydration, thereby making it an important determinant of neuronal responsiveness to changes in extracellular CO_2. Not all neurons in SC contain CA (13,25); therefore it was not surprising that approximately one-third of the neurons dialyzed with CA II were inhibited during hypercapnia. It is hypothesized that these neurons do not have endogenous CA activity and that they lack the pH regulatory mechanisms to withstand the large ensuing intracellular acidosis that occurs when loaded with CA II and exposed to hypercapnia (39).

Previous studies have shown that inhibition of CA activity will attenuate the initial and/or late neural responses to hypercapnia. Thomas (38) showed that perfusion of snail neurons with acetazolamide, and therefore inhibition of CA activity (37), attenuated the drop in intracellular pH and membrane hyperpolarization during 20 min of hypercapnia, with the greatest effect being the reduction in the initial pH responses to addition and removal of CO_2. Likewise, inhibition of CA activity will reduce the early firing rate response to CO_2 in the peripheral chemoreceptors (28) and the late depolarization response to CO_2 in cultured VLM neurons (33). Thus, washout of cytosolic CA during WCR could conceivably disrupt responses of SC neurons to short periods ($\leq$15 min) of hypercapnia.

Although whole-cell depolarizations with full and partial recoveries were observed after CA II replacement, CO_2-excited neurons were encountered less frequently compared to ICR and SUR. Moreover, CO_2-excited neurons that were loaded with CA II typically displayed excitability profiles during and after hypercapnia that differed from those measured during SUR and ICR. This suggests that other substances that participate in CO_2-induced excitation were also washed out from the neuron during WCR. The identity of these intermediate steps in the signal transduction pathway for CO_2 excitation in SC is unknown at present. Our previous ICR have shown that input resistance increases during CO_2-induced depolarization, and we have hypothesized that this corresponds to a CO_2-induced decreased potassium conductance (6,11). Exposure to the potassium channel blocker 4-aminopyridine prior to onset of hypercapnia will depolarize the neuron and abolish any additional CO_2-induced depolarization in CO_2-excited neurons in the SC (15). It remains to be determined exactly what biochemical species are involved in the reactions that transpire between the CO_2-induced intracellular acidosis and the decreased conductance. The present study, however, demonstrates that these reactions are apparently disrupted by intracellular dialysis that accompanies the conventional WCR.

Acknowledgments

We wish to thank Drs. DA Bayliss, JT Erickson, EA Gallman, WL Lawing, and DE Millhorn for their contributions to previous SUR and ICR experiments. This research was supported by NSF (BNS-9021470), NIH (1 R29 HL46308-01A1), WSU Research Challenge Award, The Parker B. Francis Foundation, and an AHA-Ohio Affiliate Undergraduate Student Fellowship to RBR.

References

1. Mitchell RA, Loeschcke HH, Massion WH, et al. Respiratory responses mediated through superficial chemosensitive areas on the medulla. J Appl Physiol 1963; 18: 523–533.
2. Mitchell Ra, Loeschcke HH, Severinghaus JW, et al. Regions of respiratory chemosensitivity on the surface of the medulla. Ann NY Acad Sci 1963; 109:661–681.
3. Schläfke ME, See WR, Loeschcke HH. Ventilatory response to alterations of H^+ ion concentration in small areas of the ventral medullary surface. Resp Physiol 1970; 10:198–212.
4. Miles R. Does low pH stimulate central chemoreceptors located near the ventral medullary surface? Brain Res 1983; 271:349–353.
5. Tojima H, Kuriyama T, Fukuda Y. Differential respiratory effects of HCO_3^- and CO_2 applied on ventral medullary surface of rats. J Appl Physiol 1991; 70:2217–2225.
6. Dean JB, Millhorn DE. CO_2-induced depolarization of neurons in nucleus tractus solitarii: Potential substrate for central chemoreceptors. In: Koepchen HP, Huopaniemi T, eds. Cardiorespiratory and Motor Coordination. Berlin: Springer-Verlag, 1992:53–59.
7. Dillon GH, Waldrop TG. In vitro responses of caudal hypothalamic neurons to hypoxia and hypercapnia. Neuroscience 1992; 51:941–950.
8. Kogo N, Arita H. In vivo study on medullary H^+-sensitive neurons. J Appl Physiol 1990; 69:1408–1412.
9. Coates LE, Li A, Nattie EE. Widespread sites of brainstem ventilatory chemoreceptors. J Appl Physiol 1993; 75:5–14.
10. Sato M, Severinghaus JW, Basbaum AI. Medullary CO_2 chemoreceptor neuron identification by c-fos immunochemistry. J Appl Physiol 1992; 73:96–100.
11. Dean JB, Lawing WL, Millhorn DE. CO_2 decreases membrane conductance and depolarizes neurons in the nucleus tractus solitarii. Exp Brain Res 1989; 76:656–661.

12. Dean JB, Bayliss DA, Erickson JT, et al. Depolarization and stimulation of neurons in nucleus tractus solitarii by carbon dioxide does not require chemical synaptic input. Neuroscience 1990; 36:207–216.
13. Neubauer JA. Carbonic anhydrase and sensory function in the central nervous system. In: Dodgson SJ, Tashian RE, Gros G, Carter ND, eds. The Carbonic Anhydrases: Cellular Physiology and Molecular Genetics. Plenum Press: New York, 1991:319–323.
14. Neubauer JA, Gonsalves SF, Chou W, et al. Chemosensitivity of medullary neurons in explant tissue cultures. Neuroscience 1991; 45:701–708.
15. Dean JB, Gallman EA, Millhorn DE. Electrophysiology and morphology of CO_2-sensitive neurons in the dorsal vagal complex (abstr). Soc Neurosci Abstr 1990; 16:1235.
16. Hamill OP, Neher E, Sakmann B, et al. Improved patch-clamp techniques for high-resolution current recording for cells and cell-free membrane patches. Pflügers Arch 1981; 391:85–100.
17. Blanton MG, Lo Turco JJ, Kriegstein AR. Whole cell recording from neurons in slices of reptilian and mammalian cerebral cortex. J Neurosci Meth 1989; 30:203–210.
18. Marty A, Neher E. Tight-seal whole-cell recordings. In: Sakmann B, Neher E, eds. Single-Channel Recording. Plenum Press: New York, 1983:107–122.
19. Neher E. The use of the patch clamp technique to study second messenger-mediated cellular events. Neuroscience 1988; 26:727–734.
20. Pusch M, Neher E. Rates of diffusional exchange between small cells and a measuring patch pipette. Pflügers Arch 1988; 411:204–211.
21. Dufy B, MacDermott A, Barker JL. Rundown of GH3 cell K^+ conductance response to TRH following patch recording can be obviated with GH3 cell extract. Biochem Biophys Res Comm 1986; 137:388–396.
22. Korn SJ, Horn R. Influence of sodium-calcium exchange on calcium current rundown and the duration of calcium-dependent chloride currents in pituitary cells, studied with whole cell and perforated patch recording. J Gen Physiol 1989; 94:789–812.
23. Malinow R, Tsien RW. Presynaptic enhancement shown by whole-cell recordings of long-term potentiation in hippocampal slices. Nature 1990; 346:177–180.
24. Horie M, Hwang TC, Gadsby DC. Pipette GTP is essential for receptor-mediated regulation of Cl^- current in dialysed myocytes from guinea-pig ventricle. J Physiol (Lond) 1992; 455:235–246.
25. Pan Y, Nauss K, Bernard DG, et al. Carbonic anhydrase in brainstem respiratory chemosensitive regions (abstr). Soc Neurosci Abst 1991; 17:201.
26. Black AMS, McCloskey DI, Torrance RW. The responses of carotid body chemoreceptors in the cat to sudden changes of hypercapnic and hypoxic stimuli. Resp Physiol 1971; 13:36–49.
27. Travis DM. Molecular CO_2 is inert on carotid chemoreceptor: Demonstration by inhibition of carbonic anhydrase. J Pharmacol Exper Ther 1971; 178:529–540.
28. Iturriaga R, Lahiri S, Mokashi A. Carbonic anhydrase and chemoreception in the cat carotid body. Am J Physiol 1991; 261:C565–C573.
29. Hanson MA, Nye PCG, Torrance RW. The location of carbonic anhydrase in relation to the blood-brain barrier at the medullary chemoreceptors of the cat. J Physiol (Lond) 1981; 320:113–125.

30. Coates LE, Li A, Nattie EE. Acetazolamide on the ventral medulla of the cat increases phrenic output and delays the ventilatory response to CO_2. J Physiol 1991; 441:433–451.
31. Horn R, Marty A. Muscarinic activation of ionic currents measured by a new whole-cell recording method. J Gen Physiol 1988; 92:145–159.
32. Rae J, Cooper K, Gates P, et al. Low access resistance perforated patch recordings using amphotericin B. J Neurosci Meth 1991; 37:15–26.
33. Martin AA, Neubauer JA. Effect of acetazolamide on CO_2 chemosensitivity of cultured medullary neurons (abstr). Soc Neurosci Abstr 1993; 19:1193.
34. Richerson GB. CO_2 modulates pacemaker neurons in the medullary raphe and parapyramidal region of the rat in vitro (abstr). Soc Neurosci Abstr 1993; 19:1403.
35. Dean JB, Gallman EA, Zhu WH, et al. Multiple effects of hypoxia on neurons in the dorsal motor nucleus (X) and nucleus tractus solitarii (NTS). Soc Neurosci Abstr 1991; 17:465.
36. Maren TH. Carbonic anhydrase: chemistry, physiology, and inhibition. Physiol Rev 1967; 47:595–781.
37. Maren TH. Use of inhibitors in physiological studies of carbonic anhydrase. Am J Physiol 1977; 232:F291–F297.
38. Thomas RC. The effect of carbon dioxide on the intracellular pH and buffering power of snail neurones. J Physiol 1976; 255:715–735.
39. Moody W. Effects of intracellular H^+ on the electrical properties of excitable cells. Annu Rev Neurosci 1984; 7:257–278.

29

Cellular and Synaptic Mechanisms Generating Respiratory Rhythm: Insights from In Vitro and Computational Studies

JEFFREY C. SMITH

National Institute of Neurological Disorders and Stroke
National Institutes of Health
Bethesda, Maryland
and University of California at Los Angeles
Los Angeles, California

GREGORY D. FUNK, STEPHEN M. JOHNSON, and JACK L. FELDMAN

University of California at Los Angeles
Los Angeles, California

I. Introduction

To understand mechanisms of respiratory rhythm generation in mammals remains a fundamental challenge in unraveling the neural control of breathing. This problem is both old and new. The origin of the rhythm of breathing has intrigued physiologists for several centuries, resulting in a general agreement that the basic oscillator lies in the lower brainstem. Recent advances in neurobiology have provided novel experimental approaches and there is a renaissance of efforts to identify basic mechanisms, particularly at the cellular level. The recent expansion of information on properties of brainstem respiratory neurons has given rise to new ideas about rhythmogenesis, but our current understanding is still limited. Fundamental questions remain in at least four main areas: *Source of the rhythm.* Where is the major locus of brainstem neurons generating the rhythm? *Mechanisms for rhythm generation.* What are the critical cellular and synaptic mechanisms? *Mechanisms for modulation of rhythm.* How is the oscillatory behavior of neurons and circuits generating the rhythm regulated? *Transmission of rhythm.* What synaptic mechanisms underlie transmission of the oscillatory drive?

In this chapter, we briefly review our recent work addressing these questions. Neural oscillators are complex, and central nervous system (CNS) processes at cellular, synaptic, and network levels must be synthesized to understand the respiratory oscillator. This poses challenges experimentally because of the difficulty of probing mechanisms in the nervous system in vivo, particularly at the cellular level. An important advance has been the development of in vitro preparations of the mammalian brainstem, especially brainstem–spinal cord and isolated medullary slice preparations from neonatal and juvenile rodents that retain functional circuits for rhythm generation. These preparations allow concurrent study at cellular, synaptic, and network levels. The bridge between the cellular and network levels, however, is problematic. Computational approaches that allow modeling with parameters of biological relevance enable synthesis of complex processes at cellular and network levels. Such modeling is a necessary adjunct to experimental studies and can provide insights into properties that cannot be investigated experimentally. We review results obtained with both in vitro experimental approaches and computational modeling. The in vitro approaches have provided novel information on sites, cellular mechanisms, and synaptic interactions; the computational studies test the plausibility of models suggested by the experimental results, and provide additional insight into mechanisms. This work is very much in progress, and we point to current controversies and significant gaps in our understanding. Recent reviews should be consulted for an overview and background (1–3). Several recent papers (4–10) should be consulted for details of experimental methodology, including discussions of the uses and limitations of the in vitro preparations as model systems.

II. Ventral Brainstem Sites of Rhythm Generation

A. Pre-Bötzinger Complex: A Candidate Site for Rhythm Generation

The caudal brainstem has long been established as the likely site for rhythmogenesis; recently interest has centered on the ventral respiratory group (VRG) of the ventrolateral medulla. The VRG is a heterogeneous cell column with electrophysiologically distinct, interconnected subregions (Fig. 1) and distinct classes of respiratory neurons. The fundamental issue is whether there is a primary locus or kernel for rhythm generation in a functionally specialized VRG subregion. Implicit in many models (e.g., Refs. 1 and 12) has been the concept of a spatially distributed oscillator involving neuronal populations throughout the VRG. However, our microsection experiments in the in vitro neonatal rat brainstem–spinal cord preparation, which have allowed systematic analysis of perturbations of rhythmogenesis with removal of defined VRG and other brainstem regions, indicate that there is a locus in a circumscribed VRG subregion (9). This subregion, which we have called the pre-Bötzinger complex (pre-BötC) (9), is sandwiched between the more rostral Bötzinger complex (BötC, containing

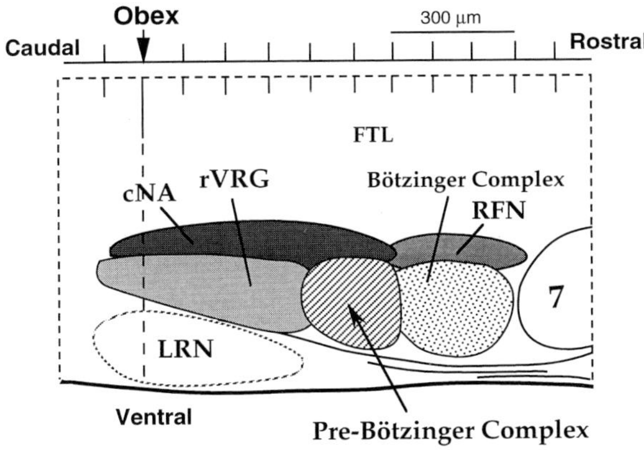

Figure 1 Sagittal view of ventral medulla showing pre-BötC and neighboring reticular formation regions. Pre-BötC is a circumscribed subregion of the classical ventral column of respiratory neurons (VRG), lying in the ventrolateral reticular formation between the caudal pole of the BötC and the rostral ventral respiratory group (rVRG). This area is at the level of the medulla where there is the transition between retrofacial nucleus (RFN) and the more caudal (semicompact) division of nucleus ambiguus (cNA). Pre-BötC is partially coextensive with cNA and the more ventral external division of NA. (See Ref. 11 for discussion of NA nomenclature.) Electrophysiologically, the pre-BötC is a transition zone between the BötC pool of expiratory neurons and inspiratory neuron pool of the rVRG, distinguished by a unique profile of respiratory interneurons proposed to be functionally specialized for rhythm generation. Other abbreviations: 7, facial nucleus; LRN, lateral reticular nucleus; FTL, lateral tegmental field. Dimensions indicated are typical for the neonatal rat. (Modified from Ref. 9.)

predominantly expiratory neurons) and the more caudal medullary pool of inspiratory neurons of the rostral VRG (rVRG) (Fig. 1). The region is sufficiently circumscribed that we have revived the notion of the existence of a *noeud vital* (9,13) introduced by French neurophysiologists during the last century (see Ref. 9). Recent studies done in vivo as well as retrospective examination of existing in vivo electrophysiological data also confirm that pre-BötC has electrophysiological and cytoarchitectonic characteristics consistent with its proposed role. Current evidence that the pre-BötC is a critical locus for rhythm generation includes the following:

1. Removal of the pre-BötC by microsection in vitro eliminates rhythmogenesis (9); disruption of rhythm generation does not occur with removal of any other brainstem or VRG region.
2. Isolated in vitro medullary slices from neonatal and juvenile rodents containing the pre-BötC retain the capacity for rhythmogenesis (9,14,15a); slices excluding the pre-BötC do not.

3. Perturbations of neuronal excitability in pre-BötC cause profound, site-specific perturbations of the rhythm (9,15,16).

4. The profiles of neuronal activity and microanatomy of the pre-BötC appear to be distinct from the rest of the VRG. The region has a unique concentration of interneurons (1,9,17–19) that have been proposed to be essential for producing respiratory phase transitions, including interneurons with spike discharge starting before and continuing into the inspiratory phase (pre-inspiratory neurons; 7,17,18) that may underlie initiation of the inspiratory phase. Moreover, there is a high ratio of propriobulbar to bulbospinal interneurons (9,20) in the pre-BötC compared to adjacent VRG regions.

5. Patterns of connectivity of pre-BötC interneurons are consistent with the notion of the pre-BötC as the origin of inspiratory drive. Projections of propriobulbar neurons in the region are targeted to medullary regions containing cranial and bulbospinal premotoneurons (1,9,20–22). Furthermore, temporal correlation analysis of excitatory interactions among medullary neurons in vivo (1,21) suggests that some pre-BötC neurons are a source of excitation to more caudally located VRG inspiratory neurons. This implies a rostrocaudal polarization of excitatory drive in the VRG cell column, consistent with pre-BötC neurons as the ultimate source of rhythmic excitation.

6. The pre-BötC contains a population of respiratory neurons with voltage-dependent, pacemaker-like properties in vitro (9). Moreover, subclasses of pre-BötC neurons continue to burst rhythmically in neonatal rat medullary slices in the absence of chemical synaptic transmission (23) consistent with the thesis of a pacemaker kernel in the pre-BötC (below).

7. The pre-BötC is a target site for second order (pulmonary stretch receptor) afferents involved in controlling inspiratory phase transitions (D. R. McCrimmon, personal communication).

The evidence is compelling that pre-BötC is the primary locus for rhythm generation in the in vitro brainstem and medullary slice preparations. Although current in vivo data are consistent with this assertion, the region has not been thoroughly explored in vivo; definitive evidence that the region is the critical locus in the in vivo nervous system is still lacking. Recent studies demonstrating that there are homologous pre-BötC regions in the neonatal and adult nervous systems from different species (18), and that neuronal perturbations in the pre-BötC cause site-specific perturbations of rhythm including apnea in the adult (e.g., 16), are beginning to provide evidence for the region's essential role in vivo. Questions remain about the cellular organization of pre-BötC, and how neurons in other regions involved in respiratory phase transitions (e.g., pontine and dorsal respiratory groups) interact with pre-BötC cells for control of rhythm. *Our hypothesis is that the pre-BötC contains the essential kernel for rhythm generation, and that it is the site where a large number of convergent inputs, including inputs from afferent circuits involved in control of phase transitions, interact with the kernel for regulation of rhythm.*

III. Cellular and Synaptic Mechanisms: The Hybrid Pacemaker-Network Model

What are the critical cellular properties and synaptic interactions underlying rhythm generation? The debate (see, e.g., Refs. 2,3,24–26) centers on whether the basic oscillator is (a) a network oscillator where rhythmogenesis is an emergent, dynamic property of interactions among synaptically coupled populations of neurons or (b) a pacemaker-driven oscillator consisting of neurons with intrinsic oscillatory properties where rhythmogenesis is explicable in terms of specific membrane currents. This issue is complex. Neuronal behavior results from interactions of cellular and synaptic properties. Neurons with intrinsic oscillatory properties can have multiple states, some of which are not oscillatory, that are regulated by synaptic inputs. Oscillatory networks can have transformational properties and dynamically reconfigure, switching from cell-driven states involving cellular oscillations to those where network interactions play dominant roles. Thus, the respiratory oscillator must be modeled as a hybrid of cellular and network properties. The fundamental question is whether there is a kernel consisting of neurons with intrinsic pacemaker properties; if so, in what states are these properties expressed, and how are they regulated by network interactions?

Our current hypothesis is that the oscillator in the pre-BötC consists of neurons with voltage-dependent pacemaker properties (9) embedded in a network (25–28). Synaptic connections provide mechanisms for control of the pacemaker oscillations, and the combination of cellular and network properties gives rise to complex network dynamic behavior, including transformational properties. We have incorporated these concepts in a formal model (27–28a) of a hybrid pacemaker-network that provides a framework for analyzing how rhythm is generated and regulated. This model has been formulated to account for the behavior of the respiratory oscillator in the highly reduced neonatal rat nervous system in vitro. We hypothesize that the basic mechanisms apply in vivo, while allowing for the possibility of transformation in different states (Refs. 26–28; see discussions below). We shall first describe the conceptual elements of the model, present some of the evidence for the model, and then consider results of computational studies.

The Model

The kernel of the oscillator is a population of synaptically coupled, bursting pacemaker neurons (Fig. 2) that generate the oscillatory drive to the rest of the network during the inspiratory phase of network activity. These neurons are bilaterally distributed in the pre-BötC and coupled by excitatory amino acid (EAA, probably glutamate)–mediated, fast synaptic connections. The pacemaker neurons receive synaptic inputs from a number of neuron populations, including distinct populations of beating (tonic) excitatory and inhibitory neurons. The beating excitatory cells provide fast EAA synaptic inputs. The tonic

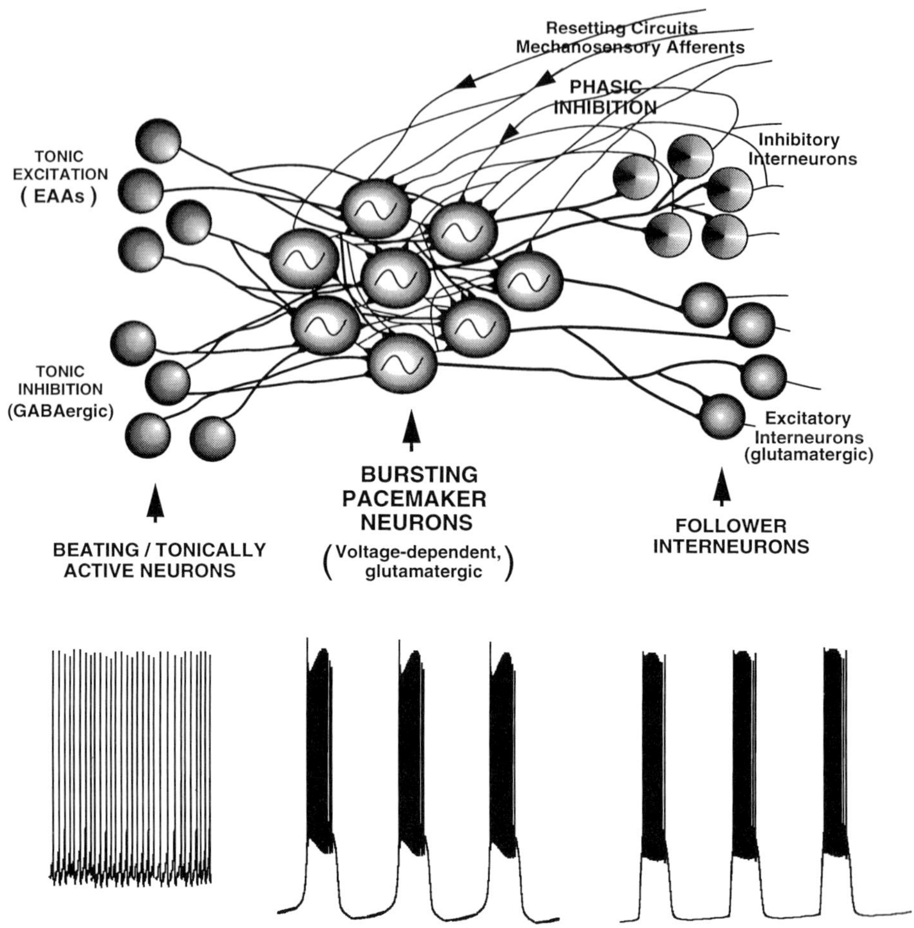

Figure 2 Simplified schematic of hybrid pacemaker-network model. The kernel of the oscillator, located in pre-BötC, consists of bilaterally distributed populations of synaptically coupled, bursting pacemaker neurons with voltage-dependent properties. Pacemaker neurons are embedded in a network and receive synaptic inputs from tonically active (beating) excitatory and inhibitory interneurons (left) and rhythmically active inhibitory interneurons (top right), which include interneurons phasically driven by mechanosensory (e.g., pulmonary stretch receptor) afferent circuits (arrows at top). Pacemaker cells provide oscillatory drive by excitatory connections to populations of follower neurons, including excitatory and inhibitory interneurons, distributed in pre-BötC and adjacent reticular formation regions. The pacemaker kernel potentially receives a large number of convergent inputs from other afferent neuronal populations (see Fig. 5) that can modulate the pacemaker neuron oscillations. Refer to text for detailed explanation.

inhibitory neurons are presumed to be GABAergic. The pacemaker cells provide rhythmic excitatory drive to populations of interneurons, including:

1. Excitatory interneurons of a polysynaptic, inspiratory drive transmission pathway(s) to premotoneurons and motoneurons. The oscillatory drive transmission is presumed to be mediated by EAAs at all synapses including the premotoneuron to motoneuron synapse.
2. Inhibitory interneurons, presumed to be GABAergic, phasically inhibit a number of neuron populations in the network, including the pacemaker cell population.

The model configuration shown in Figure 2 is simplified and does not include all active respiratory neuron populations and all of the synaptic interactions that shape the membrane potential trajectories and discharge patterns of the different classes of inspiratory and expiratory neurons in vitro (7,27a); more elaborate models that incorporate these features have been formulated (28,28a), but the principles are the same. Although neurons with the type of properties incorporated in the model have been identified in vitro, and the model can account for the behavior of the oscillator in vitro, the membrane conductances, connectivity scheme, and synaptic interactions incorporated are speculative. The simplified network configurations shown in Figures 2 through 4 represent some of the interactions and properties considered to be essential and are meant to illustrate some of the major operating principles of the oscillator, including the following:

1. *Voltage-dependent oscillatory behavior of pacemaker neurons.* The model pacemaker neurons have intrinsic membrane conductances that produce an "N-shaped" subthreshold current-voltage (I-E) relationship with a region of negative slope (Fig. 3) characteristic of cellular oscillators. The negative slope region is produced mainly by a strong, voltage-dependent inward cationic current that underlies the depolarization and burst phase of the neuronal oscillations. In the voltage range of the negative slope, the cells exhibit oscillatory bursting; the particular mix of the subthreshold currents are such that the oscillations in this voltage window are voltage-dependent and the oscillatory frequency increases as the baseline membrane potential is depolarized. *This voltage-dependence of oscillatory frequency is a basic mechanism for control of frequency.* At membrane voltages above the negative slope region, the neurons undergo a transition to a stable beating or tonic mode of action-potential generation (Fig. 3). At hyperpolarized levels, the neurons are quiescent. Thus the neurons are presumed to exhibit several distinct functional states from quiescence to oscillatory bursting to beating. Multistate oscillatory neurons are well known (e.g., Refs. 29 to 31). The functional significance of these stages for respiratory neurons, however, has not been explored. Due to this voltage-dependent behavior, the oscillatory bursting is conditional, requiring either synaptic inputs or activation of particular membrane conductances to keep the neurons in the voltage range where they generate oscillations. Thus we refer to these neurons as conditional bursting pacemakers (9,25), and we presume that many of these neurons

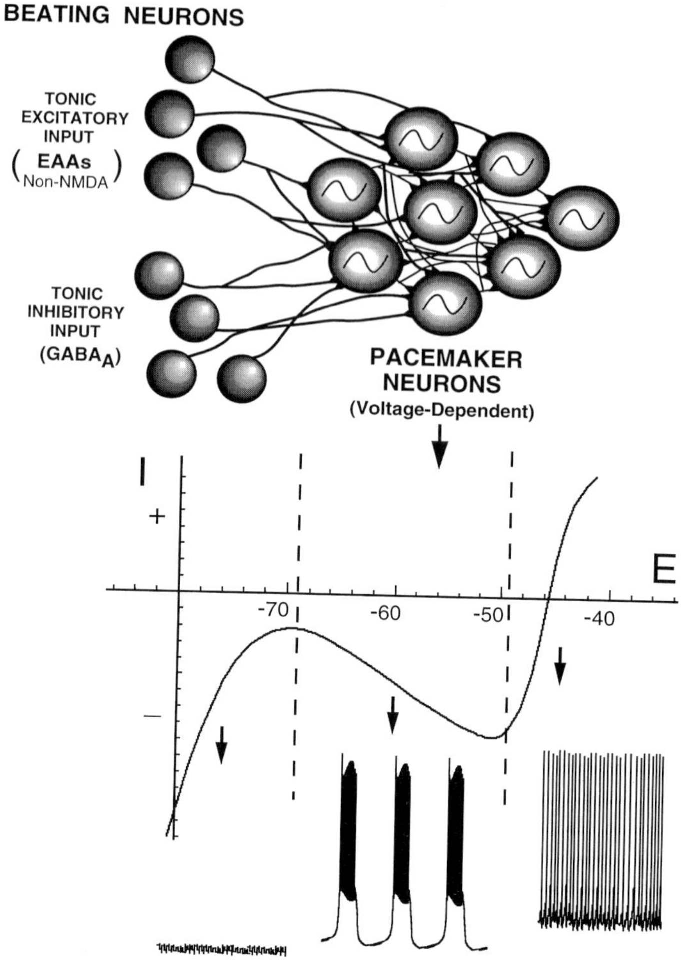

Figure 3 Schematic representation of principles of operation and functional states of pacemaker neuron kernel. Top: Pacemaker neurons (right) together with beating neurons (left) represent the most basic substrate for generation and control of oscillations. Bottom: Pacemaker neurons have multiple functional states: quiescence (lower left), oscillatory bursting (lower middle), and tonic (beating) modes of action potential generation (lower right), corresponding to different regions of the neuronal current-voltage (I-E) relationship. Middle: Oscillations occur in the voltage range with negative slope (dashed lines); oscillatory frequency in this range is voltage-dependent. Tonic synaptic inputs from beating excitatory and inhibitory neurons control oscillatory frequency and can cause transitions from oscillatory bursting to beating or quiescent states. The oscillator is presumed to operate under conditions where continuous excitatory inputs maintain voltages in or near the negative slope region. Refer to text for detailed explanation.

require at least tonic excitatory inputs to maintain oscillatory bursting (see below). In addition, the oscillatory behavior could be affected by activation/inactivation of the intrinsic conductances by a variety of signals, including modulatory neurochemical inputs or chemosensory-related signals (intra/extracellular pH or CO_2) (see "Modulation," below).

2. *Control of oscillatory behavior of pacemaker cells by tonic synaptic inputs* (Fig. 3). Tonic excitatory or inhibitory inputs from populations of beating neurons control baseline membrane potential and thus determine the functional state (quiescence, bursting, beating) of the oscillatory neurons, as well as regulating the oscillatory frequency over the voltage range for bursting. The excitatory beating neurons are presumed to have membrane conductances that maintain them in a depolarized state, spontaneously generating action potentials. Without tonic excitatory inputs, or with tonic inhibitory inputs, many of the oscillatory neurons would be hyperpolarized and would not exhibit spontaneous bursting. Thus the pacemaker neurons together with the populations of beating cells form the kernel for the generation and control of rhythm. These tonically active neurons, in addition to the oscillatory neurons, would be sensitive to modulation that would control their beating frequency and thus the level of tonic drive to pacemaker neurons.

3. *Synchronization of pacemaker neurons and initiation of inspiration.* Excitatory synaptic coupling between pacemaker neurons synchronizes their depolarization and bursting within the bilaterally distributed pacemaker cell population. The coupling is both unilateral among neighboring cells and bilateral via axon collaterals. We envision the process by which the oscillation propagates in the population and the inspiratory phase is initiated as follows. Tonic inputs maintain the pacemaker cells at a baseline voltage near the negative slope region. Due to nonuniformity in tonic inputs and neuronal excitability, there is a dispersion in the membrane potential of the pacemaker neurons, with some cells sitting at potentials in the negative slope region, which initiate bursting, while others are relatively hyperpolarized. Once initiated, the oscillation propagates through the population by mutual excitatory connections, and neurons sitting hyperpolarized to the negative slope region are pushed into the bursting regime by the phasic inputs from pacemaker cells. We presume that connectivity is such that there is a threshold phenomenon, i.e., a certain fraction of the neurons must undergo depolarization in order for the depolarizing wave and bursting to propagate through the entire population. From this scheme, we predict a spatial and temporal dispersion of onset times of spiking within the population, with a fraction of the neurons bursting early (pre-inspiratory neurons, see below) at the initiation sites within the population.

4. *Dynamic resetting of pacemaker cell oscillations by phasic inhibitory inputs* (Fig. 4). Phasic inhibitory synaptic interactions contribute to the termination of each oscillatory cycle, continually resetting the pacemaker neuron oscillation. Such inputs would arise from feedback from interneurons that receive oscillatory drive, and (in intact mammals) from inhibitory interneurons

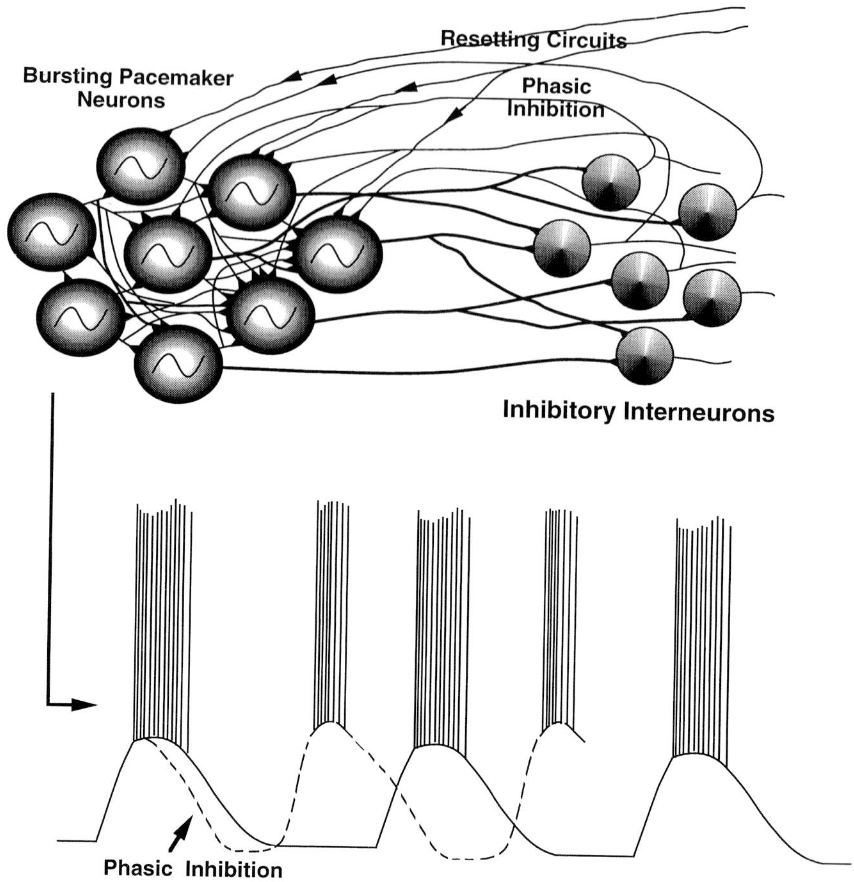

Figure 4 Schematic illustrating resetting and dynamic regulation of oscillatory frequency by phasic inhibition. Inhibitory interneurons (top right) provide phasic inputs to pacemaker neurons (top left) after temporal delays. This phasic inhibition contributes to termination of pacemaker neuron bursting (dashed line, lower left), reduces burst phase duration, and resets pacemaker oscillation each cycle. Such resetting can result in a higher steady-state oscillatory frequency.

phasically driven by pulmonary stretch receptor or other afferents that can participate in inspiratory termination. Inspiratory termination by phasic inhibition is a component of all models of the respiratory oscillator (e.g., Refs. 1,3,12,32,33). In the case of the hybrid pacemaker-network, this phasic resetting will dynamically set the oscillatory frequency higher than the endogenous frequency of the pacemaker cell population (Fig. 4). Thus the oscillatory frequency of the pacemaker neuron population is controlled by the interactions of intrinsic membrane properties, tonic synaptic inputs, and phasic inhibitory inputs.

5. *Control of pacemaker neuron oscillations by ion channel modulation.* The regulation of rhythm by tonic and phasic synaptic inputs represents

only one type of modulation (i.e., by conventional fast excitatory synaptic mechanisms). The subthreshold membrane channels/currents that underlie the pacemaker cell oscillations would be targets for modulation. Modulation of ionic channels, including by second messenger–linked processes, which alter the magnitude and time- and voltage-dependent properties of conductances, represents a fundamental mechanism for regulation of rhythm. Membrane potential shifts caused by neuromodulators could also cause transitions from oscillation to beating or quiescence.

6. *Transformational properties.* The ability of the pacemaker neurons to transform from oscillatory bursting to beating behavior can cause the mechanisms of rhythmogenesis to transform in different dynamic states. For example, with pacemaker neurons in depolarized beating states, the ability of the network to oscillate would be determined by phase-terminating inhibitory interactions. In the presence of strong phasic inhibition, stable oscillations would occur, but the oscillatory frequencies obtainable are potentially different from those with pacemaker cells in their intrinsic oscillatory mode. In the absence of strong phasic inhibition (e.g., absence of phasic afferent inputs), pacemaker and follower neurons would be tonically active, mimicking perhaps apneustic states in vivo.

Comment on Role of Synaptic Inhibition in Hybrid Pacemaker-Network

There is continuing debate about the role of synaptic inhibition in rhythm generation, particularly since pacemaker-driven oscillators do not require phasic inhibition but network oscillators generally do require such inhibition (for further discussion see Refs. 2,24–25). Synaptic inhibition has been disrupted in vitro (24,25,34) and in other experimental preparations (35) to test for involvement of pacemaker properties or inhibitory network interactions. In the hybrid pacemaker network model, tonic inhibitory inputs, interacting with tonic excitatory inputs, determine if the pacemaker neurons are bursting, beating, or quiescent. Phasic inhibitory interactions play a principal role in inspiratory phase resetting and thus regulation of the oscillations. Moreover, in the beating states, phasic inhibition is necessary for rhythmogenesis. Thus within the pacemaker network model, an essential role for phasic inhibition is state-dependent. Apart from involvement in rhythmogenesis, inhibitory interactions in the hybrid pacemaker-network model play critical roles in pattern formation (i.e., shaping the detailed patterns of neuronal discharge and membrane trajectories of different classes of respiratory neurons) (28,28a).

Synopsis of Model

Recent models of the respiratory oscillator (3,27–28a) incorporate biophysical properties of neurons and synaptic interactions, recognizing that neuronal behavior results from interactions of cellular and synaptic properties. The hybrid pacemaker-network model explicitly assumes a kernel consisting of neurons that exhibit intrinsic oscillatory behavior and that provide the rhythmic excitatory drive to the network; synaptic interactions function to synchronize, reset, and regulate the oscillations. Another recent hybrid-type model (3), which incorporates intrinsic membrane properties that could potentially give rise to cellular oscillations, assumes that the determining neurons are inhibitory rather than excitatory. In our model, inhibitory interactions are not essential for inspiratory phase termination and thus rhythmogenesis. Termination can occur via membrane conductance mechanisms. Our model, however, can also exhibit states where inhibitory interactions are required. Thus, the model should be viewed as a hybrid in two senses: (a) intrinsic oscillatory properties and synaptic interactions together are necessary for rhythmogenesis and (2) the oscillator can transform between states where the oscillatory mechanism is pacemaker-driven or dominated by network interactions.

The transformational properties give the network plasticity and are a fundamental feature of the model. The behavior of the respiratory oscillator is known to change in different brain states (e.g., sleep vs. wakefulness) or experimental conditions (e.g., intact vs. deafferented; in vivo vs. in vitro). As the afferent control signals change between these different states, the oscillator may functionally reconfigure, transforming between cell-driven and more network-like states. In vivo this plasticity may be necessary to achieve the oscillator's full range of physiological and behavioral functions. Transformation may thus provide a mechanism to extend functional capabilities.

IV. Evidence for Pacemaker Neurons

The central component of the hybrid pacemaker-network model is neurons with bursting pacemaker properties. What is the current evidence for this?

There have been speculations on possible roles of pacemaker neurons in generating respiratory rhythm (e.g., Ref. 36), which follow from the growing understanding of voltage- and ligand-dependent ionic conductances that can endow neurons with autorhythmic oscillatory properties (see Ref. 30). Until recently, direct experimental evidence has been lacking. Renewed interest in pacemaker neurons originated from results of Smith and Feldman (24,25,37), which demonstrated that conventional Cl^--dependent, ligand-gated fast synaptic inhibition, as well as slow synaptic inhibition involving specific types of K^+ conductances, was not necessary for rhythmogenesis in the in vitro brainstem-spinal cord preparation, from which it was proposed that pacemaker properties underlie rhythmogenesis in vitro (24,25,37). These results prompted us and other investigators (9,23,38–40,42) to directly test for neurons with pacemaker properties in vitro, including in the pre-BötC, providing two major pieces of evidence consistent with the pacemaker kernel hypothesis:

1. A subset of inspiratory neurons in the pre-BötC continue to burst rhythmically in isolated medullary slice preparations after block of chemical synaptic transmission with low extracellular Ca^{2+} and high Mg^{2+} (23). When synaptic transmission is intact, the majority of these cells exhibit discharge during

the inspiratory phase, with some of these inspiratory neurons exhibiting pre-inspiratory discharge up to several hundred milliseconds prior to phasic motor output. These results are consistent with the basic proposal that specialized subsets of inspiratory neurons with intrinsic pacemaker properties generate the oscillatory drive in the network during the inspiratory phase. Pre-inspiratory discharge would be predicted for some of the neurons providing the oscillatory drive (see "Synchronization," above), suggesting the involvement of pre-inspiratory neurons in rhythm generation both in vitro and in vivo (17,18).

2. A subpopulation of pre-BötC inspiratory neurons exhibit the type of intrinsic, voltage-dependent oscillatory properties in vitro (9,39,40) that were incorporated in the model. Other inspiratory neurons in pre-BötC, presumed to be follower cells in the local network, do not exhibit these properties. Their rhythmic depolarization is due to large synaptic drive inputs (7,41) characteristic of many medullary inspiratory neurons rather than due to intrinsic membrane properties.

Synopsis

Our hypothesis is that pre-BötC neurons with intrinsic, voltage-dependent bursting properties with (pre)inspiratory phase discharge are the main population of rhythm generating cells in vitro (9). These neurons exhibit a range of oscillatory frequencies, burst durations, and spike discharge profiles consistent with the respiratory motor output frequency, inspiratory phase durations, and oscillatory drive potentials/currents in the in vitro preparations (7,9). A central problem that remains in evaluating the pacemaker neuron hypothesis, however, is determining the causal relationship between neurons with intrinsic oscillatory properties and rhythm generation. This is not a trivial problem; there are currently no straightforward experimental approaches to establish such causality. The situation is complicated because several classes of respiratory neurons exhibit intrinsic oscillatory properties in vitro (9,23,38,42). Furthermore, although intrinsic conductances have been postulated to account for the behavior of respiratory interneurons in vivo (3), it remains to be determined whether these interneurons express intrinsic oscillary behavior in vivo. This technically challenging and difficult problem is complicated by the potential for transformation and the possibility that there are different mechanisms in different functional states in vivo, with some involving pacemaker behavior and others network interactions. There could also be developmental transformations of cellular/network properties, so that network (inhibitory) interactions play an increasingly vital role in rhythm generation in the adult (3). The hybrid model incorporates transformational properties, and the modeling shows that they can account for respiratory network behavior in different states, including developmental transformations (28,28a). Such models, however, remain difficult to test.

V. Evidence for Ionic Mechanisms

If our hypothesis about the central role of voltage-dependent oscillatory neurons is correct, a challenge is to define the membrane conductances underlying the cellular oscillations. Since no particular type of neuron has been shown to generate the rhythm, we can only speculate. Based on information about other types of intrinsic oscillatory cells, we can postulate that there are subthreshold currents underlying three aspects of their oscillatory behavior: an inward cationic current(s) (Na^+ and/or Ca^{2+}) responsible for the negative-slope region of the I-E curve and driving the depolarization (burst) phase; currents terminating the burst phase (e.g., K^+ currents); and currents controlling the membrane potential trajectory during the interburst phase [e.g., K^+ or mixed cationic (Na^+/K^+) currents].

We have screened for the involvement of currents that subserve rhythmic bursting in other oscillatory neurons (e.g., Refs. 30,43–46). This has been done by analyzing perturbations of the rhythm after application of channel blockers or current enhancers to the in vitro bath of en bloc medulla-spinal cord or isolated rhythmic medullary slice preparations (unpublished observations). Currents for which we tested have included:

1. *Nonspecific, cAMP-dependent cation conductances (I_C)* are responsible for maintenance of the burst phase in some invertebrate neurons (44) and proposed to be involved in bursting of some respiratory neurons in vivo (3). Attempts to enhance I_C (45) with bath application of phosphodiesterase inhibitors (e.g., isobutyl methylxanthine), membrane permeant cAMP analogues (8-bromo-cAMP), or forskolin (adenylyl cyclase stimulator), however, did not perturb the rhythm in vitro. Such results also suggest that other types of cAMP-dependent currents may play a minimal role in oscillatory burst generation in vitro, although cAMP-mediated processes could still be involved in modulation of respiratory pattern.

2. *Calcium-activated potassium conductances [$I_{K(Ca)}$]* could contribute to burst termination. Bath application of apamin and/or charybdotoxin to block the two types of $I_{K(Ca)}$ (apamin-sensitive, charybdotoxin-sensitive) (47) did not perturb rhythm in vitro.

3. *T-type or low-voltage activated calcium conductances (LVA)* are essential for the generation of neuronal oscillations in many rhythmically bursting cells (e.g., Refs. 30,31) and have been proposed to play a key role of initiating burst generation in medullary respiratory interneurons in vivo (3). Presumptive attenuation of LVA with nickel (Ni; 50 to 100 μM), however, did not perturb rhythm in the in vitro slice preparations.

At present, the involvement of Ca^{2+} conductances in rhythmogenesis is not clear. In addition to LVA, persistent Ca^{2+} currents have been proposed to be important currents underlying the burst/plateau phase of oscillatory cells (30). Observations that some neurons continue to undergo rhythmic bursting under low Ca^{2+} conditions (23,38,42) are interesting in this regard and could indicate that the main cationic current underlying the oscillatory burst phase is not Ca^{2+}.

4. *Inward-rectifying K^+ conductance (K_{ir}).* Presumptive block of K_{ir} (or a related K^+ conductance) by Rb^+ (48) profoundly affects rhythm in vitro. Increasing bath concentrations of Rb^+ progressively increase respiratory frequency and completely disrupt rhythm (49). We have proposed that several neurotransmitter pathways (see below) that modulate rhythm act via this conductance (49–51). K_{ir} or a similar K^+ current that produces a steady leakage current would contribute tonic current in the voltage regime where oscillations occur and could regulate rhythm for the type of voltage-dependent oscillator neurons we propose.

The failure of blockers/enhancers of several conductances traditionally implicated in neuronal oscillations to significantly perturb rhythm leads to our speculation that the major subthreshold currents may be voltage-dependent Na^+ and K^+ currents (see "Models of Pacemaker Neurons," below).

VI. Evidence for Control of Rhythm by Ion Channel Modulation

Modulation of intrinsic conductances underlying pacemaker oscillations would have profound effects on rhythmogenesis. A large number of transmitters have been implicated in the control of breathing (e.g., Ref. 52), but the mechanisms by which they affect neurons producing the rhythm is not yet known. There are diverse mechanisms for modulation of neuronal excitability and behavior of oscillatory networks by ion channel modulation (e.g., Refs. 53,54). We have preliminary evidence (49,50) that modulation of an inward-rectifying K^+ conductance (K_{ir}) may be an important mechanism for the control of rhythm.

Modulation of neuronal excitability by K_{ir} has been proposed to be ubiquitous in mammalian neurons (55–57), including intrinsic oscillatory neurons (43). Neurotransmitters affecting this conductance act via a G protein(s) that couples receptors to K_{ir} (55–58; Fig. 5). A number of receptors mediating inhibition, many of which have been implicated in the control of breathing (54), are G protein–coupled to K_{ir} (48,55–58), including $GABA_B$, adenosine (A1 receptors), acetylcholine (M2), norepinephrine ($\alpha 2$), serotonin (1A), dopamine (D2), opioid (μ, δ), and somatostatin receptors. Block of K_{ir} (e.g., by substance P) in some neurons produces excitation (59). [Substance P has potent stimulatory actions on respiratory rhythm generation (e.g., Ref. 60).]

We have analyzed modulation of rhythm in brainstem–spinal cord and medullary slice preparations by application of agonists at some of the receptors linked to K_{ir}, including $GABA_B$ (agonist, baclofen), μ-opioid (agonist, DAMGO), and α_2-adrenergic (agonist, clonidine). All of these agonists have inhibitory actions on rhythm generation (i.e., reduce respiratory frequency; Ref. 50), as predicted if they increase K_{ir}. These inhibitory actions are attenuated by the $G_{i/o}$-protein blocker pertussis toxin (50). We hypothesize that modulation

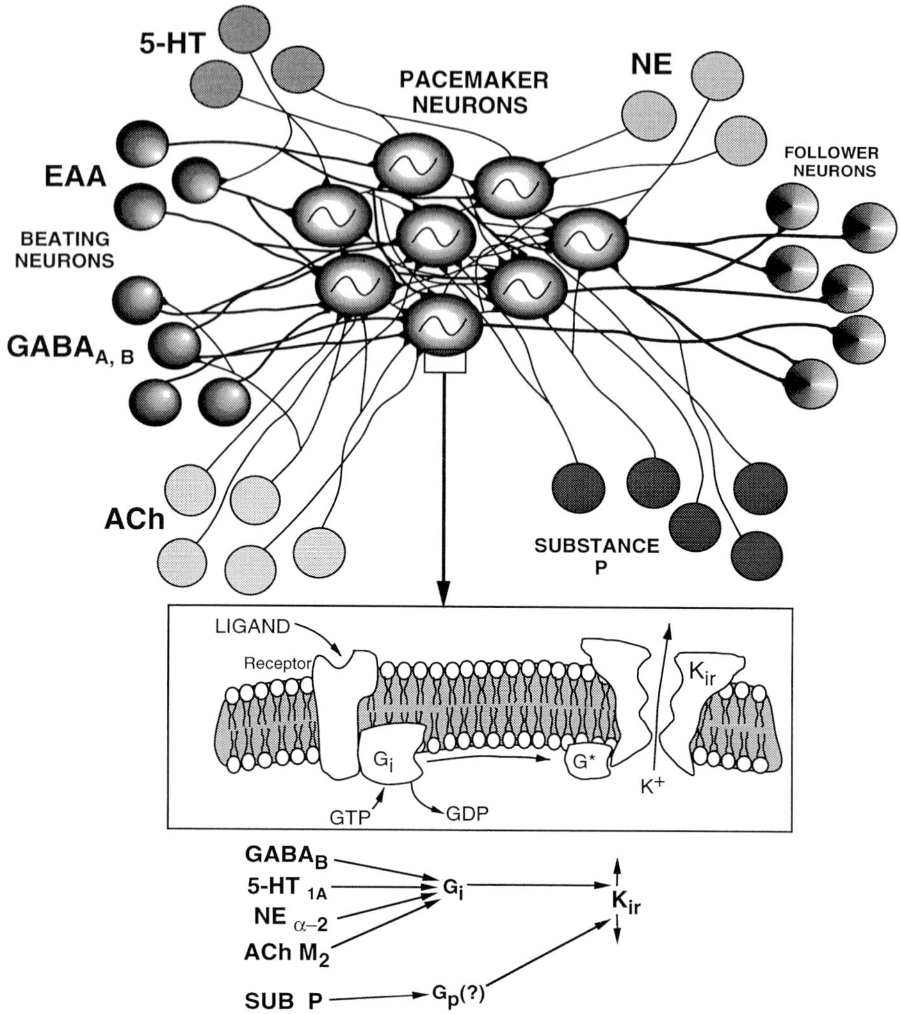

Figure 5 Neurochemical modulation of pacemaker kernel. Pacemaker neurons and beating neurons (left) potentially receive a large number of modulatory synaptic inputs, including from serotonergic (5-HT), catecholaminergic (NE), cholinergic (ACh), and peptidergic (e.g., substance P, opioid) neurons. We hypothesize that K_{ir} on pacemaker neurons is a convergent target for channel modulation by a number of neurotransmitters (bottom) whose receptors are linked to K_{ir} via G proteins (inset; after Hille, Ref. 48). G proteins (G_i, G_p) activated by receptors after transmitter binding can either increase (↑, G_i) or decrease (↓, G_p) K_{ir} (bottom). See text for explanation.

of K_{ir} via G protein–linked receptors in pre-BötC neurons represents a mechanism for control of rhythm.

VII. Evidence for Control of Rhythm by Synaptic Mechanisms

A. Excitatory Mechanisms

Excitatory synaptic interactions are obviously essential for the generation and transmission of respiratory oscillations. Endogenously released EAA, presumably glutamate, is the primary transmitter mediating fast excitatory neurotransmission for generation and transmission of oscillatory drive.

EAAs and Rhythm Generation

The principal EAA receptors [i.e., AMPA, kainic acid, NMDA, and metabotropic (AP-4) receptors; Refs. 61 and 61a] in rhythmogenesis have now been investigated in vitro (15,64). We have established that:

1. Rhythm generation is dependent on endogenously released EAA acting primarily at non-NMDA receptors (kainate and/or AMPA receptors) on pre-BötC neurons. Thus, bath application of competitive blockers of non-NMDA receptors [e.g., 6-cyano-7-nitroquinoxaline-2,3-dione (CNQX)] to medullary networks in vitro causes a dose-dependent reduction in respiratory motor discharge frequency. Microinjection of CNQX directly in the pre-BötC produces a site-specific block of rhythmogenesis in vitro (15).

2. NMDA receptor activation is not essential, at least in vitro, since NMDA channel block (e.g., with MK-801) does not cause perturbations of rhythm. The lack of involvement of NMDA receptors is of interest, since their activation can promote intrinsic oscillatory behavior of neurons and they have been proposed to play key roles in rhythmogenesis in other oscillatory motor networks. Activation of these receptors can, however, modulate respiratory rhythm (i.e., increase frequency; Ref. 64) and may contribute to inspiratory phase termination via pontine mechanisms under some conditions in vivo (66).

These observations are consistent with our hypothesis that pre-BötC pacemaker neurons require non-NMDA receptor-mediated synaptic inputs to maintain membrane potential in the range for intrinsic oscillations. Although there is currently no direct evidence, we speculate that tonic inputs onto pacemaker neurons and the mutual excitatory synaptic interactions necessary for pacemaker neuron synchronization, are mediated by postsynaptic non-NMDA receptors.

Excitatory Transmission of Inspiratory Drive

The most direct experimental evidence for involvement of EAAs is from our in vitro studies of synaptic transmission at the bulbospinal to spinal (phrenic, intercostal) motoneuron synapse (62,63,65), and more recently, the cranial

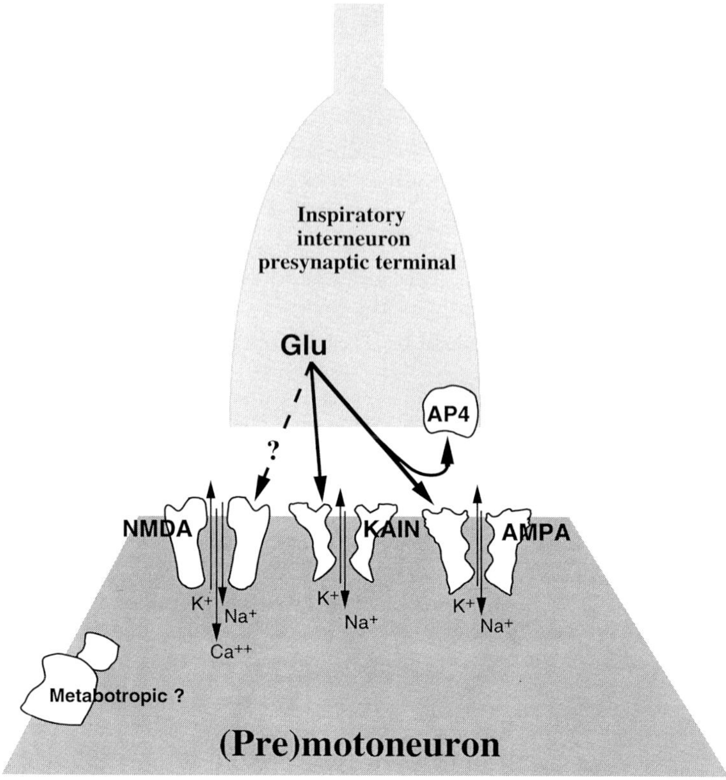

Figure 6 Model of excitatory synapse in oscillatory drive transmission pathways. Synaptically released excitatory amino acid (glutamate) interacts with one or more non-NMDA receptors (AMPA, kainate) linked to cationic channels on postsynaptic membrane of interneurons and motoneurons. NMDA receptor channels contribute only a small component of the postsynaptic inspiratory drive current. Presynaptic AP4 (metabotropic) receptors may also play a regulatory role, possibly as an autoregulatory mechanism for control of glutamate release. Modulation of synaptic transmission by AP4 receptors is potent at bulbospinal → motoneuron synapses (64–66a) but appears to play only a relatively minor role at synapses in medullary circuits (15,64). Postsynaptic metabotropic receptors may also exist (61a), but their role at synapses in respiratory drive transmission pathways has not been established. (Modified from Ref. 25.)

premotoneuron to hypoglossal motoneuron synapse (15). The accessibility of these motoneurons, as well as the ease of intracellular and axonal recording from these cells, has made them an excellent model for the analysis of synaptic mechanisms transmitting inspiratory drive. Synaptic transmission has been analyzed in detail at phrenic and hypoglossal motoneurons by measuring postsynaptic inspiratory currents generated by the endogenously released neurotransmitter. The roles of specific EAA receptor subtypes have been established, and models for excitatory transmission in the network (Fig. 6; see discussions in Ref. 2) have been developed which incorporate the following features:

1. An endogenously released EAA (glutamate) mediates inspiratory drive transmission from medullary premotoneurons to all spinal and cranial respiratory motoneurons.
2. Inspiratory drive is mediated primarily by postsynaptic non-NMDA receptors, with a relatively minor NMDA receptor component.
3. Activation of presynaptic AP4-sensitive receptors (presumably metabotropic; Ref. 66a) can potently attenuate inspiratory drive transmission to phrenic and intercostal motoneurons, but this mechanism, although present, only weakly modulates drive transmission to accessory and hypoglossal respiratory motoneurons. We have postulated that there is a degree of functional heterogeneity of receptor mechanisms in spinal and cranial synaptic pathways. A possible function of this AP4-receptor-mediated mechanism in presynaptic regulation of glutamate release has been discussed previously (25,63).

Synaptic transmission at interneuronal synapses in the drive transmission circuits has not been directly studied. NMDA and AP-4 receptor activation is not necessary for inspiratory drive transmission in medullary premotor circuits; block of these receptors in the en bloc medulla or medullary slice preparations does not block oscillatory drive transmission (15,64). We therefore presume that transmission at the interneuronal level is also mediated by non-NMDA receptors (Fig. 6).

B. Inhibitory Mechanisms

Conventional fast, chloride-dependent GABAergic and glycinergic inhibitory control of rhythm generation has been established. Agonists acting at $GABA_A$ and glycine receptors reduce respiratory frequency and inspiratory phase durations and can stop rhythmogenesis in vitro (24,25). Respiratory neurons in vitro receive phasic inhibition during the inspiratory period (41,67), indicating that there are inhibitory interneurons that receive oscillatory drive. Inhibitory interactions that can reset rhythm also are present, since activation of mechanosensory afferents can inhibit rhythmogenesis in vitro (e.g., Ref. 68) as in vivo. Removal of vagal and other afferent inputs causes perturbations of rhythm (reduce frequency) and prolongs the inspiratory phase in a manner consistent with the removal of phasic resetting inhibition (7). In addition to fast, chloride-dependent inhibition, slower inhibitory mechanisms, presumably mediated via K^+ conductances, are present (49,50).

VIII. Computational Model of Hybrid Pacemaker-Network

To test the feasibility and explore properties of the hybrid pacemaker-network model, we have developed computational models (27–28a) of bursting pacemaker neurons embedded in a network that incorporates the types of neurons and synaptic connections described above (see Fig. 2 for overview). Computational models can be helpful in understanding neural systems at all levels from membrane properties and synaptic mechanisms to complex network dynamics (see reviews in Refs. 69–71). There are well-established approaches for modeling networks of multicompartment neurons incorporating nonlinear voltage- and time-dependent membrane and synaptic properties (69–71). Recent technical developments, including the availability of neuron/network stimulation software, have made simulation of these more realistic types of models practical (e.g., Refs. 75–77). Our modeling takes a fundamentally different approach compared to other recent efforts to develop computer models of the respiratory oscillator (72–74,77a). We explicitly incorporate (a) voltage- and time-dependent membrane properties; (b) synaptic connections with time- and voltage-dependent conductances; and (c) arrays of synaptically coupled neurons, allowing spatial and temporal features of neuronal population behavior to be explored.

A. Synopsis of Simulation Approach

Neuron Models

Development of totally realistic neuron and network models is obviously not possible given the complexity of somatodendritic morphologies, the (unknown) heterogenous spatial distribution of synapses and ion channels on somatodendritic membranes, and the large numbers of interacting cells with unknown connectivity patterns and membrane conductances. Our approach has been to develop simplified models of respiratory neurons and their synaptic interactions in relatively small networks. These models contain essential features of the time- and voltage-dependent properties of cells and synapses, with network architectures that capture the basic features of the hybrid pacemaker-network. Individual neurons were modeled as one compartment (somal) or in some cases two-compartment (somal, dendritic) neurons, represented by standard equivalent electrical circuits. Somal compartments had voltage (E)- and time (t)-dependent ionic conductances [$G(E,t)$] and currents [$I(E,t)$] described by standard Hodgkin-Huxley (HH) like equations:

$$I(E,t) = G(E,t)(E\text{-}E_r),\ G(E,t) = G_{max}\cdot m^{x}(E,t)\cdot h^{Y}(E,t) \tag{1}$$

$$dm/dt = \alpha(1\text{-}m) - \beta m \tag{2}$$

$$\alpha = (a + bE)/[c + \exp(d + E/f)],\quad \beta = (a' + b'E)/[c' + \exp(d' + E/f')] \tag{3}$$

$$m_{\infty} = \alpha/(\alpha + \beta),\ \tau_m = 1/(\alpha + \beta) \tag{4}$$

where G_{max} is the maximum conductance, E_r is the current reversal potential, m is the activation parameter, m_{∞} is the steady-state activation, and τ_m is the activation time constant. The voltage- and time-dependence of m (i.e., activation) is specified by Eq. (2) and (3) above, where α and β are the rate factors (a,a′,b,b′,...,f,f′ are constants). Equations identical in form describe the voltage- and time-dependence of h, the inactivation parameter, for cases where conductances exhibit inactivation properties. These types of equations have been extensively employed to model behavior of both vertebrate and invertebrate neurons (e.g., Refs. 78 and 79; see reviews in Refs. 69 and 70).

Biophysical properties of respiratory neurons active in vitro have not yet been characterized experimentally. Parameters such as whole-cell capacitances and input conductances used in the modeling have been estimated from our whole-cell recording measurements (40,41). Beating and follower neurons have been modeled with four intrinsic membrane conductances: the voltage-dependent Na^+ and delayed rectifier K^+ conductances generating the action potential (i.e., HH Na^+ and HH K^+), leakage conductance, and an additional K^+ conductance with voltage-dependency of (in)activation modeled to give appropriate spontaneous spiking frequencies of beating cells (5 to 15 Hz) and appropriate peak spike frequencies (25 to 50 Hz) of follower neurons during rhythmic bursting (7,41). Pacemaker neurons have been modeled with a more complex set of conductances (below) that are required to produce the oscillatory behavior of some pre-BötC neurons (9,39,40). None of the parameters of the model equations has been directly measured; values for standard conductances (HH Na^+ and HH K^+) are taken from the literature; others are stipulated.

Synapse Models

Synaptic conductances, distributed on somal and dendritic compartments, were modeled by the following analytical functions: synaptic current $I_S = G_S(E\text{-}E_S)$, and $G_S = G_\infty T_S t^\alpha \exp(-\alpha t/\tau)$, where G_S is the synaptic conductance, G_∞ is the maximum synaptic conductance, E_S is the synaptic reversal potential, T_S is a constant transmitter release, and the time course of the postsynaptic conductance is specified by the above alpha function and controlled by α and τ. Parameters and synaptic reversal potentials used were consistent with conventional fast, amino acid–mediated excitatory (non-NMDA) and inhibitory (GABAergic) transmission.

Simulations

Simulations, run with the neuron/network simulation software NODUS (75) and GENESIS (76,77), have examined: (a) roles of specific types of membrane conductances in the generation and modulation of pacemaker neuron oscillations; (b) mechanisms of synchronization of pacemaker neuron bursting; (c) interactions of tonic synaptic inputs and intrinsic currents in control of pacemaker oscillations; (d) resetting of pacemaker neuron oscillations by phasic inhibition; and (e) transformational properties of the hybrid pacemaker-network. We briefly discuss several simulation results confirming the plausibility of the basic principles of operation of the hybrid pacemaker-network.

1. *Models of bursting pacemaker neurons.* Our pacemaker neuron models have incorporated up to eight membrane conductances, including HH Na^+

and HH K^+ currents, a leakage conductance, and five subthreshold conductances: (a) inward cationic (Na^+) conductance (I_{CAT}), which underlies the depolarization (burst) phase; (b,c) two voltage-dependent K^+ conductances with intermediate (I_{K1}) and long (I_{K2}) activation time constants, which interact with I_{CAT} to control burst duration, termination, and the membrane potential trajectory during the interburst interval; (d) the inward rectifying K^+ conductance (K_{ir}), which in part controls the baseline membrane potential for burst frequency modulation; (e) calcium conductance (I_{Ca}), which activates in the subthreshold potential range during membrane depolarization.

Currents (a through e) are essential for generating the subthreshold membrane potential oscillations underlying the rhythmic bursting. They represent our current best guess about the types of voltage-dependent currents required to produce the oscillatory behavior of the pacemaker cells. The steady-state, subthreshold I-E relationship of a model neuron with this set of currents, showing the negative slope region, is illustrated in Figure 7. The model neurons exhibit oscillatory bursting with voltage-dependent frequency as well as beating and quiescent states. Simulated pacemaker neurons exhibit the range of oscillatory frequencies, burst durations, and membrane potential trajectories observed in pre-BötC oscillatory neurons under current-clamp conditions (9,39,40).

2. *Control of pacemaker neuron oscillatory behavior by tonic inputs.* Simulations with single neurons or small arrays of pacemaker neurons with tonic inhibitory and/or excitatory synaptic inputs from separate populations of beating neurons (Fig. 8) confirm that the pacemaker cell oscillations can be regulated by tonic synaptic inputs. Such inputs produce shifts in baseline membrane potential and regulate oscillatory frequency in proportion to the magnitude of the synaptic currents. Strong tonic excitatory drive, which depolarize the neurons above threshold for spike generation, causes transitions from oscillatory bursting to beating. Inhibitory inputs, which hyperpolarize membrane voltage below the negative slope region, stop oscillations.

3. *Modulation of neuronal oscillations by inward rectifier K_{ir}.* Simulations of individual pacemaker neurons, in which the maximum conductance value of K_{ir} is varied, are consistent with this conductance having a strong effect on oscillatory frequency (Fig. 9). Increases in K_{ir}, mimicking the actions of inhibitory neuromodulators linked to K_{ir}, reduce oscillatory frequency and can completely stop oscillations. Reductions in K_{ir}, mimicking excitatory neuromodulators, increase oscillatory frequency.

4. *Synchronization of pacemaker neurons.* Simulations with small arrays of coupled pacemaker neurons (two populations of neurons with five neurons per population, simulating the bilaterally distributed populations of pacemaker cells) show synchronization of bursting within the pacemaker cell population by mutual excitatory synaptic connections. The simulations indicate that bursting of individual neurons can be triggered by fairly small amplitude, depolarizing synaptic inputs which activate I_{CAT}. Burst termination occurs by inactivation of I_{CAT} and activation of I_{K1} and I_{K2}. Thus the simulations support our basic

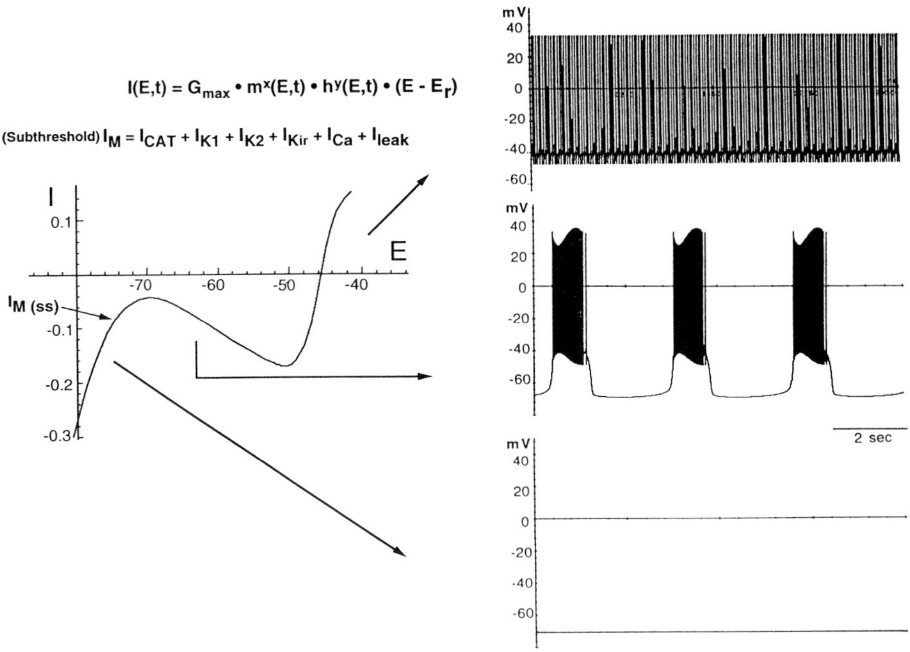

Figure 7 Example of simulated voltage-dependent oscillatory behavior (right) and subthreshold current-voltage (I-E) relationship (lower left) of model pacemaker neuron. Subthreshold I-E relationship shown represents total membrane current (I_M) resulting from summation of all currents with subthreshold activation properties (i.e., I_{CAT}, I_{K1}, I_{K2}, $I_{K_{ir}}$, I_{Ca}, and I_{leak}) under steady-state (ss) conditions. Voltage- and time-dependence of individual currents were simulated with Hodgkin-Huxley like currents [I (E,t); top]. Pacemaker neuron exhibits multiple states corresponding to different voltage regimes (arrows). Neuron exhibits bursting oscillatory behavior (middle right) under simulated current-clamp conditions with baseline potential in negative slope region of neuronal I-E relationship. Oscillatory frequency is voltage-dependent in this region, increasing at more depolarized levels of baseline membrane potential (see Fig. 8). Depolarizing baseline potential to threshold voltages for action potential generation causes model neuron to undergo transition from oscillatory bursting to stable beating behavior (upper right). Neuron is quiescent at baseline potentials more hyperpolarized than negative slope region (below −70 mV) (lower right).

hypothesis that the membrane properties underlying oscillatory bursting also promote synchronization. Furthermore, when spatial inhomogeneity of neuronal excitability and tonic synaptic input is incorporated, the simulations exhibit a dispersion of spike onset times in the population, with early and late bursting neurons, depending on the strength of synaptic coupling. The effect of connectivity patterns in the population on the distribution of activity and temporal dispersion of bursting remains to be explored.

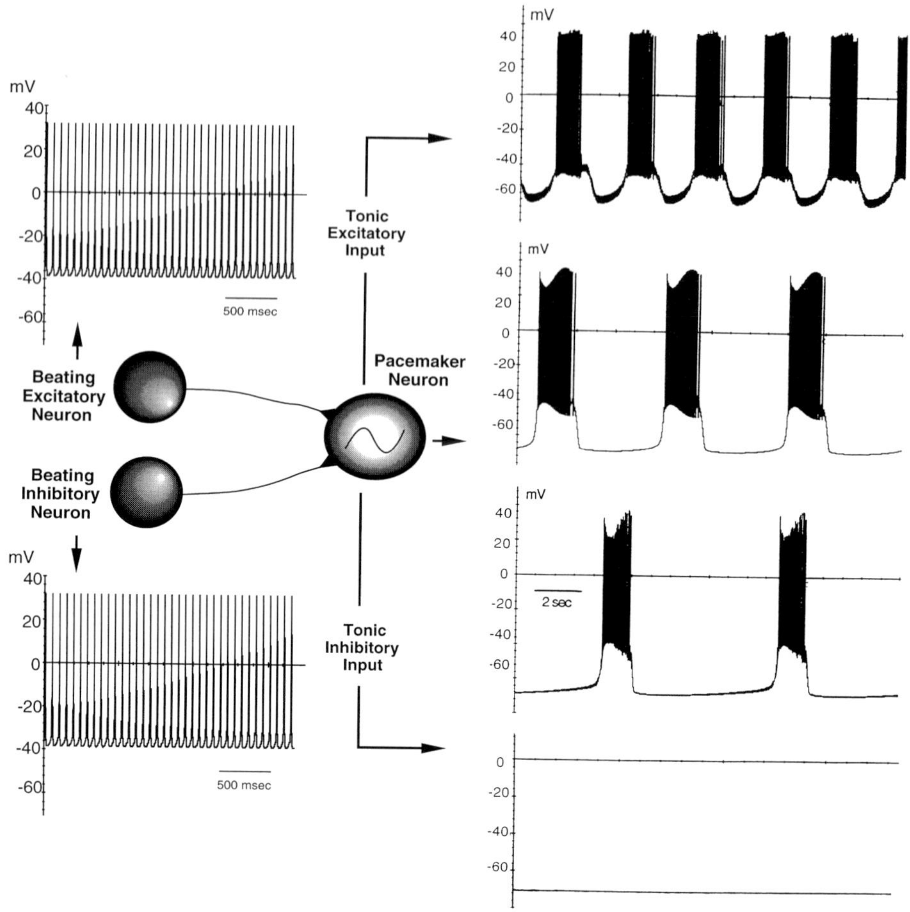

Figure 8 Simulations illustrating control of pacemaker neuron oscillations by synaptic inputs from excitatory or inhibitory beating neurons. Traces at right: Progressive shifts in baseline membrane potential with applied synaptic current cause changes in intrinsic oscillatory frequency of pacemaker neuron, reflecting voltage-dependent properties. Right middle trace: intrinsic oscillatory behavior of model neuron under simulated current clamp conditions with baseline potential in negative slope region of I-E relationship. Top right trace: Depolarizing tonic input from excitatory beating neuron increases intrinsic oscillatory frequency. Larger-amplitude excitatory inputs cause model neuron to undergo transition to beating (not shown). Bottom traces: Tonic input from beating inhibitory neuron reduces oscillatory frequency and stops oscillations at high levels of inhibitory currents (bottom right). Spiking pattern of beating neurons is shown at left.

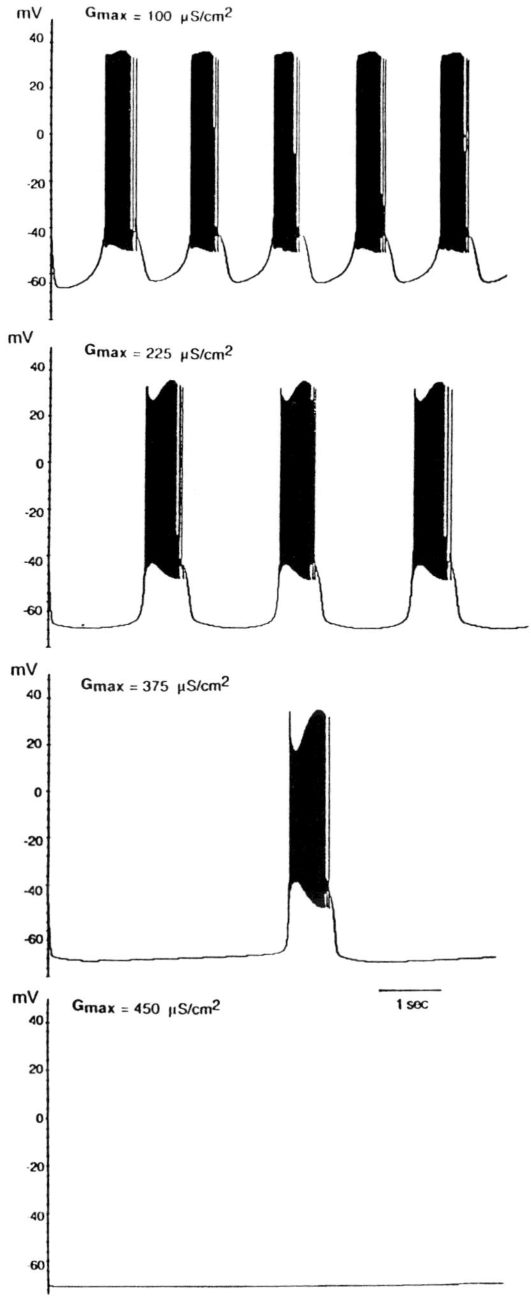

Figure 9 Modulation of model pacemaker neuron oscillations by K_{ir}. Progressive increases in maximum conductance (G_{max}) of K_{ir} reduces oscillatory frequency (middle panels) and can stop oscillations (bottom panel).

5. *Resetting of pacemaker neuron oscillations by phasic inhibition.* Phasic inhibitory synaptic inputs to the pacemaker neuron population after temporal delays ranging from tens to hundreds of milliseconds have been simulated (Fig. 10). The variable delays have been used to mimic inputs from short-delay synaptic pathways (e.g., from local interneurons) or longer-delay pathways to simulate inputs from inhibitory interneurons phasically driven by mechanosensory (e.g., pulmonary stretch receptor) afferents. Such inputs cause earlier termination of pacemaker cell bursting, reset the pacemaker cell population oscillation each cycle, and result in shorter inspiratory phase durations. This resetting mechanism results in oscillations at a frequency higher than the endogenous frequency of the pacemaker cell population without the phasic inhibition. These results are consistent with our thesis that phasic inhibitory interactions play a role in cycle-to-cycle regulation of oscillatory frequency.

6. *Transformational properties of hybrid pacemaker-network.* The behavior of the network has been simulated under conditions when pacemaker neurons undergo transitions from bursting to beating (i.e., at elevated levels of tonic excitatory inputs) and the pacemaker neurons receive phasic inhibition (Fig. 11). There are two basic network states under these conditions, depending on the strength of the inhibitory interactions: tonic network activity or stable oscillatory states. Stable oscillations are achieved at a threshold level of inhibitory input with appropriate synaptic delays. Thus, under these conditions, there is a transformation from a pacemaker-driven to a network-based mechanism for rhythmogenesis. The oscillatory behavior of the network is complex, reflecting interactions between the inhibitory currents and the intrinsic membrane currents activated as the pacemaker cells are hyperpolarized from the beating state by the inhibition. Changes in dynamic behavior that occur include (Fig. 11) (a) a shift in the network oscillations to a higher frequency range than is possible when the pacemaker cells are in their oscillatory mode and (b) a transformation of the membrane potential trajectories of pacemaker and follower cells from decrementing to augmenting trajectories. It is interesting that the oscillatory frequencies and membrane potential trajectories transform to resemble those characteristics of eupneic breathing in the neonatal rat in vivo. This may in part explain the transformations (reduction in oscillatory frequency, change in membrane potential trajectories from augmenting to decrementing) that occur when the nervous system is isolated from pulmonary afferent (inhibitory) input and transferred from in vivo to in vitro conditions (7) (see also “Resetting,” above). Recent simulations (28,28a) have also demonstrated other types of transformational behavior that may account for developmental differences in network oscillatory behavior and neuronal discharge patterns between the neonate and adult.

Synopsis

Computational studies demonstrate the plausibility of the basic principles proposed for the operation of the hybrid pacemaker-network. Given the complexity

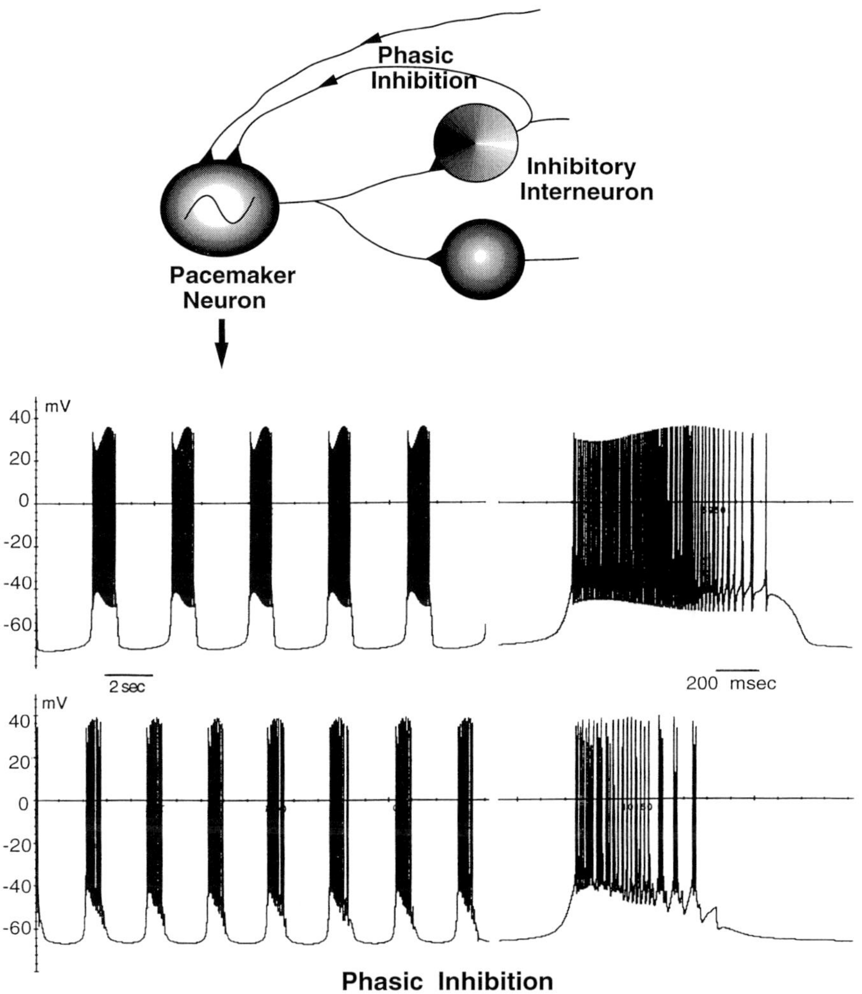

Figure 10 Simulations illustrating resetting of pacemaker neuron oscillations and frequency modulation by phasic inhibition. Simulations were done with small model circuits (top) consisting of pacemaker neuron with excitatory synaptic connections to inhibitory interneuron which provided phasic synaptic input to pacemaker neuron after specified temporal delays to mimic polysynaptic transmission delays. Traces below at left show multiple cycles of simulated oscillatory behavior; traces at right show a single burst at expanded time scale for comparison of burst patterns. Phasic inhibition reduces pacemaker neuron burst duration and increases steady-state oscillation frequency (traces at bottom left and right) compared to the case without phasic inhibitory input (upper traces). Simulated membrane potential trajectories and spike discharge patterns (rapidly peaking-decrementing profile, see traces at right) are characteristic of those in vitro.

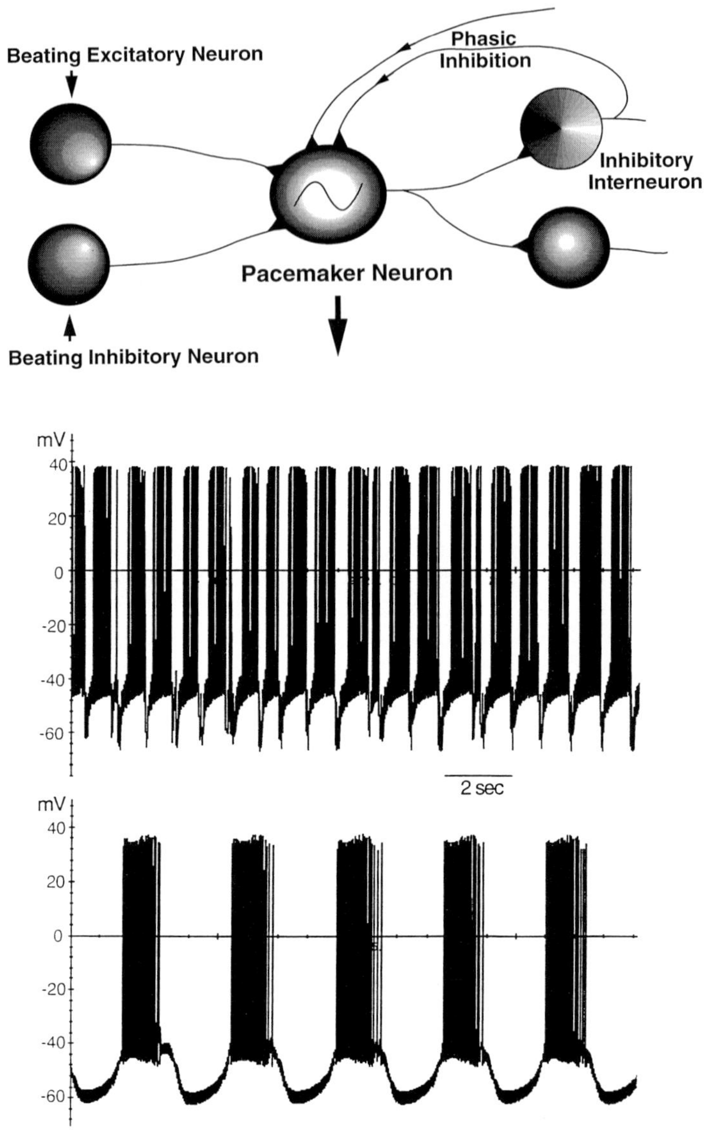

Figure 11 Simulations illustrating transformational properties of model hybrid pacemaker-network. Simulations were conducted with small network (top) consisting of pacemaker neurons, tonic beating excitatory and inhibitory neurons, and follower inhibitory and excitatory interneurons which are phasically driven by pacemaker neuron; inhibitory interneuron provides phasic inhibition to pacemaker neuron after temporal delay. Traces below show oscillatory behavior of pacemaker neuron under different network conditions. Bottom trace: behavior under conditions where pacemaker neurons are in their bursting oscillatory mode (i.e., baseline membrane voltage maintained in

and unknowns of the real system, there are obvious limitations of the model in its present form. Our modeling does, however, represent an advance in that it begins to incorporate some of the nonlinear voltage- and time-dependent membrane/synaptic properties that will be required to account for features of the real system, including intrinsic oscillatory properties. Although the majority of the model parameters used have not yet been measured, the modeling approach we have employed allows the incorporation of experimental data as they become available and close interaction between experimental and computational studies.

IX. Summary

Understanding ventral brainstem mechanisms of respiratory rhythm generation requires analysis of integrative neural function at cellular, synaptic, and network levels. A combination of biophysical studies and computational approaches will be required to solve this problem, particularly to bridge the gap in our understanding of cellular properties and network phenomena. The development of in vitro preparations that retain functional circuits for rhythm generation, coupled with advances in biophysical methodology, have provided powerful approaches for investigation of cellular and synaptic processes underlying rhythm generation. Computational approaches that allow modeling with realistic neurons and circuits enable exploration, integration, and synthesis of complex cellular and network properties. By exploiting these approaches, we have developed novel hypotheses about sites, cellular mechanisms, and network properties underlying rhythm generation, including the following:

1. The pre-BötC is the major locus of rhythm generating neurons in the mammalian brainstem.
2. Pre-BötC neurons with voltage-dependent pacemaker properties are the basic kernel for rhythm generation.
3. The respiratory oscillator is a hybrid pacemaker-network where rhythm is generated and controlled by cooperative interactions of cellular and network processes.

---

negative slope region of I-E relationship by tonic excitatory inputs). Top trace: oscillatory behavior where pacemaker neuron undergoes transformation from bursting to beating at elevated levels of tonic excitatory and weak tonic inhibitory input from beating neurons. As shown, stable oscillations of pacemaker neuron and network occur if there is strong inhibition from follower interneurons. Under these conditions, there is a transformation to a network-based mechanism for rhythmogenesis with associated changes in dynamic behavior of pacemaker neurons and network (i.e., oscillatory frequency shifts to a higher range than can be achieved with pacemakers in their intrinsic bursting mode). See text for complete explanation.

4. The hybrid oscillator has transformational properties and can exhibit cell-driven or network-based mechanisms of rhythm generation.

Further experimental and computational studies to test these hypotheses should greatly expand our understanding of rhythm generation.

Acknowledgments

This work was supported by NIH Grants HL40959, HL02204 and NS24742. G. D. Funk was supported by the Parker B. Francis Foundation. S. M. Johnson was supported by NIH HL08524. Requests for reprints should be addressed to: J. C. Smith, Laboratory of Neural Control, Bldg. 49, Rm. 3A50, National Institute of Neurological Disorders and Stroke, NIH, Bethesda, MD 20892.

References

1. Ezure K. Synaptic connections between medullary respiratory neurons and considerations on the genesis of respiratory rhythm. Prog Neurobiol 1990; 35:429–450.
2. Feldman JL, Smith JC. Neural control of respiratory pattern in mammals: An overview. In: Dempsey JA, Pack AI, eds. Regulation of Breathing. Lung Biology in Health and Disease Series, New York: Dekker, 1995:39–69.
3. Richter DW, Ballanyi K, Schwarzacher S. Mechanisms of respiratory rhythm generation. Curr Opin Neurobiol 1992; 2:788–793.
4. Berger AJ. Recent advances in respiratory neurobiology using in vitro methods. Am J Physiol (Lung Cell Mol Physiol) 1990; 3:L24–L29.
5. Feldman JL, Smith JC, Liu G. Respiratory pattern generation in mammals: In vitro en bloc analyses. Curr Opin Neurobiol 1991; 1:590–594.
6. Smith JC, Feldman JL. In vitro brainstem-spinal cord preparations for study of motor systems for mammalian respiration and locomotion. J Neurosci Meth (Special Issue: Novel Approaches to the Study of Motor Systems) 1987; 21:321–333.
7. Smith JC, Greer JJ, Liu G, Feldman JL. Neural mechanisms generating respiratory pattern in mammalian brain stem-spinal cord in vitro: I. Spatiotemporal patterns of motor and medullary neuron activity. J Neurophysiol 1990; 64:1149–1169.
8. Smith JC, Greer JJ, Ballanyi K, et al. Novel approaches to the study of cellular mechanisms generating respiratory rhythm in vitro. In: Speck DF, Dekin MS, Revelette WR, et al., eds. Respiratory Control: Central and Peripheral Mechanisms, Lexington, KY: University of Kentucky Press, 1992:34–38.
9. Smith JC, Ellenberger HH, Ballanyi K, et al. Pre-Bötzinger complex: A brainstem region that may generate respiratory rhythm in mammals. Science 1991; 254:726–729.
10. Suzue T. Respiratory rhythm generation in the in vitro brainstem-spinal cord preparation of the neonatal rat. J Physiol (Lond) 1984; 354:173–183.
11. Bieger D, Hopkins DA. Viscerotopic representation of the alimentary tract in the medulla oblongata in the rat: The nucleus ambiguus. J Comp Neurol 1987; 262:546–562.

12. Richter DW, Ballantyne D, Remmers JE. How is respiratory rhythm generated? A Model. NIPS 1986; 1:109–112.
13. Smith JC, Funk GD, Johnson SM, Feldman JL. The noeud vital of neonatal rat: Site and cellular mechanisms for respiratory rhythm generation. Proceedings of the XXXIIth Congress of the International Union of Physiological Science 1993:17.
14. Smith JC, Greer JJ, Feldman JL. Medullary slices that generate respiratory oscillations in vitro. Soc Neurosci Abstr 1990; 16:1130.
15. Funk GD, Smith JC, Feldman JL. Generation and transmission of respiratory oscillations in medullary slices: Role of excitatory amino acids. J Neurophysiol 1993; 70:1497–1515.
15a. Funk GD, Smith JC, Feldman JL. Development of thyrotropin-releasing hormone and norepinephrine potentiation of inspiratory-related hypoglossal motoneuron discharge in neonatal and juvenile mice in vitro. J Neurophysiol 1994; 72:2538–2541.
16. Ramirez JM, Pierrefiche O, Schwarzacher SW, Filloux F, McIntosh JM, Olivera BM, Richter DW. The influence of N-type calcium channel blockers on the respiratory network of adult cats. Soc Neurosci Abs 1994; 20:1755.
17. Schwarzacher SW, Smith JC, Richter DW. Respiratory neurones in the pre-Bötzinger region of cats. Pflügers Arch 1991; 418(suppl 1):R17.
18. Schwarzacher SW, Smith JC, Richter DW. The pre-Bötzinger complex in the cat. J Neurophysiol. In press.
19. Connelly CA, Dobbins EG, Feldman JL. Pre-Bötzinger complex in cats: Respiratory neuronal discharge patterns. Brain Res 1992; 590:337–340.
20. Ellenberger HH, Feldman JL. Subnuclear organization of the lateral tegmental field of the rat. I: Nucleus ambiguus and ventral respiratory group. J Comp Neurol 1990; 294:202–211.
21. Segers LS, Shannon R, Saporta S, Lindsey BG. Functional associations among simultaneously monitored lateral medullary respiratory neurons in the cat. I. Evidence for excitatory and inhibitory interactions of inspiratory neurons. J Neurophysiol 1987; 57:1078–1100.
22. Van Bibber R, Barry CL, Dobbins EG, et al. Spatial organization of premotoneurons in a local respiratory circuit studied by transsynaptic neuronal labeling with pseudorabies virus. Soc Neurosci Abstr 1992; 18:124.
23. Johnson SM, Smith JC, Funk GD, Feldman JL. Pacemaker behavior of respiratory neurons in medullary slices from neonatal rat. J Neurophysiol 1994; 72:2598–2608.
24. Smith JC, Feldman JL. Central respiratory pattern generation studied in an in vitro mammalian brainstem-spinal cord preparation. In: Seick G, Gandevia SC, Cameron WE, eds. Respiratory Muscles and their Neuromotor Control. New York: Lis, 1987:27–36.
25. Feldman JL, Smith JC. Cellular mechanisms underlying modulation of breathing pattern in mammals. Ann NY Acad Sci 1989; 563:114–130.
26. Feldman JL, Smith JC, Ellenberger HH, et al. Neurogenesis of respiratory rhythm and pattern: Emerging concepts. Am J Physiol 1990; 259:R879–R886.
27. Smith JC, Funk GD, Johnson SM, Feldman JL. A hybrid pacemaker-network model for the respiratory oscillator in mammals. Soc Neurosci Abstr 1992; 18:1279.
27a. Funk GD, Smith JC, Johnson SM, Feldman JL. Respiratory neuron populations in oscillating medullary slices from neonatal rat. Soc Neurosci Abstr 1992; 18:124.

28. Smith JC. A model for developmental transformations of the respiratory oscillator in mammals (abstr). FASEB J 1994; 8:A394.
28a. Smith JC. Computational models of the respiratory oscillator in mammals. Soc Neurosci Abstr 1994; 20:1202.
29. Chay TR, Rinzel J. Bursting, beating, and chaos in an excitable membrane model. Biophys J 1985; 47:357–366.
30. Llinás R. The intrinsic electrophysiological properties of mammalian neurons: Insights into central nervous system function. Science 1988; 242:1654–1664.
31. McCormick DA, Huguenard J. Strowbridge BW. Determination of state-dependent processing in thalamus by single neuron properties and neuromodulators. In: McKenna T, Davis J. Zornetzer SF, eds. Single Neuron Computation. San Diego, CA: Academic Press, 1992:259–290.
32. Cohen M, Feldman JL. Models of respiratory phase-switching. Fed Proc 1977; 36:2367–2376.
33. von Euler C. Brainstem mechanisms for generation and control of breathing pattern. In: Cherniack NS, Widdicombe JG, eds. Handbook of Physiology: Regulation of Respiration. Bethesda, MD: American Physiological Society, 1986: 1–68.
34. Onimaru H, Arata A, Homma I. Inhibitory synaptic inputs to the respiratory rhythm generator in the medulla isolated from newborn rats. Pflügers Arch 1990; 417:425–432.
35. Hayashi F, Lipski J. The role of inhibitory amino acids in control of respiratory motor output in an artificially perfused rat. Resp Physiol 1992; 89:47–63.
36. Feldman JL, Cleland CL. Possible roles of pacemaker neurons in mammalian respiratory rhythmogenesis. In: Carpenter DO, ed. Cellular Pacemakers. Vol II. New York: Wiley, 1982:104–128.
37. Smith JC, Feldman JL. Role of chloride-dependent synaptic inhibition in respiratory pattern generation: Studies in an in vitro mammalian brainstem-spinal cord preparation. Fed Proc 1986; 45:518.
38. Di Pasquale E, Monteau R, Hilaire G. Involvement of the rostral ventro-lateral medulla in respiratory rhythm genesis during the peri-natal period: An in vitro study in newborn and fetal rats. Dev Brain Res 1994; 78:243–252.
39. Smith JC, Ballanyi K, Richter DW, Feldman JL. Conditional bursting pacemaker neurons in a limited region of the medulla involved in respiratory rhythm generation in vitro. FASEB J 1991; 5:A734.
40. Smith JC, Ballanyi K, Richter DW, Feldman JL. Oscillatory properties of neurons in an isolated respiratory circuit analyzed by whole-cell patch-clamp techniques. Soc Neurosci Abstr 1991; 17:1580.
41. Smith JC, Ballanyi K, Richter DW. Whole-cell patch-clamp recordings from respiratory neurons in neonatal rat brainstem in vitro. Neurosci Lett 1992; 134: 153–156.
42. Onimaru H, Arata A, Homma I. Firing properties of respiratory rhythm generating neurons in the absence of synaptic transmission in rat medulla in vitro. Exp Brain Res 1989; 76:530–536.
43. Benson JA, Adams WB. The control of rhythmic neuronal firing. In: Kaczmarek L, Levitan I, eds. Neuromodulation: The Biochemical Control of Neuronal Excitability. New York: Oxford University Press, 1987:100–118.

44. Swandulla D, Lux HD. Activation of a nonspecific cation conductance by intracellular calcium elevation in bursting pacemaker neurons of Helix pomatia. J Neurophysiol 1985; 54:1430–1433.
45. Partridge LD, Swandulla D, Müller J. Modulation of calcium-activated non-specific cation currents by cyclic AMP-dependent phosphorylation in neurons of Helix. J Physiol (Lond) 1990; 429:131–145.
46. Smith SJ, Thompson SH. Slow membrane currents in bursting pacemaker neurons of tritonia. J Physiol (Lond) 1987; 382:425–448.
47. Moczydlowski E, Lucchesi K, Ravindran A. An emerging pharmacology of peptide toxins targeted against potassium channels. J Membrane Biol 1988; 105:95–111.
48. Hille B. Ionic Channels of Excitable Membranes. 2nd ed. Sunderland, MA: Sinauer, 1992.
49. Johnson SM, Smith JC, Feldman JL. Respiratory rhythm modulation by $GABA_B$, 5HT-1A, and alpha-2 adrenergic receptors: mediation via an inward rectifying K^+ conductance? Physiologist 1993; 36:A-12.
50. Johnson SM, Smith JC, Feldman JL. Modulation of respiratory rhythm in neonatal rat in vitro: role of $G_{i/o}$-protein regulated conductances. (Submitted for publication).
51. Smith JC, Funk GD, Johnson SM, Feldman JL. Neurophysiology and pharmacology of respiratory rhythm generation in mammals. Physiologist 1993; 36:A-8.
52. Dempsey JA, Olsen EB Jr, Skatrud JB. Hormones and neurochemicals in the regulation of breathing. In: Handbook of Physiology: The Respiratory System. Sec 3, Vol II, Pt 1. Bethesda, MD: Am Physiol Soc 1986:181–221.
53. Kaczmarek L, Levitan I, eds. Neuromodulation: The Biochemical Control of Neuronal Excitability. New York: Oxford University Press, 1987.
54. Marder E, ed. Neuromodulation in Circuits Underlying Behaviour. Seminars in the Neuroscience. Vol 2. Philadelphia: Saunders, 1991.
55. Nicoll RA. The coupling of neurotransmitter receptors to ion channels in the brain. Science 1990; 241:545–551.
56. Nicoll RA, Melenka RC, Kauer JA. Functional comparison of neuroreceptor subtypes in mammalian central nervous system. Physiol Rev 1990; 70:513–565.
57. North RA. Drug receptors and the inhibition of nerve cells. Br J Pharmacol 1989; 98:13–28.
58. Reuter H, Sigel E. Ionic channels: Modulation by G proteins and by phosphorylation. Curr Opin Neurobiol 1991; 1:27–31.
59. Stanfield PR, Nakijima Y, Yamaguchi K. Substance P raises neuronal membrane excitability by reducing inward rectification. Nature (Lond) 1985; 315:498–501.
60. Yamada KA, Lagercrantz J, von Euler C. Effects of substance P and TRH on ventilation and pattern of breathing in newborn rabbits. Acta Physiol Scand 1981; 113:541–543.
61. Wheal H, Thomson A, eds. Excitatory Amino Acids and Synaptic Transmission. San Diego, CA: Academic Press, 1991.
61a. Schoepp DD, Conn PJ. Metabrotropic glutamate receptors in brain function and pathology. Trends Pharmacol Sci 1993; 14:13–20.
62. McCrimmon DR, Smith JC, Feldman JL. Involvement of excitatory amino acids in neurotransmission of inspiratory drive to spinal respiratory motoneurons. J Neurosci 1989; 9:1910–1921.

63. Liu G, Feldman JL, Smith JC. Excitatory amino acid-mediated transmission of inspiratory drive to phrenic motoneurons. J Neurophysiol 1990; 64:423–437.
64. Greer JJ, Smith JC, Feldman JL. Role of excitatory amino acids in the generation and transmission of respiratory drive in neonatal rat brainstem-spinal cord in vitro. J Physiol (Lond) 1991; 437:727–749.
65. Greer JJ, Smith JC, Feldman JL. Glutamate release and presynaptic action of AP4 during inspiratory drive to phrenic motoneurons. Brain Res 1992; 576:355–357.
66. Feldman JL, Windhorst U, Anders K, Richter DW. Synaptic interaction between medullary respiratory neurons during apneusis induced by NMDA-receptor blockade in cat. J Physiol (Lond) 1992; 450:303–323.
66a. Dong X-W, Morin D, Feldman JL. Metabotropic glutamate receptors attenuate inspiratory-modulated synaptic current in phrenic motoneurons. Soc Neurosci Abstr 1994; 20:1714.
67. Onimaru H, Homma I. Whole-cell recordings from respiratory neurons in the medulla of brainstem-spinal cord preparations from neonatal rats. Pflügers Arch 1992; 420:399–406.
68. Murakoshi T, Otsuka M. Respiratory reflexes in an isolated brainstem-lung preparation of the newborn rat: Possible involvement of gamma-aminobutyric acid and glycine. Neurosci Lett 1985; 62:63–68.
69. Koch C, Segev I, eds. Methods in Neuronal Modeling: From Synapses to Networks. Cambridge, MA: MIT Press, 1989.
70. McKenna T, Davis J, Zornetzer SF, eds. Single Neuron Computation. San Diego, CA: Academic Press, 1992.
71. Bower JM, ed. Modeling the nervous system. Trends Neurosci 1992; Vol. 15.
72. Botros SM, Bruce EN. Neural network implementation of the three-phase model of respiratory rhythm generation. Biol Cybernet 1990; 63:143–153.
73. Duffin J. A model of respiratory rhythm generation. Neuroreport 1991; 2:623–626.
74. Ogilvie MD, Gottschalk A, Anders K, et al. A network model for respiratory rhythmogenesis. Am J Physiol 1992; 32:R962–R975.
75. De Schutter E. Computer software for development and simulation of compartmental models of neurons. Comput Biol Med 198:71–78.
76. Wilson M, Bhalla U, Uhley J, Bower JM. GENESIS: A system for simulating neural networks. In: Touretzky D, ed. Advances in Neural Network Information Processing Systems. San Mateo, CA: Morgan Kaufman, 1989:485–492.
77. Wilson MA, Bower JM. The simulation of large-scale neural networks. In: Koch C, Segev I, eds. Methods in Neuronal Modeling: From Synapses to Networks. Cambridge, MA: MIT Press, 1989:291–334.
77a. Balis UJ, Morris KF, Koleski J, Lindsey BG. Simulations of a ventrolateral medullary neural network for respiratory rhythmogenesis inferred from spike train cross-correlation. Biol Cybernet 1994; 70:311–327.
78. Yamada WM, Koch C, Adams PR. Multiple channels and calcium dynamics. In: Koch C, Segev I, eds. Methods in Neuronal Modeling: From Synapses to Networks. Cambridge, MA: MIT Press, 1989:97–134.
79. Golowasch J, Buchholtz F, Epstein IR, Marder E. Contribution of individual ionic currents to activity of a model stomatogastric ganglion neuron. J Neurophysiol 1992; 67:341–349.

30

Localization of Ventral Brainstem CO_2 Chemosensors by Kainic Acid Neurotoxic Lesion

ROBERT M. DOUGLAS and C. OVID TROUTH

College of Medicine
Howard University
Washington, D.C.

YUAN-HSU KANG

Naval Medical Research Institute
Bethesda, Maryland

JONAS I. ADDAE

University of the West Indies
St. Augustine, Trinidad and Tobago, West Indies

JERRY M. WRIGHT

National Institute on Alcohol Abuse and Alcoholism
National Institutes of Health
Bethesda, Maryland

I. Introduction

Physiological evidence indicates that neurons in the caudal ventrolateral medullary surface (cVMS) of rats and cats have an important role in the central regulation of both the cardiovascular and respiratory systems (1–5). Electrical stimulation (6,7) and microinjections of excitatory amino acids (EAA) (8–10) at the ventral medullary surface (VMS) elevate blood pressure and increase respiration. Gamma-amino butyric acid (GABA), EAA antagonists, and bilateral electrolytic and chemical lesions of the VMS decrease blood pressure and respiration (11–18). Immunocytochemical expression of c-*fos*, a marker for increased neuronal activity, has been used to characterize a population of carbon dioxide (CO_2)-sensitive neurons at the cVMS after 1 h of hypercapnia (19). These studies indicate that the VMS contains chemosensitive structures that are sensitive to pH and CO_2 changes. They also suggest that EAA play an important role in the central processing of cardiorespiratory afferent information. Afferent activity of these and other intracranial chemosensors at or near the VMS appear to be the dominant source of chemoreceptor drive to the respiratory controller (20–22). However, several workers have suggested that central respiratory chemosensitivity is at least partially provided by the deeper-lying respiratory centers (23,24) and that a localized and circumscribed region of central chemosensitivity other than the respiratory centers themselves may not be necessary (25,26).

Several techniques have been utilized to describe the morphological substrates subserving respiratory chemosensitivity at the ventral medulla. Schläfke and Loeschcke (27) thermally lesioned the intermediate chemosensitive zone (area S) (Fig. 1) and induced respiratory arrest in peripheral chemoreceptor-denervated cats in which the contralateral area S was coagulated. They proposed that central chemosensitivity was lost by this procedure. In a later study, Schläfke et al. (12) coagulated the chemosensitive apparatus that was believed to be operational at area S. They noted that there was diminished or no response to acidity or hypercapnia when the central chemosensory apparatus at area S was compromised. Histological examination of the coagulated tissue

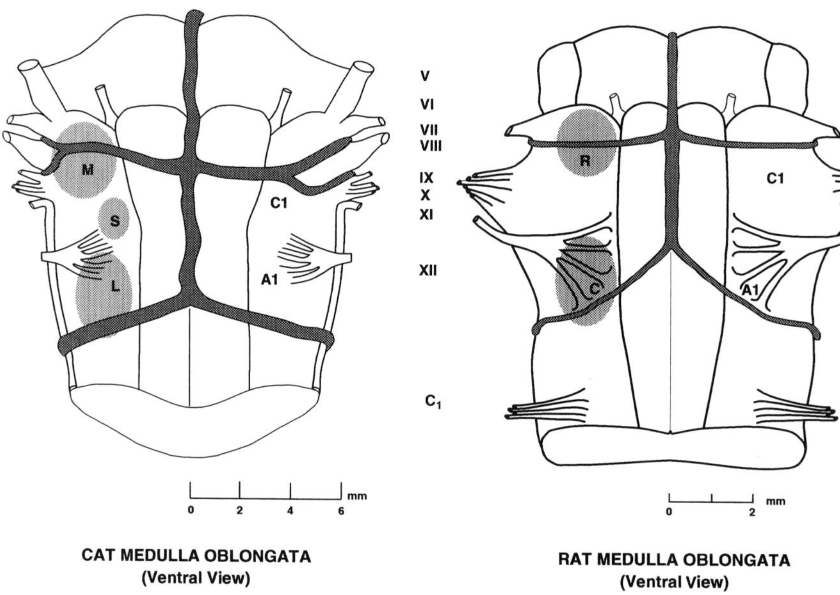

Figure 1 Ventral surface of the medulla oblongata of the cat (left) and the rat (right). In the cat, the three respiratory chemosensitive areas, "M" (Mitchell et al.) (2), "S," and "L" (Schläfke et al.) (27) are indicated on the left half of the diagram and represent the rostral, intermediate and caudal areas, respectively. In the rat, the two chemosensitive zones (Mitra et al.) (49), "R" and "C," are indicated on the left half of the diagram and represent the rostral and caudal chemosensitive zones, respectively. In both right hemimedullas, the symbols C1 and A1 represent the vasopressor and vasodepressor areas, respectively. V–XII = cranial nerves.

revealed destruction of marginal glia, superficial neurons, small blood vessels, capillaries, arcuate fibers, and other superficial fibers (12).

Berndt et al. (28) applied procaine and toxic concentrations of potassium (K^+) ions to the VMS of cats while stimulating the cVMS and the respiratory centers in peripheral chemoreceptor-denervated animals. Electrical stimulation of the cVMS caused increased ventilation, which was blocked, causing apnea following topical application of procaine to the cVMS. Electrical stimulation of the ventral respiratory group (VRG), however, reversed the procaine blockade, causing an inspiratory response. These studies indicated that the respiratory centers were still intact and functional despite the procaine-induced apnea. As with procaine, toxic concentrations of K^+ topically applied to the VMS induced apnea that was initially overridden by electrical stimulation of the VRG. However, within a few minutes, total apnea that was not overridden by electrical stimulation of the VRG ensued, indicating that K^+ had diffused to the respiratory centers and rendered them dysfunctional. These experiments demonstrated that the integrity of the VMS structures was essential to driving the respiratory centers in the absence of peripheral chemoreceptor afferents (28). Because these procedures destroyed or blocked cell bodies as well as fibers in transit, these studies failed to discriminate between surface neurons at chemosensitive sites and fibers of passage originating from other central or peripheral nervous structures. A better method had to be found.

The intraneuronal events triggered by EAA receptor stimulation have developed considerable interest over the past decade. Not only are EAA major excitatory neurotransmitters in mammalian brain but they also produce neuronal damage when present in excessive amounts. In 1978, J. T. Coyle (29), utilizing the restricted glutamate (Glu) analog kainic acid (KA), described a superior lesioning technique that destroyed cell bodies of neurons but not fibers of passage. Toxic doses of EAA are believed to destroy cell bodies of neurons but not fibers of passage (30,31). Indeed, this property provides an exquisite lesioning tool that allows for a finer localization, definition, and separation of the structures under investigation. Depending on the duration of the EAA treatment, neuronal death can be produced either by a rapid sodium- and chloride-dependent osmotic lysis (32,33) or by a more protracted calcium (Ca^{2+})-dependent process (33–35). Destabilization of ionized cytosolic calcium homeostasis and the biochemical events that are modulated by intracellular calcium (Ca_i) may also play a pivotal role in excessive Glu receptor stimulation and subsequent neuronal demise (35–38). By chemical lesioning with KA, the neuronal substrates subserving cVMS responsiveness to inspired CO_2 and topically applied acetylcholine (ACh) can be identified. We therefore applied toxic doses of KA to the VMS of rats before and after testing for the ventilatory and neuronal responses to inspired CO_2 and ACh.

In spontaneously breathing animals ($n = 8$) anesthetized with chloralose-urethane, we tested superficial neurons in the caudal chemosensitive area (area L)

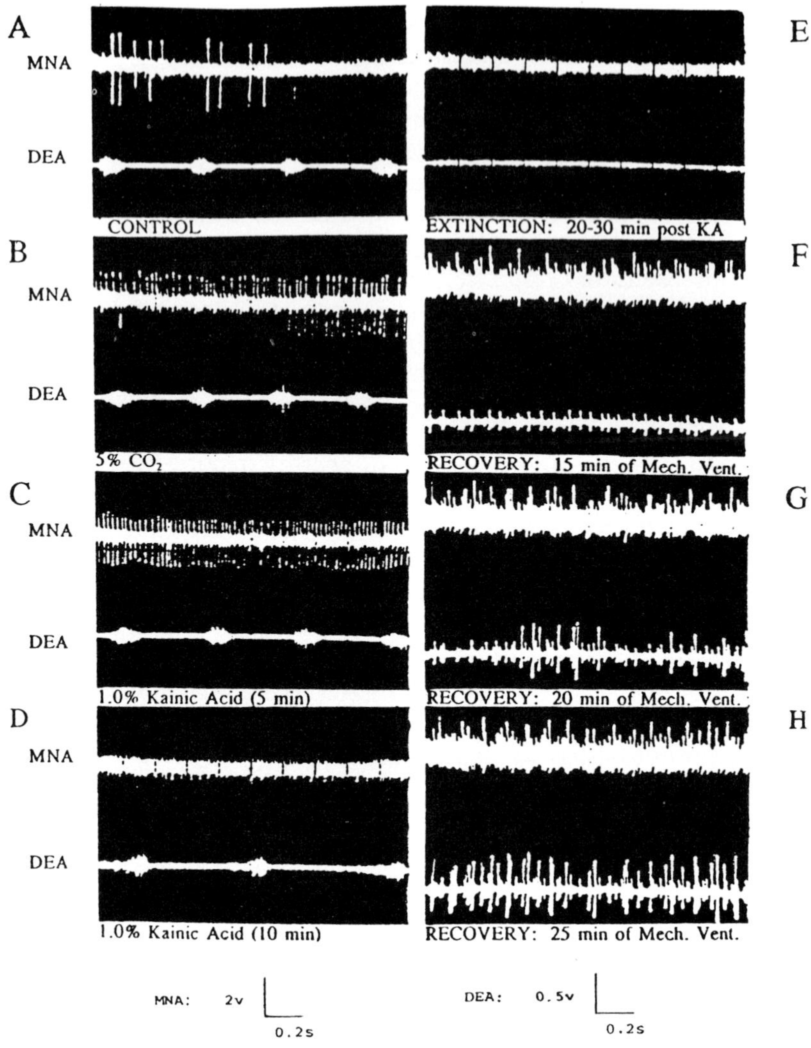

Figure 2 Nonphasic CO_2-sensitive unit discharge during the topical application of excitatory amino acids to the ventral medullary surface. Upper tracings: medullary neuronal activity (MNA). Lower tracings: diaphragmatic electromyographic activity (DEA). (A) control (MNA = 19 Hz; DEA = 48 bpm); (B) 5% $PICO_2$ (MNA = 27.9 Hz, DEA = 63 bpm); (C) 5 min post 1% kainic acid (KA) (MNA = 34.8 Hz; DEA = 53 bpm); (D) 10 min post 1% KA; (E) extinction (20 to 30 min post 1% KA), i.e., apnea and neuronal silence; (F) recovery: mechanical ventilation, 15 min postextinction; (G and H) recovery at 20 and 25 min postextinction, respectively (bpm = bursts per minute).

for their responsiveness to increased inspired CO_2 and topically applied ACh, while recording medullary neuronal activity (MNA) extracellularly and diaphragmatic electromyographic activity (DEA). Increased inspired CO_2 and ACh transiently increased MNA as well as DEA (Fig. 2A to E). Following topical application of 1% KA, there was a marked increase in MNA and DEA within 2 to 5 min. Ten to 15 min later, there were decrements in both MNA and DEA, with the animals demonstrating periodic gasping activity. At approximately 15 to 20 min after KA application, MNA subsided to extinction and DEA ceased; the animal expired. Thus KA, an excitotoxic amino acid, caused marked and prolonged increases in MNA and DEA, followed by depression to extinction and respiratory arrest.

The chemosensitive response to CO_2 and ACh must have originated from neurons at the surface and not from fibers of passage. Therefore, KA may have destroyed superficial chemosensitive VMS cell bodies, and the integrity of these chemosensitive VMS neurons in the anesthetized rat appears essential to the maintenance of the central drive of respiration (3,39–44). These results support studies of Berndt et al. (28) and Schläfke et al. (12).

II. Cardiorespiratory Effects on VMS Cardiorespiratory Control

The VMS has both respiratory and cardiovascular control elements (1,5,45, 46) and is thought to perform an important integrative function in the central regulation of the respiratory and cardiovascular systems (4). Vasopressor and vasodepressor regions have been identified within the VMS (10,16,17,47,48) and on the VMS (49–52). It has also been recognized that EAA play a significant role in the modulation of both respiratory and cardiovascular elements within and upon the ventrolateral medulla (VLM) (14,18,44,53–56). We therefore attempted to determine whether the KA-induced demise of CO_2-sensitive neurons (and elimination of CO_2 chemosensitivity) would result in dysfunction of the respiratory reflex apparatus or disruption of cardiovascular elements within the cVMS.

In these investigations, we utilized the neurotoxin KA to discriminate between cardiovascular and respiratory structures in the VMS. In spontaneously breathing animals anesthetized with chloralose-urethane, neurons in the caudal chemosensitive area (area L) were tested for their responsiveness to physiological and pharmacological stimuli before and after the topical application of 1% (47 m*M*) KA to the cVMS. As in previous experiments, 1% KA at first increased ventilation, then induced apnea within 10 to 20 min. It is possible that KA destroys only cell bodies (29); however, it is uncertain (a) whether KA causes cell death or merely metabolically depletes the cell or (b) whether fibers of passage are affected by toxic doses of EAA.

In an attempt to determine which of the two systems, respiratory or cardiovascular, was compromised in the KA-induced apneic demise of the animal, we instituted artificial ventilation with a Narco small-animal ventilator for up to 6 h immediately following the induction of the apnea by the neurotoxin KA. We speculated that the cVMS neurons in these studies may have been depolarized to the point where they were refractory to incoming stimuli, thus leading to the apnea induced by KA. After 5 to 10 min of mechanical ventilation, MNA and DEA recovered. MNA initially appeared as low -amplitude spikes that increased in frequency over the next few minutes. After 5 to 10 min, MNA increased to a maximal frequency of 84.14 ± 10.19 Hz (Fig. 2F to H). The MNA during this period of recovery is reminiscent of evoked and spontaneous epileptiform bursting described in hippocampal slice preparations (57–60) and in in vivo models of epilepsy (61–63). KA may induce activation of *N*-methyl-D-aspartate (NMDA) receptors (one of the three classic Glu receptors) which have been implicated in the generation of evoked and spontaneous burst generation (64–66).

In toxic doses, KA can directly depolarize postsynaptic elements, mobilize other EAA, induce depolarization-modulated release of the Mg^{2+} block on the NMDA channel, and markedly increase intracellular Ca^{2+} accumulation (65–72). One might assume that the KA-induced respiratory failure might in some way be associated with the toxic effects of Ca^{2+} on neurons at caudal ventral brainstem chemosensitive sites, but this must be proven. Failure of these chemosensory neurons to communicate with or modulate the activity of the central respiratory integrating centers would result in apnea. This would imply that these circumscribed superficial neuronal elements are necessary to the central drive of respiration.

During the recovery period, DEA initially demonstrated low-frequency, low-amplitude random spikes that increased in frequency and amplitude over the next 60 min to a generalized tonic state. Thereafter, DEA diminished in amplitude and frequency to the point of extinction within 3 to 4 h. The DEA indicated that the burstlike (phasic) activity reflective of phrenic nerve or diaphragmatic electromyographic activity never resumed. This result further indicates a disruption of the organization of respiratory motor output coordination and a loss of communication between the respiratory centers, per se, and the phrenic nucleus. We therefore postulate that KA renders the superficial CO_2-sensitive neuronal pool at the caudal VMS inoperable and that, in spite of the continued efficacious function of the respiratory centers and phrenic nucleus, the animal appears to undergo irreversible apnea.

After 3 to 4 h, and even though DEA had achieved extinction, MNA continued unabated until the animal expired or was sacrificed. Attempts to wean the animals from the respirator were unsuccessful, demonstrating a rapid decline in MNA, DEA, or both within 1 to 2 min. The animals expired if they were not immediately returned to the ventilator. Thus, testing for CO_2 responsiveness under these conditions has not been possible, since MNA, DEA, and ventilation could not be sustained independent of assisted breathing.

Assuming that there are cardiovascular control elements located at or near the VMS (50,73) as well as 1 to 2 mm deep (the A1 area) (10,48,74,75), it was noteworthy that superficial respiratory components were disabled by KA, yet the animal could be maintained by artificial respiration for several hours after the induction of apnea. This result implies that these cardiovascular elements were either intact or, if compromised, of insufficient influence on the putative vasomotor centers to deprive the animal of sympathetic vasomotor drive. This led to the examination of the effects of KA on the cardiovascular system.

In related experiments, we therefore examined the respiratory and blood pressure responses of rats ($n = 5$) to the topical application of EAA agonists and antagonists and KA. Respiratory responses were recorded via a low-inertia Krogh spirometer and blood pressure via a Stratham strain gauge transducer. Compared with control levels, topically applied Glu-evoked dose-related increases in mean arterial blood pressure (MABP) and tidal volume (V_T) with no apparent effect on respiratory frequency (f) (Fig. 3A). There was an apparent desensitization of the Glu response, similar to that noted by Addae and Stone (76) in the response of cortical neurons to NMDA (76–79) in that subsequent applications of Glu within 5 to 10 min of the previous applications did not induce ventilatory or blood pressure responses. Those responses, however, could again be elicited after approximately 30 min. NMDA increased MABP and V_T, and decreased f (Fig. 3B).

Conversely, the NMDA antagonists (5R,10S)-(+)-5-methyl-10,11-dihydro-5H-dibenzo[a,d]cyclohepten-5,10-imine hydrogen maleate (MK801), DL-2-amino-7-phosphono-heptanoic acid (AP7), and (±)-3-(2-carboxypiperazin-4-yl)-propyl-1-phosphonic acid (CPP) decreased both MABP and V_T, and increased f (Fig. 3C). The topical application of 1% KA produced a marked but transient increase in both MABP and V_T and a decrease in f. These experiments, as well as those of Gattie et al. (52) and of Martin and Sinclair (44) demonstrated that cells in or near the VMS contain EAA receptors both of the NMDA and non-NMDA varieties. They also appear to indicate that the EAA receptors at the VMS are involved in the timing (rhythm generation) and amplitude (pattern generation) responses of the respiratory reflex apparatus located in the caudal and rostral chemosensitive zones (80–83). A more complete examination of the EAA receptor subtypes serving cardiorespiratory responses at the VMS seems warranted.

However, as in our electrophysiological experiments, topically applied KA led to decrements in V_T and f over the next 20 to 30 min, after the initially marked increase in V_T, and subsequent apnea. If the process was uninterrupted, the animal expired, but if it was placed on a mechanical ventilator, the animal could be maintained for up to 6 h. We monitored the MABP continuously throughout this procedure and noted that, after its transient increase, MABP

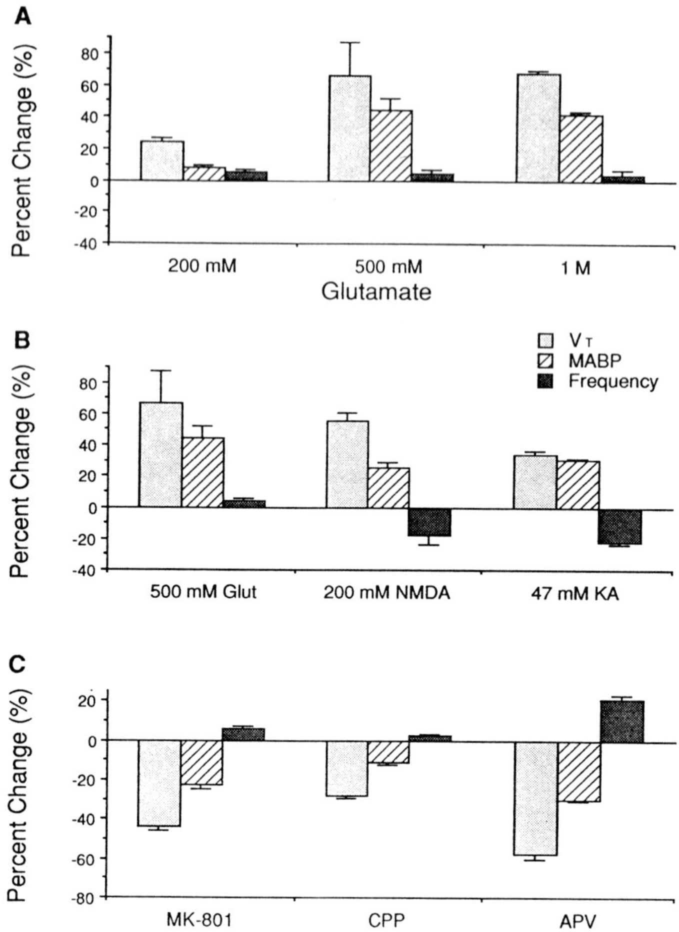

Figure 3 Effects of excitatory amino acid agonists and antagonists topically applied to the ventral medullary surface (VMS) on tidal volume (V_T), mean arterial blood pressure (MABP), and respiratory frequency (*f*) in rats. Mean percent changes ± standard error of the mean (SEM). Each animal serves as its own control. (A) Dose-related increase in V_T and MABP in response to Glu; (B) effect of NMDA (500 m*M* Glu, 200 m*M* NMDA) and non-NMDA (47 m*M* KA) receptor stimulation; (C) effect of three specific NMDA antagonists (10 m*M* MK801, 400 n*M* CPP, and 10 m*M* AP7).

returned to baseline levels and remained there when the animals were maintained via mechanical ventilation. Therefore, superficial CO_2-sensitive cells at the VMS may have some but not an overriding influence on the central cardiovascular controlling elements. A vasodepressor area (area A1) (10,84,85) whose cells contain norepinephrine has been described in the cVLM, located 1 to 1.5 mm deep to the VMS in the rat. Microinjections of EAAs in the A1 area have yielded vasodepressor responses (74), while the topical application of EAA agonists to the VMS in the same region (area L) (44,83) induces transient increases in MABP. This implies a separation or demarcation of superficial cardiorespiratory neurons and deeper-lying purely cardiovascular neurons 1 to 1.5 mm deep in the A1 area. In consideration of the rate of diffusion of small molecules such as KA (molecular weight 214.1 g), Martin and Sinclair (44) developed a mathematical model based on the solution of the one-dimensional diffusion equation and a Gallerkin finite element model and determined that a 47 m*M* (1%) KA solution applied topically to the VMS for only 2 min would achieve cytotoxic levels in a few seconds at the VMS and at deeper levels within a few minutes (44,45). KA may also be cytotoxic in nanomolar concentrations (86), which are readily achieved deep within the medullary tissue within 20 to 30 min from a topical site of application. Another aspect of the rate of diffusion of substances into brain parenchyma from a subarachnoid space or cisternal locus is the access of these molecules to the interstitial (extracellular) space via the perivascular spaces and basal laminae of penetrating arteries (87,88). Rennels et al. (88), utilizing intracisternal horseradish peroxidase perfusion (HRP), described penetrant arteries and emergent veins that were enclosed by fluid-filled perivascular spaces in contiguity with the extracellular and subarachnoid spaces, which functioned as "express routes" and have been designated as paravascular spaces. They believed that the rapid paravascular influx and spread of HRP were not simply due to diffusion but depended on the pulsations of penetrating aterioles and the convective forces generated thereby (87,88). These arguments, however, do not adequately explain the differences noted in the MABP responses to topically applied KA and KA microinjected into the A1 area. They also do not address the issue of the disruption of the respiratory apparatus at the VMS leading to apnea, while the VRG and other deeper-lying medullary neuronal pools are unaffected.

The question therefore arose as to which cells at the VMS were activated and subsequently made dysfunctional by the topical application of 1% KA without apparently affecting cardiovascular control mechanisms within the cVLM. Because calcium has been implicated in the neurotoxicity of KA and EAA (36, 89,90), we investigated the subcellular localization of calcium in cVMS cells in response to the topical application of 1% KA.

III. Ultracytochemical Localization of Calcium in Rat cVLM in Response to Topically Applied KA

The mechanism of action of KA as an excitatory amino acid in the central nervous system (CNS) appears to be the mediation of the voltage-independent portion of the synaptic response to unitary synaptic activation (91) by the opening of a voltage-independent cation channel, and the modulation of synaptic activity

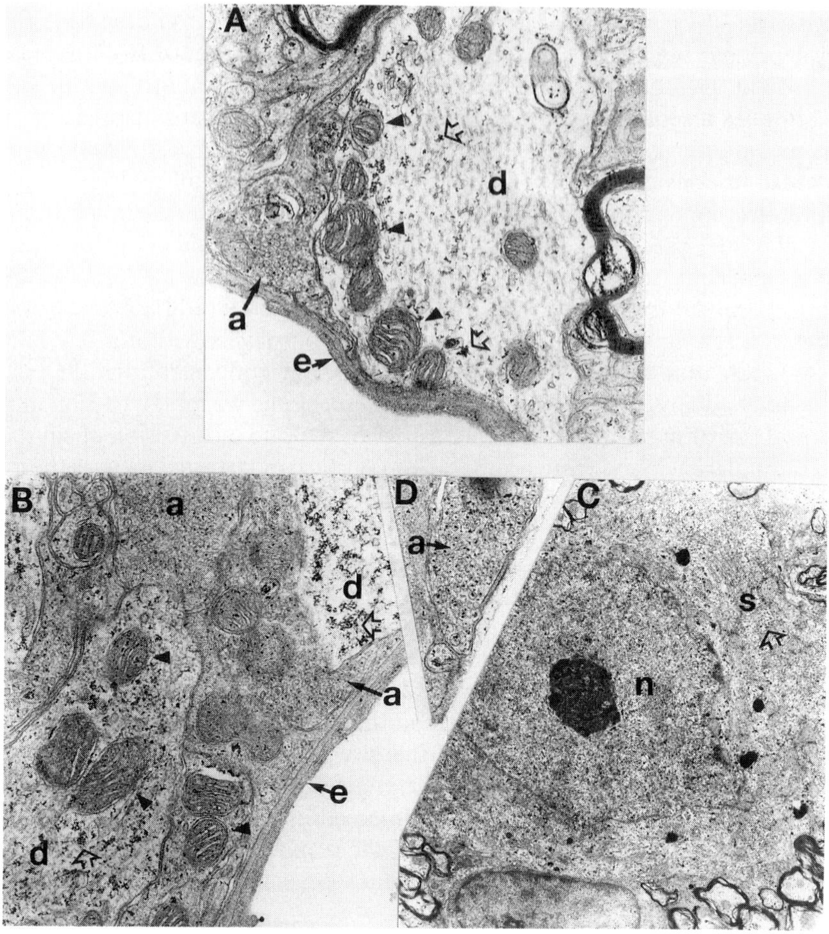

Figure 4 Control: Electron micrographs from superficial ventral brainstem at the CVLM of the normal rat processed by the oxalate-pyroantimonate method for localizing calcium. Uranyl acetate and lead citrate staining. Labels: n-nucleus, s-neuronal soma, e-endothelial wall, d-dendrite, a-axon terminal. Note fine grains or clumps of electron-dense calcium deposits (open arrows) throughout the tissue. (A) Axon terminal and dendritic process abutting blood vessel wall (neurovascular elements); numerous mitochondria (closed arrowheads) that contain variable amounts of Ca^{2+} deposit ($\times$ 21,000). (B) Dendrites and axon terminals adjacent to blood vessels ($\times$ 28,000). (C) Superficial neuron demonstrating fine nuclear and cytoplasmic deposition ($\times$ 5,200). (D) Inset: axon terminal (a) apposed to blood vessel wall with calcium deposits in clear round vesicles ($\times$ 28,000).

by calcium channels. EAA agonists depolarize cells by activation of NMDA and non-NMDA (quisqualate/AMPA, kainate) receptors that induce the influx of cationic species (Na^+, Ca^{2+}) via voltage-sensitive and voltage-independent means (91–93). EAA neurotoxicity is thought to be mediated by the influx of Na^+, Cl^-, and Ca^{2+}, leading to a sustained depolarization and the release of intracellular calcium (Ca_i) in amounts that overwhelm the Ca_i homeostatic apparatus and renders the neuron susceptible to swelling and the pejorative activation of lipases, proteases, and so on, by Ca^{2+} (36,93–95).

EAA have also been implicated in the neurotoxicity associated with several clinical conditions [such as Huntington's chorea and hypoglycemia (96–99)], and it has been postulated that, depending on the duration of EAA treatment, neuronal death can be produced either by a rapid sodium- and chloride-dependent osmotic lysis or by a more protracted calcium-dependent process. Indeed, the KA-induced increase in Ca_i is believed to be essentially caused by influx via agonist-operated and voltage-operated Ca^{2+} channels, whereas only a small contribution is generated by release from intracellular stores (95). Because EAA neuronal activation may lead to a translocation of Ca^{2+} from the extracellular fluid (ECF) and intracellular stores, we examined the distribution of cytosolic Ca^{2+} in the cVMS of KA-stimulated neurons.

We therefore utilized the combined oxalate-pyroantimonate technique (100–103) and electron microscopy to assess the role of calcium in the presumed mortal response of the chemosensory apparatus to KA stimulation and to possibly visualize and characterize these superficial elements by their altered calcium distribution. EAA receptor activation leads to increased intracellular calcium via three potential routes: voltage-gated Ca^{2+} channels, agonist-activated channels permeable to Ca^{2+} as well as monovalent ions (e.g., the NMDA channel), and by release from intracellular stores. In these studies, pledgets containing 1% KA were topically applied to the cVMS of rats ($n = 5$). The pledgets were quickly removed after a 2-min application period and the cVMS thoroughly rinsed with mCSF, pH 7.4. At the peak of the neuronal and ventilatory response to topical KA (10 min postapplication), the animals were perfused with a 90-m*M* potassium-oxalate solution of 3% glutaraldehyde and 2% paraformaldehyde, postfixed with a 2% potassium pyroantimonate–1% osmium tetroxide aqueous solution, and processed for electron microscopy. In control tissues, a discrete pattern of electron-dense Ca^{2+}-pyroantimonate deposition, similar to that described by Griffiths et al. (104) in hippocampal tissue and by Garthwaite and Garthwaite (105) in cerebellar tissue, was noted overlying some neuronal nuclei, in some dendrites, occasional mitochondria, smooth endoplasmic reticulum (SER), with sparse or no deposits in the cytosol, and none in the ECF (Fig. 5A to E). The Ca^{2+} precipitate could also be visualized intraaxonally (axoplasmic) and within synaptic terminals overlying mitochondria and vesicles. In both control and experimental tissues, the Ca^{2+}-pyroantimonate reaction prod-

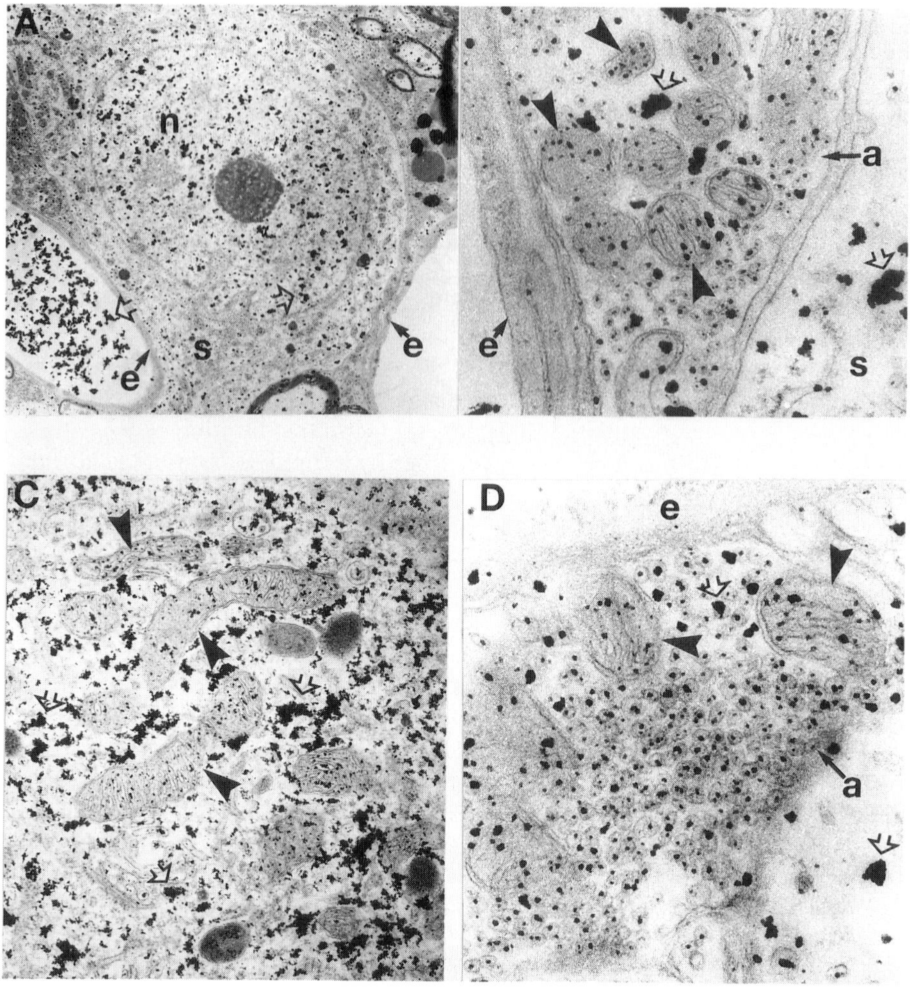

Figure 5 Effect of 1% KA applied to ventral brainstem surface. Electron micrographs in the CVLM of rats following 1% KA stimulation. Tissue prepared for calcium localization utilizing the combined oxalate-pyroantimonate technique. Labels: n-nucleus, s-neuronal soma, e-endothelial wall, d-dendrite, a-axon terminal. (A) 1% KA induces markedly increased deposition of precipitate and concomitant enlargement of precipitate size (open arrows). Cytoplasmic deposits are also increased in number and size (× 3,500); compare with Figure 4C. (B) At higher magnification (× 15,500), deposits are clearly visible in mitochondria (closed arrows), free in the cytoplasm, absent from the rough endoplasmic reticulum and within axonal terminals (vesicles and mitochondria); compare with Figure 4A and B which demonstrate reduced quantity and size of precipitates. (C) Deposition of calcium pyroantimonate within and around sometimes swollen and deranged mitochondria (closed arrows) within the soma of superficial neuron (original magnification × 35,000); compare in Figures 4 A, B, and D. (D) Axonal terminal abutting blood vessel wall with large concentrations of calcium-pyroantimonate in swollen and sometimes deranged mitochondria (× 52,000).

uct could also be found in bleb-like structures within the disrupted myelin sheaths of axons, a feature defined as a fixation artifact by Debbas et al. (106).

Medullary tissues, however, that were perfused at peak MNA following KA stimulation demonstrated a marked increase in intraneuronal Ca^{2+}-pyroantimonate deposition (Fig. 5A to E). There was a dramatic increase in cytosolic calcium deposits, indicating that ionized free calcium had been liberated within the cytosol from extra- and intracellular sources in millimolar amounts, as indicated by Simson and Spicer (107), which would be more than adequate to overwhelm the intracellular Ca^{2+} homeostatic apparatus that normally functions in the nanomolar range (104). The SER, however, which contained deposits of calcium in control tissues, essentially showed sparse or no calcium deposits in the KA-treated tissues. There were also notable increases in the number and size of precipitates overlying the nuclei, and swollen, sometimes deranged mitochondria of affected neurons; within axons and swollen dendrites; and especially within synaptic profiles where vesicular deposition appeared to be maximal. The precipitates could also be detected in the extracellular compartments and within the endothelium of medullary blood vessels traversing this region. Our results indicated that increased intracellular Ca^{2+} accumulation occurred very early in the neurotoxic sequence activated by KA and other EAA, as also recently demonstrated by Choi (108), and that it probably precedes and accompanies the cytopathology of excitotoxic-induced neuronal injury (105).

IV. Summary

This study identifies the respiratory component involved in CO_2 chemosensitivity. In spontaneously breathing rats with intact peripheral chemoreceptors, medullary neuronal activity (MNA) in the caudal chemosensitive zone on the VMS and diaphragmatic electromyographic activity (DEA) were studied in response to inspired CO_2 and topically applied ACh before and after topical application of 1% KA to the VMS. KA, an EAA analog of Glu that destroys neuronal cell bodies but not fibers of passage, caused a marked increase in MNA and DEA. This effect was followed by a gradual (over 30 min) decline in MNA and DEA, with eventual loss of activity and respiratory arrest. Within 10 to 15 min of artificial respiration, MNA and DEA recovered, however, with reduced sensitivity to inspired CO_2. It was not possible to maintain MNA and DEA for more than 2 to 3 min after weaning the animal from the respirator. In related experiments in which respiration and blood pressure were monitored utilizing a low-inertia Krogh spirometer and Statham strain gauge transducer, respectively, 1% KA produced a marked increase in V_T and MABP and a slight decrease in f. Subsequently, V_T and f gradually declined to apnea while MABP stabilized at a level slightly lower than or at control value. The animals in these experiments could be maintained on but not weaned from artificial respiration for several hours after KA application.

EAA have been implicated in the neurotoxicity associated with several clinical conditions. It has been postulated that, depending on the duration of EAA treatment, neuronal death can be produced either by rapid sodium- and chloride-dependent osmotic lysis or by a more protracted calcium-dependent process. On the other hand, destabilization of ionized cytosolic calcium homeostasis and the biomedical events that are modulated by Ca_i might play a pivotal role in excessive Glu receptor stimulation and subsequent neuronal demise. Using the combined oxalate-pyroantimonate technique, ultracytochemical localization of Ca_i was performed to assess the role of calcium in the response of the chemosensory apparatus to EAA stimulation and eventual demise. In control tissues, a discrete pattern of electron-dense Ca^{2+}-pyroantimonate deposition was noted overlying neuronal nuclei, mitochondria, and endoplasmic reticulum, with sparse deposits in the cytoplasm and none in the extracellular space. The Ca^{2+} precipitation could also be visualized intraaxonally, scattered within the myelin sheaths in bleb-like processes and within synapses overlying mitochondria and vesicles. Tissues perfused at peak MNA following KA stimulation demonstrated a dramatic increase in cytosolic calcium. The endoplasmic reticulum, however, which contained deposits of calcium in control tissues, essentially showed sparse or no calcium-pyroantimonate deposits in the KA-treated tissues. There were also notable increases in the density of precipitates overlying the nuclei and swollen mitochondria of affected neurons, within axons and swollen dendrites, and especially within synapses where vesicular deposition appeared to be maximal. The precipitate could also be detected within the extracellular compartments and within the endothelia of medullary blood vessels in this region. These results suggest that (a) KA compromises the ability of the superficial neurons to maintain the central drive to respiration; (b) KA appears to deplete the neuronal metabolic stores of superficial CO_2-sensitive neuronal elements; (c) the cardiovascular component of the VMS seems unimpaired by the topical application of KA; and (d) EAA stimulation does indeed induce a mobilization of Ca^{2+} into the cytosol, mitochondria, and nuclei of neurons from intracellular or extracellular loci. It is postulated that increased Ca_i might be involved in the neurotoxic actions of EAA agonists at the VMS.

References

1. Loeschcke HH, DeLattre J, Schläfke ME, Trouth CO. Effects on respiration and circulation of electrically stimulating the ventral surface of the medulla oblongata. Resp Physiol 1970; 10:184–197.
2. Mitchell RA, Loeschcke HH, Massion WH, Severinghaus JW. Respiratory responses mediated through superficial chemosensitive areas on the medulla. J Appl Physiol 1963; 18:523–533.
3. Trouth CO, Patrickson JW, Holloway JA, Wright LE. Neurophysiological studies on superficial medullary chemosensitive area for respiration. Brain Res 1982; 246: 47–56.

4. Millhorn DE, Eldridge FL. Role of ventrolateral medulla in regulation of respiratory and cardiovascular systems. J Appl Physiol 1986; 61:1249–1263.
5. Bruce EN, Cherniack NS. Central chemoreceptors. J Appl Physiol 1987; 62:389–402.
6. Trouth CO, Loeschcke HH, Berndt J. A superficial substrate on the ventral surface of the medulla oblongata influencing respiration. Pflügers Arch 1973; 339:135–152.
7. Ross AA, Ruggiero DA, Park DH, et al. Tonic vasomotor control by the rostral ventrolateral medulla: Effect of electrical or chemical stimulation of the area containing C1 adrenaline neurons on arterial pressure, heart rate, and plasma catecholamines and vasopressin. J Neurosci 1984; 4:474–493.
8. Gatti PJ, Norman WP, Taveira DaSilva AM, Gillis RA. Cardiorespiratory effects produced by microinjecting L-glutamic acid into medullary nuclei associated with the ventral surface of the feline medulla. Brain Res 1986; 381:281–288.
9. McAllen RM. Location of neurons with cardiovascular and respiratory actions at the ventral surface of the cat's medulla. Neuroscience 1986; 18:43–49.
10. Willette RN, Barcos PP, Krieger AJ, Sapru HN. Vasopressor and depressor areas in the rat medulla. Neuropharmacology 1983; 22:1071–1079.
11. Kumada M, Dampney RAL, Reis DJ. Profound hypotension and abolition of the vasomotor component of the cerebral ischemic response produced by restricted lesions of medulla oblongata in rabbit. Circ Res 1979; 45:63–70.
12. Schläfke ME, Kille JF, Loeschcke HH. Elimination of central chemosensitivity by coagulation of a bilateral area on the ventral medullary surface in awake cats. Pflügers Arch 1979; 378:231–241.
13. Schläfke ME, See WR, Herker-See A, Loeschcke HH. Respiratory response to hypoxia and hypercapnia after elimination of central chemosensitivity. Pflügers Arch 1979; 381:241–248.
14. McAllen RM, Neil JJ, Loewy AD. Effects of kainic acid applied to the ventral surface of the medulla oblongata on vasomotor tone, the baroreceptor reflex and hypothalamic autonomic responses. Brain Res 1982; 238:65–76.
15. Bachelard H, Gardiner SM, Bennett T. Cardiovascular responses elicited by chemical stimulation of the rostral ventrolateral medulla in conscious, unrestrained rats. J Auton Nerv Syst 1990; 31:185–190.
16. Jung R, Bruce AN, Katona PG. Cardiorespiratory responses to glutamatergic antagonists in the caudal ventrolateral medulla of rats. Brain Res 1991; 564:286–295.
17. Dillon GH, Welsh DE, Waldrop TG. Modulation of respiratory reflexes by an excitatory amino acid mechanism in the ventrolateral medulla. Resp Physiol 1991; 85:55–72.
18. Bonham AC, Jeske I. Cardiorespiratory effects of DL-homocyteic acid in caudal ventrolateral medulla. Am J Physiol 1989; 256:H688–H696.
19. Sato M, Severinghaus JW, Basbaum AI. Medullary CO_2 chemoreceptor neuron identification by c-fos immunocyto-chemistry. J Appl Physiol 1992; 73:96–100.
20. Smatresk NJ. Chemoreceptor modulation of endogenous respiratory rhythms in vertebrates. Am J Physiol 1990; 259:R887–R897.
21. Chonan T, Merrill Adams E, von Euler C, Cherniack NS. Effects of focal cooling in the ventrolateral medulla on chemoresponsiveness in dogs. Resp Physiol 1990; 80:45–54.
22. Arata A, Onimaru H, Homma I. Respiration-related neurons in the ventral medulla of newborn rats in vitro. Brain Research Bulletin 1990; 24:599–604.

23. Pappenheimer JR, Fencyl V, Heisey SR, Held D. Role of cerebral fluids in control of respiration as studied in anesthetized goats. Am J Physiol 1965; 203:436–450.
24. Walker JL, Brown AM. Unified account of variable effects of carbon dioxide on nerve cells. Science 1970; 167:1502–1504.
25. Neubauer JA, Gonsalves SF, Chou W, et al. Chemosensitivity of medullary neurons in explant tissue cultures. Neuroscience 1991; 45:701–708.
26. Dean JB, Bayliss DA, Erickson JT, et al. Depolarization and stimulation of neurons in nucleus tractus solitarii by carbon dioxide does not require chemical synaptic input. Neuroscience 1990; 36:207–216.
27. Schläfke ME, Loeschcke HH. Lokalisation eines an der Regulation von Atmung und Kreislauf beteiligten Gebietes an der ventralen Oberflache der Medulla oblongata durch Kalteblockade. Pflügers Arch 1967; 297:201–220.
28. Berndt J, Berger W, Trouth CO. Respiratory and circulatory effects of 100 mEq/L potassium or 2% procaine in the cerebrospinal fluid of cats. Pflügers Arch 1970; 321:346–363.
29. Coyle JT, Molliver ME, Kuhar MJ. In situ injection of kainic acid: A new method for selectively lesioning neuronal cell bodies while sparing axons of passage. J Comp Neurol 1978; 180:301–324.
30. Olney JW, Ho OL, Rhee V. Cytotoxic effects of acidic and sulphur containing amino acids on the infant mouse central nervous system. Exp Brain Res 1971; 14: 61–76.
31. McGeer PL, McGeer EG. Kainic acid: The neurotoxic breakthrough. CRC Crit Rev Toxicol 1982; 10:1–26.
32. Rothman SM. The neurotoxicity of excitatory amino acids is produced by passive chloride influx. J Neurosci 1985; 5:1483–1489.
33. Choi DW. Ionic dependence of glutamate toxicity. J Neurosci 1987; 7:369–379.
34. Garthwaite G, Hajos F, Garthwaite J. Ionic requirements for neurotoxic effects of excitatory amino acid analogues in rat cerebellar slices. Neuroscience 1986; 18: 437–447.
35. Manev H, Favaron M, Bertolino M, et al. Importance of sustained protein kinase C translocation and destabilization of Ca^{2+} homeostasis in glutamate-induced neuronal death. In: Guidotti A, ed. Neurotoxicology of Excitatory Amino Acids. New York, Raven Press, 1990.
36. Choi DW. Glutamate neurotoxicity in cortical cell culture is calcium dependent. Neurosci Lett 1985; 58:293–297.
37. Garthwaite G, Garthwaite J. Neurotoxicity of excitatory amino acid receptor agonists in rat cerebellar slices: Dependence on calcium concentration. Neurosci Lett 1986; 66:193–198.
38. Siman R. Role of calpain I in excitatory amino acid-induced degenerative structural changes. In: Guidotti A, ed. Neurotoxicity of Excitatory Amino Acids. New York, Raven Press, 1990.
39. Laha PK, Nayar U, China GS, Singh B. Carbon dioxide sensitivity of the central chemosensitive mechanisms: An exploration by direct stimulation in rats. Pflügers Arch 1977; 367:241–247.
40. Loeschcke HH. Review lecture: Central chemosensitivity and the reaction theory. J Physiol 1982; 332:1–24.

41. Whittaker JAC. Age related differences in the responsiveness of brainstem respiratory CO_2-chemosensory neurons. PhD dissertation, Howard University, Washington, DC, 1988.
42. Bada FJ. Identification of a naloxone-sensitive site in the brainstem influencing respiratory CO_2 chemosensitivity. PhD dissertation, Howard University, Washington, DC, 1990.
43. Nattie EE, Li A. Fluorescent location of RVLM kainate microinjections that alter the control of breathing. J Appl Physiol 1990; 68:1157–1166.
44. Martin RL, Sinclair JD. Kainic acid on the rat ventral medullary surface depresses hypoxic and hypercapnic ventilatory responses. Resp Physiol 1990; 80:55–70.
45. Nattie EE, Mills JW, Ou LC, St. John WM. Kainic acid on the rostral ventrolateral medulla inhibits phrenic output and CO_2 sensitivity. J Appl Physiol 1988; 65: 1525–1524.
46. Schläfke ME. Central chemosensitivity—A respiratory drive. Rev Physiol Biochem Pharmacol 1981; 90:171–244.
47. Bazil MK, Gordon FJ. Spinal NMDA receptors mediate pressor responses evoked from the rostral ventrolateral medulla. Am J Physiol 1991; 260:H267–H275.
48. Dampney RAL, Moon EA. Role of ventrolateral medulla in vasomotor response to cerebral ischemia. Am J Physiol 1980; 239:H349–H358.
49. Mitra J, Prabhakar NR, Overholt JL, Cherniack NS. Respiratory effects of N-methyl-D-aspartate on the ventral medullary surface. J Appl Physiol 1989; 67: 1814–1819.
50. Feldberg W. The ventral surface of the brain stem: A scarcely explored region of pharmacological sensitivity. Neuroscience 1976; 1:427–441.
51. Feldberg W, Guertzenstein PG. Vasodepressor effects obtained by drugs acting on the ventral surface of the brain stem. J Physiol 1976; 258:337–355.
52. Gatti PJ, Taviero Da Silva AM, Hamosh P, Gillis RA. Cardiorespiratory effects produced by application of L-glutamic and kainic acid to the ventral surface of the cat hindbrain. Brain Res 1985; 330:21–29.
53. Somogyi P, Minson JB, Morilak D, et al. Evidence for an excitatory amino acid pathway in the brainstem and for its involvement in cardiovascular control. Brain Res 1989; 496:401–407.
54. Abrahams TP, Hornby PJ, Walton DP, et al. An excitatory amino acid(s) in the ventrolateral medulla is (are) required for breathing to occur in the anesthetized cat. J Pharm Exp Ther 1991; 259:1388–1395.
55. Kazemi H, Hoop B. Glutamic acid and gamma amino butyric acid neurotransmitters in central control of breathing. J Appl Physiol 1991; 70:1–7.
56. Pierrefiche O, Schmid K, Foutz AS, Denavit-Saubie M. Endogenous activation of NMDA and non-NMDA glutamate receptors on respiratory neurones in cat medulla. Neuropharmacology 1991; 30:429–440.
57. Croucher MJ, Collins JF, Meldrum BS. Anticonvulsant action of excitatory amino acid antagonists. Science 1982; 216:899–901.
58. Collingridge GL, Kehl SJ, McLennan H. The antagonism of amino acid–induced excitations of rat hippocampal CA1 neurons in vitro. J Physiol 1983; 334:19–34.
59. Ryan GP, Hackman JC, Davidoff RA. Spinal seizures and excitatory amino acid–mediated synaptic transmission. Neurosci Lett 1984; 44:161–166.

60. Coan EJ, Collingridge GL. Magnesium ions block an n-methyl-d-aspartate receptor-mediated component of synaptic transmission in rat hippocampus. Neurosci Lett 1985; 53:21–26.
61. Cruzwar SJ, Meldrum BS. Protection against chemically induced seizures by 2-amino-7-phosophonoheptanoic acid. Eur J Pharmacol 1982; 85:335–338.
62. Meldrum BS, Croucher MJ, Badman G, Collins JF. Antiepileptic action of excitatory amino acid antagonists in the photosensitive baboon, Papio papio. Neurosci Lett 1983; 39:101–104.
63. Herron CE, Williamson R, Collingridge GL. A selective N-methyl-D-aspartate antagonist depresses epileptiform activity in rat hippocampal slices. Neurosci Lett 1985; 61:255–260.
64. Hablitz JJ. Conductive changes induced by DL-homocysteic acid and N-methyl-D-aspartic acid in hippocampal neurons. Brain Res 1982; 247:149–153.
65. MacDonald JF, Portietis AV, Wojtowicz JM. L-Aspartic acid induces a region of negative slope conductance in the current-voltage relationship of cultured spinal cord neurons. Brain Res 1982; 237:248–253.
66. Dingledine R, Hynes MA, King GL. Involvement of N-methyl-D-aspartate receptors in epileptiform bursting in the rat hippocampal slice. J Physiol (Lond) 1986; 380:175–189.
67. Dingledine R. N-methyl aspartate activates voltage-dependent calcium conductances in rat hippocampal pyramidal cells. J Physiol 1983; 343:385–405.
68. Mayer ML, Westbrook GL, Guthrie PB. Voltage-dependent block of Mg^{++} of NMDA responses in spinal cord neurones. Nature 1984; 309:261–263.
69. Nowak L, Bregestovski P, Ascher P. Magnesium gates glutamate-activated channels in mouse central neurones. Nature 1984; 307:462–465.
70. Davies J, Evans RH, Francis AA, Watkins JC. Excitatory amino acid receptors and synaptic excitation in mammalian central nervous system. J Physiol (Paris) 1979; 75:641–644.
71. McLennan H. The effect of decortication on excitatory amino acid sensitivity of striatal neurons. Neurosci Lett 1980; 8:313–316.
72. Ferkany JW, Coyle JT. Kainic acid selectivity stimulates the release of endogenous excitatory amino acids. J Pharm Exp Ther 1983; 225:399–406.
73. Guertzenstein PG, Silver A. Fall in blood pressure produced from discrete regions of the ventral surface of the medulla by glycine and lesions. J Physiol (Lond) 1974; 242:489–503.
74. Granata AR, Kumada M, Reis DJ. Sympathoinhibition by A1-noradrenergic neurons is mediated by neurons in the C1 area of the rostral medulla. J Auton Nerv Syst 1985; 124:387–395.
75. Granata AR, Numao Y, Kumada M, Reis DJ. A1 noradrenergic neurons tonically sympathoexcitatory neurons of C1 area in rat brainstem. Brain Res 1986; 377:127–146.
76. Addae JI, Stone TW. Interactions between topically applied excitatory amino acids on rat cerebral cortex: Discrimination by pentobarbitone. Exp Brain Res 1987; 68: 613–618.
77. Addae JI, Stone TW. Effects of topically applied excitatory amino acids on evoked potentials and single cell activity in rat cerebral cortex. Eur J Pharm 1986; 121: 337–343.

78. Mayer ML, Westbrook GL. The action of NMDA on mouse spinal neurones in culture. J Physiol 1985; 361:65–90.
79. Zorumski CF, Fishbach GD. Differentiation between glutamate receptors based on desensitization (abstr). Soc Neurosci Abst 1985; 11:105.
80. Greer JJ, Smith JC, Feldman JL. Role of excitatory amino acids in the generation and transmission of respiratory drive in neonatal rat. J Physiol 1991; 437:727–749.
81. Smith JC, Ellenberger HH, Ballanyi K, et al. Pre-Bötzinger Complex: A brainstem region that may generate respiratory rhythm in mammals. Science 1991; 254: 726–729.
82. Funk GD, Smith JC, Feldman JL. Generation and transmission of respiratory oscillations in medullary slices: Role of excitatory amino acids. J Neurophysiol 1993; 70:1497–1515.
83. Douglas RM, Bernard DG, Pan Y, et al. Cardiorespiratory effects of kainic acid on brainstem chemosensitive zones. Soc Neurosci Abstr 1992; 18:828.
84. Blessing WW, West MJ, Chalmers JP. Hypertension, bradycardia and pulmonary edema in the conscious rabbit after brainstem lesions coinciding with the A1 group of catecholamine neurons. Circ Res 1981; 49:949–958.
85. Blessing WW, Reis DJ. Inhibitory cardiovascular function of neurons in the caudal ventrolateral medulla of the rabbit: Relationship to the area containing A1 noradrenergic cells. Brain Res 1982; 253:161–171.
86. Curtis DR, Malik R. A neurophysiological analysis of the effect of kainic acid on nerve fibers and terminals in the cat spinal cord. J Physiol (Lond) 1985; 368:99–108.
87. Rennels ML, Gregory TF, Blaumanis OR, et al. Evidence for a paravascular fluid circulation in the mammalian central nervous system, provided by the rapid distribution of tracer protein throughout the brain from the subarachnoid space. Brain Res 1985; 326:47–63.
88. Rennels ML, Blaumanis OR, Grady PA. Rapid solute transport throughout the brain via paravascular fluid pathways. In: Long D, et al., eds. Advances in Neurology. Vol 52. New York, Raven Press, 1990:431–439.
89. Griffiths T, Evans MC, Meldrum BS. Temporal lobe epilepsy, excitotoxins and the mechanism of selective neuronal loss. In: Fuxe K, Roberts P, Schwarcz R, eds. Excitotoxins. Vol. 39. New York: Plenum Press, 1984.
90. Siesjö BK. Basis for the application of calcium antagonists in neurology: Calcium in the brain under physiological and pathological conditions. Exp Neurol 1990; 30:3–9.
91. Monaghan DT, Bridges RJ, Cotman CW. The excitatory amino acid receptors: Their classes, pharmacology, and distinct properties in the function of the central nervous system. Annu Rev Pharmacol Toxicol 1989; 29:365–402.
92. Hori N, French-Mullen JMH, Carpenter DO. Kainic acid responses and toxicity show pronouned Ca^{++} dependence. Brain Res 1985; 358:380–384.
93. Collingridge GL, Lester RAJ. Excitatory amino acid receptors in the vertebrate central nervous system. Pharmacol Rev 1989; 40:143–210.
94. Olney JW. Excitotoxic amino acids and neuropsychiatric disorders. Annu Rev Pharmacol Toxicol 1990; 30:47–71.
95. Frandsen A, Schousboe A. Excitatory amino acid-mediated cytotoxicity and calcium homeostasis in cultured neurons. J Neurochem 1993; 60:1202–1211.

96. McGeer EG, Olney JW, McGeer PL, eds. Kainic Acid as a Tool in Neurobiology. New York: Raven Press, 1978.
97. Fuxe K, Roberts P, Schwarcz R, eds. Excitotoxins. New York: Plenum Press, 1983.
98. Schwarcz R, Foster AC, French ED. Excitotoxic models for neurodegenerative disorders. Life Sci 1984; 35:19–32.
99. Olney JW. Excitatory amino acids and neuropsychiatric disorders. In: Hicks TP, Lodge D, McLennan H, eds. Excitatory Amino Acid Transmission. New York: Liss, 1987:217–224.
100. Borgers M, DeBrabander M, Van Reempts J, et al. Intranuclear microtubules in lung mast cells of guinea pigs in anaphylactic shock. Lab Invest 1977; 37:1–8.
101. Borgers M, Thone F, van Nueten J. The subcellular distribution of calcium and the effects of calcium-antagonists as evaluated with a combined oxalate-pyroantimonate technique. Acta Histochem 1981; 8(suppl 24):327–332.
102. Kang Y-H, McKenna T, Watson LP, et al. Cytochemical changes in hepatocytes of rats with endotoxemia or sepsis: Localization of fibronectin, calcium, and enzymes. J Histochem Cytochem 1988; 36:665–678.
103. Watson LP, Kang Y-H, Falk MC. Cytochemical properties of osteoblast cell membrane domains. J Histochem Cytochem 1989; 37:1235–1246.
104. Griffiths T, Evans MC, Meldrum BS. Intracellular calcium accumulation in rat hippocampus during seizures induced by bicuculline or L-allylglycine. Neuroscience 1983; 10:386–395.
105. Garthwaite G, Garthwaite J. Amino acid neurotoxicity: Intracellular sites of calcium accumulation associated with the onset of irreversible damage to rat cerebellar neurones in vitro. Neurosci Lett 1986; 71:53–58.
106. Debbas G, Hoffman L, Landon EJ, Hurwitz L. Electron microscope localization of calcium in vascular smooth muscle. Anat Rec 1975; 182:447–472.
107. Simson JAV, Spicer SS. Selective subcellular localization of cations with variants of the potassium (Pyro) antomonate technique. J Histochem Cytochem 1975; 23:575–598.
108. Choi DW. NMDA receptors and AMPA kainate receptors mediate parallel injury in cerebral cortical cultures subjected to oxygen-glucose deprivation (review). Progr Brain Res 1993; 96:137–143.

31

Neuronal Connections of a Ventral Brainstem Respiratory Chemosensitive Area

ANTHONY D. PAUL, ASHTON GIBBONS, ELMOND J. JOHNSON, ALECIA McLEAN, LAKEISCHA WEBB, OWEN GARDNER, BEVERLY TURNER, NADINE LYSEIGHT, and JASON TOMLIN

Oakwood College
Huntsville, Alabama

ROBERT M. DOUGLAS

College of Medicine
Howard University
Washington, D.C.

I. Introduction

Several investigations have implicated neurons at the ventral brainstem in the central chemical regulation of breathing (1–3). This chemosensitive region lies ventral to the inferior olives at the level where the hypoglossal rootlets exit the brainstem and may be reached by fluids superfusing the ventrolateral medullary surface (VMS) (Fig. 1).

Electrical stimulation of the region of maximal chemosensitivity (area L) in the cat resulted in a marked increase in ventilation (2,4). Berndt et al. (5) demonstrated that superfusion of this area with 2% procaine solution in peripheral chemoreceptor–denervated animals completely abolished the ventilatory response to electrical stimulation, causing respiratory arrest. However, electrical stimulation of the respiratory center in these animals still elicited an inspiratory response. It appears, therefore, that these superficial structures might be essential in driving the respiratory center in anesthetized cats in the absence of peripheral afferents.

If it is assumed that specialized receptors on the ventrolateral surface of the medulla oblongata are capable of sensing changes in their chemical environ-

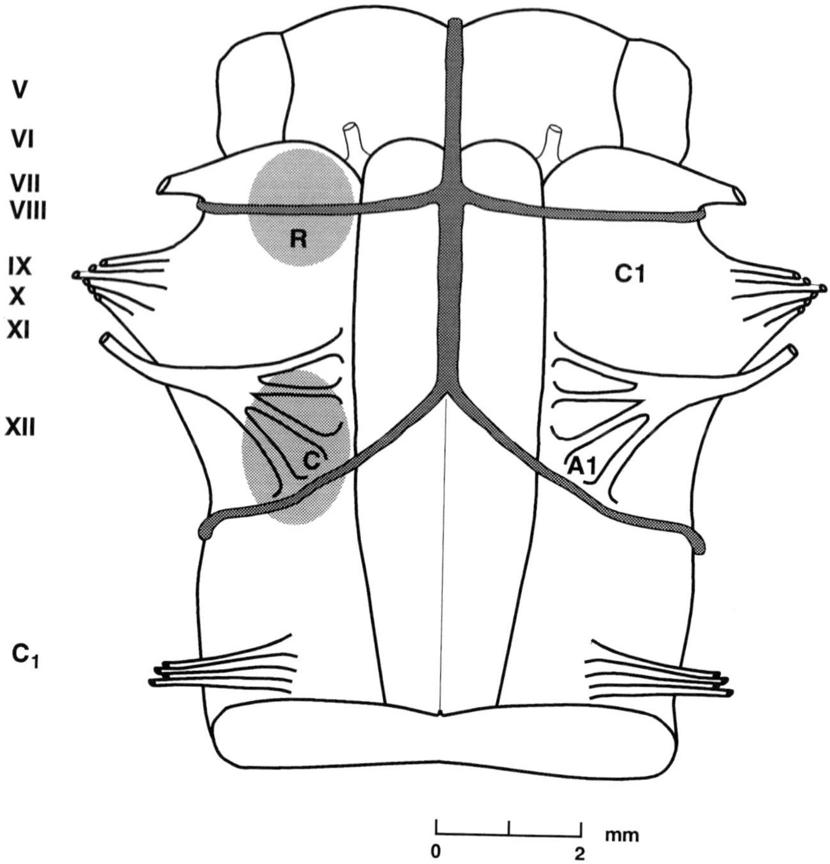

Figure 1 Ventral surface of the medulla oblongata of the rat. On the left half of the diagram, two chemosensory regions are indicated by the symbols R and C, representing the rostral and caudal chemosensitive zones, respectively. V–XII = cranial nerves.

ment, then these receptors should respond by transmitting information to the respiratory centers, thus providing the centers with the necessary input required for reflex adjustments of pulmonary ventilation.

C-fos and other immediate-early genes, in response to various physiological and pharmacological stimuli, induce concomitant nuclear binding proteins that appear to play a role in signal transduction and to function as intracellular third messengers. Sato et al. (6) described the distribution of the c-*fos* in cells of the ventral medulla oblongata that were responsive to 15% inspired CO_2. Immunocytochemical evaluation of the tissue revealed c-*fos*–positive neurons that were located within 150 μm of the VMS in the caudal ventrolateral medulla oblongata (cVLM); with a majority of cells within the first 50 μm.

In the present study, we have evaluated the neuronal connections of the caudal chemosensitive area with other respiratory control regions in the brainstem using horseradish peroxidase (HRP)-tetramethyl benzidine (TMB) reaction product as a marker for neuronal projections.

Seven rats weighing 300 to 400 g, anesthetized with chloralose urethane (200 mg/kg glucochloralose and 400 mg/kg urethane) were used. The musculi longi capitis and recti capitis ventrales, as well as the pars basalis os occipitale, the ventral portion of the atlas, and the processus odontoideus of the epistropheus were removed. The dura was then opened and reflected laterally to expose the ventral surface of the medulla oblongata. Rectal temperature was maintained constant at 37°C. The animals were then placed in the supine position with the head stabilized in a stereotaxic apparatus.

Following surgical exposure of the ventral brainstem surface, a solution of 0.02 to 0.05 μL of 4% HRP (Sigma VI) was injected into the caudal chemosensitive area on the VMS with the aid of a micromanipulator in a stereotaxic coordinate system. A 1-μL Hamilton syringe was used. At 48 h postinjection of HRP, the head of the animal was perfused and 40 μm tissue sections were processed for the TMB reaction product according to the protocol of Mesulam (7). In another series of experiments, wherever anterograde labeling was noted in the present study, microinjections of HRP were made to validate the presence of projections to that site, for example the nucleus ambiguus (NA).

In Fig. 2, the injection site was located in the cVLM overlying the exit of the hypoglossal (XII) rootlets. Only those cases where the injection was confined to the VMS were chosen for this report. A small ipsilateral projection appears to exist between the NA [or ventral respiratory group (VRG)] and the VMS. Light deposits of HRP reaction product could also be visualized in a few small, oval neurons of the ventrolateral nucleus tractus solitarius (NTS)—that is, the dorsal respiratory group (DRG), bilaterally.

Ipsilateral to the cVMS injection site, heavy labeling with HRP reaction product occurred in the nerve XII fibers traversing the medulla and in the large, multipolar neurons of the hypoglossal nucleus. This may represent uptake of HRP from the surrounding medium by nerve XII fibers. Occasional large, multipolar neurons containing HRP were also noted within the medullary reticular nucleus in the vicinity of the fibers of XII and in the lateral reticular nucleus (LRtN) (Fig. 3).

Contralateral to the injection site, HRP reaction product was detected in a few unique areas; the parvocellular region of the lateral reticular nucleus (LRt-PC), the ventrolateral and dorsolateral aspects of the inferior olivary nucleus (IO) and along the VMS. Small to medium sized neurons located in the contralateral medial and ventrolateral IO as well as superficial neurons within the thickened marginal glia (TMG) were variably labeled with HRP. A few large multipolar neurons within the contralateral LRt-PC were heavily retrogradely labeled with HRP and fibers emanating from their somata would be visualized directed toward the injection site (Fig. 4). Mesulam (7) and Kahlia

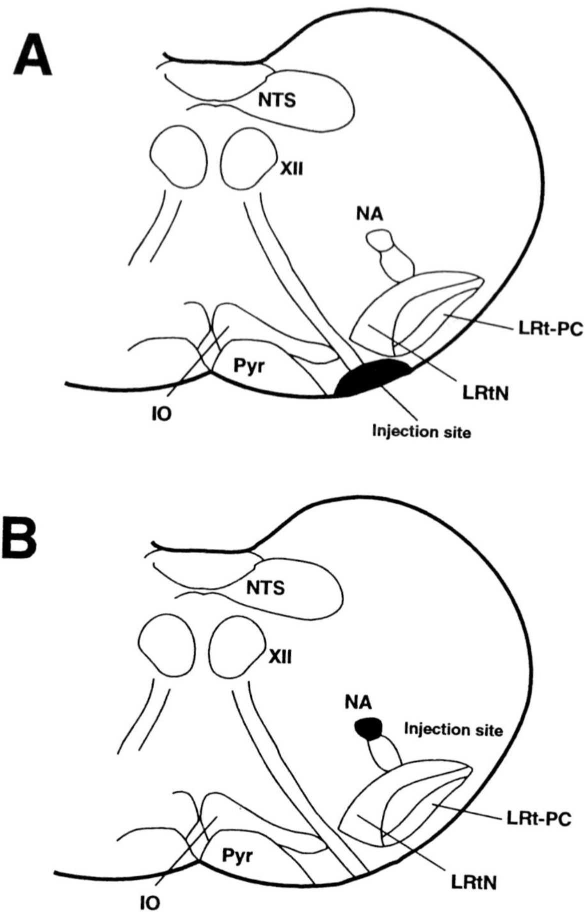

Figure 2 Diagrammatic representation of the location of HRP injection sites in the rat caudal medulla oblongata. (A) Injection site located at the VMS in the region of the hypoglossal rootlets. (B) Injection site located within the NA; one of the sites of anterograde labeling from the VMS. NTS = nucleus tractus solitarius; XII = hypoglossal nucleus; NA = nucleus ambiguus; LRt-PC = parvocellular region of the lateral reticular nucleus; LRtN = lateral reticular nucleus; Pyr = pyramidal tract; IO = inferior olivary nucleus.

and Wells (8) have demonstrated the usefulness of using TMB as the substrate for HRP histochemistry in studies where the anterograde transport of HRP is desirable. In these studies, a few tortuous fibers and axon terminal fields containing anterogradely labeled HRP could be detected within the NA, LRt-PC, and along the VMS, indicating that reciprocal connections exist between VMS,

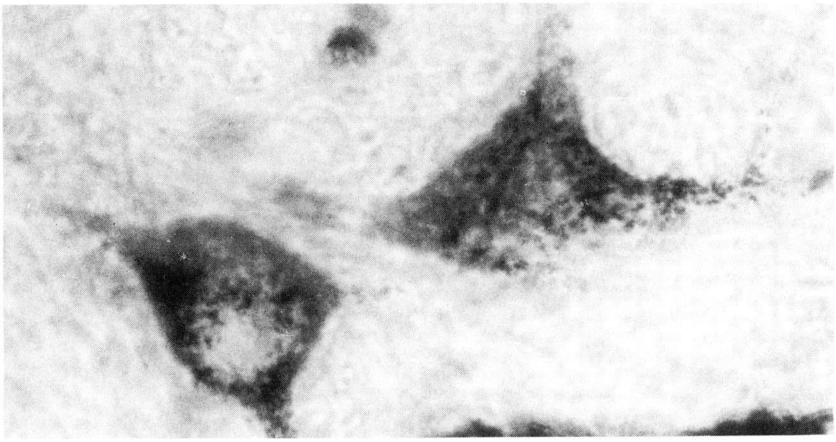

Figure 3 Light micrograph of a pair of HRP-positive neurons in the medullary reticular nucleus retrogradely labelled from a VMS injection site × 40. Neutral Red counterstain.

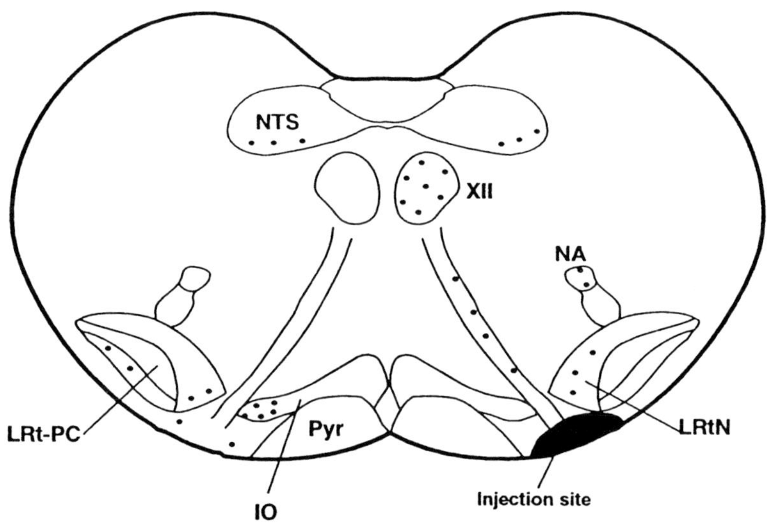

Figure 4 Diagrammatic representation of projections from various medullary subnuclei to the cVMS. Each dot represents a neuron retrogradely labeled with HRP from a VMS injection site. Ipsilateral retrograde HRP labeling was visualized in the LRtN, NA, NTS, and nerve and nucleus of XII. Contralateral retrograde HRP labeling was noted in the NTS, LRt-PC, IO, and along the VMS. LRtN = lateral reticular nucleus, NA = nucleus ambiguus, NTS = nucleus tractus solitarius, XII = hypoglossal nucleus, IO = inferior olivary nucleus, LRt-PC = parvocellular region of the lateral reticiular nucleus, Pyr = pyramidal tract, and VMS = ventral medullary surface.

ambigual, and LRt-PC neurons. Verification studies were subsequently performed by microinjection of HRP into sites demonstrating anterograde labeling. These observations agree with the reports of Norgren (9), who, with the use of tritiated proline or leucine, described projections emanating from the NA that project to the NTS. Indeed, Gallagher and Pert (10) and Walberg (11) have found that extensive reciprocal connections exist between reticular nuclei. Neurovascular elements, variably labeled with HRP, were noted in superficial regions of the medulla. These cells that are in close contact with blood vessel walls are reminiscent of the neurovascular elements described by Scheibel and Scheibel (12) and Scheibel et al. (13).

II. Summary

Regions of respiratory chemosensitivity that respond to cerebrospinal fluid pH changes and inspired carbon dioxide have been described on the VMS. In this investigation, HRP, a neuronal tracer, was utilized to ascertain the possible connections between the VMS and the deeper-lying respiratory structures. In spontaneously breathing, chloralose-urethane anesthetized rats, 0.02 to 0.05 μL of 4% HRP was microinjected into the cVMS via a 1-μL Hamilton syringe mounted on a stereotaxic coordinate system. The animals were maintained for a minimum period of 48 h to allow for the axonal transport of HRP. The injection site was located over the hypoglossal rootlets and less than 500 μm into the caudal VMS. The results reveal that the majority of neuronal connections with the caudal VMS arise (retrogradely) from (1) Ipsilaterally: lateral reticular nucleus, the nucleus ambiguus, and the nucleus tractus solitarius. HRP reaction product was also visualized within the nerve fibers and large multipolar neurons of the hypoglossal nucleus. (2) Contralaterally: IO nuclei and the parvocellular region of the LRtN (with a few neurons that have extremely long processes directed towards the midline). Neuronal connections were also identified by anterograde labeling, which indicated the presence of projections arising from the cVMS that may impinge on the respiratory centers. This investigation demonstrates that (a) superficial neuronal elements project to deeper-lying respiratory structures such as the NA and NTS; (b) bilateral connections exist between the LRt-PC nuclei; (c) possible bilateral connections between lateral inferior olivary nuclei and VMS cells, and (d) neurovascular elements may play a role in respiratory control mechanisms within the ventrolateral medulla oblongata. It is therefore possible that the lateral horns of the IO, the parvocellular regions of the LRtN, and superficial neurons within the TMG are involved in higher-order processing within the central respiratory chemoreceptor reflex pathway.

Acknowledgments

This work was supported by NIH NIGMS Grant No. S 06 GM 08016.

References

1. Mitchell RA, Loeschcke HH, Massion WH, Severinghaus JW. Respiratory responses mediated through superficial chemosensitive areas on the medulla. J Appl Physiol 1963; 18:523–533.
2. Loeschcke HH, DeLattre J, Schläfke ME, Trouth CO. Effects on respiration and circulation of electrically stimulating the ventral surface of the medulla oblongata. Resp Physiol 1970; 10:148–197.
3. Schläfke ME, See WR, Loeschcke HH. Ventilatory response to alterations of H^+-ion concentration in small areas of the ventral medullary surface. Resp Physiol 1970; 10:198–212.
4. Trouth CO, Loeschcke HH, Berndt J. A superficial substrate on the ventral surface of the medulla oblongata influencing respiration. Pflügers Arch 1973; 339: 135–152.
5. Berndt J, Berger W, Trouth CO. Respiratory and circulatory effects of 100 mEq/L potassium or 2% procaine in the cerebrospinal fluid of cats. Pflügers Arch 1970; 321:346–363.
6. Sato M, Severinghaus JW, Basbaum AI. Medullary CO_2 chemoreceptor neuron identification by c-fos immunocytochemistry. J Appl Physiol 1992; 73:96–100.
7. Mesulam MM. Tetramethyl benzidine for horseradish peroxidase neurochemistry: A non-carcinogenic blue reaction-product with superior sensitivity for visualizing neural afferents and efferents. J Histochem Cytochem 1978; 26:106–117.
8. Kahlia M, Wells RV. Brainstem projections of the aortic nerve in the cat: A study using tetramethyl benzidine as the substrate for horseradish peroxidase. Brain Res 1980; 188:23–32.
9. Norgren R. Projections from the nucleus of the solitary tract in the rat. Neuroscience 1978; 3:207–218.
10. Gallagher DW, Pert A. Afferents to brainstem nuclei (brainstem raphe, nucleus reticularis pontis caudalis and nucleus gigantocellularis) in the rat as demonstrated by microiontophoretically applied horseradish peroxidase. Brain Res 1978; 144: 257–275.
11. Walberg F. Crossed reticulo-reticular projections in the medulla, pons, and mesencephalon: An autoradiographic study in the cat. Z Anat Entwickl-Gesch 1974; 143: 127–134.
12. Scheibel ME, Scheibel AB. Does the nucleus raphe pontis have chemoreceptors or neuroendocrine functions? Soc Neurol 1974:409.
13. Scheibel ME, Tomiyasu U, Scheibel AB. "Do raphe nuclei of the reticular formation have neurosecretory or vascular sensor functions?" Exp Neurol 1975; 47:316–329.

32

Neurotransmitters in Cardiorespiratory Regulation at the Ventral Surface of the Medulla Oblongata

NANDURI R. PRABHAKAR

School of Medicine
Case Western Reserve University
Cleveland, Ohio

I. Introduction

It is well recognized that ventilatory response to CO_2 persists after denervation of arterial chemoreceptors. Pioneering work of Hans Loeschcke and his coworkers led to the identification of the ventral surface of the medulla oblongata (VMS) as an important neural substrate for mediating ventilatory response to CO_2. Since then, attention of respiratory physiologists has been focused on the VMS in elucidating its role in ventilatory homeostasis. Later studies by Feldberg, Guertzenstein and their coworkers led to the notion that VMS is also an important substrate for maintaining circulatory homeostasis. The purpose of this article is to review briefly the role of neurochemicals in the regulation of respiration and blood pressure by neural substrates in the VMS.

II. Identification of Neural Substrates in the VMS That Influence Breathing

It is well established that respiratory responses can be elicited only from certain regions of the VMS often called "chemosensitive areas." The following is a brief description of the studies that led to the identification of different areas on ventral brain stem that influence breathing.

It has long been thought that changes in blood gases, especially CO_2, affect breathing by their direct action on central neurons controlling respiration (1). In 1952, von Euler and Soderberg (2) suggested that the neural substrate responding to CO_2 is distinct from the respiratory-related neurons. Based on experiments with intraventricular perfusion of the brain, Leusen (3) suggested that chemosensitive neurons might be located superficially in the brain stem. In 1958 Loeschcke and Koepchen (4) concluded that neural substrates responding to changes in pH of the extracellular fluid (ECF) should be in close proximity to the VMS. Subsequently Mitchell, Loeschcke, and coworkers (5,6) showed that application of local anesthetics or cold mock cerebrospinal fluid (CSF) to the VMS depresses ventilation, but not when they are applied directly to the floor of the fourth ventricle. The studies of Mitchell, Loeschcke, and Schläfke have led to the identification of three areas in ventral surface of brainstem that affect breathing (6). They are located in rostral, intermediate, and caudal parts of the VMS. Electrical stimulation, application of chemicals and pledgets soaked in acidic mock-CSF to the rostral and caudal areas all increase respiration (6). In peripherally chemodenervated animals, bilateral cold blockade of intermediate areas led to apnea and attenuates or abolishes CO_2-induced increases in breathing. These studies led to the notion that the rostral and caudal areas contain neuronal substrate(s) for sensing the changes in CO_2 (or, more appropriately, changes in ECF pH); whereas the intermediate area is necessary for integration of the inputs from chemosensitive areas (6).

III. Influence of Neurochemicals on Breathing

Like other parts of the nervous system, the VMS contains several classes of neurochemicals. These include biogenic amines, amino acids, and neuropeptides. Several studies have examined their influence on ventilation. Following is a brief review of the effects of different chemicals on breathing when they are applied topically or microinjected into the chemosensitive areas of VMS.

Mitchell, Loeschcke, and collaborators (5) reported increases in respiration following local application of acetylcholine (ACh) or nicotine to the VMS. These studies were further extended by Dev and Loeschcke (7), who examined the effects of ACh on respiration in detail. These authors demonstrated that the stimulatory actions of ACh were confined to those areas of the VMS that correspond to rostral and caudal chemosensitive areas. More importantly, topical application of atropine, a muscarinic receptor antagonist to chemosensitive areas, markedly attenuated ventilatory response to CO_2. Furthermore, physostigmine applied to chemosensitive areas increased respiration (7). Other biogenic amines—such as norepinephrine, epinephrine, or serotonin—had no effect on respiration when applied to chemosensitive areas (7). From these results, it was suggested that transmission of a CO_2-induced ventilatory response requires cholinergic mechanism(s) (8). Haxhiu, Cherniack, and their collaborators (9) have shown that cholinergic agonists applied to the VMS increase the phrenic and respiratory discharge of the hypoglossal nerve. However, the effects of cholinergic agonists were greater on upper airway muscles than on phrenic nerve activity. These studies suggest that cholinergic agonists acting on chemosensitive areas of the VMS affect the output of the respiratory nerves differentially. Such differential effects could be of physiological significance in coordinating the respiratory behavior of pumping versus upper airway dilating muscles.

There is considerable evidence that amino acids affect breathing when applied to the VMS (10). In general, amino acids such as gamma-aminobutyric acid (GABA), its agonist, muscimol, taurine, and glycine inhibit breathing and depress ventilatory response to CO_2. On the other hand, application of glutamate or its agonist, *N*-methyl-D-aspartate (NMDA), stimulates breathing. When applied to the VMS, amino acids affect not only breathing but also the tone of the tracheal smooth muscle and airway secretion (11). These studies indicate that amino acids, by acting on the neurons of the VMS, can potentially influence not only pumping muscles of respiration but also upper airways.

In recent years several neuropeptides have been identified in the ventral brainstem that may function as neurotransmitters or modulators. These include tachykinin peptides, notably substance P (SP), enkephalins, somatostatin, NPY, neurotensin, and atrial natriuretic peptide (ANF) (Ref. 10 and Kumar, this volume). As elsewhere in the nervous system, some of these peptides are colocalized with other chemicals. For example, SP is colocalized with serotonin in the neurons of the rostral ventrolateral medulla (RVLM). Furthermore, Kumar and coworkers have shown that the VMS contains the enzyme neutral endopeptidase (NEP), which hydrolyzes many of the neuropeptides. Topical application of tachykinins like SP and neurokinin A (NKA) to chemosensitive areas increases phrenic nerve activity as well as the tone of the tracheal smooth muscle. Topical application of phosphoramidon, an inhibitor of NEP, potentiates respiratory response to CO_2 (12). Neurotensin also stimulates phrenic nerve activity and increases tracheal tone. The effects of neurotensin on tracheal smooth muscle appear to be mediated via parasympathetic nerves innervating the tracheal smooth muscle (13). On the other hand, peptides like somatostatin and ANF inhibit phrenic nerve activity and decrease tracheal tone (14). Interestingly, at certain doses, ANF causes a profound decrease in tracheal tone but no effect on efferent phrenic nerve activity (see Fig. 4 of Ref. 14). Thus, these studies indicate that tachykinins and neurotensin are stimulatory whereas somatostatin and ANF are inhibitory to respiration when applied to chemosensory regions of the VMS.

Thus far, studies described above demonstrate that, when applied to the VMS, neurochemicals cause pronounced changes in respiration. It is noteworthy that certain neurochemicals at given doses affect upper airway muscles more than efferent phrenic nerve activity. Likewise, pronounced changes are seen in tracheal tone without any noticeable effects on efferent phrenic nerve activity. Thus, any conclusions to be drawn about the effects of neurochemicals on respiration must take into account the heterogeneity of responses of different respiratory muscles. While the differential responses in respiratory muscles may have important physiological implications, they might also represent the different thresholds and anatomical connections from chemosensitive neurons to respiratory motoneurons.

IV. Identification of Neural Substrates in the VMS Influencing Blood Pressure

In 1972, Feldberg and Guertzenstein (15) reported that topical application of chemicals to the VMS leads to marked changes in arterial blood pressure. These authors noted that cardiovascular alterations could be elicited from two areas in the VMS that correspond closely to the caudal and rostral part of intermediate chemosensitive areas. Bilateral application of glycine or bilateral lesioning of the rostral area causes a profound fall in blood pressure (16). Application of nicotine, physostigmine, or carbachol to the caudal area also produces a fall in blood pressure (16). Later studies have shown two collections of neurons that contain catecholamines. Stimulation of these neurons results in strong vasomotor responses. One group, located more caudally, contains norepinephrine; these are often referred to as A_1 neurons (17,18). Stimulation of these neurons produces a fall in blood pressure. Some but not all A_1 neurons project to the hypothalamus and influence the secretion of arginine vasopressin (AVP). However, the A_1 group has no direct connection to the spinal cord. A second group of neurons is located more rostrally and contains epinephrine; these are called the C_1 group (17,18). Blockade of C_1 neurons by tetrodotoxin prevents the vasodepressor effects caused by the stimulation of A_1 neurons. A direct anatomical connection of C_1 neurons to the intermediolateral column of the spinal cord has been reported (19,20). These neurons constitutes vasomotor presympathetic neurons (VPN) and contain epinephrine and probably also glutaminergic cells (21). Stimulation of the rostral group of neurons raises blood pressure by producing vasoconstriction. Ablation of the rostral group of neurons lowers blood pressure and impairs the reflex response to baroreceptor stimulation (17–20). These findings have led to the notion that neurons in the RVLM provide tonic sympathetic vasomotor tone for maintaining blood pressure and that they integrate a wide variety of vasomotor reflexes (17–20).

V. Influence of Neurochemicals on Blood Pressure

Local application of carbachol or physostigmine into the RVLM produces fall in blood pressure that can be prevented by atropine. These findings led to the suggestion that cholinergic mechanisms exert a tonic inhibitory influence on RVLM neurons (22). It is well known that administration of clonidine decreases blood pressure. The hypotensive effects of clonidine were attributed to its actions on neurons in RVLM. The effects of clonidine in the VMS are mediated either by alpha-2 adrenergic receptors (see Guyenet et al., this volume) or by imidazoline receptors (see Ernsberger et al., this volume), or both. Application of epinephrine to the RVLM inhibits vasomotor activity, probably by acting on alpha-2 adrenergic receptors (23).

Microinjection of glutamate into C_1 neurons produces marked increases in blood pressure (10). Application of the inhibitory amino acid GABA to C_1 neurons lowers blood pressure. Similar injection of GABA into A_1 neurons, however, raises blood pressure (24).

Neurons in the RVLM region contain neuropeptides such as SP and neuropeptide Y (10). It has been shown that SP mediates some of the pressor effects elicited by stimulation of neurons in the RVLM (10). Application of somatostatin or ANF in the vicinity of the RVLM neurons produced a small but significant fall in blood pressure (14).

VI. Summary and Perspectives

There is little doubt that neural substrates in the VMS play important roles in the regulation of breathing and circulation. The last few decades have witnessed considerable progress in the distribution and characterization of the effects of neurochemicals on respiration and blood pressure. Several questions, however, remain unanswered. First, are there separate neurons that regulate respiration and blood pressure? Obviously, the answer to this question cannot be a simple one. Several lines of evidence, however, indicate that neurons mediating respiratory and vasomotor responses are closely intermingled and interconnected and may not be identical.

Although the effect of neurochemicals on respiration and ventilatory responses to CO_2 has been well documented, little is known regarding the contribution of these chemicals to the transmission of the CO_2 signal. Also, no information is available as to the release of neurochemicals from the chemosensitive areas of the VMS in response to graded changes in CO_2. Similar information will also be of value in clarifying the role of neurochemicals of the VMS region in thc regulation of blood pressure. Furthermore, little attention has been given to the question of whether or not the effects of these chemicals on respiration and blood pressure are secondary to changes in cerebral blood flow.

In other areas of the nervous system, there is a good deal of crosstalk among various neurotransmitters and modulators. For instance, neuropeptides have been shown to influence the release of other biogenic amines, such as acetylcholine and catecholamines. Whether such interactions among these molecules exist in the VMS is yet to be investigated. Recent studies from Trouth and Neubauer and their coinvestigators (this volume) indicate that carbonic anhydrase-containing neurons in the VMS may subserve the function of CO_2 sensing. If this is further confirmed, then it would be important to characterize the distribution of neurochemicals in this subset of carbonic anhydrase-containing VMS neurons and to assess their possible anatomical connections to respiratory and cardiovascular motoneurons.

References

1. Pitts RF. Organization of the respiratory centers. Physiol Rev 1946; 26:609–630.
2. von Euler C, Soderberg U. Medullary chemosensitive receptors. J Physiol (Lond) 1952; 118:545–554.
3. Leusen IP. Chemosensitivity of the respiratory center: Influence of CO_2 in the cerebral ventricles on respiration. Am J Physiol 1954; 176:39–44.
4. Loeschcke HH, Koepchen HP. Versuch fur Lokalisation des Angriffsortes der Atmungs- und Kreislaufwirkung von Novocain im Liquor cerebrospinalis. Pflügers Arch 1958; 266:623–641.
5. Mitchell RA, Loeschcke HH, Massion WH, Severinghaus JW. Respiratory responses mediated through superficial chemosensitive areas on the medulla. J Appl Physiol 1963; 18:523–533.
6. Schläfke ME. Central chemosensitivity: A respiratory drive. Rev Physiol Biochem Pharmacol 1981; 90:172–244.
7. Dev NB, Loeschcke HH. Topography of the respiratory and circulatory responses to acetylcholine and nicotine on the ventral surface of the medulla oblongata. Pflügers Arch 1979; 379:19–27.
8. Dev NB, Loeschcke HH. A cholinergic mechanism involved in the respiratory chemosensitivity of the medulla oblongata in the cat. Pflügers Arch 1979; 379: 29–36.
9. Haxhiu MA, Mitra J, van Lunteren E, et al. Hypoglossal and phrenic responses to cholinergic agents applied to ventral medullary surface. Am J Physiol 1984; 247: R939–R944.
10. Bruce EN, Cherniack NS. Central chemoreceptors. J Appl Physiol 1987; 62:389–402.
11. Haxhiu MA, Deal EC, Norcia MP, et al. Effect of N-methyl-D-aspartate applied to the ventral surface of the medulla on the trachea. J Appl Physiol 1987; 63:1268–1274.
12. Haxhiu MA, Deal EC, van Lunteren E, Cherniack NS. Central medullary effects of tachykinin peptides on airway tone. J Auton Nerv Syst 1989; 28:105–116.
13. Haxhiu MA, Deal EC, Trivedi RD, et al. Tracheal and phrenic responses to neurotensin applied to ventral medulla. Am J Physiol 1988; 255:R780–R786.
14. Haxhiu MA, Deal EC, van Lunteren E, Cherniack NS. Central effects of somatostatin and atrial natriuretic peptide on tracheal tone. J Appl Physiol 1993; 75:2353–2359.
15. Feldberg W, Guertzenstein PG. A vasodepressor effect of pentobarbitone sodium. J Physiol (Lond) 1972; 224:83–103.

16. Guertzenstein PG, Silver A. Fall in blood pressure produced from discrete regions of the ventral surface of the medulla by glycine and lesions. J Physiol (Lond) 1974; 242:489–503.
17. Granata AR, Kumada M, Reis DJ. Sympathoinhibition by A1-noradrenergic neurons is mediated by neurons in the C1 area of the rostral medulla. J Auton Nerv Syst 1985; 14:387–395.
18. Granata AR, Numao Y, Kumado M, Reis DJ. A1 noradrenergic neurons tonically inhibit sympathoexcitatory neurons of C1 area in rat brain stem. Brain Res 1986; 377:127–146.
19. Ross CA, Ruggiero DA, Joh TH, et al. Rostral ventrolateral medulla: Selective projections to the thoracic autonomic cell column from the region containing C1 adrenaline neurons in rat brain. J Comp Neurol 1984; 228:168–185.
20. Ross CA, Ruggiero DA, Park DH, et al. Tonic vasomotor control by the rostral ventrolateral medulla: Effect of electrical or chemical stimulation of the area containing C1 adrenaline neurons on arterial pressure, heart rate, and plasma catecholamines and vasopressin. J Neurosci 1984; 4:474–494.
21. Morrison SF, Callaway J, Milner TA, Reis DJ. Glutamate in the spinal sympathetic intermediolateral nucleus: Localization by light and electron microscopy. Brain Res 1989; 503:5–15.
22. Guertzenstein PG. Blood pressure effects obtained by drugs applied to the ventral surface of the brain stem. J Physiol (Lond) 1973; 229:395–408.
23. Guyenet PG, Stornetta RL. Inhibition of sympathetic preganglionic discharges by epinephrine and alpha-methylepinephrine. Brain Res 1982; 235:271–283.
24. Blessing WW, Reis DJ. Evidence that GABA and glycine-like inputs inhibit vasodepressor neurons in the caudal ventrolateral medulla of the rabbit. Neurosci Lett 1983; 37:57–62.

33

Age Differences in Responses of the Ventral Brainstem Respiratory CO_2-Chemosensory Area to Physiological and Pharmacological Stimulation

DAVID G. BERNARD,* C. OVID TROUTH, ROBERT M. DOUGLAS, and LUCRETIA M. SEXCIUS

College of Medicine
Howard University
Washington, D.C.

Previous studies have shown that cholinergic and opiate mechanisms participate in the regulation of ventilation at the caudal ventral medullary surface (cVMS) in the adult. In order to determine developmental differences in these mechanisms, the effects on ventilation and cVMS neuronal activity of inspired CO_2, topically applied acetylcholine (ACh), physostigmine (PHY), and β-endorphin (β-END) were investigated in spontaneously breathing adult and newborn cats. Tidal volume (V_T), respiratory frequency (f), and extracellular cVMS neuronal activity were measured.

Ventilatory Responses: Inspired CO_2, ACh, and PHY increased ventilation in the newborn kitten and adult cat; the responses decreasing with decreasing age. β-Endorphin decreased ventilation. The opiate antagonist naloxone increased the ventilatory responses to ACh, PHY, and inspired CO_2 in both age groups of cats.

cVMS Neuronal Responses: Inspired CO_2, ACh, and PHY increased the frequency of cVMS neuronal firing in the adult and newborn; however, the effect in the newborn (0 to 6 days) was small compared to that in the adult. Naloxone augmented the neuronal responses to ACh, PHY, and inspired CO_2, whereas

**Present affiliation*: Dartmouth Medical School, Lebanon, New Hampshire.

atropine reversed the CO_2-induced excitation of neuronal activity to near control values. Hexamethonium was ineffective. It is concluded that (a) there are age differences in the ventilatory and cVMS neuronal responses to inspired CO_2 and to topically applied cholinergic and opiate drugs and (b) the cVMS is a site where cholinergic-opioid interactive mechanisms may regulate central respiratory chemoreflexes in both the adult and the newborn. Neurochemical imbalance during development in these respiratory regulatory mechanisms could have pathophysiological consequences.

I. Introduction

There are functionally integrated centers in the brainstem that control breathing and vasomotor tone. These centers, which are responsive to a large variety of stimuli coming from various parts of the body, modulate cardiorespiratory activity. Critical afferent inputs to the respiratory and vasomotor centers are believed to originate from regions at the ventrolateral medullary surface (VMS) (Fig. 1) that are sensitive to CO_2 and pH. Fukuda and Loeschcke (1) identified neurons in the caudal VMS (cVMS) of rat brain slices that were responsive to

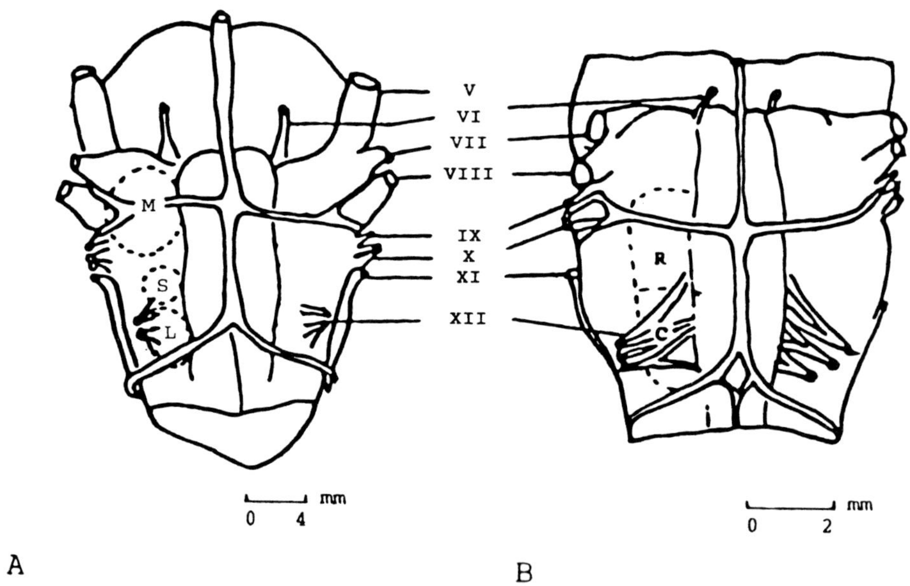

Figure 1 Schematic diagram of the ventral medullary surface of the cat and rat showing (A) the rostral, intermediate, and caudal chemosensitive regions of the cat, which are labeled M, S, and L, respectively; and (B) the rostral and caudal chemosensitive zones of the rat, which are labeled R and C, respectively. V–XII are the cranial nerves.

changes in both CO_2 and extracellular pH. They suggested that H^+-ion sensitivity was associated with cholinergic neural transmission and that this was responsible for the CO_2 chemosensitive drive to respiration.

In recent years, it has been suggested that the central CO_2 chemosensory mechanism might not be fully functional or mature at birth (2). Thus, prematurity or dysfunction of the central CO_2 chemosensory mechanism might be associated with decreased or absent chemosensitive drive to respiration. Earlier studies from our laboratory have shown that topical H^+ ion stimulation of the ventral brainstem in neonatal cats produced significantly smaller increases in activity of cVMS respiratory CO_2-sensitive neurons than in adult cats (3). In addition, direct topical application of naloxone to the VMS of cats consistently augmented the discharge frequency of cVMS neurons to hypercapnic breathing (4). These findings suggest that opioid- and ACh-sensitive sites may be located functionally at the same VMS sites where central respiratory chemoreceptors are presumed to reside (5–8). It is quite possible that cholinergic-opioid interaction might be an important factor in the central cardiorespiratory dysfunction associated with idiopathic apnea of the newborn. In this investigation, the responsiveness of the VMS chemosensory zone to pharmacological and physiological stimuli was assessed at various stages of neuronal development. The ventilatory responses to inspired CO_2, CSF-pH changes as well as cholinergic agonists and antagonists, were examined in newborn kittens and adult cats. In addition, extracellular neuronal activity of cVMS neurons was also studied in response to these pharmacological and physiological stimuli. Endogenous opiate activity was assessed by monitoring the effects of topical application of the opiate antagonist naloxone to the chemosensitive zone.

II. Methods

Newborn (age 1 to 6 days) and adult cats (age > 1 year) were anesthetized with chloralose-urethane (40 mg/kg glucochloralose and 200 mg/kg urethane) IV (adult) and IP (neonate). Rectal temperature was maintained at 37 ± 1°C by external heating or cooling. A tracheal cannula was inserted and the VMS was exposed from the pontomedullary border rostrally to about C1-C2 caudally. The heads of the spontaneously breathing animals were then stabilized in a stereotaxic apparatus (David Kopf Instruments, Inc.).

Respiratory volume and frequency were measured via the tracheal cannula connected to a low-inertia spirometer (Krogh) with an inductive transducer. The system was connected to a semiclosed circuit from which CO_2 was absorbed while O_2 was continuously replaced. Femoral arterial blood pressure was monitored with a Statham strain gauge transducer (Model P23Db). PaO_2 was maintained between 100 and 150 torr by enrichment of the inspired gas with oxygen and adjustment of its flow into the inspiratory circuit. $PECO_2$ was measured via an infrared analyzer (Beckman Medical Gas Analyzer Model LB-2) and arterial PO_2, PCO_2 and pH were measured via a clinical blood gas analyzer using polarographic electrodes (Instrumentation Laboratory pH/Blood Gas Analyzer Model 813).

Extracellular recordings of neuronal activity from the caudal chemosensitive field were made using steel microelectrodes (tip diameter 1 to 2 μm, impedance 5 to 20 MΩ) with the aid of a micromanipulator (David Kopf Instruments, Inc.) in a stereotaxic coordinate system. Using a grid system, the chemosensitive fields were mapped up to a depth of 500 μm from the surface for action potentials that changed their firing rate in response to inspired CO_2 and to changes in mock cerebrospinal fluid (mCSF) pH (pH 7.4, control; 7.0, acidic; 7.8, alkaline). Routinely, each spontaneously active cell was first tested for its responsiveness to CO_2 by increasing inspired CO_2 until an increase in ventilatory and extracellular neuronal activities were observed. Carbon dioxide was then withdrawn from the inspired air. If the cell did not decrease its firing rate within 5 min after withdrawal of CO_2, the animal was again subjected to CO_2 inhalation and a new search was initiated for another spontaneously active cell. In addition, these units were tested for their responsiveness to other modalities of sensation such as light touch, heavy pressure, noxious pinch, and joint manipulation. Chemosensitive neurons were identified in this study as those that responded to inspired CO_2 and CSF-pH changes but not to other modalities of sensation. Action potentials were amplified by a Grass P-511 preamplifier, displayed on a Tektronix 565 dual-beam oscilloscope and filmed online with a C-7 oscilloscope camera. Signal and voice protocol were recorded on magnetic tape (Ampex, PR-500) for later analysis. Diaphragmatic electromyograms (DA) were simultaneously recorded. Selected recording sites were marked by passing a direct current of 10 to 20 μA for 5 s through the extracellular recording metal microelectrode, using the microelectrode as the anode and an alligator clip attached to the skin as the cathode. Sufficient current was passed to produce an identifiable lesion. The heads of the animals were perfused (with the microelectrode in situ) via the aorta with physiological saline, followed by a 10% formaldehyde solution, and the lesions were identified in 30 μm frozen sections stained with cresyl violet or by Glees silver impregnation method.

Tidal volume (V_T), respiratory frequency (f), minute ventilation ($\dot{V}_E$) and medullary neuronal activity (MNA) were recorded. Data were calculated as percent change from control $\pm$ standard error (SEM) for each test response (V_T, f, $\dot{V}_E$, MNA, and DA), so that each unit served as its own control. Statistical significance was determined by one-way ANOVA followed by Duncan's multiple range test with the use of an IBM PS-2 Model 60 micromanipulator. Responses were considered significant for $P < 0.05$.

III. Results

A. Age Differences in the Responses to Inspired CO_2

The effect of inspired CO_2 was investigated in the neonate and compared with that of the adult cat. Inspired CO_2 resulted in a $109.52 \pm 5.58\%$ increase in V_T

in the adult cat and 29.99 ± 10.99% in the newborn. Respiratory frequency increased by 43.59 ± 4.14% in the adult and 27.68 ± 8.49% in the newborn. Thus, $\dot{V}_E$ increased by 200.20 ± 10.36% in the adult and by 65.99 ± 16.68% in the newborn. These findings indicate that the ventilatory response to inspired CO_2 decreased with decreasing age (Fig. 2).

Inspired CO_2 also produced a 22.94 ± 4.92% increase in neuronal activity in the adult and 74.00 ± 3.26% in the newborn. All recordings of VMS neuronal activity were made extracellularly from the cVMS in spontaneously breathing cats.

B. Effect of Atropine on cVMS Neuronal Responses to Inspired CO_2

Atropine (10 mg/mL) was topically applied to the VMS in spontaneously breathing animals (normocapnia) and after CO_2 breathing (hypercapnia). Atropine caused a decrease in the spontaneous activity of the cVMS neurons from 18.4 ± 1.1 Hz (control) to 11.4 ± 0.4 Hz in the adult ($n = 19$) and 9.6 ± 0.5 Hz (control) to 7.9 ± 0.4 Hz in the newborn ($n = 11$). This suggests the possibility of muscarinic involvement at these cVMS sites. Atropine applied during CO_2 breathing caused a decrease in neuronal activity from 37.86 ± 0.85 Hz peak CO_2 responses to 11.81 ± 0.34 Hz in the adult and from 12.49 ± 0.31 Hz peak CO_2 responses to 5.84 ± 0.19 Hz in the newborn. Thus atropine caused a significant decrease ($P < 0.05$) in the neuronal responses to inspired CO_2 (Fig. 3) in the adult and newborn cats.

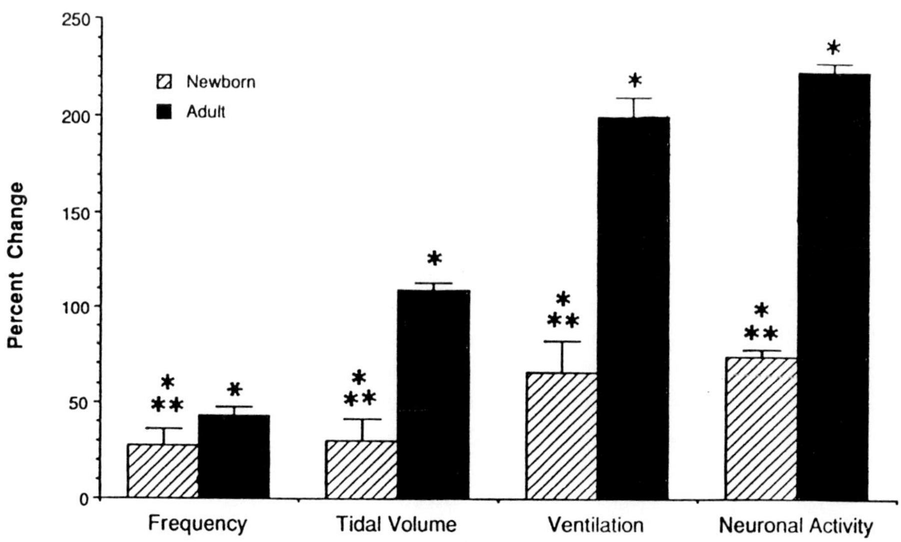

Figure 2 Effect of CO_2 on respiratory frequency, tidal volume, pulmonary ventilation, and extracellular neuronal activity in the newborn kitten and adult cat. *Significantly different from control; **significantly different from adult cat ($P < 0.05$).

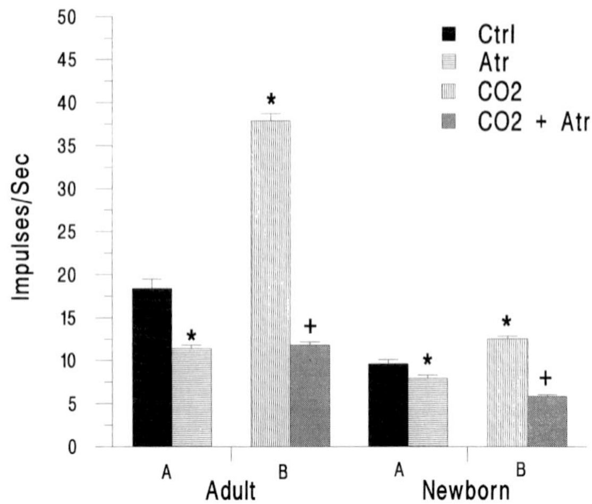

Figure 3 (A) Neuronal responses in the newborn and adult cat to atropine and (B) the effect of atropine on the CO_2 response. *Significantly different from control (ctrl); + significantly different from the CO_2 response ($P < 0.05$).

C. Age Differences in the Responses of the cVMS to Topically Applied Acetylcholine and Physostigmine

Topical application of acetylcholine (ACh) (5 mg/mL) to the cVMS resulted in an increase in V_T, f and $\dot{V}_E$ in the two age groups of cats. Tidal volume increased from 28.33 ± 2.17 mL (control) to 39.50 ± 2.98 mL in the adult and from 1.41 ± 0.21 mL (control) to 1.80 ± 0.28 mL in the newborn. Respiratory frequency increased from 18.00 ± 3.28 breaths/min (control) to 19.92 ± 3.70 breaths/min in the adult and from 21.50 ± 1.14 breaths/min (control) to 23.60 ± 1.47 breaths/min in the newborn. Minute ventilation increased from 459.25 ± 64.50 mL/min (control) to 737.25 ± 141.58 mL/min in the adult and 32.04 ± 6.29 mL/min (control) to 45.39 ± 9.31 mL/min in the newborn. As with inspired CO_2, topical application of ACh to the cVMS resulted in decreased responsiveness with decreasing age (Fig. 4).

The CO_2-sensitive neurons at the cVMS increased their firing rate in response to topical application of acetylcholine from 13.12 ± 0.57 Hz (control) to 36.28 ± 1.02 Hz in the adult and from 8.28 ± 0.22 Hz (control) to 12.45 ± 0.51 Hz in the newborn. The response in the newborn was significantly ($P <$ 0.05) reduced compared to that of the adult.

Physostigmine (PHY) also produced an increase in V_T from 27.27 ± 1.82 Hz (control) to 37.91 ± 2.58 Hz in the adult and from 1.26 ± 0.12 Hz (control) to 1.70 ± 0.17 Hz in the newborn. Furthermore, PHY produced an increase in f from 20.91 ± 2.57 breaths/min (control) to 25.09 ± 2.99 breaths/min in the adult and from 21.67 ± 1.40 breaths/min (control) to 24.11 ± 1.62 breaths/min

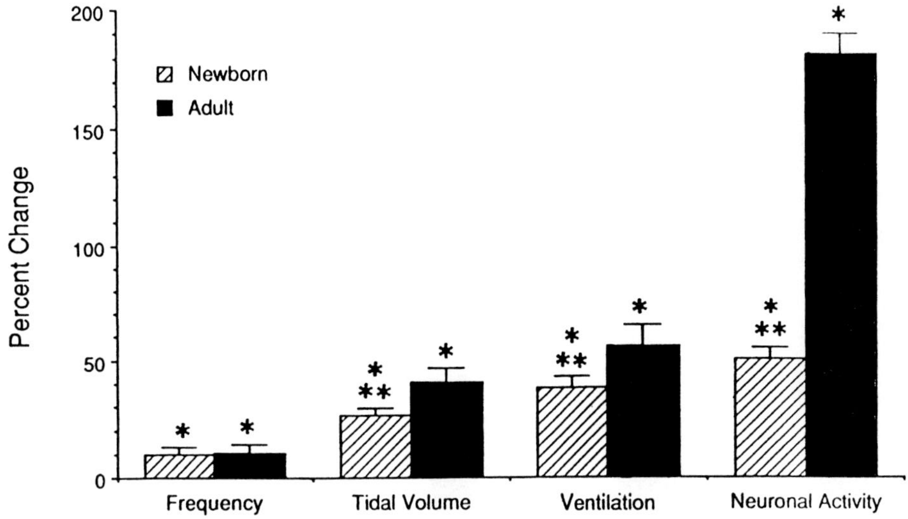

Figure 4 Effect of acetylcholine on respiratory frequency, tidal volume, pulmonary ventilation and extracellular neuronal activity in the newborn and adult cat. *Significantly different from control; **significantly different from adult cat ($P < 0.05$).

in the newborn. Minute ventilation increased from 571.82 ± 74.45 mL/min (control) to 957.00 ± 124.46 mL/min in the adult and from 27.33 ± 3.23 mL/min (control) to 40.19 ± 5.61 mL/min in the newborn following application of PHY to the cVMS.

The mean firing rate of CO_2-sensitive neurons at the cVMS increased in the two age groups of cats in response to superfusion with 5 mg/mL PHY. PHY increased the spike activity by 173.45 ± 10.28% in the adult and 53.95 ± 4.41% in the newborn. The responsiveness was smaller in the newborn compared to the adult (Fig. 5).

It was difficult to determine the amount of drug that reached the target tissues, even though application of ACh and PHY was discretely localized to a very small area of the cVMS. Drug concentrations may have been reduced by dilution with cerebrospinal fluid (CSF) and diffusion away from the site of application.

Effects of Naloxone on the CO_2, Acetylcholine, and Physostigmine

The effect of naloxone (4 mg/mL) prepared in mock CSF, pH 7.4, on the ventilatory responses to inspired CO_2, topically applied acetylcholine, and physostigmine were examined in the newborn cat and the responses compared with those of the adult. The mean ventilatory minute volume at pH 7.4 (control) in both age groups increased in response to inspired CO_2. Mean ventilatory minute

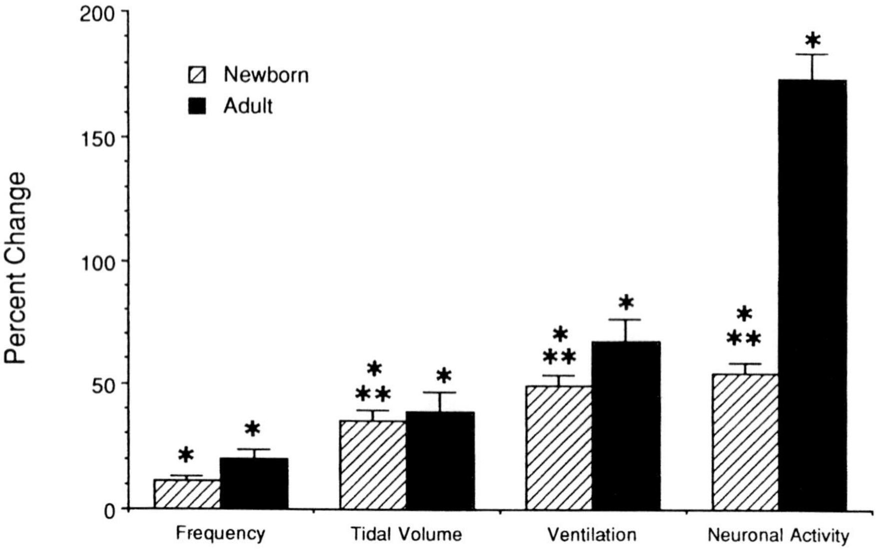

Figure 5 Effect of the anticholinesterase physostigmine on respiratory frequency, tidal volume, pulmonary ventilatory and extracellular neuronal activity in the newborn and adult cat. *Significantly different from control; **significantly different from adult ($P < 0.05$).

volume increased by 65.99 ± 16.68% (29.22 ± 2.60 mL to 48.08 ± 6.21 mL) in the neonate and by 200.20 ± 10.36% (660.53 ± 37.17 mL to 1948.30 ± 30.00 mL) in the adult (Fig. 6A).

Thus, naloxone applied topically to the cVMS augmented the ventilatory response to inspired CO_2 in newborn and adult cats. Naloxone also augmented the ventilatory responses to acetylcholine and physostigmine applied topically to the cVMS (Fig. 6B and C).

Following inspiration of CO_2, naloxone also caused an increase in spike activity of chemosensitive cVMS neurons from 222.94 ± 4.92% to 341.53 ± 9.82% in the adult and from 74.42 ± 3.26% to 137.33 ± 4.28% in the newborn (Fig. 6D). These results are indicative of an endogenous opiate involvement in the cVMS neuronal and ventilatory responses to inspired CO_2 and to cholinergic agents.

IV. Discussion

A. Age Differences in the Respiratory Responses to Inspired Carbon Dioxide

Central chemical regulation of breathing arises primarily in the lower brainstem; however, the anatomy and physiology of this system and possibly other

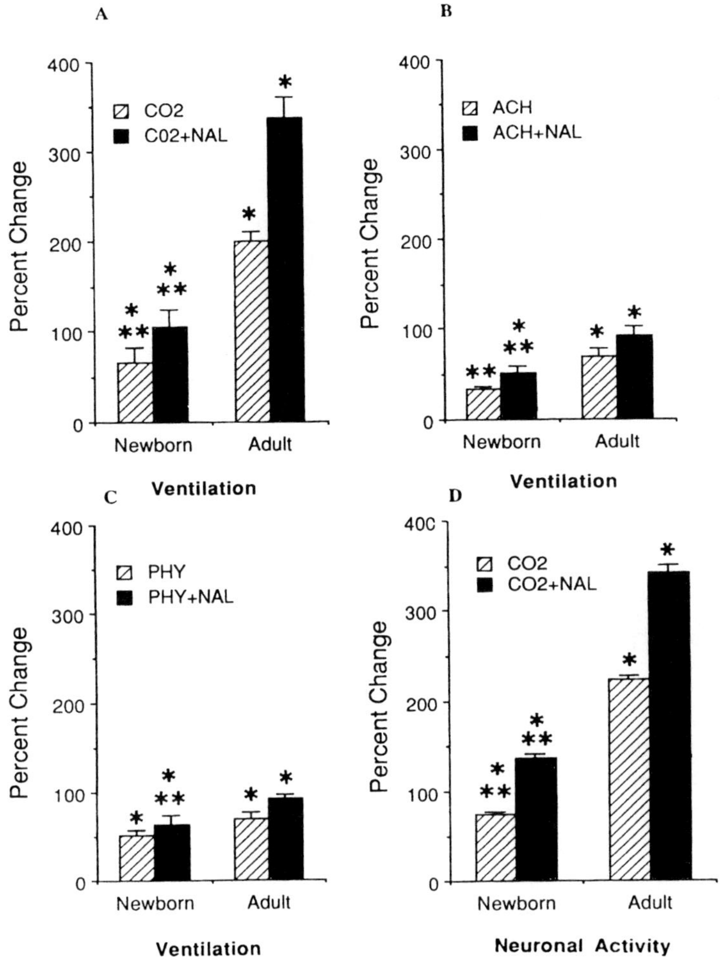

Figure 6 Effect of Nal on (A) the ventilatory response to inspired CO_2; (B) Ach; (C) PHY; (D) neuronal response to inspired CO_2. *Significantly different from control; **significantly different from adult ($P < 0.05$).

control mechanisms involved in the protection of human and nonhuman organisms against accumulation of CO_2 are not well understood. The central respiratory chemoreflex response to CO_2 has been ascribed to three acid-sensitive regions at the VMS (Fig. 1). The chemosensitive properties of these regions of the cat's VMS (Fig. 1) were identified by superfusing the ventral brainstem with acidic and alkaline mock cerebrospinal fluid while measuring either V_T or f and calculating $\dot{V}_E$ or by monitoring changes in the activity of pH-sensitive neurons at the three chemosensitive sites (5). The superficial locations of these structures were localized by electrical stimulation (6,9,10), whereas topical application of drugs directly to the VMS has been used to investigate the neuropharmacological properties of these central chemosensitive regions that drive respiration (11,12). The breathing of gas mixtures containing various concentrations of CO_2 (hypercapnic breathing) has also been used to stimulate both ventilatory and neuronal activity and is considered to be an acceptable method for studying respiratory chemosensitivity in anesthetized cats (13).

Systematic experiments conducted in our laboratory have shown that cats of all ages do not respond similarly to inspired CO_2 or to pH changes at the VMS. In longitudinal studies beginning at birth and continuing to adulthood, it was determined that the effects of pH or inspired CO_2 changes at the VMS (Fig. 1) could be organized into three age groups (3): group I: day 0 to day 6; group II: day 7 to week 6; group III: older than 7 weeks (adult).

Using the same experimental criteria in the rat, only two groups were identified: group I: day 0 to day 4 and group II: day 5 to adulthood (14). Using the age groups identified in the cat, we have investigated differences in responsiveness to inspired CO_2, cholinomimetics, and cholinolytics and to opiate antagonists. This report focuses on the differences in ventilatory and neuronal responses between adult and newborn cats.

Ventilatory Responses

The ventilatory response to inhaled carbon dioxide has been used by several investigators to assess the functional integrity of the respiratory control system (15–18), and defective neurochemical control of breathing has been implicated in the pathogenesis associated with prolonged periods of apnea often seen in premature infants (2,15,19). The primary aim of this investigation was to study age-related differences in the CO_2-chemosensory mechanism in two age groups of cats (normal neonatal and adult).

The sensitivity of the central response to inspired CO_2 was examined in the neonate and compared with that of the adult. Several workers have reported that sensitivity of the central response to CO_2 was greater in the newborn than in the fetus (15,19). In earlier studies, we have demonstrated that topical H^+ ion stimulation in neonatal cats produced less of an increase in the activity of cVMS respiratory CO_2-sensitive neurons than in adults (3). In more recent studies, we have shown that the opiate antagonist naloxone potentiates the ventilatory response to CO_2 in adults (4,20). These results together suggest that the sensitivity of the CO_2 response may be modulated not only by maturational factors but also by endogenous opiate mechanisms.

Ventilatory activity recorded in the cats of the two age groups revealed that during hypercapnic breathing, the newborn chemosensory mechanism was less sensitive to inspired CO_2 than that of the adult cat (Fig. 2). Response curves were relatively flat for newborn kittens and steeper in the adult cats. Frantz et al. (19) found that the premature infants had lower sensitivity to CO_2 than full-term infants and that the CO_2 sensitivity increased with increasing postnatal age. In a group of premature infants, Rigatto et al. (15) also reported that CO_2 sensitivity improved with gestational and postnatal age. In our study, the response to CO_2 increased with advancing postnatal age from birth to adulthood. It has been proposed by Gerhardt and Bancalari (21) that the decreased slope of the CO_2 response curve may be an important determinant of apnea in preterm infants. Therefore, prematurity and/or functional deficits in the central chemoreceptor apparatus may result in clinical conditions associated with decreased or absent chemosensitivity to the CO_2 stimulus.

Response of Brainstem-Chemosensitive Neurons

Hypotheses explaining hypercapnic hyperpnea of central origin fall into two basic groups. One group supports the notion that there are multiple sites of CO_2 chemosensitivity in the brainstem (22–24), whereas the other group supports the concept that the primary effect of inspired CO_2 is on structures located at the VMS (5–7,11,12).

The effect of CO_2 on the activity of neurons in the superficial chemosensitive area on the VMS has previously been studied by several investigators (1,8, 25). However, few systematic studies on age-related differences in the neuronal responses to inspired CO_2 have been undertaken. The current findings show that the ventral medulla oblongata contains neuronal elements in both adult and newborn cats that respond to inspired CO_2. It is quite possible that the central response to CO_2 may already begin to occur during fetal life (2). However, other reports have suggested that the central chemoreceptors may not be functionally mature at birth (2). In our studies, we have shown that the cVMS neuronal responses to inspired CO_2 decreased with decreasing age. Similar differences have been reported by Whittaker et al. (3) following superfusion of the VMS with mock CSF of varying pH. These authors identified specific milestones for development of central respiratory chemosensory neuronal responsiveness to CSF-pH changes.

We have no clear understanding of the factors regulating the decreased CO_2 response during stimulation of respiratory chemoreceptors in developing animals (26–30). One may suggest, however, that the changes in circulating progesterone and in the local production and release of acetylcholine—as well as endorphins, enkephalins, and other neuroregulators observed during maturation—could contribute to the resetting of the central chemoreflex response to CO_2 postnatally. Furthermore, developmental changes in the activity of the carbonic anhydrase (14) and intracellular signaling systems (e.g., prostaglandin, cyclic nucleotide) and changes in central sensitivity to transmitter agents could play an important role in the resetting of the central chemoreflex response to CO_2 during maturation.

Age Differences in Response to Acetylcholine and Physostigmine

Cholinergic mechanisms have been implicated in the central chemical control of respiratory activity (1,11,31,32) at the level of the cVMS. In this investigation, the topical application of the ACh to the caudal chemosensitive area (area L) of the medulla oblongata induced an increase in the frequency of action potentials recorded from superficial CO_2-sensitive neurons in cats of both age groups.

The sensitivity to ACh of the CO_2-sensitive neurons located at the cVMS was greater in adult than in newborn kittens. Several factors may be important in determining this age difference: (a) there may be a lower proportion of cholinergic neurons at the cVMS in the newborn kitten than in adult cats—this has been recently demonstrated in our laboratory by Pan et al. (14); (b) the release of ACh from presynaptic terminals due to lower ionization of calcium (33) may be inhibited in newborn kittens compared to adult cats; (c) neuronal excitation by ACh may be reduced because these neurons may not be fully developed in newborns compared to adults. In effect, maturation may produce changes that bring about an increase in sensitivity to ACh and CO_2.

The action of anticholinesterase agents topically applied to the cVMS was investigated. Application of physostigmine to the caudal chemosensitive area stimulated respiration, which suggests that either physostigmine had a direct effect on these neurons or cells located in this region may be synaptically driven by cholinergic fibers. Physostigmine may also have increased the destruction of cholinesterase, leading to the accumulation of ACh and hence enhancement of $\dot{V}_E$. The presence of cholinesterase at central synaptic regions has been demonstrated by several workers including Feldberg (34) and Krnjevic and Phillis (35). In addition, inhalation of CO_2 may increase H^+ ion concentration in the extracellular fluid of the brain, which may inactivate cholinesterase present in the surface layer. Augustinsson (36), MacIntosh and Perry (37), and Lüllmann and Peters (38) have reported that increased $[H^+]$ in the glomus cell of the carotid body leads to inactivation of cholinesterase.

Monteau et al. (39) used a brainstem–spinal cord preparation of newborn rats to analyze the central respiratory effects of ACh. They found that respiratory frequency could be decreased by muscarinic but not by nicotinic antagonists and increased by superfusates containing ACh or physostigmine, equilibrated with 5% CO_2, at pH 7.3. The present studies showed that it is possible to elicit increases in neuronal frequency in newborn and adult cats; however, the responses are different for the different age groups. In this study, application of the anti-ChE physostigmine to the caudal CO_2-chemosensitive region of the medulla produced an increase in neuronal activity and pulmonary ventilation in adult and newborn cats. The increases were significantly less ($P < 0.05$) in the neonate than in the adult.

Effects of Atropine Applied to Brainstem CO_2-Sensitive Neurons During Hypercapnic Breathing

Dev and Loeschcke (32) used atropine to block the ventilatory response to locally applied ACh on the VMS and suggested that muscarinic receptors may be involved in the respiratory drive. In rat brain slices, Fukuda and Loeschcke (1) used atropine, hexamethonium, and mecamylamine to inhibit all activity at chemosensitive VMS regions. They concluded that an intact cholinergic system on the surface layer of the VMS was necessary for normal ventilation. In this study, ACh and physostigmine applied to the cVMS increased both ventilatory and neuronal activity, whereas atropine decreased these responses. These findings therefore support the hypothesis that the cholinergic system is involved in the CO_2-induced excitation of cVMS neurons. Lipscomb and Boyarsky (23) have proposed that the respiratory center neurons themselves may be the central respiratory chemoreceptors. Central respiratory chemosensitivity has been postulated to involve cholinergic relay. Application of cholinergic agents to the chemosensitive areas induced changes in ventilation (1,4,32,40,41). VMS neurons excited by H^+ and CO_2 were also excited by ACh and physostigmine, but this is still under debate (42). The cVMS (Fig. 1) has been shown to be most sensitive to both electrical (6,10) and chemical (5) stimulation and to more specifically regulate respiration (43) than the intermediate and rostral chemosensitive regions. In the present investigation, atropine effectively blocked the CO_2-induced excitation of cVMS neurons.

Opiate Interaction in Brainstem Ventilatory Regulation

In this study naloxone was used to examine the opiate involvement at the caudal respiratory chemosensory site (cVMS). Newborn infants have been noted to have increased levels of opioid substances in their plasma as compared to their mothers or other adults (44). Opiates possess analgesic properties that may afford newborns some analgesia during the trauma of parturition. This analgesia could become less important with developmental age. It is well known that the opiates induce central respiratory depression by decreasing both tidal volume and respiratory frequency (45,46). Both pontine and medullary respiratory units seem to be involved. Might the dual role of analgesia and respiratory depression of opiates in the neonate be physiologically important?

Trouth et al. (4) recently reported that naloxone, an opiate antagonist, augments brainstem chemosensitive neuronal responses to inspired CO_2 and local pH changes at the cVMS. In our present study, naloxone augmented the ventilatory responses to topically applied ACh and physostigmine and to inspired CO_2 in adult and newborn cats. Naloxone also augmented the cVMS neuronal responses to inspired CO_2. This suggests that opiates may be involved in brainstem chemosensory control mechanisms. Several investigators have implicated the opiates in the central regulation of breathing (12,47). Apnea is produced when opioids are injected intracerebrally, and respiratory depression—produced by topical application of met-enkephalin to the VMS—is reversed by naloxone (45). Naloxone augments the ventilatory response to inspired CO_2, suggesting that endogenous opiates may have an important role in regulating the ventilatory responses to pH changes and to hypercapnia. Similar observations have been made at peripheral chemoreceptors, where met-enkephalin has been shown to inhibit naloxone-sensitive neuronal activity at the carotid body (48). Chernick and Russell (49) have shown that naloxone shortens apneic episodes in asphyxiated newborn rabbits and concluded that endogenous opiates lengthened the duration of apnea during asphyxia by inhibiting the respiratory rhythm generator. Moss and Friedman (46) found that β-endorphin, like met-enkephalin, depressed hypercapnic ventilation and is antagonized by naloxone in anesthetized dogs. It was not possible to determine why newborn kittens responded less to physiological and pharmacological stimuli than adult cats. Indeed, the decreased sensitivity of cholinergic ventral brainstem neurons to stimulation with inspired CO_2 may be referable to the effect of endogenous opioids at these sites. Our study, as well as those of Isom and Elshowihy (50) and Trouth and colleagues (4), supports the notion that an endogenous opiate mechanism may exert an important neuromodulatory effect on the ventilatory response to inspired CO_2, ACh, and PHY by influencing cVMS neuronal activity. Thus an understanding of the interactions between cholinergic and opioid mechanisms in the two age groups might provide insight into the mechanisms subserving age differences in the responsiveness of the CO_2-chemosensory structures.

Acknowledgment

Supported by MBRS Grant #5S06GM08016-22.

References

1. Fukuda Y, Loeschcke HH. A cholinergic mechanism involved in the neuronal excitation by H^+ in the respiratory chemosensitive structures of the ventral medulla oblongata of influencing respiration rats in vitro. Pflügers Arch 1979; 379:125–135.
2. Moss IR, Scarpelli EM. Generation and regulation of breathing in utero: Fetal CO_2 response test. J Appl Physiol 1979; 47:527–531.
3. Whittaker JAC, Trouth CO, Pan Y, et al. Age differences in responsiveness of brainstem chemosensitive neurons to extracellular pH changes. Life Sci 1990; 46:1699–1705.
4. Trouth CO, Bada FJ, Pan Y, et al. Naloxone application to the ventrolateral medulla enhances the respiratory response to inspired carbon dioxide. Life Sci 1991; 49: 193–200.
5. Schläfke ME, See WR, Loeschcke HH. Ventilatory response to alterations of H^+ ion concentration in small areas of the ventilatory medullary surface. Resp Physiol 1970; 10:198–212.

6. Loeschcke HH, De Lattre J, Schläfke ME, Trouth CO. Effects on respiration and circulation of electrically stimulating the ventral surface of the medulla oblongata. Resp Physiol 1970; 10:184–197.
7. Trouth CO, Loeschcke HH, Berndt J. A superficial substrate on the ventral surface of the medulla oblongata influencing respiration. Pflügers Arch 1973; 339: 135–152.
8. Trouth CO, Patrickson J, Holloway JA, Wright LE. Neurophysiological studies on superficial medullary chemosensitive area for respiration. Brain Res 1982; 246: 47–56.
9. Trouth CO. Localization in the medulla oblongata by means of electrical stimulation of superficial and deep structures influencing respiration. Pflügers Arch 1969; 307:16.
10. Trouth CO, Loeschcke HH, Berndt J. Topography of the respiratory responses to electrical stimulation in the medulla oblongata. Pflügers Arch 1973; 339:153–170.
11. Mitchell RA, Loeschcke HH, Massion WH, Severinghaus JW. Respiratory responses mediated through superficial chemosensitive areas on the medulla. J Appl Physiol 1963; 18:523–533.
12. Mitchell RA, Loeschcke HH, Severinghaus JW, et al. Regions of respiratory chemosensitivity on the surface of the medulla. Ann NY Acad Sci 1963; 109:661–681.
13. Lawson EE, Waldrop PG, Eldridge FL. Naloxone enhances respiratory output in cats. J Appl Physiol 1979; 47:1105–1111.
14. Pan Y, Douglas RM, Trouth CO. Age differences in cholinergic and catecholaminergic neurons in rat brainstem respiratory chemosensitive zones (abstr). Soc Neurosci Abstr 1992; 18:828.
15. Rigatto H, Brady JP, de la Torre Verduzco R. Chemoreceptor reflexes in preterm infants: 1. The effect of gestational and postnatal age on the ventilatory response to inhalation of 100% and 15% oxygen. Pediatrics 1975; 55:604–613.
16. Berkenbosch AAD, Bovill JG, Dahan A, et al. The ventilatory CO_2 sensitivities from Read's rebreathing method and the steady-state method are not equal in man. J Physiol (Lond) 1989; 411:367–377.
17. Berkenbosch AAD, Ward DS, Oliever CN, et al. Dynamics of ventilatory response to step changes in P_{CO_2} of blood perfusing the brain stem. J Appl Physiol 1989b; 66:2168–2173.
18. Jacobi MS, Patel CP, Saunders KB. Transient, steady-state and rebreathing responses to carbon dioxide in man, at rest and during light exercise. J Physiol (Lond) 1989; 411:85–96.
19. Frantz ID III, Alder SM, Thach BT, Taeusch WH Jr. Maturational effects on respiratory response to carbon dioxide in premature infants. J Appl Physiol 1976; 41: 41–45.
20. Bernard DG, Bada F, Pan Y, et al. Role of endogenous opiates at brainstem respiratory chemosensitive areas (abstr). Soc Neurosci Abstr 1989; 17:99.
21. Gerhardt T, Bancalari E. Ventilatory responses to CO_2 in premature infants with apnea. Pediatr Res 1979; 13:534.
22. Coates EL, Li A, Nattie EE. Widespread sites of brain stem ventilatory chemoreceptors. J Appl Physiol 1993; 75:5–14.
23. Lipscomb WT, Boyarsky LL. Neurophysiological investigations of medullary chemosensitive areas of respiration. Respir Physiol 1972; 16:362–376.

24. Millhorn DE, Eldridge EL. Role of ventrolateral medulla in regulation of respiratory and cardiovascular systems. J Appl Physiol 1986; 61:1249–1263.
25. Tojima H, Kuriyama T, Fukuda Y. Differential respiratory effects of HCO_2^- and CO_2 applied on ventral medullary surface of rats. J Appl Physiol 1991; 70:2217–2225.
26. Hutchinson EA, Percival CH, Young IM. Development of cardiovascular responses in the kitten. J Physiol 1962; 47:201–210.
27. Shineborne EA, Vapaavuori EK, Williams RL, et al. Development of baroreflex activity in unanesthetized fetal and neonatal lambs. Circ Res 1972; 31:710–718.
28. Vatner SF, Manders WT. Depressed responsiveness of the carotid sinus reflex in conscious newborn animals. Am J Physiol 1979; 237:H40–H43.
29. Walker AM, Cannata J, Dowling MH, et al. Sympathetic and parasympathetic control of heart rate in unanesthetized fetal and newborn lambs. Biol Neonate 1978; 33:135–143.
30. Blanco CE, Dawes GS, Hanson MA, McCooke HB. Carotid baroreceptors in fetal and newborn sheep. Pediatr Res 1988; 24:342–346.
31. Dev NB, Loeschcke HH. Topography of the respiratory and circulatory responses to acetylcholine and nicotine on the ventral surface of the medulla oblongata. Pflügers Arch 1979; 379:19–27.
32. Dev NB, Loeschcke HH. A cholinergic mechanism involved in the respiratory chemosensitivity of the medulla oblongata in the cat. Pflügers Arch 1979; 379:29–36.
33. Hubbard JI, Jones SF, Landau E. On the mechanism by which calcium and magnesium affect the spontaneous release of transmitter from mammalian motor nerve terminals. J Physiol (Lond) 1968; 47:201–210.
34. Feldberg W. Role of acetylcholine in the central nervous system. Br Med Bull 1950; 6:312–321.
35. Krnjevic K, Phillis JW. Pharmacological properties of acetylcholine-sensitive cells in the cerebral cortex. J Physiol 1963; 166:328–350.
36. Augustinsson KB. Cholinesterases: Study in comparative enzymology. Acta Physiol Scand 1948; 15(Suppl 52):1–182.
37. MacIntosh FC, Perry WL. Biological estimation of acetylcholine. Methods Med Res 1950; 3:76–92.
38. Lüllman H, Peters T. Influence of pH on the action of parasympathomimetic drugs. Eur J Pharmacol 1967; 2:106–112.
39. Monteau R, Morin D, Hilaire G. Acetylcholine and central chemosensitivity: In vitro study in newborn rat. Respir Physiol 1990; 81:241–254.
40. Mitchell RA. The regulation of respiration in metabolic acidosis and alkalosis. In: Brooks C, McC Kae F, Lloyd BB (eds). Cerebrospinal Fluid and Regulation of Respiration. Oxford: Blackwell, 1965:109–131.
41. Trouth CO, Millis RM, Bernard DG, et al. Cholinergic-opioid interactions at brainstem respiratory chemosensitive areas in cats. Neurotoxicology 1993; 14:459–468.
42. McHill P, Hughes E, Sinclair J. The relationship between respiratory stimulation by the superficial medullary cholinergic mechanism and by inspired carbon dioxide in anesthetized rats. J Physiol 1983; 334:42P–43P.
43. Mitra J, Prabhakar NR, Overholt JL, Cherniack NS. Respiratory and vasomotor responses to focal cooling of the ventral medullary surface of the rat. Resp Physiol 1988; 74:35–48.

44. Wardlaw SL, Stark R, Barc L, Franz AD. Plasma beta-lipotropin in human fetus at delivery: Correlation with arterial pH and PO_2. J Clin Endocrinol Metab 1979; 79:888–891.
45. Florez J, Mediavilla A. Respiratory and cardiovascular effects of met-enkephalin applied to the ventral surface of the brain stem. Brain Res 1977; 138:585–590.
46. Moss IR, Friedman E. Beta-endorphin: effects on respiratory regulation. Life Sci 1978; 23:1271–1276.
47. Gingras-Leatherman JL, McNamara MC, Hong JS, Lawson E. Development of methionine-enkephalin in microdissected areas of the rabbit brain. Brain Res 1985; 336:73–80.
48. McQueen DS, Ribeiro JA. Inhibitory actions of methionine enkephalin and morphine in the cat carotid chemoreceptors. Br J Pharmacol 1980; 71:297–305.
49. Chernick V, Russell BJ. Endogenous opiates and primary apnea (abstr). Am Rev Respir Dis 1978; 117:290.
50. Isom GE, Elshowihy RM. Naloxone-induced enhancement of carbon dioxide stimulated respiration. Life Sci 1982; 31:113–118.

34

Contribution of Carbonic Anhydrase to CO_2 Chemosensitivity in Cultured Ventral Medullary Neurons

JUDITH A. NEUBAUER and ANDREW A. MARTIN

University of Medicine and Dentistry of New Jersey–Robert Wood Johnson Medical School
New Brunswick, New Jersey

It is generally accepted that the respiratory response to CO_2/H^+ is mediated via stimulation of chemosensitive cells located bilaterally near the surface of the caudal and rostral ventrolateral medulla (1–3). Extracellular recordings from these regions have found both respiratory- and non-respiratory-related neurons that respond to changes in pH by altering their firing rate (4,5). CO_2 responses have also been described in other areas of the central nervous system (CNS)—such as the nucleus tractus solitarius (6,7), the hypothalamus (8,9), and the raphe nucleus (10)—raising the possibility of more widespread CO_2 sensitivity of neurons dispersed within the brain that may or may not have a respiratory function. Indirect evidence that the ventral lateral medullary regions are probably the central CO_2/H^+ chemosensitive sites responsible for stimulating respiration comes primarily from cooling, lesioning, and neurotoxic destruction studies which have demonstrated loss of respiratory chemosensitivity following these procedures (11–13). However, little is known about the cell phenotype or mechanism of pH transduction in these central chemoreceptors.

In vitro studies have yielded limited information regarding the characteristics of the sensing mechanism. However, several features of the pH response are noteworthy in their ability to provide insight on potential transducing mechanisms. The first is the differential respiratory response to hypercapnic and metabolic acidosis (14–16). If the extracellular pH (pH_{ex}) of the ventral medulla in the chemosensitive regions is monitored with microelectrodes, the respiratory response to changes in pH_{ex} appears to be a linear function. However, the slope of this relationship between pH and respiration is significantly less when the change in pH_{ex} is induced by metabolic acidosis rather than inhaled CO_2. That is, for equivalent changes in pH_{ex}, the respiratory response is greater to hypercapnia than to metabolic acidosis. Several suggestions have been made to explain this differential response (17,18) including different receptors for CO_2 versus HCO_3^-, a protected site (e.g., a synapse), or an intracellular site of activation.

A second insightful observation comes from studies exploring the importance of carbonic anhydrase in the central respiratory response to CO_2. Inhibition of carbonic anhydrase significantly delays and attenuates the respiratory response to CO_2 (19,20). In addition, acetazolamide displaces the relationship between the phrenic nerve response and extracellular pH to the right—that is, there is less respiratory stimulation for an equivalent change in pH_{ex} (Fig. 1).

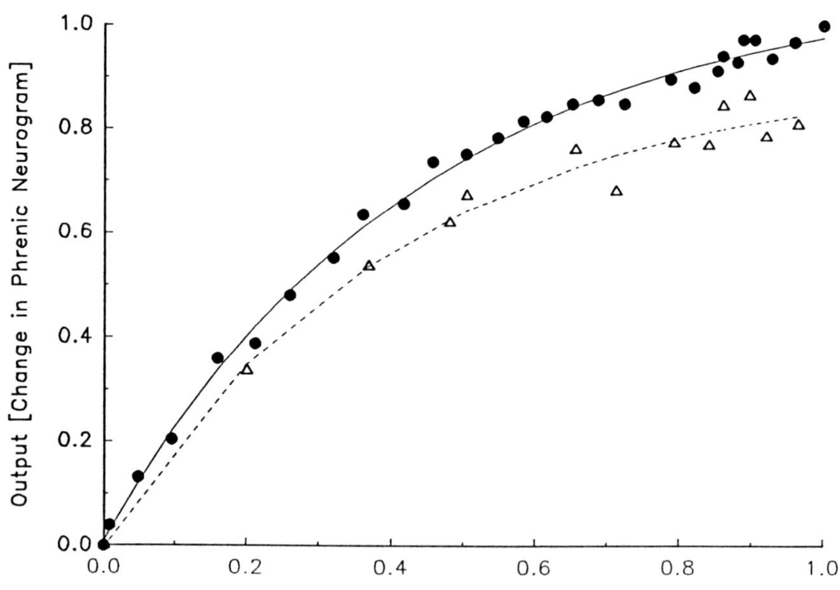

Figure 1 Phrenic neurogram and ventral medullary pH_{ex} response to a forcing function change in CO_2 before (closed circles) and after acetazolamide (open triangles).

These results suggest that extracellular pH is not the proximate stimulus of the central chemoreceptors, that an intracellular pH may be a more likely stimulus site, and that inhibition of carbonic anhydrase delays the rate of change of this intracellular pH.

Hanson et al. (19) explored the site of this effect on CO_2 more specifically by using two carbonic anhydrase inhibitors: one that crosses membranes (acetazolamide) and one that does not cross membranes (benzolamide). They found that the respiratory response to vertebral artery injections of CO_2-saturated saline was affected by acetazolamide (but not by benzolamide), suggesting that the location of the carbonic anhydrase modulation is beyond the blood-brain barrier. Furthermore, when these inhibitors were applied directly to the ventral surface of the medulla, only acetazolamide altered the response to CO_2. Taken together, these studies suggest an intracellular site for CO_2 chemosensitivity, which utilizes carbonic anhydrase to accelerate the pH change and the signal transduction process.

The lack of any anatomically distinct group of neurons on the ventral medullary surface distinguishing a chemoreceptor has made it extremely difficult to determine the exact nature of chemosensitivity in vivo. For this reason we have focused our work on developing in vitro preparations of the chemosensitive regions of the ventral medulla that retain the ability to respond to CO_2 in order to evaluate the mechanisms of CO_2 transduction. Our work has developed two types of cultures, organotypic and dissociated cell cultures, in which spontaneously active neurons retain the function of chemosensitivity (6,21,22).

Figure 2 illustrates an example of a spontaneously active ventral medullary neuron that increased its firing rate when the pH of the superfusate was reduced by increasing P_{CO_2} at constant HCO_3^-. While most of the neurons in culture were rendered less excitable when CO_2 was increased, between 25% and 35% of the neurons in culture were excited by CO_2.

In contrast, decreasing the pH of the superfusate by reducing HCO_3^- at constant P_{CO_2} did not increase the firing rate of any of the cultured neurons (Fig. 3).

Thus, we have demonstrated electrophysiologically that the neuronal function of acid sensitivity survives in culture and that these cells only respond to changes in pH induced with CO_2. But what other features characterize these neurons? Based on the previous work mentioned above, which suggested that carbonic anhydrase is dynamically involved in the respiratory response to CO_2, we were interested in whether this enzyme might be a marker for chemosensitive neurons. Although it was generally thought that carbonic anhydrase is not present in CNS neurons, subsequent work from several groups (23–25) has shown that while carbonic anhydrase is not present in peripheral motor neurons, it is present in the sensory neurons of peripheral ganglia. Thus, we hypothesized that sensory neurons involved in chemosensitivity contain carbonic anhydrase. With the exception of some suggestive evidence by Ridderstrale and

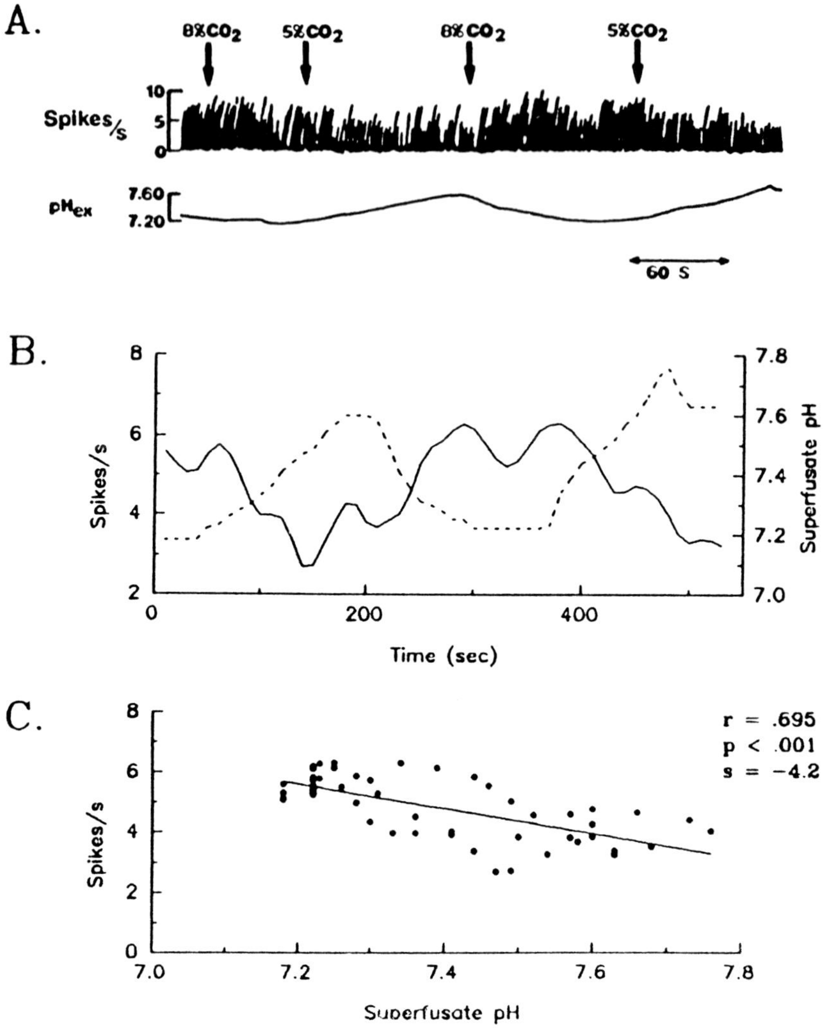

Figure 2 Representative response of a spontaneously active ventral medullary neuron in explant culture which was excited by hypercapnic acidosis. (A) Original recording traces of spike activity and superfusate pH; (B) plot of both spike activity (continuous line) and superfusate pH (dashed line) with time; (C) linear regression of spike activity against superfusate pH. Spontaneous spike activity in this neuron was greatest when pH was most acid with a mean change in spike activity of −4.2 Hz/pH unit. (From Ref. 6.)

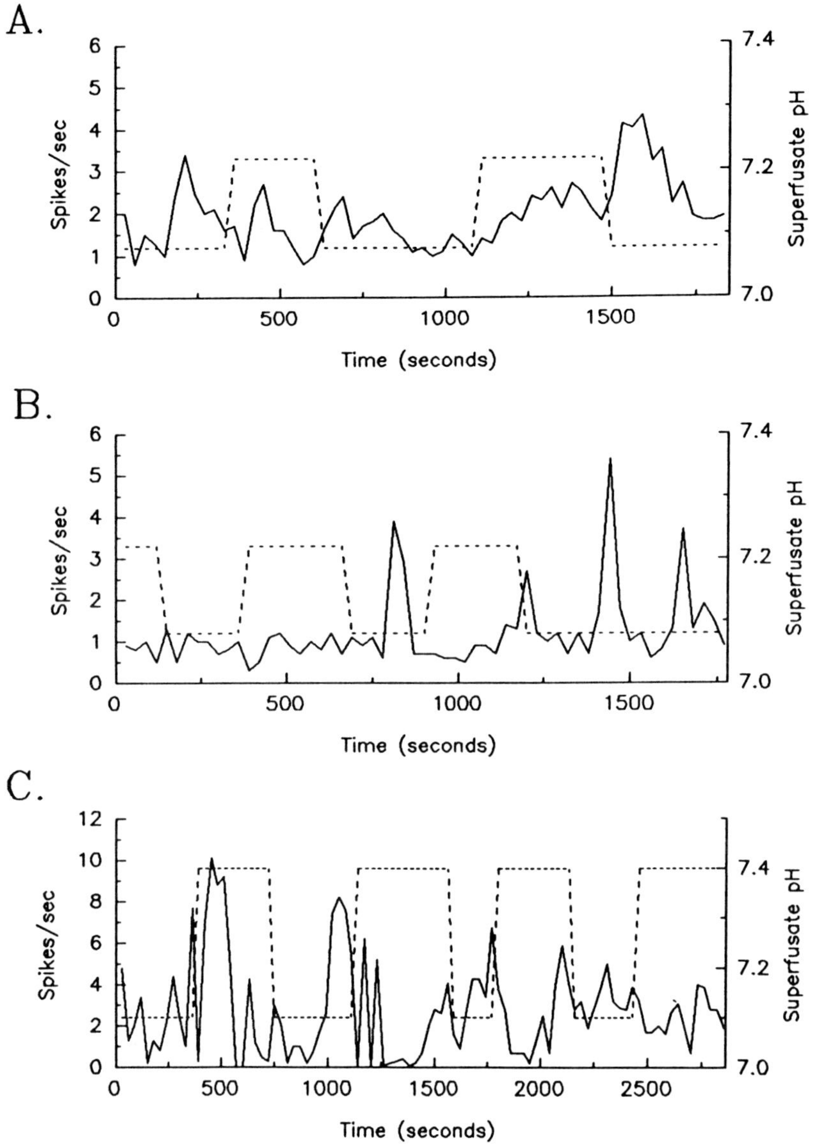

Figure 3 Spontaneous activity of ventral medullary (A) and dorsal medullary non-bursting (B) and bursting (C) neurons during alterations of the pH with changes in the HCO_3^- of the superfusate. In contrast to the responses seen with CO_2, changes in pH induced in this manner had no significant effect on the level of spontaneous neuronal activity. (From Ref. 6.)

Hanson (26) that the neuropil of some areas of the ventral medulla might be staining for carbonic anhydrase, the majority of previous attempts to localize carbonic anhydrase by histochemical techniques in neurons in tissue slices had failed primarily because of the very high background level of carbonic anhydrase in oligodendrocytes. For this reason we used the dissociated cell culture technique to enhance the sensitivity of the immunocytochemical methods, since culturing in this manner results in the growth of isolated neurons on a principally astrocytic substratum. In addition, we were interested in whether the presence of carbonic anhydrase in neurons would be specific to the ventral medulla or whether it might be present in other brain regions as well (27,28).

Dissociated cell cultures of the ventral and dorsal medulla, premotor cortex, and hypothalamus were prepared from 1-day-old rat pups. In addition, since oligodendrocytes and peripheral sensory neurons have high levels of the enzyme carbonic anhydrase, dissociated cell cultures of optic nerve and dorsal root ganglia were also prepared to serve as positive staining controls. The tissue was dissociated by subjecting it to gentle enzymatic treatment by incubation with trypsin and collagenase and dispersed mechanically by trituration. Cells were then plated on either poly-L-lysine or astrocyte-coated cover slips and incubated at 37° and 5% CO_2 in a supplemented DMEM culture medium. After 1 to 3 weeks in culture, the cultures were stained for the presence of carbonic anhydrase, using an immunocytochemical peroxidase antiperoxidase technique.

As shown in Table 1, immunocytochemical staining for carbonic anhydrase II in dissociated cell cultures of the ventral medulla revealed the presence of neurons in this region that are immunoreactive for this enzyme. However, only 25% of the ventral medullary neurons were immunoreactive for carbonic anhydrase, with the majority of neurons in this region not staining for this enzyme. In addition to the neuronal expression of carbonic anhydrase in the ventral medulla, 18% of the neurons in the dorsal medulla and 35% of the cortical neurons were immunoreactive for carbonic anhydrase; however, their staining tended to be less intense than the staining in the medulla. The greatest amount of immunoreactive neurons was observed in the hypothalamus, with 59% of

Table 1 Immunocytochemical Staining for Carbonic Anhydrase II in Dissociated Cell Cultures

	Ventral medulla	Dorsal medulla	Premotor cortex	Anterior hypothalamus	Optic nerve	Dorsal root ganglia
Positive	31	20	118	26	23	40
Negative	93	91	216	18	6	15
Percent positive	25	18	35	59	79	73

the hypothalamic neurons staining positive for carbonic anydrase. Since the hypothalamus is a brain region which subserves a number of sensory functions, including CO_2 sensitivity (8), this finding may suggest that carbonic anhydrase may be a general marker for sensory neurons or that CO_2 chemosensitivity may be a distributed function within the CNS.

The question remains, however, whether neurons immunoreactive for carbonic anhydrase are the neurons that are sensitive to changes in CO_2. To directly assess whether carbonic anhydrase is functionally important in the CO_2 response of chemosensitive neurons, we determined whether inhibition of carbonic anhydrase would alter the CO_2 chemosensitivity of dissociated rostral ventral lateral medullary neurons (22). Using the nystatin perforated patch technique (29), we studied the effect of a pulse of CO_2, which decreased the superfusate pH from 7.4 to 7.1 with a time constant of 20 s, on the repetitive firing pattern and membrane potential of ventral medullary neurons plated in a low-density dissociated culture before and after 10^{-5} *M* acetazolamide. In response to hypercarbic acidosis, we found a large, transient depolarization with excitation in 20%, a decrease in firing rate in 40%, and no change in firing rate or membrane potential in the remaining 40% of the spontaneously active neurons. Thus, differential responses of medullary neurons to CO_2 were observed in culture, suggesting that neurons retain the function of chemosensitivity in culture.

Inhibition of carbonic anhydrase with acetazolamide altered the dynamics of the responses of both CO_2-excited and CO_2-inhibited neurons.

In Figure 4, which illustrates the discharge frequency and membrane potential of a neuron excited by acidosis, prior to acetazolamide, a reduction in pH produced a rapid, transient 10-mV depolarization associated with a ninefold increase in firing rate that was followed by a sustained smaller depolarization and residual increase in firing frequency for 90 s. After acetazolamide, acidosis still produced the rapid, transient large depolarization; however, the residual depolarization and increased firing frequency were resolved within 30 s. Thus, inhibition of carbonic anhydrase reduced the duration of the sustained excitation.

In contrast, while acetazolamide altered the dynamics of the CO_2 response of inhibited neurons, it did so by prolonging the duration of inhibition. Figure 5 illustrates that prior to acetazolamide, acidosis reduced discharge rate, but this inhibition of firing frequency was resolved within 45 s. After acetazolamide, the acidosis-induced reduction in firing frequency persisted for 100 s.

These results suggest that chemosensitivity may include both excited and depressed neurons. In addition, since inhibition of carbonic anhydrase with acetazolamide slowed the recovery of depressed neurons and shortened the activation of excited neurons, the overall impact of carbonic anhydrase activity in the response to CO_2 may be to sustain activity within the chemosensitive network.

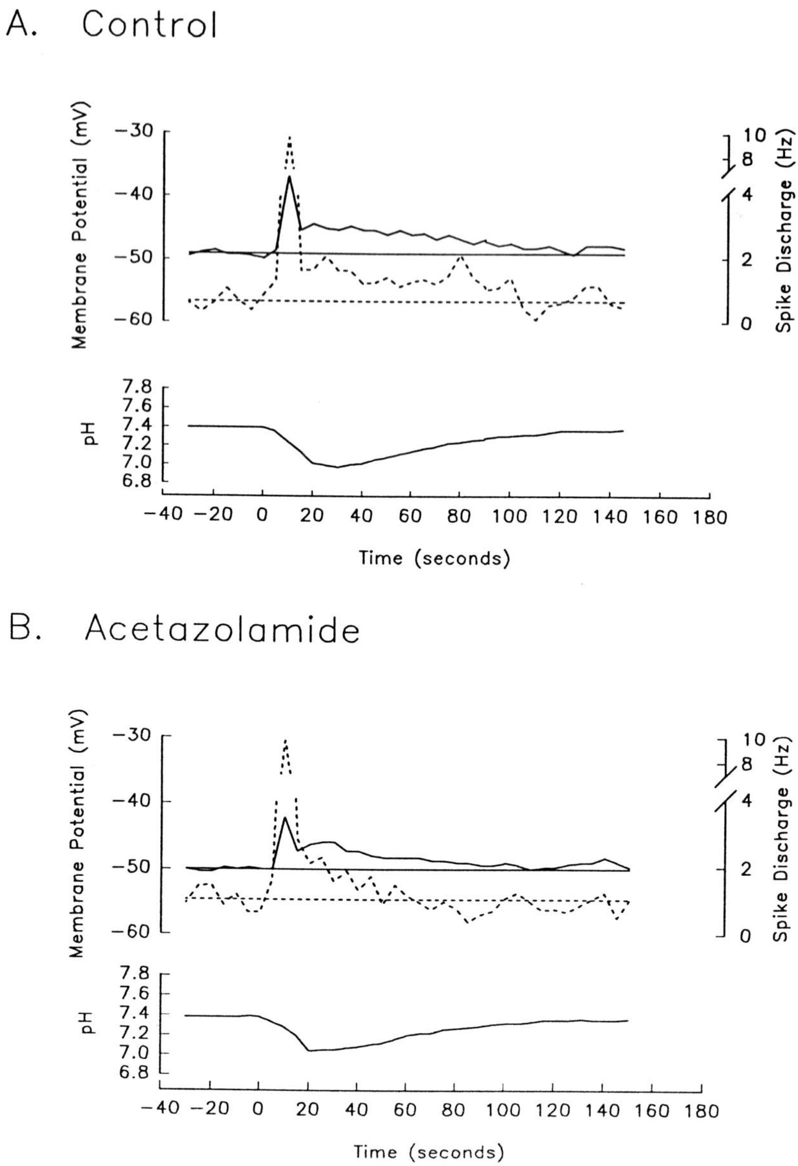

Figure 4 Representative responses of the resting membrane potential (solid line) and spike discharge (broken line) of a spontaneously active ventral medullary neuron that depolarized and increased firing frequency in response to a reduction in pH. (A) Response under control conditions; (B) Response in the presence of acetazolamide (.01 mM).

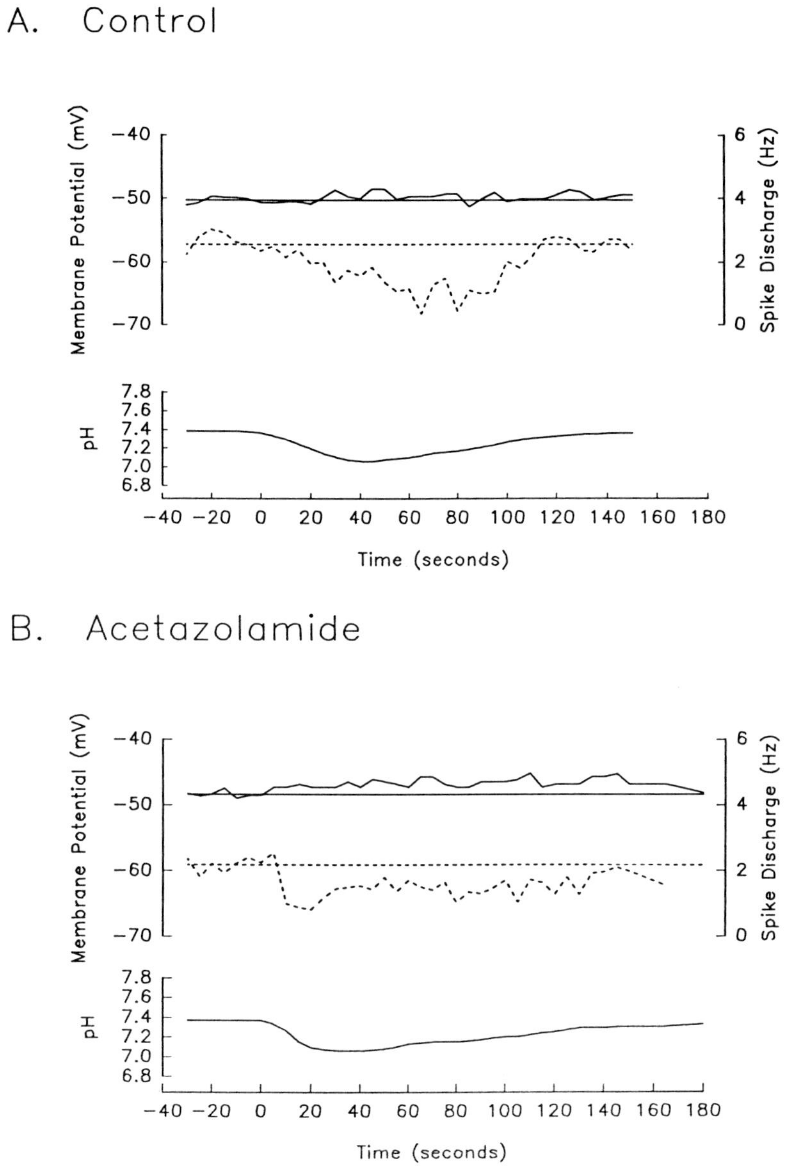

Figure 5 Representative responses of the resting membrane potential (solid line) and spike discharge (broken line) of a spontaneously active ventral medullary neuron that did not depolarize and decreased its firing frequency in response to a reduction in pH.

In summary, we have developed in vitro preparations in which a subpopulation of neurons retains its excitability to increased P_{CO_2}. In addition, we have demonstrated that a subpopulation of neurons in culture is immunoreactive for carbonic anhydrase and that inhibition of carbonic anhydrase activity alters the dynamics of the responses of CO_2-sensitive neurons, suggesting that these may be the CO_2-chemosensitive neurons. Finally, CO_2-sensitive neurons are not restricted to just the ventral medulla, further suggesting that CO_2 chemosensitivity may be a distributed function within the CNS.

References

1. Mitchell RA, Loeschcke HH, Massion WH, Severinghaus JW. Respiratory responses mediated through superficial chemosensitive areas on the medulla. J Appl Physiol 1963; 18:523–533.
2. Schläfke ME, Loeschcke HH. Lokalisation eines an der Regulation von Atmung und Kreislauf beteiligten Gebietes an der ventralen Oberflache der Medulla oblongata durch Kalteblockade. Pflügers Arch 1967; 297:201–220.
3. Schläfke ME, See WR, Loeschcke HH. Ventilatory response to alterations in H^+ ion concentration in small areas of the ventral medullary surface. Respir Physiol 1970; 10:198–212.
4. Trouth CO, Patrickson JW, Holloway JA, Wright LE. Neurophysiologic studies on superficial medullary chemosensitive area for respiration. Brain Res 1982; 246: 47–56.
5. Pokorski M. Neurophysiological studies on central chemosensor in medullary ventrolateral areas. Am J Physiol 1976; 230:1288–1295.
6. Neubauer JA, Gonsalves SF, Chou W, et al. Chemosensitivity of medullary neurons in tissue explant cultures. Neuroscience 1991; 45:701–708.
7. Dean JB, Lawing WL, Millhorn DE. CO_2 decreases membrane conductance and depolarizes neurons in the nucleus tractus solitarii. Exp Brain Res 1989; 76:656–661.
8. Tamaki Y, Nakayama T, Matsumura K. Effects of carbon dioxide inhalation on preoptic thermosensitive neurons. Pflügers Arch 1986; 407:8–13.
9. Dillon GH, Waldrop TG. In vitro responses of caudal hypothalamic neurons to hypoxia and hypercapnia. Neuroscience 1992; 15:941–950.
10. Richerson GB. CO_2 modulates pacemaker neurons in the medullary raphe and parapyramidal region of the rat in vitro (abstr). Soc Neurosci Abstr 1993; 19:1403.
11. Cherniack NS, von Euler C, Homma I, Kao FF. Graded changes in central chemoreceptor input by local temperature changes on the ventral surface of medulla. J Physiol (Lond) 1979; 287:191–211.
12. Nattie EE, Mills JW, Ou LC, St John WM. Kainic acid on the rostral ventrolateral medulla inhibits phrenic output and CO_2 sensitivity. J Appl Physiol 1988; 65:1525–1534.
13. Schläfke ME, See WR, Herker-See A, Loeschcke HH. Respiratory response to hypoxia and hypercapnia after elimination of central chemosensitivity. Pflügers Arch 1979; 381:2441–2448.

14. Eldridge FL, Kiley JP, Millhorn DE. Respiratory responses to medullary hydrogen ion in cats: Different effects of respiratory and metabolic acidosis. J Physiol (Lond) 1985; 358:258–297.
15. Shams H. Differential effects of CO_2 and H^+ as central stimuli of respiration in the cat. J Appl Physiol 1985; 58:357–364.
16. Teppema LJ, Barts PWJA, Folgering HTh, Evers JAM. Effects of respiratory and (isocapnic) metabolic arterial acid-base disturbances on medullary extracellular fluid pH and ventilation in cats. Respir Physiol 1983; 53:379–395.
17. Fukuda Y. Difference between actions of high PCO_2 and low HCO_3 on neurons in the rat medullary chemosensitive areas in vitro. Pflügers Arch 1983; 398:324–330.
18. Harada Y, Kuno M, Wang YZ. Differential effects of carbon dioxide and pH on central chemoreceptors in the rat in vitro. J Physiol (Lond) 1985; 368:679–693.
19. Hanson MA, Nye PCG, Torrance RW. The location of carbonic anhydrase in relation to the blood-brain barrier at the medullary chemoreceptors of the cat. J Physiol (Lond) 1981; 320:113–125.
20. Mishra J, Neubauer JA, Li JK-J, Edelman NH. Relationship of the dynamic characteristics of ventral medullary (Vm) pH and respiratory center output during CO_2 forcing (abstr). Fed Proc 1985; 44:1583.
21. Chou W, Neubauer JA, Geller HM, Edelman NH. Chemosensitivity of medullary and hypothalamic neurons in cell culture (abstr). Soc Neurosci Abstr 1990; 16:1061.
22. Martin AA, Neubauer JA. Effect of acetazolamide on CO_2 chemosensitivity of cultured medullary neurons (abstr). Soc Neurosci Abstr 1993; 19:1193.
23. Kazimierczak J, Sommer EW, Philippe E, Droz B. Carbonic anhydrase activity in primary sensory neurons: I. Requirements for the cytochemical localization in the dorsal root ganglion of chicken and mouse by light and electron microscopy. Cell Tissue Res 1986; 245:487–495.
24. Riley DA, Ellis S, Bain JLW. Ultrastructural cytochemical localization of carbonic anhydrase activity in rat peripheral sensory and motor nerves, dorsal root ganglia and dorsal column nuclei. Neuroscience 1984; 13:189–206.
25. Wong V, Barrett CP, Donati EJ, et al. Carbonic anhydrase activity in first-order sensory neurons of the rat. J Histochem Cytochem 1983; 31:293–300.
26. Ridderstrale Y, Hanson M. Histochemical study of the distribution of carbonic anhydrase in the cat brain. Acta Physiol Scand 1985; 124:557–564.
27. Sterbenz GC, Neubauer JA, Geller HM, Edelman NH. Carbonic anhydrase immunocytochemically localized in chemosensitive regions of the medulla (abstr). FASEB J 1988; 2:1295.
28. Neubauer JA. Carbonic anhydrase and sensory function in the central nervous system. In: Dodgson SJ, ed. The Carbonic Anhydrases: Cellular Physiology and Molecular Genetics. New York: Plenum Press, 1991:319–323.
29. Horn R, Marty A. Muscarinic activation of ionic current measured by a new whole-cell recording method. J Gen Physiol 1988; 92:145–159.

35

Age Differences in Choline Acetyltransferase, Tyrosine Hydroxylase, and Carbonic Anhydrase Neurons and Opiate Receptor Binding at Rat Ventral Brainstem

YI PAN,* C. OVID TROUTH, and ROBERT M. DOUGLAS

College of Medicine
Howard University
Washington, D.C.

PAULINE TING

Howard University Hospital
Washington, D.C.

I. Introduction

Inadequate central control of breathing has been proposed as a possible cause of idiopathic apnea in premature infants and in sudden infant death syndrome (SIDS). Gerhardt and Bancalari (1) proposed that a decreased slope of the CO_2 response curve and not impaired lung function is an important cause of apnea in preterm infants. Decreased CO_2 sensitivity has been detected in premature compared to full-term infants (2–4). Rigatto et al. (4) studied a group of premature infants and found that the CO_2 response increased with postnatal and gestational age. Krauss et al. (5) reported that premature infants demonstrated reduced ventilatory responses to increased inspired CO_2, and some infants failed to increase either tidal volume or respiratory rate during CO_2 rebreathing. Studies conducted in our laboratory (6) have shown decreased sensitivity of neurons at the ventral medullary surface (VMS) of newborn cats and rats (Fig. 1) compared with the adult in response to increased inspired CO_2 (P_ICO_2), as well as decreased cerebrospinal fluid (CSF) pH (pH 7.0), acetylcholine (ACh), and physostigmine applied topically to the VMS. The opiate antagonist naloxone augmented these responses.

**Present affiliation*: St. Louis University Hospital, St. Louis, Missouri.

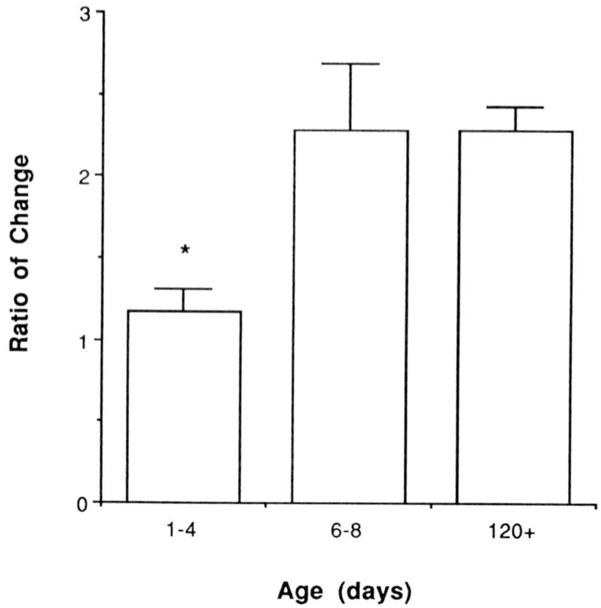

Figure 1 Graphical representation of age differences in the responses of brainstem chemosensitive neurons to inspired CO_2. Nonphasic pH-sensitive units were recorded extracellularly from the caudal respiratory chemosensitive area (cVMS) in three age groups of rats: 1 to 4 days = newborn; 6 to 8 days = young developing pup; 120+ days = adult. * = Significantly different from the adult and young developing pup groups ($P < 0.01$).

Significant contributions to the understanding of defective ventilatory control have resulted from studying the respiratory effects of the endogenous opioids. Endorphins have been shown to depress respiration (7), and antagonism of endorphin receptors by naloxone has produced respiratory stimulation in hypoxic newborn rabbits (8), hypercapnic fetal lambs (7), as well as normal adult cats (9). The specificity of the effects of naloxone on opiate receptors has been questioned, and evidence has been presented that its effects include antagonism of gamma amino butyric acid (GABA) (10). Administration of GABA to newborn cats has been used as a model for the production of apnea associated with SIDS (11). These findings suggest that the neuropharmacological manipulation of opiate levels during sleep states may in the future provide a realistic model for studying the neurochemical respiratory depression associated with SIDS. The present study, however, will characterize the development of the neurochemical respiratory drive in an anesthetized rat model. Several workers have provided evidence that cholinergic mechanisms are involved in central respiratory chemosensory control (12–17); however, the mechanisms involved in the reduced sensitivity of VMS neurons in the newborn human and animal remain unclear.

In the present study, we examined the hypothesis that the central CO_2 chemosensory structures located at the ventral brainstem might, from a functional point of view, not be fully developed at birth in normal animals and humans. Thus, prematurity and/or functional deficits in the central chemosensory mechanism might predispose to clinical conditions associated with decreased or absent chemosensitive responses to the CO_2 stimulus. We have examined the VMS for age related differences in the presence and distribution of choline acetyltransferase (ChAT), tyrosine hydroxylase (TH), and carbonic anhydrase (CA) as well as ventral brainstem opiate receptor binding in an attempt to identify and elucidate the possible mechanisms responsible for idiopathic apnea of prematurity and SIDS. We hypothesize that apnea and/or SIDS may result from defective development or dysfunction of the central cholinergic chemosensory VMS neurons regulating respiration in sleeping infants.

II. The Brainstem Respiratory Chemosensory Control Sites

Brainstem respiratory chemosensitive areas have been described on the ventrolateral medullary surface of the cat (18,19) and the rat (20–25). Three chemosensitive areas have been described (Fig. 2): a rostral (26), an intermediate, and a caudal (18,19,27), the caudal area being the most sensitive to electrical (18,28,29) and chemical (19,27) stimulation. Neuronal activity recorded at these VMS sites increased in response to increased inspired CO_2, and to CSF pH changes (19,27,30) in the cat.

Marino and Lamb (31) studied the activity of single cells in deeper regions of the medulla oblongata during CO_2 inhalation and extracellular iontophoresis of hydrogen ions in peripheral chemoreceptor–denervated cats. They concluded that cell bodies of respiratory neurons with respiratory periodicity were relatively insensitive to hydrogen ions. Further, they suggested that the paucity of hydrogen ion–sensitive cells (two) found deep within the medullary tissues did not support the notion that medullary hydrogen ion chemoreception is achieved primarily by structures located deep within the lower brainstem.

In the rat, Neubauer et al. (32) found chemosensitive neurons in cultured tissue explants from both the ventral and dorsal medulla. Fukuda and colleagues (33) compared the response of neurons in ventral and dorsal slices of the rat medulla oblongata to changing extracellular fluid pH. In the ventral medulla, they found that more than 50% of the neurons were excited by hydrogen ions, whereas in the dorsal medulla, low pH caused an inhibition of activity in most neurons and only a few were excited.

Sato et al. (34) identified sites of c-*fos* immunoreactivity at the rostral and caudal respiratory chemosensitive areas of rats breathing a CO_2-rich gas mixture (13% to 15% CO_2) in a test chamber for 1 h. Neurons at these VMS sites appeared identical to those localized by electrical (18,35) and chemical (19) stimulation in cats. According to Sato et al. (34), respiratory CO_2 chemoreceptor

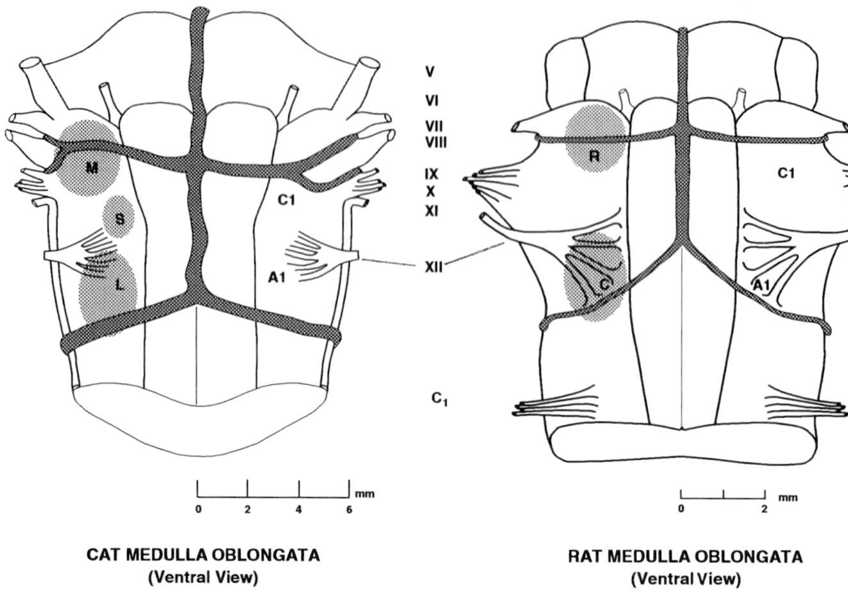

Figure 2 Ventral surface of medulla oblongata of the cat (left) and rat (right). In the cat, the three respiratory chemosensitive areas, described as "M" by Mitchell et al. (1963) and "S" and "L" by Schläfke et al. (1967 and 1970) are indicated on the left half of the diagram and represent the rostral intermediate and caudal areas, respectively. In the rat, the two chemosensitive zones described by Mitra et al. (1988) are indicated on the left half of the diagram by the symbols R and C, representing the rostral and caudal areas, respectively. In both right hemimedullas, the symbols C1 and A1 represent the vasopressor and vasodepressor areas, respectively. V–XII = cranial nerves.

neurons were identified as those which expressed c-*fos* in response to hypercapnia.

The chemosensitive areas at the VMS appear essential to the maintenance of the central drive to respiration (36). This has recently been demonstrated by topically applying kainic acid to the ventral brainstem surface in order to destroy the rat VMS neurons (23,24). Dillon et al. (25) found that the rostral VMS tonically inhibits the respiratory response to chemoreceptor stimulation and that the caudal VMS may be required for the maintenance of respiratory activity, because bilateral injection of kynurenic acid, an excitatory amino acid antagonist, caused apnea.

III. Age Differences in Brainstem Chemosensory Function

Whittaker et al. (30) studied the responses of newborn cats to CSF pH changes and compared them with the responses of the adult. The greatest sensitivity to

CSF pH changes was found in the adult group (age > 1 year). The least sensitivity was observed in the newborn group (1 to 6 days old), whereas young developing kittens (4 to 6 weeks old) exhibited an intermediate responsiveness. The chemosensory neuronal response to CSF pH changes in neonates older than 7 postnatal days was that which was characteristic for the young developing kitten group, whereas beyond the seventh week the response was one that was characteristic of the adult group. Thus, they identified specific milestones for the development of the central respiratory chemosensory responses to pH changes in the extracellular fluid (ECF).

During the perinatal developmental period, respiratory control and pattern exhibit special characteristics, such as apnea and periodic breathing; decreased sensitivity to CO_2 has been described in premature compared to full-term infants and in young compared to adult animals (2,4,5,37–39). The decreased CO_2 response may be involved in the apnea and periodic breathing seen in premature infants and in young animals. Therefore, maturational processes of the central respiratory chemosensitive apparatus may be responsible for the special characteristics of respiratory control and pattern during perinatal ontogeny.

The mechanisms of central respiratory chemoreception are unknown. Evidence suggests that CO_2 and bicarbonate ions independently influence the activity of neurons (32,40,41). According to Loeschcke (42), there is a different response to extracellular pH in metabolic acidosis from that seen in respiratory acidosis even when the ECF pH was reduced to the same extent (e.g., from 7.40 to 7.15). Arita et al. (43) assumed that the changes in ECF pH in response to inspired CO_2 may be mediated by carbonic anhydrase (CA). Developmental abnormalities in CA activity could indeed affect the chemoreceptor signal. If the hydrogen ion is the adequate stimulus to the central chemoreceptor, as Loeschcke (42) suggested, the further question is that of the location of the receptors; whether at the cell membrane, within the cytoplasm, or both. Unfortunately, there is no answer to this question at the present time.

IV. Age Differences in Distribution of Carbonic Anhydrase in Rat Ventral Brainstem

The mechanism of central chemosensitivity involving carbonic anhydrase (CA), the enzyme required for rapid hydration of CO_2, has been investigated with the aid of CA inhibitors (43–48), histochemistry, and immunocytochemistry (32, 49,50). Hanson and Ridderstrale (49) reported that a small area close to the ventral brainstem surface was characterized by a high staining of the neuropil. We investigated the location and distribution of CA at known brainstem chemosensitive sites of the newborn and adult rat utilizing histochemical and immunocytochemical techniques (Fig. 3). Nineteen adult and 26 newborn rats (1 to 15 days old) were used in the study.

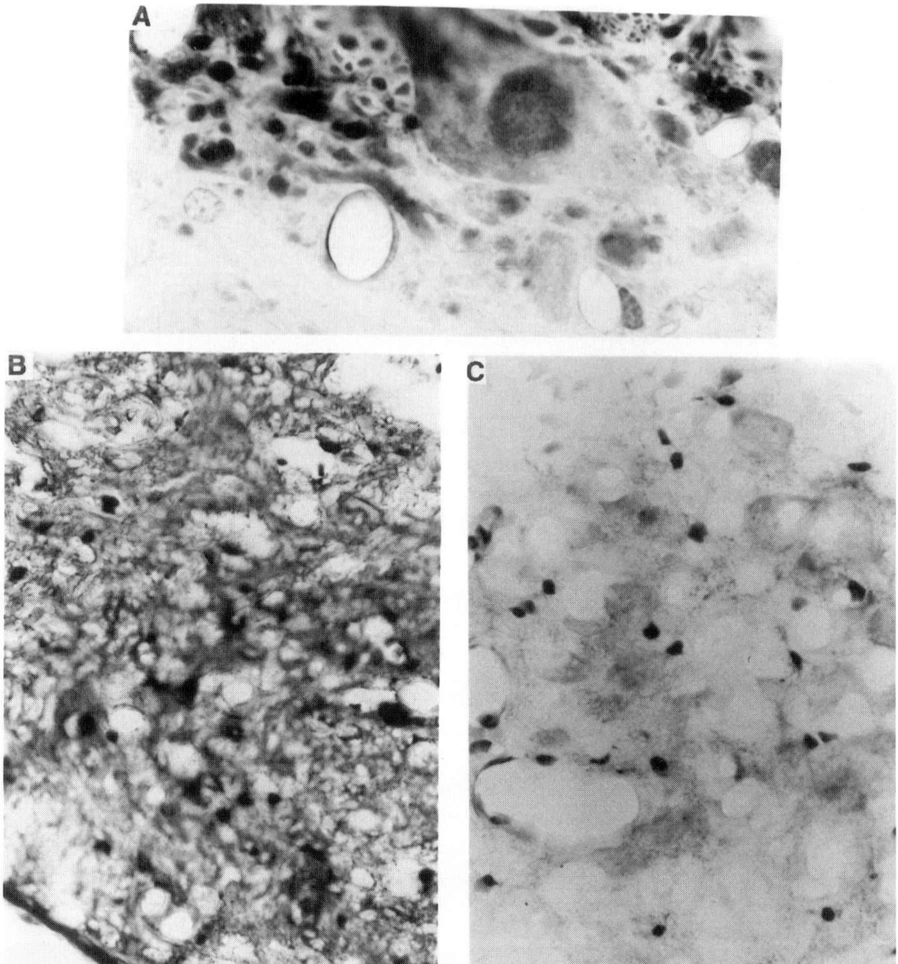

Figure 3 Histochemical and immunocytochemical localization of carbonic anhydrase (CA) in the rat caudal ventral medulla. (A) Carbonic anhydrase immunocytochemical reaction product in neurons and surrounding glia at the cVMS: adult rat, ×100. (B) Histochemical localization of CA in the adult rat cVMS. Neutral red counterstain, ×20. Positive staining of glia and neurons in the cVMS. (C) Histochemical localization of carbonic anhydrase in the newborn rat cVMS. Neutral red counterstain, ×20. Positive staining of glia. B and C were stained simultaneously.

A. Histochemical Staining

The Hansson method (50) was utilized for the CA histochemical staining. Rat brains were perfused via the aorta with 4% paraformaldehyde and 2.5% glutaraldehyde in 0.1 *M* phosphate buffer (pH 7.4), fixed 4 h at 4°C, and thoroughly rinsed. Frozen cryostat sections (10 μm) were then floated on the surface of the incubation medium (1.75 m*M* $CoSO_4$, 11.7 m*M* KH_2PO_4, 157 m*M* $NaHCO_3$, and 53 m*M* H_2SO_4) for 10 min, washed with 0.1 *M* phosphate buffer (pH 5.9), and then incubated for 2 min in 0.5% ammonium sulfide. The sections were counterstained with 1% neutral red.

Histochemical staining deposits cobalt and phosphate at the sites of carbonic anhydrase activity, which stains for soluble and membrane-bound CA. In the adult, CA was found to be ubiquitous throughout the brainstem and the CA reaction was distributed homogeneously throughout the ventral medulla. However, most of the staining occurred in the glia, which showed an intense reaction to the cobalt phosphate. At the VMS, the neurons were overlaid by a thick fibrous network of tissue that contained the CA reaction product.

Control sections from the same rat processed in the same incubation medium with the CA blocker acetazolamide (2×10^{-4} *M*) served as negative controls. Also, the 1- and 15-day-old rat medullas were incubated in the same medium together with the adult rat sections. CA-positive reactions appeared in glial and nerve fibers of the 15-day-old rat, but the staining intensity was much less than in the adult. In the 1-day-old newborn rat, however, a positive CA reaction was found in a few glia, whereas it was negative in the neurons.

B. Immunocytochemistry

Paraffin sections (4 μm) were used for the immunocytochemical staining of CA. After deparaffination, the sections were washed with phosphate buffered saline (PBS), incubated with 3% normal goat serum in PBS containing 0.3% Triton X-100 (TX) for 30 min, the incubated in a 1:200 dilution of goat anti-human CA-II serum (Green Cross Corporation, Osaka 541, Japan) at 4°C overnight. An ABC Kit and DAB Substrate Kit (Vectastain, Burlingame, CA) were used to stain the primary antibody after washing three times in PBS/0.1% TX. The sections were incubated with biotinylated anti-goat IgG for 30 min, washed with PBS/0.1% TX twice, with PBS once, and then incubated with 0.3% hydrogen peroxide solution for 30 min to abrogate endogenous peroxidase. The sections were washed twice with PBS and once with PBS/0.1% TX, then incubated with avidin-HRP for 60 min. After washing with PBS three times, the sections were incubated in 3,3′-diaminobenzidine (DAB) substrate for 5 min and counterstained with 0.5% hematoxylin.

Immunostaining revealed a significant number of CA-positive cells in the adult brainstem with positive peroxidase activity. In the CNS, CA is located primarily in glial cells (oligodendrocytes and astrocytes) throughout the brainstem

(51,52). Glial CA may be involved in the maintenance of pH and electrolyte balance in the CNS. Fukuda et al. (33) suggested that the glia at the ventral surface layer of the medulla play a role in the process of hydrogen ion–elicited changes in neuronal activity. Thus, glial elements that contain CA may be involved in chemoreception within the CNS. CA-positive neurons were located mainly in the ventral medulla, notably in the VMS-chemosensitive areas; however, they were also identified in the cochlear nuclei of the adult rat. We found some neurons at the VMS that showed a positive reaction with the anti-CA II antibody. We also found multipolar or oval-shaped neurons and some glia in the caudal ventral medulla that showed a positive CA reaction, whereas other glia were CA-negative. By comparison, the neurons in the dorsal medulla of the same section did not demonstrate CA immunoreactivity, even though CA-positive glia were seen around them. Some CA-positive neurons were found to lie very superficially at the VMS (100 μm from the surface) and to extend into the lateral reticular nuclear region. By comparison, neurons in the same VMS region of the 2- and 8-day-old rat pup medullae did not show a CA-positive reaction. The precise function of these neurons is not known; however, one might speculate that CA-positive VMS neurons are involved in the chemosensory mechanism. Two types of CA were found at the VMS: soluble CA and membrane-bound CA, and since chemoreceptors may be at the cell membrane, in the cytoplasm (42), or both, CO_2-hydrogen ion conversion could be expected to occur both on the membrane and intracellularly.

Age-related differences in CA distribution and activity have been studied by several authors. Ashby and Schuster (53) investigated the relationship between brain CA and functional maturity in rat, rabbit, cat, dog, guinea pig, and human infant. They found that there was no CA activity in the rat brain during the first 3 days after birth. Between the fifth and twelfth days a very small amount of the enzyme was found in the cerebrum. Roussel et al. (54), using newborn to 30-day-old rats and indirect immunoperoxidase techniques, indicated that the appearance of a positive reaction product to the antiserum anti-CA II was already observed in both glial types (oligodendrocytes and astrocytes) only 6 to 8 days after birth. Before this time, the positive reaction, occurring in rat brain, could be noticed only in the choroid plexus. They postulated that the appearance of CA II in oligodendrocytes and astrocytes may be interpreted as an index of the functional maturation of these cells. Kumpulainen and Korhonen (55,56) found that the brains from 1-day-old mice exhibited no CA immunoreaction, and 8-day-old mice showed a noticeable reaction only in a very small group of cells in the caudal part of the brainstem. At 15 days of age, a brown immunoperoxidase staining was evident in most parts of the brainstem. However, the earliest detectable CA II was reported to have been detected in the transitory glial cells of embryonic mouse CNS by Benjelloun et al. (57) using immunocytochemical methods.

CA enzyme activity was measured by Ghandour et al. (58). Utilizing radioimmunoassay and immunohistofluorescence techniques, they found that CA II increased considerably by the tenth postnatal day in the rat cerebellum. Saparstein et al. (59,60) observed marked changes in total brain CA activity with age. Although most CA is soluble at birth, the adult distribution of the soluble and membrane-bound forms is well established by the tenth day of age. Both forms of CA in the subcortical region of the rat are fully expressed by the sixteenth postnatal day. In mice, the activity increased with age from birth to 24 days as determined by the electrometric method (61).

From previous studies, it might be postulated that neuronal CA might be involved in the ability of neurons to transduce the CO_2 or hydrogen ion stimulus into a neuronal signal that could be the basis of respiratory CO_2 chemoreception. The presence of CA in the brain can also provide a sensitive assessment of the level of brain maturation.

Clinical and physiological findings have indicated that VMS neurons show age-related differences in their responses to inspired CO_2 and that the responsiveness of newborns is greatly reduced when compared with that of adults. One might therefore speculate that CA is essential to the integrity of the CO_2 chemosensory apparatus. Prematurity or functional deficits in the central chemoreceptor mechanism therefore would predispose to clinical conditions associated with decreased or absent chemosensitive responses to the CO_2 stimulus.

V. Age Differences in Cholinergic and Catecholaminergic Neurons in Rat Respiratory Chemosensitive Areas

The ventral brainstems of newborn and adult rats were examined by immunocytochemical techniques for the presence and distribution of cholinergic and catecholaminergic neurons. Cholinergic neurons on the VMS are sensitive to changes in the hydrogen ion concentration of the cerebrospinal fluid and to increased inspired carbon dioxide (62–64). Five adult and 9 newborn rats (1 to 15 days old) were used in the study. Alternate cryostat sections (10 μm) were incubated in primary rabbit anti-rat ChAT (Chemicon International Inc., Temecula, CA) antibodies at 1:200 dilution or 1:500 dilution of mouse anti-rat TH serum (Incstar Corporation, Stillwater, MN) overnight at 4°C. After being washed in PBS and 0.1% TX, the sections were incubated in biotinylated anti-rabbit or anti-mouse IgG, washed with PBS and 0.1% TX, incubated in 0.3% hydrogen peroxide, washed with PBS, then incubated with avidin-HRP using an ABC kit; after being washed with PBS, they were incubated in DAB substrate. ChAT- or TH-positive neurons were stained a brown color. Control sections were incubated with normal goat serum to substitute for the primary antibodies, anti-ChAT or anti-TH.

A. Choline Acetyltransferase Neurons in the Rat Medulla

Cross sections of the adult rat medulla oblongata were processed for immunocytochemical staining of ChAT. Endogenous ChAT was detected as a diffuse brown reaction product within neuronal perikarya. Multipolar or oval-shaped ChAT-positive neurons were found in the caudal chemosensitive area and in the nucleus reticularis lateralis. ChAT-positive neurons were also found to be distributed widely throughout the brainstem in the adult rat (Figs. 4 and 5). Superficial ChAT-positive neurons at the VMS were small, whereas larger, multipolar ChAT-positive neurons were observed in regions deeper than 100 μm from the VMS. By comparison, neurons in the hypoglossal nuclei at the dorsal brainstem were oval-shaped and of larger diameter. No immunocytochemical reaction was identified in the cell bodies of control sections that were processed without primary antibodies.

The brains of newborn 1-day-old rat pups that were processed simultaneously under identical conditions as those of adult rats were, by contrast, almost devoid of ChAT neurons in the VMS chemosensitive areas. Lightly positive ChAT neurons began to appear at about the fourth day in the VMS and in the hypoglossal nuclei.

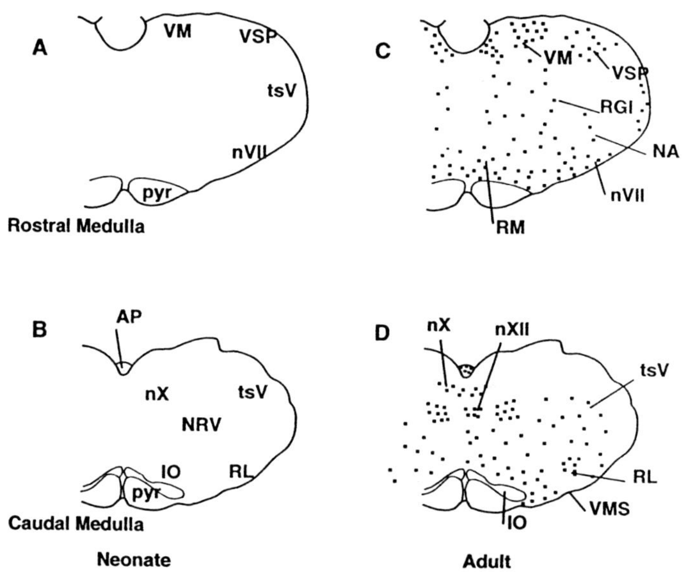

Figure 4 Distribution of choline acetyltransferase (ChAT)–positive cells in the neonatal and adult rat medulla oblongata. Dots represent the location of neurons labeled for ChAT. Note the absence of immunoreactive cells in the neonate. AP = area postrema; IO = inferior olive; NA = nucleus ambiguus; NRV = medullary reticular nucleus, Ventral part; pyr = pyramidal tract; RGI = gigantocellular reticular nucleus; RL = lateral reticular nucleus; RM = nucleus raphe magnus; tsV = nucleus of the spinal trigeminal tract; VMS = ventral medullary surface; VSP = spinal vestibular nucleus; nX = dorsal vagal nucleus; nXII = hypoglossal nucleus; nVII = facial motor nucleus.

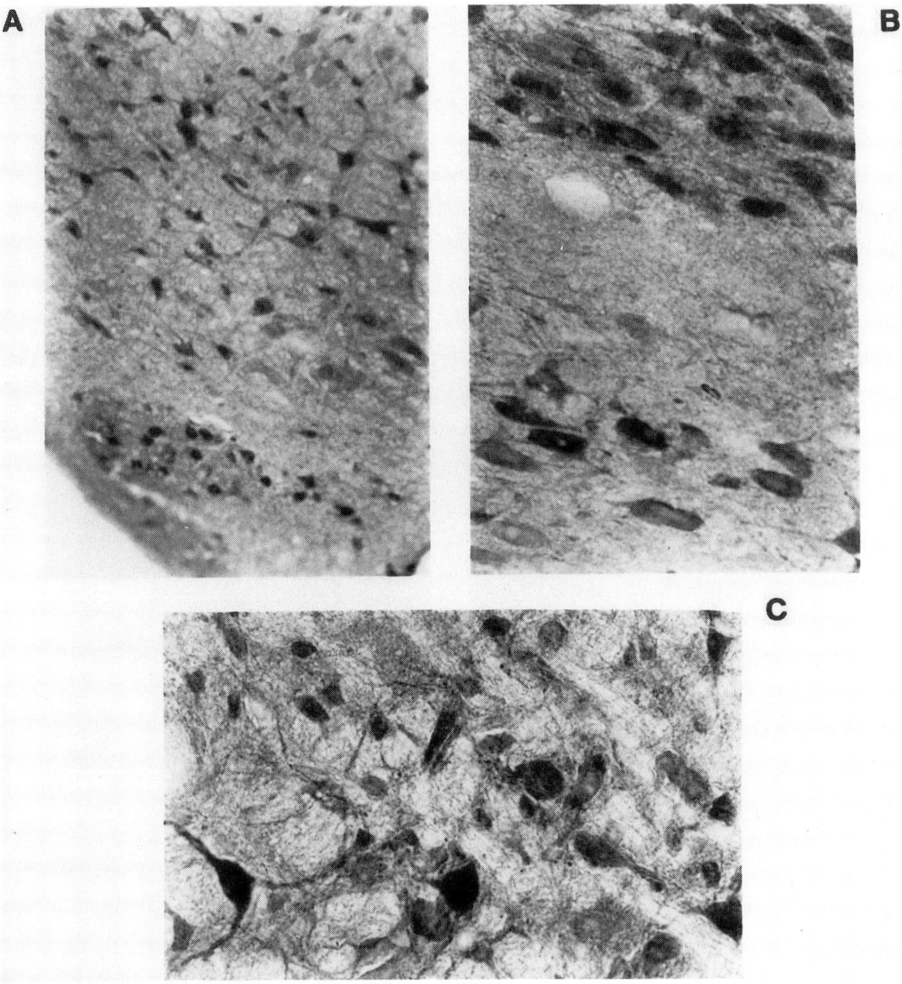

Figure 5 Distribution of choline acetyltransferase (ChAT) immunoreactivity in the adult rat caudal ventral medulla: (A) ChAT-positive neurons in the caudal chemosensitive area and the lateral reticular nucleus, ×20. (B) ChAT positive neurons in the caudal medulla and in the lateral corner of the IO, ×40. (C) ChAT-positive neurons in nucleus ambiguus.

B. Tyrosine Hydroxylase Neurons in Rat Medulla

Bipolar TH-positive neurons with fibers running parallel to the VMS were identified lateral to the hypoglossal nerve. In the areas near the nucleus reticularis lateralis, TH-positive neurons were star-shaped, with fibers running ventrally to the VMS and deeper toward the dorsal medulla. These neurons appear to be similar to the monoamine-containing neurons described by Dahlstrom and Fuxe (65) and the dopamine-B-hydroxylase (DBH)-positive neurons of Armstrong et al. (66) in the A1 area. In the caudal VMS, groups of three to four bipolar, TH-positive neurons were found fairly close together near the hypoglossal rootlets.

Multipolar or star-shaped TH-positive neurons were also found grouped together lateral to the nucleus reticularis lateralis, ventral to the nucleus ambiguus, and dorsal to the central canal in the region of the NTS and nuclei com-

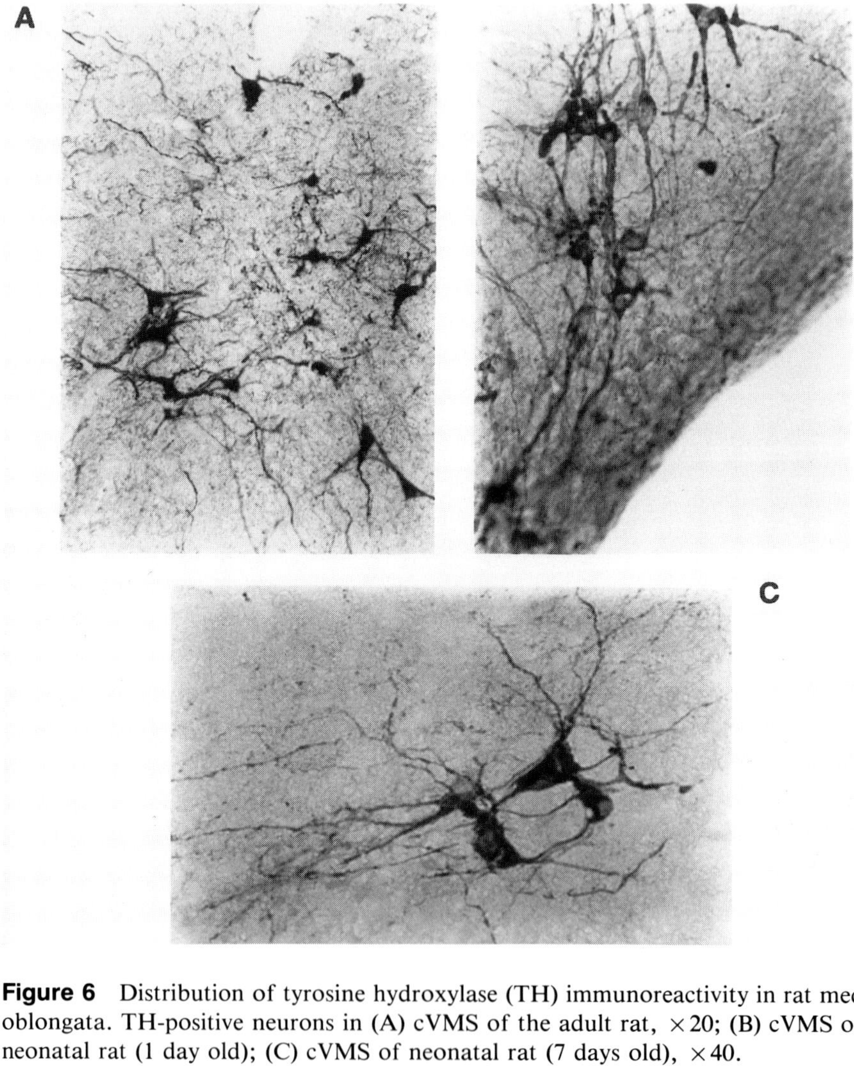

Figure 6 Distribution of tyrosine hydroxylase (TH) immunoreactivity in rat medulla oblongata. TH-positive neurons in (A) cVMS of the adult rat, ×20; (B) cVMS of the neonatal rat (1 day old); (C) cVMS of neonatal rat (7 days old), ×40.

misuralis. A few bipolar TH-positive neurons at the VMS were situated immediately lateral to the hypoglossal nerve rootlet with some TH-positive nerve fibers. TH-positive neurons have the same distribution in the newborn 1- to 4-day-old pup as in the adult (Figs. 6 and 7).

In the rat, the newborn period (1 to 4 days) is characterized by a paucity of cholinergic neurons (Fig. 4) at the VMS and reduced sensitivity to inspired carbon dioxide (Fig. 1). However, catecholaminergic neurons were already developed at the VMS in the 1- to 4-day-old pups and were similar in distribution to the adult (Figs. 6 and 7). These results suggest that age-related differences in the responsiveness to physiological stimuli might be referable to the integrity or development of cholinergic neurons.

Both cholinergic (ChAT) and catecholaminergic (TH) neurons have been demonstrated in the chemosensitive areas of the adult rat in the present investigation. Development of cholinergic neurons is believed to occur later than

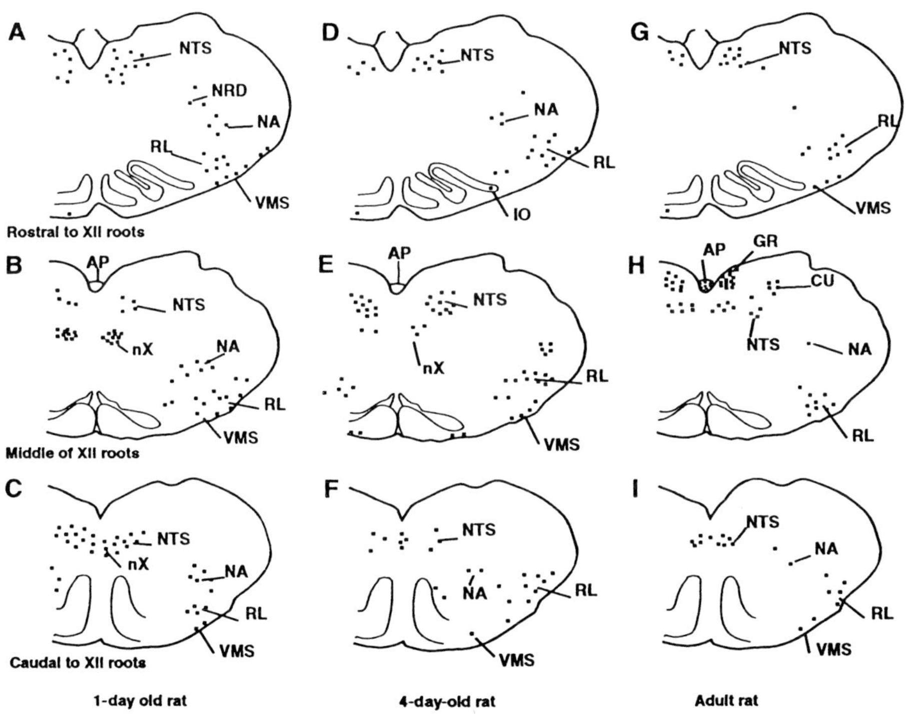

Figure 7 Distribution of tyrosine hydroxylase (TH) immunoreactive neurons in rat caudal medulla in the region of the hypoglossal rootlets. AP = area postrema; CU = cuneate nucleus; GR = gracile nucleus; IO = inferior olive; NA = nucleus ambiguus; NRD = medullary reticular nucleus, dorsal part; NTS = nucleus of the solitary tract; RL = lateral reticular nucleus; VMS = ventral medullary surface; nX = dorsal vagal nucleus.

that of catecholaminergic neurons (67). The present study demonstrates a difference in the timetable between the appearance of cholinergic and catecholaminergic neurons in the rat chemosensitive areas. It was not possible to determine whether ChAT- or TH-positive neurons described in these regions were involved in respiratory chemosensitivity, vasomotor function, or both; however, on the basis of numerous studies at the VMS, it seems that acetylcholine modulates respiratory chemoreception (13,16,17,62,63,68–72) and the catecholamines are involved in vasomotor function (73–76).

VI. Endogenous Opioids in Regulation of Breathing

A. Radioactive Receptor Binding Assay

Opiate receptors (μ, δ, and κ) were assessed by sequential displacement of overall binding of the mixed opioid agonist-antagonist. A total of 88 newborn rats (1, 3, 7, 10, and 15 days old) and 5 adult rats (older than 120 days) were sacrificed by an overdose of sodium pentobarbital (IP injection). Samples taken from the ventral part of the medulla were frozen immediately by dry ice and stored at $-70°C$. In order to reduce assay variability, the samples from the different age groups (1, 3, 7, 10, and 15 days old and older than 120 days) were assayed simultaneously under identical conditions. Samples were homogenized at 4°C in 0.32 *M* sucrose and 25 m*M* HEPES buffer (pH 7.4) and centrifuged at 4°C at 18,000 rpm (Beckman L8-*M* Ultracentrifuge) for 15 min. The pellets were resuspended in the same buffer and centrifuged again. The final concentration of the samples was 10 mg/mL. [^{3}H]-Diprenorphine (0.5 n*M*) (Amersham) was added into the samples with 10^{-6} *M* μ-opioid receptor agonist DAGO (tyr-D-ala-gly-NMe-phe-gly-ol), δ-opioid receptor antagonist ICI 174 864, κ-opioid receptor agonist U 50488H (Cambridge Research Biochemicals) or 10^{-5} *M* unselective opioid receptor antagonist naloxone. The mixtures were incubated at room temperature for 45 min, and then filtered through Whatman GF/B glass fiber paper using a Brandel Cell Harvester. The filters were transferred to scintillation vials, and 0.5 mL absolute ethanol and 4 mL Bechman EP Ready-Solv were added. The vials were counted in a Beckman LS 7800 beta scintillation counter at an efficiency of 50%. Each membrane sample protein content was determined by Peterson's modification of Lowry's method.

Data were analyzed by one-way analysis of variance (ANOVA) followed by Fisher PLSD multiple comparison test with the use of a StatView 512+ statistics program (Brainpower Inc., Calobasas, CA) on a Macintosh II computer. The changes were considered significant with $P < 0.05$.

B. Age Differences in the Distribution of Opiate Receptor Binding Sites at the Ventral Brainstem

The opioid system has been studied extensively in the brain; however, in the respiratory-related brainstem, most of the research has been focused on the nucleus tractus solitarius, which is known to be the dorsal group of the respiratory center. In this investigation, the preparation involved only the ventral medulla, which contained the ventral chemosensitive areas and the ventral group of the respiratory center. Brain opiate receptor binding at μ, δ, and κ subsites in the ventral half of medulla was measured by the radiolabeled ligand binding technique (77). Figure 8 summarizes the data on opiate receptor binding expressed in femtomoles per milligram of protein (mean $\pm$ SEM) in each age group at the ventral medulla. The total binding of unselective opiate receptor antagonist as well as binding at subsites μ and κ in the ventral medulla gradually increased after birth. During the first 7 to 10 days after birth, the total receptor binding reached a peak and then declined to adult levels. There was a significant difference between the 3- to 10-day group and the 1-day, 15-day, and adult rat groups ($P < 0.05$) in total receptor binding. The μ-receptor binding had the same pattern as the total receptors in that it reached a peak in the 7-day-old rat and then declined.

The 3- to 7-day groups were significantly different from the 1-day, 5-day, and adult rat groups ($P < 0.05$) in μ-receptor binding. The κ-receptor binding reached its highest density at 10 days. There were significant differences in κ binding between the 1- and 10-day-old pups and the 15-day-old rat ($P < 0.05$) and between the 7- to 10-day-old pups and adult rat groups ($P < 0.05$). The δ-receptor binding was low in all age groups; however, it peaked at 3 days and there was a significant difference between the 3-day-old pups and other age groups ($P < 0.05$). Of the three opiate receptor subtypes, the μ-receptor binding was the highest and revealed the same change over time as the total opiate receptor binding. It represented 72% to 75% of the total receptor binding during

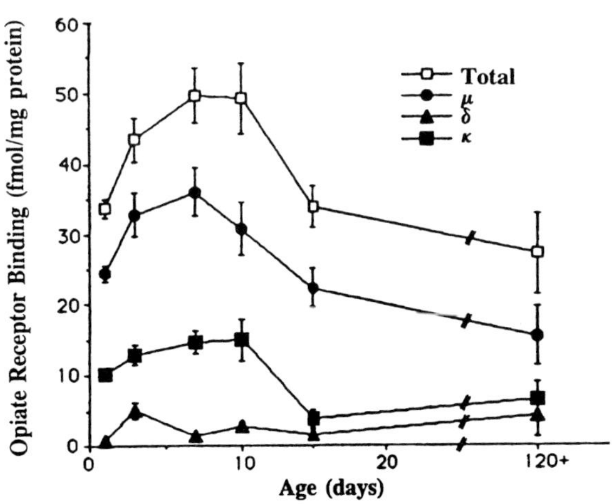

Figure 8 Opiate receptor binding with increasing age in the ventral half of the rat medulla oblongata.

the first 1 to 7 days of life and about 62% to 65% during the next 10 to 15 days; and then it decreased to 56% of total receptor binding in the adult rat. Therefore, at the ventral medulla, the highest percentage of μ-receptor binding sites were found during the newborn period. The μ receptor is the opiate receptor known to be involved in respiratory control, of which there are two subtypes, $\mu 1$ and $\mu 2$. It is believed that the $\mu 1$ receptor stimulates respiration and that the $\mu 2$ receptor depresses respiration (78). In the present study, it was not possible to differentiate between the $\mu 1$ and $\mu 2$ receptor subtypes. Leibstein et al. (79) found β-endorphin, the ligand that binds to the μ receptor, in the caudal VMS. In this laboratory, Bada (80) found that naloxone applied topically to the caudal chemosensitive area at the VMS augmented the ventilatory response to inspired CO_2, suggesting that the endogenous opiate system may be involved in the CO_2-chemosensory drive to breathing at least at the cVMS. Immunocytochemical evaluation of the cVMS by light and electron microscopy revealed the presence of β-endorphin immunoreactivity distributed widely in both the superficial type I and II neurons and the neurovascular elements (80). Also, Trouth et al. (64) found that naloxone applied topically to the cVMS during hypercapnic breathing increased minute ventilation. This effect could have been mediated through the $\mu 2$ receptor.

The highest binding of total opiate receptor was found during 3 to 10 postnatal days in this study. The μ and κ receptors had a similar binding characteristic during this period, whereas the δ-receptor binding remained low in all age groups. A similar finding has been reported by Xia and Haddad (81) using autoradiography. They demonstrated that the brainstem was essentially a μ-receptor region, the density reaching a maximum at 10 days.

The β-endorphin and enkephalin systems are also well represented in respiratory-related brain regions during perinatal development. Moss (82) measured β-endorphinlike immunoreactivity in the plasma of infants. β-Endorphin-like immunoreactivity levels in <1 to 4-day-old infants were enhanced and then fell to adult levels on the fifth day of life. They also found that β-endorphin antagonism enhanced respiratory function at rest as well as during hypercapnia and hypoxia in the newborn (1- to 4-day-old) piglet and the intermediate group (10- to 18-day-old) but not in the 25- to 42-day-old piglet (83). Another primary opioid peptide, enkephalin, was also found in high levels during development (84). Gingras-Leatherman et al. (85) microdissected the rabbit brain and found that methionine-enkephalin was higher in prenatal and early postnatal life than in the adult in nuclei involved in respiratory control (nucleus tractus solitarius, nucleus ambiguus, nucleus parabrachialis, and nucleus reticulogigantocellularis). By contrast, methionine-enkephalin levels in forebrain nuclei were lower than or equal to the adult level during development.

In the present investigation, the high binding at μ receptors in the ventral medulla has the same profile as reported by other investigators (81,83–85) during the early postnatal period. The high level of endogenous opioids functions as a strong inhibitory system in respiratory control during perinatal life. It would not be unreasonable to assume that the opioid system might be involved in some abnormalities of respiratory control, such as periodic breathing and apnea in the newborn.

VII. Morphological Structures at Ventral Brainstem Control Sites

Several studies have been undertaken to identify structures at the ventral brainstem that are involved in cardiorespiratory regulation. Among the earliest reports are those of Petrovicky (86), who described neurons embedded within the thickened marginal glia (TMG) in the cat. These findings have been confirmed by Dermietzel (87) in the intermediate area (S) and by Trouth et al. (88) in the caudal area (L). Loeschcke (89) called these TMG neurons "type II" cells. We (90) studied the VMS of the rat ultrastructurally and found type II cells within the TMG. Filiano et al. (91) considered these type II cells to be cytologically homologous to neurons of the arcuate nucleus.

Following electrical stimulation and electrolytic marking of the site of greatest sensitivity in the caudal area (L) at the VMS, Trouth et al. (92) identified a group of large multipolar neurons with maximum density at a depth 150 to 200 μm below the surface between the VMS and the ventral side of the inferior olive. Loeschcke (89) called these neurons "type I" cells. Utilizing basic principles of comparative cytoarchitecture and computer-aided reconstructions of serially sectioned brainstems, Filiano et al. (91) confirmed the existence of type I cells in the cat and equated them with the human ventrolateral medullary nucleus conterminalis (hCT). In both cats and human infants, they found that these neurons were separated into ventrolateral and dorsomedial cell clusters, the ventrolateral cell group providing the only histological characteristics that distinguishes the caudal (L) area from intermediate (S) and rostral (M) chemosensitive fields. They further postulated that the type I cell group represented the ventrolateral portion of the feline N. raphe pallidus, which has been shown to project to the ventral group of respiratory center neurons (93).

A *third group* of neurons in the chemosensitive area was found closely apposed to the walls of blood vessels that penetrated from the VMS (Fig. 9), perhaps an ideal location for sensing the blood chemistry. These cells have been designated neurovascular elements, and—like type I and II cells—are endowed with a rich array of synaptic contacts. Neurovascular relationships of this type were also observed in the rostral chemosensitive area (area M) but at greater depths (1 to 2 mm) from the ventral surface. None were found in the intermediate area. Neurons of similar shapes and sizes were also found dispersed within the marginal glia close to the blood vessels. The cells in close contact with the blood vessel walls are reminiscent of the neurovascular elements described by Scheibel and Scheibel (94) and Scheibel et al. (95), who showed specialized neurovascular relationships of subnuclei of the raphe complex of

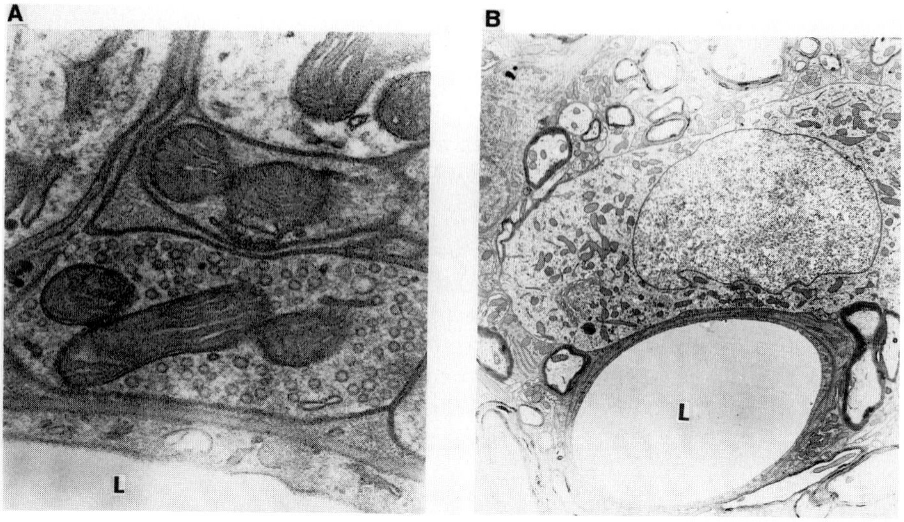

Figure 9 Electron micrographs of neurovascular elements in the cVMS of the adult rat medulla oblongata. (A) Axon terminal with mitochondria and vesicles beneath the blood vessel, abutting the basal lamina, ×30,000. (B) Cell body with large nucleus and mitochondria beneath wall of blood vessel.

the brainstem reticular substance of mice, rats, cats, monkeys, and humans. They also described neuronal cells of the nuclei raphe pontis and linealis rostralis that are characteristically organized along the walls of the raphe blood supply. They suggested that the role of these neurovascular elements might include chemosensor function, neurosecretion, and mechanoreceptor or vasomotor function. Certainly, their possible significance in a chemoreceptor role must be considered, due to the vast accumulation of data supporting chemosensitivity to extracellular pH changes in this region (13,20,21,26,96). Since the relationship between blood and the interstitial fluid surrounding the neurovascular elements reflects only a simple barrier formed by thin-walled, media-deficient intracerebral blood vessels (97), a relatively rapidly equilibrating mechanism between blood CO_2 and brainstem interstitial fluid might be postulated (89,98). On the other hand, type I and II chemosensitive neurons responding to CSF-pH changes represent a more complex system, because blood CO_2 or pH changes equilibrating the CSF pool must first overcome barriers interposed by the choroid plexus. It is also conceivable that the chemical composition of the CSF pool varies with distance from the choroid plexus (99) and with the CSF changes that must equilibrate the interstitial fluid in the vicinity of the chemosensitive neurons. In this case, a slowly equilibrating mechanism might be operative. These two systems operating together would ensure a rapid yet maintained response to blood CO_2 and acid-base changes (90,98). It is interesting that Gelfand

and Lambertsen (100) identified two functionally distinct central chemoreceptor responses to inspired CO_2 with different response rates (10 and 89 s) that were separate and distinct from peripheral chemoreceptor responses. They proposed that two anatomically distinct CO_2 chemosensory cell groups with two different blood perfusion rates might account for the central response to inhaled CO_2.

Schläfke et al. (96) systematically mapped neurons within the ventral medullary surface layer from caudal to rostral. Some neurons of the superficial layer formed part of the nucleus reticularis lateralis (in the caudal area), the nucleus paragigantocellularis (in intermediate and rostral areas) and the nucleus trapezoidalis (in the rostral area).

The ventral brainstem is highly sensitive to a variety of pharmacological agents that influence the respiratory and cardiovascular systems. Histochemical, autoradiographic, and immunocytochemical techniques have revealed the presence of many endogenous substances [e.g., ACh, serotonin, CA, TH, substance P, GABA, opiates, glutamate, neuropeptide Y, atrial naturetic peptide (ANP), brain naturetic peptide (BNP), and phenylethanol-amine-*N*-methyltransferase (PNMT)] that appear to be involved in the regulation of these two systems and perhaps a number of other bodily functions. What do these substances have to do with the chemosensor function and cardiorespiratory regulation? The apparently diffuse nature of the structures containing the reaction products identified by these techniques does not allow for easy answers regarding the specific identity of the chemosensor and its role in cardiorespiratory regulation. Thus the mechanisms underlying the basic transduction processes in central CO_2 chemoreception and cardiorespiratory regulation in the adult and newborn remain obscure mainly because the precise location of the chemoreceptors and other controlling elements in these regulatory systems await definition.

VIII. Summary

Central cardiorespiratory regulation has been ascribed to neurons located at the ventral brainstem. Age differences in the responsiveness of these systems to physiological stimuli have been reported in humans (3,4,101) and the cat (30), wherein the newborn shows decreased sensitivity to inspired CO_2 and CSF-pH changes. The primary goal of this investigation was to study age-related differences in the cardiorespiratory mechanism in normal neonatal (1- to 4-day-old), young developing (6- to 8-day-old) and adult rats ($>$ 120 days) and to assess the significance of these differences in relation to idiopathic apnea in premature infants and to sudden infant death syndrome (SIDS). We investigated the distribution of carbonic anhydrase, choline acetyl transferase, and tyrosine hydroxylase neurons and opiate receptor binding in the adult rat ventral brainstem and compared them with that of the newborn. (a) *Carbonic anhydrase (CA)*: The intensity of staining in the medulla increases with age, as shown by both histochemical and immunocytochemical methods. The CA-positive reaction product was found to be ubiquitous throughout the adult rat brainstem, particularly within the glial elements. CA-positive neurons, however, were found at the VMS within the caudal area and extending into the lateral reticular nucleus. By comparison, neurons in the same VMS region of 8-day-old and 2-day-old rat medulla did not show the CA-positive reaction. The caudal VMS is a site where CO_2 respiratory chemosensors are presumed to be localized. These results correlate well with the physiological findings, which have indicated that ventral brainstem neurons show age-related differences in responsiveness to inspired CO_2 and that the responsiveness of the newborn is greatly reduced when compared with that of the adult. One might speculate that CA is essential to the integrity of the CO_2 chemosensory function. (b) Multipolar *choline acetyltransferase* (ChAT)–positive (cholinergic) neurons identified immunocytochemically were prominent in the adult, and conspicuously absent in the newborn, appearing first on the fourth day. (c) Bipolar *tyrosine hydroxylase* (TH)–positive (catecholaminergic) neurons were found distributed throughout the ventral medulla and at the VMS by immunocytochemical techniques even in 1-day-old pups. Adrenergic and noradrenergic systems at this site may be involved in cardiovascular regulation. (d) *Opiate receptor* binding: The total binding of the unselective opiate antagonist and binding at μ and κ subsites in the ventral medulla as measured by tritiated diprenorphine gradually increased after birth, reaching a peak during the first 7 to 10 days and then declining to adult levels. However, the δ-receptor binding was low in all age groups. At the ventral medulla, the highest percentage of μ-receptor binding sites was found during the newborn period (1 to 7 days postnatal).

The newborn period is characterized by a greatly reduced sensitivity to inspired CO_2 (hypercapnia) and a paucity of CA- and cholinergic-positive neurons. This suggests that age differences in the responsiveness of the chemosensory mechanism to physiological stimuli might be referable to the integrity or development of cholinergic neurons and CA function. The observed increase in opiate receptor binding at the ventral brainstem, especially the μ subtype, might be important in modulation of cardiorespiratory function. The development of central cardiorespiratory neuronal responsiveness may therefore involve physiological adjustments associated with neurohumoral and cytoarchitectural maturation.

Acknowledgment

This work was supported by the National Institutes of Health Grant-MBRS 2-S06-RR-08016. The authors are grateful to Ms. Deepa Soni and Dr. DeLee F. Minner for copyediting and preparing the draft of the manuscript.

References

1. Gerhardt T, Bancalari E. Ventilatory response to CO_2 in premature infants with apnea. Pediatr Res 1979; 13:534.
2. Frantz ID, Adler SM, Trach BT, Taeusch WH. Maturational effects on respiratory responses to carbon dioxide in premature infants. J Appl Physiol 1976; 41:41–45.
3. Rigatto H, De la Torre Verduzco R, Cates DB. Effects of O_2 on the ventilatory response to CO_2 in preterm infants. J Appl Physiol 1975; 39:896–899.
4. Rigatto H, Bready JP, de la Torre Verduzco R. Chemoreceptor reflexes in preterm infants: 1. The effect of gestational and postnatal age on the ventilatory response to inhalation of 100% and 15% oxygen. Pediatrics 1975; 55:604–613.
5. Krauss AN, Klain DB, Waldman S, Auld AM. Ventilatory response to carbon dioxide in newborn infants. Pediatr Res 1975; 9:46–50.
6. Whittaker JAC. Age related differences in the responsiveness of brainstem respiratory CO_2-chemosensory neurons. PhD dissertation, Howard University, Washington, DC, 1988.
7. Moss IR, Friedman E. Beta-endorphin effects on respiratory regulation. Life Sci 1978; 23:1271–1276.
8. Chernick V, Madansky DL, Lawson EE. Naloxone decreases the duration of primary apnea with neonatal asphyxia. Pediatr Res 1980; 14:357–359.
9. Lawson EE, Waldrop PG, Eldridge FL. Naloxone enhances respiratory output in cats. J Appl Physiol 1979; 47:1105–1111.
10. Sawynok J, Pinsky C, La Bella FS. Minireview on the specificity of Naloxone as an opiate antagonist. Life Sci 1979; 25:1621–1631.
11. Wennergren G, Wennergren M. Respiratory effects elicited in newborn animals in the central chemoreceptors. Acta Physiol Scand 1980; 108:309–311.
12. Metz B. Brain acetylcholinesterase and respiratory reflex. Am J Physiol 1958; 192: 101–105.
13. Mitchell RA, Loeschcke HH, Severinghaus JW, et al. Regions of respiratory chemosensitivity on the surface of the medulla. Ann NY Acad Sci 1963; 109:661–681.
14. Trzebski A. Effect of stimulation of the medullary chemosensitive areas in the conscious "encephale isole" preparation in cats, in central interaction between respiratory and cardiovascular control systems. Satellite symposium, XXVIIth International Congress of Physiological Sciences, Berlin, 1977.
15. Trzebski A, Mukulski A, Przybyszewski A. Influence of the chemosensitive areas on the ventral surface of the medulla oblongata on the EEG activity in encephale isole preparations in cats. Proc Int Union Physiol Sci 1977; 13:764.
16. Dev NB, Loeschcke HH. A cholinergic mechanism involved in the respiratory chemosensitivity of the medulla oblongata in the cat. Pflügers Arch 1979; 379:29–36.
17. Dev NB, Loeschcke HH. Topography of the respiratory and circulatory responses to acetylcholine and nicotine on the ventral surface of the medulla oblongata. Pflügers Arch 1979; 379:19–27.
18. Loeschcke HH, De Lattre J, Schläfke ME, Trouth CO. Effects on respiration and circulation of electrically stimulating the ventral surface of the medulla oblongata. Resp Physiol 1970; 10:184–197.
19. Schläfke ME, See WR, Loeschcke HH. Ventilation response to alterations of H^+-ion concentration in small areas of the ventral medullary surface. Resp Physiol 1970; 10:198–212.

20. Hori T, Roth GI, Yamamoto WS. Respiratory sensitivity of rat brain-stem surface to chemical stimuli. J Appl Physiol 1970; 28:721–724.
21. Fukuda Y, Honda Y. pH-sensitive cells in the ventral-lateral surface of the rat medulla oblongata. Nature 1975; 256:317–318.
22. Laha PK, Nayar U, China GS, Singh B. Carbon dioxide sensitivity of the central chemosensitive mechanisms. Pflügers Arch 1977; 367:241–247.
23. Martin RL, Sinclair JD. Kainic acid on the rat ventral medullary surface depresses hypoxic and hypercapnic ventilatory responses. Respir Physiol 1990; 80:55–70.
24. Douglas RM, Bernard DG, Pan Y, et al. Effect of kainic acid on brainstem respiratory chemosensitive zones (abstr). Neurosci Abstr 1991; 17:788.
25. Dillon GH, Welsh DE, Waldrop TG. Modulation of respiratory reflexes by an excitatory amino acid mechanism in the ventrolateral medulla. Respir Physiol 1991; 85:55–72.
26. Mitchell RA, Loeschcke HH, Massion WH, Severinghaus JW. Respiratory responses mediated through superficial chemosensitive areas on the medulla. J Appl Physiol 1963; 18:523–533.
27. Trouth CO, Patrickson JW, Holloway JA, Wright LE. Neurophysiological studies on superficial medullary chemosensitive area for respiration. Brain Res 1982; 246: 47–56.
28. Trouth CO. Lokalisation oberflachlicher und tiefer auf die Atmung wirkender Strukturen in der Medulla oblongata durch elektrischem Reiz. Pflügers Arch 1969; 307:16–17.
29. Berndt J. A superficial substrate on the ventral surface of the medulla oblongata influencing respiration. Pflügers Arch 1973; 339:135–152.
30. Whittaker JA, Trouth CO, Pan Y, et al. Age differences in responsiveness of brainstem chemosensitive neurons to extracellular pH changes. Life Sci 1990; 46:1699–1705.
31. Marino RL, Lamb TW. Effects of CO_2 and extracellular H^+ iontophoresis on single cell activity in the cat brain stem. J Appl Physiol 1975; 38:688–695.
32. Neubauer JA, Gonsalves SF, Chou W, et al. Chemosensitivity of medullary neurons in explant tissue cultures. Neuroscience 1991; 45:701–708.
33. Fukuda Y, Honda Y, Schläfke ME, Loeschcke HH. Effect of H^+ on the membrane potential of silent cells in the ventral and dorsal surface layer of the rat medulla in vitro. Pflügers Arch 1977; 376:229–235.
34. Sato M, Severinghaus JW, Basbaum AI. Medullary CO_2 chemoreceptor neuron identification by c-fos immunocytochemistry. J Appl Physiol 1992; 73:96–100.
35. Trouth CO, Loeschcke HH, Berndt J. Topography of the respiratory responses to electrical stimulation in the medulla oblongata. Pflügers Arch 1973; 339:153–170.
36. Berndt J, Berger W, Trouth CO. Respiratory and circulatory effects of 100 mEq/l potassium or 2% procaine in the cerebrospinal fluid of cats. Pflügers Arch 1970; 321:346–363.
37. Jansen AH, Russell BJ, Chernick V. Respiratory effects of H^+ and dinitrophenol injections into the brainstem subarachnoid space of fetal lambs. Can J Physiol Pharmacol 1975; 53:726–733.
38. Gingras IL, O'Donnell KJ, Blum RF. Maternal cocaine addiction and fetal behavioral state: I. A human model for the study of sudden infant death syndrome. Med Hypoth 1990; 33:227–230.

39. Cohen G, Xu C, Henderson-Smart D. Ventilatory response of the sleeping newborn to CO_2 during normoxic rebreathing. J Appl Physiol 1991; 71:168–174.
40. Harada Y, Kuno M, Wang YZ. Differential effects of carbon dioxide and pH on central chemoreceptors in the rat in vitro. J Physiol 1985; 368:679–693.
41. Fukuda Y. Difference between actions of high pCO_2 and low $[HCO_3^-]$ on neurons in the rat medullary chemosensitive areas in vitro. Pflügers Arch 1983; 398:324–330.
42. Loeschcke HH. Central chemosensitivity and the reaction theory. J Physiol 1982; 332:1–24.
43. Arita H, Ichikawa K, Kuwana S, Kogo N. Possible locations of pH-dependent central chemoreceptors: Intramedullary regions with acidic shift of extracellular fluid pH during hypercapnia. Brain Res 1989; 485:285–293.
44. Hanson MA, Nye PCG, Torrance RW. The location of carbonic anhydrase in relation to the blood-brain barrier at the medullary chemoreceptors of the cat. J Physiol 1981; 320:113–125.
45. Adams JM, Johnson NL. Inhibiting carbonic anhydrase in brain tissue increases the respiratory response to rebreathing CO_2. Brain Res 1990; 519:23–28.
46. Leyen SA, Averill DB, Guertzenstein PG. Cardiorespiratory effects induced by acetazolamide on the ventromedullary surface of the cat. J Physiol 1990; 421:171–184.
47. Teppema LJ, Rochette F, Demedts M. Effects of acetazolamide on medullary extracellular pH and pCO_2 and on ventilation in peripherally chemodenervated cats. Pflügers Arch 1990; 415:519–525.
48. Coates EL, Li A, Nattie EE. Ventrolateral medulla (VLM) carbonic anhydrase (CA) inhibition: Changes in phrenic nerve activity (PNA) and localization of central chemoreception sites (abstr). Neurosci Abstr 1991; 17:104.
49. Hanson MA, Ridderstrale Y. Histochemical localization of carbonic anhydrase in cat brain. J Physiol 1982; 336:45.
50. Hansson HPJ. Histochemical demonstration of carbonic anhydrase activity. Histochemie 1967; 11:112–128.
51. Langui D, Delaunoy JP, Ghandour MS, Sensenbrenner M. Immunohistochemical demonstration of both carbonic anhydrase isoenzyme II and glial fibrillary acidic protein in some immature rat glial cells in primary culture. Neurosci Lett 1985; 60:151–156.
52. Cammer W, Tansey FA. Carbonic anhydrase immunostaining in astrocytes in the rat cerebral cortex. J Neurochem 1988; 50:319–322.
53. Ashby W, Schuster EM. Carbonic anhydrase in the brain of the newborn in relation to functional maturity. J Biol Chem 1950; 184:109–116.
54. Roussel G, Delaunoy J, Nussbaum J, Mandel P. Demonstration of a specific localization of carbonic anhydrase C in the glial cells of rat CNS by an immunohistochemical method. Brain Res 1979; 160:47–55.
55. Kumpulainen T, Korhonen LK. Immunohistochemical demonstration of carbonic anhydrase. Histochemistry 1978; 58:183–192.
56. Kumpulainen T, Korhonen LK. Immunohistochemical localization of carbonic anhydrase isoenzyme C in the central and peripheral nervous system of the mouse. J Histochem Cytochem 1982; 30:283–292.
57. Benjelloun S, Delaunoy JP, Gomes D, et al. Early expression of carbonic anhydrase II (CAII) in transitory glial cell of the developing murine nervous system. Dev Neurosci 1986; 8:150–159.

58. Ghandour MS, Vincendorn G, Gombos G. Carbonic anhydrase and oligodendroglia in developing rat cerebellum: A biochemical and immunohistological study. Dev Biol 1980; 77:73–83.
59. Sapirstein VS, Flynn C, Lees MB. Developmental changes in carbonic anhydrase and adenylate cyclase in quaking mice. Brain Res 1980; 185:373–383.
60. Sapirstein VS, Strocchi P, Wesolowski M, Gilbert JM. Characterization and biosynthesis of soluble and membrane-bound carbonic anhydrase in brain. J Neurochem 1983; 40:1251–1261.
61. Guillaume D, Grisar T, Vergniolle-Burette M. Glial contribution to seizure: Carbonic anhydrase activity in epileptic mammalian brain. Epilepsia 1991; 32:10–15.
62. Fukuda Y, Loeschcke HH. A cholinergic mechanism involved in the neuronal excitation by H^+ in the respiratory chemosensitive structures of the ventral medulla oblongata of rats in vitro. Pflügers Arch 1979; 379:125–135.
63. Trouth CO, Whittaker JA, Archer PW, et al. Cholinergic mechanism at brainstem respiratory chemosensitive areas. Neurotoxicology 1993; 14:459–468.
64. Trouth CO, Bada F, Pan Y, et al. Naloxone application to the ventrolateral medulla enhances the respiratory response to inspired carbon dioxide. Life Sci 1991; 49:193–200.
65. Dahlstrom A, Fuxe K. Evidence for the existence of monoamine-containing neurons in the central nervous system. Acta Physiol Scand 1964; 232(suppl):1–55.
66. Armstrong DM, Ross CA, Pickel VM, et al. Distribution of dopamine-noradrenaline-, and adrenaline-containing cell bodies in the rat medulla oblongata: Demonstrated by the immunocytochemical localization of catecholamine biosynthetic enzymes. J Comp Neurol 1982; 212:173–187.
67. Lagercrantz H, Srinivasan M. Development and function of neurotransmitter/modulator systems in the brainstem. In: Hanson MA, ed. The Fetal and Neonatal Brain Stem. New York: Cambridge University Press, 1991:1–20.
68. Fukuda Y, Honda Y, Schläfke ME, Loeschcke HH. Effect of H^+ on the membrane potential of silent cells in the ventral and dorsal surface layer of the rat medulla in vitro. Pflügers Arch 1978; 376:229–235.
69. Nattie EE, Mills JW, Ou LC. Pirenzepine prevents diethyl pyrocarbonate inhibition of central CO_2 sensitivity. J Appl Physiol 1988; 65:1962–1966.
70. Nattie EE, Li A. Ventral medulla sites of muscarinic receptor subtypes involved in cardiorespiratory control. J Appl Physiol 1990; 69:33–41.
71. Nattie EE. Central respiratory chemoreceptors. In: Haddad GG, Farber JP, eds. Developmental Neurobiology of Breathing. New York: Dekker, 1991:341–371.
72. Monteau R, Morin D, Hilaire G. Acetylcholine and central chemosensitivity: In vitro study in the newborn rat. Respir Physiol 1990; 81:241–254.
73. Ross CA, Ruggiero DA, Joh TH, Park DH. Rostral ventrolateral medulla: Selective projections to the thoracic autonomic cell column from the region containing C1 adrenaline neurons. J Comp Neurol 1984; 228:168–185.
74. Granata AR, Ruggiero DA, Park DH, et al. Brainstem area with C1 epinephrine neurons mediates baroreflex vasodepressor responses. Am J Physiol 1985; 248: H547–H567.
75. Granata AR, Kitai ST. Intracellular analysis in vivo of different barosensitive bulbospinal neurons in the rat rostral ventrolateral medulla. J Neurosci 1992; 12:1–20.
76. Ruggiero DA, Ross CA, Anwar M, et al. Distribution of neurons containing phenylethanolamine N-methyltransferase in medulla and hypothalamus of rat. J Comp Neurol 1985; 239:127–154.

77. Wood MS, Traynor JR. [3H]-Diprenorphine binding to x-sites in guinea-pig and rat brain: Evidence for apparent heterogeneity. J Neurochem 1989; 53:173–178.
78. Siren AL, Feuerstein G. The opioid system in circulatory control. N Physiol Sci 1992; 7:26–30.
79. Leibstein AG, Dermietzel R, Willemberg IM, Pauschert R. Mapping of different neuropeptides in the lower brainstem of the rat: With special reference to the ventral surface. J Auton Nerv Syst 1985; 14:299–313.
80. Bada FJ. Identification of a naloxone-sensitive site in the brainstem influencing respiratory CO_2 chemosensitivity. PhD Thesis, Howard University, Washington, DC, 1990.
81. Xia Y, Haddad GG. Ontogeny and distribution of opioid receptors in the rat brainstem. Brain Res 1991; 549:181–193.
82. Moss IR. Endorphins in fetal and neonatal breathing. In: Euler CV, Lagercrantz H, eds. Neurobiology of the Control of Breathing. New York: Raven Press, 1987: 119–124.
83. Moss IR, Runold M, Dahlin I, et al. Respiratory and neuroendocrine responses of piglets to hypoxia during postnatal development. Acta Physiol Scand 1987; 131: 533–541.
84. Bayon A, Shoemaker WJ, Bloom FE, et al. Perinatal development of the endorphin- and enkephalin-containing systems in the rat brain. Brain Res 1979; 179: 93–101.
85. Gingras-Leatherman JL, McNamara MC, Hong JS, Lawson EE. Development of methionine-enkephalin in microdissected areas of the rabbit brain. Brain Res 1985; 336:73–80.
86. Petrovicky R. Über die Glia marginalis und oberflachliche Nervenzellen im Hirnstamm der Katze. Z Anat Entw 1968; 127:221–231.
87. Dermietzel R. Central chemosensitivity, morphological studies. In: Loeschcke HH, ed. Acid-Base Homeostasis of the Brain Extracellular Fluid and Respiratory Control System. Stuttgart: PSG Publishing, 1976:52–66.
88. Trouth CO, Odek-Ogunde M, Holloway JA. Morphological observations on the superficial medullary CO_2-chemosensitive areas. Brain Res 1982; 246:35–45.
89. Loeschcke HH. Central nervous chemoreceptors. In: Respiratory Physiology. MTP International Review of Science, London: Butterworths 1974; 2:167–196.
90. Pan Y, Bada F, Bernard DG, et al. Age differences in responses of rat brainstem respiratory chemosensory neurons (abstr). Neurosci Abstr 1989; 14:627.
91. Filiano JJ, Choi JC, Kinney HC. Candidate cell populations for respiratory chemosensitive fields in the human infant medulla. J Comp Neurol 1990; 293:448–465.
92. Trouth CO, Loeschcke HH, Berndt J. Histological structures in the chemosensitive regions on the ventral surface of the cat's medulla oblongata. Pflügers Arch 1973; 339:171–183.
93. Smith JC, Morrisson DE, Ellenberger HE, et al. Brainstem projections to the major respiratory neuron populations in the medulla of the cat. J Comp Neurol 1989; 281:69–96.
94. Scheibel ME, Scheibel AB. Does this nucleus raphe pontis have chemoreceptors or neuroendocrine function? Soc Neurol 1974; 409.
95. Scheibel ME, Tomiyasu U, Scheibel A. Do raphe nuclei of the reticular formation have neurosecretory or vascular sensor functions? Exp Neurol 1975; 47:316–329.

96. Schläfke ME, Kille JF, Folgering H, et al. Breathing without central chemosensitivity. In: Umbach W, Koepchen HP, eds. Central Rhythmic and Regulation. Stuttgart: Hippokrates, 1974:97–104.
97. Frederickson RG, Low FN. Blood vessels and tissue space associated with the brain of the rat. Am J Anat 1969; 125:123–146.
98. Patrickson JW. Localization of central respiratory chemoreceptors. PhD dissertation, Howard University, Washington, DC, 1978.
99. Pappenheimer J, Miller AA, Goodrich C. Sleep promoting effects of cerebrospinal fluid from sleep-deprived goats. Proc Nat Acad Sci USA 1967; 58:513–517.
100. Gelfand R, Lambertsen CJ. Dynamic respiratory response to abrupt changes of inspired CO_2 at normal and high PO_2. J Appl Physiol 1973; 35:903–913.
101. Rigatto H, Bradt JP. Periodic breathing and apnea in preterm infants: I. Evidence for hypoventilation possibly due to central respiratory depression. Pediatrics 1972; 50:202–218.

36

Anatomic and Neurochemical Studies of the Human Ventral Medulla in Early Life

Observations Relevant to the Sudden Infant Death Syndrome

HANNAH C. KINNEY, J. J. FILIANO, A. PANIGRAPHY, and L. A. RAVA

Harvard Medical School and Children's Hospital
Boston, Massachusetts

W. F. WHITE

Pfizer, Inc.
Groton, Connecticut

I. Introduction

This report highlights work in our laboratory related to the possible role of the ventral medulla in the sudden infant death syndrome (SIDS). SIDS is the leading cause of postneonatal infant mortality in the United States today, causing approximately 7000 deaths/year (1). It is defined as the sudden death of an infant under 1 year of age that remains unexplained after a thorough case investigation, including a complete autopsy (1). Typically, SIDS is a sleep-related sudden death. Ninety percent of the deaths occur within the first 6 months of life, with a peak at 2 to 4 months. This unique and consistent age distribution suggests that, while the causes of SIDS may be heterogeneous, there is likely a final common pathway that is related to a critical period in development (2). We have proposed a triple-risk model for the pathogenesis of SIDS in which sudden death results from the intersection of three key factors: (a) an infant made vulnerable by an underlying abnormality in the control of homeostasis during sleep; (b) a critical developmental period in homeostasis and state of maturation, i.e., the first 6 months of life, as the newborn makes the transition to extrauterine life; and (c) an exogenous stressor, such as infection or overheating secondary to swaddling (3). A variety of autonomic irregularities have been reported in infants who subsequently die of SIDS, strengthening the possibility of a primary defect in autonomic and/or respiratory regulation (2). Many of these differences become apparent only during particular sleep-waking states, further suggesting central neural dysfunction (2). Maternal risk factors for SIDS (e.g., cigarette smoking during pregnancy) suggest that the intrauterine environment is suboptimal, with the implication that the pathogenesis of SIDS begins before birth (2). Subtle neurological dysfunction at birth in future SIDS victims supports the idea of pre- and/or perinatal brain injury (2).

In its broadest terms, our driving hypothesis is that SIDS, or a subset of SIDS, results from a defect in an appropriate response to a life-threatening challenge during sleep, e.g., asphyxia or hypercarbia, as the infant passes through the critical developmental period (2). A major anatomic site relevant to this hypothesis is the ventral medulla, a region critical for the integration of autonomic function, ventilation, chemoreception, and arousal (4–15). This region contains neuronal and glial cells along the surface known as the respiratory chemosensitive fields (RCF) (5,7–15). Experiments in cats and in medulla slice preparations from cats identify CO_2 and hydrogen ion responsive neuronal and glial populations at the RCF, which many investigators believe mediate ventilatory and behavioral responses to CO_2 and hydrogen ion (5,7–15). SIDS, or a subset of SIDS, may be due to a defect in neurons of the ventral surface of the medulla, resulting in an impaired response to a hypercarbic challenge during sleep that leads to a cascade of progressive hypercarbia, CO_2 necrosis, impaired airway reflexes, and ultimately, sudden death (16). The concept that a primary defect in ventral medullary surface (VMS) neurons may be involved in the pathogenesis of SIDS is supported by epidemiologic data indicating that the risk for SIDS is increased in infants sleeping prone, a position that is postulated to result in asphyxia or CO_2 rebreathing when the head is face-down (17–20). Moreover, recent clinical data indicate that some infants do sleep in the face-down position and that this position results in increased end-tidal P_{CO_2} (20). An infant with a primary VMS defect may be vulnerable to sudden death by rebreathing CO_2 in the prone position during sleep, particularly when stressed by even a minor infection and during a critical period in the maturation of respiratory and other homeostatic control systems during sleep.

In the following report, we first review anatomic evidence that the human arcuate nucleus (ARC) is homologous to a cell population in the cat VMS considered responsive to CO_2, and, second, consider evidence for a severe, developmentally related anatomic deficiency, or hypoplasia, in a subset of SIDS victims. We then provide baseline data about neurotransmitter receptor binding in the ARC and selected cardioventilatory-related nuclei of the developing human medulla that are relevant to neurochemical analysis in SIDS brainstems.

II. Anatomic Studies of the Human Infant VMS

The morphology and topographic distribution of the central chemosensors in the human medulla are not known definitively. This is because the physiological experiments required for their direct identification, namely lesion studies or stimulation by acidic solutions or electrical current, are prohibitive in humans. The only distinct neuronal population along the human VMS itself is the ARC, which historically has been considered a relay nucleus from the cerebral cortex to the cerebellum via the external arcuate fibers and inferior cerebellar peduncle (21,22). In a comparative anatomic study between the human infant and cat VMS, we found that the ARC, nucleus (n.) conterminalis, and foci of thickened marginal glia (TMG) in the human are anatomically homologous to the cat RCF (Table 1) (23). This homology is based upon cytoarchitectonic criteria (including cell size, shape, and distance from the surface) as defined in the combined neuroanatomic and physiological studies of the cat RCF of Trouth, Schläfke, and others (11–14,23), and upon three-dimensional cell distributions, as determined with the aid of computer reconstructions (23). The human n. conterminalis, located in the caudal ventrolateral medulla, is composed of large fusiform or multipolar cells that are cytologically identical to those described by Trouth et al. in the cat L field (11–13,23). In contrast to humans, however, the cells in cats do not form a distinct nucleus, but rather cluster in groups of two to three. The large, fusiform neurons described by Trouth et al. are present in ventrolateral and dorsomedial clusters extending the length of the twelfth cranial nerve (CN 12) roots, with the ventrolateral clusters matching most closely the distribution of the L field (11–13). Homologous ventrolateral neurons of the n. conterminalis and associated TMG comprise the caudal putative RCF in the human infant (Table 1). In contrast to cats, however, in humans the CN 12 extends almost the entire length of the medulla and does not demarcate the rostral boundary of the caudal RCF.

The human ARC is anatomically homologous to a second neuronal population on the cat VMS composed of smaller neurons located along the entire caudorostral length of the medulla (Table 1). In cats and humans, these neurons are small, round or oval, with indistinct Nissl substance. In cats they occur in clusters of two to four, compared to larger clusters of ten or more in humans. In humans, there are a large number of ARC neurons in the midline, but such midline cells are not shown within the experimentally determined RCF because the caudal position of the basilar artery makes this locus inaccessible in cats (23). The midline ARC is homologous to the small cells located at the ventral aspect of the cat n. raphe pallidus. In cats, the ventrolateral small round neurons and associated TMG occupy the rostral region corresponding to the S and M fields, whose distribution is marked by the roots of CN 9, 10, and 7. Homologous ARC neurons and TMG comprise the rostral putative human RCF;

Table 1 Anatomic Homology in Neuronal Cell Populations of the Ventral Medullary Surface Between the Cat and Human Infant[a]

Species	Nucleus/RCF	Neuron type	Neuron size, microns (SD)	Distance from surface of medulla, microns (SD)	Caudal-to-rostral distribution of cells
Cat	M and S RCF	Small, round	L = 16 (3) W = 11 (2)	284 (95)	Caudal medulla to caudal pons
Human	Arcuate nucleus	Small, round	L = 15 (3) W = 10 (2)	180 (83)	Caudal medulla to merge with basis pontis
Cat	L RCF	Large, fusiform	L = 31 (10) W = 11 (3)	406 (258)	From caudal to rostral extent of CN 12 roots
Human	N. conterminalis	Large, fusiform	L = 28 (6) W = 14 (3)	1003 (273)	From caudal to middle extent of medulla/CN 12 roots

Source: Ref. 23.

[a]Small, round neurons and large, fusiform neurons occasionally intermingle in caudal regions in both cat and human infant medullas. The neurons of the human n. conterminalis are deeper from the surface than large, fusiform cells (Trouth) in the cat caudal RCF; this reflects, in part, the merging of a large number of conterminalis cells with the n. raphe pallidus and medial olivary nucleus. Abbreviations: L, length; W, width; SD, standard deviation.

the landmarks of the rostral RCF are the roots of CN 9, 10, and 7, as in cats. Of note, ARC neurons extend the entire length of the human infant medulla and are probably part of the caudal as well as rostral human RCF.

A role for the ARC in vital functions is based on anatomic and physiological evidence from experimental animals and human clinicopathological correlations (15,16,24). The ablation of neurons in animals in sites homologous to the human ARC produces apnea and/or reduces blood pressure, depending on the precise site lesioned (16). Neurons in homologous positions to those of the ARC project to several nuclei related to ventilatory and autonomic regulation in the brainstem and spinal cord (16). A role for the ARC in human ventilatory control is supported by the report of its absence in an autopsied 5-month-old infant with congenital central hypoventilation syndrome (24). The human ARC is also known to project to the cerebellum (25), suggesting that it may link an autonomic/chemosensory/arousal integration zone of the ventral medulla to that structure. Further characterization of the neurochemistry, ultrastructure, connectivity, and physiological activity of the ARC and the small cells of the cat VMS is needed to confirm or refute their homology and to understand their function better.

III. An Anatomic Study of the ARC in SIDS Victims

Having identified the ARC as a putative component of the VMS cell populations related to central chemoreception, we next asked the question: Is there an anatomic deficiency of the ARC in SIDS victims? To address this question, we surveyed serially or extensively sectioned medullas of 41 SIDS victims and 27 controls in a blinded fashion. We identified two cases in which the ARC was almost totally absent; after breaking the code, we found that both cases were SIDS victims (16). Three-dimensional reconstructions and volume measurements were performed in serial sections from one of these SIDS victims and from three infant controls: the volume of the right ARC nucleus in the SIDS case was 0.7 mm^3, compared to a range of 3 to 26 mm^3 (median, 5 mm^3) in the three controls. This study suggested that ARC hypoplasia characterizes at least a subset of SIDS victims [2/41 SIDS victims (5%)] (16). Based upon the known anatomic connections of the human ARC and the connections and functions of neurons in homologous positions in experimental animals, we postulate that ARC hypoplasia may lead to sudden death in at least some SIDS victims by dyssynergy between autonomic, chemosensory, and arousal integration and the cerebellar coordination of ventilation, especially during sleep and hypercarbia in a critical developmental period (16).

A major question in the two SIDS cases with ARC hypoplasia concerns the mechanism whereby a congenital lesion in the ARC does not cause clinically apparent autonomic and ventilatory disturbances in the neonatal period but leads to death in the postneonatal period (16). The answer to this question is unknown but may be related to the same or similar mechanisms that underlie the age-related aspects of SIDS in general (2,16). Infants with ARC hypoplasia may be at particular risk for SIDS because, without the full complement of neurons and/or neuropils in the VMS cell populations, they are unable to make age-appropriate transitions in cardioventilatory and/or arousal control during the critical developmental period (2,16). Moreover, exogenous stressors may tip an already precarious balance. Of note, exogenous stressors associated with SIDS tend to cause hypercarbia, alter the demand for cardiac output and/or thermoregulation, and/or decrease arousal. These factors include infection, fever, environmentally induced hyperthermia, and bedding that traps the infant's face in an air pocket and results in rebreathing (17–20). The cause of the ARC hypoplasia in the two SIDS victims of the anatomic study is unknown. There were no other brainstem anomalies, nor was there gliosis or apparent neuronal atrophy within the ARC that would suggest a destructive or degenerative process (16). Given the timing and pattern of the formation of the human ARC from the rhombic lip (25), we speculate that the hypoplasia is restricted to rhombic lip precursors destined to differentiate into ARC neurons. Genetic or toxic insults to the regulation of ARC neuronal number during the first 16 weeks of pregnancy are considerations (16).

IV. A Baseline Study of Neurotransmitter Receptor Binding in Cardioventilatory-Related Nuclei of the Developing Human Medulla

ARC hypoplasia may characterize a small subset of SIDS victims with a putative primary defect in central chemoreception. Alternatively, this anatomic deficiency may represent the severe end of a spectrum of ARC abnormalities in a wider SIDS group, with neurochemical or molecular abnormalities (not detectable with the light microscope) at the more subtle end (16). To explore the possibility of a neurochemical deficiency in the ARC in SIDS victims, we have chosen to examine the parameter of neurotransmitter receptor binding with tissue autoradiography. A major strength of tissue receptor autoradiography is that it provides information about possible neurotransmitter abnoramlities in nuclei that are histologically unremarkable, a major consideration in the analysis of SIDS brainstems (2). This technique also allows for the localization of neurotransmitter receptor binding within precisely defined anatomic loci, even the small size of the human ARC on individual sections. In addition, it allows for analysis of all nuclei, such that an isolated defect in receptor binding in the ARC may be detected by the comparison of measurements between other nuclei as well as the ARC. Yet virtually nothing is known about neurotransmitter receptor binding in the human ARC, especially its developmental profile across early life. Thus, as a first step toward the comparative analysis of neurotransmitter receptor binding in the ARC of SIDS victims and controls, we performed a baseline study in control cases without SIDS or clinical neurological disease. We selected radioligands that bind to the receptors of neurotransmitters implicated in mediating VMS cell responses in experimental animals—e.g., muscarinic, nicotinic, opioid, and kainate (KA) receptors (10, 26–34). We also selected radioligands that bind to all or many of the subtypes of a particular receptor: given the paucity of information about neurotransmitter receptor binding in the developing human ARC in general, our initial aim is to provide baseline data about the quantitative distribution of the major subtypes combined.

A. Materials and Methods

The specimens analyzed in this study were part of a database under development in our laboratory for delineating brainstem chemical anatomy in early human life in relationship to the SIDS problem (35–40). Brainstems were obtained from individuals with postmortem intervals ≤24 h. Age is expressed as postconceptional (gestational plus postnatal) weeks. The fetal specimens at midgestation were obtained from elective abortuses autopsied in the Department of Pathology, Brigham and Women's Hospital, Boston, Massachusetts, with approval of the Clinical Protection Committee. Fetal age was determined by foot length measurements.

Quantitative receptor autoradiography was performed in frozen human fetal and infant medullas based upon methods developed in experimental animals or adult human postmortem tissues (Table 2) (35–46). Our procedures for tissue preparation and sampling have been described in detail (35). Brainstem sections from each case were incubated with different radioligands for comparison of muscarinic cholinergic, nicotinic cholinergic, opioid, serotoninergic (5HT), α_2-adrenergic, and KA receptor binding patterns. In a sample of adjacent serial-sections from each case, the following radioligands were applied: ^{3}H-quinuclidinyl benzilate (QNB) for muscarinic cholinergic receptor (mAChR) binding; ^{3}H-nicotine for nicotinic cholinergic receptor (nAChR) binding (39,45); ^{3}H-naloxone for opioid binding (35); ^{3}H-lysergic acid diethylamide for 5HT binding (38,42); ^{3}H-para-aminoclonidine (PAC) for α_2-adrenergic binding (43,44); and ^{3}H-KA for KA binding (40,46) (Table 2). Nonspecific binding was determined with appropriate displacers (Table 2). Our procedures for washing and drying the sections and generating the autoradiograms have been previously described (35,37). Sections were exposed to ^{3}H-sensitive film (LKB Ultrofilm-^{3}H) for 8 to 24 weeks (Table 2).

In this baseline study, we measured neurotransmitter receptor binding in the ARC, n. tractus solitarii (nTS), and n. paragigantocellularis lateralis (PGL). The nTS, a vagal subnucleus, is a major site of cardioventilatory reflex integration: it receives afferent information from mechano-, irritant, baro-, and peripheral and central chemoreceptors; bulbospinal neurons of the dorsal respiratory group colocalize with the nTS and drive spinal cord motor neurons

Table 2 Experimental Conditions for Neurotransmitter Receptor Binding

Receptor	Ligand	Putative binding sites	Displacer	Preincubation buffer	Incubation buffer	Incubate (min)	Expose (weeks)
Muscarinic	^{3}H-QNB 1 n*M*	m1-m5	Atropine sulfate 1 μM	NaH_2PO_4 50 n*M* NaCl 100 m*M* 30 min	NaH_2PO_4 50 n*M* NaCl 100 m*M*	120	8
Nicotinic	^{3}H-D,L-Nicotine 3.5 n*M*	$\alpha_4\beta_2$	L-Nicotine bitartrate 10 μM	None	Tris-HCl 50 m*M* $CaCl_2$ 8 m*M*	20	12
5HT	^{3}H-LSD 5 n*M*	5HT ($5HT_{1c}$) $5HT_2$	Serotonin 100 μM	TRIS-maleate 300 m*M* pH 7.4	Tris-maleate 300 m*M*, pH 7.4, ascorbic acid 0.1%	60	8
KA	^{3}H-KA 4 n*M*	High affinity: GluR5-GluR7, KA1, KA2 Low affinity: GluR1-GluR4	KA 100 μM	Tris-HCl 50 m*M* pH 7.4 30 min	Tris-HCl 50 m*M* pH 7.4 4°C	60	24
Opioid	^{3}H-Naloxone 2 n*M*	All major subtypes: mu > kappa > delta	5 μM Naloxone	Tris HCl 50 m*M* 30 min	Tris-HCl	60	10
α-2	^{3}H-PAC 1 n*M* phentolamine	High affinity: α_2	10 μm phentolamine	NaH_2PO_4 50 m*M* EDTA 10 m*M* 20 min	Tris-HCl 170 m*M* $MgCl_2$ 10 m*M*	60	12

Source: Refs. 35, 37–46, 51.

projecting to respiratory muscles (47). The nTS also projects to ventrolateral medulla sites integral to blood pressure regulation. The PGL, located in the ventrolateral reticular formation of the rostral medulla, is a putative integrative site of respiratory drive inputs and a candidate site for the rostral vasomotor control area (9). Measurements in the ARC were made in two sections from six precisely defined levels from each case; the nTS measurement was made at one midmedullary level, and PGL measurements at two rostral medulla levels. Plates in the atlas of Olszewski and Baxter (21) were used for reference. The six levels, with their atlas plate numbers (21), were: (a) caudal medulla, level of n. gracilis (Plate VIII); (b) caudal medulla, level of area postrema (Plate X); (c) midmedulla, level of n. Roller (Plate XII) (nTS measurement); (d) rostral medulla, level of n. praepositus (Plate XIV) (PGL measurement); (e) rostral medulla, level of n. vestibularis spinalis (Plate XVI) (PGL measurement); and (f) rostral medulla, level of rostral pole of principal inferior olive (Plate XVIII). In order to define the anatomic boundaries of brainstem nuclei, the tissue sections that generated the autoradiograms were stained with cresyl violet or hematoxylin and eosin and compared with the autoradiogram.

Quantitative densitometry of autoradiographs was performed with a MCID imaging system (Imaging Research Inc., Ontario). Optical densities were converted to specific activities in femtomoles/milligram (fmoles/mg) tissue with ^{3}H-standards. Receptor density was determined in specific nuclei by digitizing the nuclear boundaries directly upon the specific activity mosaic displayed on the color monitor, with superimposition of the cell-stained tissue section over the specific activity mosaic, when necessary, with software available on the imaging system. Specific activity measurements were made in a blinded fashion, without knowledge of clinical diagnosis or postconceptional age. Receptor density values from right and left paired nuclei were combined as significant differences between them were not found, and were expressed as the mean. Specific activity data were displayed as computer-generated mosaics with a linear, 15-step gray scale. The range varied depending upon the specific radioligand used. A *t*-test was used to compare radioligand binding between the fetal and infant groups.

Because tissue receptor autoradiography with tritiated ligands is subject to differential quenching by varying lipid levels in the tissue, we considered the possibility that increased lipid quench due to progression of brainstem myelination explains developmental changes in binding to the selected tritiated radioligands (37,48). To test this possibility, we corrected tritiated radioligand binding for quench using ^{3}H-succinnidyl proprionate (SP) (37,49). The brainstem sections used for ^{3}H-SP binding were adjacent to those in which binding was examined. ^{3}H-SP activity before and after delipidation was compared in the same tissue sections using the MCID system: the autoradiograms of the same tissue sections before and after delipidation were aligned with one another, and measurements of ^{3}H-SP binding were made within the identical anatomic boundaries of the selected nuclei (37). We tested the effect of quench in

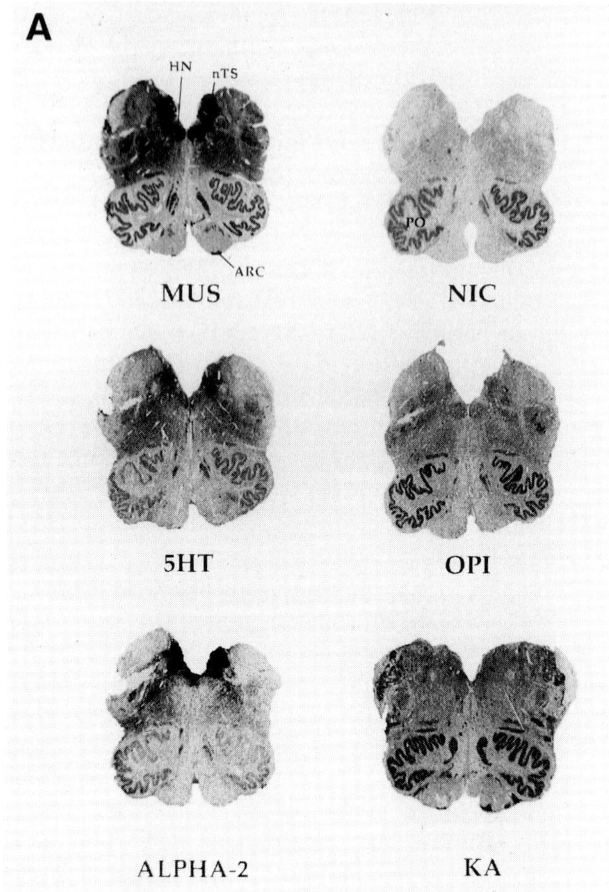

Figure 1 (A) The distribution of neurotransmitter receptor binding using six radioligands (Table 2) in the midmedulla (level of the n. Roller) and (B) rostral medulla (level of the n. prepositus) in a representative infant at 52 postconceptional weeks. As KA binding was not available in this case, it is illustrated from the medulla of a 47-postconceptional-week infant. Each image represents a specific activity mosaic in which the range of binding is illustrated by a linear 15-step gray scale. (Note that the range varies for each radioligand—e.g., 0 to 50 fmol/mg for ^{3}H-KA binding to KA receptors, 0 to 240 fmol/mg for ^{3}H-QNB binding to mAChRs). Abbreviations: MUS, muscarinic; NIC, nicotinic; 5HT, serotoninergic; OPI, opioid; alpha, α_2-adrenergic; KA, kainate; ARC, arcuate nucleus; nTS, n. tractus solitarii; PGL, n. paragigantocellularis; PO, principal inferior olive.

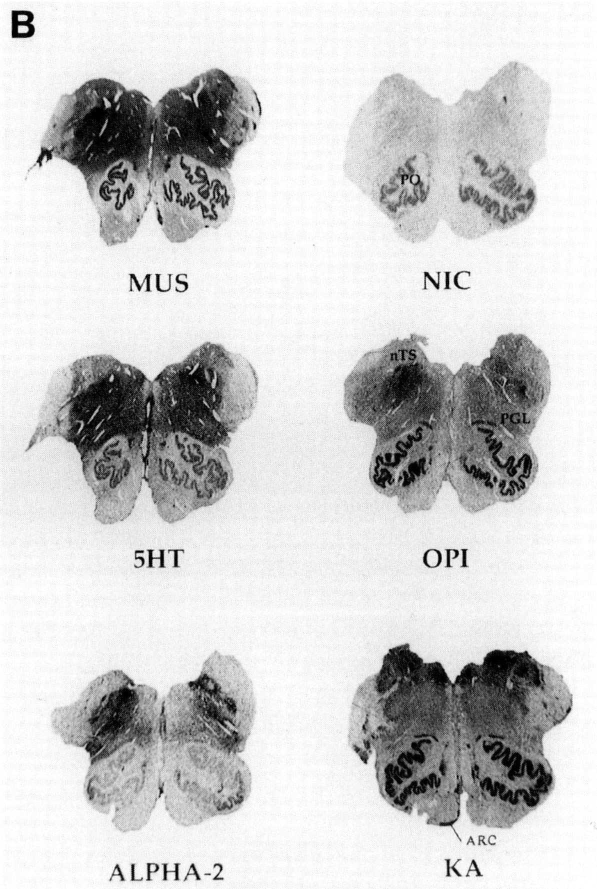

selected nuclei by dividing specific activity of binding by efficiency (^{3}H-SP binding before delipidation divided by ^{3}H-SP binding after delipidation) (37,49).

B. Results

In this baseline study, we examined radioligand binding to the selected neurotransmitter receptors in a total of 20 cases. The brainstems of 7 fetuses (19 to 26 postconceptional weeks), and 10 infants (41 to 66 postconceptional weeks) without SIDS or neurological disease were analyzed. The brainstems of 1 child and 2 adults were also analyzed as indices of maturity for comparison. The median postmortem interval was 7 h (range: 2 to 24 h). There was no apparent adverse effect of postmortem interval upon ^{3}H-ligand binding in any region sampled, as found by others analyzing the same receptors in animal and adult human studies (e.g., 35,37). The relative distribution of radioligand binding to the selected receptors was essentially the same in either of the time periods examined, i.e., fetal period or early infancy, irrespective of the cause of death, agonal conditions, or postmortem interval; thus, the cases within each of these periods are described together. The high, intermediate, and low densities that we report in this study are relative to other regions within the brainstem, but not necessarily to forebrain regions.

Distribution of Neurotransmitter Receptor Binding in the Human Infant ARC, nTS, and PGL

The distribution of radioligand binding to the selected neurotransmitter receptors is illustrated at mid- and rostral levels of the medulla in a representative infant at 52 postconceptional weeks (Fig. 1). (As KA binding was not available in this case, it is illustrated from the medulla of a 47 postconceptional week infant.) In the infant ARC, muscarinic and KA binding was intense, and 5HT binding was low-to-moderate (Fig. 1). Relative to all other brainstem nuclei, muscarinic binding in the ARC was intermediate (100 to 200 fmol/mg), compared, for example, to high binding (>200 fmol/mg) in the hypoglossal nucleus (Fig. 1) and other cranial nerve motor nuclei (39). Kainate binding was higher in the ARC (≥35 fmol/mg) and the other rhombic-lip derivatives (inferior olive and basis pontis) than in any other brainstem site (40). Muscarinic, KA, and 5HT receptor binding colocalized precisely with the cytoarchitectonic boundaries of the ARC in individual sections. The ARC was not present in every section of the medulla sampled, nor was it of uniform density in every cell cluster or at every level throughout its caudorostral distribution. These observations are consistent with previous anatomic studies by us in which we found considerable variability within and among cases in the area and density of the ARC, including between serial sections separated by as little as 100 μm (16,23). Of direct relevance to SIDS studies, this baseline variability mandates analysis in a sample of serial sections to establish significant differences in receptor binding in the ARC between SIDS and control cases.

Muscarinic, 5HT, and α-2 receptor binding was heavily concentrated in the infant nTS (Fig. 1). In contrast, KA receptor binding was low-to-moderate and nicotinic binding, negligible (Fig. 1). Throughout the infant nTS, the distribution of receptor binding for all of radioligands examined was heterogenous in both the cross-sectional and caudorostral plane. (The one exception was ^{3}H-nicotine binding, which was negligible at all levels.) ^{3}H-Naloxone binding to opioid receptors, for example, was twofold higher in the caudal than the mid-portion; rostrally it was high in the dorsolateral tegmentum, a region overlapping with and including the nTS (Fig. 1) (35). At the mid-level of the infant nTS, muscarinic, 5HT, and α_2-receptor binding was heavily concentrated within the dorsal subdivisions, with low or negligible binding in the ventral and ventrolateral subdivisions (Fig. 1). In the infant PGL, the binding was moderate (relative to other binding sites for the same radioligand within the same brainstem) for muscarinic and 5HT binding and or negligible for nicotinic, opioid, and KA binding (Fig. 1). In contrast, ^{3}H-paraclonidine binding to α_2 receptors was high in the ventrolateral reticular formation of the rostral medulla, including regions that overlap with the cytoarchitectonic boundaries of the PGL (Fig. 1).

Developmental Changes in Neurotransmitter Receptor Binding in the Human ARC, nTS, and PGL

Receptor binding to all of the six radioligands analyzed was present and regionally distributed across the medullary nuclei by midgestation, the earliest time point examined. Over the last half of gestation, dramatic changes occurred in the relative distribution and quantitative levels of receptor binding densities in the ARC, nTS, and PGL. In the fetal ARC, as in the infant ARC, binding was negligible to nicotinic, opioid, and α_2 receptors and did not distinguish the ARC from the adjacent pyramid or other medullary surface structures (Table 3). In contrast, binding to both muscarinic and KA receptors in the ARC increased (by 63% and 49%, respectively) from midgestation to early infancy (Fig. 2), whereas binding to 5HT receptors decreased (by 40%) (Table 3). The elevations in muscarinic and KA binding appeared to be transient in the infant period, with a fall to adult levels at some point thereafter (Fig. 2).

In the nTS, binding to 5HT, nicotinic, α_2, and muscarinic receptors decreased substantially from midgestation through early infancy, with the greatest decrease in 5HT and nicotinic binding (61% to 62%). In contrast, there were slight but not significant (<30%) decreases in binding to opioid and KA receptors (Table 3). In the PGL, binding decreased to varying degrees to all radioligands analyzed from midgestation to early infancy. The most substantial

Table 3 Percent Change in Receptor Binding Between Midgestation and Infancy[a]

Receptor	nTS	PGL	ARC
5HT	↓62%	↓75%	↓40%
NIC	↓61%	↓73%	NM
α_2	↓44%	↓85%	NM
MUS	↓38%	↓57%	↑63%
OPI	NS	↓58%	NM
KA	NS	↓34%	↑49%

[a]NS, not significant; NM, not measured due to visibly negligible binding in both fetal and infant periods.

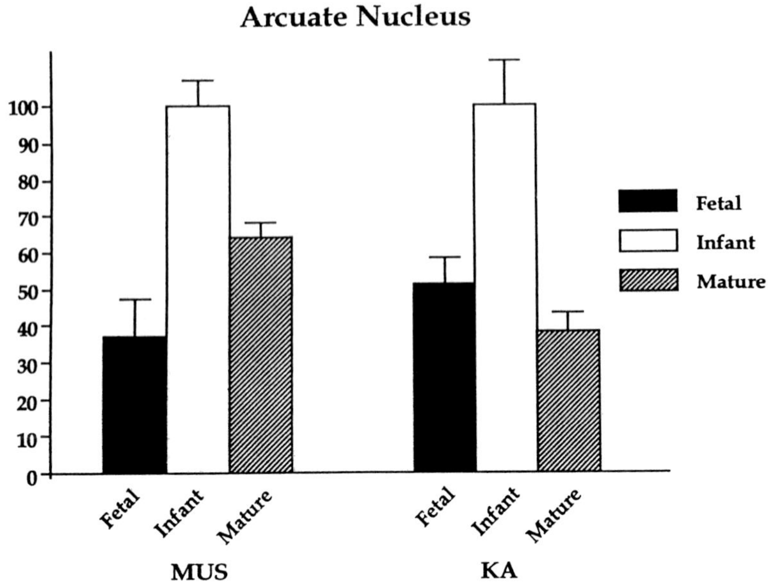

Figure 2 A comparison of the percent changes (± SEM) in the development of ^{3}H-QNB binding to muscarinic (MUS) receptors and ^{3}H-kainate (KA) binding to KA receptors in the human ARC. Within each age group, there is a minimum of three cases (see text). One hundred percent was arbitrarily determined as the age group with the highest binding.

decrease occurred in α_2 binding (85%), and the least substantial, in KA binding (34%) (Table 3). While quench did increase in reticular subnuclei between the fetal and infant periods, the change was not significant, and correction for quench did not substantially change ^{3}H-ligand binding levels in this data set (37). In contrast, there was a marked increase in quench in reticular subnuclei after early infancy to adult levels (37). In circumscribed nuclei that are not heavily intermixed with myelinated fibers, e.g., nTS, changes in lipid quenching between the fetal and infant periods were not substantial. Thus, the most dramatic quantitative changes between the fetal and infant periods of the nuclei sampled in this study were the decreases in α_2, 5HT, and nicotinic binding in the PGL (Table 3), changes that were not fully explained by progressive myelination and lipid (myelin) quenching in the reticular formation.

C. Discussion

Early human life is a critical period in the development of homeostatic control mechanisms, as the newborn adapts to independent life in the extrauterine environment (2). In the human fetus, diaphragmatic movements of breathing have been demonstrated as early as 10 gestational weeks (57); at birth, respiration switches instantaneously from sporadic movements to a smooth, rhythmic pattern. The developmental timetable of central responses to CO_2 in early human life is not completely known, but available data suggest that CO_2 responsiveness develops preterm (50) and attains adult patterns at birth or very soon thereafter. From midgestation through early infancy, dramatic changes also occur in respiratory rate, state-dependent patterning, and responses to hypoxia (2,3). While the physiological changes in human respiration are fairly well described, very little is known about the underlying neuroanatomic and neurochemical development of the central controller sites. In this baseline autoradiography study of the human medulla, we examined the developmental profile of receptor binding to several neurotransmitters implicated in mediating or modulating cardioventilation across the critical period in autonomic and state maturation. We focused upon the ARC, a candidate VMS cell population for central chemoreception (23), as well as the nTS and PGL, two nuclei considered integral to the neuronal circuitry of central cardioventilatory control (9,47). While the development of this circuitry is complex and no single parameter underlies a particular physiological change, this autoradiographic study is a step toward defining the neurochemical maturation of the circuitry and is therefore of major relevance to SIDS analysis.

The localization of cholinergic, opioid, α_2, KA, and 5HT receptor binding in the human infant nTS and PGL is consistent with animal physiological data showing the influence of the selected neurotransmitters upon cardioventilatory function via these nuclei (e.g., 35,43,44). Moreover, the receptors, cell bodies, and/or terminals of the selected neurotransmitters have been demonstrated within these two nuclei in experimental animals and/or adult humans (35,38, 42–55). Our observations in the human infant nTS, for example, correlate well with those reported in the cat and rat nTS: anatomic studies in these species show a somatotopic organization of afferents within nTS subnuclei, and a regional distribution of opioid receptors and enkephalin-positive terminals (35). By analogy to animal data, the heterogeneous distribution of opioid receptors in the human infant nTS likely reflects a precise topography of cardioventilatory terminals: the heaviest concentration of opioid receptors occurs in the caudal subdivisions, which animal data link in part to baro- and chemoreflexes (35).

The prominent localization of muscarinic and KA receptor binding in the human infant ARC is consistent with animal data showing that these receptors play major roles in mediating the physiological responses of VMS cell populations (10,26,27,29,30,32-34). The experimental applications or microinjection of many neuroactive agents (cholinergic agonists, glutamate, KA, thyroid releasing hormone, substance P, and other neuropeptides) to the VMS results in dramatic effects on ventilation. They stimulate ventilation, while muscarinic antagonists, opiates, barbiturates, and GABA inhibit it (10,28,32,34), and clonidine decreases blood pressure (30). Yet, there is evidence that application of the mAChR antagonist, atropine, and of KA to the rat VMS decreases the response to hypercarbia, suggesting that both muscarinic and KA receptors accessible to topical application are specifically involved in mediating respiratory responses to CO_2 (10,26,32,34). Indeed, a current concept for central chemoreception emphasizes the key role of cholinergic synaptic processes (10), and cholinergic cell bodies have been identified within the VMS in animals (10, 52–55). Nevertheless, it is still unclear whether the muscarinic or KA receptors are the chemoreceptors themselves, or receptors on cells or cell processes involved in the chemoreflex pathways (9). The prominent localization of muscarinic and KA receptor binding in the human ARC lends neurochemical support to the idea that this nucleus is a VMS cell population involved in central chemoreception in humans. Additional studies are needed to determine the cellular localization of the ACh- and glutamate-source neurons and specific receptor subtypes.

In contrast to prominent muscarinic and KA binding in the human infant ARC, localized opioid binding, as measured with ^{3}H-naloxone, is essentially nonexistent (35). While the precise localization of opioid receptors along the VMS is not emphasized in animal autoradiographic studies, physiological experiments indicate that opioid agonists and antagonists depress and stimulate respiration, respectively, when applied to various subregions of the VMS (28). The possibility has been raised that opioid drugs, e.g., morphine, directly affect central respiratory-related neurons by penetrating rapidly from the pial surface into the medullary parenchyma along blood vessels and nerve fibers, rather than acting on "surface" chemosensitive structures (28). This is a consideration in the human infant VMS, given the absence of distinct ^{3}H-naloxone binding sites. The opioid observations are of particular relevance to infants of opioid-addicted mothers in whom impaired central chemosensitivity to CO_2 has been demonstrated, and it has been suggested as a risk factor for sudden death in this population (56,57). Although nicotinic receptors have been anatomically identified along the cat VMS (55), and a nicotine-sensitive area has been demonstrated along the caudal VMS which overlaps with the caudal respiratory area (31), nicotinic receptor binding in the human ARC is negligible at all levels in both the fetal and infant periods. Serotoninergic binding colocalizes in the infant ARC, but few animal data are available for comparison.

The earliest appearance in the human brainstem of binding to the selected neurotransmitter receptors in this study is unknown, but our data indicate that regionalization is well established by midgestation. Between midgestation and early infancy, there are marked changes in the relative distribution and quantitative levels of receptor binding in the nTS, PGL, and ARC. In the human brainstem, almost all neuronal migrations are complete and the tegmental nuclei have achieved their basic cellular configurations by 9 to 10 gestational weeks (58). From the second trimester onward, brainstem growth is related primarily to the growth of intrinsic and extrinsic fiber systems, growth of cell bodies and dendrites, gliogenesis, and myelination (58). Neuronal migrations from the rhombic lip, however, continue to form the ARC, basis pontis, and inferior olivary complex through 24 to 26 gestational weeks (25,58). From the fourth gestational month onward, neuronal density decreases in all brainstem nuclei as nonneuronal elements proliferate. Consequently, the decreases in neurotransmitter receptor binding that we found in the nTS and PGL between the fetal and infant periods were not unexpected. Yet, a decrease in binding was not obligatory for all of the radioligands analyzed, as we have reported for ^{3}H-nicotine binding (i.e., quantitative levels remain essentially unchanged in the basis pontis and inferior olive over the second half of gestation) (37). In this study, there are no substantial decreases in opioid and KA binding in the nTS, and receptor binding not only decreases (5HT) but also increases (muscarinic and KA) in the ARC; moreover, while decreased binding occurs for all receptors analyzed in the PGL, the changes are not the same, and vary between 34% (KA) and 85% (α_2). Thus, receptor binding appears to be independently regulated during human development within specific regions. Other potential factors responsible for decreased receptor binding include a developmentally programmed loss of dendritic spines with postsynaptic receptors, loss of terminals with presynaptic receptors, loss of the receptors themselves, and/or decreased receptor binding affinity. Given that there are multiple subtypes of the receptors analyzed and that the radioligands used label more than one subtype, the decline in radioligand binding may represent differential rates of decline in expression among subtypes. Given that several neurotransmitters, including ACh, opioids, and glutamate, have been implicated as trophic factors influencing synaptogenesis and connectivity in early development, the decline in receptor binding may reflect a loss of receptors transiently elevated to mediate growth-related effects in the fetal period (35,37,59).

Overall, we found that the most dramatic changes in receptor binding to the selected neurotransmitters in the nTS and PGL occur between midgestation and early infancy, and that their relative distribution appears to be established around birth, the time of transition from dependent to independent respiratory gas exchange. Moreover, receptor binding in these two nuclei does not appear to change substantially thereafter. However, a larger sample size with multiple cases at each monthly interval is needed to detect subtle changes in the postnatal period, including the peak age range for SIDS (2 to 4 months).The dramatic increase in muscarinic and KA binding in the ARC between midgestation and early infancy suggests the possibility that this neurochemical change underlies, at least in part, chemosensory adaption to extrauterine life. Of note also is the transient elevation in KA binding in the ARC in early infancy compared to midgestation and “maturity”: given the role of glutamate receptor subtypes in mediating excitotoxicity in hypoxia-ischemia (40,60,61), we raise the possibility of a particular vulnerability of this nucleus to hypoxia-ischemia during early infancy.

Within the developmental periods examined (i.e., midgestation and early infancy), the intensity and relative distribution of binding to each of the selected radioligands were remarkably constant among cases, irrespective of the cause of death or agonal condition, suggesting a baseline pattern of binding within the time period. Nevertheless, a potential limitation of the autoradiographic study is that it is based upon an autopsy population of fetuses and infants dying of diverse causes (35,37). Although the brainstems were histologically unremarkable, it is possible that the disease or agonal condition may have adversely affected radioligand binding in a particular case such that its distribution or quantitative levels do not truly reflect those found in healthy living individuals (35, 37). In order to detect subtle abnormalities in receptor binding in the ARC, a large age-matched database is needed that accounts for wide inherent biological variability in binding and for the effects of unavoidable autopsy variables (e.g., disease and agonal states, postmortem interval) in the pediatric population. The aim of this report is to indicate the qualitative distribution of radioligand binding in the ARC, nTS, and PGL across a population of autopsied fetuses and infants and to underscore the implications of these observations for further analysis in SIDS victims.

Acknowledgments

Supported by grants from the National Sudden Infant Death Syndrome Alliance, National Institute of Child Health and Human Development (R01-HD20991), Alpha Omega Alpha Research Fellowship for Medical Students (AP), and Children's Hospital Mental Retardation Core Grant (P30-HD18655). The authors thank Mr. Timothy O'Donnell for technical help, Mr. Dixon Yun for photographic assistance, and Ms. Regena Bradeen and Ms. Sathya Reddy for help in manuscript preparation.

References

1. Willinger M, James LS, Catz C. Defining the sudden infant death syndrome (SIDS): Deliberations of an expert panel convened by the National Institute of Child Health and Human Development. Pediatr Pathol 1991; 11:677–684.
2. Kinney HC, Filiano JJ, Harper RM. The neuropathology of the sudden infant death syndrome: A review. J Neuropathol Exp Neurol 1992; 51:115–126.
3. Filiano JJ, Kinney HC. A perspective on neuropathologic findings in victims of the sudden infant death syndrome: The triple-risk model. Biol Neonate 1994; 574:1–5.
4. Dong XW, Gozal D, Rector D, Harper RM. Ventilatory CO_2-induced optical activity changes of the cat ventral medullary surface. Am J Physiol 1993; 265:494–503.
5. McAllen RM. Location of neurones with cardiovascular and respiratory function at the ventral surface of the cat's medulla. Neuroscience 1986; 18:43–49.
6. Hilton SM, Marshall JM, Timms RJ. Ventral medullary relay neurons in the pathway from the defense areas of the cat and their effect on blood pressure. J Physiol 1983; 345:149–166.
7. Issa FG, Remmers JE. Identification of a subsurface area in the ventral medulla sensitive to local changes in PCO_2. J Appl Physiol 1992; 72:439–446.

8. Loeschcke HH, DeLattre J, Schläfke ME, Trouth CO. Effects on respiration and circulation of electrically stimulating the ventral surfaces of the medulla oblongata. Resp Physiol 1970; 10:184–197.
9. Millhorn DE, Eldridge FL. Role of ventrolateral medulla in regulation of respiratory and cardiovascular systems. J Appl Physiol 1986; 61:1249–1263.
10. Nattie EE. Central respiratory chemoreceptors. In: Haddad GG, Farber JP, eds. Developmental Neurobiology of Breathing. New York: Dekker, 1991:341–371.
11. Trouth CO, Loeschcke HH, Berndt J. Histological structures in the chemosensitive regions of the ventral surface of the cat's medulla oblongata. Pflügers Arch 1973; 339:171–183.
12. Trouth CO, Loeschcke HH, Berndt J. A superficial substrate on the ventral surface of the medulla oblongata. Pflügers Arch 1973; 339:135–152.
13. Trouth CO, Odek-Ogunde M, Holloway JA. Morphological observations on superficial medullary CO_2-chemosensitive areas. Brain Res 1982; 246:35–45.
14. Schläfke ME, Kille JF, Loeschcke HH. Elimination of central chemosensitivity by coagulation of a bilateral area on the ventral medullary surface in awake cats. Pflügers Arch 1979; 378:231–241.
15. Schläfke M, Hukuhara T, See WR. Loss of central chemosensitivity, experimental studies on a clinical problem. Adv Physiol Soc 1981; 10:609–616.
16. Filiano JJ, Kinney HC. Arcuate nucleus hypoplasia in the sudden infant death syndrome. J Neuropathol Exp Neurol 1992; 51:394–403.
17. Ponsonby AL, Dwyer T, Gibbons LE, et al. Factors potentiating the risk of sudden infant death syndrome associated with the prone position. N Engl J Med 1993; 329:377–382.
18. Fleming PJ, Gilbert R, Azaz Y, et al. Interaction between bedding and sleeping position in the sudden infant death syndrome: A population based case-control study. Br Med J 1990; 301:85–89.
19. Kemp JS, Thach B. Sudden death in infants sleeping on polystryrene-filled cushions. N Engl J Med 1991; 324:1858–1864.
20. Chiodini BA, Thach BT. Impaired ventilation in infants sleeping facedown: Potential significance for sudden infant death syndrome. J Pediatr 1993; 123:686–692.
21. Olszewski J, Baxter D. Cytoarchitecture of the Human Brainstem. 2d ed. Basel: Karger, 1982.
22. Rasmussen AT, Peyton WT. Origin of the ventral external arcuate fibers and their continuity with the striae medullares of the fourth ventricle in man. J Comp Neurol 1946; 84:325–337.
23. Filiano JJ, Choi JC, Kinney HC. Candidate cell populations for respiratory chemosensitive fields in the human infant medulla. J Comp Neurol 1990; 293:448–465.
24. Folgering H, Kuyper F, Kille FJ. Primary alveolar hypoventilation (Ondine's curse syndrome) in an infant without external arcuate nucleus. Bull Eur Physiol Respir 1979; 15:659–665.
25. Essick CR. The development of nuclei pontis and nucleus arcuatus in man. Am J Anat 1912; 13:25–54.
26. Dev NB, Loeschcke HH. A cholinergic mechanism involved in the respiratory chemosensitivity of the medulla oblongata in the cat. Pflügers Arch 1979; 379:29–36.
27. Gatti PJ, Taverira Da Silva AM, Hamosh P, Gillis RA. Cardiorespiratory effects produced by application of L-glutamic and kainic acid to the ventral surface of the cat hindbrain. Brain Res 1984; 330:21–29.

28. Hurle MA, Mediavilla A, Florez J. Morphine, pentobarbital and naloxone in the ventral medullary chemosensitive areas: Differential respiratory and cardiovascular effects. J Pharmacol Exp Ther 1982; 220:642–647.
29. Lawing WL, Millhorn DE, Bayliss DA, et al. Excitatory and inhibitory effects on respiration of L-glutamate microinjected superficially into the ventral aspects of the medulla oblongata in cat. Brain Res 1987; 435:322–326.
30. Bousquet P, Guertzenstein PG. Localization of the central cardiovascular action of clonidine. Br J Pharmacol 1973; 49:573–579.
31. Feldberg W, Guertzenstein PG. Vasopressor effects obtained by drugs acting on the ventral surface of the brainstem. J Physiol Lond 1976; 258:337–355.
32. Nattie EE, Aihua L. Ventral medulla sites of muscarinic receptor subtypes involved in cardiopulmonary control. J Appl Physiol 1990; 69:33–41.
33. Nattie EE, Mills JW, Ou LC, St John WM. Kainic acid on the rostral ventrolateral medulla inhibits phrenic output and CO_2 sensitivity. J Appl Physiol 1988; 65:1525–1534.
34. Nattie EE, Wood J, Mega A, Goritiski W. Rostral ventrolateral medulla muscarinic receptor involvement in central ventilatory chemosensitivity. J Appl Physiol 1989; 66:1462–1470.
35. Kinney HC, Ottoson CK, White WF. Three-dimensional distribution of [^{3}H]naloxone binding to opiate receptors in the human fetal and infant brainstem. J Comp Neurol 1990; 291:55–78.
36. Kinney HC, Rava LA, Benowitz I. Anatomic distribution of the growth-associated protein GAP-43 in the developing human brainstem. J Neuropathol Exp Neurol 1993; 52:34–94.
37. Kinney HC, O'Donnell TJ, Kriger P, White WF. Early developmental changes in [^{3}H]nicotine binding in the human brainstem. Neuroscience 1993; 55:1127–1138.
38. White WF, Kinney HC. Distribution of serotonin receptor subtypes in human infant brainstem. X International Congress of Neuropathology, Stockholm, Sweden, September, 1986.
39. Kinney HC, Greenier V, White WF. Distribution of muscarinic cholingergic receptors in the developing human brainstem (abstr). Soc Neurosci Abstr 1990; no. 277.8.
40. Panigraphy A, Kinney HC. Developmental changes in [^{3}H]kainate binding in human brainstem sites vulnerable to perinatal hypoxia-ischemia (abstr). Soc Neurosci Abstr 1993; no. 731.9.
41. Clarke PBS, Schwartz RD, Paul SM, et al. Nicotinic binding in rat brain: Autoradiographic comparison of [^{3}H]acetylcholine, [^{3}H]nicotine, and [^{125}I]-α-bungarotoxin. J Neurosci 1985; 5:1307–1315.
42. Palacios JM, Probst A, Cortes R. The distribution of serotonin receptors in the human: High density of [^{3}H]LSD binding sites in the raphe nuclei of the brainstem. Brain Res 1983; 274:150–155.
43. Robertson HA, Leslie RA. Nonadrenergic α-2 binding sites in vagal dorsal motor nucleus and nucleus tractus solitarius: Autoradiographic localization. Can J Physiol Pharmacol 1985; 63:1190–1194.
44. Unnerstall JR, Kopajtic J, Kuhar MJ. Distribution of α-2 agonist binding sites in the rat and human central nervous system: Analysis of some functional, anatomic correlates of the pharmacologic effects of clonidine and related adrenergic agents. Brain Res Rev 1984; 7:69–101.
45. Wamsley JK, Lewis MS, Young WS, Kuhar MJ. Autoradiographic localization of muscarinic cholinergic receptors in rat brainstem. J Neurosci 1981; 1:176–191.

46. Monagahan DT, Cotman CW. Brain Res 1982; 252:91–100.
47. Feldman JL, Ellenberger HH. Central coordination of respiratory and cardiovascular control in mammals. Ann Rev Physiol 1988; 50:593–606.
48. Kinney HC, Brody BA, Kloman AS, Gilles FH. Sequence of central nervous system myelination in human infancy: II. Patterns of myelination in autopsied infants. J Neuropathol Exp Neurol 1988; 47:217–234.
49. White WF, O'Gorman S, Roe AW. Three-dimensional autoradiographic localization of quench-corrected glycine receptor specific activity in the mouse brain using ^{3}H-strychine as the ligand. J Neurosci 1990; 10:795–813.
50. Read DJC, Henderson-Smart DJ. Regulation of breathing in the newborn during different behavioral states. Ann Rev Physiol 1984; 46:675–685.
51. Cortes R, Probst A, Palacios JM. Quantitative light microscopic autoradiographic localization of cholinergic muscarinic receptors in the human brainstem. Neuroscience 1984; 12:1003–1026.
52. Eckenstein F, Sofroniew MV. Identification of central cholinergic neurons containing both choline acetyltransferase and acetylcholinesterase. J Neurosci 1983; 11:2286–2291.
53. Jones BE, Beaudet A. Distribution of acetylcholine and catecholamine neurons in the cat brain stem: A choline acetyltransferase and tyrosine hydroxylase immunohistochemical study. J Comp Neurol 1987; 261:15–23.
54. Vincent SR, Reiner PB. The immunocytochemical localization of choline acetyltransferase in the cat brain. Brain Res Bull 1987; 18:371–415.
55. Willenberg IM, Dermietzl R, Leibsten AG, Effenberger M. Mapping of cholinoceptive (nicotinoceptive) neurons in the lower brainstem: With special reference to the ventral surface of the medulla. J Auton Nerv Syst 1985; 14:287–298.
56. Chavez CJ, Ostrea EM, Stryker JC, Smialek Z. Sudden infant death syndrome among infants of drug dependent mothers. J Pediatr 1979; 95:407–409.
57. Olsten GD, Lees MH. Ventilatory response to carbon dioxide of infants following prenatal methadone exposure. J Pediatr 1980; 96:983–989.
58. Sidman RL, Rakic P. Development of the human central nervous system. In: Haymaker W, Adams RD, eds. Histology and Histopathology of the Nervous System. Vol. 1. Springfield, IL: Charles C Thomas, 1982:18–20, 83–92.
59. Hattori H, Wasterlain CG. Excitatory amino acids in the developing brain: Ontogeny, plasticity, and excitotoxicity. Pediatr Neurol 1990; 6:219–228.
60. Choi DW. Glutamate neurotoxicity and diseases of the nervous system. Neuron 1988; 1:623–634.
61. Coyle JT, Puttfarcken P. Oxidative stress, glutamate, and neurodegenerative disorders. Science 1993; 262:689–695.

37

Brainstem Mechanisms for Arousal in Fetal Sheep

JOHN M. BISSONNETTE

Oregon Health Sciences University
Portland, Oregon

I. Introduction

Respiratory drive is significantly modulated by behavioral state. The initial report which reestablished that the fetus makes breathing movements in utero (1) noted that diaphragmatic contractions were confined to periods of low-voltage electrocortical activity (LV-ECoG) considered analogous to rapid-eye-movement sleep. In the subsequent 20 years, attention has been directed to the mechanisms that regulated fetal breathing during LV-ECoG and to a certain extent to the inhibition occurring in high voltage electrocortical activity, considered to be equivalent to non-rapid-eye-movement sleep. Much less effort has been directed at understanding the mechanisms responsible for arousal in the fetus. This chapter reviews the role of oxygen and carbon dioxide perturbations on producing arousal in the fetus. In addition, it summarizes what is known about monoaminergic and cholinergic mechanisms involved in fetal arousal. Finally, some recent work focusing on the role of excitatory and inhibitory amino acid neurotransmitters in fetal arousal is discussed.

II. Definition of Arousal in the Fetus

While there is agreement that a given behavioral state in the fetus is constituted by the concurrence of a certain set of physiological measures that recur over time, exactly which measures are required to define the arousal state has been a matter of debate. Rigatto and co-authors (2) have used a window in the ewe's flank to make direct observations of the fetus. They defined fetal arousal as open eyes and purposeful movements of the head. With the constraints of this definition, no periods of arousal were observed under normal in utero conditions (2). Most authors have used biophysical criteria to define fetal arousal or wakefulness as the presence of LV-ECoG combined with active muscle (usually nuchal muscle) activity [reviewed by Szeto (3) and Jansen and Chernick (4)]. Wakefulness is a heterogenous state, and neither of these approaches addresses the question of attentive behavior in defining the level of arousal. Nevertheless, the combination of LV-ECoG and muscle activity is readily distinguished from LV-ECoG and muscle atonia, and the former certainly represents a higher level of arousal, accompanied by an elevation in blood pressure and an increase in the amplitude of fetal breathing. Using this definition, the fetus spends about 5% of the time in arousal and each episode lasts about 0.5 to 2 min (3).

III. Effect of Respiratory Gases on Arousal in Fetal Sheep

A. Oxygen

In contrast to the adult and neonate, the fetus is not aroused by hypoxia. When fetal oxygen tensions are acutely lowered during the LV-ECoG state, the immediate response is apnea followed by a change to high-voltage (HV)-ECoG (Fig. 1; see also Fig. 1 in Harding et al., Ref. 5). Since the HV-ECoG state is inhibitory for fetal breathing, apnea persists.

In contrast to the fetal response to hypoxia, an increase in arterial oxygen tension is associated with arousal (6). Hyperoxia in the sheep fetus is accomplished by ventilating the fetus in utero, and an increased level of arousal is most readily achieved by also occluding the umbilical cord (6). These experimental conditions result in a number of changes in addition to the increase in arterial oxygen tension, including an increase in lung expansion and pulmonary blood flow; loss of the umbilical circulation, placentally derived hormones, and neurochemicals; and an increase in blood pressure. The minimum increase in P_{O_2} appears to be a rise of 10 to 20 torr above the normal 20 to 25 torr (7,8). Vagotomy significantly decreases the duration of fetal arousal during hyperoxia (9), indicating the importance of pulmonary afferents.

B. Carbon Dioxide

As with hypoxia, hypercarbia does not induce arousal in fetal sheep (10,11). In these two studies, an elevation in fetal arterial CO_2 tension was accomplished

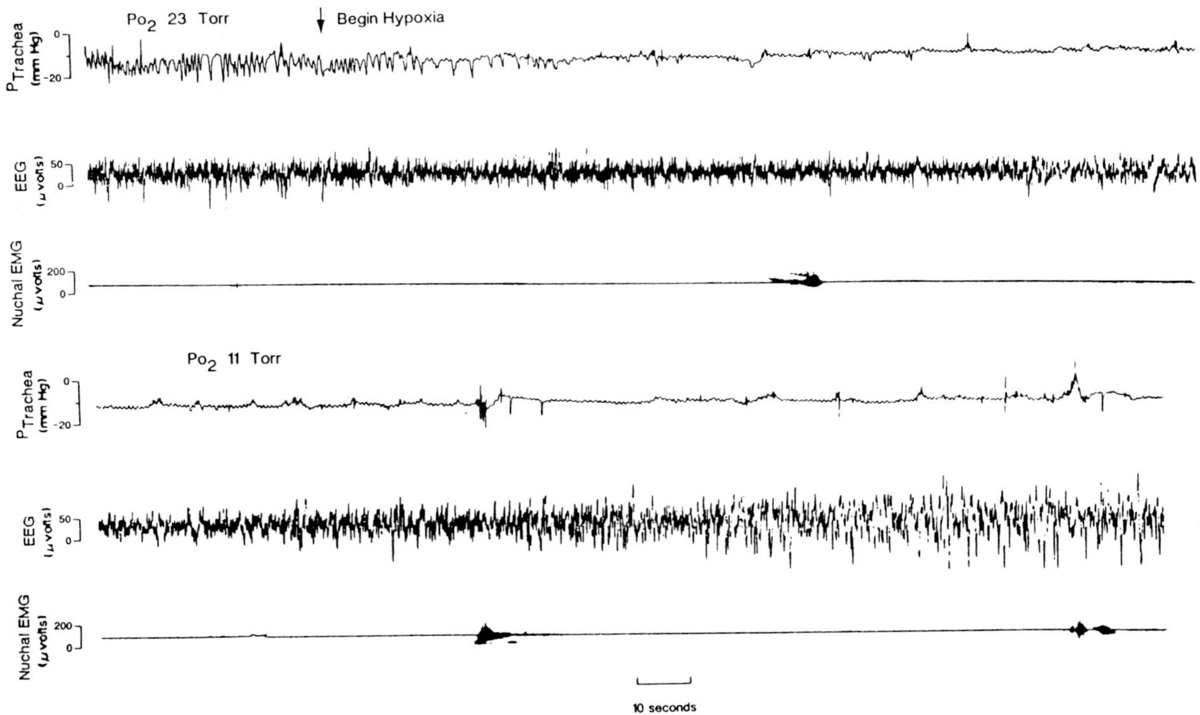

Figure 1 Effect of hypoxia on fetal behavioral state. Continuous record from 128-day gestation (term is 147 days) fetal sheep. At the arrow, the ewe began ventilation with 9% O_2, 3% CO_2, balance N_2 from a clear plastic bag placed over her head. Note rapid onset of apnea followed by a switch from low-voltage EEG to high-voltage EEG in the right half of the lower EEG trace.

by ventilating the ewe with a hypercarbic gas mixture. Hypercarbia has also been studied by ventilating the fetus in utero with high-frequency oscillation (12), but arousal was not evaluated with muscle electromyographic (EMG) recording in this study. Therefore, the combination of hypercarbia with an increase in pulmonary afferents as a stimulus for arousal in the fetus remains to be evaluated.

IV. Neurotransmitters and Arousal in Fetal Sheep

A. Cholinergic Mechanisms for Arousal

Systemic administration of the muscarinic agonist pilocarpine in the sheep fetus resulted in an increase in percent time spent in arousal (13). These studies showed a dose-dependent response, and 4.0 mg pilocarpine effected a 60% incidence of arousal. This behavioral response was blocked by atropine sulfate but not by methylatropine, which does not cross the blood-brain barrier, consistent with a central site for the muscarinic agonist. More recently the cholinergic agonist carbachol has been studied by instillation into the fourth ventricle of the sheep fetus (14). This direct approach to the medulla and caudal pons did not cause any increase in arousal (Fig. 2), indicating that the site of action for pilocarpine is more rostral in the brainstem.

Low doses of morphine and, in separate studies, thyrotropin-releasing hormone have both been shown to cause arousal in fetal sheep (15,16). In both cases, the effect is eliminated by treatment with atropine sulfate, indicating that these agents are acting through cholinergic pathways.

B. Monoaminergic Mechanisms for Arousal

Walker and his associates have made an extended study of the role for monoamines in the regulation of fetal breathing and behavioral states. Using an antibody against dopamine-β-hydroxylase, they have shown that norepinephrine or epinephrine is present in the nucleus locus ceruleus of fetal sheep at 141 to 143 days (17). This area is involved in arousal in the adult. Intravenous infusion of the inhibitor of norepinephrine uptake, desipramine, resulted in arousal in fetal sheep (18). This effect on behavioral state could be blocked with the α_1-adrenergic inhibitor prazosin. In separate studies, Joseph and Walker (18) administered 6-hydroxydopamine into the cerebrospinal fluid (CSF) of the fourth ventricle. This agent, which causes an initial displacement of transmitter stores from neurons, did not result in arousal. In additional studies, these same authors (19) infused tyramine into the fourth cerebral ventricle. Tyramine causes release of norepinephrine but, unlike 6-hydroxydopamine, does not result in release of costored neuropeptides such as corticotropin and thyrotropin-releasing hormones. This central administration of tyramine to the medulla and caudal pons did not arouse the sheep fetus. In direct studies, cisternal infusion of norepinephrine and dopamine also failed to cause arousal (19). These studies

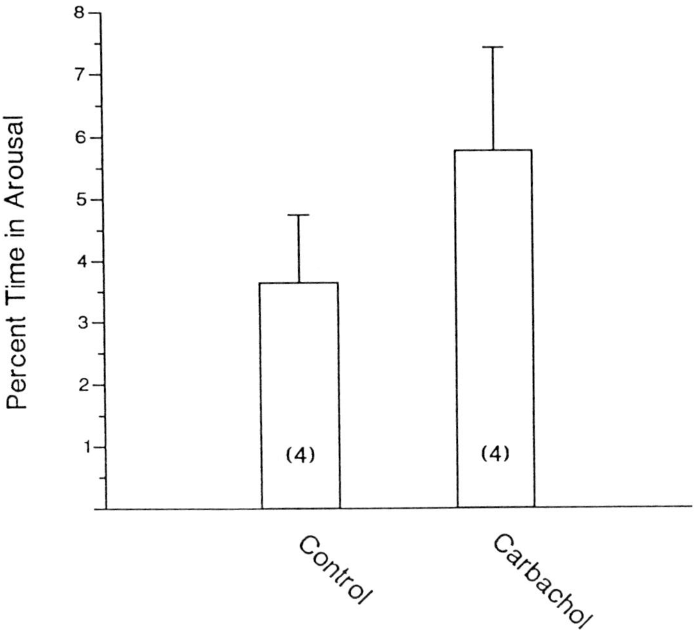

Figure 2 Effect of the cholinergic agonist carbachol on arousal in fetal sheep. Incidence of arousal during 60 min of control (left bar graph) compared to the 60 min following instillation of 1.0 μg carbachol into the CSF of the fourth ventricle. Numbers in brackets refer to the number of studies.

indicate that the adrenergic site for arousal lies in the rostral pons or more rostrally in the fetal brainstem.

C. Excitatory and Inhibitory Amino Acid Neurotransmitters and Arousal in the Fetus

In recent studies, it has been reported that blockade of the glutamate binding site on *N*-methyl-D-aspartate (NMDA) receptors in the medulla or caudal pons leads to prolonged episodes of arousal in fetal sheep (20). These experiments have been extended to show that blocking the cationic channel in NMDA receptor produces similar results (Bissonnette, Hohimer and Knopp, manuscript in preparation). Studies in an experimental model of temporal lobe epilepsy have suggested a mechanism by which this type of disinhibition could be effected (21,22). This model suggests that interneurons secreting gamma aminobutyric acid (GABA) tonically inhibit a potentially epileptogenic focus in the hippocampus. In the animals in whom neuronal hyperexcitability has been induced by prolonged afferent stimulation, these GABAergic neurons remain intact. However, they have been rendered dormant because of the loss of separate interneurons that tonically excite them. In the sheep fetus, it may be that GABAergic interneurons tonically suppress, for example, the magnocellularis reticular field, a medullary area whose neurons increase their firing rate during arousal in the adult (23). This GABA-mediated inhibition would be maintained by tonic glutamate excitement of these cells acting at NMDA receptors. Blockage of these medullary NMDA-bearing neurons would remove the GABAergic inhibition, allowing the fetus to increase its level of arousal. Further study of this proposed mechanism for arousal may lead to suggestions as to how it could be defective in situations where arousal from hypoxia and/or hypercarbia fails to take place in the neonate or adult.

References

1. Dawes GS, et al. Respiratory movements and rapid eye movement sleep in the foetal lamb. J Physiol 1972; 220:119–143.
2. Rigatto H, et al. Fetal breathing and behavior measured through a double-wall Plexiglas window in sheep. J Appl Physiol 1986; 61:160–164.
3. Szeto HH. Behavioral states and their ontogeny: Animal studies. Sem Perinatol 1992; 16:211–216.
4. Jansen AH, Chernick V. Fetal breathing and development of control of breathing. J Appl Physiol 1991; 70:1431–1446.
5. Harding R, et al. The effect of brief episodes of diminished uterine blood flow on breathing movements, sleep states and heart rate in fetal sheep. J Dev Physiol 1981; 3:231–243.
6. Baier RJ, et al. Effects of various concentrations of O_2 and umbilical cord occlusion on fetal breathing and behavior. J Appl Physiol 1990; 68:1597–1604.
7. Hasan SU, Rigaux A. Arterial oxygen tension threshold range for the onset of arousal and breathing in fetal sheep. Pediatr Res 1992; 32:342–349.
8. Alvaro R, et al. The effects of 21 or 30% O_2 plus umbilical cord occlusion on fetal breathing and behavior. J Dev Physiol 1992; 18:237–242.
9. Hasan SU, Rigaux A. Effect of bilateral vagotomy on oxygenation, arousal, and breathing movements in fetal sheep. J Appl Physiol 1992; 73:1402–1412.
10. Jansen AH, et al. Influence of sleep state on the response to hypercapnia in fetal lambs. Respir Physiol 1982; 48:125–142.
11. Rigatto H, et al. Effect of increased arterial CO_2 on fetal breathing and behavior in sheep. J Appl Physiol 1988; 64:982–987.
12. Adamson SL, et al. Umbilical cord occlusion stimulates breathing independent of blood gases and pH. J Appl Physiol 1991; 70:1796–1809.
13. Hinman DJ, Szeto HH. Cholinergic influences on sleep-wake patterns and breathing movements in the fetus. J Pharmacol Exp Ther 1988; 247:372–378.
14. Bissonnette JM, et al. A cholinergic mechanism involved in fetal breathing during the high-voltage ECoG state. Respir Physiol 1994; 96:151–162.
15. Szeto HH. Morphine-induced activation of fetal EEG is mediated via central muscarinic pathways. Am J Physiol 1991; 260:R509–R517.

16. Umans JG, et al. Effects of thyrotropin-releasing hormone in the fetal lamb. Am J Obstet Gynecol 1986; 155:1266–1271.
17. Joseph SA. Catecholamine pathways in the brainstem and control of breathing in fetal sheep. PhD thesis. Monash University, Australia, 1992:106–130.
18. Joseph SA, Walker DW. Catecholamine neurons in fetal brain. Effects on breathing movements and electrocorticogram. J Appl Physiol 1990; 69:1903–1911.
19. Joseph SA, Walker DW. Effects of intracisternal monoamines on breathing movements in fetal sheep. Am J Physiol 1993; 264:R1139–R1149.
20. Bissonnette JM, et al. Effect of N-methyl-D-aspartate blockade on the level of arousal in fetal sheep. Sleep Res 1993; 22:423.
21. Sloviter RS. Decreased hippocampal inhibition and a selective loss of interneurons in experimental epilepsy. Science 1987; 235:73–76.
22. Bekenstein JW, Lothman EW. Dormancy of inhibitory interneurons in a model of temporal lobe epilepsy. Science 1993; 259:97–100.
23. Steriade M, et al. Bulbo-thalamic neurons related to thalamocortical activation processes during paradoxical sleep. Exp Brain Res 1984; 54:463–475.

38

GABAergic Inhibition and Respiratory Rhythmicity in the Developing Opossum

JAY P. FARBER

University of Oklahoma Health Sciences Center
Oklahoma City, Oklahoma

Respiratory rhythmicity is established relatively early during the course of mammalian development. In fact, for placental mammals, respiratory movements and breathing rhythms are well known to occur prenatally. The role of specific neurotransmitters and neuromodulators in the development of breathing patterns is not well defined. Also unclear is the role of neurotransmitter systems in the apneic episodes that occur frequently in premature infants. Maturational effects related to presynaptic release of neurotransmitters and/or the postsynaptic response to neurotransmitter release may influence the generation of breathing rhythm and contribute to apneic episodes.

Apneic events can be produced in models for respiratory rhythmogenesis based on either network or pacemaker properties. Models utilizing network properties alone depend in part upon inhibitory neuronal pools to switch between inspiratory and expiratory phases (1,2). Prolongation of the respiratory cycle would result if inhibitory output from such neuronal pools were reduced. Invocation of respiratory pacemaker cells in the form of a conditional oscillator (3) could produce a similar effect, depending on the effects of inhibitory input to appropriate bursting neurons. Most simply, inhibitory input will depress rates of depolarization and might even result in complete apnea. However, a rhythm generator with mixed properties might behave in a manner similar to one depending on network properties alone (3).

The following studies from our laboratory were aimed at showing how blocking the central actions of the inhibitory neurotransmitter, gamma aminobutyric acid (GABA), affects respiratory rhythm generation in the developing opossum (*Didelphis virginiana*). The results have been interpreted with respect to the occurrence of apnea.

Data previously obtained from the opossum showed that microinjection of GABA suppressed extracellularly recorded action potentials of medullary neurons with respiratory modulation. This effect was observed in neurons from sucklings as young as approximately 3 postnatal weeks (the youngest tested) as well as adults. Further, the $GABA_A$ receptor blocking agent, bicuculline (bic), increased rates of firing of respiratory neurons in animals studied as early as the fifth postnatal week (4). Portions of the GABAergic system, therefore, seemed functional in the opossum at a relatively early developmental stage. For the purpose of comparison, it takes about 9 postnatal weeks for neurological development in the opossum to be generally comparable to that of a newborn cat or rabbit.

To examine effects of blocking central $GABA_A$ receptors on breathing in the developing opossum, solutions containing bic (0.1 to 2.0 m*M*) were placed on the exposed ventral medullary surface of 3- to 11-week-old sucklings anesthetized with a thiobarbiturate (Inactin; BYK Gulden; Konstanz, Germany). Effects on diaphragm electromyography (EMG) were examined. Bic increased the length of the respiratory cycle in sucklings at all ages tested. Typically, inspiratory duration (Ti) was also increased, while the peak of the moving average EMG was unchanged. A number of animals became apneic even at low doses (e.g., 0.1 m*M*) of bic. Removal of drug and resuscitation was often required to reestablish a breathing rhythm. The preceding effects most often occurred without tonic activation of respiratory muscles. Further, all of the preceding effects occurred below threshold for induction of seizure activity by bic (5). Substituting alpha chloralose anesthetic in a small series of animals also suppressed breathing after bic, but with the peak of moving average EMG as well as breathing rate being reduced. Another $GABA_A$ receptor antagonist, gabazine (SR 95531), produced responses similar to those of bic.

To determine possible sites of action of bic, the medulla was explored using localized microinjection techniques. Animals were tested from about the fourth through ninth postnatal week. Bic was pressure-injected into the medulla through two-barrel glass micropipettes inserted from the ventral surface. Drug concentration was 6 m*M*, and approximate injection volumes were 0.15 to 0.5 nL. The second micropipette barrel contained 2% pontamine sky blue dye in saline to mark locations of injections and to serve as control for respiratory responses. In initial studies, bic was injected in stepwise fashion at progressively greater depths in rostral as well as more caudal sites. Rostral medullary sites were generally unresponsive to bic. Effects with microinjections, after variable delay, most often occurred from locations approximately at the level of the obex. Responses from ventrolateral sites, generally in the region of the nucleus ambiguus, consistently included an increased respiratory cycle duration (Ttot) (Fig. 1); the effect was sometimes large enough to be considered an apneic response (Fig. 2). Effects on diaphragm EMG discharge were variable. Duration of response was typically over 1 min. On the basis of surface coordinates, some of the smaller animals responded to relatively medial and deep injections at the same rostrocaudal level. Values of Ttot were increased in the majority of instances, but effects were less consistent. While these locations could be close to the dorsal respiratory group, the spread of drug within the medulla may be greater in relative terms for the smaller animals. Injections from the dye-containing pipette had little or no effect on breathing.

The preceding results suggest that blockade of $GABA_A$ receptors in discrete portions of the medulla can decrease rate of breathing, sometimes to the point of apnea. The small total amount of drug injected (approx. 1 to 3 pmol) and the small injection volume argue against spread to extramedullary portions of the brain. Injection of approx. 25 pL of bic at 2 and 6 m*M* concentrations in suckling opossums occasionally produced apparent short-duration suppression of extracellularly recorded action potentials (4). This effect was preceded

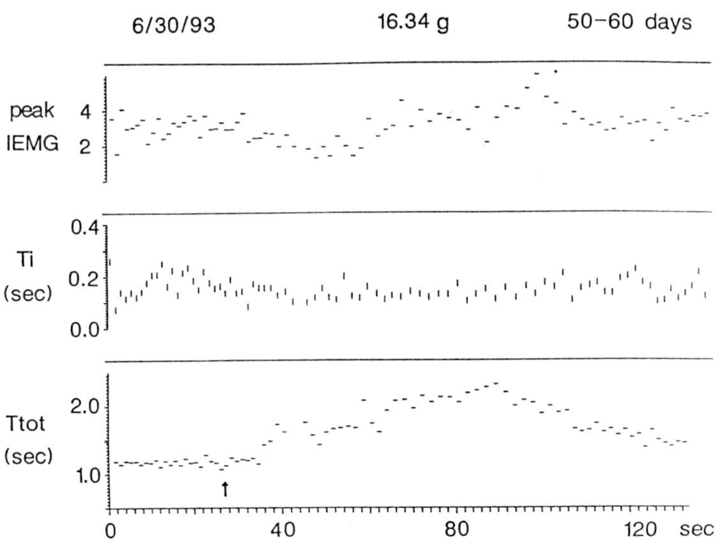

Figure 1 Breath-by-breath response to microinjection of 6 m*M* bic in an approx. 0.15-nL volume into the ventrolateral medulla. Top record shows the peak amplitude of the moving (RC) average of the diaphragm electromyogram (peak IEMG) in arbitrary units. Center record shows inspiratory time (Ti; s). Bottom record shows duration of respiratory cycles (Ttot; s). Bic is injected at the arrow (bottom record).

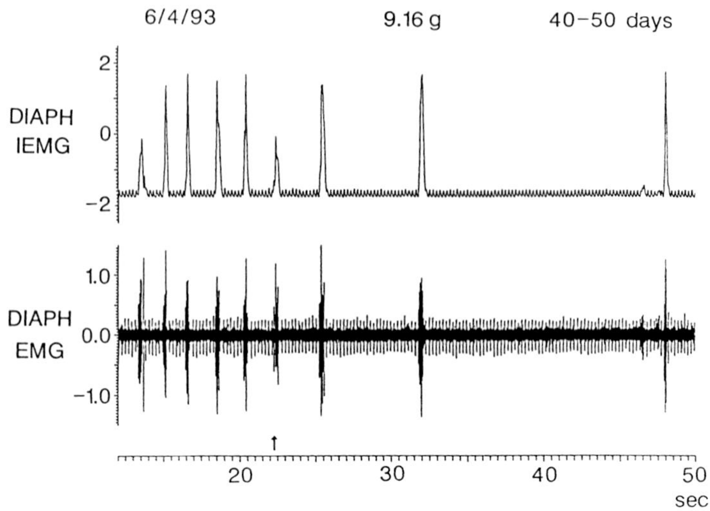

Figure 2 Occurrence of apnea after microinjection of 6 m*M* bic in approximately 0.15 nL into the ventrolateral medulla. Top trace shows the moving (RC) average of the diaphragm EMG (DIAPH IEMG); lower trace shows the raw diaphragm EMG (DIAPH EMG). Bic is injected at the arrow (below lower trace).

and followed by neuronal firing at much higher than control rates. If we assume that the preceding represents depolarization block, then it is also possible that the effects of bic on breathing pattern could arise from inability to fire action potentials as well as by blocking inhibitory effects. The relatively long duration of changes in breathing after microinjection, given the picomole quantities of bic injected, seems more consistent with cell excitation, as the drug concentration decreased over time and distance. Active sites in the present study were caudal to the pre-Bötzinger area. The pre-Bötzinger area described by Smith et al. (6) is critical for generation of breathing rhythm in the brainstem spinal cord preparation.

The present data can be compared with previous results from whole animal studies where effects on breathing of bic administration have been assessed. In the arterially perfused but intact rat, respiratory rate and inspiratory motor activity increased after bic (7). Anesthetized rabbits demonstrated decreases in both Ti and expired duration (Te), with increasing tonic inspiratory motor discharge after cerebral ventricular application of bic (8). Intravenous injection of bic in the decerebrate cat produced shortened Ti and increased inspiratory phrenic nerve activity at subconvulsant doses (9). With higher doses of bic, breathing rhythm was severely disrupted or absent (rat, cat). Even with lower doses of bic in the rabbit, bic sometimes eliminated respiratory rhythm; in those cases there was tonic activation of motor outflow to the diaphragm (8).

In vitro preparations treated with bic or withdrawal of chloride from the perfusate (to eliminate chloride channel–mediated inhibition at the postsynaptic membrane) produce a different effect. In the rat brainstem–spinal cord, bic administration or chloride removal augmented and prolonged inspiratory motor activity, but cycle length was unchanged (10). The brainstem–spinal cord of the tadpole also showed prolonged motor bursts associated with breathing after removal of extracellular chloride; this was coincident with minor prolongation of breath duration (11). In contrast to motor output affecting breathing, motor output to the gills of the tadpole was entirely suppressed by withdrawal of chloride (11). These studies have been used to support the concept of pacemaker activity for breathing that does not rely on periodic inhibition. The issue that must be addressed with respect to the present data is whether breathing rhythm in the intact preparation is dominated by network events which overwhelm the influence of a pacemaker for breathing. For example, removal of chloride from the arterial perfusate of the intact rat usually caused cessation of breathing rhythm (7).

In summary, the present results are consistent with the proposition that blockade of expiratory off-switching by GABAergic mechanisms prolongs Te and can result in apnea. Evidence from microinjection or microiontophoresis and recording studies supports the idea that GABA plays a role in the termination of discharge for populations of expiratory as well as inspiratory neurons in the medulla (4,12,13). It is of interest in this regard that intracellular recordings from respiratory neurons in the medulla of kittens showed relatively weak inhibitory postsynaptic potentials (14). This has been interpreted to suggest that inhibition was not needed to form rhythmic output from the respiratory areas of the medulla, possibly due to rhythm generation by a pacemaker. Alternatively, the present data suggest that relative weakness of inhibitory circuitry in the respiratory rhythm generator could result in apnea through suppression of phase-switching mechanisms. A third possibility may also be worthy of attention. That is, failure of inhibitory circuitry allows excessive depolarization, with apnea being derived from depolarization block of respiratory neurons involved with changing the phase of the breathing cycle.

Acknowledgment

This work was supported by HL-37318 from NIH.

References

1. Ezure K. Synaptic connections between medullary respiratory neurons and considerations on the genesis of respiratory rhythm. Prog Neurobiol 1990; 35:429–450.
2. Ogilvie MD, Gottschalk A, Anders K, et al. A network model for respiratory rhythmogenesis. Am J Physiol 1992; 263:R962–R975.

3. Feldman JL, Smith JC, Ellenberger HH, et al. Neurogenesis of respiratory rhythm and pattern: emerging concepts. Am J Physiol 1990; 259:R879–R886.
4. Farber JP. GABAergic effects on respiratory neuronal discharge during opossum development. Am J Physiol 1993; 264:R331–R336.
5. Farber JP, Hutson AM. Blockade of medullary $GABA_A$ receptors and pattern of breathing during opossum development. FASEB J 1993; 7:A399.
6. Smith JC, Ellenberger HH, Ballanyi K, et al. PreBötzinger complex; A brain stem region that may generate respiratory rhythm in mammals. Science 1991; 254:726–729.
7. Hayashi F, Lipski J. The role of inhibitory amino acids in control of respiratory motor output in an arterially perfused rat. Respir Physiol 1992; 89:47–63.
8. Schmid K, Boehmer G, Gebauer K. $GABA_A$ receptor mediated fast synaptic inhibition in the rabbit brain-stem respiratory system. Acta Physiol Scand 1991; 142: 411–420.
9. Karius DR, Ling L, Speck DF. Effects of intravenous bicuculline and strychnine on inspiratory inhibitory response in the cat. Brain Res 1992; 579:270–278.
10. Feldman JL, Smith JC. Cellular mechanisms underlying modulation of breathing pattern in mammals. Ann NY Acad Sci 1989; 563:114–130.
11. Pack AI, Galante RJ, Walker RE, et al. Comparative approach to neural control of respiration. In: Speck DF, Dekin MS, Revlette WR, Frazier DT, eds. Respiratory Control: Central and Peripheral Mechanisms. Lexington, KY: The University Press of Kentucky, 1992:52–57.
12. Champagnat J, Denavit-Saubie M, Moyanova S, Rondouin G. Involvement of amino acids in periodic inhibitions of bulbar respiratory neurons. Brain Res 1982; 237:351–365.
13. Haji A, Takeda R, Remmers JE. Evidence that glycine and GABA mediate postsynaptic inhibition of bulbar respiratory neurons in the cat. J Appl Physiol 1992; 73:2333–2342.
14. Lawson EE, Schwarzacher SW, Richter DW. Postnatal development of the medullary respiratory network in cat. In: Elsner N, Richter DW, eds. Rhythmogenesis in Neurons and Networks. Stuttgart: Thieme, 1992:69.

39

Role of the Ventral Medullary Surface in Modulating Respiratory Responses to Reflex Stimulation in Piglets

RICHARD J. MARTIN and MARTHA J. MILLER

Case Western Reserve University
and Rainbow Babies and Childrens Hospital
Cleveland, Ohio

ISMAIL A. DRESHAJ and MUSA A. HAXHIU

School of Medicine
Case Western Reserve University
Cleveland, Ohio

I. Introduction

Preterm infants, especially those with apnea, typically exhibit attenuated central chemosensitivity, as manifested by diminished ventilatory responses to CO_2 (1,2). Up to 70% of apneic episodes in such premature infants are classified as mixed apnea, during which upper airway (typically pharyngeal) obstruction accompanies a period of central apnea and may delay recovery of effective ventilation (3). It has been proposed that such mixed apnea in infants, as in adults, results from imbalance between activation of chest wall pumping and upper airway dilator muscles (4,5). There are, however, no systematically collected data on the coordination of these respiratory muscle groups when central respiratory drive is inhibited during early postnatal life.

Laryngeal stimulation is a well-documented trigger for reflex apnea, which serves to protect the lungs from aspiration. This response is most prominent during the neonatal period (6,7), and while it is assumed to be an essential protective reflex, an exaggerated response has been implicated as a cause for apnea of prematurity or sudden infant death syndrome. The mechanism responsible for the greater sensitivity of the respiratory system to the inhibitory effects of laryngeal stimulation early in development is not clear, although maturational changes in central chemosensitivity might contribute to postnatal alterations in the strength of this potent inhibitory reflex.

In order to address these issues, we developed a piglet model in which the level of central chemosensitivity could be modulated by cooling the intermediate area of the ventral medullary surface (VMS) (8,9). This technique has been previously described in adult animals (10–12). We hypothesized that during early postnatal life, (a) hypoglossal output is more sensitive than phrenic output to the withdrawal of central CO_2 chemosensitivity, (b) recovery of phrenic output precedes that of hypoglossal output as hypercapnia and hypoxia simultaneously develop under conditions of attenuated central respiratory drive, and (c) withdrawal of central chemosensitivity enhances the respiratory inhibition induced by laryngeal stimulation.

II. Effect of VMS Cooling on Hypoglossal Versus Phrenic Neural Output

All studies described here (and in the subsequent sections) were performed in 14- to 20-day-old tracheostomized piglets who were anesthetized with α-chloralose and urethane. Venous and arterial access were available for administration of anesthesia, blood pressure monitoring, and withdrawal of arterial samples. For those animals in whom respiratory neural output (ENG) was measured,

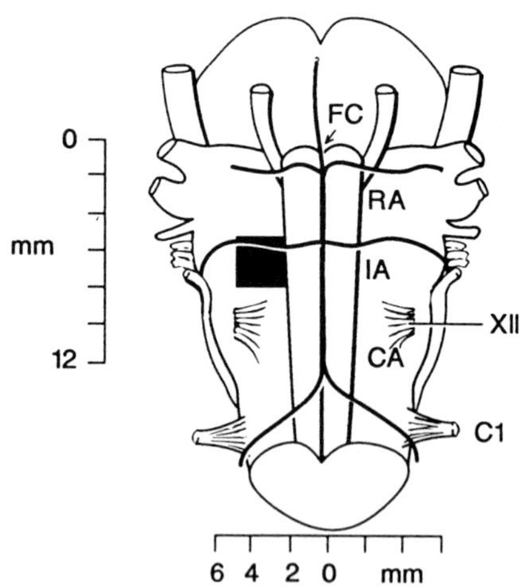

Figure 1 Ventral surface of the brainstem, with the shaded site of cooling corresponding with the intermediate area. (RA, rostral area; IA, intermediate area; CA, caudal area; FC, foramen cecum.)

paralysis was performed with gallamine and efferent electrical activity recorded from the hypoglossal nerve and the fifth or sixth root of the phrenic nerve. Bilateral vagotomy was performed in some piglets in order to eliminate the effects of lung mechanoreceptors in hypoglossal and phrenic neural output. For those animals in whom respiratory muscle activity [diaphragm electromyogram (EMG)] was

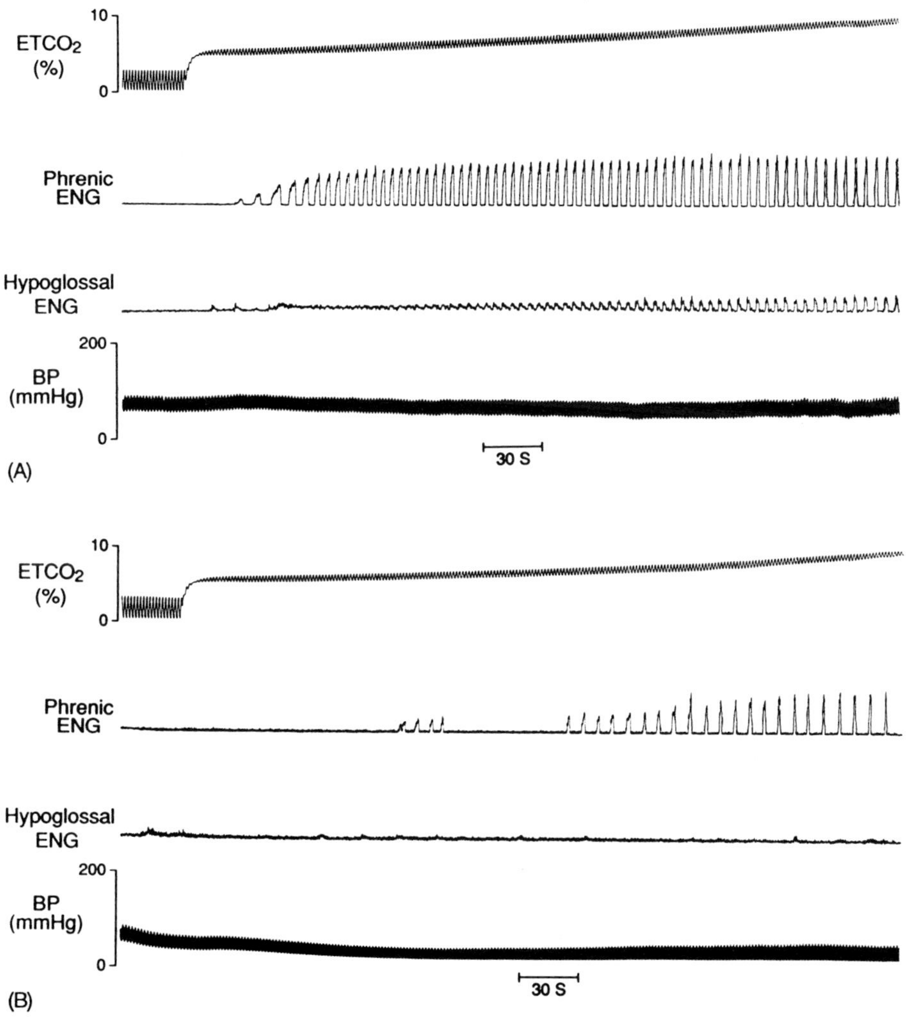

Figure 2 The phrenic and hypoglossal responses to CO_2 rebreathing in one piglet prior to (panel A) and during (panel B) cooling of the ventral medullary surface. VMS cooling caused a delayed, attenuated phrenic response and eliminated the hypoglossal response. (ET, end-tidal CO_2; BP, blood pressure.)

measured, bipolar stainless steel electrodes were implanted under direct vision into the central part of the diaphragm.

After the initial animal preparation, as indicated above for each study, a large craniotomy was performed and the dura was incised over the ventral medullary surface of the brainstem. In order to change the surface temperature of the VMS, a small liquid-cooled, two-footed probe (thermode) was used. The two thermode surfaces of 2 × 2 mm gently made contact with both sides of the ventral medulla above the exit of the hypoglossal roots (Fig. 1). A thermocouple wire was attached to the base of the thermode for recording surface medullary temperature. The temperature of the thermode was adjusted by changing the flow of either cold water (from a 5°C water bath when cooling) or warm water (from a 38°C water bath when warming).

In this series of experiments, seven piglets were hyperventilated to phrenic apnea, after which the inspired gas mixture was switched from 100% O_2 to a

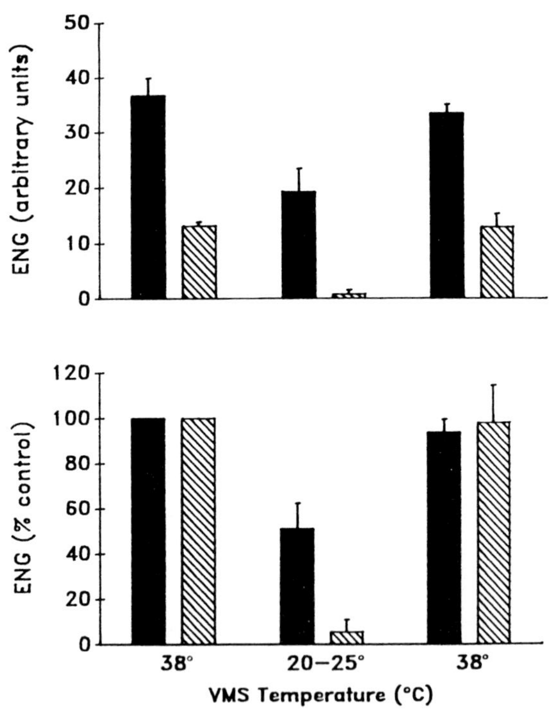

Figure 3 The effect of VMS cooling (and subsequent rewarming) on the peak phrenic (solid) and hypoglossal (striped) responses during hypercapnia. There was preferential inhibition of hypoglossal versus phrenic neural output regardless of whether ENG were expressed in arbitrary units or as a percent of control responses.

rebreathing mixture of 5% CO_2 in O_2 until expired CO_2 reached 10.5%. In each of these animals, CO_2 rebreathing was performed at VMS temperatures of 38°C and 20 to 25°C while continuously recording phrenic and hypoglossal ENG, end-tidal CO_2, and blood pressure. Responses of phrenic and hypoglossal neural output were then calculated at increasing levels of end-tidal CO_2 by averaging ENG activities over five breath epochs with every 0.5% increment in end-tidal CO_2. In three of the animals, an additional CO_2 rebreathing experiment was performed to compare the hypoglossal and phrenic responses to intermittent VMS cooling during the course of CO_2 rebreathing.

Cooling of the VMS significantly increased the CO_2 threshold for phrenic activation and decreased the peak phrenic response during CO_2 rebreathing. In each animal studied, VMS cooling caused even greater inhibition of hypoglossal than phrenic neural output. As indicated from the data in Figs. 2 and 3, VMS cooling caused elimination of the hypoglossal ENG during CO_2 rebreathing in six of the seven animals studied. When intermittent VMS cooling and rewarming were performed during hypercapnic exposure in three animals, there was greater inhibition of hypoglossal than phrenic neural output (Fig. 4).

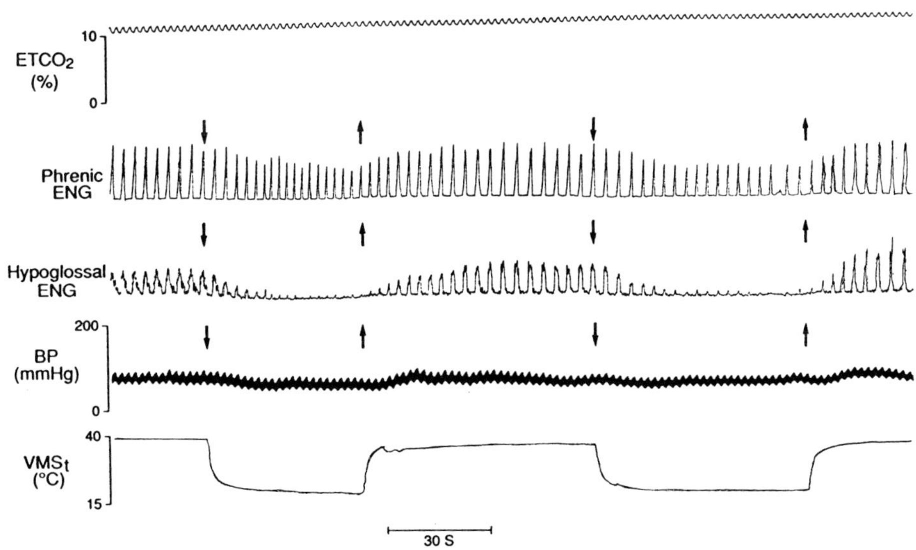

Figure 4 The phrenic and hypoglossal responses to intermittent cooling (↓) and subsequent rewarming (↑) of the VMS in one piglet during hypercapnia (as indicated by high ET CO_2). On each occasion, VMS cooling caused greater inhibition of the hypoglossal than phrenic ENG. Each episode of cooling was associated with a transient decrease in blood pressure. (From Ref. 9.)

III. Recovery of Hypoglossal Versus Phrenic Neural Output During Simultaneous Hypercapnia and Hypoxia

In these studies, 12 piglets were ventilated with a mixture of 30 to 40% inspired oxygen (balance room air) in order to attain relatively normoxic (Pa_{O_2} 106 ± 12 mm Hg) and normocapnic (Pa_{CO_2} 44 ± 6 mm Hg) conditions, as determined by blood gas evaluations (both mean ± SD). VMS cooling was then performed to 20 to 25° until phrenic apnea occurred. While VMS cooling was maintained, assisted ventilation was discontinued for approximately 60 s and a blood gas then obtained prior to resumption of assisted ventilation. This study protocol (Fig. 5) allowed us to quantify the time course of breath-by-breath recovery of phrenic and hypoglossal ENG under conditions of decreased central chemosensitivity.

There was a significant difference in the recovery of phrenic hypoglossal ENG to simulated central apnea (induced by VMS cooling) associated with development of hypercapnia and hypoxia. Recovery of respiratory ENG could be systematically analyzed in all subjects over the first seven respiratory efforts, which always occurred within 30 s of discontinuing assisted ventilation. Recovery of hypoglossal neural output was significantly attenuated when compared to that of the phrenic, as exemplified in data from Figs. 6 and 7. In 50% of the piglets, VMS cooling and cessation of assisted ventilation caused a subsequent loss of phrenic and hypoglossal neural activities, as seen in Fig. 7.

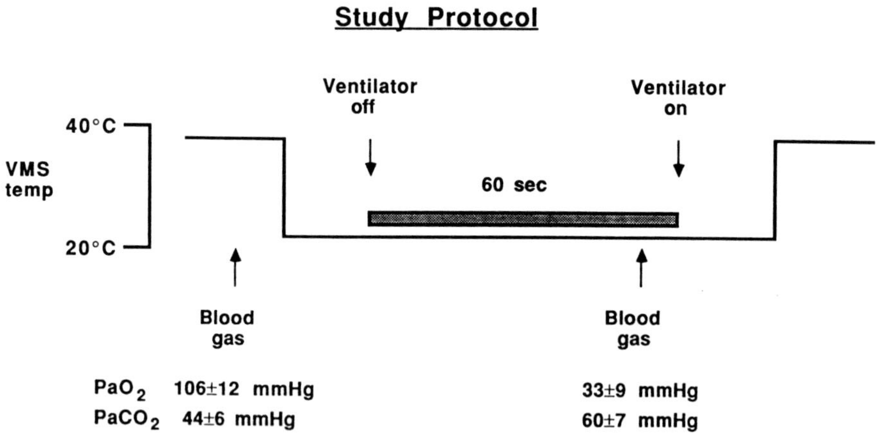

Figure 5 Schematic representation of the protocol employed to produce simultaneous hypercapnia and hypoxia during VMS cooling. The blood gas changes associated with cessation of assisted ventilation are expressed as mean ± SD. (From Ref. 9.)

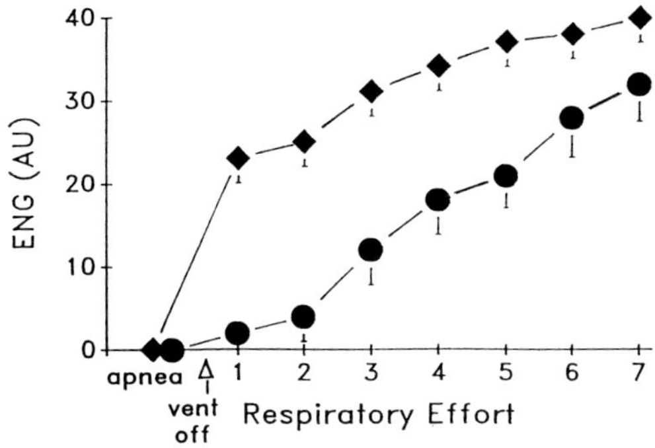

Figure 6 The recovery of phrenic and hypoglossal ENG (expressed in arbitrary units) from apnea induced by VMS cooling and cessation of assisted ventilation (↑). As respiratory neural output (efforts) reappeared, the hypoglossal response (●) was significantly attenuated when compared to the phrenic response (◆). Data are expressed as mean ±SEM. (From Ref. 9.)

IV. Effects of VMS Cooling on Respiratory Inhibition Induced by Laryngeal Stimulation

In a series of 10 piglets, the response of diaphragm EMG to superior laryngeal nerve (SLN) stimulation was evaluated at different levels of central chemosensitive drive. The superior laryngeal nerves (SLN) were isolated bilaterally and were cut at their entrance to the musculature of the larynx. SLN stimulation was performed at the minimal amplitude (threshold) that elicited a reduction (≥20%) in diaphragm EMG (DIA EMG), and the minimal amplitude (threshold) that elicited apnea (≥10 s duration) in each animal during normocapnia and with the VMS at 38°C. The same protocol for SLN stimulation was repeated during cooling of the VMS under normocapnic hyperoxic conditions and during inhalation of 7% CO_2 93% O_2 for 5 min at VMS temperatures of 38° and during cooling.

Threshold current of SLN stimulation during normocapnia caused the expected inhibition of diaphragm EMG, as seen in Fig. 8. During VMS cooling (under normocapnic conditions), SLN stimulation at the same threshold current caused significantly greater inhibition of the DIA EMG when compared to normocapnic control. In contrast, during hypercapnia, SLN stimulation caused significantly less inhibition of DIA EMG than during normocapnic control (Fig. 8). The current that induced apnea (of at least 10 s duration) was also significantly less during VMS cooling (18 ± 6 μamp) and significantly greater during hypercapnia (30 ± 10 μamp) when compared to normocapnia (24 ± 9 μamp) at a VMS temperature of 38°C (all mean ± SD). The effect of

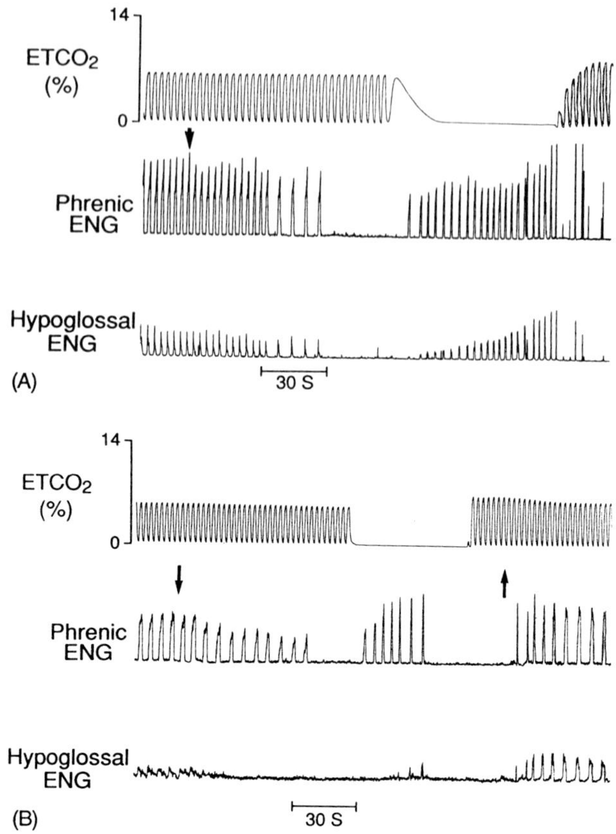

Figure 7 Representative tracings of the phrenic and hypoglossal responses to VMS cooling (↓), followed by cessation of assisted ventilation as indicated by absence of an end-tidal (ET) CO_2 tracing. The data from both animals (panels A and B) demonstrate the typically delayed recovery of hypoglossal versus phrenic activity. There was a subsequent loss of phrenic and hypoglossal activities in the animal depicted in panel B. This occurred prior to resumption of assisted ventilation (as indicated by the end-tidal CO_2 tracing). Rewarming of the VMS (↑) restored phrenic and hypoglossal neural output in this animal. (From Ref. 9.)

SLN stimulation at threshold current on DIA EMG during combined VMS cooling and hypercapnia was comparable to that during normocapnia (at a VMS temperature of 38°C), confirming that VMS cooling eliminated the effect of hypercapnia on the DIA response to SLN stimulation.

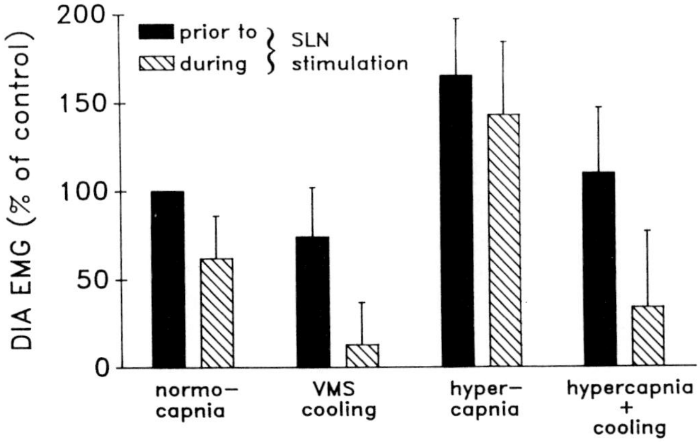

Figure 8 The response of diaphragm (DIA) EMG to superior laryngeal nerve (SLN) stimulation at threshold current performed during normocapnia, VMS cooling, hypercapnia, and combined hypercapnia and VMS cooling. EMG data (mean ± SD) are expressed as percent of control prior to SLN stimulation during normocapnia. SLN stimulation caused greater inhibition of DIA EMG during VMS cooling and less DIA inhibition during hypercapnia. DIA inhibition during the combined effects of hypercapnia and VMS cooling was comparable to that during normocapnic control conditions. (From Ref. 8.)

V. Discussion

We have performed a series of studies to characterize the role of the VMS in modulating respiratory output during early postnatal life. The technique of VMS cooling has been employed to withdraw central CO_2 chemosensitivity and evaluate the effects on chemo- and mechanoreceptor-mediated reflex responses that form an integral part of respiratory control during maturation. Within the limitations of our model, we have demonstrated that withdrawal of central chemosensitive drive causes greater inhibition of hypoglossal than phrenic neural output, delayed recovery of hypoglossal (as compared to phrenic) activity during recovery from simulated central apnea, and enhanced vulnerability to the inhibitory effects of laryngeal stimulation.

Cooling the intermediate area of the VMS in adult animal models is known to cause hypoventilation and reduce the phrenic or diaphragmatic responses to hypercapnia (10,11). It has been proposed that the intermediate area of the VMS facilitates central chemosensitive drive and acts as a relay for neural signals between the central chemoreceptors and respiratory rhythm generators (12). Experimental studies have been performed in adult animals to compare the hypoglossal and phrenic responses to VMS cooling. Studies in anesthetized vagotomized adult cats suggested that cooling of this area impaired phrenic responses to hypercapnia more than hypoglossal responses (13). Similar but more variable results were observed in anesthetized kittens (14). Haxhiu et al. subsequently reported that VMS cooling decreased hypoglossal more than phrenic activity in the same species (15). Our results in piglets support the data of Haxhiu et al. in adult cats (15), that cooling of the intermediate area of the VMS causes greater withdrawal of CO_2 sensitivity in hypoglossal than phrenic neural output. It is therefore possible that the decreased chemosensitivity associated with early postnatal maturation (1) and enhanced by apnea of prematurity (2) may preferentially inhibit neural output responsible for maintenance of upper airway patency.

Mixed apnea in spontaneously breathing infants is associated with decreased central respiratory drive, as manifested by central respiratory pauses and an attenuated DIA EMG in the obstructed inspiratory efforts (16). Subsequent resumption of airflow is associated with a high supraglottic resistance in the first breath after termination of apnea (17). We therefore sought to characterize the recovery of phrenic and hypoglossal neural output as hypercapnia and hypoxia simultaneously developed during a period of decreased central chemosensitive drive induced by VMS cooling. We observed that recovery of hypoglossal activity was significantly attenuated when compared to phrenic recovery. Such imbalance in the activity of cranial and phrenic motoneurons could compromise oropharyngeal patency and prolong recovery from central apnea by causing obstructed inspiratory efforts, as has been proposed from adult models (4,5).

Spontaneous apnea is associated with simultaneous development of hypercarbia and hypoxia. Nonetheless, previous studies have only evaluated the effects of either hypercapnia or hypoxia alone on upper airway muscle and diaphragm responses during postnatal maturation (18,19). In response to CO_2 rebreathing (under hyperoxic conditions) in healthy preterm infants, we have previously shown that genioglossus EMG exhibits a delayed response when compared to the diaphragm, consistent with a higher CO_2 threshold for activation of this upper airway muscle (18). In the studies described here, VMS cooling all but eliminated hypoglossal neural output during CO_2 rebreathing, despite $PaCO_2$ values exceeding 70 mm Hg. However, in response to cessation of assisted ventilation during VMS cooling, more modest levels of hypercapnia were consistently associated with measurable hypoglossal ENG. Therefore hypoxia, either alone or in combination with hypercapnia, was probably the stimulus for hypoglossal activation under these conditions. Subsequent disappearance of phrenic neural output during cessation of assisted ventilation in many of the piglets was probably the result of hypoxic respiratory depression, enhanced by the withdrawal of central chemosensitive drive during VMS cooling.

There is considerable evidence that laryngeal stimulation, typically induced by water in the larynx, causes apnea, and this has been most widely studied in preterm infants (6,7). While the prominence of this reflex response has been well documented in the newborn of several animal species, the mechanism for such a maturational effect is not clear (20). The strength of this inhibitory reflex during early life may not be directly related to maturational changes in laryngeal receptors or their afferent pathways but rather to changes in central integration of afferent laryngeal input. Such interaction between central respiratory drive and laryngeal-induced reflex apnea has been reported previously in piglets. Lee et al. demonstrated that the increased respiratory drive induced by aminophylline administration decreased the inhibitory effect of largyneal stimulation (21). Lawson subsequently demonstrated that in response to SLN stimulation, the length of the apnea that followed stimulus cessation was inversely related to the piglets' Pa_{CO_2} (22), consistent with our own observation that hypercapnia modulated phrenic activity during SLN stimulation. Van Vliet and Uenishi recently studied the nature of this interaction between central chemoreceptor and laryngeal afferent stimulation in decerebrated, vagotomized dogs; they proposed that increased feedback from central chemoreceptors can limit the inspiratory inhibition induced by SLN stimulation (23). Our own observation that decreased central chemosensitive drive enhances laryngeal-induced respiratory inhibition causes us to speculate that decreased central chemosensitivity may contribute to the strength of this inhibitory reflex response during early postnatal life (Fig. 9).

There are obvious limitations associated with extrapolating data from the anesthetized piglet to spontaneously breathing human infants with apnea of prematurity. Cooling the intermediate area of the VMS may attenuate functional activity of central chemosensitive structures by a different series of mechanisms than impair central chemosensitivity in apneic or immature preterm

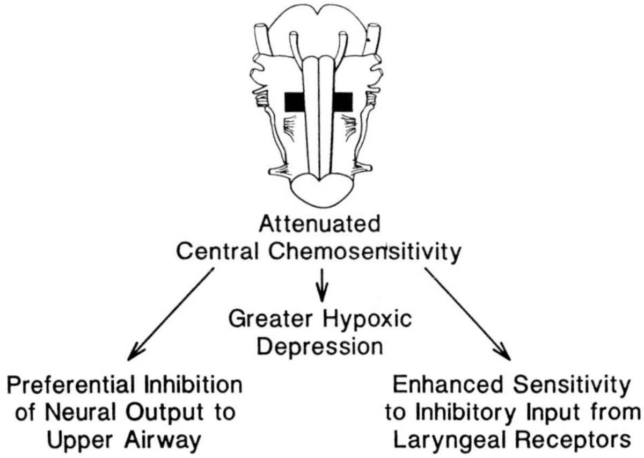

Figure 9 Hypothetical framework for some of the mechanisms whereby attenuated central chemosensitivity enhances vulnerability to neonatal apnea.

infants. Nonetheless, this technique did allow us to withdraw central chemosensitive inputs from respiratory-related neurons in the brainstem and to functionally diminish central chemosensitive drive. The result was significantly greater inhibition of hypoglossal than phrenic neural output during early postnatal life. Furthermore, simultaneous exposure to modest hypercapnia and hypoxia, under conditions of decreased central chemosensitivity, caused a delayed, attenuated hypoglossal response. These findings may be relevant to our understanding of the pathogenesis of neonatal apnea, the predisposition of infants to develop mixed apnea due to upper airway instability, and their vulnerability to respiratory inhibition induced by laryngeal stimulation.

Acknowledgment

Supported by HL25830. The authors wish to acknowledge the important contributions played by Juliann DiFiore and Marty Hughes in completing this manuscript.

References

1. Rigatto H, Brady JP, de la Torre Verduzco R. Chemoreceptor reflexes in preterm infants: II. The effect of gestational and postnatal age on the ventilatory response to inhaled carbon dioxide. Pediatrics 1975; 55:614–620.
2. Gerhardt T, Bancalari E. Apnea of prematurity: II. Respiratory reflexes. Pediatrics 1984; 74:63–66.
3. Miller MJ, Martin RJ. Pathophysiology of apnea of prematurity. In: Polin RA, Fox WW, eds. Fetal and Neonatal Physiology. Philadelphia: Saunders, 1992:872–885.
4. Remmers JE, DeGroot WJ, Sauerland EK, Anch AM. Pathogenesis of upper airway occlusion during sleep. J Appl Physiol 1978; 44:931–938.
5. Brouillette RT, Thach BT. Control of genioglossus muscle inspiratory activity. J Appl Physiol 1980; 49:801–808.
6. Perkett EA, Vaughan RL. Evidence for a laryngeal chemoreflex in some human preterm infants. Acta Pediatr Scand 1982; 71:969–972.
7. Davies AM, Koening JS, Thach BT. Upper airway chemo-reflex responses to saline and water in preterm infants. J Appl Physiol 1988; 64:1412–1420.
8. Litmanovitz I, Dreshaj I, Miller MJ, et al. Central chemosensitivity affects respiratory muscle responses to laryngeal stimulation in the piglet. J Appl Physiol. In press, 1993.
9. Martin RJ, Dreshaj I, Miller MJ, Haxhiu MA. Hypoglossal and phrenic responses to central respiratory inhibition in piglets. Respir Physiol 1994; 97:93–103.
10. Schläfke M, Loeschcke HH. Lokalisation eines an der Regulation von Atmung und Kreislauf beteiligten Gebietes an der ventralen Oberflache der Medulla oblongata durch Kalteblockade. Pflügers Arch 1967; 297:201–220.

11. Cherniack NS, von Euler C, Homma I, Kao FF. Graded changes in central chemoceptor input by local temperature changes on the ventral surface of medulla. J Physiol 1979; 287:191–211.
12. Milhorn DE, Eldridge FL. Role of ventrolateral medulla in regulation of respiratory and cardiovascular systems. J Appl Physiol 1986; 61:1249–1263.
13. Bruce EN, Mitra J, Cherniack NS. Central and peripheral chemoreceptor inputs to phrenic and hypoglossal motoneurons. J Appl Physiol 1982; 53:1504–1511.
14. Bruce EN. Hypoglossal and phrenic nerve responses to chemical stimuli in kittens. In: von Euler C, Lagercrantz H, eds. Neurobiology of the Control of Breathing. New York: Raven Press, 1985:75–80.
15. Haxhiu MA, Mitra J, van Lunteren E, et al. Responses of hypoglossal and phrenic nerves to decreased respiratory drive in cats. Respiration 1986; 50:130–138.
16. Gauda EB, Miller MJ, Carlo WA, et al. Genioglossus and diaphragm activity during obstructive apnea and airway occlusion in infants. Pediatr Res 1989; 26:583–587.
17. Miller MJ, Petrie TG, DiFiore JM. Changes in resistance and ventilatory timing that accompany apnea in premature infants. J Appl Physiol 1993; 75:720–723.
18. Carlo WA, Martin RJ, DiFiore JM. Differences in CO_2 threshold of respiratory muscles in preterm infants. J Appl Physiol 1988; 65:2434–2439.
19. Martin RJ, van Lunteren E, Haxhiu MA, Carlo WA. Upper airway muscle and diaphragm responses to hypoxia in the piglet. J Appl Physiol 1990; 68:672–677.
20. Donnelly DF. Laryngeal reflexes. In: Haddad GG, Farber JF, eds. Developmental Neurobiology of Breathing. New York: Dekker, 1991.
21. Lee JC, Stoll BJ, Downing SE. Properties of the laryngeal chemoreflex in neonatal piglets. Am J Physiol 1977; 233:R30–R36.
22. Lawson EE. Recovery from central apnea: Effect of stimulus duration and end-tidal CO_2 partial pressure. J Apply Physiol 1982; 53:105–109.
23. Van Vliet BN, Uenishi M. Antagonistic interaction of laryngeal and central chemoreceptor respiratory reflexes. J Appl Physiol 1992; 72:643–649.

40

Response of Brainstem and Cortical Neurons to O_2 Deprivation

GABRIEL G. HADDAD

Yale University School of Medicine
New Haven, Connecticut

I. Introduction

In order to understand how nerve cells function together in a network, whether at rest or during metabolic stress, it would be important to examine the physiology and biology of individual nerve cells. Although, to a certain degree, the study of isolated nerve cells might be considered artificial, since they would not be in their "natural" environment, investigations of single neurons reduce the system in such a way as to allow the examination of the *inherent* properties and responses of individual nerve cells to stimuli. Clearly, however, we have to keep in mind that the individual nerve cell response might be more difficult to observe in a network than in isolation.

One such response that we and other investigators have focused on relates to the response of neurons to lack of oxygenation in the microenvironment of nerve cells themselves. Some of the questions we have tackled are: How do nerve cells in the rat CNS respond to O_2 deprivation? Does the response depend on the type of cell deprived of O_2? Does the newborn rat have a similar response to that of the adult? Do rat nerve cells respond to O_2 deprivation as other mammalian species do, including humans? How does the response depend on cell membrane characteristics? Finally, how does the environment affect the response, especially that it may contain neuromodulators and ionic constituents that may affect the responsiveness of the cell? By trying to answer such questions, we have, in the past several years, made a number of observations and built some novel concepts and ideas regarding the cascade of events that occurs in nerve cells once they are subjected to low O_2 or total lack of O_2.

Depending on region, function and connections of nerve cells, it would be interesting to hypothesize a priori that the response to O_2 deprivation varies substantially. For example, a brainstem motoneuron that innervates a respiratory muscle, say the diaphragm or the genioglossus, may respond differently to hypoxia than an interneuron or a neuron from a different region, such as a neocortical neuron. On the one hand, from a teleological point of view, it would be important for a brainstem neuron controlling vital functions such as heart rate, blood pressure, or ventilation not to be depressed but rather to be excited when exposed to hypoxia, in spite of the potential benefits of depressing neuronal activity for cell survival when O_2 is limited. On the other hand, a transient depression in neuronal activity might be appropriate and beneficial for nerve cells whose activity may not be crucial for maintaining vital signs and immediate survival of the organism.

In this chapter, I summarize some of our recent data on the response of nerve cells to lack of O_2 and derive some important ideas and possibly some concepts. Since we have been interested in developmental issues related to oxygenation, I also detail some of the neonatal responses. Furthermore, I present data regarding some of the mechanisms involved in such responses. The remainder of this chapter is then divided into three sections. The first examines the overall response of adult and neonatal brainstem nerve cells to O_2 deprivation and the mechanisms involved in the hypoxia-induced depolarization. The second draws a parallel between brainstem and cortical neurons in their responses to O_2 lack. The third demonstrates the importance of the inherent properties of neurons in the overall response to hypoxia. Since this is becoming an area of great interest and there is already a large literature, most of this chapter is limited to describing some of our own data and conclusions that we have drawn from recent work on the effect of O_2 deprivation on neuronal function.

One of the major and important methodological questions that we dealt with is what is the nature of the reduced preparation that would be suitable for the study of these neurons. We elected, for a number of reasons, to examine these questions using either the brainstem or cortical slice preparation or freshly dissociated neurons.

II. Electrophysiological Response of Brainstem Neurons to Hypoxia

A. Adult Brainstem Responses

We have so far studied both hypoglossal (XII) and dorsovagal motoneurons (DMX). We focus in this chapter on the former, since we have now a large amount of data on the properties of these neurons and on their responses to hypoxia. We have studied so far more than 400 XII neurons and we know now that, in response to hypoxia (defined below), XII neurons depolarize, and the extent of depolarization depends on the severity of hypoxia (1,2). Over a 5-min period, XII motoneurons depolarize on average about 30 mV. The depolarization starts slowly with a latency of about 20 to 30 s and increases in rate tremendously between 1.5 and 2.5 min. Most of the depolarization occurs between 1.5 and 4.0 min, with an earlier depolarization when hypoxia is severe than when it is moderate (1,2). Furthermore, the firing frequency increases remarkably in response to injected current but abates later within the 5-min period to even below baseline levels. Apparent also is the increase in synaptic noise in the first phase of the hypoxic period (1). Other electrophysiological changes are also detected. For example, input resistance (R_N) either increases initially and drops or is attenuated directly without any increase (1). This attenuation in R_N depends on the duration and extent of hypoxia (2).

B. Neonatal Brainstem Responses

We defined neonatal neurons functionally in terms of their electrophysiological response to lack of O_2. Our strategy was to study neurons at timely intervals in the first 4 weeks of life and determine when the response to hypoxia changes from the pattern of a neonate to that of an adult (1). We found that this shift occurred between 2 and 4 weeks of age, as the pattern at 2 weeks was similar to that at 3 to 7 days of age. At 3 weeks of age, the pattern was somewhat in-between, and that at 4 weeks was very similar to that in the adult. The neonatal response consisted of a much smaller depolarization over the same hypoxia period (Fig. 1) (1). For example, neurons depolarized approximately 10 to 11 mV during the same degree of hypoxia. It was only when the hypoxic period became much longer—i.e., about 15 to 30 min—that the depolarization was about 80% of that in the adult. Therefore, the neonatal neuron had a much smaller response and resisted this depolarization. R_N did not change appreciably and the firing frequency increased somewhat without showing any subsequent decrease in discharge, as in the adult.

C. Hypoxia, Anoxia, and Ischemia

Much has been written about hypoxic or ischemic brain changes or injury. However, much confusion has also been generated with respect to the mechanisms involved in these changes. One principal issue is the nature and kind of stimulus presented to the brain or nerve cells. We and others have demonstrated that, depending on the level of O_2 deprivation, hypoxia may or may not have major effects on the brain; and the level of inspired O_2, or the level of P_{CO_2} in the arterial blood, is not a good index of brain interstitial P_{O_2}. In addition, ischemia and O_2 deprivation are very different from each other and will induce different physiological and pathophysiological responses in nerve cells.

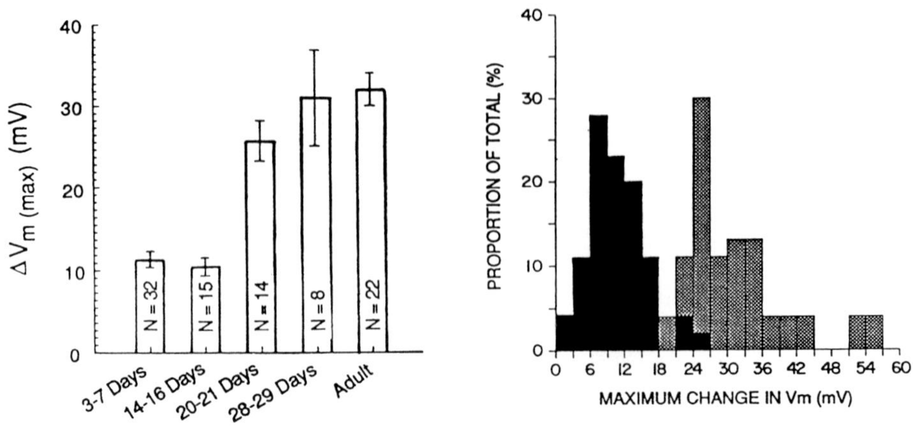

Figure 1 The anoxia-induced depolarization increases with age. Note that this depolarization reaches an adult level at about 1 month of age (left). In addition, the population of cells in the first 2 weeks of life belongs to a virtually disjointed population when compared to the population of cells at 28 to 29 days (right). (From Ref. 1.)

In order to be consistent and able to compare our results to those of others, we have measured the Po_2 tension in slices or in close approximation of dissociated cells using techniques and electrodes developed in our laboratory (3,4). We have used carbon fiber polarographic microelectrodes in both neonatal and adult brainstem slices under several experimental conditions. A number of variables were changed, including temperature, depth of measurement, brain region, animal age, tissue thickness, and ambient O_2 level (3). Interesting results were obtained, and some of these should be kept in mind when we compare animal age, different regions of the brain, and so on. We found that baseline brain tissue Po_2 was inversely proportional to distance from the surface of the slice and its temperature. In addition, white matter had a greater Po_2 than gray matter and, compared to the same region, the neonatal tissue had also a higher Po_2 than the mature one. Tissue thickness also had a major effect. For example, in 800 μm slices, Po_2 dropped to 0 at 200 to 300 μm in adult slices but not in the 400 μm slices. Furthermore, we found that graded hypoxia gave graded responses in terms of Po_2 in the slice (Fig. 2) (3).

In general, we have used the term *hypoxia* when, after reaching equilibrium, the Po_2 in the slice is about 10 to 15 torr and the term *anoxia* when the Po_2 is zero after 20 to 30 s of the start of the stimulus (3,4). Ischemia, on the other hand, is *not* equivalent to anoxia and clearly involves more than depriving the tissue of O_2 (5). This is particularly important with respect to excitatory amino acids, glucose, free fatty acids, lactic acid, and a number of other nutrients or waste products (5).

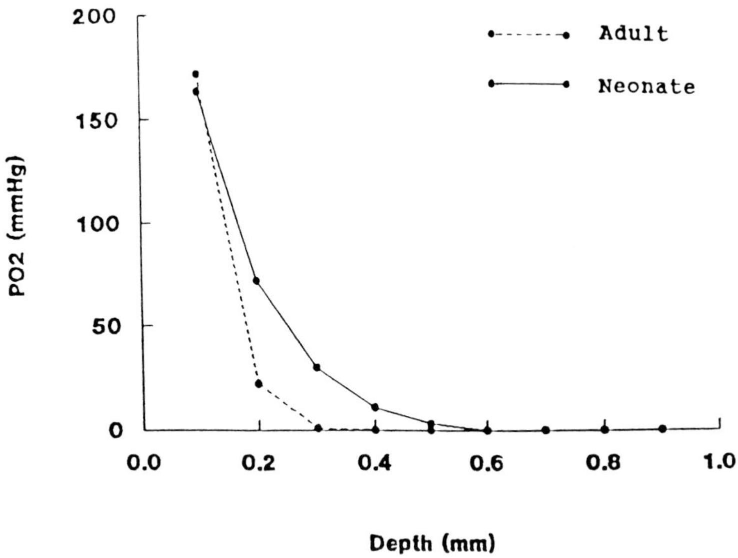

Figure 2 Importance of depth in determining Po_2 in the slice preparation. Note differences between neonate and adult. (From Ref. 3.)

D. Mechanisms of Anoxia-Induced Depolarization

A number of hypotheses with respect to the anoxia-induced depolarization were tested. First, this depolarization was not likely due to increased synaptic activity, as tetrodotoxin (TTX), high Mg^{++}/low Ca^{++}, and intracellular EGTA did not block this depolarization (1). Since K^+ increases markedly in the adult but not in the newborn in the extracellular space, one other possibility is that the anoxia-induced depolarization is related to the released K^+ (1,2,6). Indeed, K^+ increases by about 10-fold in the adult slice but by about 1 to 2 m*M* in the newborn when subjected to equally very low Po_2 (1,2,6).

Another possibility is that, in the absence of O_2, the increase in glutamate that we and others have demonstrated to occur in the extracellular space (7), depolarizes cells in the slice. However, this did not prove to be the case, as blockers of *N*-methyl-D-aspartate (NMDA) and non-NMDA receptors did not change the depolarization trajectory or pattern (8). Therefore, we believe that excitatory amino acids may have an impact on this depolarization, but on a different time scale, and potentially on the integrity of nerve cells at later stages.

One very intriguing finding in our slice work was the observation that low Na^+ (5 m*M*) in bathing solutions attenuated markedly the depolarization observed during anoxia (8). Although we do not have data on intracellular (intraneuronal) Na^+ in the slice preparation, extracellular Na^+ decreases in a parallel fashion to the depolarization seen in brainstem cells (9). We believe, therefore, that it is possible that a major component of the depolarization is based on Na^+ entry into neurons. It is important to keep in mind that a small concentration of Na^+ can increase substantially the net positive charges on the inside of the membrane; therefore the measurement of such small Na^+ change may not be easy.

Two related issues can be raised. If Na^+ gradient is decreasing, is it possible that the level of ATP is being reduced during the hypoxic period, and that this may be the reason for the collapsing Na^+ gradient? Also, is it possible that because of this collapsing gradient, the energy required to keep the remaining gradient will also be reduced? In order to determine whether the electrophysiological measures are influenced by the presence or absence of ATP, we dissociated cells, patched them, and examined their whole-cell holding potential during normoxia, hypoxia without loading the cells with ATP, and with loading the neurons patched with ATP. We found that the increase in the hyperpolarizing holding potential is significantly increased with anoxia (or cyanide), but that loading the pipette with ATP attenuated this increase in holding potential (i.e., the depolarizing potential that would have occurred had the cell not been in voltage-clamp mode) (8). It is important to note that the increase in holding potential was not eliminated during hypoxia when ATP was available, indicating that part of this depolarizing effect is not ATP-dependent. Therefore, our belief at this stage is that brainstem neurons are depolarizing in hypoxia partly because of (a) an increase in K^+ in the extracellular space and (b) an increase in Na^+ inside neurons (6,8). This observation should not be synonymous with inhibition with Na^+/K^+ ATPase, since total inhibition of this enzyme with high concentration of ouabain and strophanthidin did not block all K^+ release (half of K^+) to the extracellular space.

III. Comparative Studies: Brainstem, Hippocampus, and Neocortex

Since (a) the response of nerve cells to hypoxia depends on many factors and (b) these factors vary from one cell type to another and from one region to another, is it possible that neurons in the central nervous system (CNS) respond differently to hypoxia or anoxia? What are these factors that this response depends on? How important are these issues to injury of nerve cells when they are exposed to relatively short periods or more prolonged periods of low O_2?

Although it is clear that most mammalian cells are sensitive to O_2 deprivation, whether in the central or peripheral nervous system, whether in excitable or nonexcitable cells, it is still unclear how cells, and especially nerve cells, respond to graded O_2 deprivation and what are the reasons or the mechanisms responsible for the differential sensitivities or responsiveness to lack of O_2. We asked questions pertaining to these issues in the CNS and focused on a number of regions or neuronal types. Brainstem neurons were thus compared to neurons in the hippocampus and the neocortex for different reasons, including the facts that (a) the hippocampus is known to be extremely sensitive to anoxia, (b) the function of these cells is different in different parts of the brain, and (c) different connectivity within rostral or caudal neurons might play an important role in the sensitivity (or lack thereof) to hypoxia.

A number of electrophysiological measures were first examined in comparing the various cell types. We found both quantitative and qualitative effects that depended on the severity and duration of the stimulus and on the inherent properties of these neurons. There were three parameters that we followed while recording intracellularly from hippocampal (CA1), neocortical (NC), hypoglossal (XII), and dorsovagal (DMX) neurons in slices. These were (a) the trajectory and maximum change in V_m or membrane potential, (b) R_n or input resistance, and (c) $K^+{}_o$, or extracellular K^+ in the slice preparation. The methods used—including flow and type of bathing solutions, temperature, thickness of slices, and O_2 concentration—were all the same. XII neurons depolarized the fastest and the NC neurons the slowest with total O_2 deprivation or anoxia (Fig. 3) (1,2,5). Within 2 to 3 min of anoxia, XII neurons depolarized by more than 40 to 50 mV, while NC neurons may not have shown any depolarization by that time. Indeed, a number of these neurons tended to hyperpolarize before starting to depolarize (1,2,5). Interestingly, human temporal lobe NC neurons show a similar pattern of change in V_m as rat NC neurons (1,2,5). CA1 and DMX neurons also show a depolarization after a few minutes of anoxia, and this increased with time. Generally, however, their response was in between those of XII and NC neurons. R_n decreased in all cell types, although some XII showed, at the outset, a modest increase in R_n. The decrease in R_n was most accentuated in the DMX neurons and least in NC neurons. Interestingly also, K^+ increased in the extracellular space in all regions, but it increased much less in the neocortex than in any other region, even when O_2 was totally deprived. The reasons for this heterogeneity in response and sensitivity between the various neuronal types is not clear at the moment, but they are most likely to be dependent on the microenvironment of these neurons, the released neuromodulators, the synaptic connections and currents induced during hypoxia, and the inherent properties of these cells (1,2,5). It is interesting to note that the differences in membrane potential trajectory of these neuronal types during hypoxia does not parallel the trajectories of the increase in K^+; hence the changes in V_m cannot be explained by changes in K^+ (1,2). One other important conclusion to be drawn from these studies is the following: it seems that the dogma that neocortical cells are more sensitive to hypoxia and susceptible to injury as compared with other cell types, especially the brainstem, is fallacious. Inherently, NC cells are much more resistant to O_2 deprivation than XII or DMX or CA1 cells. This may also indicate that in the in vivo situation, the stimulus is practically never a sole depletion of O_2, and that usually the insult is more compli-

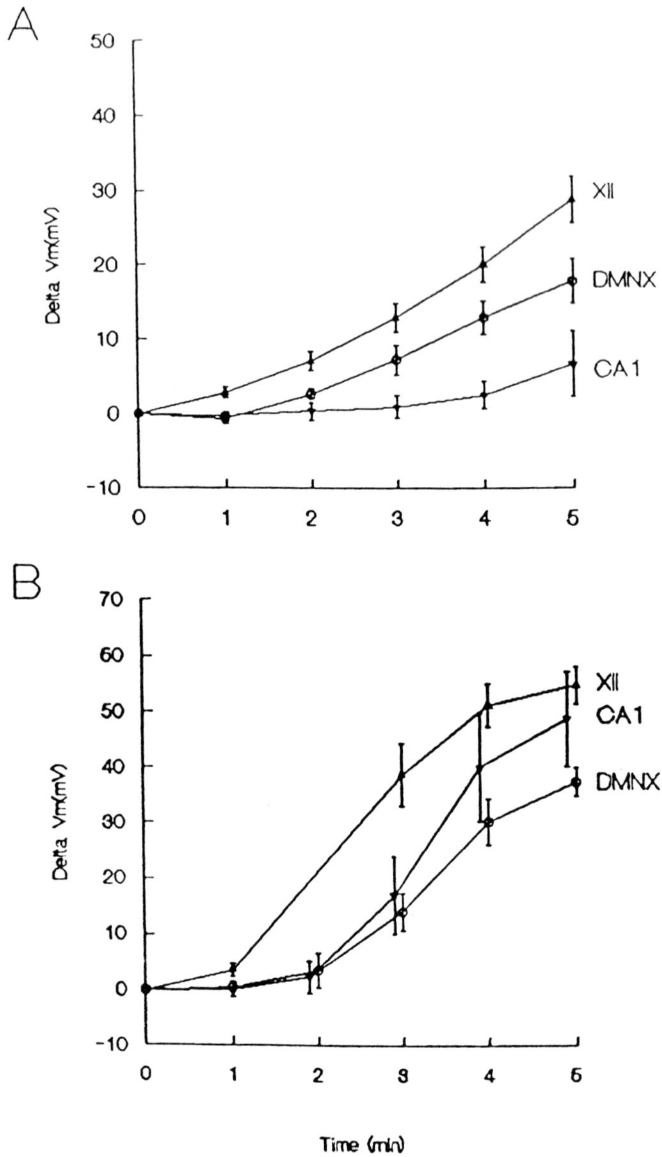

Figure 3 Average change in membrane potential in three populations of cells impaled in CA1 and brainstem neurons (DMX and XII). (A) shows trajectory before and during hypoxia and (B) shows the trajectory before and during anoxia. (From Ref. 2.)

cated, involving acid-base changes, blood flow alterations, and cellular waste products.

In other experiments, we tested the response of NC neurons in the rat as well as in human temporal cells to O_2 lack, glucose deprivation, or both. Of major interest is that the V_m trajectory had a much longer latency to fast depolarization when deprived of glucose or O_2 alone than when deprived of both stimuli (5). In addition, K^+ ion concentration was much larger outside cells when glucose and O_2 were depleted from the bathing solution than when either was administered (5). This clearly indicates the complexity of the response to ischemia when compared to that of anoxia (5).

IV. Importance of Inherent Neuronal Properties to Overall Response to Low O_2

In the past several years, we have been able to examine the importance of certain properties that nerve cells exhibit in the overall response to hypoxia. The idea is that the response of nerve cells to low O_2 is determined not only by synaptic input and connectivity but also by the membrane and other properties with which each single cell is endowed. Two examples are provided below.

A. Role of K_{ATP} Channels in Anoxia

We have recently found that neurons in some nuclei and regions in the CNS of the rat have K^+ channels that play an important role in hypoxia (6,10–12). The activation of these channels is important, since the whole response changes when they are prevented from being activated (6,10). It is important to note that although we do not know the exact relation between sulfonylurea receptors and K_{ATP} channels, evidence exists to support the idea that these receptors form a functional moiety with K_{ATP} channels (13). In XII neurons, we have demonstrated that in the presence of glibenclamide and tolbutamide, sulfonylurea receptors antagonists, the depolarization that is induced by hypoxia or anoxia is exaggerated, suggesting that the activation of these channels is helpful in dampening this depolarization (6,11). This is further supported by the fact that in the presence of these antagonists, recovery of neurons from the depolarization is jeopardized if not lost. In other types of nerve cells, such as CA1 cells, other investigators have shown that the hyperpolarization that is seen at the beginning of anoxia is eliminated when glibenclamide is used (for review, see Ref. 13). Again, this suggests that excitability is decreased when K_{ATP} channels are activated, thus decreasing O_2 consumption in neurons at a time when there is a mismatch between O_2 demand and supply. Clearly these studies provide only indirect evidence for the importance of K_{ATP} channels (6,10–11). More recently we have been able to provide more support to the idea that these channels are being activated with hypoxia and that ATP inhibits their activity (Fig. 4)

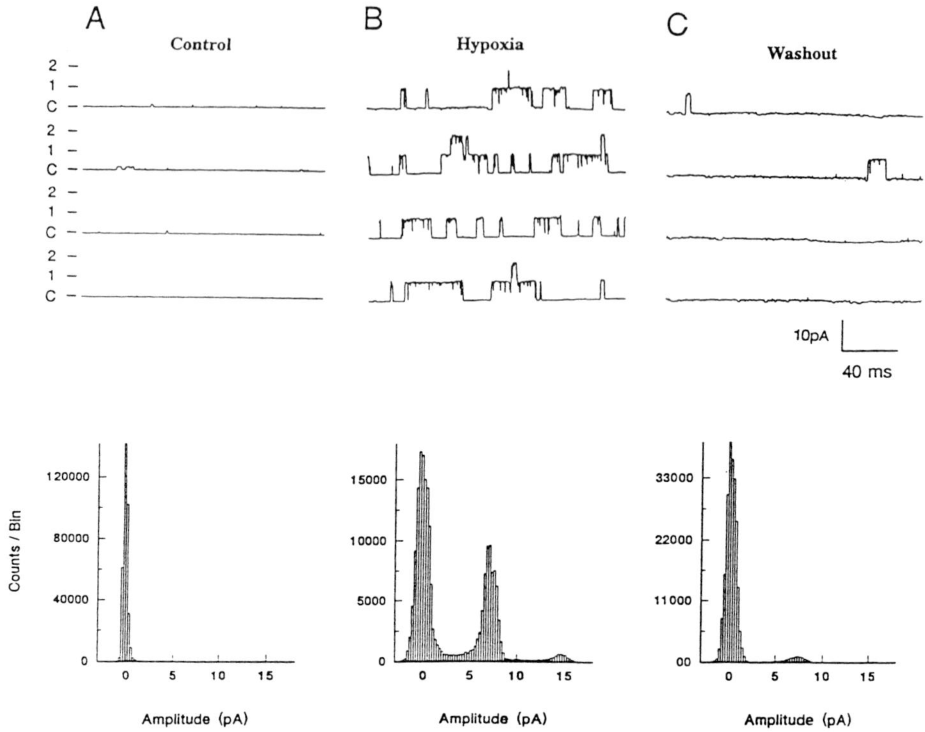

Figure 4 Single-channel recordings before, during, and after anoxia in cell-attached mode. Note that there is a large conductance channel that responds to low O_2 and increases channel probability (Po) substantially. Note that Po drops down to baseline within 1 to 2 min after control wash is administered. Lower panels show events per bin as a function of current amplitude with the interval between peaks representing the size of the current. (From Ref. 12.)

(12). It is difficult at this stage to know what is the modulator of these channels inside cells in situ when such neurons are exposed to hypoxia.

B. Role of the Voltage-Sensitive Na^+ Channels in Anoxia

We have observed that, in the human neocortical slice, neurons depress their activity during anoxia (14). This is potentially due to a number of mechanisms including (a) decreased input resistance of the cells and thus increase in electrical shunt, (b) increase in V_m (more negative), and (c) decrease in firing threshold. We have found that all of these mechanisms apply to these NC neurons (5,15). It is interesting that at the outset of anoxia, when NC cells have already decreased their excitability, NC cells have not changed in any significant way their V_m or their R_n. Hence it was assumed that their depressed activity was related to firing threshold. We tested this by dissociating neurons from the human neocortex and patching these cells to record inward currents and more specifically the current carried by Na^+ ions (14). We found that the activated voltage-sensitive Na^+ current is much smaller at all voltages tested in the presence of hypoxia than during normoxia (14,15). In addition, we found that the curve of the steady-state inactivation of the Na^+ current is shifted to the left in the absence of O_2, indicating that the steady-state activatable channels are likely to be inhibited during hypoxia (Fig. 5) (14–15). The responsible mechanisms for this steady-state inhibition are not understood, but it is interesting to note that the level of ATP in the pipette patching these neurons is an important parameter that contributes to that shift in the steady state (14). Therefore, we

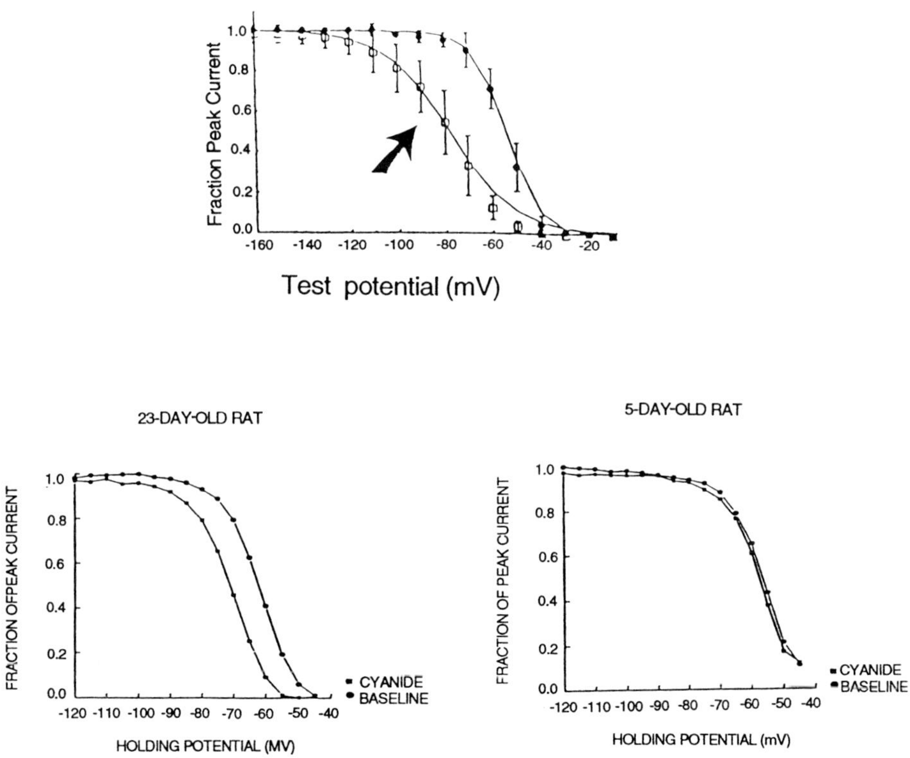

Figure 5 Top: Steady-state inactivation curve in dissociated human neurons from the temporal lobe. Note the shift in the curve to the left with cyanide or hypoxia (arrow). Bottom: Similar change in the mature rat CA1 cells (left); no such change in the CA1 in the immature CA1 (right). (From Refs. 14 and 15.)

believe that the firing threshold increases remarkably in these neocortical cells in the slice, and the basis for this is related to the decrease in the number of activatable voltage-sensitive Na^+ channels (14,15). It is important also to note that cell injury is largely eliminated if extracellular Na^+ is removed from the medium bathing the neurons during low O_2 (16).

V. Summary

Neurons in the CNS are differentially sensitive to O_2 deprivation. Age is also a major factor that enhances resistance to hypoxia by mechanisms that are not totally known. Other factors that contribute in a major way are related to the inherent properties of neurons in general and to their microenvironment. We have made major progress in the past several years in deciphering the various factors that are important in the tolerance and susceptibility to anoxia, in the sensing of O_2 depletion, and in factors that contribute to adaptation or neuronal survival. It is clear from our current knowledge that cell response and sensitivity to a stress such as hypoxia is extremely complicated and that more information must be gathered in order to model the response and the interactions between the various factors inside cells.

Acknowledgment

This work was supported by NIH grants HL39924, HD15736, and HD28940.

References

1. Haddad GG, Donnelly DF. O_2 deprivation induces a major depolarization in brain stem neurons in the adult but not in the neonatal rat. J Physiol 1990; 429:411–428.
2. Donnelly DF, Jiang C, Haddad GG. Comparative responses of brain stem and hippocampal neurons to O_2 deprivation: in vitro intracellular studies. Am J Physiol 1992; 262:L549–L554.
3. Jiang C, Agulian S, Haddad GG. O_2 tension in adult and neonatal brain slices under several experimental conditions. Brain Res 1991; 568:159–164.
4. Cummins TR, Agulian SK, Haddad GG. Oxygen tension clamp around single neurons in vitro: A computerized method for studies on O_2 deprivation. J Neurosci Methods 1993; 46:183–189.
5. Jiang C, Haddad GG. Differential responses of neocortical neurons to glucose and/or O_2 deprivation in the human and rat. J Neurophysiol 1992; 68:2165–2173.
6. Jiang C, Haddad GG. Effect of anoxia on intracellular and extracellular potassium activity in hypoglossal neurons in vitro. J Neurophysiol 1991; 66:103–111.
7. Young RSK, During MJ, Donnelly DF, et al. Effect of anoxia on excitatory amino acids in brain slices of rats and turtles: In vitro microdialysis. Am J Physiol 1993; 264:R716–R719.

8. Haddad GG, Jiang C. Mechanisms of anoxia-induced depolarization in brainstem neurons: In vitro current and voltage clamp studies in the adult rat. Brain Res 1993; 625:261–268.
9. Jiang C, Agulian S, Haddad GG. Cl^- and Na^+ homeostasis during anoxia in rat hypoglossal neurons: Intracellular and extracellular in vitro studies. J Physiol 1992; 448:697–708.
10. Jiang C, Xia Y, Haddad GG. Role of ATP-sensitive K^+ channels during anoxia: Major differences between rat (newborn and adult) and turtle neurons. J Physiol 1992; 448:599–612.
11. Xia Y, Haddad GG. Major differences in CNS sulfonylurea receptor distribution between the rat (newborn, adult) and turtle. J Comp Neurol 1991; 314:278–289.
12. Jiang C, Haddad GG. Short periods of hypoxia activate a K^+ current in central neurons. Brain Res 1993; 614:352–356.
13. Haddad GG, Jiang C. O_2 deprivation in the central nervous system: On mechanisms of neuronal response, differential sensitivity and injury. Prog Neurobiol 1993; 40:277–318.
14. Cummins TR, Jiang C, Haddad GG. Human neocortical excitability is decreased during anoxia via sodium channel modulation. J Clin Invest 1993; 91:608–615.
15. Cummins TR, Donnelly DF, Haddad GG. Effect of metabolic inhibition on the excitability of isolated hippocampal CA1 neurons: Developmental aspects. J Neurophysiol 1991; 66:1471–1482.
16. Friedman JE, Haddad GG. Removal of extracellular sodium prevents anoxia-induced injury in freshly dissociated rat CA1 hippocampal neurons. Brain Res 1993, in press.

41

Arterial Oxygen Desaturation and Respiratory Failure Associated with Sympathetic Overactivity

RICHARD M. MILLIS, C. OVID TROUTH, and SHEREÉ M. JOHNSON, and

College of Medicine
Howard University
Washington, D.C.

DON H. WOOD[†] **and OZRA DEHKORDI**

Howard University Hospital
Washington, D.C.

SAMUEL R. WRAY

University of the West Indies
Mona, Jamaica, West Indies

Previous studies have suggested that arterial oxygen desaturation (AOD) associated with sleep apnea syndromes increases plasma norepinephrine (NE) by the mechanism of excessive sympathetic nervous stimulation (SNS) and that NE inhibits ventilation by actions at the peripheral respiratory chemoreceptors. Cocaine increases plasma NE, but it also inhibits SNS by actions at the ventral medullary surface (VMS), where the central respiratory chemoreceptors are located. In the current study, we administered NE (3 to 5 μg/kg, IV) to 15 anesthetized dogs, 10 on mechanical ventilation and 5 breathing spontaneously, and in 5 mechanically ventilated cats we performed electrical stimulation of a previously described vasopressor site in the brainstem—the nucleus reticularis parvocellularis (RPC). We monitored mean arterial blood pressure (MABP), minute ventilation ($\dot{V}_E$), and arterial blood and expiratory gas tensions (Po_2, Pco_2, PN_2) continuously. To determine the sensitivity of mass spectrometric methods for measuring respiratory gas tensions, ventilation-perfusion ($\dot{V}_A/\dot{Q}$) inequality and intrapulmonary shunt were simulated by ligation or by infusion of a 50-mL bolus of venous blood into a small pulmonary artery, respectively, in 10 additional open-chested, mechanically ventilated dogs. The $\dot{V}_A/\dot{Q}$ inequality simulation increased the P_EO_2 and $Paco_2$ while decreasing Pao_2 and

[†]Deceased.

P_ECO_2 significantly. Shunt simulation decreased the PaO_2 and increased the $PaCO_2$ without changing either P_EO_2 or P_ECO_2 significantly. NE decreased both P_EO_2 and PaO_2 and increased both P_ECO_2 and $PaCO_2$ both in mechanically ventilated and spontaneously breathing dogs, with transient apnea in the latter animals. NE increased MABP, producing a marked systemic vasopressor response in all animals. NE also produced a slight decrease in the mean expiratory tidal volume with significantly increased P_EN_2 and constant P_IN_2, but these effects were too small to have been responsible for the AOD observed. The effects of direct electrical stimulation of the RPC could not be distinguished from those of NE, suggesting that a systemic vasopressor response produced by any means whatsoever might produce AOD. These findings suggest that SNS simulation by IV NE infusions in dogs inhibits ventilation and produces AOD, but that AOD cannot be prevented by mechanical ventilation. The canine model of NE infusions and the feline model of RPC electrostimulation might clarify the respiratory control mechanisms by which SNS and NE increase morbidity and mortality.

I. Introduction

Arterial oxygen desaturation (AOD) is associated with sleep apnea syndromes and might increase the plasma norepinephrine (NE) concentration by a mechanism that involves excessive stimulation of sympathetic nervous structures (1). The sympathomimetic effects of cocaine are due to its ability to inhibit neuronal reuptake of catecholamine (2), thereby causing an increase in plasma NE (3,4). However, cocaine administered directly at the brainstem decreases sympathetic activity, an effect that might be responsible for decreased vasomotor tone and cardiorespiratory failure associated with cocaine overdosage (5–7). More recent evidence that ventilation is decreased in goats by increasing the plasma NE level (8) suggests that AOD could result from a sudden increase of plasma NE and/or excessive sympathetic neural activity rather than causing it. In order to help clarify the relationships between a sudden increase of plasma NE, ventilatory failure, and AOD, we administered NE intravenously (3 μg/kg) to 15 anesthetized (Nembutal, 50 mg/kg) dogs, 5 breathing spontaneously and 10 during mechanical ventilation. To determine whether an increase in sympathetic neural activity per se could result in AOD (in the absence of pharmacologically elevating the plasma NE), we performed direct electrostimulation (1 to 4 V, 1 ms duration, 40 Hz) of a brainstem vasopressor site at the nucleus reticularis parvocellularis (RPC) in 5 cats during experimental periods of spontaneous breathing and mechanical ventilation.

II. Methods and Results

Physiological measurements of the mean arterial blood pressure were monitored from a catheter in the femoral artery, recorded by a Statham strain gauge transducer (Model P23Db), and calculated from the systolic and diastolic pressure waveforms (according to the formula 0.33 systolic + 0.67 diastolic). Minute ventilation ($\dot{V}_E$) was monitored by a Fleisch pneumotachograph and measured by a Beckman differential pressure transducer. Arterial and expiratory P_{O_2}, P_{CO_2}, and P_{N_2} values were determined by a respiratory mass spectrometer (Chemetron) by extraction of blood gases through an indwelling Silastic catheter in the abdominal aorta.

In order to validate the use of respiratory mass spectrometry for detecting AOD of various etiologies, experimental conditions of ventilation-perfusion ($\dot{V}_A/\dot{Q}$) inequality and intrapulmonary shunt were produced in anesthetized dogs subjected to thoracotomy and volume-controlled mechanical ventilation. Figure 1 shows that ligation of a pulmonary artery ($\dot{V}_A/\dot{Q}$ inequality) increased the P_EO_2 and Pa_{CO_2} and decreased the Pa_{O_2} and P_ECO_2, as predicted. Figure 2 demonstrates that injection of a 50-mL bolus of autologous venous blood directly into the pulmonary artery serving an atelectatic lobe of the lung decreased the Pa_{O_2} and increased the Pa_{CO_2}, without a change in the P_EO_2 and P_ECO_2, simulating an intrapulmonary shunt, as predicted.

A. Effects of NE on Ventilation

The effects of NE (3 μg/kg, IV) during spontaneous ventilation are demonstrated in Figure 3. A sudden systemic vasopressor response was associated

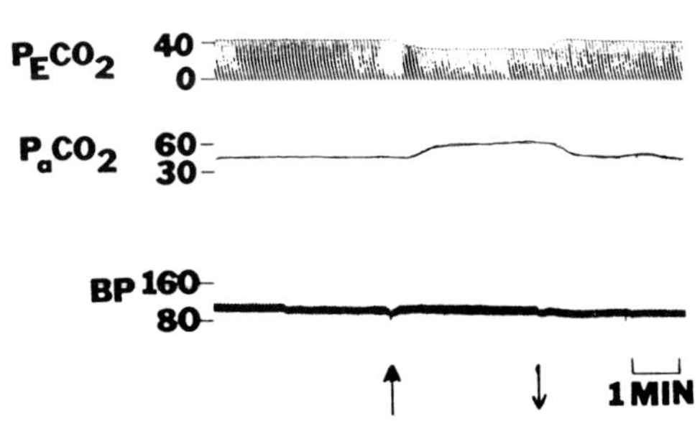

Figure 1 Responses to pulmonary artery ligation. Analog recordings of changes in arterial blood pressure (BP), mass spectrometrically measured blood gas tension (Pa_{O_2}, Pa_{CO_2}), airway gas tensions (P_EO_2, P_ECO_2) and airway flow (RESP) during volume-controlled mechanical ventilation with thoracotomy in an anesthetized dog following ligation of the left pulmonary artery (↑) and subsequent release of the ligature (↓).

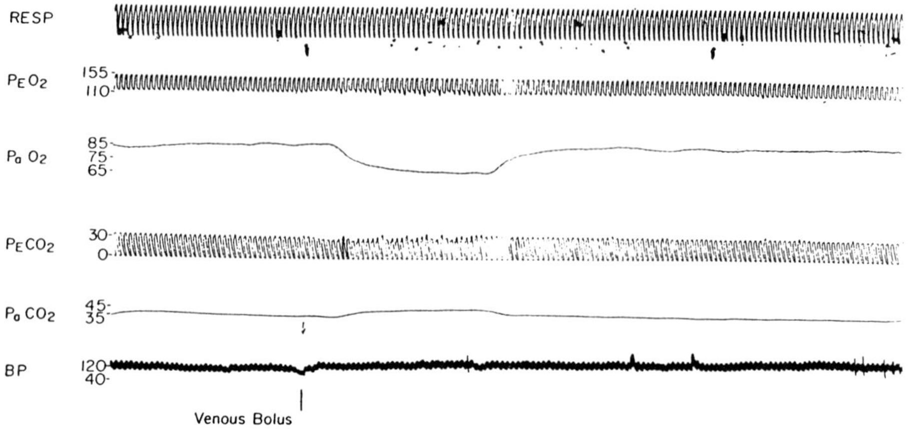

Figure 2 Simulation of shunt. Analog recordings of changes in arterial blood pressure (BP), mass spectrometrically measured blood gas tension (Pa_{O_2}, Pa_{CO_2}), airway gas tensions (P_EO_2, P_ECO_2) and airway flow (RESP) during volume-controlled mechanical ventilation with thoracotomy in an anesthetized dog subjected to injection of a 50-mL bolus of autologous venous blood into the pulmonary artery serving an atelectatic lobe of the left lung.

with transient apnea, which decreased both the P_EO_2 and PaO_2 and increased both the P_ECO_2 and Pa_{CO_2}. In order to determine whether the hypoventilation associated with increasing the plasma NE concentration could be ameliorated by mechanical ventilation, five animals were subjected to volume-controlled mechanical ventilation with an animal respirator (Harvard). Figure 4 shows that pulmonary gas exchange was nominally the same during the vasopressor response associated with administering NE intravenously to animals subjected to mechanical ventilation as that in animals breathing spontaneously. This surprising finding was not predicted and led us to the hypothesis that AOD might be associated with a large systemic vasopressor response produced by any means whatsoever (in the absence of a pharmacological vasopressor agent such as NE).

B. Effects of Brainstem Electrostimulation

We hypothesized that AOD could be produced by a systemic vasopressor response resulting from electrostimulation of a brainstem vasopressor site. Therefore, we monitored the arterial blood pressure responses following electrostimulation (1 to 4 V, 1 ms duration, 40 Hz) of the RPC of both spontaneously breathing and mechanically ventilated cats under chloralose-urethane anesthesia (40 mg/kg glucochloralose, 200 mg/kg urethane). Using fine tungsten electrodes

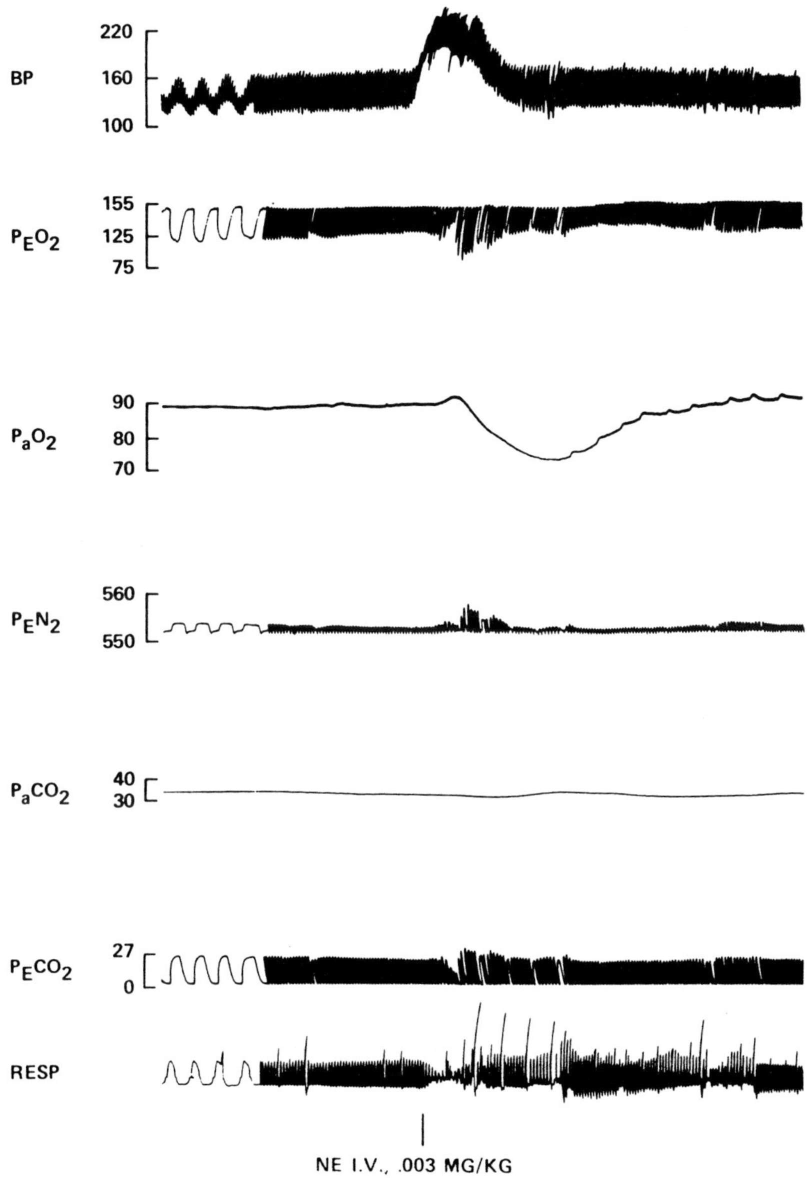

Figure 3 Responses to norepinephrine during spontaneous breathing. Analog recordings of changes in arterial blood pressure (BP), mass spectrometrically measured blood gas tension (Pa_{O_2}, Pa_{CO_2}), airway gas tensions (P_EO_2, P_ECO_2), and airway flow (RESP) in an anesthetized dog subjected to intravenous administration of norepinephrine (3 μg/kg).

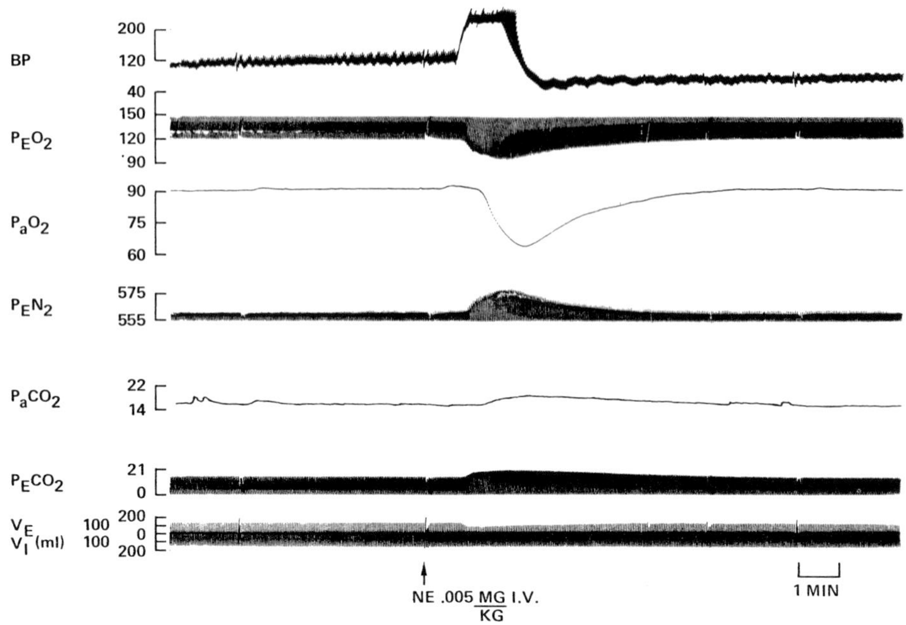

Figure 4 Responses to norepinephrine during mechanical ventilation. Analog recordings of changes in arterial blood pressure (BP), mass spectrometrically measured blood gas tension (PaO_2, $PaCO_2$), airway gas tensions (P_EO_2, P_ECO_2), inspiratory tidal volume (V_I) and expiratory tidal volume (V_E) in an anesthetized dog subjected to intravenous administration of norepinephrine (5 μg/kg).

(tip diameter 0.3 to 0.5 mm), the brainstem was electrically stimulated at the RPC. Arterial blood pressure was measured via a Statham strain gauge transducer connected to a catheter in the right femoral artery. Direct electrostimulation yielded a strong systemic vasopressor response. The ventilatory response is demonstrated in Figure 5 in spontaneously breathing cats and in Figure 6 in mechanically ventilated cats. In spontaneously breathing animals (Fig. 5), the ventilatory and respiratory gas responses to medullary electrostimulation was similar to that produced by NE (Fig. 3); it consisted of AOD with hypoventilation and transient apnea. In mechanically ventilated cats subjected to medullary electrostimulation (Fig. 6), ventilation was constant and the blood gas responses were similar to those detected in mechanically ventilated dogs exhibiting AOD following intravenous administration of NE (Fig. 4).

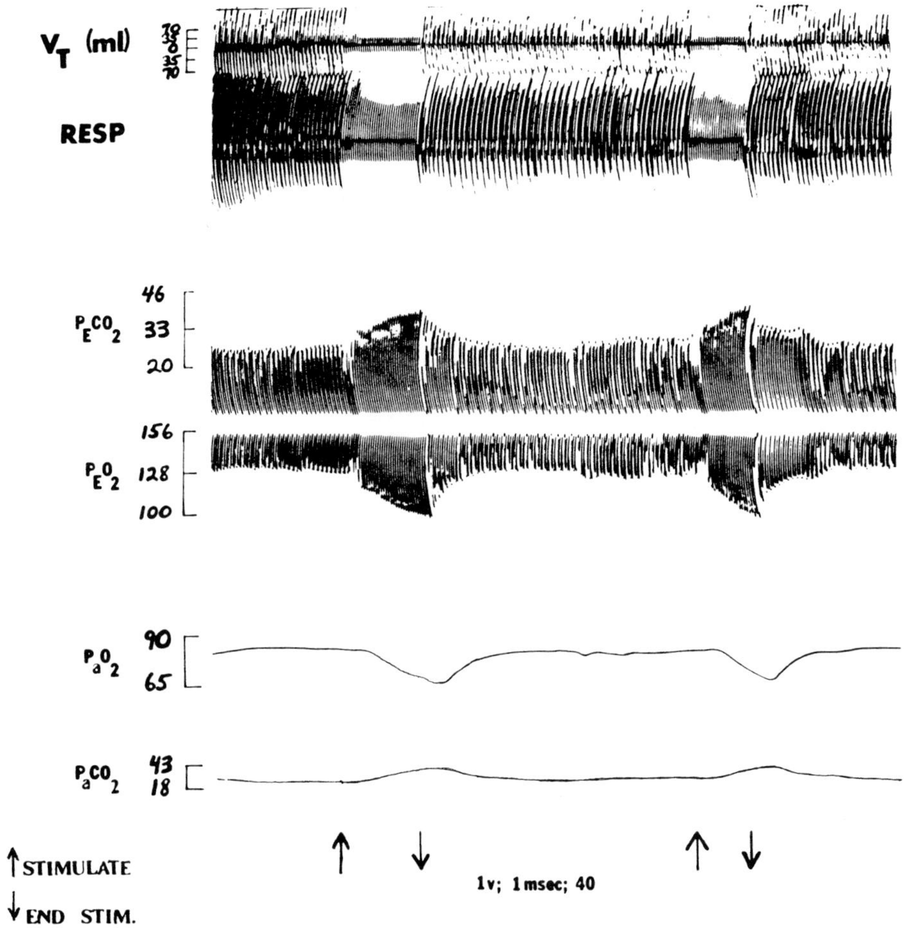

Figure 5 Responses to stimulation of the nucleus reticularis parvocellularis (RPC) during spontaneous breathing. Analog recordings of changes in arterial blood pressure (BP), mass spectrometrically measured blood gas tension (PaO_2, $PaCO_2$), airway gas tensions (P_EO_2, P_ECO_2), and airway flow (RESP), in a chloralose-urethane anesthetized cat subjected to two intervals of electrostimulation (↑, 1 to 4 V, 1 ms duration, 40 Hz) and subsequent cessation of electrostimulation (↓) of the RPC.

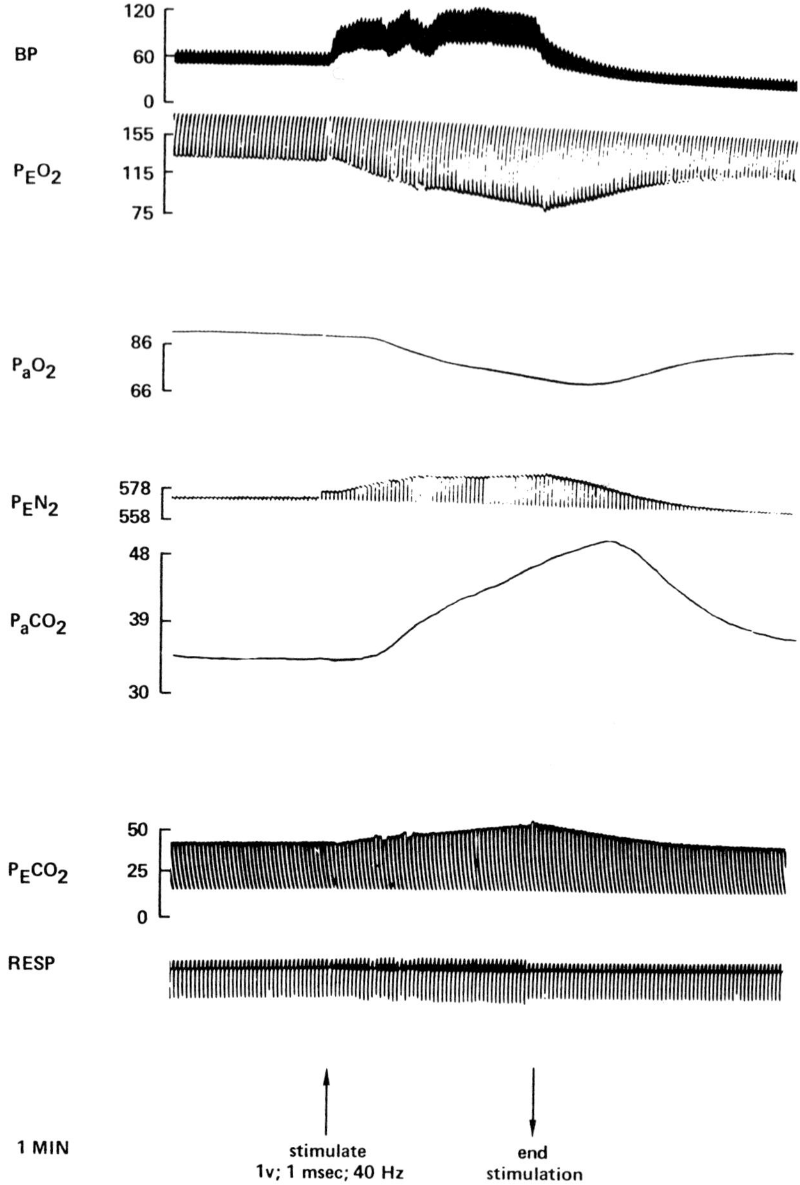

Figure 6 Responses to stimulation of the nucleus reticularis parvocellularis (RPC) during mechanical ventilation. Analog recordings of changes in arterial blood pressure (BP), mass spectrometrically measured blood gas tension (Pa_{O_2}, Pa_{CO_2}), airway gas tensions (P_EO_2, P_ECO_2), airway flow (RESP), and tidal volume (V_T) in a chloralose-urethane anesthetized cat subjected to a single interval of electrostimulation (↑, 1 to 4 V, 1 ms duration, 40 Hz) and subsequent cessation of electrostimulation (↓) of the RPC.

III. Discussion and Conclusions

The present studies have shown that sympathetic nervous overactivity stimulated by intravenous administration of NE results in hypoventilation and produces AOD in dogs. This inhibition of ventilation is a pharmacological response similar to that recently reported to have been produced by peripheral respiratory chemoreceptor responses to NE in goats (8). This model could be relevant to elucidating the neurochemical mechanisms underlying cardiorespiratory failure and AOD associated with sleep apnea syndromes or overdosage of sympathomimetics such as cocaine. We expected that the apnea and hypoventilation resulting from raising the plasma NE level in spontaneously breathing animals would be ameliorated by volume-controlled mechanical ventilation. However, the AOD persisted following intravenous administration of NE in mechanically ventilated animals. Consequently, we pursued the hypothesis that electrostimulation of the vasoactive site at the nucleus reticularis parvocellularis might be a better experimental model for sympathetic nervous overactivity because it would not require administration of a pharmacological vasopressor agent, which might have numerous diffuse sympathomimetic actions. Actions of NE on the pulmonary vasculature (9,10) have been demonstrated and could theoretically account for AOD associated with ventilation-perfusion inequality and intrapulmonary shunting that, in our validation studies, were detected mass spectrometrically. That electrostimulation of the nucleus reticularis parvocellularis inhibited ventilation and produced AOD in animals with and without mechanical ventilation demonstrates that there might be no significant difference between the sympathomimetic responses to intravenous NE and electrostimulation of a vasopressor region in anesthetized laboratory animals. These findings also suggest that intravenous administration of NE to human patients might be fraught with the danger of inhibiting ventilation or producing AOD whether or not their breathing is being supported mechanically. It is, therefore, concluded that the control mechanisms underlying the cardiorespiratory failure associated with sympathetic nervous overactivity might be elucidated using pharmacological and electrophysiological models of systemic NE administration and electrostimulation of the vasomotor center.

References

1. Eisenberg E, Zimlichman R, Lavie P. Plasma norepinephrine levels in patients with sleep apnea syndrome. N Engl J Med 1990; 322:932–933.
2. Trendelenburg U. The supersensitivity caused by cocaine. J Pharmacol Exp Ther 1959; 125:55–65.
3. Chiueh CC, Kopin IJ. Centrally mediated release by cocaine of endogenous epinephrine and norepinephrine from the sympathoadrenal medullary system of unanesthetized rats. J Pharmacol Exp Ther 1978; 205:148–154.

4. Kiritsy-Roy JA, Halter JB, Gordon SM, et al. Role of the central nervous system in hemodynamic and sympathoadrenal responses to cocaine in rats. J Pharmacol Exp Ther 1990; 255:154–160.
5. Raczkowski VFC, et al. Cocaine acts in the central nervous system to inhibit sympathetic neural activity. J Pharm Exp Ther 1991; 257:511–519.
6. Dehkordi O, et al. Central cocaine neurotoxicity at brainstem cardiorespiratory control sites. Life Sci 1993; 52:1387–1396.
7. Dehkordi O, et al. Effects of cholinomimetics on cocaine-induced hypotension and apneusis at a ventral brainstem cardiorespiratory control site. Life Sci 1994; 54:1513–1522.
8. Pizarro J, Warner MM, Ryan M, et al. Intracarotid norepinephrine infusions inhibit ventilation in goats. Resp Physiol 1992; 90:299–310.
9. Daly I de B. Intrinsic mechanisms of the lung. Q J Exp Physiol 1958; 43:2.
10. Porcelli JJ, Bergofsky EH. Adrenergic receptors in pulmonary vasoconstrictor responses to gaseous and humoral agents. J Appl Physiol 1973; 34:438.

42

Comparative Studies on the Plasticity of Central Chemosensitivity

MARIANNE E. SCHLÄFKE, THORSTEN SCHÄFER, DIETMAR SCHÄFER, and CHRISTOPH SCHÄFER

Ruhr-University Bochum
Bochum, Germany

I. Introduction

In infants with a congenital insensitivity to CO_2 (Ondine's curse syndrome, or OCS), we were unable to find any spontaneous recovery in a follow-up study over an 8-year period (1). In two children suffering from OCS, we could demonstrate that paired stimulation, using pairs of air jets and CO_2, improved ventilation, including the respiratory response to CO_2 during sleep (2). For ethical reasons, the scientific evaluation of a potentially new therapeutic measure is dependent upon basic investigations in an animal model. Because the respiratory response to CO_2 in cats can be eliminated by bilateral coagulation of a previously described intermediate area (IA) under chronic conditions (3), we used this preparation as control in order to reconfirm the lack of recovery of the respiratory response to CO_2 over a longer period. In the present approach, instead of coagulation, we applied the neurotoxin ibotenic acid (IBO) to the surface of the IA. IBO, which is known to destroy the cell soma but not the axons, spares the surrounding tissues including the vessels. Further, edema would not modify the loss of function, and the recovery from edema would not mimic the postulated return of CO_2 sensitivity. The respiratory system of this model, in which the central chemosensitive drive has been eliminated neurotoxically, is unable to respond to CO_2 or to metabolic acidosis, although it responds to hypoxia (4). The results are identical to those of the former study (3,5). The neurotoxic method, however, is superior to the method of coagulation and has better reproducibility. In addition to the method of paired stimulation described earlier, we now use the respiratory cycle as a trigger for the first stimulus; in infants, we use end-tidal P_{CO_2} ($P_{ET}CO_2$) or S_aO_2 as trigger for the stimuli.

II. Methods

A. Children

A 12-year-old girl (patient A) was hypercapnic and hypoxic during sleep but not ventilated systematically at the time she came in for a sleep study. Her ventilatory response to CO_2 at the age of 15 years corresponded to the one we will present in this study.

A 4-month-old girl (patient B) presented a negative response to CO_2 and was ventilated continuously from 2 h after birth onward. Spontaneous breathing developed only after 4 weeks of age, parallel with the procedure of paired stimulation and probably overlapping with the development of the sleep-wake cycle.

In seven children, the CO_2 response was determined in early infancy and repeatedly during childhood. The method has been described by Schäfer et al. (6). Most of these children are dependent upon artificial ventilation during sleep through a tracheostomy or other respiratory support, like nasal BiPAP, intermittent negative pressure ventilation, or phrenic nerve stimulation. In mild forms, controlled oxygen therapy has been described by which mild hypoxia is maintained (S_aO_2: 93%) and in which the oxygen supply is controlled by means of a pulse oximeter. A computer triggers the valve (7).

Since any method may force the patient to be dependent on one or several medical devices, we wanted to know whether or not the principle of the Pavlovian reflex might be useful in treating these patients. The principle of the stimulation method is presented in Figure 1 (2,8–10).

The CO_2 response curves of the two children presented here were obtained by a spirometer, from steady-state values of ventilation measured in a semi-closed system with CO_2 absorption, and from infrared absorption during hyperoxia and rebreathing of CO_2 for 7 min per CO_2 step.

B. Cats

EEG, ECG, ventilation (spirometry and induction plethysmography), end-tidal P_{CO_2} ($P_{ET}CO_2$), inspiratory F_{O_2} (F_IO_2), arterial blood pressure, and arterial blood gases were measured in four cats (2.5 to 4.8 kg) anesthetized with ketamine

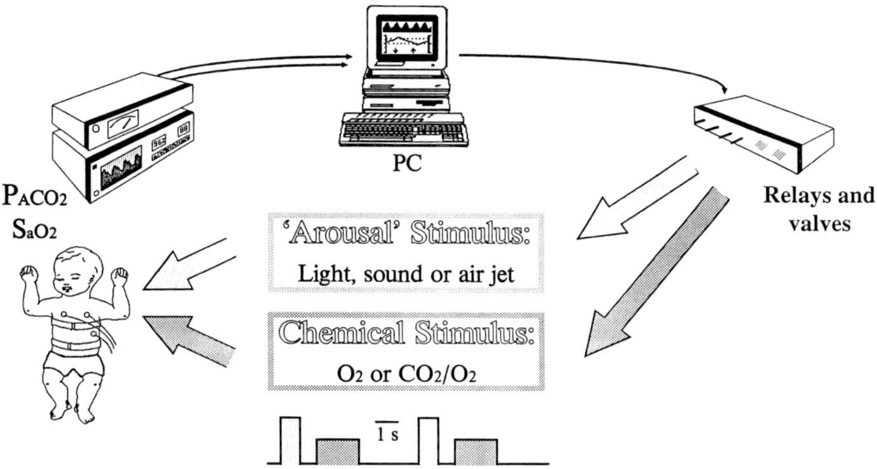

Figure 1 Principle of paired stimulation for infants triggered by $P_{ET}CO_2$ or pulse oximetry. The first stimulus is either light, sound, or an air jet; the second stimulus is CO_2 (1% to 2%). In cats, we used postinspiratory phase (spirometry) as a trigger and femoral nerve stimulation as first stimulus.

hydrochloride. Respiratory responses to CO_2 were measured by rebreathing under steady-state conditions after the surgical preparation of the ventral medullary surface (VMS). IBO (10 μg) was placed bilaterally on the IA using special tissue pledgets (1 mm^2) and removed 20 min later. Respiratory responses to CO_2 were repeatedly measured until the continuous reduction of ventilation reached steady state. The femoral nerve was prepared and cut. The rostral end was placed on a bilateral silver electrode and covered with paraffin oil.

We stimulated with 2 V, 20 Hz, and 1 ms in trains of 20 min. Within these trains, the femoral nerve was stimulated for 1 s; after a pause of 0.5 s, an increase in F_ICO_2 (0.03) followed for 2 s. These pairs were applied every 7 s. The stimulation was triggered by the respiratory cycle, so that the femoral stimulus started with the beginning of expiration. The trains were interrupted by a pause of 5 min. After every two to three trains, the CO_2 response was tested. The frontal and occipital EEG were recorded by means of silver/silver chloride electrodes, amplified by an EEG recorder (Hellige Neuroscript EE 216), and analyzed by means of fast-Fourier transformation. The EEG data were analyzed by a computer program based on the Spike 2 script language (Cambridge Electronic Design, Ltd.).

III. Results

A. Children

The comparison of CO_2 responses of seven children during the first year of life and up to 8 years of age showed no increases in ventilation with increasing PCO_2

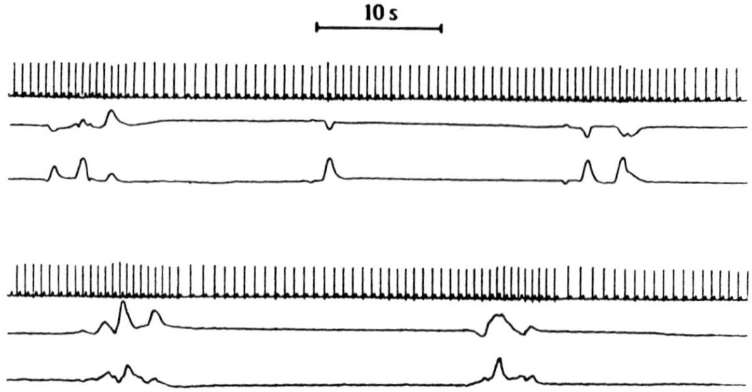

Figure 2 Tracheotomized, spontaneously breathing; recorded with induction plethymography and monitor ECG (patient A).

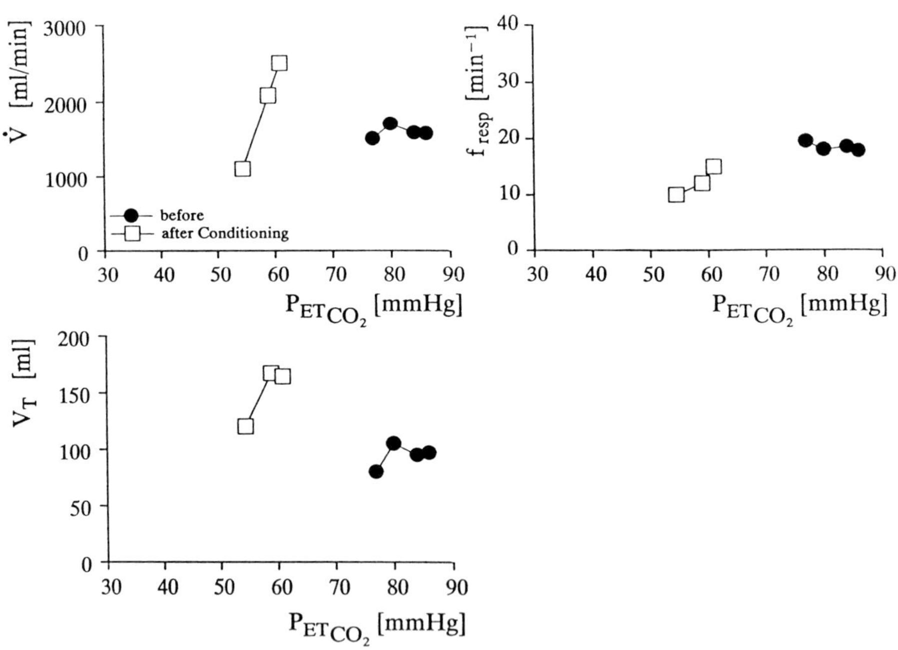

Figure 3 Respiratory response to CO_2 before and after 12 days of paired stimulation (patient A).

during sleep and wakefulness. Breathing is dependent on the peripheral chemoreceptor drive, sensory inputs, and thalamocortical activity. The threshold of the peripheral chemoreceptors varies with alveolar CO_2 partial pressure, which is the result of ventilation, but not vice versa (1). Metabolism, wakefulness, muscular exercise, sensory afferents, and the autonomic nervous system have a prominent influence on the unstable and undamped respiratory system.

Extensive periodic breathing and hypoventilation with tachy- and bradycardia has been recorded during spontaneous sleep in patient A (Fig. 2), characteristics that accompany tiredness during the daytime and typically lack dyspnea but complete insensitivity to CO_2. Figure 3 shows the response to

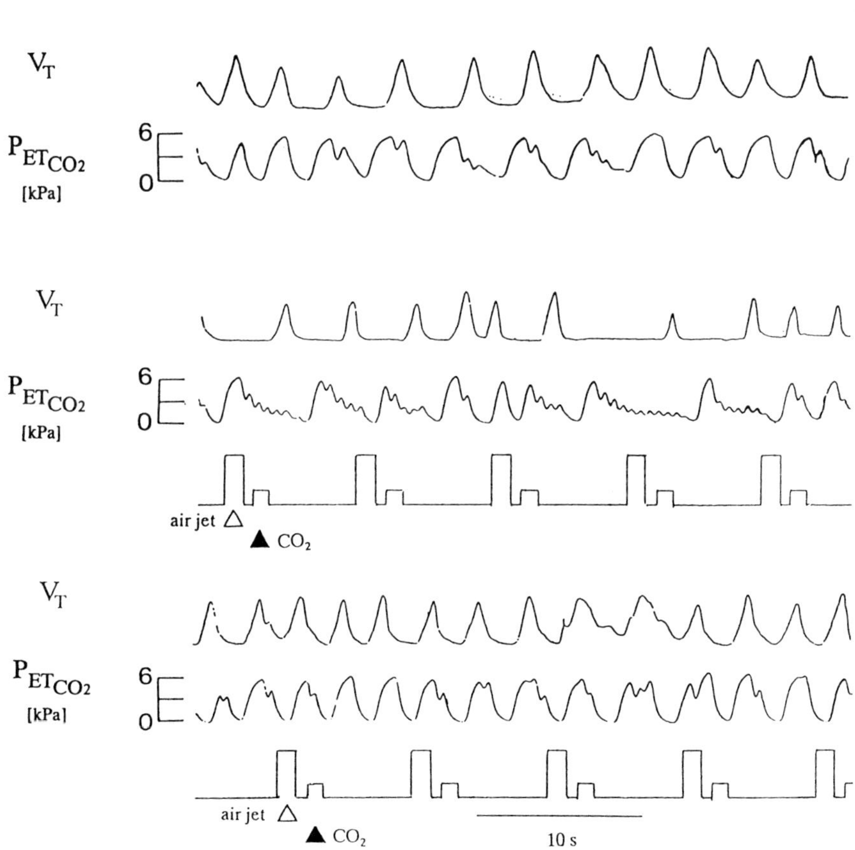

Figure 4 Original recording of ventilation (spirometry) and $P_{ET}CO_2$ during night sleep in patient B. $P_{ET}CO_2$ triggers paired stimulation and allowed spontaneous breathing during deeper sleep stages.

CO_2 before and after paired stimulation during sleep (in the same patient shown in Fig. 2). The respiratory CO_2 response curve before starting the stimulation is shifted to the extreme right and its course is flat and downsloping at P_{CO_2} values beyond 70 mm Hg. The measurement is performed during quiet sleep in hyperoxia.

Twelve days after starting the application of paired stimulation during sleep, a positive response to CO_2 during hyperoxia was measured (2). Air jets blown into the face were used as unspecific stimulus (Fig. 1).

No spontaneous breathing was found in patient B. The application of paired stimulation using air jets as nonspecific stimuli provided successful responses, causing spontaneous breathing during sleep (Fig. 4). The respiratory response to CO_2 was negative at the start of spontaneous breathing and became positive after 2 weeks of training during sleep.

For ethical reasons, the clinical work was not completed.

B. Cats

The respiratory pattern in a sleeping nonanesthetized (chronic) cat after elimination of the IA together with bradycardia is demonstrated in Figure 5. The respiratory response to CO_2 in the control phase and after coagulation of the IA is shown in Figure 6. The curve was unchanged 42 days later (in the same cat as shown in Fig. 5).

CO_2 response curves of ventilation, tidal volume, and frequency in the control 186 min after application of IBO and after two and four trains of paired stimulation are presented in Figure 7. As nonspecific stimulus, femoral nerve stimulation was used, as previously mentioned.

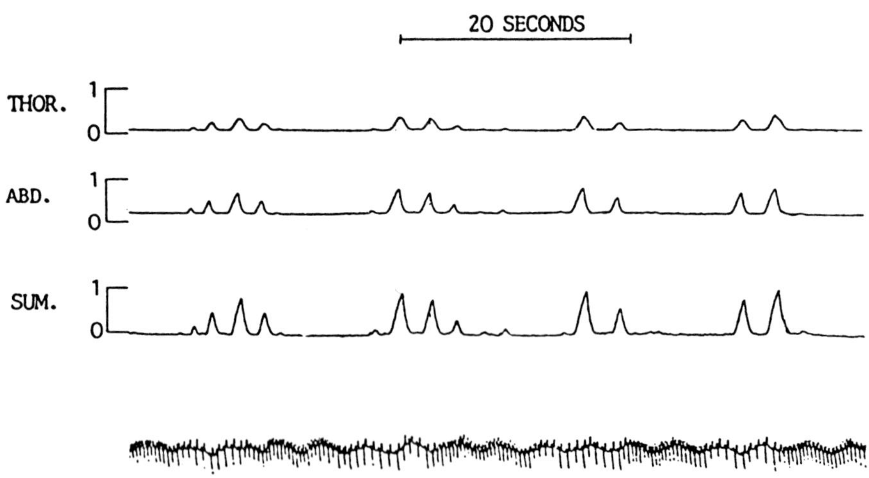

Figure 5 Periodic breathing and bradycardia arise after elimination of IA in chronic cat preparation.

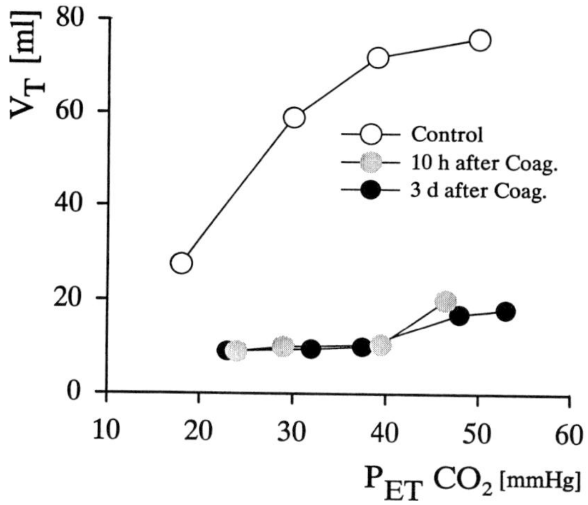

Figure 6 Respiratory response to CO_2 before and 10 h after elimination of IA. No recovery of CO_2 response 3 days later (same cat as shown in Fig. 5).

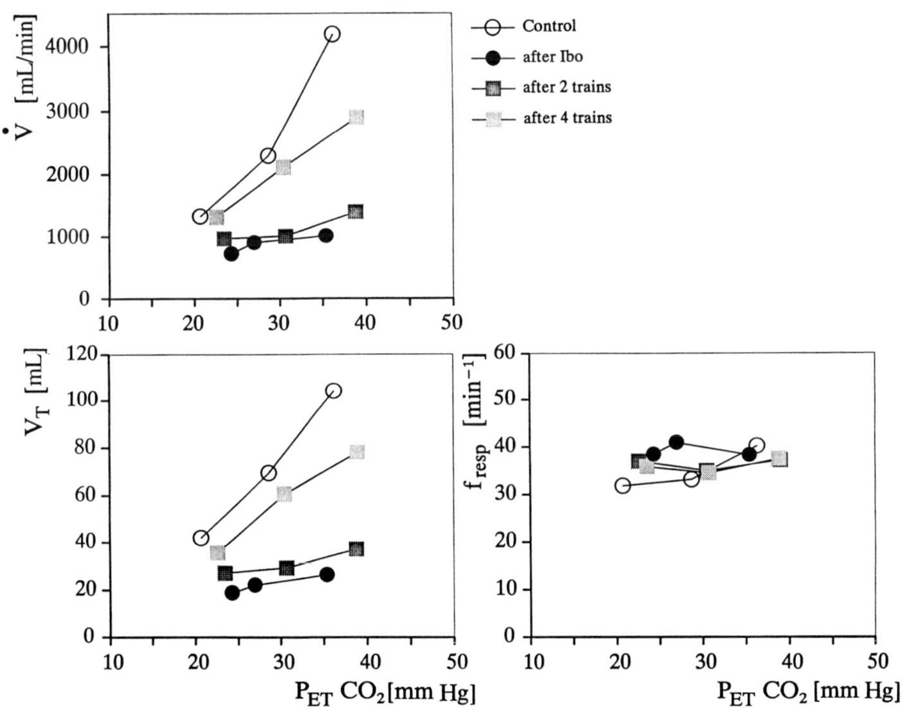

Figure 7 Respiratory response to CO_2 before and after application of IBO on surface of IA and after two and four trains of paired stimulation.

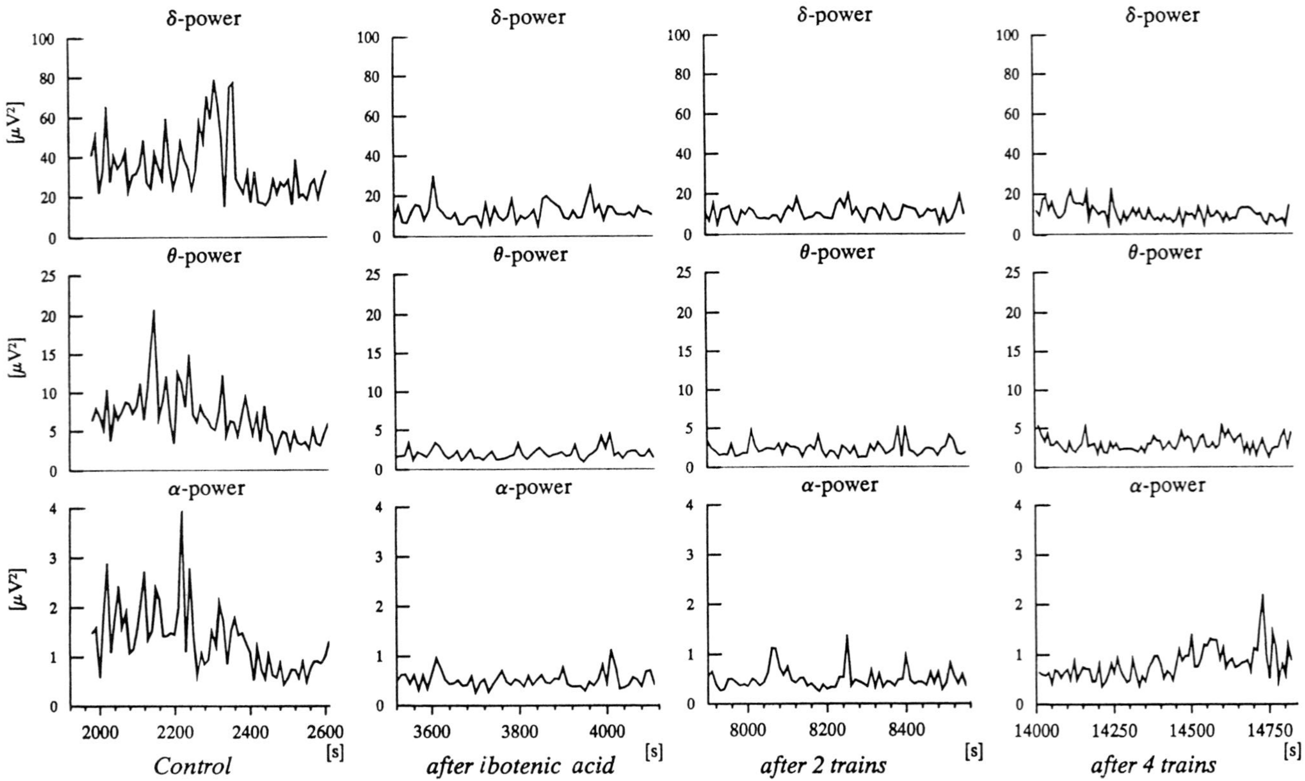

Figure 8 Elimination of IA by IBO accompanies strong reduction of delta, theta, and alpha power of EEG. The alpha power increases after paired stimulation (same cat as shown in Fig. 7).

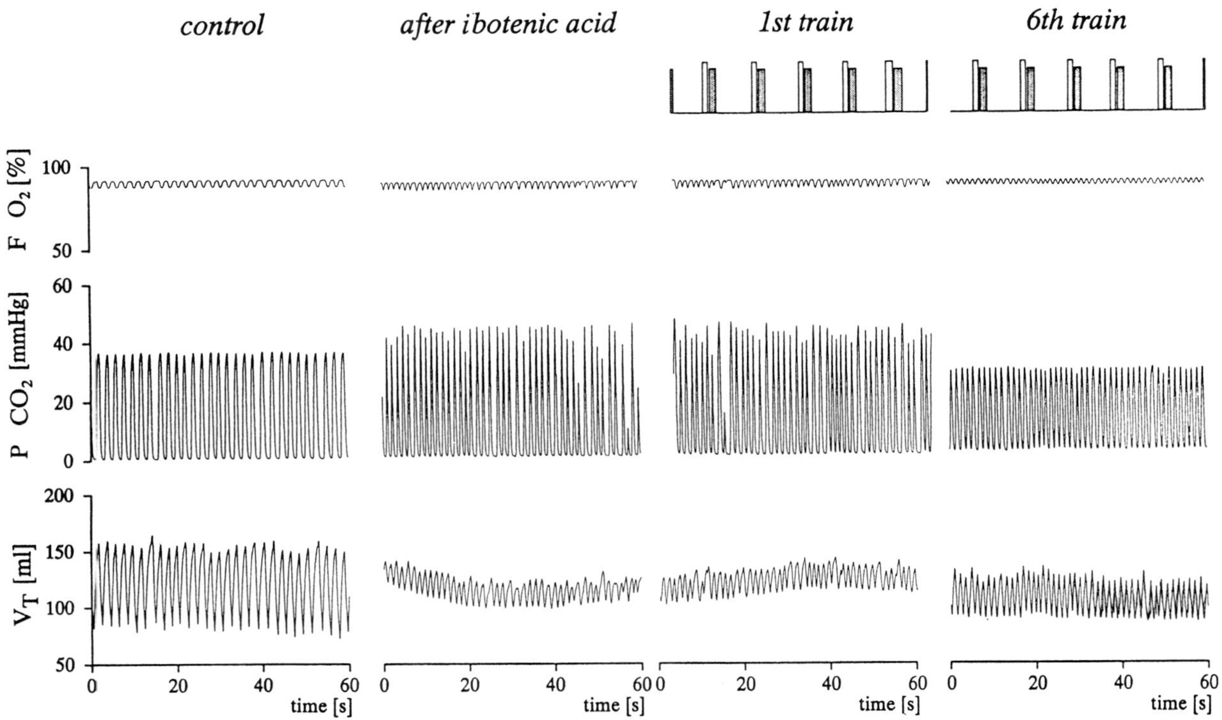

Figure 9 Respiratory patterning during hyperoxia under control conditions 90 min after local application of IBO on IA and during first and sixth trains of paired stimulation.

The delta- (0.5 to 2.5 Hz), theta- (3.0 to 7.5 Hz), and alpha-power (8.0 to 13 Hz) are shown in Figure 8 (for the same cat shown in Fig. 7). There was a strong reduction in all three power spectra after IBO had been applied. After four trains of paired stimulation, the alpha power increased.

Figure 9 gives examples of the respiratory pattern in hyperoxia during the control states 90 min after IBO, during the beginning of paired stimulation (first train), and during the fifth train of paired stimulation.

The main portion of ventilatory recovery obtained was an increase of tidal volume. This was in contrast to earlier experiments in which the recovery of ventilatory sensitivity to CO_2 was mainly a matter of the respiratory frequency response (11).

IV. Discussion

In seven children born with the diagnosis of OCS, we noted a complete insensitivity of respiration to CO_2. In a follow-up study, we were able to reconfirm the horizontal course of the CO_2 response curve up to 8 years (1,2,12,13). The patients were dependent on large-scale therapeutic measures. The alveolar P_ACO_2 showed dramatic variability. PCO_2 was unable to drive ventilation but was its result. Ventilation was driven by wakefulness, sensory inputs, reticular activity, and basically by peripheral chemoreceptors (3,5,14).

These observations required that we look for an animal model. Our hypothesis was that the VMS structures might be involved in the pathogenesis of the OCS and sudden infant death syndrome (SIDS) (15–17). Further, we demonstrated that infants within the age of increased vulnerability of the respiratory system may have a transient insensitivity to CO_2 (6,9,18,19).

The local application of IBO provided us with an adequate technique for eliminating the central chemosensitive structure more precisely. The respiratory responses to hypoxia, hypercapnia, and metabolic acidosis are similar to those obtained after coagulation of the IA and published previously (4).

Classic conditioning of the ventilatory responses to hypoxia using auditory stimuli in healthy humans has been described by Gallego and Perruchet (20). These investigators suggested that conditioning occurs in natural conditions and may play an important role in enhancing the adaptability of the organism to varying environmental conditions. On the basis of our follow-up studies in the children with OCS and in the animal model, we did not find any indication of a spontaneous adaptation to the long-term condition of respiratory acidosis but a prominent dependence of ventilation on the state of wakefulness. This has recently been documented by Shea et al. (21,22). In our study with patients and in the animal model, we showed that the principle of paired stimulation may be applied to reflex pathways within the homeostatic system.

We modified the previously described method of paired stimulation by triggering the first nonspecific stimulus within the postinspiratory phase, causing an arousal at the level of the reticular formation. This causes optimal timing of the next breath, which should be increased at the same time as the CO_2 response. The data indicated that there was a clear recovery of the CO_2 response, which in single experiments was 100% of the control. The difference with former observations, in which mainly the respiratory frequency was able to respond to CO_2 after the paired stimulation, may be attributed to the more precise method of elimination used in the present study (11).

The augmentation of the alpha activity after the paired stimulation indicates that the improvement in breathing and the return of the CO_2 response may be linked to the reticular formation. Further basic approaches are necessary to explain the observed phenomenon.

We suggest that the proposed method of paired stimulation is a way to overcome an insensitivity to CO_2 for a still unknown duration. Spontaneous recovery of the CO_2 response could not be observed, either in the children with OCS or in the animal model.

The data indicate that the respiratory system is provided with a certain plasticity and that the reticular formation plays a key role in this compensatory mechanism.

V. Conclusion

Recovery of the respiratory response to CO_2 in the animal model as well as in children with OCS does not occur spontaneously. The method of paired stimulation, using a stimulus that causes reticular arousal in combination with a short increase of inspired P_{CO_2}, can reinstall a respiratory response to CO_2 in the animal model and in children with OCS. The recovery of the alpha power parallel with the recovery of the ventilatory response to CO_2 permits the speculation that reticular activity may contribute to the effect of paired stimulation on the respiratory response to CO_2.

References

1. Schäfer T, Schläfke ME. Ventilation without CO_2-sensitivity: Ondine's curse syndrome during wakefulness and sleep. Pflügers Arch 1994; 426(suppl 6):R138.
2. Schläfke ME, Burghardt F. Training of central chemosensitivity in infants with sleep apnea. In: Schläfke ME, Koepchen HP, See WR, eds. Central Neurone Environment. Berlin: Springer-Verlag, 1983:71–81.
3. Schläfke ME, Kille JF, Loeschcke HH. Elimination of central chemosensitivity by coagulation of a bilateral area on the ventral medullary surface in awake cats. Pflügers Arch 1979; 378:231–241.
4. Schäfer D, Schläfke ME. Cardiorespiratory regulation in a model for the Ondine's curse syndrome (this volume).
5. Schläfke ME, See WR, Herker-See A, Loeschcke HH. Respiratory responses to hypoxia and hypercapnia after elimination of central chemosensitivity. Pflügers Arch 1979; 381:241–248.

6. Schäfer T, Schäfer D, Schläfke ME. Breathing, transcutaneous blood gases, and CO_2 response in SIDS siblings and control infants during sleep. J Appl Physiol 1993; 74:887–102.
7. Schäfer C, Schläfke ME. Pulsoximetrisch gesteuerte Sauerstoff-therapie bei Patienten mit mangelnder CO_2-Empfindlichkeit. Atemwegs Lungenkr 1990; 16:446–447.
8. Schläfke ME, Burghardt F, Mückenhoff K, Heimann J. Technisch unterstütztes Atemtraining bei Säuglingen mit mangelnder CO_2-Empfindlichkeit der Atmung. Biomedizinische Technik, Ergänzungsband 1981; 26:48.
9. Schläfke ME, Schäfer T, Kronberg H, et al. Transcutaneous monitoring as trigger for therapy of hypoxemia during sleep. Adv Exp Med Biol 1987; 220:95–100.
10. Schläfke ME, Schäfer T, Nebel B, et al. Development, disturbances, and training of respiratory regulation in infants. In: Peter JH, Penzel T, Podszus T, von Wichert P, eds. Sleep and Health Risk. Berlin: Springer-Verlag, 1991:476–485.
11. Burghardt F, Schläfke ME. Loss of central chemosensitivity: An animal model to overcome respiratory insufficiency. J Auton Nerv Syst 1986; Suppl:105–109.
12. Gozal D, Marcus CL, Shoseyov D, Keens TG. Peripheral chemoreceptor function in children with the congenital central hypoventilation syndrome. J Appl Physiol 1993; 74:379–387.
13. Folgering H, Kuyper F, Kille JF. Primary alveolar hypoventilation (Ondine's curse syndrome) in an infant without external arcuate nucleus: Case report. Bull Physiopathol Resp 1979; 15:659–666.
14. Schläfke ME, See WR, Massion WH, Loeschcke HH. Die Rolle "spezifischer" und unspezifischer Afferenzen fur den Antreib der Atmung, untersucht durch Reizung und Blockade von Afferenzen an der decerebrierten Katze. Pflügers Arch 1969; 312:189–205.
15. Filiano JJ, Kinney HC. Arcuate nucleus hypoplasia in the sudden infant death syndrome. J Neuropathol Exp Neurol 1992; 51:394–403.
16. Kinney HC, Filiano JJ, Panigraphy A, et al. Neurotransmitter receptor binding in cardioventilatory related nuclei of the developing human medulla oblongata (this volume).
17. Schläfke ME, Hukuhara T, Jr., See WR. Loss of central chemosensitivity: Experimental studies on a clinical problem. In: Hutas I, Debreczeni LA, eds. Advances in Physiological Science. Vol. 10. Respiration. New York: Pergamon Press, 1981: 609–616.
18. Schäfer T, Schläfke ME. Vulnerability of the respiratory system: Disposition for sudden infant death syndrome? Adv Biosci 1991; 79:127–129.
19. Schläfke ME, Schäfer T. Sleep-phase related $tcpCO_2$ in infants under closed and open loop conditions of the central pH/pCO_2 control system. In: Gaultier C, Excourrou P, Curzi-Dascalova L, eds. Sleep and Cardiorespiratory Control (Colloque INSERM). London: John Libbey, 1991:260.
20. Gallego J, Perruchet P. Classical conditioning of ventilatory responses in humans. J Appl Physiol 1991; 70:676–682.
21. Shea SA, Andres LP, Shannon D, et al. Respiratory sensations in subjects who lack a ventilatory response to CO_2. Respir Physiol 1993; 93:203–219.
22. Shea SA, Andres LP, Paydarfar D, et al. Effect of mental activity on breathing in congenital central hypoventilation syndrome. Respir Physiol 1993; 94:251–263.

43

Cardiorespiratory Regulation in a Model for Ondine's Curse Syndrome

DIETMAR SCHÄFER and MARIANNE E. SCHLÄFKE

Ruhr-University Bochum
Bochum, Germany

I. Introduction

Congenital central hypoventilation syndrome is characterized by sleep-dependent hypoventilation, the clinical manifestations of which are periodic breathing, sleep apnea, alveolar hypoventilation, and severe respiratory acidosis mainly during non-REM sleep. The syndrome was called Ondine's curse syndrome (OCS) because the automatic control function of respiratory regulation was impaired. Absence of the ventilatory response to hypercapnia has been demonstrated in OCS patients (1). The underlying pathogenetic mechanism was presumed to be defective integration of central chemoreceptor signals. Histopathological findings revealed abnormalities of the ventral surface of the medulla oblongata (VMS) in victims of OCS (2–4).

In animal experiments, two hydrogen-ion-sensitive areas have been described on the VMS (4,5). Bilateral cooling or coagulation of an area between them—the intermediate area (IA)—caused complete elimination of the central chemosensitive drive (6). Information from the rostral and caudal chemosensitive areas was presumed to converge at the IA, where it was relayed to the respiratory neuron populations. Subsequent studies also demonstrated chemosensitive neurons within the IA (7).

These concepts and hypotheses were the result of Loeschcke's comprehensive work on the central chemosensitive mechanism (8). They were challenged by investigators who proposed that chemosensitive structures driving respiration could also be localized in deeper brainstem structures in the cat, e.g., the nucleus tractus solitarius (NTS) and the locus ceruleus (9,10). However, the IA seems to play a key role, since its blockade eliminates the central chemosensitive drive completely, as demonstrated in anesthetized, decerebrate, or chronic nonanesthetized cats (4).

In order to resolve criticisms of the nonspecific methods used for elimination of central chemosensitivity, we reinvestigated the role of the IA in the control of ventilation by topically applying to the VMS the neurotoxin ibotenic acid (IBO), which is known to cause axon-sparing neuronal lesions (11). This excitatory amino acid has been described as a fairly good *N*-methyl-D-aspartate (NMDA) receptor agonist (11). NMDA receptors are involved to some extent in respiratory and cardiovascular regulation. They influence, for example, the genesis of respiratory rhythm (12), and their stimulation has been shown to augment phrenic nerve response to CO_2 (13), whereas NMDA antagonists decrease the CO_2 response (14).

We examined the effect of superficial lesions of the IA with IBO on resting ventilation, blood pressure, heart rate, and cortical activity as well as pH at the VMS. Furthermore, we tested the influence of this lesion on the steady-state responses to CO_2 and hypoxia in order to estimate the contribution of the superficial IA neurons to the central and peripheral chemosensitive mechanism. We further examined whether nonrespiratory acidosis induced by NH_4Cl or acetazolamide (ACTZ) could be compensated after IA neuronal lesions.

II. Methods

A. Anesthesia and Preparation

Nine adult cats were initially anesthetized with propionyl promazine phosphate (SC) and ketamine HCl (IM). During surgery, ketamine HCl was injected intravenously as required. After tracheostomy and removal of the larynx, the base of the skull was approached. The musculi recti capitis ventrales and musculi longi capitis were cut, the caudal basal part of the occipital bone was partially removed, and the dura mater and arachnoidea were opened. The surgical procedure was performed under oxygen breathing.

B. Recording and Data Analysis

Tidal volume (V_T), respiratory rate (*f*), end-tidal CO_2 partial pressure ($P_{ET}CO_2$), inspiratory O_2 fraction (F_IO_2), blood pressure, ECG, temporoparietal and temporooccipital EEG, and VMS pH were simultaneously and continuously measured. The data were digitized and analyzed using a personal computer. Arterial blood gas pressures (P_aCO_2, P_aO_2) and pH_a were measured in blood samples (20 μL) from the femoral artery. For statistical purposes, mean values and standard deviations were calculated at 5-min intervals for the cardiorespiratory parameters using breath-by-breath and beat-by-beat analysis and compared with the two-tailed Student's *t*-test. Unless otherwise stated, all differences refer to these mean values and were significant, with $P < 0.01$. Steady-state CO_2 responses were performed at three or more steps of $P_{ET}CO_2$. The EEG was analyzed visually and by fast-Fourier transformation.

C. Experimental Procedure

Steady-state respiratory response curves for CO_2 under hyperoxia were measured before and after lesioning of IA in each experiment. In addition, the hypoxic response was tested in three cats by lowering the F_IO_2 from 80% to 10% at different, constant $P_{ET}CO_2$ levels. The effect of nonrespiratory acidosis induced by NH_4Cl was examined in two cats. Four cats received ACTZ. In the six cats receiving NH_4Cl or ACTZ, the function of the peripheral chemoreceptors was tested qualitatively by bolus injections of 0.1% NaCN. Vagi and sinus nerves were cut at the end of four experiments.

Neurotoxic Lesion

The lesion was carried out by placing cotton pledgets (1.5 mm) soaked with 10 μL IBO solution bilaterally on IA. The concentrations used were 63.25 m*M* IBO (100 μg, 2 cats), 12.65-m*M* solution (20 μg, 1 cat), and a 6.33-m*M* solution (10 μg, 7 cats). The pledgets were removed after 5 min (63.25 m*M*) or 20 min (lower concentrations), respectively. As control, 10 μg IBO (6.33 m*M*) was applied on a field rostromedial of the IA in one experiment.

Metabolic Acidosis

The effect of NH_4Cl or ACTZ infusion was tested in two controls with intact IA. After IBO application, the administration of the drugs was performed when respiration had reached steady-state conditions again (1 to 4 h after IBO). NH_4Cl (160 mg/kg) and ACTZ (160 mg/kg) were slowly injected into the femoral vein at the rate of 0.5 mL/min.

III. Results

A. Cardiorespiratory Effects and Cortical Activity after Lesion of the IA

In two cats, tidal volume increased markedly (+33 and +64 mL) within 5 min after local applications of 100 μg IBO. This excitatory effect stopped after 17 and 12 min, respectively, and V_T slowly declined until 34 and 58 min after application of IBO, respectively, when the final apneas occurred. Both cats showed severe hypercapnia ($P_{ET}CO_2$: 74 and 93 torr, respectively) after 30 to 45 min. Systolic arterial pressure slowly declined from 110 mm Hg, starting 15 min after application of IBO, to 80 mm Hg at the thirtieth minute. No pathological patterns were observed in the EEG. The power spectra revealed an increase of power at the lower frequencies (< 8 Hz) during the excitation, decreasing thereafter.

IBO (10 and 20 μg) applied on the IA caused a small increase followed by a reduction of ventilation. Ventilation diminished to 47% to 82% of baseline 45 to 80 min after start of application, causing respiratory acidosis with an increase in $P_{ET}CO_2$ of 7 to 11 torr within 1 h. The respiratory rate, heart rate, and systolic arterial pressure exhibited changes between 1% and 22%. The EEG increased during the excitatory phase and was depressed afterwards.

All these effects were dependent on the precise application of IBO to the IA. After application of the pledgets to an area directly beneath the IA, only slight changes in f (−6%) and expiratory time (+10%) were observed. All other parameters, including the respiratory response to CO_2, showed no significant changes. An IBO lesion of the IA 275 min later caused the typical effects characterized above.

B. Hypercapnic Ventilatory Responses

After bilateral application of 10 μg IBO, a progressive reduction of the slope of the ventilatory response to inhalation of CO_2 was observed (Fig. 1).

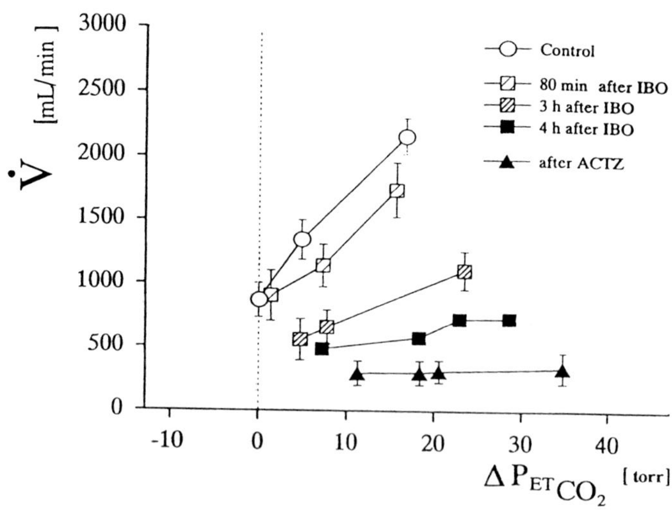

Figure 1 Ventilatory response to CO_2 (mean ± SD) in the control and several times after application of 10 μg IBO on the IA. The response was reduced after IBO and was completely abolished after infusion of ACTZ.

The tidal volume response was completely abolished in three of six cats 1 to 3 h after the lesion. The other responses showed reductions in the slopes of more than 80%. The respiratory frequency response was depressed in four experiments and slightly elevated in two cats.

C. Augmented Breaths

In order to examine the influence of the superficial lesion on the pattern of sighs, their mean V_T was measured before lesioning and after the point of maximal reduction of CO_2 sensitivity was reached. The mean V_T of sighs before lesion was 99.7 ± 24.0 mL; afterwards, it was 88.6 ± 25.6 mL. A Wilcoxon matched-pairs test failed to indicate a significant difference between these values.

D. Peripheral Chemoreceptor Function

Figure 2 shows original recordings of the respiratory parameters and the EEG after IBO lesion. Although the hypercapnic ventilatory response was almost eliminated, the hypoxic ventilatory drive and the EEG effect of mild hypoxia were still evident. In each of the three experiments, the hyperbolic ventilatory response to F_IO_2 was shifted to lower V_T values only after elimination of the

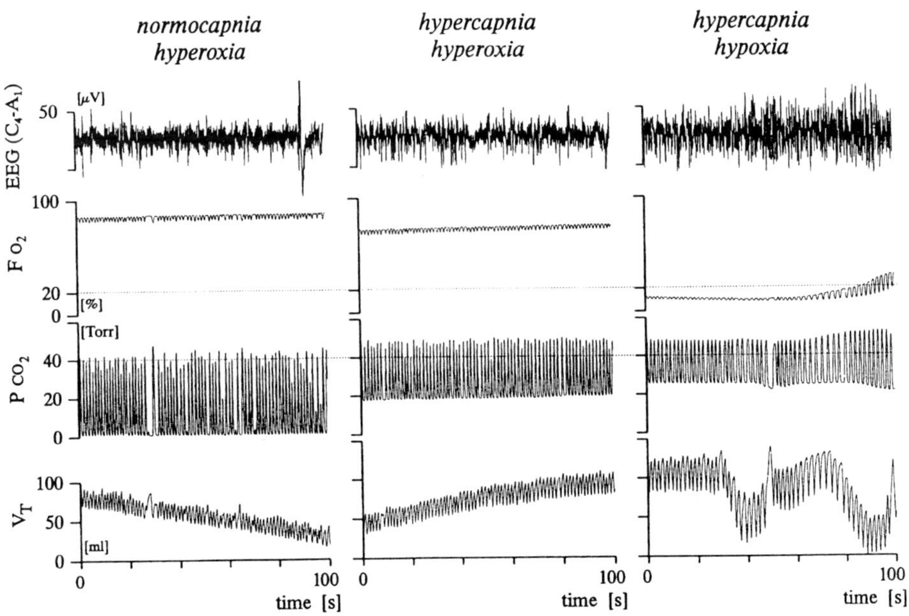

Figure 2 Original registrations of the respiratory parameters and EEG during resting ventilation and ventilatory responses to hypercapnia and hypoxia after lesion of the IA with IBO.

central chemosensitive drive. In the other experiments, the peripheral chemoreceptor function was tested by bolus injections of 0.1% NaCN. The characteristic immediate increase in V_T remained unchanged after the IBO lesion. Peripheral chemodenervation by cutting of the sinus nerves and vagi caused ventilatory arrest after IBO application.

E. Ammonium Chloride Acidosis

Intravenous infusion of NH_4Cl led to an increase in V_T and very brief acceleration of *f* in the intact animal. VMS pH showed a compensatory development of arterial blood acidification (pH_a 7.39 to 7.31) (Fig. 3).

After elimination of central chemosensitivity with IBO, NH_4Cl also caused a transient increase in *f*, whereas V_T and VMS pH rapidly decreased. Gasping and apnea occurred 5 min after the start of the NH_4Cl infusion. Arterial pH_a declined from 7.41 to 7.27 after the lesion and before acid infusion. The NH_4Cl

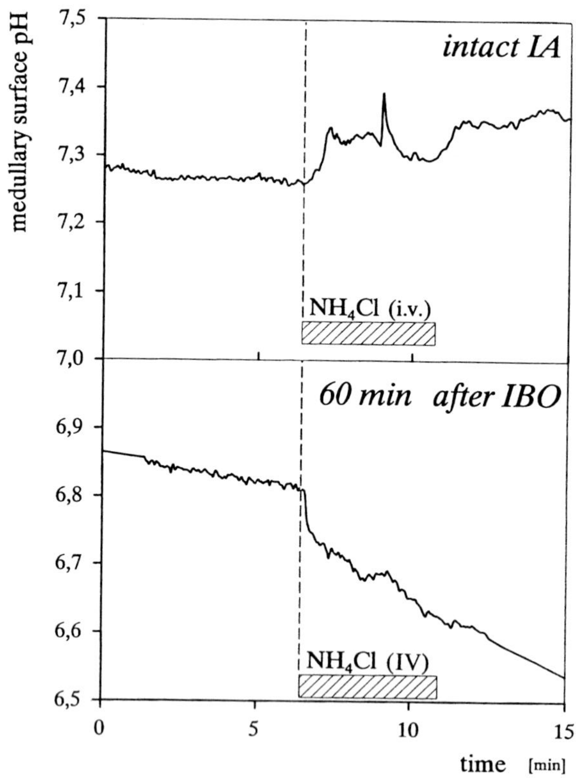

Figure 3 Medullary surface pH during infusion of ammonium chloride before and 60 min after lesion of the IA with IBO.

infusion reduced it further to pH_a 7.06, when apnea occurred. VMS pH dropped from 7.40 to 6.82 after lesioning and to 6.59 5 min after acid infusion (at the beginning of the apnea) (Fig. 3).

F. Acetazolamide

In the control, intravenous administration of 160 mg/kg ACTZ caused an increase of tidal volume by 330% within 10 min. Minimum pH_a was reached 4 h after injection, when pH_a had decreased by 0.27 units to 7.21. Up to this point, VMS pH had decreased from 7.42 to 7.09. The tidal volume response to CO_2 was shifted to lower P_aCO_2 and to very high ventilation (> 1 L/min/kg body weight). Therefore the slope was reduced. The intravenous injection of ACTZ after elimination of central chemosensitivity caused a reduction of resting ventilation by 46% resulting from decreases in V_T and f. Arterial and VMS pH diminished from 7.47 to 7.28 (pH_a) and to 7.16 (VMS pH) after the lesion. ACTZ caused further reductions to pH_a 7.04 and VMS pH 6.79 when apnea occurred.

IV. Discussion

A. Central Chemosensitive Drive of Respiration

After local application of IBO on the IA, resting ventilation decreased. The respiratory response to CO_2 was strongly reduced or completely abolished. In agreement with previous studies in which the IA was bilaterally coagulated or cooled (4), these results indicate that the neurotoxic destruction of the IA eliminates the central sensitivity for CO_2. Therefore, the criticism that arose concerning the nonspecific methods used in those experiments can be rejected with regard to their results. The findings support the notion that neurons close to the VMS in the region of the IA mediate the central chemosensitive drive. NMDA as well as non-NMDA receptors are involved in the respiratory control system located near the VMS (14). Therefore, reduction in the central chemosensitive drive could also be produced by local application of kainic acid (15,16).

B. Blood Pressure and Cortical Activity After Lesion of the IA

Bilateral application of IBO (10 μg) on the IA caused only slight changes in blood pressure and heart rate. They may be explained as being secondary to the consistent reduction in ventilation, e.g., as a decrease in respiratory modulation of sympathetic activity (17). Furthermore, each lesion was followed after a brief excitation by a general decrease of EEG power. This result may indicate a link between the central chemosensitive drive and the ascending reticular activating system at the site of the IA. Inhibitory effects on reticular neurons of local Novocain application on the VMS have been demonstrated (18) and support this hypothesis.

C. Peripheral Chemoreceptor Function after Lesion of the IA

The role of the IA for the integration of the peripheral chemoreceptors in the respiratory control system is not yet clear. Local cooling of the IA was reported not to affect the respiratory response to hypoxia (8,19,20,21,22). Other investigators found a reduction in the response to stimulation of the peripheral chemoreceptors during cooling (23) or application of kainic acid on the IA (16). In our experiments, the lesion with IBO did not affect the hypoxic drive. A peripheral chemodenervation, however, caused respiratory arrest after elimination of central chemosensitive drive, which is in agreement with previous observations by Schläfke et al. (24). The different results of Martin and Sinclair (16) may be explained by the high dose of kainic acid, the greater area of application, or distant lesions within other structures (25), e.g., the NTS. Millhorn and Eldridge (23) presumed that afferents from peripheral and central chemoreceptors do not pass through the IA but project to some second-order neurons within the medulla converging onto a common neuronal component. They observed that focal cooling of the IA blocks afferent input from the peripheral chemoreceptors, whereas in our study the neurotoxic destruction of superficial nerve cells within this region did not affect peripheral chemoreceptor function. Cooling, however, also caused more side effects concerning the cardiovascular regulation (26).

D. Nonrespiratory Acidosis

In order to examine the contribution of the peripheral chemoreceptors to the compensation of nonrespiratory acidosis, we administered NH_4Cl or ACTZ after elimination of the central chemosensitive drive. We found that NH_4Cl infusion was followed by transient augmentation of f even after the lesion, whereas the V_T response to acid infusion was eliminated, and 2 min after infusion of NH_4Cl the animal stopped breathing. The acid response of f may be explained by the thromboxane A2-mediated effects described by Shams et al. (27). An inhibition of the carbonic anhydrase by ACTZ produces a severe tissue acidosis and hypercapnia causing hyperpnea. It impairs the dynamic response of ventilation to step changes of F_ICO_2 (28), whereas the steady-state response remained unchanged (22,28). Intravenous administration reduced the peripheral chemoreceptor function (29).

In our experiments, ACTZ infusion following bilateral neurotoxic lesion of the IA caused ventilatory arrest, which might have been due to a specific inhibition of peripheral chemoreceptor function. The data from the experiments with NH_4Cl, however, indicated that the peripheral chemoreceptors were not able to compensate for a nonrespiratory acidosis after loss of central chemosensitive drive caused by IA lesion. Furthermore, additional chemosensitive sites as described by Coates et al. (9) also seem to be unable to respond to metabolic changes when the IA is lesioned. On the contrary, nonrespiratory acidosis led to an unrestrained acidification of arterial blood and VMS pH until at pH_a

7.04 ventilation stopped, which may be due to paradoxic responses of pH-sensitive neurons within the medulla oblongata (4).

V. Summary

The central chemosensitive drive is absent in patients with Ondine's curse syndrome. With regard to earlier experiments in chronic cats, we wanted to have an animal model for this syndrome in which the method of coagulation was replaced by a more precise technique. Therefore, we applied the neurotoxin ibotenic acid superficially to the IA in order to eliminate selectively neuronal somata reached from this area. This lesion caused complete elimination of the central chemosensitive drive, whereas the cardiovascular system was only slightly affected. It further led to a decrease in EEG power that may be attributed to a reduction of reticular activity. The peripheral chemoreceptor function with respect to the hypoxic ventilatory response remained intact after the lesion. Nonrespiratory acidosis, however, could not be compensated by the peripheral chemoreceptors alone and caused apnea. These data offer a model for Ondine's curse syndrome.

References

1. Gozal D, Marcus CL, Shoseyov D, Keens TG. Peripheral chemoreceptor function in children with congenital central hypoventilation syndrome. J Appl Physiol 1993; 74:379–387.
2. Folgering H, Kuyper F, Kille JF. Primary alveolar hypoventilation (Ondine's curse syndrome) in an infant without external arcuate nucleus: Case report. Bull Physiopathol Resp 1979; 15:659–666.
3. Filiano JJ, Choi JC, Kinney HC. Candidate cell populations for respiratory chemosensitive fields in the human infant medulla. J Comp Neurol 1990; 293:448–465.
4. Schläfke ME. Central chemosensitivity: A respiratory drive. Rev Physiol Biochem Pharmacol 1981; 90:172–244.
5. Schläfke ME, See WR, Loeschcke HH. Ventilatory response to alterations of H-ion concentration in small areas of the ventral medullary surface. Respir Physiol 1970; 10:198–212.
6. Schläfke ME, Kille JF, Loeschcke HH. Elimination of central chemosensitivity by coagulation of a bilateral area on the ventral medullary surface in awake cats. Pflügers Arch—Eur J Physiol 1979; 378:231–241.
7. Schläfke ME, Pokorski M, See WR, et al. Chemosensitive neurons on the ventral medullary surface. Bull Eur Physiopathol Respir 1975; 11:277–284.
8. Loeschcke HH. Central chemosensitivity and the reaction theory. J Physiol 1982; 332:1–24.
9. Coates EL, Li A, Nattie EE. Widespread sides of brain stem ventilatory chemoreceptors. J Appl Physiol 1993; 75:5–14.
10. Kogo N, Arita H. In vivo study of medullary H-sensitive neurons. J Appl Physiol 1990; 69:1408–1412.

11. Schwarcz R, Foster AC, French ED, et al. Excitotoxic models for neurodegenerative disorders. Life Sci 1984; 35:19–32.
12. Foutz AS, Champagnat J, Denavit-Saubie M. Involvement of N-methyl-D-aspartate (NMDA) receptors in respiratory rhythmogenesis. Brain Res 1989; 500:199–208.
13. Mitra J, Prabhakar NR, Overholt JL, Cherniack NS. Respiratory effects of N-methyl-D-aspartate on the ventrolateral medullary surface. J Appl Physiol 1989; 67:1814–1819.
14. Nattie EE, Gdovin M, Li A. Retrotrapezoid nucleus glutamate receptors: Control of CO_2-sensitive phrenic and sympathetic output. J Appl Physiol 1993; 74:2958–2968.
15. Nattie EE, Mills JW, Ou LC, St. John WM. Kainic acid on the rostral ventrolateral medulla inhibits phrenic output and CO_2 sensitivity. J Appl Physiol 1988; 65:1525–1534.
16. Martin RL, Sinclair JD. Kainic acid on the rat ventral medullary surface depresses hypoxic and hypercapnic ventilatory responses. Respir Physiol 1990; 80:55–70.
17. Millhorn DE, Kiley JP. Effect of graded cooling of intermediate areas on respiratory response to vagal input. Respir Physiol 1984; 58:51–64.
18. Schwanghardt F, Schroter R, Klüssendorf D, Koepchen HP. The influence of novocain block of superficial brain stem structures on discharge pattern of specific respiratory and unspecific reticular neurons. In: Umbach W, Koepchen HP, eds. Central Rhythm and Regulation. Stuttgart: Hippokrates, 1974:104–110.
19. Cherniack NS, von Euler C, Homma I, Kao FF. Graded changes in central chemoreceptor input by local temperature changes on the ventral surface of the medulla. J Physiol (Lond) 1979; 287:191–211.
20. Schläfke ME, See WR, Massion WH, Loeschcke HH. Die Rolle spezifischer und unspezifischer Afferenzen fur den Antreib der Atmung, untersucht durch Reizung und Blockade von Afferenzen an der decerebrierten Katze. Pflügers Arch 1969; 312:189–205.
21. See WR, Schläfke ME, Loeschcke HH. Role of chemical afferents in the maintenance of rhythmic respiratory movements. J Appl Physiol 1983; 54:453–459.
22. Teppema LJ, Rochette F, Demedts M. Ventilatory effects of acetazolamide in cats during hypoxemia. J Appl Physiol 1992; 72:1717–1723.
23. Millhorn DE, Eldridge FL. Role of ventrolateral medulla in regulation of respiratory and cardiovascular systems. J Appl Physiol 1986; 61:1249–1263.
24. Schläfke ME, See WR, Herker-See A, Loeschcke HH. Respiratory responses to hypoxia and hypercapnia after elimination of central chemosensitivity. Pflügers Arch 1979; 381:241–248.
25. Guldin WO, Markowitsch HJ. Epidural kainate, but not ibotenate produces lesions in local and distant regions of the brain: A comparison of the intracerebral actions of kainic acid and ibotenic acid. J Neurosci Methods 1982; 5:83–93.
26. Mitra J, Prabhakar NR, Overholt JL, Cherniack NS. Respiratory and vasomotor responses to focal cooling of the ventral medullary surface (VMS) of the rat. Respir Physiol 1988; 74:35–48.
27. Shams H, Peskar BA, Scheid P. Acid infusion elicits thromboxane A2-mediated effects on respiration and pulmonary hemodynamics in the cat. Respir Physiol 1988; 71:169–183.

28. Coates EL, Li A, Nattie EE. Acetazolamide on the ventral medulla of the cat increases phrenic output and delays the ventilatory response to CO_2. J Physiol 1991; 441:433–451.
29. Iturriaga R, Lahiri S, Mokashi A. Carbonic anhydrase and chemoreception in the cat carotid body. Am J Physiol 1991; 261:C565–C573.

44

Development of CO_2 Sensitivity During Non-REM Sleep in Infants

THORSTEN SCHÄFER and MARIANNE E. SCHLÄFKE

Ruhr-University Bochum
Bochum, Germany

The H^+/CO_2 sensitivity of the respiratory system is a major factor in maintaining the homeostasis of arterial and extracellular fluid pH and P_{CO_2}. Early human development, however, causes marked changes in the respiratory system, involving central maturation as well as changes in respiratory mechanics and airways. Phenomena in early childhood (e.g., rib cage distortion, periodic breathing, the high amount of REM sleep with muscle atonia, and increased airway resistance) tend to reduce the efficiency of respiratory efforts (1,2). We were therefore interested in the gestational and postnatal age related differences in the homeostasis of arterial P_{CO_2} and the respiratory CO_2 response during sleep in the first year of life of healthy full-term and healthy preterm infants.

The slope of the steady-state ventilatory response to inhaled CO_2 has been reported to increase 42% with decreasing prematurity from 32 to 37 weeks gestation and 62% from 2 to 27 days of postnatal age (3). Premature infants were found to have a reduced sensitivity to CO_2 when compared with term infants using the rebreathing method. Reduced sensitivity remained when all infants were compared on a weight or lung-volume basis, suggesting a maturational effect unrelated to body size (4). Occlusion pressure measurements in premature infants during breathing at steady state levels of CO_2 revealed an increase in the ventilatory response with maturation with mask pressure paralleling minute ventilation (5). These early studies, however, did not distinguish between different states of wakefulness and used invasive techniques that might have led to changes in the behavior of the respiratory system, e.g., by using face masks or intraesophageal pressure transducers.

The effects of the sleep state and postnatal maturation on steady-state CO_2 sensitivity were examined in unanesthetized premature monkeys. CO_2 sensitivity was similar between REM and non-REM sleep and increased with postnatal maturation (6). In a more detailed study on the effects of gestational and postnatal age in the premature primate, the duration of pregnancy was uncorrelated with eucapnic ventilation and CO_2 sensitivity. Postnatal development contributed to maturational changes (7). CO_2-induced changes in ventilation in healthy, sleeping infants (determined with noninvasive methods using a barometer with a sealed chamber) showed no significant differences between REM and quiet sleep between 9 days and 4 months of life. Instantaneous minute ventilation increased 17% to 33% with inhalation of 2% CO_2 without a significant age trend (8).

In a previous investigation in full-term infants, we found relative CO_2-induced changes of ventilation of 20% to 36% per torr $P_{tc}CO_2$, not correlated with age (2). However, opposite tendencies of changes in tidal volume and frequency responses with age were detected. Tidal volume ratio decreased slightly with an increase in the frequency contribution to the overall ventilatory response to CO_2.

The present cross-sectional study compares the maturation of ventilation, focusing on the PCO_2 during night sleep and the steady state CO_2 response in quiet sleep, in healthy full-term infants between 2 weeks and 18 months of life and in preterm infants near their expected birth date to the end of the first year of life.

I. Methods

A. Infant Population

Measurements of 180 healthy full-term infants were compared with 70 measurements in 48 formerly preterm infants. For the full-term infants, duration of pregnancy was 37 to 42 weeks [mean 39.8 ± 1.1 (SD) weeks], and mean birthweight was 3500 ± 450 g. For the preterm group, duration of pregnancy was 27–36 weeks [mean 31.8 ± 2.9 (SD)] weeks, and the mean birthweight 1680 ± 720 g. The young preterms were examined when they were dismissed from hospital. Most of the preterms, however, came from home. Preterms with complications, e.g., bronchopulmonary dysplasia (BPD), at the time of measurement were excluded from the analysis. Postnatal and postconceptual age data were compared.

The study had the approval of the ethics committee of the medical faculty at Ruhr University and written consent was obtained from the parents. Infants were hospitalized in the afternoon. Measurements were performed in a specially equipped sleep laboratory with a separate patients' room. They started with onset of night sleep and continued until the infants awoke in the morning.

B. Measurement Techniques

A previous study describes the polysomnographic technique in detail (2). For all infants, we continuously and simultaneously recorded respiratory movements of rib cage and abdomen using the respiratory induction plethysmography (Respitrace Systems, Ardsley, NY) calibrated with the QDC method (9); transcutaneous electrodes for PCO_2 and PO_2 measurements (TCM222, Radiometer, Copenhagen) operated at 43°C; and ECG and respiratory impedance plethysmography (Servomed, Hellige, Freiburg) for monitoring purposes. CO_2 response curves were obtained during quiet sleep without periodic breathing by elevation of inhaled CO_2 concentration to 2% in artificial air. The air, supplied by a prepared calibration gas mixture, was applied via a funnel over the infant's mouth and nose without touching them. Relative CO_2-induced changes of tidal volume, respiratory rate, and minute ventilation were calculated for the fifth minute of CO_2 inhalation and compared with the minute before the test. The difference was expressed as relative increase (%) per torr increased transcutaneous $P_{tc}CO_2$.

All tests were performed between 10 P.M. and 12 midnight. Tests with arousals leading to body movements or changes in breathing patterns indicating a change of sleep state were excluded from the analysis. In order to correlate EEG and cardiorespiratory measures for sleep states, a sleep EEG was recorded after finishing the CO_2 response tests using standard criteria (10).

II. Results

A. Influence of Age on Steady-State CO_2 Response

Full-term and preterm infants beyond 1 month after their expected birth date had similar ventilatory responses to CO_2 with an average increase in ventilation of 32% ± 2% (mean ± SEM) and 35% ± 3% per torr increased $P_{tc}CO_2$, respectively (Fig. 1). There were no age trends (for full-term infants, Spearman $r = 0.05$, $P > 0.53$; for preterms beyond 1 month, $r = -0.16$, $P > 0.23$). The group of younger preterms within the first month after their expected birth date, however, had a significantly lower ventilatory CO_2 response compared with older preterms and with the full-term infants (Figs. 1 and 2). The difference between the full-term and preterm infants within the first month of (corrected) age was due to a reduced tidal volume response to CO_2 (16% ± 4% versus 33% ± 3%, $P < 0.01$).

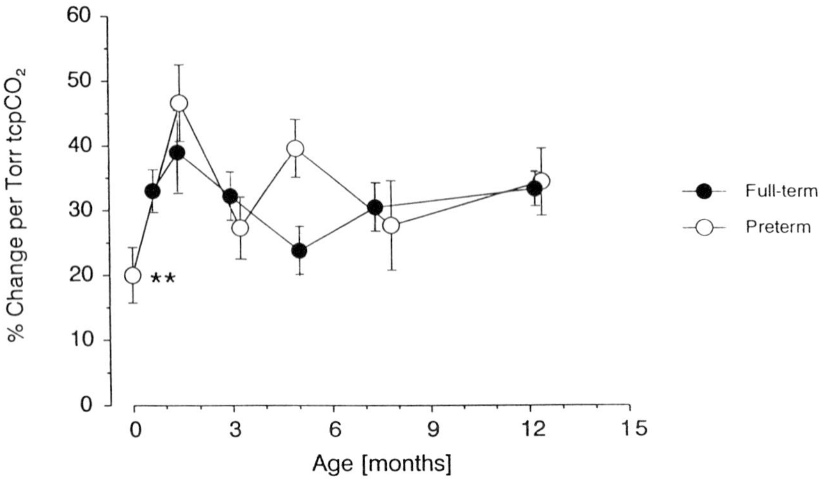

Figure 1 Relative CO_2-induced changes of ventilation per torr increased $P_{tc}CO_2$ during the first year of life in full-term and preterm infants (mean ± SEM, $^{**}P < 0.01$, t-test, compared to youngest full-term group).

B. Mean and Maximum P_{CO_2} During Sleep

Mean transcutaneously measured P_{CO_2} values did not differ in the full-term and preterm infants and did not change significantly with age (Fig. 3). Mean values were 40.9 ± 0.3 torr and 40.2 ± 0.7 torr, respectively. Mean and maximum values were closely related ($r = 0.87$, $P < 0.001$).

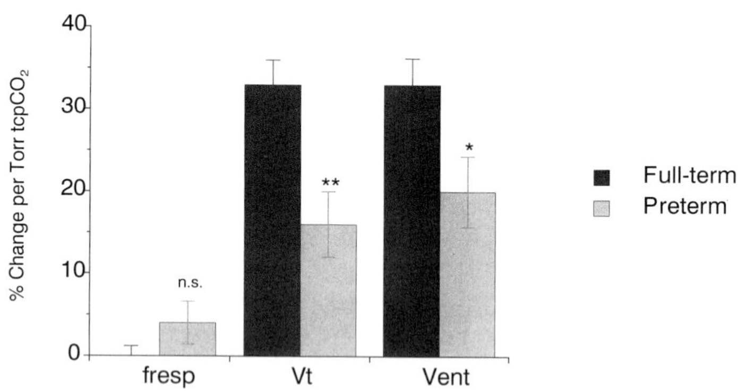

Figure 2 CO_2 response of respiratory frequency (fresp), tidal volume (Vt), and ventilation (Vent) in 1-month-old full-term and preterm infants as relative changes of control (mean ± SEM, n.s. = difference not significant, $^{*}P < 0.05$, $^{**}P < 0.01$, t-test).

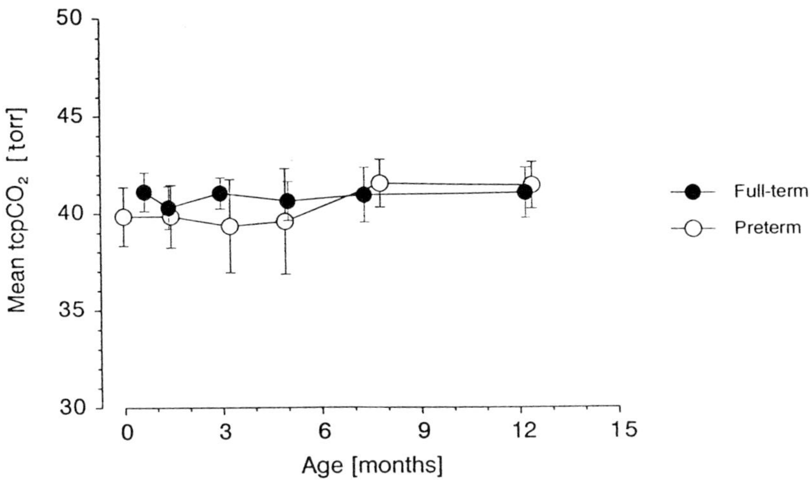

Figure 3 Mean $P_{tc}CO_2$ values during sleep in the first year of life in full-term and preterm infants (mean ± SEM).

III. Discussion

Mature ventilatory responses to CO_2 have been found previously in humans between 1 week and 4 months (8) and at 18 months (2) of life. The present investigation confirms these earlier reports with the additional observation that the same effects occur in formerly preterm infants after a certain postconceptual age. No linear age trend in preterms could be found after 30 days beyond the expected birth date. Constant mean $P_{tc}CO_2$ values during sleep are an indication for a precise regulation of carbon dioxide from birth on.

Significantly lower normoxic CO_2 responses in young preterms have been reported in humans (3–5) and monkeys (6,7) within the first few postnatal days. In our study, we found that at a postnatal age of about 2 months, corresponding to a corrected postconceptual age around the expected birth date, formerly preterm infants had a reduced CO_2 response during quiet sleep. The study design, however, does not clarify the underlying mechanisms of maturation. Changes in the mechanical properties of the respiratory system—including electromechanical coupling in respiratory muscles, maturational processes in the central nervous system that affect the H^+/CO_2-sensitive system, and recruitment of inspiratory output—may play a role. Reduced maximum airway pressures during CO_2 inhalations in preterms indicate a central maturation of respiratory drive with age (5).

Besides this central maturation of the neuronal respiratory network including the central chemosensitivity, other factors must be taken into consideration. Recently, an inhibitory respiratory sensory reflex in response to CO_2 has been described in preterm infants (11) causing short apneas (2 to 3 s) in response to 3% to 5% CO_2 inhalation and prolonged apneas at higher CO_2 concentrations. In our study, however, we used lower CO_2 concentrations and did not observe apneas at the onset of the CO_2 response tests. A reduction of CO_2 responsiveness by this reflex is unlikely.

In preterm infants at 1 month of age, different CO_2 thresholds of respiratory muscles could be found (12). During hyperoxic rebreathing, diaphragm EMG increased proportionally and linearly with tidal volume, whereas genioglossus and alae nasi EMG differed from diaphragm EMG and exhibited minimal or absent responses at low levels of hypercapnia, causing increased upper airway resistance. This phenomenon could be attributed to a reduction of the overall CO_2 response by reducing the effectiveness of the central respiratory output.

The oxygen partial pressure plays an important role in the ventilatory response to CO_2 in preterm infants. The more hypoxic the infant, the flatter was the CO_2 response curve (13), the reverse of the behavior seen in adults. We found that younger preterm infants had significantly lower $P_{tc}O_2$ values during sleep compared with full-term infants (14). In the second month of postnatal life, full-term infants had a mean $P_{tc}O_2$ of 66 torr compared to 45 torr in preterms. This relative hypoxia might already contribute to the reduction in CO_2 responsiveness. On the other hand, O_2- and CO_2-dependent changes of cerebral blood flow in preterm infants could reduce the CO_2 partial pressures at the site of the central chemosensors and thus reduce the ventilatory response if related to transcutaneous P_{CO_2}. These effects of cerebral blood flow could explain differences of overall CO_2 responses in different sleep states (15).

In the newborn piglet, the maturation of the ventilatory response to CO_2 could be attributed to a postnatal increase in the relative contribution of the peripheral chemoreceptors to the total ventilatory response and a decreasing apneic threshold (16). This effect, however, was already observed within the very first days after birth, corresponding to the time necessary for the resetting of the peripheral hypoxic sensitivity.

Disturbances in the normal maturational process of respiratory regulation during sleep have been considered as possible causes for life-threatening events and sudden infant death syndrome in infancy. We found a reduced ventilatory CO_2 response in 1-month-old infants who had a life-threatening event and needed vigorous stimulation or cardiopulmonary resuscitation (17).

We conclude that in healthy infants, CO_2 homeostasis is guaranteed throughout the first year of life despite a variety of changes in the respiratory system. In early infancy, however, there is a maturational increase of the ventilatory CO_2 response, related to postconceptual age. The data support the hypothesis that failure of the normal maturational increase in CO_2 sensitivity or intermediate suppression may play a role in the pathogenesis of SIDS.

References

1. Gaultier C. Review: Breathing and sleep during growth: Physiology and pathology. Bull Eur Physiolpathol Respir 1985; 21:55–112.
2. Schäfer T, Schäfer D, Schläfke ME. Breathing, transcutaneous blood gases, and CO_2 response in SIDS siblings and control infants during sleep. J Appl Physiol 1993; 74:88–102.
3. Rigatto H, Brady JP, Cir B, de la Torre Verduzco R. Chemoreceptor reflexes in preterm infants: II. The effect of gestational and postnatal age on the ventilatory response to inhaled carbon dioxide. Pediatrics 1975; 55:614–620.
4. Krauss AN, Klain DB, Waldman S, Auld PAM. Ventilatory response to carbon dioxide in newborn infants. Pediatr Res 1975; 9:46–50.
5. Frantz ID III, Adler SM, Thach BT, Taeusch HW Jr. Maturational effects on respiratory responses to carbon dioxide in premature infants. J Appl Physiol 1976; 41:41–45.
6. Guthrie RD, Standaert TA, Hodson WA, Woodrum DE. Sleep and maturation of eucapnic ventilation and CO_2 sensitivity in the premature primate. J Appl Physiol 1980; 48:347–354.
7. Guthrie RD, Standaert TA, Hodson WA, Woodrum DE. Development of CO_2 sensitivity: Effects of gestational age, postnatal age, and sleep state. J Appl Physiol 1981; 50:956–961.
8. Haddad GG, Leistner HL, Epstein RA, et al. CO_2-induced changes in ventilation and ventilatory pattern in normal sleeping infants. J Appl Physiol 1980; 48:684–688.
9. Sackner ME, Watson H, Belsito AS, et al. Calibration of respiratory inductive plethysmography during natural breathing. J Appl Physiol 1989; 66:410–420.
10. Anders T, Emde R, Parmelee A, eds. A manual of standardized terminology, techniques and criteria for scoring of states of sleep and wakefulness in newborn infants. Los Angeles: UCLA Brain Information Service, 1971.
11. Alvaro RE, Weintraub Z, Kwiatkowski K, et al. A respiratory sensory reflex in response to CO_2 inhibits breathing in preterm infants. J Appl Physiol 1992; 73: 1558–1563.
12. Carlo WA, Martin DJ, Difiore JM. Differences in CO_2 threshold of respiratory muscles in preterm infants. J Appl Physiol 1988; 65:2434–2439.
13. Rigatto H, de la Torre Verduzco R, Cates DB. Effects of O_2 on the ventilatory response to CO_2 in preterm infants. J Appl Physiol 1975; 39:896–899.
14. Schäfer T, Schläfke ME. Vulnerability of the respiratory system: Disposition for sudden infant death syndrome? Adv Biosci 1991; 79:127–129.
15. Parisi RA, Edelman NH, Santiago TV. Central respiratory carbon dioxide chemosensitivity does not decrease during sleep. Am Rev Respir Dis 1992; 145:832–836.
16. Wolsink JG, Berkenbosch A, DeGoede J, Olievier CN. Maturation of the ventilatory response to CO_2 in the newborn piglet. Pediatr Res 1993; 34:485–489.
17. Schläfke ME, Schäfer T, Nebel B, et al. Development, disturbances, and training of respiratory regulation in infants. In: Peter JH, Penzel T, Podszus T, von Wichert P, eds. Sleep and Health Risk. Berlin: Springer-Verlag, 1991:476–485.

45

Developmental and Pathophysiological Aspects of Respiratory Control

GABRIEL G. HADDAD

Yale University School of Medicine
New Haven, Connecticut

I. Introduction

A number of important issues have been raised concerning development and pathophysiology of ventral medullary respiratory control mechanisms. Instead of summarizing their details, I will opt for a different strategy and raise some of the aspects that need discussion or that need to be highlighted either because they are controversial or represent an advance in our understanding of respiratory control. I will raise and discuss six issues.

II. Main Issues: Development and Pathophysiology

A. Respiratory Rhythm Generation in the Neonatal Rat

New advances have been made in our understanding of respiratory rhythm generation. These have been made mostly in conjunction with a brainstem–spinal cord preparation in vitro used by a number of investigators since 1974 (1–3). One author (Smith J.) dealing with this issue presented data on some rhythmically active neurons in the pre-Bötzinger complex (1). The authors have also built a model to simulate the basis of this rhythmicity by considering a number of membrane properties and ionic channels, including Ca^{++} and K^+ channels. Though the discussion did not center on the model, it certainly focused on differences in properties and phenotypes between the newborn and the mature respiratory generating system. It was clear from the discussion that, in spite of the fact that the basic and fundamental constituents of the rhythm-generating machinery are present in the neonate—after all, the breathing rhythm is present at birth and actually much earlier in fetal life—major differences exist between the neonate and the mature infant. To name a few: (a) the patterns of respiration are well known to be different in a variety of ways; (b) the response to various types of stimuli is different; and (c) the interactions between the respiratory system and other systems including sleep, autonomic activity, and so on are different. Although such differences may prove to be based on differences not only in the central nervous system (CNS) but also in the periphery, it will be difficult to exclude the very dynamic changes that occur in the CNS in the perinatal period from having a major effect on respiration, respiratory pattern generation, and maintenance. As is well known, the rat brainstem and its neurons mature in a number of ways during the perinatal period. For example: (a) Neurons acquire functional ionic channels postnatally and increase their cytoplasmic membranes (4,5); the properties of these channels are changing. For example the properties and density Na^{++} and the Ca^{++} channels alter with age. (b) Neurons extend their dendrites and others prune some of their processes; dendritic spines disappear with age (6). (c) Neurogenesis takes place even after birth in the rat, and new neurons may be laid down in the brainstem during that period (7). With such ultrastructural and functional changes, it should not be surprising if respiratory rhythmogenesis and breathing in the immature subject were regulated very differently than in the mature one. The neonatal brainstem–spinal preparation may be useful as a model for pattern generation in general but could be misleading as a model for the pattern in the mature animal, keeping in mind some of the important observations that have been made in the past several years.

B. Developmental Responses to CO_2 and Techniques Used

Of considerable interest are the studies done in culture investigating the presence of carbonic anhydrase and in parallel studies on the excitability of neurons when exposed to CO_2. In addition, techniques involving whole-cell and nystatin patches were used to better understand the effect of CO_2 on neuronal activity in vitro. Interesting new methodology and data have been presented on the effect of CO_2 on brain activity in vivo, using light reflectance. These results raise a number of issues, but most importantly that (a) the methods used to study the effect of CO_2 are diverse, with some targeted at understanding a population response and others at single-cell properties; this will help us understand more comprehensively an age-old question that has clinical relevance; (b) it is probable that neurons are nonhomogeneous in their response to CO_2, with some showing excitation and others inhibition; (c) since neuronal excitability can depend on pH, the regulation of pH_i and pH_o becomes important, and since acid-base balance depends on interactions between neurons and glia, measurements of excitability in vitro and in vivo become crucial. One of the important controversial issues was whether the so-called CO_2 chemosensitive area on the ventral surface of the brainstem is a unique area or whether there are other such areas in the brainstem or elsewhere (8). Experiments are ongoing and answers are not yet final, but there is already evidence that cells in the CNS other than those on the ventral surface are responsive to CO_2. How this sensitivity differs from that of other cells, which cells do respond and which cells do not, what is the location of cells that respond, what is the basis for the sensitivity to CO_2, and so on are some of the questions that are currently being asked.

C. Hypoxia and Development

The effect of O_2 deprivation on the CNS in the mature and immature has been examined for a long time, but major questions still exist. One reason is that the techniques that have been used so far have not been satisfactory and novel methods had to be developed. Two such approaches were presented in this book. One of them, using in vitro biophysical techniques, examined the effect of hypoxia on single cells and how such neurons respond and adapt to low O_2 (9). The major result from this work was that neurons have some mechanisms that can be activated for the purpose of adapting to the stress, provided the stress is not overwhelming. Such a mechanism is based on changes in the biophysical properties of neurons (e.g., Na^+ channel properties) but may have their basis in the cytosol (10). Channels other than Na^+ are important also, and such experiments are being conducted in a number of laboratories. For example, the ATP-sensitive K^+ channel seems to play an important role in hypoxia in the brain. Another approach used by another group of investigators is that of the study of CNS activity to change in light reflectance. As mentioned above, this approach relies on the activity of cells in general, not only on neuronal activity. The advantage of this method is that it is rather noninvasive and it quantifies a population response. The advantage of the first approach is that molecular mechanisms at the subcellular level can be discovered. Both of these approaches will be very useful, especially in the immature or young, since such methods will be crucial for obtaining additional data that are not easy to get from the more conventional methods that have been so far used.

D. Neurotransmitters, Receptors, and Actions

A number of issues have been raised about receptor maturation in the CNS of mammals, including humans. The driving force behind this has presumably been to better characterize the newborn CNS in terms of receptor endowment and response to stress. This may include the CNS "immaturity" or "defect" in infants with sudden infant death syndrome (SIDS). By and large, it would seem (with some exceptions) that a great many receptors in the brain (including the brainstem) mature, and this is generally characterized by an increase in density rather an increase in affinity. A similar conclusion can be drawn for membrane or cellular enzymes; i.e., the number of copies increases with age. The timetable for the development of these receptors varies with the receptor species and brain region. The highest level of these receptors may not be in the adult, since some of these decline after the first 3 to 4 weeks postnatally. Good examples include opioid, GABA, and glutamate receptors (11,12) and channels such as the voltage-sensitive Na^+ channels.

The level of neurotransmitters will vary; this has been shown using nuclear magnetic resonance (NMR) and other techniques such as histochemistry, immunocytochemistry, and so on. In addition, the mechanisms that keep the neurotransmitters at a certain level in the synapse, for instance, whether by transport/uptake or enzyme degradation, will mature themselves. One major issue of considerable significance is that the response of the neonatal nerve cells to certain neurotransmitters such as gamma-aminobutyric acid (GABA) or glutamate appears to be very different from that in the adult, not only quantitatively but also in direction. At present, the reason(s) for this difference is/are clearly not simple, and these questions are not yet solved. It suffices to say, however, that differences in phenotypic responses would not be surprising, since not only neurotransmitters and receptors are changing levels and properties but also the second messenger systems and the intracellular amplification process mediating the action of these neurotransmitters may be changing as well. More data are needed in this area of research and normative data will be essential for understanding the abnormal.

E. Arousal Mechanisms and Development

One major issue raised is how to define arousal. This stems from reports that were related to responses of the fetus and the immature animal to certain neurotransmitters. Often, arousal is defined by eye openings, bodily movements, and EEG changes. However, such a definition can be problematic, since eye openings can occur during sleep, as well as bodily movements, especially in REM sleep. EEG changes can be illuminating, but most often such changes are not quantified and are used very subjectively. It would be a really important advance if we had data on the definition of arousal using multiple EEG leads and quantitative analysis. Since arousal mechanisms may be important in a number of pathophysiological states such as in SIDS, apnea of prematurity, sleep hypoventilation syndromes, and upper airway obstruction during sleep, a definition based on objective and quantitative analysis would be very valuable.

F. SIDS, Near-SIDS, and CO_2 Chemosensitivity

Previous data have not been consistent in terms of CO_2 sensitivity and SIDS. At present, no convincing evidence exists that infants at risk for SIDS (besides those infants with BPD) lack a CO_2 response. And if some do, they do not die of SIDS. The absence of the arcuate nucleus in a small ($n = 2$) population of SIDS patients compared to about 40 infants who did have the structure argues against this being a physiologically important structure. However, the partial absence or more subtle deficiencies in this nucleus may be important. At present a major effort is being made to define the neurochemistry and, in particular, the receptor chemistry of the brainstem in human brains in early infancy and the brains of babies who died of SIDS (Kinney and colleagues).

Another presentation focused on Ondine's curse or congenital hypoventilation syndrome. Using a paradigm of nonspecific stimuli associated with specific ones ($CO_2 = 2\%$ to 7%), patients with Ondine's curse seemed to have acquired a response to CO_2. Much more data will have to be generated and critically evaluated before this extraordinary occurrence can be accepted. A great many questions remain to be answered, such as relation between specificand nonspecific stimuli, paradigm used, length of time during which the response is present, nature of the disease process that generated the hypoventilation in the first place, and, of course, how this is all working, if associative "memory" really plays a role.

References

1. Smith JC, Ellenberger H, Ballanyi K, et al. Pre-Botzinger complex: A brainstem region that may generate respiratory rhythm in mammals. Science 1991; 254:726–729.
2. Feldman JL, Smith JC, Liu G. Respiratory pattern generation in mammals: In vitro en bloc analyses. Curr Opin Neurobiol 1991; 1:590–594.
3. Feldman JL, Smith JC, Ellenberger HH, et al. Neurogenesis of respiratory rhythm and pattern: Emerging concepts. Am J Physiol 1990; 259:879–886.
4. Cummins TR, Xia Y, Haddad GG. Functional properties of rat and human neocortical voltage-sensitive sodium currents. J Neurophysiol. In press.
5. Haddad GG, Getting PA. Repetitive firing properties of neurons in the ventral region of the nucleus tractus soliarius: In vitro studies in the adult and neonatal rat. J Neurophysiol 1989; 62:1213–1224.
6. Quattrochi JJ, McBride PT, Yates AJ. Brainstem immaturity in sudden infant death syndrome: A quantitative rapid golgi study of dendritic spines in 95 infants. Brain Res 1985; 325:39–48.
7. Fish I, Winick M. Cellular growth in various regions of the developing rat brain. Pediatr Res 1969; 3:407–412.
8. Dean JB, Lawing WL, Millhorn DE. CO_2 decreases membrane conductance and depolarized neurons in the nucleus tractus solitarius. Exp Brain Res 1989; 76:656–661.

9. Cummins TR, Jiang C, Haddad GG. Human neocortical excitability during anoxia via sodium channel modulation. J Clin Invest 1993; 91:608–615.
10. Haddad GG, Jiang C. O_2 deprivation in the central nervous system: On mechanisms of neuronal response, differential sensitivity and injury. Prog Neurobiol 1993; 40: 277–318.
11. Xia Y, Haddad GG. Major differences in CNS sulfonylurea receptor distribution between rat (newborn, adult) and turtle. J Comp Neurol 1991; 314:278–289.
12. Xia Y, Haddad GG. Ontogeny and distribution of $GABA_A$ receptors in the rat brainstem and rostral brain regions. Neuroscience 1992; 49:973–989.

46

Effects of Physostigmine on Cocaine-Induced Cardiorespiratory Responses at the Caudal Ventrolateral Medullary Surface of Cats

OZRA DEHKORDI and GARY C. DENNIS

Howard University Hospital
Washington, D.C.

RICHARD M. MILLIS and C. OVID TROUTH

College of Medicine
Howard University
Washington, D.C.

RUTH QUARTEY

University of Maryland
College Park, Maryland

LEVENT ERTUGRUL

University of Istanbul
Istanbul, Turkey

Previous studies have shown that cocaine possesses cholinolytic properties and that the topical administration of cocaine to cholinergic control sites at the ventral medulla of cats produces systemic hypotension with apneustic breathing. The anticholinesterase physostigmine has been effective as a "universal antidote" for overdosages of various drugs associated with central cardiorespiratory failure. We therefore studied the effect of physostigmine as a pretreatment drug for the prevention of cocaine-induced cardiorespiratory failure. Urethane (2.0 g/kg)-anesthetized cats were subjected to topical application of cocaine and physostigmine to the caudal ventrolateral medullary surface (cVMS) as follows: 125 μg, 37 mM cocaine (a) given alone; and (b) given 5 min after 10 μg, 3.6 mM physostigmine. Pretreatment with physostigmine was effective in protecting animals from cocaine-induced respiratory failure ($P<0.05$). However, its efficacy was limited to those animals that were normotensive during pretreatment. Animals that were hypotensive and had reduction in mean arterial blood pressure during physostigmine treatment experienced cardiorespiratory failure upon administration of cocaine. We believe that pretreatment with physostigmine might ameliorate cocaine-induced cardiorespiratory failure by ventral brainstem control mechanisms during normotensive states.

I. Introduction

Cardiorespiratory control regions at the ventrolateral medulla oblongata are thought to play an important role in cocaine-induced cardiorespiratory depression/failure. When applied topically to the intermediate or caudal chemosensitive areas of the medulla oblongata, cocaine produces arterial hypotension (1,2) and hypoventilation with inspiratory shift (apneusis) (2). However, the mechanisms through which cocaine acts to produce cardiorespiratory depression at the ventral medullary surface remain unclear.

Cocaine has been shown to have anticholinergic activity (3–5). On the other hand, neurons on the ventrolateral medullary surface (VMS) involved in the chemosensory response to H^+ ions stimulation are believed to be cholinergic (6–9). Therefore, the present study was undertaken to determine whether pretreatment with the cholinomimetic physostigmine, at a cholinergic cardiorespiratory control site at the caudal ventrolateral medullary surface (cVMS), could ameliorate the ventilatory failure associated with cocaine intoxication.

II. Methods

Domestic cats (2.5 to 4.4 kg) of both sexes were anesthetized with urethane (2.0 g/kg, 80% of dose titrated over 60 min). The ventral medullary surface was surgically exposed by a previously described procedure (1). Physiological measurements were made on tracheotomized animals with indwelling femoral arterial and venous catheters. Body temperature was maintained at 37 to 38°C. Arterial blood pressure (systolic, diastolic, mean) and respiratory (tidal volume, minute ventilation, and frequency) parameters were monitored by a strain gauge pressure transducer and Fleisch pneumotachograph connected to a respiratory flow transducer, respectively. The pressure and the airflow signals were logged by a physiological monitoring system (Buxco). The systolic, diastolic, mean arterial blood pressure (MABP), and heart rate (HR) were computed electronically from the pressure pulse. Tidal volume (V_T), respiratory rate (f), and minute ventilation ($\dot{V}_E$) were computed from the pneumotachographically measured airflow waveforms by a microcomputer software program (Buxco). The fractional concentration of expiratory CO_2 (F_ECO_2) was also monitored. Cocaine hydrochloride (Merck) and physostigmine sulfate (Eserine, Sigma) were dissolved in artificial cerebrospinal fluid (mock CSF, or mCSF) following the method of Schläfke et al., 1970 (10). All drugs were adjusted to physiological pH 7.4 and were applied topically to the cVMS using filter-paper pledgets. Each pledget was soaked in 5 μL of mCSF (vehicle) containing either cocaine (12.5 mg/mL, 62.5 μg per pledget), or physostigmine (1.0 mg/mL, 5 μg per pledget). The placement of the pledgets was 1.5 mm medial to the rootlets of the hypoglossal nerve.

Data were expressed as mean ± SE. Statistical significance was determined by Student's *t*-test for independent samples.

III. Results

Following surgery, a 30-min period was provided for stabilization of blood pressure and ventilation. Measurements were made during a 15-min interval in which bilateral topical application of pledgets with vehicle alone to the cVMS chemosensitive zone, medial to the hypoglossal nerve rootlets (6,11), was followed by topical application of test drugs. The experimental design is presented in Table 1. The drug protocols were as follows: 125 μg, 37 m*M* cocaine bilaterally (a) given alone in 5 cats and (b) given 5 min later, following pretreatment with 10 μg, 3.6 m*M* physostigmine in 9 cats. Cardiorespiratory parameters were evaluated during the peak responses to the topically applied test drugs and were compared to the responses detected during steady-state control periods. The time course of peak responses for all drugs was found to be within the first 5 min after their application.

A. Effects of Cocaine

Table 2 summarizes the effects of bilateral application of 125 μg cocaine to the cVMS in five cats. Administration of the mCSF vehicle produced no significant changes in either ventilation or blood pressure. Treatment with cocaine decreased both *f* and $\dot{V}_E$ significantly; the slight increase in V_T was not found to be significant. An inspiratory shift (apneusis) was regularly observed during the 5-min interval in which peak changes were identified. During hypoventilation, F_ECO_2 increased from an initial value of 5.0 ± 0.7% to 6.3 ± 0.7%. Cocaine decreased MABP significantly; HR changes were small and not statistically significant.

B. Pretreatment with Physostigmine

Cocaine produced two types of cardiorespiratory responses when applied topically to the cVMS of physostigmine-pretreated cats. The type I response, characterized by cardiorespiratory arrest, was observed in animals in which the

Table 1 Experimental Design for Cocaine (COC) Treatment and COC plus Physostigmine (COC + PHYS) Pretreatment at the Ventral Medulla

Treatments	COC ($N=5$)	PHYS ($N=9$)
Controls	mCSF	PHYS alone
Experimentals	COC alone	COC + PHYS

Table 2 Cardiorespiratory Effects of Cocaine at the cVMS ($N=5$)

	V_T mL	f bpm	$\dot{V}_E$ mL/min	F_ECO_2 %	MABP mm Hg	HR bpm
Control	18.3 ± 1.1	38.4 ± 3.0	706.0 ± 76.1	4.5 ± 0.7	150.7 ± 3.4	134.7 ± 5.7
mCSF	18.0 ± 1.0	38.0 ± 3.3	687.8 ± 80.0	5.0 ± 0.7	152.7 ± 4.5	137.0 ± 5.8
COC	19.5 ± 2.5	20.0 ± 4.1[a]	378.7 ± 72.3[a]	6.3 ± 0.7[a]	129.5 ± 6.8[a]	146.2 ± 11.8

[a]Significantly different from mCSF control (vehicle). Respiratory frequency, f, in breaths/min; heart rate, HR, in beats/min.

MABP was 130.5 ± 7.2 mm Hg during the control period. These animals experienced hypotension either before or after physostigmine pretreatment (Table 3). The type II response was characterized by no significant change in respiratory parameters (V_T, f, $\dot{V}_E$, F_ECO_2) after cocaine application (Table 4). This response is in sharp contrast to that observed in animals given cocaine either alone or after physostigmine pretreatment made them hypotensive (Fig. 1). The type II response was observed in animals in which the MABP was normal during the control period and that did not experience hypotension during physostigmine application (Table 4). The MABP of animals exhibiting the type II

Table 3 Cardiorespiratory Parameters of Physostigmine-Pretreated Cats with Cardiorespiratory Arrest After Cocaine Treatment (Type I)[a]

$N=4$	V_T mL	f bpm	$\dot{V}_E$ mL/min	F_ECO_2 %	MABP mm Hg	HR bpm
mCSF	19.8 ± 1.5	42.0 ± 3.0	840.9 ± 118.6	5.4 ± 1.0	130.5 ± 7.2[c]	197.0 ± 26.7
PHYS	17.8 ± 1.5[b]	56.0 ± 3.6[b]	1006.7 ± 122.5	5.52 ± 0.8	103.8 ± 13.4	226.0 ± 26.7

[a]Respiratory frequency, f, in breaths/min; heart rate, HR, in beats/min.
[b]Different from mCSF at $P<0.05$.
[c]Different from cats that had type II response at $P<0.05$.

Table 4 Effects of Cocaine in Physostigmine-Pretreated Normotensive Cats (Type II Response)[a]

$N=5$	V_T mL	f bpm	$\dot{V}_E$ mL/min	F_ECO_2 %	MABP mm Hg	HR bpm
mCSF	21.8 ± 1.7	36.4 ± 8.8	754.0 ± 149.9	5.2 ± 0.8	162.7 ± 10.2	156.8 ± 13.2
PHYS	17.6 ± 1.8[b]	46.4 ± 5.1	827.8 ± 141.5	6.0 ± 1.2	157.7 ± 7.5[d]	176.2 ± 10.9[b]
COC	19.6 ± 2.52	43.2 ± 5.2	810.2 ± 92.1	6.2 ± 0.9	124.0 ± 6.4[c]	216.4 ± 15.5[c]

[a]Respiratory frequency, f, in breaths/min; heart rate, HR, in beats/min.
[b]Different from mCSF vehicle at $P<0.05$.
[c]Different from physostigmine control at $P<0.05$.
[d]Different from cats that had type I response at $P<0.05$.

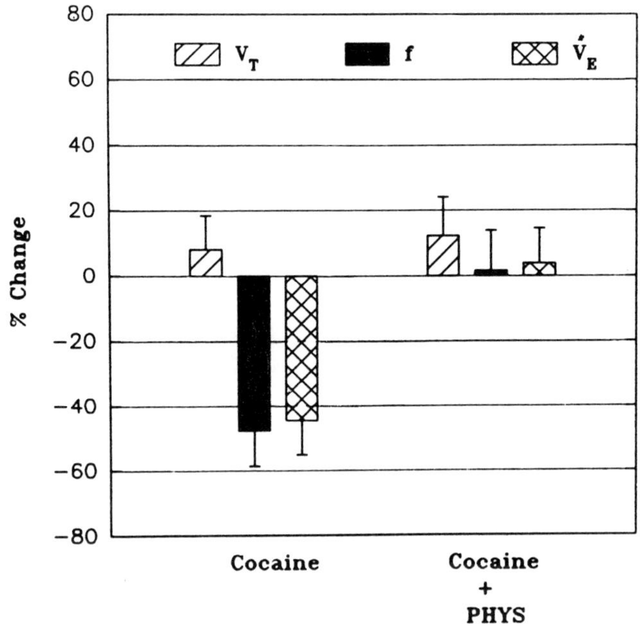

Figure 1 Ventilatory effects of physostigmine pretreatment. Percent changes from control values in tidal volume (V_T), frequency (*f*) and minute ventilation ($\dot{V}_E$) for cocaine treatment and cocaine treatment with physostigmine pretreatment (Cocaine + PHY) at the ventrolateral medullary surface.

response was significantly greater than those exhibiting the type I response ($P<0.05$). Although physostigmine protected the normotensive cats from cocaine-induced hypoventilation and apneusis, it did not protect the animals from cocaine-induced hypotension. As demonstrated in Figure 2, the hypotensive responses to cocaine in physostigmine-pretreated cats were not significantly different from those cats receiving cocaine without pretreatment.

IV. Discussion and Conclusions

In the present study, physostigmine applied directly to the VMS protected normotensive cats from cocaine-induced apneustic breathing. Although physostigmine was effective in the normotensive cats, it failed to protect the hypotensive cats from the apneustic breathing pattern and hypoventilation. Protection from cocaine toxicity has been reported in unanesthetized rats given physostigmine by the intraperitoneal route (12). In our studies, pretreatment with physostigmine produced significant prophylaxis for the hypoventilatory but not for the hypotensive effects of cocaine.

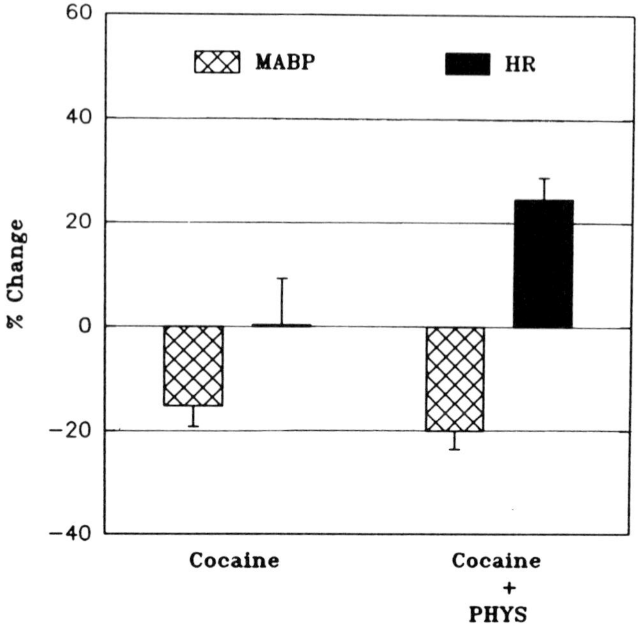

Figure 2 Cardiovascular effects of physostigmine pretreatment. Percent changes from control values in mean arterial blood pressure (MABP) and heart rate (HR) for cocaine treatment and cocaine treatment with physostigmine pretreatment (Cocaine + PHY) at the ventrolateral medullary surface.

That cholinergic mechanisms are involved in control of ventilation at the VMS is well documented (6–9,11). Anticholinesterases such as physostigmine cause acetylcholine to accumulate at synapses (13), thereby providing a mechanism for antagonizing a central cholinolytic effect. Indeed, the effect of the physostigmine-induced acetylcholine accumulation at these cVMS sites might overwhelm the cholinolytic effects of cocaine. Moreover, the finding that physostigmine protected ventilation but not blood pressure from cocaine toxicity at the cVMS suggests a fundamental difference between the interaction of cocaine with vasomotor and ventilatory control mechanisms at the VMS (14,15). Pharmacological investigations indicate that VMS respiratory mechanisms might be cholinergic, whereas the vasomotor system is adrenergic (6–9,11,16).

Physostigmine has also been used as a "universal antidote" for overdosages with ketamine (17), morphine (18), heroin (19), and antihistamines (20), all of which are considered clinical conditions not primarily of cholinergic origin but commonly associated with central respiratory depression or failure.

In conclusion, the current study demonstrates that respiratory depression/ failure associated with cocaine overdosage might be ameliorated by physostigmine directly applied to the ventral brainstem. Further studies should be performed to determine the precise mechanism by which the cholinomimetic property of physostigmine might be responsible for its protective effects on ventilatory control at the ventrolateral medullary surface.

References

1. Dehkordi O, Dennis GC, Millis RM, et al. Central cocaine neurotoxicity at brainstem cardiorespiratory control sites. Life Sci 1993; 52:1387–1396.
2. Raczkowski VFC, Hernandey YM, Erzouki HK, et al. Cocaine acts in the central nervous system to inhibit sympathetic neuron activity. J Pharmacol Exp Ther 1991; 257:511–519.
3. Sharkey J, Ritz MC, Schenden RC, et al. Cocaine inhibits muscarinic cholinergic receptors in heart and brain. J Pharmacol Exp Ther 1988; 246:1048–1052.
4. Meyer EM, Potter LT, Devane CL, et al. Effects of benzoyltropine and tropacocaine on several cholinergic processes in the rat brain. J Pharmacol Exp Ther 1990; 254:584–590.
5. Wilkerson RD. Cardiovascular effects of cocaine: Enhancement by yohimbine and atropine. J Pharmacol Exp Ther 1989; 248:57–61.
6. Mitchell RA, Loeschcke HH, Severinghaus JW, et al. Regions of respiratory chemosensitivity on the surface of the medulla. Ann NY Acad Sci 1963; 109:661–681.
7. Fukuda Y, Loeschcke HH. A cholinergic mechanism involved in the neuronal excitation by H^+ in the respiratory chemosensitive structures of the ventral medulla oblongata of rats in vitro. Pflügers Arch 1979; 379:125–135.
8. Dev NB, Loeschcke HH. Topography of the respiratory and circulatory responses to acetylcholine and nicotine on the ventral surface of the medulla oblongata. Pflügers Arch 1979; 379:19–27.
9. Dev NB, Loeschcke HH. A cholinergic mechanism involved in the respiratory chemosensitivity of the medulla oblongata in the cat. Pflügers Arch 1979; 379:29–36.
10. Schläfke ME, See WR, Loeschcke HH, et al. Ventilatory response to alternations of H ion concentration in small areas of the ventral medullary surface. Resp Physiol 1970; 10:198–212.
11. Trouth CO, Millis RM, Bernard DG, et al. Cholinergic-opioid interactions at brainstem respiratory chemosensitive areas in cats. Neurotoxicology 1993; 14:459–468.
12. Witkin JM, et al. Modulation of the lethal effects of cocaine by cholinomimetics. Life Sci 1989; 45:2295–2301.
13. McIlwain H, Bachelard HS. Biochemistry and the Central Nervous System. London: Churchill Livingstone, 1985:413–440.
14. Feldberg W, Guertzenstein PG. Vasodepressor effects obtained by drugs acting on the ventral surface of the brain stem. J Physiol (Lond) 1976; 258:337–355.
15. Millhorn DE, Eldridge FL. Role of ventrolateral medulla in regulation of respiratory and cardiovascular systems. J Appl Physiol 1986; 6:1249–1263.
16. Ross CA, Ruggiero DA, Joh TH, et al. Adrenaline synthesizing neurons in the rostral ventrolateral medulla: A possible role in tonic vasomotor control. Brain Res 1983; 273:356–361.

17. Toro-Matos A, Rendon-Platas AM, Avilla-Valdoz E, et al. Physostigmine antagonizes ketamine. Anaesth Analg 1980; 59:764–767.
18. Weinstock M, Erez E, Roll D, et al. Antagonism of the cardiovascular and respiratory depressant effects of morphine in the conscious rabbit by physostigmine. J Pharmacol Exp Ther 1981; 218:504–508.
19. Rupreht J, Dworacek B, Valkenburg M, et al. The use of physostigmine in heroin overdose. Arch Toxicol 1984; 7(suppl):467.
20. Lee JH, Turndorf H, Poppers PJ, et al. Physostigmine reversal of antihistamine induced excitement and depression. Anesthesiology 1975; 43:683–684.

47

Neurotoxic Agents at the Ventral Brainstem

C. OVID TROUTH,
RICHARD M. MILLIS,
A. JAYAM TROUTH,
and PHILLIP W. ARCHER

College of Medicine
Howard University
Washington, D.C.

W. RICHARD CAMPBELL

Morris Brown College
Atlanta, Georgia

BRENNIE E. HACKLEY, JR.
United States Army Medical Research
Institute of Chemical Defense
Aberdeen Proving Ground, Maryland

I. Introduction

The ventral brainstem might be an ideal target site for a number of toxic chemical compounds that contribute to cardiorespiratory failure and death. Although the blood-brain barrier is relatively selective, there are a few central nervous system (CNS) sites (e.g., area postrema) devoid of barrier properties. Neurotoxics that gain access to the CNS from the blood circulation might be disseminated widely through the brain not only by diffusion but also by pulsations of penetrating arterioles and the convective forces generated thereby (1,2).

II. Anticholinesterases

Acute exposure to highly toxic anticholinesterase (anti-ChE) nerve agents often results in death due to cardiorespiratory failure. Prominent among these chemical agents are diisopropylfluorophosphate (DFP), diethyl 4-nitrophenyl phosphate (paraoxon), ethyl *N*-dimethylphosphoramidocyanidate (tabun), isopropyl methylphosphonofluoridate (sarin), pinacolyl methylphosphonofluoridate (soman), tetraethyl pyrophosphate (TEPP), and *O*-ethyl-*S*-[2-(diisopropylamine)-

---

The opinions or assertions contained herein are the private views of the authors and are not to be construed as official or as the views of the Army or the Department of Defense.

ethyl]methylposphonothioate (VX). According to de Candole et al. (3), the general signs and symptoms of anti-ChE poisoning present a picture characteristic of the asphyxic state, failure of respiration being the predominant effect produced by lethal doses. Three major causes of respiratory failure due to anti-ChE agents have been proposed: (a) bronchoconstriction, laryngospasm, bronchorrhea, and associated airway obstruction, (b) neuromuscular (NM) blockade, and (c) inhibition of respiratory centers (3–7).

Failure of the respiratory control mechanism in the CNS has been suggested as the dominant factor in death by anti-ChE poisoning (8–11). Following poisoning with DFP, paraoxon, tabun, sarin, soman, or TEPP, the phrenic nerves ceased their rhythmic discharge (3,5,12). At autopsy, the diaphragm was elevated and the lungs collapsed (3). This expiratory position of the lung was consistent with inhibition of inspiratory drive. Rickett (13) observed apnea following intravenous injection of 9 × LD_{50} doses of soman in pyridostigmine-pretreated cats. Phrenic stimulation, however, caused contraction of the diaphragm, suggesting a still intact NM junction. Since nervous output from the respiratory centers is indicated by phrenic nerve output (14), quiescence of the phrenic nerve in these studies suggests central respiratory failure. The expiratory tone and forced expirations which preceded apnea in monkeys (3) could well reflect activation of expiratory neurons signaling expiration, with reciprocal inhibition of those signaling inspiration. Coincident with the forced expirations in monkeys, de Candole (3) also reported abundant discharges down the vagus nerve after inspiration had ceased. Thus, anti-ChE poisoning may be associated with loss of inspiratory drive, increased expiratory tone, and increased parasympathetic outflow. The increased parasympathetic outflow may account for some bronchospasm. In addition, it appears that following soman and other organophosphate (OP) nerve agent poisoning, ACh acts at muscarinic receptors primarily to produce brain injury and seizures (15). This muscarinic involvement is evidenced by the findings that pretreatment with antimuscarinic compounds (e.g., MK-801, atropine sulfate) appears to prevent soman-induced convulsions (16,17).

OP anti-ChE effects are considered to be irreversible but may be reduced by prior administration of carbamates, which act as reversible cholinesterase inhibitors (18–20). In the periphery, pretreatment with physostigmine or pyridostigmine protects the organism from irreversible NM blockade by soman through carbamylation of AChE. Decarbamylation restores sufficient AChE to reestablish NM function. Pretreatment protection of the enzyme is also possible with streptomycin and obidoxime (toxogonin) (21,22).

The presumption of carbamate pretreatment is that physostigmine may act centrally as a reversible anti-ChE agent to prevent transmitter breakdown at the same anatomical sites where the irreversible anti-ChE OP agents (e.g., soman) would produce their respiratory effects. Carbamate prophylaxis has been presumed to provide a means by which enough ACh may be restored by decarbamylation to prevent central respiratory failure in acute irreversible OP intoxication. Attempts have been made to determine the CNS sites of physostigmine and pyridostigmine activity that coincide with those presumed to be involved in soman lethality (central apnea). Since pyridostigmine is a quaternary ammonium compound and does not normally gain access to the brain via the blood-brain barrier, its actions are thought to be primarily peripheral (e.g., at the NM junction). Pretreatment using physostigmine primarily and pyridostigmine secondarily may cause CNS responses related to their cholinergic effects; these may also produce behavioral incapacitation, such as decreased speech and spontaneous behavior, slowed thoughts, and alteration in mood and cognition (23–25). This impairment of human performance is of immense importance and relevance today, particularly because of the recent use of pyridostigmine as a prophylactic agent for military personnel during the Gulf War and the appearance of medical symptoms referred to as the "Gulf War syndrome." These symptoms included prolonged fatigability and mood changes. Certainly changes in respiratory rate, tidal volume, and sensitivity of respiratory chemoreceptors to oxygen, CO_2, and pH are respiratory variables that may be seriously compromised during incapacitation.

Neurons at the ventral medullary surface (VMS) involved in the CO_2-chemoreflex control of respiration appear to utilize muscarinic neuroregulatory mechanisms (26,27). OP nerve agents also act at muscarinic receptor sites (16, 17). Further studies are needed to precisely define the mechanisms leading to the presumed central respiratory failure induced by the anti-ChE compounds. In this regard, we believe that the best candidate sites for anti-ChE-induced respiratory failure should be found in those respiratory control regions which, when pharmacologically blocked, cooled, or destroyed by any means whatsoever, result in apnea. The scientific literature on VMS studies is replete with examples of sites where apnea can be produced in experimental animals.

It has been proposed that ACh accumulation in areas of the pontomedullary reticular formation that activate the rapid-eye-movement (REM) stage of sleep can also inhibit respiratory muscle activity centrally (28). Interaction between cholinergic and endogenous opiate mechanisms has recently been shown to modulate the CO_2-chemoreflex response at the caudal VMS in cats (27,29), and immunocytochemical evaluation of the region by light and electron microscopy revealed the presence of beta-endorphin immunoreactivity distributed widely in the perikaryon of both the superficial "type I" neurons and neurovascular elements (30). Indeed, it has been reported that the apnea of infants and children with elevated cerebrospinal fluid beta-endorphin levels has been successfully treated with the opiate receptor antagonist naltrexone (31). In addition, (muscarinic) cholinergic-adrenergic interaction at the rostral ventrolateral medulla, nucleus tractus solitarius, and locus ceruleus has been proposed to explain the acute hypertensive responses attributed to usage of AChE inhibitors (32). The ventilatory control neurons at the VMS also appear to be responsive to kainic acid and other excitotoxic amino acids (33–35), which might play an important role in the seizure activity associated with soman intoxication (36).

III. Cocaine

Cocaine is an alkaloid of the South American shrub *Erythroxylon coca*, widely cultivated in the mountainous regions of Peru and Bolivia. It is the only known naturally occurring local anesthetic. It inhibits the reuptake of sympathetic amines (e.g., norepinephrine, dopamine) released into the synaptic cleft by adrenergic neurons (37). This potentiates sympathetically mediated vasoconstriction and hypertension. Sympathetic stimulation is believed to play a critical role in the arrhythmogenic long QT interval syndromes and ventricular arrhythmias in acute myocardial infarction (38). In humans, cocaine induces a dose-dependent increase in blood pressure and heart rate. The effects of cocaine on coronary blood flow, peripheral vascular resistance, myocardial contractility, and electrophysiological parameters have not been fully studied.

The cause of death after cocaine poisoning is attributed to cardiovascular collapse, which is often associated with generalized seizures (39). Cocaine-induced cardiorespiratory failure is multifactorial and is related to a combination of hyperthermia, hypoxia, acidosis and CNS stimulation (38). Pretreatment with agents that prevent hyperthermia and/or seizure activity has been effective in preventing a fatal outcome (40). To prevent cocaine-induced myocardial ischemia and infarction, administratiion of the calcium channel blocking agent nitrendipine has been useful and was found to produce a fourfold increase in the required lethal dose of cocaine in the rat (41). Administration of enalaprilat (an angiotensin-converting enzyme inhibitor) with diazepam (a benzodiazepine) also restored the blood pressure and cardiac rhythm to normal and also prevented myocardial lesions (42). Lethality can be modulated by the concomitant administration of cocaine and cholinomimetic agents, physostigmine in particular (43). Recent studies at the VMS show that physostigmine might inhibit the apneustic breathing pattern induced by topical application of cocaine (44). Physostigmine has been used as a "universal antidote" for overdosage with ketamine (45), morphine (46), heroin (47), and antihistamines (48), all of which are considered as clinical conditions not primarily of cholinergic origin but commonly associated with central respiratory depression or failure.

Problems of respiratory control are of long-term concern in drug-exposed infants. In a recent review, Ellis et al. (49) discussed the multitudinous problems related to fetal exposure to cocaine in utero. An increase of sudden infant death syndrome (SIDS) in infants of cocaine-using mothers has been reported by some authors (51–53), but not by others (54,55). Perry et al. (56) demonstrated a "downregulation" of placental neurotransmitter receptors in infants with intrauterine exposure to drugs of abuse. Abnormalities or depletion of receptor sites in the fetal brain interferes with normal respiratory control, causing more frequent periods of apnea (55). Woods et al. (57) demonstrated that in cocaine users significant, acute, dose-dependent increases of maternal mean arterial pressure and heart rate were associated with a marked decrease in uterine blood flow, accompanied by fetal hypoxemia and hypertension.

Mahalik et al. (58) proposed that vasoconstriction secondary to increased catecholamines, and the resulting hypoxia, was a probable contributing factor to the teratogenic potential of cocaine. Structural and functional brainstem changes in the fetus, as a consequence of norepinephrine disequilibrium secondary to maternal cocaine abuse, could alter normal maturation of cardiorespiratory control mechanisms.

Medullary stimulation by cocaine intoxication is reported to cause respiratory tachypnea marked by a rapid shallow breathing pattern. Emesis may result from stimulation of the vasomotor and vomiting centers (59). Ventilation may deteriorate to a Cheyne-Stokes pattern as the vital medullary centers are depressed (60). As hypoxemia supervenes, the patient becomes comatose and death usually results from ventricular fibrillation and asystole (61). Fatal respiratory collapse has occurred without warning in some patients restrained for delirium (62). Although the mechanism is ill defined, seizures are a common complication of acute cocaine intoxication (63). Seizures may be related to sympathomimetic effects (64) or to other complications such as cerebral infarction, intracranial hemorrhage, and CNS vasculitis (65,66). The management of acute cocaine toxicity calls for adequate support of cardiovascular and respiratory functions. The opiate antagonist naloxone has been recommended (59), as often cocaine is abused in conjunction with heroin.

Specific neuroanatomic, neurophysiological, and neurotransmitter effects of cocaine are still to be elucidated. The maturational state of the brain may play an important role in modifying the responses to cocaine administration. Further studies may provide valuable insight into the mechanisms of the cardiorespiratory problems associated with its abuse, particularly because of the widespread abuse of cocaine by adolescents, both male and female. Multiple congenital CNS defects in the offspring—cerebral infarctions, intracranial hemorrhages, cognitive problems, sleep and behavioral abnormalities—have been attributed to cocaine abuse in pregnancy. The need for prevention of dependency is great, and this can only be achieved by greater knowledge of its specific effects at various sites in the brain.

IV. Cyanide

Cyanide (hydrocyanic acid [HCN], prussic acid) is one of the most rapidly acting poisons and can be lethal within minutes of exposure not only in humans but also in animals and other forms of life, including microorganisms. Following poisoning, symptoms of salivation, nausea without vomiting, anxiety, confusion, vertigo, lower jaw stiffness, convulsions, opisthotonos, paralysis, coma, and cardiac arrhythmias may occur. Transient respiratory stimulation can be followed by respiratory failure (67). The chemical reactions in which cyanide participates might principally be responsible for the observed toxicologic effects. Cyanide complexes with metals such as iron, copper, selenium, and molybdenum, which commonly act as electron carriers, thus inhibiting the function of many metalloenzymes, including cytochrome oxidase (CCO), a mitochondrial enzyme, which contains both iron and copper and is involved in oxidation-reduction reactions and oxygen transport that are vital to cell survival.

When absorbed, cyanide reacts readily with the trivalent iron of CCO in mitochondria; cellular respiration is thus inhibited, resulting in lactic acidosis and cytotoxic hypoxia. Since utilization of oxygen is blocked, venous blood becomes rich in oxygen and can be as bright red in color as arterial blood. Respiration is increased due to oxygen lack at the peripheral chemoreceptors. There is evidence that central respiratory chemosensitive neurons at the VMS might also be stimulated by cyanide-induced local hypoxia (68–70). Exposure to cyanide is associated with a transient stage of CNS stimulation with hyperpnea and headache; thereafter, hypoxic convulsions occur and death has been attributed to failure of the respiratory centers (69).

The distribution of CCO inhibition supports the concept that brain structures are especially sensitive to the effects of cyanide. Laboratory animals nominally exhibit 35% to 80% inhibition of brain CCO (71–74) following lethal doses of cyanide. However, Pettersen and Cohen (74) observed that the levels of CCO inhibition in mouse brain were almost identical regardless of whether the animals received lethal or sublethal doses of cyanide. The heightened susceptibility of neural tissues to cyanide might merely be indicative of the tight coupling of metabolic and neural signaling events in vital control centers, whereas, in other less vital tissues, its effects are not as critical to the animal's survival.

Cyanide increases the release of neurotransmitter molecules (75). It also has direct effects on enzymes that regulate the levels of excitatory and inhibitory neurotransmitters, and it might cause an increase in the metabolic activity of specific neurons, thereby exacerbating the toxic effects of the agent. Central lesions in cyanide poisoning might result from an accumulation of intracellular free calcium, and this has been associated with tremors and seizures in mice (76,77). Recent evidence suggests that cyanide might cause release of excitatory amino acid neuroregulators (75), and intracellular accumulation of calcium ions has been associated with the experimentally induced excitotoxic lesions that produce critical dysfunction even at the ventral brainstem involving ventilatory and vasomotor control mechanisms (70). Studies of the role of excitatory amino acids in ischemic and hypoxic damage may be of relevance to the effects of cyanide on calcium levels. During cerebral ischemia, excess amounts of excitatory amino acid transmitters might be released, leading to high levels of free Ca^{2+} (78,79). NMDA receptor antagonists have been found to provide effective protection against ischemia-induced neuronal degeneration (79) and cell damage induced by hypoxia or cyanide in neuronal cell cultures (80).

Cyanide inhibits enzymes other than CCO to a lesser or greater extent. At critical respiratory or vasomotor control sites of the ventral brainstem, cyanide could produce central respiratory failure by actions either directly on the respiratory centers or on the chemosensitive drive mechanism. Glutamate decarboxylase is responsible for the conversion of glutamate, an excitatory neurotransmitter, to gamma-aminobutyric acid (GABA), an inhibitory neurotransmitter. GABA transaminase catalyzes the breakdown of released GABA. Both enzymes are inhibited by cyanide. Increases in glutamate or decreases in GABA levels may be responsible for uncontrolled neural activity. Cyanide-intoxicated animals have exhibited significant increases in glutamate levels and decreases in GABA levels in rat, mouse, and guinea pig brain at the time of onset of convulsions (81). Excitatory amino acid mechanisms have been demonstrated at the VMS (33–35) and could help explain cyanide-induced inhibition of cardiorespiratory control centers.

Acidification of the plasma resulting from inhibitioin of carbonic anhydrase by cyanide (82) might contribute to the hypoxia-induced lactic acidosis associated with cyanide intoxication. Energy-dependent ion transport systems also are involved in pH regulation. Since intracellular ATP levels are decreased because of CCO inhibition, inhibition of carbonic anhydrase might exacerbate the effects of the ATP depletion. The ventral brainstem's chemosensitive zones (27,83–86) might utilize carbonic anhydrase as a cellular transductive mechanism for neural signaling processes involved in ventilatory and vasomotor control. Consequently, inhibition of carbonic anhydrase by cyanide would be expected to contribute to a lethal control lesion at the ventral brainstem.

Effects of cyanide on guanylate cyclase have been found to be either stimulatory or inhibitory. Muscarinic acetylcholine receptor and associated ion channel activities have been shown to be regulated by cGMP, the product of guanylate cyclase activity. Interestingly, muscarinic cholinergic mechanisms seem to prominently participate in excitatory chemosensory control mechanisms at the ventral brainstem (26,87,88). This raises the possibility that inhibition of guanylate cyclase at the ventral brainstem could result in lethal central ventilatory failure. However, effects of cyanide on rat brain cGMP levels have been reported to range from increases to decreases, depending on the specific brain structures studied (89), and the precise consequence of changes of cGMP in the ventral brainstem neuropil has not been systematically investigated.

The cellular effects of cyanide poisoning suggest that cyanide exerts its toxic effects primarily by inhibiting the activity of the intramitochondrial enzyme CCO. As a result, the transfer of electrons to molecular oxygen is blocked, and critical energy-dependent functions decrease, accompanied by decrease in ATP and profound lactic acidosis, producing injuries to cellular structures. Removal of extracellular calcium or the use of specific glutamate antagonists, but not calcium channel blockers, can prevent cyanide-mediated neuronal damage (90). The precise site of the lethal cyanide-induced lesion remains uncertain. Although central respiratory failure is commonly cited as the explanation for death after cyanide poisoning, systematic studies of the effects of cyanide on central ventilatory control systems are scant.

V. Biological Agents

A. Tetrodotoxin

Tetrodotoxin is a nonprotein marine toxin (molecular weight 319) usually ingested by humans from improperly prepared puffer or globe fish. The mechanism of toxicity by which tetrodotoxin acts is blockade of fast sodium channels in the plasma membranes of excitable cells (91). It has been thought that death results from respiratory muscle paralysis (92). However, evidence that tetrodotoxin might cross the blood-brain barrier (93) has led to investigation of the central respiratory effects of tetrodotoxin. Chang et al. (94) have shown that, in addition to peripheral effects, tetrodotoxin, intraperitoneally administered to anesthetized guinea pigs, prolongs the duration of firing of neurons involved in signaling the expiratory phase of respiration. That this central effect might account for respiratory arrest is suggested by the fact that it was observed at the Bötzinger complex area of the ventral medulla, associated with the nucleus tractus solitarius and also with the ventral respiratory group of ventilatory control neurons. The Bötzinger complex is currently thought to be a primary generator of the normal respiratory rhythm by exerting powerful inhibitory and excitatory influences on inspiratory and excitatory neurons, respectively (95–100).

B. Saxitoxin

Saxitoxin is a nonprotein toxin (molecular weight 299) that also produces blockade of sodium channels (101) and can be lethal to humans after eating shellfish that have ingested a dinoflagellate plankton organism (*Gonyaulax*). Saxitoxin is also produced by a freshwater cyanobacterium (*Aphanizomenon flos-aquae*). Death has been attributed to ventilatory failure and systemic hypotension by peripheral mechanisms because it was believed that the toxin had little capacity to penetrate the blood-brain barrier. However, Borison et al. (102) have demonstrated that saxitoxin might be able to permeate the blood-brain barrier, and, more recently, Chang et al. (103) have demonstrated that, in addition to inhibition of peripheral neuromuscular transmission at the diaphragm, saxitoxin, when administered intraperitoneally to anesthetized guinea pigs, also increases the duration of firing of ventral medullary Bötzinger complex neurons. This central effect was similar to the effects of tetrodotoxin in guinea pigs (103). It was ameliorated by mechanical ventilation and mimicked by hypoxia (104).

C. Apamin

Apamin is the smallest neurotoxic polypeptide known (molecular weight 2027). It comprises about 2% by weight of the dried venom of the honeybee (*Apis mellifica*). While tetrodotoxin and saxitoxin appear to act as selective blockers of sodium channels in membranes of excitable cells, apamin produces selective blockade of potassium channels (105). Apamin-sensitive, Ca^{++}-activated potassium channels have been described in brain slices from the CA3 glial-rich region of the hippocampus subjected to kainic acid lesioning (106). The importance of neurotoxic mechanisms involving Ca^{++}-activated potassium channels has been demonstrated by the ability of intracerebroventricular injections of apamin to inhibit ethanol-induced narcosis in mice (107). Since large doses of bee venoms and ethanol are both known to produce cardiorespiratory depression/failure (108), it might be worthwhile to determine their interactions at the ventral brainstem ventilatory and vasomotor control sites, which also appear to glia-rich regions.

D. Botulinum and Alpha-Bungarotoxins

Botulinum neurotoxins comprise seven different proteins produced by the toxic anaerobic bacterium *Clostridium botulinum*, which are responsible for botulism, a potentially lethal systemic infection. Alpha-bungarotoxins are also protein toxins (molecular weight ca. 8000) in the venom of the *Elapidae* snake. Both of these toxins have been widely studied as blockers of nicotinic cholinergic receptors. Because the VMS neurons involved in chemosensory control of ventilation and vasomotor tone have been thought to utilize muscarinic neuroregulatory mechanisms (26,27,88), significant effects are not expected. However, cholinergic mechanisms of neuroregulation in the deeper-lying ventral respiratory group and Bötzinger areas have not been characterized, and central cardiorespiratory responses to these biological toxins, acting as nicotinic receptor blockers, could provide new information on the possibility of a central component to the cardiorespiratory failure associated with these agents.

References

1. Rennels ML, Gregory TE, Blaumanis OR, et al. Evidence for a paravascular fluid circulation in the mammalian central nervous system, provided by the rapid distribution of tracer protein throughout the brain from the subarachnoid space. Brain Res 1985; 326:47–63.
2. Rennels ML, Blaumanis OR, Grady PA. Rapid solute transport throughout the brain via paravascular fluid pathways. In: Long D, et al., eds. Advances in Neurology. Vol. 52. New York: Raven Press, 1990.
3. deCandole CA, Douglas WW, Lovatt Evans C, et al. The failure of respiration in death by anti-cholinesterase poisoning. Br J Pharmacol 1953; 8:466–475.
4. Krivoy WA, Hart ER, Marrazzi AS. Further analysis of the actions of DFP and curare on the respiration center. J Pharmacol Exp Ther 1951; 103–351.
5. Wright PG. An analysis of the central and peripheral components of respiratory failure produced by anticholinesterase poisoning in the rabbit. J Physiol 1954; 126: 52–70.
6. Stewart WE. The effects of sarin and atropine on the respiratory center and neuromuscular junctions of the rat. Can J Biochem Physiol 1959; 37:651–660.

7. Brimblecombe RW. Drugs acting on respiratory central cholinergic mechanism. Pharmacol Ther Br 1977; 3:65–74.
8. Ngai SR, Wang SC. Organization of central respiratory mechanism in the brain stem of the cat: Localization by stimulation and destruction. Am J Physiol 1957; 190:343–349.
9. Brown RV. The effects of intracisternal sarin and pyridine 2-aldoxime methyl methanesulphonate in anesthetized dogs. Br J Pharmacol 1960; 15:170–174.
10. Salmoiraghi GC, Bums BD. Localization and patterns of discharge of respiratory neurons in brain stem of cat. J Neurophysiol 1960; 23:2–13.
11. Rickett DL, Glenn JF, Beers ET. Central respiratory effects versus neuromuscular actions of nerve agents. Neurotoxicology 1986; 7:225–236.
12. Douglas WW, Matthews PBC. Acute tetraethyl pyrophosphate poisoning in cats and its modification by atropine or hyoscine. J Physiol 1952; 116:202–218.
13. Rickett DL. Differentiation of central and peripheral actions in soman-produced respiratory arrest. Neuroscience 1980; 6:398–390.
14. Smolders FDJ, Folgering HT, Bemardo JA. Ventilation estimated from efferent phrenic nerve activity in the paralyzed cat. Pflügers Arch 1975; 359:157–169.
15. Shih T-M. Time course effects of soman on acetylcholine and choline levels in six discrete areas of the rat brain. Psychopharmacology 1982; 78:170–175.
16. McDonough JH, Jaax NK, Crowley RA, et al. Atropine and/or diazepam therapy protects against soman-induced neural and cardiac pathology. Fund Appl Toxicol 1989; 13:256–276.
17. Braitman DJ, Sparenbork S. MK-801 protects against seizures induced by the cholinesterase inhibitor soman. Brain Res Bull 1989; 23:145–148.
18. Dirnhuber P, Green DM. Effectiveness of pyridostigmine in reversing neuromuscular blockade produced by soman. J Pharmacol 1978; 30:419–425.
19. French MC. The reversal by pyridostigmine of neuromuscular block produced by soman. J Pharm Pharmacol 1979; 31:290–294.
20. Owczarczyk H, Gajewski D, Kisielinski T. Effect of physostigmine administration on neuromuscular transmission disturbances in rats caused by dichlorophos and phospholine. Acta Physiol Pol 1980; 31:431–438.
21. Angelov A. Study of the effect of toxogonine on the neuromuscular block caused by one inhibitor of the cholinesterases (soman). Eksp Med Morfol 1969; 8:51–55.
22. Angelov A. Effect of streptomycin on the neuromuscular conductivity of cats treated with certain phosphoorganic inhibitors of cholinesterase. Eksp Med Morfol 1973; 12:106–110.
23. Risch SC, Cohen RM, Janowsky DS, et al. Mood and behavioral effects of physostigmine on humans are accompanied by elevations in plasma-endorphin and cortisol. Science 1980; 209:1545–1546.
24. Risch SC, Janowsky DS, Kalin NH, et al. Cholinergic endogeneous opioid mechanism in mood and cognition in man. Psychopharmacol Bull 1981; 17:132–137.
25. Janowsky DC, Yousef EL, et al. Acetylcholine and depression. Psychosom Med 1974; 36:248–257.
26. Fukuda Y, Loeschcke HH. A cholinergic mechanism involved in the neuronal excitation by H^+ in the respiratory chemosensitive structures of the ventral medulla oblongata of rats in vitro. Pflügers Arch 1979; 379:125–135.
27. Trouth CO, Millis RM, Bernard DG, et al. Cholinergic-opioid interactions at brainstem respiratory chemosensitive areas in cats. Neurotoxicology 1993; 14:459–468.

28. Kok A. REM sleep pathways and anticholinesterase intoxication: A mechanism for nerve agent-induced central respiratory failure. Med Hypoth 1993; 41:141–149.
29. Trouth CO, Bada FJ, Pan Y, et al. Naloxone application to the ventrolateral medulla enhances the respiratory response to inspired carbon dioxide. Life Sci 1991; 49:193–200.
30. Bada FJ. Identification of a naloxone-sensitive site in the brainstem influencing respiratory CO_2 chemosensitivity. PhD dissertation, Howard University, Washington, DC, 1990.
31. Myer et al. Naltrexone therapy of apnea in children with elevated cerebrospinal fluid β-endorphin. Ann Neurol 1990; 27:75–80.
32. Varagic VM, Prostran MS, Stepanovic S, et al. Transmitter interactions in the central cholinergic control of blood pressure regulation. Drug Metab Drug Interac 1991; 9:49–76.
33. Douglas RM, Trouth CO, Addae JI, et al. Localization of ventral brainstem CO_2 chemosensors by kainic acid neurotoxic lesion (this volume).
34. Schäfer D, Schläfke ME. Cardiorespiratory regulation in a model for Ondine's curse syndrome (this volume).
35. Martin RL, Sinclair JD. Kainic acid on the rat ventral medullary surface depresses hypoxic and hypercapnic ventilatory responses. Resp Physiol 1990; 80:55–70.
36. Wade JV, et al. Changes in extracellular amino acids during soman- and kainic acid-induced seizures. J Neurochem 1987; 49:645–650.
37. Webb R, Vanhoutte P. Cocaine and contractile responses of vascular smooth muscle from spontaneously hypertensive rats. Arch Int Pharmacodynam 1980; 253:241–256.
38. Gradman AH. Cardiac effects of cocaine: A review. Yale J Biol Med 1988; 61: 137–147.
39. Jonsson S, O'Meara M, Young J. Acute cocaine poisoning: Importance of treating seizures and acidosis. Am J Med 1983; 75:1061–1064.
40. Catravas J, Waters I. Acute cocaine intoxication in the conscious dog: Studies on the mechanism of lethality. J Pharmacol Exp Ther 1981; 217:350–356.
41. Nahas G, Trouve R. A calcium-channel blocker as antidote to the cardiac effects of cocaine intoxication. N Engl J Med 1985; 313:519–520.
42. Trouve R, Nahas G. Antidotes to lethal cocaine toxicity in the rat. Arch Pharmacodynam Ther 1990; 305:197–207.
43. Witkin JM, Goldberg SR, Katz JL, et al. Modulation of the lethal effects of cocaine by cholinomimetics. Life Sci 1989; 45:2295–2301.
44. Dehkordi O, Dennis GC, Millis RM, et al. Effects of physostigmine on cocaine-induced cardiorespiratory responses at the caudal ventrolateral medullary surface of cats (this volume).
45. Toro-Matos A, Rendon-Platas AM, Avilla-Valdoz E, et al. Physostigmine antagonizes ketamine. Anaesth Analg 1980; 59:764–767.
46. Weinstock M, Erez E, Roll D, et al. Antagonism of the cardiovascular and respiratory depressant effects of morphine in the conscious rabbit by physostigmine. J Pharmacol Exp Ther 1981; 218:504–508.
47. Rupreht J, Dworacek B, Valkenburg M, et al. The use of physostigmine in heroin overdose. Arch Toxicol (Suppl) 1984; 7:467.
48. Lee J, Turndorf H, Poppers PJ, et al. Physostigmine reversal of antihistamine-induced excitement and depression. Anesthesiology 1975; 43:683–684.

49. Ellis JE, Byrd LD, Sexson WR, Patterson-Barrett CA. In utero exposure to cocaine: A review. South Med J 1993; 86:725–731.
50. Chasnoff IJ, Burns WJ, Schnoll SH, Burns KA. Cocaine use in pregnancy. N Engl J Med 1985; 313:666–669.
51. Riley JC, Brodsky NL, Porat R. Risk for SIDS in infants with in utero cocaine exposure. A prospective study. Pediatr Res 1988; 23:454A.
52. Ward SL, Bautista D, Chan L, et al. Sudden infant death syndrome in infants of substance abusing mothers. J Pediatr 1990; 117:876–881.
53. Durand DJ, Espinoza AM, Nickerson BG. Association between prenatal cocaine exposure and sudden infant death syndrome. J Pediatr 1990; 117:909–911.
54. Bauchner H, Zuckerman B, McClain M, et al. Risk of sudden infant death syndrome among infants with in utero exposure to cocaine. J Pediatr 1988; 113:831–834.
55. Chasnoff IJ, Hunt CE, Kletter R, Kaplan D. Prenatal cocaine exposure is associated with respiratory pattern abnormalities. Am J Dis Child 1989; 143:583–587.
56. Perry BD, Pesavento DJ, Kussie PH, et al. Prenatal exposure to drugs of abuse in humans: Effects on placental neurotransmitter receptors. Neurobehav Toxicol Teratol 1984; 6:295–301.
57. Woods JR, Plessinger MA, Clarke KE. Effect of cocaine on uterine blood flow and fetal oxygenation. JAMA 1987; 257:957–961.
58. Mahalik MP, Gautieri RF, Mann DE Jr. Teratogenic potential of cocaine hydrochloride in CF-1 mice. J Pharmacol Sci 1980; 69:703–706.
59. Loper KA. Clinical toxicology of cocaine. Med Toxicol Adverse Drug Exp 1989; 4:174–185.
60. Ritchie JM, Green NM. Local anesthetics. In: Goodman LS, Gilman A, eds. The Pharmacological Basis of Therapeutics. New York: Macmillan, 1980:300–320.
61. Nanji NA, Filipenko JD. Asystole and ventricular fibrillation association with cocaine intoxication. Chest 1984; 85:132–133.
62. Sanchez-Ramos JR. Neurologic complications of drug and alcohol abuse: Psychostimulants. Neurol Clin 1993; 11:535–553.
63. Pascuale-Leon A, Dhuna A, Alfafullah I, et al. Cocaine-induced seizures. Neurology 1990; 40:404.
64. Creglar LL, Mark H. Medical complications of cocaine abuse. N Engl J Med 1986; 315:1495–1500.
65. Daras M, Tuchman AJ, Marks S. Central nervous system infarction related to cocaine abuse. Stroke 1991; 22:1320–1325.
66. Rumbaugh CL, Bergeron RT, Fang HCH, et al. Cerebral angiographic changes in the drug abuse patient. Radiology 1971; 101:335–344.
67. Hall AH, Rumack BH, Schaffer MI, LInden CN. Clinical toxicology of cyanide: North American clinical experiences. In: Ballantyne B, Marrs TC, eds. Clinical and Experimental Toxicology of Cyanides, Chap 12. Bristol, England: Wright, 1987.
68. Sun MK, Jeske IT, Reis DJ. Cyanide excites medullary sympathoexcitatory neurons in rats. Am J Physiol 1992; 262:R182–R189.
69. Mitra J, Dev NB, Romaniuk JR, et al. Cardiorespiratory changes induced by vertebral artery injection of sodium cyanide in cats. Respir Physiol 1992; 87:49–61.
70. Mitra J, Dev NB, Trivedi R, et al. Intramedullary sodium cyanide injection on respiratory and vasomotor responses in cats. Respir Physiol 1993; 93:71–82.

71. Isom GE, Way JL. Lethality of cyanide in the absence of inhibition of liver cytochrome oxidase. Biochem Pharmacol 1976; 25:605–608.
72. Isom GE, Burrows GE, Way JL, et al. Effect of oxygen on the antagonism of cyanide intoxication-cytochrome oxidase, in vivo. Toxicol Appl Pharmacol 1982; 65: 250–256..
73. Kong A, Shen A, Burrows G, et al. Effect of chlorpromazine on cyanide intoxication. Toxicol Appl Pharmacol 1983; 71:407–413.
74. Pettersen JC, Cohen SD. Antagonism of cyanide poisoning by chlorpromazine and sodium thiosulfate. Toxicol Appl Pharmacol 1985; 81:265.
75. Patel M, Ardelt BK, Yim GKW, Isom GE. Cyanide induces Ca^{2+}-dependent and independent release of glutamate from mouse brain slices. Neurosci Lett 1991; 131:42–44.
76. Johnson JD. Neurotoxic effects of cyanide and the role of calcium. Diss Abstr Int B 1986; 47:146.
77. Johnson JD, et al. Effect of pentobarbital on cyanide-induced tremors in mice and calcium accumulation in PC12 cells. Biochem Pharmacol 1987; 36:1747.
78. Winger G. PCP-NMDA connection provides hope in cerebral ischemia but new direction for antipsychotics. Trends Pharmacol Sci 1987; 8:323.
79. Kemp JA, et al. MK-801, NMDA receptors and ischemia-induced neurodegeneration. Trends Pharmacol Sci 1987; 8:414.
80. Goldberg MP, Weiss JH, Pham PC, et al. N-Methyl-D-aspartate receptors mediate hypoxic neuronal injury in cortical culture. J Pharmacol Exp Ther 1987; 243:784–791.
81. Tursky T, Sajter V. The influence of potassium cyanide poisoning on the gamma-aminobutyric acid level in rat brain. J Neurochem 1962; 9:519.
82. Taylor PW, Burgen ASV. Kinetics of carbonic anhydrase inhibitor complex formation: A comparison of anion and sulfonamide binding mechanisms. Biochemistry 1971; 10:3859.
83. Mitchell RA, Loeschcke HH, Severinghaus JW, et al. Regions of respiratory chemosensitivity on the surface of the medulla. Ann NY Acad Sci 1963; 109:661–681.
84. Schläfke ME, See WR, Loeschcke HH. Ventilatory response to alterations of H^+ ion concentration in small areas of the ventral medullary surface. Resp Physiol 1970; 10:198–212.
85. Trouth CO, Loeschcke HH, Berndt J. A superficial substrate on the ventral surface of the medulla oblongata influencing respiration. Pflügers Arch 1973; 339: 135–152.
86. Trouth CO, Loeschcke HH, Berndt J. Topography of the respiratory responses to electrical stimulation in the medulla oblongata. Pflügers Arch 1973; 339:153–170.
87. Dev NB, Loeschcke HH. Topography of the respiratory and circulatory responses to acetylcholine and nicotine on the ventral surface of the medulla oblongata. Pflügers Arch 1979; 379:19–27.
88. Dev NB, Loeschcke HH. A cholinergic mechanism involved in the respiratory chemosensitivity of the medulla oblongata in the cat. Pflügers Arch 1979; 379:29–36.
89. Persson SA, Cassel G, Sellstrom A, et al. Acute cyanide intoxication and central transmitter systems. Fund Appl Toxicol 1985; 5:S150–S159.
90. Pauwels PJ, VanAssouw HP, Leysen JE, et al. Ca^{2+}-mediated neuronal death in rat brain neuronal cultures by veratridine. Mol Pharmacol 1989; 36:525–531.

91. Kao CY. Tetrodotoxin, saxitoxin and their significance in the study of excitation phenomena. Pharmacol Rev 1966; 18:997–1049.
92. Norton S. Toxic responses of the central nervous system. In: Klaassen CD, Amdur MO, Doull J, eds. Casarett and Doull's Toxicology, 3rd ed. New York: Macmillan, 1986:378.
93. Borison HL, McCarthy LE. Respiratory and circulatory effects of saxitoxin in the cerebrospinal fluid. Br J Pharmacol 1977; 61:679–689.
94. Chang F-CT, Benton BJ, Salyer JL, et al. Respiratory and cardiovascular effects of tetrodotoxin in urethane-anesthetized guinea pigs. Brain Res 1990; 528:259–268.
95. Fedorko L, Merrill EG. Axonal projections from the rostral expiratory neurons of the Bötzinger complex to medulla and spinal cord in the cat. J Physiol 1984; 350:487–496.
96. Merrill EG, Fedorko L. Monosynaptic inhibition of phrenic motoneurons: A long descending projection from Bötzinger neurons. J Neurosci 1984; 4:2350–2353.
97. Merrill EG, Lipski J, Kubin L, et al. Origin of the expiratory inhibition of nucleus tractus solitarius inspiratory neurons. Brain Res 1983; 263:43–50.
98. Smith JC, Ellenberger HH, Bellanyi K, et al. Pre-Bötzinger complex: A brainstem region that may generate respiratory rhythm in mammals. Science 1991; 254:726–729.
99. Smith JC, et al. Cellular and synaptic mechanisms generating respiratory rhythm: Insights from in vitro and computational studies (this volume).
100. Denavit-Saubié M. Perfused isolated adult brainstem: A novel approach to study central chemosensitivity (this volume).
101. Hille B. Pharmacological modifications of the sodium channels of frog nerve. J Gen Physiol 1968; 51:199–219.
102. Borison HL, Culp WJ, Gonsalves SF, et al. Central respiratory and circulatory depression caused by intravascular saxitoxin. Br J Pharmacol 1980; 68:301–309.
103. Chang F-CT, Benton BJ, Lenz RA, et al. Central and peripheral cardio-respiratory effects of saxitoxin (STX) in urethane-anesthetized guinea pigs. Toxicon 1993; 31:645–664.
104. Benton BJ, Rivera VR, Hewetson JF, et al. Reversal of saxitonin-induced cardiorespiratory failure by a burro-raised alpha-STX antibody and oxygen therapy. Toxicol Appl Pharmacol 1994; 124:39–51.
105. Welch SP, Dunlow LD. Antinociceptive activity of intrathecally administered potassium channel openers and opioid agonists: A common mechanism of action? J Pharmacol Exp Ther 1993; 267:390–399.
106. Burnard DM, Crichton SA, MacVicar BA, et al. Electrophysiological properties of reactive glial cells in the kainate-lesioned hippocampal slice. Brain Res 1990; 510:43–52.
107. Yamamoto HA, Harris RA. Calcium-dependent 86 Rb efflux and ethanol intoxication: Studies of human red blood cells and rodent brain synaptosomes. Eur J Pharmacol 1983; 88:357–363.
108. Hughes M, Duval D, Kitabgi P, et al. Preparation of a pure monoiodo derivative of the bee venom neurotoxin apamin and its binding properties to rat brain synaptosomes. J Biol Chem 1982; 257:2762–2769.

48

A "Pneumotaxic Center" in the Ventrolateral Pons of Rats

THOMAS E. DICK and SHARON K. COLES

School of Medicine
Case Western Reserve University
Cleveland, Ohio

JOZEF S. JODKOWSKI

Institute of Physiological Sciences
Medical Academy of Warsaw
Warsaw, Poland

The "pneumotaxic center" in the pons plays a critical role in determining respiratory phase duration in a number of species. Lumsden reported in 1923 that apneustic breathing, a breathing pattern with prolonged inspiratory phase, follows transection separating the rostral pons from the medullary respiratory central pattern generator in rabbit, cat, dog, and monkey, especially in vagotomized animals (1,2). Even in humans, apneustic breathing has been associated with pontine lesions (3). However, there is a paucity of data regarding its presence in rats. Indeed, a report has questioned the existence of a pneumotaxic center in rats (4).

In other species, the rostral dorsolateral pons has been identified as the pneumotaxic center. Electrophysiological studies have characterized the medial parabrachial and Kölliker-Fuse nuclei as the specific neuronal substrates for the pneumotaxic center in human (3), cat (5–16), opossum (17), guinea pig (18), and rabbit (19,20). The physiological criteria for characterizing an area as a pneumotaxic center were derived from data obtained in lesion, stimulation, and recording studies and are listed in Table 1. For example, bilateral lesions in the dorsolateral pons led to apneustic breathing in vagotomized animals, i.e., similar to Lumsden's transection experiments (5–10,19). Small unilateral

Table 1 Criteria for a "Pneumotaxic Center"

Lesion:
1. Bilateral: Apneustic breathing in vagotomized anesthetized animal
2. Unilateral: Instability in breathing pattern

Stimulation: Response dependent on pattern of stimulation
1. Single electrical pulses
   - A. Phrenic nerve: short-latency inhibition or excitation
   - B. Hypoglossal nerve: short-latency excitation
   - C. Phase switching especially inspiration to expiration
2. Trains of electric pulses or injection of excitatory chemical
   - A. Change in amplitude in phrenic nerve activity
   - B. Change in breathing frequency

Recording: Various patterns of respiratory-modulated activity
1. Respiratory-modulated phasic activity
2. Phase-spanning activity
   - A. Peak activity near phase transition
   - B. Activity starts in one phase and ends in another
3. Respiratory-modulated tonic activity
   - A. Peak activity occurs during a phase and not associated with transition
   - B. No quiescent period

lesions destabilize respiration, as indicated by increased variability in the respiratory pattern and by increased time necessary for the respiratory cycle to return to its normal cycle length following a perturbation (11). Depending on the stimulus site, single pulses can elicit short-latency inhibition or excitation in phrenic nerve activity and can cause phase switching (12,13). Similarly, trains of stimulus pulses or microinjection of excitatory amino acids can increase or decrease breathing frequency and phrenic nerve activity (14,17). Recordings of neuronal activity have shown various patterns of respiratory-modulated activities, including phase-spanning activity and respiratory-modulated tonic activity (15,16,18).

Results from preliminary (21,22) as well as recent (23) studies have identified an area in the dorsolateral pons in rats that meets some of these criteria. In particular, microinjections of excitatory amino acids in the dorsolateral pons result in both increases and decreases in phrenic nerve activity (21,22). Potent effects were cessation of phrenic nerve activity with DL-homocysteic acid in the Kölliker-Fuse nucleus (22). In addition, bilateral lesions in the dorsolateral pons lead to apneustic breathing (23). Therefore, we hypothesized that rats are similar to other species and contain a pontine area that fulfills the criteria of a pneumotaxic center. Having identified the criteria in Table 1, we addressed our hypothesis using electrophysiological techniques, specifically (a) short-latency inhibition of phrenic nerve activity following single-pulse stimulation, (b) apneustic breathing with bilateral lesions, (c) changes in breathing frequency and/or phrenic nerve amplitude with stimulus trains and/or chemical injections, and (d) recording of respiratory modulated activity.

We used adult male rats ($n = 20$, Sprague-Dawley, 310 to 440 g). Animals were anesthetized with Equithesin (0.3 mL/100 g, IP) which is a mixture of chloral hydrate and pentobarbital (5.55 g and 1.215 g respectively per 125 mL double-distilled water). In addition to the anesthetic agent, atropine sulfate (0.5 mg/100 g, S.C.) and pancuronium bromide (0.1 mg/100 g/h, IV) were administered to reduce secretions and to paralyze the animals, respectively. Body temperature was maintained at 38°C with a heating pad. The left femoral artery and vein were cannulated to measure blood pressure and to administer drugs, respectively. A tracheostomy was performed and animals were ventilated. Animals were placed in a stereotaxic frame so that the dorsal surface of the parietal plates was horizontal. The right phrenic nerve was dissected, transected, and desheathed from a dorsal approach and the nerve was placed on a bipolar recording electrode. Phrenic nerve activity was amplified and filtered 0.1 to 3 KHz (Grass P511) and was sent to a moving averager (CWE, Inc.) where it was rectified and integrated (Paynter Filter, time constant 50 ms). The raw phrenic neural activity and the output from the moving averager were recorded on paper (Astromed DASH 8) and tape (Hewlett-Packard). In all but one animal, the vagi were transected and in three of these animals the right vagus was placed on a stimulating electrode. In the animal without vagal transection, the vagi were cut after the electrolytic lesions were produced.

Experimental protocol involved initially identifying the region of interest by determining a site from which single stimulus pulses evoked short-latency inhibition of phrenic nerve activity. Initially, tungsten electrodes (0.005 in, AM Systems) were placed stereotaxically at 9.05 to 9.25 mm caudal to bregma and 2.5 to 2.6 mm lateral to the midline. Electrode depth was referenced to the dorsal cerebral/cerebellar surface. Testing for short-latency inhibition of phrenic nerve activity began at a depth of 4500 μm and consisted of stimulating the tissue with trains of stimuli (10 Hz at 50 μA). If a short-latency inhibition was recorded at 50 μA, we then decreased the current to determine the lowest stimulus current necessary to evoke a response. We tested the tissue systematically by positioning the electrode progressively deeper in increments of 250 μm initially and 100 μm when the threshold stimulus went below 50 μA. If an electrode track was made without evoking an effect or without decreasing the threshold stimulus below 50 μA, then the electrode was repositioned.

Once an effective site was located, one of various protocols was followed. In the first protocol ($n = 13$), we lesioned the identified site bilaterally either electrolytically ($n = 8$, two of which received only unilateral lesions; the last two animals in Table 2) or chemically ($n = 5$). Electrical lesions were DC pulses, 2 to 10 mA, and 5-s pulse duration. For chemical lesions, we replaced the tungsten microelectrode with a glass micropipette containing either 1 m*M* ibotenic acid ($n = 3$) or 10 m*M* muscimol ($n = 2$) in a 2% fast green buffered physio-

Table 2 Changes in Variables of Breathing Pattern Before and After Pontine Lesion in Anesthetized, Vagotomized Rats[a]

Ti (s)		Te (s)		Ttot (s)		Resp rate (bpm)	
Before	After	Before	After	Before	After	Before	After
0.23	0.37	0.90	0.47	1.13	0.84	52.9	71.4
0.01	0.02	0.04	0.02	0.05	0.03		
0.31	3.25	0.87	1.15	1.17	4.40	51.1	13.6
0.01	0.14	0.01	0.05	0.02	0.19		
0.22	1.99	1.16	3.77	1.38	5.76	43.5	10.4
0.00	0.42	0.05	0.22	0.05	0.34		
0.36	1.33	0.69	0.92	1.05	2.25	57.0	26.6
0.00	0.08	0.02	0.03	0.02	0.11		
0.23	1.41	0.97	2.91	1.20	4.32	50.0	13.9
0.01	0.25	0.01	0.38	0.02	0.63		
0.44	0.49	1.05	0.70	1.49	1.19	40.4	57.0
0.00	0.01	0.07	0.02	0.07	0.02		
0.26	0.53	1.17	1.05	1.43	1.59	42.1	37.8
0.00	0.02	0.04	0.01	0.04	0.03		

[a]Values: mean and standard deviation of three consecutive breaths.

logical saline solution. Chemical lesions were produced by pressure injection at the stereotaxic coordinates identified by microstimulation. The amount injected was measured directly by observing the displacement of the meniscus marking its displacement, using a reticle. Phrenic nerve activity was recorded for at least 30 min following the last lesion.

In the second protocol ($n=6$), the tungsten microelectrode was replaced by a glass micropipette containing either 10 mM D-glutamate ($n=2$) or 10 mM L-glutamate ($n=4$). Both solutions were made with 2% fast green in buffered saline. The D-glutamate injections were 200 nL, which ranged from 2 to 10 times the amount of L-glutamate injections (20 to 100 nL) and served as controls.

Recently, we initiated a third protocol ($n=1$) in which neural activity in the ventrolateral pons was recorded. Activity was recorded by a glass micropipette filled with 4 mM NaCl, tip size less than 2 μm and an impedance between 10 and 25 MΩ. After recording single-unit activity, the glass micropipette was replaced with a tungsten microelectrode ($Z=5$ MΩ). A lesion was made at the site where respiratory-modulated noise was recorded via the tungsten microelectrode.

At the end of the experiment, the brainstem was removed and placed in 10% formalin and stored for at least 48 h. At least 24 h before sectioning, the tissue was blocked and placed in a 40% sucrose solution. Frozen sections (50 μm thick) were cut in the frontal plane. The sections were examined microscopically and the extent of the lesions was mapped.

We analyzed the data by measuring the duration of inspiratory phase (Ti) and the duration of the respiratory cycle (Ttot) using phrenic nerve activity as the indicator of the respiratory cycle: Ti was measured from the onset of phrenic nerve activity to its offset and Ttot was measured from the onset of one cycle to the onset of the next cycle. Measurements were made for three consecutive cycles and post-lesion measurements were made after the breathing pattern stabilized. The Ti/Ttot ratio was calculated for each animal before and after lesioning. Statistical difference was tested using the paired Student's *t*-test and was accepted if the probability for the null hypothesis was less than 0.01.

In all animals, we located a site in the ventrolateral pons between 7500 and 8500 μm below the dorsal cerebral surface from which single stimulus pulses evoked a short-latency inhibition of phrenic nerve activity (Fig. 1). The latency was measured from the onset of the stimulus pulse to the decrease of phrenic nerve activity and was less than 5 ms.

Bilateral electrolytic lesions at the sites identified by stimulation altered respiratory pattern profoundly (Fig. 2, Table 2). Duration of inspiration (Ti) increased between 1.2- and 10-fold (Table 2). In 5 animals, the duration of expiration (Te) increased also, but to a lesser extent than Ti (Table 2). In 2 animals, Te was decreased (Table 2). As a net result, the ratio of Ti to the total duration of the respiratory cycle (Ttot), an index of apneustic breathing, increased significantly in all ($n = 7$, including the two with unilateral lesions) animals (Table 2). Even after unilateral lesions, Ti was prolonged (Fig. 2B, and the last two animals in Table 2).

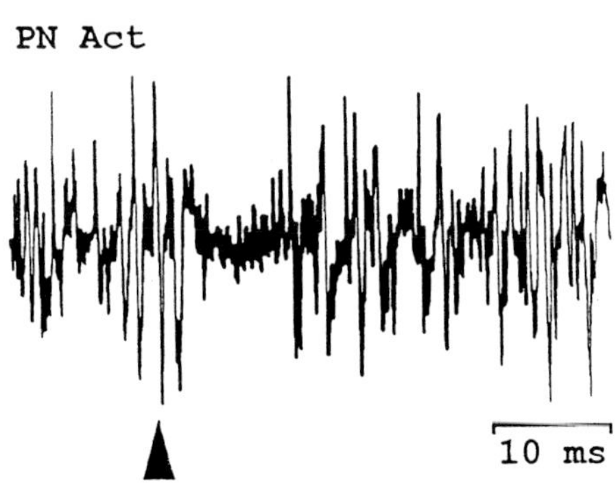

Figure 1 Short-latency inhibition of phrenic nerve activity by pontine stimulation. Latency was measured from the onset of the stimulus pulse and was 3.4 ms in this case. Stimulus pulses were delivered 8500 μm below the dorsal cerebral surface. Arrowhead = onset of the stimulus pulse (10 μA, 1 ms); PN Act = phrenic nerve activity; ms = milliseconds.

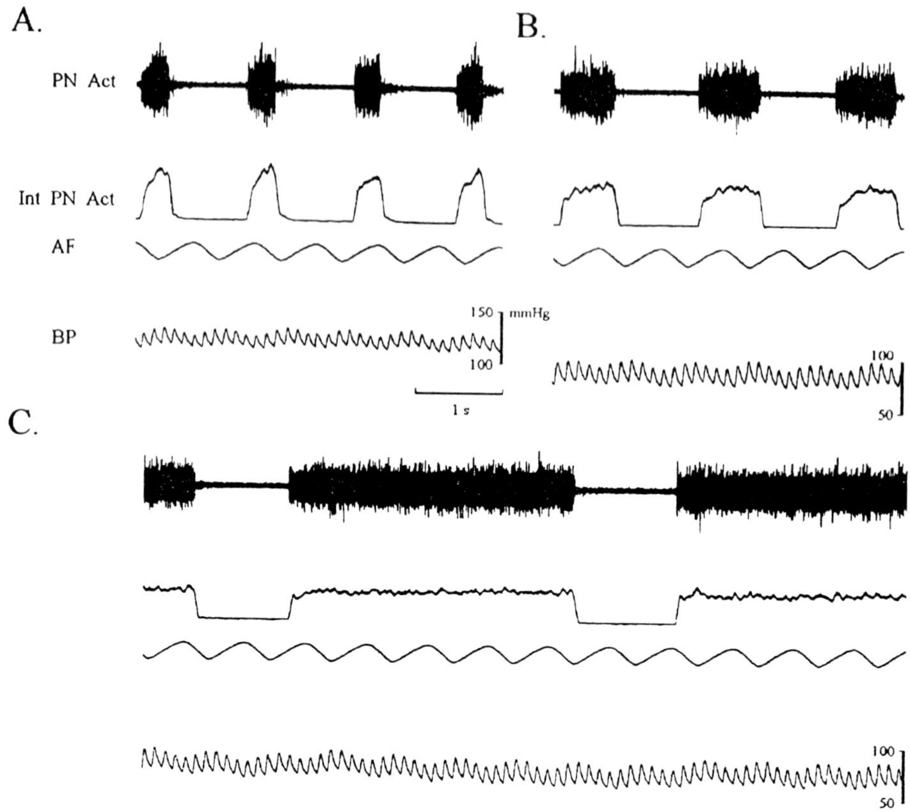

Figure 2 Effect of electrolytic lesion (10 mA, 5 s) in the ventrolateral pons of an anesthetized vagotomized rat. (A) Pons intact. (B) Stable breathing pattern after unilateral lesion. (C) Stable breathing pattern after contralateral lesion. Breathing frequency slowed due to the prolongation of inspiration (i.e., apneustic breathing). Arterial blood pressure decreased from 120 to 75 mm Hg approximately 10 s after the first lesion and remained relatively stable after the second lesion with the exception of the marked respiratory modulation. Abbreviations: PN Act = phrenic nerve activity; Int. PN Act = integrated phrenic nerve activity; AF = airflow; BP = blood pressure.

The effects of the lesions on the breathing pattern were dependent on bilateral vagal transection. When large bilateral lesions were formed in a vagally intact animal ($n = 1$), Ti was prolonged slightly, but apneusis developed more completely after vagal transection. Further, in both animals tested, continuous stimulation (10 Hz) of the proximal end of a transected vagus nerve terminated Ti and restored partially the normal breathing pattern.

Histological examination revealed that electrical lesion sites were in the ventrolateral quadrant of the pons (dotted area, Fig. 3). Lesions were lateral

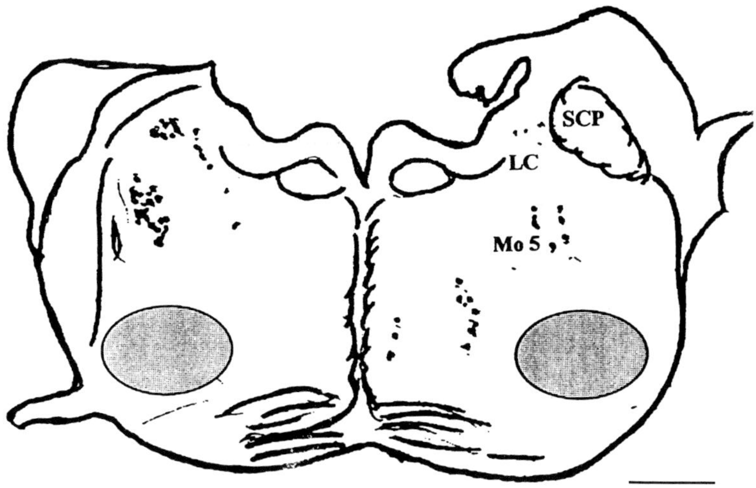

Figure 3 Drawing of coronal section of the pons. The cross-hatched areas represent damaged areas from all of the animals. On the left side, the electrode track was visible above the lesion. The distance from bregma is an approximate measure (9.05 mm caudal) and the scale bar is 11 mm. Abbreviations: LC = locus ceruleus; Mo 5 = motor nucleus of the trigeminal nerve; SCP = superior cerebellar peduncle.

to the nucleus of the trapezoid body and dorsolateral to superior olivary nucleus (the superior olivary nucleus was included in the large lesions). At its rostral end, the lesions were ventrolateral to the motor nucleus of the fifth cranial nerve (CN-V). At its caudal end, the lesion was medial to exit of the CN-VII fibers and included the A5 area. Even though the rostrocaudal extent of a single lesion was not greater than 0.5 mm, the rostrocaudal extent of all the lesions ranged from −8.7 to −10.0 mm relative to bregma.

Electrolytic stimulation and lesioning are notorious for activating both cell bodies and fibers of passage. This region of the brainstem (i.e., ventrolateral pons) contains both fibers of passage from the dorsolateral pons, the area classically defined as the pneumotaxic center, as well as cells of the A5 cell group (24,25). In addition, even in cat, axonal projections from the commissural subnuclei of the solitary tract have been shown to terminate in the ventrolateral pons (26). Therefore, we pursued these initial studies by lesioning and stimulating this area chemically as well as recording cellular activity there.

Chemical lesions consisted of both ibotenic acid (1 m*M*) and muscimol (10 m*M*) bilateral microinjections. Ibotenic acid is an excitatory amino acid agonist that causes depolarization blockade. Muscimol is a long-acting gamma-aminobutyric acid (GABA) agonist that inhibits neuronal activity directly. Both compounds act preferentially on cell bodies, block neural activity, and effectively lesion the site. Therefore, after identifying a site in the ventrolateral pons from which a short-latency inhibition could be elicited, the tungsten electrode was replaced by a glass micropipette containing either ibotenic acid or muscimol.

Ibotenic acid injections were made bilaterally near the ventrolateral pontine surface (ranging from 7800 to 8200 μm below the dorsal cerebral surface) (Fig. 4). The response to the injections was time-dependent. Initially, the amplitude of phrenic nerve activity increased, and both Ti and Te prolonged (Fig. 4B). Subsequently, phrenic nerve amplitude decreased and Ti prolonged further (Fig. 4C). Blood pressure did not change consistently. However, blood pressure increased transiently during the injection of ibotenic acid.

Muscimol injections were made near the ventral surface of the brainstem (7800 and 8000 μm below the surface). The chemical injection had both immediate (an apnea) and latent (progressive development of apneusis) effects on breathing pattern but no effect on blood pressure until the end of the experiment, at which time blood pressure decreased.

The response to excitatory amino acid injections was also time-dependent but on a shorter time scale. The immediate response to injections of L-glutamate was a termination of phrenic nerve activity, an "off switch." Subsequently,

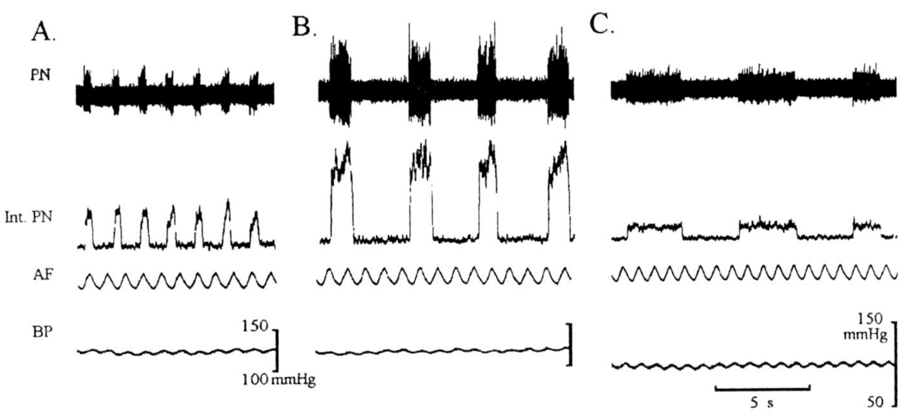

Figure 4 Effect of ibotenic acid (1 m*M*) injections in the ventrolateral pons in an anesthetized, vagotomized rat. (A) Respiration and blood pressure in vagotomized intact rat. (B) Approximately 5 min following bilateral injections of ibotenic acid (1 m*M*). The injections were approximately 15 min apart and were not of equal size (50 nL on one side and 20 nL on the contralateral side). The increase in the amplitude of phrenic nerve activity occurred after unilateral injection, but the change in timing e.g., the increase in Ti and Te, did not occur until after the contralateral injection. (C) Apneustic breathing pattern approximately 90 min after last injection. Mean arterial blood pressure is approximately 100 mm Hg. Abbreviations: PN Act = phrenic nerve activity; Int. PN = integrated phrenic nerve activity; AF = airflow; BP = blood pressure.

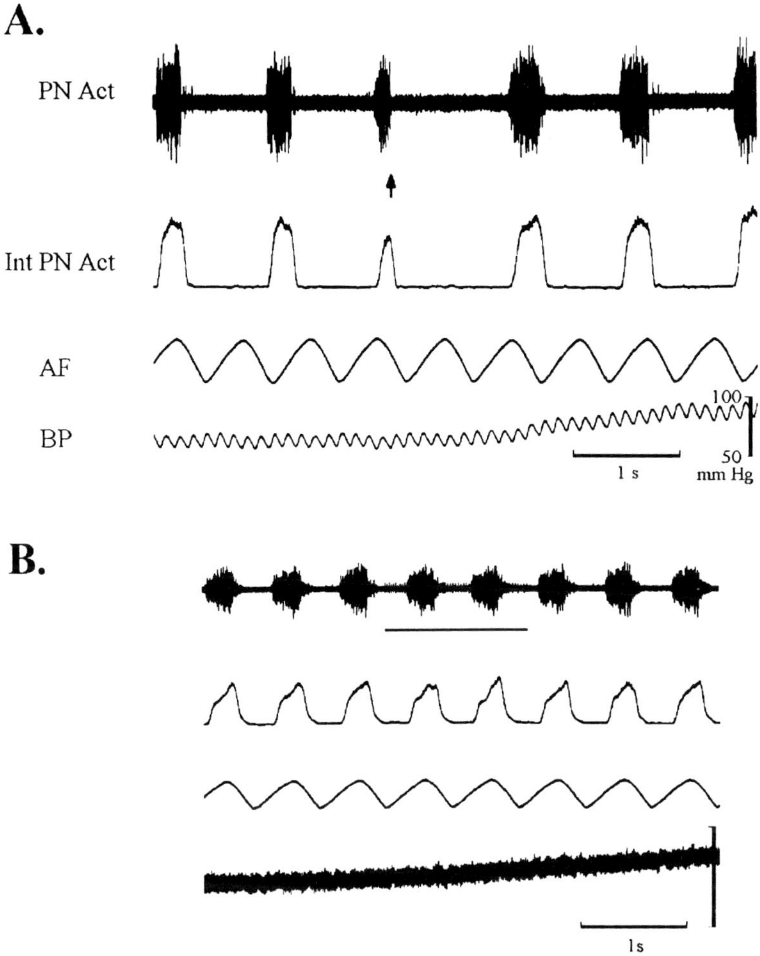

Figure 5 Effect of unilateral excitatory amino acid injection into the ventrolateral pons in rat. (A) L-glutamate (10 m*M*, 100 nL) was injected in the ventrolateral pons as a single bolus injection during inspiration. Phrenic nerve activity ceased immediately and subsequently was slightly prolonged. Te was prolonged in the shortened breath but not in the subsequent cycles. (B) D-glutamate (10 m*M*, 300 nL) was injected as a series of numerous injections (marked by bar). No change in the respiratory pattern was apparent, only a subsequent rise in blood pressure. Time (1s) and blood pressure (50 to 100 mm Hg) bars are the same in panels A and B. Abbreviations: PN Act = phrenic nerve activity; Int. PN Act = integrated phrenic nerve activity; AF = airflow; BP = blood pressure.

Ti increased transiently (Fig. 5A). Blood pressure increased as well during this subsequent increase in Ti (Fig. 5A). The respiratory response was stereospecific, indicating a dependence on an interaction between L-glutamate and its receptors. Injections of D-glutamate did not elicit a phase switching or a prolongation of Ti (Fig. 5B).

Extracellular recordings of neurons in the ventrolateral pons revealed various patterns of activity (Fig. 6). These cells were recorded with glass micropipettes filled with 4 *M* NaCl and the amplifier was AC-coupled. Unfortunately these recordings cannot distinguish between fibers of passage and cell bodies. Nevertheless, they verify a presence of highly modulated respiratory activity in the ventrolateral pons. This is in counterdistinction to respiratory-modulated sympathetic activity that has been recorded recently in the ventrolateral pons in the A5 region (23).

In conclusion, in the anesthetized, vagotomized adult rat, bilateral lesions localized specifically to ventrolateral pons resulted in apneustic breathing, i.e., a pattern in which principally Ti is lengthened. This breathing pattern was dependent on bilateral vagal transection and was partially reversed by vagal stimulation. Thus, our study confirms that a pneumotaxic center, as defined by Lumsden's classic experiments (1,2), exists in rats. Further, we believe that activity from ventrolateral pontine structures interacts with vagal input to determine respiratory phase duration in mammals.

Our physiological findings agree with observations made in a variety of species including rodents (1–3,5–10,12–23) and fulfill the criteria for a pneumotaxic center (Table 1). However, these data disagree with those of Monteau et al. (4), who reported no changes in breathing pattern following pontine transection in anaesthetized, vagotomized rats. They concluded that in adult rat pontine influences upon respiration, if any, were negligible (4). In contrast, we report apneustic breathing in rats following pontine lesions. These data contradict those of Monteau et al. (4) and support those of Wang et al. (23). We conclude that pontine influences on breathing in rats are similar to those in other mammalian species. We speculate that the failure of Monteau et al. (4) to observe changes in the breathing pattern following pontomedullary transection may be related to either incomplete transections or a strain difference among the rodent species. The transections illustrated in their report appear to be incomplete, leaving neural tissue intact in the ventral pons, which may include pathways connecting pontine and medullary respiratory networks (see Fig. 2 in Ref. 4; lines indicating transections do not go to the bottom of tissue section). In addition, a difference was reported recently between Wistar and Sprague-Dawley strains of rats regarding the breathing pattern changes following MK-801 (27). In Sprague-Dawley rats, like other animal species, apneusis followed MK-801 injections; whereas in Wistar rats, apneusis was not evident (27). We and Wang et al. (23) used Sprague-Dawley rats; Monteau et al. (4) did not identify strain.

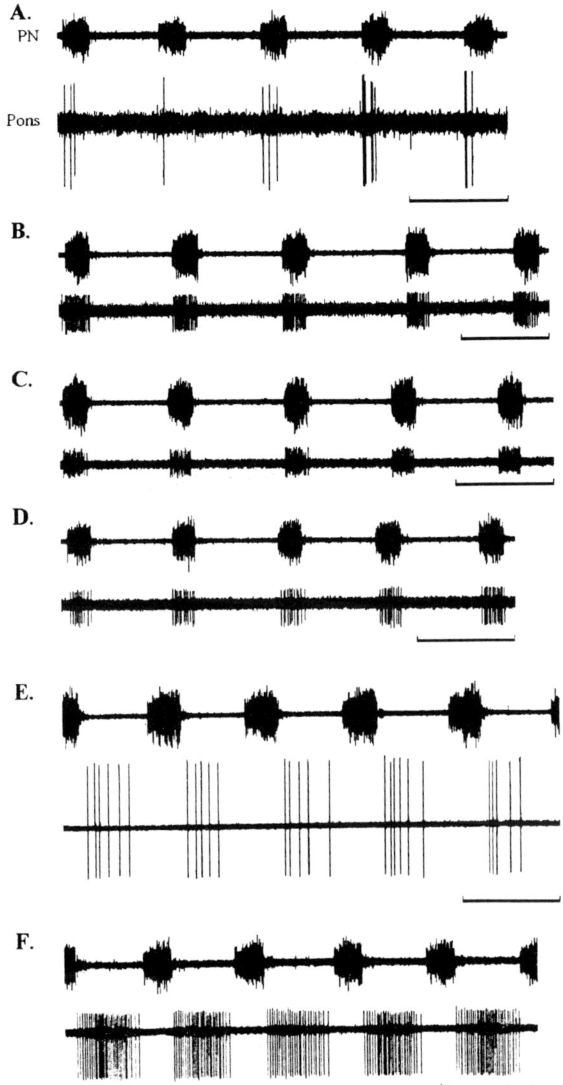

Figure 6 Extracellular recordings of highly modulated respiratory activity in the ventrolateral pons of bilaterally vagotomized rats. Tracings are paired with raw phrenic nerve activity (PN) above the pontine activity (Pons). (A) Early-inspiratory or inspiratory-decrementing activity recorded at 7400 μm below the dorsal cerebral surface. (B, C, and D) Inspiratory activities recorded 7700, 7740, and 8700 μm deep. (E) Post-inspiratory or expiratory-decrementing activity recorded 7200 μm deep. (F) Expiratory activity, 7200 μm deep. Time bar equals 1 s.

The unusual aspect of our finding is that the lesion was located in the ventrolateral pons. The lesion site was identified in vivo by determining the lowest stimulus that evoked short-latency inhibition of phrenic nerve activity. The neural elements activated by electrical stimulus pulses depend on numerous variables and fibers rather than cells that may have been preferentially activated. However, both our chemical lesioning and stimulating experiments support the hypothesis that cells in this region can modify the breathing pattern.

Electrical stimulation does provide a discrete onset time for latency measurements. The onset latency of the response recorded in the phrenic nerve activity was less than 5 ms. This short-latency response indicates a paucisynaptic if not a direct pontospinal pathway. These data are also similar to those of the cat, in which the response of phrenic nerve activity was as short (4 to 7 ms), indicating a pontospinal projection (12,13). In a recent study in which A5 neurons were antidromically activated by thoracic spinal cord stimulation, the mean response latency was greater than 25 ms (24). This discrepancy indicates multiple populations of projecting neurons from and through this area.

As indicated above, it is generally accepted that the neuronal substrate for Lumsden's pneumotaxic center is located in dorsolateral pons, specifically the medial parabrachial and Kölliker-Fuse nuclei (1–3,5–22). Indeed, previous reports indicate that this area is important in rodents as well (21–23). Only a few reports indicate that other structures in the lateral and ventrolateral pons may play any role in control of respiration, and most of these studies were performed in neonatal and fetal preparations (28–30). In an early study in the adult cat, electrolytic lesions in the ventrolateral pons in at least one cat resulted in apneustic breathing also (Fig. 8 in Ref. 9). Further, a recent study, also in the adult cat (26), indicated that second-order respiratory afferents projected to and terminated in the ventrolateral pons in addition to the dorsolateral pons. We speculated that the electrolytic lesions in the ventrolateral pons destroyed tracts between dorsolateral pontine nuclei and medullary respiratory groups and neurons in the A5 area. A potential role for neurons in the A5 group in adult rats is supported by our data from microinjections of muscimol and excitatory amino acid agonists near the ventral pontine surface.

This is the first study to report strong respiratory-modulated phasic activity in the ventrolateral quadrant in rats. These are initial recordings, and it remains unclear if they were obtained from fibers or cell bodies. Weak respiratory-modulated tonic activity has been recorded in the A5 cell group (24). The relationship between these cell populations is unclear. There were similarities and differences in both animal protocols. For instance, both utilized vagotomized, adult Sprague-Dawley rats but differed in anesthetic agents. We used Equithesin, a mixture of pentobarbital and chloral hydrate, and they induced anesthesia with halothane and maintained it with urethane. However, the cellular nature of their recordings was verified by evoking an excitatory response to locally injected L-glutamate.

Even in these initial recordings from one animal, it is apparent that different types of respiratory-modulated activity exist in the ventrolateral pons. We recorded not only from inspiratory and expiratory cells but also from early inspiratory and postinspiratory neurons. These latter two groups have been identified as representing the "kernel" of the medullary respiratory pattern generator (31). In addition, whether fibers or cells, the recorded activity could provide the neural substrate for the modulation reported by Guyenet et al. (24).

In summary, this study confirms that, in at least Sprague-Dawley rats, a pneumotaxic center exists. Further, our working hypothesis is that activity in the ventrolateral pons modulates the breathing pattern. Finally, the results in this paper and those previously published (9,26) raise the possibility that the pontine respiratory network may include structures in the ventrolateral pons even in other species.

Acknowledgments

We thank Mr. Philip Martinak for assistance in data analysis and histological processing of the tissue and Ms. Cheryl Diane Giliam-Walker for secretarial assistance in preparation of the manuscript. We acknowledge gratefully the support of the National Heart, Lung and Blood Institute, HL-42400 to TED and of the American Heart Association to SKC.

References

1. Lumsden T. Observations on the respiratory centres in the cat. J Physiol (Lond) 1923; 57:153–160.
2. Lumsden T. Observations on the respiratory centres. J Physiol (Lond) 1923; 57: 354–367.
3. Plum F, Posner JB. Diagnosis of Stupor and Coma, 2nd ed. Philadelphia: Davis, 1972.
4. Monteau R, Errchidi S, Gauthier P, et al. Pneumotaxic centre and apneustic breathing: Interspecies differences between rat and cat. Neurosci Lett 1989; 99:311–316.
5. Berger AJ, Herbert DA, Mitchell RA. Properties of apneusis produced by reversible cold block of the rostral pons. Respir Physiol 1978; 33:323–337.
6. Caille D, Vibert J-F, Hugelin A. Apneusis and apnea after parabrachial of Kölliker-Fuse N lesion: Influence of wakefulness. Respir Physiol 1981; 45:79–95.
7. von Euler C, Marttila I, Remmers JE, Trippenbach T. Effects of lesioning the parabrachial nucleus on the mechanisms for central and reflex termination of inspiration in the cat. Acta Physiol Scand 1976; 96:324–337.
8. Feldman JL, Gautier H. The interaction of pulmonary afferents and pneumotaxic center in control of respiratory pattern in cats. J Neurophysiol 1976; 39:31–44.
9. Ngai SH, Wang SC. Organization of central respiratory mechanisms in the brain stem of the cat: Localization by stimulation and destruction. Am J Physiol 1957; 190:343–349.

10. St. John WM. Characterization of the tidal volume regulating functions of the pneumotaxic center. Respir Physiol 1973; 18:64–79.
11. Oku Y, Dick TE. Phase resetting of the respiratory cycle before and after unilateral pontine lesion in cat. J Appl Physiol 1992; 72:721–730.
12. Cohen MI. Switching of the respiratory phases and evoked phrenic responses produced by rostral pontine electrical stimulation. J Physiol (Lond) 1971; 217:133–158.
13. Dick TE, Haxhiu MA, Cherniack NS. Salivary secretion elicited by activation of the parabrachial nuclei in the cat. J Auton Nerv Syst 1992; 39:19–28.
14. Denavit-Saubié M, Riche D, Champagnat J, Velutti JC. Functional and morphological consequences of kainic acid microinjections into a pontine respiratory area of the cat. Neuroscience 1981; 5:1609–1620.
15. Bianchi AL, St. John WM. Medullary axonal projections of respiratory neurons of pontile pneumotaxic center. Respir Physiol 1982; 48:357–373.
16. Cohen MI, Wang SC. Respiratory neuronal activity in pons of cats. J Neurophysiol 1959; 22:33–50.
17. Farber JP. Effects on breathing of rostral pons glutamate injection during opossum development. J Appl Physiol 1990; 69:189–195.
18. Glogowska M, Gromysz H. Respiratory activity of the pons and its influence on breathing in the guinea pig. Acta Neurobiol Exp 1988; 48:125–135.
19. Gromysz H. The effects of pontine or bulbar transection on the respiratory pattern in the rabbit. Acta Neurobiol Exp 1984; 44:239–247.
20. Hukuhara T Jr, Goto K, Takano K, Hattanmaru Y. Localization and functional organization of bulbar and pontine respiratory neurons in the rabbit. Adv Physiol Sci 1981; 10:579–586.
21. Fournier M, Feldman JL. Microinjections of excitatory amino acids into the region of the Kölliker-Fuse nucleus alter phrenic motoneuronal activity in the rat (abstr). Soc Neurosci 1988; 14:183.
22. Holmes HR, Herbert H, Saper CB. Parabrachial subnuclei in respiratory control (abstr). Soc Neurosci 1987; 13:1586.
23. Wang W, Fung M-L, St. John WM. Pontile regulation of ventilatory activity in the adult rat. J Appl Physiol 1993; 74:2801–2811.
24. Guyenet PG, Koshiya N, Huangfu D, et al. Central respiratory control of A5 and A6 pontine noradrenergic neurons. Am J Physiol 1993; 264:R1035–R1044.
25. Herbert H, Moga MM, Saper CB. Connections of the parabrachial nucleus with the nucleus of the solitary tract and medullary reticular formation in the rat. J Comp Neurol 1990; 293:540–580.
26. Otake K, Ezure K, Lipski J, Wong She RB. Projections from the commissural subnucleus of the nucleus of the solitary tract: An anterograde tracing study in the cat. J Comp Neurol 1992; 324:365–378.
27. Connelly CA, Otto-Smith MR, Feldman JL. Blockade of NMDA receptor-channels by MK-801 alters breathing in adult rats. Brain Res 1992; 596:99–110.
28. Di-Pasquale E, Morin D, Monteau R, Hilaire G. Serotonergic modulation of the respiratory rhythm generator at birth: An in vitro study in the rat. Neurosci Lett 1992; 143:91–95.
29. Hilaire G, Monteau R, Errchidi S. Possible modulation of the medullary respiratory rhythm generator by the noradrenergic A5 area: An in vitro study in the newborn rat. Brain Res 1989; 485:325–332.

30. Johnston BM, Gluckman PD. Lateral pontine lesions affect central chemosensitivity in unanesthetized fetal lambs. J Appl Physiol 1989; 67:1113–1118.
31. Smith JC, Ellenberger H, Ballanyi K, et al. Pre-Bötzinger complex: A brainstem region that may generate respiratory rhythm in mammals. Science 1991; 254:726–729.

49

Modulation of Respiratory Rhythm in the Goat by α_2-Adrenoceptors

MICHAEL S. HEDRICK, MELINDA L. RYAN, and GERALD E. BISGARD

University of Wisconsin
Madison, Wisconsin

The purpose of this study was to determine phrenic and recurrent laryngeal nerve (RLN) activities in response to intravenous administration of α_2-adrenoceptor agonists in anesthetized goats. Phrenic and RLN activities were used as indices of the inspiratory and postinspiratory (stage I expiratory) phases of the respiratory cycle, respectively. We hypothesized that prolonged phrenic apneas and respiratory arrhythmias that we have observed in awake and in anesthetized goats would be correlated with an augmentation of RLN postinspiratory activity rather than a widespread depression of breathing. Systemic infusions of the selective α_2 agonists clonidine or guanabenz resulted in phrenic apneas of varying lengths, and there was a maintained increase of tonic RLN activity throughout the length of the apnea, consistent with our hypothesis. The RLN exhibited a mixture of responses, including inspiratory, postinspiratory, and expiratory-related activities. The results suggest that α_2 agonists differentially affect spinal (phrenic) and cranial (RLN) motoneurons in a reciprocal fashion. Apneas induced by α_2 agonists are correlated with tonic postinspiratory-related activity of the RLN, which may reflect inhibition of medullary inspiratory neurons. Although the mechanism by which α_2-adrenoceptors augment postinspiratory-related RLN is unknown, we suggest that an endogenous α_2-mediated inhibitory tone, possibly noradrenergic in origin, to medullary postinspiratory neurons may be an important feature of this process. The withdrawal of an α_2-mediated inhibitory input would allow tonic activation of postinspiratory neurons which, in turn, inhibit inspiratory neurons and cause prolonged apneas. Control of postinspiratory-related activity has been implicated as an essential component in the maintenance of respiratory rhythm. Hence, our results are consistent with the three-phase model for mammalian respiratory rhythm generation.

I. Introduction

A current model for the generation of respiratory rhythm in mammals suggests that the ventilatory cycle comprises three distinct phases: inspiration, postinspiration (also called stage I expiration), and late (stage II) expiration (1,2). These three phases have been correlated with intracellular recordings from ventral brainstem neurons (1–6) and coincide with neural activities of phrenic (inspiratory), laryngeal (postinspiratory) and internal intercostal (late expiratory) motoneurons innervating muscles that control airflow during breathing (7–13).

A recent computer-generated model (14) of respiratory rhythmogenesis based on the three-phase model suggests that postinspiratory activity, in particular, has an "executive" role in the maintenance of respiratory rhythm; that is, changes in tonic drive to postinspiratory neurons exert a greater influence on respiratory rhythm than other neuronal types within the network (5,14,15). Experimental support for this concept includes studies showing that maneuvers to induce apnea, such as the stimulation of laryngeal afferents, activate medullary postinspiratory neurons and cause disturbances of respiratory rhythm (3,5,6,16).

Previous work from our laboratory indicates that intravenous administration of selective α_2-adrenergic agonists such as clonidine cause respiratory disturbances, including reductions of T_I and prolongations of T_E (apneas). These effects have been observed in awake (17) and anesthetized, artificially ventilated, vagotomized goats with peripheral chemoreceptor afferents removed, suggesting that the apneas are central in origin (18). However, it was not clear from those studies whether α_2 receptor agonists caused a widespread central depression of ventilatory effort or whether there was possibly a differential recruitment of expiratory-related activity that would be consistent with the three-phase model of respiratory rhythm described above. In the context of the network model proposed by Richter and colleagues (1,2,5), we hypothesized that apneas would be correlated with tonic activation of postinspiratory-related neurons. Efferent activity of the recurrent laryngeal nerve (RLN), which innervates vocal cord abductors and adductors, exhibits both inspiratory and postinspiratory-related activities in the cat and has been used as an index of medullary postinspiratory activity (7,11–13).

In this study we sought to determine if (a) the RLN of goats also exhibits similar respiratory-related activities compared with other species and, if so, (b) whether phrenic apneas caused by systemic administration of α_2 agonists are correlated with augmented postinspiratory-related activity of RLN. Our results suggest that pharmacological modulation of respiratory rhythm by α_2 receptors may involve changes in tonic activity to postinspiratory brainstem neurons, which is consistent with the three-phase network oscillator model.

II. Materials and Methods

A. Animal Preparation

A total of six goats were used in this study. After induction of anesthesia with sodium thiamylal (15 to 20 mg/kg) for intubation, animals were placed in dorsal recumbency under a heated blanket to maintain normal body temperature (38 to 40°C). Anesthesia was maintained with intravenous α-chloralose by continuous infusion (100 to 150 mg kg^{-1} h^{-1}). Femoral arterial and venous catheters were implanted for blood sampling and drug administration. The animals were bilaterally carotid body denervated.

The left whole recurrent laryngeal nerve (RLN) was identified immediately lateral to the trachea, cut distally, and desheathed. In three of the six animals, the contralateral vagus was also cut. The C6 phrenic nerve root was dissected free from surrounding tissue low in the neck, cut, and desheathed for neural recording. Each nerve was placed on bipolar platinum hook electrodes and immersed in mineral oil to prevent desiccation. Nerve activity was preamplified 10,000 times, filtered, and further amplified; the output was visualized on an oscilloscope (Tektronix 5113) and chart recorder (Gould TA 2000). The amplified signal was rectified and integrated to obtain a moving average of peak nerve activity.

B. Measurements

Once surgical preparations had been completed, the animals were paralyzed with a mixture of metubine iodide/pancruonium bromide and artificially ventilated. Respiratory frequency was set between 15 and 20 breaths/min and tidal volume between 300 and 500 mL, depending on the size of the animal. F_IO_2 was adjusted to maintain PaO_2 above 90 mm Hg. It was necessary to add CO_2 to the inspired gases so that breathing was above the apneic threshold for stable phrenic recording. Arterial blood gases were sampled frequently to ensure maintenance of blood gas and acid-base homeostasis during artificial ventilation.

C. Protocol

Once the surgery had been completed and the preparation stabilized, the apneic threshold for ventilation was first determined by lowering the inspired CO_2 level until phrenic activity disappeared. Pa_{CO_2} was determined at this point and $F_{I}CO_2$ was then raised until Pa_{CO_2} was approximately 5 to 8 torr above the threshold level and phrenic activity was stable. Drugs were given intravenously after stable neural activities were recorded for at least 5 min.

D. Drugs

Clonidine was dissolved in saline to obtain a stock solution (1 mg/mL), which was diluted 1:10 for IV administration. Stock solutions (1 mg/mL) of guanabenz and SKF-86466 were dissolved in dilute acetic acid and saline (1:100) before use. Clonidine and guanabenz were obtained from Sigma Chemicals (St. Louis, MO); SKF-86466 was obtained from Smith, Kline, Beecham (King of Prussia, PA).

III. Results

Phrenic and RLN activities were recorded simultaneously in all six animals. Because RLN activity was determined for whole nerve or branches contained within the main RLN trunk, we recorded from nerves bound for laryngeal adductors and abductors, and probably other upper airway muscles, known to be active during different phases of the respiratory cycle. Not surprisingly, we observed a variety of responses, which are illustrated in several neurograms (Figs. 1 to 4), that are informative for pointing out some general features of the responses that were obtained.

A. Phrenic and RLN Neural Activities

Representative neurograms showing phrenic and RLN activities from two different animals with intact vagi are shown in Figure 1. I normocapnia, RLN activity is apparent during the inspiratory phase, which clearly parallels phrenic activity. Moreover, activity occurs also in the postinspiratory phase; that is, the onset of activity starts at the end of phrenic activity, is confined to early expiration, and exhibits a decrementing profile (Fig. 1, upper panel). A second example (Fig. 1, lower panel) illustrates a different type of RLN activity that we obtained: expiratory-related activity of the RLN begins during the postinspiratory phase but is sustained throughout the entire expiratory phase.

B. Effects of α_2 Agonists on Phrenic and RLN Activities

Owing to the anatomical location of the bifurcation of the RLN and main vagal trunk, we could not perform a vagotomy on the ipsilateral side, from which

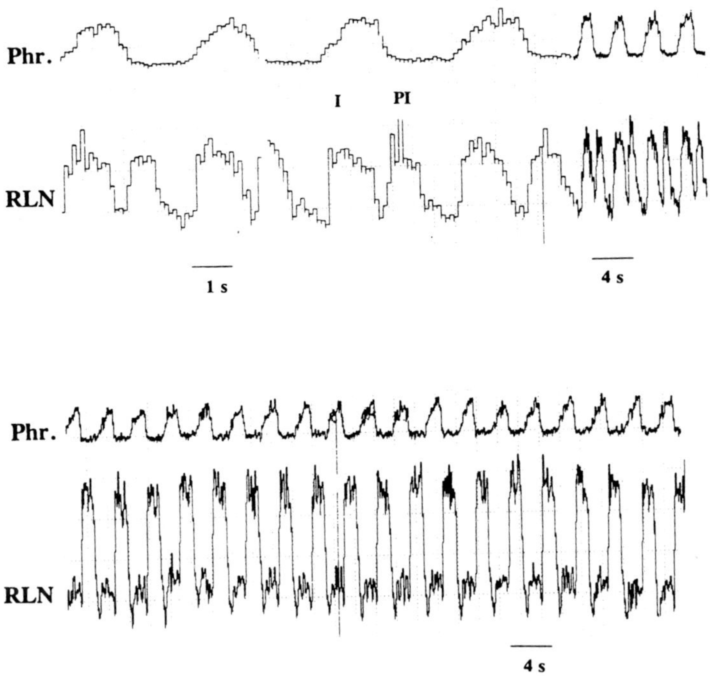

Figure 1 Examples phrenic (Phr.) and recurrent laryngeal nerve (RLN) efferent activities in two chloralose-anesthetized goats with intact vagi. In one animal (upper panel), RLN shows predominantly inspiratory (I) and postinspiratory (PI)-related activity. Inspiratory activity occurs in parallel with Phr. activity, while PI-related activity is confined to early expiration and exhibits a decrementing profile. In another animal (lower panel), RLN I-related activity is less pronounced and exhibits mainly expiratory-related activity, which spans the entire expiratory phase, with no clear PI-related component.

RLN activity was recorded. We have examined the effects of α_2 agonists on phrenic activity in animals that were bilaterally vagotomized and glomectomized (18). Figure 2 shows an example of the effects of intravenously administered clonidine (0.5 μg/kg) on phrenic activity from our previous study for comparison with vagi-intact, glomectomized goats in this study. These studies reveal no basic differences in the ventilatory responses to α_2 agonists in goats with or without intact vagi. In both groups, there is a slight yet significant reduction of T_I and a significantly increased T_E. Phrenic amplitude shows variable responses, but generally there is a trend for reductions of phrenic amplitude in both groups.

An example of phrenic and RLN responses to clonidine (0.5 μg/kg) in a vagi-intact goat is shown in Figure 3. Control phrenic and RLN activities are typically rhythmic and RLN activity is primarily confined to the expiratory

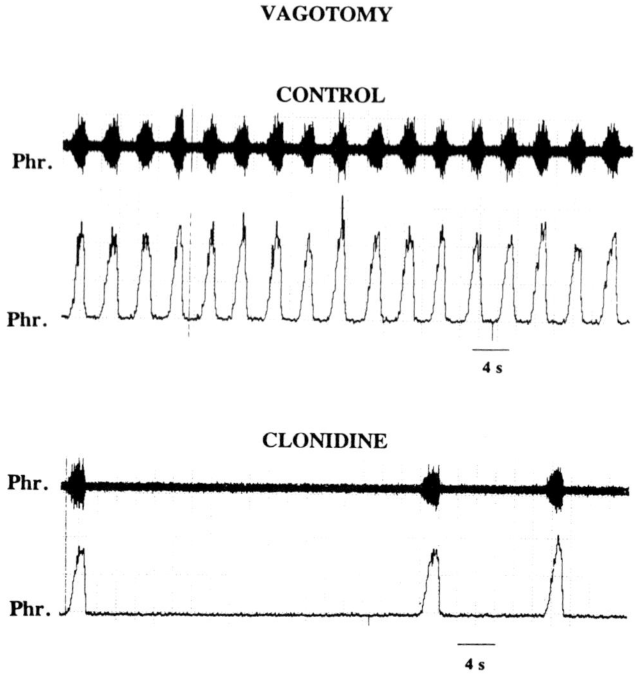

Figure 2 Effects of clonidine (0.5 μg/kg, IV) on Phr. activity in a vagotomized, anesthetized, glomectomized goat. These results were obtained in a previous study (18) and were used to compare with results obtained from vagi-intact goats in this study.

phase in this example (Fig. 3, top). Approximately 1 min after clonidine administration, there was a cessation of respiratory phrenic rhythm but a maintained increase in tonic RLN activity (Fig. 3, middle). Some 5 min after the initial dose of clonidine, phrenic activity resumed, but it was slower and more irregular (Fig. 3, bottom) than during the control period. After resumption of phrenic activity, the RLN shows reciprocal inhibition and excitation with the phrenic activity; that is, during inspiration, the RLN is inhibited and is then facilitated during the expiratory phase.

The effects of guanabenz (Fig. 4, top), another selective α_2 agonist, were similar to those of clonidine but required higher dosages (10 to 20 μg/kg). Inspiratory activity was clearly inhibited and associated with a tonic increase of RLN activity during apneas. Respiratory rhythm could be restored with SKF-86466 (19), a highly selective α_2 antagonist (Fig. 4, bottom). The antagonist returned RLN activity to a more phasic pattern with a reduction in expiratory-related activity.

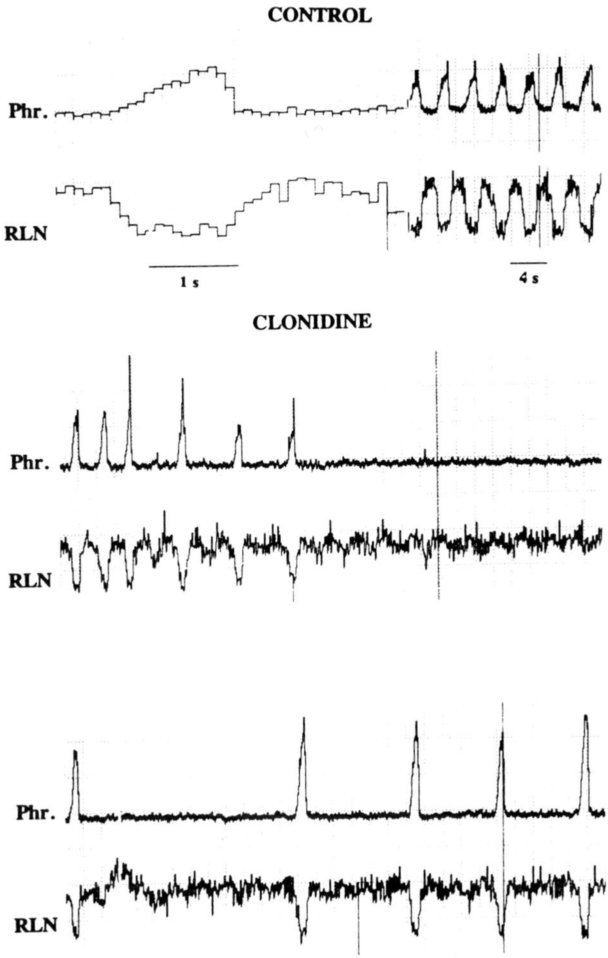

Figure 3 Effects of clonidine (0.5 μg/kg, IV) on Phr. and RLN activities in a vagi-intact goat. Control activities are shown in the top panel, followed by the response approximately 1 min (middle panel) and 5 min (bottom panel) after clonidine injection. Note the cessation of Phr. activity coinciding with tonic activation of RLN activity after 1 min. Arrhythmic breathing with prolonged apneas is apparent after 5 min, which also are correlated with tonic RLN activity.

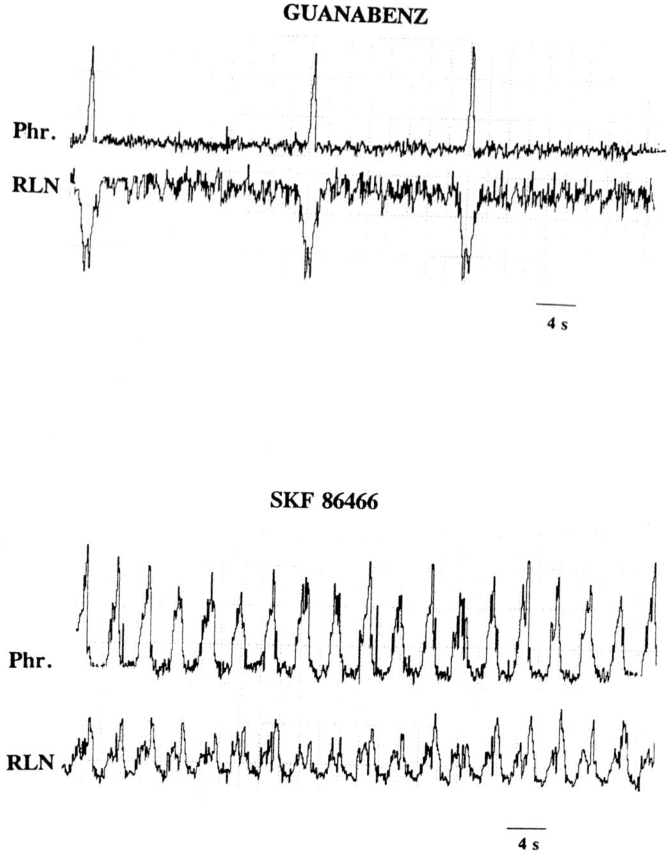

Figure 4 Effects of a different α_2 agonist, guanabenz (20 μg/kg, IV), on Phr. and RLN activities in an anesthetized goat (upper panel). Note similarity with effects of clonidine shown in Fig. 3. A selective α_2-antagonist, SKF-86466, was given to reverse the effects of guanabenz (lower panel). After the antagonist, there was a significant reduction of postinspiratory-related activity of RLN.

Changes in respiratory timing were consistent in all six goats and are summarized in Figure 5. T_I was significantly reduced after clonidine and there was a profound increase of T_E. This is consistent with our previous results using goats that were bilaterally vagotomized (18).

IV. Discussion

There are two important findings from our study: (a) RLN activity in goats exhibits inspiratory and postinspiratory-related activity—in some cases there

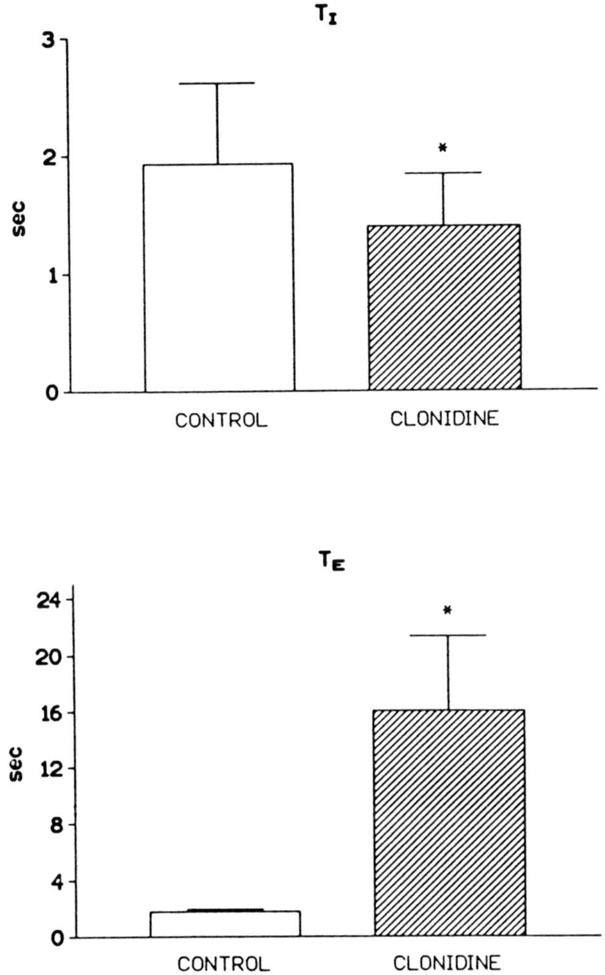

Figure 5 Summary of the effects of clonidine on inspiratory (T_I) and expiratory (T_E) durations in six anesthetized goats. There were significant reductions of T_I and increases of T_E ($P<0.05$; Wilcoxon paired sample test).

is activity that spans entirely the expiratory phase, and (b) α_2 agonists affect inspiratory and expiratory-related motoneuronal activities in a reciprocal fashion; that is, arrhythmic breathing and phrenic apneas were correlated with tonic increases of RLN activity throughout the apneic period. Our interpretation of these results is that augmented RLN activity is indicative of increased central postinspiratory activity, which may account for apneic breathing and respiratory disturbances. The data are consistent with the three-phase model for respiratory rhythmogenesis (1,2) in that central postinspiratory neuronal activity is a major determinant of respiratory rhythm.

A. Phrenic and RLN Activities

The RLN in goats exhibited a mixture of activities, which is not surprising considering that we recorded from whole nerve rather than isolating and recording from branches to specific laryngeal muscles. We therefore cannot draw firm conclusions about RLN activity as an index of any particular phase of the respiratory cycle. Zhou et al. (13) have demonstrated the importance of recording efferent activities of branches within the RLN that innervate particular laryngeal muscles. This nerve carries motor fibers innervating both vocal cord abductors (e.g., posterior cricoarytenoid, or PCA) and adductors (e.g., thyroarytenoid, or TA). Vocal cord abduction and adduction are important for regulating airflow during inspiration and early (stage I) expiration, respectively. This has been confirmed with electromyographic (EMG) data and neural recordings from PCA and TA muscles in awake humans and anesthetized animals (13,20–23).

Despite the limitation of recording whole RLN activity in this study, the results are generally consistent with several other studies in decerebrate cats (10–13). We observed both inspiratory- and expiratory-related activity in whole RLN; however, in the majority of preparations, we obtained mainly expiratory-related activity. The onset of this expiratory activity began at or immediately following peak phrenic activity, in most cases was confined to the early (postinspiratory) portion of the expiratory phase, and exhibited a decrementing profile (Fig. 1, top). Taken together, the data strongly suggest that postinspiratory-related activity is present in the RLN of anesthetized goats. In some preparations, much of this activity spanned the entire expiratory phase (Fig. 1, lower panel), which has also been noted in cats (13).

B. Effects of α_2 Agonists

Few studies have examined the effects of α_2-adrenoceptor stimulation or blockade on motor nerve activities in anesthetized animals. The general effect noted, however, is one of respiratory depression. For example, McCrimmon and Lalley (24) demonstrated that phrenic nerve activity was completely inhibited by clonidine in artificially ventilated cats. They did not observe any disruption of respiratory rhythm, but this may be related to the higher dosages used in their study (ca. 50 μg/kg) compared with this study. Higher doses of clonidine in anesthetized goats also inhibit phrenic activity (unpublished observations). Clonidine has also been shown to inhibit activity of medullary respiratory-related neurons within the ventral respiratory group (25,26). Our phrenic recordings are therefore consistent with previous studies showing that α_2 agonists cause respiratory (i.e., inspiratory) depression.

The most unique finding from our study is that clonidine and guanabenz are tonically active the RLN during the expiratory phase (Figs. 3 and 4). This is particularly surprising in light of the inhibitory effects mediated by α_2-receptor agonists in other studies. Stimulation of α_2 receptors has been shown to hyperpolarize CNS neurons (see Ref. 27). Our interpretation of the results is that α_2 agonists cause, by an as yet unidentified mechanism, an increase of tonic postinspiratory-related brainstem neuronal activity. We suggest that RLN activation is likely to be caused by a central α_2-mediated pathway rather than a peripheral reflex mechanism. The similarity of phrenic responses to α_2 agonists between vagi-intact and vagotomized goats (Figs. 2 and 3) and lack of carotid body chemoreceptor afferents rule out most feedback pathways that could result in augmented postinspiratory activity.

Several studies have shown that electrical stimulation of superior laryngeal nerve (SLN) afferents powerfully inhibits breathing and causes sustained apneas (e.g., Refs. 10,11,28); furthermore, SLN stimulation depolarizes medullary postinspiratory neurons throughout the apnea (3,5,16). This is similar to our findings of augmented RLN postinspiratory activity after administration of α_2 agonists. Because postinspiratory neurons have been shown to synaptically inhibit inspiratory and late-expiratory neurons (1,15), it is thought that postinspiratory neurons have an overall "executive" role in the brainstem respiratory oscillator network. A recent mathematical model based on this concept suggests that respiratory rhythm is highly dependent upon changes in tonic activity to postinspiratory neurons (14). The results from this study essentially support this concept, because tonic activation of RLN (presumably) postinspiratory-related activity by α_2 receptor agonists caused prolonged T_E and respiratory disturbances, as predicted by this model. Our results further imply that α_2-adrenoceptor pathways perhaps provide one source of tonic input to postinspiratory neurons (see below).

The differential and antagonistic influence of α_2 agonists on phrenic and RLN activities is similar to the differential recruitment patterns of inspiratory and expiratory motoneurons with changes in central or peripheral chemoreceptor drive. For example, cessation of phrenic activity by superior laryngeal stimulation (28) or hypocapnia (29) augments expiratory-related activity of bulbospinal or intercostal motoneurons in cats. Similarly, augmentation of phrenic activity by hypoxia inhibits abdominal expiratory nerve activity (30). The reasons for these reciprocal changes in inspiratory and expiratory activity are not entirely clear but are thought to stem primarily from differential effects of general anesthesia on central outflow from cranial and bulbospinal motoneurons to upper airway and intercostal muscles. It should be mentioned that the three-phase model of respiratory rhythmogenesis (2) predicts no overlap between inspiratory, postinspiratory, and late-expiratory phases of the respiratory cycle. Evidence from anesthetized animals generally supports this prediction (7,10–12); however, in awake dogs (31,32) and goats (33), there is clear overlap of respiratory EMG activities during the inspiratory-postinspiratory transition with central or peripheral chemoreceptor stimulation.

The neonatal rat brainstem preparation in vitro has provided additional evidence that α_2-adrenoceptors are involved in respiratory rhythmogenesis. Hilaire et al. (34) showed that the pontine A5 (noradrenergic) group inhibits the medullary rhythm generator, since medullary respiratory frequency increased after removal of the pons or injecting the A5 region with α_2 antagonists norepinephrine (NE) applied to the brainstem preparation both increases and decreases respiratory frequency (35); NE isolated to the pons consistently increases frequency, while NE applied to the medulla decreases frequency. Removing the pons also increases the respiratory frequency of the medullary oscillator (36). Clonidine mimicked the NE-induced increase in frequency when applied to the pons but had no effect when applied to the medulla (35). These results suggest a pontine inhibitory input to the medullary respiratory rhythm generator, possibly from the A5 noradrenergic group, in the neonatal rat brainstem. A significant reduction in the activity of "preinspiratory" neurons was noted in the rostral ventrolateral medulla after clonidine application (37), an effect that was blocked with forskolin, an activator of adenylate cyclase, suggesting that cAMP plays a role in mediating the effects of clonidine on respiratory rhythm. Thus, although there is evidence for α_2 receptor–mediated effects on respiratory rhythm, more studies are needed to clarify the mechanisms and pathways involved.

C. Significance of α_2-Adrenoceptors in Modulation of Respiratory Rhythm

Alpha$_2$ receptors in the CNS are generally involved in presynaptic autoregulation of NE, and stimulation of these receptors with agonists such as clonidine causes widespread reductions of NE turnover and release from neurons (38). The central effects of NE release, when mediated by α_2 receptors, have been shown to hyperpolarize neurons at pontine and medullary respiratory-related sites such as the locus ceruleus (LC) (39), the dorsal motor nucleus of the vagus (40), and the ventral respiratory group (26). Given that α_2-mediated effects cause hyperpolarization, the results from this study showing postinspiratory-related depolarization by α_2 agonists suggests one of two likely possibilities: (a) depolarization of postinspiratory neurons by a direct excitatory effect or (b) a withdrawal of an inhibitory input that unmasks a tonic excitation. We favor the second possibility, since α_2-receptor agonists are known to reduce NE release and turnover from presynaptic terminals (41). It is unknown whether NE augments or inhibits postinspiratory neurons, but a recent study on medullary inspiratory neurons of the nucleus ambiguus showed opposing effects of NE and serotonin (5-HT): NE caused inhibition while 5-HT excited inspiratory neurons (25). Alpha$_2$ agonists such as clonidine acting at presynaptic NE terminals could reduce the release of NE and thereby remove an inhibitory noradrenergic input, allowing tonic excitatory inputs to depolarize these neurons. There is evidence for serotonergic (25) and glutamatergic (42) excitation of

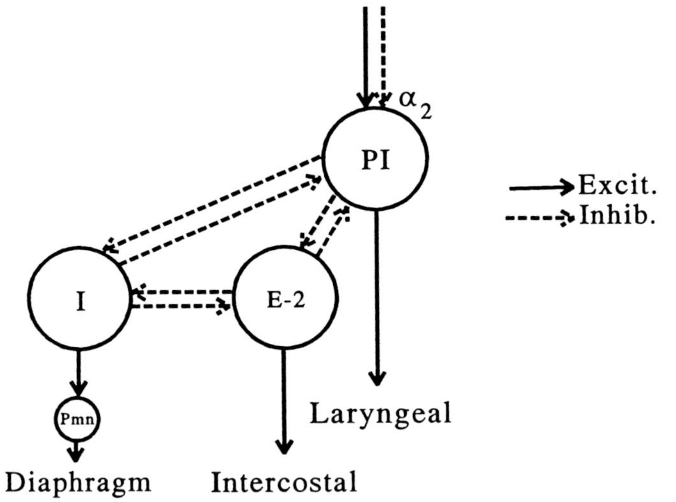

Figure 6 A hypothesis for the possible role for α_2-adrenergic receptors in the control of respiratory rhythm incorporating the three-phase network model for respiratory rhythmogenesis (1,2). Respiratory rhythm is initiated with inspiratory (I) neurons, which excite the phrenic motor nucleus (Pmn) and the diaphragm. The expiratory phase is illustrated by early (postinspiratory; PI) and late (E-2) expiratory neurons with motoneuronal outflows to laryngeal and internal intercostal muscles, respectively. I, PI, and E-2 neurons are mutually inhibitory (dashed lines) to each other and transmit excitatory (solid lines) outflow to respiratory muscles. We hypothesize that α_2-adrenergic receptors control a source of tonic inhibitory input to PI neurons (see text for details).

neurons within the nucleus ambiguus that could provide a source of tonic drive to postinspiratory neurons. Because clonidine strongly inhibits pontine A5 and LC neurons (see above) and neurons in these locations exhibit postinspiratory-related activity (43), a possible pontine noradrenergic inhibitory pathway to postinspiratory neurons in the medulla may be present. As mentioned earlier, there is evidence for this hypothesis in the neonatal rat brainstem model (34–36).

Figure 6 illustrates a hypothesis for the incorporation of α_2-adrenergic control of respiratory rhythm within the context of the three-phase network oscillator model (1,2). In this scheme, control of respiratory rhythm is achieved in part by a balance of excitatory and inhibitory inputs to medullary postinspiratory neurons. We hypothesize that α_2-adrenoceptors serve to control an inhibitory, possibly noradrenergic, pathway to medullary postinspiratory neurons. Changes in tonic drive to postinspiratory neurons constitute a key feature of the "executive" role for these neurons affecting respiratory rhythm (14). It has also been suggested that inhibitory and excitatory inputs provide the source of "conditional" drive to pacemaker cells for central rhythm generation of breathing in the neonatal rat brainstem in vitro (44,45). Whatever the mechanisms involved in the initiation and maintenance of central respiratory rhythm, α_2-receptor-mediated pathways appear to significantly contribute to its modulation.

Acknowledgments

Supported by USPHS grants HL-15473, HL-07654, and HL-08809. M. Hedrick is a NHLBI-supported postdoctoral trainee and the recipient of a National Research Service Award. The authors thank Gordon Johnson and Josue Pizarro for their excellent technical assistance.

References

1. Richter DW. Generation and maintenance of the respiratory rhythm. J Exp Biol 1982; 100:93–107.
2. Richter DW, et al. How is the respiratory rhythm generated? A model. News Physiol Sci 1986; 1:109–112.
3. Czyzyk-Krzeska MF, Lawson EE. Synaptic events in ventral respiratory neurones during apnoea induced by laryngeal nerve stimulation in neonatal pig. J Physiol (Lond) 1991; 436:131–147.
4. Lawson EE, et al. Peripheral chemoreceptor inputs to medullary inspiratory and postinspiratory neurons of cats. Pflügers Arch 1989; 414:523–533.
5. Remmers JE, et al. Reflex prolongation of stage I of expiration. Pflügers Arch 1986; 407:190–198.
6. Zheng Y, et al. Are the post-inspiratory neurons in the decerebrate rat cranial motoneurons or interneurons? Brain Res 1991; 551:256–266.
7. Dick TE, et al. Interaction between central pattern generators for breathing and swallowing in the cat. J Physiol (Lond) 1993; 465:715–730.
8. Huang Q, et al. Influence of lung volume on activities of branches of the recurrent laryngeal nerve. J Appl Physiol 1989; 67:1179–1184.
9. Hwang J-C, et al. Influence of pulmonary inflations on discharge patterns of phrenic motoneurons. J Appl Physiol 1987; 63:1421–1427.
10. Oku Y, et al. Phase-dependent dynamic responses of respiratory cycle motor activities following perturbation of the cycle in the cat. J Physiol (Lond) 1993; 461: 321–337.
11. St John WM, Zhou D. Differing control of neural activities during various portions of expiration in the cat. J Physiol (Lond) 1989; 418:189–204.
12. St John WM, Zhou D. Discharge of vagal pulmonary receptors differentially alters neural activities during various stages of expiration in the cat. J Physiol (Lond) 1990; 424:1–12.
13. Zhou D, et al. Respiratory activities of intralaryngeal branches of the recurrent laryngeal nerve. J Appl Physiol 1989; 67:1171–1178.
14. Ogilvie MD, et al. A network model of respiratory rhythmogenesis. Am J Physiol 1992; 263:R962–R975.
15. Richter DW, et al. The differential organization of medullary post-inspiratory activities. Pflügers Arch 1987; 410:420–427.

16. Lawson EE, et al. Respiratory neuronal activity during apnea and other breathing patterns induced by laryngeal stimulation. J Appl Physiol 1991; 70:2742–2749.
17. Hedrick MS, et al. Clonidine induces arrhythmic breathing in the awake goat (abstr). FASEB J 1992; 6:A1826.
18. Hedrick MS, et al. Alpha-2 adrenergic control of breathing in the anesthetized goat (abstr). FASEB J 1993; 7:A396.
19. Hieble JP, et al. Selective alpha-2 adrenoceptor blockade by SKF-86466: In vitro characterization of receptor selectivity. J Pharmacol Exp Ther 1986; 236:90–96.
20. Bartlett D Jr, et al. Laryngeal regulation of respiratory airflow. Respir Physiol 1973; 18:194–204.
21. Hutchison AA, et al. Laryngeal and diaphragmatic muscle activities and airflow patterns after birth in premature lambs. J Appl Physiol 1993; 75:121–131.
22. Kuna ST, et al. Thyroarytenoid muscle activity during wakefulness and sleep in normal adults. J Appl Physiol 1988; 65:1332–1339.
23. Kuna ST, et al. Posterior cricoarytenoid muscle activity during wakefulness and sleep in normal adults. J Appl Physiol 1990; 68:1746–1754.
24. McCrimmon DR, Lalley PM. Inhibition of respiratory neural discharges by clonidine and 5-hydroxytryptophan. J Pharmacol Exp Ther 1982; 222:771–777.
25. Arita H, Ochiishi M. Opposing effects of 5-hydroxytryptamine on two types of medullary inspiratory neurons with distinct firing patterns. J Neurophysiol 1991; 66:285–292.
26. Champagnat J, et al. Catecholaminergic depressant effects on bulbar respiratory neurons. Brain Res 1979; 160:57–68.
27. Nicoll RA, et al. Functional comparison of neurotransmitter subtypes in mammalian central nervous system. Physiol Rev 1990; 70:513–565.
28. Jodkowski JS, Berger AJ. Influences from laryngeal afferents on expiratory bulbospinal neurons and motoneurons. J Appl Physiol 1988; 64:1337–1345.
29. Sears TA, et al. Reciprocal tonic activation of inspiratory and expiratory motoneurons by chemical drives. Nature 1982; 299:728–730.
30. Fregosi RF, et al. Hypoxia inhibits abdominal expiratory nerve activity. J Appl Physiol 1987; 63:211–220.
31. Saupe KW, et al. Respiratory muscle recruitment during selective central and peripheral chemoreceptor stimulation in awake dogs. J Physiol (Lond) 1992; 448: 613–631.
32. Smith CA, et al. Differential responses of expiratory muscles to chemical stimuli in awake dogs. J Appl Physiol 1989; 66:384–391.
33. Smith CA, et al. Effects of specific carotid body and brain hypoxia on respiratory muscle control in the awake goat. J Physiol (Lond) 1993; 460:623–640.
34. Hilaire G, et al. Possible modulation of the medullary respiratory rhythm generator by the noradrenergic A5 area: An in vitro study in the newborn rat. Brain Res 1989; 485:325–332.
35. Errchidi S, et al. Noradrenergic modulation of the medullary respiratory rhythm generator in the newborn rat: An in vitro study. J Physiol Lond 1991; 443:477–498.
36. Errchidi S, et al. Permanent release of noradrenaline modulates respiratory frequency in the newborn rat: An in vitro study. J Physiol Lond 1990; 429:497–510.
37. Arata A, et al. Effects of cAMP on respiratory rhythm generation in brainstem-spinal cord preparation from newborn rat. Brain Res 1993; 605:193–199.

38. Arbilla S, Langer SZ. The regulation of neurotransmitter release by α_2 adrenoceptors in the central nervous system. In: Heal DJ, Marsden, CA, eds. The Pharmacology of Noradrenaline in the Central Nervous System. New York: Oxford University Press, 1990:141–154.
39. Andrade R, Aghajanian GK. Opiate and α_2-adrenoceptor-induced hyperpolarization of locus ceruleus neurons in brain slices: Reversal by cyclic adenosine 3′:5′-monophosphate analogues. J Neurosci 1985; 5:2359–2364.
40. Fukuda A, et al. The effects of noradrenaline on neurones in the rat dorsal motor nucleus of the vagus, in vitro. J Physiol (Lond) 1987; 393:213–231.
41. Nichols AJ. α-Adrenoceptor signal transduction mechanisms. In: Ruffolo RR, ed. α-Adrenoceptors: Molecular Biology, Biochemistry and Pharmacology. Vol. 8. Prog Basic Clin Pharmacol. Basel: Karger, 1991:44–74.
42. Reddad S, et al. Modification de l'activitie respiratoire des motoneurones larynges par application ionophoretique d'acides amines excitateurs, chez le chat (abstr). J Physiol (Paris) 1987; 82:54A.
43. Guyenet PG, et al. Central respiratory control of A5 and A6 pontine noradrenergic neurons. Am J Physiol 1993; 264:R1035–R1044.
44. Feldman JL, Smith JC. Cellular mechanisms underlying modulation of breathing pattern in mammals. Ann NY Acad Sci 1989; 563:114–130.
45. Feldman JL, et al. Neurogenesis of respiratory rhythm and pattern: Emerging concepts. Am J Physiol 1990; 259:R879–R886.

50

Multifunctional Ventral Brainstem Respiratory Neurons

Reorganization of Their Activities to Produce Vomiting

ALAN D. MILLER

Rockefeller University
New York, New York

I. Summary

Vomiting is a highly complex behavior that is generated primarily by the coordinated contraction of the major respiratory muscles. These muscles are activated in different patterns during respiration and vomiting. In particular, the diaphragm and abdominal muscles co-contract in a series of bursts of activity during both the retching and expulsion phases of vomiting. This review considers how the activity of brainstem respiratory neurons is reorganized during vomiting. Many of these neurons take an active role in generating the pattern of respiratory muscle discharge during vomiting as well as during breathing and thus can be considered to be multifunctional neurons. Other respiratory neurons are silenced during vomiting, presumably because their discharge would interfere with the unique pattern of respiratory muscle activation that occurs during vomiting. In addition, not all of the neurons in the vomiting pattern generator have yet been identified, including those that provide excitatory drive to phrenic motoneurons.

II. Introduction

Nausea and vomiting (emesis) occur frequently due to a variety of causes, including cancer chemo- and radiation therapy, surgery, pregnancy, motion and space sickness, and various diseases. The input mechanisms by which vomiting is triggered are better understood for some conditions (cancer chemotherapy, radiation) than for others (pregnancy, postsurgery, delayed emesis following chemotherapy) (1–3). Lesion studies have demonstrated that the essential coordinating circuitry for producing vomiting is located within the medulla of the brainstem. The exact organization of this coordinating circuitry is unknown and somewhat controversial. Recent studies of the distribution of cells expressing c-*fos* immunoreactivity (a marker of cell activation) in response to drug-induced emesis indicate that neurons involved in coordinating vomiting radiate from the area postrema and nucleus of the solitary tract to an arc of neurons in the lateral tegmental field implicated in somatoautonomic integration (4).

III. The Motor Act of Vomiting

Vomiting is produced primarily by intrathoracic and intraabdominal pressure changes that are generated by the coordinated action of the major respiratory muscles. These changes are accompanied by the integrated activation of the gastrointestinal tract and upper airway and postural musculature (2,5). The patterns of activity of these muscles during the retching and expulsion phases of vomiting are illustrated schematically in Figure 1. Note in particular that the diaphragm and abdominal muscles, which are activated out of phase with each other during breathing, co-contract in a series of bursts of activity during vomiting. The expulsion phase can be divided into two parts. During the first, the diaphragm and abdominal muscles co-contract, as during retching. During the second part of expulsion, phrenic discharge is inhibited while abdominal discharge continues. The periesophageal portion of the crural diaphragm relaxes during expulsion and to a lesser extent during retching, facilitating rostral movement of gastric contents. This relaxation is part of the central motor program for vomiting and does not depend on movement of vomitus within the gastrointestinal tract (6). Intercostal activity is not illustrated in Figure 1 since different patterns of discharge have been reported, presumably reflecting a complex and variable pattern of activity during vomiting. The predominant pattern observed in our laboratory in decerebrate, paralyzed cats is that the nerves to the external intercostal muscles are mainly active in phase with bursts of coactive phrenic and abdominal discharge, while those to the internal intercostals are mainly active between retching bursts and during the expulsion phase (7). The activity of the upper airway muscles during vomiting has been described in detail (8).

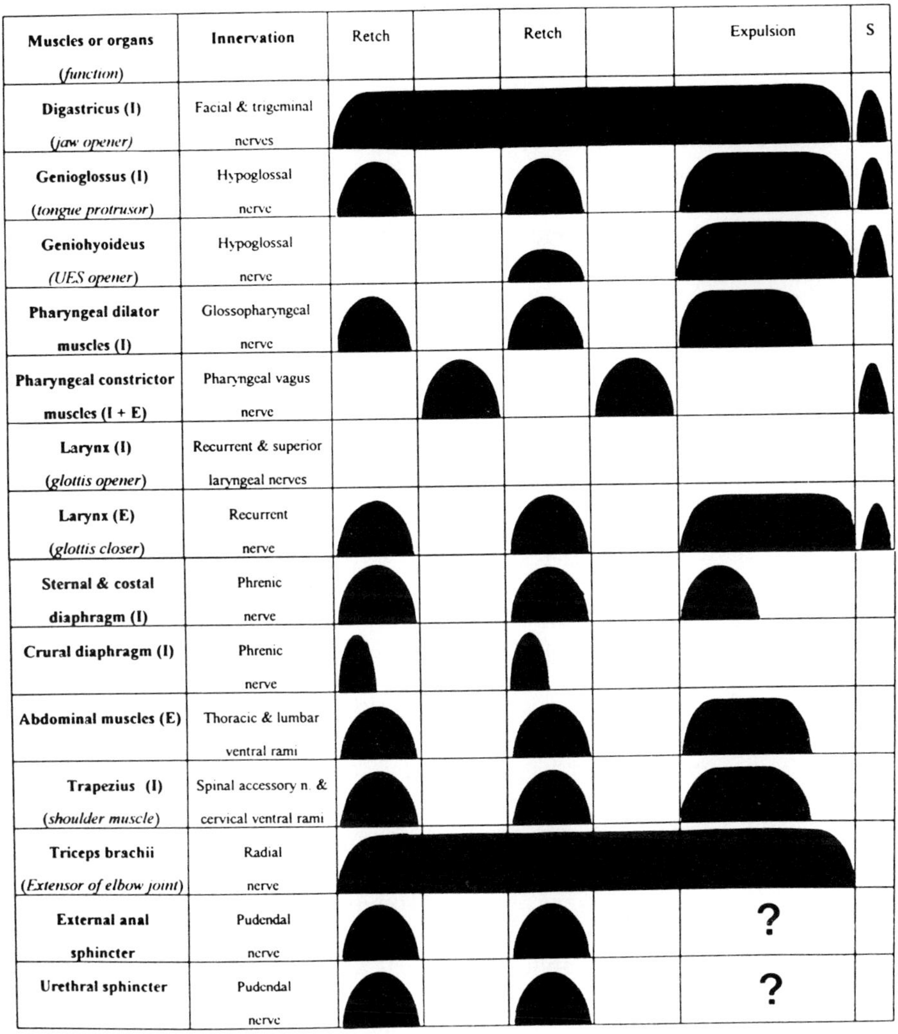

Figure 1 Schematic representation of muscle activity during vomiting. The episode illustrated consists of two retches and an expulsion, followed by swallowing (S). Timing of muscle activation is indicated in black. I and E indicate inspiratory and expiratory activities during breathing. No data indicated by "?". Abbreviations: n, nerve; UES, upper esophageal spincter. (From Ref. 5.)

IV. Activity of Medullary Respiratory Neurons During Vomiting

The contributions of brainstem respiratory neurons to the generation of respiratory muscle discharge during (fictive) vomiting are being investigated in cats that are decerebrated, paralyzed, and artificially ventilated. Fictive vomiting is identified by the same characteristic pattern of respiratory muscle nerve discharge that would generate retching and expulsion in nonparalyzed animals (9). Some types of medullary respiratory neurons are actively involved in generating the muscle discharge pattern during vomiting, while other neurons are silenced. This difference is strikingly illustrated in the firing patterns of expiratory and inspiratory medullary output neurons that convey excitatory drive to spinal respiratory motoneurons (Fig. 2). Augmenting (AUG) expiratory (E) bulbospinal neurons in the caudal ventral respiratory group (VRG) fire in two different patterns appropriate to drive abdominal or internal intercostal motoneurons during vomiting (9). One subgroup of these neurons fires in phase with abdominal and coactive phrenic discharge (E-AUG neurons in Fig. 2). Cutting the axons of these neurons by a midsagittal section between the obex and C1 abolishes abdominal discharge during both respiration and fictive vomiting (10). A second group of these E-AUG neurons fires when internal intercostal motoneurons are active between retching bursts and during expulsion (not illustrated). In contrast, most inspiratory (I) bulbospinal neurons in both the dorsal and ventral respiratory groups are actively inhibited and mainly silent during vomiting (I-AUG neurons in Fig. 2) (11,12). The medullary neurons responsible for exciting phrenic and external intercostal motoneurons during vomiting remain unidentified.

The discharge patterns of excitatory bulbospinal neurons are produced in part by inhibitory respiratory neurons with extensive medullary connections (13,14) that fire during different phases of vomiting (Fig. 2). Many inspiratory neurons with a decrementing (DEC) discharge pattern in the rostral VRG fire during coactive phrenic and abdominal bursts and thus could inhibit inspiratory bulbospinal neurons, as well as other cell types, during this phase of vomiting (15). Another subgroup of I-DEC neurons is silent throughout the vomiting episode (not illustrated). These I-DEC neurons may normally inhibit those caudal VRG E-AUG neurons that drive abdominal motoneurons; thus, their continued firing would interfere with abdominal discharge during vomiting. Another type of inhibitory neuron, augmenting expiratory neurons located in the Bötzinger complex of the rostral VRG, fires between coactive phrenic and abdominal nerve bursts and could inhibit phrenic motoneurons, inspiratory bulbospinal neurons, and other medullary neurons during this phase (BOT E-AUG neurons in Fig. 2) (16). In addition to being inhibited between retching bursts, phrenic motoneurons are also inhibited during the second phase of expulsion (17). The firing behavior of Bötzinger E-AUG neurons seems insufficient

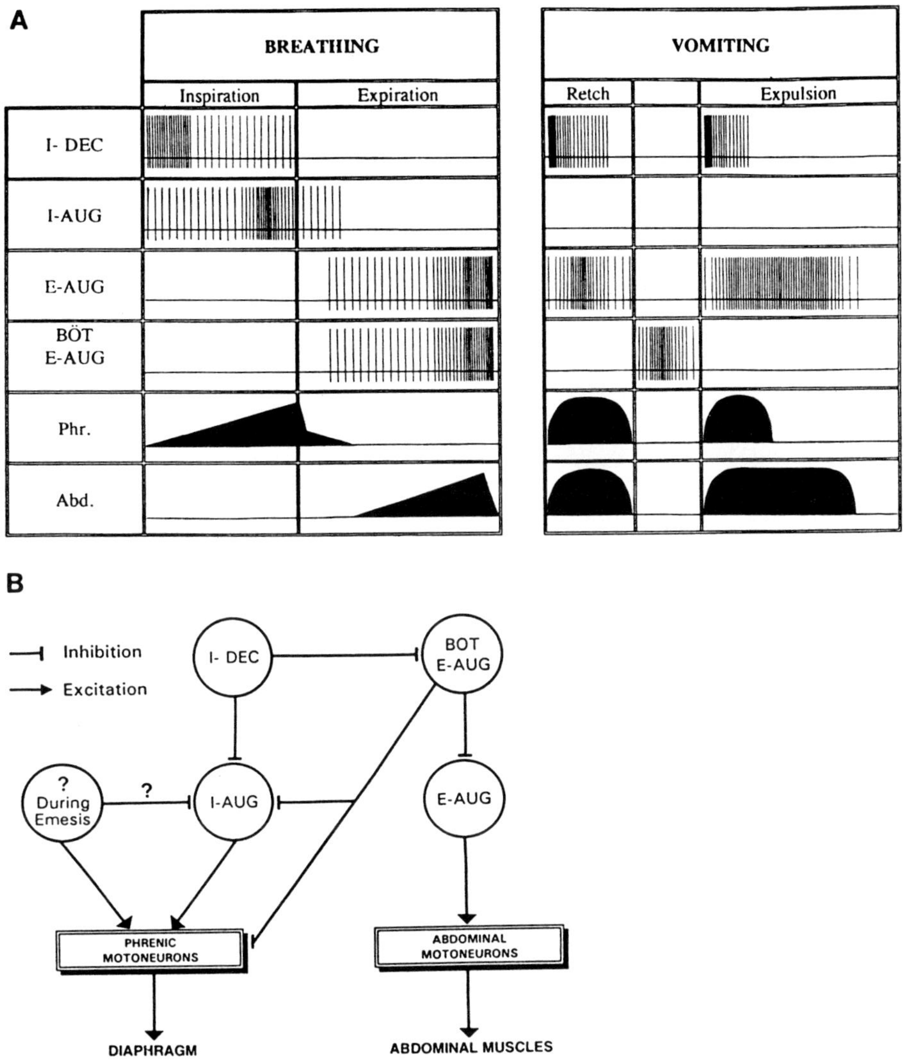

Figure 2 Simplified schematic representation of the activities of four types of medullary respiratory neurons during breathing and vomiting (A) and their connections (B). Note that the behaviors of the two expiratory (E) and two inspiratory (I) cell types are quite different during vomiting. Abbreviations: AUG, augmenting; BOT, Bötzinger; DEC, decrementing. Some cell types have been omitted from the figure for simplicity. (From Ref. 5.)

to produce this expulsion-related phrenic inhibition; thus additional yet unknown mechanisms must be involved.

Other propriobulbar cell types are mainly silenced during vomiting, presumably because their discharge would impair vomiting-related activation of spinal or cranial motoneurons (15). Included are decrementing expiratory neurons, which are known to inhibit vagal expiratory motoneurons that fire during vomiting, and most inspiratory neurons with a constant discharge pattern, some of which are known to excite inspiratory bulbospinal neurons and vagal motoneurons. The small percentage of inspiratory constant neurons that increase their discharge during vomiting may excite upper airway motoneurons (hypoglossal, glossopharyngeal) that are active during inspiration and vomiting; however, possible connections between these cell types have not been investigated.

At the level of the spinal cord, most inspiratory propriospinal neurons in the upper cervical cord (C1-C3) fire during vomiting (18). However, dysfunction of this cell group caused by the application of kainic acid has shown that these neurons are not essential for apparently normal activation of phrenic, intercostal, or abdominal motoneurons during vomiting (7). The functional significance of upper cervical inspiratory neurons remains unknown. At the level of the phrenic motor nucleus, there are respiratory-related interneurons that exhibit different responses during vomiting and presumably contribute to the output pattern observed on the motoneurons (19).

In sum, the activity of brainstem respiratory neurons during vomiting is complex. Many multifunctional neurons appear to take an active role in generating the pattern of respiratory muscle discharge during vomiting as well as breathing. These include bulbospinal VRG excitatory augmenting expiratory neurons, bulbospinal BOT inhibitory augmenting expiratory neurons, and many propriobulbar VRG inhibitory decrementing inspiratory neurons. Other respiratory neurons are mainly silenced during vomiting, presumably because their discharge would interfere with the unique pattern of respiratory muscle activation that occurs during vomiting. Included in this group are most bulbospinal excitatory augmenting inspiratory neurons and VRG propriobulbar neurons, including inhibitory decrementing expiratory neurons, most but not all excitatory constant inspiratory neurons, and a subgroup of decrementing inspiratory neurons. Still other as yet unidentified neurons are important for some components of the vomiting pattern generator, including excitation of phrenic motoneurons during retching and expulsion and inhibition of these motoneurons during the second phase of expulsion.

Acknowledgment

The author's work is supported by grant NS20585 from the National Institute of Neurological Disorders and Stroke.

References

1. Kucharczyk J, Stewart DJ, Miller AD, eds. Nausea and Vomiting: Recent Research and Clinical Advances. Boca Raton, FL: CRC Press, 1991.
2. Bianchi AL, Grélot L, Miller AD, King GL, eds. Mechanisms and Control of Emesis. Montrouge, France: Colloques INSERM/John Libbey Eurotext, 1992.
3. Sleisenger MH, ed. The Handbook of Nausea and Vomiting. Pawling, NY: Caduceus Medical Publishers Inc., 1993.
4. Miller AD, Ruggiero DA. Emetic reflex arc revealed by expression of the immediate-early gene cfos in the cat. J Neurosci 1994; 14:871–888.
5. Grélot L, Miller AD. Vomiting—Its ins and outs. News Physiol Sci 1994; 9:142–147.
6. Miller AD, Lakos SF, Tan LK. Central motor program for relaxation of periesophageal diaphragm during the expulsion phase of vomiting. Brain Res 1988; 456: 367–370.
7. Miller AD, Yates BJ. Evaluation of role of upper cervical inspiratory neurons in respiration, emesis and cough. Brain Res 1993; 606:143–147.
8. Grélot L, Barillot JC, Bianchi AL. Activity of respiratory-related oropharyngeal and laryngeal motoneurons during fictive vomiting in the decerebrate cat. Brain Res 1990; 513:101–105.
9. Miller AD, Tan LK, Suzuki I. Control of abdominal and expiratory intercostal muscle activity during vomiting: Role of ventral respiratory group expiratory neurons. J Neurophysiol 1987; 57:1854–1866.
10. Miller AD, Nonaka S. Mechanisms of abdominal muscle activation during vomiting. J Appl Physiol 1990; 69:21–25.
11. Bianchi AL, Grélot L. Converse motor output of inspiratory bulbospinal premotoneurones during vomiting. Neurosci Lett 1989; 104:298–302.
12. Miller AD, Nonaka S, Lakos SF, Tan LK. Diaphragmatic and external intercostal muscle control during vomiting: Behavior of inspiratory bulbospinal neurons. J Neurophysiol 1990; 63:31–36.
13. Ezure K. Synaptic connections between respiratory medullary neurons and considerations on the genesis of respiratory rhythm. Prog Neurobiol 1990; 35:429–450.
14. Jiang C, Lipski J. Extensive monosynaptic inhibition of ventral respiratory group neurons by augmenting neurons in the Bötzinger complex in the cat. Exp Brain Res 1990; 81:639–648.
15. Miller AD, Ezure K. Behavior of inhibitory and excitatory propriobulbar respiratory neurons during fictive vomiting. Brain Res 1992; 578:168–176.
16. Miller AD, Nonaka S. Bötzinger expiratory neurons may inhibit phrenic motoneurons and medullary inspiratory neurons during vomiting. Brain Res 1990; 521: 352–354.
17. Grélot L, Milano S, Portillo F, et al. Membrane potential changes of phrenic motoneurons during fictive vomiting, coughing, and swallowing in the decerebrate cat. J Neurophysiol 1992; 68:2110–2119.
18. Nonaka S, Miller AD. Behavior of upper cervical inspiratory propriospinal neurons during fictive vomiting. J Neurophysiol 1991; 65:1492–1500.
19. Grélot L, Milano S, Portillo F, Miller AD. Respiratory interneurons of the lower cervical (C4-C5) cord: Membrane potential changes during fictive coughing, vomiting, and swallowing in the decerebrate cat. Pflügers Arch 1993; 425:313–320.

51

Vasomotor Actions of Angiotensin II in the Ventrolateral Medulla of Spontaneously Hypertensive and Wistar Kyoto Rats

DAVID B. AVERILL

Bowman Gray School of Medicine of
Wake Forest University
Winston-Salem, North Carolina

I. Introduction

The ventrolateral medulla regulates the intrinsic activity of spinal preganglionic sympathetic neurons through a counterbalancing action of pressor and depressor circuits situated in the rostral (RVLM) and caudal (CVLM) regions of the ventral medulla, respectively (1–6). Briefly, barosensitive information is conveyed from intrinsic neurons of the nucleus of the tractus solitarius (nTS) to neurons of the CVLM. Gordon (7) has shown that *N*-methyl-D-aspartic acid (NMDA) receptors in the CVLM mediate the effect of barosensitive input to CVLM neurons. More recently, electrophysiological studies have revealed that the firing properties of CVLM neurons are consonant with the idea that these neurons are excited by baroreceptor afferent input (6,8–10). CVLM neurons project monosynaptically to pressor neurons of the RVLM (6); the effect of this projection from the CVLM is to inhibit RVLM neurons. Current evidence supports the idea that release of gamma-aminobutyric acid (GABA) from CVLM nerve terminals mediates this inhibition (11–13). Finally, it is well appreciated that pressor neurons of the RVLM project to the intermediolateral cell column of the thoracic and lumbar segments of the spinal cord and, presumably, this projection provides monosynaptic excitatory input to preganglionic sympathetic neurons.

It was within this construct of the VLM that we became interested in the potential role that Ang II might play in the regulation of cardiovascular function. High-affinity binding sites for Ang II have been demonstrated in the VLM of a number of different species (14–18). We originally showed in the cat that Ang II applied topically to the glycine-sensitive area of the ventral medulla elicited dose-dependent increases in arterial pressure (19). At the same time Allen et al. (16) found that microinjection of Ang II into the RVLM of cats produced a dose-dependent increase in blood pressure. These findings were subsequently confirmed in the rabbit by Sasaki and Dampney (20), who showed that microinjection of Ang II in the RVLM increased blood pressure, whereas microinjection of the peptide in the CVLM caused blood pressure to decrease. In addition, these investigators demonstrated that blockade of Ang II receptors in the VLM by [Sar^1,Thr^8]-Ang II produced changes in blood pressure that were directionally opposite to the effects evoked by local injection of the agonist peptide. These latter findings suggested that Ang II in the VLM could influence the tonic activity of caudal depressor and rostral pressor neurons responsible for the control of sympathetic nerve activity.

These observations then implicate Ang II in the VLM as playing a role in the regulation of cardiovascular function. However, we decided to pursue the issue of whether Ang II might influence the responsiveness of VLM neurons in the pathogenesis of hypertension. To investigate this issue, we have focused our attention on the actions of Ang II in the VLM of spontaneously hypertensive rats (SHR). The evolution of hypertension in SHR is associated with an augmented activity of the sympathetic nervous system (21–23). Another factor is enhanced activity of the brain renin-angiotensin system. Intraventricular injection of either Ang II blockers (24) or Ang I converting enzyme inhibitors (25,26) decreases the blood pressure in SHR but not that of normotensive Wistar-Kyoto (WKY) rats. Evidence also suggests that centrally administered Ang II evokes a pressor response, at least in part via enhancement of sympathetic outflow to the peripheral circulation (27,28). This chapter, therefore, will review the results of our work, which has attempted to characterize the role that Ang II acting in either the RVLM or CVLM plays in the maintenance of blood pressure in SHR and also to determine the receptor subtype at which Ang II acts in the VLM to influence the activity of vasomotor neurons.

II. Methods

A. Animal Preparation

Experiments were done in adult male SHR (13 to 16 weeks old) and age-matched WKY rats obtained from Taconic Farms (Germantown, NY). All experiments were carried out in accordance with the guiding principles in the care and use of animals as mandated by the American Physiological Society. Rats were anesthetized either with halothane (0.9% to 1.1%) and breathed a mixture of 65% room air/35% oxygen, which served as the carrier for the anesthetic agent, or with urethan (1.3 to 1.5 g/kg, IP). A femoral artery and vein were cannulated for the measurement of arterial pressure and the injection of drugs, respectively. Arterial blood gases and pH were kept within physiological limits (pH 7.35 to 7.45, Po_2 100 to 140 mm Hg, Pco_2 40 to 45 mm Hg). Body temperature was maintained at 37.5 ± 0.5°C by a heating pad.

Anesthetized rats were placed in a supine position with their heads fixed in a stereotaxic frame (David Kopf Instruments). The trachea and esophagus were transected in the lower neck and reflected rostrally. The distal trachea was cannulated to facilitate ventilation. After retraction of the bilateral longus capitis muscles, the inferior occipital bone was removed to provide a 5 × 6 mm window onto the surface of the ventral medulla oblongata. After incision and retraction of the dura, the ventral surface of the medulla was kept moist either by artificial cerebrospinal fluid (aCSF; pH 7.4) or by production of endogenous CSF. In some experiments, rats were paralyzed with *d*-tubocurarine (0.8 to 1.0 mg/kg, IV). In these experiments the tracheal cannula was connected to a ventilator (model 681, Harvard Apparatus, MA) and animals were artificially ventilated at a rate of 60 strokes/min and a tidal volume of 2.5 mL. For experiments in which sympathetic nerve activity was recorded, the abdominal plexus was exposed through a transverse incision of the lateral abdominal wall, and the inferior nerve bundle accompanying a superior mesenteric artery was placed over a bipolar silver electrode. Nerves and electrode tips were immersed in mineral oil to preserve integrity of the nerve bundle. Nerve activity was amplified and filtered (bandwidth 100 to 3000 Hz). At the end of each experiment, the noise level associated with the recording of sympathetic activity was determined following intravenous injection of hexamethonium bromide (30 mg/kg).

B. Microinjection Procedures

Microinjections were made from multibarrel micropipettes with tip diameters of 20 to 50 μm. The pipettes were made from calibrated microbore capillary glass tubing (Accu-Fill 90, Clay Adams, NJ). Tips were drawn on a micropipette puller (model PE-2, Narishige Scientific Instruments, Japan). The inner surface of the pipettes was coated with silicone (Sigmacote, Sigma Chemical Co., St. Louis, MO). Injections (50 nL) were made over a 30-s period with a hand-held syringe, as described elsewhere (29). The injected volume was measured by observing the movement of the fluid meniscus along a reticule in a microscope. Microinjections were made on one side of the brainstem unless noted otherwise.

The rostral pressor and caudal depressor areas were identified by the injection of 2 nmol of L-glutamate (L-Glu) into the RVLM or CVLM, respectively. Identification of either the rostral pressor or caudal depressor area at the time of the experiment conformed to the following modification of criteria described by Willette et al. (2): (a) the latency to the onset of change in blood pressure produced by L-Glu was no more than 5 s; (b) the response plateau occurred within 20 s after microinjection of the amino acid; and (c) the change in blood pressure was at least 30 mm Hg. In our experiments we found that the rostral pressor area was restricted to L-Glu injection at sites located 0.6 to 1.0 mm rostral to the most rostral rootlet of the hypoglossal nerve; 1.7 to 1.9 mm lateral to the midline; and 0.5 to 0.8 mm below the ventral surface. The caudal depressor area corresponded to L-Glu injection sites located between the second and third rostral rootlets of the hypoglossal nerve, 1.9 to 2.1 mm lateral to the midline, and 0.7 to 0.9 mm below the ventral surface.

Ang II (5–100 pmol), the Ang II analogue antagonist [Sar^1,Thr^8]-Ang II (100–1000 pmol), the nonpeptide Ang II antagonist losartan (10 pmol to 10 nmol), muscimol (500 pmol), and L-Glu (2 nmol) were dissolved in aCSF (in m*M*: 133.3 NaCl, 3.4 KCl, 1.3 $CaCl_2$, 1.2 $MgCl_2$, 0.6 NaH_2PO_4, 32.0 $NaHCO_3$, and 3.4 glucose). Alcian blue dye (10 nL) was injected from a separate barrel of the pipette to mark the site of injection after completing injections at a particular site.

C. Histological Analysis

At the completion of an experiment, rats were deeply anesthetized with either 4% halothane or pentobarbital sodium (50 mg, IV). The animals were then perfused transcardially with 150 mL of 0.9% NaCl followed by 150 mL of 10% phosphate-buffered formaldehyde solution. The brainstem was removed, stored overnight in 10% phosphate-buffered formaldehyde solution, and then transferred to a fixative containing 30% sucrose. Frozen brain tissue was sectioned in the coronal plane (50 μm) and stained with neutral red. Microinjection sites were identified by the deposition of Alcian blue dye and referred to standard anatomical structures of the caudal brainstem according to the atlas of Paxinos and Watson (30).

D. Statistical Analysis

Data are expressed as means ± SEM. A general linear model two-way analysis of variance followed by multiple comparisons with a Duncan's multiple range test was employed for analysis of the hemodynamic and sympathetic neural effects of Ang II or Ang II antagonist microinjection. One-way analysis of variance was also employed when appropriate. Student's *t*-test was used for comparisons of the magnitudes of blood pressure and sympathetic nerve responses to L-Glu, Ang II, and Ang II antagonist microinjection between SHR and WKY rats (31). *P* values less than 0.05 were considered statistically significant.

III. Results

A. Hemodynamic and Sympathetic Effects of Angiotensin II Microinjection in the VLM

Previously, we had shown that topical application of Ang II to the glycine-sensitive area of the cat elicited a dose-related increase in blood pressure (19). Now, we sought to investigate the potential role of this peptide in the pathogenesis of hypertension of SHR via actions of the peptide at vasomotor neurons. Thus, the first objective was to characterize the cardiovascular responses produced by microinjection of Ang II into regions of the ventrolateral medulla functionally identified as either the rostral pressor or caudal depressor area. Microinjection of either L-Glu or Ang II into the RVLM increased blood pressure, whereas injection of either substance into the CVLM decreased blood pressure. Although both Ang II and L-Glu produced directionally similar changes in blood pressure, depending upon whether the substances were injected into either the RVLM or CVLM, the time course of the cardiovascular response to Ang II differed distinctly from that produced by L-Glu injection. The hemodynamic response produced by Ang II was characteristized by a slow onset and a gradual change from baseline. In contrast, injection of L-Glu into the RVLM caused a rapidly developing and transient increase in blood pressure. Table 1 summarizes the onset times and peak latencies of the blood pressure responses to Ang II (20 and 100 pmol) and L-Glu (2nmol) microinjections into the RVLM and CVLM. The average onset time and the latency to the peak of

Table 1 Onset Times and Latencies of Blood Pressure Responses to Ang II and L-Glu Microinjection into RVLM and CVLM of Rats[a]

	Spontaneously hypertensive rats		Wistar-Kyoto rats	
	Ang II	L-Glu	Ang II	L-Glu
Rostral ventrolateral medulla				
Onset time	7.8 ± 0.5[b]	2.3 ± 0.1	9.4 ± 0.9[b]	2.6 ± 0.2
Peak latency	46.0 ± 3.3[b]	13.8 ± 0.6	42.7 ± 3.1[b]	15.4 ± 0.8
	(13)	(21)	(12)	(15)
Caudal ventrolateral medulla				
Onset time	11.5 ± 1.1[b]	2.0 ± 0.1	10.8 ± 0.8[b]	2.2 ± 0.1
Peak latency	45.3 ± 3.6[b]	12.4 ± 0.7	42.7 ± 2.4[b]	12.7 ± 0.6
	(13)	(17)	(14)	(20)

[a]Values are means ± SEM of time factors measured in seconds; no. of rats studied under different conditions is given in parentheses. Ang II, angiotensin II (20 and 100 pmol); L-Glu, L-glutamate (2 nmol).
[b]$P < 0.01$ for Ang II versus L-Glu.
Source: Ref. 46.

the blood pressure changes to Ang II were similar in SHR and WYK rats. L-Glu injected into the RVLM and CVLM evoked blood pressure responses that had similar onset times and latencies in both strains. However, Table 1 documents that the onset times and latencies to peak responses associated with L-Glu injections were significantly ($P<0.01$) shorter than those associated with Ang II injections.

Microinjections of Ang II into the RVLM elicited significant ($P<0.01$) dose-dependent increases in blood pressure that were of comparable magnitude in SHR and WYK rats (Fig. 1). Pressor responses were significant at the lowest dose of Ang II (5 pmol) and reached maximal effects at a dose of 20 pmol. Although injection of the peptide into the RVLM increased heart rate, these responses were more variable than those observed for blood pressure. Administration of L-Glu (2 nmol) into the RVLM produced comparable pressor responses in both strains. The L-Glu–evoked pressor responses were significantly ($P<0.01$) larger than the increases in blood pressure produced by Ang II injections at the same sites (Fig. 1).

Ang II microinjections into the CVLM elicited significant ($P<0.01$) dose-dependent decreases in blood pressure in both SHR and WKY rats. The depressor responses were significant at the lowest dose of the peptide (5 pmol) and reached a maximum at a dose of 20 pmol (Fig. 2). In contrast to the pressor effects produced in the RVLM, the magnitude of the decreases in blood pressure caused by Ang II was significantly greater ($P<0.01$) in SHR than in WKY rats. Decreases in blood pressure were accompanied by falls in heart rate that did not differ between SHR and WKY rats. Injection of L-Glu (2 nmol) into the CVLM elicited significantly ($P<0.01$) greater depressor responses in SHR than in WKY rats (Fig. 2).

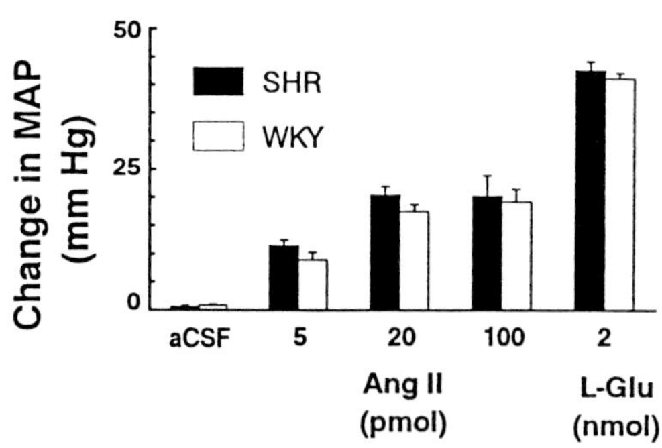

Figure 1 Changes in mean arterial pressure (MAP) in response to microinjection of artificial cerebrospinal fluid (aCSF), L-glutamate (L-Glu), and graded doses of angiotensin II (Ang II) into the rostral ventrolateral medulla of 14- to 16-week-old SHR and WKY rats. Injection volumes were 50 nL.

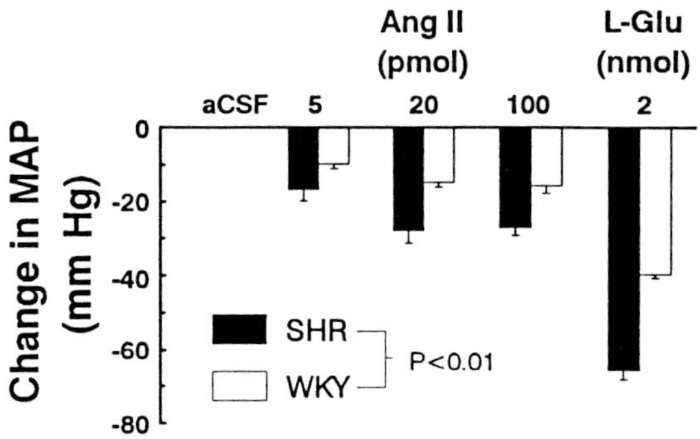

Figure 2 Changes in mean arterial pressure (MAP) in response to microinjection of artificial cerebrospinal fluid (aCSF), L-glutamate (L-Glu), and graded doses of angiotensin II (Ang II) into the caudal ventrolateral medulla of 14- to 16-week old SHR and WKY rats. Injection volumes were 50 nL. SHR showed significantly greater depressor responses than WKY rats.

Although the blood pressure changes elicited by Ang II injections into either the CVLM or RVLM were consistently smaller than the blood pressure changes produced by L-Glu injections at the same sites, these findings suggested that Ang II acted to stimulate the same vasomotor neurons which could be activated by L-Glu. Figure 3 is a composite of the locations where both Ang II and L-Glu produced either pressor or depressor responses. In both SHR and WKY rats, depressor responses were produced by injections of the peptide restricted to a region ventral to the nucleus ambiguus and dorsal to the lateral reticular nucleus, approximately at the level of the obex. In both strains, pressor responses were produced by injection of the peptide in an area encompassing the dorsolateral aspect of the lateral paragigantocellular nucleus (LPGi) and a region dorsolateral to the LPGi. This area lies at the caudal end of the facial nucleus and is consistent with the region known as the subretrofacial nucleus based upon pressor responses evoked by L-Glu injections.

Previous studies have suggested that the pressor response evoked by Ang II in the RVLM may be mediated by the sympathetic nervous system (16,19, 20). Two separate sets of experiments were conducted to establish that the pressor responses evoked by Ang II injections in the RVLM were sympathetically mediated. First, Ang II was injected into the RVLM of SHR and WKY rats before and after ganglionic blockade with hexamethonium bromide (40 mg/kg, IV). The initial injection of Ang II increased blood pressure by an average of 17 ± 4 mm Hg in SHR ($n = 3$) and 16 ± 3 mm Hg in WKY rats ($n = 5$). Delivery of hexamethonium bromide 60 min after the first injection of Ang II caused blood pressure in SHR to fall from 135 ± 3 to 65 ± 2 mm Hg in SHR and from 91 ± 4 to 49 ± 2 mm Hg in WKY rats. However, injection of Ang II (20 pmol) into the

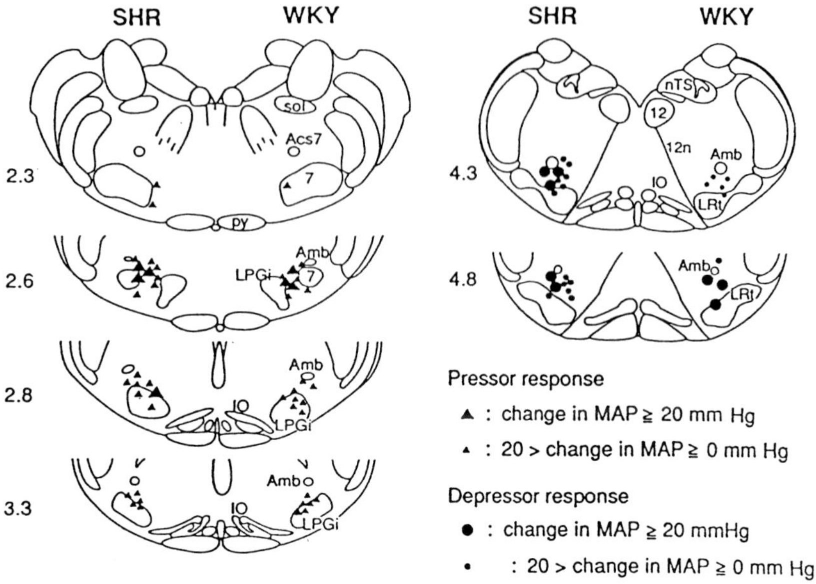

Figure 3 Sites in the ventrolateral medulla where Ang II (20 pmol) elicited either a pressor (left) or a depressor (right) response. According to the atlas of Paxinos and Watson (30), medullary sections extended from 2.30 to 4.80 mm caudal to interaural line. Abbreviations: Acs7, accessory facial nucleus; Amb, nucleus ambiguus; IO, inferior olivary nucleus; LPGi, lateral paragigantocellularis nucleus; LRt, lateral reticular nucleus; py, pyramidal tract; nTS, nucleus of tractus solitarius; sol, solitary tract; 7, facial nucleus; 12, hypoglossal nucleus; 12n, root of hypoglossal nerve; MAP, mean arterial pressure. (From Ref. 46.)

same sites after ganglionic blockade had no effect upon blood pressure or heart rate in either strain. However, ganglionic blockade was associated with a dramatic reduction in the baseline blood pressure. Therefore, in additional experiments, we assessed the effects of administration of L-Glu and Ang II into the RVLM upon splanchnic sympathetic nerve activity in SHR and WKY rats. Figure 4 shows that microinjection of Ang II (100 pmol) on average increased splanchnic sympathetic nerve activity by approximately 10% in both SHR and WKY rats. In contrast, L-Glu (2 nmol) injection at the same sites increased splanchnic sympathetic nerve activity by approximately 35% and 40% in WKY and SHR, respectively. Just as L-Glu injected into the RVLM acted as a more potent stimulus than Ang II to increase blood pressure, so did injection of the excitatory amino acid increase sympathetic nerve activity to the same relevant extent when compared to the sympathetic activation produced by Ang II injection in the RVLM. Thus, two different sets of experiments support the idea

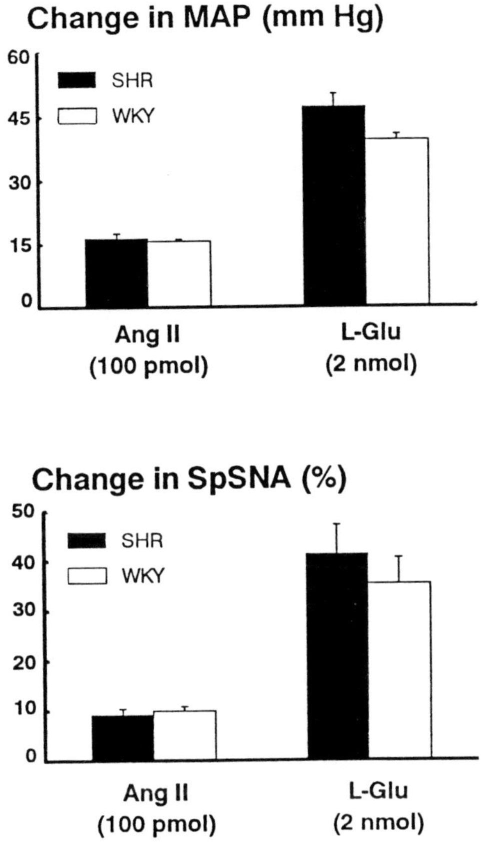

Figure 4 Changes in mean arterial pressure (MAP) and splanchnic sympathetic nerve activity (SpSNA) produced by microinjection of angiotensin II (Ang II) and L-glutamate (L-Glu) into the rostral ventrolateral medulla of SHR and WKY rats. Injection volumes were 50 nL.

that Ang II stimulates RVLM neurons, providing excitatory input to sympathetic preganglionic neurons.

B. Blockade of Angiotensin II Receptors in the Ventrolateral Medulla

The experiments reviewed above as well as those performed in cats (16,19) and rabbits (20) demonstrate that the activity of vasomotor neurons is enhanced by stimulation of Ang II receptors. These findings provided some insight into the ability of Ang II to influence the activity of vasomotor neurons. Although Ang II was less potent than L-Glu in exciting vasomotor neurons, our data suggested that this peptide could make a significant but quantitatively modest contribution to the tonic activity of a counterbalancing set of ventrolateral medullary neurons involved in the regulation of sympathetic nerve activity. Thus, it became of paramount importance to assess whether activation of Ang II receptors in the caudal depressor and rostral pressor regions by endogenous Ang II contributed to the tonic activity of the counterbalancing pools of vasomotor neurons. Investigation of this concept was pursued in both SHR and WKY rats to determine whether endogenous Ang II had a quantitatively different influence upon the tonic activity of vasomotor neurons in hypertensive rats compared to normotensive animals. To determine the physiological role that Ang II might play in the activity of vasomotor neurons in hypertensive and normotensive rats, we blocked Ang II receptors in the CVLM and RVLM with the peptide analogue antagonist [Sar^1,Thr^8]-Ang II.

Microinjection of [Sar^1,Thr^8]-Ang II into either the RVLM or CVLM produced changes in blood pressure and heart rate that were directionally opposite to the cardiovascular effects which we had previously observed for Ang II injection into each region. Injection of [Sar^1,Thr^8]-Ang II into the CVLM of SHR and WKY rats increased blood pressure and heart rate. In contrast, injection of this Ang II antagonist into the RVLM of both strains decreased blood pressure and heart rate. The pressor responses produced by [Sar^1,Thr^8]-Ang II injection in the CVLM and the depressor responses evoked by this antagonist in the RVLM developed more slowly and were more sustained than the changes in blood pressure produced by L-Glu injection in the CVLM or RVLM.

Figure 5 summarizes the changes in blood pressure produced by injection of either L-Glu (2 nmol), Ang II (20 pmol), or [Sar^1,Thr^8]-Ang II (100 pmol) into the RVLM of SHR and WKY rats. The most dramatic effect of [Sar^1,Thr^8]-Ang II injections into the RVLM was the significant ($P>0.01$) larger reduction in baseline blood pressure measured in SHR compared to WKY rats. This was in marked contrast to the repeated demonstration that Ang II elicited similar pressor responses in SHR and WKY rats and that L-Glu produced larger pressor responses than Ang II, which were of similar magnitude in SHR and WKY rats. To ascertain that the reduction in blood pressure produced by [Sar^1,Thr^8]-Ang II injection into the RVLM was caused by a reduction in sympathetic nerve activity, splanchnic sympathetic nerve activity (SpSNA) was recorded in another set of SHR and WKY rats. Injection of [Sar^1,Thr^8]-Ang II (1 nmol) into the RVLM decreased SpSNA by $21 \pm 2\%$ in SHR and by $18 \pm 4\%$ in WKY rats. These reductions in baseline SpSNA were accompanied by depressor responses in SHR (-39 ± 5 mm Hg) and WKY rats (-25 ± 4 mm Hg).

Figure 6 summarizes the changes in blood pressure produced by injection of L-Glu (2 nmol), Ang II (20 pmol), or [Sar^1,Thr^8]-Ang II (100 pmol) into the CVLM of SHR and WKY rats. When compared to WKY rats, SHR had significantly larger decreases in blood pressure in response to L-Glu ($P<0.01$) and Ang II ($P<0.05$) injections into the CVLM. In contrast, [Sar^1,Thr^8]-Ang

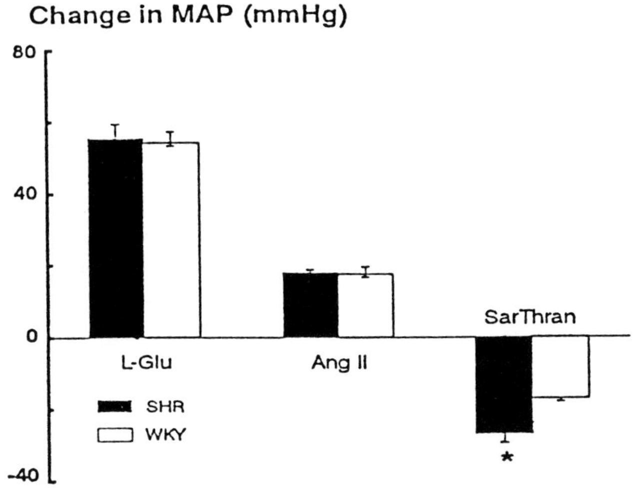

Figure 5 Changes in mean arterial pressure (MAP) produced by microinjection of L-glutamate (L-Glu), angiotensin II (Ang II), and [Sar[1],Thr[8]]-Ang II (SarThran) into the rostral ventrolateral medulla of SHR and WKY rats. Injection volumes were 50 nL. $*P<0.01$.

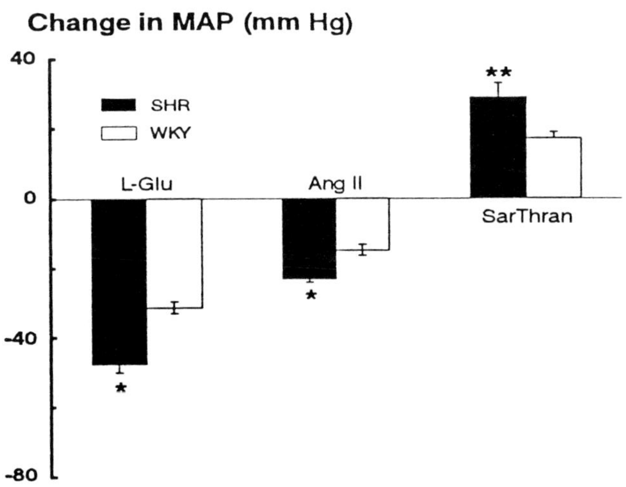

Figure 6 Changes in mean arterial pressure (MAP) produced by microinjection of L-glutamate (L-Glu), angiotensin II (Ang II), and [Sar[1],Thr[8]]-Ang II (SarThran) into the caudal ventrolateral medulla of SHR and WKY rats. Injection volumes were 50 nL. $*P<0.05$.

II injection into the CVLM increased baseline blood pressure of both strains, and this pressor response was significantly ($P<0.05$) larger in SHR.

Recent studies have shown that GABAergic projections from the CVLM provide inhibitory input to presympathetic neurons of the RVLM (11–13). We hypothesized that the increase in blood pressure produced by microinjection of [Sar^1,Thr^8]-Ang II into the CVLM was due to a reduction of GABAergic input to presympathetic neurons of the RVLM. This hypothesis was supported by the finding that [Sar^1,Thr^8]-Ang II injection into the CVLM increased SpSNA in SHR and WKY rats. However, to address more directly the concept that blockade of Ang II receptors in the CVLM reduced the tonic activity of CVLM GABAergic projections to the RVLM, we determined whether the pressor response to [Sar^1,Thr^8]-Ang II injection into the CVLM could be abolished by inhibition of RVLM neurons. Figure 7 summarizes the results of experiments in which the blood pressure changes produced by [Sar^1,Thr^8]-Ang II injection into the CVLM were determined before and after bilateral injection of the $GABA_A$ agonist muscimol into the RVLM. Injection of [Sar^1,Thr^8]-Ang II into the CVLM before delivery of muscimol into the RVLM caused pressor responses in both SHR (23 ± 1 mm Hg) and WKY rats (18 ± 2 mm Hg; $P<0.05$ SHR versus WKY).

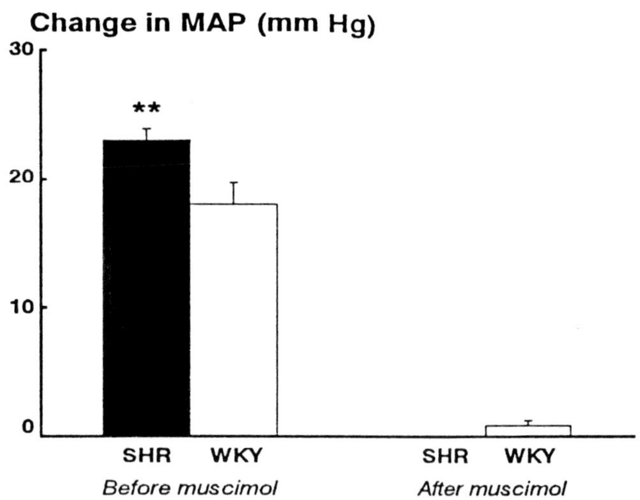

Figure 7 Changes in mean arterial pressure (MAP) produced by microinjection of [Sar^1,Thr^8]-Ang II into the caudal ventrolateral medulla before and after bilateral injection of muscimol (500 pmol each side) into the RVLM of SHR and WKY rats. The increases in MAP in response to [Sar^1,Thr^8]-Ang II injection into the caudal ventrolateral medulla after bilateral injection of muscimol into the rostral ventrolateral medulla were significantly different from the pressor responses produced by [Sar^1,Thr^8]-Ang II injection into the caudal ventrolateral medulla before muscimol injection. **$P<0.05$.

Bilateral injection of muscimol (500 pmol each side) into the RVLM decreased mean arterial pressure (MAP) from 123 ± 2 to 51 ± 3 mm Hg in SHR and from 92 ± 3 to 45 ± 2 mm Hg in WKY rats. A second injection of [Sar^1,Thr^8]-Ang II into the CVLM after bilateral injection of muscimol into the RVLM completely abolished in both strains the pressor response that previously was produced by Ang II antagonist injection into the CVLM.

C. Angiotensin II Receptor Subtypes in the Rostral Ventrolateral Medulla

The development of new agents selective for different receptor subtypes responsive to Ang II has revealed the presence of AT1 and AT2 receptors in the brainstem of rats. Recently, we showed that Ang II acts at AT1 receptors in the nTS to produce decreases in blood pressure in response to microinjection of the peptide in this dorsal medullary nucleus (32). Thus, the next objective of our studies investigating the actions of Ang II in the ventrolateral medulla of SHR and WKY rats was to determine the receptor subtype at which Ang II acts in the RVLM. To investigate the role of AT1 receptors in the RVLM, we used losartan, a 2-*n*-butyl-4-chloro-5-hydroxymethyl-1-[2′-(1H-tetrazol-5-yl) biphenyl-4-yl) methyl] imidazole potassium salt, as an antagonist selective for AT1 receptors (33).

A range of doses (0.01, 0.1, 1, and 10 nmol) of losartan was used to assess the effect of AT1 receptor blockade on the actions of Ang II in the RVLM. The lowest dose (0.01 nmol) injected into the RVLM caused a partial attenuation of the pressor response that could be elicited by Ang II (100 pmol) injection at the same site. However, higher doses (0.1, 1, and 10 nmol) of losartan attenuated by two-thirds the pressor responses that could be produced by Ang II injection in SHR and WKY rats. Since our previous studies had shown that [Sar^1,Thr^8]-Ang II could also attenuate the pressor action of Ang II injected into the RVLM, we compared the abilities of losartan and [Sar^1,Thr^8]-Ang II to block the effects of Ang II. Figure 8 illustrates the increases in blood pressure and SpSNA that were elicited by Ang II injected into the RVLM of SHR and WKY rats after prior injection of artificial cerebrospinal fluid (aCSF). When losartan (1 nmol) was injected into the RVLM immediately before injection of Ang II at the same site, the pressor response and sympathetic activation produced by Ang II were significantly ($P<0.01$) *attenuated*. In contrast, prior injection of [Sar^1,Thr^8]-Ang II into the RVLM essentially abolished the pressor response and sympathetic activation produced by Ang II injections in both SHR and WKY rats. Comparisons of the residual pressor responses produced by Ang II observed after injection of either losartan or [Sar^1,Thr^8]-Ang II revealed a significant ($P<0.05$) difference in SHR. An important feature of the doses (0.1 to 10 nmol) of losartan that significantly attenuated the actions of Ang II in the RVLM was that baseline blood pressure was not changed 5 min after injection of this Ang II antagonist. Figure 9 shows that lack of effect of

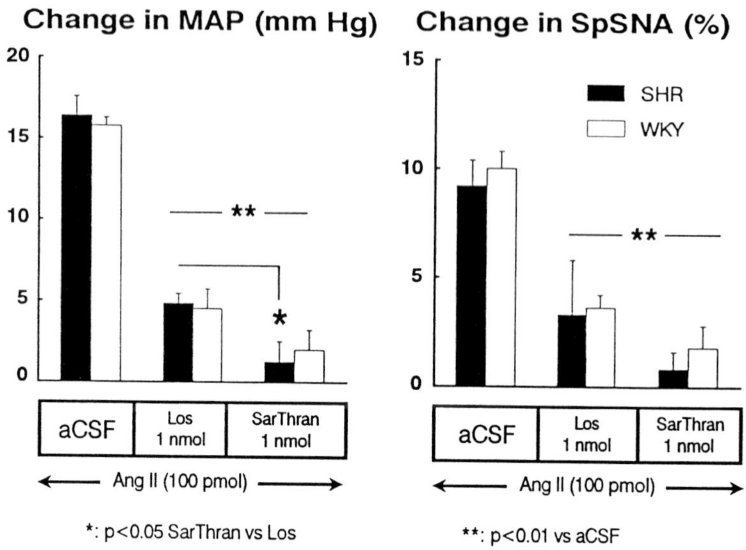

Figure 8 Changes in mean arterial pressure (MAP) and splanchnic sympathetic nerve activity (SpSNA) in response to angiotensin II (Ang II) injection into the rostral ventrolateral medulla after injection of artificial cerebrospinal fluid (aCSF), losartan (Los), or [Sar[1],Thr[8]]-Ang II (SarThran) at the same site. *$P<0.05$ for comparison of responses in SHR between losartan and [Sar[1],Thr[8]]-Ang II. **$P<0.01$ for comparison of responses produced by Ang II after either losartan or [Sar[1],Thr[8]]-Ang II to that after aCSF.

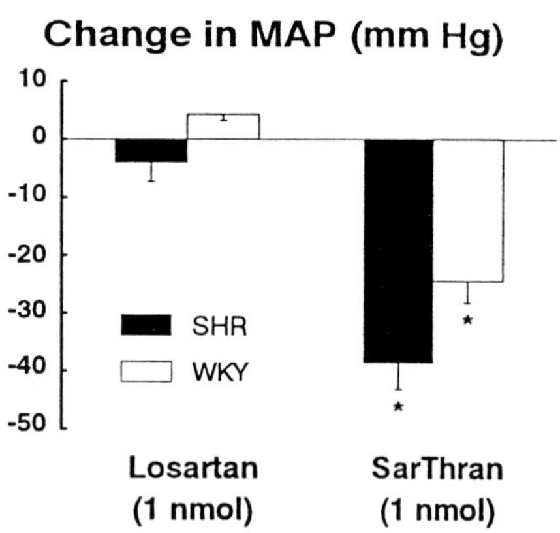

Figure 9 Changes in baseline mean arterial pressure (MAP) produced by microinjection of either losartan or [Sar[1],Thr[8]]-Ang II (SarThran) into the rostral ventrolateral medulla of SHR and WKY rats. *$P<0.01$ for comparison of responses between losartan and [Sar[1],Thr[8]]-Ang II.

losartan (1 nmol) upon baseline blood pressure was in marked contrast to the ability of [Sar^1,Thr^8]-Ang II (1 nmol) injected into the RVLM to significantly reduce baseline blood pressure. Based upon these results, we conclude that AT1 receptors play a role in mediating vasomotor actions of Ang II in the RVLM of SHR and WKY rats. However, the residual increases in blood pressure and sympathetic nerve activity produced by Ang II after losartan and the lack of effect of losartan upon baseline blood pressure suggest that Ang II may also act at non-AT1 or AT2 receptor subtypes in the RVLM.

IV. Discussion

The data presented above provide evidence that Ang II contributes to the tonic activity of vasomotor neurons of the ventrolateral medulla. Initially, we showed that this peptide acted as a neuroexcitatory agent to increase the activity of rostral pressor and caudal depressor neurons. The neuroexcitatory action of Ang II is also supported by the finding that iontophoretic application of Ang II to RVLM neurons increases the firing rate of these vasomotor neurons (34). However, the neuroexcitatory action of Ang II appeared less potent than the effects that could be evoked by L-Glu. On average, the increases in blood pressure that could be produced by peptide injection into the RVLM ranged from 32% to 48% of the magnitude that could be elicited by L-Glu injection in either SHR or WKY rats. In the CVLM, the depressor responses produced by Ang II injection ranged from 39% to 49% of the magnitude of the decreases in blood pressure evoked by L-Glu injections in either SHR or WKY rats. Another characteristic difference between the blood pressure responses evoked by Ang II and L-Glu in the VLM was that the onset times and latencies to the peak change in blood pressure were significantly longer in response to Ang II in comparison to L-Glu. The smaller changes in blood pressure and the longer time course of the changes in blood pressure evoked by Ang II might suggest that the mode of action of the peptide upon vasomotor neurons is different from that of L-Glu. Electrophysiological recordings from nTS neurons of an in vitro brainstem slice have shown that, for the same neuron, L-Glu evoked a larger change in firing rate than did Ang II, and this occurred with a shorter latency (K. L. Barnes, personal communication). These electrophysiological studies also showed that the increase in firing rate produced by Ang II was substantially more prolonged than that elicited by L-Glu. These latter findings are consistent with the notion that Ang II and L-Glu act through different neuronal mechanisms. However, one must consider that L-Glu may have neuronal effects mediated either through ligand-gated ion channels or by effects upon second-messenger systems involving changes in intracellular Ca^{2+} concentration (35). It is interesting to note that Ang II acting at AT1 receptors evokes a similar constellation of effects upon intracellular Ca^{2+} as those produced by L-Glu acting at metabotropic excitatory amino acid receptors. Thus, an important objective for future investigations will be to determine the biophysical and biochemical mechanisms by which Ang II affects excitability of vasomotor neurons.

The changes in blood pressure produced by Ang II injection into the RVLM and CVLM provide functional evidence for Ang II receptors in these regions of the rat brainstem. However, the physiological role of Ang II in these two regions of the VLM was revealed by the changes in blood pressure produced by blockade of Ang II receptors with [Sar^1,Thr^8]-Ang II. Blockade of Ang II receptors in the RVLM reduced baseline blood pressure, whereas Ang II receptor blockade in the CVLM increased baseline blood pressure. The changes in baseline blood pressure produced by local injection of [Sar^1,Thr^8]-Ang II were directionally opposite to the effects produced by injection of the agonist peptide. Qualitatively similar effects upon baseline blood pressure were observed in both SHR and WKY rats, albeit the magnitude of the changes in blood pressure was significantly larger in SHR. These data agree with our previous finding in the cat that [Sar^1,Thr^8]-Ang II applied to the ventral surface overlying the RVLM reduced blood pressure (19). Sasaki and Dampney also found that injection of [Sar^1,Thr^8]-Ang II in the VLM of rabbits produced changes in blood pressure that were in a direction opposite to those produced by the parent peptide (20). More recently, Chan et al. (36) have provided preliminary evidence that microinjection of [Sar^1,Thr^8]-Ang II reduced the firing rate of individually recorded presympathetic neurons of the RVLM. These results are taken as evidence that Ang II in the CVLM and RVLM contributes to the tonic activity of vasomotor neurons in these two regions of the VLM. In the RVLM, we have attempted to assess the relative contribution of Ang II to the tonic activity of presympathetic neurons. We compared the reduction of blood pressure produced by [Sar^1,Thr^8]-Ang II injection into the RVLM. In both SHR and WKY rats, the reduction in baseline blood pressure evoked by [Sar^1,Thr^8]-Ang II was on average 50% of the reduction in blood pressure observed when RVLM neurons were inhibited by the $GABA_A$ agonist muscimol.

The cardiovascular effects of Ang II receptor blockade in either the CVLM or RVLM appear to be the result of modulation of sympathetic outflow. First, the pressor response to [Sar^1,Thr^8]-Ang II injection into the CVLM of SHR and WKY rats was abolished after inhibition of RVLM neurons by bilateral injection of muscimol. Thus, vasomotor neurons of the RVLM are required for the action of the Ang II antagonist in the CVLM, which was to reduce inhibitory drive to RVLM vasoconstrictor neurons. This pathway agrees with the demonstration by Sasaki and Dampney (20) that injection of [Sar^1,Thr^8]-Ang II in the rabbit CVLM increased sympathetic nerve activity. We have also performed studies in Sprague-Dawley rats showing that [Sar^1,Thr^8]-Ang II injection in the CVLM caused splanchnic sympathetic nerve activity to increase (37). Second, the data reviewed above show that injection of [Sar^1,Thr^8]-Ang II into the RVLM of SHR and WKY rats reduced baseline levels of splanchnic sympathetic nerve activity. Sasaki and Dampney (20) have also shown in rabbits that injection of [Sar^1,Thr^8]-Ang II reduced renal sympathetic nerve activity.

In the last few years, agents exhibiting selectivity for different Ang II receptor subtypes have been developed; these have shown that AT1 and AT2 receptors may be associated with brain pathways involved in neural or neuroendocrine control of the cardiovascular system (32,38–41). Therefore, we determined whether Ang II acted at AT1 receptors in the RVLM. Local injection of the AT1 receptor antagonist losartan into the RVLM of SHR and WKY rats significantly attenuated the pressor responses that could be evoked by Ang II injection at the same site. However, even at the highest dose of losartan (10 nmol) utilized, there remained a residual pressor response to Ang II injection. Thus, it appeared that blockade of AT1 receptors could not abolish the neuroexcitatory action of Ang II in the RVLM. In contrast, [Sar^1,Thr^8]-Ang II at a dose of 1 nmol essentially eliminated the pressor response that could be evoked by Ang II. These findings are in contrast to a recent study by Fow et al. (32), who showed that the depressor responses produced by a low dose of Ang II (200 fmol) injected into the nTS were blocked by losartan (100 pmol). In addition, when we compared the effects of losartan and [Sar^1,Thr^8]-Ang II upon baseline blood pressure in SHR and WKY rats, we observed that losartan (1 nmol) did not reduce baseline blood pressure, whereas [Sar^1,Thr^8]-Ang II (1 nmol) caused a substantial reduction in baseline blood pressure of SHR and WKY rats. This set of data suggests that Ang II injected into the RVLM may act on non-AT1 or AT2 receptors as well as on AT1 receptors. In addition, we have the provocative finding that the actions of Ang II endogenous to the RVLM apparently are not affected by AT1 receptor blockade. Future studies need to focus on the possible role of non-AT1 receptors in the actions of Ang II or Ang II metabolites in the RVLM.

One of the goals of these experiments was to assess the role that Ang II in the VLM might have in the pathogenesis of hypertension. Since a number of studies have provided evidence that the elevated blood pressure of SHR is related to an enhanced activity of the sympathetic nervous system (21–23), we hypothesized that an action of Ang II at vasomotor neurons may be a contributing factor to the hypertension of SHR. In part, this hypothesis was based upon findings showing that central blockade of the renin-angiotensin system reduces the blood pressure of SHR (24–26). Injection of Ang II into the RVLM produced equivalent pressor responses in SHR and WKY rats, whereas injection of the peptide into the CVLM caused significantly larger decreases in blood pressure of SHR compared to WKY rats. This data set by itself did not resolve the potential role of Ang II in the VLM in the hypertension of SHR. In part, this was related to the fact that L-Glu injection in the RVLM caused equivalent pressor responses in SHR and WKY. Moreover, L-Glu injection into the CVLM evoked larger depressor responses in SHR, which were qualitatively similar to the responses evoked by Ang II. Our findings for the responses to L-Glu were in agreement with the reports of Smith and Barron (42,43). However, our data for the responses to Ang II in the RVLM did not appear consistent with the observations of Chan et al. (34). These investigators observed that application of Ang II to individually recorded RVLM neurons produced significantly larger increases in firing rate, shorter latency of augmented firing rate, and a lower current threshold for iontophoretic application of Ang II in SHR compared to WKY rats. Examination of the effects of [Sar^1,Thr^8]-Ang II injection into the VLM of SHR and WKY rats revealed that Ang II receptor blockade in the RVLM caused larger reductions in baseline blood pressure in SHR, and injection of this receptor antagonist into the CVLM produced larger increases in baseline blood pressure in SHR. The larger changes in blood pressure produced by [Sar^1,Thr^8]-Ang II injection into the RVLM and CVLM of SHR cannot be explained easily by differences in vascular reactivity between SHR and WKY because L-Glu injection in this latter set of experiments caused equivalent pressor responses in SHR and WKY rats. The question remains whether Ang II has actions in either the RVLM or CVLM that contribute to the elevated blood pressure of SHR. Our data suggest that the increased action of Ang II in the RVLM of SHR may relate to higher tonic activity of RVLM neurons in this strain. On the other hand, the action of endogenous Ang II may be augmented in the CVLM of SHR, but this enhanced activity of CVLM inhibitory neurons may be insufficient to suppress an apparent greater intrinsic pressor activity of RVLM neurons. Additional insight into this issue is provided by the work of Chan et al. (34,44,45). These investigators showed that in comparison to WKY rats, SHR have a greater proportion of RVLM neurons that fire with doublet discharges. Most intriguing is their recent finding (45) that long-term treatment of SHR with captopril not only reduced the blood pressure of SHR but also altered the firing properties of RVLM neurons such that the proportion of RVLM neurons firing in doublets or singlets was essentially the same as that determined for WKY rats. In conclusion, it will be important to determine what other factors influence the firing properties of vasomotor neurons and to identify the contribution of singlet and doublet discharging RVLM neurons in the maintenance of sympathetic nerve activity.

V. Summary

This chapter has reviewed recent work done in the author's laboratory to define the role of Ang II in the function of RVLM and CVLM in SHR and WKY rats. Data are presented demonstrating that Ang II acts as an excitatory agent in the RVLM and CVLM of both strains of rats. In addition, the results of experiments employing the receptor antagonist [Sar^1,Thr^8]-Ang II provide evidence that Ang II endogenous to both regions of the VLM contributes to the tonic activity of vasomotor neurons. Available data shows that Ang II acts at AT1 receptors in the RVLM of SHR and WKY rats. Finally, results of experiments utilizing [Sar^1,Thr^8]-Ang II as an Ang II antagonist suggest that endogenous Ang II may contribute to a greater tonic activity of CVLM and RVLM neurons of SHR.

References

1. Willette RN, Punnen S, Krieger AJ, Sapru HN. Interdependence of rostral and caudal ventrolateral medullary areas in the control of blood pressure. Brain Res 1984; 321:169–174.
2. Willette RN, Barcas PP, Krieger AJ, Sapru HN. Vasopressor and depressor areas in the rat medulla: Identification by microinjection of L-glutamate. Neuropharmacology 1983; 22:1071–1079.
3. Urbanski RW, Sapru HN. Evidence for a sympathoexcitatory pathway from the nucleus tractus solitarii to the ventrolateral medullary pressor area. J Auton Nerv Syst 1988; 23:161–174.
4. Agarwal SK, Gelsema AJ, Calaresu FR. Neurons in rostral VLM are inhibited by chemical stimulation of caudal VLM in rats. Am J Physiol 1989; 257:R265–R270.
5. Agarwal SK, Gelsema AJ, Calaresu FR. Inhibition of rostral VLM by baroreceptor activation is relayed through caudal VLM. Am J Physiol 1990; 258:R1271–R1278.
6. Agarwal SK, Calaresu FR. Monosynaptic connection from caudal to rostral ventrolateral medulla in the baroreceptor reflex pathway. Brain Res 1991; 555:70–74.
7. Gordon FJ. Aortic baroreceptor reflexes are mediated by NMDA receptors in caudal ventrolateral medulla. Am J Physiol 1987; 252:R628–R633.
8. Masuda N, Terui N, Koshiya N, Kumada M. Neurons in the caudal ventrolateral medulla mediate the arterial baroreceptor reflex by inhibiting barosensitive reticulospinal neurons in the rostral ventrolateral medulla in rabbits. J Auton Nerv Syst 1991; 34:103–118.
9. Jeske I, Morrison SF, Cravo SL, Reis DJ. Identification of baroreceptor reflex interneurons in the caudal ventrolateral medulla. Am J Physiol 1993; 264:R169–R178.
10. Terui N, Masuda N, Saeki Y, Kumada M. Activity of barosensitive neurons in the caudal ventrolateral medulla that send axonal projections to the rostral ventrolateral medulla in rabbits. Neurosci Lett 1990; 118:211–214.
11. Sun MK, Guyenet PG. GABA-mediated baroreceptor inhibition of reticulospinal neurons. Am J Physiol 1985; 249:R672–R680.
12. Urbanski RW, Sapru HN. Putative neurotransmitters involved in medullary cardiovascular regulation. J Auton Nerv Syst 1988; 25:181–193.
13. Willette RN, Barcas PP, Krieger AJ, Sapru HN. Endogenous GABAergic mechanisms in the medulla and the regulation of blood pressure. J Pharmacol Exp Ther 1984; 230:34–39.
14. Speth RC, Wamsley JK, Gehlert DR, et al. Angiotensin II receptor localization in the canine CNS. Brain Res 1985; 326:137–143.
15. Allen AM, Chai SY, Clevers J, et al. Localization and characterization of angiotensin II receptor binding and angiotensin converting enzyme in the human medulla oblongata. J Comp Neurol 1988; 269:249–264.
16. Allen AM, Dampney RAL, Mendelsohn FAO. Angiotensin receptor binding and pressor effects in cat subretrofacial nucleus. Am J Physiol 1988; 255:H1011–H1017.
17. Mendelsohn FAO, Allen AM, Clevers J, et al. Localization of angiotensin II receptor binding in rabbit brain by in vitro autoradiography. J Comp Neurol 1988; 270:372–384.
18. Rowe BP, Grove KL, Saylor DL, Speth RC. Angiotensin II receptor subtypes in the rat brain. Eur J Pharmacol 1990; 186:339–342.

19. Andreatta SH, Averill DB, Santos RAS, Ferrario CM. The ventrolateral medulla: A new site of the action of the renin-angiotensin system. Hypertension 1988; (suppl I):I-163–I-166.
20. Sasaki S, Dampney RAL. Tonic cardiovascular effects of angiotensin II in the ventrolateral medulla. Hypertension 1990; 15:274–283.
21. Judy WV, Watanabe AM, Henry DP, et al. Sympathetic nerve activity: Role in regulation of blood pressure in the spontaneously hypertensive rat. Circ Res 1976; 38 (suppl II):II-21–II-29.
22. Lundin S, Ricksten S-E, Thorén P. Renal sympathetic activity in spontaneously hypertensive rats and normotensive controls as studied by three different methods. Acta Physiol Scand 1984; 120:265–272.
23. Muratani H, Ferrario CM, Averill DB. Ventrolateral medulla in spontaneously hypertensive rats: Role of angiotensin II. Am J Physiol 1993; 264:R388–R395.
24. Mann JFE, Phillips MI, Dietz R, et al. Effects of central and peripheral angiotensin blockade in hypertensive rats. Am J Physiol 1978; 234:H629–H637.
25. Hutchinson JS, Mendelsohn FAO, Doyle AE. Blood pressure responses of conscious normotensive and spontaneously hypertensive rats to intracerebroventricular and peripheral administration of captopril. Hypertension 1980; 2:546–550.
26. Okuno T, Nagahama S, Lindheimer MD, Oparil S. Attenuation of the development of spontaneous hypertension in rats by chronic central administration of captopril. Hypertension 1983; 5(suppl II):653–662.
27. Bruner CA, Fink GD. Neurohumoral contributions to chronic angiotensin-induced hypertension. Am J Physiol 1986; 250:H52–H61.
28. Falcon JC, Phillips MI, Hoffman WE, Brody MJ. Effects of intraventricular angiotensin II mediated by the sympathetic nervous system. Am J Physiol 1978; 235: H392–H399.
29. Diz DI, Barnes KL, Ferrario CM. Hypotensive action of angiotensin II microinjected into the dorsal motor nucleus of the vagus. J Hypertens 1984; 2(suppl 3): 53–56.
30. Paxinos G, Watson C. The Rat Brain in Stereotaxic Coordinates, 2nd ed. New York: Academic Press, 1986.
31. SAS/STAT Guide for Personal Computers, 6th ed. Cary, NC: SAS Institute Inc, 1985:1–378.
32. Fow JE, Averill DB, Barnes KL. Mechanisms of angiotensin induced hypotension and bradycardia in the medial solitary tract nucleus. Am J Physiol 1994; 267:H259–H266.
33. Rhaleb NE, Roussi N, Nantel F, et al. Dup 753 is a specific antagonist for the angiotensin receptor. Hypertension 1991; 17:480–484.
34. Chan RKW, Chan YS, Wong TM. Responses of cardiovascular neurons in the rostral ventrolateral medulla of the normotensive Wistar Kyoto and spontaneously hypertensive rats to iontophoretic application of angiotensin II. Brain Res 1991; 556:145–150.
35. Conn PJ, Desai MA. Pharmacology and physiology of metabotropic glutamate receptors in mammalian central nervous system. Drug Dev Res 1991; 24:207–229.
36. Wong TM, Chan YS, Chan RKW. Neuronal and blood pressure responses to Ang II/[Sar^1,Ile^8]-Ang II microinjected into rostral ventrolateral medulla in the spontaneously hypertensive rat (abstr). Hypertension 1993; 21:586.
37. Averill DB, Muratani H, Madalin KJ, Ferrario CM. Cardiovascular actions of angiotensin II in the ventrolateral medulla. In: Kunos G, Ciriello J, eds. Central Neural

Mechanisms in Cardiovascular Regulation. New York: Springer-Verlag, 1991:104–121.
38. Tsutsumi K, Saavedra JM. Characterization and development of angiotensin II receptor subtypes (AT1 and AT2) in the brain. Am J Physiol 1991; 261:R209–R216.
39. Gehlert DR, Gackenheimer SL, Schober DA. Autoradiographic localization of subtypes of angiotensin II antagonist binding in the rat brain. Neuroscience 1991; 44:501–514.
40. Toney GM, Porter JP. Functional roles of brain AT1 and AT2 receptors in the central angiotensin II pressor response in conscious young spontaneously hypertensive rats. Dev Brain Res 1993; 71:193–199.
41. Widdop RE, Gardiner SM, Kemp PA, Bennett T. Central administration of PD 123319 or EXP-3174 inhibits effects of angiotensin II. Am J Physiol 1993; 264: H117–H125.
42. Smith JK, Barron KW. Cardiovascular effects of L-glutamate and tetrodotoxin microinjected into the rostral and caudal ventrolateral medulla in normotensive and spontaneously hypertensive rats. Brain Res 1990; 506:1–8.
43. Smith NK, Barron KW. GABAergic responses in ventrolateral medulla in spontaneously hypertensive rats. Am J Physiol 1990; 258:R450–R456.
44. Chan RKW, Chan YS, Wong TM. Electrophysiological properties of neurons in the rostral ventrolateral medulla of normotensive and spontaneously hypertensive rats. Brain Res 1991; 549:118–126.
45. Chan RKW, Chan YS, Wong TM. Effects of chronic captopril treatment on the electrical-microstimulation-induced blood pressure changes and electrophysiological properties of cardiovascular neurons in the rostral ventrolateral medulla of the spontaneously hypertensive rat. Biol Signals 1993; 2:106–116.
46. Muratani H, Averill DB, Ferrario CM. Effect of angiotensin II in the ventrolateral medulla of spontaneously hypertensive rats. Am J Physiol 1991; 260:R977–R984.

52

Perfused Isolated Adult Brainstem

A Novel Approach to Study Central Chemosensitivity

M. DENAVIT-SAUBIÉ, E. BOUDINOT, F. KATO, and M. P. MORIN-SURUN

Institut Alfred Fessard
Gif-sur-Yvette, France

THORSTEN SCHÄFER

Ruhr-University Bochum
Bochum, Germany

I. Introduction

The ventral surface of the brainstem is involved in respiratory as well as vasomotor control. Consequently it is sometimes difficult in vivo to dissociate a direct respiratory effect from indirect effects associated with changes in vasomotor tone. Furthermore, in vivo, the respiratory effect induced by an increase in P_{CO_2} and that induced by a decrease in pH cannot be differentiated. In order to address these problems, we used an in vitro brainstem preparation of the adult guinea pig (1) which excludes peripheral vasomotor influences. Furthermore, this preparation enables independent control of pH and/or P_{CO_2} by perfusion of various Krebs solutions and also provides a stable intracellular recording situation.

The isolated brainstem of newborn rats, oxygenated by bath superfusion, has been used to study newborn respiratory rhythmogenesis (2–4). Bath superfusion is also sufficient for survival of adult brain slices (5) but not intact adult brainstems, which contain the entire respiratory neuronal network. This may be explained by the large size of the adult brainstem, which cannot be sufficiently oxygenated by bath superfusion alone. In this preparation, arterial perfusion must be coupled with bath superfusion, as has been demonstrated in studies of brainstem-cerebellum connectivity (6). Using the same preparation, we have shown that the isolated adult brainstem conserves different types of neuronal rhythmicities (7) and, while perfusion of the cerebellum is not required, the vascular perfusion is indispensable to obtain a respiratory motor output (1). We routinely used this perfused adult brainstem along with the superfused neonatal rat brainstem to study the genesis and control of respiratory rhythm. However we must keep in mind that although these in vitro preparations (adult and newborn) are useful to understand some of the mechanisms involved in the rhythmogenesis, their functional neuronal networks are not entirely identical to the in vivo network.

In the present study we show that in the adult isolated brainstem, the delimitation of the neuronal structures that are irrigated by each type of perfusion (vascular or superfusion) can be used to assess their respective involvement in rhythmogenesis or central chemoreception. Such delimitation can be obtained by Po_2 measurements, neuronal recordings, and controlling of the composition of the perfusing solutions.

II. Material and Methods

A. Preparation

Data were obtained from adult guinea pigs (150 to 330 g) anesthetized with sodium pentobarbital (30 mg/kg IP), decapited at the first cervical vertebra and decerebrated at the intercollicular level. The brainstem, with the cerebellum, was rapidly removed from the cranium. During this surgical procedure, cold (4°C) Ringer's solution was dripped onto the exposed brain surfaces. The brainstem was maintained by pins, the ventral surface upward in a recording chamber filled with Krebs solution at room temperature.

The entire surface of the brainstem was completely immersed and continuously superfused at a constant rate (2 to 2.5 mL/min) in the chamber (Fig. 1A). The basilar artery was rapidly cannulated with a catheter inserted at the level of the pons, and the vascular system was perfused with Krebs solution using a peristaltic pump (1.2 to 2 mL/min). Distal ends of arteries at the first cervical segment of the spinal cord were ligated.

B. Krebs Solutions

The control Krebs solution consisted of (in m*M*): NaCl 124, KCl 5, $MgSO_4$ 2.2, KH_2PO_4 1.2, $NaHCO_3$ 26, $CaCl_2$ 2, glucose 10, and was saturated with 95% O_2 and 5% CO_2 (pH 7.35 at 27°C).

Under normal Pco_2 (5%), the pH of the solution was adjusted by changing bicarbonate concentrations as follows: acidic Krebs: $NaHCO_3$ 10 m*M* (pH 6.99 ± 0.02 SEM) and alkaline Krebs: $NaHCO_3$ 60 m*M* (pH 7.8 ± 0.005 SEM).

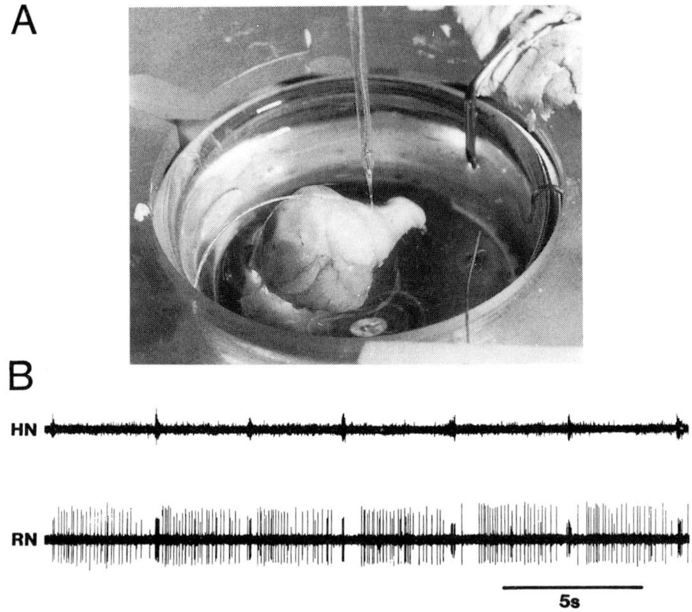

Figure 1 Respiratory-related activity of the perfused isolated preparation of adult guinea pig. (A) Photograph of the preparation in the recording chamber. The ventral surface is upward. A catheter is inserted into the basilar artery for vascular perfusion. A microelectrode is inserted through the ventral surface. The dark portion of the photograph is the nonperfused cerebellum. (B) Simultaneous recordings of rhythmic activities in the hypoglossal nerve (HN) and neurons in the ventral brainstem (RN). The multiunit recording shows two respiratory-related neuronal activities: one active during the hypoglossal burst activity (inspiratory neuron) and one active during the silent period (expiratory neuron).

The osmolarity of the Krebs solution was maintained by adjusting the NaCl concentration.

Under normal pH (7.35) adjusted with bicarbonate, increase in P_{CO_2} was obtained by increasing in the percentage of CO_2 (F_{CO_2}) bubbling the Krebs solution.

Under normal bicarbonate concentration (26 m*M*), acidic Krebs with high P_{CO_2} was obtained by increasing F_{CO_2} bubbling the Krebs.

C. Tissue P_{O_2} Measurements

Tissue P_{O_2} was measured at 100-μm intervals from the ventral to the dorsal medullary surface at the level of the ventral respiratory group (1.5 to 2.5 mm rostral to the obex and 1.5 to 2.5 mm lateral to the midline). The P_{O_2} electrodes were advanced using an electrically driven micromanipulator. In one experiment, Po_2 was measured in a superfused preparation without perfusion in the basilar artery.

Two types of oxygen-sensitive microelectrodes (Clark type, tip 20 μm, and combined needle electrodes, cathode 25 μm) connected to an oxygen meter were used. Electrodes were allowed to stabilize for at least 4 h prior to use and calibrated in Krebs solutions equilibrated with 100% N_2 and air at the same temperature as the bath solution before and after the experiments to prevent probable drifts.

D. Electrophysiological Recordings

Respiratory-related activity was recorded from the hypoglossal roots using suction electrode (Fig. 1B). Full wave neural activity was rectified and integrated using a passing RC integrating circuit with a time constant of 20 ms. Neuronal activities were recorded extracellularly through the central barrel of a multibarrel glass microelectrode filled with NaCl (4 *M*). Adjacent barrels contained methyl blue (4% in NaCl) in order to localize recording sites. The electrodes were inserted through the ventral surface in the medulla rostrally to the emergence of the hypoglossal roots and caudally to the posterior inferior cerebellar arteries. Intracellular recordings were obtained using glass micropipettes filled with KCl (3 *M*) or Kac (3 *M*) with resistance ranges of 25 to 50 mΩ and 50 to 100 MΩ, respectively.

E. Data Analysis

The periodic activity recorded from hypoglossal roots consisted of a single burst followed by low-amplitude tonic activity. These characteristics were comparable to the respiratory pattern recorded in vivo from the hypoglossal nerve; as such, the activity was considered to be respiratory-related, with the burst of activity representing the inspiratory phase. The temporal discharge pattern of neurons was classified in relation to the periodic activity of the hypoglossal nerve.

III. Results and Discussion

A. Diffusion of the Perfusate in the Tissue

The bath superfusion without vascular perfusion provided tissue oxygenation only in a superficial layer. The Po_2 decreased from a mean value of 450 mm Hg at the surface to 150 mm Hg at a depth of 200 μm and reached zero at 400 μm below the ventral surface (Fig. 2). When the superfused brainstem was also perfused through the basilar artery, the Po_2 was stable at around 200 mm Hg from 500 μm to 5000 μm, thus supplying oxygen to deep structures (Fig. 2).

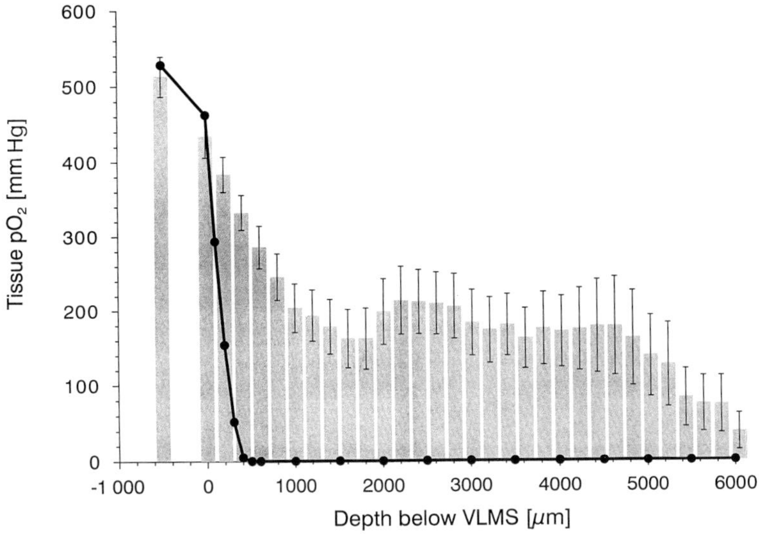

Figure 2 Tissue PO_2 profiles related to the depth from the ventral surface. The heavy line represents the PO_2 profile of one preparation only superfused by the bath. Mean values of PO_2 ($\pm$ SEM) for each step of depth obtained from 16 traces in 9 preparations perfused by basilar artery and superficial bath are presented in the histogram.

These results suggest that the absence of rhythmic respiratory activity in the hypoglossal nerve in the isolated adult brainstem maintained with only bath superfusion (1) is due to insufficient oxygenation of the respiratory network. Under bath superfusion, oxygen penetrates the superficial layer of the brainstem but cannot reach the ventral respiratory group located at 1500 μm from the surface. In superficial layers, we have identified some respiratory-related neurons (1) that may not be capable of generating a respiratory output in the absence of functional deep brainstem structures.

These results show that we can separately irrigate the superficial layers (accessed by the bath superfusion) supposedly containing the chemosensitive fields and the brainstem (accessed by the basilar perfusion) where the respiratory network is located. This dissociation is also supported by our findings that an increase of PCO_2 with acidification of the Krebs solution induces different respiratory effects depending on the route of administration (8) (bath versus arterial). The following results show that this preparation also allows a study of the specific stimuli implicated in respiratory chemosensitivity.

B. Respiratory Effects of Increased PCO_2 in the Perfusing Solution

Elevation of PCO_2 without change in pH Krebs evoked different respiratory effects depending on the routes of administration. Bath superfusion evoked an increase of the burst amplitude of the hypoglossal nerve, associated with a decrease of the burst frequency. On the contrary, perfusion of the same Krebs through the basilar artery induced an increase in the burst frequency and a decrease in the burst amplitude (Fig. 3).

C. Respiratory Effects of Changes in pH in the Perfused Solutions

In contrast to changes of P_{CO_2}, alterations in the pH of the Krebs solution evoked the same respiratory effects regardless of the perfusion routes.

Under normal P_{CO_2} (5%), a lowering pH increased the burst frequency of the hypoglossal nerve (Fig. 4). On the other hand, alkalinization of the superfusate reduced the respiratory frequency.

These results indicate that increases in P_{CO_2} without change in the pH of the Krebs have differential respiratory effects when applied to vascular perfusion and bath superfusion. This indicates that CO_2 as O_2 (9) diffusion of the bath superfusion reaches only the superficial layer of the brainstem. The respiratory effect evoked after high P_{CO_2} in the bath superfusion is consistent with the concept presented by the group of Loeschcke (10,11)—that there is P_{CO_2}

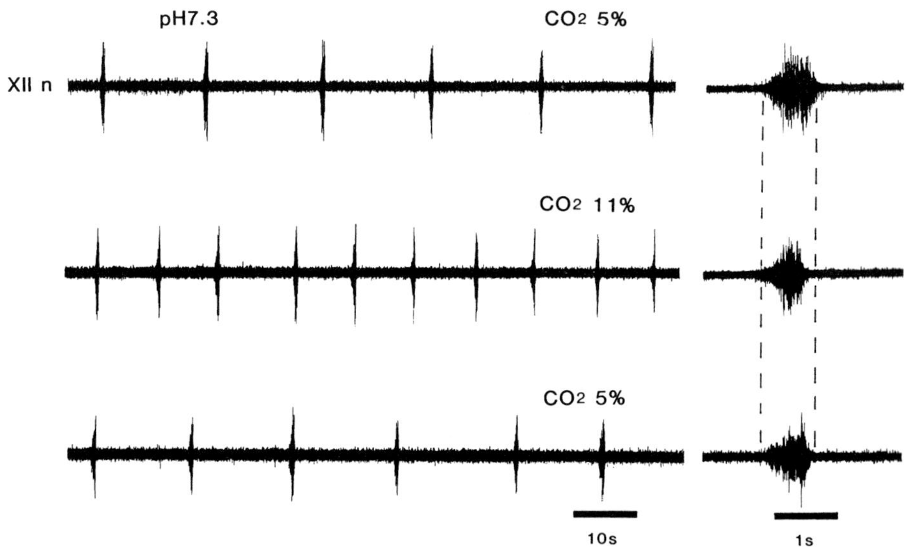

Figure 3 Increase of P_{CO_2} in basilar artery perfusate without change in pH. On the left, recordings of the hypoglossal nerve (XII n) during the control period (upper trace), the change in perfusion (middle trace), and the recovery (lower trace). On the right, recording of one burst during these different periods. In this particular case, the respiratory frequency is increased by decreasing both inspiratory and expiratory durations during change in the vascular P_{CO_2}.

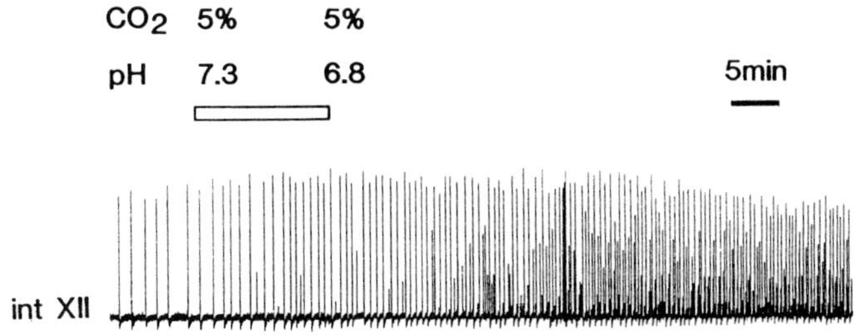

Figure 4 Effect of acidification of bath superfusion under normal P_{CO_2} on respiratory discharge of the hypoglossal nerve. Change in bath superfusion is indicated by the horizontal bar above the integrated hypoglossal nerve activity (int XII).

sensitivity in the superficial layers of the brainstem in vivo. On the other hand, similar respiratory effects are obtained when pH is modified in either one of these perfusates, suggesting that varying the pH modulates the respiratory network with a common mechanism within the brainstem and the superficial layer. This distinction suggests that different mechanisms at the neuronal level may be implicated in respiratory chemosensitive regulation.

D. Effects of Vascular Modification of pH and/or P_{CO_2} on Neuronal Discharges

A total of 49 neurons were recorded either extracellularly or intracellularly. Of these neurons, 36 were not correlated to the rhythmic burst of the hypoglossal nerve. Among these non-respiratory-related neurons, 8 evoked action potentials only when they were depolarized, while 13 were identified as respiratory-related (7 inspiratory, 4 expiratory, 2 postinspiratory).

Perfusion of acidic Krebs with high P_{CO_2}, which increases the frequency of respiratory bursts, increased the discharge frequency of most neurons recorded (Fig. 5). In this figure, the neuron presented a tonic respiratory-related discharge that allowed a possible dissociation between the neuronal chemosensitivity in the different respiratory phases. The cycle-triggered histogram of the spike distribution of this inspiratory neuron showed that spike occurrence probability during the expiratory phase was also increased, suggesting that the direct or indirect synaptic input from a chemosensitive mechanism to this neuron was modulating not only its active phase but also its inactive phase (Fig. 6).

Most of the neurons tested, respiratory or non-respiratory-related, demonstrated membrane potential hyperpolarization and a reduction in firing frequency following alkaline perfusion. In Figure 7, spikes occurring during the inspiratory burst were less sensitive to alkalinization of the solution than those

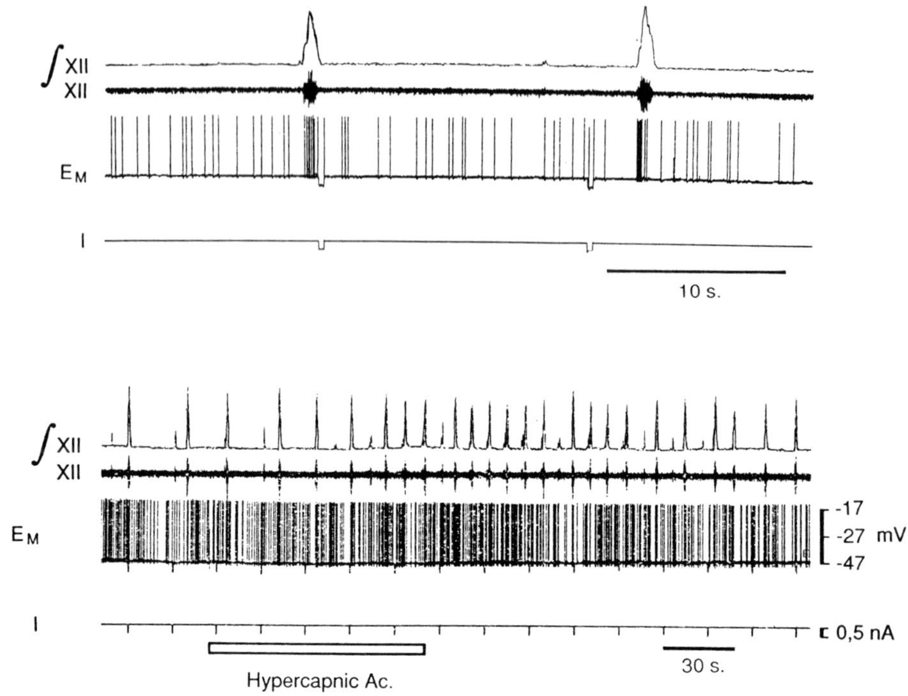

Figure 5 Simultaneous recordings of the hypoglossal nerve and an inspiratory related neuron after hypercapnic acidic Krebs perfusion through the basilar artery. Upper traces show the raw (XII) and integrated hypoglossal nerve activity (∫XII), the membrane potential of the inspiratory neuron (Em), and the current pulse (I). Lower traces represent the same recordings at a slower time base. Hypercapnic acidic vascular perfusion is indicated by the horizontal bar.

that appeared during the interburst interval. This neuronal response was associated with a reduction in the hypoglossal burst discharge frequency.

These results show that most neurons are influenced by P_{CO_2} and pH variations within the brainstem. Other studies have shown that increase in P_{CO_2} may increase, reduce, or have no effect on the discharge frequency of neurons located in the brainstem (12–15). However, the proportion of neurons that increase their discharge is particularly high in either the ventral respiratory group (16–18) or the dorsal respiratory group (18,19). All studies conducted thus far have failed to demonstrate a specific neuronal population implicated in respiratory chemosensitivity. This may be due to the extensive heterogeneity of neuronal groups within the brainstem (20) and the large dendritic arborization of these neurons extending from deep structures to the ventral surface (21). These responses to acidic Krebs with high P_{CO_2} may be the result of

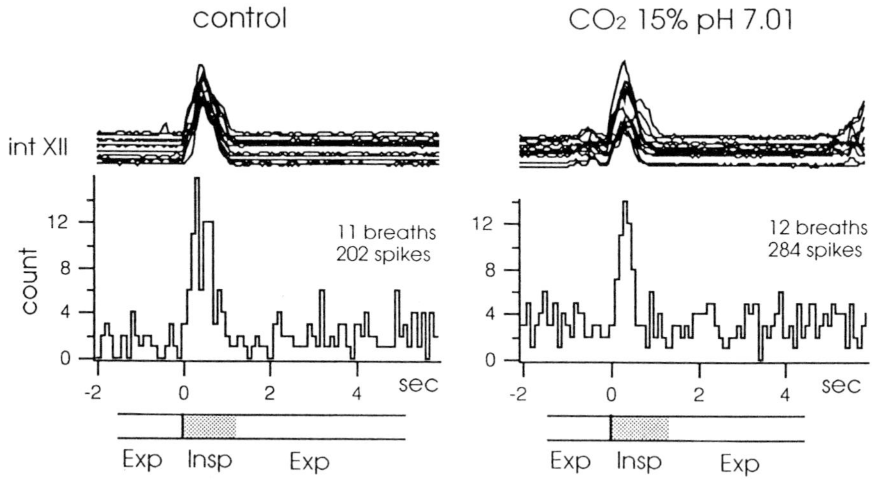

Figure 6 Cycle-triggered histogram of the spike distribution of an inspiratory (Insp) cell recorded near the paragigantocellular nucleus about 500 μm from the ventral surface. Upper trace, cycle-triggered superimposition of the integrated hypoglossal nerve activity. Bottom, cycle-triggered histogram: onset of each inspiratory activity in the hypoglossal nerve discharge was used for trigger. Abscissa, time from the trigger point; ordinates, spike counts. Bin width = 100 ms. Left, vascular perfusion with the control Krebs solution; right, vascular perfusion with high P_{CO_2} Krebs solution at maximal effect.

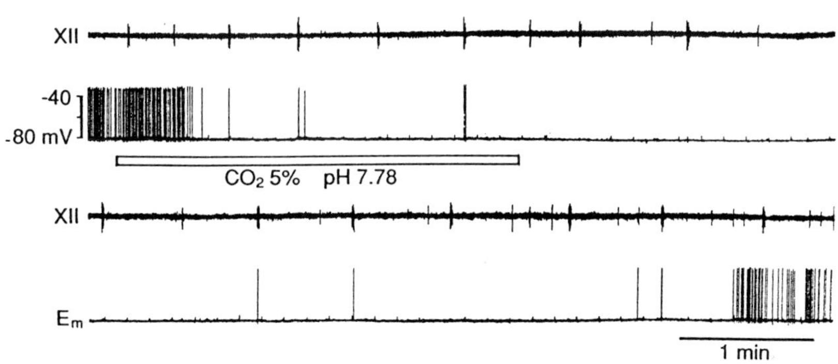

Figure 7 Simultaneous recordings of the hypoglossal nerve (XII) and the membrane potential (Em) of an inspiratory related neuron after alkalinization of the Krebs perfusing the basilar artery. Note that the last spikes before a quiescent period appeared only during the hypoglossal inspiratory burst.

synaptic inputs. In most studies, there is little evidence in favor of a direct effect of CO_2 or pH on respiratory-related neurons. In extracellular recordings, the cellular chemosensitivity cannot be assessed directly. The only direct effect has been described in the dorsal part of the brainstem in a slice preparation (19). In our intracellular recordings, respiratory-related neurons appear to be synaptically driven, as those described in vivo by Takeda and Haji (14). The analysis of single-channel activity of the neurons located in the ventral surface, which could be responsible for specific chemosensitivity, may provide important information regarding the cellular mechanism for the transduction of CO_2 and/or H^+.

IV. Conclusion

The present results have shown that only the brainstem superficial layer is oxygenated by bath superfusion, while vascular perfusion allows oxygenation in deep brainstem structures. Consequently this preparation allows for the dissociation of the chemosensitivity within superficial and deep structures of the brainstem and the evaluation of the specific role of the neurons within these different structures.

Acknowledgments

This work was supported by CNRS and DRET 92/075.

References

1. Morin-Surun MP, Boudinot E, Sarraseca H, et al. Respiratory network remains functional in a mature guinea pig brainstem isolated in vitro. Exp Brain Res 1992; 90:375–383.
2. Suzue T. Respiratory rhythm generation in the in vitro brain stem spinal cord preparation of the neonatal rat. J Physiol (Lond) 1984; 354:173–183.
3. Smith JC, Ellenberger HH, Ballanyi K, et al. Pre-Bötzinger complex: A brainstem region that may generate respiratory rhythm in mammals. Science 1991; 254:726–729.
4. Errchidi S, Hilaire G, Monteau R. Permanent release of noradrenaline modulates respiratory frequency in the newborn rat: An in vitro study. J Physiol (Lond) 1990; 429:497–510.
5. Paton JFR, Ramirez JM, Richter DW. A novel rhythmically active in vitro preparation containing the respiratory network of adult rats (abstr). XXXII Congress of the IUPS, Glasgow, Aug 1–6, 1993.
6. Llinas R, Mühlethaler M. An electrophysiological study of the in vitro, perfused brain stem-cerebellum of adult guinea-pig. J Physiol (Lond) 1988; 404:215–240.
7. Morin-Surun MP, Denavit-Saubié M. Rhythmic discharges in the perfused isolated brainstem preparation of adult guinea-pig. Neurosci Lett 1989; 101:57–61.
8. Morin-Surun MP, Boudinot E, Schäfer T, Denavit-Saubié M. Brainstem chemosensitivity: Its implication in the central respiratory regulation. Biol Neonate 1994; 65:166–170.
9. Schäfer T, Morin-Surun MP, Denavit-Saubié M. Oxygen supply and respiratory-like activity in the isolated perfused brainstem of the adult guinea pig. Brain Res 1993; 618:246–250.

10. Loeschcke HH. Central chemosensitivity and the reaction theory. J Physiol (Lond) 1982; 332:1–24.
11. Schläfke ME. Central chemosensitivity: A respiratory drive. Rev Physiol Biochem Pharmacol 1981; 90:171–244.
12. Okada Y, Mückenhoff K, Scheid P. Hypercapnia and medullary neurons in the isolated brain stem-spinal cord of the rat. Respir Physiol 1993; 93:327–336.
13. Mitchell RA, Herbert DA. The effect of carbon dioxide on the membrane potential of medullary respiratory neurons. Brain Res 1974; 75:345–349.
14. Fukuda Y, Loeschcke HH. A cholinergic mechanism involved in the neuronal excitation by H^+ in the respiratory chemosensitive structures of the ventral medulla oblongata of rats in vitro. Eur J Physiol 1979; 379:125–135.
15. Trouth CO, Patrickson JW, Holloway JA, Wright LE. Neurophysiological studies on superficial medullary chemosensitive area for respiration. Brain Res 1982; 246: 47–56.
16. Nattie EE, Fung M-L, Li A, St John WM. Responses of respiratory modulated and tonic units in the retrotrapezoid nucleus to CO_2. Respir Physiol 1993; 94:35–50.
17. Takeda R, Haji A. Synaptic response of bulbar respiratory neurons to hypercapnic stimulation in peripherally chemodenervated cats. Brain Res 1991; 561:307–317.
18. Kogo N, Arita H. In vivo study on medullary H^+-sensitive neurons. J Appl Physiol 1990; 69:1408–1412.
19. Dean JB, Bayliss DA, Erickson JT, et al. Depolarization and stimulation of neurons in nucleus tractus solitarii by carbon dioxide does not require chemical synaptic input. Neuroscience 1990; 36:207–216.
20. McCrimmon DR, Feldman JL, Speck DF, et al. Functional heterogeneity of dorsal, ventral and pontine respiratory groups revealed by micropharmacological techniques: Neurobiology of the control of breathing. New York: Raven Press, 1986: 201–208.
21. Bianchi AL, Grelot L, Iscoe S, Remmers JE. Electrophysiological properties of rostral medullary respiratory neurones in the cat: An intracellular study. J Physiol (Lond) 1988; 407:293–310.

53

Tonic Chemoreceptor Input as the Background for Respiratory Rhythm

MORTON I. COHEN

Albert Einstein College of Medicine
Bronx, New York

Many studies have shown that structures at or near the ventral medullary surface (VMS) are involved in chemoreceptor function (for reviews, cf. Refs. 1–3). The pioneering work of the Bochum group has been expanded in more recent studies, where lesions (chemical or electrolytic) or chemical stimulations have been used to explore chemoreceptive function at various brainstem sites (4,5). In addition, the VMS in the neonatal rat in vitro preparation seems to be the site of a similar chemoreceptor action (6).

To obtain more information on VMS chemoreceptor function, unit recordings in the region have been performed and the unit responses to pH and P_{CO_2} have been observed. Several studies have reported that an appreciable number of tonic (i.e., not respiratory-modulated) units increase their discharge frequency in response to lowered pH or increased P_{CO_2} (7–12). However, similar responses have also been observed for nonrespiratory units in other regions, such as the dorsal and ventral respiratory groups (7).

Recently, attention has been focused on the retrotrapezoid nucleus (RTN), a site where respiratory-modulated neurons of various types are found (13), and where chemical lesions produce marked depression or elimination of respiratory output (4). In addition, the discharges of both respiratory and nonrespiratory units in the RTN are increased by hypercapnia (8). These results suggest that the RTN may be a site of integration and transmission of central chemoreceptor input. Two qualifications of this hypothesis can be stated: (a) Because the RTN is very near the VMS, it is difficult to distinguish between an effect on VMS neurons and an effect on neurons located slightly deeper. (b) Nonrespiratory neurons excited by hypercapnia are also found in other medullary regions (7).

These observations are relevant to the following question: Does the systemic effect of chemical drive originate by actions on tonic nonrespiratory units, on respiratory-modulated units, or on both? While there is at present no definitive answer to this question, it was approached in earlier studies by observing the effect of lowered P_{CO_2} (hypocapnia) on discharge patterns of brainstem respiratory neurons and of respiratory motor outputs.

These earlier studies reported that some medullary respiratory neurons changed their discharge from a rhythmic to a tonic (nonmodulated) pattern during hypocapnia which was sufficient to produce apnea (absence of rhythmic respiratory output) (14–16). An example of this effect for an expiratory (E) neuron in the cat medulla is shown in Figure 1. It can be seen that during moderate hypocapnia (panel A), the E neuron's bursts alternate with the phrenic bursts. At a lower level of P_{CO_2} (panel B), the frequency of the cycle is markedly increased, but the pattern of alternating phrenic and E neuron bursts persists. Eventually, at the lowest P_{CO_2} level (1.5% end-tidal in panel C), rhythmicity is lost both in the phrenic and the E neuron activity, the latter showing a tonic pattern of discharge. When the inhaled CO_2 is then increased (panels D1 and D2), rhythmicity reappears: each phrenic burst (even small ones as in panel D1) is associated with interruption (silence) of the E neuron's discharge.

The phrenic and E neuron responses shown in Figure 1 are representative, respectively, of two types of response to hypocapnia that were found in a survey of brain-stem respiratory neurons in the cat (16): (a) Almost all inspiratory (I) neurons showed a reduction of firing with lowered P_{CO_2} and ultimately became silent with a sufficient degree of hypocapnia. (b) One-third of the E neurons had tonic (not respiratory-modulated) firing during hypocapnic apnea. The latter response was also shown by another category of neuron, the expiratory-inspiratory (EI) type, which during normocapnia fires with maximum frequency in late E and early I and with minimum (or zero) frequency in early E; during hypocapnia, activity appears during the previous silent period and ultimately the neuron's activity loses its respiratory rhythm. This response pattern (illustrated in Figure 14 of Ref. 16) was shown by four-fifths of the neurons of EI type.

Subsequently, other investigators (17–20) observed responses in E motor output (internal intercostal and triangularis sterni) that were similar to those of Figure 1 (tonic E activity during hypocapnic apnea and return of rhythm with raised P_{CO_2}).

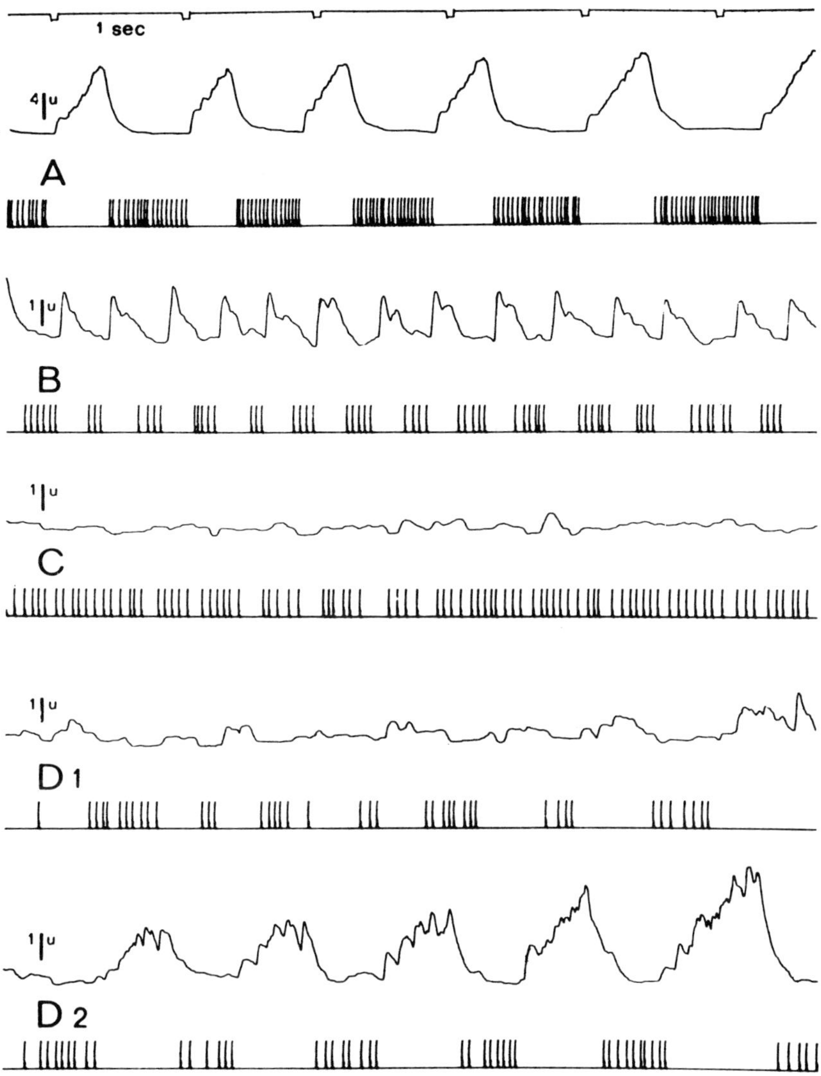

Figure 1 Effect of hypocapnia on phrenic activity and discharge of a medullary expiratory neuron. For each panel—*top trace*: phrenic (half-wave rectified and integrated with time constant of 0.1 s); *bottom trace*: standard pulses triggered from spikes of the neuron. End-tidal CO_2: (A) 3.4%; (B) 2.3%; (C) 1.5%. (A and B) Respiratory rhythm of both phrenic nerve and neuron. Note (vertical bars) that phrenic gain in panels B to D is four times that in panel A. (C) Note lack of respiratory rhythm in both signals. (D1 and D2, consecutive traces): recovery of rhythm after administration of high-CO_2 mixture starting 63 s before the trace in panel D1. (From Ref. 16.)

In addition to this "expiratory bias" during hypocapnia, in some circumstances (e.g., during hypoxia) an "inspiratory bias" was observed: diaphragmatic or phrenic activity had a tonic pattern during hypocapnic apnea (20). This pattern in the diaphragm and phrenic seems to be the usual one in the rabbit during hypocapnia (21).

Another interesting observation is that even during hypocapnic apnea, tonic (I or E) discharge is still excited by chemoreceptor input: increasing P_{CO_2} increases discharge until the point where rhythmic discharge resumes (17–20).

On the basis of these observations, qualitative models of the origin of rhythmicity were proposed (as illustrated in Fig. 18 of Ref. 16 and Fig. 12 of Ref. 18). During hypocapnic apnea, an increase of chemoreceptor input monotonically increases central I and/or E discharge, until activity in one subsystem becomes strong enough to inhibit activity in the other. Thus, rhythm arises from the background tonic excitation, which allows the discharges of I and E neurons to reach levels adequate for rhythmic reciprocal inhibition (I neurons inhibiting E neurons and E neurons inhibiting I neurons).

This role of the tonic chemoreceptor input is consistent with either a network or a conditional pacemaker model of rhythm generation (22). However, it is interesting to note that the response to increased P_{CO_2} by the in vitro preparation is quite different from the response in vivo (as shown in Fig. 3 of Ref. 6): the chemical input affects predominantly the respiratory cycle frequency, with only a minor change in the amplitude of the inspiratory burst. This suggests that in vitro, the respiratory burst has more of a pacemakerlike character than in in vivo, where network synaptic inputs may be more prominent.

Acknowledgment

Supported by N.I.H. Grant HL-27300.

References

1. Bruce EN, Cherniack NS. Central chemoreceptors. J Appl Physiol 1987; 62:389–402.
2. Loeschcke HH. Central chemosensitivity and the reaction theory. J Physiol (Lond) 1982; 332:1–24.
3. Schläfke ME. Central chemosensitivity: A respiratory drive. Rev Physiol Biochem Pharmacol 1981; 90:171–244.
4. Nattie EE, Li A, St John WM. Lesions in retrotrapezoid nucleus decrease ventilatory output in anesthetized or decerebrate cats. J Appl Physiol 1991; 71:1364–1375.
5. Coates EL, Li A, Nattie EE. Widespread sites of brain stem ventilatory chemoreceptors. J Appl Physiol 1993; 75:5–14.
6. Issa FG, Remmers JE. Identification of a subsurface area in the ventral medulla sensitive to local changes in P_{CO_2}. J Appl Physiol 1992; 72:439–446.

7. Arita H, Kogo N, Ichikawa K. Locations of medullary neurons with non-phasic discharges excited by stimulation of central and/or peripheral chemoreceptors and by activation of nociceptors in cat. Brain Res 1988; 442:1–10.
8. Nattie EE, Fung M-L, Li A, St John WM. Responses of respiratory modulated and tonic units in the retrotrapezoid nucleus to CO_2. Respir Physiol 1993; 94:35–50.
9. Pokorski M. Neurophysiological studies on central chemosensor in medullary ventrolateral areas. Am J Physiol 1976; 230:1288–1295.
10. Schläfke ME, Pokorski M, See WR, et al. Chemosensitive neurons on the ventral medullary surface. Bull Physiopathol Resp 1975; 11:277–284.
11. See WR. Untersuchungen zur Konvergenz von atemwirksamen Afferenzen and chemosensiblen Neuronen des mittleren Feldes (Area S) der ventralen Oberfläche der Medulla oblongata der Katze. Bochum: Ruhr University, 1983.
12. Trouth CO, Patrickson JW, Holloway JA, Wright LE. Neurophysiological studies on superficial medullary chemosensitive area for respiration. Brain Res 1982; 246: 47–56.
13. Connelly CA, Ellenberger HH, Feldman JL. Respiratory activity in retrotrapezoid nucleus in cat. Am J Physiol 1990; 258:L33–L44.
14. Batsel HL. Activity of bulbar respiratory neurons during passive hyperventilation. Exp Neurol 1967; 19:357–374.
15. Nesland RS, Plum F. Subtypes of medullary respiratory neurons. Exp Neurol 1965; 12:337–348.
16. Cohen MI. Discharge patterns of brain-stem respiratory neurons in relation to carbon dioxide tension. J Neurophysiol 1968; 31:142–165.
17. Bainton CR, Kirkwood PA, Sears TA. On the transmission of the stimulating effects of carbon dioxide to the muscles of respiration. J Physiol (Lond) 1978; 280: 249–272.
18. Bainton CR, Kirkwood PA. The effect of carbon dioxide on the tonic and the rhythmic discharges of expiratory bulbospinal neurones. J Physiol (Lond) 1979; 296:291–314.
19. Horner RL, Kozar LF, Phillipson EA. Tonic respiratory drive in the absence of rhythm generation in the conscious dog. J Appl Physiol 1994; 76:671–680.
20. Sears TA, Berger AJ, Phillipson EA. Reciprocal tonic activation of inspiratory and expiratory motoneurones by chemical drives. Nature 1982; 299:728–730.
21. Wyss OAM. Die tonische Innervation des Zwerchfells. Pflügers Arch 1941; 244: 712–735.
22. Smith JC, Ellenberger HH, Ballanyi K, et al. Pre-Bötzinger complex: A brainstem region that may generate respiratory rhythm in mammals. Science 1991; 254:726–729.

54

Interactions Between Area Postrema and Sympathetic or Respiratory-Related Neurons in the Rostral Ventrolateral Medulla

ANN C. BONHAM and CHRISTOPHER G. WILSON

University of California at Davis
Davis, California

I. Introduction

The rostral ventrolateral medulla contains sympathoexcitatory neurons that are considered to contribute to baseline sympathetic vasomotor tone and respiratory-related neurons, which contribute to the generation of respiratory motor pattern. The area postrema, located in the dorsomedial medulla, is the most caudal of the circumventricular organs and is best recognized as the central "chemoreceptor trigger zone" for vomiting. More recently, neurons in this structure have been shown to play an important role in regulating cardiovascular function. The studies described in this paper determine the role of area postrema neurons in modifying the discharge patterns of sympathetic or respiratory-related neurons in the rostral ventrolateral medulla in rabbits.

II. Area Postrema Interactions with Sympathetic-Related Neurons in the Rostral Ventrolateral Medulla

Sympathoexcitatory neurons in the rostral ventrolateral medulla provide a tonic excitatory drive to vasomotor preganglionic sympathetic neurons and thereby contribute to the maintenance of baseline sympathetic vasomotor tone (1–3). The neurons discharge with patterns consistently and temporally related to postganglionic sympathetic nerve activity (1–6). Many of the neurons make monosynaptic excitatory projections to preganglionic sympathetic neurons in the intermediolateral cell column (1–5,7). These sympathetic-related neurons also form the final common pathway for relaying reflex and descending changes in sympathetic nerve activity. Activation of the arterial baroreceptors (2–6,8,9), peripheral carotid chemoreceptors (10), low-pressure receptors in the heart and lung (11), nociceptors (12), and somatosensory receptors (13,14) has been shown to modulate the activity of neurons in the rostral ventrolateral medulla, as has activation of higher brain centers such as the lateral hypothalamus (15, 16). Therefore, it seems likely that most changes in sympathetic nerve activity are relayed, ultimately, by synaptic input onto these sympathetic neurons in the rostral ventrolateral medulla.

The area postrema is situated on the dorsal surface of the medulla over the fourth ventricle. In rats and rabbits, the structure spans the midline, extending from the calamus scriptorius to the obex, while in mammals other than rodents, the area postrema is bilateral (17). The cell types in the area postrema include small neurons (7 to 15 μm), flattened ependymal cells, astrocytes, and a few oligodendroglia. Because neurons in the area postrema do not have a blood-brain barrier, they are susceptible to the influences of hormones and peptides in the peripheral circulation. Previous studies have suggested that circulating substances with cardiovascular effects—including vasopressin (18, 19), endothelin (20), and angiotensin II (21)—excite neurons in the area postrema. This susceptibility to blood-borne substances led to the hypothesis that area postrema neurons, responding to humoral influences, may modulate autonomic reflex function.

A role for area postrema in regulating cardiovascular function has been demonstrated by the effects of either stimulation or destruction of area postrema neurons on sympathetic nerve activity (22,23), arterial blood pressure (24,25), and baroreflex function (18,22,26). Electrical stimulation at 10 to 40 Hz or chemical stimulation with glutamate of area postrema neurons decreases sympathetic nerve activity in rabbits without changing arterial blood pressure (22,23), decreases arterial blood pressure in rats (25,27), and increases arterial blood pressure in dogs (24). Stimulation of area postrema neurons either by circulating vasopressin or by direct background electrical excitation augments baroreflex-mediated and volume expansion–evoked inhibition of sympathetic nerve activity (18,19). Area postrema neurons are required for the progression of chronic angiotensin-induced hypertension in normotensive rats (28) and for the full expression of hypertension in spontaneously hypertensive rats (29). The neural circuitries for relaying these cardiovascular changes have not been investigated, but area postrema neurons project to a number of other brain regions with known cardiovascular functions.

Anatomical and electrophysiological studies have shown that area postrema neurons project to the medial aspect of the nucleus tractus solitarius (NTS) (29), have reciprocal connections with neurons in the lateral parabrachial nucleus (32), and also project to the rostral ventrolateral medulla (31) and to the hypothalamic paraventricular nucleus (29). Since changes in sympathetic nerve activity are relayed by sympathetic neurons in the rostral ventrolateral medulla and since area postrema neurons appear to fine-tune reflex control of sympathetic nerve activity, it seems likely that sympathetic neurons in the rostral ventrolateral medulla may be an important potential target for area postrema modulation. Accordingly, we determined, in rabbit, the effect of single electrical pulses delivered to the area postrema on the activity of sympathetic neurons in the rostral ventrolateral medulla. Sympathetic neurons exhibited discharge patterns that were synchronized to sympathetic nerve activity as determined by spike-triggered averaging; they exhibited a cardiac rhythm as determined by post R-wave analysis and were inhibited by activation of the baroreceptors produced by intravenous bolus injections of phenylephrine.

Single sequential stimuli delivered at 1 Hz to the area postrema produced an initial excitation followed by an inhibition of sympathetic neurons in the rostral ventrolateral medulla. An example is shown in Figure 1A. The area postrema–evoked increase in unit activity had an onset latency of 22 ms and peak latency of 26 ms. The excitation was followed by an inhibition that had an onset of 45 ms and lasted 123 ms. Figure 1B shows the post-R wave averages of arterial blood pressure and sympathetic nerve activity and the post-R wave histogram of unit activity. The neuron exhibited a cardiac rhythm and was most likely to discharge in late diastole ~35 ms after the R wave trigger (Fig. 1B, lowest panel). The cardiac rhythm of sympathetic nerve activity (Fig. 1B, middle panel) was delayed with respect to the cardiac rhythm of the unit (Fig. 1B, lowest panel). Figure 1C illustrates the temporal relationship between the unit activity and renal sympathetic nerve activity. Sympathetic nerve activity was locked to the unit discharge and peaked ~110 ms after the unitary spike. In contrast, sympathetic nerve activity exhibited no consistent temporal relationship to random triggers delivered with the same average frequency as that of the medullary neurons. Activation of the baroreceptors with an IV bolus injection of phenylephrine inhibited the unit and sympathetic nerve activity in parallel (Fig. 1D).

Area postrema stimuli delivered at 1 Hz also produced an excitation followed by a pronounced inhibition of averaged sympathetic nerve activity (Fig. 2). The peak increase in averaged sympathetic nerve activity occurred 190 ms after the stimuli and exceeded the mean peak for the excitatory response of the medullary sympathetic neurons to area postrema stimulation (peak latency = 29 ± 10 ms) by about 160 ms. The area postrema–evoked excitation of sympathetic nerve activity was followed by an inhibition. The postexcitatory inhibition occurred at 238 ms after the stimulus and exceeded the inhibition onset

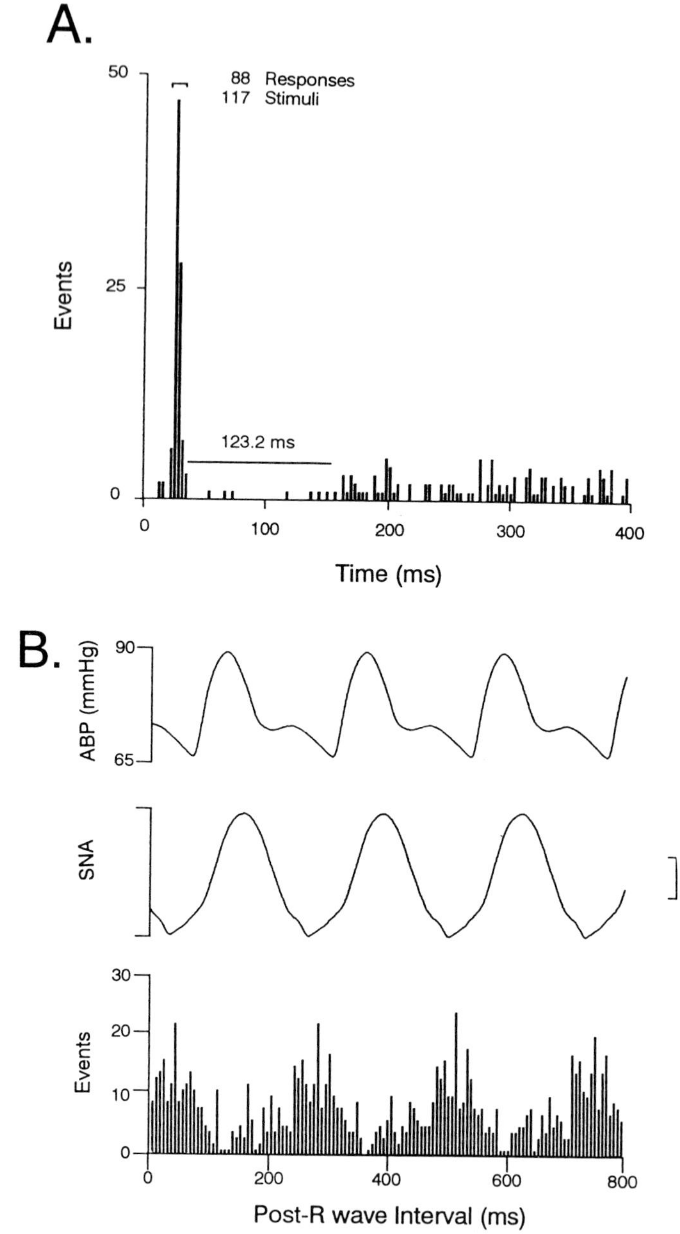

Figure 1 (A) Poststimulus time histogram of response of a sympathetic neuron in the rostral ventrolateral medulla to sequential stimuli delivered to area postrema (28 μA, 0.3 ms, 1 Hz). Bin width is 3.2 ms. (B) Post-R wave averages (500 triggers) of arterial blood pressure (ABP) and sympathetic nerve activity (SNA) and post-R wave histogram (234 triggers) of unit activity. Vertical calibration is 200 μV for SNA. Bin width is 6.4

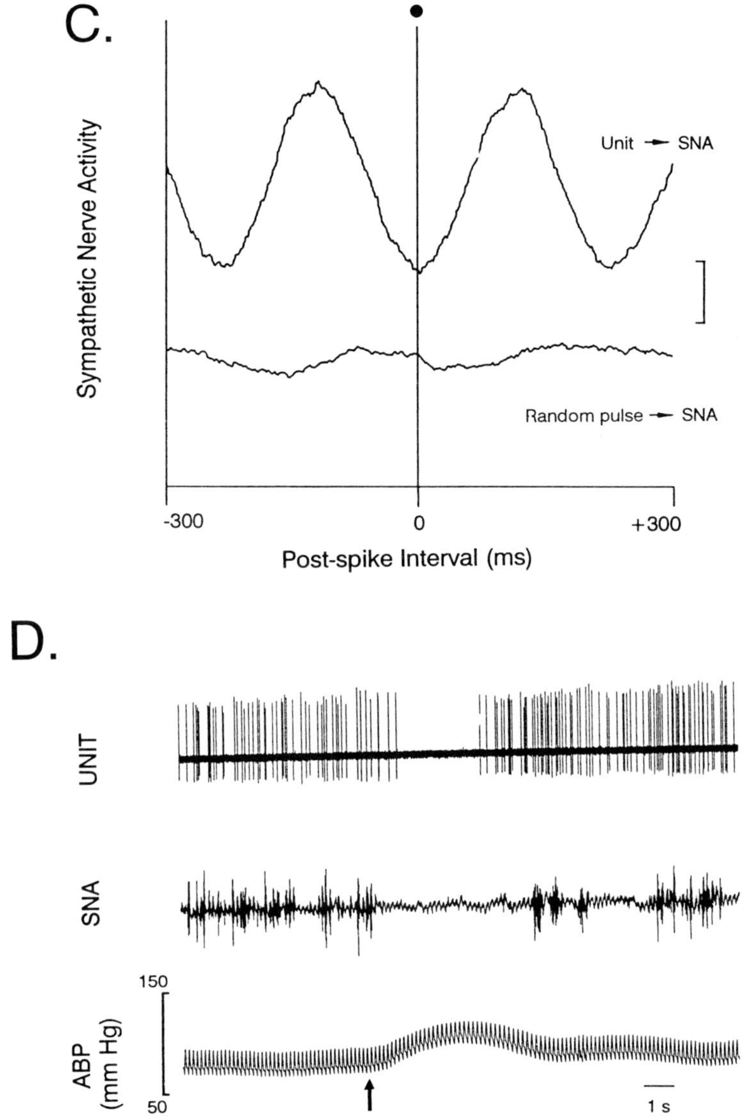

ms for the histogram. (C) Average of sympathetic nerve activity triggered by 700 spontaneous spikes of unit and 700 random pulses of the same average frequency. Peaks in upper trace exceed those in the lower by more than ninefold. Vertical calibration is 50 μV for SNA. (D) Simultaneous inhibition of medullary unit activity and sympathetic nerve discharge (SNA filtered at 1 to 1000 Hz) by IV bolus injection of phenylephrine (arrow). (From Ref. 6.)

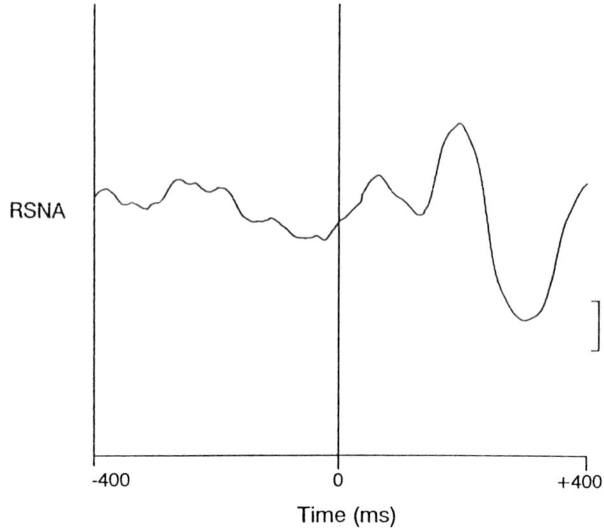

Figure 2 Average of renal sympathetic nerve activity (RSNA) triggered by 50 stimuli applied to the area postrema. Area postrema stimulation (25 μA, 0.3 ms, 1 Hz) evoked an excitatory peak in averaged RSNA followed by an inhibition.

for the sympathetic medullary neurons (onset = 53 ± 21 ms) by about 174 ms. These findings suggest that the biphasic responses of sympathetic neurons in the rostral ventrolateral medulla are transmitted to postganglionic sympathetic nerves. Sympathetic neurons were recorded from 1.5 to 2.5 mm rostral to the obex, ventral to the nucleus ambiguus at the rostral pole of the inferior olive. This distribution coincides with the electrophysiological localization of bulbospinal sympathetic neurons in the rostral ventrolateral medulla in the rabbit by Yu-Wen et al. (34) and Terui et al. (35).

Since higher-frequency stimulation (10 to 40 Hz) has been shown to decrease multifiber postganglionic sympathetic nerve activity in the rabbit (22, 23), we tested the responses of 5 medullary neurons to area postrema pulses delivered at the higher frequencies of 10 to 20 Hz for 3 s. Each pulse evoked a spike and a period of inhibition; however, after the stimulator was turned off, the cell remained inhibited for much longer (~1 to 6 s) compared with the duration of inhibition following sequential stimuli delivered at 1 Hz (~127 ms). These findings suggest a frequency-dependent response of the sympathetic nervous system to area postrema stimulation; low-frequency stimulation (1 Hz) evoked an excitatory/inhibitory response, while higher-frequency stimulation (10 to 40 Hz) resulted in a prolonged period of inhibition.

The biphasic response of the sympathetic medullary neurons to area postrema stimulation raises the possibility that the excitation and inhibition were relayed over different pathways to the rostral ventrolateral medulla. Assuming a conduction distance of 6 mm from the center of the area postrema to the recording sites in the rostral ventrolateral medulla in the rabbit and an onset latency of ~22 ms for the excitatory response, the estimated axonal conduction velocity would be ~0.3 m/s. This slow conduction velocity suggests that the impulses that elicited the excitatory responses were transmitted either directly from the area postrema to the rostral ventrolateral medulla via small, unmyelinated fibers or via a polysynaptic pathway. With an onset latency of ~53 ms for the inhibition, the calculated axonal conduction velocity is ~0.1 m/s, suggesting that the impulses that elicited the inhibition were transmitted directly via slower-conducting unmyelinated fibers or via a longer pathway than was the excitatory response.

The ability of area postrema neurons to fine-tune sympathetic responses to afferent inputs from other sources will most likely depend on the relative timing and/or the magnitude of the afferent inputs from the area postrema neurons and from the other afferent sources onto sympathetic medullary neurons. The capacity to evoke both an excitatory and inhibitory response of sympathetic medullary neurons may provide a means whereby area postrema neurons, in a time-dependent manner, can shape changes in sympathetic nerve activity by modifying the responses of sympathetic medullary neurons to a variety of other converging afferent inputs. A case in point is the ability of area postrema neurons to fine-tune baroreflex regulation of sympathetic nerve activity. Excitation of neurons in the area postrema, either by direct application of L-glutamate or by background electrical stimulation (22), augments baroreflex-mediated sympathoinhibition by increasing the gain of the curve relating decreases in sympathetic nerve activity to increases in arterial blood pressure. The baroreflex-sympathetic pathway is known to involve synapses in the NTS (33), caudal ventrolateral medulla (34,35), and the rostral ventrolateral medulla (2–4,7–9, 35). Therefore, area postrema neurons may modulate baroreflex-mediated sympathoinhibition by altering baroreceptor afferent input at any of those synapses. We have recently provided electrophysiological evidence in rabbits of time-dependent facilitatory interactions between convergent excitatory inputs from area postrema and baroreceptor afferents at the level of the NTS (36). Area postrema neurons may also augment baroreflex-mediated sympathoinhibition at the level of the rostral ventrolateral medulla; this augmentation may also depend on the timing of the inputs from the baroreceptors and area postrema neurons. For example, arrival of inhibitory volleys from the baroreceptors during the excitatory response of medullary neurons to activation of area postrema neurons might result in an occlusive interaction between the two inputs, thereby diminishing baroreceptor-mediated sympathoinhibition. On the other hand, arrival of baroreceptor input during the period of area postrema–evoked inhibition of the sympathetic medullary neurons might result in a facilitative interaction between the two inputs, thereby augmenting either the magnitude and/or the duration of baroreceptor-mediated inhibition. A model illustrating known interactions between area postrema and baroreceptor afferents in the NTS and the potential interactions between area postrema and sympathetic neurons in the rostral ventrolateral medulla is shown in Figure 3.

The second issue regarding the modulatory role for area postrema neurons of sympathetic nerve activity involves the magnitude of the input from area postrema neurons relative to that converging from other sources. Once again, the best-described interaction is between area postrema and baroreceptor afferents. Studies of baroreflex function in rabbits have suggested that area postrema–evoked decreases in renal sympathetic nerve activity are prominent only when baroreceptor afferent activity is intact (23). If tonic baroreceptor activity augments area postrema-evoked sympathoinhibition, then a steady-state increase in baroreceptor traffic to sympathetic neurons in the rostral ventrolateral medulla might augment the inhibitory responses of medullary sympathetic neurons to area postrema stimulation and diminish the excitatory responses. Accordingly,

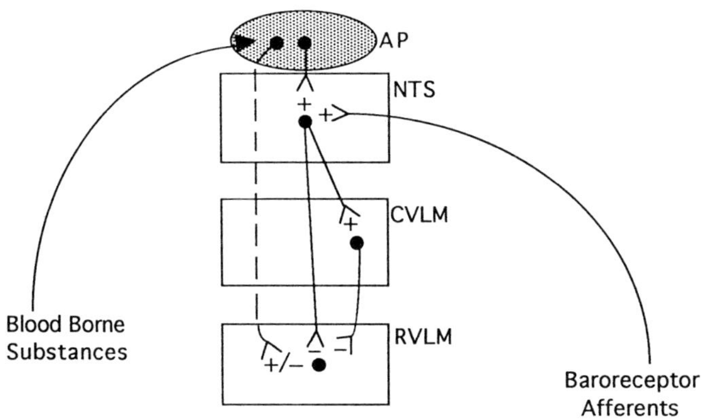

Figure 3 A model illustrating potential interactions between area postrema and medullary sympathetic neurons. Blood-borne substances excite neurons in the area postrema (AP). These AP neurons project to the nucleus tractus solitarius (NTS), where they interact in a facilitative manner with baroreceptor afferents onto neurons in the baroreflex pathway. Area postrema neurons also interact with neurons located more distally in the baroreflex-sympathetic pathway in the rostral ventrolateral medulla (RVLM). The dashed line from AP to the RVLM indicates that the number of intervening synapses is unknown. There is currently no evidence that AP neurons interact with the baroreflex-sympathetic pathway at synapses in the caudal ventrolateral medulla (CVLM). The biphasic AP input to these sympathetic neurons in the RVLM allows for area postrema neurons to modify sympathetic neuronal responses to baroreceptor input in a time-dependent manner. In a broader sense, the biphasic nature provides a mechanism whereby area postrema, responding to multiple blood-borne agents, can modulate either inhibitory or excitatory responses of medullary sympathetic neurons to afferent traffic evoked reflexly or from higher brain sites.

we performed experiments in which we compared the magnitude of the excitatory response and the duration of the inhibitory response of sympathetic medullary neurons under control conditions and during steady-state activation of the baroreceptors produced by steady-state phenylephrine infusions. An example from those findings is shown in Figure 4. During the phenylephrine infusions, the increase in the peak evoked activity was diminished and the dura-

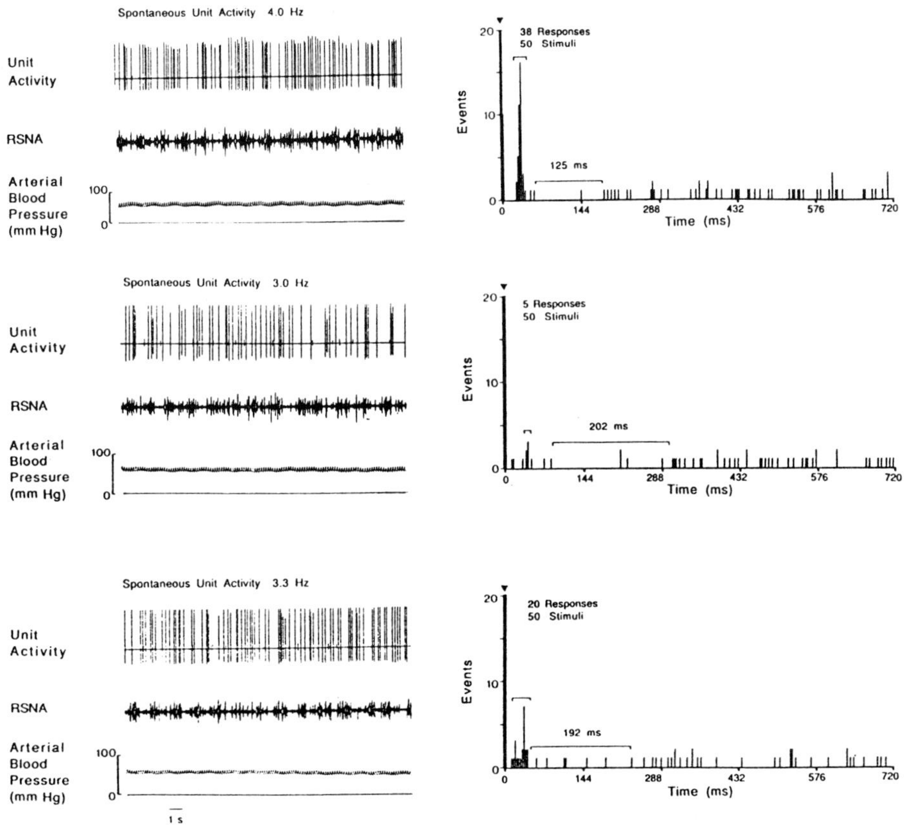

Figure 4 Example of a sympathetic neuron that was inhibited by phenylephrine infusion. Before infusion of phenylephrine (top panel), 50 area postrema stimuli (10 μA, 0.7 ms, 1 Hz) evoked 38 responses during the peak excitation and produced an inhibition that lasted 125 ms. Spontaneous activity of the neuron was 4.0 Hz. During phenylephrine infusion (middle panel), spontaneous activity was decreased to 3.0 Hz, and 50 stimuli evoked only 5 responses during the peak excitation, but they evoked a longer inhibition of 202 ms. Phenylephrine infusion was stopped and the neuron exhibited a partial recovery (bottom panel). Spontaneous activity increased to 3.3 Hz, and 50 area postrema stimuli evoked 20 responses during the peak excitation and an inhibition that lasted 192 ms. RSNA, renal sympathetic nerve activity.

tion of the spike-free period was longer. Both changes were proportional to the decrease in background neuronal activity but nonetheless suggest that increasing the magnitude of baroreceptor afferent input might favor the inhibitory response of sympathetic neurons in the rostral ventrolateral medulla to area postrema input.

III. Area Postrema Interactions with Respiratory-Related Neurons in the Rostral Ventrolateral Medulla

The rostral ventrolateral medulla also contains neurons involved in the control of respiration. The longitudinal column of respiratory cells is divided into the ventral respiratory group (VRG), which extends rostrally into the pre-Bötzinger and the Bötzinger complex. The VRG can be further subdivided into the rostral VRG (rVRG), which extends rostral to the obex in the vicinity of the nucleus ambiguus. The rVRG contains primarily inspiratory neurons with augmenting patterns (IAug), although inspiratory neurons with decrementing (IDec) and plateau patterns (IPlat) have also been recorded (37). Many neurons in the rVRG are bulbospinal, although some project to the contralateral VRG and to the dorsal respiratory group (DRG). In the rabbit, the rVRG extends from ~0.5 to 2.0 mm rostral to the obex (38). We have recorded the activity of inspiratory neurons in this region and have determined that at least some IAug, IDec, and IPlat project to the phrenic motor nucleus (39). Neurons in the VRG are thought to be involved in generating or shaping respiratory pattern. During vomiting, the discharge patterns of medullary respiratory neurons are modulated to synchronize respiratory muscles for vomiting (40); inspiratory neurons, with medullary arborizations, are excited, while bulbospinal inspiratory neurons are inhibited. Since the area postrema is regarded as the "chemoreceptor trigger zone" for vomiting, being required for the production of emesis in a variety of models (41), we determined whether area postrema stimulation might alter the discharge patterns of respiratory neurons in the rVRG. Single sequential stimuli delivered at 1 Hz to the area postrema failed to alter the discharge patterns of respiratory-related neurons in the VRG. An example is shown in Figure 5. The inspiratory neuron was inhibited by lung inflation, showing prolonged bursts and shortened interburst intervals when lung inflation was withheld and being silenced when the lungs were hyperinflated (Fig. 5A). The neuron showed no responses to area postrema stimulation (Fig. 5B). Twenty-three respiratory-related neurons were tested for responses to area postrema stimuli, and 20 of those 23 were further characterized by their responses to maintained lung inflation. None received area postrema input, but all received lung volume–related input: 9 were excited by lung inflation and discharged continuously when inflation was maintained for three cycles; 11 were either silenced or exhibited a shortened burst duration when inflation was maintained for three cycles. Neurons with respiratory-related discharge were routinely encountered just dorsal

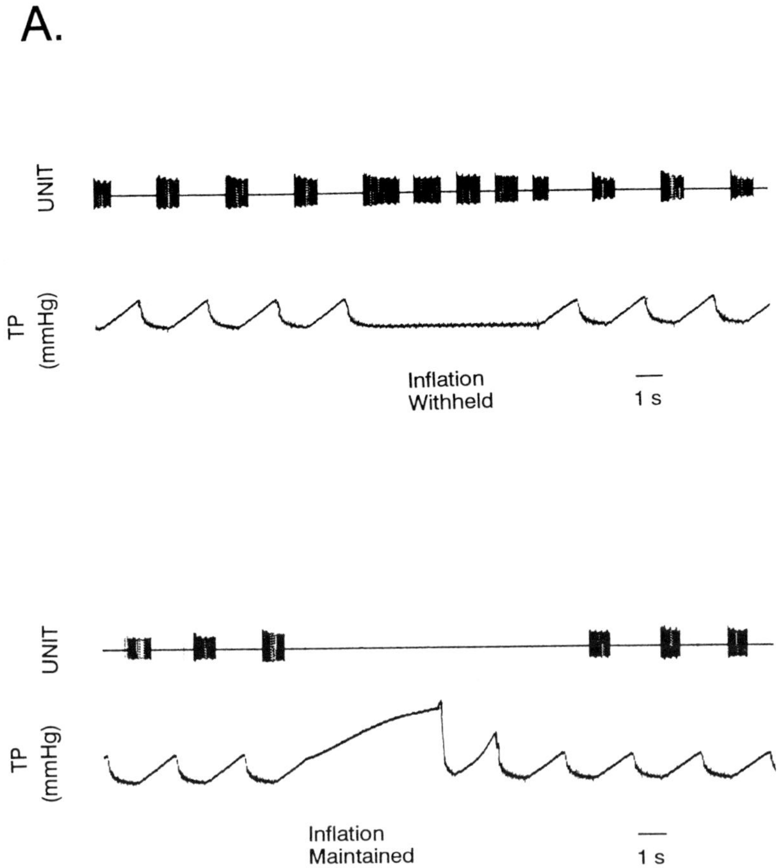

Figure 5 Example of a neuron in the rVRG that discharged with respiratory periodicity and was unaffected by area postrema stimulation. (A) Discharge pattern of the neuron in response to manipulation of lung volume. TP, tracheal pressure. (B) Poststimulus time histogram of unit response to single sequential area postrema stimuli (155 triggered sweeps; 25 μA, 0.3 ms, 1 Hz). Bin width is 3.2 ms for histogram.

and ~100 to 200 μm medial to sympathetic neurons. This region, located ~1 to 2 mm rostral to the obex, is consistent with the functionally (38) and anatomically identified (42) rVRG in rabbits. These findings suggest that low-frequency stimuli delivered to the area postrema may not alter respiratory pattern through neurons in the VRG.

IV. Summary

Area postrema neurons can provide both excitatory and inhibitory inputs to sympathetic neurons in the rostral ventrolateral medulla. The biphasic nature

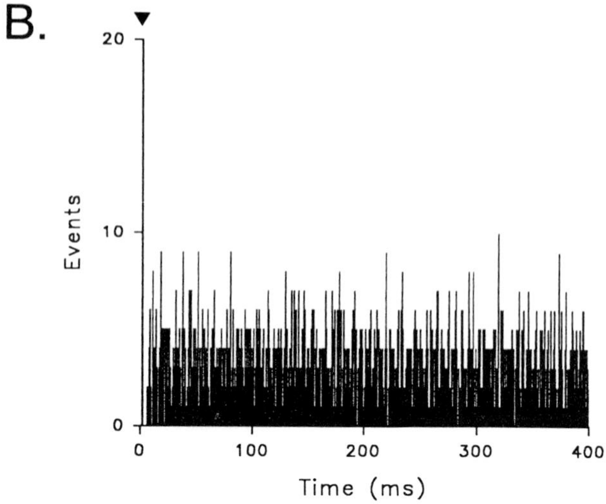

Figure 5 Continued

of the responses of sympathetic medullary neurons to area postrema stimulation may provide a mechanism whereby area postrema neurons, responding to a variety of humoral or neuronal influences, can fine-tune sympathetic nerve activity. In contrast, although the area postrema has been shown to be critical for emesis in a variety of animal models, excitation of area postrema neurons may not play a role in reorganizing the patterns of respiratory neurons in the rVRG to accommodate the vomiting reflex.

References

1. Barman SM, Gebber GL. Problems associated with the identification of brain stem neurons responsible for sympathetic nerve discharge. J Auton Nerv Syst 1981; 3: 369–377.
2. Brown DL, Guyenet PG. Electrophysiological study of cardiovascular neurons in the rostral ventrolateral medulla in rats. Circ Res 1985; 56:359–369.
3. Guyenet PG, Haselton JR, Sun M-K. Sympathoexcitatory neurons of the rostroventrolateral medulla and the origin of the sympathetic vasomotor tone. Prog Brain Res 1989; 81:105–116.
4. Morrison SF, Calloway J, Milner TA, Reis DJ. Rostral ventrolateral medulla: A source of the glutamatergic innervation of the sympathetic intermediolateral nucleus. Brain Res 1991; 562:126–135.
5. Gebber GL, Barman SM. Sympathetic-related activity of brain stem neurons in receptor-denervated cats. Am J Physiol 1981; 240(Reg Int Comp Physiol 9):R348–R355.
6. Wilson CG, Bonham AC. Area postrema excites and inhibits sympathetic-related neurons in rostral ventrolateral medulla in rabbit. Am J Physiol (Heart Circ Physiol 35): H1075–H1084, 1994.

7. Terui N, Saeki Y, Kimada M. Barosensory neurons in the ventrolateral medulla in rabbits and their responses to various afferent inputs from peripheral and central sources. Jpn J Physiol 1986; 36:1141–1164.
8. Dembowsky K, McAllen RM. Baroreceptor inhibition of subretrofacial neurons: evidence form intracellular recordings in the cat. Neurosci Lett 1990; 111:139–143.
9. Sun M-K, Guyenet PG. Arterial baroreceptor and vagal inputs to sympathoexcitatory neurons in rat medulla. Am J Physiol 1987; 252(Reg Int Comp Physiol 21): R669–R709.
10. Verberne AJM, Guyenet PG. Medullary pathway of the Bezold-Jarisch reflex in the rat. Am J Physiol 1992; 263(Reg Int Comp Physiol 32):R1195–R1202.
11. Sun M-K, Spyer KM. Responses of rostroventrolateral medulla spinal vasomotor neurones to chemoreceptor stimulation in rats. J Auton Nerv Syst 1991; 33:79–84.
12. Sun M-K, Spyer KM. Nociceptive inputs into rostral ventrolateral medulla-spinal vasomotor neurones in rats. J Physiol 1991; 436:685–700.
13. Morrison SF, Reis DJ. Reticulospinal vasomotor neurons in the RVL mediate the somatosympathetic reflex. Am J Physiol 1989; 256(Reg Int Comp Physiol 25): R1084–R1097.
14. Bauer RM, Iwamoto GA, Waldrop TG. Discharge patterns of ventrolateral medullary neurons during muscle contraction. Am J Physiol 1990; 259(Regulatory Integrative Comp. Physiol. 28):R606–R611.
15. Barman S. Descending projections of hypothalamic neurons with sympathetic nerve-related activity. J Neurophysiol 1990; 64:1019–1032.
16. Sun M-K, Guyenet PG. Hypothalamic glutamatergic input to medullary sympathoexcitatory neurons in rats. Am J Physiol 1986; 251(Reg Int Comp Physiol 20): R798–R810.
17. Morest DK. A study of the structure of the area postrema with Golgi methods. Am J Anat 1960; 107:291–303.
18. Hasser EM, Bishop VS. Reflex effect of vasopressin after blockade of V_1 receptors in the area postrema. Circ Res 1991; 67:265–271.
19. Undesser KP, Hasser EM, Haywood JR, et al. Interaction of vasopressin with the area postrema in arterial baroreflex function in conscious rabbits. Circ Res 1985; 56:410–417.
20. Ferguson AV, Smith P. Circulating endothelin influences area postrema neurons. Reg Pept 1991; 32:11–21.
21. Papas S, Smith P, Ferguson AV. Electrophysiological evidence that systemic angiotensin influences rat area postrema neurons. Am J Physiol 1990; 258(Regulatory Integrative Comp Physiol 27):R70–R76.
22. Hasser EM, Nelson DO, Haywood JR, Bishop VS. Inhibition of renal sympathetic nervous activity by area postrema stimulation in rabbits. Am J Physiol 1987; 253 (Heart Circ Physiol 22):H91–H99.
23. Hay M, Hasser EM, Undesser KP, Bishop VS. Role of baroreceptor afferents on area postrema-induced inhibition of sympathetic activity. Am J Physiol 1991; 260 (Heart Circ Physiol 29):H1353–H1258.
24. Barnes KL, Ferrario CM, Conomy JP. Comparison of the hemodynamic changes produced by electrical stimulation of the area postrema and NTS in the dog. Circ Res 1979; 45:126–143.
25. Ferguson AV, Marcus P. Area postrema stimulation induced cardiovascular changes in the rat. Am J Physiol 1988; 255(Reg Int Comp Physiol 24):R855–R860.

26. Cox BF, Hay M, Bishop VS. Neurons in area postrema mediate vasopressin-induced enhancement of the baroreflex. Am J Physiol 1990; 258(Heart Circ Physiol 27): H1943–H1946.
27. Sun M-K, Spyer KM. GABA-mediated inhibition of medullary vasomotor neurones by area postrema stimulation in rats. J Physiol 1991; 436:669–684.
28. Fink GD, Bruner CA, Mangiapane ML. Area postrema is critical for angiotensin-induced hypertension in rats. Hypertension 1987; 9:355–361.
29. Mangiapane ML, Skoog KM, Rittenhouse P, et al. Lesion of the area postrema region attenuates hypertension in spontaneously hypertensive rats. Circ Res 1989; 64:129–135.
30. Shapiro RE, Miselis R. The central neural connections of the area postrema of the rat. Comp Neurol 1985; 234:344–364.
31. Blessing WW, Hedger SC, Joh TH, Willoughby JO. Neurons in the area postrema are the only catecholamine-synthesizing cells in the medulla or pons with projections to the rostral ventrolateral medulla (C_1-area) in the rabbit. Brain Res 1987; 419:336–340.
32. Papas S, Ferguson V. Electrophysiological characterization of reciprocal connections between the parabrachial nucleus and the area postrema in the rat. Brain Res Bull 1990; 24:599–582.
33. Spyer KM. Neural organization of the baroreceptor reflex. Rev Physiol Biochem Pharmacol 1981; 88:23–124.
34. Yu-Wen L, Gieroba AJ, McAllen RM, Blessing WW. Neurons in rabbit caudal ventrolateral medulla inhibit bulbospinal barosensitive neurons in rostral medulla. Am J Physiol 1991; 261(Reg Int Comp Physiol 27):R44–R51.
35. Terui N, Masuda N, Saeki Y, Kumada M. Activity of barosensitive neurons in the caudal ventrolateral medulla that send axonal projections to the rostral ventrolateral medulla in rabbits. Neurosci Lett 1990; 118:211–214.
36. Bonham AC, Hasser EM. Area postrema and aortic or vagal afferents converge to excite neurons in nucleus tractus solitarius. Am J Physiol 1993; 264(Heart Circ 33):H1674–H1685.
37. Merill EG. The lateral respiratory neurons of the medulla: Their associations with nucleus ambiguus, nucleus retrambigualis, the spinal accessory nucleus and the spinal cord. Brain Res 1970; 24:11–28.
38. Jiang C, Shen E. Respiratory neurons in the medulla of the rabbit: distribution, discharge patterns and spinal projections. Brain Res 1991; 541:284–292.
39. Bonham AC, Ravi K, Wilson CG, Zhang Z, Kappagoda CT. Pulmonary venous congestion augments respiratory motoneuronal responses to cigarette smoke in rabbit. J Applied Physiol. In press.
40. Miller AD, Ezure K. Behavior of inhibitory and excitatory propriobulbar respiratory neurons during fictive vomiting. Brain Res 1992; 578:168–176.
41. Borison HL, Brizee KR. Morphology of emetic chemoreceptor trigger zone in cat medulla oblongata. Proc Soc Exp Biol Med 1951; 7:38–42.
42. Ellenberger HH, Vera PL, Haselton JR, et al. Brainstem projections to the phrenic nucleus: An anterograde and retrograde HRP study in the rabbit. Brain Res Bull 1990; 24:163–174.

55

The Caudal Pressor Area

A New Region of the Ventrolateral Medulla Involved in Cardiovascular Regulation

RUY R. CAMPOS, JR., O. S. POSSAS, S. L. CRAVO, O. U. LOPES, and P. G. GUERTZENSTEIN†

Escola Paulista de Medicina
São Paulo, Brazil

I. Introduction

A central question in cardiovascular regulation is: What are the mechanisms responsible for the generation and maintenance of vasomotor tone and arterial blood pressure (ABP) regulation? Evidence gathered over the last 25 years has unequivocally established that normal levels of arterial blood pressure in anesthetized animals depend upon the activity of small groups of neurons located in discrete regions of the ventrolateral medulla (VLM) (reviewed in Refs. 1 and 2).

The first conclusive evidence for a discrete localization of neurons involved in these functions was obtained by Feldberg and Guertzenstein (3). They demonstrated that several drugs known to be selectively active on cell bodies were able to evoke changes in ABP when applied topically to a restricted area of the ventral medullary surface situated just caudal to the trapezoid bodies and rostral to the roots of the twelfth cranial nerve (4). Most importantly, bilateral applications of the inhibitory amino acid glycine in this area produced a decrease in ABP equivalent to that observed immediately after spinal transection. Accordingly, stimulation of neurons in this area evoked an increase in

†Deceased.

ABP, whereas bilateral electrolytic lesions reduced ABP to spinal levels (5). This region corresponds to the area described in the pioneer work of Schläfke and Loeschcke (6). Studying regions of the VLM that are associated with respiratory control, they demonstrated that bilateral cooling of these areas produced not only respiratory arrest but also a step decrease in ABP (6).

Since then, numerous studies using the cat (7,8), rat (9–12), and rabbit (13,14) have been conducted and have confirmed and extended the original observation. In the rat, this area has been identified as the rostral ventrolateral medulla (RVLM) (12,15–17). RVLM neurons project directly to the intermediolateral cell column (16,18–20) and are responsible for the tonic excitation of sympathetic preganglionic neurons (7,21–23).

A few years after the original observations of the RVLM were made, another area with cardiovascular activity was identified in the cat's VLM by the same group (24,25). Localized caudally to the RVLM, in close proximity to the rootlets of the twelfth nerve, this area was identified as a vasodepressor area. Stimulation of this area resulted in hypotension, while its inhibition produced marked increases in ABP (9,25–27). The caudal ventrolateral medulla (CVLM) contains tonically active sympathoinhibitory neurons. Lesions or functional inactivation of this area results in increases in sympathetic nerve activity and hypertension (9,28–30). The CVLM acts through connections with the RVLM (11,28–31).

The first evidence suggesting the existence in the ventrolateral tegmentum of a third area with vasomotor activity was obtained by Feldberg and Guertzenstein (32). Working in cats under chloralose anesthesia (60 mg/kg) these authors demonstrated that topical application of leptazol on caudal sites of the VLM produced depressor responses. However, when chloralose anesthesia was increased to 90 to 120 mg/kg, a distinction between two areas in this region was detectable: a rostral area at the twelfth cranial nerve rootlets where leptazol produced its usual depressor effect, and a more caudal area, between the twelfth nerve rootlets and the first cervical nerve, where pressor effects were obtained. Furthermore, in this condition (i.e., deep anesthesia), the most caudal area was shown to be tonically active, since its inhibition by bilateral application of sodium pentobarbital produced a decrease in blood pressure. An increase in blood pressure, in response to stimulation of these caudal sites of the VLM, was later reported in anesthetized or decerebrate cats (33,34) and rats (35). However, researchers were unable to demonstrate any tonic activity from this area.

Recently, we have carried out a series of studies (36–39) to determine the functional characteristics of the caudalmost ventrolateral medulla. In addition, we have also investigated its role in the maintenance of arterial blood pressure as well as its interactions with the other two cardiovascular areas of the ventral medullary surface.

II. Methods

Experiments were performed on male adult rats (Wistar 250 to 350 g) that had been anesthetized with urethane (1.2 g/kg, IP) or on unanesthetized animals that had previously been subjected to midcollicular transection under ether anesthesia. The trachea and femoral vein were cannulated for mechanical ventilation and drug infusions, respectively. Pulsatile ABP was recorded from a cannula inserted into a femoral artery. Mean arterial pressure (MAP) was obtained by filtering the ABP signal, and heart rate (HR) was measured by a cardiotachometer that was also triggered by the ABP signal. The animals were paralyzed with gallamine (Flaxedil, 4 mg/kg, IV) and mechanically ventilated. Temperature was maintained at 37°C by a heating table. Microinjections were performed through micropipettes (OD 0.2 mm) and were placed inside guide cannulas. The length of the micropipettes and the guide cannulae was adjusted to allow only the micropipettes to be inserted into the brain tissue. The guide cannulas were directed to the desired stereotaxic position using the calamus scriptorius (CS) as a reference. Vertical positioning was obtained by slowly lowering both the micropipette and the guide cannula until a slight displacement between them was observed. Postmortem histology demonstrated that this procedure consistently permitted the placement of the micropipette tip next to the VLM surface. Microinjections, consisting of 200 nL of 0.1 *M* sodium phosphate buffer, L-glutamate (L-Glu, 50 or 100 nmol), glycine (Gly, 50 or 100 nmol), GABA (50 nmol), plus kynurenic acid (KYN, 2 nmol) or bicuculline methiodide (BIC, 0.2 nmol) were dissolved. In experiments investigating possible interactions between different areas of the VLM, a pair of micropipettes was implanted in the RVLM (2.6 mm rostral and 2.1 mm lateral to the CS) or CVLM (1.0 mm rostral and 1.8 mm lateral to the CS). In both areas, vertical positioning was obtained by lowering the micropipettes until they reached the VLM surface, as described above.

III. Localization of the Caudal Pressor Area (CPA)

Initially, to identify pressor sites within the CVLM, the VLM was mapped in terms of ABP responses to unilateral microinjections of the excitatory amino acid L-Glu (50 nmol/200 nL). The area surveyed extended from 2.0 mm caudal to 1.0 mm rostral to the CS and 1.0 to 3.0 mm lateral to the midline. Injection sites were marked and plotted in schematic maps of the VLM surface. Each animal received no more than two applications, one on each side of the brainstem. Results obtained in these experiments demonstrated that, in urethane-anesthetized rats, stimulation of a restricted area of the caudalmost portion of the VLM consistently evoked increases in ABP. These pressor responses ranged from 15 to 60 mm Hg, had a latency of 2 to 5 s, and peaked within 20 s; the

ABP remained above resting levels for up to 4 min. Heart rate responses were variable and inconstant; however, tachycardia was noted most frequently (10 to 20 bpm).

Figure 1 summarizes results obtained from 10 animals with L-Glu microinjected in several sites of the VLM. The areas from which pressor responses were obtained are shown by the circles drawn with continuous lines on the left side, while the circles drawn with interrupted lines on the right correspond to areas in which blood pressure was not affected or from which a depressor response was obtained. These depressor responses most likely resulted from stimulation of sympathoinhibitory neurons of the CVLM. The area defined by the pressor responses to L-Glu lies between the rootlets of the twelfth cranial and

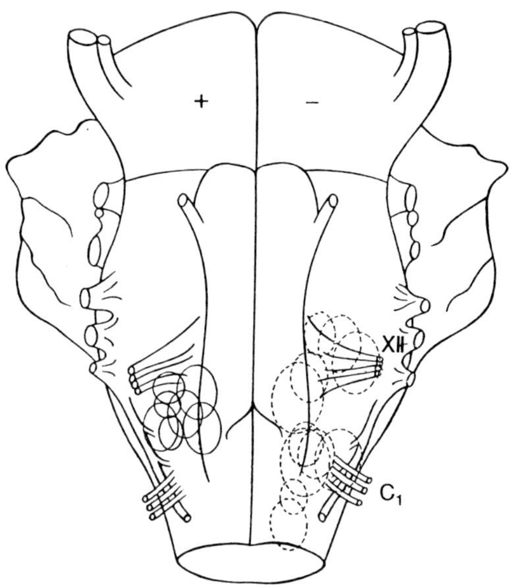

Figure 1 Diagram representing the rat's ventral medullary surface, showing localization of pressor sites in the caudal ventrolateral medulla. In urethane-anesthetized, paralyzed, and artificially ventilated rats ($n = 10$), the ventrolateral medulla was mapped out to localize sites where microinjections of L-glutamate (L-Glu, 50 nmol/200 nL) produce pressor responses. Each animal received no more than two applications, one on each side of the brainstem. Each ovoid is the result of a single application. Regardless the actual site of application, ovoids drawn on the left side (+) with continuous lines indicate sites where L-Glu produced increases in arterial blood pressure >15 mm Hg. Ovoids drawn on the right side (−) with interrupted lines indicate sites where blood pressure was unaffected or depressor responses were obtained. XII and C_1: rootlets of twelfth cranial and first cervical nerves, respectively. (From Ref. 38.)

the first cervical nerve, laterally to the pyramides, and just medial to the spinal rootlets of the eleventh cranial nerve. The actual area is probably smaller than indicated by the ovals drawn with continuous lines. This is the case, since the rostral portion of the cranialmost and the lower portion of the caudalmost circles have to be excluded from being responsible for the pressor responses obtained when microinjections were performed in this region. Furthermore, ovoids drawn at the right show that the region between the two nerves is free from interrupted ovals (i.e., no negative results were obtained from this area). This area corresponds approximately to the following stereotaxic coordinates: 0.5 to 1 mm caudal and 1.5 to 2.5 mm lateral to the CS.

Within this pressor area, increasing the L-Glu dose to 100 nmol did not cause further increases in ABP or HR responses. Bilateral applications did not result in any increase in the magnitude of pressor responses but did regularly result in their prolongation.

Since the first observation of pressor sites in the caudalmost medulla by Feldberg and Guertzenstein (32), other studies have confirmed and extended that observation. An incidental observation of pressor responses to L-Glu in caudal sites of the VLM was reported by Dembowsky et al. in cats (33). In this study, pressor responses to L-Glu stimulation were observed in about one-third of all animals. Curiously, in these animals researchers were often unable to obtain hypotension in response to stimulation of the CVLM. However, in our experience, pressor and depressor effects could be observed in the same animal and always in separate, distinct areas. A similar result (i.e., pressor and depressor sites in the same animal) was recently described in cats (34). The area defined by the pressor responses to L-Glu in our study corresponds closely to the area described previously in rats by Gordon and McCann (35). This area is designated by them as the caudal pressor area (CPA), a term that we employ in this paper. Contrary to the original observation (32), the level of anesthesia was not a relevant factor in any of these studies.

IV. Cardiovascular Responses to CPA Inhibition

In order to verify that CPA neurons have a tonic pressor activity, we studied the effects of inhibition of CPA neurons on the basal levels of ABP and HR. Unilateral microinjection of the inhibitory amino acid glycine (Gly, 50 nmol/site) into the CPA produced marked decreases in ABP, often accompanied by bradycardia. Depressor responses ranged from −15 to −45 mm Hg, and occasionally—although they are rarely reached—levels equivalent to those observed after spinal transection. As observed with L-Glu, bilateral microinjections increased duration but not magnitude of depressor responses. ABP and MAP records of Figure 2A illustrate a representative example. In a urethane-anesthetized rat, bilateral application of glycine (50 nmol/200 nL) produced a decrease in blood pressure reaching 50 mm Hg within 30 s after the injections. Blood pressure

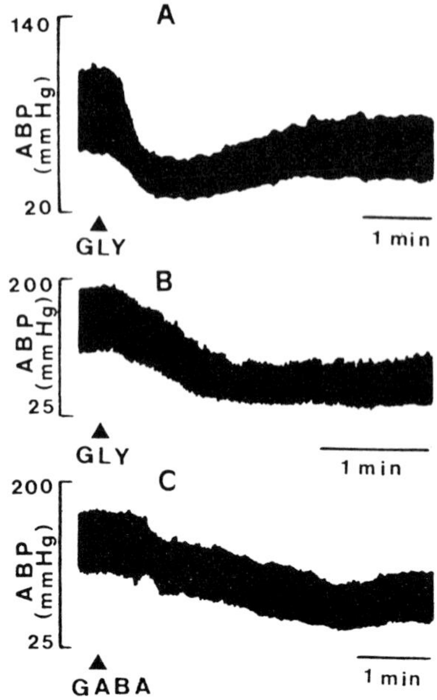

Figure 2 Effects of CPA inhibition by bilateral microinjections of glycine (Gly, 50 nmol/site) on the pulsatile arterial blood pressure (ABP, mm Hg) in a urethane-anesthetized (A) or in an unanesthetized (B) decerebrate rat. In a urethane-anesthetized rat (C), CPA was inhibited by bilateral application of GABA (50 nmol/site).

remained below basal levels for 6 min. Contrary to what was previously observed in cats by Feldberg and Guertzenstein, depressor responses to CPA inhibition were not dependent on the level of anesthesia. This was demonstrated by the fact that in unanesthetized decerebrate animals, CPA inhibition by bilateral application of glycine produced depressor responses similar to those obtained in anesthetized animals (Fig. 2B).

Results obtained in this series of experiments demonstrated that CPA neuronal activity is required for maintenance of basal levels of ABP. Although previous studies have demonstrated the presence of pressor sites in the caudalmost VLM, they have failed to demonstrate any tonic activity for this area (34,35). Since these prior studies have utilized GABA as an inhibitory agent, we tested to see if CPA activity was selectively sensitive to Gly or whether it could also be suppressed by GABA. Results obtained demonstrated that bilateral application of GABA (50 nmol/200 nL) into CPA resulted in depressor responses similar to those obtained with Gly (Fig. 2C). An explanation for this discrepancy between the present study and a previous study (35), conducted in rats, could be that previous applications were made at the level of the caudalmost rootlets of the twelfth cranial nerve. However, in the present study, applications were localized slightly more caudal to this level between the twelfth cranial and first cervical nerve rootlets. Another possible explanation is that in this study we utilized larger volumes and doses than in previous studies. Also in our study, drugs were applied immediately dorsal and spread over the ventral surface, closely resembling topical applications, and covered a larger portion of the CPA. It is possible that the combination of these factors allowed us to simultaneously inhibit a larger number of CPA neurons and to demonstrate CPA tonic activity.

Results obtained with CPA excitation or inhibition indicate that these areas exhibit the similar functional characteristics of RVLM (i.e., both are tonic sympathoexcitatory areas). Seeking to distinguish between RVLM and CPA, we compared the effects of inhibition of each of these two areas on the pressor response evoked by bilateral carotid occlusion.

V. Effects of CPA Inhibition on the Pressor Response to Carotid Occlusion

In urethane-anesthetized rats, the common carotid arteries were exposed and carefully dissected free from surrounding tissues bilaterally. Bilateral carotid occlusion was obtained by placing miniature clamps on the arteries for 1 min. The carotid occlusion pressor response (COPR) was defined as the mean increase in MAP during the occlusion, compared before and 1 min after bilateral inhibition of the CPA or RVLM.

Data in Table 1 summarize the results obtained in these experiments. Before CPA inhibition, ABP was 87 ± 6 mm Hg and COPR averaged 23 ± 3 mm Hg/s.

Table 1 Effects of CPA or RVLM Inhibition by Glycine (100 nmol/site) on the Carotid Occlusion Pressor Response (COPR)[a]

		Before inhibition		After inhibition	
Treatment	*N*	Control (MAP)	COPR (mm Hg/s)	Control (MAP)	COPR (mm Hg/s)
CPA inhibition	6	87 ± 5.9	23 ± 2.7	53 ± 5.9[b]	21 ± 4.4
RVLM inhibition	6	93 ± 9.5	28 ± 3.7	45 ± 5.3[b]	8 ± 3.0[b,c]

[a]Values are means ± SE. Comparisons of basal mean arterial blood pressure (MAP, mm Hg) and the mean increase in MAP during the carotid occlusion pressor response (COPR, mm Hg/s) before and during inhibition by bilateral microinjection of glycine into the caudal pressor area (CPA) and the rostral ventrolateral medulla (RVLM).

[b]$P < 0.05$ compared with control.

[c]$P < 0.05$ CPA compared with RVLM.

One minute after bilateral application of Gly (100 nmol/site) into the CPA, ABP decreased to 53 ± 6 mm Hg, a level significantly lower than the control, but the COPR was not modified (21 ± 4 mm Hg/s). Conversely, before RVLM inhibition, ABP was 93 ± 9 mm Hg and COPR averaged 28 ± 4 mm Hg/s, respectively. One minute after RVLM inhibition by bilateral application of Gly (100 nmol/site), ABP decreased to 45 ± 5 mm Hg (a level similar to that obtained by CPA inhibition); however, COPR was markedly reduced (8 ± 3 mm Hg/s).

Although the mechanisms involved in the carotid occlusion pressor response are not completely clear, it is accepted that this reflex is due not only to suppression of baroreceptor activity but also to stimulation of peripheral and central chemoreceptors and to cerebral ischemia. Inhibition of RVLM neurons markedly reduced the COPR, indicating that this area is involved in all neural pathways underlying these responses, as was previously shown by others (40–44). On the other hand, the maintenance of COPR during CPA inhibition indicates that, in this condition, at least one of these pathways bypasses the CPA and is thus responsible for the observed sympathoexcitation.

VI. Putative Pathways Involved in the CPA Cardiovascular Responses

To investigate the pathways underlying cardiovascular responses to CPA excitation or inhibition, we examined possible interactions between CPA and the other two cardiovascular areas of the VLM.

Initially we examined whether or not CPA pressor responses were dependent on the activity of RVLM neurons. For this purpose, two pairs of micropipettes were implanted in the RVLM and CPA of anesthetized rats, utilizing the methods described above. Pressor responses to CPA stimulation by L-Glu (50 nmol/site) were evaluated before and 1 min after RVLM inhibition by Gly (100 nmol/site). Results obtained in a typical experiment are illustrated in Fig. 3A. Bilateral stimulation of the CPA with L-Glu produced a 35-mm Hg pressor response (Fig. 3A, left). Twenty minutes later, when ABP had returned to control values, the RVLM was inhibited by bilateral application of Gly. Once the evoked hypotension was fully established (about 1 min later), stimulation of the CPA with the same dose of L-Glu produced only a 5-mm Hg rise in ABP (Fig. 3A, right). Similar results obtained in a group of six animals are summarized in Table 2. During RVLM inhibition, pressor responses to CPA stimulation were reduced from 29 ± 4 to 10 ± 4 mm Hg ($P<0.05$).

Conversely, the effects of RVLM stimulation are not inhibited by CPA blockade, as shown by a typical experiment in Fig. 3B. Initially, the RVLM was stimulated with L-Glu, producing a 60-mm Hg increase in ABP (Fig. 3B, left). After a recovery period of 20 min, CPA was inhibited by Gly (100 nmol/site) applied bilaterally; once the hypotension was established, the RVLM was

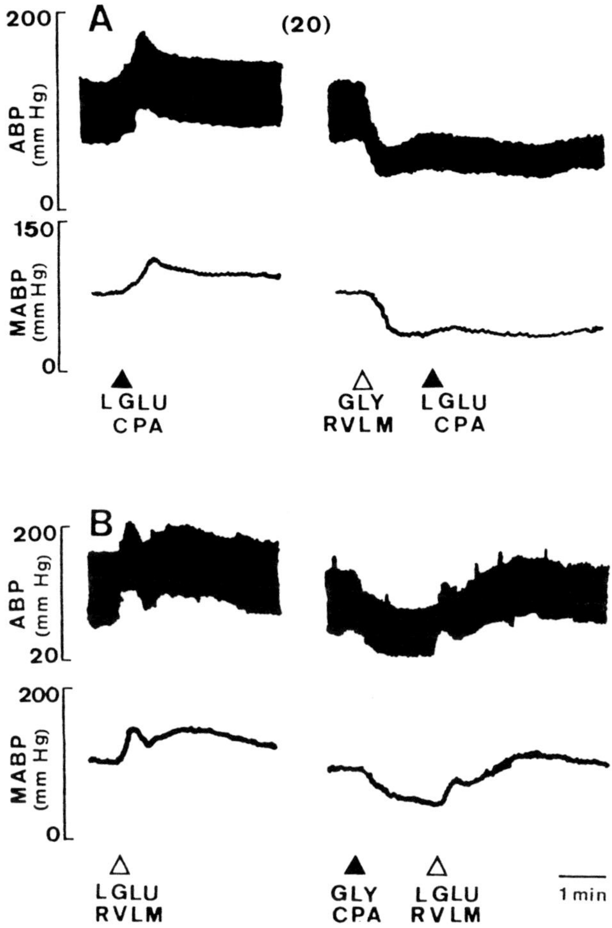

Figure 3 Interactions between the CPA and RVLM responses. (A) The CPA was stimulated by bilateral application of L-glutamate (L-Glu, filled triangles, 50 nmol/site) before (left) and during (right) RVLM inhibition by glycine applied bilaterally (Gly, open triangle, 100 nmol/site). (B) RVLM was stimulated by bilateral application of L-glutamate (open triangles, 50 nmol/site) before (left) and during (right) CPA inhibition with glycine bilaterally applied (filled triangles, 100 nmol/site). Numbers on the top of the tracings indicate time in minutes between successive records. (From Ref. 38.)

again stimulated, producing a 50-mm Hg increase in ABP (Fig. 3B, right). In a group of similar experiments ($n = 6$), pressor responses to RVLM stimulation averaged 34 ± 8 mm Hg before and 60 ± 9 mm Hg during CPA inhibition (Table 2).

These results showed that cardiovascular responses to CPA stimulation depend on the activity of sympathoexcitatory neurons in the RVLM. Pressor

Table 2 Interactions Between CPA and RVLM[a]

		Before inhibition		After inhibition	
	N	Basal (MAP)	Stim. (MAP)	Basal (MAP)	Stim. (MAP)
CPA stimulation (L-Glu, 50 nmol/site) combined with RVLM inhibition (Gly, 100 nmol/site)	6	100 ± 4.8	+29 ± 4.1	53 ± 9.1[b]	+10 ± 5.1[b]
RVLM stimulation (L-Glu, 50 nmol/site) combined with CPA inhibition (Gly, 100 nmol/site)	6	100 ± 8.5	+34 ± 8.3	63 ± 6.2[b]	+60 ± 8.7[b]

[a]Values are means ± SE. Comparisons of effects of RVLM inhibition on the pressor responses to CPA stimulation (top) and CPA inhibition on the pressor responses to RVLM stimulation (bottom).
[b]$P < 0.05$ compared with control.

responses to CPA stimulation were markedly reduced during RVLM inhibition; whereas pressor responses to RVLM stimulation were not dependent on CPA activity. They suggest that CPA effects are mediated by RVLM. Another possibility would be that CPA neurons project directly into the intermediolateral cell column, contributing to the maintenance of tonic excitation of sympathetic preganglionic neurons (SPN). In that case, the reduction of CPA pressor response during RVLM inhibition would only reflect the sudden withdrawal of an excitatory drive and a marked reduction in SPN excitability. The RVLM, however, provides a greater proportion of this drive and would have the ability to overcome the effects of CPA inhibition. This possibility, although appealing, seems unlikely, since various studies have failed to detect direct projections from this region (CPA) to the intermediolateral cell column (15,16,18,45).

We next addressed the question of whether pressor responses to CPA stimulation were due to excitation of RVLM neurons by glutamatergic synapses. In preliminary experiments, pressor responses to CPA stimulation were compared before and after microinjections of the broad-spectrum glutamate antagonist kynurenic acid (KYN, 2 nmol/site), into the RVLM bilaterally. Results obtained in these experiments (not shown) indicated that CPA pressor responses were not affected by this procedure. Curiously, in some experiments where KYN was injected into sites within the CVLM, it was shown to be effective in blocking increases in ABP during CPA stimulation. These results suggested that CPA responses were dependent on glutamatergic synapse(s) within the CVLM. These experiments, however, were somewhat disputable, since it was not possible to completely exclude the chance that these effects were only due to diffusion of the KYN from the injection sites to the CPA itself. We overcame this inconvenience by utilizing a different experimental approach. We tested not the pressor but the depressor effects of CPA inhibition during bilateral application of KYN into the CVLM.

VII. Glutamatergic Synapses Within the CVLM Are Required for the Depressor Responses Produced by CPA Inhibition

In a urethane anesthetized rat, bilateral inhibition of the CPA with Gly (100 nmol/site) produced the already described depressor response (Fig. 4A, left). After ABP had returned to control levels, kynurenic acid (KYN, 2 nmol/200 nL)

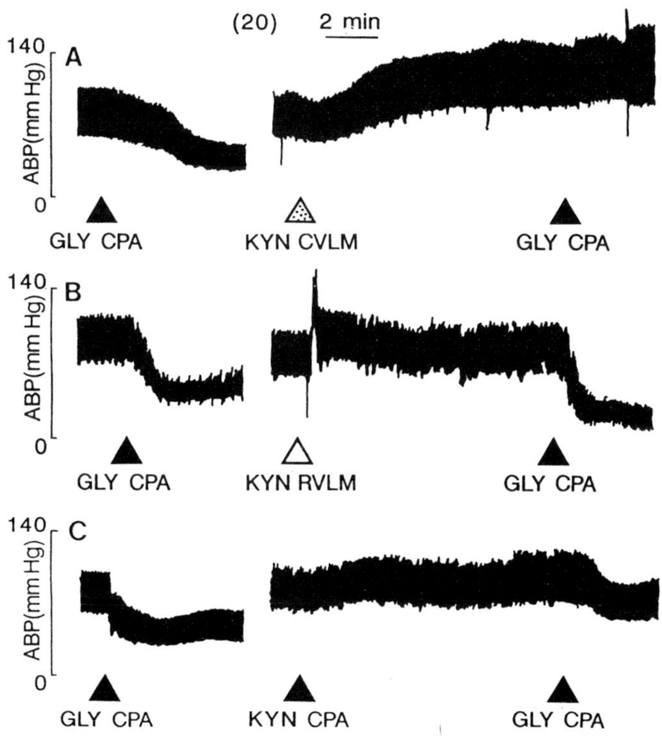

Figure 4 Effects of glutamatergic blockade on the depressor responses to CPA inhibition. In anesthetized rats, CPA was initially inhibited by bilateral microinjections of glycine (Gly, 100 nmol/site, black triangles, left). The glutamate antagonist Kynurenic acid (KYN, 4 nmol/site) was then applied to (A) the caudal ventrolateral medulla (CVLM, dotted triangle); (B) the rostral ventrolateral medulla (RVLM, open triangle) or (C) the CPA itself (black triangle), and Gly application was repeated (black triangles on the right side). Numbers on the top of the tracings indicate time in minutes between successive records.

was injected bilaterally into the CVLM. Following microinjections of KYN into the CVLM (dotted triangle, Fig. 4A, right) there was a 50-mm Hg sustained increase in arterial pressure (from 90 to 140 mm Hg). Ten minutes later, inhibition of CPA, with the same dose of Gly given previously, had no effect on arterial pressure. Results obtained in this group of experiments ($n = 6$) demonstrate that during glutamatergic blockade of the CVLM, depressor responses to CPA inhibition were reduced from -38 ± 4.3 mm Hg to only -3 ± 1.7 mm Hg (Table 3).

This same experimental design was then repeated, but now the glutamate synapses within the RVLM were blocked (Fig. 4B). Bilateral application of Gly to the CPA produced a 60 mm Hg decrease in blood pressure. After ABP had returned to basal levels, KYN was bilaterally applied to the RVLM. The application of KYN into the RVLM produced a moderate (15 to 20 mm Hg), transient increase in MAP. Ten minutes later, Gly application into the CPA was repeated, evoking a depressor response similar to that previously observed (Fig. 4B, left).

Finally, we tested the effects of KYN applied bilaterally on the CPA itself. Results are exemplified by the ABP records of Figure 4C. As observed for the RVLM, blockade of glutamate synapses within the CPA did not modify falls in ABP evoked by CPA inhibition.

Table 3 summarizes results obtained in this series of experiments. As shown in Figure 4, blockade of glutamate synapses only in the CVLM and not in the RVLM or CPA abolished depressor responses to CPA inhibition. The possibility that KYN microinjected into the CVLM spread to nearby areas and reduced depressor responses to CPA inhibition by an unspecified effect is excluded by the finding that microinjections of KYN into the RVLM or CPA itself did not modify these responses significantly.

Among the three areas examined in this study, we observed significant changes in basal levels of arterial blood pressure only when KYN was applied

Table 3 Effects of the Glutamate Antagonist Kynurenic Acid (KYN, 2 nmol/site) on Cardiovascular Responses to CPA Inhibition[a]

		Before blockade		After blockade	
Treatment KYN into	N	Control (MAP)	CPA inhib. (MAP)	Control (MAP)	CPA inhib. (MAP)
1. CVLM	6	91 ± 4.1	-38 ± 4.3	117 ± 8.5[b]	-3 ± 1.7[b]
2. RVLM	6	104 ± 4.7	-39 ± 6.7	99 ± 5.3	-40 ± 8.9
3. CPA	6	92 ± 4.8	-41 ± 3.7	82 ± 3.3	-33 ± 4.1

[a]Values are means ± SE. Comparison of effects of glutamatergic blockade by microinjections of kynurenic acid into caudal (CVLM) and rostral ventrolateral medulla (RVLM) and the caudal pressor area (CPA) on the depressor responses to CPA inhibition by glycine.
[b]$P < 0.05$ compared with control.

to the CVLM. These results are in agreement with previous studies, indicating that within the RVLM, glutamatergic synapses are mainly involved in cardiovascular reflexes rather than in tonic control of ABP (46,47). Conversely, within the CVLM, glutamatergic synapses seem to stimulate sympathoinhibitory neurons and hence contribute to maintenance of normal levels of ABP (44,48). In this area, glutamatergic synapses are also involved in mediating the baroreceptor reflex (44,49,50). A new finding in our study is the lack of significant changes in ABP after KYN application into the CPA, indicating that tonic activity of neurons in this area is not dependent on glutamatergic afferents.

In summary, results described in this section indicate that cardiovascular responses to CPA excitation or inhibition are mediated through the CVLM. However, considering that CVLM is a sympathoinhibitory area, it is difficult to imagine how CPA responses could be obtained by a direct CPA-CVLM excitatory synapse that utilizes glutamate as the neurotransmitter. It is reasonable to propose that a "change of signal" occurs within the CVLM. It might be proposed that during L-Glu application into the CPA, CPA excitation is converted into CVLM inhibition, thus leading to the observed increases in ABP. Such a mechanism would include an inhibitory component within the CVLM.

To specifically verify this hypothesis, we tested the effects of a $GABA_A$ antagonist, bicuculline, on the pressor responses to CPA stimulation. The experimental design was similar to that utilized in the preceding section: pressor responses to CPA stimulation by L-Glu (50 nmol/site) were evaluated before and after blockade of $GABA_A$ synapses by bilateral application of bicuculline (BIC, 200 pmol/site).

VIII. Effects of $GABA_A$ Receptors Blockade on the Pressor Responses to CPA Stimulation

In a urethane-anesthetized rat, bilateral stimulation of the CPA with L-Glu (50 nmol/site) produced the already described pressor response (Fig. 5A, left). After ABP had returned to control levels, BIC was bilaterally injected into CVLM. Following BIC application, a transient hypotension of approximately 25 mm Hg occurred. Ten minutes later, ABP reached a steady value (not different from the control). The CPA was again stimulated with L-Glu, but the pressor response was abolished.

Microinjections of the same dose of BIC in the RVLM had no effect on pressor responses to CPA stimulation (Fig. 5B). Bilateral application of BIC into the RVLM increased basal ABP to 115 mm Hg (from 95 mm Hg). Regardless of the higher blood pressure, CPA stimulation with L-Glu produced a 40-mm Hg increase in MAP (Fig. 5B, right), a pressor response similar to that observed during control.

We also tested the effects of BIC when injected into the CPA. As observed for the RVLM, although BIC produced a moderate increase in MAP (from

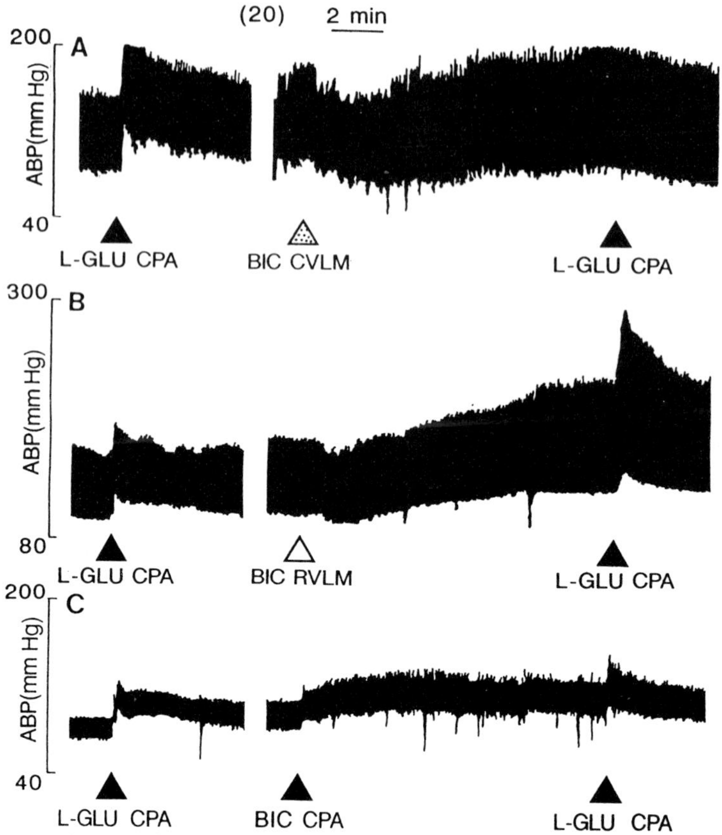

Figure 5 Effects of GABAergic blockade on the pressor responses to CPA stimulation. In anesthetized rats, CPA was initially excited by bilateral microinjections of L-glutamate glycine (L-Glu, 50 nmol/site, black triangles, left). The $GABA_A$ antagonist bicuculline (BIC, 200 pmol/site) was then applied to (A) the caudal ventrolateral medulla (CVLM, dotted triangle); (B) the rostral ventrolateral medulla (RVLM, open triangle), or (C) the CPA itself (black triangle), and L-Glu application was repeated (black triangles on the right side). Numbers on the top of the tracings indicate time in minutes between successive records.

100 to 120 mm Hg), it failed to modify the pressor responses to CPA stimulation.

Results obtained in these three groups ($n = 6$) are summarized in Table 4. As shown in the examples of Figure 5, BIC applied to the CVLM did not modify ABP basal levels but abolished the increases in blood pressure after CPA stimulation. Opposite results were obtained when BIC was microinjected into the

Table 4 Effects of the $GABA_A$ Antagonist Bicuculline (BIC, 200 pmol/site) on Cardiovascular Responses to CPA Inhibition[a]

Treatment BIC into	N	Before treatment Control (MAP)	Before treatment CPA stim. (MAP)	After treatment Control (MAP)	After treatment CPA stim. (MAP)
1. CVLM	6	103 ± 13.1	+43 ± 5.4	118 ± 11.9	+6 ± 2.7[b]
2. RVLM	6	103 ± 5.4	+35 ± 3.9	147 ± 4.7[b]	+32 ± 3.8
3. CPA	6	87 ± 5.3	+33 ± 5.1	118 ± 3.7[b]	+25 ± 3.6

[a]Values are means ± SE. Comparison of effects of GABAergic blockade by microinjections of bicuculline into caudal (CVLM) and rostral ventrolateral medulla (RVLM) and the caudal pressor area (CPA) on the pressor responses to CPA stimulation by L-glutamate.
[b]$P < 0.05$ compared with control.

RVLM or CPA: in these areas, BIC increased ABP basal levels significantly but did not modify pressor responses to CPA stimulation.

These results confirmed that CPA cardiovascular responses are mediated through the CVLM. They support the hypothesis that pressor responses to CPA stimulation result from inhibition of sympathoinhibitory neurons in the CVLM, probably through GABAergic synapses. According to this view, however, one would expect that BIC applied to the RVLM should also block CPA pressor responses. In fact, this seems to be the case. A complete blockade of pressor responses to CPA stimulation was observed in some additional experiments where the dose of BIC microinjected into the RVLM was increased to 400 pmol/site. In these experiments (not shown), following bilateral application of BIC into RVLM, a marked hypertension (MAP > 150 mm Hg) developed; in this condition, CPA stimulation by L-Glu failed to produce further increases in ABP. These experiments suggest that the CPA pressor responses observed after application of BIC (200 pmol/site) into RVLM only reflect an incomplete blockade of GABAergic synapses in this site. Previous evidence indicates that vasomotor areas within the ventrolateral medulla receive tonically active GABAergic afferents, which play a major role in the control of vasomotor tone and arterial blood pressure. Bilateral application of BIC in the RVLM produces marked increases in ABP and sympathetic nerve activity and blocks arterial baroreceptor reflexes (9,12,46,51,52). The origin of these afferents is not completely clear, but they would include GABAergic neurons within the nucleus tractus solitarius (NTS) and the CVLM (53–55). The CVLM itself seems also to be a recipient of GABAergic afferents (27,55,56). Their origin and functional role, however, remains unclear. A major new finding of this study showed that GABAergic synapses are required for modulation of CVLM sympathoinhibitory function, and that the CPA engages these afferents in the control of arterial blood pressure. Finally, at this moment, it is not possible to exclude the possibility that the CPA may be the origin of these afferents.

IX. Summary

Based on evidence reviewed in this chapter and results presented here, we conclude that the CPA is a new, distinct area in the ventrolateral tegmentum involved in cardiovascular control. The CPA has a tonic activity contributing to maintenance of normal levels of arterial blood pressure. CPA functions seem to be achieved through modulation of sympathoinhibitory neurons within the CVLM. Further investigations can provide a better comprehension of the role of the CPA in arterial blood pressure control under normal and pathological conditions.

Acknowledgments

This research is supported by the São Paulo State Research Foundation (FAPESP, grant 91/0506-0).

References

1. Chalmers J, Pilowsky PM. Brainstem and bulbospinal neurotransmitter systems in the control of blood pressure. J Hypertens 1991; 9:675–694.
2. Calaresu FR, Yardley CP. Medullary basal sympathetic tone. Annu Rev Physiol 1988; 50:511–524.
3. Feldberg W, Guertzenstein PG. A vasodepressor effect of pentobarbitone sodium. J Physiol (Lond) 1972; 224:83–103.
4. Guertzenstein PG. Blood pressure effects obtained by drugs applied to the ventral surface of the brain stem. J Physiol (Lond) 1973; 229:395–408.
5. Guertzenstein PG, Silver A. Fall in blood pressure produced from discrete regions of the ventral surface of the medulla by glycine and lesions. J Physiol (Lond) 1974; 242:489–503.
6. Schläfke M, Löschke HH. Lokalisation eines an der Regulation von Atmung und Kreislauf beteiligten Gebiets an der ventralen Oberflaeche der Medulla oblongata durch Kalteblockade. Pflüegers Arch 1967; 297:201–220.
7. McAllen RM. Identification and properties of sub-retrofacial bulbospinal neurones: A descending cardiovascular pathway in the cat. J Auton Nerv Syst 1986; 17:151–164.
8. Chai CY, Lin RH, Pan CM, et al. Pressor responses from electrical or glutamate stimulation of the dorsal or ventrolateral medulla. Am J Physiol 1988; 255:R709–R717.
9. Willette R, Krieger AJ, Barcas PB, Sapru HN. Medullary gamma-aminobutyric acid (GABA) receptors and the regulation of blood pressure in the rat. J Pharmacol Exp Ther 1983; 226:893–899.
10. Granata AR, Ruggiero DA, Park DH, et al. Lesions of epinephrine neurons in the rostral ventrolateral medulla abolish the vasodepressor components of baroreflex and cardiopulmonary reflex. Hypertension 1983; 5(suppl 5):V80–V84.

11. Willette R, Punnen S, Krieger AJ, Sapru HN. Interdependence of rostral and caudal ventrolateral medullary areas in the control of blood pressure. Brain Res 1984; 321:169–174.
12. Ross CA, Ruggiero DA, Park DH, et al. Tonic vasomotor control by the rostral ventrolateral medulla: Effect of electrical or chemical stimulation of the area containing C1 adrenaline neurons on arterial pressure, heart rate, and plasma catecholamines and vasopressin. J Neurosci 1984; 4:479–494.
13. Dampney RAL, Goodchild AK, Robertson LG, Montgomery W. Role of ventrolateral medulla in vasomotor regulation: A correlative anatomical and physiological study. Brain Res 1982; 249:223–235.
14. Farlow DM, Goodchild AK, Dampney RAL. Evidence that vasomotor neurons in the rostral ventrolateral medulla project to the spinal sympathetic outflow via the dorsomedial pressor area. Brain Res 1984; 298:313–320.
15. Ross CA, Armstrong DM, Ruggiero DA, et al. Adrenaline neurons in the rostral ventrolateral medulla innervate thoracic spinal cord: A combined immunocytochemical and retrograde transport demonstration. Neurosci Lett 1981; 25:257–262.
16. Ross CA, Ruggiero DA, Joh TH, et al. Rostral ventrolateral medulla: Selective projections to the thoracic autonomic cell column from the region containing C1 adrenaline neurons. J Comp Neurol 1984; 228:168–185.
17. Ross CA, Ruggiero DA, Reis DJ. Projections from the nucleus tractus solitarii to the rostral ventrolateral medulla. J Comp Neurol 1985; 242:511–534.
18. Amendt K, Czachurski J, Dembowsky K, Seller H. Neurons within the "chemosensitive area" on the ventral surface of the brainstem which project to the intermediolateral column. Pflügers Arch 1978; 375:289–292.
19. Morrison SF, Callaway J, Milner TA, Reis DJ. Rostral ventrolateral medulla: A source of the glutamatergic innervation of the sympathetic intermediolateral nucleus. Brain Res 1991; 562:126–135.
20. Caverson MM, Ciriello J, Calaresu FR. Direct pathway from cardiovascular neurons in the ventrolateral medulla to the region of the intermediolateral nucleus of the upper thoracic cord: Anatomical and electrophysiological investigation in the cat. J Auton Nerv Syst 1983; 9:451–475.
21. Brown DL, Guyenet PG. Cardiovascular neurons of brain stem with projections to spinal cord. Am J Physiol 1984; 247:R1009–R1016.
22. Barman SM, Gebber GL. Axonal projection patterns of ventrolateral medullospinal sympathoexcitatory neurons. J Neurophysiol 1985; 53:1551–1556.
23. Morrison SF, Milner TA, Reis DJ. Reticulospinal vasomotor neurons of the rat rostral ventrolateral medulla: Relationship to sympathetic nerve activity and the C1 adrenergic cell group. J Neurosci 1988; 8:1286–1301.
24. Bisset GW, Feldberg W, Guertzenstein PG, Rocha e Silva M Jr. Vasopressin release by nicotine: The site of action. Br J Pharmacol 1975; 54:463–474.
25. Feldberg W, Guertzenstein PG. Vasodepressor effects obtained by drugs acting in the ventral surface of the brain stem. J Physiol (Lond) 1976; 258:337–355.
26. Blessing WW, Reis DJ. Inhibitory cardiovascular function of neurons in the caudal ventrolateral medulla of the rabbit: relationship to the area containing A1 noradrenergic cells. Brain Res 1982; 253:161–171.
27. Blessing WW, Reis DJ. Evidence that GABA and glycine-like inputs inhibit vasodepressor neurons in the caudal ventrolateral medulla of the rabbit. Neurosci Lett 1983; 37:57–62.

28. Granata AR, Numao Y, Kumada M, Reis DJ. A1 noradrenergic neurons tonically inhibit sympathoexcitatory neurons of C1 area in rat brainstem. Brain Res 1986; 377:127–146.
29. Cravo SL, Morrison SF, Reis DJ. Differentiation of two cardiovascular regions within caudal ventrolateral medulla. Am J Physiol 1991; 261:R985–R994.
30. Cravo SL, Morrison SF. The caudal ventrolateral medulla is a source of tonic sympathoinhibition. Brain Res 1993; 621:133–136.
31. Agarwal SK, Gelsema AJ, Calaresu FR. Neurons in rostral VLM are inhibited by chemical stimulation of caudal VLM in rats. Am J Physiol 1989; 257:R265–R270.
32. Feldberg W, Guertzenstein PG. Blood pressure effects of leptazol applied to the ventral surface of the brain stem of cats. J Physiol (Lond) 1986; 372:445–456.
33. Dembowsky K, Czachurski J, Seller H. Some properties of the sympathoinhibition from the caudal ventrolateral medulla oblongata in the cat. In: Ciriello J, Caverson MM, Polosa C, eds. The Central Neural Organization of Cardiovascular Control. New York: Elsevier, 1989:143–158.
34. Iwamoto GA, Brtva RD, Waldrop TG. Cardiorespiratory responses to chemical stimulation of the caudalmost ventrolateral medulla in the cat. Neurosci Lett 1991; 129:86–90.
35. Gordon FJ, McCann LA. Pressor responses evoked by microinjections of L-glutamate into the caudal ventrolateral medulla of the rat. Brain Res 1988; 457:251–258.
36. Campos RR Jr, Possas OS, Cravo SL, et al. Activation of sympathetic vasomotor tone from the pressor caudal area (PCA) in ventral surface of the medulla (VSM) (abstr). Hypertens 1992; 10(suppl 4):526.
37. Silva LCS, Fontes MAP, Campagnole-Santos MJ, et al. Cardiovascular effects produced by micro-injection of angiotensin-(1-7) on vasopressor and vasodepressor sites of the ventrolateral medulla. Brain Res 1993; 613:321–325.
38. Possas OS, Campos RR Jr, Cravo SL, et al. A fall in arterial blood pressure produced by inhibition of the caudal most ventrolateral medulla: the caudal pressor area. Journal of the Autonomic Nervous System 1994; 49:235–245.
39. Campos RR Jr, Possas OS, Cravo SL, et al. Pressor caudal area a third region of ventral surface of the medulla involved in cardiovascular regulation (abstr). Inter-Amer Soc Hypertens Abstr 1993; 10:9.
40. Dampney RAL, Moon EA. Role of ventrolateral medulla in vasomotor response to cerebral ischemia. Am J Physiol 1980; 239:H349–H358.
41. Feldberg W, Rocha e Silva M Jr. Inhibition of vasopressin release to carotid occlusion by gamma-aminobutyric acid and glycine. Br J Pharmacol 1981; 72:12–24.
42. Granata AR, Ruggiero DA, Park DH, et al. Lesions of epinephrine neurons in the rostral ventrolateral medulla abolish the vasodepressor components of baroreflex and cardiopulmonary reflex. Hypertension 1983; 5(suppl 5):V80–V84.
43. Granata AR, Ruggiero DA, Park DH, et al. Brain stem area with C1 epinephrine neurons mediates baroreflex vasodepressor responses. Am J Physiol 1985; 248: H547–H567.
44. Guyenet PG, Filtz TM, Donaldson SR. Role of excitatory amino acids in rat vagal and sympathetic baroreflexes. Brain Res 1987; 407:272–284.
45. Strack AM, Sawyer WB, Platt KB, Loewy AD. CNS cell groups regulating the sympathetic outflow to adrenal gland as revealed by transneuronal cell body labeling with pseudorabies virus. Brain Res 1989; 491:274–296.

46. Sun M-K, Guyenet PG. Arterial baroreceptor and vagal inputs to sympathoexcitatory neurons in rat medulla. Am J Physiol 1987; 252:R699–R709.
47. Urbanski RW, Sapru HN. Putative neurotransmitters involved in medullary cardiovascular regulation. J Auton Nerv Syst 1988; 25:181–193.
48. Blessing WW. Baroreceptor-vasomotor reflex after N-methyl-D-aspartate receptor blockade in rabbit caudal ventrolateral medulla. J Physiol (Lond) 1989; 416:67–78.
49. Guyenet PG, Sun M-K. Role of GABA and excitatory amino acids in medullary baroreflex pathway. Organization of the autonomic nervous system: central and peripheral mechanisms. In: New York: Liss, 1987:215–225.
50. Gordon FJ. Aortic baroreceptor reflexes are mediated by NMDA receptors in caudal ventrolateral medulla. Am J Physiol 1987; 252:R628–R633.
51. Pilowsky PM, Morris MJ, Minson JB, et al. Inhibition of vasodepressor neurons in the caudal ventrolateral medulla of the rabbit increases both arterial pressure and the release of neuropeptide Y-like immunoreactivity from the spinal cord. Brain Res 1987; 420:380–384.
52. Smith JK, Barron KW. GABAergic responses in ventrolateral medulla in spontaneously hypertensive rats. Am J Physiol 1990; 258:R450–R456.
53. Meeley MP, Ruggiero DA, Ishitsuka T, Reis DJ. Intrinsic gamma-aminobutyric acid neurons in the nucleus of the solitary tract and the rostral ventrolateral medulla of the rat: An immunocytochemical and biochemical study. Neurosci Lett 1985; 58:83–89.
54. Dampney RAL, Blessing WW, Tan E. Origin of tonic GABAergic inputs to vasopressor neurons in the subretrofacial nucleus of the rabbit. J Auton Nerv Syst 1988; 24:227–239.
55. Ruggiero DA, Meeley MP, Anwar M, Reis DJ. Newly identified GABAergic neurons in regions of the ventrolateral medulla which regulate blood pressure. Brain Res 1985; 339:171–177.
56. Sved AF, Blessing WW, Reis DJ. Caudal ventrolateral medulla can alter vasopressin and arterial pressure. Brain Res Bull 1985; 14:227–232.

56

Electrophysiological and Anatomical Characterization of Barosensitive Neurons in the Rostral Ventrolateral Medulla

ANTONIO R. GRANATA

University of Tennessee, Memphis
Memphis, Tennessee

I. Introduction

The rostral ventrolateral medulla (RVLM) is an area in the ventrolateral segment of the medullary reticular formation that contains a particular population of neurons involved in the regulation of cardiovascular, respiratory, and nociceptive functions. The anatomical definition of the RVLM has been a matter of discussion, in part because the term has been used to define approximately similar medullary areas in different species; furthermore, in some studies the term has been used to define a functional area in this portion of the medulla [e.g., the vasopressor area in the RVLM (1–3)] that involves different anatomically identifiable nuclei, as, for instance, the paragigantocellular nucleus (4) and a more lateral portion called the retrofacial nucleus in the cat (5). In any case, definition of the RVLM on the basis of its anatomical cytoarchitectonic divisions has been difficult, even within the same genus (4,6,7). Nonetheless, the RVLM in the rat can be clearly defined inside the boundaries of an area limited dorsally by the nucleus ambiguus, which in its more rostral part is also called the retrofacial nucleus (8), and by the gigantocellular reticular nucleus; ventrally, the RVLM extends to the ventral medullary surface. The rostral limit of the RVLM is the facial nucleus, and the caudal one is the lateral reticular nucleus. Mediolaterally, the RVLM extends from the spinal trigeminal tract and spinal trigeminal nucleus (pars oralis) to an area containing a group of neurons immunoreactive to serotonin and substance P. In the ventral aspect (9) and more dorsally, the RVLM extends to the nucleus gigantocellularis pars ventralis.

The RVLM is interconnected with different areas in the brain and the spinal cord directly related to autonomic control. The mediocaudal nucleus of the solitary tract (NTS) at the level of the primary afferents from baro- and mechanoreceptors (10,11) densely innervates the RVLM (12,13); in turn, the RVLM innervates similar areas of the NTS. A monosynaptic connection between the NTS and the RVLM was first postulated as the main intramedullary pathway of the baro- and mechanoreceptor reflexes (14). In contrast, other reports support the idea that neurons located in the caudal extension of the ventrolateral medulla, the so-called caudal ventrolateral medulla (CVLM), could serve as a relay station between the NTS and the RVLM (15,16). However, when bilateral electrolytic lesions are placed in the CVLM in the rat or the rabbit, the vasodepressor responses elicited by electrical stimulation of the aortic depressor nerve or the central end of the vagus nerve are preserved (17,18). More recently, different reports have demonstrated that spontaneously active CVLM neurons were antidromically activated from the RVLM and excited by baroreceptor activation and by mediocaudal-NTS stimulation in the rat (19), whereas in the rabbit, these CVLM-RVLM neurons were excited by stimulation of the aortic depressor nerves and by inputs from the carotid sinus baroreceptors (20).

The central pathway of the sympathoinhibitory responses elicited by stimulation of visceral mechanoreceptor fibers in the vagus nerve is also controversial. The presence of interneurons receiving inputs from the NTS and projecting to the RVLM has been supported by some experimental evidence (21) and denied by others (22). Nevertheless, it is now accepted that both baroreceptor and visceral mechanoreceptor reflexes inhibit RVLM neurons with projections to the sympathetic preganglionic neurons in the intermediolateral column in the spinal cord, most probably through a relay station either in the CVLM or somewhere in the ventrolateral medulla (23,24).

On the other hand, it is still possible to speculate that the monosynaptic pathway between the mediocaudal-NTS and the RVLM conveys cardiovascular and pulmonary information, although its precise nature is still unknown.

The area postrema also innervates the RVLM (12,25,26); some studies in the rabbit suggest that noradrenergic neurons in the area postrema could contribute to this innervation (27), which could transmit information triggered by humoral factors unable to cross the blood-brain barrier.

The RVLM is also reciprocally connected with the lateral subdivisions of the parabrachial complex in both the rat and the cat (28,29). In addition, other studies have reported that the RVLM receives projections from, and in some cases is reciprocally connected with, the Kölliker-Fuse nucleus, the ambiguus complex, the contralateral RVLM, the nucleus raphe magnus and raphe obscurus, the central gray area, the paraventricular nucleus, the lateral hypothalamic area, the caudolateral bed nucleus stria terminalis, and the zona incerta (26,29–33). Although the aforementioned areas are all related to autonomic control, the specific function of the connections or interconnections with the RVLM is still unknown.

The RVLM has been the focus of many investigations because of its critical role in regulating the vasomotor sympathetic outflow. Different studies have demonstrated that electrolytic lesions of the RVLM (34–36), application of different inhibitory agents or neurotransmitters into an area of the ventral medullary surface overlaying the RVLM (37–41), or reducing the temperature of this medullary surface (42) causes a profound decrease in arterial pressure similar to that produced by spinal cord transection. Conversely, electrical or chemical stimulation of this area results in an important increase in arterial pressure and heart rate (7,39,43,44). The possibility that neurons located in the RVLM could tonically excite sympathetic preganglionic vasomotor neurons in the spinal cord and then play the role of a "vasomotor center" has been strongly supported by anatomical studies demonstrating a direct projection from the RVLM to the intermediolateral and intermediomedial column in the spinal cord, innervating an area that corresponds to the distribution of sympathetic preganglionic neurons (7,44–46). Furthermore, anatomical studies using electron microscopy techniques combined with anterograde tract tracing of descending medullary projections and retrograde labeling of sympathoadrenal preganglionic neurons have demonstrated the existence of synaptic inputs from the RVLM to sympathetic preganglionic neurons (47). It is important to note that the distribution of bulbospinal neurons in the RVLM is restricted to the more rostral portion of this area; the vast majority of these neurons are distributed in an area approximately 800 to 1200 μm rostral to the obex in the rat (48).

Nearly 50% of these bulbospinal neurons in the RVLM are immunoreactive to phenylethanolamine *N*-methyltransferase (PNMT), the enzyme that catalyzes the conversion of noradrenaline to adrenaline; consequently, these neurons are identical to the adrenergic group described by Höckfelt et al. (49) as the C1 adrenergic group. Because of the good correspondence between the location of the C1 adrenergic group, the bulbospinal neurons, and the medullary area critically involved in maintaining normal values of arterial pressure and heart rate, this subdivision of the RVLM has been called the C1 area (7, 18). In addition, synaptic contacts between adrenergic terminals immunoreactive to PNMT and identified sympathetic preganglionic neurons have been shown by electron microscopy. Since the RVLM is one of the more important sources of adrenaline to this part of the spinal cord, it has been assumed that C1 neurons contribute to the direct monosynaptic control of sympathetic preganglionic neurons (50). However, the possibility that these C1 adrenergic bulbospinal neurons could play the role of vasomotor neurons still poses an unresolved question. This point will be analyzed later in this study.

The critical "vasomotor" area (C1 area) involves only a *subdivision* of the nucleus paragigantocellularis lateralis in the rat (4,51); nonetheless, some investigators have called this area the "nucleus paragigantocellularis" (52). More accurately, Ross et al. (7) have anatomically defined this area of the reticular formation in the rat as the nucleus reticularis rostroventrolateralis. This area appears to be similar to the retrofacial nucleus in the cat described by Berman (8), and a seemingly similar area in the cat has been termed the "subretrofacial nucleus" by others (53).

The intermediolateral cell column also receives direct projections from the raphe pallidus, the raphe obscurus, the A5 noradrenaline-containing neurons in the ventral pons, the Kölliker-Fuse nucleus, the lateral hypothalamus, and the paraventricular nucleus in the hypothalamus. None of these pathways, however, has been demonstrated to play a role in vasomotor control.

Using extracellular recording techniques, different laboratories have identified a group of neurons in the RVLM that can be antidromically activated from the thoracic spinal cord with velocities of conduction below 10 m/s; these neurons were identified in the rat, rabbit, and cat (53–56). The recording sites, although not necessarily the neuronal location, appear to correspond to the nucleus reticularis rostroventrolateralis in the RVLM of the rat (7). The recorded neurons were spontaneously active, with a high frequency of discharge that was synchronized to the arterial pulse wave; and they were inhibited by an increase in systemic arterial pressure. In addition, the spontaneous discharge was also related to the lumbar and splanchnic sympathetic nerve discharge (55, 57). Therefore, these neurons demonstrated barosensitive properties and were considered to be components of the tonic bulbospinal sympathoexcitatory pathway.

Although some electrophysiological evidence suggests that these tonic bulbospinal neurons with barosensitive characteristics in the RVLM project to the intermediolateral column in the spinal cord (55), whether they send monosynaptic potentials to the sympathetic preganglionic neurons is still unknown.

More recent investigations have proposed that an area in the RVLM situated more medially than the classically defined "vasomotor" area—and hence also medially from the rostral pole of the C1 adrenergic neurons—could have a sympathoexcitatory function similar to that of the lateral part (58,59). Furthermore, it has been proposed that neurons in this area acting separately from the lateral counterpart could also contribute to vasomotor control, although affecting different peripheral vascular beds from those controlled by the lateral part (59–62). This medial part contains neurons that are immunoreactive to serotonin and that project to the lateral horn in the spinal cord (58,63). However, electrophysiological studies using extracellular recording techniques have not recognized barosensitive neurons with different characteristics in the RVLM, although, according to the location of the recording sites, neurons in both subdivisions of the RVLM were supposedly recorded from those studies (54,64).

We have to take into consideration at this point the technical limitations inherent in the extracellular recording technique, which cannot determine either the precise localization of the recorded neurons or the synaptic inputs received by these neurons during the cardiac cycle.

We have recently developed a technique for recording intracellular potentials from RVLM neurons in vivo (65). In this paper, we discuss the results of experiments aimed at characterizing RVLM bulbospinal neurons with barosensitive properties. The recorded neurons were labeled by intracellular injections of biocytin; this technique was combined with immunocytochemical identification of PNMT to determine whether the electrophysiologically characterized neurons were the same as the C1 adrenergic group. Finally, we have also developed an in vitro intracellular recording technique that has been used to identify similar RVLM neurons previously characterized in vivo.

II. Electrophysiological Identification of RVLM Neurons with Baroreceptor-Related Activity and Possible Involvement in Vasomotor Control

Several laboratories using an in vivo extracellular recording technique have identified a group of tonically active RVLM neurons antidromically activated from the sympathetic spinal cord and modulated by baroreceptor inputs (54–56,64). However, due to the technical limitations inherent in the extracellular recording technique, it was not possible to directly investigate the nature of these baroreceptor inputs or the transmitter produced by the bulbospinal neurons that receive these baroreceptor inputs.

We have developed a technique to record intracellular potentials in vivo and have investigated the RVLM for spontaneously active neurons with neuronal activity synchronized to the cardiac cycle. Our studies were conducted in anesthetized, paralyzed, ventilated rats (65). We investigated first at the extracellular level a group of neurons in the rostral portion of the RVLM with spontaneous action potentials occurring about every cardiac cycle, as demonstrated by simultaneous recording of the pulsatile arterial pressure and the extracellular recording (Fig. 1A). Such neurons could correspond to those described by other investigators using extracellular recording techniques (55,57). Other neurons located in the same area did not show such a clear synchronization at the extracellular level. However, simultaneous pulse-triggered histograms of the neuronal activity and averages of the pulsatile arterial pressure demonstrated a clear cardiac modulation of neuronal discharge with a conspicuous peak in the neuronal discharge in each cardiac cycle (Fig. 2). Similar types of synchronization have been reported by other investigators using an extracellular recording technique to characterize barosensitive neurons in the RVLM in the rat and the cat (53, 55).

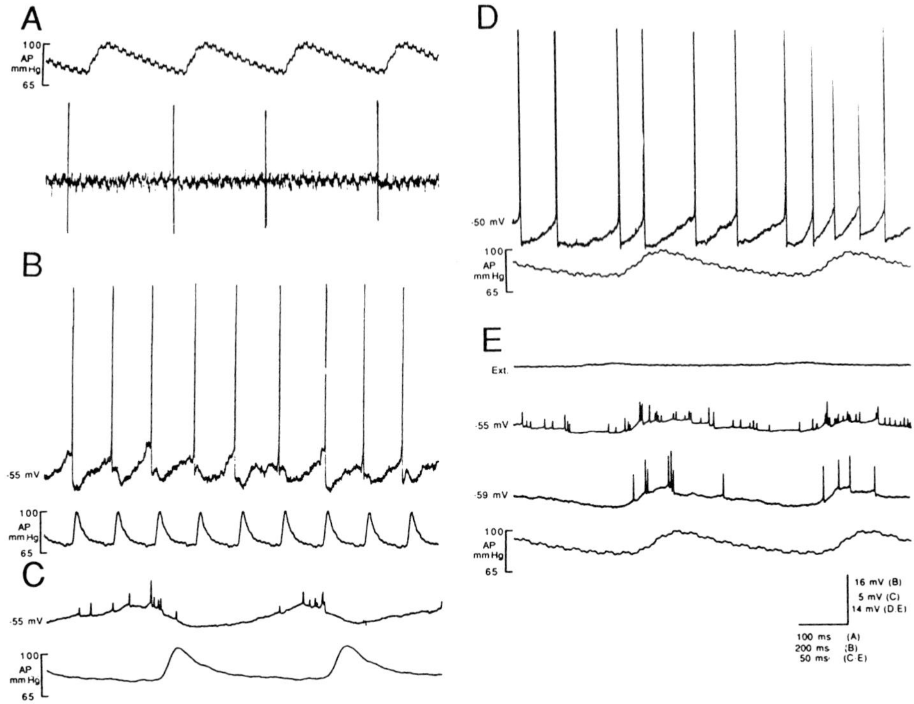

Figure 1 Extracellular and intracellular recordings from different bulbospinal RVLM neurons with cardiac-related activity. (A) Digitized record of extracellular recording of a spontaneously active RVLM neuron and arterial pressure (AP) in mm Hg. (B) Intracellular recording of the same neuron as in panel A at stable resting membrane potential (−55 mV). The neuron has recovered its pulse-related firing pattern observed in panel A. The digitized records of AP and membrane potential were simultaneously triggered by the pulsatile AP. (C) Simultaneous average of the neuronal membrane potential and pulsatile AP at resting values. The averages were triggered by the arterial pulse wave. Number of sweeps = 450. Sample period = 50 μs. (D) Simultaneous digitized records of arterial pressure (AP) (mm Hg) and intracellular recording of a different neuron at stable membrane potential (−50 mV) triggered by the arterial pulse wave. Note the increased frequency of discharge at the end of diastolic periods. (E) Simultaneous average of the AP in millimeters of mercury at resting values and the spontaneous neuronal membrane potential (upper middle trace) and the neuronal membrane potential held at −59 mV (lower middle trace). The average of the extracellular (ext.) recording is illustrated in the uppermost trace. The averages were triggered by the arterial pulse wave. Same neuron as in panel D. Number of sweeps was 550 in all cases. Sample period = 50 μs. (From Ref. 65.)

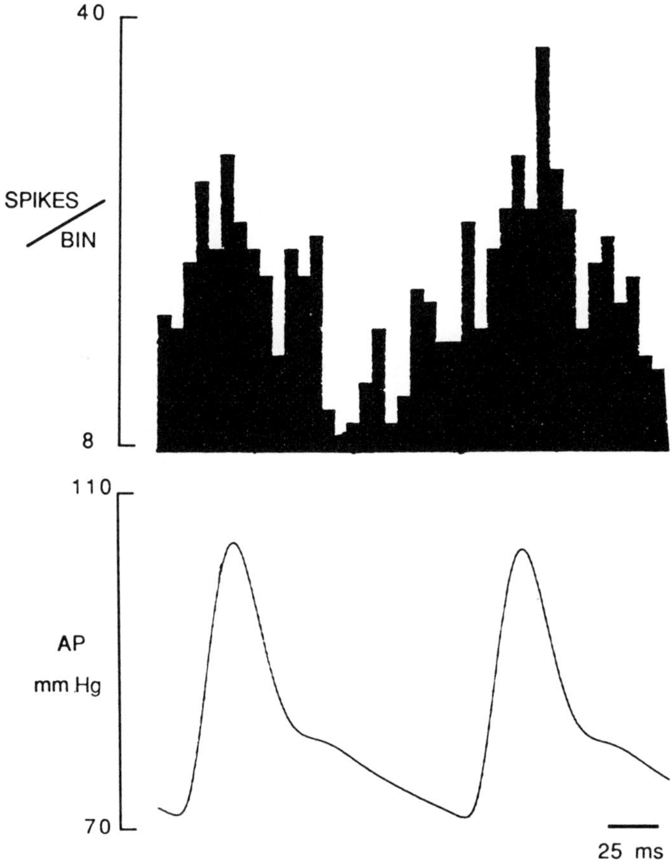

Figure 2 Pulse-triggered histogram of RVLM neuron and simultaneous average of arterial pressure. Upper and lower traces show number of spikes occurring in each time bin and the corresponding arterial pulse, respectively. Sweeps = 400; bin width = 5 ms for histogram, 1 ms for average. (From A. R. Granata, unpublished results.)

Neurons with the aforementioned characteristics were selected for intracellular analysis. During intracellular recording, some of them demonstrated a pattern of discharge similar to that found during extracellular recording; namely, spontaneous depolarizing and action potentials at practically every cardiac cycle (Fig. 1B). The simultaneous averages of pulsatile arterial pressure and neuronal membrane potential also demonstrated a clear cardiac-related rhythmic activity in this type of RVLM neuron (Fig. 1C); the average depolarization of the membrane potential correlated with the period of the pulsatile arterial pressure.

In another neuron, the cardiac-related rhythmic activity was demonstrated by a consistently faster rate of discharge at the end of the diastolic/beginning of the systolic period during spontaneous membrane potential (Fig. 1D). In these neurons, the simultaneous average of the neuronal membrane potential and pulsatile arterial pressure also demonstrated depolarization potentials occurring at the end of the diastolic/beginning of the systolic period (Fig. 1E). These potentials increased in magnitude when the neuron was hyperpolarized by passing negative current through the recording electrode (Fig. 1E). Conversely, when the neuron was depolarized, the depolarization potential decreased in magnitude. Therefore, the potentials locked to the cardiac cycle were considered excitatory postsynaptic potentials (EPSP) because of the changes in magnitude observed during hyperpolarization and depolarization.

Support for the view that these neurons with pulse-modulated activity send projections to the intermediolateral column in the spinal cord comes from electrophysiological experiments (65). We have observed that barosensitive RVLM neurons can be antidromically driven by electrical stimulation of the intermediolateral column of the ipsilateral thoracic spinal cord at the T2–T3 level (Fig. 3A to C).

The calculated conduction velocities of the descending axons from RVLM neurons characterized as barosensitive ranged from 1.2 m/s to 11.0 m/s. Other investigators have reported a second type of RVLM bulbospinal neuron with pulse-modulated activity but with slower axonal velocities of conduction. Yet in one case, these neurons, although located in the RVLM, were silent and thus incompatible with the role of tonic sympathetic nerve activity generator (55); whereas in the other case, the slowly conducting neurons were spontaneously active but located in the more medial aspect of this tonic vasomotor area—the nucleus paragigantocellularis lateralis (57).

Under our experimental conditions, we did not find spontaneously active RVLM neurons with both barosensitive properties and slow axonal velocities of conduction.

We noticed that the baroreceptor inputs were able to control the total activity of these RVLM neurons with EPSP synchronized to the cardiac cycle. The activation of the baroreceptor reflex, by increasing systemic arterial pressure with bolus intravenous injections of phenylephrine, produced hyperpolarization, a decrease in the firing frequency, and an increase in the membrane input resistance in barosensitive neurons with depolarizing potentials locked to the cardiac cycle (Fig. 4). In these experiments, the input impedance was calculated by deflections in the membrane potential evoked by constant pulses of negative current. On the other hand, the inactivation of the baroreceptor reflex by decreasing systemic arterial blood pressure by bolus intravenous injections of the peripheral vasodilator sodium nitroprusside elicited the opposite effect in these barosensitive neurons: specifically, depolarization and an increase in the firing frequency (65).

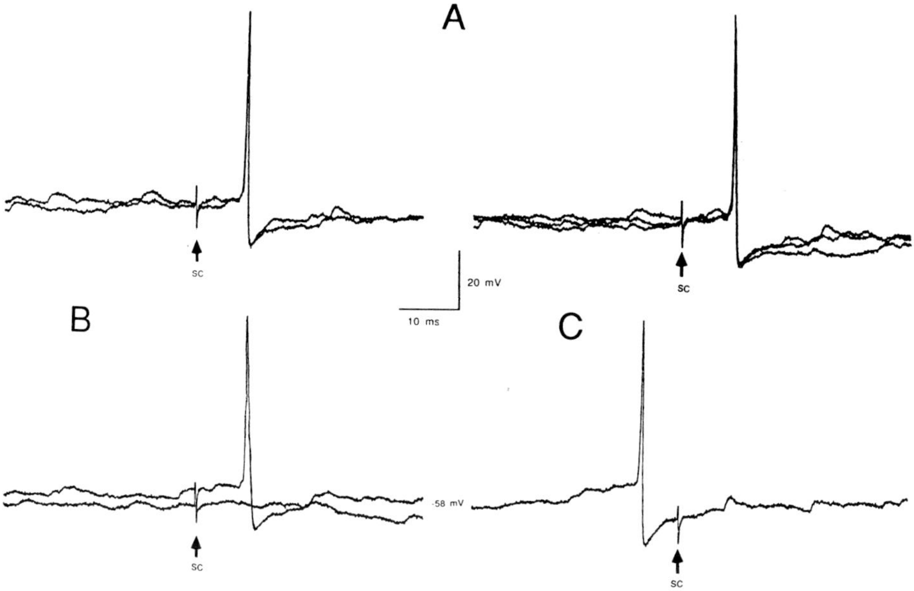

Figure 3 Antidromic activation of an RVLM neuron with pulse-modulated activity following spinal cord stimulation. (A) Electrical stimulation of the IML in the ipsilateral spinal cord (arrow) during intracellular recording produced action potentials, which demonstrated constant latency. (B) At stimulation threshold (75 μA) (arrow), spikes were all or none, and no underlying excitatory postsynaptic potentials were observed when the spike failed (the DC trace is shifted to improve visualization). (C) A spontaneous spike collided with the antidromic elicited action potential at 2× threshold of stimulation (arrow). (From Ref. 65.)

In consequence, these studies demonstrate the participation in the baroreceptor reflex of barosensitive neurons with EPSP locked to the cardiac cycle. Furthermore, they also indicate that disfacilitation is responsible for the hyperpolarization accompanied by a decrease in the firing frequency and an increase in the membrane input resistance observed following baroreceptor activation. Since it is well known that tonic sympathetic nerve activity is generated in the brainstem (66), this disfacilitation could originate in the inhibition of tonic excitatory inputs produced by other neurons in the brainstem.

In addition, these tonic excitatory inputs could be responsible for another characteristic found in the barosensitive RVLM neurons with EPSP locked to the cardiac cycle: the presence of an important ongoing and random synaptic activity observed at resting values of arterial pressure, heart rate, and neuronal membrane potential (Figs. 3 and 5).

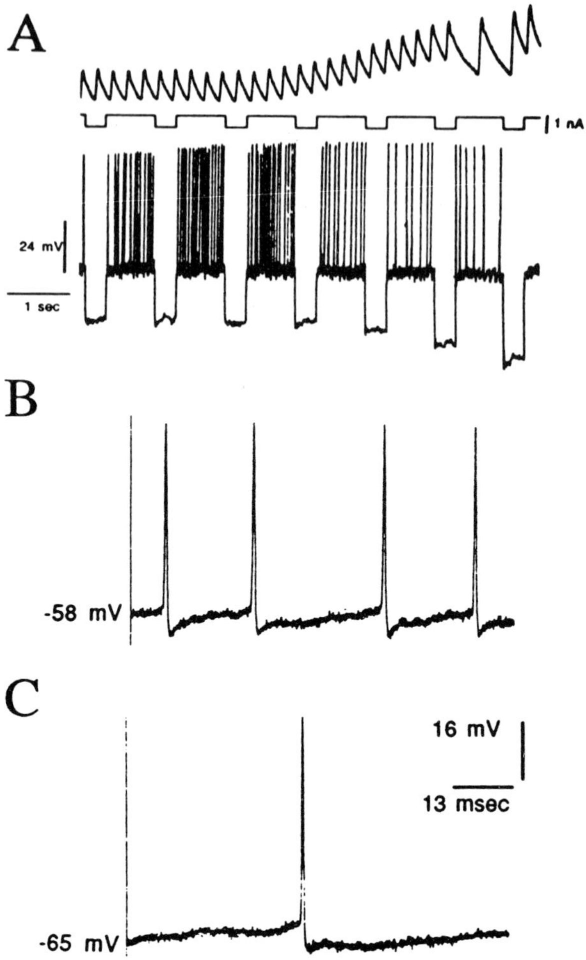

Figure 4 Inhibition of a bulbospinal RVLM neuron by phenylephrine-produced increases in systemic arterial pressure. (A) Intravenous administration of a bolus injection of phenylephrine during intracellular recording from a bulbospinal neuron in the RVLM elicited: (a) a decrease in the firing frequency; (b) an increase in the input resistance, as demonstrated by the increased amplitude of hyperpolarizing voltage deflections in response to constant-amplitude inward current pulses; and (c) hyperpolarization of the neuronal membrane potential, which can be clearly seen when the gain of the amplifier is increased: B = control; C = hyperpolarization during baroreflex activation. Note that the membrane input resistance increased in proportion to changes in arterial pressure and preceded the reflex bradycardic response. From top to bottom in panel A, systemic arterial pressure (mm Hg), control current injections, and intracellular recording (membrane potential -49mV). All spikes are attenuated in size. (From A.R. Granata, unpublished results.)

We do not know the location of these tonic excitatory neurons in the rat. However, other investigators working in the cat have postulated that sympathoexcitatory neurons located in the nucleus reticularis parvocellularis and nucleus reticularis ventralis in the lateral tegmental field drive RVLM-spinal neurons by a direct pathway from these nuclei in the lateral tegmental field to the RVLM (64,67–69).

Another characteristic that distinguishes these barosensitive neurons with EPSP locked to the cardiac cycle is that, besides firing in a single spiking mode, they have been observed to fire in a bursting pattern, with a burst occurring at the end of the diastolic/beginning of the systolic period; on several occasions the burst rode on a depolarizing potential (Fig. 5). At normal values of arterial pressure and heart rate, with the animal in a paralyzed condition and under urethane anesthesia, the bursts are composed of an arrangement of 2 to 5 spikes. Furthermore, we observed that some of these barosensitive neurons firing in a single-spiking mode switched to a bursting mode. This change in the firing mode could be the result of a relatively prolonged period of excitation, so the

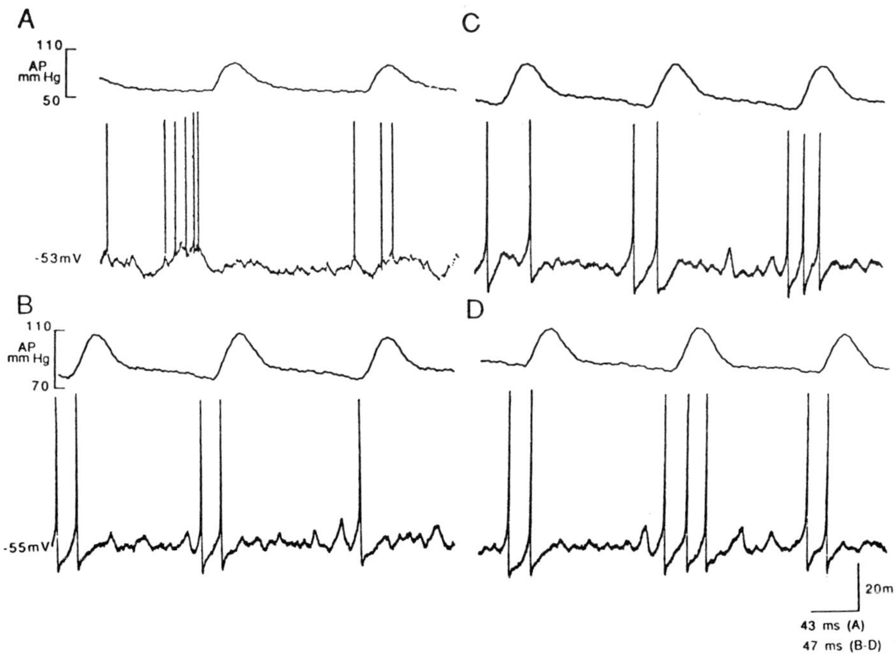

Figure 5 Spontaneously occurring burst intracellularly recorded from two barosensitive RVLM neurons. In panel A, the synchronized bursts were observed to ride on a depolarizing potential. In panels B–D (in a different neuron), the bursts or sometimes single spikes were also synchronized to the cardiac cycle. Note the important ongoing activity in this type of neuron. (From Granata, unpublished results.)

excitatory inputs received by these neurons could play even an additional role by producing a summation of spikes in the terminals of these descending neurons in the intermediolateral cell column. The summation of spikes could result in an extensive release of neurotransmitter and, hence, in large increases in postsynaptic efficacy (70,71). Therefore, the physiological role played by these barosensitive neurons could differ from that of other RVLM neurons with barosensitive properties.

Based on the aforementioned characteristics, which can be clearly distinguished from those of other RVLM-spinal neurons, we have classified these neurons as barosensitive type I neurons.

Attempts to identify and characterize barosensitive neurons in the RVLM according to the phasic baroreceptor modulatory inputs received by these neurons during each rise in systolic arterial pressure have proved difficult for some investigators lacking sufficient experience in intracellular recording from small rodents. It is very possible that the presence of conspicuous mechanical artifacts related to arterial pulsation have prevented a clear identification of these baroreceptors' inputs on barosensitive RVLM neurons. It is beyond the scope of this paper to discuss the reasons and origins of such mechanical artifacts as could appear in the hands of other investigators due to different technical deficiencies. However, we will discuss the very careful analysis followed in our investigation to eliminate the presence of prominent mechanical artifacts from our intracellular recording. First, at the end of the intracellular recording session, the electrode was withdrawn from inside the neuron to just outside it; a number of extracellular recordings were performed, and the average was calculated. We considered as valid only those intracellular recordings associated with flat averages of the corresponding extracellular control, because the flat extracellular average indicates the absence of important mechanical artifacts that could have interfered with the intracellular potentials. The possibility that flat extracellular averages could have been the result of a reduction in the recording electrode resistance due to removal of the resistance of the whole neuron by moving the electrode from the intracellular to the extracellular space is excluded. This is because the neuronal input resistance calculated as the slope of the regression line derived from the plot of the current injection versus the deflection in membrane potential ranged between 10 to 15 MΩ in these RVLM neurons; we induced similar changes in electrode resistance by passage of constant current during extracellular recording and were unable to detect fluctuations in membrane potentials locked to the cardiac cycle. Furthermore, the variability in resistance among the different electrodes used in these experiments was three to four times larger than the membrane input resistance of these neurons.

Second, we obtained intracellular potentials from a different type of neuron intermixed with barosensitive neurons in the RVLM. These different neurons can also be antidromically activated from the spinal cord; on the other hand, they show a very regular pattern of discharge at normal levels of arterial pressure and a flat average of both the neuronal membrane potentials and extracellular recordings triggered by the arterial pressure pulse. We have never seen baroreceptor modulatory activity in this type of RVLM neuron when the recording conditions and the general physiological parameters of the rat were maintained strictly under the limits of normality. Therefore if, under normal recording conditions, mechanical artifacts were unavoidable although randomly present phenomena, some neurons with the characteristics already mentioned could have demonstrated mechanical artifacts resembling baroreceptor modulatory activity. Similarly, we consistently found that the average of the neuronal membrane potential triggered by arterial pressure pulse was flat in RVLM neurons with rhythmic activity synchronized to the respiratory cycle (65).

III. Electrophysiological Characterization of RVLM Neurons with Inhibitory Postsynaptic Potentials (IPSP) Synchronized to the Cardiac Cycle

We have identified a different type of neuron in the RVLM on the basis of the pattern of discharge and responses to spinal cord stimulation. In this type of neuron, electrical stimulation of the intermediolateral column in the thoracic spinal cord evoked antidromic action potentials. Moreover, in some of these neurons, the antidromically evoked spike was preceded by a short-latency monosynaptic EPSP (65). On the other hand, the axonal conduction velocities of these bulbospinal neurons varied in a range similar to that observed in barosensitive type I neurons.

The neurons described in this group are spontaneously active, and the action potentials are produced by slow depolarizing potentials that drive the membrane potential to spike generation threshold. At spontaneous membrane potential and normal values of arterial pressure and heart rate, this type of neuron demonstrated a spontaneous firing pattern that included slower single spiking at the end of the diastolic/beginning of the systolic period, followed by single spikes firing at higher frequencies during the rest of the cardiac cycle (Fig. 6A). We observed more clearly this phasing pattern of discharge when the average of the membrane potential was triggered by spontaneous spikes (Fig. 6B). For the same neuron, at normal values of arterial pressure and heart rate, the post-R wave simultaneous average of the membrane potential and arterial pressure showed hyperpolarizing potentials occurring at the end of the diastolic/beginning of the systolic period (Fig. 6C). These hyperpolarizing potentials are inhibitory postsynaptic potentials (IPSP) because they decreased in magnitude when the neuron was hyperpolarized by passing negative current through the recording electrode (Fig. 6D). The control average of the extracellular recordings with the electrode still very close to the neuronal membrane and triggered

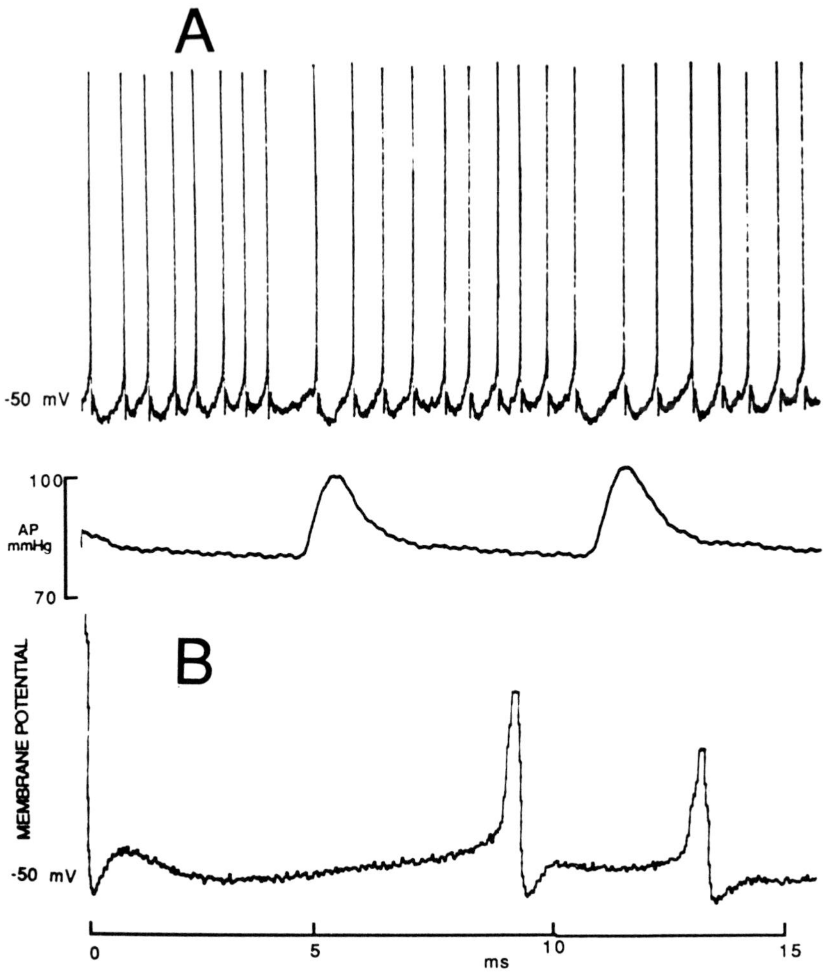

by the pulsatile arterial pressure was demonstrated to be flat (Fig. 6E). Furthermore, when the extracellular average was performed under constant injection of negative current through the recording electrode with intensities even higher than those used to drive the membrane potential during intracellular recording, only minor mechanical distortions were observed (Fig. 6E).

In these studies the IPSP were further characterized by injecting the neurons with chloride ions for several minutes. This was accomplished by passing negative current through the recording electrode filled with KCl. After the injection of chloride ions, we found that the hyperpolarization potential synchronized

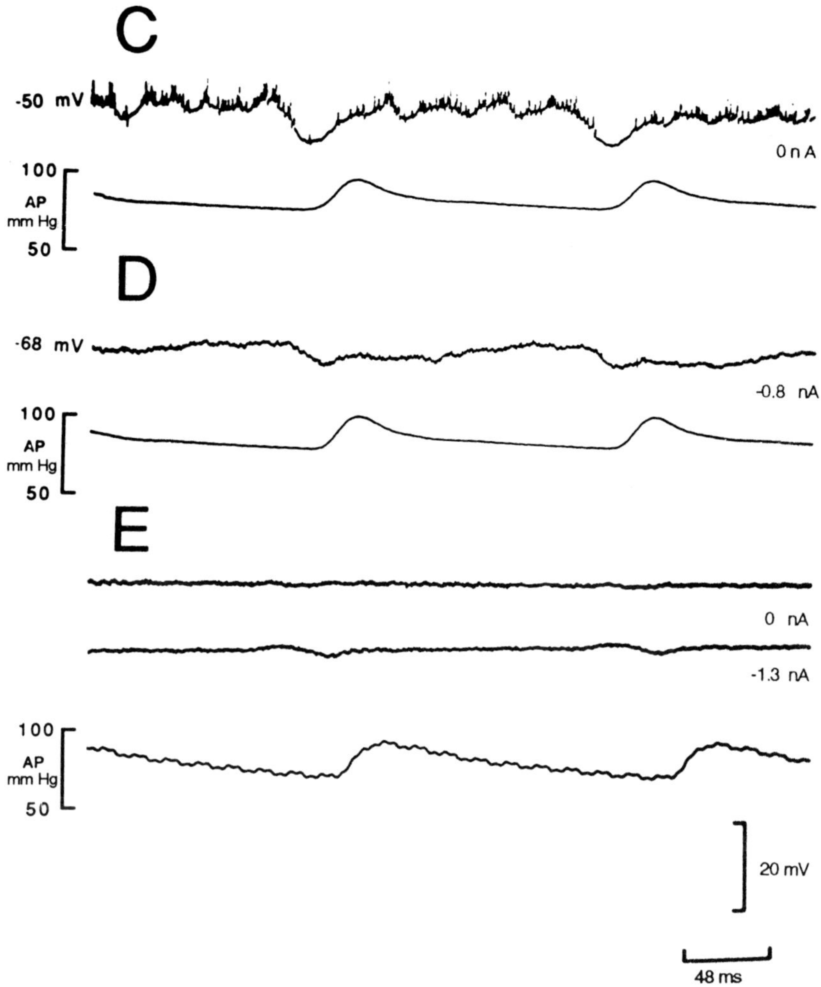

Figure 6 Intracellular recording of a RVLM bulbospinal neuron with inhibitory baroreceptor inputs. (A) Simultaneous digitized records of arterial pressure (AP) in millimeters of mercury and intracellular recording at stable membrane potential (−50 mV). The records were triggered by the arterial pulse wave. Note the decrease in the firing frequency at the end of the diastolic period. (B) Average of the resting neuronal membrane potential triggered by spontaneous spikes. Sweeps = 96; sample period = 5 μs. (C) Simultaneous averages of the neuronal membrane potential (−50 mV) and pulsatile AP in millimeters of mercury, triggered by the arterial pulse wave. Sweeps = 674; sample period = 50 μs. (D) Averages similar to those illustrated in panel C, but now at a holding membrane potential of −68 mV. (E) The simultaneous averages of AP and extracellular recordings at zero or −1.3 nA constant current injection (upper and middle traces, respectively). The averages were triggered by the arterial pulse wave. Sweeps = 700; sample period = 50 μs. Same neuron as in panels A through E. (From Ref. 65.)

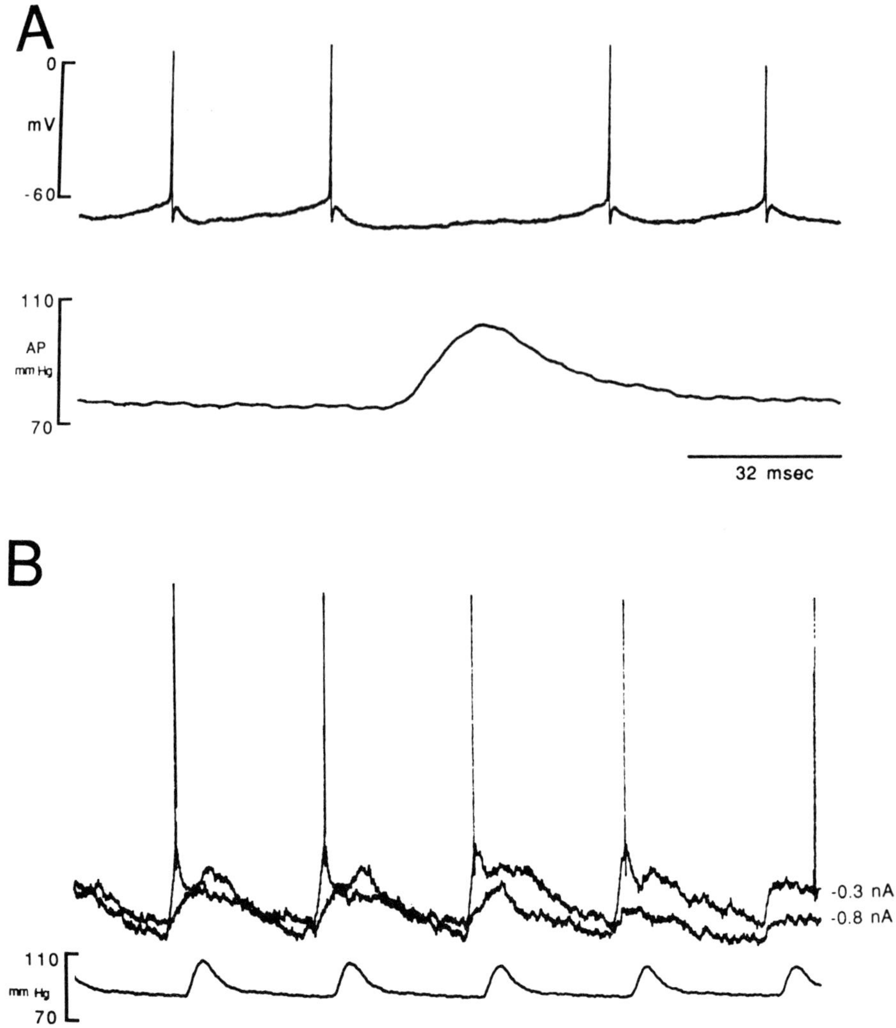

to the cardiac cycle reversed in polarity (Fig. 7B). In addition, depending on the holding membrane potential, these reversed potentials that are synchronized with the cardiac cycle also produced action potentials at every cardiac cycle (Fig. 7D). These findings, therefore, indicate that the IPSP locked to the cardiac cycle are chloride-mediated IPSP.

We discovered that the pattern of discharge of these neurons changed completely when systemic arterial pressure was constantly reduced by intravenous

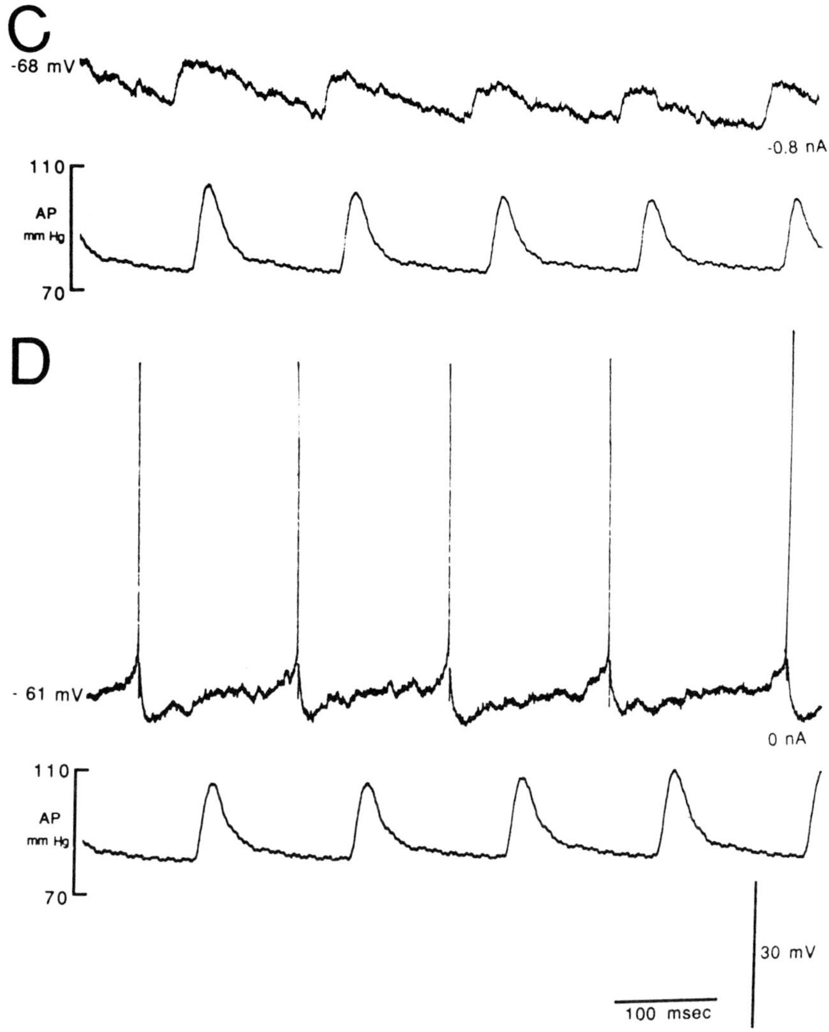

Figure 7 Intracellular recordings of a RVLM bulbospinal neuron with inhibitory baroreceptor inputs during intracellular injection of negative current. (A) Simultaneous records of arterial pressure and intracellular recording, immediately after holding the membrane potential at more negative values. The records were triggered by the arterial pulse wave. (B) Similar records as in panel A (note different time base) after intracellular injections of negative current for several minutes and holding the membrane potential with −0.3 nA or −0.8 nA negative current (microelectrode filled with KCl). (C) Simultaneous records of arterial pressure and neuronal membrane potential under similar experimental conditions as in panel B, at a holding potential of −68 mV. (D) Similar record as in panel (C), but at a membrane potential of −61 mV. Same neuron as in panels A through D. (From Ref. 65.)

infusions of sodium nitroprusside. The firing pattern during hypotension became very regular, demonstrating a pacemakerlike activity (Fig. 8A and B). The regular firing pattern was also demonstrated by the unimodal distribution of action potentials perceived when the oscilloscope was triggered by spontaneous spikes (Fig. 8C). Furthermore, the IPSP locked to the cardiac cycle present at normal values of arterial pressure and heart rate vanished during hypotension. In consequence, during hypotension, the average of the neuronal membrane potential triggered by the pulsatile arterial pressure, either at spontaneous membrane potential or during hyperpolarization (accomplished by injecting a high-intensity negative current), was flat (Figs. 8D and 9F). In addition, the regular firing pattern observed during hypotension was lost when the neurons were hyperpolarized; nevertheless, the spikes were found unrelated to the arterial pulse (Fig. 9E).

These studies indicate that these RVLM bulbospinal neurons with barosensitive properties are inhibited by chloride-dependent IPSP entrained to the

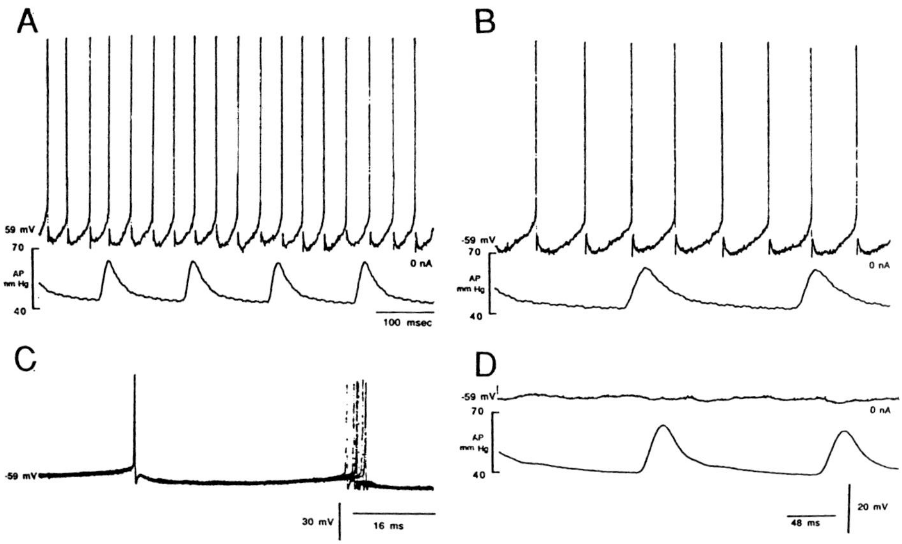

Figure 8 Intracellular recording of a RVLM bulbospinal neuron with inhibitory baroreceptor inputs during constant hypotension. (A) Simultaneous records of arterial pressure (AP) and intracellular recording at resting membrane potential during hypotension elicited by intravenous infusion of sodium nitroprusside. The records were triggered by the arterial pulse wave. Note the very regular firing pattern that took place when AP had been lowered. (B) Similar records as in panel A, but at different time base. (C) Seven superimposed oscilloscope traces triggered by spontaneous spikes. (D) Simultaneous averages of the neuronal membrane potential and pulsatile AP triggered by the arterial pulse wave during hypotension. Number of sweeps = 690; sample period = 50 μs. Experimental conditions in panel A also applicable to panels B, C, and D. (From Ref. 65.)

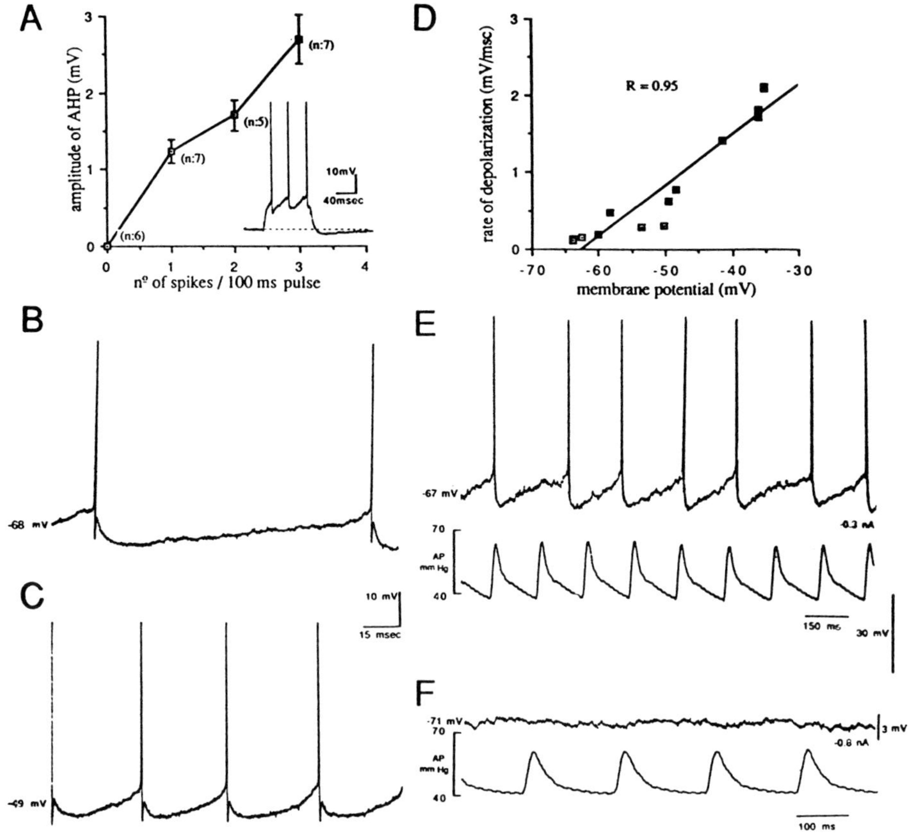

Figure 9 (A) Influence of the number of spikes elicited in a depolarizing current pulse on the amplitude of the afterhyperpolarization spike train. (B–D) Effects of changes in neuronal membrane potential on the rate of rise of the slow depolarization in RVLM neurons. (E–F) Effects of hyperpolarization and constant hypotension on a RVLM bulbospinal neuron with inhibitory baroreceptor inputs. (A) Intracellular injections of depolarizing current pulses (100 ms) with different intensities elicited to 0 to 3 spikes in RVLM neurons. At the end of the current pulse, an afterhyperpolarization was observed, as illustrated in a typical case in panel A. The amplitude of this afterhyperpolarization was increased when the number of spikes evoked by the depolarizing pulses increased. n = number of cases. Values are means ± SEM. (D) During depolarization of the neuronal membrane potential, the rate of rise (mV/ms) of the slow depolarization increased; conversely, during hyperpolarization, it decreased. The slope of the regression line = 70.0 μV/ms/mV (n = 14). Panels B and C illustrate the slow depolarization at two different membrane potentials (−58 mV in panel B, −49 mV in panel C). Note the same time base and calibration for panels B and C. (E) Simultaneous records of arterial pressure (AP) and intracellular recording on the same neuron as in Fig. 8 after intracellular injection of hyperpolarizing current for several minutes and intravenous infusion of sodium nitroprusside. The membrane potential was held at −67 mV. The records were triggered by the arterial pulse wave. (F) Simultaneous averages of the pulsatile AP with the neuronal membrane potential held at −71 mV during hypotension. The averages were triggered by the arterial pulse wave. Number of sweeps = 720; sample period = 100 μs. Panels E and F illustrate the same neuron. (From Ref. 65.)

cardiac cycle at normal values of arterial pressure. However, these modulatory baroreceptor inputs are lost during hypotension. These characteristics clearly distinguish these barosensitive neurons from other RVLM neurons with barosensitive properties. Therefore, we proposed to classify these neurons as barosensitive type II (65).

The regular pacemakerlike activity observed in barosensitive type II neurons during hypotension shows more clearly that the spontaneous action potentials of these neurons are normally preceded by slow depolarizing potentials and followed by an afterhyperpolarization (Fig. 8B). The afterhyperpolarization that follows each spike is in some cases difficult to see due to the presence of the ensuing slow depolarizing potential. However, it is known that the afterhyperpolarizing potential follows single or multiple action potentials and that the amplitude of the afterhyperpolarizing potential is proportional to the number of spikes. Consequently, we found that the amplitude of the afterhyperpolarizing potential that followed depolarizing current pulses is proportional to the number of action potentials evoked by the current pulses in these barosensitive type II neurons (Fig. 9A).

We found that the slow depolarizing potential that controls the interspike activity is voltage-dependent. The corresponding slope thus increased in magnitude with depolarization of membrane potential and, conversely, decreased during hyperpolarization (Fig. 9B and C). Further, the rate of rise of the slow depolarizing potential increases when the neuron is depolarized and decreases during hyperpolarization (Fig. 9D).

The rhythmic and regular oscillations in membrane potential and the afterhyperpolarizing potential that follows each action potential during hypotension in barosensitive type II neurons are attributes of neurons with "beating," pacemakerlike activity. Further support for this view is provided by the following observations: (a) the absence of EPSP during hyperpolarization produced by passing negative current pulses up to below spike threshold level (Fig. 10); (b) interrupting the spontaneous activity by passing short pulses of hyperpolarizing current elicited a delay in the spikes occurring after the current injection in relation to the expected time based on interspike interval determined before the injection of negative current (Fig. 11A); (c) single antidromic spikes evoked on these neurons by spinal cord stimulation also interrupted the neuronal spontaneous activity, producing a delay in the spike occurring after the antidromic spike and a "resetting" of neuronal activity similar to that observed after current injection (Fig. 11B). On the other hand, when the antidromic action potential collided with a spontaneous one, no change in the firing pattern was noted (Fig. 11C).

The presence of bulbospinal RVLM neurons with pacemakerlike properties has been demonstrated in slices of rat brainstem by other laboratories. The neurons investigated by these authors showed a regular pattern of discharge in vitro. These experimenters maintained the neurons' regular firing characteristics in vitro and did not expose underlying EPSP during hyperpolarization;

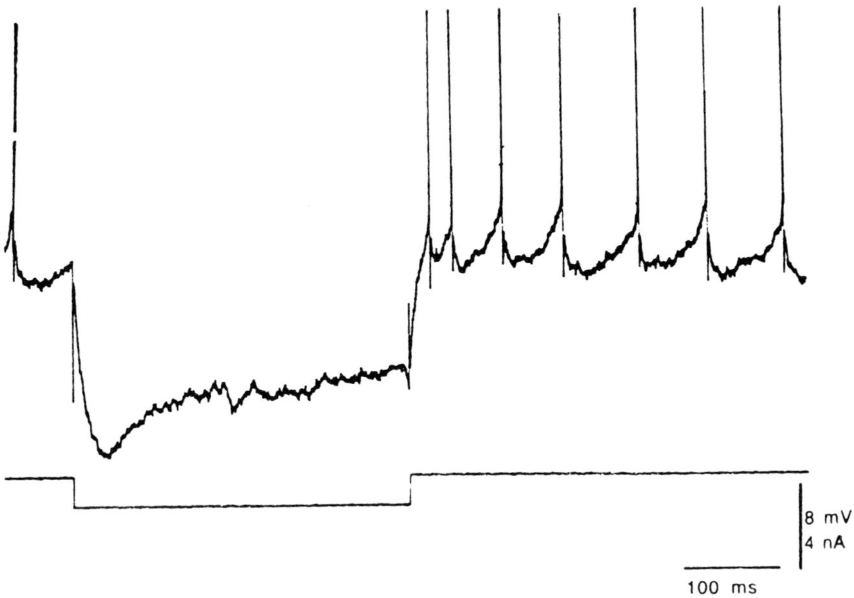

Figure 10 Responses of a physiologically identified RVLM bulbospinal neuron to intracellular current injections. At resting membrane potential (–50 mV), a hyper-polarizing current pulse produced a two-phase voltage deflection of the membrane potential. This time-dependent hang-down in the voltage deflection trace is probably due to anomalous rectification. Note that after the offset of the current pulse, the frequency of discharge was transiently increased before the neuron recovered its regular firing pattern. (From Ref. 65.)

in addition, they reset the spontaneous activity after disrupting it by hyper-polarizing current pulses (72–74). The location of these neurons appears to correspond to the nucleus reticularis rostroventrolateralis (7), the location of the neurons we characterized as barosensitive type II in our in vivo studies (65).

Bulbospinal neurons located in this part of the RVLM sent projections terminating only in the intermediolateral and intermediomedial columns. Therefore, it is likely that these neurons participate in the generation of tonic sympathetic nerve activity.

Recently, we have used a technique to record intracellular potentials from neurons in slices from the rat medulla. We identified a group of neurons in the RVLM with similar characteristics to those described for barosensitive type II neurons during hypotension. They demonstrated a very regular pattern of discharge at 32°C, the duration of action potential was in a similar range to those observed in vivo (1.2 to 1.7 ms), and they also demonstrated slow depolarizing potentials and afterhyperpolarizing potentials with features identical to those

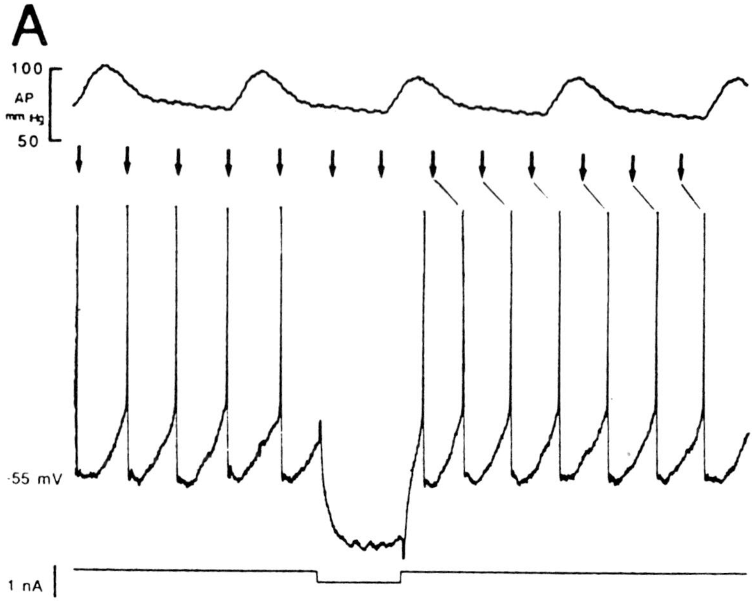

Figure 11 Resetting of pacemakerlike activity in a barosensitive RVLM neuron. (A) Hyperpolarizing current injected in a barosensitive neuron interrupted spontaneous activity, following which the pacemakerlike activity was reset. The arrows indicate the interspike interval time after the intracellular injection. (B) Similar resetting of pacemakerlike activity after interrupting neuronal activity by an antidromically evoked spike from the thoracic spinal cord. (C) The antidromic spike collided with a spontaneous spike, and this resulted in no interruption of the rhythmic discharge. Arrows (SC) = spinal cord stimulation. (From A. R. Granata, unpublished results.)

described in vivo. A representative example is shown in Fig. 12. Furthermore, when the spontaneous neuronal activity was arrested by hyperpolarizing pulses, we observed a reset of neuronal activity similar to that described in vivo (Fig. 12A).

Another important characteristic that distinguishes these type II neurons is revealed when they are hyperpolarized by pulses of negative current. A relatively fast-developing anomalous rectification that produced a decay in the hyperpolarization was observed in vivo (Fig. 10). The anomaly, first described by B. Katz in potassium-depolarized muscles (75), is a conductance that increases in magnitude when the cell is hyperpolarized and decreases under depolarization. This type of conductance is present in neurons with oscillatory behavior—for example, in the substantia nigra neurons (76) or the inferior olivary neurons (77). In these neurons, the level of anomalous rectification and

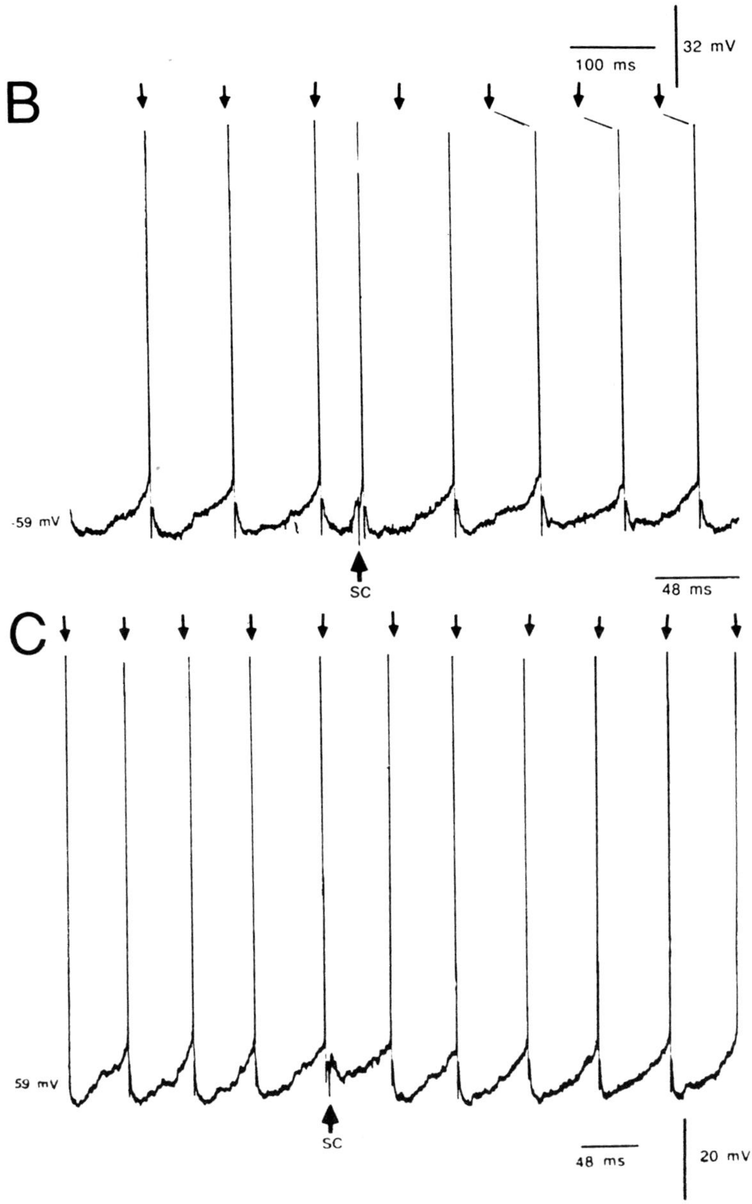
B
32 mV
100 ms
-59 mV
SC
48 ms
C
59 mV
SC
48 ms
20 mV

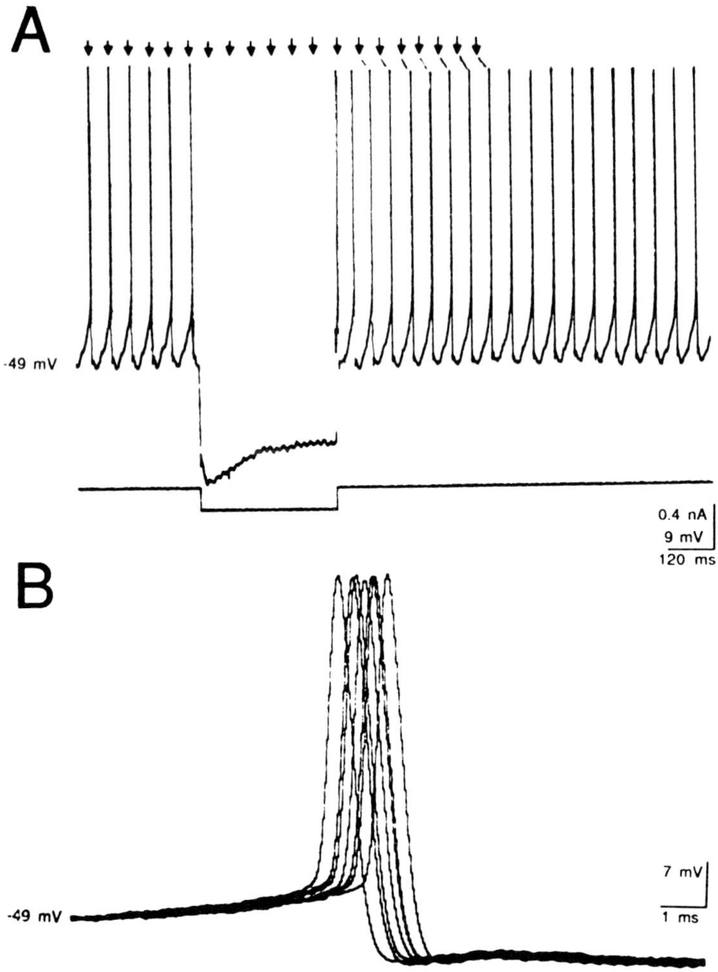

Figure 12 Intracellular recording from a regularly firing RVLM neuron from an in vitro slice preparation. (A) Hyperpolarizing current injected in the neuron interrupted spontaneous activity, following which the pacemakerlike activity was reset. The arrows indicate the interspike interval time after the intracellular injection. Note the absence of EPSP and the two-phase voltage deflection of the membrane potential during the hyperpolarizing period due to anomalous rectification. (B) Eight superimposed oscilloscope traces triggered by spontaneous spikes demonstrate the regularity of the firing. The action potential duration corresponded closely to that of the barosensitive type II neurons described in vivo. For details, see text. (From A. R. Granata, unpublished results.)

the control of the frequency of the membrane oscillation were directly related. As expected, we observed a similar type of anomalous rectification when these neurons were characterized in vitro (Fig. 12A). So far, we have found this type of conductance only in barosensitive type II neurons in vivo; hence, it has been used as an additional and important element to characterize these neurons in vitro.

On the other hand, in those pacemakerlike RVLM neurons investigated by Guyenet and collaborators (73,78), the authors did not report the presence of anomalous rectification when the RVLM pacemakerlike neurons were hyperpolarized, nor was it possible to detect any type of rectification in the membrane potential from the corresponding recordings (see Ref. 73, Fig. 2; or Ref. 78, Fig. 3). However, it is possible that, for various reasons, the authors of the referred papers did not hyperpolarize the neuronal membrane sufficiently to reveal the anomalous rectification. Therefore, we can only assume that the type of pacemakerlike neuron investigated by that group would not correspond with the barosensitive type II reported in our studies. Nevertheless, it is also possible to speculate that two different types of pacemakerlike neurons involved in vasomotor control may coexist in the RVLM.

In the rat, it was found that barosensitive RVLM spinal neurons with spontaneous activity locked to the cardiac cycle responded with a decrease in firing frequency when the aortic depressor nerve was electrically stimulated (24). Since the aortic nerves in the rat contain exclusively baroreceptor afferents, this is an additional and clear demonstration that these neurons are barosensitive, as concluded by the authors of that work.

We have observed that neurons characterized as barosensitive type II (according to the criteria already explained) also responded with hyperpolarization and a decrease in firing frequency during aortic nerve stimulation (A. R. Granata, unpublished results).

However, solely the demonstration that a neuron in the RVLM responds with inhibition to aortic nerve stimulation is not enough to prove its participation in the main baroreceptor reflex. The onset latency of the inhibitory response evoked by aortic nerve stimulation ranges from approximately 25 to 70 ms in the rat (24) to around 50 ms in the rabbit (56). It is therefore obvious that multisynaptic pathways are involved in mediating these inhibitory responses of RVLM spinal neurons triggered by electrical stimulation of the aortic nerve. Such polysynaptic pathways could involve either structures rostral to the pontomedullary border that do not participate in the main baroreceptor reflex arc or pathways revealed during electrical stimulation but not operative under physiological conditions. Consequently, the evoked response to aortic nerve stimulation is a complementary method of characterizing functional barosensitive sympathoexcitatory neurons but not a substitute for the modulatory synaptic inputs received by these neurons during each cardiac cycle.

IV. Intracellular Labeling and Double Staining for PNMT Immunoreactivity of RVLM Barosensitive Neurons

As mentioned above, there is some evidence to support the view that the C1 adrenergic neurons themselves participate in the tonic control of cardiovascular sympathetic activity and in the mediation of the baro- and mechanoreceptor reflexes. On the other hand, there are also reports supporting the idea that a similar role is played by noncatecholaminergic RVLM neurons with spinal projections (72,73). In any case, if the C1 adrenergic neurons participate in such a function, it is very doubtful that adrenaline would be the neurotransmitter released into sympathetic preganglionic neurons from C1-adrenergic neuronal terminals, as microionophoretically applied adrenaline produces depressant effects in these neurons (79,80). Furthermore, in vitro experiments have demonstrated that noradrenaline can produce both excitation and inhibition in sympathetic preganglionic neurons by activating α_1- or α_2-adrenergic receptors, respectively (81). However, it is unknown which type(s) of receptors is activated by the endogenous adrenaline released from the C1 terminals in the spinal cord.

A subpopulation of C1 adrenergic neurons also demonstrates immunoreactivity against other neurotransmitters, such as substance P, neuropeptide Y, and even excitatory amino acids (82–85). Whether these substances could act as coneurotransmitters to tonically activate sympathetic preganglionic neurons remains to be elucidated.

Overall, the C1 adrenergic neurons could be considered important candidates for vasomotor control. Because of that, we investigated whether the barosensitive neurons characterized as type I and type II correspond to the C1 adrenergic group. Consequently, the electrophysiologically characterized neurons were injected with biocytin, incubated with avidin–Texas red, and visualized under fluorescent microscopy to first determine their location in the RVLM. All neurons characterized as barosensitive in these studies were located in the RVLM in an area just caudal to the facial nerve nucleus, which extended 0.7 mm in the rostrocaudalis coordinate.

The histological sections containing the intracellularly labeled neurons were processed for PNMT or, in some cases, tyrosine hydroxylase (TH) immunoreactivity using a double-labeling immunohistofluorescence technique.

This study demonstrated that none of the neurons characterized as barosensitive type I or type II and identified by the intracellular injection of biocytin showed PNMT or TH immunoreactivity. The results of this investigation give strong support to the idea that noncatecholaminergic RVLM neurons with descending projections participate in the generation of the tonic cardiovascular sympathetic activity and in the mediation of the baroreceptor reflex.

The double-labeling technique used in this study to identify catecholaminergic neurons has been used successfully in other catecholaminergic central

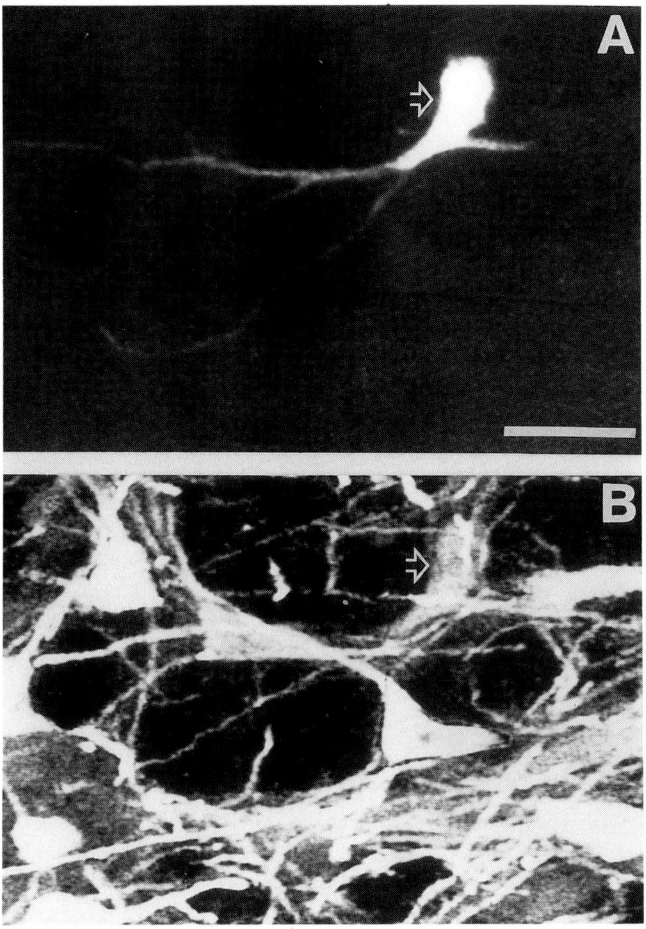

Figure 13 Photomicrograph of a spontaneously active neuron in the RVLM, intracellularly labeled with biocytin and demonstrating PNMT-immunoreactivity. In panel A, the neuron has been injected with biocytin and revealed by incubation of the tissue with avidin–Texas red. The neuron showed intense fluorescence detected with a rhodamine filter. The tissue was subsequently processed for PNMT immunocytochemical localization, using FITC-labeled secondary antibodies (B) The same field as in panel A but visualized through a blue filter. The neuron stained with biocytin in panel A (straight arrow) demonstrates PNMT-immunoreactivity in panel B (straight arrow). Scale bar: 50 μm for panels A and B. (From A. R. Granata, unpublished results.)

nervous system neurons injected with biocytin. Therefore, there is enough evidence to discard the possibility that the biocytin-avidin–Texas red histofluorescence reaction could interfere with immunohistofluorescence staining with catecholamine-synthesizing enzymes or other antibodies (86,87).

However, conclusive proof in support of this assertion has come from recent electrophysiological studies. We have obtained in vivo intracellular recordings from spontaneously active neurons in the RVLM that were identified by intracellular injections of biocytin. These neurons also demonstrated PNMT

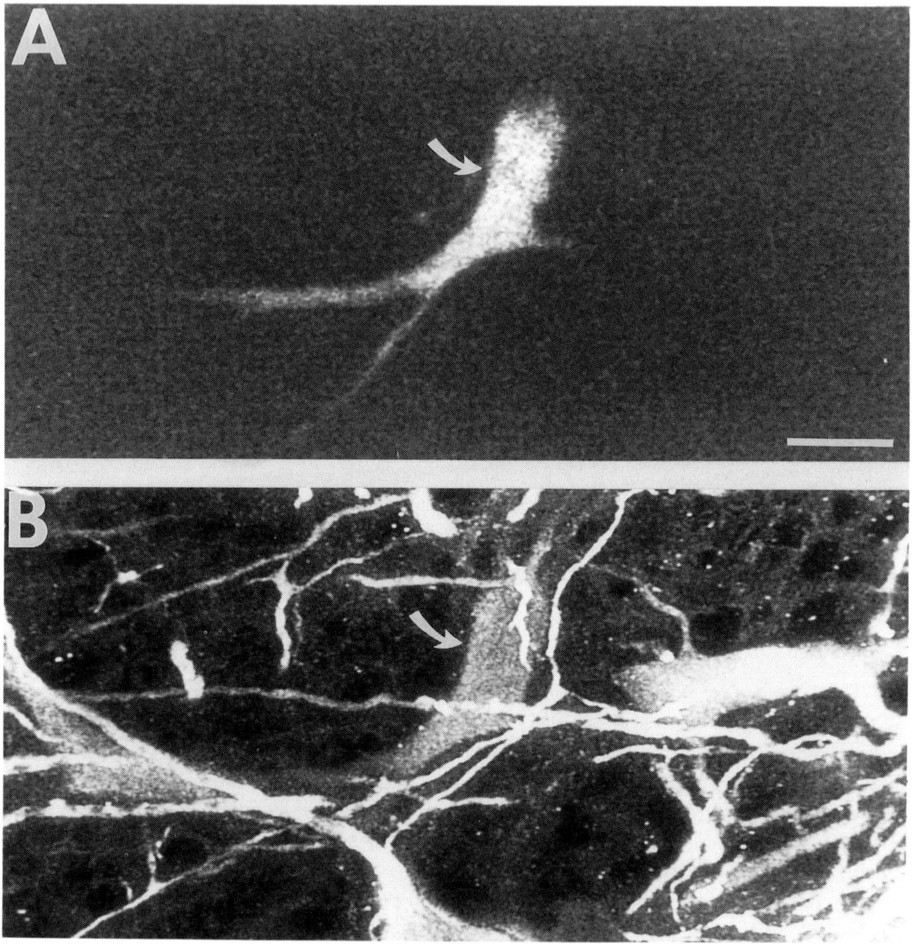

Figure 14 Photomicrograph of the same neuron shown in Figure 13 at a higher magnification. A and B as in Figure 13. Location of double-labeled neuron is indicated by curved arrows. Scale bar: 33 μm for A and B. (From A. R. Granata, unpublished results.)

immunoreactivity when they were processed by a similar double-labeling immunohistofluorescence technique (A. R. Granata, unpublished results). An example of a double-labeled neuron is shown in Figs. 13A,B; and 14A and B.

V. Summary

The RVLM is an area essential for the control and generation of tonic sympathetic nerve discharge as well as for mediating several cardiovascular reflexes, including the baro- and mechanoreceptor reflexes.

This chapter has focused on presenting information about the basic electrophysiological characteristics and synaptic inputs received by bulbospinal RVLM neurons probably involved in the tonic and reflex control of cardiovascular functions. Since it is still poorly understood how these functions are generated and integrated at the level of the RVLM, we believe that an obligatory step in resolving these questions should include a detailed analysis of the electrophysiological properties as well as the precise identification, of the transmitters contained within, and perhaps used by, RVLM neurons directly involved in such functions.

Using an in vivo intracellular recording technique, we have identified a group of RVLM neurons antidromically activated by electrical stimulation of the intermediolateral nucleus in the thoracic spinal cord. These neurons are differentiable because they receive EPSP synchronized to the cardiac cycle at the end of the diastolic/beginning of the systolic period. The baroreceptor reflex activation, elicited by increasing systemic arterial pressure, produced hyperpolarization, a decrease in the firing frequency, and an increase in the membrane input resistance of these neurons, suggesting that a disfacilitatory mechanism accounts for this effect. Inactivation of the baroreceptor reflex by decreasing systemic arterial pressure produced the opposite effect on these neurons. Because of these and other properties, we have characterized these neurons as barosensitive type I neurons.

Disfacilitation of barosensitive reticulospinal type I neurons by baroreceptor reflex activation could result from the inhibition of tonic excitatory inputs originating in other neurons, presumably located in the brainstem. These tonic excitatory inputs could be the source of the distinct ongoing and random synaptic activity typically observed in these type I neurons at resting values of arterial pressure, heart rate, and neuronal membrane potential.

Another important characteristic that demarcates these neurons is the presence of burst-firing episodes locked to the cardiac cycle. We do not know yet what mechanism is responsible for switching these neurons from a single to a burst firing mode. Nevertheless, burst-firing activity has been related with enhanced postsynaptic efficacy. Therefore, these sympathoexcitatory neurons could play a distinct role when an abrupt and important increase in sympathetic tone is needed, as, for instance, during acute “stress” situations.

We identified a different group of RVLM-spinal neurons with barosensitive properties that we named barosensitive type II neurons. These spontaneously active, non-burst-firing neurons demonstrated chloride-dependent IPSP locked to the cardiac cycle at the end of the diastolic/beginning of the systolic period; the IPSP disappeared when the baroreceptor reflex was constantly inactivated by the hypotension elicited by intravenous infusions of sodium nitroprusside. Furthermore, during hypotension, the pattern of discharge consisted of a very regular pacemakerlike activity. In barosensitive type II neurons, the spontaneous action potentials resulted from voltage-dependent, slow depolarization potentials and were followed by an afterhyperpolarization potential. These potentials, plus an anomalous rectification developed by passing hyperpolarizing current pulses, could play a role in generating the oscillatory behavior of barosensitive type II neurons. The possibility that these neurons act as a pacemaker is based on (a) the regular pattern of discharge in the absence of baroreceptor inputs, (b) the absence of EPSP during hyperpolarizing pulses below spike generation level, and (c) the "resetting" of pacemaker potentials when the neuronal activity was interrupted. We have identified the barosensitive type II neurons in an in vitro slice preparation and have confirmed the pacemakerlike properties of these neurons.

It is not yet clear whether the sympathoexcitatory barosensitivelike neurons in the RVLM [either the ones characterized as type II in these studies or the pacemaker type described by other investigators (72,73)] could play a physiological role different from that of other barosensitive-sympathoexcitatory neurons in the same area. However, it is possible to speculate that such an electrophysiological profile would make pacemakerlike neurons more suitable for maintaining the supraspinally originated ongoing synaptic activity observed in sympathetic preganglionic neurons.

Neither barosensitive type I nor barosensitive type II neurons demonstrated PNMT immunoreactivity, although in all cases they were found in the RVLM and close or relatively close to adrenergic neurons.

These results support the idea that noncatecholaminergic bulbospinal neurons in the RVLM may also play an important role in cardiovascular control.

Acknowledgments

Supported by the American Heart Association (AHA), Tennessee Affiliate, Grant TN92G13. The author would like to thank Ms. Jin Emerson-Cobb for assistance in preliminary editing.

References

1. Reis DJ, Cuenod M. Central neural regulation of carotid baroreceptor reflex in the cat. Am J Physiol 1965; 209:1267–1279.

2. Dampney RAL, Moon EA. Role of ventrolateral medulla in vasomotor response to cerebral ischemia. Am J Physiol 1980; 239:H349–H358.
3. Brown DL, Guyenet PG. Electrophysiological study of cardiovascular neurons in the rostral ventrolateral medulla. Circ Res 1985; 56:359–369.
4. Andrezik TA, Chan-Palay V, Palay SL. The nucleus paragigantocellularis lateralis in the rat: Conformation and cytology. Anat Embryol 1981; 161:355–371.
5. Taber E. The cytoarchitecture of the brain stem of the cat: I. Brain stem nuclei of cat. J Comp Neurol 1961; 116:27–69.
6. Kalia M, Fuxe K. Rat medulla oblongata: I. Cytoarchitectonic considerations. J Comp Neurol 1985; 233:285–307.
7. Ross CA, Ruggiero DA, Park DH, et al. Tonic vasomotor control by the rostral ventrolateral medulla: Effect of electrical or chemical stimulation of the area containing C1 adrenaline neurons on arterial pressure, heart rate and plasma catecholamines and vasopressin. J Neurosci 1984; 4:274–294.
8. Berman AL. The Brain Stem of the Cat: A Cytoarchitectonic Atlas with Stereotaxic Coordinates. Madison, WI: University of Wisconsin Press, 1968.
9. Loewy AD, McKellar S. Serotonergic projections from the ventral medulla to the intermediolateral cell column in the rat. Brain Res 1981; 211:146–152.
10. Ciriello J. Brainstem projections of aortic baroreceptor afferent fibers in the rat. Neurosci Lett 1983; 36:37–42.
11. Kalia M, Mesulam MM. Brainstem projections of sensory and motor components of the vagus complex in the cat: 1. The cervical vagus and nodose ganglion. J Comp Neurol 1980; 193:435–465.
12. Ross CA, Ruggiero DA, Reis DJ. Projections from the nucleus tractus solitarii to the rostral ventrolateral medulla. J Comp Neurol 1985; 242:511–534.
13. Loewy AD, Burton H. Nuclei of the solitary tract: Efferent projections to the lower brainstem and spinal cord of the cat. J Comp Neurol 1978; 181:421–450.
14. Reis DJ, Ross CA, Granata AR, Ruggiero DA. Role of C1 area of rostroventrolateral medulla in cardiovascular control. In: Buckley JP, Ferrario CM, eds. Brain Peptides and Catecholamines in Cardiovascular Regulation. New York: Raven Press, 1987:1–14.
15. Coote JH, McLeod VH. Evidence for the involvement of the baroreceptor reflex of a descending inhibitory pathway. J Physiol (Lond) 1974; 241:477–496.
16. Willette RN, Barcas PP, Krieger AJ, Sapru HN. Vasopressor and depressor areas in the rat medulla. Neuropharmacology 1983; 22:1071–1079.
17. Blessing WW, Willoughby JO. Depressor neurons in rabbit caudal medulla do not transmit the baroreceptor-vasomotor reflex. Am J Physiol 1987; 253:H777–H786.
18. Granata AR, Ruggiero DA, Park DH, et al. Brainstem area with C1 epinephrine neurons mediates baroreflex vasodepressor responses. Am J Physiol 1985; 248:H547–H567.
19. Agarwal SK, Calaresu FR. Monosynaptic connection from caudal to rostral ventrolateral medulla in the baroreceptor reflex pathway. Brain Res 1991; 555:70–74.
20. Terui N, Masuda N, Saeki Y, Kumada M. Activity of barosensitive neurons in the caudal ventrolateral medulla that send axonal projections to the rostral ventrolateral medulla in rabbits. Neurosci Lett 1990; 118:211–214.
21. Huangfu D, Guyenet PG. Sympatholytic response to stimulation of superior laryngeal nerve in rats. Am J Physiol 1991; 260:R290–R297.

22. Blessing WW, Li YW. Inhibitory vasomotor neurons in the caudal ventrolateral region of the medulla oblongata. Prog Brain Res 1989; 81:83–97.
23. Cravo SL, Morrison SF, Reis DJ. Differentiation of two cardiovascular regions within the caudal ventrolateral medulla. Am J Physiol 1991; 261:R985–R994.
24. Jeske I, Morrison SF, Cravo SL, Reis DJ. Identification of baroreceptor reflex interneurons in the caudal ventrolateral medulla. Am J Physiol 1993; 264:R169–R178.
25. Shapiro RE, Miselis RR. The central connection of the area postrema of the rat. J Comp Neurol 1985; 234:344–364.
26. Granata AR, Chang HT. Relationship of Calbindin-D 28k with afferent neurons to the rostral ventrolateral medulla in the rat. Brain Res 1994; 645:265–277.
27. Blessing WW, Hedger SC, Joh TH, Willoughby JO. Neurons in the area postrema are the only catecholamine-synthesizing cells in the medulla or pons with projections to the rostral ventrolateral medulla (C1-area) in the rabbit. Brain Res 1987; 419:336–343.
28. Fulwiler CE, Saper CB. Subnuclear organization of the efferent connections of the parabrachial nucleus in the rat. Brain Res Rev 1984; 7:229–259.
29. Lovick T. Projections from brainstem nuclei to the nucleus paragigantocellularis lateralis in the cat. J Auton Nerv Syst 1986; 16:1–11.
30. Ruggiero DA, Ross CA, Anwar M, Reis DJ. The rostral ventrolateral medulla: Immunocytochemistry of intrinsic neurons and afferent connections (abstr). Soc Neurosci Abstr 1984; 10:299.
31. Dampney RAL, Czachurski J, Dembowsky K, et al. J Auton Nerv Syst 1987; 20: 73–86.
32. Nicholas AP, Hancock MB. Evidence for projections from the rostral medullary raphe onto medullary catecholamine neurons in the rat. Neurosci Lett 1990; 108: 22–28.
33. Nicholas AP, Hancock MB. Projections from the rostral ventrolateral medulla to brainstem monoamine neurons in the rat. Neurosci Lett 1991; 122:91–95.
34. Guertzenstein PG, Silver A. Fall in blood pressure produced from discrete regions of the ventral surface of the medulla by glycine and lesions. J Physiol (Lond) 1974; 242:489–503.
35. Dampney RAL, Moon EA. Role of ventrolateral medulla in vasomotor response to cerebral ischemia. Am J Physiol 1980; 239:H349–H358.
36. Granata AR, Ruggiero DA, Park DH, et al. Lesions of epinephrine neurons in the rostral ventrolateral medulla abolish vasodepressor components of baroreflex and cardiopulmonary reflex. Hypertension 1983; 5:V80–V84.
37. Feldberg W, Guertzenstein PG. A vasodepressor effect of pentobarbitone sodium. J Physiol (Lond) 1972; 224:83–103.
38. Feldberg W, Guertzenstein PG. Vasodepressor effects obtained by drugs acting on the ventral surface of the brainstem. J Physiol (Lond) 1976; 258:337–355.
39. Benarroch EE, Granata AR, Ruggiero DA, et al. Neurons of C1 area mediate cardiovascular responses initiated from ventral medullary surface. Am J Physiol 1986; 250:R932–R945.
40. Yamada KA, McAllen RM, Loewy AD. GABA antagonists applied to the ventral surface of the medulla oblongata block the baroreceptor reflex. Brain Res 1984; 297:175–180.

41. Keeler JR, Shults CW, Chase TN, Helke CJ. The ventral surface of the medulla in the rat: Pharmacological and autoradiographic localization of GABA-induced cardiovascular effects. Brain Res 1984; 297:217–224.
42. Schläfke ME, See WR. Ventral medullary surface stimulus response in relation to ventilatory and cardiovascular effects. In: Koepchen HP, Hilton SM, Trzebski A, eds. Central Interaction between Respiratory and Cardiovascular Control Systems. Berlin: Springer-Verlag, 1980:56–64.
43. Imaizumi T, Granata AR, Benarroch EE, et al. Contributions of arginine vasopressin and the sympathetic nervous system to fulminating hypertension after destruction of neurons of caudal ventrolateral medulla in the rat. J Hypertens 1985; 3:491–501.
44. Dampney RAL, Goodchild AK, Robertson LG, Montgomery W. Role of ventrolateral medulla in vasomotor regulation: A correlative anatomical and physiological study. Brain Res 1982; 249:223–235.
45. Amendt K, Czachurski J, Dembowsky K, Seller H. Bulbospinal projections to the intermediolateral cell column: A neuroanatomical study. J Auton Nerv Syst 1979; 1:103–117.
46. Ross CA, Armstrong DM, Ruggiero DA, et al. Adrenaline neurons in the rostral ventrolateral medulla innervate thoracic spinal cord: A combined immunocytochemical and retrograde transport demonstration. Neurosci Lett 1981; 25:257–262.
47. Zagon A, Smith AD. Monosynaptic projections from the rostral ventrolateral medulla oblongata to identified sympathetic preganglionic neurons. Neuroscience 1993; 54:729–743.
48. Tucker DC, Saper CB, Ruggiero DA, Reis DJ. Organization of central adrenergic pathways: I. Relationships of ventrolateral medullary projections to the hypothalamus and spinal cord. J Comp Neurol 1987; 259:591–603.
49. Hökfelt T, Fuxe K, Goldstein M, Johansson O. Immunohistochemical evidence for the existence of adrenaline neurons in the rat brain. Brain Res 1974; 66:235–251.
50. Milner TA, Morrison SF, Abate C, Reis DJ. Phenylethanolamine N-methyltransferase containing terminals synapse directly on sympathetic preganglionic neurons in the rat. Brain Res 1988; 448:205–222.
51. Andrezik TA, Chan-Palay V, Palay SL. The nucleus paragigantocellularis lateralis in the rat: Demonstration of afferents by the retrograde transport of HRP—Conformation and cytology. Anat Embryol 1981; 161:373–390.
52. Guyenet PG, Brown DL. Nucleus paragigantocellularis lateralis and lumbar sympathetic discharge in the rat. Am J Physiol 1986; 250:R1081–R1094.
53. McAllen RM. Identification and properties of sub-retrofacial bulbospinal neurons: A descending cardiovascular pathway in the cat. J Auton Nerv Syst 1986; 17:151–164.
54. Brown DL, Guyenet PG. Cardiovascular neurons of brainstem with projections to spinal cord. Am J Physiol 1984; 247:R1009–R1016.
55. Morrison SF, Milner TA, Reis DJ. Reticulospinal vasomotor neurons of the rat rostral ventrolateral medulla: Relationship to sympathetic nerve activity and the C1 adrenergic cell group. J Neurosci 1988; 8:1286–1301.
56. Terui N, Saeki Y, Kumada M. Barosensory neurons in the ventrolateral medulla in rabbits and their responses to various afferent inputs from peripheral and central sources. Jpn J Physiol 1986; 36:1148–1164.

57. Sun MK, Guyenet PG. Medullospinal sympathoexcitatory neurons in normotensive and spontaneously hypertensive rats. Am J Physiol 1986; 250:R910–R917.
58. Minson JB, Chalmers JP, Caon AC, Renaud B. Separate areas of rat medulla oblongata with populations of serotonin- and adrenaline-containing neurons alter blood pressure after L-glutamate stimulation. J Auton Nerv Syst 1987; 19:39–50.
59. Cox BF, Brody MJ. Evidence for two functionally distinct vasomotor subregions of rostral ventrolateral medulla. Clin Exp Hypertens 1988; A10(suppl 1):11–18.
60. Cox BF, Brody MJ. Subregions of rostral ventral medulla control arterial pressure and regional hemodynamics. Am J Physiol 1989; 257:R635–R640.
61. Cox BF, Brody MJ. Mechanisms of respiration induced changes in vasomotor control exerted by rostral ventrolateral medulla. Am J Physiol 1989; 257:R626–R634.
62. Dampney RAL, McAllen RM. Differential control of sympathetic fibers supplying hindlimb skin and muscle by subretrofacial neurons in the cat. J Physiol (Lond) 1988; 395:41–56.
63. Millhorn DE, Hökfelt T, Seroogy K, et al. Immunohistochemical evidence for colocalization of γ-aminobutyric acid and serotonin in neurons of the ventral medulla oblongata projecting to the spinal cord. Brain Res 1987; 410:179–185.
64. Barman SM, Gebber GL. Axonal projection patterns of ventrolateral medullospinal sympathoexcitatory neurons. J Neurophysiol 1985; 53:1551–1566.
65. Granata AR, Kitai ST. Intracellular analysis in vivo of different barosensitive bulbospinal neurons in the rat rostral ventrolateral medulla. J Neurosci 1992; 12:1–20.
66. Alexander RS. Tonic and reflex functions of medullary sympathetic cardiovascular centers. J Neurophysiol 1946; 9:205–217.
67. Barman SM, Gebber GL. Lateral tegmental field neurons of cat medulla: A source of basal activity of ventrolateral medullospinal sympathoexcitatory neurons. J Neurophysiol 1987; 57:1410–1424.
68. Barman SM, Gebber GL. Basis for the naturally occurring activity of rostral ventrolateral medullary sympathoexcitatory neurons. Prog Brain Res 1989; 81:117–129.
69. Gebber GL, Barman SM. A physiologically based model of the brainstem generator of sympathetic nerve discharge. Prog Brain Res 1989; 81:131–139.
70. Gillary HL, Kennedy D. Pattern generation in a crustacean motoneuron. J Neurophysiol 1989; 32:595–606.
71. Gillary HL, Kennedy D. Neurotransmitter effects of impulse pattern in a crustacean motoneuron. J Neurophysiol 1989; 32:607–612.
72. Sun MK, Hackett JT, Guyenet PG. Sympathoexcitatory neurons of rostral ventrolateral medulla exhibit pacemaker properties in the presence of a glutamate-receptor antagonist. Brain Res 1988; 438:23–40.
73. Sun MK, Young BS, Hackett JT, Guyenet PG. Reticulospinal pacemaker neurons of the rat rostral ventrolateral medulla with putative sympathoexcitatory function: An intracellular study in vitro. Brain Res 1988; 442:229–239.
74. Lewis DI, Coote JH. The actions of 5-hydroxytryptamine on the membrane of putative sympatho-excitatory neurons in the rostral ventrolateral medulla of the adult rat in vitro. Brain Res 1993; 609:103–109.
75. Katz B. Les constantes electriques de la membrane du muscle. Arch Sci Physiol 1949; 2:285–299.
76. Grace AA, Bunney BS. Intracellular and extracellular electrophysiology of nigral dopaminergic neurons: 1. Identification and characterization. Neuroscience 1983; 10:301–316.

77. Yarom Y, Llinás R. Long-term modifiability of anomalous and delayed rectification in guinea pig inferior olivary neurons. J Neurosci 1987; 7:1166–1177.
78. Sun MK, Guyenet PG. Excitation of rostral medullary pacemaker neurons with putative sympathoexcitatory function by cyclic AMP and beta-adrenoceptor agonists "in vitro." Brain Res 1990; 511:30–40.
79. Coote JH. The organization of cardiovascular neurons in the spinal cord. Rev Physiol Biochem Pharmacol 1988; 110:147–185.
80. Coote JH, MacLeod VH, Fleetwood-Walker S, Gilbey MP. The response of individual sympathetic preganglionic neurons to microelectrophoretically applied endogenous monoamines. Brain Res 1981; 215:135–145.
81. Yoshimura M, Polosa C, Nishi S. Multiple actions of noradrenaline on sympathetic preganglionic neurons of the cat studied in the spinal cord slice. Prog Brain Res 1989; 81:181–190.
82. Blessing WW, Oliver JR, Hodgson AH, et al. Neuropeptide Y-like immunoreactive C1 neurons in the rostral ventrolateral medulla of the rabbit project to sympathetic preganglionic neurons in the spinal cord. J Auton Nerv Syst 1987; 18:121–129.
83. Blessing WW, Howe PRC, Joh TH, et al. Distribution of tyrosine hydroxylase and neuropeptide Y-like immunoreactivity neurons in rabbit medulla oblongata, with attention to colocalization studies, presumptive adrenaline-synthesizing perikarya, and vagal pre-ganglionic neurons. J Comp Neurol 1986; 248:285–300.
84. Lorenz RG, Saper CB, Wong DL, et al. Co-localization of substance P- and phenylethanolamine-*N*-methyltransferase-like immunoreactivity in neurons of ventrolateral medulla that project to the spinal cord: Potential role in control of vasomotor tone. Neurosci Lett 1985; 55:255–260.
85. Morrison SF, Callaway J, Milner TA, Reis DJ. Rostral ventrolateral medulla: A source of the glutamatergic innervation of the sympathetic intermediolateral nucleus. Brain Res 1991; 562:126–135.
86. Bargas J, Gallarraga E, Chang HT, Kitai ST. Electrophysiological and double-labeling: Immunohistochemical analysis of neurons in the substantia nigra zona compacta in the rat (abstr). Soc Neurosci Abstr 1988; 14:1025.
87. Kang YN, Kitai ST. Electrophysiological properties of pedunculopontine neurons and their postsynaptic responses following stimulation of substantia nigra reticulata. Brain Res 1990; 535:79–95.

57

Brainstem Pathways Mediating Limbic Forebrain Cardiovascular Responses

SERGIO B. GIANCOLA and JOHN CIRIELLO

University of Western Ontario
London, Ontario, Canada

I. Introduction

It is well recognized that the ventrolateral medulla (VLM) plays an important role in the maintenance and the reflex regulation of vasomotor tone (1). However, the contribution of limbic structures to the function of the VLM is less well understood.

Recently, the bed nucleus of the stria terminalis (BST), a limbic structure in the rostral forebrain, has been implicated in central cardiovascular regulation (2,3). The BST is thought to function as a major relay for information originating in the amygdaloid complex (4). In fact, it has recently been demonstrated that synaptic blockade in the BST attenuates the cardiovascular responses evoked by glutamate stimulation of the central nucleus of the amygdala (5). The amygdala plays an important integrative and coordinating role in a variety of behavioral, affective, visceral, and endocrine responses (6). Chemical stimulation of BST neurons has been shown to elicit depressor responses that were accompanied by decreases in heart rate (2) and increased blood flow in the hindquarters (3). The decreases in arterial pressure and heart rate have been shown to be mediated by inhibition of sympathetic vasoconstrictor fiber activity to the vasculature and cardioacceleratory fiber activity to the heart, respectively (2). In contrast, it has been reported that electrical stimulation of the BST in anesthetized and conscious spontaneously hypertensive rats elicits increases in arterial pressure (7). However, these latter observations must be interpreted with caution, as electrical stimulation also coactivates fibers of passage.

The BST is ideally located to play a major role in the control of arterial pressure, as it has direct connections with hypothalamic and brainstem structures, including the VLM (8,9), that have previously been implicated in cardiovascular regulation (7,8,10,11). In this chapter, we summarize some recent immunohistochemical and functional data suggesting that VLM neurons mediate the cardiovascular responses elicited by activation of the BST (12).

II. Fos Labeling in the Brainstem Following Stimulation of BST Depressor Sites

The induction of the nuclear phosphoprotein Fos by the activation of the protooncogene c-*fos* (13), has been used as an index of neuronal activation in central circuits following the application of a variety of different stimuli (14–20). This increase in c-*fos* activity in central neurons appears to be associated with the particular neuronal system that was being activated (19,21). This technique was used in this study to identify central structures that expressed Fos activity after microinjection of glutamate at cardiovascular responsive sites in the BST. The induction of the Fos protein was demonstrated immunohistochemically.

Localized unilateral microinjections (40 nL) of glutamate (1 *M*) into the BST of awake rats elicited decreases in mean arterial pressure (MAP) and heart rate. The MAP responses ranged from −6 to −45 mm Hg (mean, −13.3 ± 1.2 mm Hg) while the heart rate responses ranged between −5 to −30 bpm (mean, −11.1 ± 3.2 bpm). The onset of the cardiovascular responses occurred within 1 to 5 s after glutamate microinjection into the BST and the responses had a duration of approximately 1 min. A representative example of the cardiovascular responses elicited by glutamate microinjection into the BST and the location of the injection sites in these studies is shown in Figure 1. Microinjection of 0.9% saline into cardiovascular responsive sites in the BST did not elicit cardiovascular responses.

Fos-positive nuclei were observed in three main brainstem regions as a result of the activation of depressor sites in the BST: the caudal VLM, nucleus of the solitary tract (NTS), and parabrachial nucleus (PBN). Few, if any, Fos-positive nuclei were found in these structures in animals receiving repeated microinjections of physiological saline. Control animals receiving neither microinjections of glutamate nor physiological saline lacked Fos-like immunoreactivity throughout the brain.

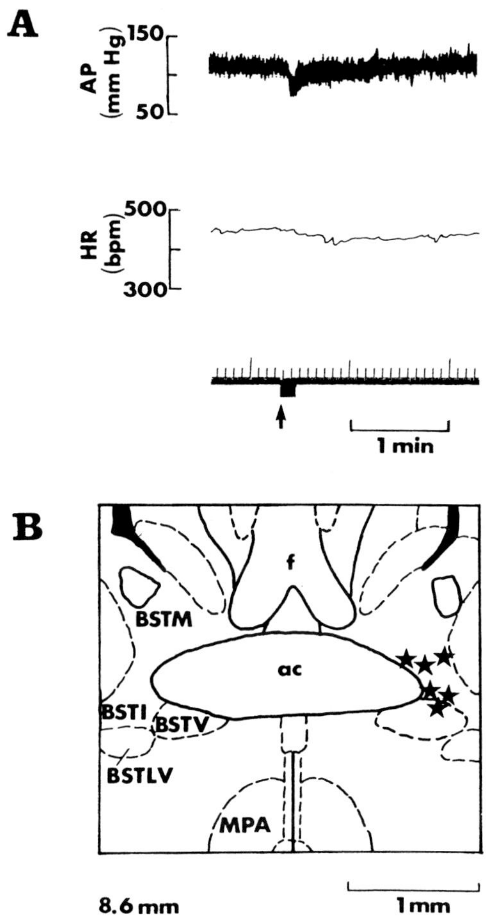

Figure 1 (A) Representative experiment showing the effect of microinjections of L-glutamate (arrow) on the arterial pressure (AP) and heart rate (HR) in an awake and freely moving rat. (B) Transverse section of the forebrain at 8.7 mm rostral to the interaural line, showing the location of the center of the glutamate injections (dots) in the bed nucleus of the stria terminalis (BST) that elicited decreases in mean AP and HR and that induced Fos immunoreactivity in brainstem autonomic nuclei. Abbreviations: ac, anterior commissure; BSTI, BST, intermediate division; BSTLV, BST, lateral division, ventral part; BSTM, BST, medial division; BSTV, BST, ventral division; f, fornix; MPA, medial preoptic area.

Within the caudal VLM, most of the Fos-labeled neurons were found in the region dorsal to the lateral reticular nucleus, dorsomedial to the spinal trigeminal tract, and ventrolateral to the nucleus ambiguus—the same region previously shown to contain the A_1-noradrenergic cell group (22,23). There were also a few scattered cells labeled in the intermediate and rostral VLM in the

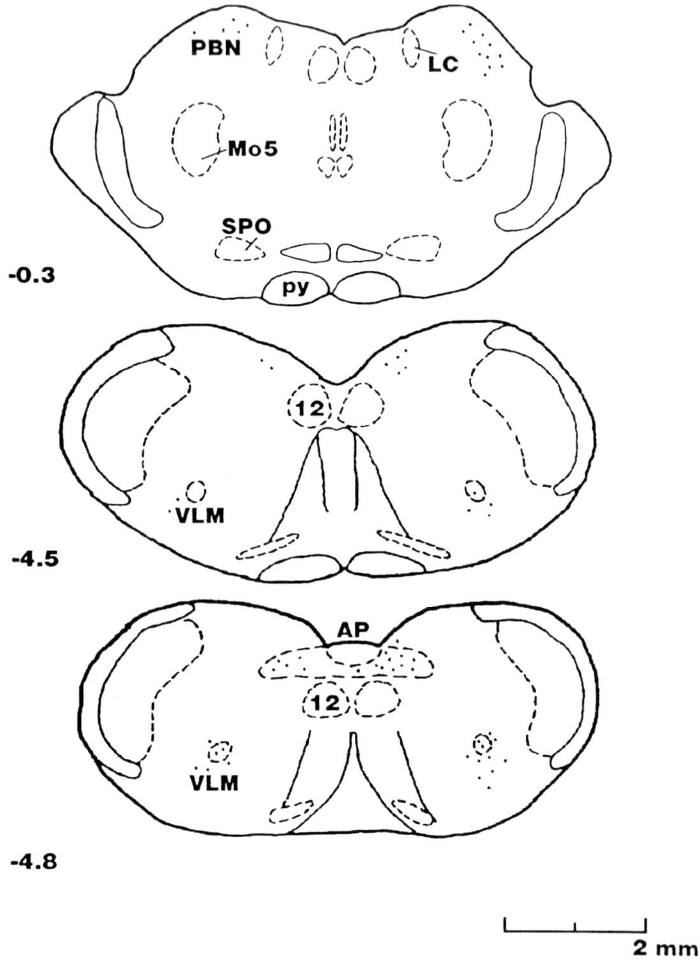

Figure 2 Distribution of *fos* immunoreactive neurons after stimulation of depressor sites in BST shown on projection drawings of the rat brainstem. Abbreviations: AP, area postrema; LC, locus ceruleus; Mo5, motor trigeminal nucleus; PBN, parabrachial nucleus; py, pyramidal tract; SPO, superior paraolivary nucleus; VLM, ventrolateral medulla; 12, hypoglossal nucleus.

region of the nucleus gigantocellularis lateralis. The labeling in the VLM was bilateral, with an ipsilateral predominance.

The distribution of Fos-labeled cells in the NTS was also similar in all animals receiving microinjections of glutamate. Most of the Fos-labeled neurons were found in the region extending from the caudal end of the area postrema to the obex and were concentrated in the medial and commissural nuclei of the NTS. These labeled neurons were observed primarily in the ipsilateral NTS.

Finally, Fos-like immunoreactivity was also observed in the PBN bilaterally. Fos-labeled neurons were most dense in the external lateral subnucleus of the PBN, adjacent to the superior cerebellar peduncle. Figure 2 schematically summarizes the extent of the Fos-immunoreactivity in these brainstem regions.

III. Effect of Brainstem Lesions on the BST Depressor Response

Previous studies have shown that activation of the BST with the excitatory amino acid glutamate (2) or its analogue D,L homocysteate (24) elicited decreases in MAP and heart rate. In addition, it was found that the depressor and bradycardiac responses elicited by BST stimulation were mediated by decreases in overall sympathetic tone (2). This evidence, combined with the Fos immunohistochemical findings summarized above, suggests that the caudal VLM, NTS, and PBN may be sites of integration for limbic autonomic responses. To test this possibility, a series of experiments was done in chloralose-anesthetized, paralyzed, and artificially ventilated rats.

In the first series, the region of the caudal VLM, NTS, and PBN that had been shown in the Fos studies to contain labeled nuclei was consecutively lesioned in order to investigate the contribution of each structure to the BST depressor response. The second series of experiments examined further the role of the caudal VLM on the BST depressor response by selectively inactivating synaptic transmission through this nucleus with the reversible pharmacological synaptic blocker cobalt chloride ($CoCl_2$).

Unilateral microinjections of glutamate (1 *M*; 20–40 nL) into the BST elicited decreases in MAP of -17 ± 2 mm Hg ($n = 8$) and heart rate of -15 ± 4 bpm ($n = 8$; Fig. 3A). These cardiovascular responses were not altered after a focal electrolytic lesion of the ipsilateral PBN (Fig. 3B). However, the magnitudes of the MAP and heart rate responses were significantly attenuated (-5 ± 1 mm Hg and 2 ± 1 bpm, respectively) 1 h after an electrolytic lesion of the ipsilateral NTS (Fig. 3C). Similarly, decreases in magnitudes were seen 1 h after an electrolytic lesion of the caudal VLM (Fig. 3D). Histological verification of the lesion sites in the medulla revealed that the largest extent of the lesion was within the NTS and caudal VLM in animals in which the MAP and heart rate responses were attenuated (Fig. 4).

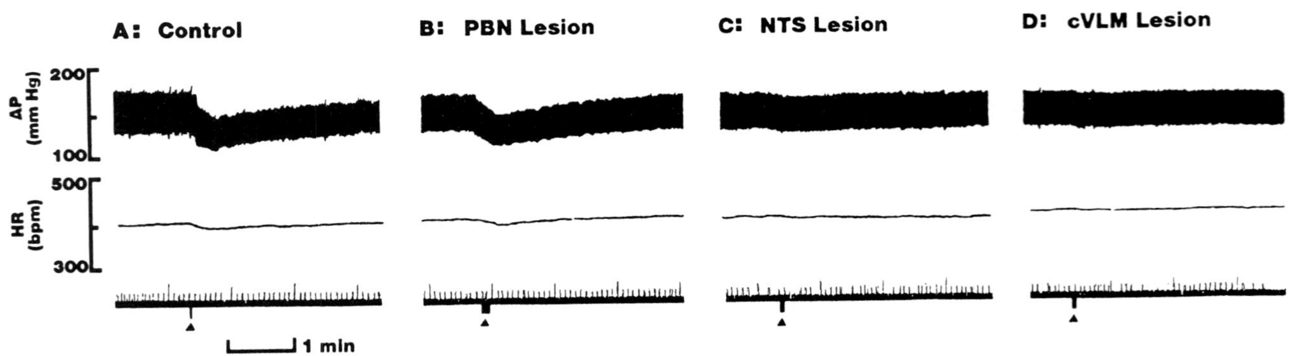

Figure 3 A representative experiment showing the effect of lesions of parabrachial nucleus (PBN), nucleus of the solitary tract (NTS), or caudal ventrolateral medulla (cVLM) on the arterial pressure (AP) and heart rate (HR) responses elicited during microinjection of L-glutamate into bed nucleus of the stria terminalis (BST). Note that after the PBN lesion (panel B), the cardiovascular responses to stimulation of BST were not altered. However, after a lesion of the ipsilateral NTS (panel C), the cardiovascular responses to stimulation of BST were significantly attenuated. An additional lesion of cVLM (panel D) abolished the cardiovascular responses to BST stimulation.

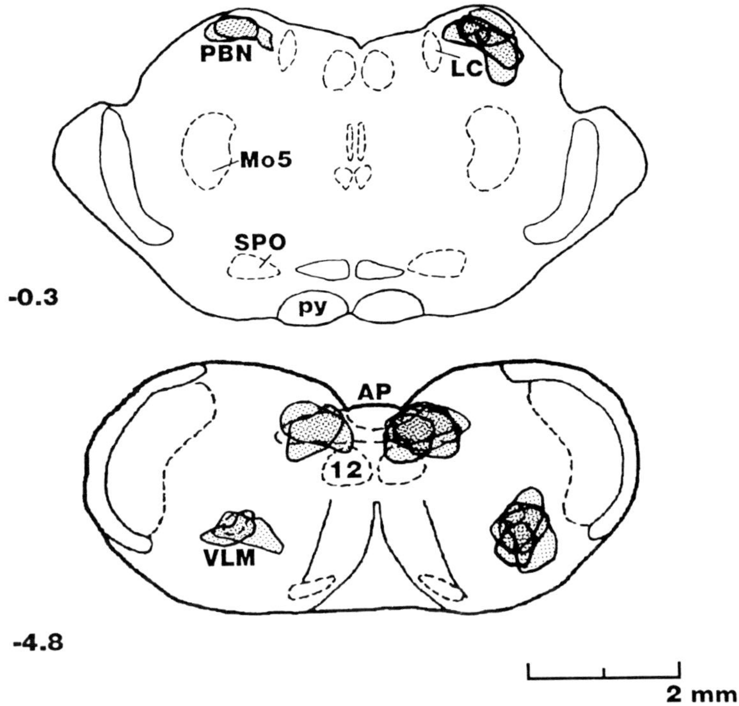

Figure 4 Transverse sections of the brainstem at −0.3 and −4.8 mm caudal to the interaural line, showing the location of electrolytic lesions (stippling) in the caudal ventrolateral medulla (cVLM), nucleus of the solitary tract (NTS), and lateral parabrachial nucleus (PBN) that attenuated the cardiovascular responses to stimulation of the ipsilateral bed nucleus of the stria terminalis. Abbreviations: AP, area postrema; LC, locus ceruleus; Mo5, motor trigeminal nucleus; PBN, parabrachial nucleus; py, pyramidal tract; SPO, superior paraolivary nucleus; VLM, ventrolateral medulla; 12, hypoglossal nucleus.

In the second series of experiments (12), injection of 0.9% saline (100 nL) into the caudal VLM did not significantly alter the cardiovascular responses elicited during stimulation of the BST (Fig. 5). However, restimulation of the BST 5 min after $CoCl_2$ was microinjected into caudal VLM resulted in an attenuation of the MAP and heart rate responses (Fig. 5). The mean decrease in MAP 5 min after $CoCl_2$ was -11 ± 3 mm Hg, or approximately 52% of control, and the mean decrease in heart rate was -4 ± 1 bpm (29% of control). Twenty min after the $CoCl_2$ microinjection, restimulation of BST elicited a decrease in MAP of -16 ± 3 mm Hg (62% of control) and a decrease in heart rate of -10 ± 3 bpm (76% of control). The magnitude of the cardiovascular

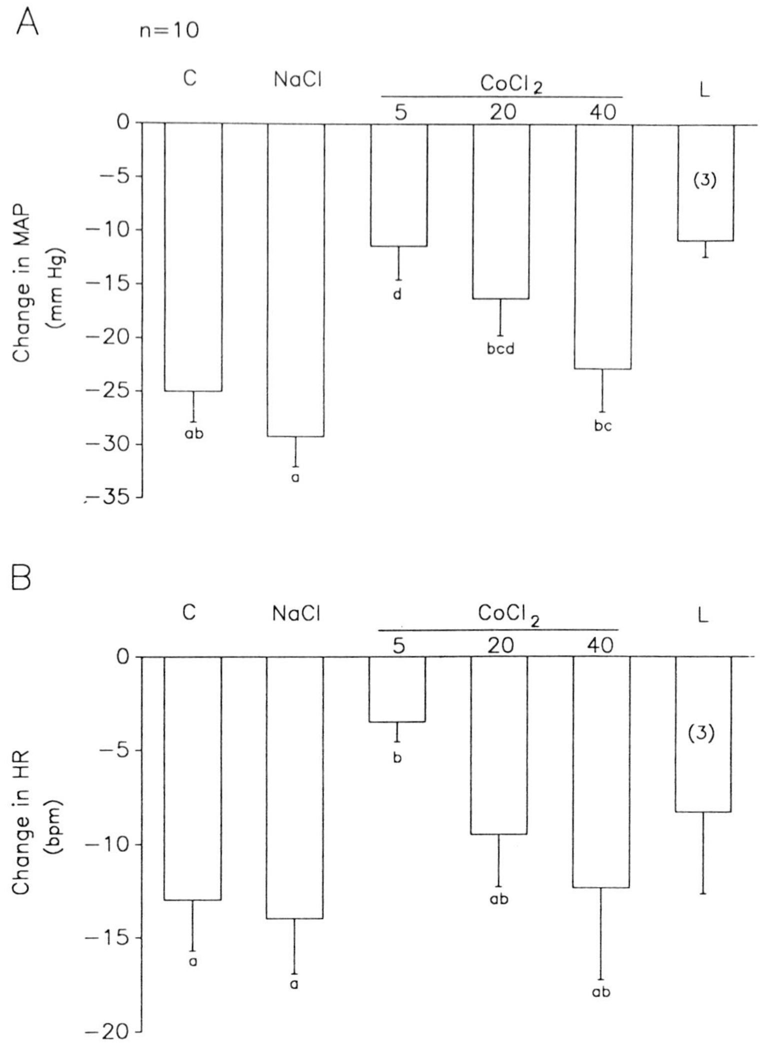

Figure 5 Histograms showing the effect of NaCl and $CoCl_2$ injections into the caudal ventrolateral medulla (cVLM) and electrolytic lesion of the cVLM on the mean arterial pressure (MAP; panel A) and heart rate (HR; panel B) responses elicited by stimulation of bed nucleus of the stria terminalis neurons (BST). C, control responses to stimulation of BST. Abbreviations: 5, 20, and 40, time in min after $CoCl_2$ injections into cVLM; L, response to BST stimulation 1 h after electrolytic lesion of cVLM. Number in parenthesis indicates number of animals. Within each histogram, any bars with letters that are different from each other are significantly different ($p < 0.05$), whereas those with similar letters are not significantly different.

responses elicited during BST stimulation returned to control values (MAP, -23 ± 4 mm Hg; heart rate, -12 ± 5 bpm), 40 min after the $CoCl_2$ was microinjected into the caudal VLM. The cardiovascular responses elicited from the BST 40 min after $CoCl_2$ microinjections into the caudal VLM were not significantly different from either control values (Fig. 5) or the response elicited from the BST 5 min after 0.9% saline was microinjected into the caudal VLM. Histological verification of the $CoCl_2$ injection sites in the medulla revealed that, in animals in which the MAP and heart rate responses were attenuated, the centers of the injection sites were within the caudal VLM. $CoCl_2$ injections that were outside the caudal VLM, in the lateral reticular nucleus or dorsal to the nucleus ambiguus, did not alter the cardiovascular responses elicited by BST stimulation.

IV. Discussion and Conclusions

In summary, these immunohistochemical and functional data have provided evidence to support the suggestion that activation of the BST with glutamate in the conscious and anesthetized rat elicits depressor responses and cardiac slowing mediated by pathways that terminate on neurons in brainstem autonomic centers, particularly the caudal VLM and the NTS.

A neuronal mapping of brainstem structures activated by stimulation of depressor sites in the BST was obtained using the c-*fos* immunohistochemical technique. However, although this method allows the visualization of a neuronal circuit activated by a specific stimulus, it can neither account for the order by which these structures were activated nor indicate whether the structures were activated via mono- or polysynaptic pathways. However, on the basis of previous anatomical and functional studies, possible pathways by which the brainstem structures in this study were activated may be suggested. Both Janssen and Ciriello (9) and Holstege et al. (8) have previously shown that BST neurons send axons to the VLM region, primarily to the region containing the A_2-noradrenergic cells. Furthermore, we have shown (12) that injection of the reversible synaptic blocker cobalt chloride ($CoCl_2$) into the caudal VLM attenuated the cardiovascular responses elicited during activation of the BST. In addition, it has also been shown that the neurons in the caudal VLM send axons to sympathoexcitatory premotor neurons in the rostral VLM (25), which, in turn, project to the intermediolateral cell column (26), the site of origin of preganglionic sympathetic axons (27). Taken together, these data suggest that caudal VLM neurons may have been activated directly by a monosynaptic pathway from BST. However, as NTS neurons were also activated and lesions of the NTS attenuated the BST cardiovascular responses, the possibility cannot be excluded that a NTS pathway to the caudal VLM also contributed to Fos induction in these neurons.

Neurons were found to be immunoreactive to Fos in the rostral aspects of the NTS and in the commissural nucleus of the NTS. As the neurons in the rostral NTS are known to function in gustatory (28) and gastric systems (29), it may be possible that neurons regulating gastric secretions (30) and motility (31) in the BST were activated along with those that evoked cardiovascular responses. Neurons in the commissural nucleus of the NTS are thought to be involved in cardiovascular regulation, as activation of these neurons elicits depressor and bradycardiac responses (32) similar to those obtained during activation of the BST. In addition, baroreceptor afferent fibers have been shown to terminate in this region (33), which in turn projects directly to the caudal VLM (34). This interpretation is consistent with our observation that lesions of the caudal NTS attenuate the BST cardiovascular responses, while an additional lesion in the caudal VLM completely eliminates these circulatory effects to BST stimulation. The functional significance of the Fos-like immunoreactivity in the PBN is somewhat more obscure. Neurons showing immunoreactivity to Fos in the PBN are unlikely to contribute to the cardiovascular responses elicited by activation of BST, as we have demonstrated that lesions of the PBN do not alter the magnitude of the BST depressor and bradycardiac responses.

Acknowledgments

The technical assistance of Z. M. Zhang is gratefully acknowledged. This work was supported by the Heart and Stroke Foundation of Ontario. J Ciriello is a Career Investigator of the Heart and Stroke Foundation of Ontario.

References

1. Ciriello J, Caverson MM, Polosa C. Function of the ventrolateral medulla in the control of the circulation. Brain Res Rev 1986; 11:359–391.
2. Ciriello J, Janssen SA. Effect of chemical stimulation of bed nucleus of the stria terminalis on arterial pressure and heart rate. Am J Physiol 1993; 34:H1516–H1522.
3. Gelsema AJ, Copeland NE, Drolet G, Bachelard H. Cardiovascular effects of neuronal activation of the extended amygdala in rats. Brain Res 1993; 626:156–166.
4. Weller KL, Smith DA. Afferent connections to the bed nucleus of the stria terminalis. Brain Res 1982; 232:255–270.
5. Roder S, Ciriello J. Contribution of the bed nucleus of the stria terminalis to the cardiovascular responses elicited by stimulation of the amygdala. J Auton Nerv Syst 1993; 45:61–75.
6., Gloor P. Anygdala. In: Handbook of Physiology: Neurophysiology. Vol II. Sec I. Washington, DC: American Physiological Society, 1960:1395–1420.
7. Phelix CF, Paull WK, Hartle DK. Immunocytochemistry (ICC) and electrical stimulation (ES) of bed nucleus of stria terminalis (BST): Connections with medullary cardiovascular (CV) regulatory centers (CRCs) in spontaneously hypertensive (SHR) and normotensive rats (abstr). Soc Neurosci Abstr 1990; 16:557.

8. Holstege G, Meiners L, Tan K. Projections of the bed nucleus of the stria terminalis to the mesencephalon, pons, and medulla oblongata in the cat. Exp Brain Res 1985; 58:370–391.
9. Janssen SA, Ciriello J. Projections from the cardiovascular responsive region of bed nucleus of the stria terminalis to medullary catecholaminergic cell groups (abstr). Soc Neurosci Abstr 1991; 17:706.
10. Gray TS, Magnuson DJ. Neuropeptide neuronal efferents from the bed nucleus of the stria terminalis and central amygdaloid nucleus to the dorsal vagal complex in the rat. J Comp Neurol 1987; 262:365–374.
11. Sawchenko PE, Swanson LW. The organization of forebrain afferents to the paraventricular and supraoptic nuclei of the rat. J Comp Neurol 1983; 218:121–144.
12. Giancola SB, Roder S, Ciriello J. Contribution of caudal ventrolateral medulla to the cardiovascular responses elicited from activation of bed nucleus of the stria terminalis. Brain Res 1993; 606:162–164.
13. Dragunow M, Faull R. The use of c-fos as a metabolic marker in neuronal pathway tracing. J of Neurosci Meth 1989; 29:261–265.
14. Dragunow M, Robertson HA, Robertson GS. Amygdala kindling and c-fos protein(s). Exp Neurol 1988; 102:261–263.
15. Dragunow M, Robertson HA. Generalized seizures induce C-fos protein(s) in mammalian neurons. Neurosci Lett 1987; 82:157–161.
16. Hunt SP, Pini A, Evan G. Induction of c-fos-like protein in spinal chord neurons following sensory stimulation. Nature 1987; 328:632–634.
17. Morgan JI, Cohen DR, Hempstead JL, Curran T. Mapping patterns of c-fos expression in the central nervous system after seizure. Science 1987; 237:192–196.
18. Rea MA. Light increases fos-related protein immunoreactivity in the rat suprachiasmatic nuclei. Brain Res Bull 1989; 23:577–581.
19. Sagar SM, Sharp FR, Curran T. Expression of c-fos protein in brain: Metabolic mapping at the cellular level. Science 1988; 240:1328–1331.
20. Hochstenbach SL, Solano-Flores LP, Ciriello J. Fos induction in brainstem neurons by intravenous hypertonic saline in the conscious rat. Neurosci Lett 1993; 158:225–228.
21. Sheng M, Greenberg ME. The regulation and function of c-fos and other immediate early genes in the nervous system. Neuron 1990; 4:477–485.
22. Armstrong DM, Ross CA, Pickel VM, et al. Distribution of dopamine-, noradrenaline- and adrenaline-containing cell bodies in the rat medulla oblongata: Demonstrated by the immunocytochemical localization of catecholamine biosynthetic enzymes. J Comp Neurol 1982; 212:173–187.
23. Kalia M, Fuxe K, Goldstein M. Rat medulla oblongata: II. dopaminergic, noradrenergic (A1 and A2) and adrenergic neurons, nerve fibers and presumptive terminal processes. J Comp Neurol 1985; 233:308–332.
24. Gelsema AJ, Calaresu FR. Chemical microstimulation of the septal area lowers arterial pressure in the rat. Am J Physiol 1987; 252:R760–R767.
25. Masuda N, Terui N, Koshiya N, Kumada M. Neurons in the caudal ventrolateral medulla mediate the arterial baroreceptor reflex by inhibiting barosensitive reticulospinal neurons in the rostral ventrolateral medulla in rabbits. J Auton Nerv Syst 1991; 34:103–118.
26. Ross CA, Ruggiero DA, Joh TH, et al. Rostral ventrolateral medulla: Selective projections to the thoracic autonomic cell column from the region containing C1 adrenaline neurons. J Comp Neurol 1984; 228:168–185.

27. Rando TA, Bowers CW, Zigmond RE. Localization of neurons in the rat spinal cord which project to the superior cervical ganglion. J Comp Neurol 1981; 196:73–83.
28. Hamilton RB, Norgren R. Central projections of gustatory nerves in the rat. J Comp Neurol 1984; 222:560–577.
29. Shapiro RE, Miselis RR. The central organization of the vagus nerve innervating the stomach of the rat. J Comp Neurol 1985; 238:473–488.
30. Henke PG. The bed nucleus of the stria terminalis and immobilization-stress: Unit activity, escape behavior and gastric pathology in rats. Behav Brain Res 1984; 11: 34–45.
31. Hermann GE, McCann MJ, Rogers RC. Activation of the bed nucleus of the stria terminalis increases gastric motility in the rat. J Auton Nerv Syst 1990; 30:123–128.
32. Nelson DO, Cohen HL, Feldman JL, McCrimmon DR. Cardiovascular function is altered by picomole injections of glutamate into rat medulla. J Neurosci 1988; 8: 1684–1693.
33. Ciriello J, Hochstenbach SL, Roder S. Central projections of baroreceptor and chemoreceptor afferent fibers in the rat. In: Barraco IRA, ed. Nucleus of the Solitary Tract. Boca Raton, FL: CRC Press, 1994:35–50.
34. Ross CA, Ruggiero DA, Reis DJ. Projections from the nucleus tractus solitarii to the rostral ventrolateral medulla. J Comp Neurol 1985; 242:511–534.

58

Vasomotor Actions of Metabotropic Glutamate Receptors in the Ventrolateral Medulla of Rats

DAVID B. AVERILL

Bowman Gray School of Medicine of Wake Forest University
Winston-Salem, North Carolina

TAKUYA TSUCHIHASHI

Kyushu University
Fukuoka City, Japan

I. Introduction

The rostral (RVLM) and caudal (CVLM) ventrolateral medulla oblongata constitute counterbalancing pools of neurons that are integral components of the baroreceptor reflex arc (1–6). Microinjection of L-glutamic acid (L-Glu) into the CVLM elicits decreases in arterial pressure and sympathetic nerve activity, whereas injection of this excitatory amino acid into the RVLM evokes increases in arterial pressure and sympathetic nerve activity (7,8). Current evidence supports the view that stimulation of CVLM neurons inhibits presympathetic neurons of the RVLM via GABAergic projections from the CVLM to the RVLM (6,9–11). On the other hand, stimulation of RVLM neurons increases sympathetic outflow via activation of RVLM projections to sympathetic preganglionic neurons residing in the intermediolateral cell column of the thoracic and lumbar segments of the spinal cord (reviewed in Ref. 12).

Two classes of excitatory amino acid or "glutamate" receptors in the mammalian central nervous system have been revealed by pharmacological and molecular biological studies. Glutamate may act at ionotropic excitatory amino acid receptors (iGluR) by actions exerted at ligand-gated ion channels. Molecular cloning studies have shown that iGluR are composed of heteromeric complexes of heterogeneous subunit proteins. This class of glutamate receptors has been further subdivided into α-amino-3-hydroxy-5-methylisoxazole-4-propionic acid (AMPA), kainic acid, and *N*-methyl-D-aspartic acid (NMDA) receptor subtypes based upon the pharmacological studies (reviewed in Ref. 13). L-glutamate may also act at metabotropic glutamate receptors (mGluR) that are coupled to GTP-binding proteins (reviewed in Refs. 14 and 15). As many as seven different mGluR have been identified by gene cloning studies (16–21). Characterization of mGluR has shown that activation of particular subtypes of mGluR may cause phosphoinositide hydrolysis and intracellular Ca^{2+} mobilization (reviewed in Refs. 22 and 23) or modulation of intracellular accumulation of cAMP (20,24–26). Although iGluR in the CVLM and RVLM have been implicated in mediating the depressor response to aortic depressor nerve stimulation (27) and the pressor response to hypothalamic stimulation (28), we sought to demonstrate that mGluR exist in the ventrolateral medulla and that selective activation of these receptors elicits cardiovascular responses.

II. Methods

A. Animal Preparations

Experiments were performed in Sprague-Dawley rats (270 to 350 g, Harlan Industries, Indianapolis, IN). All experiments were carried out in accordance with the guiding principles for the care and use of animals as mandated by the American Physiological Society. Rats were anesthetized with halothane (0.9% to 1.0%) and breathed a mixture of air and oxygen (65% and 35%, respectively). A femoral artery and vein were cannulated for the measurement of blood pressure and injection of drugs, respectively. Body temperature was maintained at 37.5 ± 0.5°C. The head of the rat was placed in a stereotaxic frame and the ventral surface of the medulla oblongata exposed (7,8). Animals were paralyzed with *d*-tubocurarine (0.8 mg/kg, IV) and artificially ventilated. Arterial blood gases and pH were maintained within physiological limits (pH 7.35–7.45, Po_2 100 to 140 mm Hg, Pco_2 40 to 45 mm Hg) by adjusting tidal volume and respiratory rate. The abdominal plexus was exposed through a transverse incision of the lateral abdominal wall, and the inferior nerve bundle accompanying a superior mesenteric artery was placed over bipolar silver electrodes. Nerve and electrodes were immersed in mineral oil to preserve integrity of the nerve fibers. Splanchnic sympathetic nerve activity (SpSNA) was amplified and filtered (bandwidth 100 to 3000 Hz). At the end of each experiment, the noise level associated with recording SpSNA was determined after intravenous injection of hexamethonium bromide (30 mg/kg).

B. Microinjection Procedures

Microinjections were made from multibarrel micropipettes with tip diameters of 20 to 50 μm. The pipettes were made from calibrated microbore capillary glass tubing (Accu-Fill 90, Clay Adams, Parsippany, NJ). Tips were drawn on a micropipette puller (model PE-2, Narishige Scientific Instruments, Japan). Injections (50 nL) were made over a 30-s period with a hand-held syringe, as described elsewhere (7,8,29). The injected volume was measured by observing the movement of the fluid meniscus along the reticle in a microscope. Microinjections were made on one side of the brainstem.

The rostral pressor and caudal depressor areas were identified by the injection of 2 nmol of L-Glu into the RVLM or CVLM, respectively. Identification of either the rostral pressor or caudal depressor area at the time of the experiment conformed to the following modification of criteria described by Willette et al. (2): (a) the latency to the onset of change in blood pressure produced by L-Glu was no more than 5 s; (b) the response plateau occurred within 20 s after microinjection of the amino acid; and (c) the change in blood pressure was at least 30 mm Hg. In our experiments, we found that the rostral pressor area (i.e., RVLM) was restricted to L-Glu injection at sites located 0.6 to 1.0 mm rostral to the most rostral rootlet of the hypoglossal nerve; 1.7 to 1.9 mm lateral to the midline; and 0.5 to 0.8 mm below the ventral surface. The caudal depressor area (i.e., CVLM) corresponded to L-Glu injection sites located between the second and third rostral rootlets of the hypoglossal nerve, 1.9 to 2.1 mm lateral to the midline, and 0.7 to 0.9 mm below the ventral surface.

All drugs for microinjection were dissolved in aCSF (in m*M*: 133.3 NaCl, 3.4 KCl, 1.3 $CaCl_2$, 1.2 $MgCl_2$, 0.6 NaH_2PO_4, 32.0 $NaHCO_3$, and 3.4 glucose, pH 7.4), which served as vehicle. Alcian blue dye (10 nL) was injected from a separate barrel of the pipette to mark the site of injection after injections at a particular site were completed.

C. Histological Analysis

At the completion of an experiment, rats were deeply anesthetized with pentobarbital sodium (50 mg, IV). The animals were then perfused transcardially with 150 mL of 0.9% NaCl followed by 150 mL of 10% phosphate-buffered formaldehyde solution. The brainstem was removed, stored overnight in 10% phosphate-buffered formaldehyde solution, and then transferred to a fixative containing 30% sucrose. Frozen brain tissue was sectioned in the coronal plane (50 μm) and stained with neutral red. Microinjection sites were identified by the deposition of Alcian blue dye and referred to standard anatomical structures of the caudal brainstem according to the atlas of Paxinos and Watson (30).

D. Experimental Protocols

Dose-Response of (1S,3R)-ACPD in RVLM and CVLM

(1S,3R)-1-aminocyclopentane-1,3-dicarboxylic acid [(1S,3R)-ACPD], which is a selective mGluR agonist, was microinjected into the RVLM or the CVLM. Doses of (1S,3R)-ACPD ranged from 100 pmol to 1 nmol for each injection. aCSF was injected as a vehicle control. At each injection site, only one dose of the drug was examined.

Effect of L-AP3 or AP7 and DNQX on (1S,3R)-ACPD Responses

The effects of (1S,3R)-ACPD (100 pmol) were examined after injection of the putative mGluR antagonist L-2-amino-3-phosphonopropionic acid (L-AP3, 2 nmol) or after injection of a solution containing the NMDA receptor antagonist 2-amino-7-phosphonoheptanoic acid (AP7, 200 pmol) and the non-NMDA receptor antagonist 6,7-dinitroquinoxaline-2,3-dione (DNQX, 200 pmol). These substances were injected into the same site of the RVLM or CVLM as subsequent injection of (1S,3R)-ACPD.

Effect of AP7 + DNQX on L-Glu, NMDA, and AMPA Responses

To assess blockade of iGluR, L-Glu (2 nmol), NMDA (20 pmol), or AMPA (5 pmol) was microinjected into the RVLM or CVLM before and 2 to 6 min after injection of AP7 + DNQX (200 pmol each) at the same site.

E. Statistical Analysis

Data are expressed as means ± SEM. One-way analysis of variance (ANOVA) followed by multiple comparisons with a Duncan's multiple range test were used for analysis of the dose response to (1S,3R)-ACPD (31). The responses evoked by (1S,3R)-ACPD in the presence of antagonists were also compared with those without antagonists by one-way ANOVA. Student's *t*-test was also used for paired observations (31). *P* values less than 0.05 were considered statistically significant.

III. Results

Baseline mean arterial pressure (MAP) and heart rate (HR) averaged 97 ± 1 mm Hg and 329 ± 4 bpm for the 62 rats included in this study. Table 1 presents

Table 1 Effect of L-Glu (2 nmol) Microinjections in the Rostral and Caudal Ventrolateral Medulla[a]

	RVLM	CVLM
Change in mean arterial pressure (mm Hg)	+35 ± 1	−41 ± 2
Change in heart rate (bpm)	+24 ± 3	−26 ± 5
Change in splanchnic SNA (%)	+78 ± 5	−58 ± 3
	(31)	(31)

[a]Values represent the mean ± SEM. Numbers in parentheses indicate number of animals in each group.

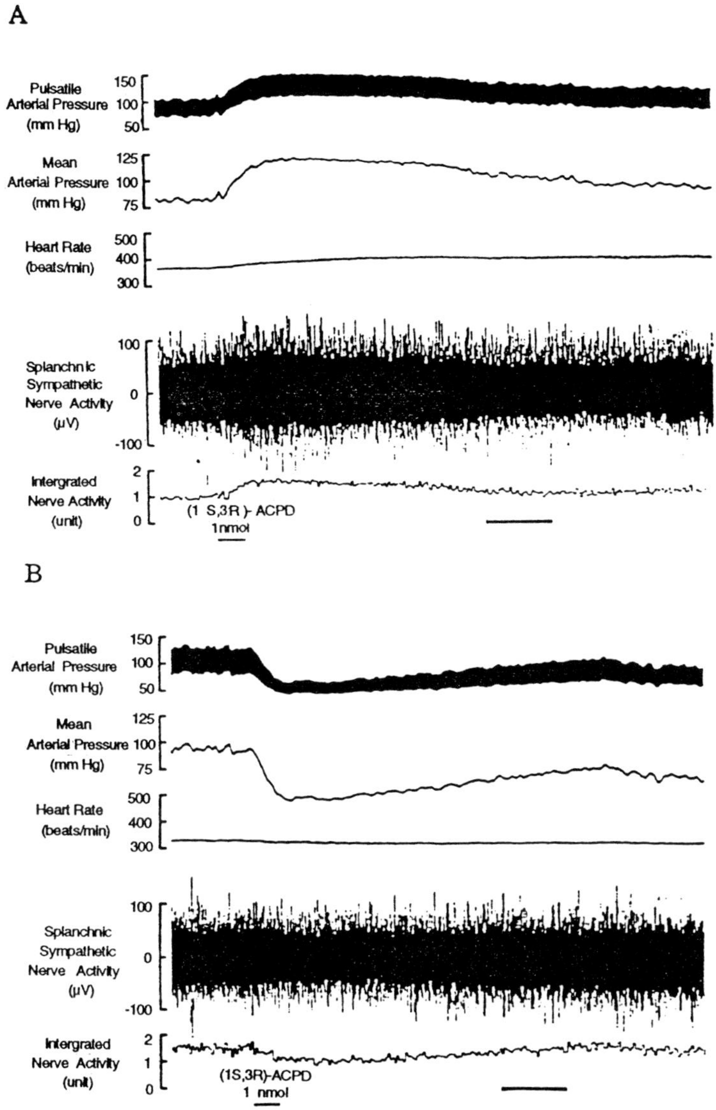

Figure 1 Polygraph tracings depict typical cardiovascular responses to microinjection of (1S,3R)-1-aminocyclopentane-1,3-dicarboxylic acid [(1S,3R)-ACPD]. Panel A: (1S,3R)-ACPD (1 nmol) microinjected into the rostral ventrolateral medulla increased mean arterial pressure, heart rate, and splanchnic sympathetic nerve activity. Panel B: (1S,3R)-ACPD (1 nmol) microinjected into the caudal ventrolateral medulla decreased mean arterial pressure, heart rate, and splanchnic sympathetic nerve activity. The horizontal bar beneath (1S,3R)-ACPD in panels A and B indicates the time of injection; the solid horizontal bar toward the right of each represents 1 minute. (From Ref. 44.)

the averaged responses of MAP, HR, and SpSNA produced by L-Glu (2 nmol) injections into the CVLM or RVLM. L-Glu injections in the RVLM produced significant increases in MAP, HR, and SpSNA, whereas injections of this amino acid in the CVLM significantly decreased MAP, HR, and SpSNA.

Figure 1 illustrates typical responses of 1 nmol of (1S,3R)-ACPD injected into the RVLM or CVLM. The predominant effects of (1S,3R)-ACPD injection into the RVLM were an increase in MAP and SpSNA and a modest increase in HR. When this mGluR agonist was injected into the CVLM, MAP and SpSNA decreased, and this was associated with a small decrease in HR. Injection of (1S,3R)-ACPD into either the RVLM or CVLM produced peak changes in MAP and SpSNA that occurred within 1 min; blood pressure and sympathetic nerve activity returned to baseline in 8 to 10 min. When two different doses of (1S,3R)-ACPD (0.1 and 1 nmol) were injected into the RVLM, the magnitude of the

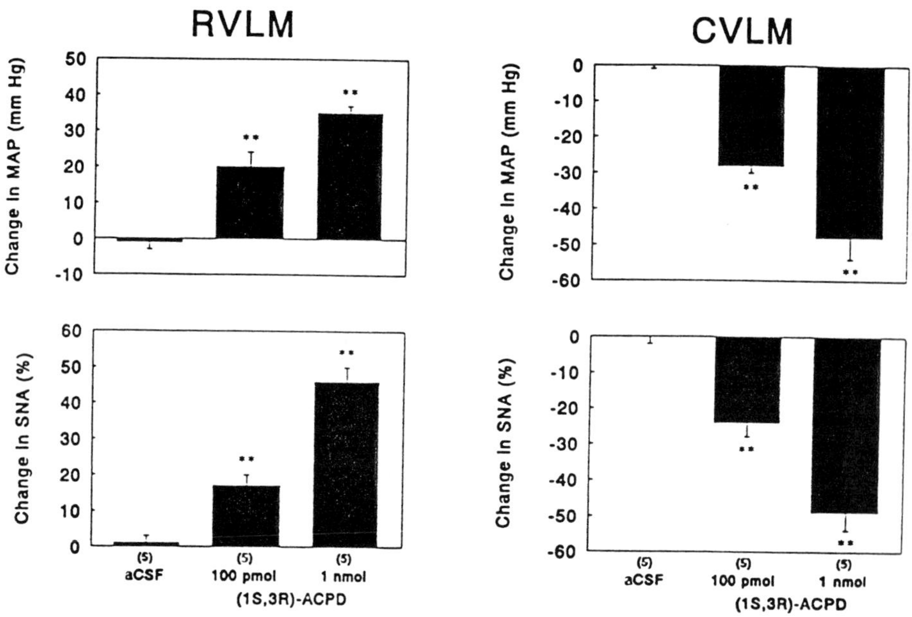

Figure 2 Dose related effects of (1S,3R)-1-aminocyclopentane-1,3-dicarboxylic acid [(1S,3R)-ACPD] microinjected into the rostral ventrolateral medulla (RVLM) or caudal ventrolateral medulla (CVLM). (1S,3R)-ACPD microinjections in the RVLM produced significant ($P<0.001$) dose-related increases in mean arterial pressure (MAP) and splanchnic sympathetic nerve activity (SNA); (1S,3R)-ACPD microinjections in the CVLM produced significant ($P<0.001$) dose-related decreases in MAP and SNA. Artificial cerebrospinal fluid (aCSF) represents the changes in MAP and SNA produced by injections of aCSF (vehicle) in either the RVLM or CVLM. **$P<0.01$ for comparison to aCSF. Number of rats for each dose is shown in parenthesis. (From Ref. 44.)

pressor response and increase in SpSNA were significantly ($P<0.001$) correlated to the dose of the agonist administered (Fig. 2). Dose-related effects in HR were also observed for injection of (1S,3R)-ACPD into the RVLM (0.1 nmol: $+7 \pm 2$ bpm, $P<0.05$ versus aCSF; 1 nmol: $+20 \pm 4$ bpm, $P<0.01$). Figure 2 also shows that the decrease in MAP and SpSNA produced by injection of (1S,3R)-ACPD (0.1 and 1 nmol) into the CVLM was significantly dose-related ($P<0.001$). The decreases in HR produced by (1S,3R)-ACPD injection into the CVLM were also dose-related (0.1 nmol: -13 ± 4 bpm; 1 nmol: -39 ± 16 bpm, $P<0.05$ versus aCSF).

One of the objectives of this study was to establish that the cardiovascular and sympathetic nerve responses produced by injections of glutamate receptor

A. Before AP7+DNQX

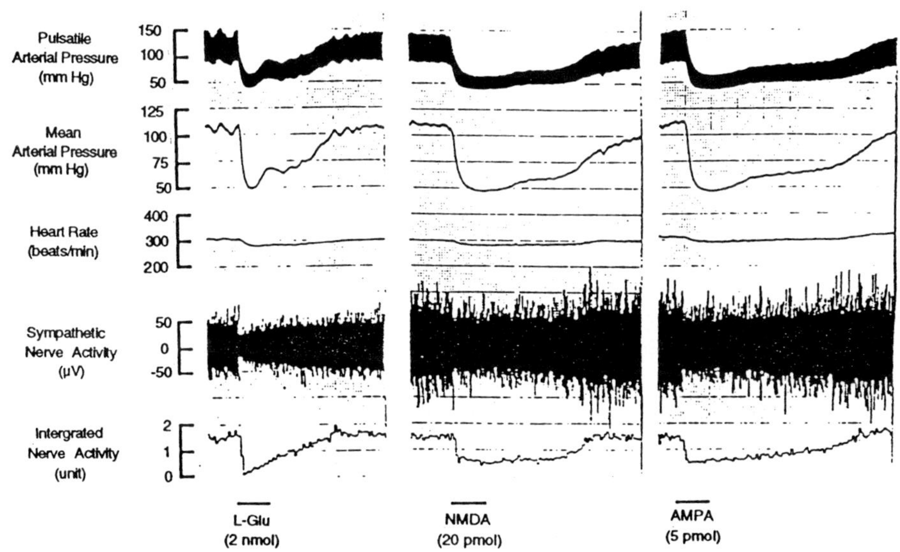

Figure 3 Effect of ionotropic glutamate receptor antagonists on responses to injections of L-glutamate (L-Glu), *N*-methyl-D-aspartic acid (NMDA), and α-amino-3-hydroxy-5-methylisoxazole-4-propionic acid (AMPA) in the caudal ventrolateral medulla (CVLM). Panel A: Decreases in blood pressure, heart rate, and splanchnic sympathetic nerve activity produced by injections of L-Glu (2 nmol), NMDA (20 pmol), and AMPA (5 pmol) into the CVLM. No ionotropic receptor antagonists had been administered before these injections. Panel B: Responses of the same cardiovascular variables when L-Glu, NMDA, and AMPA were injected into the CVLM after a prior injection of a solution containing 2-amino-7-phosphonoheptanoic acid (AP7, 200 pmol) and 6,7-dinitroquinoxaline-2,3-dione (DNQX, 200 pmol). The combination of AP7 and DNQX completely abolished the cardiovascular responses normally evoked by NMDA and AMPA, but the responses evoked by L-Glu were unaltered.

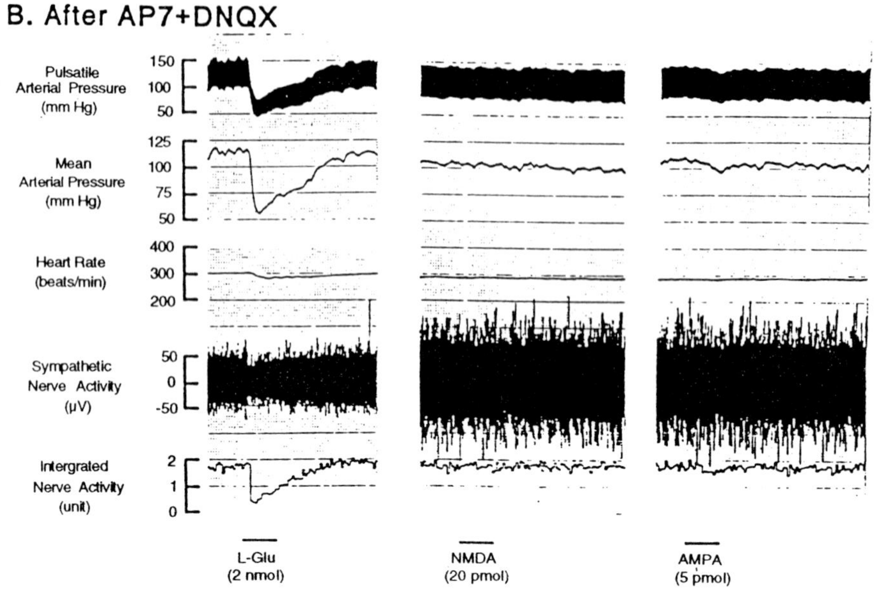

Figure 3 Continued

agonists in the VLM were due to selective stimulation of mGluR. Therefore, iGluR were blocked with a solution containing AP7 (NMDA receptor antagonist) and DNQX (non-NMDA receptor antagonist). Figure 3 is a series of chart records illustrating that injections of L-Glu (2 nmol), NMDA (20 pmol), and AMPA (5 pmol) at the same site in the CVLM each elicited depressor responses, decreases in HR, and suppression of SpSNA. However, the lower panels of Figure 3 show that when these same glutamate receptor agonists were injected at the same site after an injection of AP7 + DNQX (solution containing 200 pmol each), the blood pressure, HR, and SNA responses to NMDA and AMPA were abolished, whereas the responses to L-Glu were not altered. Figure 4 summarizes the results of this type of experiment carried out for injections of L-Glu, NMDA, and AMPA in the RVLM and CVLM. The changes in MAP and SpSNA produced by injections of NMDA or AMPA into the RVLM or CVLM were significantly ($P<0.05$ or $P<0.01$, see Fig. 4 for details) reduced by prior injection of AP7 + DNQX. In contrast, the blockade of iGluR by AP7 + DNQX did not alter the responses evoked by L-Glu injections into either the RVLM or CVLM. This finding suggested that L-Glu may have an important action at mGluR in the RVLM and CVLM. We also observed that injection of AP7 + DNQX into the RVLM did not alter baseline MAP ($-6 \pm$ mm Hg), HR (-4 ± 4 bpm), or SpSNA ($-2\% \pm 5\%$). In contrast, injection of the same solution

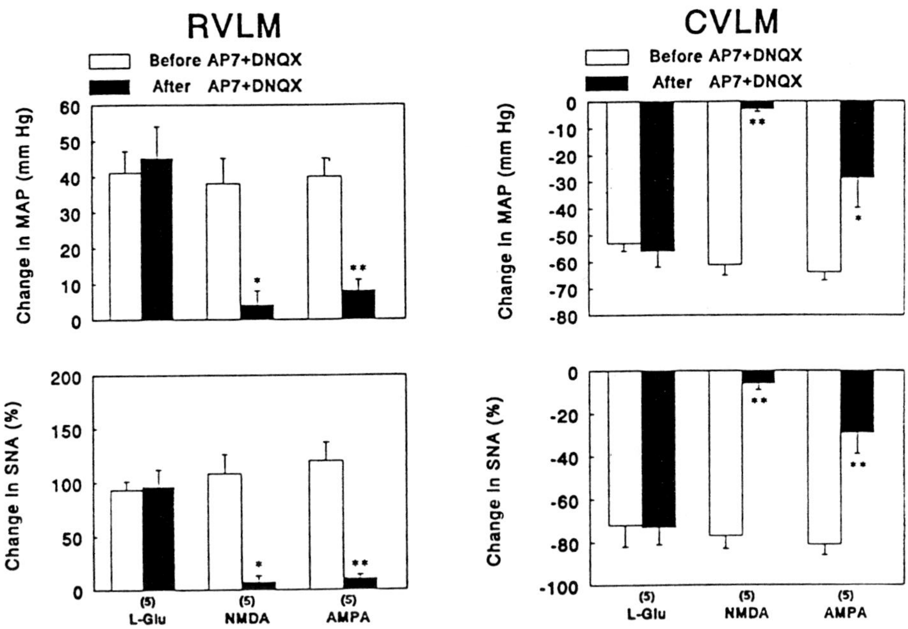

Figure 4 Bar graphs show the effects of 2-amino-7-phosphonoheptanoic acid plus 6,7-dinitroquinoxaline-2,3-dione (AP7 + DNQX, 200 pmol each) on changes in mean arterial pressure (MAP) and splanchnic sympathetic nerve activity (SNA) produced by L-glutamate (L-Glu, 2 nmol), *N*-methyl-D-aspartic acid (NMDA, 20 pmol), and α-amino-3-hydroxy-5-methylisoxazole-4-propionic acid (AMPA, 5 pmol) injected in either the rostral ventrolateral medulla (RVLM) or caudal ventrolateral medulla (CVLM). Increases or decreases in MAP and SNA produced by NMDA or AMPA were significantly reduced by AP7 + DNQX, whereas the responses elicited by L-Glu were not altered. $^*P<0.05$ and $^{**}P<0.01$ for comparisons before AP7 + DNQX. Number of rats for each agonist is shown in parentheses. (From Ref. 44.)

containing AP7 and DNQX (200 pmol each) into the CVLM significantly increased MAP (+10 ± 2 mm Hg, $P<0.005$ versus aCSF) and SpSNA (+23% ± 4%, $P<0.005$), whereas HR was not altered (+5 ± 3 bpm).

To address whether (1S,3R)-ACPD acted selectively at mGluR, we carried out experiments in which iGluR in the RVLM or CVLM were blocked with a solution containing AP7 and DNQX (200 pmol each). Figure 5 summarizes the blood pressure and SpSNA responses produced by injections of (1S,3R)-ACPD (0.1 nmol) into the RVLM or CVLM in the absence of iGluR blockade or after injections of AP7 + DNQX at the same site. The results of these experiments showed that blockade of iGluR did not alter the cardiovascular and sympathetic nerve responses evoked by (1S,3R)-ACPD injections in the RVLM or CVLM.

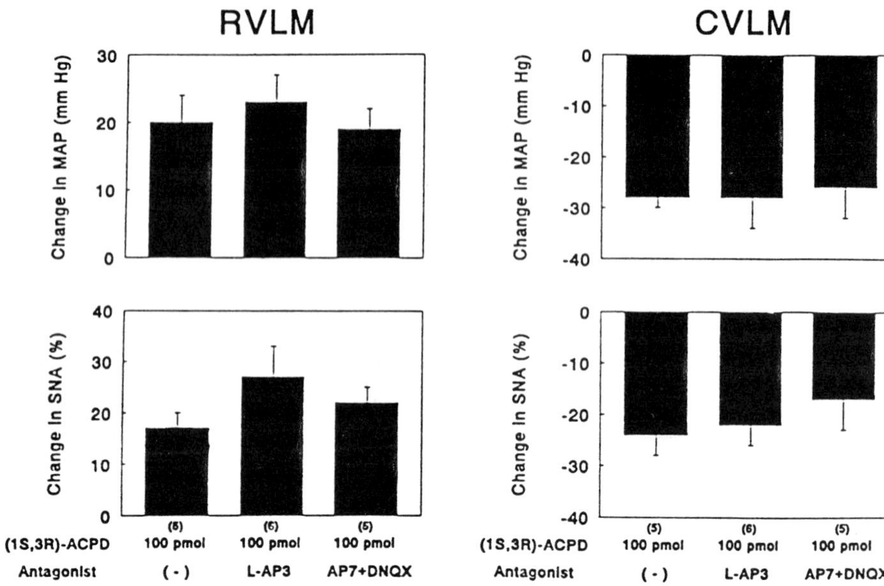

Figure 5 Bar graphs show the effects of prior injection of L-2-amino-3-phosphonopropionic acid (L-AP3, 2 nmol), 2-amino-7-phosphonoheptanoic acid (AP7, 200 pmol) plus 6,7-dinitroquinoxaline-2,3-dione (DNQX, 200 pmol), or vehicle (−) on changes in mean arterial pressure (MAP) and splanchnic sympathetic nerve activity (SNA) evoked by subsequent injection of (1S,3R)-1-aminocyclopentane-1,3-dicarboxylic acid [(1S,3R-ACPD, 100 pmol] in the rostral ventrolateral medulla (RVLM) or caudal ventrolateral medulla (CVLM). Neither L-AP3 nor AP7 + DNQX reduced the changes in MAP and SNA evoked by (1S,3R)-ACPD. Number of rats is shown in parentheses. (From Ref. 44.)

We also attempted to block the (1S,3R)-ACPD–evoked cardiovascular and sympathetic nerve responses by injection of a purported mGluR antagonist (L-AP3) in the RVLM or CVLM. L-AP3 has been reported to inhibit the phosphoinositide hydrolysis (32) and mobilization of intracellular Ca^{2+} (33) produced by *trans*-ACPD. However, injection of L-AP3 (2 nmol) in the RVLM or CVLM did not alter the blood pressure and SpSNA responses evoked by subsequent injection of (1S,3R)-ACPD (0.1 nmol) in the RVLM or CVLM (see Fig. 5).

IV. Discussion

The results of this study demonstrate the existence of mGluR in the ventrolateral medulla of rats. Moreover, activation of these glutamate receptors with the mGluR selective agonist (1S,3R)-ACPD (34–36) produced increases in MAP,

HR, and SpSNA when this agonist was injected in the RVLM and decreases in these cardiovascular variables when (1S,3R)-ACPD was injected in the CVLM. Thus, the findings of this study support the view that mGluR as well as iGluR participate in the cardiovascular responses evoked by microinjection of L-Glu in the ventrolateral medulla. Our results complement the findings of Pawloski-Dahm and Gordon (37) that *trans*-ACPD injected in the nucleus tractus solitarius (nTS) evoked depressor responses. Our findings are also in agreement with the preliminary report of McManigle et al. (38) that *trans*-ACPD injected into the CVLM of cats decreased blood pressure and HR. However, these investigators did not demonstrate that injection of *trans*-ACPD into the RVLM of cats evoked cardiovascular responses.

Two important aspects of this study were the findings that cardiovascular responses evoked by injection of L-Glu or (1S,3R)-ACPD in the RVLM or CVLM did not require stimulation of iGluR. First, blockade of iGluR with a solution containing the NMDA receptor antagonist AP7 and the non-NMDA receptor antagonist DNQX did not alter the changes in MAP and SpSNA produced by injections of (1S,3R)-ACPD in the RVLM or CVLM. These results are taken as evidence that (1S,3R)-ACPD acted selectively at mGluR in these two vasomotor regions of the VLM. Second, the cardiovascular responses produced by L-Glu injections in the RVLM or CVLM were not modified by blockade of iGluR. A similar observation has been made by Pawloski-Dahm and Gordon (37), who showed that the depressor responses produced by L-Glu injections in the nTS were not altered by injection of the broad acting iGluR antagonist kynurenic acid in the nTS. These particular findings strengthen the notion that an important action of L-Glu (at least by microinjection) is mediated by mGluR. However, these results taken alone do not permit us to establish whether L-Glu acts as a neurotransmitter in the ventrolateral medulla or nTS via actions at mGluR.

An important facet of any study investigating the action of an agonist at a particular receptor is the ability to block or attenuate the presumed receptor-mediated response by an antagonist selective for the receptor of interest. Thus, we attempted to block the cardiovascular responses of (1S,3R)-ACPD injected in the RVLM or CVLM by prior injection of L-AP3. A number of studies have demonstrated that L-AP3 can effectively prevent the phosphoinositide hydrolysis (32) and the increase in intracellular Ca^{2+} (33) evoked by *trans*-ACPD or (1S,3R)-ACPD. However, L-AP3 did not modify the cardiovascular responses that were produced by injection of (1S,3R)-ACPD or L-Glu in the RVLM or CVLM of rats. Our findings agree with the observations of Pawloski-Dahm and Gordon (37) that L-AP3 did not block the cardiovascular responses evoked by L-Glu or *trans*-ACPD injections in the nTS. It is also pertinent to point out that L-AP3 has been found ineffective in blocking mGluR mediated responses in some in vitro preparations (39,40). These results suggest that (1S,3R)-ACPD acts at mGluR subtype(s) in the ventrolateral medulla or nTS for which L-AP3 is not antagonist. This concept may be congruent with the finding that (1S,3R)-

ACPD mediated depolarization of nTS neurons does not require changes in intracellular Ca^{2+} (41). More recently, we investigated whether (RS)-4-carboxy-3-hydroxyphenylglycine or (+)-α-methyl-4-carboxyphenylglycine [(+)-MCPG] blocks the cardiovascular responses produced by (1S,3R)-ACPD in the ventrolateral medulla. (+)-MCPG has been reported to block the neuroexcitatory effects of (1S,3R)-ACPD in neonatal rat motoneurons and thalamic neurons (42). However, preliminary experiments have not shown that either agent acts as an antagonist to the cardiovascular responses evoked by (1S,3R)-ACPD in either the RVLM or CVLM.

In the last decade, mGluR have received attention as an important class of glutamate receptors in the central nervous system. Molecular cloning studies have revealed as many as seven different subtypes of mGluR (16–21). Current evidence indicates that mGluR are coupled to GTP-binding proteins and thereby influence intracellular second-messenger systems. Certain mGluR (mGluR1 and mGluR5) are coupled to a phosphoinositol hydrolysis/Ca^{2+} signal transduction cascade, whereas other receptor subtypes appear to modulate intracellular accumulation of cAMP (24,25). Recent studies have provided evidence that mGluR stimulation by (1S,3R)-ACPD potentiates the increase in cAMP produced by β-adrenergic receptor stimulation (24–26). On the other hand, receptor subtypes other than mGluR1 and mGluR5 may inhibit the cAMP cascade (20). Given the multiplicity of receptor subtypes and the differing intracellular responses evoked by particular receptor subtypes, an important advance in this field of glutamate receptors will be the development of agents that act selectively either as agonists or antagonists for specific receptor subtypes. Some progress in this regard has been made by Hayashi et al. (43), who have examined the agonist and antagonist activities of phenylglycine derivatives upon individual subtypes of mGluR expressed in Chinese hamster ovary cells. Unfortunately, the experiments reported here were performed before the more extensive investigations of compounds that may act selectively at particular receptor subtypes. Thus, our inability to block the actions of (1S,3R)-ACPD by either L-AP3 or (+)-MCPG does not permit us to establish the signal transduction mechanism(s) that may be responsible for cardiovascular responses evoked by (1S,3R)-ACPD injected in the ventrolateral medulla. More importantly, the lack of agents that clearly act as antagonists to mGluR in the ventrolateral medulla prevents us from establishing the potential physiological role of mGluR in central neural control of the cardiovascular system.

V. Summary

Results reported here provide evidence for the existence of mGluR on vasomotor neurons of the ventrolateral medulla. Stimulation of mGluR by injection of (1S,3R)-ACPD in the RVLM or CVLM elicited cardiovascular and sympathetic nerve responses that had the same pharmacological profile as that produced by injection of L-Glu at the same sites.

References

1. Willette RN, Punnen S, Krieger AJ, Sapru HN. Interdependence of rostral and caudal ventrolateral medullary areas in the control of blood pressure. Brain Res 1984; 321:169–174.
2. Willette RN, Barcas PP, Krieger AJ, Sapru HN. Vasopressor and depressor areas in the rat medulla: Identification by microinjection of L-glutamate. Neuropharmacology 1983; 22:1071–1079.
3. Urbanski RW, Sapru HN. Evidence for a sympathoexcitatory pathway from the nucleus tractus solitarii to the ventrolateral medullary pressor area. J Auton Nerv Syst 1988; 23:161–174.
4. Agarwal SK, Gelsema AJ, Calaresu FR. Neurons in rostral VLM are inhibited by chemical stimulation of caudal VLM in rats. Am J Physiol 1989; 257:R265–R270.
5. Agarwal SK, Gelsema AJ, Calaresu FR. Inhibition of rostral VLM by baroreceptor activation is relayed through caudal VLM. Am J Physiol 1990; 258:R1271–R1278.
6. Agarwal SK, Calaresu FR. Monosynaptic connection from caudal to rostral ventrolateral medulla in the baroreceptor reflex pathway. Brain Res 1991; 555:70–74.
7. Muratani H, Averill DB, Ferrario CM. Effect of angiotensin II in the ventrolateral medulla of spontaneously hypertensive rats. Am J Physiol 1991; 260:R977–R984.
8. Muratani H, Ferrario CM, Averill DB. Ventrolateral medulla in spontaneously hypertensive rats: role of angiotensin II. Am J Physiol 1993; 264:R388–R395.
9. Sun MK, Guyenet PG. GABA-mediated baroreceptor inhibition of reticulospinal neurons. Am J Physiol 1985; 249:R672–R680.
10. Urbanski RW, Sapru HN. Putative neurotransmitters involved in medullary cardiovascular regulation. J Auton Nerv Syst 1988; 25:181–193.
11. Willette RN, Barcas PP, Krieger AJ, Sapru HN. Endogenous GABAergic mechanisms in the medulla and the regulation of blood pressure. J Pharmacol Exp Ther 1984; 230:34–39.
12. Ciriello J, Caverson MM, Polosa C. Function of the ventrolateral medulla in the control of the circulation. Brain Res Rev 1986; 11:359–391.
13. Sommer B, Seeburg PH. Glutamate receptor channels: Novel properties and new clones. TIPS: Trends in Pharmacological Sciences 1993; 291–296.
14. Schoepp DD, Conn PJ. Metabotropic glutamate receptors in brain function and pathology. TIPS: Trends in Pharmacological Sciences 1993; 14:13–20.
15. Conn PJ, Desai MA. Pharmacology and physiology of metabotropic glutamate receptors in mammalian central nervous system. Drug Dev Res 1991; 24:207–229.
16. Houamed KM, Kuijper JL, Gilbert TL, et al. Cloning, expression, and gene structure of a G protein–coupled glutamate receptor from rat brain. Science 1991; 252: 1318–1321.
17. Masu M, Tanabe Y, Tsuchida K, et al. Sequence and expression of a metabotropic glutamate receptor. Nature 1991; 349:760–765.
18. Tanabe Y, Masu M, Ishii T, et al. A family of metabotropic glutamate receptors. Neuron 1992; 8:169–179.
19. Abe T, Sugihara H, Nawa H, et al. Molecular characterization of a novel metabotropic glutamate receptor mGluR5 coupled to inositol phosphate/Ca^{2+} signal transduction. J Biol Chem 1992; 267:13361–13368.

20. Okamoto N, Hori S, Akawaza C, et al. Molecular characterization of a new metabotropic glutamate receptor mGluR7 coupled to inhibitory cyclic AMP signal transduction. J Biol Chem 1994; 269:1231–1236.
21. Nakanishi S. Molecular diversity of glutamate receptors and implications for brain function. Science 1992; 258:597–603.
22. Schoepp D, Bockaert J, Sladeczek F. Pharmacological and functional characteristics of metabotropic excitatory amino acid receptors. TIPS: Trends in Pharmacological Sciences 1990; 11:508–515.
23. Schoepp DD. The biochemical pharmacology of metabotropic glutamate receptors. Bioch Soc Trans 1993; 21:97–102.
24. Winder DG, Conn PJ. Activation of metabotropic glutamate receptors in the hippocampus increases cyclic AMP accumulation. J Neurochem 1992; 59:375–378.
25. Winder DG, Conn PJ. Activation of metabotropic glutamate receptors increases cAMP accumulation in hippocampus by potentiating responses to endogenous adenosine. J Neurosci 1993; 13:38–44.
26. Gereau RW, Conn PJ. A cyclic AMP-dependent form of associative synaptic plasticity induced by coactivation of β-adrenergic receptors and metabotropic glutamate receptors in rat hippocampus. J Neurosci 1994; 14:3310–3318.
27. Gordon FJ. Aortic baroreceptor reflexes are mediated by NMDA receptors in caudal ventrolateral medulla. Am J Physiol 1987; 252:R628–R633.
28. Sun MK, Guyenet PG. Hypothalamic glutamatergic input to medullary sympathoexcitatory neurons in rats. Am J Physiol 1986; 251:R798–R810.
29. Diz DI, Barnes KL, Ferrario CM. Hypotensive action of angiotensin II microinjected into the dorsal motor nucleus of the vagus. J Hypertens 1984; 2(suppl 3): 53–56.
30. Paxinos G, Watson C. The Rat Brain in Stereotaxic Coordinates, 2d ed. New York: Academic Press, 1986.
31. SAS/STAT Guide for Personal Computers, 6th ed. Cary, NC: SAS Institute Inc, 1985:1–378.
32. Schoepp DD, Johnson BG, Smith ECR, McQuaid LA. Stereoselectivity and mode of inhibition of phosphoinositide-coupled excitatory amino acid receptors by 2-amino-3-phosphonopropionic acid. Mol Pharmacol 1990; 38:222–228.
33. Irving AJ, Schofield JG, Watkins JC, et al. 1S,3R-ACPD stimulates and L-AP3 blocks Ca^{2+} mobilization in rat cerebellar neurons. Eur J Pharmacol 1990; 186: 363–365.
34. Palmer E, Monaghan DT, Cotman CW. Trans-ACPD, a selective agonist of the phosphoinositide-coupled excitatory amino acid receptor. Eur J Pharmacol 1989; 166:585–587.
35. Desai MA, Conn PJ. Selective activation of phosphoinositide hydrolysis by a rigid analogue of glutamate. Neurosci Lett 1990; 109:157–162.
36. Schoepp DD, Johnson BG, Salhoff CR, et al. In vitro and in vivo pharmacology of trans- and cis-(+/−)-1-amino-1,3-cyclopentanedicarboxylic acid: Dissociation of metabotropic and ionotropic excitatory amino acid receptor effects. J Neurochem 1991; 56:1789–1796.
37. Pawloski-Dahm C, Gordon FJ. Evidence for a kynurenate-insensitive glutamate receptor in nucleus tractus solitarii. Am J Physiol 1992; 262:H1611–H1615.
38. McManigle JE, Berger IL, Panico WH, et al. Evidence for the presence of a metabotropic excitatory amino acid receptor in the ventrolateral medulla of the cat (abstr). FASEB J 1992; 6:A1509.

39. Kinney GA, Slater NT. Potentiation of mossy fiber-evoked EPSPs in turtle cerebellar purkinje cells by the metabotropic glutamate receptor agonist 1S,3R-ACPD. J Neurophysiol 1992; 67:1006–1008.
40. Stratton KR, Worley PF, Baraban JM. Pharmacological characterization of phosphoinositide-linked glutamate receptor excitation of hippocampal neurons. Eur J Pharmacol 1990; 186:357–361.
41. Glaum SR, Miller RJ. Metabotropic glutamate receptors mediate excitatory transmission in the nucleus of the solitary tract. J Neurosci 1992; 12:2251–2258.
42. Jane DE, Jones PLJ, Pook PCK, et al. Stereospecific antagonism by (+)-α-methyl-4-carboxyphenylglycine (MCPG) of (1S,3R)-ACPD-induced effects in neonatal rat motoneurones and rat thalamic neurones. Neuropharmacology 1993; 32:725–727.
43. Hayashi Y, Sekiyama N, Nakanishi S, et al. Analysis of agonist and antagonist activities of phenylglycine derivatives for different cloned metabotropic glutamate receptor subtypes. J Neurosci 1994; 14:3370–3377.
44. Tsuchihashi T, Averill DB. Metabotropic glutamate receptors in the ventrolateral medulla of rats. Hypertension 1993; 21:739–744.

59

Organization and Function of Ventrolateral Medullary Afferents to the Nucleus of the Solitary Tract Complex

STEFANIE RODER and JOHN CIRIELLO

University of Western Ontario
London, Ontario, Canada

I. Introduction

The ventrolateral medulla (VLM) has received considerable attention as an important integrating area for control of the circulation (1). Neurons in the VLM have been shown to be involved in the maintenance of vasomotor tone (2–4), in baroreceptor and chemoreceptor reflex mechanisms (5–11), in mediating the cerebral ischemic response and somatosympathetic reflexes (12–17), and in the control of the secretion of vasopressin (3,18–21). Most of these physiological functions have been thought to involve the catecholaminergic neurons within the VLM (3,18,22–28).

It has also been suggested that the VLM influences the gain of the baroreceptor reflex at the level of the nucleus of the solitary tract (NTS; Ref. 29), the primary site of termination of cardiovascular afferent fibers (30–33). In support of this suggestion, lesions of the caudal VLM that included the region of the A_1-noradrenergic cell group have been shown to alter baroreceptor reflex excitability (28). In addition, chemical and electrical stimulation of the VLM has been shown to elicit an increase in heart rate and in arterial pressure, during which vagal activity to the heart is inhibited (3) and sympathetic vasomotor tone is insensitive to baroreceptor inhibition (9). Furthermore, several anatomical studies have demonstrated direct projections from the VLM region to the cardiovascular areas of the NTS (34–39). However, these anatomical data have been obtained in autoradiographic or horseradish peroxidase studies, in which the injection sites were not limited to the anatomical structures studied (35, 36,38).

Additionally, although retrograde tract-tracer injections that overlap most of the NTS complex give a good indication of where the neurons that project to the NTS are located within the rostrocaudal extent of the VLM, they provide little information regarding the subnuclei in the NTS that receive the VLM projections (35,36,38–40). Retrograde tract-tracer studies must also be interpreted with caution, as fibers of passage in the injection area may have contributed to the retrograde labeling of neurons observed in the VLM. Autoradiographic studies have provided evidence for neurons within the rostral VLM that project to the ventrolateral and intermediate subnuclei of the NTS (36) and neurons within the caudal VLM that project primarily to the medial and commissural subnuclei of the NTS (35,40). In a recent study using the selective anterograde tract-tracer *Phaseolus vulgaris* leucoagglutinin (PHA-L), it was demonstrated that neurons within the rostral VLM project to the commissural and ventrolateral subnucleus of the NTS (34). Although there appears to be a discrepancy in the projections to the NTS with the use of the anterogradely transported tract tracers, it may be due to the location of the injection sites within the rostrocaudal extent of the VLM or to the fact that some neurons in the VLM do not take up and transport the tracers (41).

Injections of catecholamines into the NTS have been shown to elicit changes in arterial pressure and heart rate (42–44) and to alter baroreceptor reflex excitability (43,45,46). Therefore, the possibility exists that catecholaminergic neurons within the VLM may influence the function of NTS neurons. In support of this suggestion, the NTS has been shown to be densely innervated by catecholaminergic fibers (47–52). This catecholaminergic innervation of the NTS has been suggested to arise from neurons belonging to the A_1 and C1 catecholaminergic cell groups within the VLM (39). However, no attempt has been made to identify the subnuclei of the NTS that receive catecholaminergic inputs from VLM neurons and to determine whether a topographical organization exists within the NTS complex of projections from the A_1-noradrenergic and C1 adrenergic neurons in the VLM.

The NTS complex has also been shown to contain catecholaminergic neurons within its medial subdivision (47,50,51). These catecholaminergic neurons are also thought to play a role in cardiovascular regulation (42,53). It has been suggested that catecholaminergic inputs onto the A_2-noradrenergic neurons within the NTS may be responsible for eliciting changes in blood pressure and heart rate observed after microinjections of catecholamines into the NTS (42). This suggestion is based on the observation that some catecholaminergic neurons within the NTS receive catecholaminergic input (54), although the source of this catecholaminergic input is not known.

This chapter presents a mapping of the termination sites in the NTS complex of neurons located at different rostrocaudal levels within the VLM using the anterograde tract tracer PHA-L (55). PHA-L has the advantages of allowing the identification of the precise location of cell bodies that have incorporated the tracer for transport and permitting a clear identification of the resulting labeled axons with their presumptive terminal boutons at the light microscope level (55,56). PHA-L tract tracing was also combined in these studies with immunohistochemistry for the catecholaminergic biosynthetic enzymes tyrosine hydroxylase (TH), dopamine-beta-hydroxylase (DBH), and phenylethanolamine-*N*-methyltransferase (PNMT) to investigate the contribution of catecholaminergic neurons in the VLM to the innervation of the NTS and that of noncatecholaminergic VLM neurons to the catecholaminergic cell groups of the NTS. In addition, electrophysiological (57) and functional data are summarized, showing the effect of activation of VLM neurons on the aortic baroreceptor reflex.

II. Anatomical Organization of Afferent Projections to the NTS Complex from Neurons Within the VLM

The organization of afferent projections to the NTS complex from neurons located throughout the rostrocaudal extent of the VLM was investigated in the male Wistar rat. PHA-L injections were made within the caudal VLM that overlapped the A_1-noradrenergic neurons, within the rostral VLM that overlapped the C1 adrenergic neurons, and within the area between both the A_1-and C1 cells groups, which is known to contain a mixture of both A_1-noradrenergic and C1 adrenergic neurons (Fig. 1). This latter area will be referred to as the intermediate VLM in this chapter.

PHA-L injections within the caudal VLM (Fig. 2) resulted in a dense innervation of the commissural and medial subnuclei of the NTS caudal to area postrema (Fig. 3) bilaterally, but with an ipsilateral predominance. At the level of area postrema, the commissural and medial subnuclei of the NTS had very few PHA-L–labeled fibers. A moderate number of labeled fibers and presumptive terminal boutons were observed throughout the ventral and to a smaller extent the ventrolateral subnuclei of the NTS. The region of the lateral subnucleus of the NTS lying lateral and dorsal to the solitary tract received very few projections from caudal VLM neurons. At more rostral levels of the NTS (Fig. 2), fibers were also observed to be scattered medial to the solitary tract in the region in and around the central and medial subnuclei of the NTS. A few fibers and presumptive terminal boutons containing both PHA-L and TH and/or DBH immunoreactivity were observed in the ventral aspects of the commissural

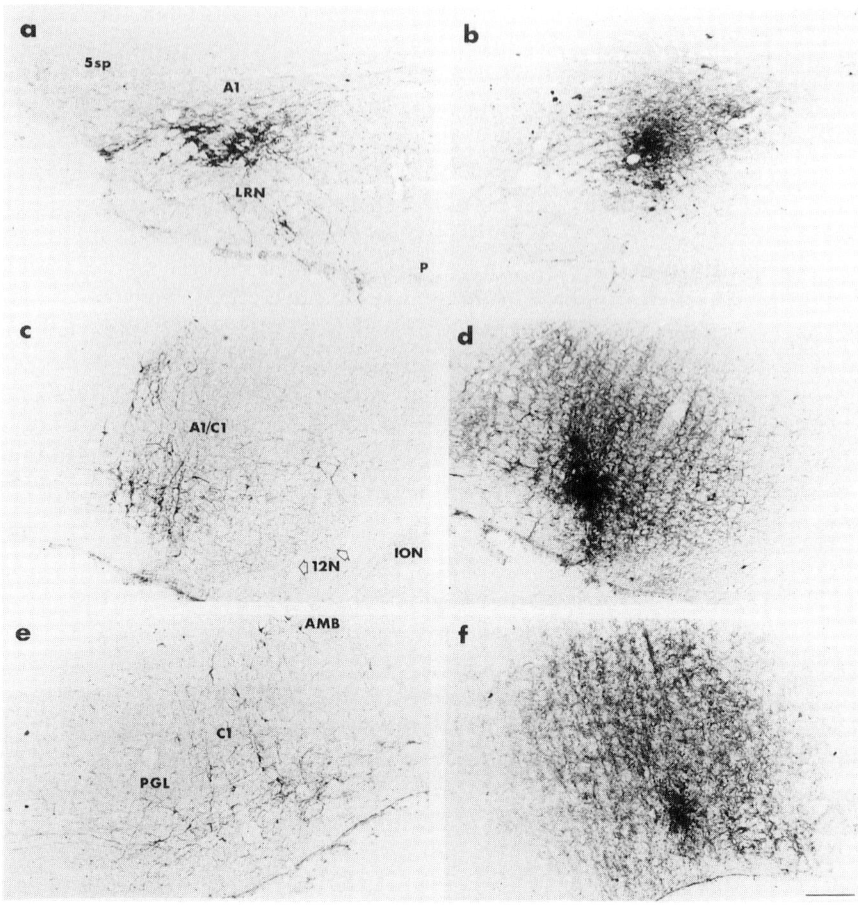

Figure 1 Bright-field photomicrographs of transverse sections of the medulla through the region of the VLM showing TH immunoreactive cells and PHA-L injection sites in the caudal (a) and intermediate (c) VLM and PNMT immunoreactive neurons (e) and a PHA-L injection (f) in the rostral VLM (f). Sections a, c, and e are adjacent to those shown in sections b, d, and f, respectively. Note that the PHA-L injection sites overlap the A_1-noradrenergic cell group (a and b), the A_1-/C1 catecholaminergic cell group (c and d) and the C1 adrenergic cell group (e and f). The calibration mark in f represents 200 μm and applies to all photomicrographs. Abbreviations: A1, A1 noradrenergic cell group; C1, C1 adrenergic cell group; AMB, nucleus ambiguus; ION, inferior olivary nucleus; LRN, lateral reticular nucleus; P, pyramidal tract, PGL, paragigantocellularis lateralis nucleus; 5sp, spinal trigeminal nucleus; 12N, hypoglossal nerve.

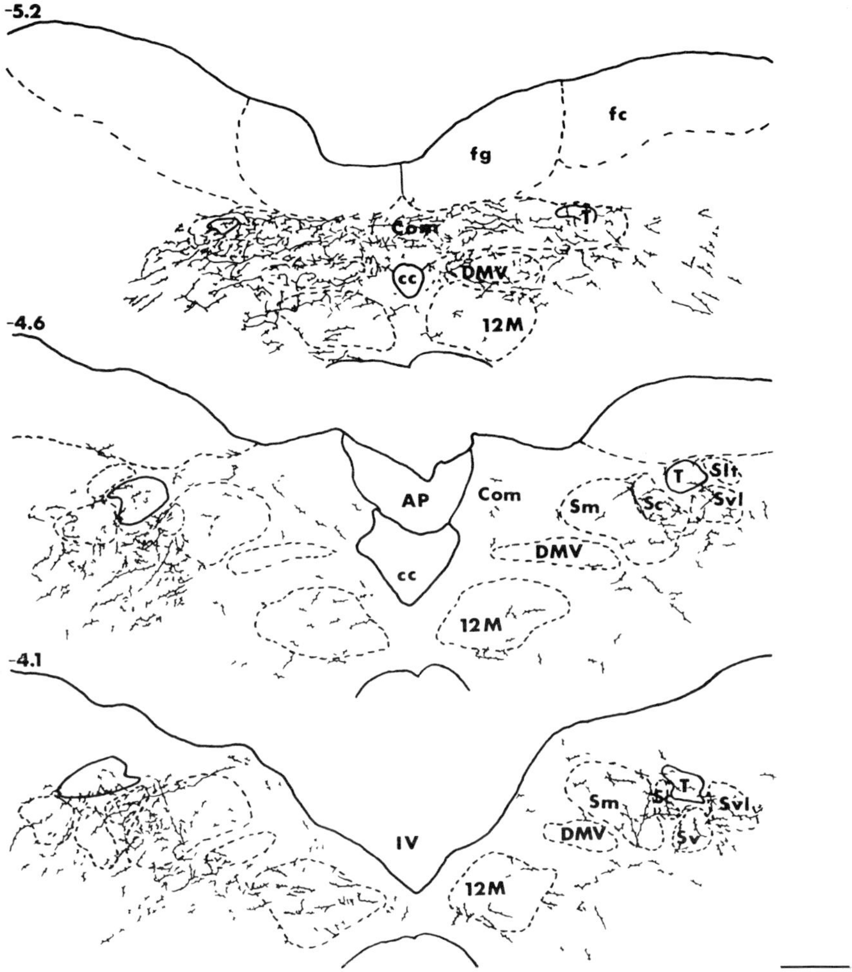

Figure 2 Series of camera lucida drawings of the rat dorsomedial medulla showing the distribution of labeled fibers and presumptive terminal boutons in the NTS complex resulting from a PHA-L injection into the caudal VLM. Calibration line represents 0.2 mm. Abbreviations: AP, area postrema; cc, central canal; Com, commissural subnucleus of nucleus of the solitary tract; DMV, dorsal motor nucleus of the vagus; fc, cuneate fasciculus; fg, gracilis fasciculus; Sm, medial subnucleus of the solitary tract; Slt, lateral subnucleus of nucleus of the solitary tract; Sv, ventral subnucleus of nucleus of the solitary tract; Svl, ventrolateral subnucleus of nucleus of the solitary tract; T, tractus solitarius; 12 M, hypoglossal nucleus; IV, fourth ventricle.

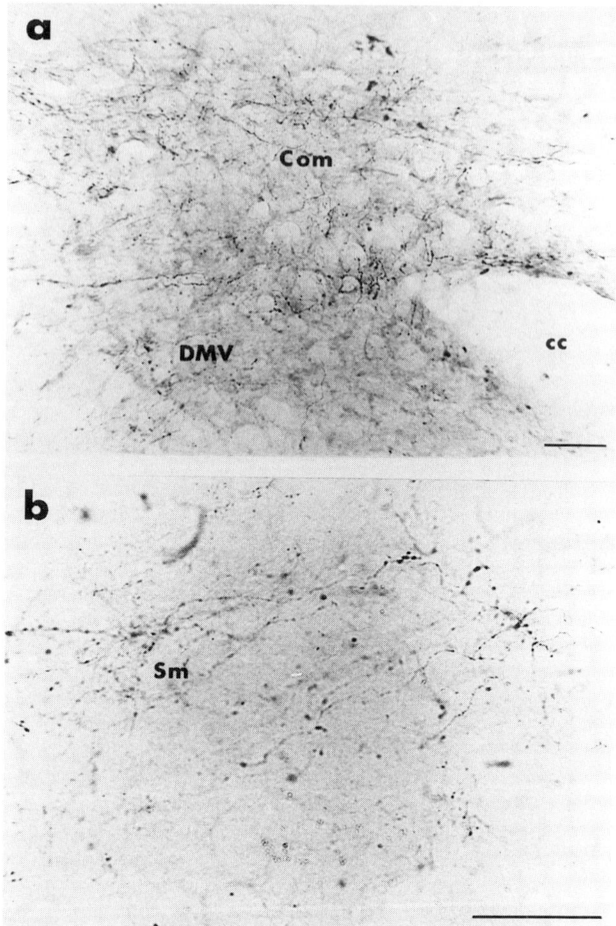

Figure 3 Bright-field photomicrographs of representative transverse sections through the NTS complex of the rat showing PHA-L labeled fibers and terminal boutons within the commissural subnucleus (Com, a) and the medial subnucleus of the nucleus (Sm; b) of the solitary tract resulting from a PHA-L injection site in the caudal VLM. The calibration marks represent 50 μm. Abbreviations: cc, central canal; Com, commissural subnucleus of nucleus of the solitary tract; DMV, dorsal motor nucleus of the vagus; sm, medial subnucleus of nucleus of the solitary tract.

nucleus (Fig. 4a and b). No other subnucleus in the NTS complex was found to contain fibers labeled for both PHA-L and TH and/or DBH.

PHA-L injections within the intermediate region of the VLM resulted in a different projection pattern in the NTS complex than that arising from the caudal VLM (Fig. 5). The commissural nucleus received few projections from

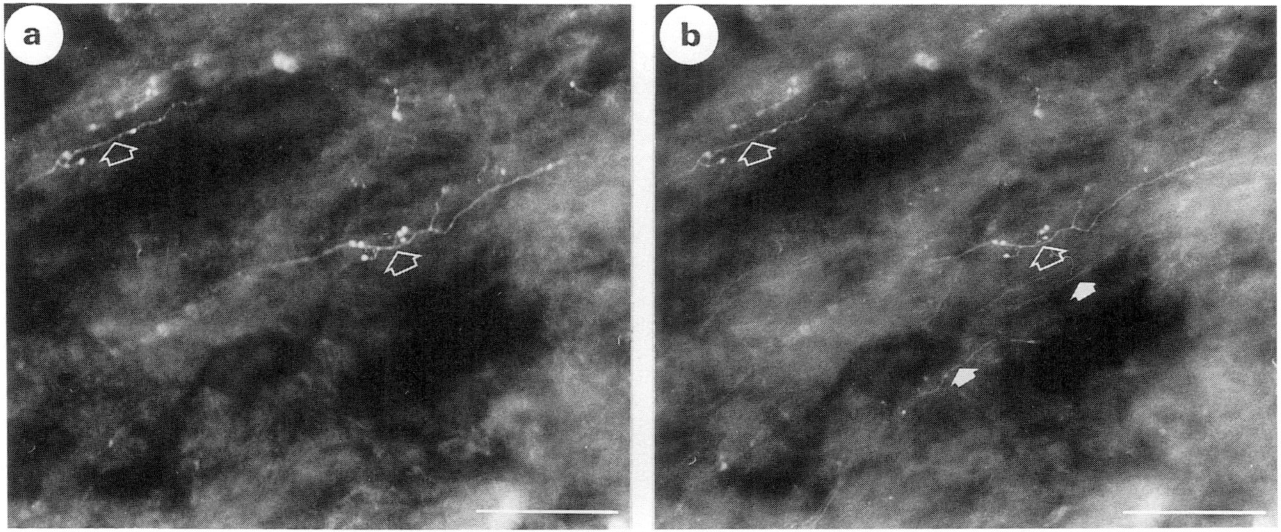

Figure 4 Photomicrographs of Texas red (a) and fluorescin (b) labeled PHA-L fiber and terminal labeling in the commissural subnucleus of nucleus of the solitary tract (A), which also contained tyrosine hydroxylase immunoreactivity (TH, B). Note that PHA-L–labeled fibers and terminal boutons (a, open arrows) were also labeled for TH (b, open arrows). Solid arrows in b point to TH-labeled fibers not immunoreactive to PHA-L (compare to a). Photomicrographs a and b were taken from the same section but under different fluorescent illuminations. Calibration mark in all photomicrographs represent 50 μm.

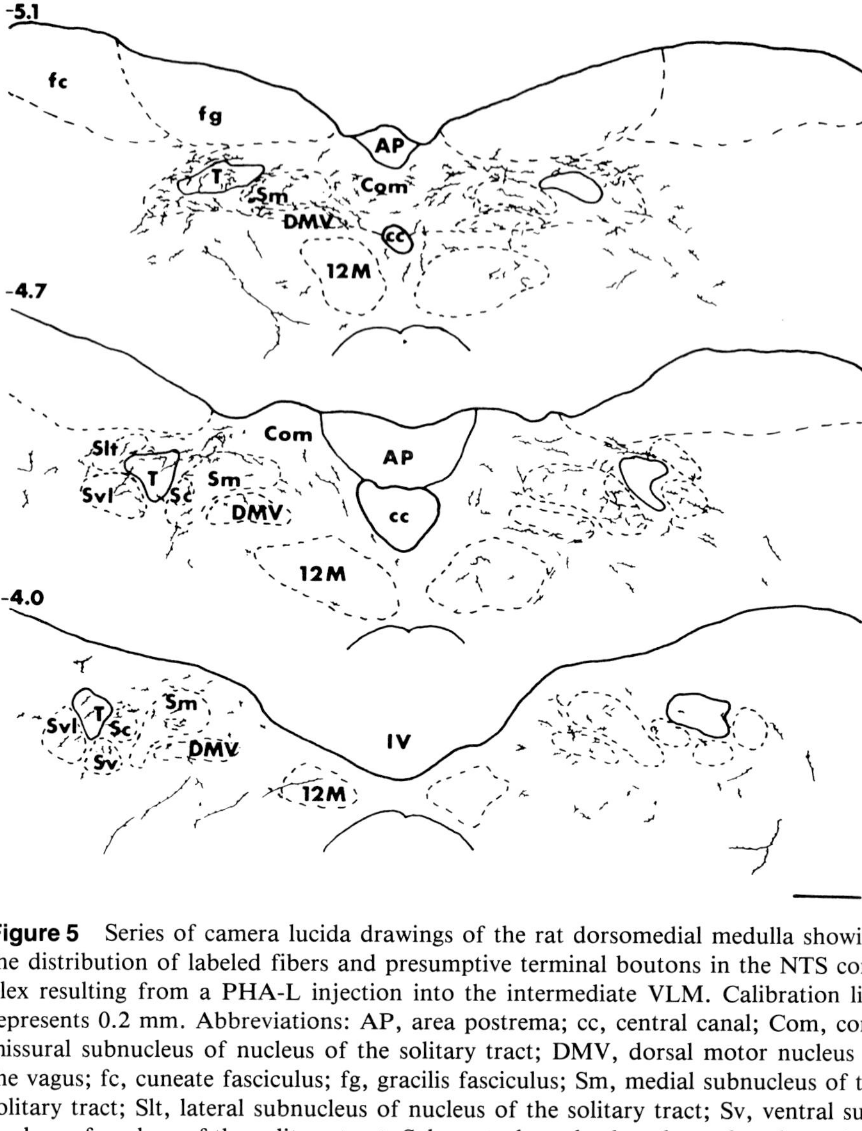

Figure 5 Series of camera lucida drawings of the rat dorsomedial medulla showing the distribution of labeled fibers and presumptive terminal boutons in the NTS complex resulting from a PHA-L injection into the intermediate VLM. Calibration line represents 0.2 mm. Abbreviations: AP, area postrema; cc, central canal; Com, commissural subnucleus of nucleus of the solitary tract; DMV, dorsal motor nucleus of the vagus; fc, cuneate fasciculus; fg, gracilis fasciculus; Sm, medial subnucleus of the solitary tract; Slt, lateral subnucleus of nucleus of the solitary tract; Sv, ventral subnucleus of nucleus of the solitary tract; Svl, ventrolateral subnucleus of nucleus of the solitary tract; T, tractus solitarius; 12 M, hypoglossal nucleus; IV, fourth ventricle.

VLM neurons in the intermediate area bilaterally. In addition, a few scattered fibers were observed around the perimeter of the medial subnucleus and encircling the central subnucleus of the NTS. Compared to projections from the caudal VLM, a denser innervation was observed to the lateral, ventral, and ventrolateral subnuclei of the caudal NTS. A small number of fibers and pre-

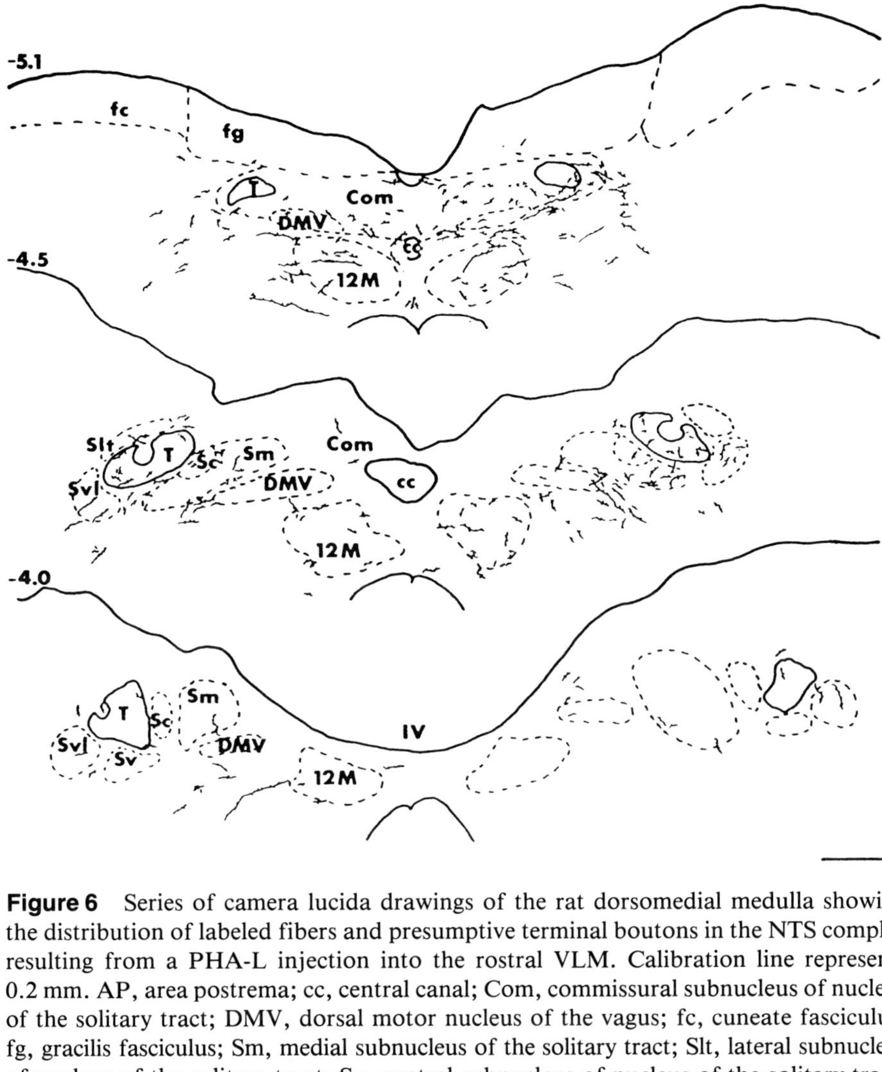

Figure 6 Series of camera lucida drawings of the rat dorsomedial medulla showing the distribution of labeled fibers and presumptive terminal boutons in the NTS complex resulting from a PHA-L injection into the rostral VLM. Calibration line represents 0.2 mm. AP, area postrema; cc, central canal; Com, commissural subnucleus of nucleus of the solitary tract; DMV, dorsal motor nucleus of the vagus; fc, cuneate fasciculus; fg, gracilis fasciculus; Sm, medial subnucleus of the solitary tract; Slt, lateral subnucleus of nucleus of the solitary tract; Sv, ventral subnucleus of nucleus of the solitary tract; Svl, ventrolateral subnucleus of nucleus of the solitary tract; T, tractus solitarius; 12 M, hypoglossal nucleus; IV, fourth ventricle.

sumptive terminal boutons in the commissural and medial subnuclei of the NTS were double labeled for PHA-L and TH and/or DBH.

PHA-L injections within the rostral VLM (Fig. 6) resulted in the least number of PHA-L labeled fibers with terminal boutons in the NTS complex compared to that after PHA-L injections in either the caudal or rostral VLM.

Few labeled fibers with terminal boutons were observed in the caudal portions of the NTS complex, along the ventral border of the commissural nucleus. In addition, few fibers with terminal boutons were scattered lateral and medial to the solitary tract at approximately the level of the obex. The lateral, ventral, and ventrolateral subnuclei of the NTS were also sparsely innervated by rostral VLM neurons. Rostral to the obex, labeled fibers were rarely observed within the NTS complex. None of the PHA-L injections that overlapped the C1 adrenergic cell group resulted in PHA-L and PNMT double-labeled fibers in the NTS complex.

III. VLM Innervation of the Catecholaminergic Neurons Within the NTS Complex

In the present study, PHA-L injections overlapping either the caudal or intermediate VLM resulted in labeled fibers with presumptive terminal boutons within both the commissural and medial subnuclei of the NTS complex. Previous anatomical studies have demonstrated that the commissural and medial subnuclei of the NTS contain noradrenergic neurons of the A_2 cell group (47, 50,51). In the present study, a few PHA-L labeled fibers with terminal boutons were observed to intermingle with TH immunoreactive neurons within the NTS at the level of the commissural subnucleus following injections of PHA-L into either the caudal or intermediate VLM (Fig. 7). At the level of the area postrema, very few PHA-L–labeled fibers with terminals were observed to intermingle with the A_2-neurons. The majority of PHA-L–labeled fibers were observed to terminate within the NTS region just lateral and dorsal to the A_2-noradrenergic cell group. PHA-L–labeled fibers with presumptive terminals were also observed rostrally within the NTS complex at the level of the C2 adrenergic neurons, although, very few of these PHA-L–labeled terminals were seen to be in close proximity to the C2 neurons. No fibers or presumptive terminal boutons immunoreactive to both PHA-L or TH and/or DBH were observed in close apposition to catecholaminergic neurons in the NTS complex.

IV. Electrophysiological Identification of Neurons Within the NTS Complex Receiving Inputs from VLM and Aortic Baroreceptors

The anatomical data presented above have provided evidence that neurons within the caudal and intermediate VLM have the greatest projection to subnuclei within the NTS complex that have previously been shown to receive inputs from baroreceptor afferent fibers carried in the aortic depressor nerve (ADN) in the rat (30–33). This observation suggests the possibility that the VLM may alter aortic baroreceptor reflex excitability at the level of the NTS.

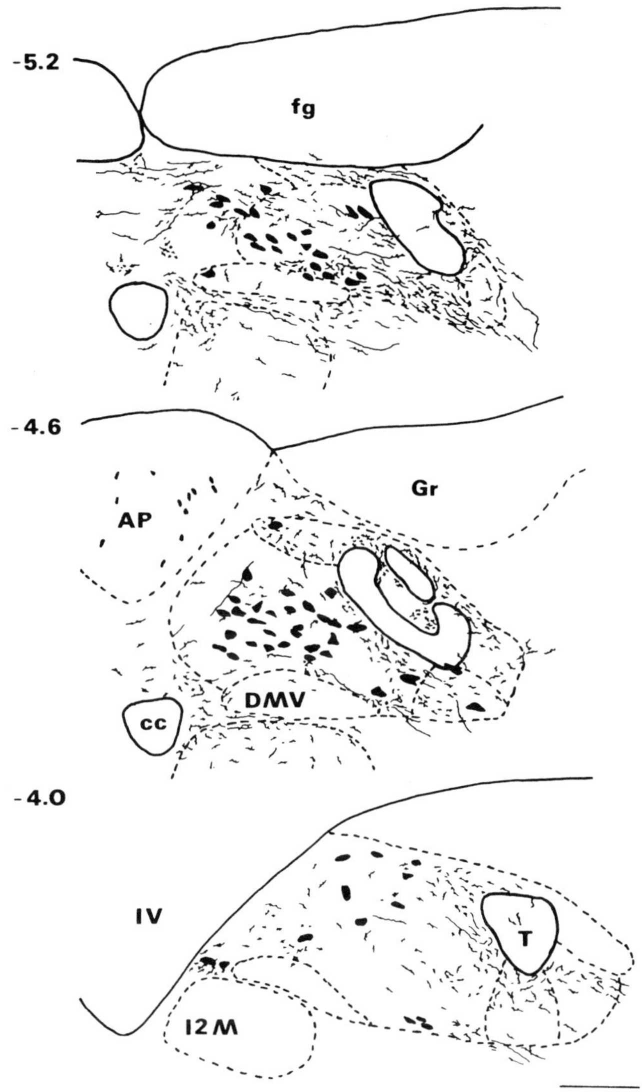

Figure 7 Series of projection drawings of the rat dorsomedial medulla showing the distribution of labeled fibers and presumptive terminal boutons resulting from a PHA-L injection that extended between the caudal and intermediate VLM in relation to the catecholaminergic neurons in the NTS complex. Calibration line represents 0.2 mm. Abbreviations: AP, area postrema; cc, central canal; DMV, dorsal motor nucleus of the vagus; fc, cuneate fasciculus; fg, gracilis fasciculus; T, tractus solitarius; 12 M, hypoglossal nucleus; IV, fourth ventricle.

To test this possibility, electrophysiological experiments were done in alpha-chloralose-anesthetized male Wistar rats to determine the effect of VLM stimulation on the response of single units in NTS to ADN stimulation (57).

Of 178 single units recorded extracellularly from histologically verified sites in the NTS region, 28% (49/178) were identified as responding orthodromically to ADN stimulation and 37% (65/178) as responding orthodromi-

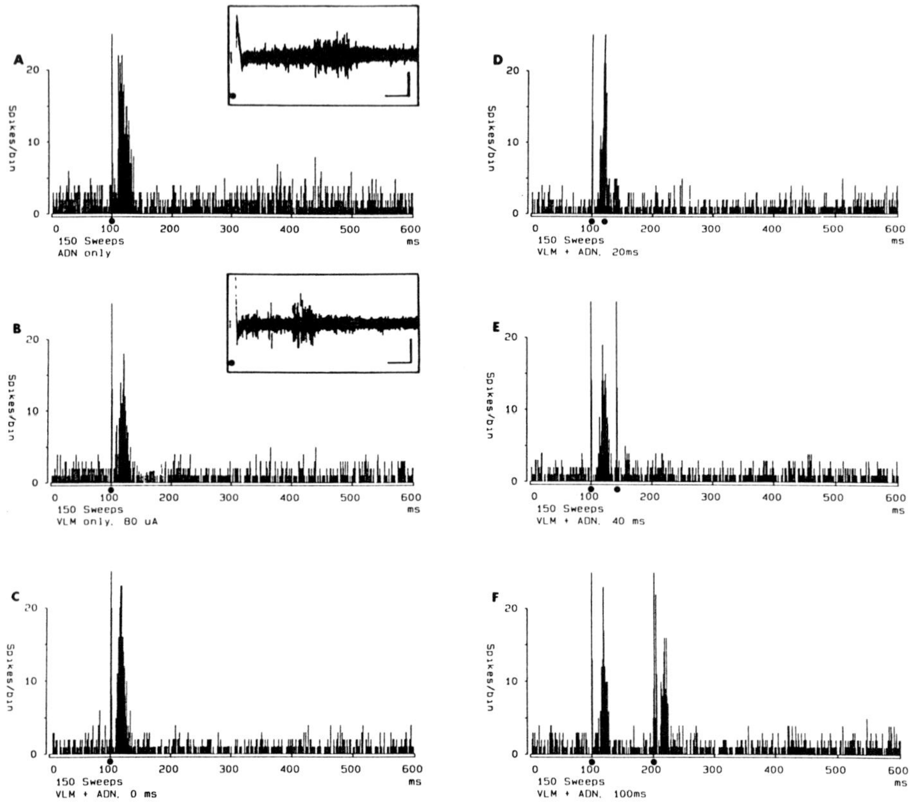

Figure 8 Peristimulus time histograms (1 ms/bin) showing the effect of applying conditioning stimuli to VLM on the firing frequency of NTS single units responding to stimulation of ADN (panels C through F). Note that the single unit increased its firing frequency during both ADN (panel A) and VLM (panel B) stimulation alone. Insets are five superimposed oscilloscope tracings of orthodromic responses of the unit to ADN (panel A) and VLM (panel B) stimulation. Calibrations marks represent 5 ms and 50 μV. Note that a conditioning stimulus applied to VLM 20 to 40 ms prior to ADN stimulation (panels D through E) reduces the response of the unit to ADN stimulation. The response to ADN stimulation begins to return to control levels approximately 100 ms after the application of the VLM stimulus. (From Ref. 57.)

cally to VLM stimulation. Twenty-nine units within the NTS responding to both ADN (mean latency, 23.9 ± 4.5 ms) and VLM (mean latency, 21.5 ± 4.4 ms) stimulation were further tested for their response to ADN stimulation in the presence of a conditioning stimulus to the VLM. Of the 29 units, 13 (45%) showed a decrease in the ADN response of approximately 60% when a conditioning stimulus to the VLM preceded the ADN test stimulus by approximately 20 to 40 ms (Fig. 8). The response began to recover by approximately 100 ms.

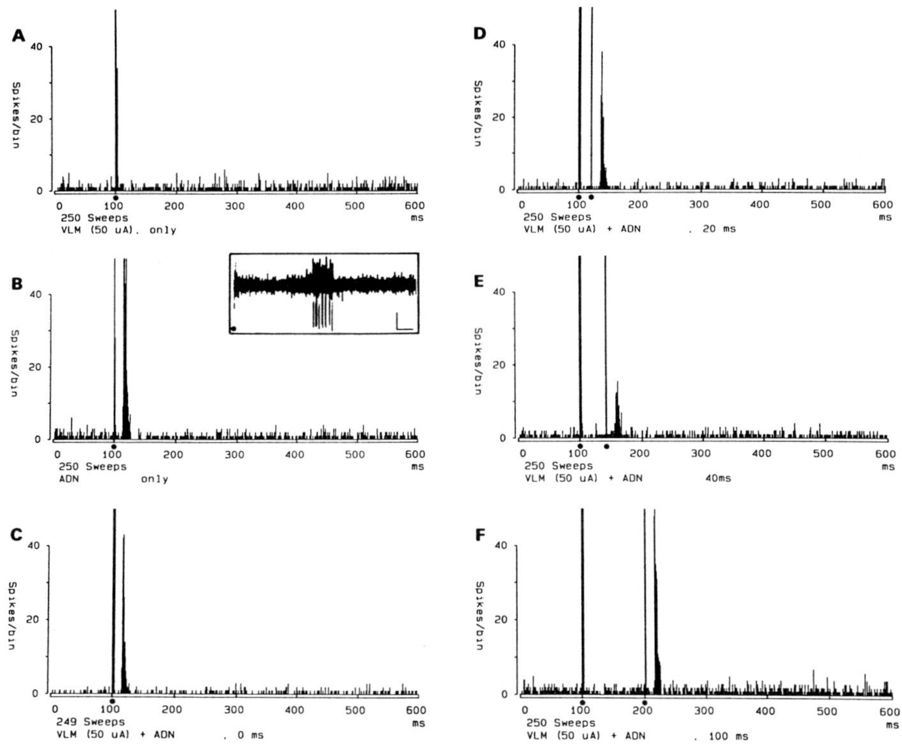

Figure 9 Peristimulus time histograms (1 ms/bin) showing the effect of conditioning stimuli to VLM on the firing frequency of an NTS single unit responding to stimulation of ADN. Note that VLM stimulation alone did not elicit a change in the firing frequency of the unit in panel A, whereas ADN stimulation excited the unit in panel B. The unit's response to ADN stimulation is reduced to less than half the original response when the conditioning stimulus applied to VLM precedes the ADN stimulation by approximately 40 ms (panel E). The response to ADN stimulation returns to control levels when the test stimulus to ADN is approximately 100 ms after the conditioning stimulus to VLM (panel F). The inset in panel B represents five superimposed oscilloscope tracings of the orthodromic response of the unit to ADN stimulation. Calibration is 100 ms and 50 μV. Stimulus was delivered at the dot. (From Ref. 57.)

An additional 13 units in the NTS that responded with excitation to stimulation of the ADN but did not alter their firing rate during stimulation of the VLM were also tested for interactions. Following a conditioning stimulus to the VLM 0 to 40 ms prior to stimulating the ADN, the ADN response in 9 of these units was reduced by approximately 40% (Fig. 9). The majority of the units that showed an interaction between the VLM and the ADN were located in and around the lateral and medial subnucleus of the NTS complex.

V. Effect of VLM Stimulation on the Vagal Component of the Aortic Baroreceptor Reflex

Glutamate stimulation (5 to 20 nL; 1 M) of 23 discrete sites in the VLM region extending from approximately the caudal pole of the lateral reticular nucleus to the nucleus paragigantocellularis lateralis failed to elicit changes in heart rate in the chloralose-anesthetized, spinal-transected, carotid sinus deafferented, paralyzed, and artificially ventilated animals. On the other hand, as shown in Figure 10A, activation of these VLM sites simultaneously with the reflex stimulation of the aortic baroreceptors after a bolus injection of phenylephrine significantly attenuated the reflex vagal bradycardia to activation of the aortic baroreceptors. The reflex vagal bradycardia during baroreceptor stimulation was attenuated by approximately 55% during the activation of VLM neurons.

In contrast, glutamate stimulation of sites ($n = 8$) in and around the external component of the nucleus ambiguus (AMB; Fig. 10B) elicited decreases in heart rate. Activation of these AMB sites during stimulation of the aortic baroreceptor reflex resulted in a decrease in heart rate that was not significantly greater than the reflex decrease to activation of aortic baroreceptors alone or stimulation of the AMB alone. However, the decrease in heart rate during the combined activation of the AMB and the aortic baroreceptor reflex was significantly smaller than the algebraic summation of the two individual responses, suggesting that the same neurons had been activated by both the glutamate injection and the reflex activation of aortic baroreceptors. The magnitude of the pressor response after the intravenous bolus injection of phenylephrine was not different during the interaction studies compared to control responses.

VI. General Discussion and Summary

The distribution of fibers and terminal labeling in the NTS complex resulting from PHA-L injections throughout the rostrocaudal extent of the VLM is in general agreement with earlier studies (34–36,40). However, several notable differences exist among these studies. The autoradiographic studies of Loewy et al. (35) and Sawchenko and Swanson (40) showed dense bilateral projections to the commissural and medial subnuclei of the NTS complex from a region corresponding in this study to the intermediate VLM. In contrast, a predominant

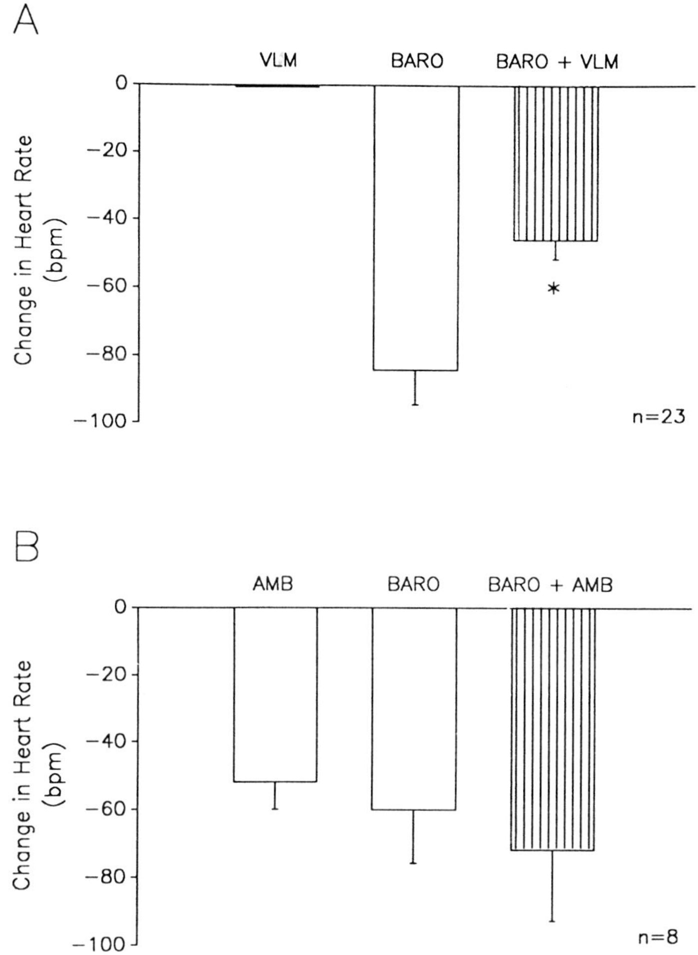

Figure 10 Histogram showing the effect of activating ventrolateral medulla (VLM; A) or nucleus ambiguus (AMB; B) neurons on the reflex vagal bradycardia to stimulation of aortic baroreceptor (BARO). Abbreviations: n, number of sites interacted; *, significantly different ($P<0.01$) from response during activation of aortic baroreceptors only.

projection was observed in this study to the ventral and ventrolateral subnuclei of the NTS and a less dense projection to the lateral subnucleus from the intermediate VLM. Few fibers were observed to the commissural and medial subnuclei. With regard to the PHA-L study done by Guyenet and Young (34), they observed dense projections to the commissural and ventrolateral subnuclei of the NTS from the rostral VLM. In this study, in agreement with the observation of McKellar and Loewy (36), a more diffuse labeling of the NTS complex was observed at the level of the obex. This labeling was mainly in the ventral, ventrolateral, lateral, and intermediate subnuclei of the NTS. Although the reasons for these discrepancies are not readily apparent, they may involve the possibility of selective uptake of the tract tracers by certain populations of VLM neurons, or the site of injection in the earlier studies may have been underestimated and have incorporated the adjacent regions of the VLM, as described in this report.

It was also observed in this study that most of the projections to the NTS complex were from non-catecholaminergic neurons in the VLM and only a small number of catecholaminergic fibers were observed in the caudal NTS after a PHA-L injection overlapping the A_1-noradrenergic cell group. Similarly, Thor and Helke (39) reported that approximately one-third of all retrogradely labeled neurons in the VLM after an injection into the NTS complex were catecholaminergic. Taken together, these data suggest that the NTS receives a moderate projection from the A_1-noradrenergic–containing cells of the caudal and intermediate VLM. However, these projections do not appear to innervate catecholaminergic cells within the NTS complex. Other transmitters contained in the pathways from the VLM to the NTS have not been identified. However, they may contain neuropeptide Y, substance P, enkephalin, serotonin, somatostatin, and/or neurotensin (27,58–60).

The data obtained in this study suggest that one of the functions of the VLM pathway to the NTS complex may be to alter the excitability of neurons receiving baroreceptor inputs. It was demonstrated that discrete activation of VLM neurons attenuated the vagal reflex bradycardia to activation of aortic baroreceptors. This effect may be the result of inhibiting NTS neurons. The excitatory response of NTS neurons to aortic nerve stimulation was found to be attenuated regardless of whether the NTS neurons were excited or not during VLM stimulation. Although the mechanisms involved in the inhibitory interactions are not known, it is likely that they involve both post- and presynaptic mechanisms (61). These data may be interpreted to suggest that during certain physiological conditions when a specific level of arterial pressure and heart rate is required, as during exercise (13), the VLM functions to alter the sensitivity of the baroreceptor reflex at the level of the NTS.

In summary, the anatomical data presented in this chapter have provided evidence to support the suggestion that different neuronal pools in the VLM project directly to specific cardiovascular areas of the NTS complex. This includes only a limited projection to the A_2-region within the NTS from the caudal and intermediate VLM. A small component of the VLM projection to the NTS complex was found to be catecholaminergic and to arise from the A_1-noradrenergic cell group. The electrophysiological data presented suggest that the projections from VLM to NTS may function to alter the vagal heart rate component of the baroreceptor reflex by inhibiting NTS neurons that receive aortic baroreceptor inputs.

Acknowledgments

This work was supported by the Heart and Stroke Foundation of Ontario. J. Ciriello is a Heart and Stroke Foundation of Ontario Career Investigator. S. Roder is the holder of an Ontario Graduate Scholarship.

References

1. Ciriello J, Caverson MM, Polosa C. Function of the ventrolateral medulla in the control of the circulation. Brain Res Rev 1986; 11:359–391.
2. Dampney RAL, Goodchild AK, Robertson LG, Montgomery W. Role of ventrolateral medulla in vasomotor regulation: A correlative anatomical and physiological study. Brain Res 1982; 249:233–235.
3. Ross CA, Ruggiero DA, Park DH, et al. Tonic vasomotor control by the rostral ventrolateral medulla: Effect of electrical or chemical stimulation of the area containing C1 adrenaline neurons on arterial pressure, heart rate, and plasma catecholamines and vasopressin. J Neurosci 1984; 4:474–494.
4. Willette RN, Barcas PP, Krieger AJ, Sapru HN. Vasopressor and depressor areas in the rat medulla: Identification by microinjection of L-glutamate. Neuropharmacology 1983; 22:1071–1079.
5. Caverson MM, Ciriello J, Calaresu FR. Chemoreceptor and baroreceptor inputs to ventrolateral medullary neurons. Am J Physiol 1984; 247:R872–R879.
6. Ciriello J, Rohlicek DV, Polosa C. Aortic baroreceptor reflex pathway: A functional mapping using [3H]2-deoxyglucose autoradiography in the rat. J Auton Nerv Syst 1983; 8:11–128.
7. Granata AR, Ruggiero DA, Park DH, et al. Lesions of epinephrine neurons in the rostral ventrolateral medulla abolish the vasodepressor components of baroreflex and cardiopulmonary reflex. Hypertension 1983; 5:V80–V84.
8. Kumada M, Terui N, Kuwaki T. Arterial baroreceptor reflex: Its central and peripheral neural mechanisms. Prog Neurobiol 1990; 35:331–361.
9. McAllen RM, Neil JJ, Loewy AD. Effects of kainic acid applied to the ventral surface of the medulla oblongata on vasomotor tone, the baroreceptor reflex and hypothalamic autonomic responses. Brain Res 1982; 238:65–76.
10. Murugaian J, Sundaram K, Krieger A, Sapru H. Electrolytic lesions in the depressor area of the ventrolateral medulla of the rat abolish depressor responses to the aortic nerve stimulation. Brain Res 1989; 499:371–377.
11. Yamada KA, McAllen RM, Loewy AD. GABA antagonists applied to the ventral surface of the medulla oblongata block the baroreceptor reflex. Brain Res 1984; 297:175–180.
12. Bauer RM, Iwamoto GA, Waldrop TG. Ventrolateral medullary neurons modulate pressor reflex to muscular contraction. Am J Physiol 1989; 257:R1154–1161.
13. Ciriello J, Calaresu FR. Lateral reticular nucleus: A site of somatic and cardiovascular integration in the cat. Am J Physiol 1977; 233:R100–109.
14. Damney RAL, Moon EA. Role of ventrolateral medulla in vasomotor response to cerebral ischemia. Am J Physiol 1980; 239:H349–H358.
15. McAllen RM. Mediation of the fastigial pressor response and a somatosympathetic reflex by ventral medullary neurones in the cat. J Physiol (Lond) 1985; 368:423–433.

16. Rohlicek CV, Polosa C. Mediation of pressor responses to cerebral ischemia by superficial ventral medullary areas. Am J Physiol 1983; 245:H962–H968.
17. Stornetta R, Morrison SL, Ruggiero DA, Reis DJ. Neurons in the rostral ventrolateral medulla mediate the somatic pressor reflex. Am J Physiol 1989; 256:R448–R462.
18. Blessing WW, Sved AF, Reis DJ. Destruction of noradrenergic neurons in rabbit brainstem elevates plasma vasopressin, causing hypertension. Science 1982; 217: 661–663.
19. Blessing WW, Willoughby JO. Inhibiting the rabbit caudal ventrolateral medulla prevents baroreceptor-initiated secretion of vasopressin. J Physiol 1982; 367:253–265.
20. Day TA, Ferguson AV, Renaud LP. Facilitatory influence of noradrenergic afferents on the excitability of rat paraventricular nucleus neurosecretory cells. J Physiol 1984; 355:237–249.
21. Day TA, Renaud LP. Electrophysiological evidence that noradrenergic afferents selectively facilitate the activity of supraoptic vasopressin neurons. Brain Res 1984; 303:233–240.
22. Ciriello J, Caverson MM, Park DH. Immunohistochemical identification of noradrenaline- and adrenaline-synthesizing neurons in the cat ventrolateral medulla. J Comp Neurol 1986; 253:216–230.
23. Chalmers JP. Brain amines and models of experimental hypertension. Circ Res 1975; 36:469–480.
24. Granata AR, Kumada M, Reis DJ. Sympathoinhibition by A_1-noradrenergic neurons is mediated by neurons in the C1 area of the rostral medulla. J Autonom Nerv Syst 1985; 14:387–395.
25. Granata AR, Ruggiero DA, Park DH, et al. Brain stem area with C1 epinephrine neurons mediates baroreflex vasodepressor responses. Am J Physiol 1985; 248: H547–H567.
26. Reis DJ, Granata AR, Joh TH, et al. Brain stem catecholamine mechanisms in tonic and reflex control of blood pressure. Hypertension 1984; 6:II-7–II-15.
27. Ruggiero DA, Cravo S, Arango V, Reis DJ. Central control of the circulation by the rostral ventrolateral reticular nucleus: Anatomical substrates. Prog Brain Res 1989; 81:49–79.
28. West MJ, Blessing WW, Chalmers J. Arterial baroreceptor reflex function in the conscious rabbit after brainstem lesions coinciding with the A1 group of catecholamine neurons. Circ Res 1981; 49:959–970.
29. Ciriello J, Caverson MM. Bidirectional cardiovascular connections between ventrolateral medulla and nucleus of the solitary tract. Brain Res 1986; 367:273–281.
30. Ciriello J. Brainstem projections of aortic baroreceptor afferent fibers in the rat. Neurosci Lett 1983; 36:37–42.
31. Ciriello J, Calaresu FR. Projections from buffer nerves to the nucleus of the solitary tract: An anatomical and electrophysiological study in the cat. J Auton Nerv Syst 1981; 3:299–310.
32. Ciriello J, Hrycyshyn AW, Calaresu FR. Horseradish peroxidase study of brain stem projections of carotid sinus and aortic depressor nerves in the cat. J Auton Nerv Syst 1981; 4:43–61.
33. Ciriello J, Hochstenbach SL, Roder S. Central projections of baroreceptor and chemoreceptor afferent fibers in the rat. In: Barraco IRA, ed. Nucleus of the Solitary Tract. Boca Raton, FL: CRC Press, 1994:35–50.

34. Guyenet PG, Young BS. Projections of nucleus paragigantocellularis lateralis to locus coeruleus and other structures in rat. Brain Res 1987; 406:171–184.
35. Loewy AD, Wallach JH, McKellar S. Efferent connections of the ventral medulla oblongata in the rat. Brain Res Rev 1981; 3:63–80.
36. McKellar S, Loewy AD. Efferent projections to the A_1-catecholamine cell group in the rat: An autoradiographic study. Brain Res 1982; 241:11–29.
37. Roder S, Ciriello J. Caudal ventrolateral medullary projections to the nucleus of the solitary tract in the cat. Neurosci Lett 1992; 134:161–164.
38. Ross CA, Ruggiero DA, Reis DJ. Projections from the nucleus tractus solitarii to the rostral ventrolateral medulla. J Comp Neurol 1985; 242:511–534.
39. Thor KB, Helke CJ. Catecholamine-synthesizing neuronal projections to the nucleus tractus solitarii of the rat. J Comp Neurol 1988; 268:264–280.
40. Sawchenko PE, Swanson LW. The organization of noradrenergic pathways from the brain stem to the paraventricular and supraoptic nuclei in the rat. Brain Res Rev 1982; 4:275–325.
41. Cowan WM, Gottlieb DI, Hendrickson AE, et al. The autoradiographic demonstration of axonal connectiions in the central nervous system. Brain Res 1972; 37: 21–51.
42. DeJong W, Zandberg P, Wijnen H, et al. Catecholamines in the nucleus tractus solitarii (NTS) and blood pressure control. In: Meyer P, Schmitt H, eds. Nervous System and Hypertension. Paris, France: Wiley-Flammarion, 1978:165–172.
43. Kubo T, Goshima Y, Hata H, Misu Y. Evidence that endogenous catecholamines are involved in α_2-adrenoceptor-mediated modulation of the aortic baroreceptor reflex in the nucleus tractus solitarii of the rat. Brain Res 1990; 526:313–317.
44. Zandberg P, DeJong W, DeWied D. Effect of catecholamine-receptor stimulating agents on blood pressure after local application in the nucleus tractus solitarii of the medulla oblongata. Eur J Pharmacol 1979; 55:43–56.
45. Feldman PD, Felder RB. α2-Adrenergic modulation of synaptic excitability in the rat nucleus tractus solitarius. Brain Res 1989; 480:190–197.
46. Feldman PD, Moises HC. Adrenergic responses of baroreceptive cells in the nucleus tractus solitarii of the rat: A microiontophoretic study. Brain Res 1987; 420:351–361.
47. Armstrong DM, Ross CA, Pickel VM, et al. Distribution of dopamine-, noradrenaline-, and adrenaline-containing cell bodies in the rat medulla oblongata: Demonstrated by the immunocytochemical localization of catecholamine biosynthetic enzymes. J Comp Neurol 1982; 212:173–187.
48. Dahlström A, Fuxe K. Evidence for the existence of monoamine-containing neurons in the central nervous system: I. Demonstration of monoamine in the cell bodies of brain stem neurons. Acta Physiol Scand 1964; 232:1–55.
49. Hökfelt T, Fuxe K, Goldstein M, Johansson O. Immunohistochemical evidence for the existence of adrenaline neurons in the rat brain. Brain Res 1974; 66:235–251.
50. Kalia M, Fuxe K, Goldstein M. Rat medulla oblongata: II. Dopaminergic, noradrenergic (A_1- and A_2-) and adrenergic neurons, nerve fibers, and presumptive terminal processes. J Comp Neurol 1985; 233:308–332.
51. Kalia M, Fuxe K, Goldstein M. Rat medulla oblongata: III. Adrenergic (C1 and C2) neurons, nerve fibers and presumptive terminal processes. J Comp Neurol 1985; 233:333–349.

52. Swanson LW, Hartman BK. The central adrenergic system. An immunofluorescence study of the location of cell bodies and their efferent connections in the rat utilizing dopamine-β-hydroxylase as a marker. J Comp Neurol 1975; 163:467–506.
53. Reis DJ, Joh TH, Nathan MA, et al. Nucleus tractus solitarii: Catecholaminergic innervation in normal and abnormal control of arterial pressure. In: Meyer P, Schmitt H, eds. Nervous System and Hypertension. Paris, France: Wiley-Flammarion, 1978:147–164.
54. Pickel VM, Chan J, Milner TA. Ultrastructural basis for interactions between central opioids and catecholamines: II. Nuclei of the solitary tracts. J Neurosci 1989; 9:2519–2535.
55. Gerfen CR, Sawchenko PE. An anterograde neuroanatomical tracing method that shows the detailed morphology of neurons, their axons and terminals: Immunohistochemical localization of an axonally transported plant lectin, Phaseolus vulgaris leucoagglutinin (PHA-L). Brain Res 1984; 290:219–238.
56. Wouterlood FG, Groenewegen HJ. Neuroanatomical tracing by use of Phaseolus vulgaris-leucoagglutinin (PHA-L): Electron microscopy of PHA-L-filled neuronal somata, dendrites, axons and axon terminals. Brain Res 1985; 326:188–191.
57. Roder S, Ciriello J. Convergence of ventrolateral medullary and aortic baroreceptor inputs in nucleus of the solitary tract. Can J Physiol Pharmacol 1993; 71:365–373.
58. Ciriello J, Caverson MM. Relation of enkephalin-like immunoreactive neurons to other neuropeptide and monoamine-containing neurons in the ventrolateral medulla. Prog Brain Res 1989; 81:3–15.
59. Ciriello J, Caverson MM, Calaresu FR, Krukoff TL. Neuropeptide and serotonin immunoreactive neurons in the cat ventrolateral medulla. Brain Res 1988; 440: 53–66.
60. Millhorn DE, Seroogy K, Hökfelt T, et al. Neurons of the ventral medulla oblongata that contain both somatostatin and enkephalin immunoreactivities project to nucleus tractus solitarii and spinal cord. Brain Res 1987; 99–108.
61. Andresen MC, Kunze DL. Nucleus tractus solitarius: Gateway to neural circulatory control. Annu Rev Physiol 1994; 56:93–116.

60

Cardiovascular Regulation
Ventral Brainstem Mechanisms

ANTONIO R. GRANATA

University of Tennessee, Memphis
Memphis, Tennessee

I. Introduction

The ventrolateral medulla (VLM) is interconnected with several structures in the brain, forming a network that is thought to be critically involved in the control of the cardiovascular sympathetic outflow. Some of these VLM pathways are organized for reflex control of arterial pressure and heart rate (HR), whereas others are suitable for integrating different and more complex functions that also affect other systems in the brain on many occasions.

The purpose of this chapter is to analyze some central and peripheral pathways to the VLM that are involved in autonomic and cardiovascular regulation with an emphasis on areas in the brain that have been considered in this book. Therefore I will concentrate on the following brain areas: (1) area postrema, (2) caudal raphe nuclei, (3) the periaqueductal central gray, and (4) the bed nucleus stria terminalis.

II. Area Postrema

The area postrema (AP) is the medullary circumventricular organ generally believed to act as a chemosensitive trigger area for circulating substances able to induce vomiting. Nonetheless, the AP is also present in the rat, a nonemetic mammal. Hence the AP is also assumed to be involved in other physiological roles, including cardiovascular regulation (1,2). As a circumventricular organ, the AP is deprived of a normal blood-brain barrier (3). Consequently, it is suitable for conveying information triggered by different hormones in the blood. For instance, the central cardiovascular effect of angiotensin II is thought to be mediated by the AP (4,5), and neurons in this area can detect changes in levels of circulating vasopressin. Furthermore, neurons in the AP project to different areas in the brain known to be involved in cardiovascular regulation (6).

Anatomical studies have shown that the AP gets afferent inputs from the paraventricular hypothalamic nucleus, the lateral parabrachial nucleus in the pons, the mediocaudal nucleus tractus solitarii (6), and the vagus nerve (7). In addition, major projection locations of AP neurons are the mediocaudal nucleus tractus solitarii and the lateral parabrachial nucleus. Besides, AP neurons project to the rostral portion of the VLM in an area where the C1 adrenergic neurons are located (the C1 area in the rostral ventrolateral medulla, RVLM), to the dorsal motor nucleus of the vagus, and to the nucleus ambiguus (6). Blessing et al. (8) found that the small catecholaminergic-containing neurons (tyrosine hydroxylase immunoreactive) in the AP of the rabbit are part of those projection neurons to the RVLM. This observation has not been confirmed in the rat (9). Nevertheless, the results of these anatomical studies indicate that AP neurons are able to directly communicate with RVLM neurons. Thus an anatomical substrate is provided for those electrophysiological results discussed in the chapter on cardiovascular regulation by Bonham and Wilson, where electrical stimulation of the AP evoked an early excitation followed by an inhibitory response on barosensitive RVLM neurons with sympathetic and cardiac rhythm (see Chap. 54).

Other studies using a single-cell recording technique from RVLM neurons also support the involvement of this area in mediating the cardiovascular effects of AP neurons (10). Electrical stimulation of the AP either increases or decreases the discharge rate of RVLM neurons (10). The results of these electrophysiological studies are particularly important because the RVLM neurons modulated by AP inputs probably play a critical role in the tonic and reflex control of sympathetic nerve activity. Therefore, the effects on RVLM neurons evoked from the AP may be transduced in a similar effect on sympathetic preganglionic neurons and eventually on postganglionic sympathetic nerves.

The actions of AP neurons on cardiovascular regulation are well documented. Electrical stimulation of the AP inhibits renal sympathetic nerve activity and enhances the baroreflex sympathoinhibitory response (11). In addition, the sympathoinhibitory effect elicited by AP stimulation is dependent on the presence of vagal and baroreceptor primary afferents into the medulla. Consequently, such sympathoinhibition is not observed in baroreceptor-denervated and vagotomized animals (12). Furthermore, chemical lesion of neurons in the AP decreases the potentiating effect of circulating vasopressin on the baroreflex sympathoinhibitory responses (13). Therefore evidence suggests that the AP potentiates the baroreflex sympathoinhibitory response. This change in the sensitivity of the baroreflex may occur at different levels of the arterial baroreceptor reflex arc. Studies performed in the rabbit demonstrated that AP neurons may increase the effectiveness of aortic depressor and vagus nerves inputs on nucleus tractus solitarii neurons (14). If these neurons in the nucleus tractus solitarii are part of the main baroreceptor reflex arc, then neurons in the RVLM that are normal components of the baroreflex pathway may also be indirectly affected by AP neurons.

Finally, since blood-borne angiotensin II reduces the baroreceptor reflex response (14), RVLM neurons involved in the baroreceptor reflex pathway may receive modulatory inputs from AP neurons that, in turn, are sensitive to changes in circulating concentrations of angiotensin II.

In summary, RVLM neurons are probably involved in mediating cardiovascular responses to stimulation of the AP. This could be a mechanism by which the AP detects the level of blood-borne vasoactive substances. In turn, these AP neurons could change or reset the activity of RVLM neurons involved in the control of circulation in response to changes in the circulating concentrations of those vasoactive substances.

III. Caudal Raphe Nuclei

Neurons in the RVLM are considered to be the main source of the tonic excitatory afferent inputs to the sympathetic preganglionic neurons in the intermediolateral cell column in the spinal cord. Ultimately, these tonic excitatory inputs are essential for maintaining normal values of arterial pressure and basal sympathetic nerve activity (for review, see Ref. 15). Other areas in the brainstem, including the caudal raphe (raphe pallidus and raphe obscurus) also innervate the intermediolateral cell column (16,17). However, a more recent investigation showed that those neurons in the raphe obscurus do not project to the intermediolateral cell column (18). Besides those spinal projections, neurons in the caudal raphe also project to other areas of the central nervous system. In particular, it is important to note those raphe projections to the RVLM where catecholaminergic neurons are located (19–21). Since both catecholaminergic and noncatecholaminergic RVLM neurons project to the intermediolateral cell column (22), neurons in the caudal raphe may evoke responses on sympathetic preganglionic neurons through a relay station in the RVLM. Electrical (23,24) or chemical (25,26) stimulation of the caudal raphe elicited pressor and depressor responses, depending on the site of stimulation in the medullary raphe. Moreover, electrical stimulation of the raphe modified sympathetic preganglionic neuron discharge (27) and sympathetic nerve activity (28,29). Finally, electrophysiological experiments demonstrated that the raphe pallidus and raphe obscurus contain neurons with spinal cord projections subserving both sympathoexcitatory and sympathoinhibitory functions (30,31). A third type of raphe neuron that likely contains serotonin and innervates the intermediolateral cell column also conveys sympathoexcitatory inputs, although these cells do not receive inputs from baroreceptors (32). Therefore, these experiments have supplied substantial information identifying basic mechanisms by which sympathetic activity can be modified by enhancing the activity of caudal raphe neurons. Nevertheless, the physiological role in cardiovascular regulation played by these raphe neurons is unknown.

The existence of a serotonin-containing pathway between the caudal raphe nuclei and the thoracic spinal cord has been demonstrated in different studies (18,33). Further, the increase in sympathetic activity elicited by stimulation of the nuclei raphe obscurus or pallidus is blocked by systemic injections of the serotonin antagonist methysergide (34). A recent study demonstrated that neurons in two cytoarchitectonically different areas of the RVLM innervate serotonin-immunoreactive neurons of the raphe obscurus and raphe pallidus (35). Therefore it is feasible that the control of cardiovascular function rendered by RVLM neurons may be accomplished by an additional indirect pathway involving serotonin-immunoreactive neurons of the medullary raphe (35). Still, it is unknown whether the bulbospinal RVLM neurons with barosensitive properties and characterized as sympathoexcitatory could send collaterals to the medullary raphe or whether, on the other hand, a different population of neurons is involved in this intramedullary pathway.

Some spinal cord projection neurons in the caudal raphe contain immunoreactive serotonin, substance P, and thyrotropin-releasing hormone (36,37). Moreover, the coexistence of these three neurotransmitters in ventral medullary neurons with projections to the intermediolateral cell column has been recently demonstrated (38). These projection neurons were found in the nucleus raphe pallidus, nucleus raphe obscurus, nucleus reticularis magnocellularis, nucleus paragigantocellularis, and parapyramidal region. Moreover, similar coexistences of these transmitters have also been found in the intermediolateral cell column (39), and the content of thyrotropin-releasing hormone in the spinal cord is diminished after systemic injection of the serotonergic neurotoxin 5,7-dihydroxytryptamine (36). Therefore, the same medullospinal neurons may use both neurochemicals as neurotransmitters. Several pieces of evidence support the idea that not only serotonin but also substance P and thyrotropin-releasing hormone are involved in cardiovascular regulation in the sympathetic spinal cord. First, physiological and pharmacological studies have demonstrated that serotonin, substance P, and thyrotropin-releasing hormone are excitatory of sympathetic preganglionic neurons (40,41). Second, intrathecal injection of substance P or thyrotropin-releasing hormone agonists produces an increase in arterial pressure and HR (42,43); on the other hand, substance P receptor antagonists produce hypotensive and bradycardic responses (44). Third, receptors for these neurotransmitters have been characterized in the intermediolateral cell column (45,46).

Besides the neurochemicals already mentioned, enkephalin is a neurotransmitter localized in the medullary raphe neurons as well as in other neurons in the RVLM with projections to the spinal cord sympathetic nuclei (47). These enkephalin-containing neurons may also participate in autonomic and cardiovascular regulation, particularly during autonomic and cardiovascular responses to stress conditions, as some studies have suggested (47–49).

Finally, recent anatomical studies have demonstrated the coexistence of gamma-aminobutyric acid (GABA) and enkephalin in the serotonergic neurons of the nucleus raphe pallidus and the nucleus raphe obscurus but have not demonstrated spinal projections of these neurons (50). The diversity of caudal raphe neurons, according to the combinations of peptides and amines contained within their neuropils and terminals in the spinal cord, could also be the result of a very complex functional organization of this medullospinal system involved in autonomic control.

IV. Periaqueductal Central Gray

A threatening stimulus generally arouses a synchronous common pattern of behavioral and autonomic responses known as the defense reaction. The biological role of the so-called defense reaction, which was very well described by W. B. Cannon in 1929 (51), can be explained as the visceral and hormonal modifications that are present during a fear and rage reaction. The behavioral changes known to happen as components of emergency reactions are determined by environmental factors. Therefore a cat confronted by a more powerful enemy will try to flee; if flight is not possible, the cat will display a defensive posture consisting of pupillary dilatation, piloerection, retraction of the ears, head retraction, arching of the back, crouching, and withdrawal, with hissing, howling, or growling. On the other hand, a cat confronting a weaker enemy will adopt a threatening posture that may ultimately be followed by attack, involving biting and hitting with the forepaws with claws unsheathed. These behavioral changes are accompanied by a cardiovascular response consisting of an increase in cardiac output, HR, and arterial pressure and vasoconstriction in splanchnic and skeletal muscle arteries, which is changed to vasodilation of skeletal muscle beds when an action such as fight or flight takes place (52–54).

S. M. Hilton and coworkers (55,56) demonstrated in the cat that neurons in the RVLM relay information for the descending pathway of the cardiovascular components of the defense reaction. The origin of the descending pathway of defense reaction was traced up to fibers of passage in the perifornical region in the hypothalamus. However, it is the midbrain periaqueductal central gray that plays a critical role in mediating the defense reaction. Selective depolarization of neurons in the midbrain periaqueductal central gray can evoke the defense reaction in either the conscious or decerebrated cat (57,58). The synchronized pattern of behavioral response is associated with an increase in arterial pressure, HR, and vasoconstriction followed by vasodilatation at the beginning of movement. There is also anatomical evidence that the increase in arterial pressure present in the defense reaction is mediated by a direct pathway from a specific division of the midbrain periaqueductal central gray to the RVLM in the cat (59). This area in the RVLM corresponds to the area where neurons involved in the tonic and reflex controls of the circulation are located (60–62). Moreover, chemical stimulation of the periaqueductal central gray at the level of the dorsolateral division in the rat increases arterial pressure and activates bulbospinal neurons in the RVLM (63). However, a different stimulation site in the periaqueductal central gray can evoke an inhibitory response on bulbospinal neurons in the RVLM that is associated with variable changes in arterial pressure (63). Therefore, a better understanding of the longitudinal organization and integration of the descending inputs from the midbrain defense regions to the RVLM will be needed to explain the complex cardiovascular responses associated with the periaqueductal central gray defensive behavior. Furthermore, the identity of the bulbospinal neurons in the RVLM that are either activated or inhibited by the periaqueductal central gray inputs is still an unresolved issue. In a recent study, at least two different types of sympathoexcitatory bulbospinal neurons in the RVLM with barosensitive properties were described (64).

In addition to the increase in arterial pressure, the cardiovascular response associated with the defense reaction implies a simultaneous increase in pulse pressure and tachycardia. These changes indicate an inhibition of the baroreceptor reflexes. Furthermore, there is evidence that the arterial baroreflexes are inhibited during the defense reaction evoked from the hypothalamus (65) and that chemical stimulation of the periaqueductal central gray decreases the baroreflex vagal bradycardiac response (66). The vascular component of the baroreceptor reflex response is also suppressed by electrical stimulation of the periaqueductal central gray, and this inhibitory pathway appears to be mediated by neurons in the parabrachial region (67). On the other hand, the direct projections to the RVLM originating in the periaqueductal central gray are apparently not involved in the modulation of the baroreflex response. However, as mentioned earlier, chemical stimulation of the periaqueductal central gray directly stimulates the barosensitive RVLM-spinal neurons (63), which are considered part of the baroreceptor reflex arc. Yet the periaqueductal central gray neurons involved in the defense reaction could additionally activate the sympathoexcitatory neurons in the RVLM by a disinhibitory mechanism involving the inhibition of the baroreceptor reflex neurons in the nucleus tractus solitarii. In fact, studies in the cat have demonstrated that stimulation of the defense area in the hypothalamus inhibits an important number of baroreceptor neurons in the nucleus tractus solitarii (68). However, the mechanisms by which periaqueductal central gray neurons inhibit the baroreceptor reflex response are certainly more complicated, and both the sympathetic and vagal components of this reflex are not equally affected by periaqueductal central gray stimulation. It is well known that activation of periaqueductal central gray perikarya causes an important inhibition of baroreflex vagal bradycardia, and yet the baroreflex hypotension is only minimally inhibited. Therefore, further investigations are needed to clearly elucidate the influence of inputs originating in the defense area on the baroreflex response.

In summary, the defense reaction is a coordinated pattern of emotional, behavioral, and visceral responses that prepare an organism to confront an emergency. It is now more widely accepted that the major structure integrating the defense reaction is the periaqueductal central gray area. Moreover, neurons in the RVLM with projections to the spinal cord receive direct projections from the periaqueductal central gray area and mediate the hypertensive response associated with the defense reaction. The inhibition of the arterial baroreflexes observed during the defense reaction could be an additional mechanism by which sympathoexcitatory neurons in the RVLM become more active, resulting in a further increase in arterial pressure.

V. Bed Nucleus Stria Terminalis

Recent studies have demonstrated that chemical stimulation of the bed nucleus stria terminalis results in changes in arterial pressure and HR (69,70). The cardiovascular responses evoked from the bed nucleus stria terminalis appear to be mediated by neurons in the caudal ventrolateral medulla (CVLM) (70). In support of this idea, anatomical studies have demonstrated that neurons in the bed nucleus stria terminalis send projections to the CVLM (71).

In addition, the bed nucleus stria terminalis could also activate other brainstem areas involved in cardiovascular regulation through a polysynaptic pathway. Using the expression of the protein c-Fos as a marker of neuronal activation, Giancola and Ciriello (Chapter 57 in this book) demonstrated that neurons in the nucleus tractus solitarii, parabrachial complex, CVLM, and RVLM contained c-Fos after chemical stimulation of the bed nucleus stria terminalis. Therefore, in addition to the CVLM, other areas in the brainstem may be involved in mediating the cardiovascular changes evoked from the bed nucleus stria terminalis. However, it is still unknown whether those additional areas receive the descending inputs from the bed nucleus stria terminalis through a relay station in the CVLM. In this regard, anatomical studies have demonstrated that the CVLM innervates both the nucleus tractus solitarii and the RVLM.

In any case, the CVLM may mediate the cardiovascular responses associated with behavioral, autonomic, and endocrine physiology, since the bed nucleus stria terminalis, as a component of the limbic system and part of the so-called extended amygdala (72), is involved in those functions.

The CVLM is also an area in the ventrolateral medulla critically involved in cardiovascular regulation. Electrolytic lesions placed in an area of the CVLM overlapping the A1 noradrenergic cell group produce a significant increase in arterial pressure (73). On the other hand, electrical or chemical stimulation of the CVLM in the rabbit (74) or rat produces a decrease in arterial pressure and HR by activating an inhibitory pathway on sympathetic vasomotor neurons in the RVLM (75). Both the vasodepressor neurons in the CVLM and the A1 noradrenergic cell group were thought to be located in the same area (the so-called A1 area) (74), and anatomical data demonstrated that neurons in the CVLM certainly project to the RVLM (76). Based mainly on very compelling pharmacological evidence, it was proposed that A1 noradrenergic neurons mediate the sympathoinhibitory response (75). However, more recent anatomical studies using double-labeling techniques demonstrated that none of the A1 adrenergic neurons projects to the RVLM (8). Therefore the origin of the very effective inhibitory catecholaminergic inputs received by the tonic sympathoexcitatory neurons in the RVLM is still a matter of discussion. These sympathoinhibitory inputs are thought to be mediated by alpha-2 adrenergic receptors.

Most of the studies dealing with the identification and characterization of the short sympathoinhibitory pathway between the CVLM and the RVLM were based on microinjections of pharmacological agents into the brain. This technique does not allow precise localization of the neurons in the CVLM responsible for the evoked effect. Therefore future investigations using more precise techniques will be needed to identify the neurons in the CVLM and the neurotransmitters responsible for the sympathoinhibitory responses.

The CVLM is also involved in the control of vasopressin release during the baroreflex. It was proposed that such an effect is mediated by a noradrenergic pathway originating in the A1 noradrenergic neurons with projections into the paraventricular and supraoptic hypothalamic nuclei where the magnocellular vasopressin-secreting neurons are located (77). This idea is supported by some experimental results. For instance, local application of noradrenaline into the supraoptic nucleus increases the secretion of vasopressin in plasma (78), and the majority of CVLM neurons with projections to the hypothalamus are noradrenergic A1 cells (79). However, the excitatory response evoked on supraoptic neurosecretory cells by stimulating the A1 area is not blocked by alpha-adrenergic antagonists, and neither endogenous nor exogenous noradrenaline acts as a fast excitatory neurotransmitter on those cells in the supraoptic nucleus (77,80,81). Therefore, the physiological role played by the A1 noradrenergic neurons in the CVLM either in controlling the sympathetic tone or in the secretion of vasopressin in the hypothalamus is still an unresolved matter.

VI. Summary and Concluding Remarks

The control of the cardiovascular system by the central nervous system involves several distinct areas in the brain that integrate afferent information from different peripheral receptors and cause, in return, reflex neurohumoral responses to effector organs that are suitable for specific physiological demands. The critical role played by the medulla oblongata in controlling the cardiovascular system has been known for more than 100 years. With the introduction of new techniques, however, substantial experimental evidence has been obtained in the last few years in support of the idea that a network of neurons in the ventrolateral medulla plays an essential role in generating the tonic cardiovascular sympathetic nerve activity and mediating several cardiovascular reflexes.

Neurons in a restricted area in the rostral portion of the ventrolateral medulla have been demonstrated to be involved in the generation of the cardiovascular sympathetic tone. Two populations of bulbospinal neurons in the RVLM have been ascribed to such function; these neurons normally supply tonic excitatory inputs to cardiovascular sympathetic preganglionic neurons in the spinal cord. The activity of these tonically discharging RVLM-spinal neurons is modulated by different peripheral and central inputs. Yet neither the nature of such inputs nor the target neurons in the RVLM have been clearly identified in many cases. Nevertheless, it is accepted that these RVLM neurons receive inhibitory inputs as a result of activating arterial baroreceptors, and they are also involved in mediating other cardiovascular reflexes such as low-pressure mechanoreceptors, somatosympathetic and chemoreceptors. Anatomical and electrophysiological studies have demonstrated that sympathoexcitatory RVLM neurons receive modulatory inputs from different areas in the brain, including the hypothalamus and the central gray, which have been associated with the defense reaction. Moreover, RVLM neurons receive important intramedullary inputs originating in the AP and medullary raphe that are probably related to the control of sympathetic activity.

In addition to the key neurons in the RVLM, other spontaneously firing neurons in the ventral medulla with projections to the spinal cord may also participate in the control of sympathetic activity. For instance, bulbospinal neurons in the nucleus raphe obscurus and nucleus raphe pallidus are known to subserve a sympathoexcitatory function. Nonetheless, their physiological role is unknown.

A small area in the CVLM containing the A1 noradrenergic neuronal group is also critically involved in cardiovascular regulation. Tonic sympathoinhibitory neurons in this area are thought to control the activity of vasomotor neurons in the RVLM. In addition, neurons in the CVLM with projections to the supraoptic and paraventricular hypothalamic nuclei mediate the baroreflex control of vasopressin secretion. The A1 noradrenergic cells are probably among these neurons. Nonetheless, the physiological role in cardiovascular regulation played by catecholaminergic neurons in the ventral medulla (A1, C1 groups) is far from being understood, and future investigations using more adequate techniques are necessary to clarify these questions.

Acknowledgments

Supported by the American Heart Association (AHA) Tennessee Affiliate, Grant TN92G13. The author would like to thank Dr. Barbara J. Kuyper for assistance in the preparation of this manuscript.

References

1. Ylitalo P, Karppanen H, Paasonen MK. Is the area postrema a control centre of blood pressure? Nature 1974; 274:58–59.
2. Barnes KL, Ferrario CM, Chernicky CL, Brosnihan KB. Participation of the area postrema in cardiovascular control of the dog. Fed Proc 1984; 43:2959–2962.
3. Wislocki GB, Putnam TJ. Note on the anatomy of the areae postremae. Anat Rec 1920; 19:281–287.
4. Joy MD, Lowe RD. Evidence that the area postrema mediates the central cardiovascular response to angiotensin II in the brain of the dog. Nature 1970; 228:1303–1304.
5. Fink GD, Bruner CA, Manglapane ML. Area postrema is critical for angiotensin-induced hypertension in rats. Hypertension 1987; 9:355–361.
6. Shapiro RE, Miselis RR. The central neural connections of the area postrema of the rat. J Comp Neurol 1985; 234:344–364.
7. Contreras RJ, Beckstead RM, Norgren R. The central projections of the trigeminal, facial, glossopharyngeal and vagus nerve: An autoradiographic study in the rat. J Auton Nerv Syst 1982; 6:303–322.
8. Blessing WW, Hedger SC, Joh TH, Willoughby JO. Neurons in the area postrema are the only catecholamine-synthesizing cells in the medulla or pons with projections to the rostral ventrolateral medulla (C1-area) in the rabbit. Brain Res 1987; 419:336–340.
9. Sun MK, Guyenet PG. Effect of clonidine and gamma-aminobutyric acid on the discharges of medullo-spinal sympathoexcitatory neurons in the rat. Brain Res 1986; 368:1–17.
10. Sun MK, Spyer KM. GABA-mediated inhibition of medullary vasomotor neurones by area postrema stimulation in rats. J Physiol (Lond) 1991; 436:669–684.
11. Hasser EM, Nelson DO, Haywood JR, Bishop VS. Inhibition of renal sympathetic nervous activity by area postrema stimulation in rabbits. Am J Physiol 1987; 253: H91–H99.
12. Hay M, Hasser EM, Undesser KP, Bishop VS. Role of baroreceptor afferents on area postrema-induced inhibition of sympathetic activity. Am J Physiol 1991; 260: H-1351–H1358.
13. Cox BF, Hay M, Bishop VS. Neurons in area postrema mediate vasopressin-induced enhancement of the baroreflex. Am J Physiol 1990; 258:H1943–H1946.

14. Bonham AC, Hasser EM. Area postrema and aortic or vagal afferents converge to excite cells in nucleus tractus solitarii. Am J Physiol 1993; 264:H1674–H1685.
15. Ciriello J, Caverson M, Polosa C. Function of the ventrolateral medulla in the control of the circulation. Brain Res Rev 1986; 11:359–391.
16. Loewy AD, McKellar S. Serotonergic projections from the ventral medulla to the intermediolateral cell column in the rat. Brain Res 1981; 211:146–152.
17. Loewy AD. Raphe pallidus and raphe obscurus projections to the intermediolateral cell column in the rat. Brain Res 1981; 222:129–133.
18. Bacon SJ, Zagon A, Smith AD. Electron microscopic evidence of a monosynaptic pathway between cells in the caudal raphe and sympathetic preganglionic neurons in the rat spinal cord. Exp Brain Res 1990; 79:589–602.
19. Bobillier P, Seguin S, Petitjean F, et al. The raphe nuclei of the cat brain stem: A topographical atlas of their efferent projections as revealed by autoradiography. Brain Res 1976; 113:440–486.
20. Basbaum AI, Clanton CH, Fields HL. Three bulbospinal pathways from the rostral medulla of the cat: An autoradiographic study of pain modulating systems. J Comp Neurol 1978; 178:209–224.
21. Nicholas AP, Hancock MB. Evidence for projections from the rostral medullary raphe onto medullary catecholamine neurons in the rat. Neurosci Lett 1990; 108: 22–28.
22. Tucker DC, Saper CB, Ruggiero DA, Reis DJ. Organization of central adrenergic pathways: I. Relationships of ventrolateral medullary projections to the hypothalamus and spinal cord. J Comp Neurol 1987; 259:591–603.
23. Adair JR, Hamilton BL, Scappaticci KA, et al. Cardiovascular responses to electrical stimulation of the medullary raphe area of the cat. Brain Res 1977; 128:141–145.
24. McCall RB. Identification of neurotransmitters which mediate cardiovascular effects elicited from the midline medulla. In: Ciriello J, ed. Organization of the Autonomic Nervous System: Central and Peripheral Mechanisms. New York: Liss, 1987:283–293.
25. Minson JB, Chalmers JP, Caon AC, Renaud B. Separate areas of rat medulla oblongata with populations of serotonin- and adrenaline-containing neurons alter blood pressure after L-glutamate stimulation. J Auton Nerv Syst 1987; 19:39–50.
26. Pilowsky PM, Kapoor V, Minson JB, et al. Spinal cord serotonin released and raised blood pressure after brain stem kainic acid injections. Brain Res 1986; 336: 354–357.
27. Cabot JB, Wild JM, Cohen DH. Raphe inhibition of sympathetic preganglionic neurons. Science 1979; 203:184–186.
28. McCall RB. Evidence for a serotonergically mediated sympathoexcitatory response to stimulation of medullary raphe nuclei. Brain Res 1984; 311:131–139.
29. Yusof APM, Coote JH. Patterns of activity in sympathetic preganglionic nerves in skeletal muscle, skin, and kidney during stimulation of the medullary raphe area of the rat. J Auton Nerv Syst 1988; 24:71–79.
30. Morrison SF, Gebber GL. Raphe neurons with sympathetic-related activity: Baroreceptor responses and spinal connections. Am J Physiol 1984; 246:R338–R348.
31. Morrison SF, Gebber GL. Axonal branching patterns and funicular trajectories of raphe spinal sympathoinhibitory neurons. J Neurophysiol 1985; 53:759–772.
32. McCall RB, Clement ME. Identification of serotonergic and sympathetic neurons in medullary raphe nuclei. Brain Res 1989; 477:172–182.

33. Skagerberg G, Björklund A. Topographic principles in spinal projections of serotonergic and non-serotonergic brain stem neurons in the rat. Neuroscience 1985; 15:445–480.
34. McCall RB. Evidence for a serotonergically mediated sympathoexcitatory response to stimulation of medullary raphe nuclei. Brain Res 1984; 311:131–140.
35. Zagon A. Innervation of serotonergic medullary raphe neurons from cells of the rostral ventrolateral medulla in rats. Neuroscience 1993; 55:849–867.
36. Helke CJ, Sayson SC, Keeler JR, Charlton CG. Thyrotropin-releasing hormone-immunoreactive neurons project from the ventral medulla to the intermediolateral cell column: Partial coexistence with serotonin. Brain Res 1986; 381:1–7.
37. Hirsch MD, Helke CJ. Bulbospinal thyrotropin-releasing hormone projections to the intermediolateral cell column: A double fluorescence immunohistochemical-retrograde tracing study in the rat. Neuroscience 1988; 25:625–637.
38. Sasek CA, Wessendorf MW, Helke CJ. Evidence for co-existence of thyrotropin-releasing hormone, substance P and serotonin in ventral medullary neurons that project to the intermediolateral cell column in the rat. Neuroscience 1990; 35:105–119.
39. Appel NM, Wessendorf MW, Elde RP. Thyrotropin-releasing hormone in spinal cord: Coexistence with serotonin and with substance P in fibers and terminals apposing identified preganglionic sympathetic neurons. Brain Res 1987; 415:137–143.
40. Backman SB, Henry JL. Effects of substance P and thyrotropin-releasing hormone on sympathetic preganglionic neurons in the upper thoracic intermediolateral nucleus of the cat. Can J Physiol Pharmacol 1984; 62:248–251.
41. Ma RC, Dun NJ. Excitation of lateral horn neurons of the neonatal rat spinal cord by 5-hydroxytryptamine. Dev Brain Res 1986; 24:89–98.
42. Helke CJ, Phillips ET. Thyrotropin-releasing hormone receptor activation in the spinal cord increases blood pressure and sympathetic tone to the vasculature and the adrenals. J Pharmacol Exp Ther 1988; 245:41–46.
43. Helke CJ, Phillips ET, O'Neill JT. Regional peripheral and CNS hemodynamic effects of intrathecal administration of a substance P receptor agonist. J Auton Nerv Syst 1987; 21:1–7.
44. Helke CJ, Phillips ET, O'Neill JT. Intrathecal administration of a substance P receptor antagonist: Studies on peripheral and central nervous system hemodynamics and on specificity of action. J Pharmacol Exp Ther 1987; 242:131–136.
45. Manaker S, Winokur A, Rostene WH, Rainbow TC. Autoradiographic localization of thyrotropin-releasing hormone receptors in the rat central nervous system. J Neurosci 1985; 5:167–174.
46. Seybold VS, Elde RP. Receptor autoradiography in thoracic spinal cord: Correlation of neurotransmitter binding sites with sympathoadrenal neurons. J Neurosci 1984; 4:2533–2542.
47. Romagnano MA, Harshbarger RJ, Hamill RW. Brain stem enkephalinergic projections to spinal autonomic nuclei. J Neurosci 1991; 11:3539–3555.
48. Marson L, Kiritsy-Roy JA, Van Loon GR. Mu-opioid peptide modulation of cardiovascular and sympathoadrenal responses to stress. Am J Physiol 1989; 257:R901–R908.
49. Nordin M, Morat P, Zainora M. The effect of endogenous opioids on blood pressure during stress. Clin Exp Pharmacol Physiol 1987; 14:303–308.

50. Kachidian P, Poulat P, Marlier L, Privat A. Immunohistochemical evidence for the coexistence of substance P, thyrotropin-releasing hormone, GABA, methionin-enkephalin, and leucin-enkephalin in the serotonergic neurons of the caudal raphe nuclei: A dual labeling in the rat. J Neurosci Res 1991; 30:521–530.
51. Cannon WB. Bodily Changes in Pain, Hunger, Fear, and Rage. 2d ed. New York: Appleton, 1929.
52. Adams DB, Bacelli G, Mancia G, Zanchetti A. Relation of cardiovascular changes in fighting to emotion and exercise. J Physiol (Lond) 1971; 212:321–335.
53. Ellison GD, Zanchetti A. Diffuse and specific activation of sympathetic cholinergic fibers in the cat. Am J Physiol 1973; 225:142–149.
54. Mancia G, Bacelli G, Zanchetti A. Hemodynamic responses to different emotional stimuli in cats: Patterns and mechanisms. Am J Physiol 1972; 223:925–933.
55. Hilton SM, Marshall JM, Timms RJ. Ventral medullary relay neurons in the pathway from the defence areas of the cat and their effect on blood pressure. J Physiol (Lond) 1983; 345:149–166.
56. Hilton SM, Smith PR. Ventral medullary neurones excited from the hypothalamic and midbrain defence areas. J Auton Nerve Syst 1984; 11:35–42.
57. Bandler R, Carrive P. Integrated defence reaction elicited by excitatory amino acid microinjection in the midbrain periaqueductal grey region of the unrestrained cat. Brain Res 1988; 439:95–106.
58. Carrive P, Dampney RAL, Bandler R. Excitation of neurones in a restricted portion of the midbrain periaqueductal gray elicits both behavioural and cardiovascular components of the defence reaction in the unanaesthetized descerebrate cat. Neurosci Lett 1887; 81:273–278.
59. Carrive P, Bandler R, Dampney RAL. Anatomical evidence that hypertension associated with the defence reaction in the cat is mediated by a direct projection from a restricted portion of the midbrain periaqueductal gray to the subretrofacial nucleus of the medulla. Brain Res 1988; 460:339–345.
60. Ross CA, Ruggiero DA, Park DH, et al. Tonic vasomotor control by the rostral ventrolateral medulla: Effect of electrical or chemical stimulation of the area containing C1 adrenaline neurons on arterial pressure, heart rate and plasma catecholamines and vasopressin. J Neurosci 1984; 4:274–294.
61. Granata AR, Ruggiero DA, Park DH, et al. Brainstem area with C1 epinephrine neurons mediates baroreflex vasodepressor responses. Am J Physiol 1985; 248: H547–H567.
62. Yamada KA, McAllen RM, Loewy AD. GABA antagonists applied to the ventral surface of the medulla oblongata block the baroreceptor reflex. Brain Res 1984; 297:175–180.
63. Lovick TA. Midbrain influences on ventrolateral medullo-spinal neurones in the rat. Exp Brain Res 1992; 90:147–152.
64. Granata AR, Kitai ST. Intracellular analysis in vivo of different barosensitive bulbospinal neurons in the rat rostral ventrolateral medulla. J Neurosci 1992; 12:1–20.
65. Coote JH, Hilton SM, Perez Gonzalez JH. Inhibition of the baroreceptor reflex on stimulation in the brain stem defence centre. J Physiol (Lond) 1979; 288:549–560.
66. Jones RO, Kirkman E, Little RA. The involvement of the midbrain periaqueductal gray in the cardiovascular response to injury in the conscious and anaesthetized rat. Exp Physiol 1990; 75:483–495.

67. Nosaka S, Murata K, Inui K, Murase S. Arterial baroreflex inhibition by midbrain periaqueductal gray in anaesthetized rats. Pflügers Arch 1993; 424:266–275.
68. Mifflin SW, Spyer KM, Withington-Wray DJ. Baroreceptor inputs to the nucleus tractus solitarii in the cat: Modulation by the hypothalamus. J Physiol (Lond) 1988; 399:369–387.
69. Gelsema AJ, Calaresu FR. Chemical microstimulation of the septal area lowers arterial pressure in the rat. Am J Physiol 1987; 252:R760–R767.
70. Giancola SB, Roder S, Ciriello J. Contribution of caudal ventrolateral medulla to the cardiovascular responses elicited by activation of bed nucleus of the stria terminalis. Brain Res 1993; 606:162–166.
71. Holstege G, Meiners L, Tan K. Projections of the bed nucleus of the stria terminalis to the mesencephalon, pons, and medulla oblongata in the cat. Exp Brain Res 1985; 58:379–391.
72. DeOlmos J, Alheid GF, Beltramino CA. Amygdala. In: Paxinos G, ed. The Rat Nervous System. Vol 1. Forebrain and Midbrain. Sydney, Australia: Academic Press, 1985:223–334.
73. Blessing WW, West MJ, Chalmers JP. Hypertension, bradycardia and pulmonary oedema in the conscious rabbit after brain stem lesions coinciding with the A1 group of catecholamine neurons. Circ Res 1981; 49:949–958.
74. Blessing WW, Reis DJ. Inhibitory cardiovascular function of neurons in the caudal ventrolateral medulla of the rabbit: Relationship to the area containing A1 noradrenergic cells. Brain Res 1982; 253:161–171.
75. Granata AR, Numao Y, Kumada M, Reis DJ. A1 noradrenergic neurons tonically inhibit sympathoexcitatory neurons of C1 area in rat brain stem. Brain Res 1986; 377:127–146.
76. Ross CA, Ruggiero DA, Reis DJ. Projections from the nucleus tractus solitarii to the rostral ventrolateral medulla. J Comp Neurol 1985; 242:511–534.
77. Randle JCR, Day TA, Jhamandas JH, et al. Neuropharmacology of supraoptic nucleus neurons: Norepinephrine and gamma-amino-butyric acid receptors. Fed Proc 1986; 45:2312–2317.
78. Blessing WW, Willoughby JO. Inhibiting the rabbit caudal ventrolateral medulla prevents baroreceptor-initiated secretion of vasopressin. J Physiol (Lond) 1985; 367:253–265.
79. Sawchenko PE, Swanson LW, Grzanna R, et al. Colocalization of neuropeptide Y immunoreactivity in brainstem catecholaminergic neurons that project to the paraventricular nucleus of the hypothalamus. J Comp Physiol 1985; 241:138–153.
80. Blessing WW, Willoughby JO. Adrenoceptor agents and baroreceptor-initiated secretion of vasopressin. In: Cowley AW, Liard JF, Ausiello DA, eds. Vasopressin: Cellular and Integrative Functions. New York: Raven Press, 1988:349–353.
81. Day TA. Control of neurosecretory vasopressin cells by noradrenergic projections of the caudal ventrolateral medulla. Prog Brain Res 1989; 81:303–317.

INDEX

D

I

K

L

R

S

W

Y

Z